Physiological Processes in Plants Under Low Temperature Stress

A. Bhattacharya

Physiological Processes in Plants Under Low Temperature Stress

Springer

A. Bhattacharya
Department of Basic Science
Indian Institute of Pulses Research
Kanpur, Uttar Pradesh, India

ISBN 978-981-16-9036-5 ISBN 978-981-16-9037-2 (eBook)
https://doi.org/10.1007/978-981-16-9037-2

This Springer imprint is published by the registered company Springer Nature Singapore Pte Ltd.
The registered company address is: 152 Beach Road, #21-01/04 Gateway East, Singapore 189721, Singapore

Preface

Despite increasing average temperature across the years, high temperatures are not the only threat and it is likely that plants, including our field crops, will also experience an increase in more devastating early spring frosts events. It has been reported that there are losses to many economically important crops just from freeze damage. Thus, rather than focusing only on yield gains and quality traits, our crop plants will need to be more flexible in response to changes in their environment so that farmers will get stable yield despite less predictable and more severe weather. To reach this goal, a better understanding on how the whole plant responds to changes in low temperature in the surrounding environment, beginning with how temperature is perceived and how the signal is transferred to nucleus. Plant exposed to a rapid and severe drop in temperature can suffer protoplasmic ice formation resulting in death. In summer, trees and herbaceous plants in northern latitudes cannot withstand freezing, but exposure to chilling temperatures induces hardening through a process called cold acclimation, and acclimated plants can survive winter temperatures far below freezing. Thus, many plants can increase their resistance to various stresses including heat, salt, and drought conditions in response to an elastic strain. The principle focus of this book is on the response of plants to unfavorable low temperatures and to low temperature stress.

Among various environmental stresses, low temperature is one of the important factors limiting the productivity and distribution of plants. It is well established that some of the molecular and physiological changes during cold acclimation are important for plant cold tolerance. The definition of chilling injury has been defined as low temperature damage in the absence of freezing. Freezing injury occurs when the external temperature drops below the freezing point of water. There are many plants, including many of our core agricultural crops that are susceptible to chilling injury and can be killed by their first experience of frost. There are also plants native to cold climates that can survive extremely low temperatures without injury. Crops have basic requirements for temperature to complete a specific phonological phase of growth or the whole life cycle. On the other hand, extremely high and low temperatures can have detrimental effects on the crop growth, development, and yield, particularly at the critical phonological phase. While some of crop plants are morphologically indeterminate, the rate of many phonological processes such as germination, initiation of flower, and development of fruiting bodies is controlled by temperature. The average daily temperature also plays an important role in determining the earliest date of sowing. Crops grown under open environments often pass through periods of abiotic stress during their life cycle. Such stresses many a time overlap so that the crop growth and productivity are adversely affected. Plants pass through a series of morphological, physiological, biochemical, and molecular changes in a quest to mitigate such adversities of the abiotic stresses. Crops experience periods of extreme low temperatures in many regions of the world and are exposed to limited water availability owing to either drought or disturbed water movement and uptake under low ambient temperatures.

Low temperature stress is an inevitable environmental factor that extensively affects plant growth and development. Cold stress is a significant abiotic stress which affects growth and development of crops, leading to loss of strength and lesions on the surface. It is known that temperature affects the main physiological and biochemical processes in plants. The physiological processes most affected are photosynthesis, growth, and respiration when temperatures are above or below the normal ranges of growth. In the same way, biochemical changes such as alterations in the viscosity, permeability, and fluidity of the cell membrane are produced at the enzymatic level. Low temperature stress decreases capacity and efficiency of photosynthesis through changes in gas exchange, pigment content, and chloroplast development and also declines chlorophyll fluorescence. Susceptible plants with cold temperatures have reduced growth, restricted use of precious varieties, and reduced yields. Plants use separate strategies to cope with stressful conditions and integrate a variety of physiological, metabolic, and molecular adaptations. Changes in the structure of proteins and lipid membranes assist to restore homeostasis of metabolites and are regarded as mechanisms by which cells feel cold temperatures. For its metabolic and physical function, the liquid state of the plasma membrane is a structural and functional asset. At low temperatures, the plasma membrane transitions from a liquid state to a stiff gel stage. Low temperature-mediated changes in the physical conformation of the membrane are mainly because of enhanced levels of unsaturated lipids, which increase the fluidity and stability of the membrane, enabling cells to adapt mechanically to cold.

Plant growth and crop yield are majorly affected by low temperature, soil water deficit, salt, and heavy metals. Physiological reactions to stress include wilting of the leaf, abscission of the leaf, decreased leaf region, and decreased water loss through transpiration. Detrimental impacts of stress can be decreased by osmotic modification, which helps with an active accumulation of solutes in the cytoplasm to maintain cell water balance. The use and application of plant growth regulators in agriculture have several practical examples. The plant growth regulators in this context are employed commercially. Commercial uses of auxins include prevention of fruit and leaf drop, promotion of flowering, thinning of fruit, induction of parthenocarpic fruit development, and rooting of cuttings for plant propagation, among others. Considerable success had been obtained in the application of plant growth regulators in some process of plant development such as flowering and fruit development as well as ripening, harvesting, and post-harvesting of fruits and vegetables. Tropical trees are generally induced to flower through environmental cues, whereas floral initiation of temperate deciduous trees is often autonomous. In the tropical evergreen tree, cool temperature is the only factor known to induce flowering, but does not ensure occurrence of floral initiation because there are important interactions with vegetative growth involvement of plant hormones. Those have been characterized in Arabidopsis, including floral initiation through photoperiod and temperature, and regulation by gibberellins and nitrate used

This book is intended to cover major effects of low temperature on various major aspects of physiological processes in plants through review articles. The book has seven review chapters all together. Under low temperature stress, the first issue faced by crop plants is lowering of seed germination percentage, slow growth and developmental processes, and changes in duration of phonological phases. The first review chapter deals with changes in seed **germination, growth, and phenology** under low temperature stress. Following poor germination and slow growth, plant water relations are disturbed under low temperature regime, and the second review chapter deals with **plant water relations** under low temperature stress. With lowering of plant water content and metabolic activities, the most affected physiological process is photosynthesis. The third review chapter deals with effects of low temperature stress on **photosynthesis and its associated processes**. Apart from water relationship and photosynthesis, biological nitrogen fixation, nitrogen uptake, and nitrogen assimilation are also affected by low temperature stress. In the fourth chapter, effects of low temperature stress **nitrogen metabolism** have been reviewed. Changes in photosynthetic and allied processes under low temperature stress causes changes in lipid metabolism in plants and the fifth chapter deals with **changes in lipid metabolism** in plants under low temperature stress. Changes in carbon, nitrogen, and lipid metabolism causes changes in growth and developmental processes due to the effects of plant growth regulators under low temperature condition and the sixth chapter deals with the **role of plant growth regulators** under low temperature conditions. Seventh chapter reviews the effects of low temperature on **dry matter partitioning and seed yield**. These review chapters address how knowledge of the physiological mechanisms of crops can contribute to reaching these goals.

These seven review chapters address knowledge of the physiological mechanisms of crops which can contribute to overcoming the adverse effects of low temperature. Suggestions and advice are most welcome for improvements in the chapters. The author is hopeful that this text will be of use to readers. Suggestions for any future edition from researchers, teachers, and students, who use the book, are welcomed. It is hoped that this book will serve the needs of students, faculty members, and researchers. The author sincerely hopes that the review chapters in this book will meet the requirements of postgraduate students as well as faculty members of plant physiology, and he will be glad to receive constructive criticism and suggestions from faculties.

Kanpur, Uttar Pradesh, India Amitav Bhattacharya

Contents

1 **Effect of Low-Temperature Stress on Germination, Growth, and Phenology of Plants: A Review** 1
1.1 Low-Temperature Stress 2
1.1.1 Chilling Injury 5
1.1.2 Freezing and Frost Injury 6
1.1.3 Physiological and Molecular Changes Under Low Temperature 7
1.2 Low Temperature and Seed Germination 10
1.2.1 Seed Priming and Germination Under Low Temperature 15
1.3 Seedling Growth Under Low Temperature 19
1.4 Root Development Under Low Temperature 23
1.5 Constraints of Environmental Temperature 26
1.5.1 Low Temperature 27
1.6 Growth and Development Under Low Temperature 30
1.6.1 Vegetative Growth 33
1.6.2 Temperature Effects on Aboveground Growth and Biomass Production 37
1.6.3 Reproductive Growth 37
1.6.4 Yield and Yield-Related Attributes 41
1.6.5 Temperature Sensitivity of Different Growth Phases 42
1.7 Response of Perennial Crops to Low-Temperature Stress 43
1.8 Mechanisms of Acclimation to Low Non-freezing Temperatures 44
1.8.1 The Role of Carbohydrates 45
1.8.2 The Role of Osmolytes 45
1.8.3 The Role of Plant Growth Hormones 47
1.8.4 Modifications to Plant Cell Membranes 48
1.8.5 The Role of Reactive Oxygen Species 49
1.8.6 Interaction(s) of Genes 50
1.8.7 Alteration of Meiotic Recombination and Somatic Mutation 54

1.9 Phenological Development 55
1.10 Temperature and Flower Induction 57
1.10.1 Effect of Temperature During Flower Maturation and Shoot Development 58
1.11 Molecular/Genetic Aspects of Flowering Response to Climate Change 64
1.12 Simulation of Phenology for Future Climate 66
References 67

2 Low Temperature Stress and Plant-Water Relationship: A Review 107
2.1 Plant and Water 108
2.2 Soil and Water 111
2.2.1 Soil Moisture Constants 111
2.2.2 Soil Temperature 114
2.2.3 Soil Water Storage and Availability to Plants 115
2.2.4 Water Transmission Characteristics 116
2.2.5 Crop Rooting Behavior and Soil Water Retention 116
2.3 Plant-Water Relations 117
2.3.1 Concept of Water Potential 121
2.3.2 Components of Water Potential 122
2.4 Water Uptake by Roots 126
2.4.1 Root Water Uptake Concepts 128
2.4.2 Water Absorption by the Roots 129
2.4.3 Root Water Uptake under Low-Temperature Conditions 130
2.4.4 Transpiration 132
2.4.5 Ascension of Water Through the Plant: Vascular System 133
2.4.6 Disruption of Water Movement 135
2.4.7 Signal Controlling Root Water Uptake Under Stress Condition 135
2.5 Plant-Water Relationship under Low-Temperature Stress 137
2.5.1 Transpiration and Transpiration Efficiency 142
2.5.2 Water-Use Efficiency 144
2.5.3 Induction of Tolerance 148
2.5.4 Low-Temperature Tolerance of Woody Plants 154
2.5.5 Approaches for Induction of Low-Temperature Stress Tolerance 155
2.5.6 Nutrient Management to Alleviate the Low-Temperature Stresses 156
2.6 Mechanism of Water Status Regulation 159
2.6.1 Morphological and Anatomical Characteristics and Water Control 160
2.6.2 Stomata Metabolism 161

2.6.3 Redox Signaling 161
2.6.4 Sugar Signaling 162
2.6.5 Control of Oxidative Stress 163
2.6.6 Control on Lipid Changes 165
2.6.7 Hormonal and Molecular Responses at Different Levels of Water Stress 165
2.6.8 Osmotic Regulation 167
2.7 Acclimation to Cold Stress 169
References 171

3 Effect of Low Temperature Stress on Photosynthesis and Allied Traits: A Review 199
3.1 Plant Metabolic Activities Under Change in Global Temperature 200
3.2 Photosynthesis Under Low Temperature 204
3.2.1 Low Temperature Stress and Chloroplast Membrane Integrity 212
3.2.2 Low Temperature Stress and Pigment System 214
3.2.3 Effect of Low Temperature on Stomata 215
3.3 Impact of Suboptimal Temperature on Photosynthesis 220
3.3.1 Photosynthesis in C4 Plants 228
3.3.2 Control of Photosynthetic Processes in C4 Plants 231
3.3.3 Thermal Acclimation of C4 Photosynthesis 233
3.4 Cellular Mechanisms Against Chilling Damage and Oxidative Stress 235
3.5 Effect of Low Temperature Stress on CO_2 Fixation 237
3.6 Effect of Low Temperature on Respiration and ATP Generation 240
3.7 Low Temperature and Photorespiratory Process 242
3.8 Cold Acclimation and Its Role in Providing Freeze Tolerance . . . 245
3.9 Low Temperature Stress-Induced Changes 248
3.9.1 Effects on Photosynthetic Pigments 248
3.9.2 Effects on Photosystems 249
3.9.3 Effects on Gas-Exchange Characteristics 251
3.9.4 Effects on Activities of Key Photosynthetic Enzymes 252
3.10 Improving Photosynthetic Capacity Through Photosynthetic Related Genes or Transcription Factors 253
3.10.1 Transfer of C4 Genes/Traits to C3 Plants 254
3.10.2 Role of Mitogen-Activated Protein Kinases (MAPKs) in Photosynthesis 255
3.11 Strategies to Ameliorate Cold Stress and Improve Photosynthesis 256
3.12 Management Strategies Against Low Temperature Stress 257
3.12.1 Selection and Breeding Strategies 257

3.12.2 Agronomic and Physiological Measures 259
3.13 Conclusion and Future Prospects 262
References 263

4 Low-Temperature Stress and Nitrogen Metabolism in Plants: A Review 299
4.1 Low-Temperature Stress and Metabolic Response in Plants 300
4.2 Nitrogen Cycle 304
4.2.1 Nitrogen Fixation 305
4.2.2 Decomposition 306
4.2.3 Nitrification 306
4.2.4 Denitrification 307
4.2.5 Biogeochemical Linkage of Nitrogen and Carbon 308
4.2.6 Soil Organic Matter 309
4.2.7 Perturbations and Nitrogen and Carbon Biogeochemistry 310
4.2.8 Organic Nitrogen 311
4.2.9 Mineralization 312
4.2.10 Biological Nitrogen Fixation 312
4.2.11 Nitrogen Fixation in Nonleguminous Plants 315
4.3 Factors Influencing Biological Nitrogen Fixation in Legumes . . . 317
4.3.1 Temperature 317
4.3.2 Soil Water Status 318
4.3.3 Nitrogen Concentration in the Root Zone 320
4.3.4 Carbon Demand for Fixation 321
4.3.5 Seasonal Regulation of Biological Nitrogen Fixation 322
4.3.6 Changes in Nitrogen Fixation with Plant Growth Stage 322
4.3.7 Influence of Nitrate on Nodulation 322
4.3.8 Influence of Nitrate on Nitrogenase Activity 323
4.4 Nitrogen Assimilation 323
4.4.1 Nitrate Reduction 324
4.4.2 Ammonia Assimilation 332
4.5 Cold Acclimation and Tolerance to Freezing Stress 335
4.5.1 Cold Acclimation 335
4.5.2 Physiological and Metabolic Responses to Low Temperature 337
4.5.3 Inducible Response 342
4.5.4 Nitrogen and Amino Acid Metabolism 343
4.5.5 Protein Metabolism 344
4.5.6 Free Amino Acids 345
4.5.7 Role of γ-Aminobutyric Acid in Cold and Frost Tolerance 348

4.5.8 Exogenous Application of Amino Acid Under Temperature Stress 349
4.5.9 Enzyme Variation 351
4.5.10 Protein Content 352
4.5.11 Genetic Analysis of Cold Acclimation 352
4.6 Protein Synthesis 353
4.6.1 Types of Proteins 353
4.6.2 Protein Synthesis 359
4.7 Chilling Stress and Potential Management Opportunities 363
4.7.1 Plant Responses to Chilling and Drought Stresses 363
4.7.2 Molecular Responses 365
4.7.3 Management Strategies 367
4.7.4 Exogenous Diethyl Aminoethyl Hexanoate Ameliorates Low-Temperature Stress 371
References 372

5 Lipid Metabolism in Plants Under Low-Temperature Stress: A Review 409
5.1 Biological Role of Lipids in Plants 410
5.2 Lipids in Plants 412
5.3 Importance of Lipid Metabolism to Cold Stress 416
5.3.1 Role of Cytosolic Lipid Droplets in Stress Response . . . 419
5.3.2 Role of Triacylglycerol in Stress Response 420
5.3.3 Lipid Antioxidant and Galactolipid Remodeling Under Low Temperature 421
5.3.4 Low-Temperature Stress and Free Fatty Acids 422
5.4 Effect of Low-Temperature Stress on Different Plants 423
5.4.1 General Effects of Chilling Temperature (Between 20 and 0 °C) on Plant Physiology 424
5.4.2 General Effects of Freezing Temperature (Below 0 °C) on Plant Physiology 424
5.4.3 Rapeseed (*Brassica napus*) 425
5.4.4 Potato (*Solanum tuberosum*) 427
5.4.5 Wheat (*Triticum aestivum*) 428
5.4.6 Rice (*Oryza sativa*) 428
5.4.7 Sorghum (*Sorghum bicolor*) 429
5.4.8 Soybean (*Glycine max*) 429
5.4.9 Food Legumes 430
5.4.10 Adzuki Beans (*Vigna angularis*) 431
5.4.11 Strawberry (*Fragaria virginiana*) 431
5.4.12 Sugarcane (*Saccharum officinarum*) 432
5.4.13 Tobacco (*Nicotiana tabacum*) 433
5.4.14 Coffee (*Coffea arabica*) 433
5.4.15 Cucumber (*Cucumis sativa*) and Fig Leaf Gourd (*Cucurbita ficifolia*) 435

5.4.16 Zoysiagrass 435
5.5 Roles of Lipid Signaling in Stress Responses 436
5.5.1 Role of Lipid as Signaling Molecule During Stress Response 437
5.5.2 Role of Phosphatidic Acid in Signaling 438
5.5.3 Role of Free Fatty Acid in Signaling 439
5.5.4 Role of Sphingolipids in Signaling 439
5.5.5 The Signal Transduction of Membrane Lipids Under Cold Stress 440
5.6 Role of Lipid in Mitigating the Low-Temperature Stress 442
5.6.1 Effects of Cold Stress on Membrane Permeability 443
5.6.2 Effects of Cold Stress on Membrane Lipid Peroxidation 444
5.6.3 Effects of Cold Stress on Membrane Lipid Component 444
5.6.4 Effect of Cold Stress on Membrane Remodeling 445
5.6.5 Effects of Cold Stress on Membrane Lipid Unsaturation 446
5.7 Mechanism of Temperature Adaptation 447
5.7.1 *cis*-Unsaturated Fatty Acids and Membrane Fluidity . . . 450
5.7.2 Low-Temperature Effect on Lipids of Cyanobacteria 451
5.7.3 Low-Temperature Effects on Lipids of Higher Plants 452
5.7.4 Low-Temperature Stress in Field and Forage Production 453
5.7.5 How Low Temperature Is Perceived? 454
5.8 Cold Stress: Symptoms, Sensing, and Strategy 455
5.8.1 Lipid Profiling Under Cold Stress 458
5.8.2 Role of Unsaturated Fatty Acids to Low-Temperature Stress 459
5.8.3 Lipid Metabolism of Horticultural Products 460
5.8.4 Characterization of Cold Stress Responses 463
5.9 Metabolic Changes During Cold Stress and Cold Acclimation . . . 464
5.9.1 Role of Triacylglycerol in Plant Stress Response 468
5.9.2 Effect of Exogenous Lipids on Cold Acclimation and Fatty Composition 472
5.10 Lipid Biosynthesis Under Low-Temperature Stress in Plants 473
5.10.1 Temperature-Induced Lipid Pathway Adjustments in Plants 475
5.10.2 Glycerolipid Response to Cold and Freezing Temperature 478
5.10.3 Triacylglycerol Biosynthesis 479
5.11 Lipid Hydrolysis Under Low-Temperature Stress 481
References 484

6 Plant Growth Hormones in Plants under Low-Temperature Stress: A Review 517
6.1 Plant Growth Hormones 518
6.2 Classes of Plant Growth Hormones 521
6.2.1 Abscisic Acid 522
6.2.2 Auxins 523
6.2.3 Gibberellins 524
6.2.4 Cytokinins 525
6.2.5 Ethylene 526
6.2.6 Brassinosteroid 526
6.2.7 Jasmonates 527
6.2.8 Salicylic Acid 528
6.2.9 Strigolactones 528
6.2.10 Polyamines 529
6.3 Importance and Use of Plant Growth Hormones 530
6.3.1 In Vitro Plant Tissue Culture 534
6.4 Physiological Responses in Plants to Low Temperature 535
6.4.1 Classification of Plants under Low Temperature 541
6.5 Role of Different Growth Hormones on Growth and Yield 543
6.6 Effect of Low Temperature on the Activity of Plant Growth Hormone 549
6.6.1 Abscisic Acid and Low-Temperature Stress 550
6.6.2 Role of Auxins under Cold Stress 551
6.6.3 Brassinosteroids and Low-Temperature Stress 552
6.6.4 Salicylic Acid and Low Temperature 552
6.6.5 Gibberellic Acids and Low Temperature 554
6.6.6 Cytokinins and Low-Temperature Stress 555
6.6.7 Jasmonic Acid and Low-Temperature Stress 556
6.6.8 Thiourea and Low-Temperature Stress 556
6.7 Role of Growth Hormones in Plants under Low Temperature 557
6.7.1 Effect on Plant Architecture 558
6.8 Mechanism of Cold Tolerance and Role of Growth Hormones 559
6.9 Metabolic Effects of Growth Hormones for Abiotic Stress Tolerance 566
6.9.1 Abscisic Acid 567
6.9.2 Auxins 568
6.9.3 Cytokinins 568
6.9.4 Ethylene 570
6.9.5 Gibberellins 571
6.9.6 Brassinosteroids 572
6.9.7 Jasmonates 573
6.9.8 Salicylic Acid 574
6.9.9 Strigolactones 576
6.9.10 Polyamines 577

6.10 Strategies for Crop Improvement through Growth Hormones under Low-Temperature Stress . . . 578
6.10.1 Compatible Organic Solutes . . . 579
6.10.2 Plant Growth Regulators . . . 582
6.10.3 Exogenous Use of Organic Compounds . . . 583
6.10.4 Agronomic Management . . . 585
1-Naphthaleneacetic Acid . . . 586
References . . . 587

7 Effect of Low Temperature on Dry Matter, Partitioning, and Seed Yield: A Review . . . 629
7.1 Dry Matter Partitioning . . . 630
7.2 Dry Matter Partitioning and Yield in Cereals . . . 633
7.3 Dry Matter Partitioning and Yield in Grain Legumes . . . 635
7.4 Dry Matter Partitioning and Yield in Grain Oilseed Crops . . . 637
7.5 Carbon Partitioning and Acclimation to Environmental Stress . . . 640
7.6 Factors Affecting Dry Matter Partitioning . . . 641
7.6.1 Energy Investment in Seeds, Fruits, and Storage Organs . . . 642
7.6.2 Breeding . . . 643
7.6.3 Determinacy . . . 644
7.6.4 Effects of Pests and Diseases . . . 645
7.6.5 Temperature Extremes During Flowering and Grain Filling . . . 645
7.6.6 Pre-/Post-anthesis Water Use . . . 646
7.6.7 Water Use, Soil Mineral Nitrogen, and Harvest Index . . . 648
7.6.8 Hormonal Control of Photosynthate Partitioning . . . 649
7.6.9 Plant Growth and Carbon Partitioning . . . 650
7.6.10 Effects of Nitrogen Partitioning . . . 651
7.7 Exploring G × E Interaction for Increasing Harvest Index in Crops . . . 652
7.8 Management Techniques to Increase Harvest Index . . . 655
7.9 Carbon Partitioning in Different Plant Parts . . . 657
7.10 Temperature Effects on Dry Matter Partitioning . . . 659
7.10.1 Low temperature (Chilling) Stress . . . 661
7.10.2 Low Root Temperature and Dry Matter Production . . . 661
7.11 Dry Matter, Grain Yield, and Harvest Index Values for Crops . . . 663
7.11.1 Wheat . . . 664
7.11.2 Barley . . . 664
7.11.3 Sorghum . . . 664
7.11.4 Rice . . . 664
7.11.5 Maize . . . 665
7.11.6 Chickpea . . . 665
7.11.7 Field Pea . . . 665

7.11.8 Faba Bean 666
7.11.9 Lentil 666
7.11.10 Lupin 666
7.11.11 *Brassica* Crops 667
7.11.12 Sunflower 668
7.12 Seed Yield 668
7.13 Low Temperature and Seed Yield 670
7.14 Factors for Yield Gaps 671
7.15 Physiological Traits Associated with Improvement in Seed Yield 672
7.15.1 Time to Flower and Plant Height 673
7.15.2 Dry Matter Production and Its Partitioning 674
7.15.3 Main Yield Components 676
7.15.4 Associations Between Yield Attributes 677
7.15.5 Yield Potential Versus Stress Adaptation: GE Interaction 678
7.16 Changing Climate and Seed Yield 679
7.17 Effect of Temperature Stress on Seed Yield 685
7.18 Physiological Avenues for Increasing Yield Potential 688
7.18.1 Increasing Radiation-Use Efficiency 688
7.18.2 Biomass Increase 691
7.18.3 Phenological Adjustment 691
7.18.4 Plant Responses to Low Temperature 692
7.18.5 Reproductive Development Under Low Temperature 692
7.18.6 Yield Components and Yield Under Low Temperature 693
7.18.7 Traits for Tolerance to Abiotic Stress 693
7.18.8 Combination of Stress 694
7.19 Engineering for Cold Stress Tolerance in Crop Plants 695
7.19.1 Role of Proteomics in Crop Stress Management 697
7.19.2 Chilling Acclimation 699
References 700

About the Author

Amitav Bhattacharya Ex Principal Scientist (Plant Physiology), Indian Institute of Pulses Research (IIPR), Kanpur, India. He earned his master's degree in Botany with specialization in Plant Physiology. He got his PhD degree in Plant Physiology from Indian Agricultural Research Institute, New Delhi, India. He served IIPR for more than 35 years and retired as Principal Scientist, and during his service period he was actively engaged in physiological research on grain legumes both in field and in laboratory. He also looked after the coordinated physiological research being conducted on different grain legume crops at different centers, spread across India. He has already published several books.

Effect of Low-Temperature Stress on Germination, Growth, and Phenology of Plants: A Review

1

Abstract

Climatic models indicated that in this century, both average temperatures and extreme events frequency will increase in the Mediterranean basin. These changes are expected to have a great impact on agriculture in general and on crop phenology in particular. Great progress has been made in understanding the responses of plants to abiotic stress. There are inherent physical, morphological, and molecular limitations to the plant's ability to respond to stress. Plant responses to abiotic stress are dynamic and complex. Among the various abiotic stresses, low temperature adversely affects germination, normal plant growth, development, and phonological events. Consistent with the increasing temperatures, crop development is expected to be faster; thus, phenological stages will be reached early, and the length of the growth period of crops with determinate cycle (i.e., cereals, grapevine, etc.) will be shorter. These impacts, together with the higher risk to have extreme climate events during sensitive phonological phases, may have strong negative effects on final yield and on yield quality. The actual impact of phonological change needs to be assessed for specific crop environment combinations, providing the basis to formulate feasible adaptation options to climate change. In other terms, the simulated changes in phenology cannot be interpreted without considering the environmental context in which a species lives. For winter crops, the effect of predicted prolonged summer drought periods and heat weaves for the next decades may be smoothed or prevented due to the faster development that will allow escaping these and then avoiding reduction in final yields. In contrast, crops, whose growing cycle takes place in summer time, are likely to experience a severe reduction of final yield as the result of increased frequency of extreme climatic events and a reduced time for biomass accumulation to yield.

A. Bhattacharya, *Physiological Processes in Plants Under Low Temperature Stress*, https://doi.org/10.1007/978-981-16-9037-2_1

Keywords

Chilling and freezing injury · Seed germination and seedling growth under low temperature · Root development in low temperature · Growth and development under low temperature · Perinea crops and low temperature · Mechanism of low-temperature acclimation · Low-temperature and phonological development · Temperature and flower initiation · Molecular and genetic aspect of flowering during climate change

1.1 Low-Temperature Stress

Temperature is one of the major environment variables affecting the growth, development, and yields of crops, especially the rate of development (Luo 2011). Crops have basic requirements for temperature to complete a specific temperature phenophase or the whole life cycle. On the other hand, extremely high and low temperatures can have detrimental effects on crop growth, development, and yield, particularly at the critical phenophase. While some of the crop plants are morphologically indeterminate, the rate of many phenological processes such as germination, floral initiation, and development of fruiting bodies is controlled by temperature (Hearn and Constable 1984). Average daily temperature also plays an important role in determining the earliest date of sowing; defining season length, which can influence both yield potential and yield quality (Bange et al. 2008; Bauer et al. 2000; Dong et al. 2006a); and determining where a particular crop can be produced sustainably. Luo et al. (2014a), working with cotton phenology under temperature regimes and using general circulation model, pointed out that there will be less impact of cold temperatures on earlier growth and potentially a longer growing season that can improve crop yield. However, there will be more incidences of hot days impacting growth and more rapid crop development in late phonological stages.

Among various environmental stresses, low temperature is one of the most important factors limiting the productivity and distribution of plants. Low temperatures, defined as low but not freezing temperatures (0–15 °C), are common in nature and can damage many plant species. In order to cope with such conditions, several plant species have the ability to increase their degree of freezing tolerance in response to low, non-freezing temperatures, a phenomenon known as cold acclimation. It is well established that some of the molecular and physiological changes that occur during cold acclimation are important for plant cold tolerance (Zhu et al. 2007). Low temperature or cold stress is another major environmental factor that often affects plant growth and crop productivity and leads to substantial crop losses (Xin and Browse 2000; Sanghera et al. 2011). Chilling stress results from temperatures cool enough to produce injury without forming ice crystals in plant tissues, whereas freezing stress results in ice formation within plant tissues. Plants differ in their tolerance to chilling (0–15 °C) and freezing (<0 °C) temperatures. Both chilling and freezing stresses are together termed low temperature or cold stress: the damage due to cold stress can range from chilling injury and freezing injury to suffocation and heaving. In general, plants from temperate climatic regions are considered to be chilling-tolerant to variable degrees, and their freezing tolerance can be increased by exposing to cold, but non-freezing, temperatures; this process is known as cold acclimation. However, generally, the plants of tropical and subtropical origins are sensitive to chilling stress and lack this mechanism of cold acclimation (Sanghera et al. 2011). Low temperature may affect several aspects of crop growth, viz., survival, cell division, photosynthesis, water transport, growth, and finally crop yield.

Temperature is the main factor that determines the geographic distribution of organisms, in the context of both latitudinal and altitudinal gradients of thermal niche occupation (Hochachka and Somero 2002). The thermal range of the biosphere is broad and varies from +400 °C of the hydrothermal winds operating at the bottom of the oceans (Jupp and Schultz 2000) to −94.3 °C of the surface air of the Dome Argus, Antarctica (Scambos 2017). No organism has the capacity to withstand the full range of biosphere temperatures. Thus, life is limited to a much narrower temperature range, from −20 to +122 °C (Deming 2002; Fitter and Hay 2002). Specifically, the existence of plants is limited to an approximate thermal range of −10 to +60 °C, defined by the freezing point of intracellular water and the temperature of protein denaturation (Fitter and Hay 2002; Źróbek-Sokolnik 2012; Taiz and Zeiger 2015). This survival range can be exemplified by the woody trees of regions, such as Alaska, northern Canada, Europe, and Asia, which are highly adapted to intracellular water freezing (Taiz and Zeiger 2015), and by plants from arid regions such as cacti and agaves, which tolerate high diurnal temperatures that exceed 60 °C (Nobel 2003). The importance of temperature as a physical factor on the distribution of organisms is a consequence of its direct influence on molecular (DNA, proteins) or supramolecular (membranes, chromosomes) structures, which results from merely a thermodynamic effect (Ruelland and Zachowski 2010; Knight and Knight 2012). These changes are usually fast; therefore, changes in the ambient temperature can be quickly detected by cell organelles, triggering specific pathways of biochemical and molecular responses in each of these cell compartments and making up an integrated cell response to temperature changes (Ruelland and Zachowski 2010).

Minimum survival temperatures vary according to the natural environment, acclimation state, and growth form of the plant and, in many cases, may vary among different organs and tissues within the plant (Larcher 2003). Tissues of chilling-intolerant lowland tropical plants that never experience subfreezing temperatures can be killed at temperatures between 0 and 10 °C, while tissues that can survive these temperatures but not freezing temperatures are referred to as chilling-tolerant. Plants from regions with episodic or persistent seasonal temperatures below 0 °C are usually described using pairs of the words *frost*, *freezing*, or *cold* and *tolerant*, *hardy*, or *resistant*, with the terms *tolerance*, *hardiness*, and *resistance* used to describe the phenomenon of survival at low temperature. Low-temperature tolerance can be classified as follows:

- Moderate low temperature, −20 to –40 °C
- Intermediate low temperature, −40 to –60 °C
- Extreme low temperature, <–60 °C

More systematic exploration of low-temperature tolerance requires quantitative estimates of minimum survival temperatures for whole plants or plant tissues. These can be assessed by a wide variety of methods. Typically, whole plants or plant parts are exposed to a range of subfreezing temperatures in a temperature-controlled chamber, although some studies have applied low-temperature treatments to intact plants in the field (Taschler et al. 2004). Low-temperature stress results in various observable or measurable symptoms of injury, including death of whole plants, visible necrosis of specific tissues and organs, or less obvious cellular symptoms that can be detected by vital staining, osmotic responsiveness, chlorophyll fluorescence, or measurement of relative electrolyte leakage in affected tissues. The latter gives a useful measure of injury because a general symptom of cellular injury is a loss of semipermeability of the plasma membrane, which then results in the release of intracellular electrolytes (Steponkus 1984). These kinds of measurements are often used to determine a minimum survival temperature or construct temperature response curves and interpolate the temperature, resulting in 50% plant or tissue death, LT_{50}. Under natural conditions, trees may be subject to more complex environmental conditions than those imposed in laboratory tests, such as prolonged low temperature, repeated freezing and thawing, solar warming followed by rapid cooling, or light stress, so that laboratory estimates may not correspond to minimum survival temperatures in the field. In general, laboratory estimates of LT_{50} or minimum survival temperature are somewhat to well below the minimum temperatures encountered in the sampling location or natural range of the species in question. When different methods are directly compared, they often give generally similar estimates of LT_{50} (Burr et al. 1990), and LT_{50} values based on the same or similar methods can be compared among different tissues and species or track relative changes in low-temperature tolerance over time.

Low- and high-temperature stress are becoming the major researchable issues for agricultural scientists worldwide due to the challenges imposed by changing climate (Shah et al. 2011). Based on the ability to occupy thermal niches, organisms can be classified into psychrophiles, which live and reproduce at temperatures below +15 °C, some of which maintain metabolic activities at temperatures up to −20 °C; mesophiles, which live comfortably between +15 and +40 °C; and thermophiles, which have their best performance from +50 to +60 °C (moderate thermophiles). The term *hyperthermophiles* (or extreme thermophiles) has been used for organisms with optimal growth rates above +80 °C (Counts et al. 2017; Dalmaso et al. 2015; Pikuta et al. 2007). This classification is based mainly on studies of microorganisms that live in environments with extremely low and high temperatures (e.g., Antarctica and geothermal regions, respectively). Therefore, the thermal limits of each class may vary according to the temperatures that different groups of organisms can tolerate (Nievola et al. 2017).

Crops grown under open environments often pass through periods of abiotic stress during their life cycle. Such stresses many a time overlap, so crop growth and productivity is adversely affected. Plants pass through a series of morphological, physiological, biochemical, and molecular changes in a quest to mitigate such adversities of the abiotic stresses (Hussain et al. 2018). A lot of work reports the adaptive responses of crop plants to different abiotic stresses, wherein emphasis has been laid on individual stress factors (Jongdee et al. 2002; Chinnusamy et al. 2007; Nejad et al. 2010; Basu et al. 2016; Wang et al. 2016; Anjum et al. 2017). Crops experience periods of extreme low temperatures in many regions of the world (Ruelland et al. 2009; Wang et al. 2016) and are exposed to limited water availability owing to either drought or disturbed water movement and uptake under low ambient temperatures (Shinozaki et al. 2003; Zhang et al. 2004; Beck et al. 2007). Studies show that primarily chilling and drought pose a similar impact on stomatal development and leaf growth; nevertheless, the mechanisms of drought-induced changes in some physiological processes are quite different than those induced by chilling (Deng et al. 2012). Plants may exhibit common molecular and physiological responses on exposure chilling and drought (shared response); others may be specific to a given stress factor (Atkinson et al. 2013; Sewelam et al. 2014). In general, chilling stress thermodynamically declines the kinetics of several physiological as well as metabolic processes occurring in plants (Ruelland et al. 2009; Hussain et al. 2016). It severely reduces the rate and uniformity of germination, hampers seedling vigor, and delays ontogenetic plant development (Cruz and Milach 2004; Oliver et al. 2007), resulting in severe crop yield losses (Ruelland et al. 2009).

1.1.1 Chilling Injury

Chilling injury occurs when chilling-sensitive plant species are stored at temperatures above the freezing point of tissues but lower than 15 °C. Chilling-sensitive plants are those that are sensitive to chilling and suffer damage when exposed to it. Chilling temperatures effects on plants in temperate climates lead to a reduction or complete crop failure due to either direct damage or delayed maturation (Lukatkin et al. 2012). Even a small drop in temperature can cause visible damage to chilling-sensitive plants. The most noticeable visual symptoms of chilling injury in herbaceous plants are leaf and hypocotyl wilting (Frenkel and Erez 1996), which often precedes the appearance of infiltration (water-saturated areas) (McMahon et al. 1994); the appearance of surface pits and large cavities (Cabrera et al. 1992; Frenkel and Erez 1996); discoloration of leaves and internal tissues (Yoshida et al. 1996a; Tsuda et al. 2003); accelerated aging and rupture of chilled tissues; slow, incomplete, or uneven ripening (Dodds and Ludford 1990); deterioration of the structure and flavor (Ventura and Mendlinger 1999); increased susceptibility to decay (Cabrera et al. 1992); and, in the case of prolonged chilling, leaf necrosis and plant death (Frenkel and Erez 1996). Potential symptoms of chilling injury are surface lesions; water-soaking of tissues; water loss; desiccation or shrivelling; internal discoloration; tissue breakdown; failure of fruit to ripen, or uneven or slow ripening; accelerated senescence and ethylene production; shortened storage or shelf life; compositional changes; loss of growth or sprouting capability; and wilting and increased decay due to leakage of plant metabolites, which encourage growth of microorganisms, especially fungi (Lukatkin et al. 2012). Seeds of chilling-sensitive plants do not germinate at temperatures below 10–15 °C (Ismail et al. 1997) and by this parameter can be divided into two main groups (Markowski 1988). The seeds of the first group (representatives, Solanaceae and pumpkin) are not damaged during imbibitions at chilling temperatures. With temperature increase they grow normally, but initiation of root growth leads to underdeveloped root tip tissue, tissue necrosis after the root tip, and damage to the cortex or stele (Jennings and Saltveit 1994). The second group includes plants whose seeds are particularly sensitive to low temperatures during imbibitions and may not germinate at low temperatures: beans, soybeans, chickpeas, corn, and cotton (Zemetra and Cuany 1991). A characteristic effect of chilling temperatures on chilling-sensitive plants is growth slowing, more pronounced in susceptible species and varieties in comparison with the tolerant species (Venema et al. 1999a). In addition, there is a delayed development and lengthening of the growing season (Skrudlik and Kościelniak 1996). At the same time, flowering rate and fruit and seed filling are reduced (Skrudlik and Kościelniak 1996; Lejeune and Bernier 1996).

Chilling injuries occur above the freezing point (0 °C), and plants of tropical and subtropical origin are most susceptible. Injured foliage appears purple or reddish and sometimes wilted. Chilling injury can be obvious or invisible. Chilling can delay crop blooming, cause direct damage, or reduce plant vigor. Chilling injury happens often in tropical and subtropical plants grown in subtropical areas but can happen in native, temperate forest plants as well, depending on critical temperatures, duration of low temperature, temperature changes, age, hydration status of plants, and time of year. Plants can drop damaged leaves, become wilted, produce misshapen new growth, display discolored foliage, or even have whole or portions of the plants dying. The damage may also be unseen by the naked eye but manifest later as delayed blooming or stunted growth.

1.1.2 Freezing and Frost Injury

Freezing and frost injuries occur at or below the freezing point (0 °C). Freeze damage happens during an advection freeze when an air mass with temperatures below freezing moves into an area and displaces warmer air, causing the temperature of plants to become low enough for ice crystals to form inside their cells. Frost damage happens during a radiation freeze. This occurs on clear nights without wind when plants radiate more heat to the atmosphere than they received. This creates a temperature inversion where cold air close to the ground gets trapped by moist, warmer air above it. When the air temperature at ground level nears or drops below freezing, the plant's temperature becomes colder than the surrounding air temperature, causing ice crystals to form on the surface of the foliage, stems, and flowers. During both frost and freeze injury, ice crystals form in plant tissues, dehydrating cells and disrupting/rupturing plant membranes. This causes physical damage as expanding rigid ice crystals puncture cellular membranes and dehydration damage as all water within plant tissues freezes, rendering it inaccessible for photosynthesis. Once physical damage has occurred, it manifests itself as brown to yellow necrosis, often leading to portions of the plant or the entire plant dying. Cold hardy species are less susceptible to freezing and frost injury.

1.1.3 Physiological and Molecular Changes Under Low Temperature

Plants respond and survive under stress conditions by bringing changes at the molecular and cellular levels as well as at the biochemical and physiological levels (Xin and Browse 2000). Low-temperature stress inhibits seedling establishment, effecting early growth stages of rice and resulting in poor crop maturation. In order to gain stable rice production, cold tolerance at the seedling stage is an important character. One of the most effective ways to avoid the low-temperature damage is to develop cold-tolerant genotype (Lou et al. 2007). Mineral nutrition acquisition and assimilation are strongly influenced by both high- and low-temperature stress in plants (Rivero et al. 2006). Some essential nutrients such as nitrogen, sulfur, phosphorus, magnesium, and calcium are structurally important for the proteins, nucleic acids, chlorophylls, certain secondary metabolites, and defense-related micro- and macromolecules, while others have both structural and functional roles (Epstein and Bloom 2005; Taiz and Zeiger 2006).

As a result of exposure to low temperatures, many physiological and biochemical cell functions have been correlated with visible symptoms (wilting, chlorosis, or necrosis). Often these adverse effects are accompanied by changes in cell membrane structure and lipid composition (Matteucci et al. 2011), cellular leakage of electrolytes and amino acids, a diversion of electron flow to alternate pathways (Seo et al. 2010), alterations in protoplasmic streaming, and redistribution of intracellular calcium ions (Knight et al. 1998). It also involves changes in protein content and enzyme activities (Ruelland and Zachowski 2010) as well as ultrastructural changes in a wide range of cell components, including plastids, thylakoid membranes and the phosphorylation of thylakoid proteins, and mitochondria (Zhang et al. 2011a, b). Brief exposures to low temperatures may only cause transitory changes, and plants generally survive. However, prolonged exposure to stress causes plant necrosis or death. To overcome stresses generated by exposure to low non-freezing temperatures, plants can trigger a cascade of events that cause changes in gene expression and thus induce biochemical and physiological modifications that enhance their tolerance (Zhu et al. 2007). This phenomenon is known as chilling or cold acclimation.

In addition to reduced plant growth and productivity (Tommasini et al. 2008), the nutrient uptake behavior of plants is also impaired under chilling temperature (Xu and Huang 2006) and soil water-deficit stress (Hu et al. 2007). Temperature affects the physiochemical and microbial processes in soils, which may modify the plant-nutrient relationships (Yan et al. 2012). Poor root system of plants under chilling stress reduces the uptake of several nutrients, including nitrogen, phosphorus, and potassium (Domisch et al. 2002; Yan et al. 2012). Under chilling stress, shoot growth of maize seedlings was reduced by the direct effect of low temperature on shoot meristems and by restricted supply of nutrients via roots (Hund et al. 2007). Farooq et al. (2009b) concluded that reduced root length, low hydraulic conductance, poor root branching, and thicker root axis under chilling stress lead to reduced mineral nutrient uptake in plants. However, the variations in such effects may arise based on plant species, stress period, physiological plant growth stage, or the frequency of changing the nutrient solutions.

Low temperatures have an effect on mineral nutrition of plants. Absorption of ions by roots is difficult, as well as their movement in the aboveground parts of plants. The distribution of nutrients between the plant organs is disrupted, with general decrease in nutrient content in the plant. Chilling of plants leads to a decrease in the activity of nitrate reductase, reduction in the nitrogen incorporation in the amino acids and proteins, and a drop in the proportion of organic phosphorus and an increase in inorganic phosphorus content (Zia et al. 1994), which is a consequence of a breach of phosphorylation and enhanced decomposition of organic phosphorus compounds. Mechanisms to reduce the absorption of nutrients by chilling temperatures include depression of respiration and/or oxidative phosphorylation, impaired enzymatic transport systems associated with conformational protein changes in membranes, changes in membrane potential, reduction of supply of ATP to H^+-transporting ATPase, as well as lowering the permeability coefficients for ions.

Chilling injury is thought to be caused by loss of membrane fluidity, detected as membrane phase changes below the critical temperatures in sensitive taxa. Biochemical consequences include altered reaction rate of membrane-bound enzymes, ion leakage, and loss of compartmentalization. The resulting metabolic imbalances lead to cellular injury. Chilling may produce lesions in the thylakoid membranes of the leaf chloroplasts and result in lowered photosynthesis. Most tropical plants are susceptible to chilling injury, whereas temperate zone plants are less often affected. The economic losses caused by chilling injury to temperate-zone horticultural crops are difficult to estimate, but are not thought to be great. Fruit of some horticultural crops, e.g., apples, tomatoes, and cucumber, suffer breakdown disorders in or following low-temperature storage that are attributed to chilling injury. Some vegetable transplants are also believed to be prone to chilling injury at low temperature if not properly hardened when transplanted from the greenhouse. Freezing injury, which affects survival and production of many plants, is of great economic importance to temperate zone horticulture and will be dealt with exclusively in the remainder of this review. Various types of freezing injury result from untimely frosts in the spring and autumn or from extremely low temperatures in midwinter. Some susceptible plants, such as corn and beans, are killed the moment they freeze (1–4 °C), whereas some hardy woody perennials can tolerate temperatures as low as –60 °C when they are fully acclimated. Most horticultural crops are damaged at temperatures between these extremes. The hardiness level varies with taxa, plant organ, and season.

The major adverse effect of cold stress in plants has been seen in terms of plasma membrane damage. This has been documented due to cold stress-induced dehydration (Steponkus et al. 1993). The plasma membrane is made up of lipids and proteins. Lipids in the plasma membrane are made up of two kinds of fatty acids: unsaturated and saturated fatty acids. Unsaturated fatty acids have one or more double bonds between two carbon atoms, whereas saturated fatty acids are fully saturated with hydrogen atoms. Lipids containing more saturated fatty acids solidify faster and at temperatures higher than those containing unsaturated fatty acids. Therefore, the relative proportion of these two types of fatty acids in the lipids of the plasma membrane determines the fluidity of the membrane (Steponkus et al. 1993). At the transition temperature, a membrane changes from a semi-fluid state into a semicrystalline state. Cold-sensitive plants usually have a higher proportion of saturated fatty acids in their plasma membrane. Therefore, cold-sensitive plants have a higher transition temperature. On the contrary, cold-resistant plants have a higher proportion of unsaturated fatty acids and hence a lower transition temperature. The agricultural crops which can withstand even during the freezing temperatures of late spring or early fall frost can be used more successfully for cultivation during cold stress. Therefore, selection of low-temperature-tolerant crops is very important for the sustainability of agriculture.

It has been noticed that cold-induced ice formation is the real cause of plant damage. Ice formation in plant tissues during cold stress leads to dehydration. Ice is formed in the apoplastic space of a plant tissue because that has relatively lower solute concentration. It is known that the vapor pressure of ice is much lower than water at any given temperature. Therefore, ice formation in the apoplast establishes a vapor pressure gradient between the apoplast and surrounding cells. As a result of this gradient, the cytoplasmic water migrates down the gradient from the cell cytosol to the apoplastic space. This adds to existing ice crystals in the apoplastic space and causes a mechanical strain on the cell wall and plasma membrane, leading to cell rupture (Olien and Smith 1977; Uemura and Steponkus 1997). In addition to the well-established harmful effect of cold stress alterations in lipid composition of the biomembranes, affecting their fluidity (Welti et al. 2002), certain additional factors may also contribute to damage induced by cold stress. This includes synthesis and accumulation of compatible solutes, synthesis of cold acclimation-induced proteins (Shinozaki and Yamaguchi-Shinozaki 2000), changes in the carbohydrate metabolism (Frankow-Lindberg 2001), and the boosting of the radical scavenging potential of the cells (Hernández-Nistal et al. 2002; Baek and Skinner 2003). Taken together, cold stress results in loss of membrane integrity, leading to solute leakage. Further, cold stress disrupts the integrity of intracellular organelles, leading to the loss of compartmentalization. Exposure of plants to cold stress also causes reduction and impairing of photosynthesis, protein assembly, and general metabolic processes. Recently, attempts have also been directed toward analyzing the effect of cold stress on the whole plant metabolome. Metabolic profiling of *Arabidopsis* plants revealed that cold acclimation increases 75% of the 434 analyzed metabolites (Cook et al.

2004; Kaplan et al. 2004). However, metabolite profiles in *Arabidopsis* do not appear to correlate with cold acclimation capacity (Hannah et al. 2006). The role of such metabolites in plants has been known as osmoprotectants. In addition to their role as osmoprotectants and osmolytes, certain metabolites induced during cold acclimation might act as signals for reconfiguring gene expression (Yadav 2010). For example, cold stress induces the accumulation of proline, a well-known osmoprotectant. Cold stress affects virtually all aspects of cellular function in plants. One of the major influences of cold stress-induced dehydration is membrane disintegration. Such changes caused by cold stress adversely affect the growth and development of plants.

Oxidative stress is crucial in relation to chilling-induced injuries in plants (Srivalli et al. 2003; Hussain et al. 2016). Chilling stress exacerbates reactive oxygen species production in plants' cell. Excess production and the accumulation of reactive oxygen species causes oxidative damage at cellular level, disrupts cellular membranes, and leads to enzyme inactivation, protein degradation, and ionic imbalance in plants (Baier et al. 2005; Tarchoune et al. 2010). The reactive oxygen species disturb the cellular macromolecules, including DNA, and hence may cause deletion of bases due to alkylation and oxidation, which are linked with various physiological and biochemical disorders in plants (Tuteja and Tuteja 2001). Plants possess a highly efficient and sophisticated antioxidative defense system to control the overproduction of reactive oxygen species (Hussain et al. 2016). The reactive oxygen species-induced damages and disruption of cellular homeostasis are alleviated by the action of different enzymatic (e.g., catalase, superoxide dismutase, peroxidase, glutathione reductase, glutathione peroxidase) and nonenzymatic (e.g., ascorbic acid, carotenoids, α-tocopherol, and glutathione content) antioxidants (Gill and Tuteja 2010; Chen et al. 2016; Hussain et al. 2016). In plants, the levels of reactive oxygen species are regulated by their production rate as well as the extent of their detoxification by enzymatic and/or nonenzymatic antioxidant systems. Mechanism of reactive oxygen species production and its scavenging by high antioxidative capacity has been associated with tolerance of plants to abiotic stresses (Gill and Tuteja 2010).

1.2 Low Temperature and Seed Germination

Germination is the process by which a plant grows from a seed. Germination can be thought of in a general sense as anything expanding into greater being from a small existence or germ, a method that is commonly used by many seed germination projects. Germination is the growth of a plant contained within a seed; it results in the formation of the seedling. The seed of a vascular plant is a small package produced in a fruit or cone after the union of male and female sex cells. All fully developed seeds contain an embryo and, in most plant species, some store of food reserves, wrapped in a seed coat. Some plants produce varying numbers of seeds that lack embryos; these are called empty seeds, which never germinate. Dormant seeds are ripe seeds that do not germinate because they are subject to external environmental conditions that prevent the initiation of metabolic processes and cell growth.

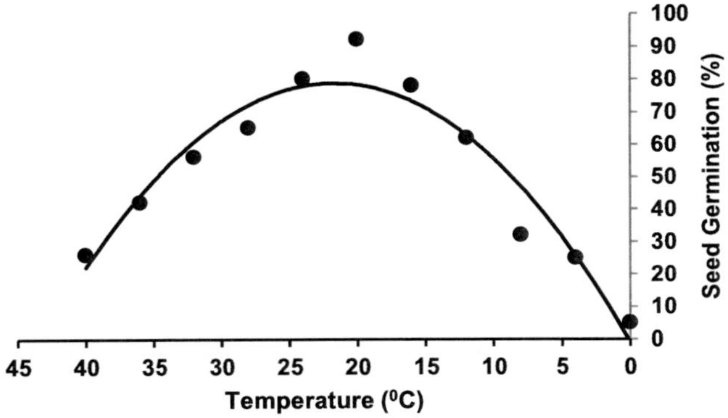

Fig. 1.1 Effect of temperature on seed germination

Since seed germination is one of the most critical stages in the life cycle of plants, the differences in germination time have important effects on plant growth, competitiveness, and reproduction, particularly for annuals in arid areas (Kos and Poschlod 2010; Akiyama and Agren 2013; Lu et al. 2016).

Low temperature is one of the main abiotic stresses affecting the growth, development, and spatial distribution of plants (Sun and Huang 2011) and tends to reduce and delay the germination rate of seeds and even causes germination failure (Wang et al. 2008). However, most desert plants such as *Salsola affinis* (Wei et al. 2007; Luo et al. 2014b), *Salsola korshinskyi* (Li et al. 2012), *Lepidium perfoliatum* (Yang et al. 2015), *Haloxylon ammodendron* (Tian 2014), *Anabasis elatior* (Han et al. 2011), and *Anabasis aphylla* (Peng et al. 2018) are highly adaptable to cold environments and can thus germinate under such conditions. Seed germination is sensitive to environmental condition. Seed germination can occur only in response to specific combinations of environmental cues present in the field such as temperature regime, precipitation, or light (Donohue et al. 2010), although after dormancy is fully broken, seeds of many species can germinate over a wide range of conditions (Yi et al. 2019). Temperature requirement for dormancy break and germination have been the focus of much seed ecology research (Baskin and Baskin 2014; Finch-Savage and Leubner-Metzger 2006). Under proper conditions, the seed begins to germinate, and the embryonic tissues resume growth, developing toward a seedling. Seed germination depends on both internal and external conditions. The most important external factors include temperature, water, oxygen, and sometimes light or darkness (Raven et al. 2005). Various plants require different variables for successful seed germination. Often this depends on the individual seed variety and is closely linked to the ecological conditions of a plant's natural habitat. For some seeds, their future germination response is affected by environmental conditions during seed formation; most often these responses are types of seed dormancy.

Cold stress generally results in poor germination (Fig. 1.1), stunted seedlings, yellowing of leaves, withering, and reduced tillering. The effects of cold stress at the reproductive stage of plants delay heading and result in pollen sterility, which is

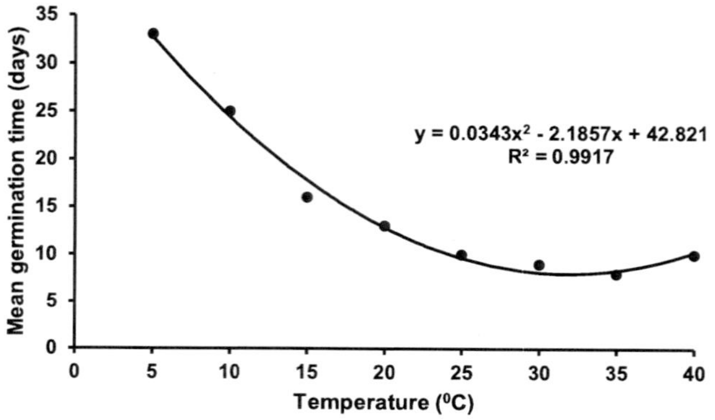

Fig. 1.2 Effect of temperature on the time of germination

thought to be one of the key factors responsible for the reduction in grain yield of crops (Suzuki et al. 2008). Appropriate time for sowing wheat will change in accordance to environmental as well as soil temperature (Rashed Mohassel and Koochaki 2000). Wheat seeds germinate well in temperatures above 4 °C and budding occurs sooner. Minimum temperature for the bud production was about 2 °C and near 0 °C in mid-winter wheat, with the best and most suitable temperatures ranging from 8 to 10 °C and the maximum temperature ranging from 20 to 22 °C (Khodabandeh 2003). Germination stage is sensitive to soil temperature because the absorption seed needs water by enzymatic activity or is breathing (Voorhees et al. 1981). Germination and growth before emergence normally is controlled by soil temperature (Hegarty 1973). According to Macduff and Wild (1986), 3–9 °C is the optimum temperature for turnip root growth in terms of length and number. Barley root length increased with increasing temperature in the thermal range of 3–25 °C after 20 days of germination (Zabihi-e-Mahmoodabad et al. 2011). Abbasal-Ani and Hay (1983) reported that root systems of seedlings of barley, oats, rye, and wheat are shorter at low temperatures (5 °C) than at high temperatures (15 and 25 °C) and the rate of growth is also fast. Davidson (1969) has reported that temperature has an effect on the growth allocation between shoots and roots. Chilling injury is a serious problem during germination and early seedling growth in many plant species. The optimum temperature range for germination of rice seed lies between 20 and 35 °C, with 10 °C being the minimum critical value below which rice does not germinate (Yoshida 1981). There are many reports on positive correlation between germination at low temperature and root development at an early stage and between the germination and seedling establishment (Nahar et al. 2012). Angadi et al. (2000a, b) observed that the number of days to 50% germination in *Brassica napus* was only 3 days at 8 °C, which was nearly 13 days at 2 °C. It has been demonstrated that under low-temperature regimes, the rate of seed germination is low, or, in other words, the number of days taken for seed germination is higher under low temperature as compared to normal temperature (Fig. 1.2). There have been reports about the low rate of germination. This low-temperature effect was more pronounced in *Brassica rapa*, because at 2 °C, emergence was less than 50%, even after 20 days of sowing (Angadi et al. 2000a, b). Buriro et al. (2011) reported that the increase in temperature significantly enhanced germination and related traits in wheat cultivars. All the wheat varieties germinated well (80–97%) sown at 10–30 °C. The maximum seed germination vigor index occurred at 20–30 °C, and these temperature regimes were identified as optimum for wheat seed germination. The delay in germination percentage and the reduced germination percentages were observed in *Gossypium hirsutum* at low temperature below 20 °C (Krzyzanowski and Delouche 2011). In *Triticum aestivum*, the germination is drastically hampered at temperature below 8–10 °C (Zabihi-e-Mahmoodabad et al. 2011). The optimum germination temperature differs by species. Seed germination of *Hosta sieboldiana* occurs over a range of temperature, from 10 to 30 °C, but the germination rate is most rapid at 25–30 °C (Kanazawa et al. 2015). The optimum germination temperature of *Hosta plantaginea* is 25 °C, but the germination percentage is very low (<25%, Oh et al. 2003).

Rice seed germination and some physiological parameters during seed germination are affected by low temperature (Xiaochuang et al. 2017; Kuk et al. 2003; Wang et al. 2018). Low temperature (19–20 °C) results in a reduction in photosynthetic parameters and affects antioxidant enzyme activity and membrane lipid peroxidation (Rao et al. 2019). A previous study suggested that the difference in high air temperature and soil temperature could affect carbohydrate metabolism in plants (Xu and Huang 2000). Increasing the root-zone temperature has a positive effect on plant development (Beauchamp and Lathwell 1967). Root growth is affected by various soil temperatures (McMichael and Burke 1998), and root temperature significantly affects root growth and function (Arai-Sanoh et al. 2010). Additionally, soil temperature significantly affects root vigor, root proline content, and malondialdehyde concentration (Bai et al. 2017). Soil temperature also affects photosynthesis (Wu et al. 2012). Thus, high or low temperature during the day or at night in the air or in the soil as well as high- or low-temperature irrigated water for rice plants could result in different growth responses in rice plants (Li et al. 2019).

In chickpea (*Cicer arietinum* L.), cold temperature (<15 °C) at emergence reduces seed germination with low vigor. In sensitive genotypes, cold temperature causes whole plant necrosis and plant death (Devasirvatham et al. 2015). Genotypic differences in chickpea was identified in Australia, and field screening for cold tolerance during the late vegetative stage was done by exposing them to −7.4 °C for 3 weeks (Malhotra and Singh 1991), and cold-tolerant genotypes were identified, which showed improved establishment under low temperatures (Wery et al. 1994). Seeds from different species and even seeds from the same plant germinate over a wide range of temperatures. Seeds often have temperature range within which they will germinate, and they will not do so above or below this range. Many seeds germinate at temperatures slightly above 16–24 °C, while others germinate just above freezing, and still others germinate only in response to alternations in temperature between warm and cool. Some seeds germinate when the soil is cool (between −2 and 4 °C), and some when the soil is warm (24–32 °C). Some seeds require exposure to cold temperatures (vernalization) to break dormancy. Some seeds in a dormant state will not germinate even if conditions are favorable. Seeds that are dependent on temperature to end dormancy have a type of physiological dormancy. For example, seeds requiring the cold of winter are inhibited from germinating until they take in water in the fall and experience cooler temperatures. Four degrees Celsius is cool enough to end dormancy for most cool dormant seeds, but some groups, especially within the family Ranunculaceae and others, need conditions cooler than –5 °C. Some seeds will only germinate after hot temperatures during a forest fire, which cracks their seed coats; this is a type of physical dormancy.

During the early germination process of soybean, low temperatures can significantly reduce the rate of imbibition, the ability of embryo tissue to expand, and mitochondrial respiration (Vertucci and Leopold 1983). The rate of hypocotyl elongation significantly decreases with decreasing temperature below 30 °C (Hatfield and Egli 1974). Interestingly, the aftereffects of low temperatures during the seedling stage can substantially extend the vegetative growth rate and increase the number of axillary branches, the rate of dry weight per plant, and pod setting (Skrudlik and Kościelniak 1996), whereas the effects of high temperatures are mostly studied and considered damaging on the reproductive growth and yield potential in soybean, especially under the conventional soybean production system of the United States (Ashlock et al. 2000; Tacarindua et al. 2013; Hatfield and Prueger 2015). Many argue that the success of the early soybean production system of the United States was due to continuously increasing global air temperatures over the years (Tacarindua et al. 2013; Egli and Cornelius 2009; Thuzar et al. 2010), and they emphasize the importance of determining heat/cold tolerance among the available soybean cultivars during early growth stages (Alsajri et al. 2019).

Temperature and photoperiod predominantly affect morphological and physiological growth and development of soybean plant among other environmental variables (Setiyono et al. 2007). While the phenological response to temperature can primarily determine soybean variety selection for cultivation in a given geographical location during early growth stages with little interaction of photoperiod (Setiyono et al. 2007), photoperiod modifies the response to temperature with changing geographical locations and therefore serves as a basis for classifying the cultivars by maturity group (Heatherly and Elmore 2004). Studies in the past have determined genotypic variability in phenological responses to temperatures for the traits such as germination, plant height, node number, net photosynthesis, leaf area, and fruit number per plant by either varying the planting date in the field (Heatherly 2005; Salmerón et al. 2014, 2015; Bruns 2011; Grichar and Biles 2014) or utilizing controlled environment facilities (Tacarindua et al. 2013; Singh et al. 2017a, b; Cober et al. 2014; Kurosaki and Yumoto 2003; Ewuebker et al. 2001). Therefore, soybean cultivars' tolerance to low temperatures within or across maturity groups at constant photoperiods can be best achieved by utilizing controlled environment facilities.

Most common annual vegetables have optimal germination temperatures between 24 and 32 °C, though many species (radishes or spinach) can germinate at significantly lower temperatures, as low as 4 °C, thus allowing them to be grown from seeds in cooler climates. Suboptimal temperatures lead to lower success rates and longer germination periods. Some field crops are extremely sensitive to chilling and drought, particularly during germination and early phases of seedling development. Each seed requires optimum temperature and soil moisture for germination. Chilling stress severely impairs the germination and often lowers the seedling vigor (Kang and Saltveit 2002; Wang et al. 2016), delays plant development, and ultimately causes severe yield losses (Cruz and Milach 2004; Oliver et al. 2007; Ruelland et al. 2009). Chilling stress is known to thermodynamically limit the kinetics of various physiological as well as metabolic functions in plants (Ruelland et al. 2009). Chilling stress evokes shoot water deficit due to its thermophilic nature and reduced root water uptake (Stewart et al. 1990).

Although germination responses to climate has been investigated at population and community levels (Cao et al. 2018; Cavieres and Arroyo 2000; Giménez-Benavides et al. 2005; Liu et al. 2011, 2018; Shimono and Kudo 2005; Wagner and Simons 2009), knowledge of differential seed germination responses within and between species in a community is important in predicting community composition under future climatic conditions. Plants have evolved strategies involving both predictive germination and optimization of their fitness such that seeds germinate in the current environment and others remain dormant and thus hedge their bet against unpredictable conditions that are not favorable for seedling establishment (Gremer et al. 2016).

1.2.1 Seed Priming and Germination Under Low Temperature

Chilling stress is a critical factor that determines the geographical distribution of many field crop species and their productivity (Hussain et al. 2016). Chilling stress during seed germination causes severe problems around the world; moreover, due to global climate change, increased frequencies of extreme weather events, such as unexpected and sudden chilling events occurring after seed sowing, often have negative effects on maize seed germination and shoot growth (Cao et al. 2019). Chilling stress thermodynamically constrains the kinetics of many physiological and metabolic processes in plants. It has been widely reported that the deleterious effects of chilling stress on crops are caused by the induction of oxidative stress. Seed priming as a pre-sowing seed treatment has been proven to be an approach for achieving rapid and uniform emergence effectively, practically, and easily as well as for improving seed vigor and seedling viability under unfavorable environmental conditions (Marta et al. 2016; Espanany et al. 2016).

With the development of intensive agriculture worldwide, transplant cultivation has become a limiting factor. Direct seeded cultivation is an alternative to transplanting. Seed priming is a useful treatment, applied prior to planting, which partially hydrates the seeds to a point of germination process initiation, followed by drying, which prevents radicle emergence. Seed can be primed by either uncontrolled hydration, hydropriming (Casenave and Toselli 2007; Ghassemi-Golezani et al. 2010), or controlled hydration methods (Chen and Arora 2013; Ibrahim 2016), which include osmotic priming, solid matrix priming, and hormonal priming (Basra et al. 2006; Foti et al. 2008). Seed priming improves seed performance with a faster and synchronized germination under optimal and suboptimal conditions (Sivritepe et al. 2003). Partial hydration is sufficient for the occurrence of numerous physiological processes typical of the first stages of imbibition (pre-germinative metabolism) (Paparella et al. 2015).

There are different priming methods depending on the way in which seed hydration is controlled; the most common are pure water and re-drying to original moisture content (hydropriming), osmotic agents that limit water uptake (osmopriming), and a solid matrix (matric priming) (Ibrahim 2016). Although hydropriming is a simple method, the main disadvantage is the difficulty of limiting the abrupt absorption of water (Debbarma and Das 2017; Shukla et al. 2018). Osmopriming, which involves the use of osmotic agents or salts, is an easy, cheap, and effective method (Ibrahim 2016). Osmopriming involves the use of adverse osmotic solutions, like polyethylene glycol, or different salts such as chlorides, sulfates, nitrates, etc. The effects of seed priming with potassium nitrate have been previously studied in different plant species (Amjad et al. 2007; Shim et al. 2008; Eskandari and Kazemi 2011). Primed seeds show increased germination rate and seedling establishment at suboptimal conditions (Jafar et al. 2012). The significance of various seed priming strategies at low temperature has previously been studied. For example, polyethylene glycol osmopriming improved emergence of tomato seeds (Amooaghaie et al. 2010), seed priming with potassium nitrate and sodium chloride decreased time to 50% germination and increased final germination of sorghum (Tiryaki and Buyukcingil 2009), and priming with sodium selenite enhanced bitter gourd seed germination under suboptimal temperature conditions (Chen and Sung 2001). It had been shown that treatment with 0.5% potassium nitrate solution had the most improving effect. This treatment reduced mean germination time and time for 50% germination and increased germination energy at low and mixed temperature. Priming showed no positive effect on the final germination, and it was concluded that seed priming with potassium nitrate could be used to achieve better seedling establishment at low-temperature conditions in maize (Čanak et al. 2016).

In order to maintain biodiversity and restoration of tropical and temperate forests with native species, it is necessary to assure more than 90% seed germination under natural conditions (Bozzano et al. 2014; Montagnini and Finney 2011; Rogers and Montalvo 2004). Nevertheless, the success of these projects relies heavily on the proper selection of native species. Restoration practices based on native species are urgently required to face the effects of deforestation and climate change (Peraza-Villarreal et al. 2020). Performance of seeds and seedlings can be increased by applying pretreatment to seeds and seedlings before transplanting them to the field (Benítez-Rodríguez et al. 2014; Singh and Chaturvedi 2018). Seed priming (hydration followed by dehydration) has been successfully implemented to increase seed and seedling performance of agricultural plants and also for some native plants (Benech-Arnold and Sánchez 2004; Pedrero-López et al. 2016; Sánchez et al. 2001, 2006; Schmidt 2000). The priming treatments involve two steps: (a) seed hydration in water (hydropriming), osmotic solutions of NaCl (osmopriming), or other solutes, in a solid matrix such as vermiculite (matrix priming) or in the soil (natural priming; Benítez-Rodríguez et al. 2014; Gamboa de Buen et al. 2006; González-Zertuche et al. 2001), and (b) seed dehydration, in which all priming treatments ought to end before radicle protrusion; thus, after hydration, seeds are dehydrated up to their initial water content. During hydration, important advances in the germination process take place, such as mobilization of reserves and the repair of DNA, RNA, cell membranes, organelles, and others (Bray 1995). Seeds maintain these changes during storage, and, consequently, primed seeds germinate fast and synchronically. Hydropriming and natural priming have been applied successfully to orthodox seeds (long-lived; Orozco-Segovia et al. 2014). It has been demonstrated that priming treatments can be an option to improve restoration practices (Benítez-Rodríguez et al. 2014; Orozco-Segovia et al. 2014; Pedrero-López et al. 2016; Singh and Chaturvedi 2018).

The effects of incorporating salicylic acid into a potassium nitrate priming solution on the germination and emergence of eggplant seeds at 15 °C were investigated (Zhang et al. 2011a, b). Seeds were primed into 3% potassium nitrate containing 0.05, 0.1, 0.5, or 1 mM salicylic acid for 6 days, respectively. After priming, seeds were either immediately used for germination and emergence test at 15 °C or stored at 4 °C for 1 month and then for the germination test. The primed eggplant seeds in general improved the final germination percentage, germination rate, and germination synchrony at 15 °C compared with non-priming seeds. Priming seeds in 3% solution of potassium nitrate supplement with 0.1 mM salicylic acid resulted in the best priming effect compared with other priming treatments and non-priming treatment, and all of the beneficial priming effects remained after being stored at 4 °C for 30 days. These results indicated that priming seeds in 3% potassium nitrate solution containing 0.1 mM salicylic acid could be used as an effective method to improve low-temperature performance of eggplant seeds and subsequent seedling growth.

Low-temperature stress during germination and early seedling growth is an important constraint of crop production. The effects of seed priming with 0.25%, 0.50%, and 0.75% (w/v) chitosan (linear polysaccharide composed of randomly distributed β-(1→4)-linked D-glucosamine and *N*-acetyl-D-glucosamine) solutions at 15 °C on growth and physiological changes were investigated using two maize (*Zea mays* L.) inbred lines, chilling-tolerant and chilling-sensitive. While seed priming with chitosan had no significant effect on germination percentage under low-temperature stress, it enhanced germination index, reduced the mean germination time, and increased shoot height, root length, and shoot and root dry weights in both maize lines. The decline of malondialdehyde content and relative permeability of the plasma membrane and the increase of the concentrations of soluble sugars and proline, peroxidase activity, and catalase activity were detected both in the chilling-sensitive and chilling-tolerant maize seedlings after priming with the three concentrations of chitosan. Priming with 0.50% chitosan for about 60, approximately 64, hours seemed to have the best effects. Thus, it suggests that seed priming with chitosan may improve the speed of germination of maize seed and benefit for seedling growth under low-temperature stress (Guan et al. 2009).

Melatonin (*N*-acetyl-5-methoxytryptamine), an indoleamine, is a highly conserved molecule occurring in evolutionarily distant organisms and has proven to be an abiotic antistress agent in plants (Li et al. 2016; Chen et al. 2009; Arnao and Josefa 2006). Several studies have demonstrated that lower dose of melatonin ($<10\ \mu M$) can stimulate the growth of maize plants and the germination of cucumber seeds under cold stress (Posmyk et al. 2009; Kolodziejczyk et al. 2016; Simlat et al. 2020). In cucumber, seed priming with melatonin promotes lateral root formation and seed germination under water-stress conditions (Zhang et al. 2013). In addition, melatonin acts as the first line of defense and as an internal sensor of oxidative stress in plants. For instance, exogenous applications of melatonin can increase photosynthetic carbon assimilation by improving plant antioxidant defense of organelles under drought or low-temperature stress in barley (Li et al. 2016). Melatonin can also increase the chance of survival by enhancing starch metabolism and energy supplies in response to damage caused by environmental stressors from heavy metals and temperature fluctuations (Zhang et al. 2015a; Liu et al. 2015). Thus, improved performance of primed seeds in terms of germination and seedling growth under chilling stress might be attributed to an enhanced antioxidant system and increased starch metabolism.

Direct sown rice seeds may be subjected to low temperatures, drought, flooding, and other adverse conditions. Embryonic cell membrane damage results in severe extracellular solute infiltration, lower seed moisture and vitality, and accumulation of reactive oxygen species. All these factors reduce the seed germination rate (Du et al. 2019). At low temperatures, however, antioxidants such as peroxidase, catalase, and glutathione peroxidase in the seeds scavenge free radicals. In this way, these enzymes protect the plants by alleviating low-temperature chilling effects. Seed initiation can increase the antioxidant enzyme content and vigor of seeds. Seed priming can also enhance antioxidant levels and the activity of oxidative defensive enzymes (Roqueiro et al. 2012; Kaklewski et al. 2008). Certain trace elements are also known as seed initiators. When selenium is used as a seed initiator, it increases seed activity and vitality and serves as a human dietary source of this essential micronutrient (Tapiero et al. 2003). Selenium is absorbed during rice germination and improves germination quality (Khaliq et al. 2015; Lazo-Vélez et al. 2015), root activity (Hartikainen et al. 2000), and seedling growth and vitality (Teklić et al. 2013). Selenium also increases the antioxidant and defensive substance contents and activities in plants (Roqueiro et al. 2012), provides energy for germination, and accelerates phosphorylation during membrane formation (Kaklewski et al. 2008). Selenium improves stress resistance and antioxidant and enzyme activity in plants. Although selenium can improve seed germination and seedling quality, sodium selenate is rarely used as a seed initiator. Sodium selenite is known to initiate and prime seeds (Kaklewski et al. 2008).

1.3 Seedling Growth Under Low Temperature

Seed germination is also related to seed traits. One of the most studied regeneration traits is seed mass, which varies considerably among and within species (Leishman et al. 2000; Moles et al. 2005). Species with small seeds have high temporal variation in germination, and large seeds produce larger seedlings that perform better than smaller ones under stressful conditions (Kitajima and Fenner 2000). Seedlings from large seeds usually can benefit from early emergence by maximizing development of deep root systems and thus defend aboveground climatic stress before the onset of the unfavorable season (Brunner et al. 2015).

The abiotic root-soil environment is seldom optimum for continuous growth and can have a profound influence on root anatomical characteristics, root developmental patterns, and the seedling physiology of roots. Root systems in relation to stress tolerance have been reviewed by Miller (1986), who discussed the effects of water, oxygen, mechanical impedance, and temperature stress on mature plants. Latimer (1991) reviewed vegetable seedling growth responses to mechanical stress. Adverse environments that include the interaction of factors such as nonoptimum temperature, deficiency or imbalance of nutrients, drought, and waterlogging have received less attention in seedlings. The establishment of seedling is a critical process to plant growth, especially under adverse environmental conditions (Bohnert et al. 1995). Seedlings adapt to stressful environments through different mechanisms, including changes in morphological and developmental patterns as well as physiological and biochemical processes. Adaptation is associated with maintaining osmotic homeostasis by metabolic adjustments that lead to the accumulation of metabolically compatible compounds such as soluble sugar, malondialdehyde, and proline. It also includes modification of related enzyme activity and cell membrane stability (Liang et al. 2009; Chinnusamy et al. 2007). Moreover, it is well known that plant structural modifications and growth pattern adjustments are useful indices of the consequences of stress environment (Rauf 2008; Ushio et al. 2008). Therefore, plants that have adapted to low temperatures have mediators of these substances. The effects of water or low-temperature stress on the change of physiological indexes have been reported (Yang and Lin 2008; Cao et al. 2009).

At different times during the life cycle, rice plant is differentially sensitive to temperature stress. Hence, the critically low and high temperatures, normally below 20 °C and above 30 °C, vary from one growth stage to another (Dubey et al. 2018). Each species requires a range of temperatures for seed germination and seedling establishment (Bradbeer 1988). Since seedling development stages are sensitive to environmental changes (Shimono and Kudo 2003), seedlings may need to be established in the first growing season to survive in the following season (Maruta 1994). In rice, a temperature of 22 °C or below is considered subnormal for seedling growth. Seedling growth may be reasonably good up to 35 °C, above which it declines sharply. Nishiyama (1977) reported that the critical minimum temperature for shoot elongation ranged from 7 to 16 °C and from 12 to 16 °C for root elongation. The critical minimum for elongation of both shoot and root is, hence, about 10 °C. A moderate increase in temperature speeds up leaf emergence, and temperature is a

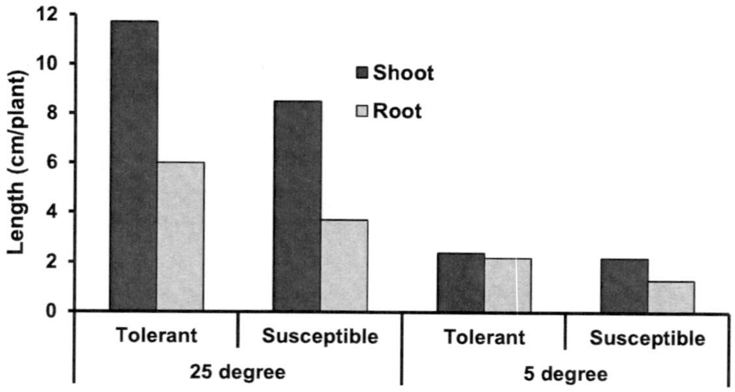

Fig. 1.3 Effect of low temperature on shoot and root length of susceptible and tolerant genotypes of *Brassica napus* at 25 days after germination under 25 and 5 °C

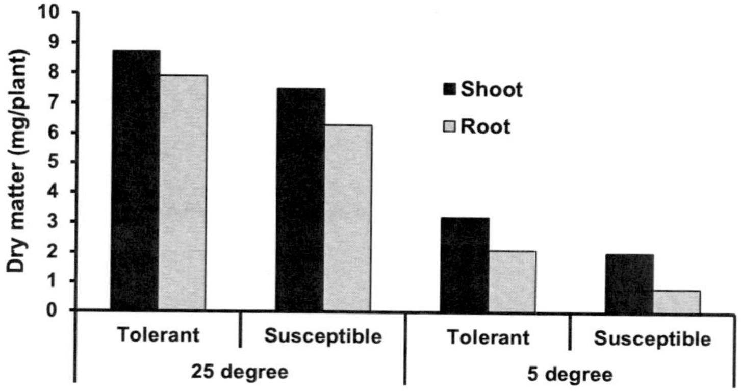

Fig. 1.4 Effect of low temperature on shoot and root length of susceptible and tolerant genotypes of *Brassica napus* at 25 days after germination under 25 and 5 °C

principal environmental determinant of leaf appearance in rice (Gao et al. 1992). The phyllochron concept, which is defined as the time interval between the appearances of successive leaf tips (Klepper et al. 1982), is used to predict the appearance of individual leaves, expressed in thermal time, with units of degree-days. The leaf appearance rate is not constant with time when rice plants are grown at constant temperature (Yin et al. 1996).

The effect of temperature on early seedling growth has been the subject of many studies. Parameters of growth are elongation rate of the primary root or the shoot at a certain stage of seedling development. Low temperature is a limiting factor of seed germination and seedling growth. With low-temperature-tolerant and susceptible genotypes, Fu et al. (2017) observed reduction in lengths and dry weights of shoot and root at 25 days after germination under 5 °C as compared to 25 °C (Figs. 1.3 and 1.4).

The optimum temperature for root and shoot elongation of maize seedlings is 30–34 °C. Many reports describe differential response of genotypes to low temperatures. Genetic variations in percentage of germination, percentage of emergence, and time to emergence at low temperature were reported by Eagles and Hardacre (1979) and Eagles and Brooking (1981). It has been reported that with the best materials, the rate of the growth processes is extremely slow at temperatures around 10 °C. Several researchers have shown that optimum temperature for germination and seedling growth depends on the species. Generally, temperate crops germinate between 0 and 35 °C, whereas tropical crops germinate between 10 and 45 °C. Within species, optimal temperature varies significantly between the genotypes. Although the growth and development of root systems are influenced by the environment, they are also under genetic control.

Reported cardinal temperatures for cotton growth and development are in the range of 12–15 °C for minimum temperature and 20–30 °C for optimum temperature (Reddy et al. 1991). Reduction in growth and development processes, chilling injuries, and delayed maturation of cotton can be attributed to low temperatures (Bange and Milroy 2004). Cotton seedlings exposed to chilling temperatures early in the season take longer to develop, have a slower biomass accumulation, and could halt seedling emergence (Christiansen and Thomas 1969). Reddy et al. (2017) have advocated the functional relationships between temperature and cotton seed emergence, and pre-squaring above- and belowground growth and developmental responses will be useful in assisting management and in improving functionality of many cotton models for on-farm and research and policy decisions. Seed germination, emergence, and early-season vigor are important aspects of cotton development, which play an important role in early-season canopy development and light interception. Cotton seedlings typically grow very slowly early in the season because most of the biomass is partitioned to root growth during this period (Hodges et al. 1993; Snider and Oosterhuis 2015). This slow growth trait combined with low soil and air temperatures during the early season (Hodges et al. 1993; Reddy et al. 1995) may lead to poor stand establishments. Although some information on cotton root growth and soil environment is available (McMichael and Quisenberry 1993), functional algorithms between many root traits and temperature from emergence and squaring are lacking for developing models for management. There have been many studies describing cotton growth and development at various stages of growth. In many of those studies, the temperature effects have been imposed following seedling emergence (Reddy et al. 1992a, b, c, 1993, 1997). Soil temperature is an important factor that impacts root growth and development. For most crop plants, the response of germination rate and seedling growth to temperature has recently been of great interest (Vassilevska-Ivanova and Tcekova 2002). In general, root growth tends to increase with the increase in soil temperature until the optimum temperature is reached and decrease when temperature exceeds the optimum temperature.

Low temperature and cold and wet soils during early spring represent major constraints for the cultivation of tropical crops, bearing the risk of poor germination, impaired seedling establishment, and reduced nutrient acquisition due to limited root growth and activity. This is frequently associated with poor vegetative plant development, reduced stress resistance, and finally reduced crop yield (Imran et al. 2013), although under favorable conditions, short cold periods can be tolerated and later compensated until final harvest (Saeidnejad et al. 2012). In plant species, such as maize, with optimum temperatures of 25–30 °C for germination and plant growth, even moderately low soil temperatures <15 °C are already detrimental (Kaspar and Bland 1992). Fu et al. (2017) pointed out that extreme temperatures give rise to an oxidative damage in plants. Cold is a severe environmental stress that disrupts the metabolic balance of cells, resulting in excessive accumulation of reactive oxygen species produced as a result of oxidative damage, which includes hydrogen peroxide (H_2O_2), singlet oxygen ($^1O_2$), superoxide ($O_2^{-\bullet}$), and the hydrogen radical ($HO^\bullet$) (Mittler 2002). To alleviate oxidative damage, plants have evolved protective enzymatic and nonenzymatic defense systems to detoxify reactive oxygen species and reduce oxidative stress. These strategies include the production of antioxidant enzymes, lipid-soluble antioxidants, and water-soluble reductants (Mittler et al. 2004; Duan et al. 2012). Temperature is one of the most critical factors affecting the germination of seeds (Verma et al. 2010). Earlier studies also demonstrated that variations in germination of different species were related to altitude (Giménez-Benavides and Milla 2013; Mondoni et al. 2011). If germination patterns change in response to habitat, it could be expected that seeds from higher elevations would germinate better under low-temperature conditions (Cavieres and Arroyo 2000).

Due to root growth limitation and slow diffusion rates, plant availability and acquisition of sparingly soluble nutrients, such as phosphorus, potassium, iron, zinc, manganese, and copper, are particularly affected by low soil temperatures (Wan et al. 2001). Therefore, placement of these nutrients close to the seedling roots or as seed treatments is among the most widespread practical measures to counteract detrimental effects of low soil temperatures on seedling establishment and early growth of maize (Imran et al. 2013; Bradáčová et al. 2016; Nkebiwe et al. 2016). Meanwhile, starter applications of ammonium phosphates by shallow subsurface placement below the seeds (Nkebiwe et al. 2016) belong to the standard fertilization strategies employed for maize production systems in temperate climates. Micronutrients with stress-protective functions, such as zinc and manganese, are frequently applied as foliar sprays. However, this is not possible in the seedling stage, and usually, the formulations are more expensive than soil fertilizers. More cost-effective placement strategies including seed dressings are increasingly employed to promote stress resistance, early growth, and crop establishment (Farooq et al. 2012; Bradáčová et al. 2016). Similar stress-protective effects have been recorded also for amendments with silicon (Liang et al. 2015). Many studies have highlighted a role of silicon in the suppression of oxidative damage in various plant species under a wide range of stress conditions. Protective effects have been reported against low-temperature stress in wheat (Liang et al. 2008) and cucumber (Liu et al. 2009).

In spite of the importance of the root architecture that determines efficient acquisition of soil nitrogen and water, understanding of root growth and development is minimal. Root system architecture has been referred to as an integrative result of lateral root initiation, morphogenesis, emergence, and growth (Dubrovsky and Forde 2012) and thus provides key traits that could be used to screen cultivars for survival under low-temperature conditions. Soil temperature affects how a plant root system permeates the soil. Root system development is a function of growth and development which is temperature-dependent (Kaspar and Bland 1992). Little information is available describing the effects of temperature on cotton root morphology and root-related traits at early growth stages. Root systems of tropical or subtropical origin species are sensitive to chilling temperatures. For example, root growth decreased when squash (*Cucurbita pepo* L.) and cucumber (*Cucumis sativus* L.) seedlings were exposed to 2 and 6 °C for 48 h, and when seedlings were returned to 26 °C, irreversible damage (inability to resume growth) occurred after 96 h (Reyes and Jennings 1994). Short-term cyclic cold stresses (2 °C) before transplanting generally stimulated seedling height, leaf area, and shoot and root growth but decreased chlorophyll content in tomato seedlings (Dufault and Melton 1990). Temperature between 10 and 13 °C suppressed tomato root growth (Hardenburg et al. 1986). However, Leskovar and Cantliffe (1991) reported that stem length and root fresh and dry weights were not suppressed when containerized tomato transplants were stored at 5 °C for 8 days. When containerized tomato seedlings were transplanted and mulched in beds with clear polyethylene, seedling root extension rates between 1 and 2 weeks after transplanting were significantly correlated with soil temperature during the first week (Wien et al. 1993).

Temperature directly influences the nutrient status of seedlings by affecting nutrient absorption and transport from the roots to the shoots. Seedling root elongation of beans, cucumber, eggplant (*Solanum melongena* L.), pea, pepper, radish (*Raphanus sativus* L.), and spinach (*Spinacia oleracea* L.) was optimum when constant in the 16.7–21.1 °C range, and a synergistic effect with phosphorus nutrition occurred on beans, cucumber, eggplant, and pepper, but not on radish, spinach, and pea (Wilcox and Pfeiffer 1990). Enhanced early root growth and nutrition uptake, particularly phosphorus, characterized tomato transplants grown with clear polyethylene mulch where soil temperatures were 5–7 °C higher than soil not mulched (Wien et al. 1993). Papadopoulos and Tiessen (1987) reported that leaf tissue accumulation of nutrients at high root temperatures (27 °C vs. 21 °C) was greater for phosphorus than for nitrogen, with minimum or no effect on the accumulation of potassium, calcium, and magnesium.

1.4 Root Development Under Low Temperature

Raven and Edwards (2001) defined root as "roots are axial multicellular structures of sporophytes of vascular plants which usually occurs underground, have strictly apical elongation growth, and generally have gravitropic responses which range from positive gravitropism to diagravitropism, combined with negative phototropism." The apical meristem of one (lower vascular plants) to many (all seed plants) diving cells produces a root cap acropetally and initials of stele, cortex, and epidermis basipetally. The branching of roots involves the endogenous origin of new root apical meristems in the pericycle (Raven and Edwards 2001). The most conserved functions of roots present in extant plants are anchorage to substrate and uptake of water and mineral nutrients. The evolution of multicellular organs such as roots was necessary for successful colonization of land by early plants (Pires and Dolan 2012).

It is widely acknowledged that soil temperature is one of the primary factors affecting plant growth. In fact, soil temperature could be even more important than air temperature, because it usually lags behind (although it tracks) air temperature in spring (Pregitzer et al. 2000). As a consequence, soil temperature occasionally may not meet the requirements for root formation, although air temperatures are favorable. Cold soils could thus determine the growth limit of plants at high latitudes or altitudes, and this limit should occur at lower altitudes and latitudes in trees compared with low-stature plants, because the crowns and canopy shade and thus cool their rooting zones (Körner 1998; Körner et al. 2003). In a global comparison of root-zone temperatures at high-elevation climatic tree lines, a mean of 6.7 °C (±0.8) was found (Körner and Paulsen 2004).

In experiments where all the potential growth-limiting factors, except for low soil temperature, have been adjusted to non-limiting levels, a supposedly uniform root-zone temperature of 5 °C has been found to inhibit root initiation and further lengthening in seedlings of *Picea abies* (Vapaavuori et al. 1992), *Pinus cembra* (Turner and Streule 1983), *Pinus sylvestris* (Vapaavuori et al. 1992; Iivonen et al. 1999; Domisch et al. 2001), *Pinus taeda* (Bilan 1967), *Larix decidua* (Tranquillini 1973), and *Populus tremuloides* (Wan et al. 1999; Landhausser et al. 2001). Optimal root growth has been found to occur only above 15 °C in *Pinus abies*, *Larix decidua*, *Pseudotsuga menziesii*, *Betula verrucosa*, *Quercus robur*, *Carpinus betulus*, and *Pinus nigra* (Lyr 1996). Tyron and Chapin III (1983) reported that roots of *Picea mariana* seedlings grown in sand, in pots, and in the glasshouse were able to elongate at 5 °C, but they did not specify rates or the experimental setup they used to manipulate soil temperature. The actual temperatures across the full soil/pot profile were not measured in any of these studies; thus, periodic topsoil warming under irradiance cannot be excluded. In cold substrates (<5 °C), root growth in *Pinus sylvestris* has been found to be constrained by plasma membrane H^+-ATPase transport, which has multiple functions in cell growth (Ryyppö et al. 1998; Iivonen et al. 1999). Neither external nutrient supply nor the nutrient demand of the shoot increased plasma membrane-ATPase activity at root-zone temperatures below 13 °C, demonstrating a direct thermal inhibition of active transport in these processes in *Pinus sylvestris* root cells, under cold soil conditions. In addition, it was demonstrated that root growth is correlated with an increase in the activity of the plasma membrane-ATPase.

The expansion of a plant's root system depends not only on the amount of roots produced by the plant and their arrangement but also on the cost incurred in the construction of a given root length (specific root length, m/g). Few studies have explored the influence of low soil temperature on specific root length. Chapin III (1974) found a smaller root diameter in cold-adapted populations of marsh plants compared with warm-adapted plants. Körner and Renhardt (1987) have reported that alpine forbs have thinner roots and a higher specific root length compared with low-altitude forbs. It has been hypothesized that higher specific root length, and thus more efficient investment of carbon for soil exploration, would compensate for the slower nutrient and water absorption in cold habitats (Chapin III 1974) or for poorer mycorrhization (Körner 1998). However, thinner roots are not as advantageous when other root functions are considered. Thicker roots have a greater ability to penetrate soil and provide a better anchorage for the plant (Hutchings and John 2003), as well as being more efficient in water and nutrient transport. Under controlled conditions, most root studies have been conducted using thermostatic water baths to cool down the growth substrate of plants. The temperature inside each bath is set to a constant value such that all roots should experience the same thermal regime, independent of their position in the root system. While temporal variation in temperature has been simulated (Halter et al. 1997; Iivonen et al. 1999; Domisch et al. 2002), Alvarez-Uria and Körner (2007) could not find any analysis of root growth responses to temperature changes similar to natural soil temperature profiles.

Plants have developed various mechanisms to tolerate water or salt stress, including an altered shoot-root ratio. Thus, it is known that soil water deficit may cause a greater inhibition of shoot growth compared with root growth, and, in some cases, the absolute root biomass in drying soil may increase when compared to well-watered soils (Santos et al. 2007). However, it is known that the degree of the responses of roots to environmental stresses may vary considerably within a family, a genus, and even a species (Francini and Sebastiani 2010; Kachout et al. 2011; Rahnama et al. 2011). The exposure of root systems to extreme high and low temperatures during the growing season negatively affects root growth. The roots of container-grown plants may be subjected to extreme temperatures not normally found under field conditions, which alter root growth (Mathers et al. 2007; Miralles et al. 2009). Exposure to low temperatures is one of the most common abiotic stresses of plants (Wu and Zou 2010). If the temperature of the soil drops significantly from optimum values, then the structure and function of root systems may be altered. In addition, lower temperatures reduced the shoot length and the shoot-root ratio and increased the percentage of brown roots in *Lotus creticus* (Franco et al. 2001). Nevertheless, compared with moderate temperatures, low temperature strongly reduced root length, root surface area, and root volume but increased the shoot-root dry matter ratios in *Citrus tangerine* seedlings (Wu and Zou 2010). This was also the case with *Lactuca sativa* (He et al. 2009) and *Myrtus communis* (Miralles et al. 2009). Furthermore, the exposure of roots to cold reduced hydraulic conductivity and can result in a marked water stress in the whole plant (Maurel et al. 2010), as Lee et al. (2004) also reported for *Cucumis sativus* plants.

Architecture of roots is increasingly studied in crops such as rice, corn, and cotton in identifying responses involved in stress tolerance during seedling growth (Singh et al. 2017a, b, 2018; Wijewardana et al. 2015). However, little is known about the soybean root system under stressful conditions (Khaled et al. 2011). Early assessments of whole root systems without breaking off the finer parts were nearly impractical in the past (Khan et al. 2016). For this reason, previous studies mostly screened cultivars for abiotic stress tolerance based on above-growth traits, like height, leaf area, and node numbers (Salmerón et al. 2014, 2015; Khan et al. 2016). However, the introduction of root phenotype systems, like hydroponics, gels, wax-petroleum layers, and WinRHIZO root scanner, has offered plant and soil scientists to evaluate root system architecture traits with minimal destruction (Singh et al. 2017b; Reddy et al. 2017; Brand et al. 2016). Studies have successfully exploited the above technologies to define the relationship between temperature stress tolerance and root traits, including root length, diameter, thickness, surface area, and lateral root numbers (Singh et al. 2018; Wijewardana et al. 2015; Brand et al. 2016). Further, differences in correlation between root and shoot traits to different abiotic stresses were also found during the seedling stage (Singh et al. 2018). Therefore, combined analysis of above- and belowground growth and developmental traits is important in identifying cultivars for abiotic stress tolerance.

1.5 Constraints of Environmental Temperature

It has been accepted that our climate is changing due to increased "greenhouse gas atmospheric concentrations," and this change is expected to have an important impact on different economic sectors (e.g., agriculture, forestry, energy consumptions, tourism, etc.) (Hanson et al. 2006). In particular, for agriculture, such a change in climate may have significant impacts on crop growth and yield, since these are largely determined by the weather conditions during the growing season.

Abiotic stresses are the main cause of extensive agricultural production losses worldwide (Barnabás et al. 2008; Athar and Ashraf 2009). Among abiotic stresses, drought, salinity, and extreme temperatures are the major environmental constraints that modern agriculture has to cope with. It has been estimated that they may be responsible for over 50% yield reduction in major crop plants. However, severity of losses depends on the plant development stage at which the stress occurs and its intensity and duration (Bray et al. 2000; Ashraf et al. 2008; Atteya 2003; Monneveux et al. 2006; Lafitte et al. 2007). Plant responses toward abiotic stresses have been the subject of various researches since decades. Research efforts have expanded from model plants to various crop plants like wheat, barley, rice, maize, and other economically important species. The development of the "omics" technologies (e.g., genomics, transcriptomics, proteomics, metabolomics) has revolutionized plant science research and has enabled more holistic study of interactions among biological components using models and/or networks to integrate genes, metabolites, proteins, regulatory elements, fluxes, and others; this methodology is defined as systems biology (Yuan et al. 2008). Present research permitted to explain functions of many key genes, proteins, metabolites, and molecular networks involved in plant responses to salinity, drought, heat, cold, heavy metals, and other abiotic stresses. It has been reported that under environmental stress during the flowering and fruiting period, abscission of flowers and flower buds may occur (Abdulmalik et al. 2012) in sweet peeper (*Capsicum annuum* L.), and this can result in serious yield decrease in pepper production (Minges and Warholic 1973). The abscission can be caused by several factors such as extreme temperature, lack of moisture, or low light conditions (Wien 1990). According to Dagdelen et al. (2004), yield of pepper plant is seriously reduced when pepper is exposed to moisture stress

Table 1.1 Threshold high temperatures for some crop plants

Crop plants	Threshold temperature (°C)	Growth stage	References
Wheat	26	Post-anthesis	Stone and Nicolás (1994)
Rice	34	Grain yield	Peng et al. (2004)
Corn	38	Grain filling	Thompson (1986)
Sorghum	36	Panicle emergence	Prasad et al. (2006a)
Cotton	45	Reproductive	Rehman et al. (2004)
Pearl millet	35	Seedling	Ashraf and Hafeez (2004)
Brassica	29	Flowering	Morrison and Stewart (2002)
Groundnut	34	Pollen production	Vara Prasad et al. (2000)
Cool season pulses	25	Flowering	Siddique et al. (1999)
Cowpea	41	Flowering	Patel and Hall (1990)
Tomato	30	Emergence	Adams et al. (2001)
Broccoli	35	Flower initiation	Björkman and Pearson (1998)
Canola	28	Vegetative growth	Aksouh-Harradj et al. (2006)

during flowering and fruit formation stage. Abiotic stresses such as drought, salinity, and high and low temperatures negatively affect plant growth and development, causing cellular water deficit, cell membrane injury, loss of enzyme activities, and other defects and resulting in severe reductions of crop yields (Zhu 2002; Yamaguchi-Shinozaki and Shinozaki 2006).

Temperature influences photosynthesis, water and nutrient absorption, transpiration, respiration, and enzyme activity. These factors govern germination, flowering, pollen viability, fruit set, rates of maturation and senescence, yield, quality, harvest duration, and shelf life. Different plants have different threshold temperature (Table 1.1). However, for most crop species, optimum temperatures usually range around 25 °C. High temperature is a crucial problem in growing good crops of high-temperature-sensitive vegetables, including tomato. It impairs different morphological criteria (Warrag 1999). Temperature requirements (usually based on night temperature) of plants are given below by the cardinal values and derived range for "effective growth" (growth range) and "optimum growth" (optimum range) that Krug (1991) has used for major vegetables.

1.5.1 Low Temperature

While temperature and other abiotic stresses are clearly limiting factors for crops cultivated on marginal lands, crop productivity everywhere is often at the mercy of random environmental fluctuations. Low temperature of the field environment restricted the vegetative growth and delayed all the phenological stages in comparison with control plants. It has been shown that chickpea (*Cicer arietinum* L.) lacks cold tolerance and is sensitive to chilling temperatures (>8 °C), especially at its reproductive phase (Bakht et al. 2006). The reproductive structures can withstand temperature of 8 °C minimum to 22 °C maximum during the coldest period. The most advantageous temperature range for normal flowering, fertilization, and seed set is 10–14 °C average minimum temperature and 25–31 °C average maximum temperature. Temperature within the chilling range can limit the growth and vigor of chickpea at all phenological stages but is considered most damaging to yield at reproductive stage. In India, chickpea is grown as a winter-season crop. The northern parts of the Indian subcontinent and southern Australia are the most affected regions due to chilling, as the temperature is below 15 °C at flowering (Clark 2001; Bakht et al. 2006; Berger 2007). During the reproductive phase, low temperature is detrimental to normal flowering and pod development, which causes prolonged reproductive phase, floral abortion, poor pollen germination, impaired ovule development, failure in pod set, and reduction in seed filling that drastically affects crop productivity (Nayyar et al. 2005a). The reproductive phase, especially in early maturing chickpea genotypes, coincides with low temperature of the surroundings, causing abortion of flowers and impairment in pod set (Kumar et al. 2005). A deep investigation of the development of male and female components is required to prove the sensitivity of the developmental stage and actual cause of reproductive failure.

Hot- or cold-temperature stresses can be detrimental to all phases of plant development. As the majority of our food supply is a product of sexual reproduction in flowering plants, understanding how different plants cope with stress during their reproductive (gametophytic) phase is critical to managing the future of agricultural productivity (Zinn et al. 2010). During the short time surrounding fertilization, even a single cold night can be fatal to reproductive success for many plant species. While temperature stress has been extensively studied and reviewed (Iba 2002; Yamaguchi-Shinozaki and Shinozaki 2006; Chinnusamy et al. 2007; Kotak et al. 2007; Wahid et al. 2007), most of the literature emphasizes insights from experimentally accessible tissues, such as leaves and roots. By comparison, studies on sexual reproduction are often more difficult because gamete development and fertilization are complex processes that occur during a short window of time and are predominantly hidden within tissues of the flower. Nevertheless, sexual reproduction has been long recognized as being highly stress-sensitive, with reproductive stress tolerance often a limiting trait in crop plant productivity (Charles and Harris 1972; Herrero and Johnson 1980). This aspect of stress sensitivity has been reviewed by Barnabás et al. (2008), Hedhly et al. (2008), and Thakur et al. (2010). Plants have evolved complex mechanisms to cope with daily changes in the environment. At a molecular level, this is illustrated by the thousands of transcriptional changes observed in seedlings, leaves, roots, and pollen as plants reprogram cellular processes to adapt to hot or cold temperatures (Fowler and Thomashow 2002; Kreps et al. 2002; Busch et al. 2005: Larkindale and Vierling 2008; Frank et al. 2009). One developmental program that provides a very robust stress tolerance is the production of a dehydrated embryo that can remain dormant for long periods of time inside a seed. Similarly, pollen development often includes the production of dehydrated pollen grains, which are also relatively stress-tolerant. While these dormancy programs are used in special circumstances, actively growing plants utilize less extreme measures for dealing with daily changes in the environment. These daily responses to changing environments have been extensively studied in root and leaf tissues, providing a framework of general principles for investigating stress tolerance in all plant cells, including cells involved in reproduction.

The impact of temperature stress is a complex function of intensity, duration, and rate of temperature change (Wahid et al. 2007; Thakur et al. 2010). Low temperature can alter multiple aspects of cellular physiology. For example, temperature can dramatically change membrane fluidity, nucleic acid, and protein structures, as well as metabolite and osmolyte concentrations (Wang et al. 2003; Howarth 2005; Chinnusamy et al. 2007). Low-temperature stresses induce the production of reactive oxygen species, which at elevated concentrations will result in oxidative damage and, potentially, cell death (Apel and Hirt 2004).

While many plants can withstand highly stressful conditions, they nevertheless require adequate time to sense and adapt to new environments. Understanding how plants sense and respond to specific combinations of stress is a challenge (Suzuki and Mittler 2006). For example, increased levels of reactive oxygen species can function to trigger transcriptional changes in both hot and cold stress conditions. Hormone production and signaling may also change during temperature stress (Kotak et al. 2007). For example, abscisic acid, salicylic acid, and ethylene signaling are activated in response to high-temperature stress. Abscisic acid signaling can also play a significant role in establishing cold tolerance (Thomashow 1999). In addition, the phytohormones ethylene, gibberellins, and auxin are also thought to be involved in cold stress signaling (Lee et al. 2005).

Low temperatures also modulate leaf growth. Plants grown under chilling conditions developed smaller leaves than those grown under control or heat conditions. Growth under low temperatures often results in significant alterations in leaf morphology. The most noticeable effect is a reduction in specific leaf area (the ratio of leaf area to leaf dry mass) (Armstrong et al. 2007). It is also remarkable that under cold conditions, the leaves become thicker than those observed under control or high temperature (Rodríguez et al. 2015). It has been previously reported that plants grown under chilling conditions show reduced leaf expansion and increased mesophyll thickness (Williams and Black 1993). Since temperature affects leaf size, it could affect also the leaf shape. Juvenile leaves of *Brassica oleracea* have an oval shape. Although the leaves of varieties follow this general rule, under control conditions, the Acephala group developed leaves slightly longer than its width, whereas the opposite behavior was observed in the leaves of the Capitata group. There was no effect of temperature in the leaf shape of the Capitata group, whereas the leaves of the Acephala group developed under thermal stress become longer than its width. At the autotrophic stage, plant growth strongly depends on the capacity of the photosynthetic apparatus to fix carbon. Photosynthesis is one of the most affected cellular reactions by environmental changes; concretely the PSII activity is especially sensitive to thermal stress (Ruelland and Zachowski 2010).

The effects of various abiotic stresses on structural and functional aspects of reproductive phase have been reported in soybean (*Glycine max*) (Gass et al. 1996; Kokubun et al. 2001), rice (*Oryza sativa*) (Imin et al. 2004), pea (*Pisum sativum*) (Guilioni et al. 1997), and chickpea (Kaur et al. 2008). In rice, low temperature during the reproductive stage has been found to cause degeneration of spikelets, incomplete panicle exsertion, and increase in spikelet sterility (Terres 1991). Later, Takeoka et al. (1992) has identified that booting stage in rice is highly susceptible to chilling and the damage at this particular stage of reproductive development causes around 30–40% yield reduction in the temperate grown rice worldwide (Nishiyama 1995; Andaya and Mackill 2003a). Cold stress induces flower abscission in soybean (Gass et al. 1996), pigeon pea (*Cajanus cajan*) (Sandhu et al. 2007), and chickpea (Nayyar et al. 2005a), but the actual cause of reproductive failure is unclear.

Pollen sterility is a common cause of low fruit set under cold temperatures at flowering in rice, sorghum (*Sorghum bicolor*), strawberry (*Fragaria* sp.), and tomato (*Lycopersicon lycopersicum*) (Lin and Peterson 1975; Toriyama and Hinata 1984). Low temperatures may cause deformation of flowers or floral parts (Rylski 1979), leading to functional sterility or the formation of deformed fruit. It was observed that pod set failure and deformation in fruit could be associated with abnormal ovary development (Rylski 1979). Fertilization process, subsequent embryo development, and seed filling are potential targets of cold stress. It was observed that under cold conditions, the development of male gametophyte is impaired, which leads to smaller pollen grains and an increase in non-viable pollen (Gudkova 1980). Imin et al. (2004), while studying the effects of early cold stress on the maturation of rice anthers, demonstrated for the first time that cold-temperature stress at the young microspore stage enhances and induces partial degradation of proteins in the rice anthers at the trinucleate stage. Chickpea can experience yield losses up to 50% due to cold stress (Saxena and Johansen 1990), depending upon the severity of stress and sensitivity of the genotype. Hence, incorporation of cold tolerance in chickpea cultivars is an important prerequisite for winter sowing of the crop (Singh et al. 1990; Bakht et al. 2006).

1.6 Growth and Development Under Low Temperature

Chilling stress, damage caused by low but above freezing temperatures, has been recognized as a unique environmental impact on crop plant growth. Susceptibility to chilling injury imposes late plantings or prevents the cultivation of many crops in regions where temperatures can drop below the optimal growth temperatures for individual plant species. Early planting could contribute to a longer growing season, a more effective utilization of late spring and early summer rainfall, and an enhanced yield potential (Ercoli et al. 2004).

Plants exposed to chilling temperatures suffer injury, exhibiting reduction or stop of growth. This phenomenon is initially reversible but ultimately irreversible and can result in cell death. Chilling adversely affects a wide range of processes, including crop germination and establishment, photosynthesis, flowering, plant growth, and grain yield (Ting et al. 1991; Staggenborg and Vanderlip 1996; Verheul et al. 1995). The degree of chilling injury can vary with plant species, stage of crop development, and the conditions of irradiance and mineral nutrition during chilling stress. In chilling-sensitive species, it has been recognized that chilling causes greater injury under illuminated conditions, owing to photooxidative damage determined by the increase of the formation of harmful active oxygen species (Long 1983; Wise 1995; Aroca et al. 2001). The degree of dysfunction observed after chilling of tomato plants depended on mineral nutrition, affecting the production of root-borne phytohormones (Starck et al. 1997), and plants grown with limited supply of mineral nutrients were more susceptible to chilling (Starck et al. 2000). In spinach, nitrogen deficiency greatly decreased the rate of acclimation to low temperature (Martindale and Leegood 1997). Chilling injury is time dependent, and changes in physiological activity precede the development of visual symptoms of injury. Rates of respiration and ion leakage increases at chilling temperature, while photosynthetic activity and carbohydrate metabolism decrease (Karnok and Beard 1983; White and Schmidt 1989; Sonoike 1999). Chilling temperatures cause starch accumulation in chloroplasts of some plant species, inhibiting the photosynthetic process because of feedback inhibition of photosynthetic enzymes (West 1970; Rogers et al. 1977). Photoinhibition of photosynthesis is enhanced at chilling temperatures, and one of the major reasons for this enhancement is the decreased utilization of excitation energy in carbon metabolism at chilling temperature (Sonoike 1999). Chilling damage depends not only on conditions during chilling but also on the environment before and after chilling, such as temperature, light conditions, and water status of the plants prior to chilling and the light conditions after chilling (Starck et al. 1994; Szalai et al. 1996; Smeets and Wehner 1997). Sorghum is one of the most economically important plants grown in Italy, but, originating from the semiarid tropics, it is susceptible to cold stress and often expresses poor early-season vigor and reduced competitive ability against weeds, owing to low temperatures after sowing. The effects of chilling stress on sorghum have been studied in relation to chilling-induced declines in photosynthetic capacity. However, there are many situations, i.e., cloudy conditions or darkness, where the photosynthetic apparatus of sorghum may not be affected by low temperatures (Taylor and Rowley 1971). Reduced growth may result from more direct effects of temperature on growing tissues or from the reduction of photosynthate flow through the phloem. According to Wardlaw and Bagnall (1981), although low temperatures may reduce the carrying capacity of the phloem of sorghum, this is unlikely to be an important factor in regulating plant growth at low temperatures. Lang and Minchin (1986) in sequential chill-warm cycles demonstrated that the translocation process was sensitive only to the first cycle, subsequent cycles having no effect owing to sorghum adaptation mechanisms.

The rate of plant growth and development is dependent upon the temperature surrounding the plant, and each species has a specific temperature range represented by a minimum, maximum, and optimum (Hatfield and Prueger 2015). Growth is an irreversible increase in volume, size, or weight, which includes the phases of cell

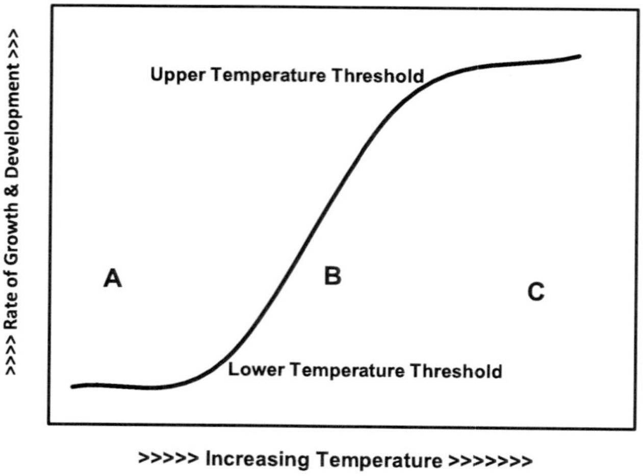

Fig. 1.5 Typical relationship between rate of plant growth and development and temperature. Growth and development ceases when temperature declines below the lower-temperature thresholds (A) or increases above the upper-temperature threshold (C). Growth and development increases rapidly when temperature falls between lower- and upper-temperature thresholds (B)

division, cell elongation, and differentiation. Both cell division and cell enlargement are affected under environmental stress, owing to impaired enzyme activities, loss of turgor, and decreased energy supply (Farooq et al. 2009a; Taiz and Zeiger 2006). Heat units can be used as a management tool for more efficient timing of irrigation and nutrient inputs to a crop. The thermal environment impacts the development of all crop systems. Plants will develop over a range of temperatures, which is defined by the lower and upper temperature thresholds for growth (Fig. 1.5). Heat unit systems take into account the elapsed time that local temperatures fall within the set upper- and lower-temperature thresholds and thereby provide an estimate of the expected rate of development for the crop. Heat unit systems have largely replaced days after planting in crop phenology models because they take into account day-to-day fluctuations in temperature. Phenology models describe how crop growth and development are impacted by weather and climate and provide an effective way to standardize crop growth and development among different years and across many locations (Baskerville and Emin 1969).

Chilling stress is known to cause significant reductions in plant growth and development. Various studies have reported the negative effects of chilling stress on the growth of rice (Hussain et al. 2016; Wang et al. 2016), cotton (Zhao et al. 2012), tomato (Starck et al. 2000), potato (Svensson et al. 2002), muskmelons (Wang et al. 2004), and sugarcane (Thakur et al. 2010; Anjum et al. 2011; Zhu et al. 2013). It has been reported that chilling stress severely damaged the shoot and root growth attributes, compared with control (Hussain et al. 2018). Further, the interactive damage caused by chilling and drought was found to be additive for all growth attributes. Chilling stress may cause necrotic lesions on leaves, delay leaf development, prolong cell cycle with decreased cell production, induce wilting, and increase susceptibility to pathogens and diseases (Korkmaz and Dufault 2001; Rymen et al. 2007). Decreased rate of both cell division and elongation causes smaller leaf area (Jouyban et al. 2013), while slow leaf initiation rate leads to reduced number of leaves under periods of chilling stress (Lukatkin et al. 2012; Jouyban et al. 2013). Leaf growth under drought stress is mainly restricted by decreased leaf water potential (Hsiao and Xu 2000), while interrupted water flow from xylem to other elongating cells and reduced turgor pressure under severe drought stress lead to reduced cell elongation and smaller leaf area in crop plants (Nonami 1998; Schuppler et al. 1998; Taiz and Zeiger 2006).

Chilling stress restricts root growth by decreasing root length, biomass, and morphology (Cutforth et al. 1986), which ultimately reduces volume of the root system for exploring the nutrients and water (Richner et al. 1996). Nevertheless, the reported effects of drought stress on root growth have been diverse. For instance, drought stress increased the root length of *Catharanthus roseus* (Jaleel et al. 2008) and sunflower (Tahir et al. 2002), but did not affect the root growth of wheat and maize (Sacks et al. 1997). Increase in root length under drought conditions has been attributed to increased abscisic acid content in roots (Manivannan et al. 2007). Wullschleger et al. (2005) stated that root dry biomass of *Populus* species was reduced under mild and severe drought stress. In general, varieties with deep and prolific root systems were better able to sustain growth and yield under water-deficit conditions (Babu et al. 2001).

1.6.1 Vegetative Growth

Cold temperature can be distinguished from freezing in terms of the range of temperatures that cause the related damages (Iba 2002). Many species of tropical or subtropical origin are injured or killed by non-freezing low temperatures and exhibit various symptoms of chilling injury, such as chlorosis, necrosis, or growth retardation. In contrast, chilling-tolerant species are able to grow at such low temperatures (Sanghera et al. 2011). Plant vegetative growth is greatly affected by environmental changes especially during seed germination and seedling development (Shinozaki and Yamaguchi-Shinozaki 2007). In contrast to the many studies on the effect of drought on vegetative development, there have been few reports about the effect of drought on reproductive growth. Crops such as wheat (*Triticum aestivum*) and rice (*Oryza sativa*) show partial male sterility under low temperature, leading to significant reduction of grain production (Sheoran and Saini 1996; Lalonde et al. 1997; Saini 1997). Cold temperature generally encourages prolonged vegetative growth in chickpea. It has been established that temperature is the main determinant for flower initiation in most environments, although some authors have linked flower initiation in chickpea to photothermal response (Roberts et al. 1985). In northern NSW, Australia, flower initiation can commence at $<15\ ^{\circ}C$, although the occurrence of flower abortion will likely to be high (Jenkins and Brill 2011).

The degree of injury in plants usually depends on the time of occurrence (growth stage), severity of chilling, and low-temperature duration (Li et al. 1981). Low temperature has the potential to affect growth and development of plants during any developmental stage, from germination to grain filling (Ye et al. 2009). However, Yoshida (1981) showed that sensitivity to cold varies between stages and rice plants have a lower threshold temperature (10–13 °C) for cold damage during the early stages of development (germination and vegetative), what makes them less sensitive to cold than during the reproductive stage, which has a higher threshold temperature for damage (18–20 °C). Despite these differences among stages, cool weather and cold irrigation water can damage the plant during any developmental stage, such as germination, seedling, vegetative, reproductive, and grain maturity (Andaya and Mackill 2003b; Ji et al. 2008; Xu et al. 2008).

Low-temperature stress inhibits various metabolic reactions, thus preventing the expression of full genetic potential of plants, which is expressed by different phenotypic symptoms (Chinnusamy et al. 2007). Some of the common low-temperature injuries in plants are reduced seedling growth, seedling discoloration, leaf yellowing, leaf whitening, white specks, white bands, withering after transplanting, a reduced rate of tillering, stunting, and so on (Nahar et al. 2012). According to Angadi et al. (2000a, b), temperatures below 10 °C result in slower and reduced growth and premature stem elongation in *Brassica napus*, *Brassica rapa*, and *Raphanus sativus*. It is well reported that plants at their seedling stage are very much sensitive to cold stress. At the early stage of plant growth, various phenotypic symptoms in response to chilling stress are surface lesions, chlorosis, necrosis, desiccation, tissue breakdown and water-soaked appearance of tissues, reduced leaf expansion, and wilting (Solanke and Sharma 2008; Jiang et al. 2002). Nahar et al. (2009a) observed variations in symptom for cold injury in rice including stunted growth, yellowing of leaves, abnormal number of tiller, malformed grain, abnormal colours in grain. Generally, exposure to cold temperature affects crop growth and development in two ways concurrently. First, developmental events in the shoot apex are affected, which directly determine the differentiation of the panicle and hence potential yield and spikelet fertility resulting in fewer grains. Second, photosynthesis is impaired, which reduces growth and results in indirect yield loss because there is less carbohydrate available for grain production (Takeoka et al. 1992). The cellular structures and components are also damaged due to the cold, which have been studied for a long time. The development process and ontogeny of the organelles may be disrupted by chilling stress (Kratsch and Wise 2000). Nahar et al. (2009a, b) observed cold stress induced morphological symptoms like stunted plant, bushy plants, early maturity, and yellowing of leaves. Modifications of cellular components include swelling of the mitochondria, plastids, and thylakoid lamellae, vesiculation of thylakoid, accumulation of lipid droplets and eventually disorganization of the entire plastid (Murphy and Wilson 1981), reduction in ribosomal numbers, dilation of endoplasmic reticulum, vesiculation of cytoplasmic membranes, condensation of nuclear chromatin, invagination of plasmalemma, and vacuolation of membranous vesicles (Kratsch and Wise 2000; Ishikawa 1996), which were noticed during chilling stress. Without these in extreme cases, chilling results in accelerated senescence and eventually plant death (Sharma et al. 2005).

Wolfe (1991) used chilling-sensitive species of *Phaseolus vulgaris* (L.) and *Zea mays* (L.) and chilling-sensitive species of *Spinacia oleracea* (L.) in growth chambers under warm- (28/18 °C day/night cycle) and cool (18/12 °C)-temperature regimes and made growth analysis to evaluate leaf area and biomass partitioning during early autotrophic growth. It was reported that advantages exhibited by *Pisum sativum and Spinacia oleracea* were a relatively minor delay in emergence, a higher net assimilation rate, and less than 10% reduction in relative growth rate at 18/12 °C. Chilling-sensitive species showed 40–50% reductions in relative growth rate, and for *Phaseolus vulgaris,* this was most associated with lower specific leaf area and photosynthetic area at 18/12 °C. A significantly lower specific leaf area at a cooler temperature was also observed in *Zea mays*, but factors controlling net assimilation rate dominated growth response for this species. Visibly thicker leaves and increased leaf density (i.e., lower specific leaf area) were observed in all species when grown at 18/12 °C.

Rice is a cold-sensitive plant that originated from tropical or subtropical zones. When low temperature occurs during the reproductive stages, it can cause serious yield and yield component losses (Farrell et al. 2006a). The optimum temperature for rice cultivation is between 25 and 35 °C, and in temperate regions, rice growth is aided by a limited period that favors its growth (De Los Reyes et al. 2003). Exposure to cold temperature affects all phonological stages of rice and lower grain production and yield, too. Low temperature during vegetative stage can cause slow growth and reduce seedling vigor (Ali et al. 2006), reduce the number of seedlings, reduce tillering (Shimono et al. 2002), increase plant mortality (Farrell et al. 2006a; Baruah et al. 2009; Fujino et al. 2004), and increase growth period (Alvarado and Hernaiz 2007), and in the reproductive stage, it can cause panicle sterility and lower grain production and yield (Shimono et al. 2002). Critical stages for cold damage include germination, booting, flowering, and filling stages (Satake 1976b).

The symptoms of cold sensitivity and damage vary according to the growth stage of the rice plant (Yoshida 1981). In the germination stage, the most common symptoms of cold-temperature damage are delayed and lower percentage of germination (da Cruz and Milach 2000). At the vegetative stage, chilling damage is expressed through yellowing of the leaves, lower stature, and decreased tillering of rice plants (Ghadirnezhad and Fallah 2014). When cold coincides with the reproductive stage of the rice plant, sterility of the spikelets is the most common symptom of injury, but incomplete panicle exsertion and spikelet abortion may also occur (Sataka and Hayase 1970). Spikelet sterility may result from pollen abortion due to cold during microsporogenesis, when pollen grains are being formed, at the booting stage (Mackill et al. 1996). Low temperatures can also result in abnormalities at anthesis, such as cessation of anther development, nonripening of pollen, nonemergence of anthers from spikelets, partial or no anther dehiscence, pollen grains remaining in anther loculi, little or no pollen shedding, and failure of pollen to germinate after reaching stigmas (Ito et al. 1970). During the grain filling stage, chilling temperature may cause delayed and incomplete grain maturation (Ye et al. 2009).

The after effects of continuous and variable chilling temperatures acting periodically for 7 days at seedling stage (day/night, 2/2 °C, 5/5 °C, 15/2 °C, 15/5 °C, 15/15 °C; and control 22/18 °C) on growth, development, and final yield of two soybean cultivar was studied (Skrudlik and Kościelniak 1996). The temperature decrease lengthened the duration of the vegetation period before flowering for both cultivars, which allowed compensation for the reduced as an effect of the chilling rate of daily increments in the leaf area and the dry weight of the aboveground parts of the plants. The most important symptom of chilling treatment was a considerable increase in the number of axillary branches after flowering, which caused an increase in the dry weight of plants. The subsequent reaction of the growth of leaves to chilling at seedling stage was, after flowering, genotype dependent. Delay in plant growth induced by chilling allowed for the avoidance of low temperatures during flowering in this experiment. This chilling during flowering was one of the reasons of worse pod setting and a reduction in the number of seeds per pod on control plants.

Seasonal changes in temperature promote many developmental processes (i.e., flowering, germination, or grain filling) (Wheeler et al. 1996; Roberts and Summerfield 1987). However, above or below certain thresholds, temperatures limit geographical distribution and productivity of many important crops. Contrary to metazoans, plants do not have specialized cell types that allow perception of temperature fluctuations. The mechanisms through which plants perceive temperature have been proposed to be similar for both high and low temperatures, although the intracellular signaling and physiological response differ between both stimuli (Ruelland and Zachowski 2010). The first structure that responds to temperature fluctuations is the plasma membrane. Both high and low temperatures cause changes in the fluidity of this structure, which activates an intracellular signal cascade (Örvar et al. 2000). Associated with the plasma membrane, the cytoskeleton is another sensor of temperature fluctuations. Exposure of plants to growth-limiting temperatures induces the depolymerization of microtubules and microfilaments (Müller et al. 2007). These two structures are intimately involved in cell morphogenesis (Fowler and Quatrano 1997), and its rearrangement may explain variations of the leaf shape in plant growth under extreme temperatures (Campitelli and Stinchcombe 2013).

Crops with a biennial life cycle are exposed to seasonal temperature variations ranging from below 0 °C to more than 40 °C. Among biennial plants, *Brassica oleracea* L. stands out as one of the most important species in the world from an economical point of view. Human selection has led to the development of a range of cultivars within this species, in which different organs are used for human or livestock consumption (Gomez-Campo and Prakash 1999). This wide diversity may be reflected in different mechanisms to respond to thermal stress among the different cultivars of this species. Rodríguez et al. (2015) has pointed out that although high temperatures significantly reduced the dry weight of seedlings, in general terms, low temperature had a higher impact in *Brassica oleracea* physiology than high temperature.

1.6.2 Temperature Effects on Aboveground Growth and Biomass Production

The effects of growth temperature on perennials are wide-ranging, influencing many aspects of development, including leaf morphology, petiole length, and stem thickness (Samach 2012). Early-spring plants start their aboveground development under particular ecological conditions, i.e., comparatively low air and soil temperatures and high insolation. However, these parameters are not essential environmental factors for the completion of their short life cycle in spring (De Hertogh and le Nard 1993). Once the leaf apex has emerged early in spring, the rapid growth of the whole plant starts. This could be guaranteed by the fact that the chloroplast structure is formed during underground growth in darkness, prior to the appearance of photosynthesis, which allows the plant to reach a maximum photosynthetic rate rapidly (Mamushina et al. 2002, 2008; Mamushina and Zubkova 1996).

Relatively high temperatures (18/14 °C day/night temperature) during the growth period lead to faster growth in geophytes and earlier flowering. However, in this temperature range, the plants have smaller flowers and exhibit a shorter duration of vegetation. Moreover, a warmer spring always results in a lower bulb yield than for plants grown at lower temperature. In a cooler temperature (12/8 °C), bulb growth and leaf activity are maintained for a longer period (De Hertogh and le Nard 1993; Gandin et al. 2011; Lapointe and Lerat 2006; Badri et al. 2007; Lundmark et al. 2009). Shorter growth time can also be explained by early activation of different sucrose-cleaving enzymes in the bulbs grown under warmer temperatures, which results in the faster shift from elongation to cell maturation and to the earlier cessation of starch accumulation (Gandin et al. 2011; Lundmark et al. 2009). Additionally, the rate of carbon assimilation at warmer temperatures rapidly becomes too high for the capacity to incorporate carbon into the bulb, leading to feedback inhibition of the photosynthetic rate. Since the size of the bulb is considered to be one of the major parameters that influence the quality of flowers, low-temperature conditions are more favorable during the aboveground growth of early-spring plants. This will promote the production of larger underground organs and flowers in the next spring.

1.6.3 Reproductive Growth

Low-temperature stresses have several major effects on reproductive tissues that contribute to poor seed set yield:

- Early or delayed flowering
- Asynchrony of male and female reproductive development

- Defects in parental tissue
- Defects in male and female gametes

While the focus of this review is on pollen, some brief examples of temperature effects on other aspects of reproductive development will be provided first.

Low-temperature stress can trigger either early or delayed flowering, depending on the species and other environmental conditions. One important modifier is the photoperiod, which provides seasonal information in which stress can be interpreted in the appropriate context (Putterill et al. 2004; Craufurd and Wheeler 2009). Nevertheless, moderate heat stress will often accelerate flowering, which may cause reproduction to occur before plants accumulate adequate resources (i.e., biomass) for allocation to developing seeds. Mechanisms of heat-stimulated bolting and flowering in *Arabidopsis thaliana* are being uncovered (Balasubramanian et al. 2006; Tonsor et al. 2008). Cold temperatures will typically delay flowering, which may cause seeds to develop later in the growing season under suboptimal temperatures.

Temperature stress can sometimes have different effects on male and female structures, thereby creating asynchrony between male and female reproductive development (Herrero 2003; Hedhly et al. 2008). In maize (*Zea mays*), floral asynchrony is a significant problem under conditions of combined stress from heat and water deficit (Barnabás et al. 2008). A third category of temperature stress effects includes defects in the structure and function of parental tissues (i.e., corollas, carpels, and stamens). Low-temperature stress can reduce the number, decrease the size, and cause deformity of floral organs (Takeoka et al. 1991; Morrison and Stewart 2002). For example, a high frequency of flower abortion in response to low temperatures is well documented for *Cicer arietinum* L. (Croser et al. 2003).

Reproductive phases of the life cycle of plants are more vulnerable to cold stress (Nishiyama 1995). The reproductive phase subjected to low-temperature stress is influenced differently in each of its subphases. During the development of male gametophytes, low temperature causes disruption of meiosis, tapetal hypertrophy, stunted development of pollen grain, anther protein degradation, pollen sterility, and pollen tube deformation. In female gametophyte development, its effects are characterized by reduced style and ovary length, disruption of meiosis, reduced stigma receptivity, callose deposition in style, damage to embryo sac components, and arrest of the fertilization process. At flowering, low temperature may cause delayed flowering, bud abscission, and sterile or distorted flowers, while at grain filling, the source-sink relation is altered, kernel filling rate is reduced, and ultimately small-sized, unfilled, or aborted seeds are produced (Jiang et al. 2002; Thakur et al. 2010). Flower buds of *Simmondsia chinensis* can be damaged or killed by temperatures of -2 to -5.5 °C (Nelson and Palzkill 1993). Farrell et al. (2006a) observed that cold water significantly reduced panicle emergence in rice. Low temperatures affect not only normal heading but also panicle exsertion and prevent the normal elongation of internodes of rice. Cold stress (11 °C) caused abnormality in panicle initiation process, delayed heading, incomplete panicle emergence, sterile and malformed spikelets, and spikelet degeneration symptoms in rice genotypes

(Nahar et al. 2009a, b). In canola, low, but non-freezing, temperatures prior to flowering slow the rate of plant development, delay flowering, slow the rate of flower opening, and reduce the amount of pollen shed, and in severe frost, pod abortion is evident in many species (Miller et al. 2001). The effect of low temperature prevailing at the early stages of plant growth is sometimes associated with the premature flowering in Chinese cabbage (Kalisz and Cebula 2001).

There are relatively few examples in which the effects of temperature stress on female reproductive organs have been investigated. For cold stress, a study on chickpeas showed reduced ovule size, reduced ovule viability, missing embryo sacs, and impaired pistil function in temperature-sensitive cultivars (Srinivasan et al. 1999). In rice (*Oryza sativa*) and maize, evidence suggests that ovule/female fertility is fairly tolerant of moderate cold stress (Hayase et al. 1969; Dupuis and Dumas 1990). More is known of the effects of temperature stress on male reproductive structures (Barnabás et al. 2008; Thakur et al. 2010). Cold temperatures can induce pollen sterility, which may be due to disruption of sugar metabolism in the tapetum, ultimately abolishing starch accumulation (i.e., energy reserves) in the pollen grains (Oliver et al. 2005). Evidence suggests that cold-induced disruption of anther sugar transport and corresponding pollen sterility is signaled by abscisic acid, in part by downregulating expression of cell wall invertase and monosaccharide transporters (Oliver et al. 2007).

In the final category of temperature-induced reproductive effects, temperature stress may directly affect the development of male and female gametes. The effects of temperature stress on male gametes are well documented for numerous plant species. Pollen maturation, viability, germination ability, and pollen tube growth can be negatively affected by heat (Dupuis and Dumas 1990; Peet et al. 1998; Prasad et al. 1999; Aloni et al. 2001; Young et al. 2004). Cold stress can disrupt mitosis I and II, thus preventing rice microspores from maturing into tricellular pollen grains (Sataka and Hayase 1970). Cold temperatures inhibited pollen germination and shortened pollen tubes in *Trifolium repens* and chickpea (Jakobsen and Martens 1994; Srinivasan et al. 1999).

Dupuis and Dumas (1990) studied the response of maize male and female reproductive tissues to high- and low-temperature stresses by treating male and female tissues separately to 4 h of control temperature (28 °C) or temperature stress (4 °C or 40 °C) and then using them for in vitro fertilization with a complementary reproductive organ. No significant difference in fertilization was found between male maize spikelets treated with control (28 °C), cold (4 °C), or heat (40 °C) temperatures and then crossed with control pollen at 28 °C. Fertilization rates from crosses using pollen submitted to cold (4 °C) stress were not significantly different from the control. However, fertilization was unsuccessful when using pollen submitted to heat stress prior to application to the spikelet, suggesting that maize pollen maturation is highly sensitive to heat stress in a manner similar to the tomato example described above. In addition, in vitro fertilization was unsuccessful when pollinated spikelets were treated to 6 h of heat stress immediately following application of pollen grown under control temperatures. This indicates that the heat stress sensitivity of pollen extends beyond the early events of pollen grain maturation.

The young microspore stage is the most susceptible stage to cold injury in rice plants (Sataka and Hayase 1970). The young microspore stage occurs approximately 10–12 days prior to heading (Sataka and Hayase 1970; Heenan 1984). The threshold temperature that induces spikelet sterility at the young microspore stage differs between cultivars and depends on factors such as the duration of low-temperature exposure and growing conditions. The critical air temperature that induces cold damage depends on the cultivar (Satake 1976a). The duration of low temperature is also an essential element determining the extent of cold damage. Spikelet sterility increases with the increase in duration of low temperature during the young microspore stage (Hayase et al. 1969; Ito 1976, 1978; Nishiyama 1978; Heenan 1984).

Temperature stress during flowering is considered a major challenge to yield in many chickpea production areas. Mean daily temperatures at or exceeding 15 °C can cause flower abortion (Clarke and Siddique 1998). Temperatures of less than 10 °C during flowering induce flower shedding, low pod set, and ultimately poor seed set. Poor pollen viability and germination are the main reasons for low pod set (Savithri et al. 1980). Cold temperature also reduces partitioning of assimilates to the vegetative parts, resulting in reduced harvest index. This reduction in harvest index is more common in south Asia and Australia than in other production areas (Siddique and Sedgley 1986; Saxena 1990). Poor pod set in chickpea can occur due to the failure of male or female floral parts, or both. Low temperatures at flowering can affect anther dehiscence. Mean daily temperatures of 15 °C can also reduce anther dehiscence and pollen load on stigma (Srinivasan et al. 1999). However, at a similar temperature, pollen viability and pollen germination on the stigma were higher in the cold-tolerant lines than in the cold-sensitive cultivars (Srinivasan et al. 1999). Pollen function was clearly more sensitive to temperature change than pistil function (esterase activity). Clarke and Siddique (2004) showed that pollen viability and pollen germination on the stigma were the primary reasons why pod set in chickpea was reduced during low-temperature stress. Pollen sensitivity to low temperatures was identified at 5 and 9 days before anthesis (Clarke 2001). Clarke and Siddique (2004) and Srinivasan et al. (1999) also observed that low temperature did not affect the pistil function, i.e., esterase activity. However, pollen tube growth on the styles of sensitive genotypes was retarded due to cold temperatures. Short pollen tube length at low temperature (15 °C) in in vitro germination tests was observed at 20 and 25 °C by Savithri et al. (1980). Ultimately, pod set is reduced by low temperature as observed by Srinivasan et al. (1999). They found that pod set was reduced at a low-temperature regime of 15/5 °C compared with the control (25/15 °C).

Like other crops, rice also shows response to low-temperature stress during different developmental stages (Zeng et al. 2017). The reproductive stage in rice is a critical period. Farrell et al. (2001, 2006b) reported that rice was susceptible to abundant low-temperature-induced damage during the young microspore stage, and 12 h of exposure to low temperature can cause spikelet sterility. Low-temperature-induced spikelet sterility during the microsporogenesis stage is considered genotype dependent, which for cold-tolerant genotypes was a nighttime air temperature of 15 °C for 4 consecutive days and for cold-sensitive genotypes was a nighttime air temperature between 17 and 19 °C (Satake 1969).

1.6.4 Yield and Yield-Related Attributes

Suboptimal temperatures reduce the growth and metabolic activities, leading to reduction in agronomic and yield attributes of the crops. Rizza et al. (1994) reported that drought and chilling stress severely impaired barley productivity. The chilling stress in chickpea at reproductive stage led to flower abortion and poor pod setting (Clarke and Siddique 2004; Nayyar et al. 2005a). In temperate growing areas, cold temperature is responsible for 30–40% reduction in rice yields because chilling stress resulted in degeneration of spikelets, panicle deformation, and poor spikelet fertility (Andaya and Mackill 2003a). Low temperature stress during reproductive developmental stages of grain crops may induce flower abscission, ovule abortion, pollen sterility, pollen tube distortion, and poor fruit set, thus resulting in reduced final yields (Thakur et al. 2010). Whaley et al. (2004) found that chilling stress during stem elongation reduced the internode extension, caused spikelet death, and lowered biomass accumulation and grain yield in wheat. Low temperature at jointing and booting stages of wheat significantly decreased the number of productive tillers per plant (Li et al. 2015) and negatively affected the development of young spikes (Thakur et al. 2010). It can be said that chilling temperature adversely affects yield and yield components of crop plants. Reductions in these traits could be the outcome of stress-induced modifications in metabolic and physiological activities and corresponding negative implication for plant reproductive organs with reduced productivity. Onset of chilling in harmful to plants; nevertheless, plant reproductive stage is perhaps the most sensitive stage to these stress factors than pre- and post-reproductive stages.

Low-temperature-induced spikelet sterility, in particular during the reproductive period, could be related to the physiological response mechanisms in the anther, pollen, and stigma. Earlier studies have indicated that short anther dehiscence, poor pollen grains, and low pollen germination on stigmas were all highly correlated with spikelet sterility under low-temperature-induced stress (Matsui et al. 2001; Prasad et al. 2006a). Moreover, Satake (1976b) considered that the main reason for this could be attributed to low-temperature-induced damage of male organs, such as pollen, not of female organs, such as the stigma. Hence, if pollen grains were deficient at anthesis, they could impede pollination, leading to spikelet sterility. These results might be similar to high-temperature stress in rice. A larger anther and longer stigma (Suzuki 1981, 1982) were found to contribute to the increase in the basal pore length and found to improve low-temperature tolerance during the reproductive stage, and larger anthers and longer stigmas also ultimately benefit successful pollination (Matsui and Kagata 2003). In addition, panicle exsertion (Rang et al. 2011) and peduncle length (He et al. 2009) have been reported to be responsible for enhanced spikelet sterility under water stress, as has high temperature (Rang et al. 2011), but similar results have not been documented under low-temperature stress.

In most rice genotypes, flowering and anthesis undergo a 5-day period with most spikelets at anthesis; the similar patterns of flowering have been reported in *indica* and *japonica* subspecies (Nishiyama and Blanco 1980; Prasad et al. 2006b). Studies have shown that flowering (containing anthesis and fertilization), as a developmental stage in rice, in particular at the booting stage (microsporogenesis), is considered most susceptible to low temperature (Farrell et al. 2006a; Maite and Ellis 2015). The influences on spikelet flowering in rice have been verified under high temperature (Rang et al. 2011). The climatic effects on the time of day and the duration of flowering in rice also differ (Julia and Dingkuhn 2012). Jagadish et al. (2007) analyzed the flowering patterns and the time of day of spikelets reaching anthesis under high temperature in *indica* and *japonica* rice.

Post-anthesis temperature stress, particularly after commencement of pod set, can cause significant pod abortion and decreased grain filling. In chickpea, cold stress decreased the rate and duration of grain filling and produced smaller seeds (Nayyar et al. 2007; Kaur et al. 2008). Chickpea average seed weight and size decreased by 41% and 24% at 13/5 °C, respectively, as compared to 28/17 °C, owing largely to a reduction in seed filling duration from 20 days (non-stressed) to 14 days (cold stressed) (Nayyar et al. 2005b). Similarly, low post-anthesis temperature reduced yield by 1.3 t/ha in Australia, during 2009 (Moore et al. 2010). In a controlled environment study, Wang et al. (2006) reported a grain yield reduction of 33–39% for post-anthesis heat stress compared with pre-anthesis heat stress. This was possibly due to poor remobilization of photosynthates to the grain, thus lowering seed weight and seed number per plant (Wang et al. 2006). Both temperature stress extremes influenced seed shape and seed coat color. Generally, temperature stress reduces cotyledon cell number, cell expansion, grain filling rate, and ultimately seed weight (Munier-Jolain and Ney 1998). Temperature stress can influence grain filling by altering the concentration of hormones, particularly abscisic acid and enzymes, in plant tissue. As discussed earlier, abscisic acid plays a significant role in both cold and heat stress tolerance in chickpea (Nayyar et al. 2005a; Kumar et al. 2012a) and is generally downregulated under stress. Exogenous application of abscisic acid increased tolerance to cold stress by improving survival rate through the reproductive stage (Nayyar et al. 2005a). The exogenous application of abscisic acid decreased electrolyte leakage and increased pollen viability, germination, flower retention, pod set, seed size, and grain yield (Kumar et al. 2008). Similarly, Kumar et al. (2012b) showed that abscisic acid treatment reduced oxidative injury in chickpea under high temperature.

1.6.5 Temperature Sensitivity of Different Growth Phases

The rate of development of plants is generally responsive to photoperiod for only part of their life cycle (i.e., between emergence and flowering), though some post-flowering processes, such as the rate of flower initiation, are affected by photoperiod and can therefore influence the duration of the seed filling period (Summerfield et al. 1998). In contrast, temperature affects the rate of development throughout the life cycle (Roberts et al. 1996). Three distinct stages of pre-flowering development can be identified in plants, namely, the pre-inductive or juvenile, inductive, and post-inductive phases. Photoperiod only affects the duration of the inductive phase (Yin et al. 1997b), and in photoperiod-sensitive cultivars, the effects of photoperiod are usually substantial and are the major determinant of flowering time. In most annual crop species examined, the duration of the pre-inductive or juvenile phase is short days, not weeks: soybean (Collinson et al. 1992) and lentil (Roberts et al. 1986). A long juvenile trait has also been identified in soybean (Hinson 1989).

Typically, the response to temperature of many plant processes, including the rate of development (inverse of days to flower), is described in terms of cardinal temperatures (base or minimum, T_b; optimum, T_o; and maximum or lethal, T_m temperature) and the thermal sum ($\boldsymbol{\theta}$) or rate (inverse of the duration). This response can be quantified by simple bilinear or broken stick models with a sharply defined T_o or by linear models with a flat-top/plateau response that have a maximum rate of development over a range of temperatures (Craufurd and Wheeler 2009). Curvilinear quadratic or beta-function models (Yin et al. 1997a) that have a near optimal rate of development over a range of temperatures have also been used. This basic response to temperature is affected by photoperiod in photoperiod-sensitive genotypes and therefore must be determined under short-day or inductive photoperiod (Roberts and Summerfield 1987; Yan and Wallace 1998).

1.7 Response of Perennial Crops to Low-Temperature Stress

Like annual crops, perennial crops are also sensitive to extreme temperature. Fruits and nut trees are important crop plants which often face extreme temperature stress-induced damages. Every fruit tree species has a range of optimum temperatures (Table 1.2) above or below which the growth and yield are markedly reduced (Hasanuzzaman et al. 2013). The mean temperature range for the optimum growth of most tropical fruits is about 24–30 °C (Whiley et al. 1989; Dinesh and Reddy

Table 1.2 Ranges of optimum temperature for seed germination in some perineal crops

Tree species	Optimum temperature (°C)	References
Mango—*Mangifera indica*	24–27	Davanport (2009)
Lychee—*Litchi chinensis*	25–35	Tindall (1994)
Guava—*Psidium guajava*	23–28	Verheij and Coronel (1992)
Jackfruit—*Artocarpus heterophyllus*	16–28	Haq (2006)
Longan—*Dimocarpus longan*	20–25	Verheij and Coronel (1992)
Durian—*Durio zibethinus*	24–30	Verheij and Coronel (1992)
Rambutan—*Nephelium lappaceum*	24–32	Verheij and Coronel (1992)
Banana—*Musa* spp.	20–35	Viktorova (1983)
Grapes—*Vitis vinifera*	10–35	White et al. (2006)
Coconut—*Cocos nucifera*	20–32	Louis and Annappan (1980)
Cashew—*Anacardium occidentale*	30–35	Bezerra et al. (2007)
Wild jujube—*Ziziphus lotus*	35	Maraghni et al. (2010)

2012). Mango (*Mangifera indica*) tree can tolerate high temperature up to 48 °C only for a certain period of time (Mukherjee and Litz 2009) but has only partial tolerance to low temperature. Schaffer et al. (2009) observed that monoembryonic mango cultivars tend to be more low-temperature-tolerant than polyembryonic cultivars. However, several studies have shown that low temperature promotes reproductive morphogenesis in mango. Dinesh and Reddy (2012) studied the responses of fruit trees to temperature and observed differential responses to temperature in different fruit species and observed genotypic differences toward the response of low temperature. Tindall (1994) reported that rambutan is adapted to warm tropical climates of approximately 22–30 °C and is sensitive to temperatures below 10 °C. In mangosteen, low temperature (below 20 °C) markedly slowed down the overall growth of the tree. Low temperatures during floral development in fruit trees result in fewer healthy flowers, but high temperatures during floral development cause dryness and lead to sterility. Larcher (1995) showed that banana, papaya, mango, grape, and orange are sensitive to low temperature and lose their quality and productivity. Aslamarz et al. (2009) studied several walnut cultivars and genotypes and found that temperature had a great influence on the performance of trees and high-temperature regimes result in the best-quality fruits but extremely high temperatures for extended periods of time are known to cause damage.

1.8 Mechanisms of Acclimation to Low Non-freezing Temperatures

Cold acclimation is a process whereby plants increase their freezing tolerance after exposure to a short period of low but non-freezing cold temperatures. Whereas cold acclimation can be induced at 12 °C, vernalization requires lower temperatures, suggesting that cold acclimation and vernalization pathways are independent (Bond et al. 2011). Exposure to long-term cold promotes flowering in vernalization-sensitive *Arabidopsis* plants via the vernalization pathway. In contrast, exposure to short-term cold or overexpression of cold-responsive genes delays flowering by activating the transcription factor *FLOWERING LOCUS C* (Seo et al. 2009; Jung et al. 2012). The primary mechanisms involved in cold acclimation are related to a number of processes, including molecular and physiological modifications occurring in plant membranes, the accumulation of cytosolic calcium ion, increased levels of reactive oxygen species and the activation of reactive oxygen species scavenger systems, changes in the expression of cold-related genes and transcription factors, alterations in protein and sugar synthesis, proline accumulation, and biochemical changes that affect photosynthesis (Bhattacharya 2018). The injuries caused by low temperature are generally categorized into chilling stress (temperatures above 0 °C) and freezing stress (temperatures below 0 °C) (Li et al. 2018; Huang et al. 2014). Chilling stress often leads to growth and photosynthesis reduction, leaf wilting and chlorosis or even necrosis, cellular membrane damages, as well as oxidative stress in plants (Suzuki and Mittler 2006). In addition, temperatures below 0 °C caused freezing stress, which resulted in the formation of ice crystals within the cell, mechanical damages, as well as metabolic dysfunction in plants (Xin and Browse 2000). However, plants have evolved complex mechanisms to tolerate chilling and freezing stresses, such as accumulation of carbohydrates and proteins, resulting in large quantities of soluble sugars, amino acids, and cold-induced stress-related proteins (Trischuk et al. 2014), as well as hormone homeostasis inducing the gene expression that functions to stabilize membranes against freezing-induced injury (Thomashow 1999; Fowler and Thomashow 2002).

1.8.1 The Role of Carbohydrates

The acquisition of freezing tolerance is called cold acclimation, which is generally initiated after a period of decreases in above 0 °C temperatures and photoperiod (Levitt 1980). Growth inhibition resulting from low temperature reduced the capacity for energy utilization, which leads to feedback inhibition of photosynthesis (Huang et al. 2014). Differences in the capacity to minimize photoinhibition and recover photosynthesis during cold acclimation have been shown to contribute to intra- and interspecific differences in freezing tolerance (Pocock et al. 2001). Carbohydrates are synthesized from a complex series of reactions involving photosynthesis in plants (Huang et al. 2014). Changes in the three major types of water-soluble carbohydrates, monosaccharides (glucose and fructose), disaccharides (sucrose, trehalose), and fructans, have often been reported in plants subjected to low temperature (Theocharis et al. 2012a). Those sugars may serve as typical osmoprotectant, cellular membrane stabilizer, scavengers of reactive oxygen species, and signaling molecules in plants to tolerant chilling and freezing stress (Theocharis et al. 2012a). Many studies have showed that higher accumulation of sugars, or individual sugar fractions such as starch (Zhang et al. 2006), fructans (Dionne et al. 2001), and sucrose (Zhang et al. 2006), during cold acclimation is associated with great cold tolerance in turfgrass (Cai et al. 2004; Hoffman et al. 2010). Sucrose consists of the monosaccharides glucose and fructose, and it plays an important role in growth, development, storage, signaling, and stress acclimation (Janska et al. 2010). Leaf starch potentially contributed carbon to carbohydrate accumulation in response to low temperature (Sicher 2011). Fructans often accumulated during cold acclimation in cereals and grasses, which serve to stabilize cell membranes (Livingston III and Heyer 2009). Furthermore, trehalose, a nonreducing disaccharide of glucose, has been implicated in modulating levels of freezing tolerance and may also be involved in starch accumulation (Theocharis et al. 2012a).

1.8.2 The Role of Osmolytes

Osmolytes are compounds soluble in the solution within a cell, or in the surrounding fluid, which play a crucial role in maintaining fluid balance and cell volume. For instance, when a cell swells due to external osmotic pressure, membrane channels open and allow efflux of osmolytes, which carry water with them, restoring normal cell volume (Khan et al. 2017). These osmolytes allow feasible water uptake by the plants (Flowers 2004; Ashraf and Foolad 2007) and scavenge free radicals (Okuma et al. 2004; Banu et al. 2010). Osmolytes like reducing and non reducing sugars and proline, are abundant, while glycine betaine is present in very low concentration or even absent in some plants that are not capable to accumulate it (Maqsood et al. 2006; Ahmed et al. 2010). These osmolytes protect plants from stresses by detoxification of reactive oxygen species, stabilizing proteins and enzymes, cellular osmotic adjustment, and protection of membrane integrity (Verbruggen and Hermans 2008). Accumulation of sugars plays an important role during cold acclimation, which is frequently observed in the leaves of *Arabidopsis*, tomato, and petunia (Venema et al. 1999a, b; Usadel et al. 2008; Nägele et al. 2011; Bauerfeind et al. 2015). Glucose and fructose could be transported from the apoplast into the cytosol at the expense of energy, where they play a protective role in the cell. In addition to the protective role, under natural conditions, the content of soluble sugars increases during the onset of winter time when plants are exposed to cold, whereas soluble sugars decline in spring when plants are deacclimating (Siminovitch 1981). A recent study on *Petunia hybrida* indicates that cold-induced sugar accumulation in source leaves could have a protective role against freezing temperatures, which impedes carbohydrate transport and utilization in the growth sinks (Bauerfeind et al. 2015). Moreover, plants have evolved various mechanisms to overcome the changes during acclimation to cold-like accumulation of soluble sugars (Wanner and Junttila 1999), amino acids (Shao et al. 2006a), and glycine betaine (Kishitani et al. 1994) and alteration in cryobehavior and membrane composition (Örvar et al. 2000; Shao et al. 2008). These play a pivotal role in the protection of cells from low-temperature injury. During cold acclimation, sugar level increases under artificial conditions (Shao et al. 2006b, 2008), which could be due to improved transport mediated by sucrose synthase or invertase (Sauer 2007; Iordachescu and Imai 2008). Moreover, glycine betaine occurs abundantly in chloroplast and plays a pivotal role in the protection of thylakoid membrane, which maintains proper efficiency of photosystem II (Robinson and Jones 1986; Genard et al. 1991). Glycine betaine accumulates during stress conditions in many crop plants such as spinach, barley, wheat, sugar beet, and sorghum (Rhodes and Hanson 1993; Yang et al. 2003). Sensitive genotypes accumulate less glycine betaine than tolerant genotypes in response to stress. Some plants such as *Arabidopsis*, mustard, tobacco, and rice do not produce glycine betaine under non-stress or stress conditions (Rhodes and Hanson 1993). Exogenous application of glycine betaine to non-accumulating or low-accumulating plants could help to reduce the adverse effects of environmental stress (Yang and Lu 2005). Furthermore, exogenous application of glycine betaine improved freezing tolerance in wheat plants (Allard et al. 1998), as well as growth of tobacco and *Arabidopsis* under low-temperature stress (Somersalo et al. 1996; WeiBing and Rajashekar 2001). The transport rate of glycine betaine increases under chilling or osmotic stress, and the addition of glycine betaine to the growth medium conferred tolerance to both salt and chilling stress in the cells of *Listeria monocytogenes* (Ko et al. 1994). Moreover, proline widely occurs in higher plants and accumulates in larger amounts in response to stress (Ozturk and Demir 2002; Hsu et al. 2003; Kavi Kishore et al.

2005; Ashraf and Foolad 2007). It was considered as an inert compatible osmolyte that protects subcellular structure and macromolecules under osmotic stress (Kavi Kishore et al. 2005). The proline acts as an osmolyte, stabilizes subcellular stress, buffers redox potentials, and scavenges free radicals (Szabados and Savouré 2010). Proline plays varied roles in plant stress, such as a mediator of osmotic adjustment, inducer of osmotic-related genes, stabilizer of proteins and membranes, and scavenger of reactive oxygen species (Verbruggen and Hermans 2008; Szabados and Savouré 2010; Theocharis et al. 2012b). It maintains proper $NAD^+/NADP^+$ ratio and alleviates cytoplasmic acidosis (Hare and Cress 1997). Proline has been shown to function as a molecular chaperon which protects protein integrity and enhances the activity of different enzymes. The most chilling-sensitive plants that accumulate proline under low-temperature stress conditions do not acquire cold tolerance (Kushad and Yelenosky 1987), unless a high concentration of proline was applied during stress (Xin and Li 1993).

Low- and high-temperature stress can reduce growth, development, yield, and quality, which causes valuable crop loss in terms of productivity of tomato (Khan et al. 2015a). Low-temperature stress causes profound morphological and physiological changes that adversely affect plant growth, development, and ultimately yield and quality (Sayyari et al. 2013; Khan et al. 2015a). Tomato crop is quite sensitive to cold stress and shows reduction in overall growth, yield, and quality contributing parameters (Shi et al. 2012; Khan et al. 2015a). Low-temperature stress adversely affects the reproductive phase, leading to increased flower abortion and fruit set. Plants grown in cold stress have significantly reduced the number of fruits and fruit yield per plant, and the rate of reduction in fruit number has been reported to be in proportion to the level of cold stress in tomato (Khan et al. 2015a). At harvest, fruits developed on plants grown in low-temperature stress possess significantly lower values for lycopene and β-carotene content. This loss is increased further with decrease in the level of low-temperature stress. Phenolic compound-induced alleviation of low-temperature stress has been reported in tomato (Orabi et al. 2015), cucumber (Zhang et al. 2015b), and watermelon (Sayyari et al. 2013). The application of phenolic compounds assists in improving tolerance against low-temperature stress in plants (Meena et al. 2018).

1.8.3 The Role of Plant Growth Hormones

Plant hormones may serve as signals in regulating plant growth, development, and adaptation to environmental stress (Strivastava 2002). For abscisic acid, they can affect the diverse processes, such as leaf senescence, seed dormancy, and germination as well as cell division and elongation (Finkelstein and Gibson 2002). In addition, it has also been reported to be involved in the responses to abiotic stresses including chilling and freezing in plants (Finkelstein and Gibson 2002). Cytokinins are essentially involved in various plant developmental processes including cell division and enlargement, chloroplast biogenesis, nutrient mobilization, and leaf senescence (Ryu and Cho 2015), as well as facilitate the responses to delay both stomatal closure and leaf senescence under abiotic stresses (Ryu and Cho 2015). Auxin such as indole-3-acetic acid can promote root initiation and also delay plant senescence (Zhang et al. 2009). Bioactive gibberellins are involved in plant growth and development such as leaf expansion, stem elongation, and flowering (Ryu and Cho 2015). Jasmonic acid is a critical signaling molecule for diverse developmental processes and defense responses in plants (Llanes et al. 2016) and has been shown to induce a response similar to that of abscisic acid for alleviating chilling injury (Hu et al. 2013). Salicylic acid has also been demonstrated to be involved in the responses of plants to a broad range of abiotic stresses, including low temperature (Khan et al. 2015b).

1.8.4 Modifications to Plant Cell Membranes

Membranes are a primary site of cold-induced injury. Several studies have demonstrated that membrane rigidification, coupled with cytoskeletal rearrangements, calcium influxes, and the activation of mitogen-activated protein kinase cascades, triggers low-temperature responses (Sangwan et al. 2002). The lipid composition of the plasma membrane and chloroplast envelopes in acclimated plants changes such that the threshold temperature for membrane damage is lowered relative to that for non-acclimated plants (Uemura and Steponkus 1999). This is achieved by increasing the cold-adapted membranes' unsaturated fatty acid content, which makes them more fluid (Vogg et al. 1998). The process of cold acclimation promotes the stabilization of membranes, which prevents damage leading to cell death. The acclimation process also activates mechanisms that protect membrane fluidity by ensuring the optimal activity of associated enzymes (Matteucci et al. 2011). Photosynthesis and its related pigments at the physiological level are strongly affected by exposure to cold. The cessation of growth resulting from cold stress reduces the capacity for energy utilization, causing feedback inhibition of photosynthesis (Ruelland and Zachowski 2010). In cold-acclimated winter annuals, photosynthetic activity is maintained by increases in the abundance and activity of several Calvin cycle enzymes (Goulas et al. 2006). This recovery is associated with elevated levels of thylakoid plastoquinone A and a concomitant rise in the apparent size of the intersystem electron donor pool to PSI (Baena-Gonzalez et al. 2001). Consequently, non-photochemical quenching increases in cold-stressed leaves in parallel with increased zeaxanthin levels to compensate for the reduced electron consumption by photosynthesis. Ruelland and Zachowski (2010) reported that energy dissipation via non-photochemical quenching and electron transport was not only enhanced following cold acclimation but also contributed to protection from oxidative damage:

- Although xanthophylls are not considered photosynthetic pigments per se, xanthophylls (notably, violaxanthin, antheraxanthin, and zeaxanthin) help in protecting the photosystems, and their abundance increases at low temperatures (Ivanov et al. 2006). Xanthophylls have structural roles and act as natural antioxidants, quenching triplet chlorophyll and singlet oxygen, which are potentially harmful to the chloroplast (Passarini et al. 2009; Han et al. 2010). It has also been postulated that unbound zeaxanthin and other carotenoids may also stabilize thylakoid membranes against putative peroxidative damage and heat stress (Laugier et al. 2010).

- Flavonoids accumulate in leaves and stems in response to low temperatures. They are synthesized via the phenylpropanoid pathway, which is controlled by key enzymes, including phenylalanine ammonia-lyase and chalcone synthase (Sharma et al. 2007). It has been reported that cold stress induces transcriptomic modifications that increase flavonoid biosynthesis, including reactions involved in anthocyanin biosynthesis and the metabolic pathways that supply it (Crifo et al. 2011).

1.8.5 The Role of Reactive Oxygen Species

Reactive oxygen species have been reported to be involved in processes leading to plant stress acclimation (Suzuki et al. 2011). This finding indicates that reactive oxygen species are not simply toxic by-products of metabolism but act as signaling molecules that modulate the expression of various genes, including those encoding antioxidant enzymes and modulators of H_2O_2 production (Suzuki et al. 2011). In addition, low-temperature stress has been reported to cause significant increases in the levels of the soluble nonenzymatic antioxidants ascorbate and glutathione, as well as the activity of the main NADPH-generating dehydrogenases (Airaki et al. 2011).

Chilling or cold stress leads to reduced growth and leaf expansion (Sowinski et al. 2005; Rymen et al. 2007), wilting (Bagnall et al. 1983), and chlorosis (Yoshida et al. 1996b) in plants and may even lead to necrosis (tissue death) of plants (Yadav 2010). Rice is sensitive to chilling stress, and its persistence leads to poor germination, stunted seedling, yellowing or withering, and decreased tillering (Mukhopadhyay et al. 2004). Effect of low temperature, both directly through metabolic inhibition (decreased metabolism, freezing the cell membrane, and loss of membrane (Jewell et al. 2010)) and indirectly through the reaction caused by cold osmotic reactions (dehydration and freezing due to cold), prevent full expression of the genetic potential of plants. The destruction of the plasma membrane of plant cells and pigments has been reported as the major side effect of cold stress in plants (Steponkus et al. 1993). It has generally been noticed that plants native to warm habitats, such as soybean, corn, and cotton, exhibit symptoms of injury upon exposure to low non-freezing temperatures or hypothermia (Jenabiyan et al. 2014). The degree of damage, however, depends on the sensitivity of plants to cold stress and may vary from one plant to another. Angadi et al. (2000a, b) observed that temperatures below 10 °C negatively influenced the growth and stem elongation in *Brassica napus*, *Brassica rapa*, and *Raphanus sativus* plants. Also Rivero et al. (2001) reported that low-temperature stress decreased plant biomass and antioxidant enzyme activity in tomato and watermelon plants. Genotype response to different environments has a great deal of importance in selecting the process of these lines against abiotic stress tolerance (Hazer Jaribi et al. 2013). Soybean is sensitive to temperatures below 15 °C, which is reflected in the metabolism, growth, development, and yield changes (Balestrasse et al. 2010). According to the findings of Hume and Jackson (1981), a single night of dark chilling, with a minimum temperature of 8 °C, can reduce plant growth and prevent pod formation in soybean. The damage will be more severe if cold stress occurs with high light intensity. Furthermore, the combination of high light intensities and low temperatures, such as those experienced on cold but sunny mornings in spring, can cause irreversible damage to young soybean seedlings (Balestrasse et al. 2010). Moosavi (2007) on forage sorghum and Mohammadi-Nikpoor and Keshavarzi (1995) on safflower reported an increase in plant height with the decrease of light intensity.

The accumulation of osmolytes is also related to low-temperature tolerance. Free proline, soluble proteins, and soluble sugar are among the major osmolytes. Accumulation of proline was reported in many plant species under diverse abiotic stress conditions, and its role in plant stress tolerance was widely investigated and discussed (Delauney and Verma 1993). The amount of soluble protein increased under low-temperature conditions (Bravo et al. 1999), and their abundance is correlated to the degree of freezing tolerance (Karimzadeh et al. 2006).

1.8.6 Interaction(s) of Genes

Genes expressed under stress conditions may increase tolerance to growth retardation, cold, and high salt stress in transgenic rice and *Arabidopsis* (Ito et al. 2006). It has been determined that in comparison with other cereals like wheat and barley, rice is more sensitive to cold stress (Wen et al. 2002). Low temperature stress activate many genes, which either act as protectants toward stress-induced damage or regulate the expression of other genes and transduce signal (Shinozaki et al. 2003). Davletova et al. (2011) reported that, in contrast to many stress-specific signaling and regulatory genes, Zat12, which is a zinc finger protein, responds to various abiotic and biotic stresses and plays an important role in abiotic stress signaling in *Arabidopsis*. Similarly, MYB transcription factors play a central role in plant growth and respond to stress conditions (Yang et al. 2012). OsMYB2 encodes a MYB transcription factor (stress-responsive element), which is involved in the regulation of various pathways, leading to tolerance of rice to dehydration, stress, salt, and cold. Low-temperature exposure often induces a variety of biochemical, physiological, and enzymatic changes in plant, which can result in an acclimation response (Hughes and Dunn 1996).

Low temperature strongly suppresses many proteins and genes of rice seedlings but also induces several genes and proteins (Christie et al. 1991; Minhas and Grover 1999; Sung et al. 2003). These proteins have been considered to be involved in the adaptation to low-temperature stress conditions (Guy 1990; Thomashow 1999; Sung et al. 2003). Low temperature has been found to rapidly induce gene expression of alcohol dehydrogenase in rice seedlings (Christie et al. 1991; Minhas and Grover

1999). Alcohol dehydrogenase is an enzyme involved in ethanolic fermentation and is the most studied enzyme in relation to anaerobiosis (Kennedy et al. 1992; Ricard et al. 1994; Tadege et al. 1999). However, the biological role of alcohol dehydrogenase is so far apparent only under anaerobiosis, and the activity of ethanolic fermentation and ethanol concentration in the seedlings under low-temperature conditions has not yet been determined (Naguchi 2007).

Cold temperatures can induce pollen sterility, which may be due to disruption of sugar metabolism in the tapetum, ultimately abolishing starch accumulation (i.e., energy reserves) in the pollen grains (Oliver et al. 2005). Evidence suggests that, cold-induced disruption of anther sugar transport and corresponding pollen sterility is signaled by abscisic acid, in part by downregulating expression of cell wall invertase and monosaccharide transporters (Oliver et al. 2007). Cold stress can disrupt mitosis I and II, thus preventing rice microspores from maturing into tricellular pollen grains (Sataka and Hayase 1970). Cold temperatures inhibited pollen germination and shortened pollen tubes in *Trifolium repens* and chickpea (Srinivasan et al. 1999).

High temperatures cause poor anther dehiscence characterized by tight closure of the locules, which was shown to reduce pollen dispersal in rice and tomato (*Solanum lycopersicum*) (Matsui and Omasa 2002; Sato et al. 2002). Cold temperatures can induce pollen sterility, which may be due to disruption of sugar metabolism in the tapetum, ultimately abolishing starch accumulation (i.e. energy reserves) in the pollen grains (Oliver et al. 2005). Evidence suggests that, cold-induced disruption of anther sugar transport and corresponding pollen sterility is signaled by abscisic acid, in part by downregulating expression of cell wall invertase and monosaccharide transporters (Oliver et al. 2007). Heat stress reduces carbohydrate deposition in pollen grains (Jain et al. 2007). The preexisting carbohydrate reserves within the pollen grains fuel tube growth, but pollens later switch to using carbohydrates provided by the transmitting tract of the style (Herrero and Arbeloa 1989). Cotton (*Gossypium hirsutum*) flowers exposed to moderately high temperatures showed reduced carbohydrate reserves (particularly sucrose) and ATP production in the pistils, which correlated with a heat stress-induced decline in net photosynthesis (Snider et al. 2009).

Under low temperature, plasma membrane is thought to be the primary site of injury because of its central role in the regulation of various cellular processes (Takahashi et al. 2013). Low temperature stress result in peroxidation of the membrane lipids and the accumulation of malondialdehyde, which binds to some proteins and destructs the membrane system (Chen 1991) and therefore has been considered as an important indicator of membrane system injuries (Fan et al. 2012). Some studies showed that more malondialdehyde will be generated in plants with less tolerance to stress (Zhai et al. 2013). Plants under abiotic stress have evolved a defense system against oxidative stress by increasing the activities of reactive oxygen species-scavenging enzymes, such as superoxide dismutase, catalase, and peroxidase (Miller et al. 2010). Superoxide dismutase plays a crucial role in antioxidant defense because it catalyzes the dismutation of O_2^- into H_2O_2, and catalase and peroxidase reduces the H_2O_2 to H_2O and protects superoxide dismutase from oxidation (Tian et al. 2012). Therefore, the activities of these enzymes may be connected with plant tolerance to stress.

Classical genetic studies have demonstrated that the ability of plants to cold acclimate is a quantitative trait involving the action of many genes with small additive effects (Thomashow 1990). Many approaches have been made to determine the nature of genes with roles in freezing tolerance: the isolation and characterization of genes induced during cold acclimation, the isolation and characterization of mutants affected in freezing tolerance, and transcriptome analysis. Most molecular studies on plant responses to cold stress are focused on the mechanism of cold acclimation rather than on chilling tolerance. Nevertheless, recent evidence indicates that some of the molecular changes that occur during cold acclimation are also important for chilling tolerance (Gong et al. 2002; Dong et al. 2006b). It appears that chilling tolerance that is exhibited by temperate plants is not entirely constitutive and that at least part of it is developed during exposure to chilling temperatures (Jan et al. 2009). Numerous physiological and molecular changes occur during cold acclimation, and among them, the transcriptional activation and repression of genes by low temperature are of central importance (Thomashow 1999). Reprogramming of gene expression results in the accumulation not only of protective proteins but also of hundreds or more of metabolites, some of which are known to have protective effects.

Temperature-induced change in membrane fluidity is one of the immediate consequences in plants during temperature stresses and might represent a potential site of perception and/or injury (Horvath et al. 1998; Örvar et al. 2000). The importance of proper membrane fluidity in temperature tolerance has been delineated by mutation analysis and transgenic and physiological studies. At low temperature, greater membrane lipid unsaturation appears to be crucial for optimum membrane function. The accumulation of sucrose and other simple sugars that typically occurs with cold acclimation also seems likely to contribute to the stabilization of membranes as these molecules can protect membranes against freeze-induced damage in vitro (Strauss and Hauser 1986; Anchordoguy et al. 1987). In addition, there is emerging evidence that certain novel hydrophilic and late embryogenesis abundant polypeptides also participate in the stabilization of membranes against freeze-induced injury. These hydrophilic and late embryogenesis abundant polypeptides are predicted to contain regions capable of forming amphipathic α-helices, which are shown to have strong effect on intrinsic curvature of monolayers and their propensity to form hexagonal II phase. They are said to defer their formation at lower temperatures (Epand et al. 1995). An additional hypothesis suggests that the extensive water-binding capacity of these hydrophilic proteins might provide a protective environment in the proximity of membranes during freezing and result in membrane stabilization. Although freezing injury is thought to result primarily from membrane lesions caused by cellular dehydration, additional factors may also contribute to freezing-induced cellular damage. There is evidence that freeze-induced production of reactive oxygen species contributes to membrane damage and that intercellular ice can form adhesions with cell walls and membranes and cause cell rupture (Olien and Smith 1977). In addition, there is evidence that protein denaturation occurs in plants at low temperature (Guy et al. 1998), which could potentially result in cellular damage. In these cases, the enhancement of antioxidative mechanisms (Aroca et al. 2003), increased levels of sugars in the apoplastic space (Livingston and Henson 1998), and the induction of genes encoding molecular chaperones (Guy and Li 1998), respectively, could have protective effects.

Cold stress can be just as lethal as heat stress. When a cell freezes, the water inside it expands as it turns to ice. This can cause the cell membrane to rupture and lead to cell death. Plants respond to cold temperatures by activating metabolic pathways that protect their cells from cold and freezing conditions. One protection strategy is to accumulate sugars, which decreases the temperature at which ice forms, similar to the effect of putting salt on roads. Another is to produce proteins that stabilize membranes to help them resist rupture. Transcription factors play an important role in activating metabolic pathways in response to changing environmental conditions; these are the proteins that regulate the expression of genes and determine when they are turned on or off to produce the proteins they encode. There are hundreds—even thousands—of different transcription factors that have specialized roles in plant growth and development by regulating different sets of genes at different times. Previously, research has shown that three DREB1-type transcription factors act as master switches in cold-inducible gene expression in a number of plant species. Dehydration-responsive element-binding (DREB1) genes that encode these transcription factors are themselves induced by cold; thus, cold triggers the production of the DREB1 transcription factors, which then act to turn on the expression of a large number of cold-induced genes encoding proteins that increase cold stress tolerance in plants. Other transcription factors known as CAMTAs (calmodulin-binding transcription activator) contribute to the cold-induced activation of the DREB1 transcription factors, but how cold activates any of these transcription factors is still unknown. The work of Kidokoro et al. (2017) revealed that plants recognize cold stress as two different signals. One signal is caused by a rapid temperature drop, and this signaling pathway functions during both day and night. The other signal is caused by both rapid and more gradual temperature decreases, but this signaling pathway functions only during the day. CAMTA transcription factors induce DREB1 gene expression in response to a rapid temperature decrease during day and night, whereas another type of transcription factor involved in circadian rhythms regulates DREB1 expression in response to both rapid and slow temperature decreases only during the day. The presence of these two signaling pathways suggests that two different sensor systems might function in the recognition of cold stress signals in plants. Identifying the components of these sensor systems and how they work to allow plants develop cold tolerance can help to breed plants that are less susceptible to cold damage.

In bacteria, nucleic acid-binding cold shock proteins accumulate at cold temperatures and function as transcription antiterminators or translational enhancers by destabilizing RNA secondary structure (Jones and Inouye 1994). Some cold shock proteins, domain-containing proteins, in plants are upregulated by cold stress and might function as RNA chaperones in the regulation of translation (Nakaminami et al. 2006; Kim et al. 2007). A different cold-responsive nucleic acid-binding protein, a zinc finger containing glycine-rich RNA-binding protein from *Arabidopsis* designated at RZ-1a, is also upregulated by cold stress, and genetic analysis supports its function in freezing tolerance (Kim et al. 2005). Another group of RNA chaperones, RNA helicases, are involved in every step of RNA metabolism. In cyanobacteria, cold-induced DEAD-box RNA helicases (RNA helicases of the DEAD-box family) are important players in RNA metabolism in most living organisms. Despite high conservation between DEAD-box proteins, they participate in many different processes: Linder and Jankowsky (2011) suggested to unwind cold-stabilized secondary structure in the 5′ untranslated region of RNA during cold stress (Yu and Owttrim 2000).

1.8.7 Alteration of Meiotic Recombination and Somatic Mutation

When exposed to an abiotic stress, plants can draw upon an extensive repertoire of strategies, genetic and epigenetic, that potentially enable them to better endure that stress, either during their own lifetimes or those of their progeny (Jorgensen 2004; Mirouze and Paszkowski 2011; Kinoshita and Seki 2014). Within developmental timeframes, responses to stress mobilize epigenetic changes, the most common of which involves the unsilencing of genes via the demethylation of DNA (Mirouze and Paszkowski 2011). These epigenetic revisions are often redeployed if the plant experiences stress again at some later point (Mirouze and Paszkowski 2011; Kinoshita and Seki 2014). Within evolutionary timeframes, stress has the potential to induce genuine, stably inherited genetic change (Bilichak and Kovalchuk 2016). Many forms of stress have been shown to increase the rates of meiotic recombination and various somatic mutations (Boyko et al. 2005; Molinier et al. 2006; Yao and Kovalchuk 2011; Avramova 2015), the latter being of relevance because plant gametes do not develop in isolation from the soma. It has even been speculated that plants have evolved the ability to selectively induce mutations in genes associated with a beneficial response to the initiating stress (McClintock 1984; Jorgensen 2004). In contrast, after studying the transgenerational effects of a wide range of stresses on rates of intrachromosomal recombination, it was concluded that responses in the progeny of stressed plants were "rare and stochastic" and that it is the genetic variation arising from such stochasticity which is the main or sole source of any evolutionary benefits (Pecinka et al. 2009).

One of the most important varieties of natural stress that can promote genetic variability in plants is extreme temperature. Nevertheless, the effects of heat and cold on the rates of meiotic recombination and somatic mutations have only been partially described, and the findings have not been consistent (Saini et al. 2017). Meiotic recombination has been the most studied type of change, either directly or via rates of chiasmata formation, i.e., range of species. Heat and cold have variously been reported to increase rates (Barth et al. 2000; Francis et al. 2007), decrease rates (Swanson 1942), or have no effect (Jensen 1981). For somatic mutations, where rates have been studied in the progeny of stressed parental plants, there is a similar story of variability. Microsatellite instability, particularly, and base substitution rates also have been reported to respond in a mixed manner to cold stress (Yao and Kovalchuk 2011). For intrachromosomal recombination, one study reported increased rates in response to cold in *Arabidopsis* (Yao and Kovalchuk 2011), while another, using two lines of *Arabidopsis*, found that cold had no effect and that heat affected only one line, which caused rates to decrease (Pecinka et al. 2009).

1.9 Phenological Development

Plants vary tremendously in their physiological behavior over the course of their life cycles. As plants change physiologically and morphologically through their various stages of growth, water and nutritional requirements will change considerably as well. Flowering is a critical life history trait that ensures seed production required for the survival of species. Over recent years, enormous scientific progress has been made to understand the molecular basis of this trait. Briefly, flowering, or, more broadly, the transition from vegetative to reproductive growth, requires genetic and epigenetic reprogramming and reallocation of metabolic and biochemical resources throughout the plant. According to the review articles by Andrés and Coupland (2012) and Blümel et al. (2014), flowering is regulated by an elaborate network of genetic pathways responsive to endogenous and environmental stimuli. This ensures that the transition to reproduction coincides with favorable conditions. In *Arabidopsis thaliana*, the vernalization (cooling of seed during germination in order to accelerate flowering when it is planted), photoperiod, circadian clock, sugar budget, age, thermosensory, autonomous, and gibberellin pathways converge on a few floral integrator genes that promote flowering (Kazan and Lyons 2016).

Environmental cues regulate the timing of critical life history transitions in plants, such as the shift from vegetative growth to reproduction (Anderson et al. 2011). Environmental signals can:

- Promote simultaneous flowering among conspecifics, thereby increasing the probability of outcrossing and decreasing interspecific hybridization.
- Prevent individuals from flowering during adverse periods.
- Increase the probability that seeds develop, are dispersed, and germinate under optimal conditions (Rathcke and Lacey 1985; Simpson and Dean 2002; Engelmann and Purugganan 2006; Inouye 2008).

Plants show a strong flowering time response to photoperiod, temperature, and the duration of winter (vernalization) (Rathcke and Lacey 1985; Kudoh et al. 1995; Michaels and Amasino 2000; Weinig et al. 2002; Lempe et al. 2005; Balasubramanian et al. 2006; Li et al. 2006; Sherry et al. 2007; Kim et al. 2009; Wilczek et al. 2009), and natural selection often favors different flowering times in contrasting environments (Stinchcombe et al. 2004; Engelmann and Purugganan 2006; Hall and Willis 2006; Scarcelli et al. 2007; Verhoeven et al. 2008; Sandring and Ågren 2009). The onset of flowering has been extensively studied at the physiological and molecular levels in model organisms (Turck et al. 2008; Ehrenreich et al. 2009) and agronomic species (Nakagawa et al. 2005; Buckler et al. 2009).

In winter cereals such as wheat and barley that require vernalization (long exposure to non-freezing temperatures), the vernalization pathway is mainly mediated via the flowering activator VRN-1 (Diallo et al. 2012). Vernalization triggers epigenetic modifications at FLC in *Arabidopsis* or at VRN-1 in cereals to promote flowering. An epigenetic "memory" of cold is imprinted on these loci so that they can promote flowering when favorable conditions (longer days and warmer temperatures) are experienced (Sheldon et al. 2009; Diallo et al. 2012). However, although winter cereals display freezing tolerance during vegetative development, they can still be sensitive to cold temperatures during flowering. Interestingly, VRN-1 has a role in suppressing the expression of genes (Dhillon et al. 2010).

The research on *Arabidopsis* has greatly illuminated the current understanding of the processes involved in flowering. However, it is also becoming evident that plants have evolved diverse mechanisms to regulate flowering. For instance, transcription factor *FLOWERING LOCUS C* (FLC) seems to be specific to *Brassica* species, and an FLC homologue does not seem to exist in cereals (Cockram et al. 2007). Rather than the flowering activator, the key vernalization response gene in cereals (Greenup et al. 2009; Diallo et al. 2012) is involved. Varieties within crop species also have different photoperiod sensitivities that have arisen via adaptation to different growth environments or through breeding (Coles et al. 2010; Gómez-Ariza et al. 2015). Stress-regulated flowering is not formally recognized as a floral transition pathway per se. However, a number of studies suggest that both biotic and abiotic stress factors play key roles in controlling the transition to flowering.

It can be speculated that plants have the ability to sense the strength of stress factors and respond appropriately. It also appears that the effect of stress-induced flowering, which can be influenced by day length, is dependent on the timing of the stress and the plant species. Under gradual and mild stress, flowering is delayed so that reproductive processes can resume at a later time. However, under terminal stress where physical damage to tissue is expected to occur, flowering is accelerated to ensure reproduction, even if this comes at the expense of yield. Indeed, in crop plants, the correct timing of flowering is critical for adaptation to specific environments and closely correlates with grain yield (Gao et al. 2014). If flowering occurs prematurely under stressful environments, seed set and grain filling may be compromised. If flowering is delayed, the plant risks succumbing to terminal stress before producing any seed. The association between stress and flowering time can lead to undesirable consequences. For instance, heterologous expression of a MYB transcription factor from chrysanthemum in *Arabidopsis* enhances drought and salinity tolerance but can also lead to delays in flowering (Shan et al. 2012). Therefore, enhanced understanding of the links between these processes is essential for engineering of stress tolerance in crops. A challenge for future agriculture in a changing climate is to predict which combination of traits to breed into crops in order to maximize development while avoiding stress-induced yield penalties.

1.10 Temperature and Flower Induction

Flower induction is a complex systemic process regulated by numerous genes, promoters, and inhibitors, which are triggered by various signals, such as photoperiod and temperature (Corbesier and Coupland 2003; Horvath 2009). Floral initiation is the key developmental step that marks the process of sexual reproduction in plants. The process is proceeded by flower induction, i.e., the processing and translocation of a flower-inducing chemical signal that is perceived by the vegetative bud meristems to render them floral. The process is usually triggered by an environmental signal but can also be endogenously controlled. Floral initiation marks the stage of no return when the first morphological changes are taking place at the shoot apical meristem. During the following stage of flower differentiation, the successive whorls of flower organs are then laid down at the shoot apex. Collectively, these events constitute the processes of flower bud formation. The main research focuses on the determination of the optimal temperature for flower initiation, not on physiological studies of floral stimulus transition (Khodorova and Boitel-Conti 2013). In most geophytes, the most important factor controlling their flowering is the seasonal thermoperiodicity, whereas the effects of light on flower induction are commonly less important. Most species do not show a preference for photoperiod, although in a small number of bulbous plants that respond to both temperature and day length (e.g., *Allium, Colchicum*), flowering is generally favored by short days (Wiebe 1994; Ofir and Kigel 2006). In different species, flower induction and initiation may take place at different times of the year. For those in which these processes begin during the summer months (e.g., *Tulipa, Narcissus, Hyacinthus, Crocus*, some *Iris*), they are supported by relatively high temperatures. The temperature optimum for induction and early floral organogenesis also varies between species from 9 to 25 °C. The physiological requirements are probably determined by the ecological origin of the species. For instance, those originating from temperate zones (e.g., *Allium, Iris*) usually require a very short period of relatively high temperature (about 20 °C) for flower induction, but this should be followed by a lower temperature for optimal flower meristem differentiation. Species originating from Irano-Turanian regions (e.g., *Tulipa, Narcissus*) require a relatively high temperature for flower induction, initiation, and early differentiation. The further the temperature is from the optimal range, the lower is the rate of floral meristem development. For all studied species, low temperature suppresses flower induction and initiation and leads to the formation of vegetative meristem (De Hertogh and le Nard 1993; Noy-Porat et al. 2009; Kim et al. 2007a, b; Niedziela et al. 2008). Once flower differentiation is completed, most bulbs require a period of low temperature (4–9 °C) from the beginning of autumn, which enables stem elongation and anthesis in spring.

Flowering is a complex process that is controlled by a number of internal and environmental factors (Lang 1965). In tree crops such as apple, the process is particularly complicated for several reasons. First, most woody plants, including apple, have a long juvenile phase, during which they cannot be induced to flower by any means (Zimmerman 1972). Second, the annual apple shoot must develop a critical number of leaves (16–20 depending on cultivar) before it may respond to environmental factors and undergo floral initiation (McLaughlin and Green 1991). This means that flower bud formation may in fact be a two-step process: first, a vegetative growth process to produce the critical number of nodes and, second, the floral initiation process proper (Heide et al. 2020). Third, there is a time span of more than 10 months between floral initiation and anthesis (blooming) in apple trees (Verheij 1996), during which time the tree is going through induction and breaking of dormancy and fruits are developing in parallel with flower initiation and differentiation. These are processes that are known to interfere with flower bud formation (Chan and Cain 1967). Finally, only a fraction of the buds of an apple tree becomes floral, and buds on dwarf shoots (spur buds) and buds on vigorously growing extension shoots have different seasonal timing of flower bud formation (Rivero et al. 2017). In addition, because of its size, the apple tree does not lend itself readily for experimentation in controlled environments.

Environmental conditions are known to influence flower bud formation in apple, but experimental results have often been inconsistent and highly variable (Hanke et al. 2007). Shading has long been known to strongly reduce flowering in apple trees (Tromp 1984). Although Tromp (1984) found that flower bud formation was better in 14-h long-day than in 8-h short-day conditions, it was concluded that the result was due to differences in daily light integral, which was almost twice as large in the long-day treatment as in the short-day treatment. It was also demonstrated by Heide and Prestrud (2005) that vegetative growth cessation and dormancy induction are likewise unaffected by photoperiod in apple but rather controlled by low temperature. In controlled environments, Tromp (1976) found that shoot growth of apple trees was reduced, while flower bud formation was stimulated at 17 °C or 19 °C compared with 24 °C when applied from full bloom. Raising the temperature from 17 to 24 °C late in the experimental period also reduced flower bud formation.

1.10.1 Effect of Temperature During Flower Maturation and Shoot Development

Despite the apparent dormant state of bulbs during the autumn and winter months, active developmental processes continue to take place. All these processes require energy, carbohydrate partitioning, and water uptake, which can only be supplied by the underground organ source (De Hertogh and le Nard 1993; Kamenetsky and Okubo 2013). The mobilization and transport of the reserves in bulbs seem to be affected by temperature conditions (De Hertogh and le Nard 1993; Kamenetsky and Okubo 2013).

1.10.1.1 Water Status

During the storage period, growing buds are large recipients of water. Water redistribution in bulbs and its transfer to the buds plays an important role in the complete development of flowers and is temperature-dependent. Indeed, in tulips, formed flower buds are aborted after storage in the wrong temperature conditions (17 °C). The water content in these flower buds is almost half of that in normal flowers of bulbs stored at optimal temperature (5 °C), thus suggesting that at least one of the possible reasons for bud abortion may be its insufficient water supply (Van Kilsdonk et al. 2002). It has been reported that during storage, low temperature induces water transfer from lateral scales to central ones and also enhances the subsequent transfer of water from the basal plate and scales to the developing bud (Zemah et al. 1999; Van der Toorn et al. 2000; Van Kilsdonk et al. 2002). Moreover, low-temperature storage leads to a selective expression of the aquaporin γTIP gene in stalks after planting (Balk and de Boer 1999). Since aquaporins facilitate water transport during cell enlargement and contribute greatly to cell growth (Eisenbarth and Weig 2005: Ehlert et al. 2009), the absence of γTIP gene expression might also play a role in water deficiency in buds and the inhibition of stalk growth in plants stored at ambient temperature.

1.10.1.2 Respiration

Although flower bud development begins at a slow rate during the autumn-winter period, the growth of leaves and flowers takes place in spring. This active growth requires energy to supply, for example, ATP for enzyme working and sucrose loading to phloem. These energy demands might be provided by a temporarily increased respiration. Temperature does not appear to affect the respiratory activity of bulbs during storage, but, after planting, precooled bulbs show significantly higher respiration and energy production than those stored at ambient temperature (about +18 °C), (Samach and Wigge 2005). The respiration is higher both in the bulb scales and in developing shoots (Kannerworff and van der Plas 1994; Lee et al. 2002), probably due to the increased number of mitochondria following the cold treatment (Khodorova et al. 2007). The respiratory metabolism is also affected by temperature during the epigenous growth of geophytes. Growing at warmer temperatures leads to a rapid increase in photosynthetic and photorespiratory rates, which results in faster starch accumulation and starch saturation in underground organs. The limit of the capacity for carbon accumulation in the sink organ provokes the feedback inhibition of the photosynthetic rate and, consequently, faster leaf senescence in comparison with plants growing at a lower temperature (Gandin et al. 2011).

1.10.1.3 Carbohydrate Distribution

It is well known that low temperature leads to the intensive activity of starch hydrolysis and the sucrose synthetic pathway (Stitt and Hurry 2002; Guy et al. 2008), although it seems that starch hydrolysis in tulips and *Corydalis* is not influenced by temperature conditions during storage. Starch degradation also occurs in peripheral scales more intensively than in central ones, independently of storage temperature (Van der Toorn et al. 2000; Kamenetsky et al. 2003; Khodorova et al. 2010). Since geophytes do not photosynthesize during the autumn-winter period, starch hydrolysis is the only source of soluble sugars. Thus, the dynamics of sugar accumulation in bulbs generally follow those of starch degradation. During storage, the total sugar concentration in bulbs is not influenced by differences in the surrounding temperature (Khodorova et al. 2010), and the amount of sucrose is always greater than that of reducing sugars (Ranwala and Miller 2008a). The activities of sucrose-cleaving enzymes, sucrose synthase, and invertases remain at low levels (Balk and de Boer 1999). However, even if the accumulation of sugars during storage is unlikely to be dependent on temperature, cold treatment has a considerable influence on sugar transport. For example, in tulip and *Corydalis* bulbs, low-temperature storage leads to a transfer of nutrients from the storage organ to the bud, which is suppressed under a higher-temperature regime (Khodorova et al. 2010; Khodorova 2011).

Since the plants deprived of low-temperature storage generally demonstrate only slight growth after planting, lower sugar concentrations in their bulbs and shoots are expected. Indeed, by the time of anthesis in tulip, the sugar content (hexoses, in particular) in stalks of correctly stored bulbs may be up to tenfold higher than in bulbs stored at ambient temperature (Lambrechts et al. 1994). In fact, low-temperature storage leads to an elevated expression of invertase genes and high activities of these enzymes in growing stalks, which maximizes hexose production (Balk and de Boer 1999; Lambrechts and Kollöfel 1993; Ranwala and Miller 2008b). There is no doubt that the accumulated carbohydrate reserve participates in shoot elongation. Since hexoses, particularly glucose, are known to contribute to cell division and elongation (Weschke et al. 2003; Sreenivasulu et al. 2004), it has been suggested that the slow growth of shoots in plants deprived of low temperature is due to an insufficient supply of sugars. Sucrose-cleaving enzymes (invertases and sucrose-synthase) are activated when there is a demand for metabolic carbohydrate and energy (Salerno and Curatti 2003; Roitsch and Gonzalez 2004). Apparently, without low-temperature treatment, there is no induction of carbohydrate demand in shoots and no activation of invertases and sucrose-synthase (Lambrechts et al. 1994; Ranwala and Miller 2008b). At the same time, sucrose-cleaving enzymes play an important role in the control of plant development (Koch 2004), and the low activities of these enzymes in the stalks of bulbs stored at higher temperature (Lambrechts and Kollöfel 1993) may also prevent the initiation of stalk development and the subsequent storage of carbohydrates.

1.10.1.4 Endogenous Plant Growth Regulators

The accumulations of endogenous abscisic acid and cytokinins in bulbs are not temperature-dependent processes (Letham et al. 2003), whereas the production of auxin and gibberellin is affected by the surrounding temperature and probably plays one of the leading roles in geophyte growth regulation.

Gibberellins

Gibberellic acid is known for its role in the elongation of axial organs (stems, petioles, and inflorescences), flower development (Hsu et al. 2008; Rieu et al. 2008), and vernalization (Rebers et al. 1994). For bulbous plants, it has been reported that the amount of gibberellic acid correlates directly with shoot elongation rate and the presence/absence of cold treatment. After planting, when shoot elongation starts, the concentration of gibberellic acid generally increases rapidly in the internodes and basal plates of precooled plants, whereas non-cooled ones exhibit low gibberellic acid levels in these organs (Rebers et al. 1995, 1994; Ranwala and Miller 2008b; Rietveld et al. 2000). It has been shown that de novo biosynthesis of gibberellic acid is a prerequisite factor for shoot elongation in bulbous plants, which begins once shoot elongation has started. Biosynthesis of gibberellic acid is also crucial for sprouting. The suppression of gibberellic acid biosynthesis, even in cold-treated bulbs, leads to low stem growth (Ranwala and Miller 2008a, b).

Auxins

The role of auxins as factors involved in stem elongation and growth regulation in different plant species is widely discussed (Perrot-Rechenmann 2010). In bulbous plants, auxin is reported to be the main hormonal factor involved in the induction of tulip stalk elongation (Rietveld et al. 2000). The primary site of auxin biosynthesis is the flower bud. The removal of floral buds causes weaker growth of the stalk, even in cooled bulbs, whereas the application of the exogenous auxin at the site of the removed buds induces normal elongation of the stem (De Hertogh and le Nard 1993; Okubo and Uemoto 1985). Auxin is the only known factor that shows a correlation between cold treatment requirement and the induction of stem and flower growth in bulbs. Rietveld et al. (2000) have shown that, upon cooling, tulip stalks show a higher responsiveness and sensitivity to auxin, which increases gradually with the length of cold treatment. In the absence of cold treatment, plants are not sensitive to auxin signals. It has been suggested that this fluctuation in auxin response is the main factor responsible for the different growth behavior in vivo of cold-treated and untreated tulip bulbs (Khodorova and Boitel-Conti 2013). The response to auxin in stalks is not sufficient for their elongation without cold treatment (Rietveld et al. 2000). It can also be mentioned that auxin not only is required for the induction of shoot elongation but is also involved in it. Inhibition of auxin transport reduces stem elongation and results in a decrease in the quantity of gibberellins (Okubo and Uemoto 1985).

In different plant species, both auxin and gibberellic acid have central roles in temperature-controlled elongation responses (Heggie and Halliday 2005). Moreover, the concentration of auxin may regulate key genes encoding enzymes for gibberellic acid biosynthesis (Ozga et al. 2003; Ross et al. 2003; Demason 2005; Björkland et al. 2007). This joint action of two hormones guarantees the growth of shoots and leaves: auxin promotes gibberellic acid biosynthesis, thus acting as a mobile factor, whereas gibberellic acid is the actual effector of growth (Ozga et al. 2003). Since stalk elongation in bulbous plants has also been shown to be due to the parallel activity of auxin and gibberellins, it was suggested that auxin signals not only play a role in the induction of shoot growth but also probably affect the onset of gibberellic acid biosynthesis. The accumulation of the latter in growing shoots induces the remobilization of all reserves in bulbs in order to supply nutrients for growth. For instance, accumulation of gibberellic acid leads to an enhanced expression of invertase genes, which establishes a large sink of sucrose transported from the bulb and provides hexoses necessary for shoot elongation. A normal growing shoot also induces an increase in respiration and water flux.

1.10.1.5 Effect of Cropping System

The assessment of cropping systems' response to a changing climate plays an important role for the evaluation of near-future economic assets, and the study of crop phenology response was indicated as a key stage for a better formulation of adaptation policies and options (Duchêne and Schneider 2005; Wolfe et al. 2005; Sadras and Monzon 2006). Phenology is in fact the most important attribute involved in the final yield assessment and consequently in the adaptation of crops to the changing environment. Both the timing of phenological stages and the relative duration of the pre- and post-flowering phases are in fact critical determinants of yield (Sadras and Connor 1991). The activity that is most demanding for a crop (i.e., reproductive phase) should take place at the time of optimal conditions (i.e., temperature and rainfall) (Visser and Both 2005), whereas its duration should be as long as possible for optimal biomass partitioning to the fruit (Bindi et al. 1992).

The impact of climate change on the crop phenology of typical Mediterranean annual (sunflower, maize, potato, and durum wheat) and perennial (grapevine and olive tree) crops were reviewed (Osborne et al. 2000; Moriondo and Bindi 2005, 2006; Moriondo et al. 2011; Bindi and Moriondo 2005). The use of crop growth and phenology simulation models coupled to the outputs of a general circulation model has been extensively. Phenology and yield of annual crops were simulated using the CropSyst model (Stockle et al. 2003), whereas the phenology of grapevine and olive was reproduced using the grapevine model developed by Bindi et al. (1997) and growing degree-day accumulation models (Osborne et al. 2000; Moriondo et al. 2001), respectively. The general circulation model data were directly used as model inputs or downscaled at higher-resolution dynamically regional circulation model (RCMs) or empirically downscaling (ED), for a better climate simulation at local scale.

Currently, relatively little is known about the extent to which the plant's epigenetic "memory" of winter affects stress responses on a global scale. Expression studies in cereals suggest that a small number of stress-related genes may acquire a memory of vernalization. While many genes differentially regulated during vernalization return to basal levels upon transfer to ambient growth conditions, several stress-related genes remain up- or downregulated for a week after a return to ambient temperatures (Greenup et al. 2011; Huan et al. 2013). A correlation between the length of vernalization and defense gene expression and disease resistance has also been demonstrated in cereal crops (White and Jenkyn 1995; Gaudet et al. 2011), suggesting that prolonged exposure to cold plays developmental as well as defensive roles. Together, these examples indicate the multiplicity and complexity of molecular events associated with stress tolerance and flowering in plants.

Since plant phenology is largely temperature dependent, there was an overall trend for earlier occurrence of key phenological events and consequent shortening of crop growing cycle for winter and summer annual crops, grapevine, and olive tree (Osborne et al. 2000; Bindi and Moriondo 2005; Moriondo and Bindi 2007). More interestingly, when the effect of higher temperatures on the single phenological phase duration was investigated, an asymmetric impact on the length of vegetative and reproductive phases was evident both in annual and perennial crops, resulting in large fitness consequences. In winter crops, for instance, higher temperatures as simulated for the future scenarios shortened the pre-flowering phase, and this allowed grain filling periods to take place in cooler periods that favor a longer time period for biomass partitioning to grains (Sadras and Monzon 2006). Additionally, a shorter vegetative phase allowed the crop to escape the summer drought, resulting finally in an increased final yield when compared to the present period (including the CO_2 effect, data not shown). In contrast, the grain filling period was significantly reduced for a typical summer crop. The same trend was simulated for grapevine, where higher temperatures, advancing bud break phase shifted post-bud break development to earlier, relatively cooler conditions that cancelled out the increasing temperature trend. As a consequence, higher temperatures did not decrease the length of bud break-anthesis phase, whereas the duration of anthesis-maturation phase was largely shortened with respect to the present period.

Phenology, or the relationship between climate and the sequence and timing of developmental events or stages, provides a foundation for understanding crop development and growth (McMaster et al. 2011). Flowering is a life history trait determined by plant genotype, the environment, and the interaction between them (Rathcke and Lacey 1985). Synchronized flowering can increase the effective population size, reduce genetic structure, and maximize outcrossing (Primack 1985). In contrast, asynchronous flowering can impede gene flow (Primack 1985; Cruzan et al. 1994) and reduce reproductive success (Rathcke and Lacey 1985; Fox 1990a). In addition to genetic mechanisms, flowering is also affected by environmental factors such as photoperiod (Shitaka and Hirose 1998), temperature (Price and Waser 1998; Sandvik and Totland 2000), and water stress (Fox 1990b, 2001; Bram and Quinn 2000; Stanton et al. 2000).

Circadian clocks are 24-h biological oscillators, which generally enable organisms to anticipate predictable, daily changes that are linked to the rotation of the Earth. Such clocks have been found in a wide range of organisms, including cyanobacteria, fungi, insects, and mammals (Young and Kay 2001). The timing of flowering, a critical stage of development in the life cycle of most plants when seed number is determined, is important for adaptation to both the abiotic stresses of temperature and the water-deficit constraints (Curtis 1968) within the growing season. For example, in many annual crops, brief episodes of hot temperatures (>32–36 °C) can greatly reduce seed set, and hence crop yield, if they coincide with a brief critical period of only 1–3 days around the time of flowering (Vara Prasad et al. 2000; Wheeler et al. 2000; Jagadish et al. 2008). Therefore, the moderation of crop development will be critical to the impacts of climate change on yield in two ways: through determining the season length and hence the availability of radiation, water, and nutrient resources for growth and by affecting the exposure of the crop to climate extremes. Adaptation to moderate changes in climate that influence temperature, season length, and planting dates, as well as the occurrence of abiotic stress, can be achieved by selecting varieties with appropriate flowering times and crop durations (Richards 2006). Farmers and plant breeders have very successfully selected/manipulated life cycle duration and phenology to maximize the range of environments in which crops grow as well as their yield

(Roberts et al. 1996) at least for current climates. A major challenge for crop improvement is how to plan for future climate change. The genetic and environmental moderation of the timing of flowering is therefore central to the responses described above. The timing of flowering within a season is largely determined by responses to temperature and photoperiod, and in whole plants at suboptimal temperatures, these quantitative responses are reasonably well understood (Wallace and Yan 1998). However, responses to temperature and photoperiod at supra-optimal temperatures are poorly understood—though clearly, these will become more important as the frequency of high-temperature events increases under projected climate change. In recent years, molecular biology has also greatly contributed to our understanding of flowering gene pathways (Baurle and Dean 2006; Tsuji et al. 2008), although the effects of temperature and temperature-photoperiod interactions on these pathways have not been studied.

Plant species distributions are determined by the response of populations to regional climates. It has been pointed out that common ragweed (*Ambrosia artemisiifolia*) is an invasive wind-pollinated annual plant causing considerable health and economic problems (Chauvel et al. 2006; Fumanal et al. 2007; Essl et al. 2009; Šikoparija et al. 2009). Climate-dependent phenological models suggest that the distribution of *Ambrosia artemisiifolia* in Europe is limited by low temperatures in the North, where plants are prevented from completing reproduction by autumn frosts, which kill adult plants (Chapman et al. 2014), and drought in the South, which inhibits seed germination and seedling emergence (Storkey et al. 2014). Skálová et al. (2015) have stated that temperature is the main determinant of common ragweed's distribution and identify nutrient availability as a factor that results in the realized niche being smaller than the fundamental niche; both of these need to be taken into account when predicting the future spread of *Ambrosia artemisiifolia*.

1.11 Molecular/Genetic Aspects of Flowering Response to Climate Change

Little is known about the molecular mechanisms that control flowering times in response to ambient temperature, other than for the vernalization response, even in *Arabidopsis* (Baurle and Dean 2006). Similarly, little is known about the response of flowering pathway genes to elevated CO_2. Flowering time (usually measured by leaf number) in wild-type *Arabidopsis* responds to temperature in a similar manner to other plants, occurring sooner at warm (up to ~27 °C) than at cool (16 °C) temperatures (Blazquez et al. 2003; Thingnaes et al. 2003; Balasubramanian et al. 2006). Comparisons of mutants of the photoperiod, gibberellin, and autonomous pathways have shown that ambient temperature (growth temperature) is sensed through genetic pathways involving *FCA* and *FVA* and integrated through *FT* (FLOWERING LOCUS T) (Blazquez et al. 2003; Lee et al. 2007) and that temperature leads to a photoperiod-independent activation of *FT* (Balasubramanian et al. 2006). Temperature also affects responses of phytochrome mutants (Blazquez et al.

2003; Halliday et al. 2003); for example, *PHYA* (Phytochrome A) is not able to promote flowering at cooler temperatures, a response that might contribute to P × T interaction. However, while warm/short days (27 °C/8 h) apparently induce flowering at the same time as cool/long days (16 °C/16 h) in *Arabidopsis* (Balasubramanian et al. 2006), temperature cannot usually compensate for non-inductive photoperiod in most crop plants.

In *Arabidopsis*, Springer et al. (2008) examined the response to elevated CO_2 in two genotypes: one selected for high seed number at elevated CO_2 and one a random control. They found that flowering was not affected by CO_2 in the control genotype, and unsurprisingly the downregulation of *FLC* (Flowering Locus C) and the upregulation of *SOC1* and *LFY* (Leafy) were similar at ambient and elevated CO_2. However, in the adapted genotype, flowering was delayed by 7–9 days at elevated CO_2, and this was associated with no downregulation of *FLC* over the course of the experiment and, as a result, later upregulation of *SOC1 and LFY expression.* Elevated CO_2, as expected, increased plant size, and relative increases in the adapted and control genotypes at ambient and elevated CO_2 were proportional to the delay in flowering. It has been suggested that higher sucrose levels may act as a signal to influence flowering. However, it may be that selection for high seed number at elevated CO_2 was associated with later flowering and hence larger plant size and yield.

Models predicting flowering time based on flowering pathway genes have been proposed for *Arabidopsis* (Welch et al. 2003, 2004; Van Oosterom et al. 2004), and these models do predict flowering time in constant photoperiod and temperature environments. Furthermore, these simple gene network models can generate important physiological parameters, such as critical photoperiods. However, as stated above, none of these models currently quantify basic (suboptimal) temperature responses and certainly not supra-optimal temperatures; nor do they model temperature-photoperiod interactions.

The amount of water consumed by ornamental plants depends on the particular species and cultivar, the cultivation system, and the plant growing season. It has been estimated that, on average, 100–350 kg of water is needed to produce 1 kg of plant dry matter (Jiménez and Caballero 1990). Water-deficit affects negatively the process of flowering in many plant species by reducing the fertility of newly formed flowers (Slawinska and Obendorf 2001). The growth of salt-treated plants is often limited by the ability of roots to extract water from the soil and transport it to the shoot due to the osmotic component of salinity (Rodríguez et al. 1997). Sepaskhah and Yarami (2009) indicated that saffron (*Crocus sativus* L.) flower and corm were the most and the least sensitive organ to soil water depletion, respectively. Furthermore, Shillo et al. (2002) reported two bulb species, namely, *Hippeastrum × hybridum* Hort. and *Ornithogalum arabicum* L., which were very sensitive to salinity; and the degree of damage was correlated to the salinity level, and this response was expressed as weight reduction in all plant organs. Al-Humaid and Moftah (2006) indicated that plant biomass, number of leaves, length, and weight of marketable inflorescences and bulb yield of tuberose (*Polianthes tuberosa* L.) were significantly reduced by water deficit (Bahadoran and Salehi 2015).

1.12 Simulation of Phenology for Future Climate

Most commonly documented impact of the climate change on crops is the shortening of development stages in a warmer climate and the resulting changes in areas of crop suitability. In many studies, this response dominates the impact of climate change on yield. Many simulation studies of changes in areas of crop suitability use simple thermal time relationships to represent the rate of development (Fischer et al. 2005), and only a few model the effects of supra-optimal temperatures on the rate of development (Wheeler 2008). Given that it is known that photoperiod sensitivity of duration to flowering is a key determinant of crop adaptation to climate and that many crop varieties are photoperiod-sensitive to some extent, it seems important to explore how photoperiod sensitivity may affect the response of crop phenology to temperature warming. For this, the photothermal model of flowering time was used to simulate duration from sowing to tassel initiation in maize in the current climate at one maize growing location and at a range of prescribed temperature increases. Specifically, the aim was to determine whether photoperiod sensitivity affected the response of duration from sowing to tassel initiation to increases in mean temperature and to a more variable temperature regime.

One hundred years of current climate at Zaria, Nigeria (11.1°N, 7.7°E), were generated using the MarkSim weather generator (Jones and Thornton 2000). Another set of 100 years was generated with an increase in temperature variability that was simulated by increasing the diurnal temperature range by 50% compared with the current climate simulations. Temperature increases of +1, +2, to +6 °C were added separately to the daily mean temperature of these two sets of current climates to provide a total of 14 sets of 100 years of climate. The photoperiod on each day was calculated from standard astronomical day length equations (Keisling 1982). A sowing date of 15 May, the average date of the start of the rains at this location (Craufurd and Qi 2001), was used in all simulations. Time to tassel initiation was simulated using the photothermal time model with the appropriate temperature and photoperiod sensitivity for maize (PT model) and with the same temperature sensitivity alone (temperature model, i.e., no response to photoperiod). The mean and coefficient of variation of duration from sowing to tassel initiation was calculated for each set of 100 years of simulations.

Mean duration from sowing to tassel initiation became progressively shorter with an increase in mean temperature until an apparent optimum, beyond which duration progressively increased, indicated that the current mean temperatures at Zaria are below the optimum. Photoperiod sensitivity (PT model) delayed tassel initiation by 6 days in the current climate (zero change in temperature) and changed the response to temperature warming. Most notably, the apparent optimum for this development stage increased from +2 °C without photoperiod sensitivity (T model) to +4 °C with photoperiod sensitivity. The interannual variability in the duration from sowing to tassel initiation was similar between the simulations with and without photoperiod sensitivity until +2 °C, beyond which interannual variability in development was greater at a given temperature warming in the absence of photoperiod sensitivity. Also, interannual variability in development was greater when the diurnal temperature variability was increased. To explore this further, the change with temperature in the coefficient of variation of duration to tassel initiation between the climate simulations with and without extra temperature variability was examined. Again, photoperiod sensitivity affected the simulated response to temperature warming. The increase in variability of the timing of tassel initiation due to increased temperature variability at +3 °C and warmer was reduced with photoperiod sensitivity.

References

Abbasal-Ani MK, Hay RKM (1983) The influence of growing temperature on the growth and morphology of cereal seedling root systems. J Exp Bot 34:1720–1730

Abdulmalik MM, Olarewaju JD, Usman IS, Ibrahim A (2012) Effects of moisture stress on flowering and fruit set in sweet pepper (Capsicum annum L.) cultivars. Prod Agric Technol 8(1):191–198

Adams SR, Cockshull KE, Cave CRJ (2001) Effect of temperature on the growth and development of tomato fruits. Ann Bot 88:869–877

Ahmed CB, Rouina BB, Sensoy S, Boukhriss M, Abdullah FB (2010) Exogenous proline effects on photosynthetic performance and anti-oxidant defense system of young olive tree. J Agric Food Chem 58:4216–4222

Airaki M, Leterrier M, Mateos RM, Valderrama R, Chaki M, Barroso JB, Del Rio LA, Palma JM, Corpas FJ (2011) Metabolism of reactive oxygen species and reactive nitrogen species in pepper (Capsicum annuum L.) plants under low temperature stress. Plant Cell Environ 35:281–295

Akiyama R, Agren J (2013) Conflicting selection on the timing of germination in a natural population of Arabidopsis thaliana. J Evol Biol 27:193–199

Aksouh-Harradj NM, Campbell LC, Mailer RJ (2006) Canola response to high and moderately high temperature stresses during seed maturation. Can J Plant Sci 86(4):967–980

Al-Humaid AI, Moftah AE (2006) Response of vegetative and reproductive parameters of water stressed tuberose plant to vapor grad and kaolin antitranspirants. King Saud Univ Reposit 18: 127–139

Ali MG, Naylor REL, Matthews S (2006) Distinguishing the effects of genotype and seed physiological age on low temperature tolerance of rice (Oryza sativa L.). Exp Agric 42(3): 337–349

Allard F, Houde M, Krol M, Ivanov A, Huner NPA, Sarhan F (1998) Betaine improves freezing tolerance in wheat. Plant Cell Physiol 39:1194–1202

Aloni B, Peet M, Pharr M, Karni L (2001) The effect of high temperature and high atmospheric CO_2 on carbohydrate changes in bell pepper (Capsicum annuum) pollen in relation to its germination. Physiol Plant 112:505–512

Alsajri FA, Singh B, Wijewardana C, Irby JT, Gao W, Reddy KR (2019) Evaluating soybean cultivars for low- and high-temperature tolerance during the seedling growth stage. Agronomy 9(13):1–20

Alvarado R, Hernaiz S (2007) Antecedentesgeneralessobre el arroz en Chile. In: Alvarado R (ed) Arroz Manejo Tecnológico, Boletín INIA, vol 162. Instituto de Investigaciones Agropecuarias INIA, Centro Regional de Investigación Quilamapu, Chillán, Chile, p 179

Alvarez-Uria P, Körner C (2007) Low temperature limits of root growth in deciduous and evergreen temperate tree species. Funct Ecol 21(2):211–218

Amjad M, Ziaf K, Iqbal Q, Ahnad L, Riaz MA, Saqib ZA (2007) Effect of seed priming on seed vigor and salt tolerance in hot pepper. Pak J Agric Sci 44(3):408–416

Amooaghaie R, Nikzad K, Shareghi B (2010) The effect of priming on emergence and biochemical changes of tomato seeds under suboptimal temperatures. Seed Sci Technol 38:508–512

Anchordoguy TJ, Rudolph AS, Carpenter JF, Crowe JH (1987) Modes of interaction of cryoprotectants with membrane phospholipids during freezing. Cryobiology 24(4):324–331

Andaya VC, Mackill DJ (2003a) Mapping of QTLs associated with cold tolerance during the vegetative stage in rice. J Exp Bot 54:2579–2585

Andaya VC, Mackill DJ (2003b) QTLs conferring cold tolerance at the booting stage of rice using recombinant inbred lines from a japonica × indica cross. Theor Appl Genet 106:1084–1090

Anderson JT, Lee CR, Mitchell-Olds T (2011) Life history QTLs and natural selection on flowering time in Boechera stricta, a perennial relative of Arabidopsis. Evolution 65(3):71–787

Andrés F, Coupland G (2012) The genetic basis of flowering responses to seasonal cues. Nat Rev Genet 13:627–639

Angadi SV, Cutforth HW, Miller PR, McConkey BG, Entz H, Brandt SA, Volkmar KM (2000a) Seeding management to reduce temperature stress in Brassica species during reproductive growth. Can J Plant Sci 80(4):693–701

Angadi SV, Cutforth HW, McConkey BG (2000b) Seeding management to reduce temperature stress in Brassica species. Saskatchewan Soils Crops Proc

Anjum SA, Xie Y, Wang LC, Saleem MF, Man C, Lei W (2011) Morphological, physiological and biochemical responses of plants to drought stress. Afr J Agric Res 6:2026–2032

Anjum SA, Ashraf U, Tanveer M, Khan I, Hussain S, Shahzad B, Zohaib A, Abbas F, Saleem MF, Ali I, Wang LC (2017) Drought induced changes in growth, osmolyte accumulation and antioxidant metabolism of three maize hybrids. Front Plant Sci 8:69

Apel K, Hirt H (2004) Reactive oxygen species: metabolism, oxidative stress, and signal transduction. Annu Rev Plant Biol 55:373–399

Arai-Sanoh Y, Ishimaru T, Ohsumi A, Kondo M (2010) Effects of soil temperature on growth and root function in rice. Plant Prod Sci 13:235–242

Armstrong AF, Wardlaw KD, Atkin OK (2007) Assessing the relationship between respiratory acclimation to the cold and photosystem II redox poise in Arabidopsis thaliana. Plant Cell Environ 30:1513–1522

Arnao MB, Josefa HR (2006) The physiological function of melatonin in plants. Plant Signal Behav 1:89–95

Aroca R, Irigoyen JJ, Sanchez-Diaz M (2001) Photosynthetic characteristics and protective mechanisms against oxidative stress during chilling and subsequent recovery in two maize varieties differing in chilling sensitivity. Plant Sci 161(4):719–726

Aroca R, Irigoyen JJ, Sánchez-Diaz M (2003) Drought enhances maize chilling tolerance. II. Photosynthetic traits and protective mechanisms against oxidative stress. Physiol Plant 117(4):540–549

Ashlock LO, Klerk R, Huitink G, Keisling T, Vories ED (2000) Planting practices, Arkansas soybean handbook MPV, vol 197. University Arkansas Cooperative Extension Service, Little Rock, AR, pp 35–49

Ashraf M, Foolad MR (2007) Roles of glycine betaine and proline in improving plant abiotic stress resistance. Environ Exp Bot 59(2):206–216

Ashraf M, Hafeez M (2004) Thermotolerance of pearl millet and maize at early growth stages: growth and nutrient relations. Biol Plant 48:81–86

Ashraf M, Athar HR, Harris PJC, Kwon TR (2008) Some prospective strategies for improving crop salt tolerance. Adv Agron 97:45–110

Aslamarz AA, Vahdati K, Rahemi M (2009) Estimation of chilling and heat requirements of some Persian walnut cultivars and genotypes. HortScience 44(3):697–701

Athar HR, Ashraf M (2009) Strategies for crop improvement against salinity and drought stress: an overview. In: Ashraf M, Ozturk M, Athar HR (eds) Salinity and water stress: improving crop efficiency, vol 44, 1st edn. Springer, New York, pp 1–16

Atkinson NJ, Lilley CJ, Urwin PE (2013) Identification of genes involved in the response to simultaneous biotic and abiotic stress. Plant Physiol 162:2028–2041

Atteya AM (2003) Alteration of water relations and yield of corn genotypes in response to drought stress. Bulg J Plant Physiol 29:63–76

Avramova Z (2015) Transcriptional "memory"of a stress: transient chromatin and memory (epigenetic) marks at stress-response genes. Plant J 83(1):149–159

Babu RC, Shashidhar HE, Lilley JM, Thanh ND, Ray JD, Sadasivam S, Sarkarung S, O'Toole JC, Nguyen HT (2001) Variation in root penetration ability, osmotic adjustment and dehydration tolerance among accessions of rice adapted to rainfed lowland and upland ecosystems. Plant Breed 120:233–238

Badri MA, Minchin PE, Lapointe L (2007) Effects of temperature on the growth of spring ephemerals: Crocus vernus. Physiol Plant 130:67–76

Baek KH, Skinner DZ (2003) Alteration of antioxidant enzyme gene expression during cold acclimation of near-isogenic wheat lines. Plant Sci 165:1221–1227

Baena-Gonzalez E, Gray JC, Tyystjarvi E, Aro EM, Maenpaa P (2001) Abnormal regulation of photosynthetic electron transport in a chloroplast ycf9 inactivation mutant. J Biol Chem 276: 20795–20802

Bagnall D, Wolfe JOE, King RW (1983) Chill-induced wilting and hydraulic recovery in mung bean plants. Plant Cell Environ 6(6):457–464

Bahadoran M, Salehi H (2015) Growth and flowering of two tuberose (Polianthes tuberosa L.) cultivars under deficit irrigation by saline water. JAgricSciTechnol 17(2):415–426

Bai R, Chen L, Zhang X, Wei G, Wei C (2017) Effect of salinity and soil temperature on the growth and physiology of drip-irrigated rice seedlings. Arch Agron Soil Sci 63:513–524

Baier M, Kandlbinder A, Golldack D, Dietz K (2005) Oxidative stress and ozone: perception, signalling and response. Plant Cell Environ 28:1012–1020

Bakht J, Bano A, Dominy P (2006) The role of abscisic acid and low temperature in chickpea (Cicer arietinum) cold tolerance. II. Effects on plasma membrane structure and function. J Exp Bot 57: 707–715

Balasubramanian S, Sureshkumar S, Lempe J, Weigel D (2006) Potent induction of Arabidopsis thaliana flowering by elevated growth temperature. PLoS Genet 2:980–989

Balestrasse KB, Tomaro ML, Batlle A, Noriega GO (2010) The role of 5 aminolevulinic acid in the response to cold stress in soybean plants. Phytochemistry 71:2038–2045

Balk PA, de Boer AD (1999) Rapid stalk elongation in tulip (Tulipa gesneriana L. cv. Apeldoorn) and the combined action of cold-induced invertase and the water-channel protein γTIP. Planta 209:346–354

Bange MP, Milroy SP (2004) Impact of short term exposure to cold temperatures on early development of cotton (Gossypium hirsutum L.). Aust J Agric Res 55:655–664

Bange MP, Caton SJ, Milroy SP (2008) Managing yields of high fruit retention in transgenic cotton (Gossypium hirsutam L.) using sowing date. Aust J Agric Res 59:733–471

Banu NA, Hoque A, Watanabe-Sugimoto M, Islam M, Uraji M, Matsuoka K, Nakamura Y, Murata Y (2010) Proline and glycine betaine ameliorated NaCl stress via scavenging of hydrogen peroxide and methylglyoxal but not superoxide or nitric oxide in tobacco cultured cells. Biosci Biotechnol Biochem 74:2043–2049

Barnabás B, Jager K, Feher A (2008) The effect of drought and heat stress on reproductive processes in cereals. Plant Cell Environ 31:11–38

Barth S, Melchinger AE, Devezi-Savula B, Lübberstedt T (2000) A high-throughput system for genome-wide measurement of genetic recombination in Arabidopsis thaliana based on transgenic markers. Funct Integr Genomics 1:200–206

Baruah AR, Ishigo-Oka N, Adachi M, Oguma Y, Tokizono Y, Onishi K, Sano Y (2009) Cold tolerance at the early growth stage in wild and cultivated rice. Euphytica 165(3):459–470

Baskerville GL, Emin P (1969) Rapid estimation of heat accumulation from maximum and minimum temperatures. Ecology 50(3):514–517

Baskin CC, Baskin JM (2014) Seeds: ecology, biogeography, and evolution of dormancy and germination, 2nd edn. Elsevier, Academic Press, San Diego, CA

Basra SMA, Farooq M, Wahid A, Khan MB (2006) Rice seed invigoration by hormonal and vitamin priming. Seed Sci Technol 34:775–780

Basu S, Ramegowda V, Kumar A, Pereira A (2016) Plant adaptation to drought stress. Version 1. 5: F1000 Faculty Rev-1554
Bauer PJ, Frederick JR, Bradow JM, Sadler EJ, Evans DE (2000) Canopy photosynthesis and fiber properties of normal and late planted cotton. Agron J 92:518–523
Bauerfeind MA, Winkelmann T, Franken P, Druege U (2015) Transcriptome, carbohydrate, and phytohormone analysis of Petunia hybrida reveals a complex disturbance of plant functional integrity under mild chilling stress. Front Plant Sci 6:583
Baurle I, Dean C (2006) The timing of developmental transitions in plants. Cell 125(4):655–664
Beauchamp EG, Lathwell DJ (1967) Root-zone temperature effects on the early development of maize. Plant Soil 26:224–234
Beck EH, Fettig S, Knake C, Hartig K, Bhattarai T (2007) Specific and unspecific responses of plants to cold and drought stress. J Biosci 32:501–510
Benech-Arnold RL, Sánchez RA (2004) Handbook of seed physiology: applications to agriculture. The Haworth Press, New York
Benítez-Rodríguez L, Gamboa-deBuen A, Sánchez-Coronado ME, Alvarado-López S, Soriano D, Méndez I, Vázquez-Santana S, Carabias-Lillo J, Mendoza A, Orozco-Segovia A (2014) Effects of seed burial on germination, protein mobilization and seedling survival in Dodonaea viscosa. Plant Biol 16(4):732–739
Berger JD (2007) Eco-geographic and evolutionary approaches to improving adaptation of autumn-sown chickpea (Cicer arietinum L.) to terminal drought: the search for reproductive chilling tolerance. Field Crop Res 104:112–122
Bezerra MA, de Lacerda CF, Filho EG, de Abreu CE, Prisco JT (2007) Physiology of cashew plants grown under adverse conditions. Braz J Plant Physiol 19(4):449–461
Bhattacharya A (2018) Abiotic stress and physiological traits in plants. In: Abiotic stress and plant physiology, Metabolic activities, vol 1. New India Publishing Agency, New Delhi
Bilan MV (1967) Effect of low temperature on root elongation in loblolly pine seedlings. In: Proc. 14th Int. Union Foreset Res. Oog. Cong., München, Part 4, Section 23, pp 74–82
Bilichak A, Kovalchuk I (2016) Transgenerational response to stress in plants and its application for breeding. J Exp Bot 67:2081–2092
Bindi M, Moriondo M (2005) Impact of a 2 °C global temperature rise on the Mediterranean region: agriculture analysis assessment. In: Climate change impacts in the Mediterranean resulting from a 2 °C global temperature rise, A WWF Report
Bindi M, Ferrini F, Miglietta F (1992) Climatic change and the shift in the cultivated area of olive trees. J Agric Mediterranea 22:41–44
Bindi M, Miglietta F, Gozzini B, Orlandini S, Seghi L (1997) A simple model for simulation of growth in grapevine (Vitis vinifera L.). I. Model development. Vitis 36:67–71
Björkland S, Antti H, Uddestrand I, Moritz T, Sundberg B (2007) Cross-talk between gibberellin and auxin in development of poplar wood: gibberellin stimulates polar auxin transport and has a common transcriptome with auxin. Plant J 52:499–511
Björkman T, Pearson KJ (1998) High temperature arrest of inflorescence development in broccoli (Brassica oleracea var. italica L.). J Exp Bot 49(318):101–106
Blazquez M, Hoon Ahn J, Weigel D (2003) A thermosensory pathway controlling flowering time in Arabidopsis thaliana. Nat Genet 33(2):168–171
Blümel M, Dally N, Jung C (2014) Flowering time regulation in crops—what did we learn from Arabidopsis? Curr Opin Biotechnol 32:121–129
Bohnert HJ, Nelson DE, Jensen RG (1995) Adaptations to environmental stresses. Plant Cell 7(7): 1099–1111
Bond DM, Dennis ES, Finnegan EJ (2011) The low temperature response pathways for cold acclimation and vernalization are independent. Plant Cell Environ 34:1737–1748
Boyko A, Filkowski J, Kovalchuk I (2005) Homologous recombination in plants is temperature and day-length dependent. Mutat Res 572(1–2):73–83

Bozzano M, Jalonen R, Thomas E, Boshier D, Gallo L, Cavers S, Bordács S, Smith P, Loo J (2014) Genetic considerations in ecosystem restoration using native tree species. State of the World's Forest Genetic Resources—Thematic Study. FAO and Biodiversity International, Rome

Bradáčová K, Weber NF, Moradtalab N, Asim M, Imran M, Weinmann M, Neumann G (2016) Micronutrients (Zn/Mn), seaweed extracts, and plant growth–promoting bacteria as cold–stress protectants in maize. Chem Biol Technol Agric 3:19

Bradbeer JW (1988) Seed dormancy and germination. Chapman and Hall, New York, pp 27–54

Bram MR, Quinn JA (2000) Sex expression, sex-specific traits, and the effects of salinity on growth and reproduction of Amaranthus cannabinus (Amaranthaceae), a dioecious annual. Am J Bot 87:1609–1618

Brand D, Wijewardana C, Gao W, Reddy KR (2016) Interactive effects of carbon dioxide, low temperature, and ultraviolet-B radiation on cotton seedling root and shoot morphology and growth. Front Earth Sci 10:607–620

Bravo LA, Close TJ, Corcuera LJ, Guy CL (1999) Characterization of an 80 kDa dehydrin-like protein in barley responsive to cold acclimation. Physiol Plant 106(2):177–183

Bray CM (1995) Biochemical processes during osmopriming of seeds. Water relations in seed germination. In: Kigel J (ed) Seed development and germination. Marcel Dekker, New York, pp 767–789

Bray EA, Bailey-Serres J, Weretilnyk E (2000) Responses to abiotic stress. In: Buchanan BB, Gruissen W, Jones RL (eds) Biochemistry and molecular biology of plants, 1st edn. Wiley, Rockville, pp 1158–1249

Brunner I, Herzog C, Dawes MA, Arend M, Sperisen C (2015) How tree roots respond to drought. Front Plant Sci 6:547

Bruns HA (2011) Planting date, rate, and twin-row vs. single-row soybean in the Mid-South. Agron J 103:1308–1313

Buckler ES, Holland JB, Bradbury PJ, Acharya CB, Brown PJ, Browne C, Ersoz E, Flint-Garcia S, Garcia A, Glaubitz JC, Goodman MM, Harjes C, Guill K, Kroon DE, Larsson S, Lepak NK, Li HH, Mitchell SE, Pressoir G, Peiffer JA, Rosas MO, Rocheford TR, Romay MC, Romero S, Salvo S, Villeda HS, da Silva HS, Sun Q, Tian F, Upadyayula N, Ware D, Yates H, Yu JM, Zhang ZW, Kresovich S, McMullen MD (2009) The genetic architecture of maize flowering time. Science 325:714–718

Buriro M, Oad FC, Keerio MI, Tunio S, Gandahi AW, Hassan SWU, Oad SM (2011) Wheat seed germination under the influence of temperature regimes. Sarhad J Agric 27:539–543

Burr KE, Tinus RW, King RM (1990) Comparison of three cold hardiness tests for conifer seedlings. Tree Physiol 6:351–369

Busch W, Wunderlich M, Schoffl F (2005) Identification of novel heat shock factor dependent genes and biochemical pathways in Arabidopsis thaliana. Plant J 41(1):1–14

Cabrera RM, Saltveit ME, Owens K (1992) Cucumber cultivars differ in their response to chilling temperatures. J Am Soc Hortic Sci 117(5):802–807

Cai Q, Wang S, Cui Z, Sun J, Ishii Y (2004) Changes in freezing tolerance and its relationship with the contents of carbohydrates and proline in overwintering centipede grass (Eremochloa ophiuroides (Munro) Hack.). Plant Prod Sci 7:421–426

Campitelli BE, Stinchcombe JR (2013) Testing potential selective agents acting on leaf shape in Ipomoea hederacea: predictions based on an adaptive leaf shape cline. Ecol Evol 3:2409–2423

Čanak P, Mirosavljević M, Ćirić M, Vujošević B, Kešelj J, Stanisaljević D, Mitrović B (2016) Seed priming as a method for improving maize seed germination parameters at low temperatures. Ratar Povrt 53(3):106–110

Cao JH, Li XB, Lin WF, Xie GS, Chen JM (2009) Survey on resistance of 12 species oil palm cultivated in the field. Chin J Trop Agric 29:1–6

Cao S, Liu K, Du G, Baskin JM, Baskin CC, Bu H, Qi W (2018) Seedling emergence of 144 subalpine meadow plants: effects of phylogeny, life cycle type and seed mass. Seed Sci Res 28:93–99

Cao Q, Li G, Cui Z, Yang F, Jiang X, Diallo L, Kong F (2019) Seed priming with melatonin improves the seed germination of waxy maize under chilling stress via promoting the antioxidant system and starch metabolism. Sci Rep 9:15044
Casenave EC, Toselli ME (2007) Hydropriming as a pre-treatment for cotton germination under thermal and water stress conditions. Seed Sci Technol 35:88–98
Cavieres LA, Arroyo MTK (2000) Seed germination response to cold stratification period and thermal regime in Phacelia secunda (Hydrophyllaceae), altitudinal variation in the Mediterranean Andes of central Chile. Plant Ecol 149(1):1–8
Challinor A, Wheeler TR (2008) Crop yield reduction in the tropics under climate change: processes and uncertainties. Agric Forest Meteor 148(3):343–356
Chan BG, Cain JC (1967) The effect of seed formation on subsequent flowering. Proc Am Soc Hortic Sci 91:63–68
Chapin FS III (1974) Morphological and physiological mechanisms of temperature compensation in phosphate absorption along a latitudinal gradient. Ecology 55:1180–1198
Chapman DS, Haynes T, Beal S, Essl F, Bullock JM (2014) Phenology predicts the native and invasive range limits of common ragweed. Glob Chang Biol 20:192–202
Charles WB, Harris RE (1972) Tomato fruit-set at high and low-temperatures. Can J Plant Sci 52: 497–506
Chauvel B, Dessaint F, Cardinal-Legrand C, Bretagnolle F (2006) The historical spread of Ambrosia artemisiifolia L. in France from herbarium records. J Biogeogr 33:665–673
Chen S (1991) Injury of membrane lipid peroxidation to plant cell. Plant Physiol Commun 27:84–90
Chen K, Arora R (2013) Priming memory invokes seed stress-tolerance. Environ Exp Bot 94:33–45
Chen CC, Sung JM (2001) Priming bitter gourd seeds with selenium solution enhances germinability and antioxidative responses under sub-optimal temperature. Physiol Plant 111:9–16
Chen Q, Qi WB, Reiter RJ, Wei W, Wang BM (2009) Exogenously applied melatonin stimulates root growth and raises endogenous indoleacetic acid in roots of etiolated seedlings of Brassica juncea. J Plant Physiol 166:324–328
Chen W, Guo C, Hussain S, Zhu B, Deng F, Xue Y, Geng M, Wu L (2016) Role of xylo-oligosaccharides in protection against salinity-induced adversities in Chinese cabbage. Environ Sci Pollut Res 23(2):1254–1264
Chinnusamy V, Zhu J, Zhu JK (2007) Cold stress regulation of gene expression in plants. Trend Plant Sci 12:444–451
Christiansen MN, Thomas RO (1969) Season-long effects of chilling treatments applied to germinating cottonseed. Crop Sci 9:672–673
Christie PJ, Hahn M, Walbot V (1991) Low-temperature accumulation of alcohol dehydrogenase-1 mRNA and protein activity in maize and rice seedlings. Plant Physiol 95:699–706
Clark H (2001) Improving tolerance to low temperature in chickpea. Presented at the 4th European conference on grain legumes, Cracow, Poland
Clarke HJ (2001) Improving tolerance to low temperature in chickpea. In: 4th European conference on grain legumes. Towards the sustainable production of healthy food, feed and novel products, Cracow, Poland, pp 34–35
Clarke HJ, Siddique KHM (1998) Growth and development. In: Loss S, Brandon N, Siddique KHM (eds) The chickpea. Western Australia Agriculture Bulletin 1326. Department of Western Australia Agriculture, Australia
Clarke HJ, Siddique KHM (2004) Response of chickpea genotypes to low temperature stress during reproductive development. Field Crop Res 90:320–334
Cober ER, Curtis DF, Stewart DW, Morrison MJ (2014) Quantifying the effects of photoperiod, temperature and daily irradiance on flowering time of soybean isolines. Plan Theory 3:476–497
Cockram J, Jones H, Leigh FJ, O'Sullivan D, Powell W, Laurie DA, Greenland AJ (2007) Control of flowering time in temperate cereals: genes, domestication, and sustainable productivity. J Exp Bot 58:1231–1244

Coles ND, McMullen MD, Balint-Kurti PJ, Pratt RC, Holland JB (2010) Genetic control of photoperiod sensitivity in maize revealed by joint multiple population analysis. Genetics 184: 799–812

Collinson ST, Ellis RH, Summerfield RJ, Roberts EH (1992) Durations of the photoperiod-sensitive and photoperiod-insensitive phases of development to flowering in four cultivars of rice (Oryza sativa L.). Ann Bot 70:339–346

Cook D, Fowler S, Fiehn O, Thomashow MF (2004) A prominent role for the CBF cold response pathway in configuring the low temperature metabolome of Arabidopsis. Proc Natl Acad Sci U S A 101:15243–15248

Corbesier L, Coupland G (2003) The quest for florigen: review of recent progress. J Exp Bot 57: 3395–3403

Counts JA, Zeldes BM, Lee LL, Straub CT, Adams MWW, Kelly RM (2017) Physiological, metabolic and biotechnological features of extremely thermophilic microorganisms. Wiley Interdiscip Rev Syst Biol Med 9(3):e1377

Craufurd PQ, Qi A (2001) Photothermal adaptation of sorghum (Sorghum bicolor) in Nigeria. Agric For Meteor 108:199–211

Craufurd PQ, Wheeler TR (2009) Climate change and the flowering time of annual crops. J Exp Bot 60(9):2529–2539

Crifo T, Puglisi I, Petrone G, Recupero GR, Lo Piero AR (2011) Expression analysis in response to low temperature stress in blood oranges: implication of the flavonoid biosynthetic pathway. Gene 478:1–9

Croser JS, Clarke HJ, Siddique KHM, Khan TN (2003) Low-temperature stress: implications for chickpea (Cicer arietinum L.) improvement. Crit Rev Plant Sci 22:185–219

Cruz R, Milach S (2004) Cold tolerance at the germination stage of rice: methods of evaluation and characterization of genotypes. Sci Agric 61:1–8

Cruzan MB, Hamrick JL, Arnold ML, Bennett BD (1994) Mating system variation in hybridizing irises: effects of phenology and floral densities on family outcrossing rates. Heredity 72:95–105

Curtis DL (1968) The relation between the date of heading of Nigerian sorghums and the duration of the growing season. J Appl Ecol 5(1):215–226

Cutforth HW, Shaykewich CF, Cho CM (1986) Effect of soil water and temperature on corn (Zea mays L.) root growth during emergence. Can J Soil Sci 66:51–58

da Cruz RP, Milach SCK (2000) Breeding for cold tolerance in irrigated rice. Ciência Rural 30:909–917

Dagdelen N, Yilmaz E, Sezg F, Gurbuz T (2004) Effects of water stress at different growth stages on processing pepper (Capsisum annuum cv. Kapija) yield, water use and quality characteristic. Pak J Biol Sci 7(12):2167–2172

Dalmaso G, Ferreira D, Vermelho A (2015) Marine extremophiles: a source of hydrolases for biotechnological applications. Mar Drugs 13(4):1925–1965

Davanport TL (2009) Reproductive physiology. In: Litz RE (ed) The mango: botany, production and uses, 2nd edn. CABI, London

Davidson RL (1969) Effect of root/leaf temperature differentials on root/soot ratios in some pasture grasses and clover. Ann Bot 33(3):561–569

Davletova S, Schlauch K, Coutu J, Mittler R (2011) The zinc-finger protein Zat12 plays a central role in reactive oxygen and abiotic stress signaling in arabidopsis. Plant Physiol 139:847–856

De Hertogh AA, le Nard M (1993) The physiology of flower bulbs: a comprehensive treatise on the physiology and utilization of ornamental flowering bulbous and tuberous plants. Elsevier Science Publishers, Amsterdam, p 812

De Los Reyes BG, Myers SJ, McGrath JM (2003) Differential induction of glyoxylate cycle enzymes by stress as a marker for seedling vigor in sugar beet (Beta vulgaris). Mol Genet Genomics 269(5):692–698

Debbarma M, Das SP (2017) Priming of seed: enhancing growth and development. Int J Curr Microbiol Appl Sci 6(12):2390–2396

Delauney A, Verma DPS (1993) Proline biosynthesis and osmo regulation in plants. Plant J 4:215–223

Demason DA (2005) Auxin-cytokinin and auxin-gibberellin interactions during morphogenesis of the compound leaves of pea (Pisum sativum). Planta 222:151–166

Deming JW (2002) Psychrophiles and polar regions. Curr Opin Microbiol 5(3):301–309

Deng B, Du W, Liu C, Sun W, Tian S, Dong H (2012) Antioxidant response to drought, cold and nutrient stress in two ploidy levels of tobacco plants: low resource requirement confers polytolerance in polyploids? Plant Growth Regul 66:37–47

Devasirvatham V, Gaur PM, Raju TN, Trethowan RM, Tan DKY (2015) Field response of chickpea (Cicer arietinum L.) to high temperature. Field Crop Res 172:59–71

Dhillon T, Pearce SP, Stockinger EJ, Distelfeld A, Li C, Knox AK, Vashegyi I, Vágújfalvi A, Galiba G, Dubcovsky J (2010) Regulation of freezing tolerance and flowering in temperate cereals: the VRN-1 connection. Plant Physiol 153:1846–1858

Diallo AO, Ali-Benali MA, Badawi M, Houde M, Sarhan F (2012) Expression of vernalization responsive genes in wheat is associated with histone H3 trimethylation. Mol Genet Genomics 287:575–590

Dinesh MR, Reddy BMC (2012) Physiological basis of growth and fruit yield characteristics of tropical and sub-tropical fruits to temperature. In: Sthapit B, Rao R, Sthapit S (eds) Tropical fruit tress species and climate change. Biodiversity International, Rome, pp 45–70

Dionne J, Castonguay Y, Nadeau P, Desjardins Y (2001) Freezing tolerance and carbohydrate changes during cold acclimation of green-type annual bluegrass (Poa annua L.) ecotypes. Crop Sci 41:443–451

Dodds GT, Ludford PM (1990) Surface topology of chilling injury of tomato fruit. HortSci 25(11): 1416–1419

Domisch T, Finér L, Lehto T (2001) Effects of soil temperature on biomass and carhydrate allocation in Scots pine (Pinus sylvestris) seedlings at the beginning of the growing season. Tree Physiol 21:465–472

Domisch T, Finér L, Lehto T (2002) Growth, carbohydrate and nutrient allocation of scots pine seedlings after exposure to simulated low soil temperature in spring. Plant Soil 246(1):75–86

Dong H, Li W, Tang W, Li Z, Zhang D, Niu Y (2006a) Yield qyality and leaf senescence of cotton grown at varying planting dates and densities in the Yellow River Valley of China. Field Crop Res 98:106–115

Dong CH, Xiangyang HU, Tang W, Zheng X, Kim YS, Lee B, Zhu JK (2006b) A putative Arabidopsis nucleoporin, AtNUP160, is critical for RNA export and required for plant tolerance to cold stress. Mol Cell Biol 26(24):9533–9543

Donohue K, Casas RRD, Burghardt L, Kovach K, Willis CG (2010) Germination, post-germination adaptation, and species ecological ranges. Annu Rev Ecol System 41:293–319

Du B, Luo H, He L, Zhang L, Liu Y, Mo Z, Pan S, Tian H, Duan M, Tang X (2019) Rice seed priming with sodium selenate: effects on germination, seedling growth, and biochemical attributes. Sci Rep 9:4311

Duan M, Ma NN, Li D, Deng YS, Kong FY, Lv W, Meng QW (2012) Antisense-mediated suppression of tomato thylakoidal ascorbate peroxidase influences anti-oxidant network during chilling stress. Plant Physiol Biochem 58:37–45

Dubey AN, Verma S, Goswami S, Devedee AK (2018) Effect of temperature on different growth stages and physiological process of rice crop-a review. Bull Environ Pharmacol Life Sci 7(11): 129–136

Dubrovsky JG, Forde BG (2012) Quantitative analysis of lateral root development: pitfalls and how to avoid them. Plant Cell 24:4–14

Duchêne E, Schneider C (2005) Grapevine and climatic changes: a glance at the situation in Alsace. Agron Sustain Dev 25(1):93–99

Dufault RJ, Melton RR (1990) Cyclic cold stresses before transplanting influence tomato seedling growth, but not fruit earliness, fresh market yield, or quality. J Am Soc Hortic Sci 115:559–563

Dupuis I, Dumas C (1990) Influence of temperature stress on in vitro fertilization and heat-shock protein-synthesis in maize (Zea mays L.) reproductive tissues. Plant Physiol 94:665–670
Eagles HA, Brooking IR (1981) Populations of maize with more rapid and reliable seedling emergence than Cornbelt dents at low temperature. Euphytica 30:755–763
Eagles HA, Hardacre AK (1979) Genetic variation in maize (Zea mays L.) for germination and emergence at 10°C. Euphytica 28:287–295
Egli DB, Cornelius PL (2009) A regional analysis of the response of soybean yield to planting date. Agron J 101:330–335
Ehlert C, Maurel C, Tardieu F, Simonneau T (2009) Aquaporin-mediated reduction in maize root hydraulic conductivity impacts cell turgor and leaf elongation even without changing transpiration. Plant Physiol 150:1093–1104
Ehrenreich IM, Hanzawa Y, Chou L, Roe JL, Kover PX, Purugganan MD (2009) Candidate gene association mapping of Arabidopsis flowering time. Genetics 183:325–335
Eisenbarth D, Weig AR (2005) Dynamics of aquaporins and water relations during hypocotyl elongation in Ricinus communis L. seedlings. J Exp Bot 56:1831–1842
Engelmann K, Purugganan MD (2006) The molecular evolutionary ecology of plant development: flowering time in Arabidopsis thaliana. Adv Bot Res 44:507–526
Epand RM, Shai Y, Segrest JP, Anantharamaiah GM (1995) Mechanisms for the modulation of membrane bilayer properties by amphipathic helical peptides. Biopolymers 37(5):319–338
Epstein E, Bloom AJ (2005) Mineral nutrion of plants: principles and prospectives, 2nd edn. Sinauer, UK
Ercoli L, Mariotti M, Masoni A, Arduini I (2004) Growth response of sorghum plants to chilling temperature and duration of exposure. Eur J Agron 21(1):93–103
Eskandari H, Kazemi K (2011) Effect of seed priming on germination properties and seedling establishment of cowpea (Vigna sinensis). Notulae Scientia Biologicae 3(4):113–116
Espanany A, Fallah S, Tadayyon A (2016) Seed priming improves seed germination and reduces oxidative stress in black cumin (Nigella sativa) in presence of cadmium. Ind Crop Prod 79:195–204
Essl F, Dullinger S, Kleinbauer I (2009) Changes in the spatio-temporal patterns and habitat preferences of Ambrosia artemisiifolia during its invasion of Austria. Preslia 81:119–133
Ewuebker F, Mullen RE, Koehler K (2001) Flooding and temperature effects on soybean germination. Crop Sci 41:1857–1861
Fan WJ, Zhang M, Zhang HX, Zhang P (2012) Improved tolerance to various abiotic stresses in transgenic sweet potato (Ipomoea batatas) expressing spinach betaine aldehyde dehydrogenase. PLoS One 7(5):e37344
Farooq M, Wahid A, Kobayashi NDF, Basra SMA (2009a) Plant drought stress: effects, mechanisms and management. Sustain Agric 29:185–212
Farooq M, Aziz T, Wahid A, Lee DJ, Siddique KHM (2009b) Chilling tolerance in maize: agronomic and physiological approaches. Crop Pasture Sci 60(6):501–516
Farooq M, Wahid A, Siddique KHM (2012) Micronutrient application through seed treatments - a review. J Soil Sci Plant Nutr 12:125–142
Farrell TC, Williams RL, Fukai S (2001) The cost of low temperature to the NSW rice industry. In: Proceeding of the 10th Australian agronomy conference, Aust. Soc. Agron., Hobart, TAS
Farrell TC, Fukai S, Williams RL (2006a) Minimizing cold damage during reproductive development among temperate rice genotypes. I. Avoiding low temperature with the use of appropriate sowing time and photoperiod-sensitive varieties. Aust J Agric Res 57:75–88
Farrell TC, Fox KM, Williams RL, Fukai S (2006b) Genotypic variation for cold tolerance during reproductive development in rice: screening with cold air and cold water. Field Crop Res 98: 178–194
Finch-Savage WE, Leubner-Metzger G (2006) Seed dormancy and the control of germination. New Phytol 171:501–523
Finkelstein RR, Gibson SI (2002) ABA and sugar interactions regulating development: cross-talk or voices in a crowd? Cur Opin Plant Biol 5:26–32

Fischer G, Shah M, Tubiello FN, van Velhuizen H (2005) Socioeconomic and climate change impacts on agriculture: an integrated assessment, 1990–2080. Philos Trans R Soc B Biol Sci 360(1463):2067–2083

Fitter A, Hay R (2002) Environmental physiology of plants, 3rd edn. Academic, London

Flowers TJ (2004) Improving crop salt tolerance. J Exp Bot 55:307–319

Foti R, Abureni K, Tigere A, Gotosa J, Gere J (2008) The efficacy of different seed priming osmotica on the establishment of maize (Zea mays L.) caryopses. J Arid Environ 72:1127–1130

Fowler JE, Quatrano RS (1997) Plant cell morphogenesis: plasma membrane interactions with the cytoskeleton and cell wall. Annu Rev Cell Dev Biol 13:697–743

Fowler S, Thomashow MF (2002) Arabidopsis transcriptome profiling indicates that multiple regulatory pathways are activated during cold acclimation in addition to the CBF cold response pathway. Plant Cell 14:1675–1690

Fox GA (1990a) Components of flowering time variation in a desert annual. Evolution 44:1404–1423

Fox GA (1990b) Drought and the evolution of flowering time in desert annuals. Am J Bot 77(11): 1508–1518

Fox GA (2001) Failure-time analysis: studying times-to-events and rates at which events occur. In: Scheiner SM, Gurevitch J (eds) Design and analysis of ecological experiments. Oxford University Press, Oxford, UK

Francini A, Sebastiani L (2010) Copper effects on Prunus persica in two different grafting combinations (P. persica, P. amygdalus and P. cerasifera). J Plant Nutr 33:1338–1352

Francis KE, Lam SY, Harrison BD, Bey AL, Berchowitz LE, Copenhaver GP (2007) Pollen tetrad-based visual assay for meiotic recombination in Arabidopsis. Proc Natl Acad Sci U S A 104: 3913–3918

Franco JA, Bañón S, Fernández JA, Leskovar DI (2001) Effect of nursery regimes and establishment irrigation on root development of Lotus creticus seedlings following transplanting. J Hortic Sci Biotechnol 76:174–179

Frank G, Pressman E, Ophir R, Althan L, Shaked R, Freedman M, Shen S, Firon N (2009) Transcriptional profiling of maturing tomato (Solanum lycopersicum L.) microspores reveals the involvement of heat shock proteins, ROS scavengers, hormones and sugars in the heat stress esponse. J Exp Bot 60:3891–3908

Frankow-Lindberg BE (2001) Adaptation to winter stress in nine white clover populations: changes in non-structural carbohydrates during exposure to simulated winter conditions and 'spring' regrowth potential. Ann Bot 88:745–751

Frenkel C, Erez A (1996) Induction of chilling tolerance in cucumber (Cucumis sativus) seedlings by endogenous and applied ethanol. Physiol Plant 96(4):593–600

Fu JJ, Liu J, Yang LY, Miao YJ, Xu YF (2017) Effects of low temperature on seed germination, early seedling growth and antioxidant systems of the wild Elymus nutans Griseb. J Agric Sci Technol 19:1113–1125

Fujino K, Sekiguchi H, Sato T, Kiuchi H, Nonoue Y, Takeuchi Y, Ando T, Lin SY, Yano M (2004) Mapping of quantitative trait loci controlling low-temperature germinability in rice (Oryza sativa L.). Theor Appl Genet 108:794–799

Fumanal B, Chauvel B, Bretagnolle F (2007) Estimation of pollen and seed production of common ragweed in France. Annu Agric Environ Med 14:233–236

Gamboa de Buen A, Cruz-Ortega R, Martínez-Barajas E, Sánchez-Coronado ME, Orozco-Segovia A (2006) Natural priming as an important metabolic event in the life story of Wigandia urens (Hydrophyllaceae) seeds. Physiol Plant 128:520–530

Gandin A, Gutjahr S, Dizengremel P, Lapointe L (2011) Source-Sink imbalance increases with growth temperature in the spring geophyte Erythronium americanum. J Exp Bot 62:3467–3479

Gao L, Jun Z, Huang Y, Zhang L (1992) Rice clock model—a computer model to simulate rice development. Agric For Meteorol 60:1–16

Gao H, Jin M, Zheng XM, Chen J, Yuan D, Xin Y, Wang M, Huang D, Zhang Z, Zhou K, Sheng P, Ma J, Ma W, Deng H, Jiang L, Liu S, Wang H, Wu C, Yuan L, Wan J (2014) Days to heading

7, a major quantitative locus determining photoperiod sensitivity and regional adaptation in rice. Proc Natl Acad Sci U S A 111(46):16337–16342
Gass T, Schori A, Fossati A, Soldati A, Stamp AP (1996) Cold tolerance of soybean (Glycine max (L.) Merr.) during the reproductive phase. Eur J Agron 5:71–88
Gaudet DA, Wang Y, Frick M, Puchalski B, Penniket C, Ouellet T, Robert L, Singh J, Laroche A (2011) Low temperature induced defence gene expression in winter wheat in relation to resistance to snow moulds and other wheat diseases. Plant Sci 180(1):99–110
Genard H, Le Saos J, Hillard J, Tremolieres A, Boucaud J (1991) Effect of salinity on lipid composition, glycine betaine content and photosynthetic activity in chloroplasts of Suaeda maritime. Plant Physiol Biochem 29:421–427
Ghadirnezhad R, Fallah A (2014) Temperature effect on yield and yield components of different rice cultivars in flowering stage. Int J Agron 2014:1–4
Ghassemi-Golezani K, Chadordooz-Jeddi A, Nasrollahzadeh S, Moghaddam M (2010) Effects of hydro-priming duration on seedling vigour and grain yield of pinto bean (Phaseolus vulgaris L.) cultivars. Notulae Botanicae Horti Agrobotanici Cluj-Napoca 38(1):109–113
Gill SS, Tuteja N (2010) Reactive oxygen species and antioxidant machinery in abiotic stress tolerance in crop plants. Plant Physiol Biochem 48:909–930
Giménez-Benavides L, Milla R (2013) Comparative germination ecology of two altitudinal vicariant Saxifraga species endemic to the North of Spain. Plant Biol 15(3):593–600
Giménez-Benavides L, Escudero A, Pérez-García F (2005) Seed germination of high mountain Mediterranean species: altitudinal, interpopulation and interannual variability. Ecol Res 20: 433–444
Gómez-Ariza J, Galbiati F, Goretti D, Brambilla V, Shrestha R, Pappolla A, Courtois B, Fornara F (2015) Loss of floral repressor function adapts rice to higher latitudes in Europe. J Exp Bot 66: 2027–2039
Gomez-Campo C, Prakash S (1999) Origin and domestication. In: Gomez-Campo C (ed) Biology of Brassica Coenospecies. Elsevier Science B.V, Amsterdam, pp 33–58
Gong Z, Lee H, Xiong L, Jagendorf A, Stevenson B, Zhu JK (2002) RNA helicase-like protein as an early regulator of transcription factors for plant chilling and freezing tolerance. Proc Natl Acad Sci U S A 99(17):11507–11512
González-Zertuche L, Vázquez-Yanes C, Gamboa A, Sánchez-Coronado ME, Aguilera P, Orozco-Segovia A (2001) Natural priming of Wigandia urens seeds during burial: effects on germination, growth and protein expression. Seed Sci Res 11:27–34
Goulas E, Schubert M, Kieselbach T, Kleczkowski LA, Gardestrom P, Schroder W, Hurry V (2006) The chloroplast lumen and stromal proteomes of Arabidopsis thaliana show differential sensitivity to short- and long-term exposure to low temperature. Plant J 47:720–734
Greenup A, Peacock WJ, Dennis ES, Trevaskis B (2009) The molecular biology of seasonal flowering-responses in Arabidopsis and the cereals. Ann Bot 103:1165–1172
Greenup AG, Sasani S, Oliver SN, Walford SA, Millar AA, Trevaskis B (2011) Transcriptome analysis of the vernalization response in barley (Hordeum vulgare) seedlings. PLoS One 6: e17900
Gremer JR, Kimball S, Venable DL (2016) Within and among year germination in Sonoran Desert winter annuals: bet hedging and predictive germination in a variable environment. Ecol Lett 19: 1209–1218
Grichar WJ, Biles SP (2014) Response of soybean to early-season planting dates along the upper Texas Gulf Coast. Int J Agron 4:190–195
Guan YJ, Hu J, Wang XJ, Shao CX (2009) Seed priming with chitosan improves maize germination and seedling growth in relation to physiological changes under low temperature stress. J Zhejiang Univ Sci B 10(6):427–433
Gudkova TI (1980) Physiological and cytological studies of the reasons for pollen sterility in spring wheat under low temperature. Nauch Tr Linengar S Kh In Ta 394:103–108

Guilioni L, Wery J, Tardieu F (1997) Heat stress-induced abortion of buds and flowers in pea. Is sensitivity linked to organ age or to relations between reproductive organs? Ann Bot 80:159–168

Guy CL (1990) Cold acclimation and freezing stress tolerance. Role of protein metabolism. Annu Rev Plant Physiol Plant Mol Biol 41:187–223

Guy C, Li QB (1998) The organization and evoluion of the spinach stress 70 molecular chaperone gene family. Plant Cell 10(4):539–556

Guy C, Haskell D, Li QB (1998) Association of proteins with the stress 70 molecular chaperones at low temperature: evidence for the existence of cold labile proteins in spinach. Cryobiology 36(4):301–314

Guy C, Kaplan F, Kopka J, Selbig J, Hincha DK (2008) Metabolomics of temperature stress. Physiol Plant 132:220–235

Hall MC, Willis JH (2006) Divergent selection on flowering time contributes to local adaptation in Mimulus guttatus populations. Evolution 60:2466–2477

Halliday K, Salter M, Thingnaes E, Whitelam G (2003) Phytochrome control of flowering is temperature sensitive and correlates with expression of the floral integrator FT. Plant J 33(5): 875–885

Halter R, Sands R, Ashton DH, Nabiar EKS (1997) Root growth of subalpine and montane Eucalyptus seedlings at low soil temperatures. Trees 12:35–41

Han H, Gao S, Li B, Dong XC, Feng HL, Meng QW (2010) Overexpression of violaxanthin de-epoxidase gene alleviates photoinhibition of PSII and PSI in tomato during high light and chilling stress. J Plant Physiol 167:176–183

Han JX, Wei Y, Yan C, An SZ (2011) The vivipary characteristic of Anabasis elatior and its ecological adaptation. Acta Ecol Sin 31(10):2662–2668

Hanke M-V, Flachowsky H, Peil A, Hättasch C (2007) No flower no fruit - genetic potentials to trigger flowering in fruit trees. Genes Genomes Genomics 1(1):1–20

Hannah MA, Wiese D, Freund S, Fiehn O, Heyer AG, Hincha DK (2006) Natural genetic variation of freezing tolerance in Arabidopsis. Plant Physiol 142:98–112

Hanson C, Palutikof J, Perry AH (2006) Modelling the impact of climate extremes. Clim Res 31:1–133

Haq N (2006) Fruits for the future 10. Jackfruit (Artocarpus heterophyllus). Southampton Centre for Underutilised Crops, Southampton, p 192

Hardenburg RE, Watada AE, Wang CY (1986) The commercial storage of fruits, vegetables, and florist and nursery stocks. U.S. Department of Agriculture, Washington, DC, p 66

Hare PD, Cress WA (1997) Metabolic implications of stress-induced proline accumulation in plants. Plant Growth Regul 21:79–102

Hartikainen H, Xue T, Piironen V (2000) Selenium as an anti-oxidant and pro-oxidant in ryegrass. Plant Soil 225:193–200

Hasanuzzaman M, Nahar K, Fujita M (2013) Extreme temperature responses, oxidative stress and antioxidant defense in plants. In: Vahdati K, Leslie C (eds) Abiotic stress – plant response and application in agriculture, Chapter No. 6. InTech, Rijeka, Croatia, pp 169–205

Hatfield JL, Egli DB (1974) Effect of temperature on the rate of soybean hypocotyl elongation and field emergence. Crop Sci 14:423–426

Hatfield JL, Prueger JH (2015) Temperature extremes: effect on plant growth and development. Weather Clim Extremes 10(A):4–10

Hayase H, Satake T, Nishiyama I, Ito N (1969) Male sterility caused by cooling treatment at the meiotic stage in rice plants. II. The most sensitive stage to cooling and the fertilising ability of pistils. Proc Crop Sci Soc Jpn 38:706–711

Hazer Jaribi E, Jabarov K, Sobouri A, Kaandi R (2013) Evaluation of genotypic x environments interactions using path coefficienr analysis in soybean. Caspian J Appl Sci Res 2(7):46–54

He XQ, Du C, Shao Z, Li Q (2009) Effect of salt and water stress on seed germination of Dianthus Chinensis L. Acad Conf Hortic Sci Technol Proc:60–62

Hearn AB, Constable GA (1984) Cotton. In: Goldsworthy PR, Fisher NM (eds) The physiology of tropical crops. Wiley, Chichester, UK, pp 495–527
Heatherly LG (2005) Midsouthern USA soybean yield affected by maturity group and planting date. Crop Manag 4(1):1–18
Heatherly LG, Elmore RW (2004) Managing inputs for peak production. In: Boerma HR, Specht JE (eds) Soybeans: improvement, production, and uses, agronomy monograph. American Society of Agronomy, Crop Science Society of America and Soil Science Society of America, Madison, WI, pp 451–536
Hedhly A, Hormaza JI, Herrero M (2008) Global warming and plant sexual reproduction. Trends Plant Sci 14:30–36
Heenan DP (1984) Low temperature induced floret sterility in the rice cultivars Calrose and Inga as influenced by nitrogen supply. Aust J Exp Agric Anim Husb 24:255–259
Hegarty TW (1973) Temperature relations of germination in the field. In: Heydecker W (ed) Seed ecology, vol 4. Butterworths, UK, pp 11–31
Heggie L, Halliday KJ (2005) The highs and lows of plant life: temperature and light interactions in development. Int J Dev Biol 49:675–687
Heide OM, Prestrud AK (2005) Low temperature, but not photoperiod, controls growth and dormancy induction in apple and pear. Tree Physiol 25:109–114
Heide OM, Tivero R, Sønsteby A (2020) Temperature control of shoot and floral initiation in apple (Malus x domestica Borkh). CABI Agric Biosci 1:8
Hernández-Nistal J, Dopico B, Labrador E (2002) Cold and salt stress regulates the expression and activity of a chickpea cytosolic Cu/Zn superoxide dismutase. Plant Sci 163:507–514
Herrero M (2003) Male and female synchrony and the regulation of mating in flowering plants. Philos Trans R Soc London Ser B Biol Sci 358:1019–1024
Herrero M, Arbeloa A (1989) Influence of the pistil on pollen-tube kinetics in peach (Prunus persica). Am J Bot 76(10):1441–1447
Herrero MP, Johnson RR (1980) High-temperature stress and pollen viability of maize. Crop Sci 20:796–800
Hinson K (1989) Use of a long juvenile trait in cultivar development. In: Pascale AJ (ed) Proceedings of the world soybean research conference IV. Asociacion Argentina de la Soja, Buenos Aires, Argentina, pp 983–1000
Hochachka PW, Somero GN (2002) Biochemical adaptation: mechanism and process in physiological evolution. Oxford University Press, New York
Hodges HF, Reddy KR, McKinion JM, Reddy VR (1993) Temperature effects on cotton. Bull. 990. Mississippi Agric. & Forestry Exp. Stn., Mississippi State Univ., Mississippi State
Hoffman L, DaCosta M, Ebdon JS, Watkins E (2010) Physiological changes during cold acclimation of perennial ryegrass accessions differing in freeze tolerance. Crop Sci 50:1037–1047
Horvath D (2009) Common mechanisms regulate flowering and dormancy. Plant Sci 177:523–531
Horvath I, Glatz A, Vervasovszki V, Török Z, Páli T, Balogh G, Kovács E, Nádasdi L, Benkö S, Joó F, Vigh L (1998) Membrane physical state controls the signaling mechanism of the heat shock response in Synechocystis PCC 6803: identification of hsp17 as a 'fluidity gene'. Proc Natl Acad Sci U S A 95(7):3513–3518
Howarth CJ (2005) Genetic improvements of tolerance to high temperature. In: Ashraf M, Harris PJC (eds) Abiotic stresses: plant resistance through breeding and molecular approaches. Haworth Press Inc, New York, pp 277–300
Hsiao TC, Xu LK (2000) Sensitivity of growth of root versus leaves to water stress; biophysical analysis and relation to water transport. J Exp Bot 51:1595–1616
Hsu SY, Hsu YT, Kao CH (2003) The effect of polyethylene glycol on proline accumulation in rice leaves. Biol Plant 46:73–78
Hsu YF, Tzeng JD, Liu MC, Yei FL, Chung MC, Wang CS (2008) Identification of anther-specific/predominant genes regulated by gibberellin during development of lily anthers. J Plant Physiol 165:553–563

Hu Y, Burucs Z, Tucher SV, Schmidhalter U (2007) Short-term effects of drought and salinity on mineral nutrient distribution along growing leaves of maize seedlings. Environ Exp Bot 60(2): 268–275

Hu Y, Jiang L, Wang F, Yu D (2013) Jasmonate regulates the INDUCER OF CBF EXPRESSION-C-REPEAT BINDING FACTOR/DRE BINDING FACTOR1 cascade and freezing tolerance in Arabidopsis. Plant Cell 25(8):2907–2924

Huan Q, Mao Z, Zhang J, Xu Y, Chong K (2013) Transcriptome wide analysis of vernalization reveals conserved and species-specific mechanisms in Brachypodium. J Integr Plant Biol 55: 696–709

Huang BR, DaCosta M, Jiang YW (2014) Research advances in mechanisms of turfgrass tolerance to abiotic stresses: from physiology to molecular biology. Crit Rev Plant Sci 33(2–3):141–189

Hughes MA, Dunn MA (1996) The molecular physiology of plant acclimation to temperature. J Exp Bot 47:291–305

Hume DJ, Jackson AKH (1981) Frost tolerance in soybeans. Crop Sci 21:689–692

Hund A, Richner W, Soldati A, Fracheboud Y, Stamp P (2007) Root morphology and photosynthetic performance of maize inbred lines at low temperature. Eur J Agron 27(1):52–61

Hussain S, Khan F, Cao W, Wu L, Geng M (2016) Seed priming alters the production and detoxification of reactive oxygen intermediates in rice seedlings grown under sub-optimal temperature and nutrient supply. Front Plant Sci 7:439

Hussain HA, Hussain S, Khaliq A, Ashraf U, Anjum SA, Men S, Wang L (2018) Chilling and drought stresses in crop plants: implications, cross talk, and potential management opportunities. Front Plant Sci 9:393

Hutchings MJ, John EA (2003) Distribution of roots in soil, and root foraging activity. In: Kroon H, Visser EJW (eds) Root ecology, Ecological studies, vol 168. Springer, Berlin, pp 33–60

Iba K (2002) Acclimative response to temperature stress in higher plants: approaches of gene engineering for temperature tolerance. Annu Rev Plant Biol 53:225–245

Ibrahim EA (2016) Seed priming to alleviate salinity stress in germinating seeds. J Plant Physiol 192:38–46. https://doi.org/10.1016/j.jplph.2

Iivonen S, Rikala R, Ryyppö A, Vapaavuori E (1999) Responses of Scots pine (Pinus sylvestris) seedlings grown in different nutrient regimes to changing root zone temperature in spring. Tree Physiol 19:951–958

Imin N, Kerim T, Rolfe BG, Weinman JJ (2004) Effect of early cold stress on maturation of rice anthers. Proteomics 4:1873–1882

Imran M, Mahmood A, Römheld V, Neumann G (2013) Nutrient seed priming improves seedling development of maize exposed to low root zone temperatures during early growth. Eur J Agron 49:141–148

Inouye DW (2008) Effects of climate change on phenology, frost damage, and floral abundance of montane wildflowers. Ecology 89:353–362

Iordachescu M, Imai R (2008) Trehalose biosynthesis in response to abiotic stresses. J Int Plant Biol 50:1223–1229

Ishikawa HA (1996) Ultrastructural features of chilling injury: injured cells and the early events during chilling of suspension cultured mung bean cells. Am J Bot 83:825–835

Ismail AM, Hall AE, Close TJ (1997) Chilling tolerance during emergence of cowpea associated with a dehydrin and slow electrolyte leakage. Crop Sci 37(4):1270–1277

Ito N (1976) Male sterility caused by cooling treatment at the young microspore stage in rice plants. XV. Effect of moderate cooling before or after the critical stage on the sterility induced by cooling at the critical stage. Proc Crop Sci Soc Jpn 45:558–562

Ito N (1978) Male sterility caused by cooling treatment at the young microspore stage in rice plants. XVI. Changes in carbohydrates, nitrogenous and phosphorus compounds in rice anthers after cooling treatment. Jpn J Crop Sci 47:318–323

Ito N, Hayase H, Satake T, Nishiyama I (1970) Male sterility caused by cooling treatment at the meiotic stage in rice plants. III. Male abnormalities at anthesis. Proc Crop Sci Soc Jpn 39:60–64

Ito Y, Nakanomyo I, Motose H, Iwamoto K, Sawa S, Dohmae N, Fukuda H (2006) Dodeca-CLE peptides as suppressors of plant stem cell differentiation. Science 313:842–845
Ivanov AG, Sane PV, Krol M, Gray GR, Balseris A, Savitch LV, Oquist G, Huner NP (2006) Acclimation to temperature and irradiance modulates PSII charge recombination. FEBS Lett 580:2797–2802
Jafar MZ, Farooq M, Cheema MA, Afzal I, Basra SMA, Wahid MA, Aziz T, Shahid M (2012) Improving the performance of wheat by seed priming under saline conditions. J Agron Crop Sci 198:38–45
Jagadish SVK, Craufurd PQ, Wheeler TR (2007) High temperature stress and spikelet fertility in rice (Oryza sativa L.). J Exp Bot 58:1627–1635
Jagadish SVK, Craufurd PQ, Wheeler TR (2008) Phenotyping parents of mapping populations of rice for heat tolerance during anthesis. Crop Sci 48(3):1140–1146
Jain M, Prasad PVV, Boote KJ, Hartwell AL, Chourey PS (2007) Effects of season-long high temperature growth conditions on sugar-to-starch metabolism in developing microspores of grain sorghum (Sorghum bicolor L. Moench). Planta 227(1):67–79
Jakobsen HB, Martens H (1994) Influence of temperature and aging of ovules and pollen on reproductive success in Trifolium repens L. Ann Bot 74:493–501
Jaleel CA, Gopi R, Sankar B, Gomathinayagam M, Panneerselvam R (2008) Differential responses in water use efficiency in two varieties of Catharanthus reseus under drought stress. C R Biol 331:42–47
Jan N, Mahboob-ul-Hussain, Khurshid IA (2009) Cold resistance in plants: a mystery unresolved. Electron J Biotechnol 12(3):1–15
Janska A, Marsik P, Zelenkova S, Ovesna J (2010) Cold stress and acclimation-what is important for metabolic adjustment? Plant Biol 12:395–405
Jenabiyan M, Pirdashi H, Yaghoubiab Y (2014) The combined effect of cold and light intensity stress on some morphological and physiological parameters in two soybean (Glycin max L.) cultivars. Int J Biosci 5(3):189–197
Jenkins L, Brill R (2011) Chickpea time of sowing trial. DAN00156: Northern Pulse Agronomy Project. NSW DPI, Australia
Jennings P, Saltveit ME (1994) Temperature effects on imbibition and germination of cucumber (Cucumis sativus) seeds. J Am Soc Hortic Sci 119(3):464
Jensen J (1981) Effect of temperature on genetic recombination in barley. Hereditas 94:215–218
Jewell MC, Campbell BC, Ian D (2010) Transgenic plants for abiotic stress resistance. In: Kole C, Michler CH, Abbott AG, Hall TC (eds) Transgenic crop plant. Springer, Berlin, pp 67–132
Ji SL, Jiang L, Wang YH, Liu SJ, Liu X, Zhai HQ, Yoshimura A, Wan JM (2008) QTL and epistasis for low temperature germinability in rice. Acta Agronom Sin 34:551–556
Jiang QW, Kiyoharu O, Ryozo I (2002) Two novel mitogen-activated protein signaling components, OsMEK1 and OsMAP1, are involved in a moderate low-temperature signaling pathway in Rice. Plant Physiol 129:1880–1891
Jiménez R, Caballero M (1990) El riego y la Fertilización. In: Jiménez R, Caballero M (eds) El Caltivo Industrial de Plantas en Maceta. Ediciones de Horticultura, Reus, Spain, pp 101–122
Jones PG, Inouye M (1994) The coldshock response-a hot topic. Mol Microbiol 11(5):811–818
Jones PG, Thornton PK (2000) MarkSim. Software to generate daily weather data for Latin America and Africa. Agron J 92(3):445–453
Jongdee B, Fukai S, Cooper M (2002) Leaf water potential and osmotic adjustment as physiological traits to improve drought tolerance in rice. Field Crop Res 76:153–163
Jorgensen RA (2004) Restructuring the genome in response to adaptive challenge: McClintock's bold conjecture revisited. Cold Spring Harb Symp Quant Biol 69:349–354
Jouyban Z, Hasanzade R, Sharafi S (2013) Chilling stress in plants. Int J Agric Crop Sci 5:2961
Julia C, Dingkuhn M (2012) Variation in time of day of anthesis in rice in different climatic environments. Eur J Agron 43:166–174
Jung JH, Seo PJ, Park CM (2012) The E3 ubiquitin ligase HOS1 regulates Arabidopsis flowering by mediating CONSTANS degradation under cold stress. J Biol Chem 287:43277–43287

Jupp T, Schultz A (2000) A thermodynamic explanation for black smoker temperature. Nature 403(6772):880–883

Kachout SS, Mansoura AB, Hamza KJ, Leclerc JC, Rejeb MN, Ouerghi Z (2011) Effects of water stress on plant growth in Atriplex hortensis L. J Hortic Sci Biotechnol 86:101–106

Kaklewski K, Nowak J, Ligocki M (2008) Effects of selenium content in green parts of plants on the amount of ATP and ascorbate-glutathione cycle enzyme activity at various growth stages of wheat and oilseed rape. J Plant Physiol 165:1011–1022

Kalisz A, Cebula S (2001) Direct plant covering and soil mulching in the spring production of some Chinese cabbage cultivars. Effect of temperature on premature bolting. Folia Horticulturae 13: 13–22

Kamenetsky R, Okubo H (2013) Ornamental geophytes: from basic science to sustainable production. CRC Press, Boca Raton, FL, p 578

Kamenetsky R, Zemah H, Ranwala AP, Vergelt F, Ranwala NK, Miller WB, van As H, Bendel P (2003) Water status and carbohydrate pools in tulip bulbs during dormancy release. New Phytol 158:109–118

Kanazawa Y, Xia K, Tan H, Suzuki K (2015) Seed storage longevity of Hosta sieboldiana (Asparagaceae). South Afr J Bot 98:6–9

Kang HM, Saltveit ME (2002) Chilling tolerance of maize cucumber and rice seedling leaves and roots are differentially affected by salicylic acid. Physiol Plant 115:571–576

Kannerworff WA, van der Plas LHW (1994) Respiration of bulb scale fragments of tulip after storage at 5°C. Plant Sci 104:31–38

Kaplan F, Kopka J, Haskell DW, Zhao W, Schiller KC, Gatzke N, Sung DY, Guy CL (2004) Exploring the temperature-stress metabolome of Arabidopsis. Plant Physiol 136:4159–4168

Karimzadeh G, Sharifi-Sirchi GR, Jalali-Javaran M, Dehghani H, Francis D (2006) Soluble proteins induced by low temperature treatment in the leaves of spring and winter wheat cultivars. Pak J Bot 38:1015–1026

Karnok KJ, Beard JB (1983) Effects of gibberellic acid on the CO_2 exchange rates of bermuda-grass and St. Augustinegrass when exposed to chilling temperature. Crop Sci 23(3):514–517

Kaspar TC, Bland WL (1992) Soil temperature and root growth. Soil Sci 154:290–299

Kaur G, Kumar S, Nayyar H, Upadhyaya HD (2008) Cold stress injury during the pod-filling phase in chickpea (Cicer arietinum L.). Effects on quantitative and qualitative components of seeds. J Agron Crop Sci 194:457–464

Kavi Kishore PB, Sangam S, Amrutha RN, Laxmi PS, Naidu KR, Rao KRSS, Rao S, Reddy KJ, Theriappan P, Sreenivasulu N (2005) Regulation of proline biosynthesis, degradation, uptake and transport in higher plants: its implications in plant growth and abiotic stress tolerance. Curr Sci 88:424–438

Kazan K, Lyons R (2016) The link between flowering time and stress tolerance. J Exp Bot 67(1): 47–60

Keisling TC (1982) Calculation of the length of day. Agron J 74(4):758–759

Kennedy RA, Rumpho ME, Fox TC (1992) Anaerobic metabolism in plants. Plant Physiol 100:1–6

Khaled O, Washington E, Lage P, Kantartzi SK, Lightfoot DA, Kassem MA (2011) Comparison of early and conventional soybean production systems for yield and other agronomic traits. Atlas J Plant Biol 1:1–5

Khaliq A, Aslam F, Matloob A, Hussain S, Geng M, Wahid A, Rehman H (2015) Seed priming with selenium: consequences for emergence, seedling growth, and biochemical attributes of rice. Biol Trace Elem Res 166:236–244

Khan TA, Fariduddin Q, Yusuf M (2015a) Lycopersicon esculantum under low temperature stress: an approach toward enhanced antioxidants and yield. Environ Sci Pollut Res Int 22(18): 14178–14188

Khan MI, Fatma M, Per TS, Anjum NA, Khan NA (2015b) Salicylic acid-induced abiotic stress tolerance and underlying mechanisms in plants. Front Plant Sci 6:462

Khan MA, Gemenet DC, Villordon A (2016) Root system architecture and abiotic stress tolerance: current knowledge in root and tuber crops. Front Plant Sci 7:1584

Khan TA, Fariduddin Q, Yusuf M (2017) Low temperature stress – is phytohormones application a remedy? Environ Sci Pollut Res 24:21574–21590

Khodabandeh N (2003) Cereals, 7th edn. Tehran University Press, Tehran, pp 78–111

Khodorova NV (2011) A study of adaptation to cold in a geophyte species (Corydalis bracteata (Steph.) Pers, Fumariaceae DC.) and an approach of secondary metabolism during plant development. PhD Thesis, Jules Verne University of Picardy, Amiens, France. p 181

Khodorova MV, Boitel-Conti M (2013) The role of temperature in the growth and flowering of geophytes. Plan Theory 2(4):699–711

Khodorova NV, Koteyeva NK, Miroslavov EA (2007) Ultrastructural changes in phloem parenchyma cells in Corydalis bracteata (Fumariaceae) growing outdoors and indoors. Botanicheskii Z 92:1011–1023

Khodorova NV, Miroslavov EA, Shavarda AL, Laberche JC, Boitel-Conti M (2010) Bud development in corydalis (Corydalis bracteata) requires low temperature: a study of developmental and carbohydrate changes. Ann Bot 105:891–903

Kidokoro S, Yoneda K, Takasaki H, Takahashi F, Shinozaki K, Yamaguchi-Shinozaki K (2017) Different cold-signaling pathways function in the responses to rapid and gradual decreases in temperature. Plant Cell 29:760–774

Kim YO, Kim JS, Kang H (2005) Cold-inducible zinc finger-containing glycine-rich RNA binding protein contributes to the enhancement of freezing tolerance in Arabidopsis thaliana. Plant J 42(6):890–900

Kim JS, Park SJ, Kwak KJ, Kim YO, Kim JY, Song JJ, Boseung G, Jung CH, Kang H (2007) Cold shock domain proteins and glycine-rich RNA-binding proteins from Arabidopsis thaliana can promote the cold adaptation process in Escherichia coli. Nucleic Acids Res 35(2):506–516

Kim SH, Niedziela CE, Nelson PV, de Hertogh AA, Swallow WH, Mingis NC (2007a) Growth and development of Lilium longiflorum "Nellie White" during bulb production under controlled environments: II. Effects of shifting day/night temperature regimes on scale bulblets. Sci Hortic 112:89–94

Kim SH, Niedziela CE, Nelson PV, de Hertogh AA, Swallow WH, Mingis NC (2007b) Growth and development of Lilium longiflorum "Nellie White" during bulb production under controlled environments: I. Effects of constant, variable and greenhouse day/night temperature regimes on scale and stem bulblets. Sci Hortic 112:95–98

Kim DH, Doyle MR, Sung S, Amasino RM (2009) Vernalization: winter and the timing of flowering in plants. Annu Rev Cell Dev Biol 25:277–299

Kinoshita T, Seki M (2014) Epigenetic memory for stress response and adaptation in plants. Plant Cell Physiol 55:1859–1863

Kishitani S, Watanabe K, Yasuda S, Arakawa K, Takabe T (1994) Accumulation of glycinebetaine during cold acclimation and freezing tolerance in leaves of winter and spring barley plants. Plant Cell Environ 17:89–95

Kitajima K, Fenner M (2000) Ecology of seedling regeneration. In: Fenner M (ed) Seeds: ecology of regeneration in plant communities, 2nd edn. CABI, Wallingford, UK, pp 331–359

Klepper B, Rickman RW, Peterson CM (1982) Quantitative characterization of vegetative development in small cereal grains. Agron J 74:789–792

Knight MR, Knight H (2012) Low-temperature perception leading to gene expression and cold tolerance in higher plants. New Phytol 195(4):737–751

Knight H, Brandt S, Knight MR (1998) A history of stress alters drought calcium signalling pathways in Arabidopsis. Plant J 16:681–687

Ko R, Smith LT, Smith GM (1994) Glycine betaine confers enhanced osmotolerance and cryotolerance in Listeria monocytogenes. J Bacteriol 176:426–431

Koch K (2004) Sucrose metabolism: regulatory mechanism and pivotal roles in sugar sensing and plant development. Curr Opin Plant Biol 7:235–246

Kokubun M, Shimada S, Takahashi M (2001) Flower abortion caused by pre-anthesis water deficit is not attributed to impairment of pollen in soybean. Crop Sci 41:1517–1521

Kolodziejczyk I, Dzitko K, Szewczyk R, Posmyk MM (2016) Exogenous melatonin improves corn (Zea mays L.) embryo proteome in seeds subjected to chilling stress. J Plant Physiol 193:47–56
Korkmaz A, Dufault RJ (2001) Developmental consequences of cold temperature stress at transplanting on seedling and field growth and yield. I. Watermelon. J Am Soc Hortic Sci 126:404–409
Körner C (1998) A re-assessment of high elevation treeline positions and their explanation. Oecologia 115:445–459
Körner, Ch. And Paulsen, J. 2004. A world-wide study of high altitude treeline temperatures. J Biogeogr, 31: 713–732
Körner C, Renhardt U (1987) Dry-matter partitioning and root length leaf-area ratios in herbaceous perennial plants with diverse altitudinal distribution. Oecologia 74:411–418
Körner C, Paulsen J, Pelaez-Riedl S (2003) A bioclimatic characterisation of Europe's alpine areas. In: Nagy L, Grabherr G, Körner C, Thompson DBA (eds) Alpine biodiversity in Europe. Springer, Berlin, pp 13–28
Kos M, Poschlod P (2010) Why wait? Trait and habitat correlates of variation in germination speed among Kalahari annuals. Oecologia 162:549–559
Kotak S, Larkindale J, Lee U, von Koskull-Doring P, Vierling E, Scharf KD (2007) Complexity of the heat stress response in plants. Curr Opin Plant Biol 10:310–316
Kratsch HA, Wise RR (2000) The ultrastructure of chilling stress. Plant Cell Environ 23:337–350
Kreps JA, Wu YJ, Chang HS, Zhu T, Wang X, Harper JF (2002) Transcriptome changes for Arabidopsis in response to salt, osmotic, and cold stress. Plant Physiol 130:2129–2141
Krug H (1991) Gemüseproduktion. Verlag Paul Parey, Berlin/Hamburg, Germany, pp 335–341
Krzyzanowski FC, Delouche JC (2011) Germination of cotton seed in relation to temperature. Revista Brasileira de Sementes 33:543–548
Kudoh H, Ishiguri Y, Kawano S (1995) Phenotypic plasticity in Cardamine flexuosa – variation among populations in plastic response to chilling treatments and photoperiods. Oecologia 103: 148–156
Kuk YI, Shin JS, Burgos NR, Hwang TE, Han O, Cho BH, Jung S, Guh JO (2003) Antioxidative enzymes offer protection from chilling damage in rice plants. Crop Sci 43:2109–2117
Kumar S, Nayyar H, Bains TS, Kaur G (2005) Low temperature effects on early maturing chickpea genotype ICCV-96029. Int Chickpea a Pigeonpea Newsl 12:19–22
Kumar S, Kaur G, Nayyar H (2008) Exogenous application of abscisic acid improves cold tolerance in chickpea (Cicer arietinum L.). J Agron Crop Sci 194:449–456
Kumar S, Thakur P, Kaushal N, Malik JA, Gaur P, Nayyar H (2012a) Effect of varying high temperatures during reproductive growth on reproductive function, oxidative stress and seed yield in chickpea genotypes differing in heat sensitivity. Arch Agron Soil Sci 59:823–843
Kumar S, Kaushal N, Nayyar H, Gaur PM (2012b) Abscisic acid induces heat tolerance in chickpea (Cicer arietinum L.) seedlings by facilitated accumulation of osmoprotectants. Acta Phys Plant 34:1651–1658
Kurosaki H, Yumoto S (2003) Effects of low temperature and shading during flowering on the yield components in soybeans. Plant Prod Sci 6:17–23
Kushad MM, Yelenosky G (1987) Evaluation of polyamine and proline levels during low temperature acclimation of citrus. Plant Physiol 84:692–695
Lafitte HR, Yongsheng G, Yan S, Li ZK (2007) Whole plant responses, key processes, and adaptation to drought stress: the case of rice. J Exp Bot 58:169–175
Lalonde S, Beebe DU, Saini HS (1997) Early signs of disruption of wheat anther development associated with the induction of male sterility by meiotic-stage water deficit. Sex Plant Reprod 10:40–48
Lambrechts H, Kollöfel C (1993) Soluble and insoluble invertase activity in elongating Tulipa gesneriana flower stalk. Physiol Plant 89:830–834
Lambrechts H, Rook F, Kollöfel C (1994) Carbohydrate status of tulip bulbs during cold-induced flower stalk elongation and flowering. Plant Physiol 104:515–520

Landhausser SM, DesRochers A, Lieffers VJ (2001) A comparison of growth and physiology in Picea glauca and Populus tremuloides at different soil temperatures. Can J For Res 31:1922–1929

Lang A (1965) Physiology of flower initiation. In: Ruhland W (ed) Encyclopedia, Plant Physiol, vol 15. Springer, Berlin, pp 1380–1536

Lang A, Minchin PEH (1986) Phylogenetic distribution and mechanism of translocation inhibition by chilling. J Exp Bot 37(176):389–398

Lapointe L, Lerat S (2006) Annual growth of the spring ephemeral Erythronium americanum as a function of temperature and mycorrhizal status. Can J Bot 84:39–48

Larcher W (1995) Physiological plant ecology-ecophysiology and stress physiology of functional groups. Springer, Berlin

Larcher W (2003) Physiological plant ecology: ecophysiology and stress physiology of funtional groups. Springer, Berlin

Larkindale J, Vierling E (2008) Core genome responses involved in acclimation to high temperature. Plant Physiol 146:748–761

Latimer JG (1991) Mechanical conditioning for control of growth and quality of vegetable transplants. Hortic Sci 26:1456–1461

Laugier E, Tarrago L, Vieira Dos Santos C, Eymery F, Havaux M, Rey P (2010) Arabidopsis thaliana plastidial methionine sulfoxide reductases B, MSRBs, account for most leaf peptide MSR activity and are essential for growth under environmental constraints through a role in the preservation of photosystem antennae. Plant J 61:271–282

Lazo-Vélez MA, Chávez-Santoscoy A, Serna-Saldivar SO (2015) Selenium-enriched breads and their benefits in human nutrition and health as affected by agronomic, milling, and baking factors. Cereal Chem 92:134–144

Lee AK, Suh JK, Roh MS (2002) Flowering and changes in respiration in Asiatic hybrid lilies as influenced by bulb vernalization. Sci Hort 123:366–371

Lee SH, Singh AP, Chung GC (2004) Rapid accumulation of hydrogen peroxide in cucumber roots due to exposure to low temperature appears to mediate decreases in water transport. J Exp Bot 55(403):1733–1741

Lee BH, Henderson DA, Zhu JK (2005) The Arabidopsis cold-responsive transcriptome and its regulation by ICE1. Plant Cell 17:3155–3175

Lee J, Yoo S, Park S, Hwang I, Lee J, Ahn J (2007) Role of SVP in the control of flowering time by ambient temperature in Arabidopsis. Genes Dev 21(4):397–402

Leishman MR, Wright IJ, Moles AT, Westoby M, Fenner M (2000) The evolutionary ecology of seed size. Trends Ecol Evol 7:368–372

Lejeune P, Bernier G (1996) Effect of environment on the early steps of ear initiation in maize (Zea mays L.). Plant Cell Environ 19(2):217–224

Lempe J, Balasubramanian S, Sureshkumar S, Singh A, Schmid M, Weigel D (2005) Diversity of flowering responses in wild Arabidopsis thaliana strains. Public Lib Sci Genet 1:e6

Leskovar DI, Cantliffe DJ (1991) Tomato transplant morphology affected by handling and storage. Hortic Sci 26:1377–1379

Letham DS, Smith HG, Willcocks DA (2003) Cytokinin metabolism in Narcissus bulbs: chilling promotes acetylation of zeatin riboside. Funct Plant Biol 30:525–532

Levitt J (1980) Responses of plants to environmental stresses, Chilling, freezing and high temperature stresses, vol 1, 2nd edn. Academic, New York

Li TG, Visperas RM, Vergara BS (1981) Correlation of cold tolerance at different growth stages in rice. Acta Bot Sin 23:203–207

Li Y, Roycewicz P, Smith E, Borevitz J (2006) Genetics of local adaptation in the laboratory: flowering time quantitative trait loci under geographic and seasonal conditions in Arabidopsis. PLoS One 1:e105

Li KL, Liu T, Li YG, Shao LP, Zhang R, Song ZL, Liu HF (2012) Seed polymorphism and germination of Salsola korshinskyi drob. J Shihezi Univ Nat Sci 30(3):276–281

Li XN, Pu HC, Liu FL, Zhou Q, Cai J, Dai T, Cao W, Jiang D (2015) Winter wheat photosynthesis and grain yield responses to spring freeze. Agron J 107:1002–1010
Li X, Tan DX, Jiang D, Liu F (2016) Melatonin enhances cold tolerance in drought-primed wild-type and abscisic acid-deficient mutant barley. J Pineal Res 61:328–339
Li S, Yang Y, Zhang Q, Liu N, Xu Q, Hu L (2018) Differential physiological and metabolic response to low temperature in two zoysiagrass genotypes native to high and low latitude. PLoS One 13(6):e0198885
Li S, Jiang H, Wang J, Wang Y, Pan S, Tian H, Duan M, Wang S, Tang X, Mo Z (2019) Responses of plant growth, physiological, gas exchange parameters of super and non-super rice to rhizosphere temperature at the tillering stage. Sci Rep 9:10618
Liang Y, Zhu J, Li Z, Chu G, Yangfang D, Zhang J (2008) Role of silicon in enhancing resistance to freezing stress in two contrasting winter wheat cultivars. Environ Exp Bot 64(3):286–294
Liang LH, Mei X, Lin F, Xia J, Liu SJ, Wang JH (2009) Effect of low temperature stress on tissue structure and physiological index of cashew young leaves. Ecol Environ Sci 18:317–320
Liang Y, Nikolic M, Bélanger R, Gong H, Song A (2015) Silicon–mediated tolerance to drought and low–temperature stress, silicon in agriculture. Springer Science + Business Media, Dordrecht
Lin SS, Peterson ML (1975) Low temperature induced floret sterility in rice. Crop Sci 15:657–660
Linder P, Jankowsky E (2011) From unwinding to clamping—the DEAD box RNA helicase family. Nat Rev Mol Cell Biol 12:505–516
Liu JJ, Lin SH, Xu PL, Wang XJ, Bai JG (2009) Effects of exogenous silicon on the activities of antioxidant enzymes and lipid peroxidation in chilling-stressed cucumber leaves. Agric Sci Chin 8(9):1075–1086
Liu K, Baskin J, Baskin C, Bu H, Liu M, Liu W, Du G (2011) Effect of storage conditions on germination of seeds of 489 species from high elevation grasslands of the eastern Tibet Plateau and some implications for climate change. Am J Bot 98:12–19
Liu J, Wang W, Wang L, Sun Y (2015) Exogenous melatonin improves seedling health index and drought tolerance in tomato. Plant Growth Regul 77:317–326
Liu K, Cao S, Du G, Baskin JM, Baskin CC, Bu H, Qi W, Ting L (2018) Linking seed germination and plant height: a case study of a wetland community on the eastern Tibet Plateau. Plant Biol 20:886–893
Livingston DP, Henson CA (1998) Apoplastic sugars, fructans, fructan exohydrolase, and invertase in winter oat: responses to second-phase cold hardening. Plant Physiol 116(1):403–408
Livingston HD III, Heyer A (2009) Fructan and its relationship to abiotic stress tolerance in plants. Cell Mol Life Sci 66:2007–2023
Llanes A, Andrade A, Alenabo S, Luna V (2016) Alterations of endogenous hormonal levels in plants under drought and salinity. Am J Plant Sci 7:1357–1371
Long SP (1983) C4 photosynthesis at low temperatures. Plant Cell Environ 6(4):345–363
Lou Q, Chen L, Sun Z, Xing Y, Li J, Xu X, Mei M, Luo J, Luo ML (2007) A major QTL associated with cold tolerance at seedling stage in rice (Oryza sativa L.). Euphytica 158:87–94
Louis H, Annappan RS (1980) Environmental effects on yield of coconut palm. Indian Coconut J 12:1–4
Lu JJ, Tan DY, Baskin CC, Baskin JM (2016) Effects of germination season on life history traits and on transgenerational plasticity in seed dormancy in a cold desert annual. Sci Rep 6:250076
Lukatkin AS, Brazaityte A, Bobinas C, Duchovskis P (2012) Chilling injury in chilling-sensitive plants: a review. Agriculture 99(2):111–124
Lundmark M, Hurry V, Lapointe L (2009) Low temperature maximizes growth of Crocus vernus (L.) Hill via changes in carbon partitioning and corm development. J Exp Bot 60:2203–2213
Luo Q (2011) Temperature thresholds and crop production: a review. Clim Chang 109(3):583–598
Luo Q, Bange M, Clancy L (2014a) Cotton crop phenology in a new temperature regime. Ecol Model 285:22–29

Luo N, Liu T, Liu HF, Liu ZC, Tang JG (2014b) Difference of seed germination characteristics between Salsola affinis C.A. Mey and Salsola brachiata Pall in Junggar Basin. J Shihezi Univ Nat Sci 32(5):608–614

Lyr H (1996) Effect of the root temperature on growth parameters of various European tree species. Ann For Sci 53:317–323

Macduff JH, Wild A (1986) Effects of temperature on parameters of root growth relevant to nutrient uptake: measurements on oilseed rap and barley grown in flowing nutrient solution. Plant Soil 94:321–332

Mackill DJ, Coffman WR, Garrity DP (1996) Rainfed lowland rice improvement. International Rice Research Institute, Manila, p 242

Maite ME, Ellis RH (2015) Temporal sensitivities of rice seed development from spikelet fertility to viable mature seed to extreme-temperature. Crop Sci 55:354–364

Malhotra RS, Singh KB (1991) Gene action for cold tolerance in chickpea. Theor Appl Genet 82: 598–601

Mamushina NS, Zubkova EK (1996) Effect of temperature on potential photosynthesis and photosynthetic carbon metabolism in C_3 plants with different seasonal patterns of development. Russ J Plant Physiol 43:313–318

Mamushina NS, Voznesenskaya EV, Zubkova EK, Maslova TG, Miroslavov EA (2002) Structural and functional changes of mesophyll cells during leaf growth in two species of spring ephemers. Russ J Plant Physiol 49:171–178

Mamushina NS, Zubkova EK, Yudina OS (2008) Metabolism of ^{14}C-Glycine as a substrate for photorespiration of the leaf at different developmental stages of ephemeroides. Russ J Plant Physiol 55:41–47

Manivannan P, Jaleel CA, Sankar B, Kishorekumar A, Somasundaram R, Lakshmanan GMA, Panneerselvam R (2007) Growth, biochemical modifications and proline metabolism in Helianthus annuus L. as induced by drought stress. Colloids Surf B Biointerfaces 59(2): 141–149

Maqsood A, Shahbaz M, Akram NA (2006) Influence of exogenously applied glycine betaine on growth and gas exchange characteristics of maize (Zea mays L.). Pak J Agric Sci 43:36–41

Maraghni M, Gorai M, Neffati M (2010) Seed germination at different temperatures and water stress levels, and seedling emergence from different depths of Ziziphus lotus. South Afr J Bot 76(3):453–459

Markowski A (1988) Sensitivity of different species of field crops to chilling temperature. II. Germination, growth and injuries of seedlings. Acta Physiol Plant 10(3):275–283

Marta B, Szafranska K, Posmyk MM (2016) Exogenous melatonin improves antioxidant defense in cucumber seeds (Cucumis sativus L.) germinated under chilling stress. Front Plant Sci 7:575

Martindale W, Leegood RC (1997) Acclimation of photosynthesis to low temperature in Spinacia oleracea L. II. Effects of nitrogen supply. J Exp Bot 48(10):481–487

Maruta E (1994) Seedling establishment of Polygonum cuspidatum and Polygonum weyrichii var. alpinum at high altitudes of Mt. Fuji. Ecol Res 9:205–213

Mathers HM, Lowe SB, Scagel C, Struve DK, Case LT (2007) Abiotic factors influencing root growth of woody nursery plants in containers. Hort Technol 17:151–162

Matsui T, Kagata H (2003) Characteristics of floral organs related to reliable self pollination in rice (Oryza sativa L.). Ann Bot 91:473–477

Matsui T, Omasa K (2002) Rice (Oryza sativa L.) cultivars tolerant to high temperature at flowering: anther characteristics. Ann Bot 89(6):683–687

Matsui T, Omasa K, Horie T (2001) The difference in sterility due to high temperatures during the flowering period among japonica rice varieties. Plant Prod Sci 4:90–93

Matteucci M, D'Angeli S, Errico S, Lamanna R, Perrotta G, Altamura MM (2011) Cold affects the transcription of fatty acid desaturases and oil quality in the fruit of Olea europaea L. genotypes with different cold hardiness. J Exp Bot 62:3403–3420

Maurel C, Simonneau T, Sutka M (2010) The significance of roots as hydraulic rheostats. J Exp Bot 61:3191–3198

McClintock B (1984) The significance of responses of the genome to challenge. Science 226:792–801

McLaughlin JW, Green DW (1991) Fruit and hormones influence flowering of apple. I. Effect of cultivar. J Am Soc Hortic Sci 116:446–449

McMahon MJ, Permit AJ, Arnold JE (1994) Effects of chilling on Episcia and Dieffenbachia. J Am Soc Hortic Sci 119(1):80–83

McMaster GS, Edmunds DA, Wilhelm WW, Nielsen DC, Prasad PVV, Ascough JC (2011) PhenologyMMS: a program to simulate crop phenological responses to water stress. Comput Electron Agric 77:118–125

McMichael BL, Burke JJ (1998) Soil temperature and root growth. Hortic Sci 33:947–951

McMichael BL, Quisenberry JE (1993) The impact of the soil environment on the growth of root systems. Environ Exp Bot 33:53–61

Meena YK, Khurana DS, Kaur N, Singh K (2018) Towards enhanced low temperature stress tolerance in tomato: an approach. J Environ Biol 39:529–535

Michaels SD, Amasino RM (2000) Memories of winter: vernalization and the competence to flower. Plant Cell Environ 23:1145–1153

Miller DE (1986) Root systems in relation to stress tolerance. Hortic Sci 21:963–970

Miller P, Lanier W, Brandt S (2001) Using growing degree days to predict plant stages. Montana State University Extension Service, Montana

Miller G, Suzuki N, Citci-Yilmaz S, Mittler R (2010) Reactive oxygen species homeostasis and signalling during drought and salinity stresses. Plant Cell Environ 33(4):453–467

Minges PA, Warholic D (1973) Pepper problems in New York. In: Proceedings of National pepper conference. Packers Intl., St. Charles, pp 1–14

Minhas D, Grover A (1999) Transcript levels of genes encoding various glycolytic and fermentation enzymes change in response to abiotic stresses. Plant Sci 146:41–51

Miralles J, Nortes PA, Sánchez-Blanco MJ, Martínezsánchez JJ, Bañón S (2009) Above-ground and pot-in pot production systems for Myrtus communis L. Trans ASABE 52:93–101

Mirouze M, Paszkowski J (2011) Epigenetic contribution to stress adaptation in plants. Curr Opin Plant Biol 14:267–274

Mittler R (2002) Oxidative Stress, antioxidants and stress tolerance. Trend Plant Sci 7(9):405–410

Mittler R, Vanderauwera S, Gollery M, Van Breusegem F (2004) Reactive oxygen gene network of plants. Trend Plant Sci 9(10):490–498

Mohammadi-Nikpoor A, Keshavarzi A (1995) Study of effect of sowing date and plant density on yield and yield components of safflower in Mashad, Iran. M.Sc. Thesis on Agriculture, Ferdowsi University of Mashad, Mashad, Iran

Moles AT, Ackerly DD, Webb CO, Tweddle JC, Dickie JB, Pitman AJ, Westoby M (2005) Factors that shape seed mass evolution. Proc Natl Acad Sci U S A 102:10540–10544

Molinier J, Ries G, Zipfel C, Hohn B (2006) Transgeneration memory of stress in plants. Nature 442:1046–1049

Mondoni A, Probert RJ, Rossi G, Vegini E, Hay FR (2011) Seeds of alpine plants are short lived: implications for long-term conservation. Ann Bot Lond 107(1):171–179

Monneveux P, Sánchez C, Beck D, Edmeades GO (2006) Drought tolerance improvement in tropical maize source populations: evidence of progress. Crop Sci 46:180–191

Montagnini F, Finney C (2011) Restoring degraded landscapes with native species in Latin America. Nova Science Publishers, Hauppauge, New York

Moore K, Jenkins L, Hertel K, Callaghan G (2010) Chickpea in 2010: low temperature effects in 2009. GRDC update paper. www.grdc.com.au

Moosavi SGHR (2007) Effect of plant density and planting pattern on agronomic traits and yield of forage sorghum. Final report of agricultural design in Islamic Azad University, Birjand Branch, Iran

Moriondo M, Bindi M (2005) Cambiamenti climatici e ambiente vitivinicolo: impatto e strategie di adattamento. Rapporto finale per ARSIA

Moriondo M, Bindi M (2006) Comparison of temperatures simulated by GCMs, RCMs and statistical downscaling: potential application in studies of future crop development. Clim Res 30:149–160

Moriondo M, Bindi M (2007) Impact of climate change on the phenology of typical Mediterranean crops. Italian J Agrometereol 12(3):5–12

Moriondo M, Orlandini S, De Nuntiis P, Mandrioli P (2001) Effect of agrometeorological parameters on the phenology of pollen emission and production of olive trees (Olea europaea L.). Aerobiologia 7:225–232

Moriondo M, Giannakopoulos C, Bindi M (2011) The role of climate extremes in crop yield climate change impact assessments. Clim Chang 104:679–701

Morrison MJ, Stewart DW (2002) Heat stress during flowering in summer brassica. Crop Sci 42: 797–803

Mukherjee SK, Litz RE (2009) Introduction: botany and importance. In: Litz RE (ed) The mango: botany, production and uses, 2nd edn. CAB International, London, pp 1–18

Mukhopadhyay A, Vij S, Tyagi AK (2004) Over expression of a zinc-finger protein gene from rice confers tolerance to cold, dehydration, and salt stress in transgenic tobacco. Proc Natl Acad Sci U S A 101:6309–6314

Müller J, Menzel D, Šamaj J (2007) Cell-type-specific disruption and recovery of the cytoskeleton in Arabidopsis thaliana epidermal root cells upon heat shock stress. Protoplasma 230:231–242

Munier-Jolain NG, Ney B (1998) Seed growth rate in grain legumes II. Seed growth rate depends on cotyledon cell number. J Exp Bot 49:1971–1976

Murphy C, Wilson JM (1981) Ultrastructural features of chilling injury in Episcia reptans. Plant Cell Environ 4:261–265

Nägele T, Kandel BA, Frana S, Meissner M, Heyer AG (2011) A systems biology approach for the analysis of carbohydrate dynamics during acclimation to low temperature in Arabidopsis thaliana. FEBS J 278:506–518

Naguchi HK (2007) Submergence acclimation to low-temperature stress in rice roots. Plant Prod Sci 10(3):297–302

Nahar K, Hasanuzzaman M, Majumder RR (2009a) Effect of low temperature stress in transplanted Aman Rice Varieties mediated by different transplanting dates. Acad J Plant Sci 2:132–138

Nahar K, Biswas JK, Shamsuzzaman AMM, Hasanuzzaman M, Barman HN (2009b) Screening of indica rice (Oryza sativa L.) genotypes against low temperature stress. Bot Res Int 2:295–303

Nahar K, Biswas JK, Shamsuzzaman AMM (2012) Cold stress tolerance in rice plant: screening of genotypes based on morphophysiological traits. LAP Lambert Academic, Berlin

Nakagawa H, Yamagishi J, Miyamoto N, Motoyama M, Yano M, Nemoto K (2005) Flowering response of rice to photoperiod and temperature: a QTL analysis using a phenological model. Theor Appl Genet 110:778–786

Nakaminami K, Karlson DT, Imai R (2006) Functional conservation of cold shock domains in bacteria and higher plants. Proc Natl Acad Sci U S A 103(26):10122–10127

Nayyar H, Bains T, Kumar S (2005a) Low temperature induced flower abortion in chickpea: relationship to abscisic acid and cryoprotectents in reproductive organs. Environ Exp Bot 53: 39–47

Nayyar H, Bains TS, Kumar S (2005b) Chilling effects during seed filling on accumulation of seed reserves and grain yield. J Sci Food Agric 85:1925–1930

Nayyar H, Kaur G, Kumar S, Upadhyaya HD (2007) Low temperature effects during seed filling on chickpea genotypes (Cicer arietinum L.): probing mechanisms affecting seeds reserves and yield. J Agron Crop Sci 193:336–344

Nejad TS, Bakhshande A, Nasab SB, Payande K (2010) Effect of drought stress on corn root growth. Rep Opin 2:47–52

Nelson JM, Palzkill DA (1993) Irrigation effects on growth, cold tolerance of flower buds, and seed yield of jojoba. In: Janick J, Simon JE (eds) New crops. Wiley, New York, pp 360–362

Niedziela CE, Kim SH, Nelson PV, de Hertogh AA (2008) Effects of N-P-K deficiency and temperature regime on the growth and development of Lilium longiflorum "Nellie White" during bulb production under phytotron conditions. Sci Hortic 116:430–436

Nievola CC, Carvalho CP, Carvalho V, Rodrigues E (2017) Rapid responses of plants to temperature changes. Temperature (Austin) 4(4):371–405

Nishiyama I (1977) Decrease in germination activity of rice seeds due to excessive desiccation in storage. Jpn J Crop Sci 46:111–118

Nishiyama I (1978) Male sterility caused by cooling treatment at the young microspore stage in rice plants. XVIII. Some enzyme activities in anthers during and after the cooling. Jpn J Crop Sci 47: 551–556

Nishiyama I (1995) Damage due to extreme temperatures. In: Matsuo T, Kumazawa K, Ishii R, Ishihara H, Hirata H (eds) Science of the rice plant. Food and Agriculture Policy Research Center, Tokyo, Japan, pp 769–812

Nishiyama I, Blanco L (1980) Avoidance of high temperature sterility by flower opening in the early morning. Jpn Agric Res Q 14:116–117

Nkebiwe PM, Weinmann M, Bar-Tal A, Müller T (2016) Fertilizer placement to improve crop nutrient acquisition and yield: a review and meta-analysis. Field Crop Res 196:389–401

Nobel PS (2003) Environmental biology of agaves and cacti. Cambridge University Press, Cambridge

Nonami H (1998) Plant water relations and control of cell elongation at low water potentials. J Plant Res 111:373–382

Noy-Porat T, Flaishman MA, Eshel A, Sandler-Ziv D, Kamenetsky R (2009) Florogenesis of the Mediterranean geophyte Narcissus tazetta and temperature requirements for flower initiation and differentiation. Sci Hortic 120:138–142

Ofir M, Kigel J (2006) Opposite effects of daylength and temperature on flowering and summer dormancy of Poa bulbosa. Ann Bot 97:659–666

Oh HS, Kim HH, Moon SJ, Kwon SJ, Lee CH (2003) Effect of temperature and priming treatment on seed germination of Hosta plantaginea. J Korean Soc Hortic Sci 44:267–270. (In Korean with English abstract)

Okubo H, Uemoto S (1985) Changes in endogenous gibberellin and auxin activities during first internode elongation in tulip flower stalk. Plant Cell Physiol 26:709–719

Okuma E, Murakami Y, Shimoishi Y, Tada M, Murata Y (2004) Efects of exogenous application of proline and betaine on the growth of tobacco cultured cells under saline conditions. Soil Sci Plant Nutr 50:1301–1305

Olien CR, Smith MN (1977) Ice adhesions in relation to freeze stress. Plant Physiol 60:499–503

Oliver SN, Van Dongen JT, Alfred SC, Mamun EA, Zhao X, Saini HS, Fernandes SF, Blanchard CL, Sutton BG, Geigenberger P, Dennis ES, Dolferus R (2005) Cold-induced repression of the rice anther-specific cell wall invertase gene OSINV4 is correlated with sucrose accumulation and pollen sterility. Plant Cell Environ 28(12):1534–1551

Oliver SN, Dennis ES, Dolferus R (2007) ABA regulates apoplastic sugar transport and is a potential signal for cold-induced pollen sterility in rice. Plant Cell Physiol 48:1319–1330

Orabi SA, Dawood MJ, Saleem SR (2015) Comparative study between the physiological role of hydrogen peroxide and salicylic acid in alleviating low temperature stress on tomato plants grown under sand-ponic culture. Sci Agric 9:49–59

Orozco-Segovia A, Sánchez-Coronado ME, Martínez-Villegas JA, Pedrero-López LV, Becerra-Vázquez A, Rosete-Rodríguez A, Peraza-Villarreal H (2014) Ecofisiología de semillas de plantas tropicales: El acondicionamiento mátrico una herramienta útil para germinar especies nativas, útiles para la restauración y conservación de especies. In: dos Santos Silva TR, do Nascimento Moura CW, de Lima LC, Ribeiro dos Santos FS (eds) Botânica na America Latina: Conhecimento, interação e difusão. Memorias del LXV Congresso Latinoamericano de Botânica. Sociedad Latinoamericana de Botánica, Salvador, Brazil, pp 194–203

Örvar BL, Sangwan V, Omann F, Dhindsa RS (2000) Early steps in cold sensing by plant cells: the role of actin cytoskeleton and membrane fluidity. Plant J 23:785–794

Osborne CP, Chuine I, Viner D, Woodword FI (2000) Olive phenology as a sensitive indicator of future climatic warming in the Mediterranean. Plant Cell Environ 23:701–710

Ozga JA, Yu J, Reinecke DM (2003) Pollination-, development-, and auxin-specific regulation of gibberellin 3bet-hydroxylase gene expression in pea fruits and seeds. Plant Physiol 131:1137–1146

Ozturk L, Demir Y (2002) In vivo and in vitro protective role of proline. Plant Growth Regul 38: 259–264

Papadopoulos AP, Tiessen H (1987) Root and air temperature effects on the elemental composition of tomato. J Am Soc Hortic Sci 112:988–993

Paparella S, Araújo SS, Rossi G, Wijayasinghe M, Carbonera D, Balestrazzi A (2015) Seed priming: state of the art and new perspectives. Plant Cell Rep 34:1281–1293

Passarini F, Wientjes E, Hienerwadel R, Croce R (2009) Molecular basis of light harvesting and photoprotection in CP24: unique features of the most recent antenna complex. J Biol Chem 284: 29536–29546

Patel PN, Hall AE (1990) Genotypic variation and classification of cowpea for reproductive responses to high temperatures under long photoperiods. Crop Sci 30:614–621

Pecinka A, Rosa M, Schikora A, Berlinger M, Hirt H, Luschnig C, Scheid OM (2009) Transgenerational stress memory is not a general response in Arabidopsis. PLoS One 4(4): e5202

Pedrero-López LV, Rosete-Rodríguez A, Sánchez-Coronado ME, Mendoza-Hernández PE, Orozco-Segovia A (2016) Effects of hydropriming treatments on the invigoration of aged Dodonaea viscosa seeds and water-holding polymer on the improvement of seedling growth in a lava field. Restor Ecol 24:61–70

Peet MM, Sato S, Gardner RG (1998) Comparing heat stress effects on male-fertile and male-sterile tomatoes. Plant Cell Environ 21:225–231

Peng S, Huang J, Sheehy JE, Lanza RC, Visperas RM, Zhong X, Centeno GS, Khush GS, Cassman KG (2004) Rice yields decline with higher night temperatures from global warming. Proc Natl Acad Sci U S A 101(27):9971–9975

Peng MW, Wang M, Jiang P, Chang YL, Chu GM (2018) The impact of low temperature on seed germination of two desert species in Junggar Basin of China. Appl Ecol Environ Res 16(5): 5771–5780

Peraza-Villarreal H, Sánchez-Coronado M, Linding-Cisneros R, Tinoco-Ojanguren C, Velázquez-Rosas N, Cámara-Cabrales L, Orozco-Segovia A (2020) Seed priming effects on germination and seedling establishment of useful tropical trees for ecological restoration. Trop Conserv Sci 11(1):1–15

Perrot-Rechenmann C (2010) Cellular responses to auxin: division versus expansion. Cold Spring Harb Perspect Biol 2:a001446

Pikuta EV, Hoover RB, Tang J (2007) Microbial extremophiles at the limits of life. Crit Rev Microbiol 33(3):183–209

Pires ND, Dolan L (2012) Morphological evolution in land plants: new designs with old genes. Philos Trans R Soc B Biol Sci 367(1588):508–518

Pocock TH, Hurry VM, Savitch LV, Huner NPA (2001) Susceptibility to low-temperature photoinhibition and the acquisition of freezing tolerance in winter and spring wheat: the role of growth temperature and irradiance. Physiol Plant 113:499–506

Posmyk MM, Balabusta M, Wieczorek M, Sliwinska E, Janas KM (2009) Melatonin applied to cucumber (Cucumis sativus L.) seeds improves germination during chilling stress. J Pineal Res 46:214–223

Prasad PVV, Craufurd PQ, Summerfield RJ (1999) Fruit number in relation to pollen production and viability in groundnut exposed to short episodes of heat stress. Ann Bot 84:381–386

Prasad PVV, Boote KJ, Allen LH Jr (2006a) Adverse high temperature effects on pollen viability, seed-set, seed yield and harvest index of grain-sorghum [Sorghum bicolor (L.) Moench] are more severe at elevated carbon dioxide due to high tissue temperature. Agric For Meteorol 139: 237–251

Prasad PVV, Boote KJ, Allen LH, Sheehy JE, Thomas JMG (2006b) Species, ecotype and cultivar differences in spikelet fertility and harvest index of rice in response to high temperature stress. Field Crop Res 95:398–411
Pregitzer KS, King JS, Andrew JB, Brown E (2000) Responses of tree fine roots to temperature. New Phytol 147:105–115
Price MV, Waser NM (1998) Effects of experimental warming on plant reproductive phenology in a subalpine meadow. Ecology 79:1261–1271
Primack RB (1985) Patterns of flowering phenology in communities, populations, individuals, and single flowers. In: White J (ed) The population structure of vegetation. Dr. W. Junk, Dordrecht, The Netherlands, pp 571–593
Putterill J, Laurie R, Macknight R (2004) It's time to flower: the genetic control of flowering time. BioEssays 26:363–373
Rahnama A, Munns R, Poustini K, Watt M (2011) A screening method to identify genetic variation in root growth responses to a salinity gradient. J Exp Bot 62:69–77
Rang ZW, Jagadish SVK, Zhou QM, Craufurd PQ, Heuer S (2011) Effect of high temperature and water stress on pollen germination and spikelet fertility in rice. Environ Exp Bot 70:58–65
Ranwala AP, Miller WB (2008a) Analysis of nonstructural carbohydrates in storage organs of 30 ornamental geophytes by high-performance anion-exchange chromatography with pulsed amperometric detection. New Phytol 180:421–433
Ranwala AP, Miller WB (2008b) Gibberellin-mediated changes in carbohydrate metabolism during flower stalk elongation in tulips. Plant Growth Regul 55:241–248
Rao GS, Ashraf U, Kong LL, Mo ZW, Xiao LZ, Zhong KY, Fahad R, Tang X (2019) Low soil temperature and drought conditions at flowering stage affect physiology and pollen traits of rice. J Integr Agric 18(8):1859–1870
Rashed Mohassel MH, Koochaki A (2000) Principles and operation of dryfarming. (Translation), 6th edn. Jahad Daneshgahi of Mashhad Press, Tehran, pp 142–143
Rathcke B, Lacey EP (1985) Phenological patterns of terrestrial plants. Annu Rev Ecol Syst 16: 179–214
Rauf S (2008) Breeding sunflower (Helianthus annuus L.) for drought tolerance. Commun Biometry Crop Sci 3:29–44
Raven JA, Edwards D (2001) Roots: evolutionary origins and biogeochemical significance. J Exp Bot 52(Root Special Issue):381–401
Raven PH, Ray F, Evert S, Eichhorn E (2005) Biology of plants, 7th edn. W.H. Freeman and Company Publishers, New York, pp 504–508
Rebers M, Romeijn G, Knegt E, van der Plas LHW (1994) Effects of exogenous gibberellins and paclobutrazol on floral stalk growth of tulip sprouts isolated from cooled and non-cooled tulip bulbs. Physiol Plant 92:661–667
Rebers M, Vermeer E, Knegt E, Shelton CJ, Plas LHW (1995) Gibberellin levels and cold-induced floral stalk elongation in tulip. Physiol Plant 94:687–691
Reddy VR, Reddy KR, Baker DN (1991) Temperature effects on growth and development of cotton during the fruiting period. Agron J 83:211–217
Reddy KR, Hodges HF, McKinion JM, Wall GW (1992a) Temperature effects on Pima cotton growth and development. Agron J 84:237–243
Reddy KR, Hodges HF, Reddy VR (1992b) Temperature effects on cotton fruit retention. Agron J 84:26–30
Reddy KR, Reddy VR, Hodges HF (1992c) Temperature effects on early season cotton growth and development. Agron J 84:229–237
Reddy KR, Hodges HF, McKinion JM (1993) A temperature model for cotton phenology. Biotronics 22:47–59
Reddy KR, Hodges HF, McKinion JM (1995) Cotton crop responses to a changing environment. In: Rosenzweig C (ed) Climate change and agriculture: analysis of potential international impacts, ASA Spec. Publ, vol 59. ASA, Madison, pp 3–30

Reddy KR, Hodges HF, McKinion JM (1997) Crop modeling and applications: a cotton example. Adv Agron 59(225):290

Reddy KR, Brand D, Wijewardana C, Gao W (2017) Temperature effects on cotton seedling emergence, growth, and development. Agron J 109(4):1379–1387

Rehman H, Malik SA, Saleem M (2004) Heat tolerance of upland cotton during the fruiting stage evaluated during cellular membrane thermostability. Field Crop Res 85:149–158

Reyes E, Jennings PH (1994) Response of cucumber (Cucumis sativus L.) and squash (Cucurbita pepo L. var. melopepo) roots to chilling stress during early stages of seedling development. J Am Soc Hortic Sci 119:964–970

Rhodes D, Hanson AD (1993) Quaternary ammonium and tertiary sulfonium compounds in higher-plants. Annu Rev Plant Physiol Plant Mol Biol 44:357–384

Ricard B, Couée I, Raymond P, Saglio PH, Saint-Ges V, Pradet A (1994) Plant metabolism under hypoxia and anoxia. Plant Physiol Biochem 32:1–10

Richards RA (2006) Physiological traits used in the breeding of new cultivars for water-scarce environments. Agric Water Manag 80(1–3):197–211

Richner W, Soldati A, Stamp P (1996) Shoot-to-root relations in field grown maize seedlings. Agron J 88:56–61

Rietveld PL, Wilkinson C, Franssen HM, Balk PA, van der Plas LHV, Weisbeek PJ, Douwe de Boer A (2000) Low temperature sensing in tulip (Tulipa gesneriana L.) is mediated through an increased response to auxin. J Exp Bot 51:587–594

Rieu I, Ruiz-Rivero O, Fernandez-Garcia N, Griffiths J, Powers SJ, Gong F, Linhartova T, Eriksson S, Nilsson O, Thomas SG, Phillips AL, Hedden P (2008) The gibberellin biosynthetic genes AtGA20ox1 and AtGA20ox2 act, partially redundantly, to promote growth and development throughout the Arabidopsis life cycle. Plant J 53:488–504

Rivero RM, Ruiz JM, García PC, Lopez-Lefebre LR, Sánchez E, Romero L (2001) Resistance to cold and heat stress: accumulation of phenolic compounds in tomato and watermelon plants. Plant Sci 160:315–321

Rivero H, Huertas M, Francini R, Vila L, Darre LE (2006) Concentrations of As, Cd, Co, Cr, Cu, Fe, Hg, K, Mg, Mn, Mo, Na, Ni, Pb and Zn in Uruguayan rice determined by atomic absorption spectrometry. Atomic Spectrosc 27:48–55

Rivero R, Sønsteby A, Heide OM, Måge F, Remberg SF (2017) Flowering phenology and the interrelations between phenological stages in apple trees (Malus domestica Borgh.) as influenced by the Nordic climate. Acta Agric Scand B Soil Plant Sci 67:292–302

Rizza F, Crosatti C, Stanca AM, Cattivelli L (1994) Studies for assessing the influence of hardening on cold tolerance of barley genotypes. Euphytica 75:131–138

Roberts EH, Summerfield RJ (1987) Measurement and prediction of flowering in annual crops. In: Atherton JG (ed) Manipulation of flowering. Butterworth, London, pp 17–50

Roberts EH, Hadley P, Summerfield RJ (1985) Effects of temperature and photoperiod on flowering in chickpeas (Cicer arietinum L.). Ann Bot 55:881–892

Roberts EH, Summerfield RJ, Muehlbauer FJ, Short RW (1986) Flowering in lentil (Lens culinaris Medic.): the duration of the photoperiodic inductive phase as a function of accumulated day length above the critical photoperiod. Ann Bot 58:235–248

Roberts EH, Qi A, Ellis R, Summerfield RJ, Lawn RJ, Shanmugasundaram S (1996) Use of field observations to characterise genotypic flowering responses to photoperiod and temperature: a soyabean exemplar. Theor Appl Genet 93(4):519–533

Robinson SP, Jones GP (1986) Accumulation of glycine betaine in chloroplasts provides osmotic adjustment during salt stress. Aust J Plant Physiol 13:659–668

Rodríguez P, Dell AJ, Morales D, Sánchez-Blanco MJ, Alarcón JJ (1997) Effect of salinity on growth, shoot water balance, and root hydraulic conductivity in tomato plants. J Agron Crop Sci 128:439–444

Rodríguez VM, Soengas P, Alsono-Villaverde V, Sotelo T, Cartea ME, Velasco P (2015) Effect of temperature stress on the early vegetative development of Brassica oleracea L. BMC Plant Biol 15:145

Rogers DL, Montalvo AM (2004) Genetically appropriate choices for plant materials to maintain biological diversity. Report to the USDA Forest Service, Rocky Mountain Region, Lakewood, CO, USA. University of California, Berkeley
Rogers RA, Dunn JH, Nelson CJ (1977) Photosynthesis and cold hardening in zoysia and bermudagrass. Crop Sci 17(3):727–732
Roitsch T, Gonzalez MC (2004) Function and regulation of plant invertases: sweet sensations. Trends Plant Sci 9:606–613
Roqueiro G, Maldonado S, Mdel CR, Maroder H (2012) Fluctuation of oxidative stress indicators in Salix nigra seeds during priming. J Exp Bot 63:3631–3642
Ross JJ, Davidson SE, Wolbang CM, Bayly-Stark E, Smith JJ, Reid JB (2003) Developmental regulation of the gibberellin pathway in pea shoots. Funct Plant Biol 30:83–89
Ruelland E, Zachowski A (2010) How plants sense temperature. Environ Exp Bot 69:225–232
Ruelland E, Vaultier MN, Zachowski A, Hurry V (2009) Cold signalling and cold acclimation in plants. Adv Bot Res 49:35–150
Rylski I (1979) Fruit set and development of seeded and seedless tomato fruits under diverse regimes of temperature and pollination. J Am Soc Hortic Sci 104:835–839
Rymen B, Fiorani F, Kartal F, Vandepoele K, Inzé D, Beemster GTS (2007) Cold nights impair leaf growth and cell cycle progression in maize through transcriptional changes of cell cycle genes. Plant Physiol 143:1429–1438
Ryu H, Cho YG (2015) Plant hormones in salt stress tolerance. J Plant Biol 58:147–155
Ryyppö A, Repo T, Vapaavuori E (1998) Development of freezing tolerant in roots and shoots of Scot pine seedlings at non-freezing temperature. Can J For Res 28(4):557–565
Sacks MM, Silk WK, Burman P (1997) Effect of water stress on cortical cell division rates within the apical meristem of primary roots of maize. Plant Physiol 114:519–527
Sadras VO, Connor DJ (1991) Physiological basis of the response of harvest index to the fraction of water transpired after anthesis: a simple model to estimate harvest index for determinate species. Field Crop Res 26:227–239
Sadras VO, Monzon JP (2006) Modelled wheat phenology captures rising temperature trends: shortened time to flowering and maturity in Australia and Argentina. Field Crop Res 99:136–146
Saeidnejad AH, Pouramir F, Naghizedek M (2012) Improving chilling tolerance of maize seedlings under cold conditions by spermine application. Not Sci Biol 4:110–117
Saini HS (1997) Effects of water stress on male gametophyte development in plants. Sex Plant Reprod 10:67–73
Saini R, Singh AK, Dhanpal S, Saeed TH, Hyde GJ, Baskar R (2017) Brief temperature stress during reproductive stages alters meiotic recombination and somatic mutation rates in the progeny of Arabidopsis. BMC Plant Biol 17:103
Salerno GL, Curatti L (2003) Origin of sucrose metabolism in higher plants: when, how and why? Trends Plant Sci 8:63–69
Salmerón M, Gbur EE, Bourland FM, Buehring NW, Earnest L, Fritschi FB, Golden BR, Hathcoat D, Lofton J, Miller TD (2014) Soybean maturity group choices for early and late plantings in the Midsouth. Agron J 106:1893–1901
Salmerón M, Gbur EE, Bourland FM, Earnest L, Golden BR, Purcell LC (2015) Soybean maturity group choices for maximizing radiation interception across planting dates in the Midsouth United States. Agron J 107:2132–2142
Samach A (2012) Control of flowering, Chapter 25. In: Altman A (ed) Plant biotechnology and agriculture. Prospects for the 21st century. Academic, London, p 387
Samach A, Wigge PA (2005) Ambient temperature perception in plants. Curr Opin Plant Biol 8: 483–486
Sánchez A, Orta R, Muñoz BC (2001) Pretratamientos pregerminativos de hidratación-deshidratación de las semillas y sus efectos en plantas de interés agrícola. Agronomía Costarricense 25:67–91

Sánchez JA, Muñoz BC, Hernández L, Montejo L, Suárez AG, Torres-Arias Y (2006) Pretratamientos robustecedores de semillas para mejorar la emergencia y el crecimiento de Trichospermum mexicanum, árbol tropical pionero. Agronomía Costarricense 30:7–26

Sandhu JS, Gupta SK, Singh S, Dua RP (2007) Genetic variability for cold tolerance in pigeonpea. J SAT Agric Res 5:1–3

Sandring S, Ågren J (2009) Pollinator-mediated selection on floral display and flowering time in the perennial herb Arabidopsis lyrata. Evolution 63:1292–1300

Sandvik SM, Totland O (2000) Short-term effects of simulated environmental changes on phenology, reproduction, and growth in the late-flowering snowbed herb Saxifraga stellaris L. Ecoscience 7:201–213

Sanghera GS, Wani SH, Hussain W, Singh NB (2011) Engineering cold stress tolerance in crop plants. Curr Genomics 12:30–43

Sangwan V, Orvar BL, Beyerly J, Hirt H, Dhindsa RS (2002) Opposite changes in membrane fluidity mimic cold and heat stress activation of distinct plant MAP kinase pathways. Plant J 31: 629–638

Santos TP, Lopes CM, Rodrigues ML, Souza CR, Silva JR, Maroco JP, Pereira S, Chaves MM (2007) Effects of deficit irrigation strategies on cluster microclimate for improving fruit composition of Moscatel field-grown grapevines. Sci Hortic 112:321–330

Sataka T, Hayase H (1970) Male sterility caused by cooling treatment at the young microspore stage in rice plants. V. Estimation of pollen developmental stage and the most sensitive stage to coolness. Proc Crop Sci Soc Jpn 39:468–473

Satake T (1969) Research on cold injury of paddy rice plants in Japan. Jpn Agric Res Q 4:5–10

Satake T (1976a) Sterile-type cool injury in paddy rice plants. In: Climate and rice. International Rice Research Institute, Los Baños, Philippines, pp 281–300

Satake T (1976b) Determination of the most sensitive stage to sterile-type cool injury in rice plants. Res Bull Hokkaido Natl Agric Exp Stn 113:1–44

Sato S, Peet MM, Thomas JF (2002) Determining critical pre- and post-anthesis periods and physiological processes in Lycopersicon esculentum Mill. exposed to moderately elevated temperatures. J Exp Bot 53(371):1187–1195

Sauer N (2007) Molecular physiology of higher plant sucrose transporters. FEBS Lett 581:2309–2317

Savithri KS, Ganapathy PS, Sinha SK (1980) Sensitivity to low temperature in pollen germination and fruit-set in Cicer arietinum L. J Exp Bot 31:475–481

Saxena MC (1990) Problems and potentials in chickpea production in the nineties. In: Van Rheenen HA, Saxena MC (eds) Chickpea in the nineties. Proceedings of the 2nd international workshop on chickpea improvement. ICRISAT, Patancheru, India, pp 13–25

Saxena NP, Johansen C (1990) Realized yield potential in chickpea and physiological considerations for further genetic improvement. In: Sinha SK, Sane PV, Bhargava SC, Agarwal PK (eds) Proceedings of international congress on plant physiology. Indian Soc. Plant Physiol. & Biochem., New Delhi, India, pp 279–288

Sayyari M, Ghanbari F, Fatahi S, Bavandpour F (2013) Chilling tolerance improving of watermelon seedling by salicylic acid seed and foliar application. Not Sci Biol 1:67–73

Scambos T (2017) NASA-USGS Landsat 8 Satellite pinpoints coldest spots on Earth. http://www.nasa.gov/content/goddard/nasa-usgs-landsat-8-satellite-pinpoints-coldest-spots-on-earth

Scarcelli N, Cheverud JM, Schaal BA, Kover PX (2007) Antagonistic pleiotropic effects reduce the potential adaptive value of the FRIGIDA locus. Proc Natl Acad Sci U S A 104:16986–16991

Schaffer B, Urban L, Lu P, Whiley AW (2009) Ecophysiology. In: Litz RE (ed) The mango botany, production and uses. CABI, London

Schmidt LH (2000) Guide to handling of tropical and subtropical forest seed. University of Copenhagen, Danida Forest Seed Centre, Copenhagen, Denmark

Schuppler U, He P, John PCL, Munns R (1998) Effect of water stress on cell division and cell-division-cycle 2-like cell-cycle kinase activity in wheat leaves. Plant Physiol 117:667–678

Seo E, Lee H, Jeon J, Park H, Kim J, Noh YS, Lee I (2009) Crosstalk between cold response and flowering in Arabidopsis is mediated through the flowering-time gene SOC1 and its upstream negative regulator FLC. Plant Cell 21:3185–3197

Seo PJ, Kim MJ, Park JY, Kim SY, Jeon J, Lee YH, Kim J, Park CM (2010) Cold activation of a plasma membrane-tethered NAC transcription factor induces a pathogen resistance response in Arabidopsis. Plant J 61:661–671

Sepaskhah AR, Yarami N (2009) Interaction effects of irrigation regime and salinity on flower yield and growth of saffron. J Hortic Sci Biotechnol 84(2):216–222

Setiyono TD, Weiss A, Specht J, Bastidas AM, Cassman KG, Dobermann A (2007) Understanding and modeling the effect of temperature and daylength on soybean phenology under high-yield conditions. Field Crop Res 100:257–271

Sewelam N, Oshima Y, Mitsuda N, Ohme-Takagi M (2014) A step towards understanding plant responses to multiple environmental stresses: a genome-wide study. Plant Cell Environ 37: 2024–2035

Shah F, Huang J, Cui K, Nie L, Shah T, Chen C, Wang K (2011) Impact of high-temperature stress on rice plant and its traits related to tolerance. J Agric Sci 149:545–556

Shan H, Chen S, Jiang J, Chen F, Chen Y, Gu C, Li P, Song A, Zhu X, Gao H, Zhou G, Li T, Yang X (2012) Heterologous expression of the chrysanthemum R2R3-MYB transcription factor CmMYB2 enhances drought and salinity tolerance, increases hypersensitivity to ABA and delays flowering in Arabidopsis thaliana. Mol Biotechnol 51(2):160–173

Shao HB, Liang ZS, Shao MA (2006a) Osmotic regulation of 10 wheat (Triticum aestivum L.) genotypes at soil water deficits. Colloid Surf B Biointerfaces 47:132–139

Shao HB, Chen XY, Chu LY, Zhao XN, Wu GY, Yong B (2006b) Investigation on the relationship of proline with wheat anti drought under soil water deficits. Colloids Surf B Biointerfaces 53(1): 113–119

Shao HB, Chu LY, Shao MA, Zhao CX (2008) Advances in functional regulation mechanisms of plant aquaporins: their diversity, gene expression, localization, structure and roles in plant soil-water relations. Mol Membr Biol 25(3):1–12

Sharma P, Sharma N, Deswal R (2005) The molecular biology of the low-temperature response in plants. BioEssays 27:1048–1059

Sharma N, Cram D, Huebert T, Zhou N, Parkin IA (2007) Exploiting the wild crucifer Thlaspi arvense to identify conserved and novel genes expressed during a plant's response to cold stress. Plant Mol Biol 63:171–184

Sheldon CC, Finnegan JE, Peacock WJ, Dennis ES (2009) Repression by vernalization in Arabidopsis. Plant J 59:488–498

Sheoran IS, Saini HS (1996) Drought-induced male sterility in rice: changes in carbohydrate levels and enzyme activityies associated with the inhibition of starch accumulation in pollen. Sex Plant Reprod 9:161–169

Sherry RA, Zhou XH, Gu SR, Arnone JA, Schimel DS, Verbug PS, Wallace LL, Luo YQ (2007) Divergence of reproductive phenology under climate warming. Proc Natl Acad Sci U S A 104: 198–202

Shi XM, Jin F, Huang YT, Du XW, Li CM, Wang M, Shao H, Jin MJ, Wang J (2012) Simultaneous determination of five plant growth regulators in fruit by modified quick, easy, cheap, effective, rugged and safe (QuEChERS) extraction and liquid chromatographytandem mass spectrometry. J Agric Food Chem 60:60–65

Shillo R, Ding M, Pasternak D, Zaccai M (2002) Cultivation of cut flower and bulb species with saline water. Sci Hortic 92(1):41–54

Shim SI, Moon JC, Jang CS, Raymer P, Kim W (2008) Effect of potassium nitrate priming on seed germination of seashore paspalum. HortScience 43(7):2259–2262

Shimono Y, Kudo G (2003) Intraspecific variations in seedling emergence and survival of Potentilla matsumurae (Rosaceae) between alpine fellfield and snowbed habitats. Ann Bot 91:21–29

Shimono Y, Kudo G (2005) Comparisons of germination traits of alpine plants between fellfield and snowbed habitats. Ecol Res 20:189–197

Shimono H, Hasegawa T, Iwama K (2002) Response of growth and grain yield in paddy rice to cool water at different growth stages. Field Crop Res 73(2–3):67–79

Shinozaki K, Yamaguchi-Shinozaki K (2000) Molecular responses to dehydration and low temperature: differences and cross-talk between two stress signaling pathways. Curr Opin Plant Biol 3: 217–223

Shinozaki K, Yamaguchi-Shinozaki K (2007) Gene networks involved in drought stress response and tolerance. J Exp Bot 58(2):221–227

Shinozaki K, Yamaguchi-Shinozaki K, Seki M (2003) Regulatory network of gene expression in the drought and cold stress responses. Curr Opin Plant Biol 6:410–417

Shitaka Y, Hirose T (1998) Effects of shift in flowering time on the reproductive output of Xanthium canadense in a seasonal environment. Oecologia 114:361–367

Shukla N, Kuntal H, Shanker A, Sharma S (2018) Hydro-priming methods for initiation of metabolic process and synchronization of germination in mung bean (Vigna radiata L.) seeds. J Crop Sci Biotechnol 21:137–146

Sicher R (2011) Carbon partitioning and the impact of starch deficiency on the initial response of Arabidopsis to chilling temperatures. Plant Sci 181:167–176

Siddique KHM, Sedgley RH (1986) Chickpea (Cicer arietinum L.), a potential grain legume for southwestern Australia: seasonal growth and yield. Crop Pasture Sci 37:245–261

Siddique KHM, Loss SP, Regan KL, Jettner RL (1999) Adaptation and seed yield of cool season grain legumes in Mediterranean environments of south-western Australia. Aust J Agric Res 50: 375–387

Šikoparija B, Smith M, Skjøth CA, Radišić P, Milkovska S, Šimić S, Brandt J (2009) The Pannonian plain as a source of Ambrosia pollen in the Balkans. Int J Biometeor 53:263–272

Siminovitch D (1981) Common and disparate elements in the processes of adaptation of herbaceous and woody plants to freezing: a perspective. Cryobiology 18:166–185

Simlat M, Szewczyk A, Ptak A (2020) Melatonin significantly influences seed germination and seedling growth of Stevia rebaudiana Bertoni. PLoS One 6:e5009

Simpson GG, Dean C (2002) Flowering - Arabidopsis, the rosetta stone of flowering time? Science 296:285–289

Singh JS, Chaturvedi RK (2018) Tropical dry deciduous forest: research trends and emerging features. Springer Nature Singapore, Singapore

Singh KB, Malhotra RS, Saxena MC (1990) Sources for tolerance to cold in Cicer species. Crop Sci 30:1136–1138

Singh B, Reddy KR, Redoña ED, Walker T (2017a) Developing a screening tool for osmotic stress tolerance classification of rice cultivars based on in vitro seed germination. Crop Sci 57:387–394

Singh B, Reddy KR, Redoña ED, Walker T (2017b) Screening of rice cultivars for morpho-physiological responses to early-season soil moisture stress. Rice Sci 24:322–335

Singh B, Norvell E, Wijewardana C, Wallace T, Chastain D, Reddy KR (2018) Assessing morphological characteristics of elite cotton lines from different breeding programmes for low temperature and drought tolerance. J Agron Crop Sci 204:467–476

Sivritepe N, Sivritepe HO, Eris A (2003) The effects of NaCl priming on salt tolerance in melon seedlings grown under saline conditions. Sci Hortic 97:229–237

Skálová H, Moravcová L, Dixon AFG, Kindlmann P, Pyšek P (2015) Effect of temperature and nutrients on the growth and development of seedlings of an invasive plant. AoB Plants 7:plv044

Skrudlik G, Kościelniak J (1996) Effects of low temperature treatment at seedling stage on soybean growth, development and final yield. J Agron Crop Sci 176(2):111–117

Slawinska J, Obendorf RL (2001) Buckwheat seed set in plant and during in vitro inflorescence culture: evaluation of temperature and water deficient stress. Seed Sci Res 11:223–233

Smeets L, Wehner TC (1997) Environmental effects on genetic variation of chilling resistance in cucumber. Euphytica 97(2):217–225

Snider JL, Oosterhuis DM (2015) Physiology. In: Fang D, Percy R (eds) Cotton, Agron. Monograph No. 57, 2nd edn. ASA, CSSA, and SSSA, Madison, pp 339–400

Snider JL, Oosterhuis DM, Skulman BW, Kawakami EM (2009) Heat stress-induced limitations to reproductive success in Gossypium hirsutum. Physiol Plant 137(2):125–138

Solanke AU, Sharma AK (2008) Signal transduction during cold stress in plants. Physiol Mol Biol Plants 14:69–79

Somersalo S, Kyei-Boahen S, Pehu E (1996) Exogenous glycine betaine application as a possibility to increase low temperature tolerance of crop plants. Nordisk Jordbruksforskning 78:10

Sonoike K (1999) The different roles of chilling temperatures in the photoinhibition of photosystem I and photosystem II. J Photochem Photobiol B Biol 48(2–3):136–141

Sowinski P, Rudzinska-Langwald A, Adamczyk J, Kubica I, Fronk J (2005) Recovery of maize seedling growth, development and photosynthetic efficiency after initial growth at low temperature. J Plant Physiol 162:67–80

Springer CJ, Orozco R, Kelly JK, Ward JK (2008) Elevated CO_2 influence the expression of floral initiation genes in Arabidopsis thaliana. New Phytol 178(1):63–67

Sreenivasulu N, Altschmied L, Radchuk V, Gubatz S, Wobus U, Weschke W (2004) Transcript profiles and deduced changes of metabolic pathways in maternal and filial tissues of developing barley grains. Plant J 37:539–553

Srinivasan A, Saxena NP, Johansen C (1999) Cold tolerance during early reproductive growth of chickpea (*Cicer arietinum* L.): genetic variation in gamete development and function. Field Crop Res 60(3):209–222

Srivalli B, Sharma G, Khana-Chopra R (2003) Antioxidative defense system in an upland rice cultivar subjected to increasing intensity of water stress followed by recovery. Physiol Plant 119:503–512

Staggenborg S, Vanderlip RL (1996) Sorghum grain yield reduction caused by duration and timing of frwwzing temperature. Agron J 88(3):473–477

Stanton ML, Roy BA, Thiede DA (2000) Evolution in stressful environments. I. Phenotypic variability, phenotypic selection, and response to selection in five distinct environmental stresses. Evolution 54:93–111

Starck Z, Choluj D, Niemyska B (1994) Effect of preceding temperature and subsequent condition on response of tomato plants to chilling. Acta Physiol Plant 16(4):329–336

Starck Z, Bogdan J, Choluj D (1997) Effect of phosphorus and potassium supply on tomato plant response to chilling stress. In: Crop development for the cool and wet regions of europe, COST, vol 814, pp 49–56

Starck Z, Niemyska B, Bogdan J, Tawalbeh RNA (2000) Response of tomato plants to chilling stress in association with nutrient or phosphorus starvation. Plant Soil 226:99–106

Steponkus PL (1984) Role of the plasma membrane in freezing injury and cold acclimation. Annu Rev Plant Physiol 35:543–584

Steponkus PL, Uemura M, Webb MS (1993) A contrast of the cryostability of the plasma membrane of winter rye and spring oat-two species that widely differ in their freezing tolerance and plasma membrane lipid composition. In: Steponkus PL (ed) Advances in low-temperature biology, vol 2. JAI Press, London, pp 211–312

Stewart CR, Martin BA, Reding L, Cerwick S (1990) Seedling growth, mitochondrial characteristics, and alternative respiratory capacity of corn genotypes differing in cold tolerance. Plant Physiol 92:761–766

Stinchcombe JR, Weinig C, Ungerer MC, Olsen KM, Mays C, Halldorsdottir SS, Purugganan MD, Schmitt J (2004) A latitudinal cline in flowering time in Arabidopsis thaliana modulated by the flowering time gene FRIGIDA. Proc Natl Acad Sci U S A 101:4712–4717

Stitt M, Hurry V (2002) A plant for all seasons: alterations in photosynthetic carbon metabolism during cold acclimation in Arabidopsis. Curr Opin Plant Biol 5:199–206

Stockle CO, Donatelli M, Nelson R (2003) CropSyst, a cropping systems simulation model. Eur J Agron 18:289–307

Stone PJ, Nicolás ME (1994) Wheat cultivars vary widely in their responses of grain yield and quality to short periods of post-anthesis heat stress. Aust J Plant Physiol 21:887–900

Storkey J, Stratonovitch P, Chapman DS, Vidotto F, Semenov MA (2014) A process-based approach to predicting the effect of climate change on the distribution of an invasive allergenic plant in Europe. PLoS One 9:e88156

Strauss G, Hauser H (1986) Stabilization of lipid bilayer vesicles by sucrose during freezing. Proc Natl Acad Sci U S A 83(8):2422–2426

Strivastava LM (2002) Plant growth and developent: hormones and environment. Academic, San Diego, CA

Summerfield RJ, Asumadu H, Ellis RH, Qi A (1998) Characterization of the photoperiod response of post-flowering development in maturity isolines of soybean [Glycine max (L.) Merrill] 'Clatk'. Ann Bot 82:775–771

Sun W, Huang Y (2011) Global warming over the period 1961–2008 did not increase high-temperature stress but did reduce low-temperature stress in irrigated rice across China. Agric For Meteorol 151(9):1193–1201

Sung DY, Kaplan F, Lee KJ, Guy CL (2003) Acquired tolerance to temperature extremes. Trend Plant Sci 8:179–187

Suzuki S (1981) Cold tolerance in rice plants with special reference to the floral characters. I. Varietal differences in anther and stigma lengths and effects of planting densities on these characters. Jpn J Breed 31:57–64

Suzuki S (1982) Cold tolerance with rice plants with special reference to the floral characters. II. Relations between floral characters and the degree of cold tolerance in segregating generations. Jpn J Breed 32:9–16

Suzuki N, Mittler R (2006) Reactive oxygen species and temperature stresses: a delicate balance between signalling and destruction. Physiol Plant 126:45–51

Suzuki K, Nagasuga K, Okada M (2008) The chilling injury induced by high root temperature in the leaves of rice seedlings. Plant Cell Physiol 49:433–442

Suzuki N, Koussevitzky S, Mittler R, Miller G (2011) ROS and redox signaling in the response of plants to abiotic stress. Plant Cell Environ 35:259–270

Svensson J, Ismail A, Palva ET, Close TJ (2002) Dehydrins. In: Storey KB, Storey JB (eds) Cell and molecular responses to stress. Elsevier, Amsterdam, pp 155–171

Swanson CP (1942) Meiotic coiling in Tradescantia. Bot Gaz 103:457–474

Szabados L, Savouré A (2010) Proline: a multifunctional amino acid. Trends Plant Sci 15:89–97

Szalai G, Janda T, Paldi E, Szigeti Z (1996) Role of light in the development of post-chilling symptoms in maize. J Plant Physiol 148(3–4):378–383

Tacarindua CRP, Shiraiwa T, Homma K, Kumagai E, Sameshima R (2013) The effects of increased temperature on crop growth and yield of soybean grown in a temperature gradient chamber. Field Crop Res 154:74–81

Tadege M, Dupuis I, Kuhlemeier C (1999) Ethanolic fermentation, new functions for an old pathway. Trend Plant Sci 4:320–325

Tahir MHN, Imran M, Hussain MK (2002) Evaluation of sunflower (Helianthus annuus L.) inbred lines for drought tolerance. Int J Agric Biol 4(3):398–400

Taiz L, Zeiger E (2006) Stress physiology. In: Taiz L, Zeiger E (eds) Plant physiology. Sinauer Associates, Inc, Sunderland, pp 671–681

Taiz L, Zeiger E (2015) Plant physiology and development, 6th edn. Sinauer Associates, Inc, Sunderland

Takahashi D, Li B, Nakayama T, Kawamura Y, Uemura M (2013) Plant plasma membrane proteomics for improving cold tolerance. Front Plant Sci 4:90

Takeoka Y, Hiroi K, Kitano H, Wada T (1991) Pistil hyperplasia in rice spikelets as affected by heat-stress. Sex Plant Reprod 4:39–43

Takeoka Y, Mamun AA, Wada T, Kaufman PB (1992) Reproductive adaptation of rice to environmental stress. Jap. Sci. Soc. Press, Elsevier, Tokyo, Japan, p 237

Tapiero H, Townsend DM, Tew KD (2003) The antioxidant role of selenium and seleno-compounds. Biomed Pharmacother 57:134–144

Tarchoune I, Sgherric C, Izzo R, Lachaal M, Ouerghi Z, Navari-Izzo F (2010) Antioxidative response of Ocimum basilicum to sodium chloride and sodium sulphate salinization. Plant Physiol Biochem 48:772–777

Taschler D, Beikircher B, Neuner G (2004) Frost resistance and ice nucleation in leaves of five woody timberline species measured in situ during shoot expansion. Tree Physiol 24:331–337

Taylor AO, Rowley JA (1971) Plants under climatic stress. I. Low temperature, high light effects on photosynthesis. Plant Physiol 47(5):713–718

Teklić T, Lončarić Z, Kovačević V, Singh BR (2013) Metallic trace elements in cereal grain – a review: how much metal do we eat? Food Ener Secur 2:81–95

Terres AL (1991) Melhoramento de arroz irrigado para tolerância ao frio no Rio Grande do Sul Brasil. In: Reunión Sobre Mejoramiento De Arroz En El Cono Sur: Goiânia. IICA Procisur, Trabajos, Montevideo, pp 91–103

Thakur P, Kumar S, Malik JA, Berger JD, Nayyar H (2010) An overview: cold stress effects on reproductive development in grain crops. Environ Exp Bot 67(3):429–443

Theocharis A, Clément C, Barka EA (2012a) Physiological and molecular changes in plants grown at low temperatures. Planta 235:1091–1105

Theocharis A, Bordiec S, Fernandez O, Paquis S, Dhondt-Cordelier S, Baillieul F, Clement C, Ait BE (2012b) Burkholderia phytofirmans PsJN primes Vitis vinifera L. and confers a better tolerance to low non-freezing temperatures. Mol Plant-Microbe Interact 25:241–249

Thingnaes E, Torre S, Ernsten A, Moe R (2003) Day and night temperature responses in Arabidopsis: effects of gibberellin and auxin content, cell size, morphology and flowering time. Ann Bot 92(4):601–612

Thomashow MF (1990) Molecular genetics of cold acclimation in higher plants. Adv Genet 28:99–131

Thomashow MF (1999) Plant cold acclimation: freezing tolerance genes and regulatory mechanisms. Annu Rev Plant Physiol Plant Mol Biol 50:571–599

Thompson LM (1986) Climatic change, weather variability and corn production. Agron J 78:649–653

Thuzar M, Puteh AB, Abdullah NAP, Lassim MBM, Jusoff K (2010) The effects of temperature stress on the quality and yield of soya bean [Glycine max (L.) Merrill]. J Agric Sci 2:172–179

Tian Y (2014) The lethal factors and survival strategy in Haloxylon ammodendron germination and ecesis process. PhD Thesis, Xinjiang University

Tian Z, Wang F, Zhang W, Liu C, Zhao X (2012) Antioxidant mechanism and lipid peroxidation patterns in leaves and petals of marigold in response to drought stress. Hortic Environ Biotechnol 53(3):183–192

Tindall HD (1994) Rambutan cultivation. Food and Agriculture Organization of the United Nations FAO, Rome. 978-9-25103-325-8

Ting CS, Owens TG, Wolfe DW (1991) Seedling growth and chilling stress effect on photosynthesis in chilling-sensitive and chilling tolerant cultivars of Zea mays. J Plant Physiol 137(5): 559–564

Tiryaki I, Buyukcingil Y (2009) Seed priming combined with plant hormones: influence on germination and seedling emergence of sorghum at low temperature. Seed Sci Technol 37: 303–315

Tommasini L, Svensson JT, Rodriguez EM, Wahid A, Malatrasi M, Kato K, Wanamaker S, Resnik J, Close TJ (2008) Dehydrin gene expression provides an indicator of low temperature and drought stress: transcriptome-based analysis of barley (Hordeum vulgare L.). Funct Integr Genomics 8(4):387–405

Tonsor SJ, Scott C, Boumaza I, Liss TR, Brodsky JL, Vierling E (2008) Heat shock protein 101 effects in A. thaliana: genetic variation, fitness and pleiotropy in controlled temperature conditions. Mol Ecol 17:1614–1626

Toriyama K, Hinata K (1984) Anther respiratory activity and chilling resistance in rice. Plant Cell Physiol 25:1215–1217

Tranquillini W (1973) Der Wasserhaushalt junger Forstpflanzen nach dem Versetzen und seine Beeinflussbarkeit. Cbl ges Forstw 90:46–52

Trischuk RG, Schilling BS, Low NH, Gray GR, Gusta LV (2014) Cold acclimation, de-acclimation and re-acclimation of spring canola, winter canola and winter wheat: the role of carbohydrates, cold-induced stress proteins and vernalization. Environ Exp Bot 106:156–163

Tromp J (1976) Flower-bud formation and shoot growth in apple as affected by temperature. Sci Hortic 5:331–338

Tromp J (1984) Flower-bud formation in apple as affected by air and root temperature, air humidity, light intensity, and day length. Acta Hortic 149:39–47

Tsuda H, Niimura Y, Katoh T (2003) Chill injury in Saintpaulia leaf with special reference to leaf spot formation. J Agric Sci 47(4):283–289

Tsuji H, Tamaki S, Komiya R (2008) Florigen and the photoperiodic control of flowering in rice. Rice 1(1):25–35

Turck F, Fornara F, Coupland G (2008) Regulation and identity of florigen: FLOWERING LOCUS T moves center stage. Annu Rev Plant Biol 59:573–594

Turner H, Streule A (1983) Wurzelwachstum und Sprossentwicklung junger Koniferen im Stressklima der alpinen Waldgrenze, mit Berücksichtigung von Mikroklima, Photosynthese und Stoffproduktion. In: Böhm W, Kutschera L, Lichtenegger E (eds) Wruzelökolgie und ihre Nutzanwendung. Irdning Bundesanstalt für Alpenländische Landwirtschaft, Gumpenstein

Tuteja N, Tuteja R (2001) Unravelling DNA repair in human: molecular mechanisms and consequences of repair defect. Crit Rev Biochem Mol Biol 36:261–290

Tyron P, Chapin FS III (1983) Temperature control over root growth and root biomass in taiga forest trees. Can J For Res 13:827–833

Uemura M, Steponkus PL (1997) Effect of cold acclimation on membrane lipid composition and freeze induced membrane destabilization. In: Plant cold hardiness, molecular biology, biochemistry and physiology. Plenum, New York, pp 171–179

Uemura M, Steponkus PL (1999) Cold acclimation in plants: relationship between the lipid composition and the cryostability of the plasma membrane. J Plant Res 112:245–254

Usadel B, Bläsing OE, Gibon YPF, Höhne M, Günter M (2008) Multilevel genomic analysis of the response of transcripts, enzyme activities and metabolites in Arabidopsis rosettes to a progressive decrease of temperature in the non-freezing range. Plant Cell Environ 31:518–547

Ushio A, Mae T, Makino A (2008) Effects of temperature on photosynthesis and plant growth in the assimilation shoots of a rose. Soil Sci Plant Nutr 54(2):253–258

Van der Toorn A, Zemah H, van As H, Bendel P, Kamenetsky R (2000) Developmental changes and water status in tulip bulbs during storage: visualization by NMR imaging. J Exp Bot 51: 1277–1287

Van Kilsdonk MG, Nicolay MG, Franssen JM, Kolloffel C (2002) Bud abortion in tulip bulbs studied by magnetic resonance imaging. J Exp Bot 53:1603–1611

Van Oosterom EJ, Hammer G, Chapman SC (2004) Can transition to flowering be modelled dynamically from the gene level. In: 4th International crop science congress, Brisbane, Australia, p 7

Vapaavuori EM, Rikala R, Ryyppö A (1992) Effects of root temperature on growth and photosynthesis in conifer seedlings during shoot elongation. Tree Physiol 10:217–230

Vara Prasad PV, Craufurd PQ, Summerfield RJ, Wheeler TR (2000) Effect of short epidosed of heat stress on flower production and fruit set of groundnut (Arachis hypogea L.). J Exp Bot 51:777–784

Vassilevska-Ivanova R, Tcekova Z (2002) Effect of temperature on seed germination and seedling growth of sunflower (Helianthus annuus L.). Biol Physiol Des Plantes 55(10):67

Venema JH, Posthumus F, de Vries M, van Hasselt PR (1999a) Differential response of domestic and wild Lycopersicon species to chilling under low light: growth, carbohydrate content, photosynthesis and the xanthophyll cycle. Physiol Plant 105(1):81–88

Venema JH, Posthumus F, van Hasselt PR (1999b) Impact of suboptimal temperature on growth, photosynthesis, leaf pigments and carbohydrates of domestic and high-altitude wild Lycopersicon species. J Plant Physiol 155:711–718
Ventura Y, Mendlinger S (1999) Effects of suboptimal low temperature on yield, fruit appearance and quality in muskmelon (Cucumis melo L.) cultivars. J Hortic Sci Biotechnol 74(5):602–607
Verbruggen N, Hermans C (2008) Proline accumulation implants: a review. Amino Acids 35:753–759
Verheij FA (1996) Morphological and physiological aspects of the early phases of flower bud formation of apple. PhD Thesis, Wageningen Univ. Wageningen, The Netherlands, p 148
Verheij EWM, Coronel RE (1992) Edible fruits and nuts, vol 2. Plant Resources of South-East Asia Prosea Foundation, Bogor, pp 128–131
Verheul MJ, Hasselt PRV, Stamp P (1995) Comparison of maize inbred lines differing in low temperature tolerance effect of acclimation at suboptimal temperature on chloroplast functioning. Ann Bot 76(1):7–14
Verhoeven KJF, Poorter H, Nevo E, Biere A (2008) Habitat-specific natural selection at a flowering-time QTL is a main driver of local adaptation in two wild barley populations. Mol Ecol 17:3416–3424
Verma SK, Kumar B, Ram G, Singh HP, Lal RK (2010) Varietal effect on germination parameters at controlled and uncontrolled temperature in Palmarosa (Cymbopogon martinii). Ind Crop Prod 32(3):696–699
Vertucci CW, Leopold AC (1983) Dynamics of imbibition by soybean embryos. Plant Physiol 72: 190–193
Viktorova MK (1983) Plant growing in the tropics and subtropics. Mir Publisher, Moscow, pp 214–218
Visser ME, Both C (2005) Shifts in phenology due to global climate change: the need for a yardstick. Proc R Soc B 272:2561–2569
Vogg G, Heim R, Gotschy B, Beck E, Hansen J (1998) Frost hardening and photosynthetic performance of Scots pine (Pinus sylvestris L.). II. Seasonal changes in the fluidity of thylakoid membranes. Planta 204:201–206
Voorhees WB, Allmares RR, Johnson CE (1981) Alleviating temperature stress. In: Arkin GF, Taylor HM (eds) Modifying the root environment to reduce crop stress. Am. Soc. Agric. Eng., St. Joseph, MI, pp 217–266
Wagner I, Simons AM (2009) Divergent norms of reaction to temperature in germination characteristics among populations of the arctic alpine annual, Koenigia islandica. Arctic Antarctic Alpine Res 41:388–395
Wahid A, Gelani S, Ashraf M, Foolad MR (2007) Heat tolerance in plants: an overview. Environ Exp Bot 61:199–223
Wallace DH, Yan W (1998) Plant breeding and whole-system physiology. Improving adaptation, maturity and yield. CABI, Wallingford, UK
Wan XC, Landhausser SM, Zwiazek JJ, Lieffers VJ (1999) Root water flow and growth of aspen (Populus tremuloides) at low root temperatures. Tree Physiol 19:879–884
Wan X, Zwiazek JJ, Lieffers VJ, Landhäusser M (2001) Hydraulic conductance in aspen (Populus tremuloides) seedlings exposed to low root temperatures. Tree Physiol 21:691–696
Wang WX, Vinocur B, Altman A (2003) Plant responses to drought, salinity and extreme temperatures: towards genetic engineering for stress tolerance. Planta 218:1–14
Wang B, Vinocur O, Altman AS (2004) Role of plant heat-shock proteins and molecular chaperones in the abiotic stress response. Trend Plant Sci 9:244–252
Wang J, Gan YT, Clarke F, McDonald CL (2006) Response of chickpea yield to high temperature stress during reproductive development. Crop Sci 46:2171–2178
Wang Y, Zhang XM, Li L, Li HX (2008) Seed on response of germination of four species of Salsola L. to main ecology factors. Seed 27(8):58–63

Wang W, Chen Q, Hussain S, Mei J, Dong H, Peng S, Huang J, Cui K, Nie L (2016) Pre-sowing seed treatments in direct-seeded early rice: consequences for emergence, seedling growth and associated metabolic events under chilling stress. Sci Rep 6:19637

Wang Y, Cui Y, Hu G, Wang X, Chen H, Shi Q, Xiang J, Zhang Y, Zhu D, Zhang Y (2018) Reduced bioactive gibberellin content in rice seeds under low temperature leads to decreased sugar consumption and low seed germination rates. Plant Physiol Biochem 133:1–10

Wanner L, Junttila O (1999) Cold-induced freezing tolerance in Arabidopsis. Plant Physiol 120: 391–399

Wardlaw IF, Bagnall D (1981) Phloem transport and the regulation of growth of Sorghum bicolor (Moench) at low temperatures. Plant Physiol 68(2):411–414

Warrag MOA (1999) Flowering and fruiting of tomato (Lycopersicum esculentum Mill) in Qassin Saudi Arabia during summer. J King Saud Univ Agric Sci 3:241–250

Wei Y, Dong M, Huang ZY (2007) Seed polymorphism, dormancy, and germination of Salsola affinis (Chenopodiaceae), a dominant desert annual inhabiting the Junggar Basin of Xinjiang, China. Aust J Bot 55(4):1–7

WeiBing X, Rajashekar CB (2001) Glycine betaine involvement in freezing tolerance and water stress in Arabidopsis thaliana. Environ Exp Bot 46:21–28

Weinig C, Ungerer MC, Dorn LA, Kane NC, Toyonaga Y, Halldorsdottir SS, Mackay TFC, Purugganan MD, Schmitt J (2002) Novel loci control variation in reproductive timing in Arabidopsis thaliana in natural environments. Genetics 162:1875–1884

Welch S, Roe J, Dong Z (2003) A genetic neural network of flowering time control in Arabidopsis thaliana. Agron J 95(1):71–81

Welch S, Dong Z, Roe J (2004) Modelling gene networks controlling transition to flowering in Aabidopsis. In: 4th International crop science congress, Brisbane, Australia, p 20

Welti R, Li W, Li M, Sang Y, Biesiada H, Zhou HE, Rajashekar CB, Williams TD, Wang X (2002) Profiling membrane lipids in plant stress responses. Role of phospholipase Da in freezing induced lipid changes in Arabidopsis. J Biol Chem 277:31994–32002

Wen J, Li J, Guo Y (2002) Use carbonyl content of proteins in evaluating antioxidative health food. J Food Hygien 14:13–17

Wery J, Silim SN, Knights EJ, Malhotra RS, Cousin R (1994) Screening techniques and sources of tolerance to extremes of moisture and air temperature in cool season food legumes. Euphytica 73:73–83

Weschke W, Panitz R, Gubatz S, Wang Q, Radchuk R, Weber H, Wobus U (2003) The role of invertases and hexose transporters in controlling sugar ratios in maternal and filial tissues in barley caryopses during early development. Plant J 3:395–341

West SH (1970) Biochemical mechanisms of photosynthesis and growth depression in Digitaria decumbens when exposed to low temperatures. Proc IX Int Grassl Congr 11:514–517

Whaley JM, Kirby EJM, Spink JH, Foulkes MJ, Sparkes DL (2004) Frost damage to winter wheat in the UK: the effect of plant population density. Eur J Agron 21:105–115

Wheeler TR, Hong TD, Ellis RH, Batts GR, Morison JIL, Hadley P (1996) The duration and rate of grain growth, and harvest index, of wheat (Triticum aestivum L.) in response to temperature and CO_2. J Exp Bot 47:623–630

Wheeler TR, Craufurd P, Elis RH, Porter JR (2000) Temperature variability and yield of annual crops. Agric EcoSyst Environ 82(1–3):159–167

Whiley AW, Rasmussen TS, Saranah JB, Wolstenholme BN (1989) Effect of temperature on growth, dry matter production and starch accumulation in ten mango (Mangifera indica L.) cultivars. J Hortic Sci Biotechnol 64(6):753–765

White N, Jenkyn JF (1995) Effects of sowing date and vernalization on the growth of winter barley and its resistance to powdery mildew (Erysiphe graminis f. sp. hordei). Ann Appl Biol 126(2): 269–283

White RH, Schmidt RE (1989) Bermudagrass response to chilling temperatures as influenced by iron and benzyladenine. Crop Sci 29(3):768–773

White MA, Diffenbaugh NS, Jones GV, Pal JS, Giorgi F (2006) Extreme heat reduces and shifts United States premium wine production in the 21st Century. Proc Natl Acad Sci U S A 103: 11217–11222

Wiebe HJ (1994) Effects of temperature and daylength on bolting of leek (Allium porrum L.). Sci Hortic 59:177–185

Wien HC (1990) Screening pepper cultivars for resistance to flower abscission: a comparison of techniques. Hortic Sci 25:1634–1636

Wien HC, Minotti PL, Grubinger VP (1993) Polyethylene mulch stimulates early root growth and nutrient uptake of transplanted tomatoes. J Am Soc Hortic Sci 118:207–211

Wijewardana C, Hock M, Henry B, Reddy KR (2015) Screening corn hybrids for cold tolerance using morphological traits for early-season seeding. Crop Sci 55:851–867

Wilcox GE, Pfeiffer CL (1990) Temperature effect on seed germination, seedling root development and growth of several vegetables. J Plant Nutr 13:1393–1403

Wilczek AM, Roe JL, Knapp MC, Cooper MD, Lopez-Gallego C, Martin LJ, Muir CD, Sim S, Walker A, Anderson J, Egan JF, Moyers BT, Petipas R, Giakountis A, Charbit E, Coupland G, Welch SM, Schmitt J (2009) Effects of genetic perturbation on seasonal life history plasticity. Science 323:930–934

Williams DG, Black RA (1993) Phenotypic variation in contrasting temperature environments: growth and photosynthesis in Pennisetum setaceum from different altitudes on Hawaii. Funct Ecol 7:623–633

Wise RR (1995) Chilling-enhanced photooxidation: the production, action and study of reactive oxygen species produced during chilling in the light. Photosynth Res 45(2):79–97

Wolfe DW (1991) Low temperature effects on early vegetative growth, leaf gas exchange and water potential of chilling-sensitive and chilling-tolerant crop species. Ann Bot 67(3):205–212

Wolfe DW, Schwartz MD, Lakso AN, Otsuki Y, Pool RM, Shaulis NJ (2005) Climate change and shifts in spring phenology of three horticultural woody perennials in northeastern USA. Int J Biometeorol 49:303–309

Wu QS, Zou YN (2010) Beneficial roles of Arbuscular mycorrhizas in citrus seedlings at temperature stress. Sci Hortic 125:289–293

Wu SH, Jansson PE, Kolari P (2012) The role of air and soil temperature in the seasonality of photosynthesis and transpiration in a boreal Scots pine ecosystem. Agric For Meteor 156:85–103

Wullschleger SD, Yin TM, DiFazio SP, Tschaplinski TJ, Gunter LE, Davis MF, Tuskan GA (2005) Phenotypic variation in growth and biomass distribution for two advanced-generation pedigrees of hybrid poplar. Can J For Res 35(8):1779–1789

Xiaochuang C, Chu Z, Lianfeng Z, Junhua Z, Hussain S, Lianghuan W, Qianyu J (2017) Glycine increases cold tolerance in rice via the regulation of N uptake, physiological characteristics, and photosynthesis. Plant Physiol Biochem 112:251–260

Xin Z, Browse J (2000) Cold confort farm: the acclimation of plants to freezing temperatures. Plant Cell Environ 23(9):893–902

Xin Z, Li PH (1993) Relationship between proline and abscisic acid in the induction of chilling tolerance in maize suspension-cultured cells. Plant Physiol 103(2):607–613

Xu Q, Huang B (2000) Effects of differential air and soil temperature on carbohydrate metabolism in creeping bentgrass. Crop Sci 40:1368–1374

Xu Q, Huang B (2006) Seasonal changes in root metabolic activity and nitrogen uptake for two cultivars of creeping bentgrass. Hortic Sci 41(3):822–826

Xu LM, Zhou L, Zeng YW, Wang FM, Zhang HL, Shen SQ, Li ZC (2008) Identification and mapping of quantitative trait loci for cold tolerance at the booting stage in a japonica rice near isogenic line. Plant Sci 174(3):340–347

Yadav SK (2010) Cold stress tolerance mechanisms in plants. A review. Agron Sustain Dev 30: 515–527

Yamaguchi-Shinozaki K, Shinozaki K (2006) Transcriptional regulatory networks in cellular responses and tolerance to dehydration and cold stresses. Annu Rev Plant Biol 57:781–803

Yan WK, Wallace DH (1998) Simulation and prediction of plant phenology for five crops based on photoperiod x temperature interaction. Ann Bot 81:705–716

Yan Q, Duan Z, Jingdong M, Xun L, Fei D (2012) Effects of root-zone temperature and N, P, and K supplies on nutrient uptake of cucumber (Cucumis sativus L.) seedlings in hydroponics. Soil Sci Plant Nutr 58(6):707–717

Yang HG, Lin WF (2008) Effect of low temperature stress on some physiological characteristics of oil palm (Elaeis guineensis Jacq.). Seedlings 29:326–332

Yang X, Lu C (2005) Photosynthesis is improved by exogenous glycine betaine in salt-stressed maize plants. Physiol Plant 124:343–352

Yang WJ, Rich PJ, Axtell JD, Wood KV, Bonham CC, Ejeta G, Mickelbart MV, Rhodes D (2003) Genotypic variation for glycine betaine in sorghum. Crop Sci 43:162–169

Yang A, Dai X, Zhang WH (2012) A R2R3-type MYB gene, OsMYB2, is involved in salt, cold, and dehydration tolerance in rice. J Exp Bot 63:2541–2556

Yang N, Zhao HP, Ge FW, Li YH, Zeng WJ, Zhao HX (2015) Physiological response of two lepidium species to low temperature stress during seed germination. Arid Zone Res 32(4): 760–765

Yao Y, Kovalchuk I (2011) Abiotic stress leads to somatic and heritable changes in homologous recombination frequency, point mutation frequency and microsatellite stability in Arabidopsis plants. Mutat Res 707:61–66

Ye H, Du H, Tang N, Li X, Xiong L (2009) Identification and expression profiling analysis of TIFY family genes involved in stress and phytohormone responses in rice. Plant Mol Biol 71:291–305

Yi F, Wang Z, Baskin CC, Baskin JM, Ye R, Sun H, Zhang Y, Ye X, Liu G, Yang X, Huang Z (2019) Seed germination responses to seasonal temperature and drought stress are species-specific but not related to seed size in a desert steppe: implications for effect of climate change on community structure. Ecol Evol 9(4):2149–2159

Yin X, Kropff MJ, Ellis RH (1996) Rice flowering in response to diurnal temperature amplitude. Field Crop Res 48:1–9

Yin X, Kropff MJ, Nakagawa H, Horie T, Goudriaan J (1997a) A model for photothermal responses of flowering in rice. II. Model evaluation. Field Crop Res 51:201–211

Yin X, Kropff M, Ynalvez MA (1997b) Photoperiodically sensitive and insensitive phases of preflowering development in rice. Crop Sci 37:182–190

Yoshida S (1981) Fundamentals of rice crop science. International Rice Research Institute, Los Baños, pp 1–63

Yoshida R, Kanno A, Kameya T (1996a) Cool temperature-induced chlorosis in rice plants. II. Effects of cool temperature on the expression of plastid-encoded genes during shoot growth in darkness. Plant Physiol 112:585–590

Yoshida R, Kanno A, Sato T, Kaтeya T (1996b) Cool temperature-induced chlorosis in rice plants. Relationship between the induction and a disturbance of etioplast development. Plant Physiol 110(3):997–1005

Young MW, Kay SA (2001) Time Zones: a comparative genetics of circadian clocks. Nat Rev Genet 2:702–715

Young LW, Wilen RW, Bonham-Smith PC (2004) High temperature stress of Brassica napus during flowering reduces micro- and megagametophyte fertility, induces fruit abortion, and disrupts seed production. J Exp Bot 55:485–495

Yu E, Owttrim GW (2000) Characterization of the cold stress-induced cyanobacterial DEAD-box protein CrhC as an RNA helicase. Nucleic Acids Res 28(20):3926–3934

Yuan JS, Galbraith DW, Dai SY, Griffin P, Stewart CN Jr (2008) Plant systems biology comes of age. Trends Plant Sci 13:165–171

Zabihi-e-Mahmoodabad R, Jamaati-e-Somarin S, Khayatnezhad M, Gholamin R (2011) Effect of cold stress on germination and growth of wheat cultivars. Adv Environ Bot 5(1):94–97

Zemah H, Bendel P, Rabinowitch HD, Kamenetsky R (1999) Visualization of morphological structure and water status during storage of Allium aflatunense bulbs by NMR imaging. Plant Sci 147:65–73

Zemetra RS, Cuany RL (1991) Variation among inbreds for seed response to low temperatures in maize (Zea mays L.). Maydica 36(1):17–23
Zeng Y, Zhang Y, Xiang J, Pan X, Zhu D (2017) Effects of low temperature stress on spikelet-related parameters during anthesis in Indica–Japonica hybrid rice. Front Plant Sci 8:1350
Zhai F-F, Han L, Liu J-X, Qian Y-Q, Ju G-S, Li W, Zhang S-W, Sun Z-Y (2013) Assessing cold resistance of mutagenic strains of perennial ryegrass under artificial low-temperature stress. Acta Prata Sin 22:268–279
Zhang M, Duan L, Zhai Z, Li J, Tian X, Wang B, He Z, Li Z (2004) Effects of plant growth regulators on water deficit-induced yield loss in soybean. In: Proceedings of the 4th international crop science congress, Brisbane, QLD
Zhang XZ, Ervin EH, LaBranche AJ (2006) Metabolic defense responses of seeded bermuda grass during acclimation to freezing stress. Crop Sci 46:2598–2605
Zhang XZ, Ervin EH, Evanylo GK, Haering KC (2009) Impact of biosolids on hormone metabolism in drought-stresses tall fescue. Crop Sci 49:364–370
Zhang S, Jiang H, Peng S, Korpelainen H, Li C (2011a) Sex-related differences in morphological, physiological, and ultrastructural responses of Populus cathayana to chilling. J Exp Bot 62:675–686
Zhang Y, Liu H, Shen S, Zhang X (2011b) Improvement of eggplant seed germination and seedling emergence at low temperature by seed priming with incorporation SA into KNO_3 solution. Front Agric China 5:534–537
Zhang N, Zhao B, Zhang HJ, Weeda S, Yang C, Yang ZC, Ren S, Guo YD (2013) Melatonin promotes water-stress tolerance, lateral root formation, and seed germination in cucumber (Cucumis sativus L.). J Pineal Res 54(1):15–23
Zhang N, Sun Q, Zhang H, Cao Y, Weeda S, Ren S, Guo YD (2015a) Roles of melatonin in abiotic stress resistance in plants. J Exp Bot 66(3):647–656
Zhang Y, Zhang M, Yang H (2015b) Postharvest chitosan-g-salicylic acid application alleviates chilling injury and preserves cucumber fruit quality during cold storage. Food Chem 174:558–563
Zhao J, Li S, Jiang T, Liu Z, Zhang W (2012) Chilling stress - the key predisposing factor for causing Alternaria alternata infection and leading to cotton (Gossypium hirsutum L.) leaf senescence. PLoS One 7:e36126
Zhu JK (2002) Salt and drought stress signal transduction in plants. Annu Rev Plant Biol 53:247–273
Zhu JH, Dong CH, Zhu JK (2007) Interplay between cold-responsive gene regulation metabolism and RNA processing during plant cold acclimation. Curr Opin Plant Biol 10:290–295
Zhu JJ, Li YR, Liao JX (2013) Involvement of anthocyanins in the resistance to chilling-induced oxidative stress in Saccharum officinarum L. leaves. Plant Physiol Biochem 73:427–443
Zia MS, Salim M, Aslam M, Gill MA, Rahmatullah (1994) Effect of low temperature of irrigation water on rice growth and nutrient uptake. J Agric Sci 173(1):22–31
Zimmerman RH (1972) Juvenility and flowering in woody plants: a review. HortScience 10:447–455
Zinn KE, Tunc-Ozdemir M, Harper JF (2010) Temperature stress and plant sexual reproduction: uncovering the weakest links. J Exp Bot 61(7):1959–1968
Źróbek-Sokolnik A (2012) Temperature stress and responses of plants. In: Ahmad P, Prasad MNV (eds) Environmental adaptations and stress tolerance of plants in the era of climate change. Springer, New York, pp 113–134

Low Temperature Stress and Plant-Water Relationship: A Review

2

Abstract

A common effect of several abiotic stresses is to cause tissue dehydration. Such dehydration is caused by the imbalance between root water uptake and leaf transpiration. Under some specific stress conditions, regulation of root water uptake is more crucial to overcome stress injury than regulation of leaf transpiration. The rate of root water uptake of a given plant is the result of its root hydraulic characteristics, which are ultimately regulated by aquaporin activity and, to some extent, by suberin deposition. Plant-water relations concern how plants control the hydration of their cells, including the collection of water from the soil, its transport within the plant, and its loss by evaporation from the leaves. Flow of water through the plant and soil over macroscopic distances is driven by gradients in hydrostatic pressure. Over microscopic distances (e.g., across semi-permeable membranes), it is driven by gradients in water potential. Evaporation of water from leaves is primarily controlled by stomata and, if not made good by the flow of water from the soil through the plant to the leaves, results in the plants' wilting. Resistances to this flow are still not well understood. Plant water deficit is initiated as the crop demand for water exceeds the supply. The capacity of plants to meet the demand and thus avoid water deficit depends on their "hydraulic machinery." This machinery determines the ability to transport sufficient amount of water from the soil to the atmosphere via the stomata in order to provide for transpiration, transpirational cooling, and carbon assimilation. Water is transported by way of the soil-plant-atmosphere continuum, and it is largely controlled by the resistances in the continuum as determined by root, stem, leaf, stomata, and cuticular hydraulic resistances. The gross effects of deficient and of excessive soil moisture on plant growth are well known, but controversy has existed for many years around the question whether the so-called available moisture is equally available for plant growth or available only with such increasing difficulty that plant growth functions are retarded before the wilting

A. Bhattacharya, *Physiological Processes in Plants Under Low Temperature Stress*, https://doi.org/10.1007/978-981-16-9037-2_2

point is reached. Various measurable aspects of plant growth do not respond in the same manner to increasing moisture stress. Crops will respond to irrigations, although measured soil moisture stress is quite low.

Keywords

Low-temperature stress and plant-water relation · Soil moisture storage and availability · Plant water transmission characteristics · Concept of water potential · Water uptake by root · Leaf water transpiration · Water-use efficiency · Induction of low-temperature tolerance · Mechanism of water status regulation · Acclimation to cold stress

2.1 Plant and Water

Every plant has requirement for physical factors, viz., light, temperature, air, water, etc., and for certain chemical factors, e.g., inorganic nutrients and organic nutrients in optimum amount and in available form at the right time as its nutrition for its proper growth and development. Any deviation from this optimum condition of any factor essential for its growth will lead to aberrant change in physiological processes, and due to this, the plant body will experience tension, and this state can be referred to as plant under stress. Temperature on earth changes along with season, and most plants are specially adopted to grow over a limited range of temperatures (Smith et al. 1999). Both very high or low temperature cause severe changes in the plant metabolic machinery. Both these conditions are stressful for the growth and disturb the plants' physiology and biochemical process (Dwivedi and Raghvendra 2004). All plant species perform their metabolic processes and complete the life cycle under an optimum temperature range. Among the various climatic factors, temperature is vital for the growth of autotrophic plants, besides light, water, and atmospheric CO_2. Temperature in a true sense dictates the rate at which metabolic activity proceeds in plants, and it must be present at a certain level of warmth in plants for productive metabolism (Ziegler 1998). Any change in temperature affects plants greatly so stress imposed by temperature has an important implication for agriculture, for example, it has been conjectured that a 1 °C decrease in the world mean temperature would bring a great amount of changes in metabolic processes and metabolic products (Kumar et al. 2018). Different scientists have defined stresses differently depending on their mode of studies. Lichtenthaler (2006) defined stress as any substance or condition which stops a plant's metabolism, development, or growth. Stress was also defined by Mahmood (2002) as any element that decreases plant reproduction and development below the genotype's potential. However, plant response to stress is very complex, including morphological, physiological, biochemical, and molecular changes. Some plants have morphological adaptations in order to avoid undesirable conditions, and others can alter their physiology metabolism, gene expression, and various developmental activities to tolerate the effect of specific stresses (Rihan et al. 2017).

When the environmental temperature is considerably lower than the optimal growth temperatures, then it is said to be low-temperature stress. In majorly of plants, this temperature is usually below 10 °C the low temperature. It has been decided where there is formation of ice within the tissues, causing water deficit inside the plant, would be freezing temperature (Pearce 1999). Low temperature has a negative effect on the metabolism with a corresponding reduction in crop quality and quantity. Chilling and freezing resistances are complex phenomena, as most plants are capable of hardening to cold, i.e., acquisition of increased resistance upon exposure to low temperature. The effect of freezing temperature is irreversible as they cause loss of cell proteins and disturbed plasma membrane integrity. When exposed to temperature lower than 12–13 °C, tropical plants are more susceptible to chilling than those growing in cold regions. When exposed to temperature higher than 12–13 °C. Low temperatures can damage plants, both by a chilling effect, leading to physiological and developmental abnormalities, and by freezing, causing cellular damage directly or via cellular dehydration. Lyons (1973) described many symptoms of low-temperature injury. Some physiological processes such as flowering in rice are extremely sensitive to low temperatures, and damage may occur at temperatures as high as 20 °C. Commonly visible symptoms of low-temperature injury to the leaves include wilting, bleaching due to photooxidation of pigments, waterlogging of the intercellular spaces, browning, and eventually leaf necrosis and plant death (Levitt 1980; Witt and Barfield 1982). Dudal (1976) estimated that 15% of arable land is affected by freezing temperature stress. Low temperatures can reduce crop yields in several ways. Chilling and freezing injury can directly affect crop growth by causing physical damage or by interfering with normal biochemical and physiological functions, thus reducing yield. More subtly, low temperatures reduce potential agricultural productivity by limiting the crop or varieties that can be grown in a particular area, with cold-tolerant species/cultivars often not those with the highest potential yield. Low-temperature exposure can be both a daily and seasonal factor to which plants must adapt, including, in some regions, prolonged periods of low temperature lasting many months.

Water, being considered as a universal solvent, occupies 75% of our planet in the form of oceans. Water is also found in the atmosphere in the form of hydrospheric mantle. The evaporation of water from the surface of the ocean, formation of clouds, and raining are a natural cycle that evolved during the course of evolution of this planet. In the course of chemical evolution, the birth of life has chosen H_2O as the medium of biochemical activities. Thus, water has become the mother of life or "solvent of life." Cells of all organisms are made up of 90% or more water. All other components are either dissolved or suspended in water to form protoplasm, which is often referred to as the physical basis of life. It acts as a medium for all biochemical reaction that takes place in the cell and also acts as a medium of transportation from one region to another region. Water is a remarkable compound made up of hydrogen and oxygen (2:1), and it has high specific heat, high heat of vaporization, and high heat of fusion and expansion (colligative properties). Water because of its bipolar nature acts as a universal solvent for it dissolves more substances than any other solvent. Electrolytes and non-electrolytes like sugars and proteins dissolve very well.

Even some hydrophobic lipid molecules show some solubility in water. Water acts as a good buffer against changes in the hydrogen ion concentration (pH). This is because of its ionization property. Certain xerophytes use water as a buffer system against high temperature. Water also exhibits viscosity and adhesive properties. Because of hydrogen bonds, water molecules are attracted toward each other, and they are held to each other with considerable force. This force of attraction is called cohesive force. Thus, water possesses a high tensile strength. If this water is confined in very narrow columns of dimensions of xylem vessels, its tensile and cohesive forces reach very high values (1000–1200 g). And this force is very helpful in the ascent of sap. Water is of great importance in osmoregulation, particularly in the maintenance of turgidity of cells, opening and closing of stomata, and growth of the plant body. Water is an important substrate in photosynthesis, for it provides reducing power in CO_2 fixation; water is also used in breaking or making chemical bonds of polypeptides, polynucleotides, carbohydrates, etc.

Well-watered plants are turgid. Their cells, which are enclosed in a strong but slightly elastic wall, are distended by an internal pressure that may be as high as 1 MPa, five times the pressure in a car tire and ten times the pressure of the atmosphere. Plants perform best when they are turgid. Many of the structures of higher plants serve to maintain their cells sufficiently hydrated to function—to grow, to photosynthesize, and to respire—even though most of these cells are in the shoots of the plants and so not only are remote from the supply of water in the soil but are also exposed to a dry environment. A well-hydrated leaf may transpire several times its own volume of water during a day. Water evaporates from wet cell walls into the internal gas spaces of the leaf. It then flows away as vapor, largely through the stomata, which are variable pores in the surface of the leaf. The loss is an unavoidable consequence of the stomata being open, as they must be to allow carbon dioxide to enter the leaf. The relative humidity inside a leaf is typically greater than 99%, and thus there is usually a large difference of absolute humidity across the stomata that induce the rapid diffusion of water vapor out of the leaf. Although a leaf may lose much water by evaporation, its net loss of water is usually small. Evaporation from cell walls creates in them a large suction that replenishes water by drawing it from the soil, principally not only via the plant's vascular system but also through the flow across cells in the roots and leaves.

Crops experience periods of extreme low temperatures in many regions of the world (Ruelland et al. 2009; Wang et al. 2016a) and are exposed to limited water availability owing to either drought or disturbed water movement and uptake under low ambient temperatures (Shinozaki et al. 2003; Zhang et al. 2004; Beck et al. 2007). Such an exposure of plants to chilling and drought simultaneously hampers plant growth and hence is detrimental to productivity (Tommasini et al. 2008). Studies show that primarily chilling and drought pose a similar impact on stomatal development and leaf growth; nevertheless, the mechanisms of drought-induced changes in some physiological processes are quite different than those induced by chilling (Deng et al. 2012). Plants may exhibit common molecular and physiological responses on exposure to chilling and drought (shared response); others may be specific to a given stress factor (Atkinson et al. 2013; Sewelam et al. 2014). In general, chilling stress thermodynamically declines the kinetics of several physiological as well as metabolic processes occurring in plants (Ruelland et al. 2009; Hussain et al. 2016). It severely reduces the rate and uniformity of germination, hampers seedling vigor, and delays ontogenic plant development (Cruz and Milach 2004; Oliver et al. 2007), resulting in severe crop yield losses (Ruelland et al. 2009). Hussain et al. (2018) reviewed the recent observations on plant responses to chilling and drought stresses on shared as well as unique basis and tried to find a common thread potentially underlying these responses.

2.2 Soil and Water

Soil has been defined as a three-dimensional, dynamic, natural body occurring on the surface of the earth that is a medium for plant growth and whose characteristics have resulted from the forces of climate and living organisms acting upon parents material, as modified by relief, over a period of time (Tamhane et al. 1970). To serve as a favorable medium for plant growth, soil must store and supply water and nutrients in addition to the anchorage it provides. Soil is a porous medium; therefore, due to the geometry of the pore spaces between the soil particles and the nature of the surfaces, soil has the capacity to hold water. This property of soil enables it to retain precipitation or irrigation water in the root zone to be used by plants over time. The amount of water held depends upon the porosity and pore size distribution and the capillary pressure of water in the soil. The force by which the water is held by soil (soil water suction/tension) is the force that the plant roots have to overcome for extracting water retained by soil. It is expressed in the units of pressure. Earlier, it was generally expressed in bars (or atmospheres) but is now expressed in pascals (Pa). The relationship between the various units is given below:

$$1 \text{ bar} = 1020 \text{ cm of water column*}$$
$$1 \text{ atm} = 1030 \text{ cm of water column}$$
$$1 \text{ bar} = 10^5 \text{ Pa} = 100 \text{ kPa} = 0.1 \text{ MPa}$$

*Pressure exerted by a column of water 1020 cm high

Conventionally, the soil water potential is expressed in terms of hydraulic head, i.e., soil water potential expressed in terms of height of water column. For example, if the soil water tension is 1 bar (soil water potential of −1 bar), in terms of water column height, it will be 1020 cm. For the sake of convenience, 1 bar is considered equivalent to 1 atm and approximately 1000 cm of water column height.

2.2.1 Soil Moisture Constants

There are many terms that are associated with the water that is contained or retained by the soil. The common ones are saturation, field capacity, permanent wilting point,

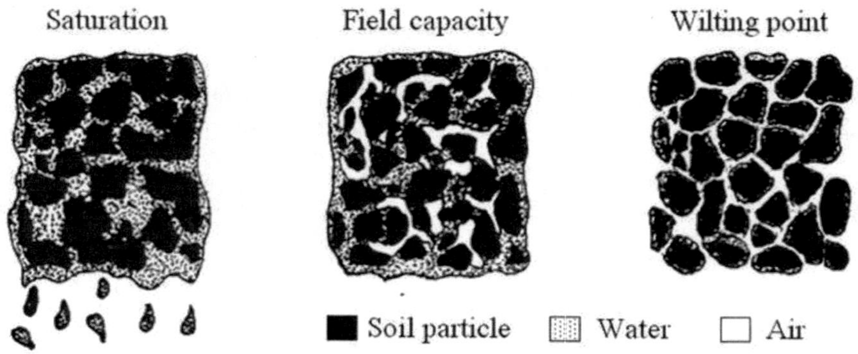

Fig. 2.1 Volumes of water and air associated with soil solids at different moisture levels

and hygroscopic coefficient. The water held by the soil at these points is shown schematically in Fig. 2.1.

2.2.1.1 Saturation

Soil is a porous medium, and when all the pores of the soil are filled with water, it is referred to as saturated soil. Saturation percentage or the water held by soil can be determined by bringing the soil core/soil sample in contact with water column at a height corresponding to the midpoint of the sample and determining the moisture held by soil or by using the formula for porosity:

$$P = \{1 - \mathrm{BD}/\mathrm{PD}\} \times 100 \quad (1)$$

where P is the porosity in percent, BD is the bulk density (g cm^{-3}), and PD is the particle density (g cm^{-3}) of soil. PD can be generally assumed to be 2.65 (g cm^{-3}) for mineral soils. It can also be estimated by determining the moisture content of the core of which saturated hydraulic conductivity has been determined without allowing any water to drain out of the sample.

The water potential in soil affects water reservoir and its availability for plants; hence, it has a large impact on plant growth and production. Furthermore, the soil water content exerts a great influence on some physical and chemical properties of soil, such as the oxygen content, which interferes with root breathing, microbial activity, and soil chemical status. Soil water potential is directly dependent on soil physical characteristics and varies with time and space, depending on soil water balance. That balance is determined by input (rain, irrigation) and output of the soil (drainage, evaporation, and root absorption). It is important to emphasize that behavior of water into soil differs from that in a pot. That is, soil water interacts with the matrix and solutes, and it is under pressure or tension, resulting in various energy states, relative to free water (Kirkham 2005). With regard to the physiological aspect, it is important to point out that the water content in soil is associated with field capacity, the permanent wilting point, and the available water content.

The term “field capacity” corresponds to the maximum water content that a given soil can retain by capillarity, after saturation and gravity drainage (Fig. 2.1), and it is conventionally estimated as the water content when the matrix potential is −0.03 MPa (−0.3 bar). There is no real and unique value for accurately characterizing the field capacity of a given soil. Furthermore, the continuous drainage can induce an overestimation of the water consumption of the plant. Despite these uncertainties, the term field capacity is still useful for a qualitative understanding—rather than a quantitative understanding—of the water behavior of a particular soil, providing an estimate of the maximum limit of water accumulation. In general, clay soils or those with higher content of organic matter (upper to 5% of organic matter) present a higher soil water holding capacity (average field capacity ranging from 35 to 40%vol). In contrast, sandy soils have a lower water holding capacity, and field capacity typically ranges from 10 to 15%vol.

The wilting point is also known as the permanent wilting point and can be defined as the amount of water per unit weight (or volume) of soil that is so tightly retained by the soil matrix that roots are unable to absorb causing the wilting of plant. In other words, it corresponds to the water potential of soil under which plants cannot maintain turgor pressure, even if a series of defense mechanisms have been triggered. Similar with field capacity, soil water content at wilting point is not a unique and precise value although it conventionally measures at −1.5 MPa (−15 bar). The wilting point is influenced by the physical and chemical characteristics of soil, but also by the plant species considered. Conventionally, the wilting point is estimated as the water content when the matrix potential of the soil is −1.5 MPa (−15 bar). Some plant species can absorb soil water at a potential much smaller than this limit. For example, olive trees can set a water potential gradient between dry soil (−3 MPa) and leaf (−7 MPa) (Dichio et al. 2006). Similarly, Larrea divaricata may absorb water at −6.0 MPa soil water potential (Kirkham 2005). One desert plant (Larrea tridentata) can survive with soil water potentials up to −11.5 MPa, maintaining the photosynthetic activity of leaves within the range between −5 and −8 MPa (Fitter and Hay 2002). At the permanent wilting point, the water potential of soil tends to be less than or equal to the osmotic potential of the plant, which is extremely low in plants adapted to dry environments.

The water content in the soil is essential for calculating the available water content for the plants. The available water content is calculated considering the soil volume explored by roots and the percentage of water content determined as the difference between field capacity and wilting point. With the reduction of soil water potential, there is also a reduction in its hydraulic conductivity (i.e., water moves slowly in the soil), limiting the water absorption capacity of the roots. In this scene—and for a majority of crops—the yields are reduced if the water content in the soil approaches the wilting point. Thus, the available water content should be considered as a relative value, and, for the same soil water potential, it may have different proportions of accessibility, depending on the ability of each species to exploit or capture available water (Chavarria and dos Santos 2012).

2.2.1.2 Hygroscopic Coefficient

It is the water held by the soil at 31 bars (3.1 MPa) soil moisture tension and can be determined using a pressure membrane assembly meant for higher pressures. It can also be approximated by allowing an oven dry soil to absorb moisture from the air till constant weight is attained. The water between saturation and field capacity is called gravitational water. It flows out under the influence of gravity and is considered unavailable to plants. The water between field capacity and hygroscopic coefficient is referred to as capillary water as it is bound to the soil by capillary forces. The water held by the soil between field capacity and permanent wilting point is termed as available water for plants, while the water held between permanent wilting point and hygroscopic coefficient is considered unavailable to plants. Some plants are able to survive even at moisture contents below permanent wilting point for short periods. Field capacity can also be determined in the laboratory with this equipment using an applied pressure of 1/3 bar for clayey soils and 0.1–0.2 bar for sandy to sandy loam soils. Field capacity is considered as the upper limit of available water to plants and the permanent wilting point as the lower limit. The relationship between the amounts of water held in a soil and the force by which it is held (capillary pressure or suction or tension) is depicted in the form of a curve commonly referred to as the soil water characteristic or release curve or simply the pF curve.

2.2.2 Soil Temperature

Soil temperature is the function of heat flux in the soil as well as heat exchanges between the soil and atmosphere (Elias et al. 2004). It is also defined as the function of the internal energy of the soil (Geiger et al. 2003). The transfer of heat in the soil (Zhao et al. 2007a) and the latent heat exchanges at the surface (Nwankwo and Ogugurue 2012) are the primary causes of variations in soil temperature. The main source of soil temperature is solar radiation. Soil temperature is measured with a thermometer. Soil temperature varies seasonally and daily which may result from changes in radiant energy and energy changes taking place through soil surface (Chiemeka 2010). Onwuka (2018) divided these factors into two: the amount of heat made available to the soil surface and the amount of heat dissipated from soil surface down the profile. The factors that affect the amount of heat supplied at the soil surface include soil color (Chiemeka 2010), mulching (Martias and Musil 2012), solar radiation (Geiger et al. 2003), slope of land surface (Elizaberashivili et al. 2010), vegetative cover (Decker et al. 2003), organic matter content (Abu-Hamdeh and Reeder 2000), and evaporation (Lu et al. 2009).

Soil temperature has a range of 10°–28°, and it influences soil respiration by increasing the activity of extracellular enzymes that degrade polymeric organic matter in soils (Conant et al. 2008), increase microbial retake of soluble substrates (Allison et al. 2010), and increase microbial respiration rates (Wallenstein et al. 2010). Increase in soil temperature increases the soil nitrogen mineralization rates through the increase in microbial activity and increase in the decomposition of organic matter in the soil (Yan and Hangwen 2014). Soil temperature below freezing point decreases mineralization by inhibiting microbial activity and decreasing diffusion of soluble substrates in the soil (Kaiser et al. 2007). Most soil microorganisms require temperatures between 10 and 35.6 °C for their activities. Soil microbial activities decrease with low soil temperatures, and at freezing point, most activities cease (Allison 2005). Water uptake decreases with low temperature. This is due to the increased viscosity and decreased absorption rate of water at low temperature (Toselli et al. 1999). Decreased water uptake reduces the rate of photosynthesis. Increase in soil temperature improves root growth because of the increase in metabolic activity of root cells and the development of lateral roots (Repo et al. 2004). Low soil temperature results in reduced tissue nutrient concentrations and as such decreases root growth (Balland and Arp 2005; Bronick and Lal 2005; Buchas 2001; Conyers et al. 2012; McMichael and Burke 1998; Pradel and Pieri 2000; Puhe 2003).

2.2.3 Soil Water Storage and Availability to Plants

The soil water storage capacity is generally expressed in terms of available water capacity. The total available water in soil is regarded as the difference between soil moisture content of "field capacity" and "wilting point."

It is customary to express available water capacity in terms of depth dimensions so that it can be considered with irrigation, evaporation rate, rainfall, and irrigation water quantity, which are generally expressed in terms of depth. The available water capacity of specific layers, generally available as percent by weight, can be expressed as "cm" if bulk density values and depth of soil layer of the root zone are also determined as illustrated below:

$$D_{\mathrm{w}} = \{(\mathrm{FC} - \mathrm{PWP}) \times \mathrm{BD} \times D_{\mathrm{s}}\}/100 \qquad (2)$$

where D_{w} is the depth of water contained in a specific soil layer (cm); D_{s}, thickness of the soil layer (cm); BD, bulk density of the particular soil layer (g cm^{-3}); FC, soil water content at field capacity (% by wt); and PWP, soil water content at permanent wilting point (% by wt). The value of "D_{w}" is computed for each soil layer and then added up for the entire profile.

However, the values of bulk density, field capacity, and permanent wilting point vary from layer to layer within a soil profile, soil to soil, and location to location even under the same major soil group for the simple reason that there will be a variation in soil depth and available water capacity limits also. Therefore, as already emphasized earlier, the actual determination has to be made for the specific soil under consideration.

Different views regarding availability of water to plants appear apparently contrasting; however, they can be reconciled if consideration is given to the water transmission characteristics of soils, crop rooting depth and density, and evaporative potential of atmosphere. In general, a greater soil water depletion can be tolerated by plants under the conditions of low evaporativity, deep and dense rooting of plants, and medium- to fine-textured soils. In contrast, under the conditions of high evaporativity, sparse rooting, and coarse-textured soils, only a small depletion of soil water can be tolerated by plants, and frequent light irrigations will be needed to maximize crop growth. Irrigation schedule needs to be optimized in terms of time and amount of irrigation on the basis of permissible water depletion for a given combination of crop, soil, and climate. Under limited water supply situations, crop water deficit may be planned by identifying optimal and suboptimal water deficit sequences.

2.2.4 Water Transmission Characteristics

Hydrophysical properties of soil, viz., moisture retention characteristics and hydraulic conductivity, greatly influence the availability of water to plant roots. As suction increases, i.e., moisture content decreases, hydraulic conductivity decreases more steeply in coarse-textured soils than in fine-textured soils. Thus, not only is the amount of moisture retained in a coarse-textured soil less (Table 2.1) but also the movement of moisture from the bulk soil to the site of moisture absorption, i.e., the root surface is much slower as compared with the fine-textured soil.

2.2.5 Crop Rooting Behavior and Soil Water Retention

Crop rooting behavior influences the soil water availability to a very large extent. In general, the greater the proliferation and density of roots, the lesser the sensitivity of the crop to soil water depletion. Moreover, compared with the roots in the upper layers, the lower-lying roots, although less abundant, are more effective in water uptake because of being younger and less crowded and often growing in a better soil moisture regime. The root completely fills the soil pore under adequate soil moisture conditions, while, under severe stress, shrinkage of cortical cells of root results in root shrinkage and, thus, significantly reduces soil-root contact (Fig. 2.2).

Agriculturists have pursued better crop adaptations to soil water deficits in systems that depend on rainfed water inputs; under these circumstances, crop water uptake ability plays a pivotal role in crop yield (Kato and Okami 2011). Across the soil-plant-atmosphere continuum, hydraulic water flow is induced by the gradient between soil water potential (Ψ_{soil}) and leaf water potential (Ψ_{leaf}). Plant

Table 2.1 Available water capacity of soils of different textural classes

Textured classes	Available water (cm/cm of soil)
Loamy sand	0.074
Sandy loam	0.146
Loam	0.191
Silt loam	0.234
Sandy clay loam	0.209
Silty clay loam	0.204

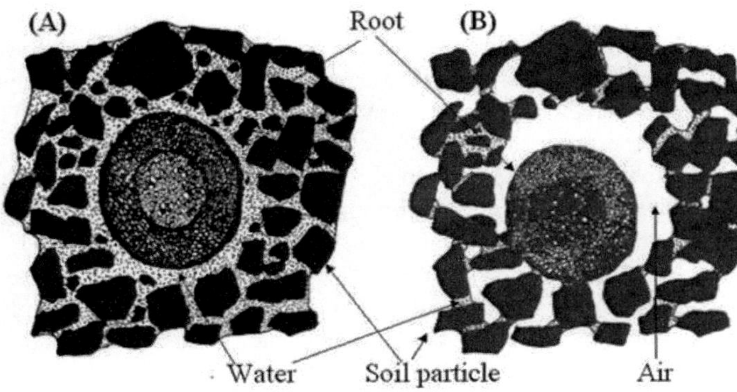

Fig. 2.2 Cross-section of root surrounded by soil. (**a**) Adequate soil moisture. (**b**) Severe water deficit stress

water uptake is determined by Ψ_{soil}, Ψ_{leaf}, and the hydraulic conductance of water transport from roots to leaves (whole-plant hydraulic conductivity, K_{pa}; Boyer 1996). K_{pa} in intensely transpiring plants is largely determined by hydraulic conductance of the root system, which can be used to estimate water uptake ability of roots (Steudle and Peterson 1998). Consequently, numerous studies have determined responses of crops to soil drying through measurements of root-system hydraulic conductance normalized by root length or root surface area (root hydraulic conductivity, L_p; Steudle 2000).

Agriculture consumes 70% of the freshwater resource, but less water is becoming available for irrigation owing to the global climate change and competition from urbanization and industrial development (Pennisi 2008). In particular, increasing physical water scarcity is a main constraint for irrigated *Oryza sativa* production (Peng et al. 2009). To cope with water shortage, water-saving rice cultivation technologies are being developed (Bouman et al. 2007). Rice has many traits derived from its semi-aquatic ancestor (e.g., shallow root system), but at the same time, it has an ability to adapt to soil water deficit (Gowda et al. 2011).

2.3 Plant-Water Relations

Water is the most abundant constituent of all physiologically active plant cells. Leaves, for example, have water contents which lie mostly within a range of 55–85% of their fresh weight. Other relatively succulent parts of plants contain approximately the same proportion of water, and even such largely nonliving tissues as wood may be 30–60% water on a fresh-weight basis. The smallest water contents in living parts of plants occur mostly in dormant structures, such as mature seeds and spores. The great bulk of the water in any plant constitutes a unit system. This water is not in a static condition. Rather it is part of a hydrodynamic system, which in terrestrial plants involves the absorption of water from the soil, its translocation throughout the plant, and its loss to the environment, principally in the process known as transpiration. The typical mature, vacuolate plant cell constitutes a tiny osmotic system, and this idea is central to any concept of cellular water dynamics. Although the cell walls of most living plant cells are quite freely permeable to water and solutes, the cytoplasmic layer that lines the cell wall is more permeable to some substances than to others. In some kinds of plant cells, movement of water occurs principally by the process of imbibition rather than osmosis. The swelling of dry seeds when immersed in water is a familiar example of this process.

Plant-water relations concern how plants control the hydration of their cells, including the collection of water from the soil, its transport within the plant, and its loss by evaporation from the leaves. The water status of plants is usually expressed as "water potential," which has units of pressure, is always negative, and in simple form is the algebraic sum of the hydrostatic pressure and the osmotic pressure of water. Flow of water through the plant and soil over macroscopic distances is driven by gradients in hydrostatic pressure. Over microscopic distances (e.g., across semi-permeable membranes), it is driven by gradients in water potential. Evaporation of water from leaves is primarily controlled by stomata and, if not made good by the flow of water from the soil through the plant to the leaves, results in the plants' wilting.

Various gases diffuse into and out of physiologically active plants. Those gases of greatest physiological significance are carbon dioxide, oxygen, and water vapor. The great bulk of the gaseous exchanges between a plant and its environment occur through tiny pores in the epidermis that are called stomates. Although stomates occur on many aerial parts of plants, they are most characteristic of, and occur in greatest abundance in, leaves. The term transpiration is used to designate the process whereby water vapor is lost from plants. Although basically an evaporation process, transpiration is complicated by other physical and physiological conditions prevailing in the plant. Whereas loss of water vapor can occur from any part of the plant which is exposed to the atmosphere, the great bulk of all transpiration occurs from the leaves. There are two kinds of foliar transpiration:

- Stomatal transpiration, in which water vapor loss occurs through the stomates.
- Cuticular transpiration, which occurs directly from the outside surface of epidermal walls through the cuticle. In most species, 90% or more of all foliar transpiration is of the stomatal type.

Transpiration is a necessary consequence of the relation of water to the anatomy of the plant and especially to the anatomy of the leaves. Terrestrial green plants are dependent upon atmospheric carbon dioxide for their survival. In terrestrial vascular plants, the principal carbon dioxide-absorbing surfaces are the moist mesophyll cell walls which bound the intercellular spaces in leaves. Ingress of carbon dioxide into these spaces occurs mostly by diffusion through open stomates. When the stomates are open, outward diffusion of water vapor unavoidably occurs, and such stomatal transpiration accounts for most of the water vapor loss from plants. Although transpiration is thus, in effect, an incidental phenomenon, it frequently has marked indirect effects on other physiological processes which occur in the plant because of its effects on the internal water relations of the plant.

In terrestrial rooted plants, practically all of the water which enters a plant is absorbed from the soil by the roots. The water thus absorbed is translocated to all parts of the plant. The mechanism of the "ascent of sap" (all translocated water contains at least traces of solutes) in plants, especially tall trees, was one of the first processes to excite the interest of plant physiologists. The upward movement of water in plants occurs in the xylem, which, in the larger roots, trunks, and branches of trees and shrubs, is identical with the wood. In the trunks or larger branches of most kinds of trees, however, sap movement is restricted to a few of the outermost annual layers of wood.

Root pressure is generally considered to be one of the mechanisms of upward transport of water in plants. While it is undoubtedly true that root pressure does account for some upward movement of water in certain species of plants at some seasons, various considerations indicate that it can be only a secondary mechanism of water transport. Upward translocation of water (actually a very dilute sap) is engendered by an increase in the negativity of water potential in the cells of apical organs of plants. Such increases in the negativity of water potentials occur most commonly in the mesophyll cells of leaves as a result of transpiration.

The successively smaller branches of the root system of any plant terminate ultimately in the root tips, of which there may be thousands and often millions on a single plant. Most absorption of water occurs in the root tip regions and especially in the root hair zone. Older portions of most roots become covered with cutinized or suberized layers through which only very limited quantities of water can pass. Whenever the water potential in the peripheral root cells is less than that of the soil water, movement of water from the soil into the root cells occurs. There is some evidence that, under conditions of marked internal water stress, the tension generated in the xylem ducts will be propagated across the root to the peripheral cells. If this occurs, water potentials of greater negativity could develop in peripheral root cells than would otherwise be possible. The absorption mechanism would operate in fundamentally the same way whether or not the water in the root cells passed into a state of tension. The process just described, often called passive absorption, accounts for most of the absorption of water by terrestrial plants.

The phenomenon of root pressure represents another mechanism of the absorption of water. This mechanism is localized in the roots and is often called active absorption. Water absorption of this type only occurs when the rate of transpiration is low and the soil is relatively moist. Although the xylem sap is a relatively dilute solution, its osmotic pressure is usually great enough to generate more negative water potential than usually exists in the soil water when the soil is relatively moist. A gradient of water potentials can thus be established, increasing in negativity across the epidermis, cortex, and other root tissues, along which the water can move laterally from the soil to the xylem.

Plants are sessile organisms that cannot escape from environmental constraints, and, as a result, they have evolved numerous adaptive responses to cope with environmental stresses. Most environmental limitations share common effects and responses such as a reduction in growth and photosynthesis, oxidative damage, hormonal changes, and the accumulation of numerous stress-related proteins. These changes are usually the result of tissue dehydration (Kacperska 2004; Dobra et al. 2010). Tissue dehydration occurs when there is an imbalance between root water uptake and leaf transpiration (Aroca et al. 2001; Jackson et al. 2003). When leaves begin to dehydrate, plants generally start closing their stomata; however, under some environmental limitations or in specific plant genotypes, modification of root water uptake capacity plays a more important role compared with stomatal closure in avoiding stress-induced growth reduction (Matsuo et al. 2009). Water deficit occurs in tissues under soil water deficit, low and high temperature, salt stress, or flooding stress conditions (Aroca et al. 2001, 2007; Wahid and Close 2007). Under some environmental conditions, dehydration is the first signal that induces the plant to respond (Jia et al. 2002; Christmann et al. 2007), and the importance of the hydration state of tissues in the response of plant to different stresses is well supported by experimental evidence (Aroca et al. 2001; Bouchabke-Coussa et al. 2008; Matsuo et al. 2009). The importance of root water uptake capacity in coping with several environmental limitations is supported by the work of Aroca et al. (2001) who found that two maize genotypes differing in chilling sensitivity also differed in the response of their root water uptake rate to chilling stress. Hence, while the tolerant genotype kept its leaf transpiration and root water uptake rates unchanged during chilling periods, the sensitive genotype became dehydrated because it decreased first its root water uptake rate and then its leaf transpiration. Also, Matsuo et al. (2009), comparing three rice genotypes, found a good correlation between root water uptake capacity, estimated as root hydraulic conductivity, and shoot dry weight under water-limited conditions. In the same way, Hattori et al. (2008) found that the ameliorative effect of silicon application in the response of sorghum plants to osmotic stress was mainly caused by an improvement of root water uptake capacity, avoiding water imbalance caused by the osmotic stress imposed.

Relative water content, leaf water potential, transpiration rate, stomatal conductance, and leaf and canopy temperature are the important attributes describing plant-water relations (Farooq et al. 2009a). Sales et al. (2013) reported severe reductions in leaf water potential of sugarcane under the simultaneous occurrence of cold and drought stresses, compared with individual stress factor. Generally, the chilling-sensitive plants exhibit water-stress symptoms under low temperatures, which are triggered by reduced root hydraulic conductivity followed by severe decline in leaf water potential and loss of turgor pressure (Aroca et al. 2003). McWilliam et al. (1982) and Lee et al. (1993) also reported that decline in root hydraulic conductivity and loss of stomatal control caused chilling-induced wilting in plants. Aroca et al. (2003) stated that low temperature-induced reduction in vapor-pressure deficit between the leaf surface and the atmosphere decreased transpiration rate and root water uptake drastically. Chilling stress caused shoot water deficits in maize plants by significant reductions in root water uptake instead of leaf transpiration (Janowiak and Markowski 1994). Likewise, Javier et al. (1997) reported that under chilling stress, root water uptake decreased in drought-hardened plants similar to non-drought-hardened plants, while transpiration rate was lower in drought-hardened plants only.

Jensen and Taylor (1961) have pointed out that low temperatures cause a decrease in the absorption and movement of water in plants. This reduction is less in species native to cool environments than in species which normally grow in warm ones (Kramer 1942). This has been attributed to the combined effects of decreased permeability of the root membranes and the increased viscosity of the water. Kramer (1940) observed that water flow through root systems increased as temperature was increased to 35 °C, the highest temperature studied. Application of kinetic theory to similar biological systems has been made with success (Johnson et al. 1954). This approach can reveal much concerning factors that limit the rate of biological processes that are not now well known.

2.3.1 Concept of Water Potential

The energy status of water is expressed in terms of water potential, which is a measure of chemical potential of water. Water potential can be defined as the free energy of water available (without temperature change) to do work. Simply water potential is the potential energy of water in a system relative to pure water in the same temperature and pressure. It measures the tendency of water to move through soil plant atmosphere continuum.

The chemical potential of the water depends on the mean free energy per molecule and the concentration of water molecules, i.e., on the mole fraction of the water. The degree to which the presence of solute reduces the chemical potential of the water in the solution below that of pure free water can be expressed as:

$$\mu_{\mathrm{w}} - \mu^{2}{}_{\mathrm{w}} = RT \ \ln \ N_{\mathrm{w}} \tag{3}$$

where μ_{w} is the chemical potential of water in the solution, $\mu^{2}{}_{\mathrm{w}}$ is the chemical potential of pure water at the same temperature and pressure (ergs per mole), R is the gas constant, T is the absolute temperature in °K, and N_{w} is the mole fraction of water.

For use with ionic solutions, the mole fraction is replaced by the activity of water, a_{w}, and, for general use, by the relative vapor pressure, *e/e°*.

$$\mu_{\mathrm{w}} - \mu_{\mathrm{w}} = RT \ \ln \ \left(e/e^{\circ}\right) \tag{4}$$

- *When the vapor pressure of the water in the system under consideration is the same as that of pure free water, ln e/e° is zero, and the potential difference is also zero. Thus,* ***pure free water is defined as having a potential of zero****.*
- *When the vapor pressure of the system is less than that of pure water, ln e/e° is a negative number; hence, the potential of the system is less than that of pure free water and is expressed as a negative number.*

Since the expression of chemical potential in units of ergs per mole is inconvenient in discussions of cell water relations, units of energy per unit of volume was chosen. The measurements are compatible with pressure units, which can be obtained by dividing both sides of the above equation by the partial molal volume of water, V_w (cm^3/mol). The resultant term is called the water potential, Ψ_w. The symbol for water potential is the Greek letter Psi (Ψ).

$$\Psi_w = \left(\mu_w - \mu^2{}_w\right)/V_w = \left(RT/V_w\right) \ln \left(e/e^{\circ}\right) \quad (5)$$

Water potential = the chemical potential of water/the partial molar volume.
$= J\ mol^{-1}/L\ mol^{-1}$.
$= N \times m\ mol^{-1}/m^3\ mol^{-1}$.
$= N\ m^{-2} = MPa$.
1 MPa = 10 Bars = 10 Atm. (1 Atm = 760 Mm hg = 14. 7 Lbs. Sq. in^{-1}*)*

Plant water potential expresses the chemical potential of water in the plant system relative to the chemical potential of pure water at the same temperature and pressure. However, the following points need to be noted carefully:

- Water always moves from high water potential to low water potential.
- The addition of solute decreases water potential.
- Pressure increases water potential.
- In cells, water moves by osmosis to areas where water potential is lower.

2.3.2 Components of Water Potential

The total water potential (Ψ) of a system can be partitioned into different component potentials.

$$\Psi = \Psi_s + \Psi_p + \Psi_m + \Psi_s \quad (6)$$

where Ψ_p, Ψ_s, Ψ_m, and Ψ_s are referred to as pressure, solute/osmotic, matric, and gravitational potential, respectively.

2.3.2.1 Solute/Osmotic Potential (Ψ_s)

It arises due to the presence of dissolved solutes in water. Since presence of solutes reduces the potential of water to below that of pure water, the osmotic potential is below zero or a negative value. Thus, Ψ_s of a system is determined by the osmolality, i.e., the total concentration of dissolved particles in a solution without regard for the particle size, density, configuration, or electrical charge. The vapor pressure of water in a solution is lowered in proportion to the extent to which the mole fraction of water in the solution is decreased by adding solute. Hence, when water is separated from a solution by a membrane permeable to water but impermeable to the solute, water will move across the membrane along a gradient of decreasing vapor pressure or chemical potential into the solution until the vapor pressures of solution and pure water become equal. The pressure that must be applied to the solution to prevent the movement of water into the solution when separated by semi-permeable membrane is termed the osmotic pressure. The negative value of osmotic pressure is equal to osmotic potential.

The Van't Hoff equation can be used to calculate the osmotic pressure of solutions:

$$\pi = (n_s/V)RT \tag{7}$$

where π is the osmotic pressure in pascals; V, the volume of solvent in liters; n_s, the moles of solute; R, the gas constant (0.00832 L MPa/degree mol at 273°K); and T, the temperature in °K.

For 1 mol of solute in 1 L of solvent at 273 °K (0 °C), $\pi = 22.7 \times 10^5$ Pa or 2.27 MPa (22.4 atm or 22.7 bars). In the case of soil, osmotic potential plays a significant role where high concentrations of salts occur, for example, in saline, soils or waters with high salinity are being used, as that influences the uptake of water by plant. In soil, osmotic potential is given by:

$$\Psi_s = -0.36 \times \text{EC} \tag{8}$$

where EC is the electrical conductivity (dS m^{-1}). Osmotic potential is generally ignored as far as movement of water within the soil is considered.

Osmotic adjustment is generally thought to be the major mechanism that helps to maintain cell turgor in evergreen plant species when the water potential decreases, sustaining water uptake and plant metabolic activity during drought (Turner and Jones 1980). Osmotic adjustment occurs when the osmotic potential of cells is lowered through active solute accumulation rather than passively as a result of tissue dehydration. Under cold conditions, this is believed to decrease the freezing temperature of the cell sap reducing the probability of ice formation. Additionally, increases in compatible solutes such as proline, other osmotically active amino acids, and soluble sugars protect the structural integrity of cell membranes from cellular dehydration during extracellular freezing (Thomas and James 1993; Xin and Browse 2000; Kosová et al. 2007). Osmotic adjustment is the decrease of osmotic potential owing to the net accumulation of solutes in response to water deficit conditions (Zhang et al. 1999). Under osmotic stress, plants maintain the water relations by accumulating various osmotically active molecules/ions such as soluble sugars, organic acids, proline, glycinebetaine, potassium, calcium, and chloride ions. Osmoregulation/osmotic adjustment helps plants to sustain growth and photosynthetic rate and assimilate partitioning under stressful conditions (Subbarao et al. 2000). Farooq et al. (2009a) stated that osmotic adjustment is directly related to the accumulation of osmolytes in different concentrations, which largely depends on stress severity, plant type, and the growth stage at which the stress is imposed (Shao et al. 2006). As chilling stress has been reported to disturb plant osmotic balance and cause dehydration stress (Farooq et al. 2009a; Wang et al. 2016a), therefore, under chilling and drought stress, osmotic adjustment assists the plants to maintain higher turgor potential (Zhang et al. 1999; Chaves et al. 2009). Morgan et al. (Morgan 1990) reported that osmotic adjustment is crucial in postponing the dehydration stress under water-limited conditions. Greater accumulation of osmolytes is generally linked with drought tolerance (Blum 2005) because these facilitate in the maintenance of leaf turgor to improve conductance of stomata for efficient CO_2 intake (Kiani et al. 2007) and promote the water uptake ability of roots (Chimenti et al. 2006). Mercado et al. (1997) reported that protein contents in the tissues of chilling-sensitive plants are usually decreased with chilling stress, mainly because of reduced synthesis which, in turn, increases the level of free amino acids particularly proline (Jouve et al. 1993). Endogenous concentrations of proline were linked with chilling tolerance capability in maize (Songstad et al. 1990) and drought tolerance in rice (Hsu et al. 2003) and pea (Alexieva et al. 2001). Proline mainly accumulates in younger leaves and is recognized as the most important solute against abiotic stresses (Perez-Perez et al. 2009). Most of the studies have shown that osmotic potential of plants is more affected by drought as compared to chilling stress (Farooq et al. 2009b).

Changes in tissue elasticity, which modify the relationship between turgor pressure and cell volume, also play an important role in the cell resistance to dehydration (Scholz et al. 2012). Similar to the drought effect, freezing-induced cell dehydration results in the collapse of the cell wall around the shrinking protoplasm and, depending on the properties of cell components, may lead to different degrees of frost injuries (Levitt 1980). When the tissue elasticity is high, the cell wall deforms readily, and thus a loss of water will cause a large change in volume but a small change in turgor, because the cell wall shrinks and continues to squeeze the cytoplasm. On the other hand, when the tissues are rigid, a large loss of water causes little change in the volume of the cell, but a rapid decrease in turgor that allows the cell to avoid further water loss. In addition to its effects on dehydration avoidance, the cell wall mechanical properties contribute to the formation of an effective barrier against the propagation of extracellular ice in cell walls and to the avoidance of intracellular freezing (Solecka et al. 2008). Deposition of extensin, a glycoprotein, and other structural components on cell wall during cold acclimatization is known to provide greater structural rigidity, preventing cell contraction and collapse during freezing (Weiser et al. 1990; Cavender-Bares 2005).

2.3.2.2 Pressure Potential (Ψ_p)

Pressure potential is equivalent to the hydrostatic pressure exerted on the cell wall by plasma membrane due to water uptake-induced inflation of protoplast, in other words, the hydrostatic pressure to which water in a liquid phase is subjected. It was known formerly as wall pressure or turgor pressure. In a turgid plant cell, pressure potential is usually positive, but the pressure potential of xylem in a transpiring plant, which is under considerable tension, will be negative. When a plant cell is having sufficient water, it inflates the cell membrane and hence develops a pressure against the cell wall. At incipient plasmolysis when the cell faces water deficit, turgor potential will be zero. Immediately after cell division, plant cells take up water and expand. Cell expansion ceases once cell wall is fully developed. Hence, the final volume of the cell, and thus size and shape of plant organs, depends upon the water availability. Water deficit at any period of plant growth thus may result in smaller organs such as small leaves, fruits, and short plant. Loss of cell turgor leads to drooping of shoots, leaves, etc. Pressure potential may be positive or negative; the higher the pressure, the greater the potential energy in a system, and vice versa. Therefore, a positive Ψ_p (compression) increases Ψ_{total}, while a negative Ψ_p (tension) decreases Ψ_{total}. Positive pressure inside cells is contained by the cell wall, producing turgor pressure in a plant. Turgor pressure ensures that a plant can maintain its shape. A plant's leaves wilt when the turgor pressure decreases and revive when the plant has been watered. Pressure potentials are typically around 0.6–0.8 MPa, but can reach as high as 1.5 MPa in a well-watered plant. As a comparison, most automobile tires are kept at a pressure of 30–34 psi or about 0.207–0.234 MPa. Water is lost from the leaves via transpiration (approaching $\Psi_p = 0$ MPa at the wilting point) and restored by uptake via the roots. Plants can also regulate Ψ_p by opening and closing the stomata. Stomatal openings allow water to evaporate from the leaf, reducing Ψ_p and Ψ_{total}. This increases water potential between the water in the petiole (base of the leaf) and in the leaf, thereby encouraging water to flow from the petiole into the leaf.

2.3.2.3 Matric Potential (Ψ_m)

Interaction of water with matrix (capillary or electrostatic) reduces the potential of water to below that of pure water, and hence it is a negative value. Matric potential is the amount of water bound to the matrix of a plant via hydrogen bonds and is always negative to zero. In a dry system, it can be as low as −2 MPa in a dry seed or as high as zero in a water-saturated system. Every plant cell has a cellulosic cell wall, which is hydrophilic and provides a matrix for water adhesion, hence the name matric potential. The binding of water to a matrix always removes or consumes potential energy from the system. Matric potential is similar to solute potential because the hydrogen bonds remove energy from the total system. However, in solute potential, the other components are soluble, hydrophilic solute molecules, whereas in matric potential, the other components are insoluble, hydrophilic molecules of the plant cell wall. Matric cannot be manipulated by the plant and is typically ignored in well-watered roots, stems, and leaves. Normally matric potential is a negligible component of plant water potential, but it is the most important component of soil water potential. Soil water movement essentially occurs in reference to the hydraulic gradient expressed as change in soil matric potential per unit length.

2.3.2.4 Gravitational Potential (Ψ_g)

Gravity potential is always negative or zero in a plant with no height. Without height, there is no potential energy in the system. The force of gravity pulls water downward to the soil, which reduces the total amount of potential energy in the water in the plant (Ψ_{total}). The taller the plant, the taller the water column, and the more influential the gravitational potential becomes. On a cellular scale and in short plants, this effect is negligible and easily ignored. However, over the height of a tall tree like a giant coastal redwood, the plant must overcome an extra 1 MPa of resistance because of the gravitational pull of -0.1 MPa m^{-1}. It arises due to gravitational forces acting on water. It may be positive or negative depending upon whether the system is above or below the reference point. The contribution of gravitational potential to plant water potential is negligible except for tall trees as the contribution of gravitational potential to the water potential is about 0.01 MPa per meter height. Among these components, the major components of water potential in plants are the turgor potential and osmotic potential, while in soil, the major component of water potential is the matric and gravitational potential.

2.4 Water Uptake by Roots

The maximum water uptake is mediated by 20–200 mm root region from the growing meristem, and it contains a large amount of root hairs, which increases the surface area and, thus, the extraction of soil moisture from a larger soil volume. Root system continuously grows and extracts water and nutrients from the soil. Soil water enters the root apoplast and moves mainly through apoplastic pathway till it reaches the Casparian strip. At this point, water enters the symplastic pathway. A cross-section of a cereal plant root showing types of cells in root and pathway of water and nutrient transport is depicted in Fig. 2.3.

Thus, at this point of water transport in roots, maximum resistance occurs. Water channel proteins called aquaporins, which are located on the plasma membrane, mediate entry of water through plasma membrane into the symplast. Once water crosses the Casparian strip, water can move in both symplast and apoplast till it reaches the apoplastic xylem in the stele. Through the xylem, water reaches the

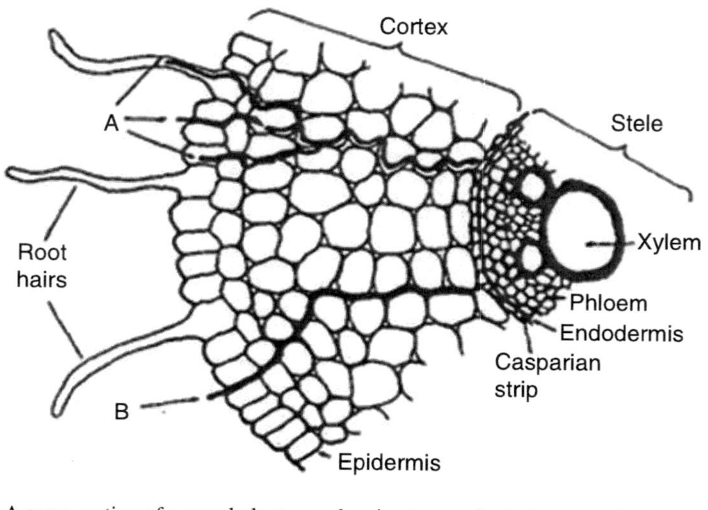

Fig. 2.3 A cross-section of a cereal plant root showing types of cells in root and pathway of water and nutrient transport. (**a**) Symplasmic and (**b**) apoplasmic pathway

leaves, where it can take either symplastic or apoplastic pathway until it reaches the stomatal cavity.

Ascent of sap or movement of water from root to leaf is explained by the cohesion-tension theory. It relies on the physical properties of water, on mechanisms of liquid transport, and on the anatomical features of the xylem, the sap conducting system. Water within the whole plant forms a continuous network of liquid columns (due to hydrogen bonding of water molecules with each other) from the absorbing surfaces of roots to the evaporating surfaces. About 99% of these water columns consist of xylem vessels and tracheids, and the remaining 1% is constituted by the wall and cytoplasm of living cells. Surface tension of water at the evaporating surface provides the driving force for the movement of water. Due to surface tension and the small radius of the curvature of the capillary menisci at the evaporative surfaces, water potential of adjacent cells is lowered by evaporation. Since water column is continuous in vascular pathway, it transfers the variations of water potential throughout the plant instantaneously. From the stomatal cavity, water vapor evaporates to the atmosphere through stomatal pores present in the epidermis. This loss of water through evaporation from the stomata contained in the leaves of plants is called transpiration. The rate of transpiration is controlled by the energy gradient of water or the water potential gradient.

The pathway of movement of water from soil to atmosphere through plant is called soil-plant-atmosphere continuum (SPAC), an analogy to Ohm's law of flow of electrical current through a system. Water uptake and movement in the plant are driven by the energy status of water present in the SPAC. Water moves from high-energy status, i.e., high water potential (in soil), to low-energy status, i.e., low water potential (in atm). Water flow in the SPAC depends on water potential gradients and resistances according to the following equations, valid for steady-state conditions:

$$\mathrm{Jv} = \text{Potential gradient/resistance} = \Delta\Psi/R \tag{9}$$

where Jv is the rate of water flow or transpiration; $\Delta\Psi$, water potential gradient; and R, resistance to water flow. The rate of water flow from soil to root depends upon water potential gradient between root and soil and the resistance in the soil.

This can be expressed as:

$$\mathrm{Jv} = (\Psi_{\mathrm{soil}} - \Psi_{\mathrm{root}})/R_{\mathrm{soil}} \tag{10}$$

Similarly, the rates of water flow from root to xylem and xylem to leaf and leaf to air are expressed as:

$$\mathrm{Jv} = \left(\Psi_{\mathrm{root}} - \Psi_{\mathrm{xylem}}\right)/R_{\mathrm{root}} \tag{11}$$

$$\mathrm{Jv} = \left(\Psi_{\mathrm{xylem}} - \Psi_{\mathrm{leaf}}\right)/\left(R_{\mathrm{xylem\ to\ leaf}}\right) \tag{12}$$

$$\mathrm{Jv} = (\Psi_{\mathrm{leaf}} - \Psi_{\mathrm{air}})/(R_{\mathrm{leaf\ to\ air}}) \tag{13}$$

Hence, the rate of transpiration or the rate of flow of water from soil to air through plant system depends upon the water potential gradient between soil and air and the resistances to water movement in the various paths of soil and plant. Thus, the rate of transpiration (Jv) can be expressed as:

$$\mathrm{Jv} = (\Psi_{\mathrm{soil}} - \Psi_{\mathrm{air}})/R_{(\mathrm{soil+root+xylem+leaf})} \tag{14}$$

In reality, steady-state conditions seldom exist, and water potential in plants quickly decreases after sunrise and recovers during the night if the soil has adequate moisture.

2.4.1 Root Water Uptake Concepts

Two main forces regulate the root water uptake rate, namely, osmotic and hydrostatic forces. Hydrostatic force is generated by the transpiration stream, whereas osmotic force is generated by the root pressure. Root water transport is divided into radial and axial transport. The axial transport consists of the water moving along the xylem vessels to aerial parts, and it does not contribute in a major way to the resistance of water transport through the whole plant (Knipfer and Fricke 2011). In woody plants, this axial transport can be an important determinant of resistance because of cavitation events (Dalla-Salda et al. 2009). On the other hand, radial water flow from the soil solution to the root xylem vessels has high resistance to root water transport and involves three dynamically exchangeable paths (Steudle and Peterson 1998). The apoplastic path comprises water moving through the pores between the fibrils of the cell wall and through the intercellular spaces. The symplastic path consists of water moving through the cytoplasm and through plasmodesmata between cells. It has been assumed that under transpiring conditions, the main route for radial water transport was the apoplastic path, and under conditions where transpiration is reduced, the main route would be the cell-to-cell path (Steudle and Peterson 1998; Javot and Maurel 2002). However, Knipfer and Fricke (2010, 2011) and Fritz and Ehwald (2011) demonstrated that in some species and under some conditions, water flowing by the cell-to-cell path could account for almost the whole radial root water transport even under transpiring conditions. Role of root morphology and anatomy in the overall root water transport capacity cannot be underestimated (Bramley et al. 2009). Independent of the radial water pathway that predominates under specific environmental conditions, the role of aquaporins in root water uptake has been abundantly documented (Kaldenhoff et al. 1998; Javot et al. 2003; Postaire et al. 2010).

Aquaporins first described in plants by Maurel et al. (1993) are membrane intrinsic proteins found in all living organisms (Agre et al. 1993) that facilitate the passage of water by forming a proteinaceous pore in the membrane. Osmotic gradients drive water transport through aquaporins. In plants, aquaporins are divided into five families:

- Plasma membrane intrinsic proteins (PIPs).
- Tonoplast intrinsic proteins (TIPs).
- Nodulin26-like intrinsic proteins (NIPs).
- Small and basic intrinsic proteins (SPIs).
- Uncharacterized intrinsic proteins (XIPs).

The number of aquaporin genes present in a plant species is very high, ranging from 30 to >70 (Maurel et al. 2008; Park et al. 2010). Localization of each aquaporin protein inside the plant cells varies among cell membranes. At the same time, the membrane aquaporin location in the cell is dynamic since some transport of them from internal membranes to the plasma membrane has been found (Zelazny et al. 2007; Wudick et al. 2009).

To estimate the root water uptake capacity of a single root or of the whole root system, hydraulic conductance measurements are undertaken, and a direct correlation between hydraulic conductance values and root water uptake rates has been observed (Nardini and Pitt 1999; Li et al. 2005a). In this sense, the contribution of aquaporins to hydraulic conductance values has been tested by several approaches. The involvement of aquaporins in hydraulic conductance regulation was assayed by the inhibition of hydraulic conductance by mercurial reagents (Maggio and Joly 1995; Carvajal et al. 1996), since several plant aquaporins have a cysteine residue sensitive to mercury (Daniels et al. 1996). More conclusive approaches involved the use of transgenic plants in which the expression of some aquaporins was inhibited. Siefritz et al. (2002) and Postaire et al. (2010) found that a specific plasma membrane aquaporin of tobacco and *Arabidopsis*, respectively, regulated hydraulic conductance under hydrostatic forces. On the other hand, again Siefritz et al. (2002) and Javot et al. (2003) found that a specific plasma membrane intrinsic aquaporin of tobacco and *Arabidopsis*, respectively, regulated hydraulic conductance under osmotic forces. These findings clearly support the involvement of plasma membrane intrinsic aquaporin in the regulation of hydraulic conductance under both osmotic and hydrostatic forces and, therefore, their involvement in the whole root water uptake rate (Aroca et al. 2012).

2.4.2 Water Absorption by the Roots

Essentially all of the water used by land plants is absorbed from the soil by roots. A root system consists of a complex network of individual roots that vary in age along their length. Roots grow from their tips and initially produce thin and non-woody fine roots. Fine roots are the most permeable portion of a root system and are thought to have the greatest ability to absorb water, particularly in herbaceous (i.e., non-woody) plants (McCully 1999). Fine roots can be covered by root hairs that significantly increase the absorptive surface area and improve contact between roots and the soil. Some plants also improve water uptake by establishing symbiotic relationships with mycorrhizal fungi, which functionally increase the total absorptive surface area of the root system (McElrone et al. 2013). As the roots absorb water, there is a reduction in the water potential in the rhizosphere. This process establishes a water potential gradient between the rhizosphere and a neighboring region of the soil which presents a higher water potential and which coordinates the water movement toward the roots of a transpiring plant. This water movement in the soil occurs mainly through mass flow due to the fact that the water-filled micropores of the soil are interconnected. Therefore, water flows from soil to root at a rate depending on the water potential gradient between soil and plant which is affected by plant water need, hydraulic conductivity of the soil, soil type, and soil water content. Sandy soils have higher conductivity due to greater porosity, but they also retain less water in relation to clay soils or soils rich in organic matter. At field capacity, water is initially removed from the center of the largest pores (spaces $\geq$50 nm, that are too large to have any significant capillary force) between the soil particles, maintaining the water next to the particles due to adhesive forces. The reduction in water content causes a drastic decrease in soil hydraulic conductivity, because the water is replaced by air in the spaces between the soil particles. Thus, the water movement in the soil is limited to the periphery of soil pores, which can promote restrictions in the hydraulic conductivity to the root surface and reach the permanent wilting point.

2.4.3 Root Water Uptake under Low-Temperature Conditions

It is well established that low-temperature conditions cause leaf dehydration in sensitive plants (Nagasuga et al. 2011). As for other stresses, this leaf dehydration is caused by the imbalance between water lost by leaf transpiration and water uptake by roots (Vernieri et al. 2001). Hence, although the transpiration rate decreases under low-temperature conditions because of a decrease in the vapor pressure difference between the leaf surface and the atmosphere (Aroca et al. 2003), stomata of the sensitive plants remain open, while those of the tolerant plants close more rapidly (Aroca et al. 2003; Bloom et al. 2004). At the same time, root water uptake also decreases drastically when the temperature goes down, because of the decrease in the vapor-pressure deficit (Aroca et al. 2003) and the increase in the viscosity of water (Bloom et al. 2004). The reduction of hydraulic conductance under low-temperature conditions not only has a physical explanation but also has biological metabolic causes. In some species or in plants grown under particular conditions before the low-temperature treatment is applied, a break point in the Arrhenius plot experiments was observed (Murai-Hatano et al. 2008). These break points are attributed to membrane injury, most probably caused by a phase transition of membrane lipids (Nihsida and Murata 1996).

Several low-temperature-tolerant plant species are able to recover their hydraulic conductance after a prolonged time (a few days) of exposure to low temperature (Aroca et al. 2003, 2005; Nagasuga et al. 2011). It was hypothesized that this recovery could be caused by an enhancement of aquaporin activity. Aroca et al.

(2005) found that a maize low-temperature-tolerant genotype accumulated plasma membrane intrinsic protein (PIP) in its roots after 3 days of exposure to 5 °C, concomitantly with an increase in the phosphorylation state of the PIP2 proteins. This plasma membrane intrinsic protein accumulation was correlated with even higher hydraulic conductance values of the low-temperature-treated maize plants. Aquaporin phosphorylation has been shown to enhance aquaporin activity (Maurel et al. 1995) by keeping aquaporin pores in the open state (Daniels and Yeager 2005; Tornroth-Horsefield et al. 2006). Aroca et al. (2005) also found the same plasma membrane intrinsic protein aquaporin response in a low-temperature-sensitive maize genotype, which kept its hydraulic conductance at very low levels during low-temperature stress. So, the response of plasma membrane intrinsic protein aquaporins to low-temperature stress was not the only explanation for the different hydraulic conductance behavior between the sensitive and the tolerant maize genotypes. Aroca et al. (2005) also found that the root membranes of the sensitive genotype were injured by the low-temperature treatment. Hence, to tolerate low-temperature stress, it was not sufficient to accumulate more plasma membrane intrinsic proteins in the root membranes. The tolerant genotype also possessed robust mechanisms against membrane injury caused by low-temperature conditions.

In agreement with the accumulation of plasma membrane intrinsic protein in roots during low-temperature periods, other researchers also found that some plants grown at low temperatures had higher hydraulic conductance than plants grown at higher temperatures, when hydraulic conductance was measured at the higher temperatures (Wan et al. 1999; Lee et al. 2005b). Also, Matsumoto et al. (2009) found that rice plants overexpressing one plasma membrane intrinsic protein1 aquaporin were more tolerant to low-temperature stress than untransformed plants. So, it seems that accumulation of plasma membrane intrinsic protein is crucial to tolerate low-temperature stress. Indeed, Lee et al. (2005b) found that a low-temperature-sensitive plant species (*Cucumis sativus*) accumulated more suberin in both root exo- and endodermis than a tolerant species (*Cucurbita ficifolia*). This higher suberin accumulation was correlated with lower hydraulic conductance values under low-temperature conditions of the sensitive species. Moreover, Lee et al. (2005a) also observed a more pronounced and constant increase in the double bond index of the root plasma membrane in *Cucurbita ficifolia* plants than in that of *Cucumis sativus* under low-temperature conditions. Such an increase was caused by raising the amount of linolenic acid and by diminishing the amount of stearic acid. The increase in the double bond index could be correlated with the higher hydraulic conductance observed in *Cucurbita ficifolia* than in *Cucumis sativus* under low-temperature conditions (Lee et al. 2005b; López-Pérez et al. 2009).

There are many studies on the response of rice (*Oryza sativa* L.) plants to low temperature that includes the temperatures of air and water (soil) in the root zone. However, it is important to investigate the effects of air and water temperatures on rice plants because they have great effect on various growth processes individually and the flooded water temperature often differs from the air temperature (Nagasuga et al. 2011). The magnitude and direction of the differences vary with weather conditions and the growth stage of rice (Shimono et al. 2007a, b; Kuwagata et al.

2008). In addition, successive water supply to the paddy field and excessive water drainage from the soil often decrease the flooded water temperature even on sunny days. Quantitative investigation on the effect of the temperature in the root zone should contribute not only to understanding the mechanism of rice plant resistance to low temperature more clearly but also to the improvement of rice cultivation in cool climate area. During the vegetative period, a lower flooded water temperature reduces the rate of tillering, delays leaf number, and reduces leaf elongation (Takamura et al. 1960; Matsushima et al. 1964; Shimono et al. 2002). Grain yield can also be reduced by physiological changes, such as nutrient absorption by root system, that occur during the vegetative period (Shimazaki et al. 1963). Since the low temperature of flooded water greatly affects biomass and grain yield.

Several studies have been conducted to identify the underlying mechanisms (Takamura et al. 1960; Matsushima et al. 1964; Satake 1976; Shimono et al. 2002). Leaf growth, which is important for dry matter production, is strongly affected by environmental factors such as temperature (Sato 1972; Shimono et al. 2002). The reduction of dry matter production by the low temperature of flooded water might be a direct effect of low temperature on the shoot growth. On the other hand, Shimono et al. (2004) indicated that the water content of plant and bleeding from the plant after shoot excision were lower under a low flooded water temperature condition, suggesting that the low temperature of flooded water inhibits water uptake by the root and thereby reduces dry matter production. It is important to evaluate the effect of a low root temperature on water uptake separately from that on other processes to better understand the mechanism of rice plant resistance to low temperature. It has been shown that low root temperature inhibits water uptake by roots (Kramer and Boyer 1995). Since the water movement throughout a plant is strongly regulated by that in the root system, the inhibition of water uptake by low root temperature should affect plant water content and, as a result, dry matter production. Water uptake is controlled by both the water permeability of the root surface and the root volume. Low root temperature inhibits root growth, although there are wide differences among species in the optimal temperature for root growth (Kramer and Boyer 1995). However, there are few studies on the relationship between low root temperature and root volume in rice plants. On the other hand, Murai-Hatano et al. (2008) reported that root hydraulic conductivity in rice, which relates to its permeability, decreased with decreasing root temperature and that hydraulic conductivity decreased dramatically at temperatures below 15 °C. Such phenomenon has been reported in several other plant species (Clarkson 1976; Markhart et al. 1979; Wan et al. 2001; Lee et al. 2004b). It is necessary to clarify which of these factors is mainly affected by a low root temperature.

2.4.4 Transpiration

About 99% of the water taken up by the plant is lost to the atmosphere through transpiration. Evaporation of water from leaf surface occurs through two pathways: (1) direct evaporation from the outer epidermal cell walls through the cuticle to the atmosphere and (2) from the cell wall to the stomatal cavity and then to the atmosphere through the stomata. Depending upon the thickness and wax load, the water loss through the cuticle varies. Majority of the water is lost in transpiration through the stomatal pathway. Water deficit stress develops when the water removed by the plant from the soil is not replenished through irrigation/rainfall. When water uptake is less than the rate of transpiration, cellular water deficit occurs. The leaf water potential has to be maintained at a critical value in order to maintain cell turgidity and open stomata to support photosynthesis. The sensitivity of the plant processes to water potential changes varies greatly. Hsiao (1973) identified cell expansion as the most sensitive plant process to cellular water deficits. However, wide genetic variation in plant processes for water deficit stress tolerance is found within and across species.

- For a given genotype, the transpiration rate is regulated by the stomatal opening and closing, depending upon the water availability. The rate of transpiration varies widely in different plant species due to the differences in root architecture, leaf area, leaf structure, leaf number, cuticle thickness, and wax load and stomatal characters. Since water vapor is lost to the atmosphere and CO_2 is taken inside the leaf for photosynthesis through the stomata, any reduction in transpiration through the stomata will also reduce the rate of photosynthesis. Hence, crop yield can be expressed as:

$$\text{Yield} = T \times \text{WUE} \times \text{HI} \tag{15}$$

where T is the total seasonal crop transpiration; WUE, crop water-use efficiency; and HI, crop harvest index (the ratio of economic yield to total biomass).

2.4.5 Ascension of Water Through the Plant: Vascular System

The presence of plants outside the water environment—among other factors—has been related to the evolution of the vascular system, which allows for the speedy upward movement of water to meet the demand of transpiration from the leaves. Water supply through cells by diffusion (difference in chemical gradient) alone is not able to maintain the hydration of a perspiring canopy plant. The water flows from the roots to the shoot of the plant through the xylem. The general mechanism to explain this upward movement of water is the cohesion-tension theory. Later, the cohesion theory has been questioned as a result of assessments of tension in xylem vessels, which do not present a direct relation with tension values measured on leaves through pressure chambers. Furthermore, it is assumed that the hydrophobic interaction between the internal walls of the xylem and the sap composition (lipids, proteins, polysaccharides, etc.) prevents the development of a tensile strength larger than 1 MPa (Zimmermann et al. 2004), which is smaller than the estimated tension of rising water in a 30-m-high tree (3 MPa). However, despite these questions, many studies argue that the fundamentals of the cohesion-tension theory are still valid for explaining the water flow in the soil-plant-atmosphere continuum (Richter 2001; Steudle 2001; Cochard 2002; Tyree 2003).

Once in the xylem tissue, water moves easily over long distances in these open tubes. There are two kinds of conducting elements (i.e., transport tubes) found in the xylem: (1) tracheids and (2) vessels. Tracheids are smaller than vessels in both diameter and length and taper at each end. Vessels consist of individual cells, or "vessel elements," stacked end to end to form continuous open tubes, which are also called xylem conduits. Vessels have diameters approximately that of a human hair and lengths typically measuring about 5 cm although some plant species contain vessels as long as 10 m. Xylem conduits begin as a series of living cells, but as they mature, the cells commit suicide (referred to as programmed cell death), undergoing an ordered deconstruction where they lose their cellular contents and form hollow tubes. Along with the water conducting tubes, xylem tissue contains fibers which provide structural support and living metabolically active parenchyma cells that are important for storage of carbohydrates, maintenance of flow within a conduit, and radial transport of water and solutes. When water reaches the end of a conduit or passes laterally to an adjacent one, it must cross through pits in the conduit cell walls. Bordered pits are cavities in the thick secondary cell walls of both vessels and tracheids that are essential components in the water transport system of higher plants. The pit membrane, consisting of a modified primary cell wall and middle lamella, lies at the center of each pit and allows water to pass between xylem conduits while limiting the spread of air bubbles (i.e., embolism) and xylem-dwelling pathogens. Thus, pit membranes function as safety valves in the plant water transport system. Averaged across a wide range of species, pits account for >50% of total xylem hydraulic resistance. The structure of pits varies dramatically across species, with large differences evident in the amount of conduit wall area covered by pits, and in the porosity and thickness of pit membranes.

After traveling from the roots to stems through the xylem, water enters leaves via petiole (i.e., the leaf stalk) xylem that branches off from that in the stem. Petiole xylem leads into the mid-rib (the main thick vein in leaves), which then branch into progressively smaller veins that contain tracheids and are embedded in the leaf mesophyll. In dicots, minor veins account for the vast majority of total vein length, and the bulk of transpired water is drawn out of minor veins (Sack and Holbrook 2006; Sack and Tyree 2005). Vein arrangement, density, and redundancy are important for distributing water evenly across a leaf and may buffer the delivery system against damage (i.e., disease lesions, herbivory, air bubble spread). Once water leaves the xylem, it moves across the bundle sheath cells surrounding the veins. It is still unclear the exact path water follows once it passes out of the xylem through the bundle sheath cells and into the mesophyll cells, but is likely dominated by the apoplastic pathway during transpiration (Sack and Holbrook 2006).

2.4.6 Disruption of Water Movement

Water transport can be disrupted at many points along the SPAC resulting from both biotic and abiotic factors. Root pathogens (both bacteria and fungi) can destroy the absorptive surface area in the soil, and similarly foliar pathogens can eliminate evaporative leaf surfaces, alter stomatal function, or disrupt the integrity of the cuticle. Other organisms (i.e., insects and nematodes) can cause similar disruption of above- and belowground plant parts involved in water transport. Biotic factors responsible for ceasing flow in xylem conduits include pathogenic organisms and their by-products that plug conduits; plant-derived gels and gums produced in response to pathogen invasion; and tyloses, which are outgrowths produced by living plant cells surrounding a vessel to seal it off after wounding or pathogen invasion. Abiotic factors can be equally disruptive to flow at various points along the water transport pathway. During drought, roots shrink and lose contact with water adhering to soil particles—a process that can also be beneficial by limiting water loss by roots to drying soils (i.e., water can flow in reverse and leak out of roots being pulled by drying soil). Under severe plant dehydration, some pine needle conduits can actually collapse as the xylem tensions increase.

Water moving through plants is considered meta-stable because at a certain point the water column breaks when tension becomes excessive—a phenomenon referred to as cavitation. After cavitation occurs, a gas bubble (i.e., embolism) can form and fill the conduit, effectively blocking water movement. Both sub-zero temperatures and drought can cause embolisms. Freezing can induce embolism because air is forced out of solution when liquid water turns to ice. Drought also induces embolism because as plants become drier, tension in the water column increases. There is a critical point where the tension exceeds the pressure required to pull air from an empty conduit to a filled conduit across a pit membrane—this aspiration is known as air seeding. An air seed creates a void in the water, and the tension causes the void to expand and break the continuous column. Air seeding thresholds are set by the maximum pore diameter found in the pit membranes of a given conduit.

2.4.7 Signal Controlling Root Water Uptake Under Stress Condition

It is well known that abscisic acid modifies root hydraulic properties. The most common effect is an increase in root hydraulic conductance when abscisic acid is added to the root medium (Aroca 2006; Mahdieh and Mostajeran 2009; Ruiz-Lozano et al. 2009). Nevertheless, in some cases, no effect (Wan and Zwiazek 2001) or even a decrease in root hydraulic conductance (Aroca et al. 2003) was observed. The most robust evidence about the role of abscisic acid in regulating root hydraulic conductance comes from studies of abscisic acid-deficient plants or abscisic acid-overproducing plants. Thus, abscisic acid-deficient tomato plants have lower whole-plant hydraulic conductance, as well as a lower root exudation rate (Nagel et al. 1994). Also, Thompson et al. (2007) found that tomato plants overproducing abscisic acid had higher root hydraulic conductance values than wild-type plants. Abscisic acid is known to accumulate in root cells exposed to abiotic stresses such as drought (Simonneau et al. 1998), salt (Jia et al. 2002), low temperature (Aroca et al. 2003), or flooding (Olivella et al. 2000). In the case of flooding stress, a decrease of root abscisic acid contents has also been observed (Rodríguez-Gamir et al. 2011). Hence, it is possible that abscisic acid may regulate, at least in part, root hydraulic conductance changes under stress conditions. However, abscisic acid addition and drought stress usually have opposite effects on root hydraulic conductance. Thus, while abscisic acid usually increases root hydraulic conductance, drought inhibits it (Aroca 2006). This apparent contradiction could possibly be explained by assuming that exogenous application of abscisic acid to the roots causes an excess of abscisic acid or a different cellular location of the abscisic acid (Aroca et al. 2003). Nevertheless, under some specific drought conditions, an increase in root hydraulic conductance values has been observed (Singh and Sale 2000; Siemens and Zwiazek 2003), although in these studies abscisic acid levels were not determined.

Wan (2010) found that external application of abscisic acid inhibited the negative effect of NaCl on cell root hydraulic conductance and explained this by suggesting that abscisic acid promoted cell osmotic adjustment. However, no direct measurements supporting that osmotic adjustment were presented. Exogenous abscisic acid also modulates root plasma membrane aquaporin expression and protein abundance (Jang et al. 2004; Aroca 2006; Aroca et al. 2008b; Mahdieh and Mostajeran 2009; Ruiz-Lozano et al. 2009). When expression of four plasma membrane aquaporin genes was studied in abscisic acid-deficient tomato plants, it was found that under well-watered conditions, two plasma membrane aquaporin genes remained unchanged, and two other plasma membrane aquaporin genes were upregulated by abscisic acid depletion. In contrast, under drought conditions, three plasma membrane aquaporin genes were downregulated, and only one gene was upregulated (Aroca et al. 2008a). However, no protein data are available from this study. Hence, it is clear that abscisic acid is involved in the response of root hydraulic conductance and plasma membrane aquaporins to different abiotic stresses, although the mechanism behind such regulation is still far from being understood.

Ethylene, another stress-related plant hormone, could also play a role in the regulation of root water uptake under several abiotic stress conditions. It is known that ethylene levels vary under different abiotic stress conditions such as drought (Liang 2003), salt (Quinet et al. 2010), or flooding (Huang et al. 1997). For example, Kamaluddin and Zwiazek (2002) and Islam et al. (2003) found an enhancement of root hydraulic conductance values by an exogenous application of ethylene in several tree species. In contrast, Li et al. (2009a) found a negative effect of ethylene application on root hydraulic conductance values in *Medicago falcata* plants. This contrasting result could be caused by different concentration, time of exposure, or, simply, the different plant species used. Qaderi et al. (2007) found that canola (*Brassica napus*) plants with low levels of ethylene production have lower root hydraulic conductance values under non-stressed conditions. Another stress-related plant hormone that regulates root hydraulic conductance is salicylic acid. Boursiac et al. (2008) found that exogenous salicylic acid inhibited root hydraulic conductance in *Arabidopsis* plants by causing invaginations of plasma membranes containing plasma membrane aquaporin proteins, as occurs under salt stress conditions. Moreover, Volobueva et al. (2004) found that salicylic acid treatment inhibited symplastic water transport in maize roots. Apart from the above results, no more information is available about how other plant hormones could regulate root water uptake under abiotic stress conditions.

Other groups of molecules that may govern root water uptake under abiotic stress conditions are the reactive oxygen species. Reactive oxygen species are generated in root tissues under different abiotic stress conditions (Blokhina et al. 2001; Aroca et al. 2005; Katsuhara et al. 2005; Bian and Jiang 2009). Among reactive oxygen species, hydrogen peroxide (H_2O_2) is the most studied molecule in relation to root water uptake properties (Lee et al. 2004a; Aroca et al. 2005; Ye and Steudle 2006; Rhee et al. 2007; Boursiac et al. 2008; Benabdellah et al. 2009). In the earlier studies of Lee et al. (2004a) and Aroca et al. (2005), it was shown that exogenous application of H_2O_2 inhibited cell and root hydraulic conductance, respectively. Several mechanisms including membrane damage (Aroca et al. 2005), closure of aquaporin pores (Ye and Steudle 2006), or formation of plasma membrane vesicles containing plasma membrane aquaporin (Boursiac et al. 2008) have been postulated. However, in specific plant species or genotypes (more precisely, plants tolerant to low-temperature conditions), no effect of exogenous H_2O_2 on root hydraulic conductance was observed (Aroca et al. 2005; Rhee et al. 2007). In addition, Benabdellah et al. (2009) found that low concentrations of exogenous H_2O_2 increased root hydraulic conductance in *Phaseolus vulgaris* plants. Consequently, reactive oxygen species may serve as signals in controlling root hydraulic conductance behavior under abiotic stress conditions, stimulating or inhibiting root hydraulic conductance depending on their cellular concentration. This hypothesis is also supported by the fact that the root hydraulic conductance response to abscisic acid is modified by the addition of antioxidants (Aroca 2006). Obviously, more research is needed in order to ascertain how reactive oxygen species may be involved in root hydraulic conductance regulation under several stress conditions.

2.5 Plant-Water Relationship under Low-Temperature Stress

Chilling stress results from temperatures cool enough to produce injury without forming ice crystals in plant tissues, whereas freezing stress results in ice formation within plant tissues. Plants differ in their tolerance to chilling (0–15 °C) and freezing (<0 °C) temperatures. Both chilling and freezing stresses are together termed low-temperature or cold stress: the damage due to cold stress can range from chilling injury and freezing injury to suffocation and heaving. Low-temperature stress is a major environmental factor that often affects plant growth and crop productivity and leads to substantial crop losses (Xin and Browse 2000; Sanghera et al. 2011). In general, plants from temperate climatic regions are considered to be chilling tolerant to variable degrees, and their freezing tolerance can be increased by exposing to cold, but non-freezing, temperatures; this process is known as cold acclimation. However, generally, the plants of tropical and subtropical origins are sensitive to chilling stress and lack this mechanism of cold acclimation (Sanghera et al. 2011). Low temperature may affect several aspects of crop growth, viz., survival, cell division, photosynthesis, water transport, growth, and finally crop yield.

Several physiological and biochemical changes distinguish plant genotypes during and after cold acclimation (Guy 1990; Koster and Lynch 1992; Hughes and Dunn 1996; Grossi et al. 1998; Thomashow 1999; Stitt and Hurry 2002). One such distinction among genotypes is tissue water content that decreases more rapidly in certain genotypes than others during the cool autumn season (Chen and Gusta 1978; Fu et al. 1999). Martin (1927) observed from field results that northern selections contained less tissue water in late autumn than did those selected farther south where winter freezing conditions are less severe. Nass (1983) observed similar differences in plant water content among winter wheat and rye (*Secale cereale* L.) cultivars. The reduction of tissue water content in response to low temperature is believed to be due to an impedance of water uptake by the roots (Limin and Fowler 1985). Reduced root hydraulic conductance occurs shortly after plants are transferred from warm to cooler temperatures. Bolger et al. (1992) observed this effect within a few hours after cotton (*Gossypium hirsutum* L.) plants were transferred from 30 to 18 °C. Similarly, Malone (1993) observed a decrease in hydraulic conductance of spring wheat roots within minutes after plants were transferred from 22 to 8 °C. The effect of cool temperatures on lowering water content in cultivars could result from (1) inhibited plant water uptake or (2) plant water loss exceeding uptake. It was hypothesized that less water would be removed from soil at 2 °C. This hypothesis was tested by comparing foliage water content and plant water uptake of a northern and a southern cultivar at 2 °C and a photosynthetic photon flux density (PPFD) at or near light compensation (Kenefick et al. 2002).

It had been observed that good water conditions in winter are conducive to the safe overwintering of forage (Naydenova et al. 2013). When the soil water content is high, the surface temperature changes gradually; soil freezing and thawing is not very intense, and encountering sudden temperature rises and cooling or cryogenic stress is relatively light for alfalfa, and thus the overwinter survival rate is high (Sun et al. 2004). The loss of agricultural and livestock production caused by cryogenic and drought disasters is a global problem. Low-temperature damage in the Northern United States and Western Canada, where alfalfa is planted in large areas, has led to severe losses in yield (Keane et al. 2018). The northern China area has cold winters, with a dry and windy climate, which causes some crops to suffer from cryogenic and drought stress from winter to early spring. Mild cases include damaged branches and leaves, while the severe cases cause the death of the entire plant. Environmental stresses in early winter (including cold and dryness) and the following spring (including drought and cold damage caused by drastic changes in temperature) are the factors causing the death of winter crops (Yu et al. 2015). In 2002, nearly 10,000 hm^2 of alfalfa in northern China suffered varying degrees of damage due to low temperatures (Sun et al. 2004). In areas where the soil is often frozen in winter, watering before the soil freezes is called “freezing irrigation.” After freezing, a frozen soil layer forms around the root system to keep the temperature in the root zone relatively stable and prevent sudden changes in the outside temperature from harming the plants. After thawing in early spring, timely watering is known as "thawing irrigation," which can also play the same protective role of meeting crop needs for water in the turning green period (Bingzhe et al. 2010), thereby improving the rate of returning green (Mikić et al. 2015). Wu et al. (2020) have reported that alfalfa absorbed soil water when the soil water content was high enough in the root layer when under high water volume freezing irrigation (irrigation in early winter when soil is freezing) but not under low and medium water volume freezing irrigation. Alfalfa gradually began to absorb soil water on the third day after thawing irrigation (irrigation in late winter when the soil is thawing) and showed different water uptake characteristics under low, medium, and high water volume. Thawing irrigation also accelerated the regeneration of alfalfa.

Earlier studies, however, failed to establish how winter water recharge occurred. Instead of taking water from the soil, trees could, for short periods, rely on water stored in the stem or extract snow melt water through the leaves or shoot bark (Cowling and Kedrowski 1980; Boyce et al. 1991). Boyce and Lucero (2002) reported that large saplings of *Picea engelmannii* Parry used water stored in the sapwood for transpiration, whereas smaller saplings seemed to rely more on soil water uptake. In boreal climates, tree stems and branches freeze in wintertime (Sakai and Larcher 1987), which lowers xylem conductance, making it more difficult to use the water stored in the stems. However, Owston et al. (1972) showed water movement in stems of *Pinus contorta* Dougl. and *Abies magnifica* A. Murr. even when air temperatures dropped below –16 °C every night of their measurement period. Stem diameter variation gives information on soil water uptake by trees. Stem diameter fluctuates daily because of transpiration-induced tension changes in the stem sap (Irvine and Grace 1997; Perämäki et al. 2001). The fluctuation follows the pattern of transpiration, and the daily amplitude reflects the balance between transpiration rate and water uptake from soil (Perämäki et al. 2001). Lack of transpiration or the absence of available soil water results in zero amplitude. Sevanto et al. (2001) showed that drought results in an overall decreasing trend in stem diameter, whereas an increase in free water in the soil (e.g., after rainfall) makes the stem swell (Sevanto et al. 2006a, b).

Among abiotic stresses, temperature extremes are very critical for determining the agricultural production. Plants exhibit a maximum rate of growth and development at an optimum temperature or over a diurnal range of temperatures. Deviation of ambient temperature from optimal value, causes physiological, biochemical, metabolic, and molecular changes occur within plants. At the extremes of the natural temperature range of a plant, the degree of physiological, cellular, metabolic, and molecular dysfunction becomes so severe that it leads to death. Plants feel stress under high- as well as low-temperature exposure. Mesophilic temperate plants possess inducible temperature-stress tolerance. During high-temperature exposure, stress tolerance can be induced by exposure to short-term elevated temperature, and this is known as acquired thermotolerance (Kotak et al. 2007), while at lower temperatures, stress tolerance can be induced by exposure to reduced temperature and is known as chilling tolerance and/or cold acclimation. Chilling tolerance is the ability of a plant to tolerate low temperatures (0–15 °C) without injury or damage (Somerville 1995), while cold acclimation is an enhanced tolerance to the physical and physiochemical vagaries of freezing stress (Thomashow 1999). Both cold acclimation and chilling tolerance involve an array of biochemical, molecular, and metabolic processes (Thomashow 1999; Larkindale et al. 2005; Kotak et al. 2007; Zhu et al. 2007). Exposure of plants to temperature stress leads to the modification of metabolism in two ways. Firstly, plants try to adjust their cellular metabolism that altered due to rising or falling of temperatures. Temperature stress changes the structure, catalytic properties, and function of enzymes (Kubien et al. 2003) and membrane metabolite transporters. Interestingly, regulatory mechanisms of plants become active and function to restore normal metabolite levels and, most importantly, metabolic fluxes (Schwender et al. 2004; Fernie et al. 2005). Secondly, the modifications of metabolism in response to temperature stress are mainly linked to enhanced tolerance mechanisms. Many metabolites thought to have important properties that could contribute to induced stress tolerance have long been linked to stress responses (Thomashow 1999; Nayyar et al. 2005).

Each plant has an optimum set of temperatures for its proper growth and development. It has generally been noticed that plants native to warm habitats exhibit symptoms of injury upon exposure to low non-freezing temperatures (Lynch 1990). For example, plants such as maize, soybean, cotton, tomato, and banana show signs of injury upon exposure to temperatures below 10–15 °C (Hopkins 1999). However, the appearance of injury symptoms depends upon the sensitivity of a plant to cold stress and varies from plant to plant. Cold stress-induced injury in plants may appear after 48–72 h of stress exposure. Plants exposed to cold stress show various phenotypic symptoms that include reduced leaf expansion, wilting, and chlorosis (yellowing of leaves) and may lead to necrosis (death of tissue). Cold stress also severely affects the reproductive development of plants, and this has been seen in rice plants at the time of anthesis (floral opening), which leads to sterility in flowers (Jiang et al. 2002).

Chilling of sensitive plants affects all components of water regime and causes loss of water, resulting in strong wilting (Vernieri et al. 1991; Boese et al. 1997; Bloom et al. 2004). It is based on two main factors: rapid decline in the ability of roots to absorb water and transport it to the shoots (Bolger et al. 1992) and reduced ability to close stomata in response to subsequent water deficit (Wilkinson et al. 2001; Bloom et al. 2004). Insufficient water supply provokes rapid drop in water potential of leaves during the first hours of cooling (Boese et al. 1997). The degree of chilling damage of plants can be reduced by means of preventing the disturbance of the water regime (Vernieri et al. 1991; Boese et al. 1997).

Cell membrane plays major roles in water and nutrient movements within and outward the cell. Intra- and extracellular water and nutrient movements are inhibited due to low temperature due to membrane damage under low temperature. There can be of two types of abnormalities during low-temperature stress. Cold damages the membrane that makes the membrane permeable to undesired nutrients and ions and causes ion leakage; another is cell membrane, and cell wall can be ruptured by the cold which is also responsible for disrupting cellular homeostasis by destroying both intra- and extracellular nutrient and water movements (Salinas 2002; Mahajan and Tuteja 2005). Severe dehydration may also occur due to freezing of cell constituents, solutes, and water (Yadav 2010). Available literature states that when temperatures drop below 0 °C, the ice formation generally begins in the intracellular spaces because the intracellular fluid has a higher freezing point as compared to the other suborganelles of cell (Thomashow 1999; Yadav 2010). Low and freezing temperatures also lead to cellular dehydration and reduce water and nutrient uptake and conduction by the roots in some plants, thus causing osmotic stress (Chinnusamy et al. 2007). Yadav (2010) stated that dehydration during cold occurs mainly due to reduction in water uptake by roots and is a hindrance to closure of stomata. The success or failure of a seedling in the field is strongly related to the development of its root system under cold stress (Enns et al. 2006). In the root of cucumber, it was found that chilling caused injury to the cortical cells and further long-time exposure increased the density of cytoplasm and damage the endoplasmic reticulum (Lee et al. 2002). Chilling-sensitive plants exposed to low temperature usually show water-stress symptoms due to decreased root hydraulic conductance and decreased leaf water and turgor potentials (Aroca et al. 2003). Freezing-induced increase in water viscosity is partly accounted for an initial decrease in root hydraulic conductance (Matzner and Comstock 2001). During cold stress, another phenomenon is common with the imbalanced water movement. The altered metabolic functions include production of more enzymes, isozymes though they help in maintaining more catalytic activity to cope with the low-temperature stress (Hurry et al. 1994).

The survival of plants in freezing temperatures depends on their capacity for cold acclimation (McKhann et al. 2008). It is very important to study the frost damage mechanism and to breed cold-tolerant varieties since the average minimum temperature is below 0 °C in about 64% of the earth's land area and it is below −10 °C in about 48% (Deane 1994). Frost technically means the formation of ice crystals, and this includes a phase change from vapor or liquid to solid. Ice formation (crystallization) consists of two main steps: (a) nucleation and (b) crystal growth. Nucleation is the process when the molecules start to gather, forming clusters at the nanometer scale. However, these clusters are required to grow to a specific size in order to form stable nuclei, and at this point, the atoms arrange in a defined manner to form the crystal structure. Subsequently, the nuclei keep growing to form ice and cause damage effects on plant cells (Sakai and Larcher 1987).

Research about the mechanism of ice initiation in plants has greatly decreased recently because of its essential role in specific aspects of cold hardiness (Gusta et al. 2009; Aryal and Neuner 2010; Hacker et al. 2011; Walters Jr et al. 2011). Ice formation may take place in either the intracellular or extracellular spaces depending on the cooling conditions (Guy 1990). Intercellular freezing can happen when cooling rates are high or after a significant supercooling. Supercooling is defined as when liquid water cools below the melting temperature (0 °C) without any ice formation. The supercooling point is the minimum low temperature reached before ice formation. Extracellular freezing takes place in the spaces between cells in water transporting elements (xylem and phloem) or on the external surfaces of the plant. Ice will grow and spread from the initial formation point (nucleation site) through the extracellular spaces in contiguous water. This process will continue until equilibrium of water potential is achieved resulting in dehydration effects on the plant cells (Wisniewski and Fuller 1999). Several forms of membrane damage can result as a consequence of dehydration caused by freezing effects. This could include expression-induced lysis, lamellar-to-hexagonal-phase transitions, and additional jump lesions. It has been demonstrated that protein denaturation caused by the effect of low temperature could cause cellular damage (Guy et al. 1998). Takahashi et al. (2013) reported that the alternation of plasma membrane compositions and functions during cold acclimation is one of the most important adaptation mechanisms for plant cold tolerance.

Generally, plants can be classified into three different classes according to their low-temperature tolerance (Stushnoff et al. 1984):

- The first group includes frost-tender plants that are sensitive to chilling injury and can be killed by short periods of exposure to temperature just below freezing point. Beans (*Phaseolus lunatus*), corn (*Zea mays*), rice (*Oryza sativa*), and tomatoes (*Solanum lycopersicum*) are examples of plants in this category.
- Low-temperature acclimation of plants in the second group allows them to tolerate the presence of extracellular ice in their tissues. Their frost resistance ranges from the broad-leafed summer annuals, which are killed at temperatures slightly below freezing point, to perennial grasses that can survive exposure to −40 °C. As temperatures decrease, the outward migration of intracellular water to the growing extracellular ice crystal causes dehydration stress that will eventually result in irreversible damage to the plasma membrane.
- The final group is made up of very cold hardy plants that are predominantly temperate woody species. Like the plants in the previous group, their lower limits of cold tolerance are dependent on the stage of acclimation, the rate and the degree of the temperature decline, and the genetic capability of tissues to accommodate extracellular freezing and the accompanying dehydration stress.

2.5.1 Transpiration and Transpiration Efficiency

Mass flow of liquid water from the roots to the leaves is driven in part by capillary action, but primarily driven by water potential differences. If the water potential in the ambient air is lower than the water potential in the leaf airspace of the stomatal pore, water vapor will travel down the gradient and move from the leaf airspace to the atmosphere. This movement lowers the water potential in the leaf airspace and causes evaporation of liquid water from the mesophyll cell walls. This evaporation increases the tension on the water menisci in the cell walls and decreases their radius and thus the tension that is exerted on the water in the cells. Because of the cohesive properties of water, the tension travels through the leaf cells to the leaf and stem xylem where a momentary negative pressure is created as water is pulled up by the xylem from the roots. As evaporation occurs at the leaf surface, the properties of adhesion and cohesion work in tandem to pull water molecules from the roots, through the xylem tissue, and out of the plant through the stomata. In taller plants and trees, the force of gravity can only be overcome by the decrease in hydrostatic (water) pressure in the upper parts of the plants due to the diffusion of water out of stomata into the atmosphere. Water is absorbed at the roots by osmosis, and any dissolved mineral nutrients travel with it through the xylem.

In leaf mesophyll, there is an extensive system of intercellular spaces which correspond to the internal surface of water contact with the air. There is a potential water potential gradient in the interface between the cell walls and the intercellular spaces, mobilizing the water by the cell walls from the final extremities of the xylem bundles. This water flow by the cell walls occurs in an analogous way to the principles of water movement through the soil matrix. As the leaf cuticle represents a barrier to water outlet—allowing on average only 5% of water permeability—the water vapor moves from leaf intercellular spaces to the atmosphere predominantly through stomatal diffusion. This process of the loss of water vapor by the leaves is called transpiration and corresponds to the majority (90%) of the volume of water absorbed by plants. In the soil-plant-atmosphere continuum of water flow, there are two major factors determining the water potential of a plant: (1) the water potential of the soil, which characterizes the water supply, and (2) transpiration, which defines the loss of water. The plant, which is an intermediate in this process, may regulate the water potential gradient between the soil and the atmosphere primarily through the regulation of stomatal conductance.

The stomata have a quick and fine control of the water relations of a plant, coordinating the control of the water potential gradient between the leaf and the air. In this interface of the leaf with the environment, it is important to note that small changes in the relative humidity of the air are reflected in major changes in the water potential gradient, which requires a stomatal control so as to maintain the water stability of the plant. A simple variation from 100% to 99% of the relative humidity already corresponds to a decrease of −1.36 MPa in the water potential of the air. This decrease becomes more evident in the water potential of the air in average (80%) and in extreme (50%) conditions of relative humidity, which, respectively, provide values of −14 and −93.6 MPa at 20 °C (Nobel 2009). Throughout this water route between the leaf and the air, there are two components that can exert resistance to the diffusion process: (1) stomatal resistance, which is coordinated by the stomatal opening, and (2) resistance of the air boundary layer, which is located closest to the leaf surface and is directly influenced by wind speed.

Transpiration efficiency has been defined as the total dry matter produced per unit of water transpired and is an important crop characteristic, especially when water resources are becoming scarcer in many regions (Vadez et al. 2014). The transpiration efficiency is actually quite complex, as it depends on both genetic and environmental components [$TE = k_d/(e^*_a - e)_d$], where the term in parentheses reflects the vapor-pressure deficit and k_d is a coefficient that reflects the CO_2 concentration in the stomatal chamber (Sinclair 2012). Hence, transpiration efficiency shows an inverse relationship with the vapor-pressure deficit (Tanner and Sinclair 1983). Modeling study showed that setting a maximum transpiration rate during the mid-day period, when vapor-pressure deficit is highest, would improve the yield of sorghum under water-limited conditions and increase transpiration efficiency (Sinclair et al. 2005). This genetic trait would confer the capacity to alter the effective vapor-pressure deficit value that drives transpiration efficiency. Large genotypic variations in the restriction of transpiration under a high vapor-pressure deficit have been reported in several legume species, such as soybean (Fletcher et al. 2007), chickpea (Zaman-Allah et al. 2011), cowpea (Belko et al. 2012), and peanut (Devi et al. 2010), as well as in cereals such as sorghum (Gholipoor et al. 2010), pearl millet (Kholova et al. 2010), and wheat (Schoppach and Sadok 2012).

2.5.2 Water-Use Efficiency

The water requirement of a crop is the quantity of water needed for normal growth, development, and yield and may be supplied by precipitation or by irrigation or by both. Water is needed mainly to meet the demand of evaporation, transpiration, and metabolic needs of the plants. The water requirement of a crop is dependent on factors such as variety, growth stage, duration, plant population, and growing season besides soil texture, structure, and depth and climatic factor such as temperature, relative humidity, and wind velocity. Some crops like cotton, groundnut, and pulses prefer stress during early vegetative growth to suppress excess vegetative growth (Sinha et al. 1985). In many crops, the initial establishment and flowering stages are highly sensitive to drought conditions that results in the poor performance of the root system and also shedding of flowers, in addition to lodging at maturity in some crops. Water-use efficiency, the amount of carbon gain relative to water use (Leakey et al. 2019), has been studied as a key target for crop improvement for at least a century. There are considerable differences between plant species to produce a unit dry matter per unit amount of water used resulting in widely varying values of water-use efficiency (Fischer and Turner 1978). Thus, genetic variation in water-use efficiency assumes considerable importance, especially in situations when water is limiting. Transpiration efficiency is a component of water-use efficiency, being aerial dry matter production per unit of water transpired by the crop. A significant increase in water-use efficiency could be obtained by agronomic practices such as sub-surface irrigation and mulches and plant attributes like canopy structure and rate of evaporative losses from soil (Rebetzke and Richards 1999; Richards and Luckacs 2002). There are various morphological, anatomical, phonological, physiological, and biochemical parameters, which enable crop plants to regulate the water required for the production of dry matter and yield. These include stomatal behavior, canopy structure, leaf movement and surface reflectance, leaf area, and root systems (Subbarao and Johansen 2002; Chaerle et al. 2005; Blum 2009). At present, only some of these traits have been studied at the molecular level.

In agriculture, water-use efficiency has been slowly improving due to use of genotypes with increased water-use efficiency (Condon et al. 2004) and due to adoption of innovative irrigation practices such as drip irrigation and deficit irrigation (Pereira et al. 2006; Costa et al. 2007). Drip irrigation improves water-use efficiency by reducing runoff and evaporation losses (Khanna-Chopra and Singh 2011). Deficit irrigation aims to optimize water productivity in crops by deliberately allowing crops to sustain some degree of water deficit and yield reduction to an extent with significant reduction of irrigation water. Variation in transpiration efficiency in crop plants can result from changes in water vapor flux through stomata as in wheat (Condon et al. 1990) or change in photosynthetic capacity as in groundnut (Rao et al. 1995). A significant negative correlation has been shown between photosynthetic capacity and specific leaf area, thus forming a basis of variation in transpiration efficiency. The C4 plants exhibit certain advantages compared to C3 plants such as reduced photorespiration and increase nitrogen-use and water-use efficiency. The advent of genetic engineering in crop plants has resulted in attempts to induce C4 characteristics in C3 plant such as tobacco, potato, and rice by genetic manipulation (Matsuoka et al. 2001). When stomata open and CO_2 is absorbed, water is inevitably lost. The balance between two processes is maintained by the stomatal aperture (Woodward 2002). Stomata respond directly to ambient humidity, thereby strongly influencing plant transpiration efficiency. The efficiency and speed of stomatal movement to fluctuating water availability are major factors in maximizing photosynthesis and water-use efficiency. Since C4 species can maintain high photosynthetic rates even when stomatal conductance is low compared with their C3 counterparts, it follows that C4 photosynthesis promotes higher water-use efficiencies than are found in C3 species (Huxman and Monson 2003; Kocacinar et al. 2008; Osborne and Sack 2012). Experiments have shown that instead of a gradual improvement in water-use efficiency from C3 species, across the intermediate range, and to a full C4 pathway, the increase in water-use efficiency resembles a threshold effect: Type I and II intermediates have water-use efficiency on a par with C3 species, while C4-like species have a high water-use efficiency akin to C4 plants (Kocacinar et al. 2008; Vogan and Sage 2011). Way et al. (2014) have pointed out that C4 photosynthesis evolved independently numerous times, probably in response to declining atmospheric CO_2 concentrations, but also to high temperatures and aridity, which enhance water losses through transpiration. Here, the environmental factors controlling stomatal behavior of leaf-level carbon and water exchange were examined across the evolutionary continuum from C3 to C4 photosynthesis at current (400 $\mu mol\ mol^{-1}$) and low (280 $\mu mol\ mol^{-1}$) atmospheric CO_2 conditions.

There are several direct and indirect pathways by which low soil temperatures may impair plant metabolism and growth (Göbel et al. 2019), first through reductions in root growth and decreased root water/nutrient uptake and second indirectly through reduced photosynthetic carbon gain due to stomatal closure in the course of cold-induced drought stress (DeLucia et al. 1992; Shimono et al. 2004; Alvarez-Uria and Körner 2007). Sachs (1875) observed that plants from warmer climates wilted when exposed to cold soil of 3–5 °C, while plants from cooler climates did not. Numerous examples of plants reducing transpiration rate at soil temperatures below 10 °C are given by Kramer (1969), but in most cases, it remained unclear whether physical constraints (increased viscosity of water) or physiological limitations (such as altered root membrane permeability) were the main causes of reduced water uptake. Some of these soil temperature effects on plant metabolism and growth seem to take place in partial or full independence of air temperature. It is still not very clear that whether low soil temperature is as influential on plant metabolism as low air temperature. Some studies have found that, under certain conditions, soil temperature may have an even larger influence on growth than air temperature, as it influences root physiology more directly. For example, Xu and Huang (2000) found high soil temperatures to be more detrimental than high air temperatures for the growth of the grass *Agrostis palustris*. The photosynthesis of *Picea rubens* saplings was more affected by soil temperature than by air temperature (Schwarz et al. 1997). Kato et al. (2006) identified soil temperature to be more important for the net ecosystem exchange of CO_2 of grasslands on the Tibetan Plateau than air temperature. Göbel et al. (2019) pointed out that soil frost (−2 °C) had a strong negative effect on gas exchange and stomatal conductance in all three species, independent of the elevation of origin. Leaf dark respiration was stimulated by soil frost in *Deschampsia flexuosa*, but not in *Kobresia pygmaea*, which also had a lower temperature optimum of photosynthesis. Soil cooling from 15 to 5 °C did not significantly reduce stomatal conductance and gas exchange in any of the species, and it was concluded that these are able to maintain a relatively high root water uptake in cold, non-frozen soil, but the high-alpine *Kobresia pygmaea* seems to be especially well adapted to warm shoot-cold root episodes, as it has a higher photosynthetic activity at 10 than 20 °C air temperature and does not upregulate leaf dark respiration upon soil freezing, as was observed in the grasses from warmer climates.

One of the plant damages caused by chilling is the reduction of photosynthetic rates, since the cold affects both electronic transport in the thylakoids and carbon fixation (Allen and Ort 2001). In addition, cold can inhibit photosynthesis by the reduced availability of free phosphate in the chloroplast due to the low utilization rate and, consequently, the accumulation of triose phosphate, reducing phosphate availability for photosynthesis (Huner et al. 1993; Taiz and Zeiger 2015). In plants susceptible to low temperatures, photosynthesis is interrupted just above the freezing point mainly due to the susceptibility of thylakoids to cold, while in tolerant plants, photosynthesis can be maintained even at negative temperatures until the freezing point of cellular fluids (Larcher 2004). Photosynthetic damage can occur within a few hours of cold exposure, as demonstrated for rubber plants (*Hevea brasiliensis* Muell. Arg.), which presented injuries after 4 h of exposure to 10 °C (Mai et al. 2009). However, these plants showed to be tolerant to cold, and after their transference to 28 °C, they returned to their previous values of photosynthetic rates in less than 24 h. In chicory plants (*Cichorium intybus* L.), the lowest maximum photosynthetic rate found in plants maintained at 4 °C was related to changes in the activity of enzymes involved in CO_2 fixation and the availability of this gas for photosynthesis (Devacht et al. 2009). A reduction in the content of photosynthetic pigments induced by cold may also contribute to the decrease in photosynthesis. For example, cold-sensitive genotypes of corn (*Zea mays* L.) grown at 14/12 °C (day/night) showed reductions in the content of the photosynthetic pigments and, consequently, in the photosynthetic rates when compared to tolerant genotypes (Haldimann 1999).

To increase the crop yield, either the transpiration or water-use efficiency must be increased, keeping harvest index at its maximum value for each species. Since water is a scarce resource, it is imperative to enhance the water-use efficiency of crop plants. High water-use efficiency is crucial for crop production under water-limited environments (Bhattacharya 2019). Water-use efficiency is influenced by the rate of transpiration and photosynthesis. C3 plants, such as rice, wheat, etc., fix CO_2 in photosynthesis through the Calvin cycle into a three-carbon compound. C4 plants, such as maize, sugarcane, sorghum, etc., first fix CO_2 in photosynthesis through the Hatch-Slack cycle into a four-carbon compound, which is then exported to specialized buddle sheath cells, where CO_2 is released and converted into sugars by the Calvin cycle. In general, water-use efficiency of C4 plants is higher than the C3 plants, mainly due to minimal or negligible amount of photorespiration in C4 plants. Nevertheless, significant variation exists in water-use efficiency within each of these categories of plants.

Another damage related to cold is the change in the cell membranes' fluidity. Under cold exposure, membranes become rigid, while under heat, they become more fluid. Chilling may cause a reduction in membrane fluidity (Nievola et al. 2017), due to the transition of the lipid components from a fluid-crystalline state to a solid-gel state, leading to dysfunctions such as increased permeability, water and intracellular solute loss, and inactivation of transport channels (Buchanan et al. 2015; Iba 2002). The membrane lipids can move within one of the layers and rotate around their own axis, and these movements are thermodynamically directed and temperature dependent (Ruelland and Zachowski 2010). These movements are reduced when temperature decreases, making the membrane more rigid. In most chilling-sensitive plants, the temperature at which the transition phase begins is close to 10 °C. In addition, differences in the structure of membrane lipids can influence chilling tolerance. Sensitive plants usually have a higher content of saturated fatty acid residues in the lipids, while the more resistant ones show an increase in the desaturation of these molecules (Ahmad and Prasad 2012).

In order to control the increase of reactive oxygen species and thus contribute to cooling tolerance, several species use the antioxidant system. For example, in watermelon plants (*Citrullus lanatus* (Thomb.) Mansf. cv. dulce maravilla) cultivated at 10 °C for 30 days, the lowest cold tolerance was related to the higher H_2O_2 content and reduction in the activity of ascorbate peroxidase, catalase, and glutathione reductase when compared to plants grown at 35 °C (Rivero et al. 2002). Interestingly, the comparison of cold tolerance in leaves of female and male plants of the Chinese *Populus cathayana* Rehd. showed an increase in H_2O_2 and ascorbate peroxidase values in both sexes after cultivation at 4 °C for 14 days, but superoxide dismutase and glutathione reductase levels were higher in male plants, suggesting that male plants of this species are more tolerant to chilling than females (Zhang et al. 2011). The increase in the enzyme activities in response to cold may be related to adjustments in gene expression. It has been reported that rubber plants grown at 10 °C have a 2.5-fold increase in ascorbate peroxidase gene expression after 24 h and the activity levels of this enzyme also increased after 1 h of cold exposure, demonstrating that responses of the antioxidant system may occur rapidly (Mai et al. 2009).

Sugars can also be related to antioxidant defense (Demidchik 2015; Couée et al. 2006). The antioxidant role of glucose might be related to its function in the pentose phosphate pathway, increasing the production of NADPH, one of the major cofactors of the reactive oxygen species scavenging enzymes, and thus contributing to decrease the formation of this molecule in the cells (Couée et al. 2006). Fructose also plays a relevant role as an antioxidant, since this sugar was twice more effective than glucose in removing $\dot{O}_2^-$, and therefore, it was associated with a key role in defense during chilling (Bogdanović et al. 2008). In cold-acclimated plants, raffinose might be involved in the stabilization of PSII (Knaupp et al. 2011). Membrane stabilization is an important mechanism of cold resistance, and carbohydrates play a key role in this process (Tarkowski and Van den Ende 2015). Sucrose, fructose, and oligosaccharides from the raffinose family can be inserted between the lipid polar groups decreasing membrane permeability (Hincha et al. 2003), which might be increased due to the membrane damage induced by cold. Exposure to cold also leads to a reduction in plant growth, as demonstrated for *Cistus albidus* L. and *Quercus ilex* L (Oliveira and Peñuelas 2004). This growth decrease was associated with the reduction of the starch/soluble sugars ratio, suggesting that photoassimilation under these conditions was deficient for maintaining growth since the stock components were remobilized and not re-stocked (Oliveira and Peñuelas 2004). Different varieties of chicory also showed a growth reduction when transferred from 16 to 4 °C, which was associated with changes in the photosynthesis during this period (Devacht et al. 2009).

2.5.3 Induction of Tolerance

The impact of cooling on plants depends on the chilling rate, exposure time, and other associated stresses (Buchanan et al. 2015). Depending on the tolerance of the species, the damage caused by low temperatures (chilling injury) may occur in a few hours or days (Shinozaki et al. 2015), determining the degree of sensitivity of a given species. When there is a sudden cold, the injury can occur after a few minutes and is called a "cold shock" (Phadtare et al. 1999; Levitt 1980).

Rate of metabolic processes (biochemical processes) deceases gradually with a decrease in temperature and may cease under severe stresses (Nievola et al. 2017). Cold temperature stress (0 to −10 °C) has broad-spectrum effects on cellular components and metabolic processes of plants. Cold temperature extremes impose stresses of variable severity depending on the intensity and duration of the stress. Several studies indicate that the membrane systems of the cell are the primary site of freezing injury in plants (Levitt 1980; Steponkus 1984) and freezing-induced membrane damage results primarily from the severe dehydration associated with freezing (Steponkus 1984; Steponkus et al. 1993). As temperatures drop below 0 °C, ice formation is generally initiated in the intercellular spaces in the extracellular fluid having a higher freezing point (lower solute concentration) than the intracellular fluid (Jan and Andrabi 2009). Because the chemical potential of ice is less than that of liquid water at a given temperature, the formation of extracellular ice results in a drop in water potential outside the cell. Consequently, there is movement of unfrozen water down the chemical potential gradient from inside the cell to the intercellular spaces. At 10 °C, more than 90% of the osmotically active water typically moves out of the cells, and the osmotic potential of the remaining unfrozen intracellular and intercellular fluid is greater than 5 osmolar. Multiple forms of membrane damage can occur as a consequence of freezing-induced cellular dehydration including expansion-induced lysis, lamellar-to-hexagonal II phase transitions, and fracture jump lesions (Steponkus et al. 1993). Waraich et al. (2012) have pointed out that low-temperature stress during reproductive development induces flower abscission, pollen sterility, pollen tube distortion, ovule abortion, and reduced fruit set, which ultimately lowers yield. A number of approaches are being used to alleviate the effect of temperature stress in crop plants. Proper plant nutrition is one of the good strategies to alleviate the temperature stress and in crop plants. Low-temperature-induced change in membrane fluidity is one of the immediate consequences in plants during low-temperature stress and might represent a potential site of perception and/or injury (Horváth et al. 1998; Orvar et al. 2000). It is well documented that freezing-induced production of reactive oxygen species contributes to membrane damage and that intercellular ice can form adhesions with cell walls and membranes and cause cell rupture (Olien and Smith 1977). There is also an evidence that protein denaturation occurs in plants at low temperature (Guy et al. 1998) which could potentially result in cellular damage. Though the primary site of cold injury and hence electrolyte leakage is the plasma membrane (Bhandari and Nayyar 2014) in all the plants, however, in nature, the effects in response to various stresses may be specific or non-specific, i.e., they may overlap, e.g., effects of cold and draught stress may overlap. To alleviate the harmful effects of cold stress, plants may accumulate various cryoprotectants.

There are two types of the low-temperature stress which affect plant growth and development, namely, chilling (0–10 °C) and freezing, which occurs when the temperature falls below zero (Song et al. 2015). With the effects of global warming and drastic weather changes, the low temperature has become one of the major forms of abiotic stress with enormous effects on plant growth and development (Hasanuzzaman et al. 2013). The low temperature of the field environment restricted the plant's vegetative growth and delayed all the phonological stages in comparison to plants under controlled environment (Barrero-Gil et al. 2016). Apart from this, it led to some vegetative aberrations like chlorosis, necrosis of leaf tips, and curling of whole leaves. Reproductive stage damage involved the abscission of juvenile buds, flower abortion of balls and in extreme conditions that causes plant death (Bolek 2010). Plants have developed several response mechanisms to various forms of abiotic stressors, including chilling and freezing stresses. The most common phenomenon is acclimation, which mainly occurs through enhanced tolerance to environmental stress after priming treatment (Shi et al. 2018). The plant stress response is regulated at the molecular level (Iba 2002) and induces several changes in biochemical pathways and physiological processes of stressed plants (Bartwal et al. 2013).

Generally, plants from temperate regions, including *Arabidopsis,* winter wheat, barley, and oilseed rape, show an increase in freezing tolerance in response to low non-freezing temperatures, a phenomenon known as cold acclimation (Miura and Furumoto 2013). By contrast, some important crops such as rice, maize, tomato, and cotton, which originated from the tropical and subtropical climates, cannot acclimatize to extreme temperatures and are therefore sensitive to chilling stress (Zhao et al. 2015).

Chilling-sensitive plants exposed to low temperatures usually show water-stress symptoms due to decreased root hydraulic conductance and leaf water and turgor potentials (Aroca et al. 2003). Cold stress effects on crop plant have been studied since long time in many economically important crops (Farooq et al. 2009a; Nahar et al. 2009a, 2009b; Wang et al. 2009) among which some of are chilling sensitive and unable to survive cold temperatures. Low temperature affects the plants in every stage of life starting from germination up to maturity. Low-temperature stress inhibits various metabolic reactions, thus preventing the expression of full genetic potential of plants which is expressed by different phenotypic symptoms (Chinnusamy et al. 2007). Some of the common low-temperature injury in plants are reduced seedling growth, seedling discoloration, leaf yellowing, leaf whitening, white specks, white bands, withering after transplanting, a reduced rate of tillering, stunting, and so on (Nahar et al. 2012). According to Angadi et al. (2000), temperatures below 10 °C result in slower and reduced growth and premature stem elongation in *Brassica napus, Brassica rapa*, and *Raphanus sativus*. It is well reported that plants at their seedling stage are very much sensitive to cold stress. At the early stage of plant growth, various phenotypic symptoms in response to chilling stress are surface lesions, chlorosis, necrosis, desiccation, tissue breakdown, and water-soaked appearance of tissues, reduced leaf expansion, and wilting (Solanke and Sharma 2008; Jiang et al. 2002). Nahar et al. (2009b) observed cold injury symptom in rice including stunted growth, yellowing of leaves, abnormal number of tiller, malformed grain, and abnormal colors in grain.

Cold stress, which includes both chilling and freezing injuries, modifies gene expression and plant metabolism with consequent effects on many biological functions (Janmohammadi et al. 2015). There are a disruption of plasmatic membrane (Steponkus 1984), changes of protein synthesis (Duman and Wisniewski 2014) and limited breathing, photosynthesis, and carbon fixation (Liu et al. 2012). Freeze damage is not caused by the actual freeze temperature but by the formation of ice crystals (Beck et al. 2007). The formation of ice crystals in the extracellular space thus causes the drainage of water from the protoplast into the extracellular space and the subsequent dehydration of cells. Although dehydration causes the fading of tissues, it prevents the adverse formation of ice in the protoplasts (Scott 2008). Despite this, the dehydration of protoplasts remains the main cause of freeze damage. The two distinct strategies taken by plants to combat low-temperature stress are avoidance and tolerance. Stress avoidance entails preventing the freezing of sensitive tissues. A more elaborate avoidance strategy involves supercooling, in which endogenous ice nucleation is prevented by inhibiting the formation of ice nucleators (Janská et al. 2010). Balbuena et al. (2011) studied sunflower cold acclimation changes in freezing; susceptible and tolerant lines were investigated using a label-free comparative proteomic approach. They stated that cold-responsive proteins were mostly involved in metabolism, protein synthesis, and energy and defense processes and that tolerant lines have different proteome responses to cold acclimation. Tetreault et al. (2016) examined variation in cold acclimation capacity and freezing tolerance of the perennial sunflower species. Freezing tolerance was the highest in plants from the northernmost latitude under both non-cold-acclimated and cold-acclimated experimental conditions. The plants from all populations retained the ability to increase freezing tolerance through the process of cold acclimation.

Generally, exposure to cold temperature affects crop growth and development in two ways concurrently. First, developmental events in the shoot apex are affected which directly determine the differentiation of the panicle and hence potential yield and spikelet fertility resulting in fewer grains (Hasanuzzaman et al. 2013). Second, photosynthesis is impaired which reduces growth and results in indirect yield loss because there is less carbohydrate available for grain production (Takeoka et al. 1992). The cellular structures and components are also damaged due to the cold which have been studied for a long time. The development process and ontogeny of the organelles may be disrupted by chilling stress (Kratsch and Wise 2000). Nahar et al. (2009a, b) observed cold stress-induced morphological symptoms like stunted plant, bushy plants, early maturity, and yellowing of leaves. Modifications of cellular components include swelling of mitochondria, plastids, and thylakoid lamellae, vesiculation of thylakoid, accumulation of lipid droplets, and eventually the disorganization of the entire plastid (Murphy and Wilson 1981); reduction in ribosomal numbers, dilation of the endoplasmic reticulum, vesiculation of cytoplasmic membranes, condensation of nuclear chromatin, invagination of plasmalemma, and vacuolation of membranous vesicles (Kratsch and Wise 2000; Ishikawa 1996) were noticed during chilling stress. Without these in extreme cases, chilling results in accelerated senescence and eventually plant death (Sharma et al. 2005).

Extreme temperature injuries caused either by cold or by high temperature first attack on the cell membrane. There are many studies where cellular membranes have been shown as the primary site of freezing injury in plants (Steponkus et al. 1998; Thomashow 2001). Cell membrane is damaged in two ways, viz., disruption of protein lipid structure, protein denaturation, and precipitation of solutes that indulges the membrane permeability. Due to low-temperature stress, the fatty acids become unsaturated, and the lipid protein ratios of the membrane become altered which ultimately affect the membrane fluidity and structure as well (Wang and Li 2006). The flexible liquid-crystalline phase is converted into a solid-gel phase, thereby affecting the cellular function in different ways, viz., increased membrane permeability increases ion leakage, allows the entrance of undesirable anions and cations into the cell, obstructs the exchange of essential ions, hampers the osmosis and diffusion processes, etc. All the phenomena are responsible for disrupting cellular homeostasis (Farooq et al. 2009a). Conversion of cellular water into ice is a major reason for cell rupture in cold stress. At first, the ice formation occurs in apoplast having low solute concentration; this creates a vapor pressure between cytoplasm and apoplast and results in the migration of unfrozen cytoplasmic or cytosol water to the apoplast. This water gives a pressure which is the cause of enlargement of existing ice crystals and the pressure toward the cell wall and cell membrane which leads to cell rupture (Mckersie and Bowley 1997; Olien and Smith 1977). Cell membrane plays major roles in water and nutrient movements within and outward the cell. Intra- and extracellular water and nutrient movements are inhibited due to low temperature due to membrane damage. There can be of two types of abnormalities during low-temperature stress. Cold damages the membrane that makes the membrane permeable to undesired nutrients and ions and causes ion leakage; another is cell membrane, and cell wall can be ruptured by the cold which is also responsible for disrupting cellular homeostasis by destroying both intra- and extracellular nutrient and water movements (Salinas 2002; Mahajan and Tuteja 2005). Severe dehydration may also occur due to freezing of cell constituents, solutes, and water (Yadav 2010). Available literature states that when temperatures drop below 0 °C, the ice formation generally begins in the intracellular spaces because the intracellular fluid has a higher freezing point as compared to the other suborganelles of cell (Yadav 2010; Thomashow 1999). Low and freezing temperatures also lead to cellular dehydration and reduce water and nutrient uptake and conduction by the roots in some plants, thus causing osmotic stress (Chinnusamy et al. 2007). Yadav (2010) stated that dehydration during cold occurs mainly due to reduction in water uptake by roots and is a hindrance to closure of stomata. The success or failure of a seedling in the field is strongly related to the development of its root system under cold stress (Enns et al. 2006). In the root of cucumber, it was found that chilling caused injury to the cortical cells and further long-time exposure increased the density of cytoplasm and damage the endoplasmic reticulum (Lee et al. 2002). Chilling-sensitive plants exposed to low temperature usually show water-stress symptoms due to decreased root hydraulic conductance and decreased leaf water and turgor potentials (Aroca et al. 2003). Freezing-induced increase in water viscosity is partly accounted for an initial decrease in root hydraulic conductance (Matzner and Comstock 2001). During cold stress, another phenomenon is common with the imbalanced water movement. Altered metabolic functions are e.g., production of more enzymes, isozymes though they help in maintaining more catalytic activity to cope with the low-temperature stress (Hurry et al. 1994).

Trees at the seedling stage are physiologically and morphologically very sensitive to cold stress which can produce cell damage and reduce growth or even the tree's chances of survival (Yildiz et al. 2014). Therefore, the ability to cope with low temperatures is an important parameter driving the distribution of trees such as conifers in different locations (Bannister and Neuner 2001). At the molecular level, cold stress alters membrane fluidity and consequently membrane permeability. Plants respond to cold stress by inducing physiological and molecular changes, including the plant metabolic profile. These changes may play an advantageous role in cell cryoprotection during cold stress or prior to freezing temperatures (Thomashow 1999). However, seedlings' resistance to cold temperatures and the metabolic phenotype are likely to differ according to the plant genetic structure, the genetic origin, or the provenance (Kreyling et al. 2010), and this difference may affect the plant's ability to adapt to cold stress (Alonso-Blanco et al. 2005). For instance, differential sensitivity to winter cold was reported by Colombo et al. (1995) in Douglas fir (*Pseudotsuga menziesii*) tissues from different seedling provenances. In a similar manner, provenance significantly affected stem and needle hardiness of Scotch pine (*Pinus sylvestris*) (Bannister and Neuner 2001). At the molecular level, there are several descriptions of genes whose role is important in cold tolerance or acclimation (Thomashow 1999) or even genes whose overexpression may increase cold tolerance (Porcel et al. 2018).

Táibi et al. (2018) evaluated several physiological parameters both under control conditions and under cold stress. Under control growth conditions, we only observed minor differences among the tested seed sources. However, cold stress conditions generated several differences in the physiological and molecular responses of *Pinus halepensis* seedlings. Water potential (Ψ_w) of the cold-tolerant seed sources exhibited a significant decrease under the cold stress, as compared to the cold-sensitive ones. In addition, cold-tolerant seed sources exhibited higher stomatal conductance, transpiration, and net photosynthesis. Also, maximal efficiency of PSII, comprised between 0.74 and 0.78 for both treatments, was higher for cold-tolerant seedlings. Even though the quantum yield of non-cyclic electron transport (ΦPSII) was slightly higher in cold-sensitive seedlings under controlled conditions, cold-tolerant seedlings manifested higher values under cold stress. However, most of the differences, although statistically significant, were small. Cold stress is known to decrease membrane fluidity which can induce membrane damage and concomitantly water loss and oxidation, as the loss of fluidity affects the different electronic transport chains and increases the production of reactive oxygen species.

When plants are exposed to any form of stress, the internal equilibrium dynamism rapidly changes, leading to excessive production of reactive oxygen species, which causes extensive damage to various cell structures (Cai et al. 2019). The response of plants to low-temperature stress is a highly complex process involving multiple levels of regulation. When plants are exposed to cold stress, a series of physiological and biochemical alterations at the molecular levels are induced to increase plant cold tolerance and survival under cold stress conditions. These changes include increased accumulation of reactive oxygen species scavenging enzymes, rapid alterations in malondialdehyde content, modifications in lipid compositions such as an increase in ultrastructured fatty acid content which affects cell membrane fluidity, changes in protein and carbohydrate composition, and accumulation of anti-freeze and antioxidative substances such as soluble sugars, proline, superoxide dismutase, peroxidase, and catalase among others (Hussain et al. 2016). After decades of intensive studies, cold-responsive system in plants at the transcriptional level is becoming clear (Wang et al. 2017).

Cotton (*Gossypium* spp.), originally from tropical and subtropical climates, is sensitive to low temperature during whole growth period, but is especially affected at the germination and seedling stage (Shan et al. 2007). The genus *Gossypium* includes 46 diploid ($2n = 2x = 26$) and 6 tetraploid ($2n = 2x = 52$) species, all the diploid cotton species subsequently diversified to produce 8 groups A–G and K (Chakravarthy et al. 2012). All tetraploid cotton species originated through polyploidization between the A-genome and the D-genome species (Renny-Byfield et al. 2016). The A-genome species have experienced a long period of artificial selection for their high-quality fiber, whereas the D-genome species under natural selection maintained abundant genetic diversity and a number of desirable characters, such as fiber quality and resistance to salt, heat, drought, cold insect, and diseases (Kirungu et al. 2018). *Gossypium thurberi*, a wild diploid cotton species, is endemic to the desert sands of the southern part of Arizona in America and has adapted to the local temperature climate. The species is highly tolerant to low temperature and can survive low temperature as low as −6 °C (Liang et al. 1996). Researchers have identified some significant genes related to chilling tolerance in cotton, such as *GhDREB1* (Shan et al. 2007), *GhTIP1* (Li et al. 2009b), GhMT3a (Xue et al. 2009), phospholipase Dα (Kargiotidou et al. 2010), *GbCBF1* (Guo et al. 2011), *GbPATP* (Liu et al. 2015), and *GhTPS11* (Wang et al. 2016c). The overexpression of *DREB/CBF* genes has been found to confer cold tolerance to transgenic plants (Kang et al. 2013). Moreover, several *CBL* genes have been identified in cotton, through blast search of the CBL proteins from various dicotyledonous plants such as reported cotyledons' *CBF* sequences of *Arabidopsis* (Fowler et al. 2005), *Brassica napus* (Jing et al. 2008), and soybean (Li et al. 2005b; Kidokoro et al. 2015), among other plants, in order to maximize the *CBF* genes' discovery. These studies have proven to be useful clues for understanding the mechanism of cold tolerance.

2.5.4 Low-Temperature Tolerance of Woody Plants

Much of the literature on low-temperature tolerance in plants is focused on herbaceous crop species such as cereal grasses, potato, alfalfa, and *Arabidopsis*, with more limited work on woody plants including fruit and ornamental species and some important forest species (e.g., Gusta et al. 2009; Wisniewski and Gusta 2014). These species may be subject to stress, injury, or death due to low-temperature stress when grown under marginal or changing temperature regimes. Studies have yielded considerable insight into the molecular and biophysical mechanisms and functional genomics of tolerance of temperatures as low as −50 °C. However, relatively little attention has been paid to plants that naturally survive some of the lowest temperatures on earth (Strimbeck et al. 2015).

In the taiga forest regions of Siberia and Canada, temperatures range from record lows of around −64 °C to record highs of 36 °C, thereby spanning a full 100 °C. The mean monthly temperatures for December, January, and February are all around -40 °C. Many of the plants and animals in this extreme environment overwinter under the protection of snow or in the soil. In contrast, evergreen pine (*Pinus*), spruce (*Picea*), and fir (*Abies*) species, along with deciduous larch (*Larix*) and a few angiosperm tree and shrub species, remain exposed above the snow and survive the extreme cold, variable light, dry conditions, and high winds of the regions' winters. Similarly, plants in arctic regions and habitats where there is little snow may be exposed to extreme low temperature and other stresses for months at time.

Exposed plants in these harsh environments likely employ mechanisms of low-temperature tolerance that go well beyond those of the well-studied species. An understanding of how these plants survive can contribute to crop improvement and technologies for dry and frozen preservation of foods, drugs, and other biological materials (e.g., Langis and Steponkus 1990; Tada et al. 1990; Siow et al. 2007). Because the phenology of dormancy and low-temperature tolerance may be affected by global warming (Kramer et al. 2000), which is generally expected to be greatest in winter and at higher latitudes, results of this work may also be applied in understanding and predicting the implications of global warming for individual tree species and forest ecosystems, especially in boreal regions.

2.5.5 Approaches for Induction of Low-Temperature Stress Tolerance

The importance of proper membrane fluidity in low-temperature tolerance has been delineated by mutation analysis and transgenic and physiological studies. At low temperature, greater membrane lipid unsaturation appears to be crucial for optimum membrane function. The plants have several mechanisms or approaches to cope with low-temperature stress (Steponkus et al. 1993). Cold acclimation is a key approach to stabilize membranes against freezing injury. It prevents expansion-induced lyses and the formation of hexagonal II phase lipids in rye and other plants (Steponkus et al. 1993). Multiple mechanisms appear to be involved in this stabilization. The best documented are changes in lipid composition (Steponkus et al. 1993). Similarly, the accumulation of sucrose and other simple sugars that typically occurs with cold acclimation also seems likely to contribute to the stabilization of membranes as these molecules can protect membranes against freezing-induced damage in vitro (Strauss and Hauser 1986; Anchordoguy et al. 1987). In addition, there is emerging evidence that certain novel hydrophilic and late embryogenesis abundant proteins also participate in the stabilization of membranes against freezing-induced injury (Epand et al. 1995). These hydrophilic and late embryogenesis abundant proteins are predicted to contain regions capable of forming amphipathic a-helices which are shown to have strong effect on intrinsic curvature of monolayers and their propensity to form hexagonal II phase. They are said to defer their formation at lower temperatures (Epand et al. 1995). Another mechanism through which plants can cope with the low-temperature stress might be the extensive water binding capacity of hydrophilic proteins which provide a protective environment in the proximity of stabilization. Although freezing injury is thought to result primarily from membrane lesions caused by cellular dehydration, additional factors may also contribute to freezing-induced cellular damage (Jan and Andrabi 2009). The enhancement of antioxidative mechanisms (Aroca et al. 2003), increased levels of sugars in the apoplastic space (Livingston and Henson 1998), and the induction of genes encoding molecular chaperones (Guy and Li 1998), respectively, could have protective effects to reduce the freezing-induced cellular damage.

2.5.6 Nutrient Management to Alleviate the Low-Temperature Stresses

Inadequate and unbalanced supply of mineral nutrients and impaired soil fertility are particular problems, causing decreases in global food production, especially in the developing countries. It is estimated that around 60% of cultivated soils have growth-limiting problems associated with mineral-nutrient deficiencies and toxicities (Cakmak 2002). Adequate nutrition is essential for the integrity of plant structure and key physiological processes such as nitrogen and magnesium is structural part of chlorophyll needed for photosynthesis, phosphorus is needed for energy production and storage, is a structural part of nucleic acids, potassium is needed for osmotic regulation and activation of enzymes (Waraich et al. 2011).

A well-nourished plant is expected to produce more biomass per unit of transpired water than a low-nourished plant. Studies have suggested that plant nutrients are not only required for better plant growth and development but also helpful to improve agricultural water-use efficiency (Waraich et al. 2011). A number of reports are available indicating the role of nutrients in alleviating various abiotic stresses, for example, silicon has beneficial effects on increased salinity tolerance in wheat (Tahir et al. 2011) and potassium for increased salinity tolerance (Munns 2005).

2.5.6.1 Nitrogen

Nitrogen plays a very crucial role in temperature stress tolerance. Nitrogen fertilization has been reported to mitigate the adverse effects of abiotic stresses (Waraich et al. 2011). Nitrogen in the form of nitric oxide is a highly reactive, free radical with a broad spectrum of regulatory functions in many physiological processes, such as seed germination, leaf expansion, cell senescence, ethylene emission, stomatal closure, and programmed cell death, and a signal molecular mediating response to abiotic and biotic stresses such as drought stress, salinity, UV-B radiation, and both high- and low-temperature stress (Zhao et al. 2007b; Yang et al. 2006; Crawford and Guo 2005; Zhang et al. 2006). Nitric oxide may protect plant against stress by acting as an antioxidant directly scavenging the reactive oxygen species generated under high- or low-temperature stress (Wendehenne et al. 2001). Some reports revealed that nitric oxide acts as a signal in the inducement of thermotolerance in plant by activating active oxygen scavenging enzymes (Yang et al. 2006; Song et al. 2008). In addition, Uchida et al. (2002) reported that northern blot analysis demonstrated that nitric oxide protected the chloroplast against oxidative damage under heat stress by inducing the expression of gene encoding small heat shock protein 26 (HSP26).

2.5.6.2 Potassium

Potassium plays a crucial role in the survival of crop plants under environmental limitations. Potassium is essential for many physiological processes, such as photosynthesis, translocation of photosynthates into sink organs, maintenance of turgidity, and activation of enzymes under stress conditions (Mengel and Kirkby 2001). Potassium deficiency causes an increase in NADPH-dependent O_2 generation in root cells which indicates that increased reactive oxygen species production during both photosynthetic electron transport and NADPH-oxidizing enzyme reactions may be involved in membrane damage and chlorophyll degradation in potassium-deficient plant (Waraich et al. 2011).

Plants have developed a wide range of adaptive/resistance mechanisms to maintain productivity and ensure survival under a variety of environmental stresses like drought, chilling, frost stresses, and high-temperature stress. Low-temperature stress affects membrane lipid fluidity and thus may alter membrane structure (Marschner 1995). Low temperature also affects photosynthetic electron transport, stomatal conductance, Rubisco activity, and CO_2 fixation in plants due to the conversion of O_2 to reactive oxygen species (Huner et al. 1998; Foyer et al. 2002). Under low supply of potassium, chilling- or frost-induced photooxidative damage can be exacerbated causing more decreases in plant growth and yield. Potassium supply in high amounts can provide protection against oxidative damage caused by chilling or frost. Kafkafi (1990) reported that increasing potassium concentration in irrigation water provided important protection against stem damage from low night temperatures in carnation plants. Similarly, Grewal and Singh (1980) reported that in potato plants, decreases in yield and increases in leaf damage induced by frost under field conditions could be alleviated by high application of potassium fertilizer. Similar results have also been reported by Hakerlerler et al. (1997), who observed that potassium supply enhanced total plant yield by 2.4-fold, 1.9-fold, and 1.7-fold in tomato, pepper, and eggplant, respectively, depending on the source of potassium fertilizer.

2.5.6.3 Calcium

Generally, plant genotypes that tolerate low-temperature stress are able to maintain high leaf water potential by closing their stomata and preventing transpirational water loss (Wilkinson et al. 2001). Calcium has been shown to be an essential requirement for chilling-induced stomatal closure in chilling-tolerant genotypes. Increasing the calcium supply induces stomatal closure, and this effect is most distinct in plants grown at low temperatures. It is also believed that abscisic acid-induced stomatal closure is partially mediated by calcium released from internal guard cell stores or the apoplast (Wilkinson et al. 2001), and this function seems to make calcium a major contributing factor to chilling tolerance and protection of leaves from dehydration. Calcium is considered to play a role in mediating stress response during cold injury, recovery from injury, and acclimation to cold stress (Palta 2000). It has been suggested that calcium is necessary for recovery from low-temperature stress by activating the plasma membrane enzyme ATPase which is required to pump back the nutrients that were lost in cell damage (Palta 2000). Since dehydration is the common denominator, calcium also has a role to play in freeze injury tolerance. Calcium has a very prominent role in the maintenance of cell structure. It activates the plasma membrane enzyme ATPase which pumps back the nutrients lost during cell membrane damage due to calcium deficiency and recovers the plant from cold injury. Calcium also plays a role as calmodulin which controls the plant metabolic activities and enhances the plant growth under low-temperature stress condition (Waraich et al. 2011).

2.5.6.4 Magnesium

Magnesium is involved in numerous physiological and biochemical processes in plants affecting growth and development (Waraich et al. 2011). It plays an essential role in photosynthesis and many other metabolic processes. It is well documented that magnesium plays an important function in the electron transport chain of the chloroplast. Magnesium plays a role to transfer energy from photosystem II to nicotinamide adenine dinucleotide phosphate ($NADP^+$) and protect thylakoid membrane which in turn reduces the accumulation of excitation energy and oxidative damage (Halliwell 1987). Magnesium increased the activities of antioxidative enzymes and the concentration of antioxidant molecules in bean (Cakmak and Marschner 1992; Cakmak 1994), *Mentha pulegium* (Candan and Tarhan 2003), maize (Tewari et al. 2004), pepper (Anza et al. 2005), and mulberry (Tewari et al. 2006). Moreover, magnesium-deficient plants have also shown to accumulate significantly higher amount of malondialdehyde, a general indicator of lipid peroxidation (Candan and Tarhan 2003; Tewari et al. 2004). Magnesium increases the root growth and root surface area which helps to increase uptake of water and nutrients by root. Magnesium being a constituent of chlorophyll increases the amount of sucrose and enhances the transport of sucrose from leaves to roots (Waraich et al. 2011). Magnesium improves carbohydrate translocation by increasing phloem export and reduces reactive oxygen species generation and photooxidative damage to chloroplast under temperature stress (high or low) conditions. Maintenance of chloroplast structure by improving magnesium nutrition enhances the photosynthetic rate under temperature stress which in turn improves the productivity (Waraich et al. 2011).

2.5.6.5 Boron

Low-temperature stress inhibits the growth and development of plants (Xu et al. 2008). High- or low-temperature stress induces the production of reactive oxygen species such as superoxide radical (O_2^-) and hydrogen peroxide (Xu et al. 2008). The accumulation of reactive oxygen species damages membrane lipids and can lead to the death of plant cells (Molassiotis et al. 2006). Boron can increase the antioxidant activities of plants and thereby alleviate reactive oxygen species damage induced by high- or low-temperature stress. Boron nutrition improves sugar transport in the plant body which helps to improve seed germination and seed grain formation. This in turn improves the yield by improving the temperature stress (Waraich et al. 2011). Boron application also improves the carbohydrate metabolism and decreases the phenolic compounds in leaves. This in turn reduces the production of reactive oxygen species and enhances the photosynthetic rate and reduces the cell damage (Waraich et al. 2011).

2.5.6.6 Manganese

Manganese has no direct role in temperature stress alleviation. It can reduce the adverse effects of temperature stress indirectly by enhancing the photosynthetic rate and nitrogen metabolism in the plant body. Manganese nutrition reduces the interveinal chlorosis and brown necrotic spots on leaves and reduces premature leaf drop. Manganese is also reported to be involve in the activation of many enzymes in plant systems, mostly in oxidation-reduction, decarboxylation, and hydrolytic reactions (Marschner 1995); hence, it may play a role in the detoxification of reactive oxygen species. It has been reported that manganese has a crucial role in the diminution of the production of oxygen free radicals and increases the antioxidative compounds and enzymatic activities (Aktas et al. 2005; Turhan et al. 2006; Aloni et al. 2008) under low-temperature stress.

2.5.6.7 Selenium

Selenium deficiency is usually associated with increased lipid peroxidation which alters the integrity of cell membranes and, consequently, affects cell functions (Valko et al. 2005). Studies have shown that selenium at low concentrations can protect plants from several types of abiotic stresses (Hawrylak-Nowak et al. 2010; Valadabadi et al. 2010). High- or low-temperature stress can promote accumulation of reactive oxygen species in the chloroplasts and decrease the antioxidant activity. A decrease in antioxidant enzyme activity is noticed during leaf senescence (Srivalli and Khanna-Chopra 2004). Reports are there on lettuce (*Lactuca sativa* L.) and ryegrass (*Lolium perenne* L.) showing that although selenium is toxic at high concentrations, it can exert beneficial effects on plants at low concentrations. Xue et al. (2001) reported that plants grown under high- or low-temperature stress showed less senescence-related oxidative stress and maintained green leaf color for a longer period when treated with selenium.

2.6 Mechanism of Water Status Regulation

Plant water deficits may occur as a consequence of a seasonal decline in soil water availability, developing in the long term, or may result from drought spells. An increased evaporative demand of the atmosphere, occurring mostly on a daily basis, affects total carbon gain by the crops, even irrigated ones. Genotypes native from climates with marked seasonality are able to acclimate to the fluctuating environmental conditions, enhancing their efficiency for those conditions (Pereira and Chaves 1993, 1995). In the case of slowly developing water deficits, plants may also escape dehydration by shortening their life cycle. In the case of rapid dehydration, oxidative stress developing as a secondary effect is potentially very damaging to the photosynthetic machinery (Ort 2001). The capacity for energy dissipation (Flexas and Medrano 2002) and metabolic protection (induced or constitutive) against the damaging effects of reactive oxygen species (Foyer and Noctor 2003) is a key element for the success of plants under drought. Tissue tolerance to severe dehydration is not common in higher plants, including crops, but does arise in species native from extremely dry environments (Ingram and Bartels 1996). Understanding the mechanisms underlying those different responses can support the design of new management tools and genotypes for modern precision agriculture.

It is well known that a major effect of decreased water availability is diminished leaf carbon fixation due to stomatal closure, which may start at moderate plant water deficits. At the whole plant level, total carbon uptake is further reduced due to the concomitant or even earlier inhibition of growth. It has been shown that cell division and expansion are directly inhibited by water stress (Zhu 2001). Slower growth has been suggested as an adaptive feature for plant survival under stress, because it allows plants to divert assimilates and energy, otherwise used for shoot growth, into protective molecules to fight stress (Zhu 2002) and/or to maintain root growth, improving water acquisition (Chaves et al. 2003). This feature may be relevant for crops intended for drought-prone areas, but inconvenient for regions where only mild and sporadic stress is likely to occur (Chaves and Oiveira 2004).

2.6.1 Morphological and Anatomical Characteristics and Water Control

Every time plants open their stomata to allow the influx of carbon dioxide, they lose water by diffusion. The organization of the mesophyll and leaf dimensions is modified by water restrictions and can provide a strategy for plants to affect the stomatal conductance and CO_2 diffusion (Evans et al. 2009). Plants which need greater efficiency in water use (μmol CO_2 fixed/H_2O transpired) have a strategic demand to retain higher CO_2 concentrations inside. Characteristics of leaf anatomy related to water deficit are reduction of thickness, higher cell density, and smaller intercellular spaces, all of which try to mitigate the problems of excessive water loss (Chartzoulakis et al. 2002). The reduction in cell size and, consequently, the reduction of tissues are associated with the turgor decrease of cells (Ogbonnaya et al. 1998). However, smaller cells may stay more turgid when compared to larger cells, having better capacity to tolerate conditions of water restriction (Burghardt et al. 2008). These changes in cell size result in an increased internal surface to of CO_2 exchange per leaf area, seeking to maintain the photosynthetic rate with reduced stomatal conductance (Ennajeh et al. 2010).

Leaf cuticle presents variations in anatomical position and chemical composition. Chemically, these cuticles are characterized by two specific groups of lipids: (1) cutin, which forms the support structure of the membrane, and (2) waxes deposited on the external surface—called epicuticular waxes—and also strongly dispersed within the matrix of the cutin called cuticular waxes (Devine et al. 1993). The most important function of epicuticular waxes is to avoid leaf water loss to the environment. In addition, they reduce leaf nutrient loss and prevent excessive solar radiation, pathogenic microorganisms, cooling, wind damage, and physical abrasion (Mendgen 1996; Barnes and Cardoso-Vilhena 1996; Kerstiens 1996).

2.6.2 Stomata Metabolism

Stomatal complexes are present in green stems, flowers, and fruits, averaging 30–400 per square millimeter. The stomata liberate into the atmosphere around 95% of the water that exits from the leaves and causes 90% of gas exchange, while only 5% of the water output is through the cuticle. There are two types of stomata morphology, kidney-shaped and grass-like, sized between 5 and 15 μm wide and 20 μm long. The stomata work as a hydraulic valve regulated by water. When turgid, the stomatal pore opens, and when flaccid due to water loss, it closes. In the early morning, light in the blue band (440–490 nm) indicates the arrival of solar radiation and possibility of photosynthetic activity. Thus, the plant opens the stomatal pores to allow influx of carbon dioxide. The first theory about stomatal opening control was called the "starch-sugar hypothesis," and it was widely accepted during the early 1940s. This hypothesis suggested that the hydrolysis of starch in soluble sugars decreased the osmotic potential of cell, promoting water absorption and stomatal opening. However, that theory lost strength as advances were made by studies of potassium movement in the guard cells. Currently, three modes covering the osmoregulation processes of guard cells are accepted: (1) the influx of potassium and chloride through proton pump activation and the synthesis of malate from starch breakdown; (2) sucrose synthesis by starch hydrolysis; and (3) sucrose synthesis by carbon dioxide fixation due to photosynthetic activity.

2.6.3 Redox Signaling

Redox signals are early warnings, exerting control over the energy balance of a leaf. Redox signaling molecules include some key electron carriers, such as the plastoquinone pool, or electron acceptors (e.g., ferredoxin/thioredoxin system) as well as reactive oxygen species (e.g., hydrogen peroxide). The plastoquinone redox state was shown to control gene transcription of photosystem reaction centers of cyanobacteria and chloroplasts (Allen 1993). In particular, a reduced plastoquinone pool activates the transcription of the PSI reaction center, whereas an oxidized pool activates the transcription of the PSII reaction center (Li and Sherman 2000). The intracellular concentrations of reactive oxygen species are controlled by the plant detoxifying system, which includes ascorbate and glutathione pools. Accumulating evidence suggests that these compounds are implicated in redox signal transduction, acting as secondary messengers in hormonal-mediated events (Foyer and Noctor 2003), namely, stomatal movements (Pei et al. 2000).

Hydrogen peroxide and other redox compounds play an important role in the stress perception of the apoplast, which acts as a bridge between the environment and the symplast. It has been observed that hydrogen peroxide is transported from the apoplast to the cytosol through the aquaporins, suggesting that the regulation of signal transduction can also occur via the modulation of transport systems (Pastori and Foyer 2002). Nitric oxide, a reactive nitrogen species, acts as a signaling molecule, in particular by mediating the effects of hormones and other primary signaling molecules in response to environmental stimuli. It may act by increasing cell sensitivity to these molecules (Neill et al. 2003). Nitric oxide was shown to play a role as an intermediate of abscisic acid effects on guard cells (Hetherington 2001; Neill et al. 2003). Nitric oxide may be also involved in stress perception by the apoplast, since this compartment can be a major site of its synthesis. It is also likely that both nitric oxide and hydrogen peroxide are synthesized in parallel and act in a concerted way in a number of physiological responses, including stomatal responses to the environmental stresses. The balance between nitric oxide and hydrogen peroxide also seems to play a role in some critical cellular responses, including programmed cell death (Chaves et al. 2009).

Because nitrite can act as a precursor of nitric oxide, nitrate reductase-dependent nitric oxide production is now receiving much attention. Since the activity of nitrate reductase is highly regulated by the environment (including nitrate supply, light, temperature, CO_2, and cytosolic pH), this may be reflected in nitric oxide production and regulatory functions, such as those exerted on stomatal aperture (Garcia-Mata and Lamattina 2003). It was also suggested that nitric oxide might operate over long distances, acting, for example, as root signal via nitrite coming from the roots to the shoot via the xylem stream. It would then produce nitric oxide in the guard cells. This evidence suggests that besides the role of nitrate reductase in the coordination of carbon to nitrogen metabolism, this enzyme might also participate in the regulation of stomatal response to abscisic acid and other stress factors.

2.6.4 Sugar Signaling

The carbohydrate status of the leaf, which is altered in quantity and quality by water deficits, may act as a metabolic signal in the response to environmental stress (Koch 1996; Jang and Sheen 1997; Chaves et al. 2003). The signaling role of sugars under this context is not totally clear. In general, drought can lead either to increased (under moderate stress) or to constant (under intense stress) concentration of soluble sugars in leaves, in spite of lowered carbon assimilation, because growth and export are also inhibited. Under very severe dehydration, soluble sugars may decrease (Pinheiro et al. 2001). However, starch synthesis is, in general, strongly depressed, even under moderate water deficits (Chaves 1991).

An increase in acid invertase activity was observed in leaves of droughted plants, coinciding with the rapid accumulation of glucose and fructose in maize leaves (Trouverie et al. 2003) and with the accumulation of glucose, fructose, and sucrose, in both leaf blades and petiole of lupins (Pinheiro et al. 2001). The trend of changes observed in sucrose of the leaf petioles is anti-parallel to the changes in leaf blades, suggesting that, under severe stress, leaves are increasing export (Pinheiro et al. 2001). Interestingly, the activity of acid vacuolar invertase was highly correlated with xylem sap abscisic acid concentration (Trouverie et al. 2003). Molecular analysis indicated that abscisic acid is a powerful enhancer of the vacuolar invertase activity and expression (Trouverie et al. 2003). There is also the indication of a direct glucose control of abscisic acid biosynthesis. An increase in the transcription of several genes of abscisic acid synthesis by glucose was observed in *Arabidopsis* seedlings (Cheng et al. 2002). Modulation of the expression of abscisic acid signaling genes by glucose and abscisic acid was also reported. Other evidence indicates that CO_2, light, water, and other environmental signals can be integrated and perceived as sugar signals (Pego et al. 2000), suggesting that different signal types may be perceived by the same receptor or that the signal pathways converge downstream (Ho et al. 2001). On the other hand, sugars travelling in the xylem of droughted plants or sugars that might increase dramatically in the apoplast of guard cells under high light are likely to exert an important influence on stomatal sensitivity to abscisic acid (Wilkinson and Davies 2002).

Crosstalk between the sugar and plant hormone pathways, namely, those of abscisic acid and ethylene (Pego et al. 2000; Leon and Sheen 2003), was also revealed. It was shown, for example, glucose and abscisic acid at high concentrations act in synergy to inhibit growth, whereas at low concentrations they can promote growth. On the other hand, it was demonstrated that the glucose inhibition of growth could be overcome by ethylene, although, in general, this hormone acts as a growth inhibitor (Leon and Sheen 2003). Responses and interactions appear to be dependent both on concentrations and on the particular tissue; an example of the latter is the opposite effect of abscisic acid on the growth of shoot and root (Sharp 2002).

Sugars are also involved in the control of the expression of different genes related to biotic stress and lipid and nitrogen metabolism (Koch 1996; Jang and Sheen 1997). They also affect the expression of genes encoding photosynthesis via a complex and branched pathway. Depletion of sugars triggers an increase in photosynthetic activity, presumably due to a de-repression of sugar controls on transcription, and an accumulation of sugars, due to a lower consumption of photoassimilates, have the opposite effect (Pego et al. 2000).

2.6.5 Control of Oxidative Stress

The coffee crop is confined to the inter-tropical zone, from 20–25°N in Hawaii down to 24°S in Brazil, mainly due to ecological factors related to temperature and humidity (Smith 1989). Both low positive and high temperatures can remarkably affect coffee yields, although this is not always easily perceived. Nonetheless, within the referred zone, the strongest climatic limitations of coffee are frost and drought. Frost easily damages and eventually completely kills the coffee tree, as indeed occurs from time to time either in southern Brazil or, closer to the equator, at altitudes around 2000 m and above. Drought episodes are much more frequent than frosts, so drought is considered the major environmental stress affecting coffee production in most coffee-growing countries. In some marginal regions with no irrigation, coffee yields may decrease as much as 80% in very dry years.

In order to deal with oxidative stress, plants developed a complex and integrated network of protection mechanisms that avoid the production and/or control the action of the produced reactive oxygen species. According to its action, these mechanisms may be divided into those that (DaMatta and Ramalho 2006):

- *Promote the reduction of energy capture*—a highly efficient way of avoiding a chloroplast energy overcharge is reducing energy absorption by leaves. This may be achieved through leaf movements and rolling, production of reflecting surfaces, movement of chloroplasts to the opposite side of the cell to light incidence, decreases in chl content, etc. Such mechanisms may decrease the absorption of incident radiation by up to 40% (Lawlor 2001; Karpinski et al. 2002).
- *Increase the photochemical use or the non-photochemical dissipation ability of excess energy*—improved photochemical use of energy can be achieved through the reinforcement of thylakoid electron transport components, Rubisco, etc., allowing the maintenance of thylakoid reduction state (Xu et al. 1999; Kornyeyev et al. 2001). The processes decrease the production of highly reactive compounds and prevent the over-reduction of the electron transport chain and the over-acidification of the thylakoid lumen, which sensitize PSII to photodamage (Müller et al. 2001; Mano 2002).
- *Control the production and the level of highly reactive compounds*—an overproduction of reactive oxygen species could occur under environmental stress conditions, requiring a complementary line of defense to efficiently remove them. That is the function of this pathway which involves the action of several scavengers, including enzymes (ascorbate peroxidase, monodehydroascorbate reductase, dehydroascorbate reductase, glutathione reductase) and hydrophilic (ascorbate and reduced glutathione) and lipophilic (zeaxanthin and α-tocopherol) molecules (Foyer et al. 1994; Krause 1994; Alscher et al. 1997; Adams III and Barker 1998; Munné-Bosch and Alegre 2002; Perl-Treves and Perl 2002).
- *Promote long-term structural changes*—a consolidated acclimation should also involve long-term structural changes at the membrane level, namely, quantitative and qualitative changes in lipids and protein components. The ability to maintain de novo synthesis, to repair and replace affected cellular components, is of utmost importance. Also, the maintenance of fluidity of membranes is essential for drought and cold tolerance and can be achieved through changes in the balance of lipid classes and fatty acid saturation (Nishida and Murata 1996; Routaboul et al. 2000; Xin and Browse 2000; Iba 2002). These changes in fatty acid unsaturation levels could be achieved by preferential synthesis of palmitic acid (C16:0) and a decrease in linolenic acid (C18:3).

The role of oxidative stress control for the expression of cold acclimation ability could be inferred from the comparison between several coffee genotypes with different cold tolerance, being cv. Apoatã and *Coffea dewevrei* more sensitive and Icatu more tolerant, as already mentioned. Icatu showed a lower increase in membrane selectivity loss and in peroxidative degradation, which could be related to the maintenance of (or increase in) the global fatty acid saturation level of membrane lipids and the reinforcement of the antioxidative system, responsible for maintaining an organized and functional membrane structure less sensitive to reactive oxygen species attack on the double bonds of unsaturated fatty acid (Campos et al. 2003;

Ramalho et al. 2003). On the other hand, it is reasonable to admit that with temperatures close to 0 °C, the antioxidative mechanisms based on enzymes will be severely suppressed in vivo (Geiger et al. 1992; Adams III and Demmig-Adams 1995).

2.6.6 Control on Lipid Changes

Changes in the lipid classes concerning cold tolerance are also important, namely, those in the galactolipids monogalactosyldiacylglycerol (MGDG) and digalactosyldiacylglycerol (DGDG), which are the main lipid components of thylakoids, representing up to about 60% (Harwood 1997; Xin and Browse 2000). Reduction in the ratio between monogalactosyldiacylglycerol and digalactosyldiacylglycerol due to a preferential digalactosyldiacylglycerol synthesis have been related to higher tolerance to cold (Hurry et al. 1998; Iba 2002; Karpinski et al. 2002) and drought (Gigon et al. 2004).

However, DaMatta and Ramalho (2006) working with coffee pointed out that such changes were insufficient to increase cold tolerance, as found in *Coffea dewevrei* and *Coffea canephora* cv. Apoatã and in the hybrid Piatã (*Coffea arabica* x *Coffea dewevrei*) which showed decreases of about 14%, 27%, and 41%, respectively, in the ratio between galactolipids monogalactosyldiacylglycerol and digalactosyldiacylglycerol, but proved to be cold sensitive, and it was suggested that changes in the ratio are not decisive for low-temperature acclimation in coffee (Campos et al. 2003). Moreover, trans-Δ^3-hexadecenoic acid has been implicated in the stabilization/association of light harvesting complex from photosystems with the reaction centers, as well as in the process of replacement of damaged D1 protein (Moon et al. 1995; Harwood 1997; Iba 2002). On this evidence, it was suggested that the fatty acid constitution and the degree of saturation of phosphatidylglycerol in chloroplast membranes play an important role in cold tolerance in coffee. However, it should be kept in mind that an equilibrium should be reached between changes that allow a higher membrane tolerance to peroxidation but will turn the membrane more rigid at low temperatures (higher fatty acid saturation level) and those that promote membrane fluidity but allow reactive oxygen species attack (lower fatty acid saturation level).

2.6.7 Hormonal and Molecular Responses at Different Levels of Water Stress

There are several strategies for the adaptation to dry environments that can be considered as a tool for progress in overcoming limiting conditions on growth and production. Generally, some plants can accumulate water to delay or escape such water-limited condition, while others can deal with the limitations through decreased metabolic activity (Bartels 2005). The effect of water restriction depends on the degree and duration of the stress, the stage of plant development, the genotypic ability of the species, and environmental interactions. Due to climate changes, several studies have looked to understand the biochemical and molecular basis of water stress (Yokota et al. 2006; Xoconostle-Cázares et al. 2010). Water deficit can influence a plant on several levels, with cell expansion and growth being the first processes to respond to water limitation. With the gradual increase of stress, other processes are also affected, such as photosynthesis and allocation of assimilates. At the cellular level, membranes and proteins can be damaged with the increment of reactive oxygen species or peroxidation. These responses are common with other abiotic stresses, such as the effects of salinity and low temperatures, resulting in the synthesis of a similar group of proteins (Artlip and Wisniewski 2002).

When the plant water potential is influenced by the water restriction of soil or saline stress, the stomata must respond quickly to avoid water deficit. Signaling between the root and stomata is carried out by abscisic acid, which has been considered to be a "plant hormone of stress" due to its participation in the signaling networks of other factors of stress (Artlip and Wisniewski 2002). Abscisic acid is synthesized from carotenoid by the synthesizing enzyme of abscisic acid (zeaxanthin epoxidase, 9-cis-epoxycarotenoid dioxygenase, and aldehyde oxidase) which is induced in the root apex or in the parenchyma cells of vascular bundles by water or saline stress. After the synthesis of abscisic acid in the roots, it is transported through the xylem to the leaves alone or conjugated with glucose (the latter being more appropriate for transport over long distances). The proportion of each form of transport (alone or conjugated) is variable between species (Sauter et al. 2002). Once in the leaf, conjugated abscisic acid is hydrolyzed into its free form by the apoplastic enzyme β-D-glucosidase, inducing stomatal closure through a signaling system in the guard cells of chloroplasts (Yokota et al. 2006).

Other studies also point out that the leaves must act as sensors of relative humidity in order to avoid desiccation. This is clear from observations of the extremely quick closure of stomata with increments in the vapor pressure gradient between the leaf and the air, even when there is adequate water availability in the soil (Assmann et al. 2000). In addition, there is genetic evidence that leaf sensitivity to relative humidity is related to abscisic acid metabolism, exerting a hormonal effect over a short distance. Abscisic acid has been related to quantitative and qualitative variations in the gene expression and protein synthesis stimulated by water stress (Artlip and Wisniewski 2002). Meanwhile, it is notable that some of the proteins that are de novo synthesized do not appear in the responses to the application of abscisic acid and these signals/response routes to the water stress have abscisic acid-dependent and abscisic acid-independent routes (Yokota et al. 2006).

Salicylic acid plays a key role in providing tolerance against high- as well as low-temperature stress. Dat et al. (1998) reported that foliar spray of lower concentrations of salicylic acid in mustard increased the H_2O_2 level and also reduced the catalase activity when accompanied with hardening (45 °C for 1 h), thereby increasing the potential of plants to withstand the heat stress. Larkindale and Huang (2004) reported that the pre-treatment with salicylic acid in *Agrostis stolonifera* had no effect on peroxidase activity, whereas the catalase activity declined, compared to control. A similar response was observed in potato plantlets, raised from the cultures, supplemented with lower concentrations of acetyl salicylic acid (Lopez-Delgado et al. 1998). Besides providing tolerance to the plants against high-temperature stress, exogenous application of salicylic acid also induces resistance against the low-temperature stress (chilling or cold stress). An enhanced cold tolerance in maize plants, grown in hydroponic solutions, supplemented with 0.5 mM of salicylic acid was observed by Janda et al. (1997, 1999). Salicylic acid application reduced electrolyte leakage and catalase activity with a concomitant enhancement in the activities of glutathione reductase and guaiacol peroxidase. Salicylic acid may exert deleterious effects on plants under normal growth conditions. Janda et al. (2000) observed a decline in net photosynthetic rate, stomatal conductance, and transpiration rate in maize plants after 1 day of salicylic acid treatment under normal growth conditions. Kang and Saltveit (2002) reported that electrolyte leakage due to low-temperature stress in the leaves of maize, cucumber, and rice plants can be significantly reduced by the application of lower concentrations of salicylic acid. Similar reports also indicate that exogenous salicylic acid application alleviates the damaging effects of low temperatures in rice and wheat (Szalai et al. 2000; Tasgin et al. 2003), bean (Senaratna et al. 2000), and banana (Kang et al. 2003a) by activating various antioxidant enzymes in maize (Janda et al. 1999, 2000) and banana (Kang et al. 2003b).

2.6.8 Osmotic Regulation

Osmotic regulation process of stomatal opening occurs by the movement of solutes—called osmolytes—thereby influencing water movement between the cells. This water movement has several purposes in relation to cell hydration inside plants, such as the stomatal opening and higher water absorption via root system. Several compounds that work in the osmotic regulation of plants are known: carbohydrates (sucrose, sorbitol, mannitol, glycerol, arabinitol, pinitol) (Hare et al. 1998), nitrogen compounds (proteins, betaine, glutamate, aspartate, glycine, choline, putrescine, 4-gamma-aminobutyric acid) (Kinnersley and Turano 2000), organic acids (malate and oxalate), and so on (Sairam and Tyagi 2004). From the standpoint of thermodynamics, there are concepts relating to water potential that are elementary in the explication of water movement. In turgid cells, the water potential (Ψ_w) is composed of pressure (Ψ_ρ) and osmotic ($\Psi\pi$) potentials. The greatest possibility of interference in the Ψ_w of plant cells is by varying the amount of internal solutes. Thus, osmotic regulation is a process in which variation of the amount of solutes—and consequently in osmotic potential—interferes in water movement. The higher the concentration of solutes in a solution, the greater will be the disorder of the system (entropy). Furthermore, this condition leads to more negative osmotic potential and, consequently, to a more negative water potential. The water inside the plant flows from the sites with higher water potentials to those with lower ones. Thus, an increase in solute concentration in the cell favors the entry of water.

Among the solutes, the best-studied is potassium (K^+), the lack of which is related to tissue dehydration. However, most references point out that the main role of potassium within the osmoregulation process is the regulation of stomata opening (Hsiao and Läuchli 1986). It also acts in water absorption by the roots, in transpiratory control, and in the cells' ability to resist low temperatures (Grewal and Singh 1980). In particular, under conditions of potassium deficit, sugars from starch hydrolysis become more relevant in osmotic regulation than in the process of stomatal opening (Poffenroth et al. 1992). The proline amino acid also stands out as an osmotic regulator and is linked to stress by both water deficit and salinity (Molinari et al. 2004; Zhu et al. 2005). Proline is a well-established protective molecule and has been reported to impart protection against cold stress to many plants by not only maintaining its osmolarity but also acting as molecular chaperon, thereby stabilizing the structure and hence functioning of crucial proteins and enzymes (Diehl et al. 2013). Proline also protects the plant by maintaining the double-stranded structure of genetic material, i.e., DNA (Law and Jacobsen 2010), and by upregulating the antioxidative stress machinery (Rejeb et al. 2014). Thus, if exogenously supplied, proline can well mitigate the harmful effects of cold stress as reported by Kumar et al. (2011) in cold-stressed chickpea. Proline-supplemented chickpea plants exhibited improved pollen viability, pollen germination, and pollen tube growth along with better flower and pod retention. Similarly, proline conferred appreciable cold tolerance to rapeseed plants as well (Jonytiene et al. 2012). Nevertheless, proline does not only exert the function of osmoregulation in plant cells during periods of water deficit, but it can also protect against the activity of free radicals, regulate the pH in the cytoplasm, protect against the denaturation of macromolecules, and also act as a source of carbon and nitrogen under conditions of stress (Sharma and Dietz 2006; Sivakumar et al. 2000; Díaz et al. 1999). Other compounds with importance in osmotic regulation are glycinebetaine and trehalose. Glycinebetaine, like proline, does not just exert or exercise its function in osmotic potential. These compounds also act in the reduction of reactive oxygen species produced under conditions of stress in plants. Glycinebetaine is a quaternary ammonium compound and has been proposed to be an effective cryoprotectant as it protects the cold-stressed plant plants from photoinhibition-induced damage to photosynthetic machinery (Giri 2011) by protecting the various components of PSII (Kurepin et al. 2017) and also as a molecular chaperone. Exogenously supply of glycinebetaine improved the performance of tomato plants by Karabudak et al. (2014). Similarly, glycinebetaine-supplemented sweet pepper plants also performed better than untreated one under cold stress (Wang et al. 2016b) (which corroborated similar reports on wheat (Zhang et al. 2010) and other grain crops (Thakur et al. 2010)). Trehalose, an alpha-linked disaccharide, has been proposed to be playing protective roles under various abiotic stresses including cold and freezing (Lordachescu and Imai 2008). There are numerous reports regarding trehalose levels going up under low temperatures rice in *Arabidopsis* (Veyres et al. 2008) and rice (Pan et al. 2012). Thus, exogenously supplied trehalose thus can well protect not only cold-stressed plants as proved in *Arabidopsis* (Hanhong et al. 2005) but also yeast (Hirasawa et al. 2001) and even of *Cryptococcus laurentii* and *Rhodotorula glutinis* (Li et al. 2008). The protective effects of trehalose have been credited to its membrane-stabilizing ability and depression in phase transition temperature of biomembranes and remain amorphous even under completely dehydrated conditions (Sundaramurthi et al. 2010).

With the process of tissue dehydration, there is production of reactive oxygen species, such as the singlet oxygen ($^1O_2$), the superoxide radical (O^{2-}), the hydrogen peroxide (H_2O_2), and the hydroxyl radical (OH^-) (Liu and Huang 2000). Reactive oxygen species are produced from oxygen metabolism and play an important role as indicators in the stress process of water deficit. Among these are oxygen ions, peroxides, and free radicals, and these compounds will cause oxidative stress in cells and prejudice their operation.

2.7 Acclimation to Cold Stress

In temperate or boreal climates, low to subzero temperature is an environmental factor influencing plant life cycle by affecting seed germination, plant growth, and the time for flowering. Whereas low-temperature stress (<10 °C) suppresses the growth rate and leaf expansion, freezing temperatures (<0 °C) lead to a combination of drought stress syndromes, mechanical wounding, and finally plant death (Xin and Browse 2000; Penfield 2008; Guo et al. 2018). Therefore, native plants have developed a strategy, known as cold acclimation (Thomashow 1999), to prepare themselves in low temperatures for upcoming freezing temperatures. This process includes the formation and accumulation of cryoprotectants like soluble sugars (Guy et al. 1992), prolines (Verbruggen et al. 1993; Gilmour et al. 2000), flavonoids or anthocyanin (Schulz et al. 2015, 2016), changes in lipid and protein compositions of cellular membranes (Uemura et al. 1995; Miki et al. 2019), and major changes in the plant transcriptome and proteome (Thomashow 1999; Miki et al. 2019; Hoermiller et al. 2017; Zuther et al. 2019).

In *Arabidopsis*, numerous changes in gene expression accompanying the cold acclimation process are initialized through the C-REPEAT BINDING FACTOR (CBF)-dependent cold pathway. Here, cold-activated CBF transcription factors (CBF1–CBF3) are inducing cold response genes, which are, for example, responsible for the biosynthesis of protectants, and COLD-REGULATED (COR) genes (Gilmour et al. 1998). Well-described examples for COR genes are COR15A and COR15B, which are strongly upregulated in a CBF-dependent manner and which enhance the freezing resistance by stabilizing the chloroplast membranes when constitutively (over)expressed (Thalhammer et al. 2014; Artus et al. 1996). CHALCONE SYNTHASE (CHS) is another cold-inducible gene (Leyva et al. 1995), encoding a key regulatory enzyme of the anthocyanin biosynthetic pathway, but it is not part of the CBF regulon (Fowler and Thomashow 2002; Park et al. 2015, 2018). Given the fact that only ~12% of the cold-regulated genes are regulated by CBFs (Park et al. 2015, 2018), one has to assume that also other transcription factors are of importance for plant cold acclimation. For example, recently the transcription factor genes BZR1 (Li et al. 2017) and ERF105 (Bolt et al. 2017) have been shown to be important for regulating freezing tolerance and cold acclimation in *Arabidopsis*. Several plant hormones are also involved in regulating the response to cold, in particular abscisic acid (Shi and Yang 2014; Lee and Seo 2015) and gibberellin (Achard et al. 2008), but also growth hormones such as cytokinin (Jeon et al. 2010) although its role is debated (Cortleven et al. 2019).

It is important to note that not only long-term cold treatment activates acclimation and an improved response to a second cold stress but also cold treatments as short as 24 h (van Buer et al. 2016). This modification of a future stress by a previous stress characterizes priming (Baier et al. 2019; Conrath et al. 2015). Priming has been described as a resource-saving strategy overcoming the limitations of constitutive cold acclimation (Xin and Browse 2000). Under conditions of variable temperatures, it might be favorable if plants would remember cold stress to be prepared to respond effectively to an eventual future cold stress without investing resources into continued maintenance of the acclimated state. In this way, they may use more of the available resources for growth (Hilker et al. 2016). A variety of mechanisms have been shown to be involved in priming stress responses, ranging from lasting changes in metabolism to modification of transcriptions factors and chromatin changes (Hilker et al. 2016; Conrath et al. 2015). Priming of the response to cold stress has been shown to be regulated by tAPX (van Buer et al. 2016) and to be associated with partly accession-specific changes of transcripts and metabolites (Zuther et al. 2015, 2019; Byun et al. 2014). However, genes important for cold memory have as yet not been described, while several genes are known that play a role in maintaining a primed state in response to elevated temperatures. These include the transcription factor genes HEAT SHOCK FACTOR A2 (HSFA2) (Lämke et al. 2016) and FORGETTER1 (FGT1) (van Buer et al. 2019).

Leuendrof et al. (2020) studied the impact of discontinuous or sustained cold stress (4 °C) on in vitro grown *Arabidopsis thaliana* seedlings of different age and their ability to get primed and respond differently to a later triggering stress. It was observed that cold treatment of 7-day-old seedlings induced the expression of cold response genes but did not cause a significantly enhanced freezing resistance. The competence to increase the freezing resistance in response to cold was associated with the formation of true leaves. Discontinuous exposure to cold only during the night led to a stepwise modest increase in freezing tolerance provided that the intermittent phase at ambient temperature was less than 32 hours. Seedlings exposed to sustained cold treatment developed a higher freezing tolerance which was further increased in response to a triggering stress during 3 days after the priming treatment had ended indicating cold memory. Interestingly, in all scenarios, the primed state was lost as soon as the freezing tolerance had reached the level of naïve plants indicating that an effective memory was associated with an altered physiological state.

Cellular and metabolic changes that occur during cold acclimation include increased levels of sugars, soluble proteins, proline, and organic acids as well as the appearance of new isoforms of proteins and altered lipid membrane composition (Hughes and Dunn 1990). There is an increasing body of evidence to show that many of these biochemical and physiological changes are regulated by low temperature through changes in gene expression and in recent years a number of low-temperature-responsive genes have been cloned from a range of both dicotyledon and monocotyledon species. The ability to acclimate to frost tolerance is under genetic control, and a low temperature of 6 °C to 2 °C, which will cold acclimate genetically competent plants, such as winter cereals, will damage a chill-sensitive species, such as maize. However, not all chill-tolerant plants are able to acclimate for frost tolerance, and many spring cereal cultivars can be classified in this group. In addition to acclimation for frost tolerance, a low positive temperature treatment may also acclimate chill-tolerant plants for growth at low temperatures (Hughes and Pearce 1988). In species and cultivars which will cold acclimate, the low-temperature treatment may also alter the growth habit and vernalize the plants. Further, there is often an interaction between low temperature and other environmental factors, particularly day length and water status (Pan et al. 1994).

References

Abu-Hamdeh NH, Reeder RC (2000) Soil thermal conductivity effects of density, moisture, salt concentration and organic matter. J Am Soil Sci Soc 64(4):1285–1290

Achard P, Gong F, Cheminant S, Alioua M, Hedden P, Genschik P (2008) The cold-inducible CBF1 factor-dependent signaling pathway modulates the accumulation of the growth-repressing DELLA proteins via its effect on gibberellin metabolism. Plant Cell 20(8):2117–2129

Adams WW III, Barker DH (1998) Seasonal changes in xanthophyll cycle-dependent energy dissipation in *Yucca glauca* Nuttall. Plant Cell Environ 21:501–511

Adams WW III, Demmig-Adams B (1995) The xanthophyll cycle and sustained thermal energy dissipation in *Vinca minor* and *Euonymus kiautschovicus* in winter. Plant Cell Environ 18:117–127

Agre P, Sasaki S, Chrispeels MJ (1993) Aquaporins: a family of water channel proteins. Am J Phys 265:F461

Ahmad P, Prasad MNV (2012) Environmental adaptations and stress tolerance of plants in the era of climate change. Springer, New York, Dordrecht, Heidelberg, London

Aktas H, Karni L, Chang DC, Turhan E, Bar-Tal A, Aloni B (2005) The suppression of salinity; associated oxygen radicals production, in pepper *(Capsicum annuum* L.) fruit, by manganese, zinc and calcium in relation to its sensitivity to blossom end rot. Physiol Plant 123:67–74

Alexieva V, Sergiev I, Mapelli S, Karanov E (2001) The effect of drought and ultraviolet radiation on growth and stress markers in pea and wheat. Plant Cell Environ 24:1337–1344

Allen J (1993) Control of gene expression by redox potential and the requirement for chloroplast and mitochondrial genomes. J Theo Biol 165:609–631

Allen DJ, Ort DR (2001) Impacts of chilling temperatures on photosynthesis in warm-climate plants. Trends Plant Sci 6(1):36–42

Allison SO (2005) Cheaters, diffusion and nutrient content decomposition by microbial enzymes in spatially structured environments. Ecol Tell 8(6):626–635

Allison D, Wallenstein MD, Bradford MA (2010) Soil carbon response to warming depends on microbial physiology. Nat Geosci 3:336–340

Aloni B, Karni L, Deventurero G, Turhan E, Aktas H (2008) Changes in ascorbic acid concentration, ascorbate oxidase activity, and apoplastic pH in relation to fruit development in pepper *(Capsicum annuum* L.) and the occurrence of blossom end rot. J Hortic Sci Biotechnol 83:100–105

Alonso-Blanco C, Gomez-Mena C, Llorente F, Koornneef M, Salinas J, Martínez-Zapater JM (2005) Genetic and molecular analyses of natural variation indicate CBF2 as a candidate gene for underlying a freezing tolerance quantitative trait locus in *Arabidopsis*. Plant Physiol 139: 1304–1312

Alscher RG, Donahue JL, Cramer C (1997) Reactive oxygen species and antioxidants: relationships in green cells. Physiol Plant 100:224–233
Alvarez-Uria P, Körner C (2007) Low temperature limits of root growth in deciduous forest and evergreen temperate tree species. Funct Ecol 21(2):211–218
Anchordoguy TJ, Alan SR, Carpenter JF, Crowe JH (1987) Modes of interaction of cryoprotectants with membrane phospholipids during freezing. Cryobiology 24:324–331
Angadi SV, Cutforth HW, Miller PR (2000) Seeding management to reduce temperature stress in *brassica* species. In: Saskatchewan soils and crops proceedings. Semiarid Prairie Agricultural Research Centre, Saskatchewan, Canada
Anza M, Riga P, Garbisu C (2005) Time course of antioxidant responses of *Capsicum annuum* subjected to progressive magnesium deficiency. Ann Appl Biol 146:123–134
Aroca R (2006) Exogenous catalase and ascorbate modify the effects of abscisic acid (ABA) on root hydraulic properties in *Phaseolus vulgaris* L. plants. J Plant Growth Regul 25:10–17
Aroca R, Tognoni F, Irigoyen JJ, Sánchez-Díaz M, Pardossi A (2001) Different root low temperature response of two maize genotypes differing in chilling sensitivity. Plant Physiol Biochem 39: 1067–1073
Aroca R, Vernieri P, Irigoyen JJ, Sánchez-Díaz M, Tognoni F, Pardossi A (2003) Involvement of abscisic acid in leaf and root of maize (*Zea mays* L.) in avoiding chilling-induced water stress. Plant Sci 165:671–679
Aroca R, Amodeo G, Fernández-Illescas S, Herman EM, Chaumont F, Chrispeel MJ (2005) The role of aquaporins and membrane damage in chilling and hydrogen peroxide induced changes in the hydraulic conductance of maize roots. Plant Physiol 137:341–353
Aroca R, Porcel R, Ruiz-Lozano JM (2007) How does arbuscular mycorrhizal symbiosis regulate root hydraulic properties and plasma membrane aquaporins in *Phaseolus vulgaris* under drought, cold or salinity stresses? New Phytol 173:808–816
Aroca R, Alguacil MM, Vernieri P, Ruiz-Lozano JM (2008a) Plant responses to drought stress and exogenous ABA application are modulated differently by mycorrhization in tomato and an ABA-deficient mutant (sitiens). Microb Ecol 56:704–709
Aroca R, Vernieri P, Ruiz-Lozano JM (2008b) Mycorrhizal and non-mycorrhizal *Lactuca sativa* plants exhibit contrasting responses to exogenous ABA during drought stress and recovery. J Exp Bot 59:2029–2041
Aroca R, Porcel R, Ruiz-Lozano JM (2012) Regulation of root water uptake under abiotic stress conditions. J Exp Bot 63(1):43–57
Artlip TS, Wisniewski ME (2002) Induction of proteins in response to biotic and abiotic stresses. In: Pessarakli M (ed) Handbook of plant and crop physiology. Marcel Dekker Inc., New York, pp 657–679
Artus NN, Uemura M, Steponkus PL, Gilmour SJ, Lin C, Thomashow MF (1996) Constitutive expression of the cold-regulated *Arabidopsis thaliana* COR15a gene affects both chloroplast and protoplast freezing tolerance. Proc Natl Acad Sci U S A 93(23):13404–13409
Aryal B, Neuner G (2010) Leaf wettability decreases along an extreme altitudinal gradient. Oecologia 162:1–9
Assmann SM, Snyder JA, Lee YRJ (2000) ABA-deficient (aba1) and ABA-insensitive (abi1-1, abi2-1) mutants of Arabidopsis have a wild-type stomatal response to humidity. Plant Cell Environ 23(4):387–395
Atkinson NJ, Lilley CJ, Urwin PE (2013) Identification of genes involved in the response to simultaneous biotic and abiotic stress. Plant Physiol 162:2028–2041
Baier M, Bittner A, Prescher A, van Buer J (2019) Preparing plants for improved cold tolerance by priming. Plant Cell Environ 42:782–800
Balbuena TS, Salas JJ, Martínez-Force E, Garcés R, Thelen JJ (2011) Proteome analysis of cold acclimation in sunflower. J Proteome Res 10:2330–2346
Balland V, Arp PA (2005) Modeling soil thermal conductivities over a wide range of conditions. J Environ Eng Sci 4(6):549–558

Bannister P, Neuner G (2001) Frost resistance and distribution of conifers. In: Bigras FJ, Colombo SJ (eds) Conifer cold hardiness. Kluwer Academic Publishers, Dordrecht, pp 3–21

Barnes JD, Cardoso-Vilhena J (1996) Interactions between electromagnetic radiation and cuticle. In: Kerstiens G (ed) Plant cuticles: an integrated functional approach. Bios Scientific Publishers, Lancaster, pp 157–174

Barrero-Gil J, Huertas R, Rambla JL, Granell A, Salinas J (2016) Tomato plants increase their tolerance to low temperature in a chilling acclimation process entailing comprehensive transcriptional and metabolic adjustments. Plant Cell Environ 39(10):2303–2318

Bartels D (2005) Desiccation tolerance studied in the resurrection plant *Craterostigma plantagineum*. Integ Compara Biol 45(5):696–701

Bartwal A, Mall R, Lohani P, Guru SK, Arora S (2013) Role of secondary metabolites and brassinosteroids in plant defense against environmental stresses. J Plant Growth Regul 32(1): 216–232

Beck EH, Fettig S, Knake C, Hartig K, Bhattarai T (2007) Specific and unspecific responses of plants to cold and drought stress. J Biosci 32:501–510

Belko N, Zaman-Allah M, Cisse N, Diop NN, Zombre G, Ehlers JD, Vadez V (2012) Lower soil moisture threshold for transpiration decline under water deficit correlates with lower canopy conductance and higher transpiration efficiency in drought-tolerant cowpea. Funct Plant Biol 39:306–322

Benabdellah, K., Ruiz-Lozano, J.M. and Aroca, R. 2009. Hydrogen peroxide effects on root hydraulic properties and plasma membrane aquaporin regulation in *Phaseolus vulgaris*. Plant Mol Biol, 70: 647–661

Bhandari K, Nayyar H (2014) Low temperature stress in plants: an overview of roles of cryoprotectants in defense, vol 1, 3rd edn. Springer, New York, pp 193–265

Bhattacharya A (2019) Water use efficiency under changing climatic condition. In: Changing climate and resource efficiency in plants. Elsevier, Academic Press, USE, USE, pp 111–180

Bian SM, Jiang YW (2009) Reactive oxygen species, antioxidant enzyme activities and gene expression patterns in leaves and roots of Kentucky bluegrass in response to drought stress and recovery. Scientia Hort 120:264–270

Bingzhe F, Fugui M, Ruifang L, Duowen S, Xiangwei H (2010) Effect of covering soil and irrigation on overwintering rate and yield of *Silphium perfoliatum* L. Chin J Grassl 32:106–109

Blokhina OB, Chirkova TV, Fagerstedt KV (2001) Anoxic stress leads to hydrogen peroxide formation in plant cells. J Exp Bot 52:1179–1190

Bloom AJ, Zwieniecki MA, Passioura JB, Randall LB, Holbrook NM, St Clair DA (2004) Water relations under root chilling in a sensitive and tolerant tomato species. Plant Cell Environ 27: 971–979

Blum A (2005) Drought resistance, water-use efficiency, and yield potential—are they compatible, dissonant, or mutually exclusive? Aust J Agric Res 56:1159–1168

Blum A (2009) Effective use of water (EUW) and not water-use efficiency (WUE) is the target of crop yield improvement under drought stress. Field Crop Res 112:119–123

Boese SR, Wolfe DW, Melkonian JJ (1997) Elevated CO_2 mitigates chilling-induced water stress and photosynthetic reduction during chilling. Plant Cell Environ 20(5):625–632

Bogdanović J, Mojović M, Milosavić N, Mitrović A, Vucinić Z, Spasojević I (2008) Role of fructose in the adaptation of plants to cold-induced oxidative stress. Eur Biophys J 37(7): 1241–1246

Bolek Y (2010) Genetic variability among cotton genotypes for cold tolerance. Field Crops Res 119(1):59–67

Bolger TP, Upchurch DR, McMichael BL (1992) Temperature effects on cotton root hydraulic conductance. J Environ Exp Bot 32(1):49–54

Bolt S, Zuther E, Zintl S, Hincha DK, Schmülling T (2017) ERF105 is a transcription factor gene of *Arabidopsis thaliana* required for freezing tolerance and cold acclimation. Plant Cell Environ 40:108–120

Bouchabke-Coussa O, Quashie ML, Seoane-Redongo J, Fortabat M-N, Gery C, Yu A, Linderme D, Trouverie J, Granier F, Téoulé E, Durand-Tardif M (2008) *ESKIMO1* is a key gene involved in water economy as well as cold acclimation and salt tolerance. BMC Plant Biol 8:125

Bouman BAM, Humphreys E, Tuong TP, Barker R (2007) Rice and water. Adv Agron 92:187–237

Boursiac Y, Boudet J, Postaire O, Luu DT, Tournaire-Roux C, Maurel C (2008) Stimulus-induced downregulation of root water transport involves reactive oxygen species-activated cell signaling and plasma membrane internalization. Plant J 56:207–218

Boyce RL, Lucero S (2002) Role of roots in winter water relations of Engelmann spruce saplings. Tree Physiol 19(13):893–898

Boyce RL, Friendland AJ, Webb ET, Herrick GT (1991) Modeling the effect of winter climate on high elevation red spruce shoot water contents. For Sci 37(6):1567–1580

Boyer JS (1996) Advances in drought tolerance in plants. Adv Agron 56:187–218

Bramley H, Turner NC, Turner DW, Tyerman SD (2009) Roles of morphology, anatomy and aquaporins in determining contrasting hydraulic behavior of roots. Plant Physiol 150:348–364

Bronick CJ, Lal R (2005) Soil structure and management: a review. Geoderma 124(1–2):3–22

Buchanan BB, Gruissem W, Jones RL (2015) Biochemistry & molecular biology of plants, 2nd edn. Wiley Blackwell, West Sussex, UK

Buchas GD (2001) Soil temperature regimes. In: Smith KA et al (eds) Soil and environmental analysis: physical methods. Marcel Dekkar, New York, pp 539–594

Burghardt M, Burghardt A, Gall J, Rosenberger C, Riederer M (2008) Ecophysiological adaptations of water relations of *Teucrium chamaedrys* L. to the hot and dry climate of xeric limestone sites in Franconia (*southern Germany*). Flora 203:3–13

Byun YJ, Koo MY, Joo HJ, Ha-Lee YM, Lee DH (2014) Comparative analysis of gene expression under cold acclimation, deacclimation and reacclimation in *Arabidopsis*. Physiol Plant 152: 256–274

Cai X, Magwanga RO, Xu Y, Zhou Z, Wang X, Hou Y, Wang Y, Zhang Y, Liu F, Wang K (2019) Comparative transcriptome, physiological and biochemical analyses reveal response mechanism mediated by *CBF4* and *ICE2* in enhancing cold stress tolerance in *Gossypium thurberi*. AoB Plants 11(6):plz045

Cakmak I (1994) Activity of ascorbate-dependent H_2O_2-scavenging enzymes and leaf chlorosis are enhanced in magnesium and potassium-deficient leaves, but not in phosphorus-deficient leaves. J Exp Bot 45:1259–1266

Cakmak I (2002) Plant nutrition research: priorities to meet human needs for food in sustainable ways. Plant Soil 247:3–24

Cakmak I, Marschner H (1992) Magnesium deficiency and high light intensity enhance activities of superoxide dismutase, ascorbate peroxidase and glutathione reductase in bean leaves. Plant Physiol 98:1222–1227

Campos PS, Quartin V, Ramalho JC, Nunes MA (2003) Electrolyte leakage and lipid degradation account for cold sensitivity in leaves of *Coffea* sp. plants. J Plant Physiol 160:283–292

Candan N, Tarhan L (2003) Relationship among chlorophyll-carotenoid content, antioxidant enzyme activities and lipid peroxidation levels by Mg^{2+} deficiency in the *Mentha pulegium* leaves. Plant Physiol Biochem 41:35–40

Carvajal M, Cooke DT, Clarkson DT (1996) Responses of wheat plants to nutrient deprivation may involve the regulation of water-channel function. Planta 199:372–381

Cavender-Bares J (2005) Impacts of freezing on long distance transport in woody plants. In: Holbrook NM, Zweiniecki MA (eds) Vascular transport in plants. Elsevier, Boston, pp 401–424

Chaerle L, Saibo N, van der Straeten D (2005) Tuning to pores: towards engineering plants for improved water use efficiency. Trends Biotech 23:308–315

Chakravarthy VSK, Reddy TP, Reddy VD, Rao KV (2012) Current status of genetic engineering in cotton (*Gossypium hirsutum* L): an assessment. Crit Rev Biotechnol 8551:1–18

Chartzoulakis K, Patakasb A, Kofidisc G, Bosabalidisc A, Nastoub A (2002) Water stress affects leaf anatomy, gas exchange, water relations and growth of two avocado cultivars. Sci Hortic 95: 39–50

Chavarria G, dos Santos HP (2012) Plant water relations: absorption, transport and control mechanism. In: Montanaro G (ed) Advances in selected plant physiology aspects. InTech, Croatia, pp 105–132

Chaves MM (1991) Effects of water deficits on carbon assimilation. J Exp Bot 42(1):1–16

Chaves MM, Oiveira MM (2004) Mechanisms underlying plant resilience to water deficits: prospects for water-saving agriculture. J Exp Bot 55(407):2365–2384

Chaves MM, Maroco JP, Pereira JS (2003) Understanding plant response to drought: from genes to the whole plant. Funct Plant Biol 30:239–264

Chaves MM, Flexas J, Pinheiro C (2009) Photosynthesis under drought and salt stress: regulation mechanisms from whole plant to cell. Ann Bot 103(4):551–560

Chen PM, Gusta LV (1978) The role of water in cold hardiness of winter cereals. In: Li PH, Sakai A (eds) Plant cold hardiness and freezing stress: mechanisms and crop implications. Academic Press, New York, pp 165–173

Cheng W-H, Endo A, Zhou L, Penney J, Chen H-C, Arroyo A, Leon P, Nambara E, Asami T, Seo M, Koshiba T, Sheen J (2002) A unique short-chain dehydrogenase / reductase in Arabidopsis glucose signalling and abscisic acid biosynthesis and functions. Plant Cell 14(11):2732–2743

Chiemeka IU (2010) Soil temperature profile of Uturu, Nigeria. Pacific J Sci Tech 11(1):478–482

Chimenti CA, Marcantonio M, Hall AJ (2006) Divergent selection for osmotic adjustment results in improved drought tolerance in maize (*Zea mays* L.) in both early growth and flowering phases. Field Crops Res 95:305–315

Chinnusamy V, Zhu J, Zhu JK (2007) Cold stress regulation of gene expression in plants. Trends Plant Sci 12:444–451

Christmann A, Weiler EW, Steudle E, Grill E (2007) A hydraulic signal in root-to-shoot signalling of water shortage. Plant J 52:167–174

Clarkson, D.T. 1976. The influence of temperature on the exudation of xylem sap from detached root systems of rye (*Secale cereale*) and barley (*Hordeum vulgare*). Planta, 132(3): 297–304

Cochard H (2002) A technique for measuring xylem hydraulic conductance under high negative pressure. Plant Cell Environ 25(6):815–819

Colombo SJ, Zhao S, Blumwald E (1995) Frost hardiness gradients in shoots and roots of *Picea mariana* seedlings. Scand J For Res 10:32–36

Conant RT, Drijber RA, Haddix MC (2008) Sensitivity of organic matter decomposition to warming varies with its quality. Glob Chang Biol 14(4):868–877

Condon AG, Farquhr GD, Richards RA (1990) Genetic variation in carbon isotope discrimination and transpiration efficiency in wheat. Leaf gas exchange and whole plant studies. Ayst J Plant Physiol 17:9–22

Condon AG, Richards RA, Rebetzke GJ, Farquar GD (2004) Breeding for high water-use efficiency. J Exp Bot 55:2447–2460

Conrath U, Beckers GJ, Langenbach CJ, Jaskiewicz MR (2015) Priming for enhanced defense. Annu Rev Phytopathol 53:97–119

Conyers, M., Newton, P., Condon, J.R, Poile, G., Mele, P. and Ash, G.J. 2012. Long term trials end with a quasi–equilibrium between C, N, and pH and implication for C sequestration. Soil Res 50(7): 527–535

Cortleven A, Leuendorf JE, Frank M, Pezzetta D, Bolt S, Schmülling T (2019) Cytokinin action in response to abiotic and biotic stresses in plants. Plant Cell Environ 42(3):998–1018

Costa JM, Ortudo MF, Chaves MM (2007) Deficit irrigation as a strategy to save water physiology and potential application to horticulture. J Int Plant Biol 49(10):1421–1434

Couée I, Sulmon C, Gouesbet G, El Amrani A (2006) Involvement of soluble sugars in reactive oxygen species balance and responses to oxidative stress in plants. J Exp Bot 57(3):449–459

Cowling JE, Kedrowski RA (1980) Winter water relations of native and introduced evergreens in interior Alaska, USA. Can J Bot 58(1):94–99

Crawford NM, Guo FQ (2005) New insights into nitric oxide metabolism and regulatory functions. Trends Plant Sci 10:195–200

Cruz R, Milach S (2004) Cold tolerance at the germination stage of rice: methods of evaluation and characterization of genotypes. Sci Agric 61:1–8

Dalla-Salda G, Martínez-Meier A, Cochard H, Rozenberg P (2009) Variation of wood density and hydraulic properties of Douglas-fir (*Pseudotsuga menziesii* (Mirb.) Franco) clones related to a heat and drought wave in France. Forest Ecol Manage 257:182–189

DaMatta FM, Ramalho JDC (2006) Impacts of drought and temperature stress on coffee physiology and production: a review. Braz J Plant Physiol 18(1):55–81

Daniels MJ, Yeager M (2005) Phosphorylation of aquaporin PvTIP3;1 defined by mass spectrometry and molecular modeling. Biochemistry 44:14443–14454

Daniels MJ, Chaumont F, Mirkov TE, Chrispeels MJ (1996) Characterization of a new vacuolar membrane aquaporin sensitive to mercury at a unique site. Plant Cell 8:587–599

Dat JF, Lopez-Delgado H, Foyer CH, Scott IM (1998) Parallel changes in H_2O_2 and seedlings. Plant Physiol 116:1351–1357

Deane C (1994) Whole plant and tissue culture studies on frost tolerance in cauliflower. Ph. D. Thesis. The National University of Ireland

Decker KT, Wang D, Waite C, Scherbatskoy T (2003) Snow removal and ambient air temperature in northern Vermont. J Am Soil Sci Soc 67(4):4234–1242

DeLucia EH, Heckathorn SA, Day TA (1992) Effects of soil temperature on growth, biomass allocation and resource acquisition of *Andropogon gerardii* Vitman. New Phytol 120(4): 543–549

Demidchik V (2015) Mechanisms of oxidative stress in plants: from classical chemistry to cell biology. Environ Exp Bot 109:212–228

Deng B, Du W, Liu C, Sun W, Tian S, Dong H (2012) Antioxidant response to drought, cold and nutrient stress in two ploidy levels of tobacco plants: low resource requirement confers polytolerance in polyploids? Plant Growth Regul 66:37–47

Devacht S, Lootens P, Roldán-Ruiz I, Carlier L, Baert J, Van Waes J, Van Bockstaele E (2009) Influence of low temperatures on the growth and photosynthetic activity of industrial chicory *Cichorium intybus* L. partim. Photosynthetica 47(3):372–380

Devi MJ, Sinclair TR, Vadez V (2010) Genotypic variation in peanut for transpiration response to vapor pressure deficit. Crop Sci 50:191–196

Devine M, Duke OS, Fedtke C (1993) Physiology of herbicide action. P.T.R Prentice–Hall, Englewood Cliffs, NJ

Díaz P, Borsani O, Monza J (1999) Acumulación de prolina en plantas en respuesta al estrés osmotico. Agrociencia 3(1):1–10

Dichio B, Xiloyannis C, Sofo A, Montanaro G (2006) Osmotic regulation in leaves and roots of olive tree (*Olea europaea* L.) during water deficit and rewatering. Tree Physiol 26:179–185

Diehl RC, Guinn EJ, Capp MW, Tsodikov OV, Record MT Jr (2013) Quantifying additive interactions of the osmolyte proline with individual functional groups of proteins: comparisons with urea and glycine betaine, interpretation of m-values. Biochemistry 52(35):5997–6010

Dobra J, Motyka V, Dobrev P, Malbeck J, Prasil IT, Haisel D, Gaudinova A, Havlova M, Gubis J, Vankova R (2010) Comparison of hormonal responses to heat, drought and combined stress in tobacco plants with elevated proline contents. J Plant Physiol 167:1360–1370

Dudal R (1976) Inventory of major soils of the world with special reference to mineral stress. In: Wright MJ (ed) Plant adaption to mineral stress in problem soils. Cornell Univ. Agricultural Experiment Station, Ithaca, New York, pp 3–23

Duman JG, Wisniewski MJ (2014) The use of antifreeze proteins for frost protection in sensitive crop plants. Environ Exp Bot 106:60–69

Dwivedi P, Raghvendra AS (2004) Modulation of osmatic stress effects on photosynthesis and respiration by temperature in mesophyll protoplast of pea. Ind J Exp Biol 42:1208–1211

Elias EA, Cichota R, Torraiaini HH, van Lier QD (2004) Analytical and soil temperature model: correction for temporal variation of daily amplitude. Am Soil Sci Soc J 68(3):784–788

Elizaberashivili ES, Urshadze TF, Elizaberashivili M, Elizaberashivili SE, Schafer M (2010) Temperature regimes of some types in Georgia. Eurasian Soil Sci 43(4):427–435

Ennajeh M, Vadel AM, Cochard H, Khemir H (2010) Comparative impacts of water stress on the leaf anatomy of a drought-resistant and a drought-sensitive olive cultivar. J Hort Sci Biotech 85(4):289–294

Enns LC, Mccully ME, Canny MJ (2006) Branch roots of young maize seedlings, their production, growth, and phloem supply from the primary root. Funct Plant Biol 33:391–399

Epand RM, Shai Y, Segrest JP, Anantharamaiah G, M. (1995) Mechanisms for the modulation of membrane bilayer properties by amphipathic helical peptides. Biopolymers 37:319–338

Evans JR, Kaldenhoff R, Genty B, Terashima I (2009) Resistances along the CO_2 diffusion pathway inside leaves. J Exp Bot 60:2235–2248

Farooq M, Aziz T, Wahid A, Lee DJ, Siddique KHM (2009a) Chilling tolerance in maize: agronomic and physiological approaches. Crop Pasture Sci 60:501–516

Farooq M, Wahid A, Kobayashi N, Fujita D, Basra SMA (2009b) Plant drought stress: effects, mechanisms and management. Agron Sustain Dev 29:185–212

Fernie AR, Geigenberger P, Stitt M (2005) Flux an important, but neglected, component of functional genomics. Curr Opin Plant Biol 8:174–182

Fischer RA, Turner NC (1978) Plant productivity in arid and semiarid zones. Rev Plant Physiol 29: 277–317

Fitter A, Hay R (2002) Environmental physiology of plants. Academic Press, San Diego, CA

Fletcher AL, Sinclair TR, Allen LH (2007) Transpiration responses to vapor pressure deficit in well watered 'slow-wilting' and commercial soybean. Environ Exp Bot 61:145–151

Flexas J, Medrano H (2002) Energy dissipation in C_3 plants under drought. Funct Plant Biol. 29: 1209–1215

Fowler S, Thomashow MF (2002) Arabidopsis transcriptome profiling indicates that multiple regulatory pathways are activated during cold acclimation in addition to the CBF cold response pathway. Plant Cell 14:1675–1690

Fowler SG, Cook D, Thomashow MF (2005) Low temperature induction of *Arabidopsis* CBF1 is gated by the circadian clock. Plant Physiol 137:961–968

Foyer CH, Noctor G (2003) Redox sensing and signalling associated with reactive oxygen in chloroplasts, peroxisomes and mitochondria. Physiol Plant 119:355–364

Foyer CH, Lelandais M, Kunert KJ (1994) Photooxidative stress in plants. Physiol Plant 92:696–717

Foyer CH, Vanacker H, Gomez LD, Harbinson J (2002) Regulation of photosynthesis and antioxidant metabolism in maize leaves at optimal and chilling temperatures: review. Plant Physiol Biochem 40:659–668

Fritz M, Ehwald R (2011) Mannitol permeation and radial flow of water in maize roots. New Phytol 189:210–217

Fu P, Wilen RW, Wu GH, Robertson AJ, Gusta L (1999) Dehydrin gene expression and leaf water potential differs between spring and winter cereals during cold acclimation. J Plant Physiol 156(3):394–400

Garcia-Mata C, Lamattina L (2003) Abscisic acid, nitric oxide and stomatal closure: is nitrate reductase one of the missing links? Trends Plant Sci 8:20–26

Geiger DR, Servaites JC, Shieh W-J (1992) Balance in the source-sink system: a factor in crop productivity. In: Baker NR, Thomas H (eds) Crop photosynthesis: spatial and temporal determinants. Elsevier Science Publishers, Amsterdam, pp 155–176

Geiger R, Aron RN, Todhunter P (2003) The climate near the ground. Rownaan and Little Field Publishers, Inc., Lanham, pp 42–50

Gholipoor M, Prasad PVV, Mutava RN, Sinclair TR (2010) Genetic variability of transpiration response to vapor pressure deficit among sorghum genotypes. Field Crops Res. 119:85–90

Gigon A, Matos AR, Laffray D, Zuily-Fodil Y, Pham-Thi AT (2004) Effect of drought stress on lipid metabolism in the leaves of *Arabidopsis thaliana* (ecotype Columbia). Ann Bot 94:345–351

Gilmour SJ, Zarka DJ, Stockinger EJ, Salazar MP, Houghton JM, Thomashow MF (1998) Low temperature regulation of the Arabidopsis CBF family of AP2 transcriptional activators as an early step in cold-induced COR gene expression. Plant J 16:433–442

Gilmour SJ, Sebolt AM, Salazar MP, Everard JD, Thomashow MF (2000) Overexpression of the Arabidopsis CBF3 transcriptional activator mimics multiple biochemical changes associated with cold acclimation. Plant Physiol 124:1854–1865

Giri J (2011) Glycine betaine and abiotic stress tolerance in plants. Plant Signal Behav 6:1746–1751

Göbel L, Coners H, Hertel D, Willinghöfer S, Leuschner C (2019) The role of low soil temperature for photosynthesis and stomatal conductance of three graminoids from different elevations. Front Plant Sci 10:330

Gowda VRP, Henry A, Yamauchi A, Shashidhar HE, Serraj R (2011) Root biology and genetic improvement for drought avoidance in rice. Field Crops Res 122:1–3

Grewal JS, Singh SN (1980) Effect of potassium nutrition on frost damage and yield of potato plants on alluvial soils of the Punjab (India). Plant Soil 57:105–110

Grossi M, Giorni E, Rizza F, Stanca AM, Cattivelli L (1998) Wild and cultivated barleys show differences in the expression pattern of a cold-regulated gene family under different light and temperature conditions. Plant Mol Biol 38(6):1061–1069

Guo HM, Li ZC, Zhang H, Xin YZ, Cheng HM (2011) Cloning of cotton CBF gene for cold tolerance and its expression in transgenic tobacco. Acta Agron Sinica 37:286–293

Guo X, Liu D, Chong K (2018) Cold signaling in plants: insights into mechanisms and regulation. J Integr Plant Biol 60:745–756

Gusta LV, Wisniewski M, Tanino KK (2009) In: plant cold hardiness: from the laboratory to the field. Centre for Biosciences and Agriculture International, Oxfordshire

Guy CL (1990) Cold acclimation and freezing stress tolerance: role of protein metabolism. Annu Rev Plant Biol 41:187–223

Guy C, Li QB (1998) The organization and evolution of the spinach stress 70 molecular chaperone gene family. Plant Cell 10:539–556

Guy CL, Huber JL, Huber SC (1992) Sucrose phosphate synthase and sucrose accumulation at low temperature. Plant Physiol 100:502–508

Guy C, Haskell D, Li QB (1998) Association of proteins with the stress 70 molecular chaperones at low temperature: evidence for the existence of cold labile proteins in spinach. Cryobiology 36: 301–314

Hacker J, Ladinig U, Wagner J, Neuner G (2011) Inflorescences of alpine cushion plants freeze autonomously and may survive subzero temperatures by supercooling. Plant Sci 180:149–156

Hakerlerler H, Oktay M, Eryüce N, Yagmur B (1997) Effect of potassium sources on the chilling tolerance of some vegetable seedlings grown in hotbeds. In: Johnston AE (ed) Food security in the WANA region, the essential need for balanced fertilization. International Potash Institute, Basel, pp 317–327

Haldimann P (1999) How do changes in temperature during growth affect leaf pigment composition and photosynthesis in *Zea mays* genotypes differing in sensitivity to low temperature? J Exp Bot 50(333):543–550

Halliwell B (1987) Oxidative damage, lipid peroxidation and antioxidant protection in chloroplasts. Chem Phys Lipids 44:327–340

Hanhong B, Eliot H, Richard S (2005) Exogenous trehalose promotes non-structural carbohydrate accumulation and induces chemical detoxification and stress response proteins in *Arabidopsis thaliana* grown in liquid culture. Plant Sci 168:1293–1301

Hare PD, Cress WA, Staden JV (1998) Dissecting the roles of osmolyte accumulation during stress. Plant Cell Environ 21:535–553

Harwood JL (1997) Plant lipid metabolism. In: Dey PM, Harborne JB (eds) Plant biochemistry. Academic Press, San Diego, pp 237–271

Hasanuzzaman, M., Nahar, K. and Fukita, M. 2013. Extreme temperature responses, oxidative stress and antioxidant defense in plants. In: K. Vahdati, and C. Leslie, (ed.).,Abiotic stress—

plant responses and applications in agriculture, chapter no. 6. INTECH, DOI: https://doi.org/10.5772/54833
Hattori T, Sonobe K, Araki H, Inanaga S, An P, Morita S (2008) Silicon application by sorghum through the alleviation of stress induced increase in hydraulic resistance. J Plant Nutr 31:1482–1495
Hawrylak-Nowak B, Matraszek R, Szymanska M (2010) Selenium modifies the effect of short-term chilling stress on cucumber plants. Biol Trace Elem Res 138(1–3):307–315
Hetherington AM (2001) Guard cell signaling. Cell 107:711–714
Hilker M, Schwachtje J, Baier M, Balazadeh S, Bäurle I, Geiselhardt S, Hincha DK, Kunze R, Mueller-Roeber B, Rillig MC, Rolff J, Romeis T, Schmülling T, Steppuhn A, van Dongen J, Whitcomb SJ, Wurst S, Zuther E, Kopka J (2016) Priming and memory of stress responses in organisms lacking a nervous system. Biol Rev Camb Philos Soc 91(4):1118–1133
Hincha DK, Zuther E, Heyer AG (2003) The preservation of liposomes by raffinose family oligosaccharides during drying is mediated by effects on fusion and lipid phase transitions. Biochim Biophys Acta Biomembr 1612(2):172–177
Hirasawa R, Yokoigawa K, Isobe Y, Kawai H (2001) Improving the freeze tolerance of baker's yeast by loading with trehalose. Biosci Biotechnol Biochem 65:522–526
Ho SL, Chao YC, Tong WF, Yu SM (2001) Sugar co-ordinately and differentially regulates growth- and stress-related gene expression via a complex signal transduction network and multiple control mechanisms. Plant Physiol 125:877–890
Hoermiller II, Naegela T, Augustin H, Weckwerth W, Hever AG (2017) Subcellular reprogramming of metabolism during cold acclimation in *Arabidopsis thaliana*. Plant Cell Environ 40(5):602–610
Hopkins WG (1999) The physiology of plants under stress. In: Introduction to plant physiology, 2nd edn. Wiley, New York, pp 451–475
Horváth I, Glatz A, Varvasovszki V, Torok Z, Pali T, Balogh G, Kovacs E, Nadasdi L, Benko S, Joo F, Vigh L (1998) Membrane physical state controls the signaling mechanism of the heat shock response in Synechocystis PCC 6803: identification of hsp17 as a 'fluidity gene'. Proc Natl Acad Sci U S A 95(7):3513–3518
Hsiao T (1973) Plant responses to water stress. Ann Rev Plant Physiol 24:519–570
Hsiao TC, Läuchli A (1986) Role of potassium in plant–water relations. In: Tinker B, Läuchli A (eds) Advances in plant nutrition, vol 2. Praeger Scientific, New York, pp 281–311
Hsu SY, Hsu YT, Kao CH (2003) The effect of polyethylene glycol on proline accumulation in rice leaves. Biol Plant 46:73–78
Huang B, Johnson JW, Box JE, NeSmith DS (1997) Root characteristics and hormone activity of wheat in response to hypoxia and ethylene. Crop Sci 37:812–818
Hughes MA, Dunn MA (1990) The effect of temperature on plant growth and development. Biotechnol Genet Eng Rev 8:161–188
Hughes MA, Dunn MA (1996) The molecular biology of plant acclimation to low temperature. J Exp Bot 47(296):291–305
Hughes MA, Pearce RS (1988) Low temperature treatment of barley plants causes altered gene expression in shoot meristems. J Exp Bot 39:1461–1467
Huner NPA, Öquist G, Hurry VM, Krol M, Falk S, Griffith M (1993) Photosynthesis, photoinhibition and low temperature acclimation in cold tolerant plants. Photosynth Res 37(1):19–39
Huner NPA, Oquist G, Sarhan F (1998) Energy balance and acclimation to light and cold. Trends Plant Sci 3:224–230
Hurry VM, Malmberg G, Gardestorm P, Oquist G (1994) Effects of a short term shift to low temperature and of long term cold hardening on photosynthesis and ribulose-1, 5-bisphosphate carboxylase/oxygenase and sucrose phosphate activity in leaves of winter rye (*Secale cereale* L.). Plant Physiol 106:983–990

Hurry V, Huner N, Selstam E, Gardeström P, Öquist G (1998) Photosynthesis at low growth temperatures. In: Raghavendra AS (ed) Photosynthesis. A comprehensive treatise. Cambridge University Press, Cambridge, pp 238–249

Hussain S, Khan F, Cao W, Wu L, Geng M (2016) Seed priming alters the production and detoxification of reactive oxygen intermediates in rice seedlings grown under sub-optimal temperature and nutrient supply. Front Plant Sci 7:439

Hussain HA, Hussain S, Khaliq A, Ashraf U, Anjum SA, Men S, Wang L (2018) Chilling and drought stresses in crop plants: implications, cross talk, and potential management opportunities. Front Plant Sci 9:393

Huxman TE, Monson RK (2003) Stomatal responses of C3, C3–C4 and C4 *Flaveria* species to light and intercellular CO_2 concentration: implications for the evolution of stomatal behavior. Plant Cell Environ 26:313–322

Iba K (2002) Acclimative response to temperature stress in higher plants: approaches of gene engineering for temperature tolerance. Annu Rev Plant Biol 53(1):225–245

Ingram J, Bartels D (1996) The molecular basis of dehydration tolerance in plants. Ann Rev Plant Physiol Plant Mol Biol 47:377–403

Irvine J, Grace J (1997) Continuous measurements of water tensions in the xylem of trees based on elastic properties of wood. Planta 202(4):455–461

Ishikawa HA (1996) Ultrastructural features of chilling injury: injured cells and the early events during chilling of suspension cultured mung bean cells. Am J Bot 83:825–835

Islam MA, MacDonald SE, Zwiazek JJ (2003) Responses of black spruce (*Picea mariana*) and tamarack (*Larix laricina*) to flooding and ethylene. Tree Physiol 23:545–552

Jackson MB, Saker LR, Crisp CM, Else MA, Janowiak F (2003) Ionic and pH signaling from roots to shoots of flooded tomato plants in relation to stomatal closure. Plant Soil 253:103–113

Jan NHM, Andrabi KI (2009) Cold resistance in plants: a mystery unresolved. Electr J Biotech 12: 1–15

Janda T, Szalai G, Tari I, Paldi E (1997) Exogenous salicylic acid has an effect on chilling symptoms in maize (*Zea mays* L.) plants. In: Sowinski P, Zagdanska B, Aniol A, Klaus P (eds) Crop development for cool and wet European climate. ECSP-EEC-EAEC, Brussels, Belgium, pp 179–187

Janda T, Szalai G, Tari I, Paldi E (1999) Hydroponic treatment with salicylic acid decreases the effect of chilling injury in maize (*Zea mays* L.) plants. Planta 208:175–180

Janda T, Szalai S, Antunovics ZS, Horvath E, Paldi E (2000) Effect of benzoic acid and aspirin on chilling tolerance and photosynthesis in young maize plants. Maydica 45:29–33

Jang JC, Sheen J (1997) Sugar sensing in higher plants. Trends Plant Sci 2:208–214

Jang JY, Kim DG, Kim YO, Kim JS, Kang HS (2004) An expression analysis of a gene family encoding plasma membrane aquaporins in response to abiotic stresses in *Arabidopsis thaliana*. Plant Mol Biol 54:713–725

Janmohammadi M, Zolla L, Rinalducci S (2015) Low temperature tolerance in plants changes at the protein level. Phytochemistry 117:76–89

Janowiak F, Markowski A (1994) Changes in leaf water relations and injuries in maize seedlings induced by different chilling conditions. J Agron Crop Sci 172:19–28

Janská A, Maršík P, Zelenková S, Ovesná J (2010) Cold stress and acclimation—what is important for metabolic adjustment? Plant Biol 12:395–405

Javier PDJ, José IJ, Manuel SD (1997) Chilling of drought-hardened and non-hardened plants of different chilling-sensitive maize lines changes in water relations and aba contents. Plant Sci 122:71–79

Javot H, Maurel C (2002) The role of aquaporins in root water uptake. Ann Bot 90:301–313

Javot H, Lauvergeat V, Santoni V, Martin-Laurent F, Güclü J, Vinh J, Heyes J, Franck KI, Schäffner AR, Bouchez D, Maurel C (2003) Role of a single aquaporin isoform in root water uptake. Plant Cell 15:509–522

Jensen RD, Taylor SA (1961) Effect of temperature on water transport through plants. Plant Physiol 36(5):639–642

Jeon J, Kim NY, Kim S, Kang NY, Novák O, Ku S-J, Cho C, Lee DJ, Lee E-J, Strand M, Kim J (2010) A subset of cytokinin two-component signaling system plays a role in cold temperature stress response in Arabidopsis. J Biol Chem 285:23371–23386

Jia WS, Wang YQ, Zhang SQ, Zhang JH (2002) Salt-stress-induced ABA accumulation is more sensitively triggered in roots than in shoots. J Exp Bot 53:2201–2206

Jiang QW, Kiyoharu O, Ryozo I (2002) Two novel mitogen-activated protein signaling components, OsMEK1 and OsMAP1, are involved in a moderate low-temperature signaling pathway in rice. Plant Physiol 129:1880–1891

Jing Z, Bo Z, XiaoFen J, RiHe P, YuShan Q, Zhen Z (2008) Cloning and analysis of CBF transcription factors from *Brassica napus* cv. Huyou 15. Plant Physiol Commun 44:1106–1112

Johnson FH, Eyring H, Polissar MJ (1954) The kinetic basis of molecular biology. John Wiley and Sons, New York

Jonytiene, V., Burbulis, N., Kupriene, R. and Blinstrubiene, A. 2012. Effect of exogenous proline and de-acclimation treatment on cold tolerance in *Brassica napus* shoots cultured *in-vitro*. J Food Agric Environ, 10(1): 327–330

Jouve L, Engelmann F, Noirot M, Charrier A (1993) Evaluation of biochemical markers (sugar, proline, malondialdehyde and ethylene) for cold sensitivity in microcuttings of two coffee species. Plant Sci 91:109–116

Kacperska A (2004) Sensor types in signal transduction pathways in plant cells responding to abiotic stressors: do they depend on stress intensity? Physiol Plant 122:159–168

Kafkafi U (1990) Impact of potassium in relieving plants from climatic and soil-induced stresses. In: Johnston AE (ed) Food security in the WANA region, the essential need for balanced fertilization. International Potash Institute, Basel, pp 317–327

Kaiser C, Meyer H, Biasi C, Rusalimova O, Barsukov P, Richter A (2007) Conservation of soil organic matter through cryoturbation in arctic soils in Siberia. J Appl Soil Ecoli 112(G2): 173–183

Kaldenhoff R, Grote K, Zhu JJ, Zimmermann U (1998) Significance of plasmalemma aquaporins for water transport in *Arabidopsis thaliana*. Plant J 14:121–128

Kamaluddin M, Zwiazek JJ (2002) Ethylene enhances water transport in hypoxic aspen. Plant Physiol 128:962–969

Kang HM, Saltveit ME (2002) Chilling tolerance of maize, cucumber and rice seedling leaves and roots are differentially affected by salicylic acid. Physiol Plant 115:571–576

Kang GZ, Wang CH, Sun GC, Wang ZX (2003a) Salicylic acid changes activities of H_2O_2-metabolizing enzymes and increases the chilling tolerance of banana seedlings. Environ Exp Bot 50:9–15

Kang GZ, Wang ZX, Sun GC (2003b) Participation of H_2O_2 in enhancement of cold chilling by salicylic acid in banana seedlings. Acta Bot Sin 45:567–573

Kang J, Zhang H, Sun T, Shi Y, Wang J, Zhang B, Wang Z, Zhou Y, Gu H (2013) Natural variation of C-repeat-binding factor (CBFs) genes is a major cause of divergence in freezing tolerance among a group of *Arabidopsis thaliana* populations along the Yangtze River in China. New Phytol 199:1069–1080

Karabudak T, Bor M, Özdemir F, Türkan I (2014) Glycine betaine protects tomato (*Solanum lycopersicum*) plants at low temperature by inducing fatty acid desaturase7 and lipoxygenase gene expression. Mol Biol Rep 41(3):1401–1410

Kargiotidou A, Kappas I, Tsaftaris A, Galanopoulou D, Farmaki T (2010) Cold acclimation and low temperature resistance in cotton: *Gossypium hirsutum* phospholipase dalpha isoforms are differentially regulated by temperature and light. J Exp Bot 61:2991–3002

Karpinski S, Wingsle G, Karpinska B, Hällgren J-E (2002) Low-temperature stress and antioxidant defense mechanisms in higher plants. In: Inze D, van Montagu M (eds) Oxidative stress in plants. Taylor & Francis, London, pp 69–103

Kato Y, Okami M (2011) Root morphology, hydraulic conductivity and plant water relations of high-yielding rice grown under aerobic conditions. Ann Bot 108(3):575–583

Kato T, Tang Y, Gu S, Hirota M, Du M, Li Y, Zhao X (2006) Temperature and biomass influences on interannual changes in CO_2 exchange in an alpine meadow on the Qinghai-Tibetan plateau. Glob Chang Biol 12(7):1285–1298

Katsuhara M, Otsuka T, Ezaki B (2005) Salt stress-induced lipid peroxidation is reduced by glutathione S-transferase, but this reduction of lipid peroxides is not enough for recovery of growth in Arabidopsis. Plant Sci 169:369–373

Keane RE, Mahalovich MF, Bollenbacher BL, Manning ME, Loehman RA, Jain TB, Holsinger LM, Larson AJ (2018) 2018. In: Halofsky JE, Peterson DL (eds) Effects of climate change on forest vegetation in the northern Rockies, vol 63. Springer International Publishing, Cham, Switzerland, pp 59–95

Kenefick DG, Koepke JA, Sutton F (2002) Plant water uptake by hard red winter wheat (*Triticum aestivum* L.) genotypes at 2°C and low light intensity. BMC Plant Biol 2:8.1

Kerstiens G (1996) Diffusion of water vapour and gases across cuticles and through stomatal pores presumed closed. In: Kerstiens G (ed) Plant cuticles: an integrated functional approach. Bios Scientific Publishers, Lancaster, pp 121–134

Khanna-Chopra R, Singh S (2011) Approaches to increase water use efficiency in horticultural and grain crops—an overview. Plant Stress 5(Special Issue 1):52–63

Kholova J, Hash CT, Kumar PL, Yadav RS, Kocova M, Vadez V (2010) Terminal drought-tolerant pearl millet *Pennisetum glaucum* (L.) R. Br. Have high leaf ABA and limit transpiration at high vapour pressure deficit. J Exp Bot 61:1431–1440

Kiani SP, Talia P, Maury P, Grieu P, Heinz R, Perrault A, Nishinakamasu V, Hopp E, Gentzbittel L, Paniego N, Sarrafi A (2007) Genetic analysis of plant water status and osmotic adjustment in recombinant inbred lines of sunflower under two water treatments. Plant Sci 172(4):773–787

Kidokoro S, Watanabe K, Ohori T, Moriwaki T, Maruyama K, Mizoi J, Htwe NMPS, Fujita Y, Sekita S, Shinozaki K, Yamaguchi-Shinozaki K (2015) Soybean DREB1/CBF-type transcription factors function in heat and drought as well as cold stress-responsive gene expression. Plant J 81(3):505–518

Kinnersley AM, Turano FJ (2000) Gamma aminobutyric acid (GABA) and plant response to stress. Crit Rev Plant Sci 19:479–509

Kirkham MB (2005) Principles of soil and plant water relations. Elsevier Academic Press, San Diego, CA

Kirungu JN, Deng Y, Cai X, Magwanga RO, Zhou Z, Wang X, Wang Y, Zhang Z, Wang K, Liu F (2018) Simple sequence repeat (SSR) genetic linkage map of D genome diploid cotton derived from an interspecific cross between *Gossypium davidsonii* and *Gossypium klotzschianum*. Int J Mol Sci 19(1):E204

Knaupp M, Mishra KB, Nedbal L, Heyer AG (2011) Evidence for a role of raffinose in stabilizing photosystem II during freeze – thaw cycles. Planta 234(3):477–486

Knipfer T, Fricke W (2010) Root pressure and solute reflection coefficient close to unity exclude a purely apoplastic pathway of radial water transport in barley (*Hordeum vulgare*). New Phytol 187:159–170

Knipfer T, Fricke W (2011) Water uptake by seminal and adventitious roots in relation to whole-plant water flow in barley (*Hordeum vulgare* L.). J Exp Bot 62:717–733

Kocacinar F, McKown AD, Sage TL, Sage RF (2008) Photosynthetic pathway influences xylem structure and function in *Flaveria* (Asteraceae). Plant Cell Environ. 31:1363–1376

Koch KE (1996) Carbohydrate-modulated gene expression in plants. Ann Rev Plant Physiol Plant Mol Biol 47:509–540

Kornyeyev D, Logan BA, Payton P, Allen RD, Holaday AS (2001) Enhanced photochemical light utilization and decreased chilling-induced photoinhibition of photosystem II in cotton overexpressing genes encoding chloroplast-targeted antioxidant enzymes. Physiol Plant 113: 323–331

Kosová K, Vitamvas P, Prasil IT (2007) The role of dehydrins in plant response to cold. Biol Plant 51:601–617

Koster KL, Lynch DV (1992) Solute accumulation and compartmentation during the cold acclimation of *puma* rye. Plant Physiol 98(1):108–113

Kotak S, Larkindale J, Lee U, von Koskull-Doring P, Vierling E, Scharf KD (2007) Complexity of the heat stress response in plants. Curr Opin Plant Biol 10:310–316

Kramer PJ (1940) Root resistance as a cause of decreased water absorption by plants at low temperatures. Plant Physiol 15(1):63–79

Kramer PJ (1942) Species difference with respect to water absorption at low soil temperatures. Am J Bot 29(10):828–832

Kramer PJ (1969) Plant and soil water relations. In: A modern synthesis. McGraw Hill, New York, NY

Kramer PJ, Boyer JS (1995) In water relations of plants and soils. Academic Press, San Diego

Kramer K, Leinonen I, Loustau D (2000) The importance of phenology for the evaluation of impact of climate change on growth of boreal, temperate and Mediterranean forests ecosystems: an overview. Int J Biometeorol 44:67–75

Kratsch HA, Wise RR (2000) The ultrastructure of chilling stress. Plant Cell Environ. 23:337–350

Krause GH (1994) Photoinhibition induced by low temperatures. In: Baker NR, Bowyer JR (eds) Photoinhibition of photosynthesis—from molecular mechanisms to the field. Bios Scientific Publishers, Oxford, pp 331–348

Kreyling J, Wiesenberg GLB, Thiel D, Wohlfart C, Huber G, Walter J, Jentsch A, Konnert M, Beierkuhnlein C (2010) Cold hardiness of *Pinus nigra* Arnold as influenced by geographic origin, warming, and extreme summer drought. Env Exp Bot 78:99–108

Kubien DS, von Caemmerer S, Furbank RT, Sage RF (2003) C4 photosynthesis at low temperature. A study using transgenic plants with reduced amounts of rubisco. Plant Physiol 132:1577–1585

Kumar S, Malik J, Thakur P, Kaistha S, Sharma KD, Upadhyaya HD, Berger JD, Nayyar H (2011) Growth and metabolic responses of contrasting chickpea (*Cicer arietinum*) genotypes to chilling stress at reproductive phase. Acta Physiol Plant 33:779–787

Kumar R, Chaurasiya PC, Singh RN, Singh S (2018) A review report: low temperature stress for crop production. Int J Pure App Biosci 6(2):575–598

Kurepin LV, Ivanov A, Zaman M, Pharis RP, Hurry V, Hüner NP (2017) Interaction of glycine betaine and plant hormones: protection of the photosynthetic apparatus during abiotic stress. Springer International Publishing, Cham, pp 185–202

Kuwagata T, Hamasaki T, Watanabe T (2008) Modeling water temperature in a rice paddy for agro-environmental research. Agri Forest Meteorol 148(11):1754–1766

Lämke J, Brzezinka K, Altmann S, Bäurle I (2016) A hit-and-run heat shock factor governs sustained histone methylation and transcriptional stress memory. EMBO J 35:162–175

Langis R, Steponkus P (1990) Cryopreservation of rye protoplasts by vitrification. Plant Physiol 92: 666–671

Larcher W (2004) Ecofisiologia vegetal. RiMa, São Carlos

Larkindale J, Huang B (2004) Thermotolerance and antioxidant systems in *Agrostis stolonifera:* involvement of salicylic acid, abscisic acid, calcium, hydrogen peroxide, and ethylene. J Plant Physiol 161:405–413

Larkindale J, Hall JD, Knight MR, Vierling E (2005) Heat stress phenotypes of *Arabidopsis* mutants implicate multiple signaling pathways in the acquisition of thermotolerance. Plant Physiol 138:882–888

Law JA, Jacobsen SE (2010) Establishing, maintaining and modifying DNA methylation patterns in plants and animals. Nat Rev Genet 11(3):204–220

Lawlor DW (2001) Photosynthesis, 3rd edn. Bios Scientific Publishers, Oxford

Leakey ADB, Ferguson JN, Pignon CP, Wu A, Jin Z, Hammer GL, Lobel DB (2019) Water use efficiency as a constraint and target for improving the resilience and productivity of C3 and C4 crops. Ann Rev Plant Biol 70:781–808

Lee HG, Seo PJ (2015) The MYB96-HHP module integrates cold and abscisic acid signaling to activate the CBF-COR pathway in *Arabidopsis*. Plant J 82:962–977

Lee TM, Lur HS, Chu C (1993) Role of abscisic acid in chilling tolerance of rice (*Oryza sativa* L.) seedlings. Endogenous abscisic acid levels. Plant Cell Environ 16:481–490

Lee, H., Guo, Y., Ohta, M., Xiong, L., Stevenson, B. and Zhu, J.K. 2002. *LOS2*, a genetic locus required for cold responsive transcription encodes a bi-functional enolase. EMBO J 21: 2692–2702

Lee SH, Singh AP, Chung GC (2004a) Rapid accumulation of hydrogen peroxide in cucumber roots due to exposure to low temperature appears to mediate decreases in water transport. J Exp Bot 55:1733–1741

Lee SH, Singh AP, Chung GC, Ahn SJ, Noh EK, Steudle E (2004b) Exposure of roots of cucumber (*Cucumis sativus*) to low temperature severely reduces root pressure, hydraulic conductivity and active transport of nutrients. Physiol Plant 120(3):413–420

Lee SH, Ahn SJ, Im YJ, Cho K, Chung GH, Cho BH, Han O (2005a) Differential impact of low temperature on fatty acid unsaturation and lipoxygenase activity in figleaf gourd and cucumber roots. Biochem Biophys Res Comm 330:1194–1198

Lee SH, Chung GC, Steudle E (2005b) Gating of aquaporins by low temperature in roots of chilling-sensitive cucumber and chilling tolerant figleaf gourd. J Exp Bot 56:985–995

Leon P, Sheen J (2003) Sugar and hormone connections. Trends Plant Sci 8:110–116

Leuendrof JE, Frank M, Schmülling T (2020) Acclimation, priming and memory in the response of *Arabidopsis thaliana* seedlings to cold stress. Sci Rep 10:689

Levitt J (1980) responses of plants to environmental stress, Chilling freezing, and high temperature stresses, vol 1, 2nd edn. Academic Press

Leyva A, Jarillo JA, Salinas J, Martinez-Zapater JM (1995) Low temperature induces the accumulation of phenylalanine ammonia-lyase and chalcone synthase mRNAs of *Arabidopsis thaliana* in a light-dependent manner. Plant Physiol 108:39–46

Li H, Sherman LA (2000) A redox-responsive regulator of photosynthesis gene expression in the cyanobacterium *Synechocystis* sp. strain PCC 6803. J Bacteriol 182:4268–4277

Li Y, Xu H, Cohen S (2005a) Long-term hydraulic acclimation to soil texture and radiation load in cotton. Plant Cell Environ 28:492–499

Li XP, Tian AG, Luo GZ, Gong ZZ, Zhang JS, Chen SY (2005b) Soybean DRE-binding transcription factors that are responsive to abiotic stresses. Theo App Genet 110:1355–1362

Li BQ, Zhou ZW, Tian SP (2008) Combined effects of endo- and exogenous trehalose on stress tolerance and biocontrol efficacy of two antagonistic yeasts. Biol Control 46:187–193

Li DD, Wu YJ, Ruan XM, Li B, Zhu L, Wang H, Li XB (2009a) Expressions of three cotton genes encoding the PIP proteins are regulated in root development and in response to stresses. Plant Cell Rep 28:291–300

Li DD, Tai FJ, Zhang ZT, Li Y, Zheng Y, Wu YF, Li XB (2009b) A cotton gene encodes a tonoplast aquaporin that is involved in cell tolerance to cold stress. Gene 438:26–32

Li H, Ye K, Shi Y, Cheng J, Yang S (2017) BZR1 positively regulates freezing tolerance via CBF-dependent and CBF-independent pathways in Arabidopsis. Mol Plant 10(4):545–559

Liang WS (2003) Drought stress increases both cyanogenesis and beta-cyanoalanine synthase activity in tobacco. Plant Sci 165:1109–1115

Liang Z, Jiang R, Zhong W, Zhao G, Li A, Feng H, Li Z, Sui S (1996) Studies on the trispecific hybrid (*Gossypium barbadense* - *G. thurberi* - *G. hirsutum*) and new variety breeding. Acta Agron Sinica 22:673–680

Lichtenthaler HK (2006) The stress concept in plants: an introduction. Ann N Y Acad Sci 851:187–198

Limin AE, Fowler DB (1985) Cold-hardiness of sequential winter wheat tissue segments to differing temperature regimes. Crop Sci 25(5):838–843

Liu X, Huang B (2000) Heat stress injury in relation to membrane lipid peroxidation in creeping bentgrass. Crop Sci 40(1):503–510

Liu YF, Qi MF, Li TL (2012) Photosynthesis, photoinhibition, and antioxidant system in tomato leaves stressed by low night temperature and their subsequent recovery. Plant Sci 196:8–17

Liu T, Guo S, Lian Z, Chen F, Yang Y, Chen T, Ling X, Liu A, Wang R, Zhang B (2015) A P4-ATPase gene GbPATP of cotton confers chilling tolerance in plants. Plant Cell Physiol. 56: 549–557

Livingston DP, Henson CA (1998) Apoplastic sugars, fructans, fructan exohydrolase, and invertase in winter oat: responses to second-phase cold hardening. Plant Physiol 116:403–408

Lopez-Delgado H, Dat JF, Foyer CH, Scott IM (1998) Induction of thermotolerance in potato microplants by acetylsalicylic acid and H_2O_2. J Exp Bot 49:713–720

López-Pérez L, Martínez-Ballesta MC, Maurel C, Carvajal M (2009) Changes in plasma membrane lipids, aquaporins and proton pump of broccoli roots, as an adaptation mechanisms to salinity. Phytochemistry 70:492–500

Lordachescu M, Imai R (2008) Trehalose biosynthesis in response to abiotic stress. J Integr Plant Biol 50(10):1223–1229

Lu S, Ren T, Gong Y, Horton R (2009) Am improved model for predicting soil thermal conductivity from water content at room temperature. Am J Soil Sci Soc 71(1):8–14

Lynch DV (1990) Chilling injury in plants: the relevance of membrane lipids. In: Katterman F (ed) Environmental injury to plants. Academic Press, New York, pp 17–34

Lyons JM (1973) Chilling injury in plants. Annu Rev Plant Physiol 24:445–466

Maggio A, Joly RJ (1995) Effects of mercury-chloride on the hydraulic conductivity of tomato root systems. Evidence for a channel mediated water pathway. Plant Physiol 109:331–335

Mahajan S, Tuteja N (2005) Cold, salinity and drought stresses: an overview. Arch Biochem Biophys 444:139–158

Mahdieh M, Mostajeran A (2009) Abscisic acid regulates root hydraulic conductance via aquaporin expression modulation in Nicotiana tabacum. J Plant Physiol 166:1993–2003

Mahmood, R. 2002. *In vitro* effect of salt on the vigor of potato (*Solanum tuberosum* L.) plantlets. Int J Biotechnol 1: 73–77

Mai J, Herbette S, Vandame M, Kositsup B, Kasemsap P, Eric C, Julien JL, Améglio T, Roeckel-Drevet P (2009) Effect of chilling on photosynthesis and antioxidant enzymes in *Hevea brasiliensis* Muell. Arg Trees 23(4):863–874

Malone M (1993) Rapid inhibition of leaf growth by root cooling in wheat: kinetics and mechanism. J Environ Exp Bot 44(268):1663–1669

Mano J (2002) Early events in environmental stresses in plants - induction mechanisms of oxidative stress. In: Inzé D, van Montagu M (eds) Oxidative stress in plants. Taylor Francis, London, pp 217–245

Markhart AH, Fiscuc EL, Naylor AW, Kramer PJ (1979) Effect of temperature on water and ion transport in soybean and broccoli systems. Plant Physiol 64(1):83–87

Marschner H (1995) Mineral nutrition of higher plants, 2nd edn. Academic Press, New York

Martias AD, Musil S (2012) Temperature and thermal diffusivity within a range of land soil near Oragle, Arizona. J Arizona-Nevada Acad Sci 44(1):15–21

Martin JH (1927) Comparative studies of winter hardiness in wheat. J Agric Res 35:493–535

Matsumoto T, Lian HL, Su WA, Tanaka D, Liu CW, Iwasaki I, Kitagawa Y (2009) Role of the aquaporin PIP1 subfamily in the chilling tolerance of rice. Plant and Cell Physiol 50:216–229

Matsuo N, Ozawa K, Mochizuki T (2009) Genotypic differences in root hydraulic conductance of rice (*Oryza sativa* L.) in response to water regimes. Plant Soil 316:25–34

Matsuoka M, Furbank RT, Fukayama H, Miao M (2001) Molecular engineering of C_4. Ann Rev Plant Physiol Plant Mol Biol 52:297–314

Matsushima S, Tanaka T, Hoshino T (1964) Analysis of yield determining process and its application to yield-prediction and culture improvement of lowland rice. LXXI. Combined effects of air-temperatures and water-temperatures at different stage of growth on the growth and morphological characteristics of rice plants. Proc Crop Sci Soc Jpn 33:135–140

Matzner S, Comstock J (2001) The temperature dependence of shoot hydraulic resistance implications for stomatal behaviour and hydraulic limitation. Plant Cell Environ 24:1299–1307

Maurel C, Reizer J, Schroeder JI, Chrispeels MJ (1993) The vacuolar membrane protein gamma-TIP creates water-specific channels in *Xenopus* oocytes. EMBO J 12:2241–2247

Maurel C, Kado RT, Guern J, Chrispeels MJ (1995) Phosphorylation regulates the water channel activity of the seed specific aquaporin alpha-TIP. EMBO J 14:3028–3035
Maurel C, Verdoucq L, Luu DT, Santoni V (2008) Plant aquaporins: membrane channels with multiple integrated functions. Ann Rev Plant Biol 59:595–624
McCully ME (1999) Roots in soil: unearthing the complexities of roots and their rhizospheres. Ann Rev Plant Physiol Plant Mol Biol 50:695–718
McElrone AJ, Choat B, Gambetta GA, Brodersen CR (2013) Water uptake and transport in vascular plants. Nat Educ Knowl 4(5):6
Mckersie BD, Bowley SR (1997) Active oxygen and freezing tolerance in transgenic plants. In: Li PH, Chen THH (eds) Plant cold hardiness, molecular biology, biochemistry and physiology. Plenum Press, New York, pp 203–214
McKhann H, Gery C, Berard A, Leveque S, Zuther E, Hincha D, DeMita S, Brunel D, Teoule E (2008) Natural variation in CBF gene sequence, gene expression and freezing tolerance in the Versailles core collection of *Arabidopsis thaliana*. BMC Plant Biol 8:105
McMichael BC, Burke JJ (1998) Soil temperature and root growth. Hort Sci 33(6):947–951
McWilliam JR, Kramer PJ, Musser RL (1982) Temperature-induced water stress in chilling-sensitive plants. Aust J Plant Physiol 9:343–352
Mendgen K (1996) Fungal attachment and penetration. In: Kerstiens G (ed) Plant cuticles: an integrated functional approach. Bios Scientific Publishers, Lancaster, pp 175–188
Mengel K, Kirkby EA (2001) Principles of plant nutrition, 5th edn. Kluwer Academic Publishers, Dordrecht
Mercado JA, Reid MS, Valpuesta V, Quesada MA (1997) Metabolic changes and susceptibility to chilling stress in *Capsicum annuum* plants grown at suboptimal temperature. Aust J Plant Physiol 24:759–767
Miki Y, Takahashi D, Kawamura Y, Uemura M (2019) Temporal proteomics of *Arabidopsis* plasma membrane during cold- and de-acclimation. J Proteome 197:71–81
Mikić A, Čupina B, Rubiales D, Mihailović V, Šarūnaitė L, Fustec J, Antanasović S, Krstić Đ, Bedoussac L, Zorić L, Đordević V, Perić V, Srebrić M (2015) Models, developments, and perspectives of mutual legume intercropping. Adv Agron 130:337–419
Miura K, Furumoto T (2013) Cold signaling and cold response in plants. Int J Mol Sci 14(3): 5312–5337
Molassiotis A, Sotiropoulos T, Tanou G, Diamantidis G, Therios I (2006) Boron induced oxidative damage and antioxidant and nucleolytic responses in shoot tips culture of the apple root-stock EM9 (*Malus domestica* Borkh). Environ Exp Bot 56:54–62
Molinari, H.B.C., Marur, C.J., Bespalhok Filho, J.C., Kobayashi, A.K., Pileggi, M., Leite Jr, R.P., Pereira, L.F.P. and Vieira, L.G.E. 2004. Osmotic adjustment in transgenic citrus rootstock *Carrizo citrange* (*Citrus sinensis* Osb x *Poncirus trifoliata* L. Raf) overproducing proline. Plant Sci 167: 1375–1381
Moon BY, Higashi S, Gombos Z, Murata N (1995) Unsaturation of the membrane lipids of chloroplasts stabilizes the photosynthetic machinery against low-temperature photoinhibition in transgenic tobacco plants. Proc Natl Acad Sci U S A 92:6219–6223
Morgan PW (1990) Effects of abiotic stresses on plant hormone systems. In: Alscher RG, Cumming JR (eds) Stress responses in plants: adaptation and acclimation mechanisms. Wiley-Liss Inc., New York, NY, pp 113–146
Müller P, Li X-P, Niyogi KK (2001) Non-photochemical quenching. A response to excess light energy. Plant Physiol 155:1558–1566
Munné-Bosch S, Alegre L (2002) The function of tocopherols and tocotrienols in plants. Crit Rev Plant Sci 21:31–57
Munns R (2005) Genes and salt tolerance: bringing them together. Tansley review. New Phytol 167:645–656
Murai-Hatano M, Kuwagata T, Sakurai J, Nonami H, Ahmed A, Nagasuga K, Matsunami T, Fukushi K, Maeshima M, Okada M (2008) Effect of low root temperature on hydraulic conductivity of rice plants and the possible role of aquaporins. Plant Cell Physiol 49:1294–1305

Murphy C, Wilson JM (1981) Ultrastructural features of chilling injury in *Episcia reptans*. Plant Cell Environ 4:261–265
Nagasuga K, Murai-Hatano M, Kuwagata T (2011) Effects of low root temperature on dry matter production and root water uptake in rice plants. Plant Prod Sci 14(1):22–29
Nagel OW, Konings H, Lambers H (1994) Growth-rate, plant development and water relations of the ABA-deficient tomato mutant sitiens. Physiol Plant 92:102–108
Nahar K, Hasanuzzaman M, Majumder RR (2009a) Effect of low temperature stress in transplanted Aman Rice varieties mediated by different transplanting dates. Acad J Pl Sci 2:132–138
Nahar K, Biswas JK, Shamsuzzaman AMM, Hasanuzzaman M, Barman HN (2009b) Screening of indica rice (*Oryza sativa* L.) genotypes against low temperature stress. Bot Res Int 2:295–303
Nahar K, Biswas JK, Shamsuzzaman AMM (2012) Cold stress tolerance in rice plant: screening of genotypes based on morphophysiological traits. Lambert Academic Publishing, Saarbrücken, Germany
Nardini A, Pitt F (1999) Drought resistance of *Quercus pubescens* as a function of root hydraulic conductance, xylem embolism and hydraulic architecture. New Phytol 143:485–493
Nass HG (1983) Relationship of plant water content and winter hardiness of winter wheat and fall rye grown in Atlantic Canada. Can J Plant Sci 63(1):67–71
Naydenova G, Hristova T, Aleksiev Y (2013) Objectives and approaches in the breeding of perennial legumes for use in temporary pasture lands. Biotechnol Anim Husb 29(2):233–250
Nayyar H, Chander K, Kumar S, Bains T (2005) Glycine betaine mitigates cold stress damage in chickpea. Agron Sustain Dev 25:381–388
Neill SJ, Desikan R, Hancock JT (2003) Nitric oxide signalling in plants. New Phytol 159:11–35
Nievola CC, Carvalho CP, Carvalho V, Rodrigues E (2017) Rapid responses of plants to temperature changes. Temperature (Austin) 4(4):371–405
Nihsida I, Murata N (1996) Chilling sensitivity in plants and cyanobacteria: the crucial contribution of membranes lipids. Ann Rev Plant Physiol Plant Mol Biol 47:541–568
Nishida I, Murata N (1996) Chilling sensitivity in plants and cyanobacteria: the crucial contribution of membrane lipids. Annu Rev Plant Physiol Plant Mol Biol 47:541–568
Nobel PS (2009) Physicochemical and environmental plant physiology, 4th edn. Elsevier, Academic Press, London
Nwankwo C, Ogugurue D (2012) An investigation of temperature variation at soil depth in peanuts of southern Nigeria. Am J Environ Eng 2(5):142–147
Ogbonnaya CI, Nwalozie MC, Roy-Macauley H, Annerose DJM (1998) Growth and water relations of Kenaf (*Hibiscus cannabinus* L.) under water deficit on a sandy soil. Indus Crops Products 8:65–76
Olien CR, Smith MN (1977) Ice adhesions in relation to freeze stress. Plant Physiol 60(4):499–503
Oliveira G, Peñuelas J (2004) Effects of winter cold stress on photosynthesis and photochemical efficiency of PSII of the Mediterranean *Cistus albidus* L. and *Quercus ilex* L. Plant Ecol 175: 179–191
Olivella C, Biel C, Vendrell M, Savé R (2000) Hormonal and physiological responses of *Gerbera jamesonii* to flooding stress. HortScience 35:222–225
Oliver SN, Dennis ES, Dolferus R (2007) ABA regulates apoplastic sugar transport and is a potential signal for cold-induced pollen sterility in rice. Plant Cell Physiol 48:1319–1330
Onwuka BM (2018) Effects of soil temperature on some soil properties and plant growth. Adv Plant Agric Res 8(1):34–37
Ort DR (2001) When there is too much light. Plant Physiol 125:29–32
Orvar BL, Sangwan V, Omann F, Dhindsra RS (2000) Early steps in cold sensing by plant cells: the role of actin cytoskeleton and membrane fluidity. Plant J 23:785–794
Osborne CP, Sack L (2012) Evolution of C4 plants: a new hypothesis for an interaction of CO_2 and water relations mediated by plant hydraulics. Philoso. Transac Royal Soc B: Biol Sci 367:583–600
Owston PW, Smith JL, Halverson HG (1972) Seasonal water movement in tree stems. For Res 18(4):267–272

Palta JP (2000) Stress interactions at the cellular and membrane levels. Hort Sci 25(11):1377
Pan A, Hayes PM, Chen F, Chen THH, Blake T, Wright S, Karsai I, Beds Z (1994) Genetic analysis of the components of winter hardiness in barley (*Hordeum vulgare* L.). Theo Appl Genet 89: 900–910
Pan J, Zhang M, Kong X, Xing X, Liu Y, Zhou Y, Liu Y, Sun L, Li D (2012) ZmMPK17, a novel maize group D MAP kinase gene, is involved in multiple stress responses. Planta 235(4): 661–676
Park W, Scheffler BE, Bauer PJ, Campbell BT (2010) Identification of the family of aquaporin genes and their expression in upland cotton (*Gossypium hirsutum* L.). BMC Plant Biol 10:142
Park S, Lee CM, Deherty CJ, Kim Y, Thomashow MF (2015) Regulation of the Arabidopsis CBF regulon by a complex low-temperature regulatory network. Plant J 82(2):193–207
Park S, Gilmour SJ, Grumet R, Thomashow MF (2018) CBF-dependent and CBF-independent regulatory pathways contribute to the differences in freezing tolerance and cold-regulated gene expression of two Arabidopsis ecotypes locally adapted to sites in Sweden and Italy. PLoS One 13:e0207723
Pastori GM, Foyer CH (2002) Common components, networks, and pathways of cross-tolerance to stress. The central role of 'redox' and abscisic acid-mediated controls. Plant Physiol 129:460–468
Pearce RS (1999) Molecular analysis of acclimation to cold. Plant Grow Regul 29:47–76
Pego JV, Kortstee AJ, Huijser C, Smeekens SCM (2000) Photosynthesis, sugars and the regulation of gene expression. J Exp Bot 51:407–416
Pei ZM, Murata Y, Benning G, Thomine S, Klusener B, Allen GJ, Grill E, Schroeder JI (2000) Calcium channels activated by hydrogen peroxide mediate abscisic acid signalling in guard cells. Nature 406:731–734
Penfield S (2008) Temperature perception and signal transduction in plants. New Phytol 179:615–628
Peng S, Tang Q, Zou Y (2009) Current status and challenges of rice production in China. Plant Prod Sci 12(1):3–8
Pennisi E (2008) The blue revolution, drop by drop, gene by gene. Science 320:171–173
Perämäki M, Nikinmaa E, Sevanto S, Ilvesniemi H, Siivola E, Hari P, Vesala T (2001) Tree stem diameter variations and transpiration in scots pine: an analysis using a dynamic sap flow model. Tree Physiol 21(12–13):889–897
Pereira JS, Chaves MM (1993) Plant water deficits in Mediterranean ecosystems. In: Smith JAC, Griffiths H (eds) Plant responses to water deficits—from cell to community. Bios Scientific Publishers, Oxford, pp 237–248
Pereira JS, Chaves MM (1995) Plant responses to drought under climate change in Mediterranean-type ecosystems. In: Moreno JM, Oechel WC (eds) Global change and mediterranean-type ecosystems. Springer-Verlag, Berlin, Heidelberg, New York, pp 140–160
Pereira JS, Chaves MM, Cladeira MC, Correia AV (2006) Water availability and productivity. In: Morison J, Croft M (eds) Plant growth and climate change. Blackwell Publishers, London, pp 118–145
Perez-Perez JG, Robles JM, Tovar JC, Botia P (2009) Response to drought and salt stress of lemon 'Fino 49' under field conditions: water relations, osmotic adjustment and gas exchange. Sci Hortic 122:83–90
Perl-Treves R, Perl A (2002) Oxidative stress: An introduction. In: Inzé D, van Montagu M (eds) Oxidative stress in plants. Taylor Francis, London, pp 1–32
Phadtare S, Alsina J, Inouye M (1999) Cold-shock response and cold-shock proteins. Curr Opin Microbiol 2(2):175–180
Pinheiro C, Chaves MM, Ricardo CP (2001) Alterations in carbon and nitrogen metabolism induced by water deficit in the stems and leaves of *Lupinus albus* L. J Exp Bot 52:1063–1070
Poffenroth M, Green DB, Tallman G (1992) Sugar concentrations in guard cells of *Vicia faba* illuminated with red or blue light. Plant Physiol 98:1460–1471

Porcel R, Bustamante A, Ros R, Serrano R, Mulet JM (2018) BvCOLD1: a novel aquaporin from sugar beet (*Beta vulgaris* L.) involved in boron homeostasis and abiotic stress. Plant Cell Environ 2018:1–14

Postaire O, Tournaire-Roux C, Grondin A, Boursiac Y, Morillon R, Shäffner AR, Maurel C (2010) A PIP1 aquaporin contributes to hydrostatic pressure-induced water transport in both the root and rosette of *Arabidopsis*. Plant Physiol 152:1418–1430

Pradel E, Pieri P (2000) Influence of glass layer in vineyard and soil temperature. Aust J Grape Wine Res 6(1):59–67

Puhe J (2003) Growth and development of the root system of Norway spruce (*Picea abies*) in forest stands-a review. For Ecol Manag 175(1–3):253–273

Qaderi MM, Islam MA, Reid DM, Shah S (2007) Do low-ethylene-producing transgenic canola (*Brassica napus*) plants expressing the ACC deaminase gene differ from wild-type plants in response to UVB radiation? Can J Bot 85:148–159

Quinet M, Ndayiragije A, Lefevre I, Lambillotte B, Dupont-Gillain CC, Lutts S (2010) Putrescine differently influences the effect of salt stress on polyamine metabolism and ethylene synthesis in rice cultivars differing in salt resistance. J Exp Bot 61:2719–2733

Ramalho JC, Quartin V, Fahl JI, Carelli ML, Leitão AE, Nunes MA (2003) Cold acclimation ability of photosynthesis among species of the tropical *Coffea* genus. Plant Biol 5:631–641

Rao RCN, Udaykumar M, Farquhar GD, Talwar HS, Prasad TG (1995) Variation in carbon isotope discrimination and its relationship to specific leaf weight and Ribose-1,s-bisphosphate carboxylase content in groundnut genotypes. Aust J Plant Physiol 22:545–551

Rebetzke GJ, Richards RA (1999) Genetic improvement of early vigour in wheat. Aust J Agric Res 50:291–301

Rejeb KB, Abdelly C, Savouré A (2014) How reactive oxygen species and proline face stress together. Plant Physiol Biochem 80:278–284

Renny-Byfield S, Page JT, Udall JA, Sanders WS, Peterson DG, Arick MAII, Grover CE, Wendel JF (2016) Independent domestication of two old world cotton species. Genome Biol Evol 8: 1940–1947

Repo TI, Leinonen AR, Finer L (2004) Effect of soil temperature on bid phenology, chlorophyll fluorescence, carbohydrate content and cold hardiness of Norway spruce seedlings. Physiol Plant 121(1):93–100

Rhee JY, Lee SH, Singh AP, Chung GC, Ahn SJ (2007) Detoxification of hydrogen peroxide maintains the water transport activity in figleaf gourd (*Cucurbita ficifolia*) root system exposed to low temperature. Physiol Plant 130:177–184

Richards RA, Luckacs Z (2002) Seedling vigour in wheat-sources for variation for genetic and agronomic improvement. Aust J Agric Res 42:111–121

Richter H (2001) The cohesion theory debate continues: the pitfalls of cryobiology. Trends Plant Sci 6(10):456–457

Rihan HZ, Al-Issawi M, Fuller MP (2017) Advances in physiological and molecular aspects of plant cold tolerance. J Plant Interact 12(1):143–157

Rivero RM, Ruiz JM, García PC, López-Lefebre LR, Sánchez E, Romero L (2002) Response of oxidative metabolism in watermelon plants subjected to cold stress. Funct Plant Biol 29(5):643

Rodríguez-Gamir J, Ancillo G, González-Mas MC, Primo-Millo E, Iglesias DJ, Forner-Giner MA (2011) Root signalling and modulation of stomatal closure in flooded citrus seedlings. Plant Physiol Biochem 49:636–645

Routaboul J-M, Fischer S, Browse J (2000) Trienoic fatty acids are required to maintain chloroplast function at low temperature. Plant Physiol 124:1697–1705

Ruelland E, Zachowski A (2010) How plants sense temperature. Environ Exp Bot 69(3):225–232

Ruelland E, Vaultier MN, Zachowski A, Hurry V (2009) Cold signalling and cold acclimation in plants. Adv Bot Res 49:35–150

Ruiz-Lozano JM, Alguacil MM, Bárzana G, Vernieri P, Aroca R (2009) Exogenous ABA accentuates the differences in root hydraulic properties between mycorrhizal and non mycorrhizal maize plants through regulation of PIP aquaporins. Plant Mol Biol 70:565–579

Sachs J (1875) Geschichte Der Botanik Vom 16. Jahrhundert Bis 1860. Oldenbourg, München
Sack L, Holbrook NM (2006) Leaf hydraulics. Ann Rev Plant Biol 57:361–381
Sack L, Tyree MT (2005) Leaf hydraulics and its implications in plant structure and function. In: Holbrook NM, Zwieniecki MA (eds) Vascular transport in plants. Elsevier, Academic Press, San Diego, CA, pp 93–114
Sairam RK, Tyagi A (2004) Physiology and molecular biology of salinity stress tolerance in plants. Curr Sci 86:407–421
Sakai A, Larcher W (1987) Frost survival of plants. Ecol Stud 62:157–165
Sales CRG, Ribeiro RV, Silveira JAG, Machado EC, Martins MO, Lagôa AMMA (2013) Superoxide dismutase and ascorbate peroxidase improve the recovery of photosynthesis in sugarcane plants subjected to water deficit and low substrate temperature. Plant Physiol Biochem 73:326–336
Salinas J (2002) Molecular mechanisms of signal transduction in cold acclimation. In: Scheel D, Wasternack C (eds) Plant signal transduction. Oxford University Press, London, pp 116–139
Sanghera GS, Wani SH, Hussain W, Singh NB (2011) Engineering cold stress tolerance in crop plants. Curr Genomics 12:30–43
Satake T (1976) Sterility-type cool injury to rice plants in Japan. In: Takahashi K, Yoshino M (eds) Climatic change and food production. Tokyo University, Tokyo, Japan, pp 246–254
Sato K (1972) Growth responses of rice plant to environmental conditions. I. the effects of air-temperatures on the growth at vegetative stage. Proc Crop Sci Soc Jpn 41:388–393
Sauter A, Dietz KJ, Hartung W (2002) A possible tress physiological role of abscisic acid conjugates in root-to-shoot signalling. Plant Cell Environ 25(2):223–228
Scholz FG, Bucci SJ, Arias N, Meinzer FC, Goldstein G (2012) Osmotic and elastic adjustments in cold desert shrubs differing in rooting depth: coping with drought and subzero temperatures. Oecologia 170:885–897
Schoppach R, Sadok W (2012) Differential sensitivities of transpiration to evaporative demand and soil water deficit among wheat elite cultivars indicate different strategies for drought tolerance. Environ Exp Bot 84:1–10
Schulz E, Tohge T, Zuther E, Fernie AR, Hincha DK (2015) Natural variation in flavonol and anthocyanin metabolism during cold acclimation in *Arabidopsis thaliana* accessions. Plant Cell Environ 38:1658–1672
Schulz E, Tohge T, Zuther E, Fernie AR, Hincha DK (2016) Flavonoids are determinants of freezing tolerance and cold acclimation in *Arabidopsis thaliana*. Sci Rep, 6: e34027
Schwarz P, Fahey T, Dawson T (1997) Seasonal air and soil temperature effects on photosynthesis in red spruce (*Picea rubens*) saplings. Tree Physiol 17(3):187–194
Schwender J, Ohlrogge J, Shachar-Hill Y (2004) Understanding flux in plant metabolic networks. Curr Opin Plant Biol 7:309–317
Scott P (2008) Physiology and behaviour. John Wiley and Sons, New York, p 305
Senaratna T, Touchell D, Bunn E, Dixon K (2000) Acetyl salicylic acid (aspirin) and salicylic acid induce multiple stress tolerance in bean and tomato plants. Plant Growth Regul 30:157–161
Sevanto S, Vesala T, Perämäki M, Nikinmaa E (2001) Xylem diameter changes as an indicator of stand-level evapotranspiration. Boreal Environ Res 6(1):45–52
Sevanto S, Suni T, Pumpanen J, Grönholm T, Kolari P, Nikinmaa E, Hari P, Vesala T (2006a) Wintertime photosynthesis and water uptake in a boreal forest. Tree Physiol 26(6):749–757
Sevanto S, Suni T, Pumpanen J, Grönholm T, Kolari P, Nikinmaa E, Hari P, Vesala T (2006b) Wintertime photosynthesis and water uptake in a boreal forest. Tree Physiol 26(6):749–757
Sewelam N, Oshima Y, Mitsuda N, Ohme-Takagi M (2014) A step towards understanding plant responses to multiple environmental stresses: a genome-wide study. Plant Cell Environ 37: 2024–2035
Shan DP, Huang JG, Yang YT, Guo YH, Wu CA, Yang GD, Gao Z, Zheng CC (2007) Cotton GhDREB1 increases plant tolerance to low temperature and is negatively regulated by gibberellic acid. New Phytol 176:70–81

Shao HB, Chen XY, Chu LY, Zhao XN, Wu GY, Yong B (2006) Investigation on the relationship of proline with wheat anti drought under soil water deficits. Colloids Surf B Bio Interfaces 53: 113–119

Sharma SS, Dietz KJ (2006) The significance of amino acids and amino acid-derived molecules in plant responses and adaptation to heavy metal stress. J Exp Bot 57:711–726

Sharma P, Sharma N, Deswal R (2005) The molecular biology of the low-temperature response in plants. BioEssays 27:1048–1059

Sharp RE (2002) Interaction with ethylene: changing views on the role of abscisic acid in root and shoot growth responses to water stress. Plant Cell Environ. 25:211–222

Shi Y, Yang S (2014) In: Zhang D-P (ed) Abscisic acid: metabolism, transport and signaling. Springer, Netherland, pp 337–363

Shi Y, Ding Y, Yang S (2018) Molecular regulation of CBF signaling in cold acclimation. Trends Plant Sci 23(7):623–637

Shimazaki Y, Satake T, Watanabe K (1963) Studies on the growth and nutrient absorption of the rice plant reared with cool water irrigation during its early growth stage. Res Bull Hokkaido Natl Agric Exp Stn 80:1–12

Shimono H, Hasegawa T, Iwama K (2002) Response of growth and grain yield in paddy rice to cool water at different growth stages. Field Crops Res 73(1):67–79

Shimono H, Hasegawa T, Fujimura S, Iwama K (2004) Responses of leaf photosynthesis and plant water status to low water temperature at different growth stages. Field Crops Res 89(1):71–83

Shimono H, Hasegawa T, Iwama K (2007a) Modeling the effects of water temperature on rice growth and yield under a cool climate. I Model development. Agron J 99(5):1327–1337

Shimono H, Hasegawa T, Kuwagata T, Iwama K (2007b) Modeling the effects of water temperature on rice growth and yield under a cool climate II. Model application. Agron J 99(5):1338–1344

Shinozaki K, Yamaguchi-Shinozaki K, Seki M (2003) Regulatory network of gene expression in the drought and cold stress responses. Curr Opin Plant Biol 6:410–417

Shinozaki K, Uemura M, Bailey-Serres J, Bray EA, Weretilnyk E (2015) Responses to abiotic stress. In: Buchanan BB, Gruissem W, Vickers K, Jones RL (eds) Biochemistry and molecular biology of plants. John Wiley & Sons, Hoboken

Siefritz F, Tyree MT, Lovisolo C, Schubert A, Kaldenhoff R (2002) PIP1 plasma membrane aquaporin in tobacco: from cellular effects to function in plants. Plant Cell 14:869–876

Siemens JA, Zwiazek JJ (2003) Effects of water deficit stress and recovery on the root water relations of trembling aspen (*Populus tremuloides*) seedlings. Plant Sci 165:113–120

Simonneau T, Barrieu P, Tardieu F (1998) Accumulation rate of ABA in detached maize roots correlates with root water potential regardless of age and branching order. Plant Cell Environ 21: 1113–1122

Sinclair TR (2012) Is transpiration efficiency a viable plant trait in breeding for crop improvement? Funct Plant Biol 39:359–365

Sinclair TR, Hammer GL, van Oosterom EJ (2005) Potential yield and water-use efficiency benefits in sorghum from limited maximum transpiration rate. Funct Plant Biol 32:945–952

Singh DK, Sale PWG (2000) Growth and potential conductivity of white clover roots in dry soil with increasing phosphorous supply and defoliation frequency. Agron J 92:868–874

Sinha SK, Agarwal PK, Khanna-Chopra R (1985) Irrigation in India: a physiological and phonological approach to water management in grain crops. In: Hillel D (ed) Advances in irrigation science. Academic Press, New York, pp 130–212

Siow LF, Rades T, Lim MH (2007) Characterizing the freezing behavior of liposomes as a tool to understand the cryopreservation procedures. Cryobiology 55:210–221

Sivakumar P, Sharmila P, Saradhi PP (2000) Proline alleviates salt-stress induced enhancement in rubisco oxygenase activity. Biochem Biophys Res Commun 279(1):512–515

Smith AW (1989) Introduction. In: Clarke RJ, Macrae R (eds) Coffee, Volume 1: Chemistry. Elsevier, London, pp 1–41

Smith BN, Jones AR, Hanson LD, Criddle RS (1999) Growth, respiration rate, and efficiency response to temperature. In: Handbook of plant and crop stress, 2nd edn. Marcel Dekker, New York, pp 417–440

Solanke AU, Sharma AK (2008) Signal transduction during cold stress in plants. Physiol Mol Biol Plants 14(1–2):69–79

Solecka D, Zebrowski J, Kacperska A (2008) Are pectins involved in cold acclimation and de-acclimation of winter oil-seed rape plants? Ann Bot (Lond) 101:521–530

Somerville C (1995) Direct tests of the role of membrane lipid composition in low-temperature-induced photoinhibition and chilling sensitivity in plants and cyanobacteria. Proc Natl Acad Sci U S A 92:6215–6218

Song L, Ding W, Shen J, Zhang Z, Bi Y, Zhang L (2008) Nitric oxide mediates abscisic acid induced thermotolerance in the calluses from two ecotypes of reed under heat stress. Plant Sci 175:826–832

Song G, Wang M, Zeng B, Zhang J, Jiang C, Hu Q, Geng G, Tang C (2015) Anther response to high-temperature stress during development and pollen thermotolerance heterosis as revealed by pollen tube growth and in vitro pollen vigor analysis in upland cotton. Planta 241(5):1271–1285

Songstad DD, Duncan DR, Widholm JM (1990) Proline and polyamine involvement in chilling tolerance of maize suspension cultures. J Exp Bot 41:289–294

Srivalli B, Khanna-Chopra R (2004) The developing reproductive 'sink' induces oxidative stress to mediate nitrogen mobilization during monocarpic senescence in wheat. Biochem Biophys Res Commun 325:198–202

Steponkus PL (1984) Role of the plasma membrane in freezing injury and cold acclimation. Ann Rev Plant Physiol 35:543–584

Steponkus PL, Uemura M, Webb MS (1993) A contrast of the cryostability of the plasma membrane of winter rye and spring oat-two species that widely differ in their freezing tolerance and plasma membrane lipid composition. In: Steponkus PL (ed) Advances in low-temperature biology, vol 2. JAI Press, London, pp 211–312

Steponkus PL, Uemura, M, Joseph, RA, Gilmour SJ, Thomashow MF (1998) Mode of action of the COR15a gene on the freezing tolerance of *Arabidopsis thaliana*. Proc Nat Acad Sci U S A, 95: 14570–14575

Steudle E (2000) Water uptake by plant roots: an integration of views. Plant Soil 226:45–56

Steudle E (2001) The cohesion-tension mechanism and the acquisition of water by plant roots. Ann Rev Plant Physiol Plant Mol Biol 52:847–875

Steudle E, Peterson CA (1998) How does water get though roots? J Exp Bot 49:775–788

Stitt M, Hurry V (2002) A plant for all seasons: alterations in photosynthetic carbon metabolism during cold acclimation in Arabidopsis. Curr Opin Plant Biol 5(3):199–206

Strauss G, Hauser H (1986) Stabilization of lipid bilayer vesicles by sucrose during freezing. Proc Natl Acad Sci U S A 83(8):2422–2426

Strimbeck GR, Schaberg PG, Fossdal CG, Schröder WP, Kjellsen TD (2015) Extreme low temperature tolerance in woody plants. Front Plant Sci 19:884

Stushnoff C, Flower DB, Brule-Babel A (1984) Breeding and selection for resistance to low-temperature. In: Vose PB (ed) Plant breeding—a contemporary basis. Elsevier, pp 115–136

Subbarao GV, Johansen C (2002) Transpiration efficiency: avenues for genetic improvement. In: Pessarakdi M (ed) Handbook for crop and plant physiology. Marcel Dekker Inc., New York, pp 835–855

Subbarao GV, Nam NH, Chauhan YS, Johansen C (2000) Osmotic adjustment, water relations and carbohydrate remobilization in pigeon pea under water deficits. J Plant Physiol 157:651–659

Sun Q, Wang Y, Hou X (2004) Alfalfa winter survival research summary. Pratacultural Sci 21:21–25

Sundaramurthi P, Patapoff TW, Suryanarayanan R (2010) Crystallization of trehalose in frozen solutions and its phase behavior during drying. Pharma Res 27:2374–2383

Szalai G, Tari I, Janda T, Pestenacz A, Paldi E (2000) Effects of cold acclimation and salicylic acid on changes in ACC and MACC contents in maize during chilling. Biol Plant 43:637–640

Tada N, Sato M, Yamanoi J, Mizorogi T, Kasai K, Ogawa S (1990) Cryopreservation of mouse spermatozoa in the presence of raffinose and glycerol. J Reprod Fertil 89:511–516

Tahir MA, Aziz T, Rahmatullah. (2011) Silicon induced growth and yield enhancement in two wheat genotypes differing in salinity tolerance. Commun Soil Sci Plant Anal 42:395–407

Táibi K, Del Campo AD, Vilagrosa A, Ballésa JM, López-Gresa MP, López-Nicolás JM, Mulet JM (2018) Distinctive physiological and molecular responses to cold stress among cold-tolerant and cold-sensitive *Pinus halepensis* seed sources. BMC Plant Biol 18:236

Taiz L, Zeiger E (2015) Plant physiology and development, 6th edn. Sinauer Associates, Inc.

Takahashi D, Li B, Nakayama T, Kawamura Y, Uemura M (2013) Plant plasma membrane proteomics for improving cold tolerance. Front Plant Sci 4:90

Takamura Y, Takeuchi S, Hasegawa H (1960) Studies on the effects of soil temperature upon the growth of crop plants. III. Soil temperature and leaf emergence of rice plant. VIII. Relation of temperature of several parts of rice plant to the rate of leaf emergence. Proc Crop Sci Soc Jpn 29: 195–198

Takeoka Y, Mamun AA, Wada T, Kaufman BP (1992) Reproductive adaptation of rice to environmental stress. Elsevier Science Publisher, Amsterdam, The Netherland. pp. 8.10

Tamhane RV, Motiramani DP, Bali YP, Donahue RL (1970) Soils: their chemistry and fertility in tropical Asia. Prentice-Hall of India Pvt. Ltd., New Delhi, p 475

Tanner CB, Sinclair TR (1983) Efficient water use in crop production: research or research? In: Taylor HM, Jordan WR, Sinclair TR (eds) Limitations to efficient water use in crop production. ASA, CSSA and SSSA, Madison. WI, pp 1–27

Tarkowski ŁP, Van den Ende W (2015) Cold tolerance triggered by soluble sugars: a multifaceted countermeasure. Front Plant Sci 6:203

Tasgin E, Atici O, Nalbantoglu B (2003) Effects of salicylic acid and cold on freezing tolerance in winter wheat leaves. Plant Growth Regul 41:231–236

Tetreault, H.M., Kawakami, T., Ungerer, M.C. and Levy, C. 2016. Low temperature tolerance in the perennial sunflower *Helianthus maximiliani*. Am Midland Nat, 175: 91–102

Tewari RK, Kumar P, Tewari N, Srivastava S, Sharma PN (2004) Macronutrient deficiencies and differential antioxidant responses-influence on the activity and expression of superoxide dismutase in maize. Plant Sci 166:687–694

Tewari RK, Kumar P, Sharma PN (2006) Magnesium deficiency induced oxidative stress and antioxidant responses in mulberry plants. Sci Hortic 108:7–14

Thakur P, Kumar S, Malik JA, Berger JD, Nayyar H (2010) Cold stress effects on reproductive development in grain crops: an overview. Environ Exp Bot 67:429–443

Thalhammer A, Bryant G, Sulpice R, Hincha DK (2014) Disordered cold regulated15 proteins protect chloroplast membranes during freezing through binding and folding, but do not stabilize chloroplast enzymes in vivo. Plant Physiol 166(1):190–201

Thomas H, James AR (1993) Freezing tolerance and solute changes in contrasting genotypes of *Lolium perenne* L. acclimated to cold and drought. Ann Bot 72:249–254

Thomashow MF (1999) Plant cold acclimation: freezing tolerance genes and regulatory mechanisms. Annu Rev Plant Physiol Plant Mol Biol 50:571–599

Thomashow MF (2001) So what's new in the field of plant cold acclimation? Lots! Plant Physiol 125:89–93

Thompson AJ, Andrews J, Mullholland BJ, McKee JMT, Hilton HW, Horridge JS, Farquhar GD, Smeeton RC, Smillie IRA, Black CR, Taylor IB (2007) Overproduction of abscisic acid in tomato increases transpiration efficiency and root hydraulic conductivity and influences leaf expansion. Plant Physiol 143:1905–1917

Tommasini L, Svensson JT, Rodriguez EM, Wahid A, Malatrasi M, Kato K, Wanamaker S, Rosnik J, Close TJ (2008) Dehydrin gene expression provides an indicator of low temperature and drought stress: transcriptome-based analysis of barley (*Hordeum vulgare L.*). Funct Integr Genomics 8(4): 387–405

Tornroth-Horsefield S, Wang Y, Hedfalk K, Johanson U, Karlsson M, Tajkhorshid E, Neutze R, Kjellbom P (2006) Structural mechanism of plant aquaporin gating. Nature 439:688–694

Toselli M, Flore JA, Maragoni B, Masia A (1999) Effect of root-zone temperature on nitrogen accumulation by non-breeding apple trees. J Hort Sci Biotech 74:118–124
Trouverie J, Thevenot C, Rocher J-P, Sotta B, Prioul J-L (2003) The role of abscisic acid in the response of a specific vacuolar invertase to water stress in the adult maize leaf. J Exp Bot 54(390):2177–2186
Turhan E, Karni L, Aktas H, Deventurero G, Chang DC, Bar-Tal A, Aloni B (2006) Apoplastic antioxidants in pepper (*Capsicum annuum* L.) fruit and their relationship to blossom-end rot. J Hortic Sci Biotechnol 81:661–667
Turner NC, Jones MN (1980) Turgot maintenance by osmotic adjustment: a review and evaluation. In: Turner NC, Kramer PJ (eds) Adaptation of plants to water and high temperature stress. Wiley, New York, pp 84–104
Tyree MT (2003) Plant hydraulics: the ascent of water. Nature 423(6943):923–923
Uchida A, Jagendorf AT, Hibino T, Takabe T, takabe, T. (2002) Effects of hydrogen peroxide and nitric oxide on both salt and heat stress tolerance in rice. Plant Sci 163(3):515–523
Uemura M, Joseph RA, Steponkus PL (1995) Cold acclimation of *Arabidopsis thaliana* (effect on plasma membrane lipid composition and freeze-induced lesions). Plant Physiol 109(1):15–30
Vadez V, Kholova J, Medina S, Kakkera A, Anderberg H (2014) Transpiration efficiency: new insights into an old story. J Exp Bot 65(21):6141–6153
Valadabadi SA, Shiranirad AH, Farahani HA (2010) Ecophysiological influences of zeolite and selenium on water deficit stress tolerance in different rapeseed cultivars. J Ecol Nat Environ 2: 154–159
Valko M, Morris H, Cronin MT (2005) Metals, toxicity and oxidative stress. Curr Med Chem 12: 1161–1208
van Buer, J., Cvetkovic, J. and Baier, M. 2016. Cold regulation of plastid ascorbate peroxidases serves as a priming hub controlling ROS signaling in *Arabidopsis thaliana*. BMC Plant Biol 16: 163
van Buer J, Prescher A, Baier M (2019) Cold-priming of chloroplast ROS signalling is developmentally regulated and is locally controlled at the thylakoid membrane. Sci Rep 9:3022
Verbruggen N, Villarroel R, van Montagu M (1993) Osmoregulation of a pyrroline-5-carboxylate reductase gene in *Arabidopsis thaliana*. Plant Physiol 103:771–781
Vernieri P, Pardossi A, Tognoni F (1991) Influence of chilling and drought on water relations and abscisic acid accumulation in bean. Aust J Plant Physiol 18(1):25–35
Vernieri P, Lenzi A, Figaro M, Tognoni F, Pardossi A (2001) How roots contribute to the ability of *Phaseolus vulgaris* L. to cope with chilling-induced water stress. J Exp Bot 52:2199–2206
Veyres N, Aono M, Sangwan-Norreel BS, Sangwan RS (2008) The *Arabidopsis* sweetie mutant is affected in carbohydrate metabolism and defective in the control of growth, development and senescence. Plant J 55:665–686
Volobueva OV, Velikanov GA, Baluska F (2004) Regulation of intercellular water exchange in various zones of maize root under stresses. Rus J Plant Physiol 51:676–683
Vogan PJ, Sage RF (2011) Water-use efficiency and nitrogen-use efficiency of C3–C4 intermediate species of *Flaveria* Juss. (Asteraceae). Plant Cell Environ 34:1415–1430
Wahid A, Close TJ (2007) Expression of dehydrins under heat stress and their relationship with water relations of sugarcane leaves. Biol Plant 51:104–109
Wallenstein M, Allison SD, Ermakovich J, Steinweg JM, Sinsabaugh RL (2010) Control of temperature sensitivity of soil enzymes, a key driver of in-situ enzyme rates. In: Shukla G, Verma A (eds) Soil enzymology-Chapter 13. Springer-Verlag, pp 245–258
Walters KR Jr, Serianni AS, Voituron Y, Sformo T, Barnes BM, Duman JG (2011) A thermal hysteresis-producing xylomannan glycolipid antifreeze associated with cold tolerance is found in diverse taxa. J Comp Physiol B 181:631–640
Wan XC, Landhäusser SM, Zwiazek JJ, Lieffers VJ (1999) Root water flow and growth of aspen (*Populus tremuloides*) at low root temperatures. Tree Physiol 19:879–884
Wan XC, Zwiazek JJ (2001) Root water flow and leaf stomatal conductance in aspen (*Populus tremuloides*) seedlings treated with abscisic acid. Planta 213:741–747

Wan X, Zwiazek JJ, Lieffers VJ, Landhausser SM (2001) Hydraulic conductance in aspen (*Populus tremuloides*) seedlings exposed to low root temperature. Tree Physiol 21(10):691–696

Wan XC (2010) Osmotic effects of NaCl on cell hydraulic conductivity of corn roots. Acta Biochim Biophys Sin 42:351–357

Wang LJ, Li SH (2006) Salicylic acid-induced heat or cold tolerance in relation to Ca^{2+} homeostasis and antioxidant systems in young grape plants. Plant Sci 170:685–694

Wang Z, Wang J, Wang F, Bao Y, Wu Y, Zhang H (2009) Genetic control of germination ability under cold stress in rice. Rice Sci 16(3):173–180

Wang W, Chen Q, Hussain S, Mei J, Dong H, Peng S, Huang J, Cui K, Nie L (2016a) Pre-sowing seed treatments in direct-seeded early rice: consequences for emergence, seedling growth and associated metabolic events under chilling stress. Sci Rep 6:19637

Wang Q, Ding T, Zuo J, Gao L, Fan L (2016b) Amelioration of postharvest chilling injury in sweet pepper by glycine betaine. Postharvest Biol Technol 112:114–120

Wang CL, Zhang SC, Qi SD, Zheng CC, Wu CA (2016c) Delayed germination of *Arabidopsis* seeds under chilling stress by overexpressing an abiotic stress inducible GhTPS11. Gene 575: 206–212

Wang DZ, Jin YN, Ding XH, Wang WJ, Zhai SS, Bai LP, Guo ZF (2017) Gene regulation and signal transduction in the ICE-CBF-COR signaling pathway during cold stress in plants. Biochem Mosc 82:1103–1117

Waraich EA, Ahmad R, Ashraf MY, Saifullah, Ahmad M (2011) Improving agricultural water use efficiency by nutrient management in crop plants. Acta Agric Scand Sect B: Plant Soil Sci 61(4): 291–304

Waraich EA, Ahmad R, Halim A, Aziz T (2012) Alleviation of temperature stress by nutrient management in crop plants: a review. J Soil Sci Plant Nutr 12(2):221–244

Way DA, Katul GG, Manzoni S, Vico G (2014) Increasing water use efficiency along the C_3 to C_4 evolutionary pathway: a stomatal optimization perspective. J Exp Bot 65(13):3683–3693

Weiser RL, Wallner SJ, Waddell JW (1990) Cell wall and extensin mRNA changes during cold acclimation of pea seedlings. Plant Physiol 93:1021–1026

Wilkinson S, Clephan AL, Davies WJ (2001) Rapid low temperature induced stomatal closure occurs in cold-tolerant Commelina communis leaves but not in cold-sensitive tobacco-leaves, via a mechanism that involves apoplastic calcium but not abscisic acid. Plant Physiol 126(4): 1566–1578

Wilkinson S, Davies WJ (2002) ABA-based chemical signalling: the co-ordination of responses to stress in plants. Plant Cell Environ 25:195–210

Wisniewski M, Fuller M (1999) Ice nucleation and deep supercooling in plants: new insights using infrared thermography. In: Cold-adapted organisms: ecology, physiology, enzymology and molecular biology. Springer-Verlag, Berlin, pp 105–118

Wisniewski M, Gusta LV (2014) The biology of cold hardiness: adaptive strategies preface. Environ Exp Bot 106:1–3

Witt M, Barfield BJ (1982) Environmental stress and plant productivity. In: Rechcigl Jr (ed) Handbook of agricultural productivity. CRC Press, Boca Raton, FL, pp 347–374

Wendehenne D, Pugin A, Klessig DF, Durner J (2001) Nitric oxide: comparative synthesis and signaling in animal and plant cells. Trends Plant Sci 6:177–183

Woodward FI (2002) Potential impacts of global elevated CO_2 concentrations on plants. Plant Biol 5:207–211

Wu J, Li H, Niu J, Liu T, Zheng H, Xu X, Miao S (2020) Water uptake patterns of alfalfa under winter irrigation in cold and arid grassland. Water 12:1093

Wudick MM, Luu DT, Maurel C (2009) A look inside: localization patterns and functions of intracellular plant aquaporins. New Phytol 184:28–302

Xin Z, Browse J (2000) Cold comfort farm: the acclimation of plants to freezing temperatures. Plant Cell Environ 23:893–902

Xoconostle-Cázares B, Ramírez-Ortega FA, Flores-Elenes L, Ruiz-Medrano R (2010) Drought tolerance in crop plants. Am J Plant Physiol 5(5):241–256

Xu CC, Jeon YA, Lee C-H (1999) Relative contributions of photochemical and non-photochemical routes to excitation energy dissipation in rice and barley illuminated at a chilling temperature. Physiol Plant 107:447–453

Xu Q, Huang B (2000) Growth and physiological responses of creeping bentgrass to changes in air and soil temperatures. Crop Sci 40(5):1363–1368

Xu PL, Guo YK, Bai JG, Shang L, Wang XJ (2008) Effects of long-term chilling on ultrastructure and antioxidant activity in leaves of two cucumber cultivars under low light. Physiol Plant 132: 467–478

Xue T, Hartikainen H, Piironen V (2001) Antioxidative and growth-promoting effect of selenium on senescing lettuce. Plant Soil 237:55–61

Xue T, Li X, Zhu W, Wu C, Yang G, Zheng C (2009) Cotton metallothionein GhMT3a, reactive oxygen species scavenger, increased tolerance against abiotic stress in transgenic tobacco and yeast. J Exp Bot 60:339–349

Yadav SK (2010) Cold stress tolerance mechanisms in plants. A review. Agron Sust Develop 30: 515–527

Yan L, Hangwen X (2014) Effect of soil temperature, flooding and organic matter addition in N_2O emission from a soil of Hongze lake wetland, China. Sci World J 2014:272684

Yang JD, Yun JY, Zhang TH, Zhao HL (2006) Presoaking with nitric oxide donor SNP alleviates heat shock damages in mung bean leaf discs. Bot Stud 47:129–136

Ye Q, Steudle E (2006) Oxidative gating of water channels (aquaporins) in corn roots plant. Cell Environ 29:459–470

Yildiz D, Nzokou P, Deligoz A, Koc I, Genc M (2014) Chemical and physiological responses of four Turkish red pine (*Pinus brutia* ten.) provenances to cold temperature treatments. Eur J Forest Res 133:809

Yokota A, Takahara K, Akashi K (2006) Water stress. In: Rao KVM, Raghavendra AS, Reddy KJ (eds) Physiology and molecular biology of stress tolerance in plants. Springer, Netherlands, pp 15–39

Yu H, Liu H, Wang J (2015) Effects of cover and irrigation on winter surviving rate, soil temperature and soil moisture of algonquin alfalfa. Chin J Grassl 37:107–111

Zaman-Allah M, Jenkinson DM, Vadez V (2011) Chickpea genotypes contrasting for seed yield under terminal drought stress in the field differ for traits related to the control of water use. Funct Plant Biol 38:270–281

Zelazny E, Borst JW, Muylaert M, Batoko H, Hemminga MA, Chaumont F (2007) FRET imaging in living maize cells reveals that plasma membrane aquaporins interact to regulate their subcellular localization. Proc Nat Acad Sci U S A 104:12359–12364

Zhang J, Nguyen HT, Blum A (1999) Genetic analysis of osmotic adjustment in crop plants. J Exp Bot 50:291–302

Zhang M, Duan L, Zhai Z, Li J, Tian X, Wang B, Zhong H, Zhao L (2004) Effects of plant growth regulators on water deficit-induced yield loss in soybean. In: Proceedings of the 4th International Crop Science Congress, Brisbane, QLD, pp. 575.

Zhang YY, Wang LL, Liu YL, Zhang Q, Wei QP, Zhang WH (2006) Nitric oxide enhances salt tolerance in maize seedlings through increasing activities of proton-pump and Na1/H1 anti-port in the tonoplast. Planta 224:545–555

Zhang XY, Liang C, Wang GP, Luo Y, Wang W (2010) The protection of wheat plasma membrane under cold stress by glycine betaine overproduction. Biol Plant 54(1):83–88

Zhang S, Jiang H, Peng S, Korpelainen H, Li C (2011) Sex-related differences in morphological, physiological, and ultrastructural responses of *Populus cathayana* to chilling. J Exp Bot 62(2): 675–686

Zhao YM, Xao Mong MA, Wang L (2007a) Variations of soil temperature and soil moisture in northern Tibet plateau. J Glaciol Geocryol 29(4):578–583

Zhao MG, Tian QY, Zhang WH (2007b) Nitric oxide synthase-dependent nitric oxide production is associated with salt tolerance in Arabidopsis. Plant Physiol 144:206–217

Zhao C, Lang Z, Zhu JK (2015) Cold responsive gene transcription becomes more complex. Trends Plant Sci 20(8):466–468
Zhu JK (2001) Cell signalling under salt, water and cold stresses. Cur Opin Plant Biol 4:401–406
Zhu JK (2002) Salt and drought stress signal transduction in plants. Ann Rev Plant Biol 53:247–273
Zhu X, Gong H, Chen G, Wang S, Zhang C (2005) Different solute levels in two spring wheat cultivars induced by progressive field water stress at different developmental stages. J Arid Environ 62:1–14
Zhu J, Dong CH, Zhu JK (2007) Interplay between cold-responsive gene regulation, metabolism and RNA processing during plant cold acclimation. Curr Opin Plant Biol 10:290–295
Ziegler P (1998) Environmental stress and stress responses plants. In: The co-action between living system and the planet. University of Geneva, Switzerland, pp 67–100
Zimmermann U, Schneider H, Wegner LH, Haase A (2004) Water ascent in tall trees: does evolution of land plants rely on a highly metastable state? New Phytol 162(3):575–615
Zuther E, Juszczak I, Lee YP, Baier M, Hincha DK (2015) Time-dependent deacclimation after cold acclimation in Arabidopsis thaliana accessions. Sci Rep 5:12199
Zuther E, Schaarschmidt S, Fischer A, Erban A, Pagter M, Mubeen U, Giavalisco P, Kopka J, Sprenger H, Hincha DK (2019) Molecular signatures associated with increased freezing tolerance due to low temperature memory in *Arabidopsis*. Plant Cell Environ 42(3):854–873

Effect of Low Temperature Stress on Photosynthesis and Allied Traits: A Review

3

Abstract

Environmental stress is an inescapable reality for most plants growing in natural settings. Conditions of sub- or supraoptimal temperatures, water deficit, water logging, salinity, and pollution can have dramatic effects on plant growth and development and, in agricultural settings, yield. Chilling and freezing to low temperatures leads to disturbances in all physiological processes—water regime, mineral nutrition, photosynthesis, respiration, and metabolism. Inactivation of metabolism, observed at chilling of chilling-sensitive plants, is a complex function of both temperature and duration of exposure. Freezing tolerance is essential for perennial plants, and ability to adapt to extreme temperature is crucial for their survival in many environments. Response of plants to low temperature exposure is associated with a change in the rate of gene transcription of a number of low molecular weight proteins. In C3 species, photosynthesis is classically considered to be limited by the capacities of ribulose 1·5-bisphosphate carboxylase/oxygenase (Rubisco), ribulose bisphosphate (RuBP) regeneration, or inorganic phosphate regeneration. Using both theoretical and empirical evidence, we describe the temperature response of instantaneous net CO_2 assimilation rate in terms of these limitations, and evaluate possible limitations on A at elevated temperatures arising from heat-induced lability of Rubisco activase. In C3 plants, Rubisco capacity is the predominant limitation on net CO_2 assimilation across a wide range of temperatures at low CO_2 (<300 μbar) while at elevated CO_2, the limitation shifts to inorganic phosphate regeneration capacity at suboptimal temperatures, and either electron transport capacity or Rubisco activase capacity at supraoptimal temperatures. In C4 plants, Rubisco capacity limits net CO_2 assimilation below 20 °C in chilling-tolerant species, but the control over net CO_2 assimilation at elevated temperature remains uncertain. Acclimation of C3 photosynthesis to suboptimal growth temperature is commonly associated with a disproportional enhancement of the inorganic phosphate regeneration capacity. In

A. Bhattacharya, *Physiological Processes in Plants Under Low Temperature Stress*, https://doi.org/10.1007/978-981-16-9037-2_3

many C4 species from warm habitats, acclimation to cooler growth conditions increases levels of Rubisco and C4 cycle enzymes which then enhance net CO_2 assimilation below the thermal optimum. By contrast, few C4 species adapted to cooler habitats increase Rubisco content during acclimation to reduced growth temperature; as a result, A changes little at suboptimal temperatures. Global change is likely to cause a widespread shift in patterns of photosynthetic limitation in higher plants. Limitations in electron transport and Rubisco activase capacity should be more common in the warmer, high CO_2 conditions expected by the end of the century.

Keywords

Low temperature and photosynthesis · Pigment system · Temperature and chloroplast membrane integrity · Low temperature and chlorophyll · Suboptimal temperature and photosynthesis in C3 and C4 plants · Cellular mechanism against chilling temperature and oxidative stress · CO_2 fixation under low temperature · ATP generation · Photorespiratory process · Cold acclimation · Improving photosynthetic capacity · Strategies to ameliorate cold stress and improve photosynthesis · Management strategies against cold stress

3.1 Plant Metabolic Activities Under Change in Global Temperature

The global population is forecast to increase by 2 billion and reach 9.8 billion people in 2050. This means that it is mandatory to raise crop productivity by 70% to meet the projected demand by the middle of the century (Alexandratos and Bruinsma 2012). Food quality should also be taken into account because it must provide all the essential nutrients to maintain human health. Alongside the increase in population, changes in the diet, particularly due to the higher meat consumption in some developing countries, may exacerbate the demand for feed crops. Abiotic environmental stresses are considered as major limitations threatening worldwide food security and have a great impact on global crop production. It has been proven that the impact of these changes on natural systems and human health has already been harmful. It is reported that the global mean land and ocean surface temperatures have increased by 0.8 °C during the period 1888–2012 (IPCC 2001, 2014). Further, although some degree of uncertainty exists in how the global surface temperature will rise, worldwide averaged surface temperature is foreseen to increase by 1.4–5.8 °C by 2100 (IPCC 2001). In addition, extreme heat events have been detected since 1950, and it is assumed that such extreme events are going to occur often in the future (IPCC 2007, 2014). In fact, heat waves are projected to be more intense and longer-lasting, while cold episodes are projected to decrease significantly. Such changes are projected to occur almost everywhere (Meehl et al. 2007).

One of the most variable conditions in the field is temperature, and relatively severe frost, caused by temperatures below −20 °C, can be expected to occur over

42% of the earth's surface (Larcher 1995). Low temperature is therefore a major determinant of the geographical distribution and productivity of plant species. Exacerbating this problem, plants from high latitudes and high altitudes are faced with short growing seasons and the need to grow at low temperatures for prolonged periods to extend the growing season. Thus, the capacity for active photosynthesis during prolonged exposure to low growth temperatures is essential in determining their successful site occupancy and subsequent productivity. At low temperatures, between 0 and 10 °C the enzymes that carry out photosynthesis do not work efficiently, and this decreases the photosynthetic rate. This leads to a decrease in glucose production and will result in stunted growth. For plants inside a greenhouse, the installation of a greenhouse heater and thermostat prevents this from occurring. At medium temperatures, between 10 and 20 °C, the photosynthetic enzymes work at their optimum levels, so photosynthesis rates gauge high. Depending on the particular plant in question, set the greenhouse thermostat to a temperature within this range for best results. At these optimum temperatures, the limiting factor becomes the diffusion of carbon dioxide into the leaves. At temperatures above 20 °C, the rate of photosynthesis decreases because the enzymes do not work as efficiently at this temperature. This is despite the increase of carbon dioxide diffusion into leaves. At a temperature above 40 °C the enzymes that carry out photosynthesis lose their shape and functionality, and the photosynthetic rate declines rapidly.

Despite the importance of low temperatures in determining agricultural productivity and ecological diversity at higher latitudes and altitudes, relatively little is known about either the short-term or long-term effects of cold on the underlying biochemical responses of plant energy metabolism, processes that contribute to plant growth (Hurry et al. 2002). Temperature may be the main environmental factor determining the natural latitudinal and altitudinal distribution of plants. Temperature on earth changes along attitudinal and attitudinal clines, and also with season, most plants are specially adopted to grow over a limited range of temperatures (Smith et al. 1999). Abiotic stress is one of the most important factors responsible for the global loss of crop productivity, decreasing average yield by more than 55% (Bray et al. 2000) for major crops. Improvement in agricultural practices is vital to address global food demand, where stress-tolerant crops have provided promising solution to the farmers. Extremes of temperature limit plant survival and its reproduction, and therefore temperature is required as one of the strongest determinants of plant growth and metabolism (Smith et al. 1999). The extremes of temperature either very high or very low cause severe alteration in the plant metabolism machinery. Both these conditions are stressful for the growth and disturb the plants' physiology and biochemical process (Dwivedi and Raghvendra 2004).

All plant species perform their metabolic functions and complete the life cycle under optimum temperature range. Among the various climatic factors, vital for the growth of autotrophic plants, temperature is the crucial one, besides light, water, atmospheric CO_2, minerals, and oxygen. Temperature in a true sense dictates the rate at which metabolism's activity proceeds in plants, and it must be present at a certain level of warmth in plants for productive metabolism (Ziegler 1998). It is known that temperature affects the main physiological and biochemical processes in plants. The physiological processes most affected are photosynthesis, growth, and respiration when temperatures are above or below the normal ranges of growth (Hasanuzzaman et al. 2013). In the same way, biochemical changes such as alterations in the viscosity, permeability, and fluidity of the cell membrane are produced (Hasanuzzaman et al. 2013), and, at the enzymatic level, studies have reported changes or inhibition of Ribulose-1,5-bisphosphate carboxylase/oxygenase, phosphoenolpyruvate carboxylase, pyruvate phosphate dikinase, adenosine triphosphate synthase, and antioxidant enzymes, among others (Hasanuzzaman et al. 2013; Song et al. 2014). Temperatures below and above the optimum affect metabolism differently. As temperatures rise above the optimum, reaction rates will increase up to the point where physical destruction of metabolic components begins to occur and reaction rates begin to decrease. Between the optimum and this physically destructive temperature, high temperature thermal stress is generally reversible. As temperatures decline below the optimum, reaction rates decrease in a largely reversible manner, though below some temperature, oxidative stress and changes in membrane structure can result in nonreversible damage that requires repair before optimal metabolism can resume. Whereas severe or lethal low temperature stresses are virtually guaranteed in most cotton production environments, severe and lethal high temperature stresses, although entirely possible, generally are limited to water deficit conditions (Holaday et al. 2016). Temperature stress is becoming the major concern for plant scientists worldwide due to the frequent climatic change that adversely influences the sustainable agricultural practices around the globe (Watanabe and Kume 2009; Shah et al. 2011). Therefore, low temperature stress is an inevitable environmental factor that extensively affects plant growth and development. Environmental stresses can reduce growth, quality, yield, and productivity (Thakur et al. 2010; Zhang et al. 2013), which cause significant crop losses (Wang et al. 2003; Vinocur and Altman 2005). Low temperature stress decreases capacity and efficiency of photosynthesis through changes in gas exchange, pigment content, and chloroplast development and also declines chlorophyll fluorescence (Farooq et al. 2009c). There are many studies which indicated that cellular membranes are the primary site of chilling injury in plants (Thomashow 2001; Zhang and Tian 2009). It is damaged in two ways, viz., disruption of protein lipid structure, protein denaturation, and leakage of solutes that indicate the membrane instability. Due to low temperature stress, the fatty acids become unsaturated, and the ratio of lipid-protein in the membrane is altered, ultimately affecting the membrane fluidity as well as structure (Wang and Li 2006). The membrane stability also leads to leakage of free electrons and generation of reactive oxygen species. Differential activation of antioxidant system is also assumed to confer tolerance to chilling stress in various crop plants.

Any change in temperature affect plants greatly, so stress imposed by temperature has important implication for agriculture; for example, it has been conjectured that a 1 °C decrease in the world mean temperature. Temperature can be categorized into two categories, viz., low temperature and high temperature. Low temperature stress is when the temperatures are considerably lower than the optimal growth temperature resulting in chilling temperature stress in plant. In majority of plants this temperature is usually below 10 °C; the low temperature, i.e., lower than 10 °C, can be further subdivided into two main categories. This is decided by the fact whether these is formation of ice within the tissues or nat. Both these conditions also cause water deficit inside the plant (Pearce 1999). Low temperature and frost have a negative effect on the metabolism with a corresponding reduction in crop quality and quantity. Chilling resistance and freeing resistance are complex phenomenon, as most plants are capable of hardening to cold, i.e., acquisition of increased resistance upon exposure to low temperature. This results in irreversible and proper honor loss of cell proteins and disturbed plasma membrane integrity. Tropical plants are more susceptible to chilling than those growing in cold regions, when exposed to temperature higher than 12–13 °C.

Low temperature during winter is a significant factor determining geographical distribution of temperate crop and perennial grass species, and this could be contributed to the negative impact of low temperature on photosynthetic apparatus functioning (Kalaji et al. 2011; Borawska-Jarmułowicz et al. 2014) and plants' capability for acclimation (Harrison et al. 1997). Chlorophyll fluorescence measurements allow evaluating physiological status of plants based on detection changes in some photosystem II components, electron transport chain components, and light-dependent photochemical reactions. It is a very sensitive tool that allows sensing changes in general bioenergetic status of plants' photosynthetic apparatus (Borawska-Jarmułowicz et al. 2014).

An altered nitrate uptake and its metabolism are observed under low temperature. Reduction in nitrogen uptake (Chandra and Pareak 1990), decreased nitrogen partitioning in young shoots induced remobilization of nitrogen from older to younger leaves and inhibition in nitrogen assimilation process at low temperatures. Low temperature treatment modulates the effect of hyper-osmoticum stress on photosynthesis and respiration and results in increased participation of alternative pathway (Dwivedi and Raghvendra 2004). Low temperature stress also manifests increased oxidative damage in plants. There is enhancement of antioxidant defense mechanism of the plant under low temperature stress. A major response of the plant under low temperature stress is enhanced activities and content of each of the as scavenging enzyme and ran enzymatic antioxidant (Shah et al. 2001). As both chilling and freezing temperature cause water deficit inside the cells the compatible solute manital, which alleviates water stress has been shown to protect chloroplastic, phosphokinase from hydroxyl radical of damage under stressfully condition (Shen et al. 1997).

Change in membrane fluidity has been well documented upon exposure of plant to low temperatures. Chloroplast membranes are also susceptible to these changes. A reduction in chloroplast membrane content and its disorganization has been reported in spinach leaves under chilling stress. Swelling of chloroplast and accumulation of starch grains within the chloroplast has also been reported in plants upon exposure to low temperature. Low temperature enhanced the distribution of excitation energy of PS I and altered the amount transferred from chlorophyll "b" to chlorophyll "a" in maize. The most commonly accumulated sugar in plant exposed to low temperature is sucrose. It gets accumulated to about tenfold in spinach leaves with an increase in sucrose phosphate syntheses activity upon exposure to cold temperature (Guy et al. 1992a). Over all photosynthetic CO_2 assimilation decreases to 50% under low temperature stress (Nie et al. 1992).

3.2 Photosynthesis Under Low Temperature

Most plants, irrespective of origin, can photosynthesize and grow in the temperature range of 10–35 °C, and sudden temperature changes within these limits cause no irreversible damage to the photosynthetic apparatus. However, to occupy a habitat with prevailing temperature below or above this range, plants have adapted morphological, physiological, biochemical, and biophysical properties to maintain functional integrity. Examples of such adaptations are provided by plants native to cold, arctic, or hot, desert regions. The physiological processes resulting to growth are genetically controlled, and plants native to cool climates usually develop poorly in warm climates and vice versa. The responses of net photosynthesis to contrasting temperature conditions have frequently been studied in different groups of plants. Low temperature is a relative expression. Plants of tropical or subtropical origin may incur irreversible damage to photosynthesis by temperature below 10 °C, whereas plants from cooler climates may photosynthesize and develop normally at temperature down to around 10 °C.

Photosynthesis is a crucial physiological process which actively determines the survival, growth, development, and yield of the plant system (Banerjee and Roychoudhury 2018a). It is well-known that photosynthesis determines plant growth and development by incorporating atmospheric carbon into carbohydrates (Ribeiro et al. 2006). The effects of both instantaneous and growth temperatures on photosynthetic machinery have attracted attention of plant physiologists who are interested in studying the physiological responses of citrus plants to temperature changes (Ribeiro et al. 2003, 2004). Temperature effects on photosynthesis may occur through an increased oxygenase activity of ribulose–1,5–bisphosphate carboxylase/ oxygenase (Rubisco) (Laisk and Edwards 1998; Ribeiro et al. 2004) as well as impaired photochemical activity due to membrane injury and damage of electron transport chain components (Berry and Björkman 1980). In addition to the inhibition of Rubisco activase (Law and Crafts-Brandner 1999), high temperature may also limit photosynthesis by stomatal closure under photorespiring condition (Ribeiro et al. 2004). As a result, environmental stresses like cold, heat, salinity, drought, and heavy metal toxicity primarily deteriorate the overall efficiency of the photosynthetic machinery. Certain plant species have developed a number of novel mechanisms to robustly tackle the suboptimal conditions. Such evolutionary modifications aid in the gradual adaptation of plants to moderate or extreme conditions of abiotic stresses (Hirayama and Shinozaki 2010). Low temperature or cold is a major type of environmental stressor which limits plant survival, ecosystem abundance, and even productivity. Temperatures ranging between 0 and 15 °C are responsible for cold stress. Freezing stress is caused by still lower temperatures up to −15 °C (Xin and Browse 2001). The mechanisms of tolerance against cold and freezing stress are distinctly different. Cold stress is the major factor which inhibits agricultural production in high altitudes and hilly terrains (Banerjee et al. 2017). Crops of tropical and subtropical origin are incapable of adapting to cold stress. Such plant species might be able to cope up with temporary and transitory alterations induced by cold stress. However, they are incapable of surviving under prolonged cold stress. The stressed plants are manifested with physiological symptoms like chlorosis, wilting, and necrosis (Zhu et al. 2007). In this chapter, we have highlighted the various deteriorative effects of cold stress on photosynthesis of plants.

Photosynthesis is the process converting light energy to chemical energy and storing the energy in the bonds of sugars and other organic compounds. Carbohydrates are synthesized from water and carbon dioxide, and oxygen is released as a by-product. In plants and algae, the process of photosynthesis occurs in specialized organelles known as chloroplasts which contain the photosynthetic apparatus. Photosynthesis mainly occurs in mesophyll cells in plant leaves. Photosynthesis is one of the major metabolic processes which is sensitive to environmental stresses, particularly temperature. It is important to regulate the energy flow to optimize carbon fixation and prevent light-induced damage. Photosynthesis consists of two major processes: the light-dependent electron transport chain and the light-independent carbon fixation cycle or the Calvin-Benson-Bassham cycle. Plants are subjected to several harsh environmental stresses that adversely affect growth, metabolism, and yield. Soil water deficit, salinity, low and high temperatures, flood, pollutants, and radiation are the important stress factors limiting the productivity of crops (Lawlor 2002). Several abiotic (light, temperature, water availability, nutrients, and soil structure) factors affect the growth in higher plants (Lichtenthaler 1996, 1998). Although the plant growth is controlled by a multitude of physiological, biochemical, and molecular processes, photosynthesis is a key phenomenon, which contributes substantially to the plant growth and development. The chemical energy expended in a number of metabolic processes is, in fact, derived from the process of photosynthesis, which is capable of converting light energy into a usable chemical form of energy. This key process occurs in all green plants, whether lower or higher, occurring in oceans or on land as well as in photosynthetic bacteria (Taiz and Zeiger 2010; Pan et al. 2012). However, stressful environments, including drought, salinity, and unfavorable temperatures, considerably hamper the process of photosynthesis in most plants by altering the ultrastructure of the organelles and concentration of various pigments and metabolites including enzymes involved in this process as well as stomatal regulation. Environmental stresses have a direct impact on the photosynthetic apparatus, essentially by disrupting all major components of photosynthesis including the thylakoid electron transport, the carbon reduction cycle, and the stomatal control of the CO_2 supply, together with an increased accumulation of carbohydrates, peroxidative destruction of lipids, and disturbance of water balance (Allen and Ort 2001).

At the physiological level, photosynthesis is one of the first biochemical processes affected on exposure to high or low temperature stress, which in turn severely hampers the growth and productivity (Fig. 3.1). Low temperatures may disturb the

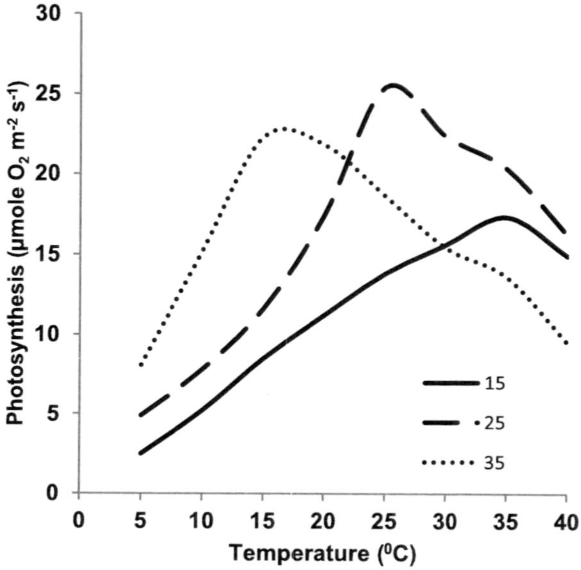

Fig. 3.1 Temperature dependence of photosynthetic rates in leaves grown at three different temperatures (15°, 25°, and 35 °C)

key organs of photosynthesis, including chloroplast and thylakoid membranes, causes swelling of plastids and thylakoid lamellae, vesiculation of thylakoid, accumulation of lipid drops, and ultimately disorganization of entire plastid (Kratsch and Wise 2000; Ishikawa 1996). Low temperature also disrupts the systems including electron transport, carbon cycle metabolism, and stomatal conductivity. Among the photosynthetic apparatus, photosystem II is the primary target of damage under low temperature stress. Moreover, low temperature reduces the activities of stromal and carbon assimilation enzymes like Calvin cycle enzyme and ATP synthase, and restricts RuBisCO regeneration and limits the photophosphorylation (Allen and Ort 2001). Another impact of low temperature exposure is the decline of carbon export from leaves which results in the accumulation of soluble carbohydrates (Strand et al. 2003). Yordanova and Popova (2007) stated that exposure of wheat plants to a low temperature (3 °C) for 48 h and 72 h resulted in decreased levels of chlorophyll, CO_2 assimilation, and transpiration rates. Photosynthesis is strongly reduced below 18 °C (Ramalho et al. 2003), while temperatures around 4 °C dramatically depress photosynthetic performance (Silva et al. 2004). The decline of photosynthetic capacity in low temperature is related to a decrease in the quantum efficiency of photosystem II and the activities of PS I, the ATP synthase, and the stromal enzymes of the carbon reduction cycle (Allen and Ort 2001). Partelli et al. (2009) showed that coffee plant resulted in 30% reduction in chlorophyll a, 27% reduction in chlorophyll b, and 29% reduction of total chlorophyll when the day/night temperature decreases from 25/20° to 13/8 °C. For carotenoids, there are 86% reduction of α-carotene, 57% reduction in β-carotene, 68% reduction in α/β-carotene ratio, 32% reduction in lutein, but 21% increase in zeaxanthin. In *Oryza sativa*, the total chlorophyll content was reduced by

50% due to exposure to low temperature (15/10 °C) for 2 weeks (Aghaee et al. 2011). In a study, Reda and Mandoura (2011) reported that even at low temperature stress of 3 °C the enzyme chlorophylase is still activated and led to a decline in chlorophyll in *Triticum aestivum* plant. The altered growth resulting from cold stress reduces the capacity for energy utilization and causing the feedback inhibition of photosynthesis (Ruelland and Zachowski 2010). In cold-acclimated winter annual plants, photosynthetic activity is maintained by an increase in the abundance and activity of specific enzymes of Calvin cycle (Goulas et al. 2006). This increase could be associated with higher levels of thylakoid plastoquinone-A along with improvement in electron donor pool to PS I (Baena-Gonzalez et al. 2001). Consequently, non-photochemical quenching increases in cold-stressed leaves in parallel with increased levels of zeaxanthin to compensate for the reduced electron consumption in photosynthesis. Zeaxanthin protects the PS II reaction centers from over-excitation (Krol et al. 1999). Moreover, Ruelland and Zachowski (2010) reported that energy dissipation via non-photochemical quenching and electron transport was not only enhanced following cold acclimation but also contributed to protection from oxidative damage. However, the recovery of the activation state of Rubisco, following CO_2 reduction at low temperature, is consistent with a regulatory feedback reflecting limitations in the RuBP regeneration capacity. The possible reasons for the low temperature-induced suppression of photosynthetic traits could be due to decreased transport of carbohydrate from the leaves, stomatal resistance, destruction of the photosynthetic apparatus, and water splitting complex of PS I (Khan et al. 2017). This results in the inhibition of electron transport, uncoupling of electron transfer and energy storage, and loss in the activity and synthesis of key enzymes of Calvin cycle and C4 cycle (Venema et al. 1999a, b; Van Heerden et al. 2003; Garstka et al. 2007). Similar changes in photosynthesis and related traits under chilling stress are also reported by various other workers (Johnsen et al. 2007; DeRidder and Crafts-Brandner 2008; Suzuki et al. 2008).

The typical response of leaf net photosynthesis to temperature can be described by a peaked surface (Fitter and Hay 2002), with low photosynthesis at cool temperatures, increasing to a maximum rate at optimal temperatures and then decreasing again at very high temperatures. This peaked temperature response has been described many times in the literature for a wide range of species (Battaglia et al. 1996; Fitter and Hay 2002). The net photosynthesis and temperate responses are highly variable among studies (Lin et al. 2012). This variability may be the result of both acclimation to temperature, which involves short- to long-term changes at organismal level, and adaptation to temperature, which involves evolutionary changes in different environments. However, the mechanisms underlying variability in net photosynthesis to temperature responses are not well understood, and acclimation and adaptation are therefore not easily predictable.

There are three primary sets of processes that control the net photosynthesis to temperature response, namely, biochemical, respiratory, and stomatal processes. Much of the effort to date to understand variability in net photosynthesis to temperature responses has focused on biochemical processes. The two major biochemical processes thought to limit photosynthesis are the carboxylation of ribulose-1,5-

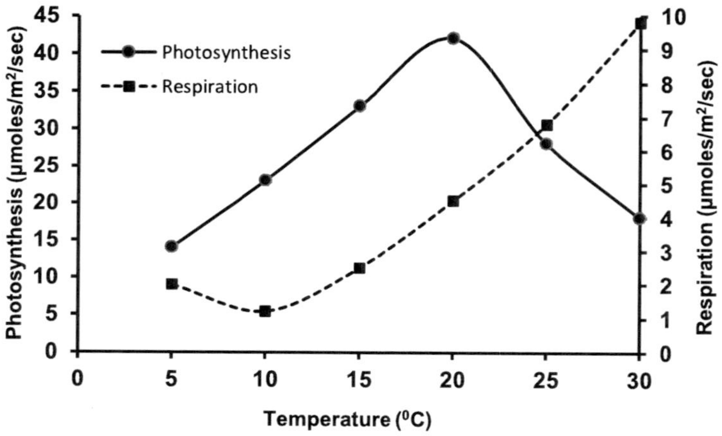

Fig. 3.2 Effect of different temperatures on rates of photosynthesis (O_2 evolution) and dark respiration in leaves of *Cicer arietinum* L

bisphosphate (RuBP) and electron transport photochemistry for the regeneration of RuBP in the Calvin cycle (Farquhar et al. 1980). A third biochemical process, triose-phosphate utilization, can also limit photosynthesis at high internal CO_2 concentrations or with chilling (Sage and Kubien 2007). The biochemical model of C3 photosynthesis proposed by Farquhar et al. (1980) has been widely applied to describe the temperature dependence of component biochemical processes underlying leaf photosynthesis (Dreyer et al. 2001; Medlyn et al. 2002b; Kattge and Knorr 2007). Each of the biochemical processes is temperature dependent (Leuning 2002; Medlyn et al. 2002a). The temperature dependences of these biochemical limitations have frequently been described in the form of mathematical models based on the Arrhenius equation for enzyme activity as a function of temperature (for RuBisCo carboxylation rate), and a peaked function incorporating effects of enzyme deactivation and conformational changes with warming (Harley et al. 1992; Medlyn et al. 2002a). In chickpea (*Cicer arietinum* L.) photosynthetic O_2 evolution showed an increasing trend up to 40 °C, when decreases were probably caused by heat stress (Fig. 3.2).

This means that one O_2 molecule is released while 19 molecules of CO_2 are assimilated under non-photorespiratory condition. On the other hand, the lowest photosynthesis/respiration ratio was found at 50 °C (Fig. 3.3), indicating heat damage to plant metabolism. Respiration is less sensitive than photosynthesis to heat stress, being abruptly inhibited at near-lethal temperature (Hällgren et al. 1991). In fact, a temperature increase of 25 °C (from 25 to 50 °C) caused an increase of 4.8 times in leaf respiration under non-photorespiring conditions (Fig. 3.2). Since respiratory and photosynthetic processes are integrated and related (Raghavendra and Padmasree 2003), one could argue that a higher relationship between photosynthesis and respiration indicates that a higher amount of O_2 is released in relation to its

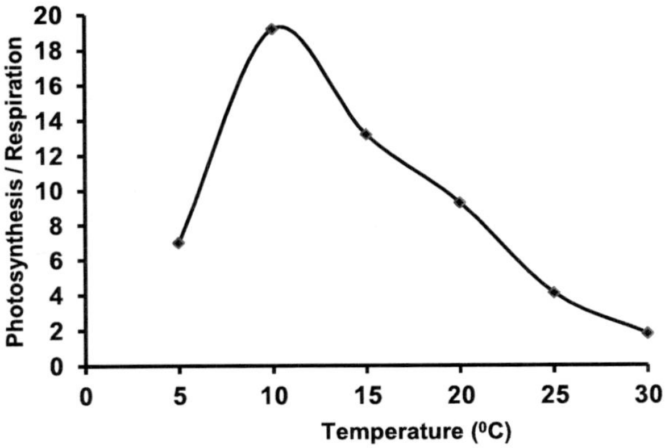

Fig. 3.3 Effect of different temperatures on ratios between rates of photosynthesis and dark respiration in leaves of *Cicer arietinum* L

consumption, which means high efficiency of atmospheric CO_2 gain. The highest photosynthesis/respiration ratios were verified at 30 and 35 °C, being around 19 (Fig. 3.3).

Differences in the net photosynthesis to temperature response among species or growth conditions are commonly attributed to differences in the temperature dependence of these biochemical processes (Walcroft et al. 1997; Hikosaka et al. 2006). Higher plants vary considerably in temperature sensitivities of Rubisco carboxylation (Medlyn et al. 2002a; Kattge and Knorr 2007), which may potentially indicate adaptation in these responses. In addition to the effects of temperature on photosynthetic biochemistry, effects on respiratory and stomatal processes may also be important. Net leaf photosynthesis is given by gross photosynthesis less leaf respiration. Leaf respiration is also a temperature-dependent process and contributes to the overall net photosynthesis to temperature response. It has been suggested that the ratio of gross photosynthesis to dark respiration is lower in warmer versus cooler conditions (Atkin and Tjoelker 2003). It has therefore been hypothesized that increases in respiration with temperature may become increasingly responsible for decreased net photosynthesis under progressively higher temperatures (Kirschbaum 1999). Hence examination of mitochondrial respiration as a component of net photosynthesis is important to consider in the overall temperature response of C3 plants. Stomatal regulation of the internal CO_2 concentration is the third key process determining the overall temperature response.

The role of stomatal conductance in regulating net photosynthesis is well known. Stomata may respond to temperature itself (Hendrickson et al. 2004a; Mott and Peak 2010), as well as the increase in vapor pressure deficit, that tends to accompany increasing temperature (Leuning 1995; Day 2000). Stomatal regulation can affect the net photosynthesis to temperature response in two major ways. First, stomatal aperture depends on the value of vapor pressure deficit during measurements. Additionally, vapor pressure deficit is determined by water vapor content of the air, which can vary significantly among seasons and environments. In warm environments or in summer, the water vapor content of the air tends to be higher than in cool environments or in winter, giving a higher dewpoint and, for a given temperature, a lower vapor pressure deficit. The difference in vapor pressure deficit at a given temperature will change net photosynthesis to temperature response through its effect on stomatal conductance. There is considerable variation among woody plant species in sensitivity to vapor pressure deficit, and some studies suggest that this sensitivity may be dependent on growth temperature (Lloyd and Farquhar 1994; Medlyn et al. 2011). As such, there may also be acclimation of stomatal sensitivity to vapor pressure deficit that affects the net photosynthesis to temperature response measured in different seasons. Both aspects—stomatal sensitivity to vapor pressure deficit and stomatal acclimation to vapor pressure deficit—can contribute to the role of stomatal regulation in the overall net photosynthesis to temperature response.

The ability of crop plants to acclimate to different environments is directly or indirectly associated with their ability to acclimate at the level of photosynthesis, which in turn affects biochemical and physiological processes and, consequently, the growth and yield of the whole plant (Chandra 2003). Chilling stresses alter the normal rates of photosynthesis and other gas exchange attributes in crop plants. While working on sugarcane, Sales et al. (2013) found that drought stress alone or in combination with chilling stress severely affected the leaf gas exchange, photochemical activity, CO_2 assimilation, generation of energetic pressure at the PS II level, electron transport rate, and transpiration. Under chilling stress, low photosynthetic rates in crop plants have been attributed owing to poor stomatal and mesophyll conductance of CO_2, impaired chloroplastic development, restricted metabolite transport, decreased quantum efficiency, and the quantum yield for CO_2 assimilation (Nie et al. 1995; Haldiman 1998; Sowiński et al. 2005). Jung and Steffen (1997) argued that chilling stress reduces photosynthetic efficiencies and enhances photoinhibition process due to over excitation of thylakoid membranes and subsequent impairment of photosynthetic function. Moreover, chilling slows down the rate of CO_2 fixation and restricts the $NADP^+$ supplement available to accept electrons from the electron transport chain (Wise 1995). On the other hand, drought stress reduces photosynthesis by inhibiting leaf area and decreasing the rate of photosynthesis per unit leaf area (Wahid and Rasul 2005; Farooq et al. 2009b). Drought stress slows down carbon fixation rate by directly restricting metabolism or by limiting the CO_2 entry into the leaf (Apel and Hirt 2004; Farooq et al. 2009b). Imbalance between capturing and utilization of light (Foyer and Noctor 2000), decrease in Rubisco activity (Bota et al. 2004), alterations in the photosynthetic pigments (Anjum et al. 2003), and damage to photosynthetic apparatus (Fu and Huang 2001) are some of the major reasons for drought-induced reductions in photosynthesis. However, the drought-induced limitations of photosynthesis through metabolic distortions are more complex than stomatal limitations, and these mainly occur through reduced synthesis of photosynthetic pigments (Reddy et al. 2004). During drought stress, limited intercellular CO_2 concentration leads to the accumulation of reduced photosynthetic electron transport components that can potentially reduce the molecular oxygen, resulting in the production of reactive oxygen species, which are deleterious to photosynthetic apparatus (Basu et al. 2016). Although chilling and drought stresses have some unique effects on photosynthetic functions of plants, yet both these stresses are detrimental for plant photosynthetic apparatus and light harvesting mechanism.

Chilling-sensitive plants are those that suffer a dramatic and often slowly reversible, i.e., damaging, inhibition of plant processes at low temperatures usually below 15 °C (Berry and Björkman 1980). Photosynthetic and growth processes of chilling-tolerant plant species are also restricted by low temperatures but are able to recover relatively rapidly upon return to warmer conditions (Wolfe 1991). Chilling temperatures can limit photosynthesis via stomatal closure, inhibition of thylakoid electron transport and photophosphorylation, inhibition of key enzymes in sucrose and starch biosynthesis, Rubisco inactivation, and limitation of sink strength and phloem loading (Allen and Ort 2001). When low temperature events physically and metabolically restrict the demand for carbon, supply exceeds demand and photosynthesis is down regulated to correct this imbalance. Photosynthesis modeling has been used to demonstrate end-product limitation of photosynthesis in grapevines under both drought and warm conditions (Maroco et al. 2002), but has not been investigated under low temperature conditions (Hendrickson et al. 2004a). During and after chilling, the rate of photosynthesis in the leaves of chilling-sensitive plants decreased, and this is more related to decreasing temperature and lengthening of chilling period and persisted for a long time after transfer of chilled plants in the heat (Allen and Ort 2001; Boese et al. 1997; Li et al. 2004; Sonoike 1998; Strauss et al. 2007; Van Heerden et al. 2003). The physiological reasons for the suppression of photosynthesis are the inhibition of phloem transport of carbohydrates from the leaves, stomata limitation, destruction of the photosynthetic apparatus, damage to water- splitting complex of photosystem one, inhibiting electron transport, and uncoupling of electron transfer and energy storage, changes in the activity and inhibition of synthesis of key enzymes of the Calvin cycle and C4-way (Garstka et al. 2007; Kingston-Smith and Foyer 2000; Smith et al. 1999; Terashima et al. 1998; Van Heerden et al. 2003; Venema et al. 1999a; Yoshida et al. 1996). Cold-sensitive crop species have smaller temperature homeostasis of leaf photosynthesis than cold-tolerant species (Yamori et al. 2009). Chilling of sensitive plants in light had much stronger effects on the photosynthetic apparatus than chilling in the dark (Szalai et al. 1997; Alam and Jacob 2002). It is considered that a disturbance of photosynthesis due to the light chilling is largely a result of photoinhibition and photo-oxidation occurring in the chilling-sensitive plants (but not cold-resistant), as a result of the excess energy of excitation obtained by photosynthetic apparatus. Photoinhibition of photosynthesis is the lowering of photosynthetic activity under excessive illumination during chilling (Wang et al. 2008). It increases with decreasing temperature and increasing light intensity (Janda et al. 1999). Primary site photoinhibition is the PS II. However, it was discovered that photoinhibition occurs at relatively low light and low temperature, and the main site of damage is photosystem (Sonoike 1996, 1999). Decrease of photosynthesis at chilling temperatures may be a consequence of photo oxidative damage to the photosystems in the membranes of chloroplasts, which is manifested by increased lipid per oxidation, degradation of chlorophyll, carotene, and xanthophylls (Fryer et al. 1998;

Kingston-Smith and Foyer 2000). It was caused by activated oxygen species and was associated with reduced antioxidant activity of tissues (Alam and Jacob 2002; Leipner et al. 1997, 2000; Terashima et al. 1998). The inactivation of metabolism is a complex function of both temperature value and duration of its effects (Breidenbach et al. 1990). It is difficult to distinguish between metabolic changes in chilled plants, occurring as a result of chilling damage or preceding it. So, protein content in tissues of chilling-sensitive plants is usually reduced with chilling, mainly due to a sharp decrease in synthesis (Mercado et al. 1997; Levitt 1980). As a result of inhibition of protein synthesis is the increase in the level of free amino acids (Kanda 1998), especially proline (Jouve et al. 1993; Duncan and Widholm 1991), accumulation of which is considered as the element of the mechanism of cold hardening. Low temperatures reduce the activity of many enzymes (Guy 1990). The reasons for this may be the dissociation of multimeric enzymes, protein-lipid and hydrophobic interactions disorders, reversible changes in kinetic properties of enzymes, and allosteric regulation (Matsuo et al. 1994). Keeping the chilling-sensitive plants at low temperature, the concentration of soluble sugars increased, and starch content decreased significantly in all organs (Jouve et al. 1993). Changes in the level of carbohydrates caused by the chilling are associated with impaired respiration, photosynthesis, and the activity of enzymes of carbohydrate metabolism (Ebrahim et al. 1998). Various physiological functions are not equally sensitive to cooling. Physiological dysfunction induced by low temperatures can be converted (or function restored) if the tissue is returned to normal temperature before the appearance of damage. Thus, temperatures below critical trigger the disturbances of physiological functions, but these disturbances do not lead to visible manifestations of injury or to changes in the rate of growth and development, since disturbances of the physiological processes are reversible until they become stable. Irreversible damage arising from prolonged chilling may be caused by the accumulation of toxic metabolites (Graham and Patterson 1982).

3.2.1 Low Temperature Stress and Chloroplast Membrane Integrity

Loss of membrane fluidity is the primary effect of cold stress in plants and cyanobacteria. The cold adapted plant species increase the content of unsaturated fatty acids leading to elevated membrane fluidity at low temperature. The non-acclimated plants on the contrary cannot modify the lipid content and hence are prone to membrane damage (Vogg et al. 1998). Rupture in the chloroplast membrane negatively affects thylakoid structure, chlorophyll content, photosynthetic enzyme activities, and electron transport (Banerjee and Roychoudhury 2018a). Paul et al. (1992) highlighted these changes along with abnormal stomatal closure to be largely responsible for reduced photosynthetic efficiencies in plants growing at low temperature. The osmoprotectants like soluble sugars, proline, and polyamines protect the chloroplast membrane from cold-induced injuries by stabilizing the macromolecular structure (Shao et al. 2006; Banerjee and Roychoudhury 2016). Li et al. (2018a, b) revealed that short-term chilling stress in young tea plants resulted in a significant decrease in maximal photosynthetic efficiency, disintegration of membrane systems, and oxidative damages to the photosynthetic apparatus.

In model systems, it has been demonstrated that carotenoids, as integrated membrane constituents, may provide an alternative for the modification of the membrane structure and for the regulation of the membrane dynamics (Havaux 1998). These apolar compounds serve as protective pigments against high-light stress in photosynthetic organisms (Demmig-Adams and Adams 1996). A carotenoid molecule has a rod-like structure, which often terminates with polar groups. The length of a typical carotenoid matches the thickness of the hydrophobic membrane core (Frank et al. 1999). The molecular architecture of these pigments is responsible for their localization and orientation in the membranes and for their effects on the membrane properties (Gruszecki and Sielewiesiuk 1991; Strzalka and Gruszecki 1994; Gruszecki et al. 1999). Studies on a mixture of liposomes and carotenoids demonstrated that formation of the hydroxy derivatives of these compounds not only increased the rigidity of the lipid phase, but also blocked the penetration of protons through the membrane (Berglund et al. 1999) and significantly decreased the oxygen diffusion-concentration product in the hydrocarbon region of model membranes (Subczynski et al. 1991).

However, the question of whether carotenoid molecules can be present separately in the lipid phase of photosynthetic membranes or whether they are exclusively attached to protein complexes of the photosynthetic apparatus has not been clarified, although indirect evidence points to their existence in lipid bilayers. Havaux and Niyogi (1999) found that, besides their known roles in protection against high-light stress, the xanthophylls in higher plants can also protect the highly unsaturated glycerolipids of the photosynthetic membranes against peroxidation. A similar protective effect was found in a model system (Woodall et al. 1997). It has also been demonstrated that the relative abundance of xanthophylls in the thylakoid membranes of higher plants leads to an enhanced thermostability of the lipid phase in photosynthetic membranes by changing their fluidity (Tardy and Havaux 1997; Havaux and Gruszecki 1993). The use of xanthophyll molecules as internal markers of phase properties in biomembranes likewise suggests the presence of xanthophyll molecules in the lipid phases of these membranes (Gombos and Vigh 1986). To address the problem of the structural aspects of carotenoids in the protection against heat and light stresses, studies were conducted using cyanobacteria (Várkonyi et al. 2002). The cyanobacteria have a well-studied lipid desaturase system (Gombos and Murata 1998), their carotenoid composition is rather simple (Hirschberg and Chamovitz 1994), and their carotenoid content responds to high-light stress. The induction of carotenoid synthesis by low temperature has also been detected in some cyanobacterial strains (Celerin et al. 1998; Miskiewicz et al. 2000).

3.2.2 Low Temperature Stress and Pigment System

The initial photosynthetic reactions are catalyzed by PS I and PS II, which trap light energy and convert it into redox potential energy (Ensminger et al. 2006). Low temperature induces the loss of photosynthetic capacity by triggering PS II activity, thus limiting photon density (Ball et al. 1991). Cold stress rapidly affects the equilibrium in the chloroplasts. The chlorophyll biosynthesis is affected resulting in imbalances in PS II and the antenna complexes (Ensminger et al. 2006). As an urge to increase photon capture, plants exposed to cold usually accumulate higher concentration of accessory pigments like chlorophyll b and carotenoids compared to that of chlorophyll a. The carotenoids and chlorophyll remain associated with the light harvesting complex "a" and light harvesting complex b, respectively. Carotenoids are efficient antioxidants which protect the families of light harvesting polypeptides from cold-induced injuries (Banerjee and Roychoudhury 2016). The xanthophyll pool constituted by violaxanthin, antheraxanthin, and zeaxanthin was found to increase in the stressed plants (Krol et al. 1988). This is probably due to the antioxidant capacity of xanthophylls, independent of non-photochemical quenching (Banerjee and Roychoudhury 2015). In addition, xanthophylls act as the starting material for synthesis of the universal stress phytohormone, abscisic acid. The endogenous levels of abscisic acid have been found to be stimulated in plants growing at low temperature (Gusta et al. 2005). The synergistic application of cold and high intensity of light enhanced the accumulation of total xanthophylls. The antheraxanthin and zeaxanthin fractions increased fourfold at 5 °C compared to that at normal temperature (Savitch et al. 2002). The *Arabidopsis* seedlings exposed to high light intensity at low temperature exhibited increased lipid peroxidation and chlorophyll bleaching leading to photoinhibition. On exposure to cold temperatures, the winter conifers organize their photosynthetic apparatus by decreasing the number of functional PS II reaction centers. A steep loss of the light harvesting chlorophyll occurs. A large thylakoid protein complex is generated and further associated with the light harvesting complex II, PS I and PS II (Ensminger et al. 2006). The PS I in *Arabidopsis* plants were more affected compared to PS II under low light treatments at chilling temperature (Zhang and Scheller 2004). Gombos et al. (1994a) reported that low temperature stress drastically slowed down the replenishing of D1 proteins (core reaction center of the PS II) leading to photoinhibition at photosystem II. It has been observed that the equilibrium maintained between degraded D1 and newly synthesized D1 protein determines the extent of photosystem II dependent photoinhibition (Aro et al. 1993). Cold stress delays this recovery process, which requires newly synthesized D1 to be inserted into the thylakoids and finally incorporated in the photosystem I complex (Banerjee and Roychoudhury 2018a).

Photosynthesis consists of vectorial transfer of electrons from water in the lumen to nicotinamide adenine dinucleotide phosphate ($NADP^+$) in the stroma mediated by a set of redox carriers. Chlororespiration involves electron transfer reactions from stromal reductants to oxygen via the plastoquinone pool (Rumeau et al. 2007). The chlororespiratory components respond to abiotic stresses like drought, heat, and high light. The level of nucleus-encoded, plastid-localized terminal oxidase (PTOX) increased in the alpine plant species and *Arabidopsis* seedlings under high light intensity at low temperature (Laureau et al. 2011; Ivanov et al. 2012). However, the content of the plastid encoded NADH dehydrogenase complex remained unchanged during cold stress (Ivanov et al. 2012). Paredes and Quiles (2015) observed that the *Hibiscus rosa-sinensis* L. plants were sensitive to cold stress due to the reduced photochemical efficiency of PS II and decrease in the electron transport capacity. Exposure to low temperature resulted in an increase in electron donation by NADPH and ferredoxin to plastoquinone. These observations suggested that the cyclic electron flow confers photosynthetic protection. The cyclic flow did not increase in plants where only the stem and not the root were exposed to cold. As a result, due to insufficient dissipation of excess photons, the PS II exhibited significant damage. The chlororespiratory enzymes like PTOX and NADH dehydrogenase complex increased in plants only where the stem was cooled, whereas their levels remained unchanged in the sets where the entire plant was exposed to stress. Thus, chlororespiration participates in the protection of PS II during cold stress in the absence of an increased cyclic electron flow around PS I (Paredes and Quiles 2015). Zhou et al. (2017) investigated the effects of low temperature in the cold-acclimatized *Rhododendron chrysanthum* L. using a proteomic approach. Among 153 identified photosynthesis-related proteins, seven including the Rubisco large subunit were found to increase under cold stress compared to the control seedlings.

3.2.3 Effect of Low Temperature on Stomata

Reduced air and leaf temperature will usually reduce evaporative demand. However, cool roots reduce hydraulic conductivity and substantially inhibit water uptake from the soil. Owing to the high specific heat capacity of the soil, transient root chills are far less common in the field than low shoot temperatures. As well as not reflecting a natural chill, experiments that chill whole potted plants have to be interpreted with caution because of the ease with which unrealistic drought stress can be induced. This is because the soil warms up after a chill much more slowly than do leaves and the surrounding air, and therefore, unless humidity is maintained close to saturation, the evaporative demand will increase faster than the water supply.

Stomata play an important role in regulating photosynthesis and transpiration. Stomatal density is the number of stomata per unit leaf area, whereas the stomatal index is a function of both the number of stomata and the size of the epidermal cells. Stomatal density and index are ecophysiological parameters that can be affected by environmental factors such as drought, humidity, light, heavy metals, and CO_2 concentration (Royer 2001; Banon et al. 2004; Vezza et al. 2018). Numerous studies have shown that stomatal size, density, and pattern of stomatal distribution in leaves are controlled by long-term temperatures effects (Maherali et al. 2002; Franks and Farquhar 2007; Zheng et al. 2013; Hill et al. 2014; Yan et al. 2017; Wu et al. 2018). Photosynthetic capacity is closely linked to stomatal density (Xu et al. 2016). Wu et al. (2018) reported that warming had a different effect on two cultivars of *Syzygium*. High temperature significantly decreased stomatal size in *Syzygium rehderianum* and stomatal density in *Syzygium superba*. Warming also decreased photosynthesis in *Syzygium superba*, but had no significant effect on photosynthesis in *Syzygium rehderianum* (Wu et al. 2018). Wu et al. (2018) reported that a reduction in stomatal density had a negative effect on photosynthesis, while maintaining stomatal conductance and the photosynthetic rate was probably due to a reduction in stomatal size. Accordingly, investigating stomatal responses in the leaves of different *Stevia* cultivars experiencing low temperature may be important for gas exchange and photosynthetic capacity (Hajihashemi et al. 2018).

In many chilling-sensitive species, such as cucumber, tomato, bean, cotton, and soybean, low temperatures can cause stomata to appear locked open and unable to respond normally to leaf water deficit (Martin et al. 1981; Bagnall et al. 1983; Eamus 1987; Guye and Wilson 1987). Such species will quickly wilt if not warmed under extremely high humidity, especially if their roots are still cool. By contrast, when more realistic, near saturation conditions with warm roots are used, stomatal closure following a chill can be observed in these and other species. There are two potential causes of chill-induced stomatal closure. First, a direct inhibition of mesophyll photosynthesis could lead to a rise in internal CO_2 concentration, which in turn induces stomatal closure. Alternatively, stomata themselves might be the primary target of the chill, and their closure could lead to a reduction in internal CO_2 concentration, precipitating a decline in photosynthesis. Approaches that distinguish between these two different mechanisms primarily involve the determination of the dependence of CO_2 uptake on internal CO_2 concentration. Such assessments of stomatal limitation of photosynthesis frequently attribute the chill-induced inhibition of CO_2 fixation to a combination of stomatal and non-stomatal effects (Martin et al. 1981; Flexas et al. 1999; Allen et al. 2000; Bongi and Long 1987). However, because interactions between components of photosynthesis appear to be mediated by internal CO_2 concentration, identification of the extent to which stomata limit photosynthesis remains problematic. The regulation of leaf stomatal conductance is a key phenomenon in plants as it is vital for both a prevention of desiccation and CO_2 acquisition (Dodd 2003; Medici et al. 2007). The decrease in photosynthetic rate under stressful conditions (salinity, drought, and extreme temperatures) is normally attributed to suppression in the mesophyll conductance and the stomata closure at moderate and severe stress (Flexas et al. 2004; Chaves et al. 2009).

There are reports on the effect of cold stress on stomatal apparatus and epidermal cells (Boese and Huner 1990; Ameglio et al. 2001). It has been reported (Hajihashemi et al. 2018) that the cold stress significantly affected leaf and stem anatomy in the *Stevia* cultivars, increasing the stomatal number, size, density, and index, and decreasing epidermal cell density. In spinach, stomatal number and epidermal cell density did not change when exposed to low temperature (5 °C) stress (Boese and Huner 1990). In *Stevia*, the increase in stomatal index and density could increase water loss and reduce leaf water potential, which may explain the observed reductions in water use efficiency, intercellular CO_2, and net photosynthesis in the nine tested cultivars. Some plants exhibit low leaf water potentials in response to chilling stress, which may limit photosynthetic activity (Pardossi et al. 1992; Wilkinson et al. 2001). Cold stress increased the width of derma, epiderma, xylem, phloem, and sclerenchyma, which was followed by an increase in stem diameter in all *Stevia* cultivars. Equiza et al. (2001) suggested that increased epidermal thickness in response to cold stress is an adaptive reaction to increase freezing tolerance. Sclerenchyma is one of the main supporting tissues of growing plant organs (Leroux 2012); its increase in response to cold stress may have increased stem hardness in the *Stevia* cultivars. The increase in xylem vessels in response to cold stress may increase water transport from roots to leaves to prevent reductions in water potential. The diameter and frequency of xylem vessels are critical determinants of water conductance (Lovisolo and Schubert 1998).

3.2.3.1 Low Temperature and Chlorophyll

Chlorophyll is one of the major chloroplast components for photosynthesis, and relative chlorophyll content has a positive relationship with photosynthetic rate. The decrease in chlorophyll content under drought stress has been considered a typical symptom of oxidative stress and may be the result of pigment photo-oxidation and chlorophyll degradation. Photosynthetic pigments are important to plants mainly for harvesting light and production of reducing powers. Both the chlorophyll a and b are prone to soil dehydration (Farooq et al. 2009b). Decreased or unchanged chlorophyll level during drought stress has been reported in many species, depending on the duration and severity of drought (Kyparissis et al. 1995; Zhang and Kirkham 1996). Drought stress caused a large decline in the chlorophyll a content, the chlorophyll b content, and the total chlorophyll content in different sunflower varieties (Manivannan et al. 2007).

In the process of photosynthesis (Fig. 3.4), two key events occur mandatorily; light reactions, in which light energy is converted into ATP and NADPH and oxygen is released, and dark reactions, in which CO_2 is fixed into carbohydrates by utilizing

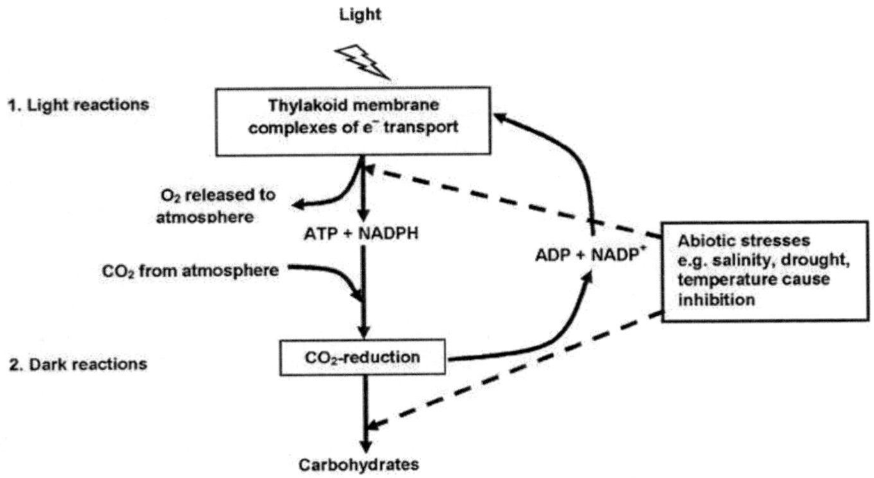

Fig. 3.4 Light and dark reaction of photosynthesis

the products of light reactions, ATP and NADPH (Lawlor 2001; Taiz and Zeiger 2010; Dulai et al. 2011). There are two main pathways of CO_2 fixation, C3 and C4. Plants have been categorized into C3, C4, or C3-C4 intermediate plants depending on the spatial distribution of these two pathways within leaf tissues or as crassulacean acid metabolism plants with a temporal distribution (Doubnerová and Ryšlavá 2011; Freschi and Mercier 2012). Plants possessing the different types of photosynthetic mechanisms are adapted to specific climatic zones. For example, C3 plants, representing over 95% of the Earth plant species, thrive well in cool and wet climates, with usually low light intensity. In contrast, C4 plants occur in hot and dry climatic conditions with usually high light intensity. Generally, C4 and crassulacean acid metabolism plants are the best adapted to arid environments, because they have higher water-use efficiency than that of C3 plants. C4 plants have higher photosynthetic efficiency than C3 plants, namely, in arid, hot, and under high light conditions, because they possess an additional carbon fixation pathway and characteristic anatomy to limit photorespiration. Furthermore, crassulacean acid metabolism plants can effectively save metabolic energy and water during harsh environmental conditions by closing their stomata during the day (Taiz and Zeiger 2010). Different growth and development related processes depend on the interplay of intracellular organelles. The chloroplast is the key site for photosynthesis, in which both light and dark reactions of photosynthesis take place. However, this organelle is highly sensitive to different stressful environments such as salinity, drought, extremes of temperature, flooding, varying light intensity, and ultraviolet radiation, and it plays a premier role in the modulation of stress responses (Biswal et al. 2008; Saravanavel et al. 2011). All these stresses reduce the photosynthetic rate by stress-induced stomatal or non-stomatal limitations (Saibo et al. 2009; Rahnama et al. 2010). For example, soil water deficit stress, particularly at its mild intensity, can inhibit leaf photosynthesis and stomatal conductance in most green plants (Medrano et al. 2002).

Chlorophyll fluorescence is a direct tool for detecting photosystem II efficiency, as the ratio of F_v to maximal fluorescence emission (F_v/F_m) (Logan et al. 2007; Adams et al. 2013a, 2014). Although photoinhibition reduces efficiency of photosynthetic electron transport, it can be regarded as a means of protection of the photosynthetic apparatus in response to environmental stress; its key characteristics are a reduction in F_v/F_m and dissipation of light energy as heat. Despite the reduction in photosynthetic capacity, it is often accompanied by increased sugar accumulation, which is a typical stress response in many plants (Adams et al. 2006, 2013a, 2014; Logan et al. 2014). In *Vicia minor*, a reduction in photosynthetic rate and photosystem II efficiency in response to low temperature was accompanied by soluble sugar accumulation (Adams et al. 2013a, 2014). Characterization of the efficiency of PS I and II is another tool that reflects the efficiency of both PS I and II in different conditions (Živčák et al. 2008). Indications of photosynthetic apparatus responses to environmental stress can also be based on variations in intercellular CO_2 concentration, photosynthetic rate, or water use efficiency (Adams et al. 2013a, 2014; Cohu et al. 2014).

The depression in photosynthetic performance of maize leaves at low temperatures is associated with reductions in both the maximum quantum yield of photochemistry (Φ_{max}) (Stirling et al. 1991) and the light-saturated rate of carbon dioxide assimilation (A_{sat}) (Long et al. 1983; Baker 1994). These depressions in photosynthetic activity of maize leaves have been ascribed to chill-induced photoinhibition of PS II (Baker et al. 1988; Ortiz-Lopez et al. 1990). Photoinhibition of PS II, however, is unlikely to be the sole cause for the large reduction in photosynthetic productivity at low temperatures. Nie and Baker (1991) demonstrated that an effect on chloroplast biogenesis could also contribute to the reduction in photosynthesis at low temperatures. They observed that exposure of leaves to low temperatures resulted in the preferential loss of some thylakoid polypeptides which have been correlated with a loss of activity of PS I and II. Growth temperature was observed to have negligible effects on the content of the Light Harvesting Complexes (20–26 kDa), LHCII (27–28 kDa), and the extrinsic water splitting polypeptide (33 kDa), which are encoded by the nuclear genome. On the other hand, thylakoid polypeptides which are encoded by the chloroplast genome, namely, chlorophyll a protein (65–70 kDa), α and β subunits of the coupling factor (58 and 57 kDa), cytochrome f (33 kDa), D1 protein (32 kDa), PS I subunit (22 kDa), and cytochrome b6/f subunit IV (17 kDa), were found to decrease as a proportion of the total thylakoid protein with decreasing growth temperatures (Nie and Baker 1991; Bredenkamp et al. 1992; Hahn and Walbot 1989). Most of the studies on the effect of low temperature on the development of the thylakoid membrane focused mainly on one maize cultivar (Val and Baker 1989; Ortiz-Lopez et al. 1990; Nie and Baker 1991; Stirling et al. 1991; Bredenkamp et al. 1992). Moseki (2004) studied the effects of low temperatures on the development of the thylakoid membrane which were investigated in a range of *Zea mays* L. cultivars, by examining the polypeptide profiles isolated from plants grown at suboptimal temperatures in controlled environment cabinets and reported that the six cultivars examined suggest that thylakoid polypeptide profiles may be used as a criterion for screening low temperature tolerance in maize lines.

3.2.3.2 Regulation of Sucrose-Starch Accumulation Under Low Temperature

Short-term low temperature stress results in an inhibition of sucrose biosynthesis which leads to restriction in phosphate recycling and photophosphorylation (Labate and Leegood 1988; Labate et al. 1990). This occurs because the maximum capacity for triose phosphate utilization is temperature dependent (Sage and Sharkey 1987) and the requirement of the chloroplast for phosphate is increased at low temperature (Leegood and Walker 1983). Restriction in triose phosphate utilization may trigger feedback mechanisms that reduce the rate of photosynthetic electron transport, limit ATP supply, and thus, reduce the rate of photosynthetic carbon assimilation (Sharkey 1990; Pammenter et al. 1993). In contrast to low temperature stress, cold acclimation of cereals results in an increase in photosynthetic capacity (Hurry and Huner 1991; Hurry et al. 1994). The increased CO_2 assimilation rates following cold acclimation have been reported to be associated with increase in the activities of

Rubisco, stromal and cytosolic fructose-1,6-bisphosphate (Hurry et al. 1995), as well as sucrose-phosphate synthase (Guy et al. 1992b; Holaday et al. 1992). However, adjustment of CO_2 assimilation rates and concomitant adjustment in the sucrose biosynthetic pathway are species and cultivar dependent (Holaday et al. 1992). In fact, winter wheat cultivars can be distinguished from spring by their ability to adjust their photosynthetic capacity to keep quinine in oxidized state, and thus an increased resistance to photoinhibition (Öquist et al. 1993a; Hurry et al. 1994). Furthermore, the capacity to adjust photosynthetic capacity and the redox state of quinine has been correlated with the freezing tolerance of cereals (Öquist et al. 1993a). It has been proposed that cold-acclimated spring cultivars exhibit the capacity to adjust CO_2 assimilation and hence the redox state of quinine because of an ability to overcome the potential restriction of triose phosphate utilization imposed by low temperature (Hurry et al. 1994). It has been hypothesized (Savitch et al. 1997) that cold acclimated spring should exhibit characteristics of feedback-limited photosynthesis, that is, reduced rates of CO_2 assimilation and electron transport due to restriction of triose phosphate utilization and cod acclaimed winter wheat should exhibit no restriction.

3.3 Impact of Suboptimal Temperature on Photosynthesis

Photosynthesis is the principal process of capturing light energy to form carbohydrates and is sensitive to low temperature (Adams and Demmig-Adams 2004; Ensminger et al. 2006; Fan et al. 2015; Stewart et al. 2016). Among various abiotic stresses, low temperature is one of the most important factors limiting the crop productivity and distribution of plants. Low temperatures are defined as low but not freezing temperatures from 0 °C to −15 °C; they are common in nature and can damage many plant species. Chilling temperatures range from 0 °C to 15 °C, whereas freezing temperature starts from 0 °C and further. Chilling and freezing temperatures are different from each other and have different tolerance mechanisms. Low temperature stress is a major factor that limits the agricultural productivity in hilly areas and high altitude areas especially (Xin and Browse 2001). Plants of tropical and subtropical origins are sensitive to chilling temperature (0–10 °C) and are incapable of cold acclimation. Whereas plants belonging to temperate climatic regions are considered to be chilling tolerant with variable degrees of temperatures and can increase their freezing tolerance by being exposed to chilling, nonfreezing temperatures; this is known as cold acclimation (Zhu et al. 2007). Brief exposure of plants to low temperature stress may cause necrosis, wilting, and chlorosis and finally leads to death of plants. Photosynthesis and growth processes of chilling-tolerant plant species are restricted by low temperature but are able to recover relatively rapidly returning to warmer condition (Wolfe 1991). Grapevines which are often grown in cool climates for premium quality wine production are frequently exposed to chilling temperature well below 20 °C (Gladstone 1992). Studies have shown that both non-acclimated and acclimated grapevines are sensitive to chilling temperature (Hendrickson et al. 2004a).

In many habitats, crop and native plants are periodically exposed to low temperatures; this can significantly reduce their productivity (Fryer et al. 1995; Greaves 1996; Haldiman 1998). Among the many physiological mechanisms responsible for this is the potential for damage to the photosynthetic apparatus caused by high light (Baker 1994; Jones and Demmers-Derks 1999). The capacity of plants to tolerate low temperatures, and to adapt to them, varies among plant species and cultivars. The ability of French bean plants to tolerate short periods of high irradiance (simulating natural sunflecks) under conditions when utilization of the reducing power generated by photosynthetic electron transport in carbohydrate production and growth is limited by low temperatures is investigated in this paper. Absorption of excess light can be deleterious because it leads to increased production of singlet oxygen and reduced reactive oxygen species; these react with, and damage, proteins and other components of photosystems leading to true photoinhibitory damage. Critical conditions of low temperature are commonly experienced by arctic or alpine species in their short summer (Jones and Demmers-Derks 1999) and by crop species in temperate zones on cold spring mornings. In such situations, controlled dissipation of excess excitation energy absorbed by the chlorophylls is essential to protect the photosynthetic apparatus. French bean was chosen for the present study because there is a substantial background literature on temperature/photoinhibition responses of this species, which provides a good comparative basis for the present extension to realistic sunfleck treatments (Tsonev et al. 2003). Furthermore, it is an important crop whose growth in temperate regions is significantly restricted by sensitivity to low temperatures (Nishida and Murata 1996). This sensitivity is compounded by photoinhibition at high light. Nevertheless, the sensitivity can be modulated, and there is good evidence that *Phaseolus* can undergo substantial hardening or acclimatization to low temperatures (El-Saht 1998). There is also genetic variability in tolerance to low temperatures (Melo et al. 1997).

Factors that enhance photoinhibition at low temperature include reduced utilization of energy in photosynthetic carbon metabolism; restricted rates of D1 protein turnover; decreased xanthophyll cycle activity; and inadequate rates of removal of active oxygen species (Krause 1994a). The effect of oxygen on the separate processes of photoinhibition and photosynthetic down-regulation is poorly understood because of its dual role (Krause 1994b). On the one hand, it has protective effects through photorespiration and the Mehler reaction (Osmond and Grace 1995), although Ruuska et al. (2000) have suggested that the Mehler reaction may have less effect than once thought. On the other hand, oxygen promotes photoinhibition mainly by triggering D1 degradation and by increasing the probability of forming reactive oxygen species such as the superoxide anion radical (Brestic et al. 1995; Wiese et al. 1998).

Another important aspect of down-regulation is the degree to which acclimatization to cold may enhance protective mechanisms. Acclimatization of the photosynthetic apparatus to low temperature and to high light is well documented, but the mechanisms are still unclear. These include increased capacity of carbon assimilation at low temperature and increased activities of scavenger systems. Öquist et al. (1993a, b) reported that irrespective of temperature (0–25 °C) or the state of cold-hardiness, the susceptibility of photosynthesis in cereals to photoinhibition is controlled by the redox state of quinine A pool, the primary photosystem II acceptor. These authors suggested that the primary reason for photoinhibition at low temperature is an increase in the proportion of closed reaction centers due to slower rates of CO_2 assimilation at low temperature (Tsonev et al. 2003).

The genus *Coffea* has approximately 100 species, with commercial relevance for *Coffea arabica* and *Coffea canephora* (Davis et al. 2006). Brazil is the world's largest coffee producer and exporter. Low temperatures interfere with the photosynthetic process in several ways (Partelli et al. 2009). They lower stomatal conductance, photochemical efficiency of the photosystem II, thylakoid electron transport rate, enzyme activity and carbon metabolism, as well as the photosynthetic pigment complex systems (Suzuki et al. 2008) and membrane lipids (Campos et al. 2003). Coffee is particularly sensitive to cold, especially *Coffea arabica, Coffea canephora*, and *Coffea dewevrei* (DaMatta et al. 1997; Ramalho et al. 2003), which are responsible for over 99% of the world's coffee production. Previous works showed that photosynthesis is strongly reduced below 18 °C (Ramalho et al. 2003), while temperatures around 4 °C dramatically depress photosynthetic performance and yield (DaMatta et al. 1997; Silva et al. 2004). However, a gradual exposure to low positive temperatures highlighted the possibility of photosynthetic cold acclimation in some coffee genotypes (Ramalho et al. 2003), which was related to membrane stability (Campos et al. 2003). Oxidative stress often occurs when plants remain under adverse environmental conditions (drought, high irradiance, extreme temperatures, and nutritional stresses) due to changes in the light energy capture balance and in its use (Demmig-Adams et al. 1995; Ramalho et al. 2003). In fact, when the energy trapped by the photosynthetic pigments exceeds consumption requirements for carbon assimilation, increased production of highly reactive molecules of chlorophyll (^{3}Chl) and oxygen ($^1O_2$,O_2^-, H_2O_2 and OH) may occur, leading to damages in the photosystems, in enzymes, in membrane lipids, and in DNA (Suzuki et al. 2008). Therefore, the xanthophyll cycle is an important photoprotective mechanism when excess of light energy occurs, since it performs thermal energy dissipation, preventing overproduction of highly reactive molecules of chlorophyll and oxygen (Ma et al. 2003; Ramalho et al. 2003; Cai et al. 2007).

As a result of exposure to low temperature, many physiological and morphological changes have been correlated with visible symptoms like wilting, chlorosis, and necrosis that lead to death of plants (Adam and Murthy 2014). Plants ultimately show marked changes in biochemical and cell functions such as the following:

Biomembrane lipid composition:

- Small molecular accumulation (Gilmour et al. 2000; Yamaguchi-Shinozaki and Shinozaki 2006).
- Amino acid (Wanner and Junttila 1999; Shao et al. 2006).
- Redistribution of intercellular calcium ions (Knight et al. 1998).
- Cellular leakage of electrolytes.

- Diversion of electron flow to alternate pathways (Seo et al. 2010).
- Enzyme activities (Ruelland and Zachowski 2010).

When a plant is exposed to a suboptimal temperature after acclimation to an optimum or higher temperature, net carbon assimilation declines immediately, whether the plant is able to acclimate eventually to lower temperatures or not (Hendrickson et al. 2004b; Pérez et al. 2001). Factors from the restriction of stomatal conductance to leaf cell biochemical effects can be involved in this decline in net carbon assimilation. The stomatal effects are evident even when the temperature of the roots of the plants is maintained at an optimum temperature to maintain water absorption while the shoots are being chilled (Allen and Ort 2001; Bunce 2000; Perera et al. 1995). However, biochemical factors often are more constraining than stomatal aperture, especially when the temperature is far below the optimum. Hendrickson et al. (2004b) indicated that as the temperature falls from optimum to approximately 15 °C, changes in net carbon assimilation for grape involve Rubisco deactivation. Other studies have noted similar negative effects of chilling on Rubisco activation, either after an extended period of chilling (Holaday et al. 1992) or after night chilling (Allen et al. 2000). The deactivation of Rubisco could be related to a reduction in the availability of ATP from electron transport in the chloroplast. This ATP is needed by Rubisco activase in the process of removing inhibitory sugar phosphates from the active site of Rubisco, thus allowing Rubisco to be activated (Byrd et al. 1995). Acclimation to chilling by chilling-tolerant species involves an increase in the activation state of Rubisco and an increase in the total extractable activity of the enzyme (Campbell et al. 2007; Holaday et al. 1992; Hurry et al. 2000; Pérez et al. 2001).

With chilling, especially below 15 °C, net carbon assimilation is constrained, at least in the short term, by the decreased ability of the Calvin-Benson cycle to regenerate ribulose 1,5 bisphosphate (RuBP) for Rubisco (Hendrickson et al. 2004b; Labate and Leegood 1988; Leegood and Furbank 1986; Perera et al. 1995; Silim et al. 2010). In such situations, CO_2 assimilation by C3 plants becomes insensitive to changes in O_2 concentration and net carbon assimilation saturates at a low intercellular CO_2 concentration, basically eliminating any constraint of stomatal conductance on net carbon assimilation that might occur at slightly suboptimal temperatures. Studies with chloroplasts (Leegood and Walker 1983; Mächler et al. 1984) and leaves (Hendrickson et al. 2004b; Labate and Leegood 1988; Leegood and Furbank 1986; Silim et al. 2010) indicated that such exposure to a decline in temperature causes a phosphate limitation to net carbon assimilation. Allowing excised leaves to accumulate phosphate prior to the exposure to chilling temperatures alleviates this limitation (Hendrickson et al. 2004b). Phosphate is required for ATP synthesis in the chloroplast to drive the regeneration of RuBP in the Calvin-Benson cycle net carbon assimilation slowing of the rate of the release of phosphate from phosphorylated sugars could restrict the amount of phosphate available in the chloroplast for ATP synthesis. The synthesis of ADP-glucose for starch synthesis by ADP-glucose pyrophosphorylase (AGPase) in the chloroplast is one such process (Hendriks et al. 2003). This enzyme uses glucose-1-phosphate as substrate along with ATP producing ADP-glucose and pyrophosphate, which can be cleaved to release phosphate. However, for spinach and wheat acclimation to low temperature, AGPase activity does not increase (Martindale and Leegood 1997; Savitch et al. 1997), suggesting that this pathway to release phosphate is not of major importance during chilling for these species. In contrast, cotton subjected to 15 °C does increase the AGPase activity (Perera et al. 1995). Another process that releases phosphate is sucrose synthesis in the cytosol. The phosphate is shuttled into the chloroplast by the triose-phosphate translocator of the inner chloroplast envelope. Acclimation to chilling exposure increases the extractable activity of sucrose phosphate synthase and cytosolic fructose bisphosphatase (Guy et al. 1992a; Holaday et al. 1992; Hurry et al. 2000; Lundmark et al. 2006; Pérez et al. 2001; Savitch et al. 1997) that are critical in sucrose synthesis. Reducing the expression for sucrose phosphate synthase reduces the ability of *Arabidopsis thaliana* (L.) Heynh. to acclimate to chilling (Strand et al. 2003). Also, in concert with an increase in sucrose synthesis, triose-phosphate exchange for phosphate appears to be an important process to upregulate during chilling (Lundmark et al. 2006). These changes during acclimation could improve phosphate cycling to the chloroplast.

Another major biochemical problem that could restrict RuBP regeneration is inactivation of the stromal fructose bisphosphatase and sedoheptulose bisphosphatase in the Calvin-Benson cycle (Allen and Ort 2001; Holaday et al. 1992; Payton et al. 1997; Pérez et al. 2001; Sassenrath et al. 1990). These enzymes are activated in the light by the reduction of disulfide bonds utilizing reducing power ultimately from the light-absorbing reactions (Scheibe 1990). During chilling, the critical sulfhydryl groups of stromal fructose bisphosphatase and sedoheptulose bisphosphatase become more oxidized, which alters the kinetic properties of the enzymes. These enzymes in chilling-tolerant plants tend to deactivate to a lesser degree in the short term (Pérez et al. 2001). The ability to acclimate to chilling conditions and improve net carbon assimilation involves at least full activation of stromal fructose bisphosphatase (Holaday et al. 1992) and an increase in the extractable activity of both stromal fructose bisphosphatase and sedoheptulose bisphosphatase (Holaday et al. 1992; Hurry et al. 2000; Pérez et al. 2001). Despite the apparent sensitivity of photosynthesis to chilling, photosynthate (carbohydrate) accumulates in the leaf with time at low temperature (Campbell et al. 2007; Hurry et al. 2000; Paul et al. 1992; Pérez et al. 2001; Venema et al. 1999a). Over several days, an increase in leaf carbohydrate is known to cause a downregulation of genes for certain photosynthetic proteins of the Calvin-Benson cycle and light absorption (Koch 1996; Paul and Foyer 2001; Strand et al. 1997). Thus, plants of chilling-sensitive species likely would suffer further reductions in net carbon assimilation with long exposures to chilling if the export of carbohydrate from leaves could not be improved. Although this carbohydrate accumulation occurs for most plants, whether they are able to acclimate to chilling or not, it does not appear to affect negatively gene expression for chilling-tolerant plants, such as *Arabidopsis thaliana*, during the acclimation process (Strand et al. 1997). In fact, the gene expression for many critical enzymes is increased as part of the acclimation process, consistent with the change in the extractable activities of key enzymes of carbon metabolism.

Interestingly, during acclimation to low temperature, respiration rates reach a maximum before rates of net carbon assimilation do (Campbell et al. 2007), suggesting that the metabolic function is aimed at improving the processes that utilize carbohydrate before carbohydrate synthesis increases.

Photosynthesis in some plant has been shown to be sensitive to reductions in temperature below optimum (DeRidder and Crafts-Brandner 2008; Königer and Winter 1993; Payton et al. 1997; Perera et al. 1995; Winter and Königer 1991). Even a temperature of 20 to 21 °C can cause a substantial reduction of net carbon assimilation in the short term (DeRidder and Crafts-Brandner 2008; Winter and Königer 1991). In addition, growth at 20 °C reduces leaf area, further reducing the total carbon gain (Winter and Königer 1991). When researchers utilize plants in pots for their analysis, a large decrease in stomatal conductance often is observed in the short term, possibly due to the chilling of the root system that reduces water availability to the leaves (DeRidder and Crafts-Brandner 2008; Perera et al. 1995). Keeping the roots at optimum temperature during the chilling period maintained stomatal conductance and improved net carbon assimilation in one experiment (DeRidder and Crafts-Brandner 2008) but had no effect on net carbon assimilation in another experiment (Perera et al. 1995). These effects of chilling in the light on gas-exchange parameters for cotton, especially on stomatal conductance, can extend into a subsequent period of optimum temperature, as might occur in the morning following a cool night (Zhang et al. 2012). However, there appears to be genotypic variation in the sensitivity to a period of chilling and the rapidity with which photosynthesis recovers at optimum temperature (Wu et al. 2012).

Analyses of the relative rates of photosystem II electron transport with those of CO_2 assimilation frequently, but not always (Kingston-Smith et al. 1997), imply that chilling leads to an increase in alternative (i.e., non-CO_2) electron sinks (Flexas et al. 1999; Allen et al. 2000; Fryer et al. 1998). It has been argued (Osmond and Grace 1995) that exploiting oxygen as a terminal electron acceptor, in both Rubisco oxygenase photorespiration and the Mehler–ascorbate-peroxidase reaction, protects plants from photodamage in bright light. Thylakoid electron transport is intrinsically liable to produce active oxygen species. To prevent calamitous damage to component proteins and lipids, plants have numerous antioxidant systems. It is becoming increasingly clear that the regeneration of these antioxidants, such as ascorbate, can be an important and variable electron sink. The Mehler–ascorbate-peroxidase (water–water) cycle has been recently reviewed (Asada et al. 1998). The absence of a method to quantify the size of this sink means that it is currently unclear whether ascorbate regeneration is a major additional sink for electrons observed following a chill. In maize, the activities of the enzymes involved have been reported to rise when grown under cool conditions in the field (Fryer et al. 1998) but to decline following a short chill under controlled conditions (Jahnke et al. 1991). Interference with this process of antioxidant regeneration in response to a light chill might be a major cause of the observed inhibition of photosynthesis. In addition to the direct effects of this oxidative potential, light chill-induced oxidative stress can lead to a change in the redox state of the stroma. Inhibition of Benson-Calvin cycle enzymes was found to closely mirror the chilling-induced inhibition of photosynthesis in tomato (Sassenrath and Ort 1990; Sassenrath et al. 1990). The pool sizes of fructose-1,6-biphosphate and sedoheptulose-1,7-bisphosphate increased in tomato leaves during chilling, and these increases appeared to be caused by direct inhibition of fructose-1,6-biphosphatase and sedoheptulose-1,7-bisphosphatase activities. These are two regulatory enzymes of the Benson-Calvin cycle whose activity is regulated via thioredoxin-mediated reduction of specific sulfhydryl groups. A further possible candidate for a chill-induced alternative electron sink is cyclic electron transport. It has been argued that such a cycle operates around PS I and the cytochrome b6-f complex using a ferredoxin–plastoquinone oxidoreductase to regenerate reduced plastoquinone (plastoquinol) from ferredoxin (Cleland and Bendall 1992), and contributes to the ΔpH required to engage photoprotection (Heber et al. 1995). A chill-induced increase in non-linear electron transport should be visible as a relative stimulation of the quantum efficiency of PS I versus PS II electron transport. The absence of a substantial shift in this relationship at low temperatures (Kingston-Smith et al. 1999) suggests that chilling does not induce a significant increase in cyclic electron transport (Allen and Ort 2001).

In addition to a reduction in stomatal conductance with chilling, biochemical problems have been noted for cotton photosynthesis in a growth chamber experiment under a moderate photon flux density (450–540 μmol m^{-2} s^{-1}) and a constant 15 °C, but primarily after 4–8 days of chilling (Perera et al. 1995). Thus, a brief period of moderate chilling with a moderate photon flux density, such as occurs during a cool morning, might not cause substantial biochemical disruptions to carbon assimilation. Initially, Rubisco activity (not activated in vivo) remained high, and the total extractable activity and the activation state of stromal fructose bisphosphatase rose, likely factors in keeping net carbon assimilation from declining more than 13% at 15 °C the first day the plant is exposed to 15 °C. The lack of a substantial decrease in the O_2 sensitivity of net carbon assimilation until day 4 suggests that there was no major problem, initially, with RuBP regeneration. Thus, with respect to the early responses to 15 °C and a moderate photon flux density, aspects of cotton photosynthesis are insensitive to this temperature and mimicked many of the acclimation responses of chilling-tolerant plants. However, exposures to a much lower temperature in full sun (35 min at 5 °C and 1700 μmol photons m^{-2} s^{-1}) can cause the activation state of stromal fructose bisphosphatase in cotton leaves to decline (Payton et al. 1997). Perera et al. (1995) showed that, with a long-term (8 days) exposure to 15 °C, net carbon assimilation did diminish for cotton by nearly 50%. The Rubisco initial activity declined approximately 25%, and the stromal fructose bisphosphatase activity and activation state declined during this period. One possible cause of this considerable further decline in net carbon assimilation and changes in enzyme activity after day 1 of the experiment could have been the maintenance of high leaf sucrose and hexose (glucose plus fructose) contents that could have signaled a downregulation of gene expression for photosynthetic enzymes. Also, although increases in ADP-glucose pyrophosphorylase, cytosolic stromal fructose bisphosphatase, and glucose-1-phosphate uridylyltransferase occurred with chilling, sucrose phosphate synthase activity declined, suggesting that eventually, net carbon assimilation became increasingly constrained by the phosphate supply to the chloroplast, consistent with the decline in the O_2 sensitivity of net carbon assimilation to 4% from 33% in the 8-day period. However, it is also possible that there were increasing problems of maintaining electron transport and the redox state of the chloroplast.

When more light energy is absorbed than can be utilized by the Calvin-Benson cycle during chilling, problems with components of the light-absorption apparatus and oxidative state of the chloroplast can develop (Kornyeyev et al. 2001; Payton et al. 2001; Perera et al. 1995). In such a situation, photo-inactivation of photosystem II and photosystem I might occur, and redox-sensitive enzymes, such as stromal fructose bisphosphatase and sedoheptulose bisphosphatase, might inactivate. The extent of inactivated PS II complexes can be estimated from chlorophyll a fluorescence analysis (F_v/F_m) after leaves or leaf discs have been allowed to remain in the dark for 30–60 min. With time in moderate to high photon flux density during chilling (10–15 °C), F_v/F_m declines for cotton leaves, indicating an inactivation or damage to PS II complexes (Kornyeyev et al. 2001; Perera et al. 1995). The inactivated PS II complexes might not recover quickly once the leaves warm to an optimum temperature. For the cotton plants subjected to 15 °C, the complete recovery of F_v/F_m values required 2 days at 28 °C (Perera et al. 1995). The following section addresses this problem of oxidative stress in cotton leaves during chilling, what mechanisms the plants employ to reduce the negative effects on photosynthesis, and the potential that cotton has to subdue its occurrence in the field.

Each plant has its unique set of temperature requirements, which are optimum for its proper growth and development net carbon assimilation set of temperature conditions, which are optimum for one plant may be stressful for another plant. Many plants, especially those, which are native to warm habitat, exhibit symptoms of injury when exposed to low non-freezing temperatures (Lynch 1990). These plants including maize (*Zea mays*), soybean (*Glycine max*), cotton (*Gossypium hirsutum*), tomato (*Lycopersicon esculentum*), and banana (*Musa* sp.) are in particular sensitive to temperatures below 10–15 °C and exhibit signs of injury (see Lynch 1990; Guy 1990; Hopkins 1999). The symptoms of stress-induced injury in these plants appear from 48 to 72 h; however, this duration varies from plant to plant and also depends upon the sensitivity of a plant to cold stress. Various phenotypic symptoms in response to chilling stress include reduced leaf expansion, wilting, and chlorosis (yellowing of leaves) and may lead to necrosis (death of tissue). Chilling also severely hampers the reproductive development of plants; for example, exposure of rice plants to chilling temperature at the time of anthesis (floral opening) leads to sterility in flowers (Jiang et al. 2002).

The major malicious effect of freezing is that it induces severe membrane damage (Steponkus et al. 1993). This damage is largely due to the acute dehydration associated with freezing. Membrane lipids are primarily composed of two kinds of fatty acids unsaturated as well as saturated fatty acids. Unsaturated fatty acids have one or more double bonds between two carbon atoms (−CH=CH−), whereas saturated fatty acids are fully saturated with hydrogen atoms ($-CH_2-CH_2-$). It is a well-known fact that lipids containing saturated fatty acids solidify at temperatures higher than those containing unsaturated fatty acids. Therefore, the relative proportion of unsaturated fatty acids in the membrane strongly influences the fluidity of the membrane (Steponkus et al. 1993). The temperature at which a membrane changes from semi-fluid state to a semi-crystalline state is known as the transition temperature. Chilling-sensitive plants usually have a higher proportion of saturated fatty acids and, therefore, a higher transition temperature. Chilling-resistant species on the other hand are marked by higher proportion of unsaturated fatty acids and correspondingly a lower transition temperature (Mahajan and Tuteja 2005). It has been suggested that reduction in net photosynthesis under low temperature stress is due to disruption of almost all major components of photosynthesis (Allen and Ort 2001; Dahal et al. 2012). In maize (*Zea mays* L.) and oats (*Avena sativa* L.), the efficiency of excitation capture by photosystem II reaction centers and the quantum yield of electron transport were higher in tolerant genotypes than that in sensitive varieties (Fracheboud et al. 1999; Rizza et al. 2001). The performance index and the chill factor index were higher in the tolerant genotypes of soybean [*Glycine max* (L.) Merr.] under chilling stress (Strauss et al. 2006).

The success of many crops rests on their ability to withstand the freezing temperature of late spring or early autumn frost. Therefore tolerance to freezing temperatures is particularly important for the sustainability of agricultural crops. As understanding the basics of a disease is essential for its cure, in the same way understanding of low freezing induces its injurious effects on plants is essential for the development of frost-tolerant crops. The real cause of freeze-induced injury to plants is the ice formation rather than low temperatures. It is noteworthy to mention here that dehydrated tissues such as seeds and fungal spores can survive at very low temperatures without any symptoms of injury. Even cryopreservation is a common method for storage of seeds and other biological materials, which is based on the fact that water essentially solidifies without the formation of ice crystals. Ice formation in plants begins in the apoplastic space as it has relatively lower solute concentration. As the vapor pressure of ice is much lower than water at any given temperature, ice formation in the apoplast establishes a vapor pressure gradient between the apoplast and surrounding cells. The unfrozen cytoplasmic water migrates down the gradient from the cell cytosol to the apoplast, which contributes to the enlargement of existing ice crystals and causes a mechanical strain on the cell wall and plasma membrane leading to cell rupture (McKersie and Bowley 1997; Olien and Smith 1997). Freeze-induced cellular dehydration results in multiple forms of membrane damage including expansion-induced-cell lyses and fracture lesions (McKersie and Bowley 1997; Uemura and Steponkus 1997) and lamellar-to-hexagonal-II phase transition. Although freeze exerts its effect largely by membrane damage due to severe cellular dehydration, certain additional factors may also contribute to damage induced by freeze.

3.3.1 Photosynthesis in C4 Plants

C4 plants are rare in the cool climates characteristic of high latitudes and elevations, but the reasons for this are unclear (Kubien et al. 2003). Unlike C3 higher plants, C4 species are generally not found in cold environments, with only a few species occurring in the cool-temperate climate zone. Since the C4 mechanism has evolved independently from C3 ancestors on many occasions, it is assumed that absence from cold climates reflects a fundamental barrier within the photosynthetic mechanism (Naidu and Long 2004), rather than physiological intolerance of low temperature in any other process within C4 species (Long 1999; Sage 1999; Sage et al. 1999). *Miscanthus x giganteus* (Hodkinson and Renvoize 2001) is a C4 grass species that is uniquely cold-tolerant (Beale and Long 1995; Bullard et al. 1995; Beale et al. 1996). In contrast to other closely related C4 species utilizing the NADP-malic enzyme pathway, such as sugarcane (*Saccharum officinarum*), sorghum (*Sorghum vulgare*), and maize (*Zea mays*), *Miscanthus x giganteus* is capable of developing photosynthetically competent leaves under chilling temperatures (<10 °C) early in the growing season (Beale et al. 1996) and maintains relatively high photosynthetic rates across a range of measurement temperatures when grown at either 14 °C or 25 °C (Naidu et al. 2003). *Miscanthus x giganteus* is also unique among other low-temperature-tolerant C4 genera such as *Muhlenbergia* and *Flaveria* (Sage et al. 1999) in its ability to achieve high efficiencies of energy conversion and accumulate large amounts of biomass at low temperatures (Beale and Long 1995).

The reason for the relative lack of C4 plants in cool regions remains unclear. The lower quantum yield of photosynthesis in C4 versus C3 species at low temperatures has been proposed to account for the differences in the biogeography of the two pathways (Ehleringer 1978). However, differences in quantum yield only effect carbon uptake when light is limiting, and in plant canopies CO_2 assimilation is a function of both light-saturated and light-limited photosynthetic rates (Ehleringer 1978; Long 1999). Alternatively, one or more of the carbon-concentrating reactions of the mesophyll might be inherently chilling sensitive. Pyruvate orthophosphate dikinase and phospho*enol*pyruvate carboxylase can both dissociate below 8 °C–12 °C in vitro (Edwards et al. 1985; Krall and Edwards 1993). However, considerable species and ecotypic variability has been noted, and these steps are not fundamentally prone to failure at low temperatures in vivo (Leegood and Edwards 1996; Matsuba et al. 1997).

The possibility that the bundle sheath reactions may limit C4 photosynthesis at suboptimal temperatures has received less attention. Björkman and Pearcy (1971) found that the activation energy of photosynthesis was similar to that of Rubisco in several warm-climate C3 and C4 species. Variation in Rubisco activity was correlated with differences in the photosynthetic capacity of *Atriplex lentiformis* (Torr.) Wats. grown at different temperatures (Pearcy 1977a). In *Bouteloua gracilis* Lag. and *Muhlenbergia montanum* (Nutt.) A.S. Hitch., the light-saturated net photosynthetic rate and the maximal in vitro Rubisco activity are equivalent below 17 °C and 22 °C, respectively (Pittermann and Sage 2000, 2001). These findings indicate that Rubisco capacity can limit the rate of C4 photosynthesis below approximately 20 °C.

C4 photosynthetic rate is limited by different processes depending upon external environmental conditions. At low light, the product of maximum quantum yield of CO_2 uptake and leaf absorptance determines photosynthetic rate (Long et al. 1993).

Additionally, the diversion of absorbed energy into alternative energy sinks, such as active oxygen species, reduces photosynthetic rate under low light conditions (Ort and Baker 2002). In saturating light, metabolic control of C4 photosynthesis varies according to the internal CO_2 concentration. At low internal CO_2 concentration, as may result at low stomatal conductances, phosphoenol-pyruvate carboxylase activity will control photosynthetic rate, while at saturating internal CO_2 concentration, photosynthesis can be limited by a variety of processes, mainly phosphoenol-pyruvate carboxylase regeneration via pyruvate orthophosphate dikinase capacity and maximal ribulose bisphosphate carboxylase/oxygenase activity (von Caemmerer 2000). While ribulose 1,5-bisphosphate (RuBP) regeneration (limited by whole-chain electron transport) may also affect photosynthesis at high light and high internal CO_2 concentration, metabolic control over C4 photosynthesis has been shown to reside mainly with three major C4 photosynthetic enzymes: Rubisco, phosphoenol-pyruvate carboxylase, and phosphoenol-pyruvate dikinase (Usuda et al. 1985; Matsuoka et al. 2001). Analysis of the response of photosynthesis to internal CO_2 concentration suggests that in healthy leaves growing in the current ambient CO_2 concentration, internal CO_2 concentration will be saturating, and control will therefore reside either with phosphoenol-pyruvate regeneration or Rubisco, rather than with phosphoenol-pyruvate carboxylase. This has been supported by analysis of control coefficients in antisense and co-sense-transformed *Flaveria*, a C4 dicot (Furbank et al. 1997).

Under low temperature conditions, it remains unclear whether C4 photosynthesis is controlled through maintenance of high carboxylation efficiency, high capacity for phosphoenol-pyruvate regeneration, high rates of Rubisco activity, or some combination of these. Studies have demonstrated reduced activity of Rubisco (Kingston-Smith et al. 1997; Pittermann and Sage 2000, 2001), phosphoenol-pyruvate dikinase (Du et al. 1999), or phosphoenol-pyruvate carboxylase (Kingston-Smith et al. 1997; Chinthapalli et al. 2003) upon chilling. Kubien et al. (2003) showed that transgenic reduction in the activity of Rubisco in *Flaveria bidentis* correlated closely with the reduction in light-saturated photosynthesis at low temperature. However, maintenance of enzyme activity of both phosphoenol-pyruvate carboxylase and Rubisco corresponded to maintenance of photosynthetic capacity in the semicold-tolerant C4 NADP-malic enzyme grass *Paspalum dilatatum* Poir. in response to growth at low temperature (Cavaco et al. 2003).

Do stomata play a role in decreased photosynthetic efficiency of C4 plants at low temperature? For *Zea mays* the results appear equivocal. At chilling temperatures, stomatal conductance can increase (Rodriguez and Davies 1982), which explains why the relative water content of leaves drops during chilling even though the leaf–air water vapor pressure deficit is low (Aguilera et al. 1999). However, leaves subjected to a transient low temperature treatment showed a Ci on return to warm temperatures that was ca. 20% below that of controls (Aguilera et al. 1999), suggesting that growth at low temperature has the potential to lower stomatal conductance to an extent that photosynthesis could be affected. Lee et al. (2002) saw an almost 50% reduction in stomatal conductance due to cold temperature, averaged across 49 maize cultivars. If stomata close in response to low temperature, then internal CO_2 concentration may be decreased such that phoshoenol-pyruvate carboxylase exerts metabolic control over photosynthetic rate (Collatz et al. 1992).

High conversion efficiencies for whole stands in the field depend on maximizing light-limited photosynthesis which depends on the maximum quantum yield of CO_2 assimilation and leaf absorptance (Long et al. 1994), as well as the rate of light-saturated photosynthesis. In the natural environment, significant amounts of carbon gain occur under light-limited conditions in which photosynthetic rate is independent of the activities of enzymes of carbon metabolism, but depends on the efficiency of light absorption and transduction of the absorbed energy into ATP and NADPH. Loss of pigments and development of non-photochemical and alternative photochemical sinks for absorbed energy will lower light-limited photosynthetic rate (Long 1999).

Zea mays leaves that grow at suboptimal temperatures show impaired development of the photosynthetic apparatus and a depressed photosynthetic capacity (Nie and Baker 1991; Nie et al. 1995), which is evident as a significantly decreased maximum quantum yield of photosynthesis (Long et al. 1983; Baker et al. 1989, 1990). Reduced susceptibility to photoinhibition and low-temperature impairment of development is a potential mechanism of cold tolerance in *Miscanthus x giganteus* and would be evidenced by an ability to maintain a greater quantum yield at low temperature than *Zea mays*. Although low, photorespiration is a competing alternate photochemical energy pathway that can operate within C4 species and is therefore a potential factor that may decrease efficiency (Maroco et al. 1997, 2000; Laisk and Edwards 1998). Also, *Zea mays* grown at low temperatures has been shown to develop large alternative sinks for electron transport, possibly associated with the Mehler reaction (Andrews et al. 1995; Fryer et al. 1998). Siebke et al. (2003) reported that non-photorespiratory light-dependent O_2 uptake at high CO_2 concentrations was, on average, 18% of gross O_2 evolution for several NAD-malic enzyme and NADP-malic enzyme grass species, suggesting a high capacity for the Mehler reaction in C4 grasses. These alternative energy sinks, both non-photochemical (Leipner et al. 2000) and photochemical (Wingler et al. 2000; Ort and Baker 2002), can depress the efficiency of light-limited photosynthesis, although they may play an important role in protecting against photodamage under chilling conditions.

3.3.2 Control of Photosynthetic Processes in C4 Plants

In general, photosynthesis can function without harm between 0 and 30 °C in cold adapted plants that are active in winter and early spring or grow at high altitude and latitude (Larcher 2003). In plants from equitable habitat (e.g., warm season crops), photosynthesis operates safely between 7 and 40 °C, and in plants from hot environments, such as tropical species and summer species in the desert, photosynthesis operates between 15 and 45 °C with no apparent problem (Bunce 2000). In all cases, photosynthesis shows an optimum temperature that roughly corresponds to the middle of the non-harmful range, and drops off with increasing slope as temperatures rise above the thermal optimum. With changes in growth conditions, the thermal optimum can shift, typically by one-third to one-half the number of degrees as the shift in growth temperature (Berry and Björkman 1980). Some species, particularly those in more equitable habits, can acclimate to temperature change as indicated by shifts in the thermal optimum and enhanced assimilation rates at the new growth temperature (Mawson et al. 1986; Atkin et al. 2006). Other species, such as specialists for extreme environments, show less potential to acclimate, and exhibit similar thermal responses in warm or cool growing conditions (Atkin et al. 2006). Eventually, at temperatures near the extreme end of the functional range, injury occurs, and the rate of photosynthesis is irreversibly impaired.

Over the past 30 years, numerous ideas have been proposed to explain the biochemical controls over C3 and C4 photosynthesis, and sophisticated theoretical models have been developed to test the possibilities (von Caemmerer and Furbank 1999; von Caemmerer 2000). As a result, a good understanding of the limitations over photosynthesis at the thermal optimum in C3 plants (von Caemmerer 2000); however, the biochemical controls over the rate of photosynthesis at non-optimal temperatures are less clear. Advances in model parameterization and increased experimental attention have in recent years improved our understanding of the mechanisms controlling photosynthetic responses to both high and low temperatures (Bernacchi et al. 2001, 2002, 2003; Medlyn et al. 2002; Salvucci and Crafts-Brandner 2004a). These advances occur at a time when society is increasingly concerned about global climatic change, and at present it is the first wave of responses to climatic change in the form of range shifts, phenological adjustments, and population collapses (Jump et al. 2006; Menzel et al. 2006; Mouthon and Daufresne 2006). To be able to predict the effects of climatic change, and to adapt agricultural systems and land management to a warmer world, it is imperative to understand how temperature affects photosynthetic carbon gain.

The rate of net CO_2 assimilation in the leaves of C3 plants is generally characterized as being under the control of three distinct processes: the capacity of ribulose 1·5-bisphosphate carboxylase/oxygenase (Rubisco) to consume ribulose bisphosphate (RuBP) (RuBP-saturated photosynthesis or Rubisco-limited photosynthesis), the capacity of the Calvin cycle and the thylakoid reactions to regenerate RuBP (RuBP regeneration-limited photosynthesis), and the capacity of starch and sucrose synthesis to consume triose phosphates and regenerate inorganic phosphate for photophosphorylation (inorganic phosphate regeneration-limited photosynthesis or triose phosphate use-limited photosynthesis) (Harley and Sharkey 1991; von Caemmerer 2000). The Rubisco capacity to consume RuBP is generally the predominant limitation on net CO_2 assimilation at light saturation and CO_2 levels below the current ambient partial pressure of 380 mbar. Because of this, the initial slope of the response of net photosynthesis to intercellular CO_2 partial pressure (the net CO_2 assimilation/internal CO_2 concentration response) is determined by Rubisco capacity at light saturation (Hikosaka et al. 2006). RuBP regeneration capacity (as affected by the rate of light harvesting) becomes limiting at subsaturating light intensities at all levels of CO_2 (Sage et al. 1990). At saturating light intensities, moderate temperature (25–30 °C), and elevated CO_2, either the capacity of RuBP regeneration or the inorganic phosphate regeneration capacity limits net CO_2 assimilation (Sharkey 1985a). RuBP regeneration capacity at light saturation generally reflects limitations in electron transport capacity, a feature that allows for the estimation of electron transport capacity in vivo using gas exchange measurements at high CO_2 (von Caemmerer 2000; Cen and Sage 2005). Characteristic responses of net CO_2 assimilation to CO_2 and O_2 variation in C3 leaves indicate whether net CO_2 assimilation is limited by the RuBP regeneration capacity versus the inorganic phosphate regeneration capacity. When net CO_2 assimilation is controlled by the RuBP regeneration capacity, it is stimulated by an increase in CO_2 or a reduction in O_2 supply under conditions where photorespiration occurs (Sharkey 1988). This is normally the case above 5 °C, and below a CO_2 level of 500 mbar (cooler temperatures, 35 °C) (Ehleringer et al. 1991). The sensitivity of RuBP regeneration-limited net CO_2 assimilation to CO_2 and O_2 reflects the competition between these gases for RuBP; as CO_2 levels increase, oxygenase activity is increasingly inhibited and net CO_2 assimilation rises with decreasing slope to a plateau that begins at the CO_2 level where photorespiration becomes nil. By contrast, when net CO_2 assimilation is limited by the inorganic phosphate regeneration capacity, there is little if any stimulation following CO_2 enrichment or O_2 reduction at CO_2 levels where photorespiration is present (Sharkey 1985a, b, 1988). This lack of CO_2 or O_2 sensitivity occurs because inorganic phosphate is used as fast as it becomes available, a situation that is not altered by reducing photorespiration (Sharkey 1985b). In practice, the easiest approach to detect an inorganic phosphate regeneration limitation is to model the predicted sensitivity of net CO_2 assimilation to variation in O_2 and CO_2, and compare this with observed sensitivities (Sharkey 1988). Oxygen sensitivity of net CO_2 assimilation can be defined as (A21–A2) / A2 where A21 is net CO_2 uptake at current O_2 levels (21% O_2) and A2 is net CO_2 assimilation at 2% O_2 where photorespiration is nil (Sage and Sharkey 1987). Carbon dioxide sensitivity, defined as $(A_{air}–A_{sat})/A_{sat}$, compares net CO_2 assimilation in a lower CO_2 atmosphere (A_{air}) with CO_2- saturated net CO_2 assimilation (Sage 1990). If RuBP regeneration capacity is limiting, the observed O_2 or CO_2 sensitivity will be equivalent to the respective theoretical sensitivity; if the Pi regeneration capacity is limiting, the ratio will be well below the theoretical value, and will often be near zero or negative (Sage and Kubien 2007).

3.3.3 Thermal Acclimation of C4 Photosynthesis

C4 photosynthesis shows three general patterns of thermal acclimation. The first is observed in generalist species such as found in warm desert *Atriplex* shrubs that are active year-round. Here, the breadth of the thermal response and the values of net CO_2 assimilation at the thermal optimum change a little with variation in growth temperature; instead, the thermal optimum shifts in the direction of the growth temperature (Osmond et al. 1980). The second response has largely been observed in C4 species from cool environments [e.g., *Atriplex confertifolia*, *Bouteloua gracilis*, *Miscanthus*, *Muhlenbergia* sp., *Paspalum dilatatum* (Sage 2002; Cavaco et al. 2003; Naidu et al. 2003)], but can occur in summer active weeds [*Atriplex rosea* (Björkman and Pearcy 1971)]. Here, the thermal response of net CO_2 assimilation changes little below measurement temperatures of 20 °C following growth in a cooler environment, while it changes at elevated temperature. In the third type of response, net CO_2 assimilation changes at all measurement temperatures following thermal acclimation; species with this type of response often increase net CO_2 assimilation following transfer to cooler conditions that are not stressful [e.g., warm-season grasses and *Atriplex sabulosa* (Dwyer et al. 2007)]. In contrast to these acclimation responses, there is also the response of chilling-sensitive C4 plants such as maize to low growth temperature. These species show a loss in photosynthetic capacity, membrane integrity is impaired, and enzyme activity is reduced at all measurement temperatures following prolonged chilling (Long 1999; Naidu et al. 2003). These are considered to be stress responses that lead to a loss of vigor and possibly death. The mechanistic explanation for acclimation in C4 photosynthesis remains unclear, although recent progress in our understanding of the biochemical controls over C4 photosynthesis allows for some partial evaluations. Because Rubisco imposes a strong limitation on net CO_2 assimilation at low temperature, the most obvious means of thermal acclimation to suboptimal temperatures would be to enhance Rubisco content. Studies to date indicate some species do acclimate in this manner, while numerous others do not. Acclimation involving a change in Rubisco content occurs in the warm season grasses *Panicum coloratum* and *Cenchrus ciliaris*, and the warm-season dicot *Flaveria bidentis* grown at 25/20 and 35/30 °C; plants grown at the cooler temperatures had 30–40% more Rubisco than plants grown at the warmer temperatures, which was associated with a rise in net CO_2 assimilation below 25 °C (Dwyer et al. 2007). In C4 *Atriplex lentiformis* grown at 23 °C days, Rubisco content was more than double the levels measured in plants grown at 43 °C, and this was associated with a near doubling of measured net CO_2 assimilation at 10 °C (Pearcy 1977a). In maize grown at 19 versus 31 °C, Rubisco activity was by 21% greater while pyruvate orthophosphate dikinase and NADP–malic enzyme activities were 64–100% greater in the 19 °C-grown plants (Ward 1987). These responses are distinct from responses of warm-season C4 grasses in chilling conditions (10–15 °C), where Rubisco and other protein levels gradually decline following chilling exposure (Du et al. 1999; Naidu and Long 2004). In chilling-tolerant, cool-season C4 grasses, there is no change in Rubisco content nor in net CO_2 assimilation below 20 °C, indicating minimal acclimation capacity to cool conditions (Pittermann and Sage 2001; Cavaco et al. 2003; Naidu et al. 2003; Kubien and Sage 2004a). Instead, acclimation responses are mostly observed at the thermal optimum. Cool and warm-grown plants of the boreal C4 grass *Muhlenbergia glomerata* exhibited identical net CO_2 assimilation/transpiration responses below 23 °C, while the warm-grown grass had 50% higher net CO_2 assimilation at the thermal optimum and a 5 °C higher thermal optimum (Kubien and Sage 2004b). *Miscanthus x giganteus*, a cool-tolerant C4 grass, shows the same net CO_2 assimilation/transpiration response below 20 °C in warm- and cool-grown plants, and an enhancement in net CO_2 assimilation at the thermal optimum in the warm-grown plants (Naidu et al. 2003). Cool-grown *Paspalum dilitatum* also showed no change in the net CO_2 assimilation/transpiration response and Rubisco content below 20 °C compared to warm-grown plants, but had a higher net CO_2 assimilation at the thermal optimum and less heat tolerance (Cavaco et al. 2003). The higher rate at elevated temperature in the cool-grown *Paspalum dilitatum* was associated with greater total protein in the leaves; however, phosphor enol-pyruvate carboxylase activity was unchanged by the treatment (Cavaco et al. 2003). The most comprehensive C4 acclimation study is by Dwyer et al. (2007) with three warm-season C4 species (*Cenchrus ciliaris*, *Flaveria bidentis*, and *Panicum coloratum*) grown at 25/20 and 35/20 °C. In these species, net CO_2 assimilation was reduced at the thermal optimum, and the thermal optimum for net CO_2 assimilation was about 1–3 °C higher, in the warmer-grown plants (Dwyer et al. 2007). In *Panicum coloratum* and *Cenchrus ciliaris*, net CO_2 assimilation showed a pronounced decline above 35 °C in the plants grown at 25/20 but not 35/30 °C. In the cooler-grown plants, the increase in net CO_2 assimilation at the thermal optimum was associated with increased Rubisco capacity, a rise in carbonic anhydrase activity, a rise in leaf nitrogen content, and the production of leaves with higher leaf mass per area. The two grass species also had greater cytochrome-f content in the cool-relative to the warm-grown plants, indicating that there may have been an increase in electron transport capacity that supported an increase in net CO_2 assimilation at the thermal optimum. In all three species phosphoenol-pyruvate carboxylase activity, photosystem II content, and chlorophyll content were unchanged. Dwyer et al. (2007) did not examine Rubisco activation state or activase lability.

3.4 Cellular Mechanisms Against Chilling Damage and Oxidative Stress

Numerous studies suggested that photooxidative damage by reactive oxygen intermediates is a major cause in the decrease in net carbon assimilation following exposure to a variety of environmental stresses, including excess light and suboptimal temperature (Allen 1995; Alscher et al. 1997; Bowler et al. 1992; Königer and Winter 1993; Payton et al. 1997, 2001; Smirnoff 1995). Earlier studies in cotton showed significant impacts of low temperature on net carbon assimilation in short-term acute exposures (35 min at 5 °C and a photon flux density of 1700 μmol m^{-2} s^{-1}) (Payton et al. 1997) or more moderate stress (20 °C and full sun for 1 day) (Königer and Winter 1993). Both of these studies showed rapid decreases in net carbon assimilation as well as significant lag times in recovery of net carbon assimilation after re-exposure to optimal temperatures.

Reactive oxygen intermediates are generated in cellular reactions, such as electron transport in mitochondria and chloroplasts, where electron transfer to molecular oxygen can occur. Under normal conditions, oxidative damage is minimal due to the metabolism of active oxygen species by both enzymatic and nonenzymatic mechanisms. However, during exposure to environmental stress, the mechanisms that normally scavenge toxic oxygen species could become overwhelmed, resulting in membrane damage, pigment bleaching, enzyme inactivation, and decreased net carbon assimilation or photoinhibition (Allen 1995; Bowler et al. 1992; Foyer et al. 1994; Smirnoff 1995; Van Camp et al. 1996). The damage caused by reactive oxygen intermediates and their products is known as oxidative stress.

Chloroplasts are highly susceptible to oxygen toxicity for many reasons:

- They have a high internal O_2 concentration during exposure to light.
- They contain several molecules, such as reduced ferredoxin, that can reduce O_2 to superoxide anions (O_2^-).
- Lipids in the chloroplast membranes have a high percentage of unsaturated fatty acid tails that are susceptible to peroxidation.
- Excited chlorophyll can generate singlet oxygen resulting in protein damage (Halliwell and Gutteridge 1989).

During photosynthesis, the photochemical oxidation of H_2O generates O_2 and electrons. Noncyclic electron flow occurs when photosystem II transfers these electrons to photosystem I via intermediate electron acceptors and results in the reduction of ferredoxin, which is then used to reduce $NADP^+$ on the stromal side of the thylakoids. A proton gradient across the thylakoid membrane is established as protons are released in the chloroplast lumen during electron transport. This proton gradient is used by a thylakoid-bound ATP synthetase to generate ATP by photophosphorylation of ADP and is also a regulator of electron transport rate and photosystem II photochemistry (Foyer 1996). When NADPH/NADP ratios increase, formation of O_2^- via pseudocyclic electron flow to O_2 could serve to maintain electron flow through photochemistry and protect the transthylakoid gradient (Osmond and Grace 1995; Schreiber and Neubauer 1990). However, the O_2^- and subsequently formed H_2O_2 could react in the presence of Fe^{2+} or Fe^{3+} to form highly toxic hydroxyl radicals (OH^-) and must be scavenged to prevent cellular and subcellular damage (Bowler et al. 1992; Halliwell and Gutteridge 1989). Additionally, excited chlorophyll can transfer energy to O_2 and generate singlet oxygen if the energy is not dissipated by transfer to reaction centers or lost as fluorescence (Demmig-Adams and Adams 1996). Aside from the damaging effects of OH, H_2O_2 can inactivate several enzymes in the Calvin-Benson cycle, notably the bisphosphatases, as mentioned previously (Buchanan 1980; Kaiser 1979). Furthermore, increasing concentrations of H_2O_2 can cause damage to metal-containing enzymes, including superoxide dismutase and ascorbate peroxidase (Bunkelmann and Trelease 1996; Ishikawa et al. 1996). Thus, the primary role of the reactive oxygen intermediates scavenging mechanism is to remove O_2 and H_2O_2 to prevent the formation of OH and prevent uncontrolled fluctuations in redox state of the cell.

For chilling-sensitive plant species, decreased photosynthesis in response to low temperature and high light results from stomatal limitations, decreased activity of enzymes involved in carbon assimilation, and inorganic phosphate limitation to ATP synthesis (Holaday et al. 1992; Labate and Leegood 1988; Perera et al. 1995; Sassenrath et al. 1990). A considerable amount of attention has been focused on the role of the xanthophyll cycle in nonphotochemical dissipation of excess energy (reviewed by Demmig-Adams and Adams 1996). One postulate is that non-photochemical quenching involves the direct transfer of energy from excited chlorophyll a to de-epoxidized xanthophyll pigments, zeaxanthin and antheraxanthin, located in the light-harvesting antenna complexes of PS II and PS I (Demmig-Adams and Adams 1996). Non-photochemical quenching decreases the efficiency of PS II, which slows energy entry into the electron transport chain to rates at which the products of electron transport (e.g., NADPH, ATP) can be used by the leaf. Although it has been suggested that other mechanisms also might contribute to non-photochemical quenching of chlorophyll fluorescence (Adams et al. 1990; Johnson et al. 1993), this xanthophyll-cycle activity is thought to account for most of the non-photochemical quenching. Additionally, Sonoike (1996) characterized a novel type of photoinhibition at PS I in chilling-sensitive plants in which reactive oxygen intermediates damage the Fe-S centers on the acceptor side of P700. This inactivation of PS I could lead to conditions where electron transport intermediates downstream from PS II remain reduced, increasing rates of PS II inactivation (Sonoike 1996). Grace and Logan (1996) showed that despite increases in the xanthophyll cycle pool size and in the rate of non-photochemical quenching, the rate of O_2 photoreduction increased in plants exposed to increasing growth photon flux density. They concluded that increases in leaf antioxidants, specifically superoxide dismutase, ascorbate peroxidase, glutathione reductase, ascorbate, and glutathione, protect against this potential oxidative stress. Although several isozymes of superoxide dismutase, ascorbate peroxidase, and glutathione reductase exist in most subcellular compartments, the majority of their activity is concentrated in the chloroplast (Edwards et al. 1994; Foyer 1993; Gillham and Dodge 1986). Superoxide dismutase is a major scavenger of O_2^-, and its enzymatic activity results in the production of H_2O_2. In the chloroplast, the H_2O_2 is eliminated by ascorbate peroxidase (stromal and thylakoid-bound) via the peroxidation of reduced ascorbate yielding monodehydroascorbate and H_2O. The monodehydroascorbate radicals can disproportionate spontaneously into reduced ascorbate and dehydroascorbate or be reduced by ferredoxin or by monodehydro-ascorbate reductase using NAD(P)H. The reduction of dehydroascorbate to reduced ascorbate by dehydroascorbate reductase is dependent on reduced glutathione which is, in turn, maintained by the NADPH$^-$dependent activity of glutathione reductase. Additionally, reduced glutathion plays a role in enzymatic detoxification of other electrophilic compounds in reactions catalyzed by glutathione S-transferases. Reduced ascorbate and reduced glutathione can also directly serve as reducing agents for nonenzymatic reduction of peroxides and free radicals (Creissen et al. 1996; Foyer et al. 1994).

3.5 Effect of Low Temperature Stress on CO_2 Fixation

Low temperatures reduce CO_2 fixation, which in turn reduces the availability of substrates for photosynthetic electron transport ($NADP^+$) and photophosporylation (ADP), resulting in feedback inhibition of electron transport (Stachula 2015). Low temperatures also directly reduce the rate of photosynthetic electron transport by increasing thylakoid membrane viscosity, caused by alteration in biophysical properties of thylakoid lipids (Huner et al. 1998), restricting the diffusion of plastoquinone. Thus, when plants are exposed to low temperature an imbalance is created when chlorophyll antenna complexes trap more energy that can be processed biochemically (Huner et al. 1993; Ensminger et al. 2006; Wilson et al. 2006). The result of plastoquinone being reduced and of the reduced availability of electron acceptors ($NADP^+$) at photosystem I is the synthesis of reactive oxygen species (Møller et al. 2007). Consequently, cold treatment increases excitation pressure, induces photoinhibition (Powles and Björkman 1982; Hurry and Huner 1991, 1992; Huner et al. 1998), and increases the production of reactive oxygen species (Møller et al. 2007). Low temperatures also attenuate the activity of scavenging enzymes, and this further promotes the overaccumulation of reactive oxygen species such as superoxide (O_2^-), hydrogen peroxide, and the hydroxyl radical (Møller et al. 2007) and to harmful effects such as lipid peroxidation, carbonylation of proteins, formation of disulfide and dityrosine bridges, modification of amino acids in polypeptide chains, DNA damage, as well as modifications of the structure of pigments (Gill and Tuteja 2010). The other effect of low temperature on photosynthesis is connected with Calvin cycle, where activity of enzymes involved in carbon fixation slows down. One of the consequences of thylakoid membranes being in an over-energized state is photodamage, caused initially by the increased formation of reactive oxygen species. Superoxide is the major cause of photodamage to PS I in the cold, whereas hydrogen peroxide and superoxide are the most damaging forms for PS II (Tjus 2001). The major site of photoinhibition in PS II is the inactivation of the D1 protein (Aro et al. 1993). Recovery from PS II photoinhibition is strongly temperature dependent, and low temperature will decrease the rate of repair of PS II (Gombos et al. 1994b; Kalisz et al. 2016). Failure to dissipate or avoid accumulating excess excitation energy leads to photooxidative damage to the photosynthetic apparatus, which is often manifested as bleaching, chlorosis, or bronzing of leaves (Niyogi 2000).

Cold stress suppresses sucrose production in the cytosol resulting in the increase of phosphorus associated intermediates. Thus, phosphate cycling between the chloroplast and cytosol is inhibited due to utilization of the available inorganic phosphate sink (Hurry et al. 2000). As a result, the rate of ATP biosynthesis is retarded, and this impedes the generation of ribulose 1,5--bisphosphate required for CO_2 fixation. Gecin (2013) studied the relationship between carbon flux rate, photosynthesis, and low temperature in *Hordeum vulgare* L. seedlings. It was revealed that short-term cold stress at 5 °C reduced the biochemical reaction and sink rates by consuming less energy along with decreased absorption and production of CO_2. Ramalho et al. (2003) reported elevated export of carbon in the leaves of cold stressed winter wheat seedlings compared to the control sets. The stressed plants exhibited increased activities of enzymes involved in the Calvin cycle and sugar metabolism. Proper channelization of metabolites toward the synthesis of non-enzymatic osmoprotectants and enzymatic antioxidants ensures protection of crucial enzymes against cold-induced injuries. The cold-acclimatized winter wheat plants have evolved such important adaptations in order to tackle long-term cold stress during prolonged winter seasons (Banerjee and Roychoudhury 2018b). Chen et al. (2015)

identified the relations between abiotic stresses and Rubisco activase (a nuclear gene encoding a chloroplast protein involved in photosynthesis). Bioinformatic analysis of the cis acting elements in the Rubisco activase promoter revealed a possible induction by multiple abiotic stresses like cold, heat, salt, and drought. Thus, apart from acting as an activation enzyme of Rubisco, Rubisco activase is also a multiple responder to abiotic stresses (Chen et al. 2015; Sehrawat et al. 2013). Stitt and Hurry et al. (2002) concluded that the photosynthetic decline in plants exposed to low temperature is mainly due to degradation of Rubisco. Cold-tolerant herbaceous plants have been found to increase the availabilities of adenylates and phosphorylated intermediates along with an increase in the chloroplastidic inorganic phosphorus content. This probably elevated the ribulose 1,5-bisphosphate (RuBP) regeneration capacity of the plants during cold stress (Stitt and Hurry 2002). Oliveira and Penuelas (2004) examined the effects of natural Mediterranean winter conditions on the potted *Cistus albidus* L. and *Quercus ilex* L. plants. The stressed plants exhibited lower light, CO_2_saturated CO_2 assimilation rates, and apparent quantum yield compared to those grown in normal greenhouse conditions. In spite of decreased leaf absorptance, leaf damage, and plant mortality, *Citrus albidus* was found to better adapt to cold stress than *Quercus ilex*. However, *Quercus ilex* maintained low rates of net assimilation along with higher leaf and plant survival compared to those in *Citrus albidus* (Oliveira and Penuelas 2004). Bilska-Kos et al. (2018) found strong inhibition of net CO_2 assimilation *Zea mays* L. lines exposed to cold stress. The stressed plants developed thicker leaves and mesophyll layers. Low temperature of 2 °C deteriorated plant growth, chlorophyll content, net photosynthetic rate, stomatal conductance, intercellular CO_2, and transpiration rate in zoysia grass (Li et al. 2018a, b). In another study, it was found that 3 weeks of cold acclimation in *Lolium multiflorum* Lam. and *Festuca arundinacea* Schreb. resulted in stable maintenance of photochemical processes along with slightly reduced CO_2 assimilation and significantly decreased stomatal conductance (Augustyniak et al. 2018).

Low temperature affects different aspects of photosynthesis such as sucrose synthesis in the cytosol leading to accumulation of phosphorus-related intermediates. This results in the depletion of the available inorganic phosphate and in the decreased cycling of inorganic phosphate between the cytosol and chloroplast (Hurry et al. 2000). This in turn impedes synthesis of the ATP necessary for the regeneration of ribulose 1–5 bis-phosphate to support CO_2 fixation. Carbohydrate generation has been reported to have greater instantaneous low temperature sensitivity than other components of photosynthesis (Leegood and Edwards 1996). Pollock and Lloyed (1987) have shown that starch synthesis is more sensitive to low temperature than sucrose and fructose synthesis in cold-tolerant species. Growth of winter rye at 5 °C results in an increased capacity for CO_2-saturated O_2 evolution, dampening of the fluorescent induction, and reductions in the lag time to reach steady-state photosynthesis. Low temperature stress inhibits sucrose synthesis in the cytosol coursing decreased inorganic phosphate cycling between cytosol and chloroplast (Hurry et al. 2000). Concomitant with the upgradation of carbon metabolism, cold-tolerant winter wheat also exhibits stimulation of carbon export from the leaf compared to cold stressed plants. Cold-induced tolerance includes increase in enzyme activities (e.g., from the Calvin cycle and sugar metabolism), reinforcement of energy dissipation mechanism, and antioxidative molecules and qualitative and quantitative changes in lipids and protein membrane composition and protects the cell structure by keeping metabolic pathway regulation mechanism intact (Ramalho et al. 2003; Quartin et al. 2004). In contrast, growth and development at low temperature increased the expression and subsequent activity of Calvin cycle enzymes (Hurry et al. 2000). Decline in photosynthesis after chilling during dark and light photoperiods attributes loss in Rubisco activity. It has been suggested that chilling damages the Rubisco protein itself. In addition cold-tolerant herbaceous plants grown at cold temperatures exhibit an increase in inorganic phosphorus availability in the chloroplast (Stitt and Hurry 2002) as well as in adenylates and phosphorylated intermediates and in the capacity for the regeneration of RUBP (Hurry et al. 1994).

3.6 Effect of Low Temperature on Respiration and ATP Generation

Plant vitality may be compromised if demand for non-structural carbohydrates exceeds supply over prolonged periods of time. Such a non-structural carbohydrate imbalance may result in the reduction of growth, loss of reproductive capacity, or delayed recovery from stress. Under extreme circumstances, excessive non-structural carbohydrates depletion may kill the plants (Adams et al. 2013b; Sevanto et al. 2014; Dickman et al. 2015). During periods of limited photosynthesis, like winter dormancy, plants as well as trees depend solely on stored non-structural carbohydrates to maintain basic metabolic functions, produce defensive compounds, and retain cell turgor (McDowell et al. 2008; Sevanto 2014). With amassing threats to plants due to climate change, knowing how environmental factors affect non-structural carbohydrates demand is critical for agricultural and forest management, modeling climate impacts, and better understanding plant dormancy biology.

Cellular respiration is the primary energy generator for all cellular activities and, therefore, constitutes a major sink for non-structural carbohydrates, especially when growth is limited. As an enzymatic process, respirational activity is intimately linked to the environment via temperature. Thus, like many other biological processes, its thermal kinetics can be represented by the Arrhenius equation and the thermal coefficient,

$$Q_{10} = (R_2/R_1)^{[10/(T_2-T_1)]}$$

where R_1 and R_2 are respiration rates and T_1 and T_2 are the temperatures at which respiration is measured. Respirational Q_{10} varies between 1.3 and 2.7 for many biological systems, e.g., single enzymes in the respiratory pathway (Brooks and Farquhar 1985), mitochondria (Klikoff 1966), soil microbes (Pietikäinen et al. 2005), various plants (Carey et al. 1997; Atkin et al. 2005; Edwards and Hanson

1996), fungi (Pietikäinen et al. 2005), and whole ecosystems (Flanagan and Johnson 2005). In each of these systems, Q_{10} is considered constant across the wide range of inhabitable temperatures. However, plants are known to have a variable Q_{10} (Atkin et al. 2005), which is mainly associated with high temperatures and attributed to substrate availability, enzyme denaturation (Atkin and Tjoelker 2003), and/or temperature acclimation (Noguchi et al. 2015).

The consequence of keeping plants at chilling temperatures is a change in respiratory rate. There is evidence of its decline, occurring as a result of destruction of the mitochondria structure, the general lowering of kinetic energy, and the inhibition of some enzymes (Prasad et al. 1994; Munro et al. 2004; Lawrence and Holaday 2000). Others have reported that an increase in respiratory activity during chilling and prolonged elevation of the respiration rate after cold exposure may indicate irreversible metabolic dysfunction and accumulation of incompletely oxidized intermediates (Steward et al. 1990; Yadegari et al. 2008). The mechanism of stimulation is unknown, but it is possible to assume that it was the result of uncoupling of oxidative phosphorylation. It is also possible that the increased respiration reflects a reaction to the transfer of plants from chilling temperatures to the higher temperatures (Zauralov and Lukatkin 1997). As a result of decreased respiration and increased consumption of energy-rich phosphates at chilling temperatures is a reduction of ATP levels (Takeda et al. 1995; Lawrence and Holaday 2000). Cold-tolerant crop species have greater temperature homeostasis of leaf respiration than cold-sensitive species (Yamori et al. 2009). Chilling reduces the cytochrome path of the electron transport in seedlings (Prasad et al. 1994; Reyes and Jennings 1997) and enhances alternative respiratory pathways (Purvis and Shewfelt 1993; Ordentlich et al. 1991; Gonzalez-Meler et al. 1999). Perhaps these alternative pathways play an important role in plant adaptation to chilling (Steward et al. 1990). They are triggered at the chilling period and increase with decreasing temperature (Ordentlich et al. 1991). These alternative pathways induced by chilling caused a decrease in superoxide generated in mitochondria (Purvis and Shewfelt 1993; Hu et al. 2008). Armstrong et al. (2008) analyzed temperature-dependent sensitivity of leaf respiration in Arabidopsis during cold acclimation and suggested the importance of an alternative oxidation pathway in this process. Moreover, Talts et al. (2004) observed that cold-treated plants often display higher rates of respiration. Heidarvand et al. (2017) pointed out that mitochondria from low temperature-tolerant species respond to low temperature through remodeling their composition that changes the structural and functional properties of the organelles. In this review, we discuss physiological aspects of mitochondrial respiration rate that are affected by low temperature, as well as changes in the abundance of respiratory components under low temperature. The latter includes components of the phosphorylating and non-phosphorylating pathways and adjustments of mitochondrial membrane composition.

It is well known that living cells generate heat during the metabolic processes, involving hydrolysis of ATP and mitochondrial respiration because of partial uncoupling of oxidative phosphorylation (Skulachev 1999). Respiration of plant cells are also affected by temperature (Rasion 1980), and there is a linear increase in

O_2 uptake with temperature (Buetow 1962). In addition to short-term responses, species grown at low temperatures often show higher rates of respiration than species grown at higher temperatures when measured at the same temperature (Collier and Cummins 1990). The increased rate of respiration at low temperatures in mitochondria involves two energy dissipating systems, alternative oxidase pathway (Purvis and Shewfelt 1993) and plant uncoupling mitochondrial protein (Vercesi 2001). This stimulation of respiration was reported as adaptive behavior of plants grown at cold conditions compared with related species from warmer conditions (McNulty and Cummins 1987). This stimulation of respiration causes a decrease in the yield of ATP synthase which is linked to an increase in heat dissipation (Calegario et al. 2003). These systems are also involved in removal of reactive oxygen species, because they do not generate a proton electrochemical gradient (Møller 2001). These systems are also implicated in the removal of reactive oxygen species, because they do not generate a proton electrochemical gradient (Møller 2001). An important function of alternate oxidase pathway is to modulate respiration and to maintain the metabolic and energetic homeostasis during the disturbance of respiration process by the adverse growth condition or stress (Moore et al. 2008; Sieger et al. 2005; Fiorani et al. 2005). Under phosphate limitation, the cytochrome pathway, which is coupled with ATP generation, is inhibited because of the lack of phosphate (Plaxton 2004), therefore impairing electron transport and hindering the oxidation of NADH to NAD^+. NAD^+ is required for the continuation of tricarboxylic acid cycle, which supplies carbon skeletons for biosynthesis (Millenaar and Lambers 2003). In addition, alternate oxidase system also provides a flexible way for plants to regulate energetic requirements. For instance, at high phosphate potential (high ATP/ADP ratio) or high NADPH status, alternative pathway is activated to bring down the energy level (Jarmuszkiewicz et al. 2001). Although the electron flow from reduced ubiquinone to AOX is not coupled with ATP generation, the electron transfer in the upstream complex I produces proton motive force, which means that when the cytochrome pathway is inhibited, alternate oxidase pathway supports ATP production, albeit with low efficiency, by maintaining the electron flow to oxygen and allowing energy production through respiration to sustain plant growth (Vanlerberghe et al. 1997).

3.7 Low Temperature and Photorespiratory Process

RuBP oxygenase-carboxylase (rubisco), a key enzyme in photosynthesis, is the molecular equivalent of a good friend with a bad habit. In the process of carbon fixation, rubisco incorporates carbon dioxide into an organic molecule during the first stage of the Calvin cycle. Rubisco is so important to plants that it makes up 30% or more of the soluble protein in a typical plant leaf. But rubisco also has a major flaw: instead of always using CO_2 as a substrate, it sometimes picks up O_2 instead (Fig. 3.5). This side reaction initiates a pathway called photorespiration, which, rather than fixing carbon, actually leads to the loss of already-fixed carbon as

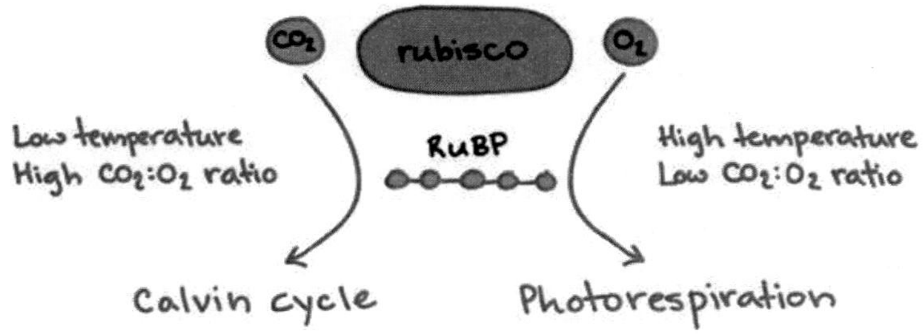

Fig. 3.5 Rubisco using either CO_2 or O_2 as substrate

CO_2. Photorespiration wastes energy and decreases sugar synthesis, so when rubisco initiates this pathway, it's committing a serious molecular *faux pas*.

Rubisco can bind to either carbon dioxide or oxygen depending on environmental conditions. Binding to carbon dioxide and initiation of the Clavin cycle is favored at low temperatures and at a high carbon dioxide-to-oxygen ratio. Binding to oxygen and the initiation of photorespiration is favored at high temperatures and a low carbon dioxide-to-oxygen ratio. Two key factors are the relative concentrations CO_2 and O_2, and the temperature. When a plant has its stomata open, CO_2 diffuses in, and O_2 and water vapor diffuse out, and photorespiration is minimized. However, when a plant closes its stomata, O_2 from photosynthesis builds up inside the leaf. Under these conditions, photorespiration increases due to the higher ratio of O_2–CO_2.

Photorespiration begins in the chloroplast, when rubisco attaches O_2 to RuBP in its oxygenase reaction. Two molecules are produced: a three-carbon compound, 3-phospho glyceric acid, and a two-carbon compound, phosphoglycolate. 3-Phosphoglyceric acid is a normal intermediate of the Calvin cycle, but phosphoglycolate cannot enter the cycle. To recover some of the lost carbon, plants put phosphoglycolate through a series of reactions that involve transport between various organelles. Three-fourths of the carbon that enters this pathway as phosphoglycolate is recovered, while one-fourth is lost as CO_2 (Fig. 3.6).

Characteristic responses of net CO_2 assimilation to CO_2 and O_2 variation in C_3 leaves indicate whether net CO_2 assimilation is limited by the RuBP regeneration capacity versus the phosphate regeneration capacity. When net CO_2 assimilation is controlled by the RuBP regeneration capacity, it is stimulated by an increase in CO_2 or a reduction in O_2 supply under conditions where photorespiration occurs (Sharkey 1988). This is normally the case above 5 °C and below a CO_2 level of 500 μ bar (cooler temperatures, <15 °C) to over 1000 μ bar CO_2 (warmer temperatures, >35 °C) (Ehleringer et al. 1991). The sensitivity of RuBP regeneration-limited net CO_2 *assimilation* to CO_2 and O_2 reflects the competition between these gases for RuBP; as CO_2 levels increase, oxygenase activity is increasingly inhibited and net CO_2 *assimilation* rises with decreasing slope to a plateau that begins at the CO_2 level where photorespiration becomes nil. By contrast, when net CO_2 assimilation is limited by the phosphate regeneration capacity, there is little if any stimulation following CO_2 enrichment or O_2 reduction at CO_2 levels where photorespiration is

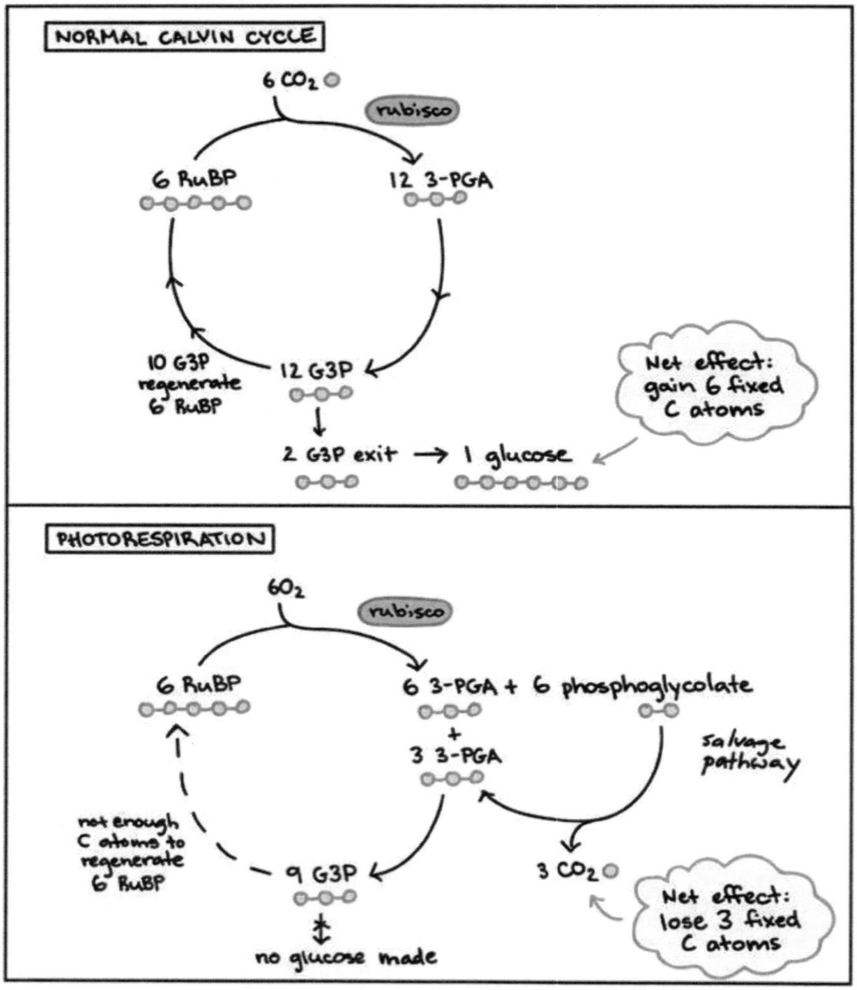

Fig. 3.6 Comparison of Calvin cycle and photorespiration pathways

present (Sharkey 1985a, b, 1988). This lack of CO_2 or O_2 sensitivity occurs because phosphate is used as fast as it becomes available, a situation that is not altered by reducing photorespiration (Sharkey 1985b). In practice, the easiest approach to detect a phosphate regeneration limitation is to model the predicted sensitivity of net CO_2 assimilation to variation in O_2 and CO_2, and compare this with observed sensitivities (Sharkey 1988). If RuBP regeneration capacity is limiting, the observed O_2 or CO_2 sensitivity will be equivalent to the respective theoretical sensitivity; if the phosphate regeneration capacity is limiting, the ratio will be well below the theoretical value, and will often be near zero or negative (Cowling and Sage 1998; Sun et al. 1999; Hendrickson et al. 2004b; Kubien and Sage 2004a).

Low temperature can enhance O_2 uptake processes due to increased O_2 solubility, which could enhance photorespiration (Flexas et al. 1999). The chilling-induced physiological imbalance leads to elevated levels of reactive oxygen species (Suzuki and Mittler 2006; Einset et al. 2007). To counter increased reactive oxygen species levels, antioxidant machinery is induced.

During cold stress, catalase is induced, as observed in two chilling-tolerant rice cultivars (Guo et al. 2006). Further, using transcriptome analysis, Cheng et al. (2007) showed that a reactive oxygen species mediated regulatory module functions as an early component of the chilling stress response pathway in rice, including upregulation of genes such as catalase, indicating the contribution of photorespiration in cold stress situations (Voss et al. 2013). Similarly, Hoshida et al. (2000) found that transgenic rice over-expressing chloroplastic glutamine synthetase shows increased tolerance to chilling stress by virtue of increased capacity for photorespiration. Photorespiratory glycolate oxidase activity is affected by low temperatures (Streb et al. 2005); short-term (4 h) chilling stress in the evergreen plant *Pachysandra terminalis* activates photorespiration, as evidenced by upregulated expression of glycolate oxidase and hydroxyl pyruvate reductase genes (Zhou et al. 2006).

3.8 Cold Acclimation and Its Role in Providing Freeze Tolerance

Cold acclimation can be defined as the modifications of anatomy, physiology, and metabolism that occur in response to sub-optimum temperatures that minimize irreversible freeze-damage and improve plant fitness. This process implies massive changes in gene expression (Thomashow 2010; Calzadilla et al. 2016), enzymatic activities (Barrero-Gil and Salinas 2013), proteomic changes (Amme et al. 2006), and subcellular reprogramming of metabolism (Hoermiller et al. 2017). Although the mechanisms behind cold acclimation are not fully understood, it is known that photosynthetic responses to cold but non-freezing temperature can be important. Photosystems, electron transport chain, as well as the Calvin-Benson cycle and other carbon metabolism reactions are involved in the production of carbohydrates during photosynthesis. This complex process is finely regulated through different signals, allowing its homeostasis under different environmental conditions. However, its deregulation reduces CO_2 fixation capacity and induces photoinhibition (Gururani et al. 2015). In general terms, the response to photoinhibition involves different photoprotective mechanisms which avoid a redox imbalance into the electron transport chain, chloroplast reactive oxygen species generation, and protein photodamage (Goh et al. 2012).

The primary function of cold acclimation is to stabilize the membranes against freeze injury. Acclimation results in increase in proportion of unsaturated fatty acids and thereby a drop in transition temperature (Williams et al. 1988). It functions to prevent the expansion-induced lyses and formation of hexagonal two phase lipids in rye and other plants (Steponkus et al. 1993; Uemura and Steponkus 1997). Cold acclimation results in physical and biochemical restructuring of cell membranes through changes in the lipid composition and induction of other non-enzymatic proteins that alter the freezing point of water. Addition of solutes decreases the freezing point of water to a more negative value, thus preventing ice formation.

The acclimation potential of photosynthesis to temperature greatly varies with the plant species and ecotypes. Although a shift in the optimum temperature for photosynthesis is generally less than one-half of that in the growth temperature (Berry and Björkman 1980), several plants show dramatic changes in the temperature-response curve of photosynthesis. The optimum temperature for photosynthesis in winter wheat (*Triticum aestivum* L.) grown at different seasons of the year increased with increase in the mean air temperature at a rate of about 3 °C increase for each 4 °C increase in the growth temperature (Yamasaki et al. 2002; Sawada 1970). A 15^0 °C increase in the growth temperature resulted in a 15 °C increase in optimum temperature for photosynthesis in *Pinus taeda* (Strain et al. 1976), and acclimation of *Saxifraga cernua* to a 10 °C higher temperature was accompanied with about a 10 °C upward shift in the optimum temperature (Mawson et al. 1986).

Several mechanisms for thermal acclimation of photosynthesis have been proposed. Plants grown at low temperatures had higher levels of ribulose-1,5-bisphosphate carboxylase and other enzymes, which are involved in carbon metabolism compared with plants grown at high temperatures (Maruyama et al. 1990; Holaday et al. 1992; Hurry et al. 1995; Strand et al. 1999). Growth at low temperatures also resulted in higher leaves of cytosolic fructose 1,6-bisphosphatase and sucrose-phosphate synthases, which regenerate orthophosphate during sucrose synthesis (Crespi et al. 1991; Holaday et al. 1992; Makino et al. 1994; Hurry et al. 1995). Photosynthetic acclimation to low temperature was, therefore, suggested to involve an increase in the capacity of enzymatic reactions that limit photosynthesis at low temperature. However, a downward shift in the optimum temperature of photosynthesis, which results in not only enhanced photosynthetic performance at low temperatures but also reduced photosynthetic performance at high temperatures, cannot be explained only in terms of quantitative increases in the level of enzymes. Growth at high temperatures resulted in an increase in the threshold temperature above which irreversible heat inactivation of photosynthesis occurs (Pearcy 1977b; Badger et al. 1982). Acclimation to high temperatures was, therefore, related to increased heat-tolerance of the photosynthetic apparatus (Berry and Björkman 1980; Badger et al. 1982). The extent of thermal stabilization was, however, limited and not large enough to account for an upward shift in the optimum temperature for photosynthesis in the plants with a high acclimation potential.

Farquhar and von Caemmerer (1982) reported that a shift in the optimum temperature for photosynthesis also results from a change in balance between the carboxylation capacity of ribulose-1,5-bisphosphate (which is determined from the initial slope of the response curve of photosynthetic rate to intercellular CO_2 concentration) and the regeneration capacity of ribulose-1,5-bisphosphate (which is estimated from rate of CO_2-saturated photosynthesis). Experiments showed that growth temperature affects not only the ratio of carboxylation capacity to the regeneration capacity of ribulose-1,5-bisphosphate but also the temperature dependence of the two capacities (Hikosaka et al. 1999; Bunce 2000). Changes in temperature dependence of photosynthesis in eight plants were more closely related to changes in temperature dependence of the two capacities than to changes in their ratio (Bunce 2000). Ribulose-1,5-bisphosphate regeneration is mediated by electron transport whose capacity and temperature dependence are influenced by growth temperature. At a low temperature, plants grown at low temperatures showed higher rates of photosynthetic electron transport than those grown at high temperatures (Huner 1985; Mitchell and Barber 1986; Mawson and Cummins 1989). An increase in growth temperature also led to an upward shift in optimum temperature for electron transport, although the extent of shift varied greatly with the plant species (Badger et al. 1982; Mawson and Cummins 1989).

Under adverse environmental conditions that restrict the rate of photosynthesis, photon capture can exceed the rate at which the energy can be used, resulting in production of reactive oxygen species and cell damage (Demmig-Adams and Adams 1992a; Tikkanen et al. 2012). It has been shown that low temperatures are directly related to reactive oxygen species increase (Ensminger et al. 2006; Grimaud et al. 2013), and a positive correlation has been observed between detoxification capacity of these molecules and cold stress tolerance (Lee et al. 2007; Yoshimura et al. 2004). In addition, it has been suggested that an efficient electron sink has the potential to counterbalance the excess of light energy absorption (Murchie and Niyogi 2011). In consequence, an increase in carbon assimilation could be strongly relevant in the photosynthetic acclimation capacity and in the avoidance of chloroplast photo-oxidative stress in plants (Adams et al. 2013a).

Low temperatures induce a number of alterations in cellular components, including the extent of unsaturated fatty acids (Cossins 1994), the composition of glycerolipids (Lynch and Thompson Jr. 1982), changes in protein and carbohydrate composition, and the activation of ion channels (Knight et al. 1996). Accumulation of sucrose and other simple sugars that occurs with cold acclimation also contributes to the stabilization of membrane as these molecules can protect membranes against freeze damage. Freezing tolerance is a multigenic trait. Low temperatures activate a number of cold-inducible genes (Jones and Inouye 1994), such as those that encode dehydrins, lipid transfer proteins, translation elongation factors, and the late-embryogenesis-abundant proteins (Nishida and Murata 1996). Moreover, intercellular ice formation can cause a mechanical strain on cell wall and membrane leading to cell rupture (McKersie and Bowley 1997; Olien and Smith 1997). There is also substantiation that protein denaturation occurs in plants at low temperature which could also result in cellular damage (Guy et al. 1998).

Global proteomic analysis of rice leaves subjected progressively to low temperature (Cui et al. 2005), and chloroplast proteome analysis against drought stress and recovery in tomato (Tamburino et al. 2017) lead to the conclusion that changes in chloroplast proteins occur in the response and acclimation to abiotic stress. Thus, in the present study we aimed to evaluate the role of chloroplast in the *Lotus japonicus* low temperature response, by studying chloroplast stress-responsive proteins under cold stress using a proteomic approach. In addition, confocal microscopy and flow cytometry have been used to identify and quantify chloroplastic reactive oxygen species generation. Calzadilla et al. (2019) hypothesize that different chloroplast strategies are involved in plant tolerance to cold stress.

Overall, cold acclimation results in protection and stabilization of the integrity of cellular membranes, enhancement of the antioxidative mechanisms, increased intercellular sugar levels, as well as accumulation of other cryoprotectants including polyamines that protect the intracellular proteins by inducing the genes encoding molecular chaperones (Guy and Li 1998). All these modifications help the plant to withstand and surpass the severe dehydration associated with freezing stress.

3.9 Low Temperature Stress-Induced Changes

3.9.1 Effects on Photosynthetic Pigments

Tewari and Tripathy (1998) found that chlorophyll synthesis under the low temperature (7 °C)- and high temperature (42 °C) stressed cucumber seedlings was affected by 90 and 60%, respectively. The suppression in chlorophyll biosynthesis was found to be partially due to the inhibition in 5-aminolevulinic acid biosynthesis under both low (78%) and high-temperature (70%) regimes. Furthermore, biosynthesis of protochlorophyllide in low- and high-temperature-stressed seedlings was impaired by 90 and 70%, respectively. In hexaploid wheat seedlings, protochlorophyllide synthesis, porphobilinogen deaminase, and protochlorophyllide oxido-reductase were affected similarly to that of cucumber, which suggests that temperature stress has generally a similar effect on enzymes involved in chlorophyll biosynthesis in both wheat and cucumber. Photosystem II has been long believed to be a prominent heat-sensitive component of photosynthesis (Schrader et al. 2004), but it can perform normal functioning up to 45 °C (Gombos et al. 1994a). However, there are some reports showing that moderately high temperature (35–45 °C) can induce the cyclic transport of electrons and thylakoid membranes become leaky (Sharkey 2005). Similarly, permeability of the thylakoid membranes is one of the most heat-sensitive components of the photosynthetic apparatus (Havaux 1996), which could be counteracted by zeaxanthin. An increased stability of thylakoid membranes was observed at mild heat treatment of potato (35 °C for 2 h), which indicated that de-epoxidized xanthophylls maintained thylakoids and thylakoid membranes against heat-induced disorganization (Havaux 1996; Brugnoli et al. 1998). The deactivation of ribulose-1,5-bisphosphate carboxylase/oxygenase at mild heat stress could be attributed to the deleterious effects of heat on chloroplast reactions (Sharkey 2005; Velikova et al. 2012).

Photosynthesis is sensitive to changes in temperature, but a reversible decline has been observed under mild temperature stress, while a permanent impairment was found after an exposure to severe heat stress (Wang et al. 2010; Centritto et al. 2011). At high temperatures, PS II and stroma become oxidized, and a significant reduction takes place in PS I. In addition, evidence supports the existence of considerable cyclic electron flow at high temperature, which suggests that maintenance of an energy gradient across the thylakoid membrane as well as adenosine triphosphate homeostasis might be involved in the prevention of irreversible impairment under high-temperature regimes (Sharkey and Zhang 2010).

Altogether, different stressful environments, including salinity, drought, and heat, cause generally a considerable reduction in contents of important photosynthetic pigments, particularly chlorophyll. This reduction may occur due to stress-induced impairment in pigment biosynthetic pathways or in pigment degradation. However, the extent of these phenomena depends on the species, variety, duration of plant exposure, and tolerance of the stress. Reduction in photosynthetic pigments, whether through the impairment in pigment biosynthesis or destruction of pigments, may lead to the impairment in electron transport and hence reduced photosynthetic capacity in most plants (Brini 2017).

3.9.2 Effects on Photosystems

Excess irradiance may be harmful for plants that are unable to balance the absorbed/utilized energy ratio (Huner et al. 1998). This may be even worse when plants are exposed simultaneously to high light and low temperatures which decrease carbon and other enzymatic assimilation processes, creating a greater imbalance because light absorption is largely temperature insensitive (Huner et al. 1998). However, cold acclimation decreases susceptibility to photoinhibition (Krause 1994) by causing several metabolic alterations and producing changes at the chloroplast level that may restore the energy balance (Bravo et al. 2007). Widely accepted hypothesis is that cold acclimation may improve the ability of plants to maintain metabolism at low temperature by keeping Q_A, the primary quinone electron acceptor, more oxidized (Huner et al. 1998; Melis 1999). Other plants use different strategies. For instance, cold-acclimated *Pinus contorta* L. partially loses PS II reaction centers, reduces needle chlorophyll per unit area, and reduces its daily carbon gain (Savitch et al. 2002). All these changes are accompanied by an increased and sustained capacity for heat dissipation through non-photochemical quenching. It is also known that xanthophyll levels increase during cold acclimation and the half-time to develop qE (energy-dependent quenching of chlorophyll florescence) decreases (Krause 1994). All these cold acclimation-induced changes may help to restore the energy balance and hence reduce the incidence of low-temperature-induced photodamage.

Colobanthus quitensis (Kunth) Bartl. Caryophyllaceae extends from the Maritime Antarctic and along the Andes Mountains to Ecuador, with one site in Mexico (Lewis Smith 2003). It usually grows above 2500 m above sea level, in the Andes Mountains. The Antarctic *Colobanthus quitensis* plants have been described as morphologically and physiologically adapted to succeed in these cold environments (Mantovani and Vieira 2000; Pérez-Torres et al. 2004). For instance, its photosynthetic machinery is well adapted to low temperature (Xiong et al. 1999, 2000). The photosynthetic responses of a *Colobanthus quitensis* population from the Andes of

Central Chile have been studied in the field (Casanova-Katny et al. 2006), and ecotypic differentiation of the Antarctic and Andes populations has been proposed (Gianoli et al. 2004; Sierra-Almeida et al. 2007).

Antarctic and Andean environments can differ significantly in solar radiation and temperature during the summer. Photon flux density in the Antarctic summer can go up to 1500 μmol photons m^{-2} s^{-1} on sunny days; however, sunny days are infrequent, accounting for less than 20% in summer (Xiong et al. 1999; Xiong and Day 2001). Average temperature in the Antarctic summer is about 3 °C (Alberdi et al. 2002). In the Andes mountains, photon flux density in summer is frequently as high as 2500 μmol photons m^{-2} s^{-1}, and the average temperature at 2600 m above sea level asl is about 13 °C (Cavieres and Arroyo 1999). These differences in photon flux density and temperature regime during the growing season could have resulted in selection for altered plasticity of the photosynthetic apparatus to cope with high light and low temperature in Andes and Antarctic ecotypes, respectively. In order to test this hypothesis, the two ecotypes of *Colobanthus quitensis* described above (Gianoli et al. 2004) were grown, one from Antarctica (sea level) and the other from the Andes (2700 m above sea level), under the same controlled laboratory conditions. The purpose of this study was to assess the effect of cold acclimation on the capacity of these ecotypes of *Colobanthus quitensis* to cope with excess irradiance and low temperature-induced photoinhibition.

Photosynthetic pigments present in the photosystems are believed to be damaged by stress factors resulting in a reduced light-absorbing efficiency of both photosystems (PSI and PSII) and hence a reduced photosynthetic capacity (Geissler et al. 2009; Zhang et al. 2011). Light energy absorbed by chlorophyll is transformed into chlorophyll fluorescence (Maxwell and Johnson 2000). Despite the fact that the extent of chlorophyll fluorescence does not comprise more than 1–2% of total light absorbed by the chlorophyll, its measurement is convenient and noninvasive. It gives a valuable insight into exploitation of the excitation energy by PS II, and indirectly by the other protein complexes of the thylakoid membranes (Roháček 2002), particularly in plants exposed to stressful conditions. Xanthophylls are carotenoids located in light-harvesting antenna complexes of almost all photosynthetic organisms; they play an important role in light harvesting, photoprotection, and assemblage of light harvesting complex (Latowski et al. 2004; Misra et al. 2006). Zeaxanthin (one of xanthophylls) is formed from violaxanthin by violaxanthin de-epoxidase; it plays a key role in minimizing the overexcitation in higher plants (Kuczyńska et al. 2012). In the xanthophyll cycle, the interconversion of two carotinoids, violaxanthin and zeaxanthin, takes place, and it has a substantial role in photoprotection of plants. Due to its considerable importance, it is a promising target for genetic engineering to enhance stress tolerance in plants. The over expression of the *chyB* gene, which is responsible for encoding β-carotene hydroxylase one of enzymes in zeaxanthin biosynthetic pathway), induced twofold enhancement in the pool size of the xanthophylls cycle in *Arabidopsis thaliana* (Davison et al. 2002). The dissipation of excess light energy as heat within light harvesting complex protects photosynthetic apparatus against the oxidative damage. The xanthophyll cycle, particularly the de-epoxidation of violaxanthin to zeaxanthin through antheraxanthin, was reported to play an important role in energy dissipation at high light intensity (Demmig-Adams and Adams 1992b; Misra et al. 2006). The dissipation of excitation energy is determined as nonphotochemical quenching of chlorophyll fluorescence during photosynthetic electron transport, which significantly correlates with contents of zeaxanthin and antheraxanthin produced during the xanthophyll cycle (Niyogi et al. 1997).

Although the photochemical efficiency of PS II was similar in both cabbage and kidney bean seedlings, the de-epoxidation state of violaxanthin increased sixfold in kidney bean, while no significant change was observed in cabbage under saline treatments. Similarly, nonphotochemical quenching increased in kidney bean, but decreased in cabbage. This showed that pigments involved in the xanthophyll cycle influenced nonphotochemical quenching in both cabbage and kidney. Thus, the increase in the de-epoxidation state of violaxanthin in salt-stressed kidney plants may be a signal to shield the pigment-protein complexes from salt-induced photodamage (Misra et al. 2006). During nonphotochemical quenching, the light harvesting complex of PS II undergo conformational changes, thereby generating a modification in pigment interactions causing the development of energy traps. Thus, nonphotochemical quenching plays a key role in the protection of PS II from photodamage. Nonphotochemical quenching is considered as an indicator of excess excitation energy (Ruban et al. 2002; Parida et al. 2007).

3.9.3 Effects on Gas-Exchange Characteristics

Carbon dioxide exchange characteristics have been regarded an important indicator of the growth of plants, because of their direct link to net productivity (Ashraf 2004; Piao et al. 2008). However, the effect of any stress on photosynthesis could be caused by stomatal, nonstomatal, or both factors (Athar and Ashraf 2005; Saibo et al. 2009); nevertheless, the extent of stress-induced stomatal or nonstomatal regulation depends on the species. Different environmental stresses including chilling temperature are known to depress gas-exchange characteristics to a varying extent thereby affecting overall photosynthetic capacity of most plants. For example, Lawlor and Cornic (2002) reported that the leaf net CO_2 assimilation rate of higher plants decreased substantially as the leaf water potential and relative water content decreased. The control of water loss through stomatal regulation has been recognized as an early plant response to stress (Jia and Zhang 2008; Harb et al. 2010). As stress continues, the stomata closure occurs for longer periods during the day. This, in turn, leads to the reduced carbon assimilation rate and water loss, resulting in maintenance of the carbon assimilation at the cost of low water availability (Brock and Galen 2005; Sausen and Rosa 2010; Pan et al. 2011). Stomatal limitation was generally considered to be the major factor of reduced photosynthesis under environmental stress condition (Galmés et al. 2007; Bousba et al. 2009).

In view of a number of reports, it is evident now that CO_2 uptake and photosynthetic yield decline considerably or even cease, when temperature deviate from optimum value (Islam 2011; Xie et al. 2011). Temperature stress, both high and low temperature, can damage the ultrastructure of the thylakoid membrane, thereby modulating its biochemical properties (Wang et al. 2010). High temperature has been reported to inactivate the oxygen evolving complex of PS II (Zhang et al. 2001; Toth et al. 2011). Photosynthetic capacity (rate at which leaves are able to fix carbon during photosynthesis and it is typically expressed as the amount of carbon dioxide that is fixed per meter squared per second, e.g., as μmol m^{-2} s^{-1}) has also been considered as one of the potential indicators of temperature stress tolerance in different plant species, because the gas-exchange characteristic showed a positive association with the drought tolerance in different species or cultivars within the species in several studies, e.g., tomato (Camejo et al. 2005; Singh et al. 2005) and melon (Kitroongruang et al. 1992).

3.9.4 Effects on Activities of Key Photosynthetic Enzymes

One of the most prominent effects of various stresses is the stomata closure, which leads to a lower concentration of intercellular CO_2, which in turn causes deactivation of Rubisco as well as other enzymes such as sucrose phosphate synthase and nitrate reductase (Chaves et al. 2009; Mumm et al. 2011). The increase in the amount of Rubisco can be beneficial for the survival of plants under harsh environmental conditions, because a positive relationship of leaf Rubisco contents with net photosynthetic rate has been reported in most C3 plants (Taub 2010; Makino 2011).

Fructose-1,6-bisphosphatase has been considered as one of the potential enzymes that can cause the decline of photosynthetic activity under stressful conditions, because it is involved in the regeneration of the Rubisco substrate, ribulose-1,5-bisphophate (RuBP). Ghosh et al. (2001) reported that the activity of fructose-1,6-bisphosphatase was positively associated with the different stress tolerance of the two rice species, because it was inhibited in the salt-sensitive species (e.g., *Oryza sativa*), but it remained unaffected in the wild, salt-tolerant species (*Porteresia coarctata*) under saline regimes. Salt stress, drought, and chilling temperature stress can also adversely affect levels and activities of different enzymes involved in the mechanism of photosynthesis, thereby impairing the synthesis of carbohydrates as well as water-use efficiency of plants, which in turn adversely affects the overall plant yield (Kumar and Singh 2009; Gill et al. 2011).

Change in temperature has been reported to reduce the solubility of CO_2 relative to O_2 within the leaf tissue resulting in reduced availability of CO_2 as a substrate for CO_2 concentrating mechanisms (Salvucci and Crafts-Brandner 2004b; Li et al. 2007). This in turn significantly affects the activities of key photosynthetic enzymes. At temperatures above the thermal optimum, Rubisco activase, the key regulatory enzyme of Rubisco, is reported to dissociate in both C3 and C4 species, causing a reduction in the photosynthetic capacity (Raines 2011; Sage and Zhu 2011). C3 plants require a stabilization of Rubisco activase to acclimate themselves to high temperatures (Kurek et al. 2007; Kumar et al. 2009). C4 plants are generally more tolerant to elevated temperatures than C3 plants. However, photosynthesis in C4 plants is also perturbed by high temperatures. For example, as well as high-temperature induced impairment in Rubisco activase functioning and the activation state in C4 plants, the inhibition occurs also in electron transport, phosphoenol pyruvate carboxylation, and phosphoenol pyruvate regeneration at elevated temperatures (Raines 2006; Kelly 2006).

It is evident that the activities of enzymes involved in C3 and C4 photosynthetic pathways are altered to a varying extent under stressful environments. However, it depends on the type of species, stomatal and nonstomatal factors, as well as their interaction, how far the changes in the activities of these enzymes affect photosynthetic capacity. Furthermore, a number of enzymes of both CO_2 concentrating mechanisms are regulated by light (Taiz and Zeiger 2010; Doubnerová and Ryšlavá 2011), so the light intensity and its duration in addition to the intensity of additional stress play also a significant role in the regulation of these enzymes. It has also been observed that under most stresses the balance between different enzymes is also perturbed. However, an optimal balance of different enzymes is necessary for the normal functioning of the photosynthetic pathways.

3.10 Improving Photosynthetic Capacity Through Photosynthetic Related Genes or Transcription Factors

Two approaches are currently used to improve the photosynthetic capacity in different plant species. Firstly, an effort is devoted to developing transgenic C3 plants with over-expression of ribulose-1,5-bisphosphate carboxylase/oxygenase genes or other enzymes of the C3 pathway. Secondly, attempts are underway to transfer the genes of a key C4 photosynthetic pathway to C3 plants. Considerable progress has been made during the last two decades in engineering the photosynthetic genes using advanced protocols of the recombinant DNA technology. Using the first strategy, Feng et al. (2007) produced a transgenic line of japonica rice by overexpressing *OsSbp* cDNA from an indica rice cultivar. This transgenic line showed beside an enhanced activity of the C3 enzyme sedoheptulose-l,7-bisphosphatase (SBPase, EC 3.1.3.37), which is involved in the regeneration of RuBP, also the enhanced photosynthetic capacity as well as the increased tolerance to salt stress. They suggested that the enhanced rate of photosynthesis under salt stress might have been due to the enhanced supply of RuBP to ribulose-1,5-bisphosphate carboxylase/oxygenase by the activation of SBPase. In another study, Tanaka et al. (2005) assessed the role of the overexpression of chloroplastic glutamine synthetase, in stress tolerance of rice. They showed that the overexpression of glutamine synthetase significantly reduced the tissue Na^+ content under stress regimes, suggesting the putative role of this enzyme in stress tolerance of rice. Similar to this study, Kozaki and Takeba (1996) showed enhanced protection from photoinhibition of transgenic tobacco plants overexpressing glutamine synthetase. Karaba et al. (2007) found a significant improvement in water use efficiency and in biomass production due to improved photosynthetic assimilation and decreased transpiration rate in transgenic rice plants under salt- and drought stress conditions by overexpressing *Arabidopsis* HARDY (*HRD*) gene. The expression of the pea *ABR17* (ABA-responsive17) cDNA, a member of pathogenesis-related proteins (PR10), in *Arabidopsis thaliana* showed the improved rate of germination under salt, cold temperature, or both conditions. Furthermore, the transgenic *Arabidopsis* plants showed the enhanced tolerance to freezing temperature, suggesting the potential utility of the *ABR17* gene to engineer multiple stress tolerance due to *ABR17*-mediated increased photosynthesis (Srivastava et al. 2004).

3.10.1 Transfer of C4 Genes/Traits to C3 Plants

Several attempts have been made during the past three decades to improve capacity of C3 plants by transferring C4 traits to C3 plants. Initially, classical hybridization of C4 plants with those of C3 was done, but most of the resultant C3-C4 hybrids were infertile (Brown and Bouton 1993). However, with the advent of recombinant DNA technology, it is possible now to transfer and express substantially the enzymes of C4 pathways in appropriate places within the leaves of C3 plants (Miyao 2003; Begonia and Begonia 2007; Kajala et al. 2012). For example, Häusler et al. (2002) reported that the first cDNAs of the C4-phosphoenolpyruvate carboxylase sequences for cloning were derived from maize (Izui et al. 1986) and *Flaveria trinervia* (Poetsch et al. 1991). Later on, a tobacco transgenic line was developed overexpressing maize phosphoenolpyruvate carboxylase gene under the control of its own promoter (Hudspeth et al. 1992) or by using the constitutively expressing *Cauliflower mosaic virus* (CaMV) 35S promoter (Kogami et al. 1994). However, in all these transgenic lines, although the activity of C4-phosphoenolpyruvate carboxylase increased twofold compared with that in nontransformed plants, the transgenic and nontransformed lines did not differ significantly in the net CO_2 assimilation rate. In another study, Ku et al. (1999) transferred the complete maize phosphoenolpyruvate carboxylase gene to rice plants; the activity of phosphoenolpyruvate carboxylase in the transgenic rice increased about 110-fold compared with the nontransgenic plants and threefold the maize activity. The incorporation of the intact maize C4-*Pdk* gene into rice was also reported to be effective in overproducing phosphopyruvate dikinase about 40-fold over that of wild-type rice plants (Fukayama et al. 2001).

Similarly, the introduction of the maize C4-specific NADP-malic enzyme cDNA to rice plants enhanced the activity of NADP-malic enzyme to 30- or 70-fold in rice leaves compared with the non-transgenic rice (Takeuchi et al. 2000; Tsuchida et al. 2001). Although Matsuoka et al. (2001) have reported that a number of C3 plants overexpressing C4 phosphoenolpyruvate carboxylase have shown enhanced photosynthetic capacity, the success is not so significant, when the phosphoenolpyruvate carboxylase gene is transferred to a phylogenetically distant plant species. For example, the intact, maize C4-specific phosphoenolpyruvate carboxylase gene was transferred to tobacco plants, but it was not fully expressed in the tobacco leaves (Hudspeth et al. 1992). This was ascribed to incorrect transcription initiation. In addition, incorrect splicing may take place, when genes from monocots are transferred to dicots (Ruan et al. 2012). Furthermore, the conversion of C3 to C4 plant requires a rearrangement or de novo formation of anatomical structures within the leaves to localize the photosynthetic enzymes in appropriate leaf tissues. Such promising anatomical changes, parallel to C4 leaf anatomy, have not been achieved so far. Therefore, the manipulation of genes responsible for the development of C4 specific structures in C3 plants is essential rather than only engineering the genes of key C4 enzymes. Thus, this is one of the major causes of the little success in achieving the desired goal in terms of developing C3 transgenic lines with C4 trait. In addition, although photosynthesis-related genes have been transferred from C4 to C3 plants and their photosynthetic capacity has been examined under nonstress conditions, no attempts have been reported to test the effectiveness of this approach under stressful conditions.

3.10.2 Role of Mitogen-Activated Protein Kinases (MAPKs) in Photosynthesis

Mitogen-activated protein kinases respond to a variety of extracellular stimuli including mitogens, proinflammatory cytokines, osmotic stress, and heat shock, and they pass on information from sensors to cellular processes in all eukaryotes (Nakagami et al. 2005; Diédhiou et al. 2008; Inagaki et al. 2008). They are known to regulate a multitude of cellular processes, such as mitosis, differentiation, gene expression, and cell death or survival (Zhang and Klessig 2001). MAP kinases are widespread in eukaryotes. They are effectively involved in the signal transduction of various metabolic processes. MAPK cascades comprise minimally three protein kinases, a MAP kinase kinase kinase (MAPKKK), a MAP kinase kinase (MAPKK), and a MAP kinase (MAPK), which function in concert. Activation of each MAPK occurs when it undergoes phosphorylation. Following the activation, mitogen-activated protein kinases move from the cytoplasm into the nucleus, wherein they phosphorylate many transcription factors leading to alterations in the gene expression (Whitmarsh and Davis 2000; Janknecht 2003; Zhang et al. 2011). There is strong evidence now that MAP kinases are involved in the mechanisms of tolerance to a variety of biotic and abiotic stresses in plants (Huang et al. 2015; Zhang et al. 2011). For example, increased expression of genes encoding a MAP kinase module is strong evidence of the involvement of this enzyme cascade in plants exposed to low temperature, salt, or drought stress (Mizoguchi et al. 1996; Dai et al. 2007; Kant et al. 2007). In *Arabidopsis*, the MAPK kinase 2 (MKK2) has been reported to be activated by both cold and salt stress (Teige et al. 2004). Similarly, mitogen-activated protein kinases have been reported to be activated by osmotic stresses in *Medicago sativa* and tobacco (Jonak et al. 2002) and by salt stress in rice (Diédhiou et al. 2008). Shou et al. (2004) have shown that the tobacco MAPKK (mitogen activated protein kinase kinase) and NPK1 (nitrogen activated protein kinase) expressed constitutively in maize resulted in a maize transgenic line with enhanced drought tolerance trait measured in terms of kernel mass and *net photosynthetic rate*. The enhanced drought tolerance of the transgenic line of maize was shown to be strongly associated with a markedly high rate of photosynthesis, suggesting the active role of NPK1 in the mechanism of protection of photosynthesis machinery from water stress damage.

It is evident that development of transgenic lines of C3 plants overexpressing C4 photosynthetic enzymes is a meaningful approach to improve photosynthetic capacity of C3 plants and to bring it to the level of C4 plants. However, the extent, to which C4 enzymes transferred to C3 plants play a role in effectively fixing CO_2, depends on a number of factors including the localization of introduced C4 enzymes within the leaf tissues and coordination of C4 enzymes with already CO_2 fixing pathways or other allied pathways operative in C3 plants. It has been observed that although overexpression of a single C4 enzyme can modulate the photosynthetic metabolism in C3 plants, in most cases, it does not have significant effects on photosynthesis. Furthermore, little success of attaining photosynthetic capacity in C3 plants equivalent to that of C4 plants by transferring C4 genes to C3 plants could have been due to non-transformation of C3 leaf anatomical structure to that of C4 leaf. This may be the main reason of a low effectiveness of the attempts to bring C3 photosynthetic capacity at par with that of C4 through genetic engineering. Thus, alternatively, efforts should be made to improve the efficiency of CO_2 concentrating processes in C3 plants by enhancing the activities of key enzymes through genetic manipulation.

3.11 Strategies to Ameliorate Cold Stress and Improve Photosynthesis

Cold stress usually deteriorates the overall photosynthetic apparatus. Exhaustive research has revealed some experimental strategies which can be utilized to ameliorate the cold-induced damages of this important physiological process (Adam and Murthy 2014). Priming of the seeds or stressed seedlings with chemical agents can stimulate photosynthesis even under suboptimal conditions (Roychoudhury and Banerjee 2016). Exogenous application of melatonin at 100 μM concentration reduced the damage to photosynthetic apparatus in cold stressed *Lycopersicon esculentum* L. leaves (Yang et al. 2018). The treated seedlings exhibited increased electron transfer rates and quantum yields of PS I and PS II at low temperature (Yang et al. 2018). In another study, Li et al. (2016) deciphered the relations between melatonin and abscisic acid in a process of cold tolerance induced by drought priming. It was found that exogenous melatonin application in the drought primed seeds of barley induced the endogenous antioxidant and abscisic acid levels. The treated seedlings exhibited sustained photosynthetic electron transport during cold stress (Li et al. 2016). Cold priming of the winter wheat seeds at 5.2 °C stimulated the reactive oxygen species scavenging systems (Li et al. 2014). The trapped energy flux, electron transport, and photosynthetic rates increased in the primed plants grown at low temperature (Li et al. 2014). Cvetkovic et al. (2017) also reported the positive correlation between cold priming and stabilization of PS II activity in *Arabidopsis*. Priming with the compatible solute, glycinebetain essentially conferred photosynthetic protection in plants exposed to cold stress (Roychoudhury and

Banerjee 2016; Giri 2011). Transgenic rice plants overexpressing choline oxidase net carbon assimilation exhibited increased photosynthetic rate when subjected to cold stress (Kathuria et al. 2009). Hur et al. (2004) showed that knockout of the proline synthesizing gene, Pyrroline-5-carboxylate synthetase 2, severely retarded the photosynthetic process and growth in rice at 4 °C. Thus, Pyrroline-5-carboxylate synthetase 2 proved to be the crucial gene for conferring cold tolerance in rice. Treatment of the medicinal herb, *Stevia rebaudiana* Bertoni, with polyamine supplement improved the photochemical efficiency of PS II along with an increase in the photosynthetic pigments during cold stress. The treated plants also accumulated higher quantities of proline and glycinebetain during low temperature stress compared to the non-treated sets (Peynevandi et al. 2018). Overexpression of glycerol-3-phosphate acyltransferase involved in cellular fatty acid saturation conferred protection to the photosynthetic apparatus in transgenic tobacco plants exposed to low temperature (Murata et al. 1992). The transgenic tobacco plants overexpressing superoxide dismutase efficiently scavenged the cold-induced generation of reactive oxygen species and also exhibited 20% higher photosynthetic activity compared to the wild-type plants (Sen Gupta et al. 1993). Pilon-Smits et al. (1995) overexpressed the dehydrin cold regulated in *Arabidopsis* plants and observed enhanced freezing tolerance of the chloroplasts and protoplasts (Banerjee and Roychoudhury 2020).

3.12 Management Strategies Against Low Temperature Stress

Chilling stresses have been shown to have adverse effects on the growth and productivity of crop plants (Liu et al. 1998), and plant tolerance to this stress might have certain protective mechanisms in shared (Stockinger et al. 1997). Plants often develop a cascade of strategies through physiological, biochemical, and molecular modifications at cell level to cope with these stresses. Studies have shown that mechanisms of plant tolerance to chilling stress can be regulated through developing tolerant plant genotypes, genetic modifications, seed treatments, application of plant growth regulators, and compatible solutes and use of plant mineral nutrients.

3.12.1 Selection and Breeding Strategies

Integration of conventional, molecular, and omic-based techniques can successfully be employed to develop tolerant plant genotypes (Maqbool et al. 2017). Conventional breeding has been instrumental in developing chilling tolerant cultivars in various crops (Jha et al. 2017). Various rice varieties, e.g., "Koshihikari" in Japan, "Silewah" in Indonesia, and "Padi Labou Alumbis" in Malaysia, were released for chilling stress tolerance by conventional breeding method (Ahmad et al. 2014a). Based on both open air and controlled conditions, Rodriguez et al. (2007) reported that "EP80 × Puenteareas" population in maize was an important source of low temperature tolerance. Implementation of the modern omics approaches and identification of QTLs/genes for chilling tolerance can significantly support crop improvement strategies aimed to develop high yielding cultivars under low-temperature condition (Ahmad et al. 2014a; Jha et al. 2017). In rice, selection using different parameters led to the development of low temperature tolerant genotypes such as "HSC55," "M103," and "Jyoudeki" based on low spikelet sterility (Farrell et al. 2006; Ye et al. 2009). Suh et al. (2010) reported that phenotyping selection and simple sequence repeat makers' identification methods are helpful in screening of plant genotypes against chilling stress. They found three QTLs responsible for seed setting percentage under chilling stress by using a recombinant inbred line population developed by tropical japonica × temperate japonica.

Under stress conditions, estimating final grain yield is a simple and easy method for screening tolerant crops. Promising tolerant genotypes can be identified by direct selection of grain yield under stress conditions (Verulkar et al. 2010). Hence, conventional breeding approach includes the selection of crop plants for its yield and related attributes under stress conditions (Ahmad et al. 2014a). In general, breeding strategies have been employed to manipulate the genetic makeup of crops for enhancing their tolerance against stress (Maqbool et al. 2017); however, marker-assisted selection proved better than classical breeding for screening traits associated with stress tolerance (Ahmad et al. 2014a) and enhancing stress tolerance in plants (Jongdee et al. 2002). The molecular basis of drought tolerance is generally based on QTL analysis and segregation mapping (Ali et al. 2017). The identification of QTLs associated with drought tolerance is an important tool for marker-assisted selection of plants with desired characters (Farooq et al. 2009b). Marker-assisted recurrent selection involves the accumulation of more superior alleles responsible for drought tolerance in plants (Varshney et al. 2012). A large number of QTLs have already been identified for several stress tolerance traits; nevertheless, epigenetic quantitative trait loci and epistatic QTLs discovered in the future can be used for molecular breeding (Gupta et al. 2017). Sequential or simultaneous transfer of QTLs at different developmental stages of rice may lead to cultivars with increased stress tolerance capacities (Lang et al. 2013). Overexpression of late embryogenesis abundant gene has been reported to increase stress tolerance in rice (Xiao et al. 2007). Furthermore, while discussing the constraints in breeding approaches, Ouk et al. (2006) reported that low heritability of grain yield and lack of effective selection criteria are the main hindrances to develop stress-tolerant cultivars. Similarly, Kumar et al. (2008) reported that many previous trials on stress tolerance failed to impose severe stress; therefore true and precise stress-tolerant lines were not selected. Although conventional breeding methods for stress tolerance have been used for a long time, nevertheless, advancements in genomics and molecular breeding approaches would have the significant role in the development of stress-tolerant cultivars (Kumar et al. 2015). Moreover, selection of crop cultivars for their unique responses to chilling stress might be helpful for breeding programs to develop stress-tolerant plant types.

3.12.2 Agronomic and Physiological Measures

Plants respond to chilling stress via series of agronomic and physiological processes. Exogenous application of compatible solutes, plant growth hormones, plant mineral nutrient, and seed priming can be effective in alleviating the adverse effects of drought and chilling stresses.

3.12.2.1 Application of Compatible Solutes

Compatible solutes protect the plants from osmotic stress by not posing any detrimental effects on enzymes, membranes, and other macromolecules even at higher concentrations (Cechin et al. 2006; Kiani et al. 2007). Compatible solutes include glycinebetaine, soluble sugars, proline, sugar alcohols, trehalose, and some organic acids (Kiani et al. 2007). Many of the osmolytes are organic solutes in nature while some of the essential ions such as K^+ (Yokoi et al. 2002). Among various osmolytes, proline has been found to have multiple roles in plant stress regulation and tolerance potential (Yamada et al. 2005; Farooq et al. 2009a, b). Proline and polyamines have been reported to impart chilling tolerance in maize (Songstad et al. 1990). Proline retains the potential to relieve low temperature injury in cold-sensitive plants. Although most of the chilling-sensitive plants also accumulate proline, the levels of accumulation are not enough to induce cold tolerance; therefore, exogenous application of proline can be a potential option to enhance its level in plants (Hussain et al. 2018).

Soluble sugars and sugar alcohols serve as osomoregulators, signalling molecules, and cryoprotectants in plants (Ruelland et al. 2009) and helps in cleansing stress-generated reactive oxygen species (Van den Ende and Valluru 2009). Chilling tolerance has been observed by increase of total soluble sugars in many plant species (Farooq et al. 2009a; Wang et al. 2016). Soluble sugars reduce osmotic potential and enhance water absorption under water limited conditions, and thus improve stress tolerance in plants (Farooq et al. 2009b). Exogenous application of glycinebetaine play critical roles in abiotic stress tolerance in plants by modifying antioxidant activities, membrane integrity, osmotic adjustment, and reactive oxygen species detoxification (Park et al. 2006; Xing and Rajashekar 2001; Farooq et al. 2008). Xing and Rajashekar (2001) reported that glycinebetaine accumulation was positively related to the cold stress tolerance in plants. Glycinebetaine improved the growth of *Solanum tuberosum* plants under low temperature conditions (Somersalo et al. 1996) and freezing tolerance in *Arabidopsis thaliana* (Xing and Rajashekar 2001). Its foliar application was helpful in low temperature tolerance of several plant species (Xing and Rajashekar 2001; Farooq et al. 2009b). In short, application of compatible solutes such as glycinebetaine, soluble sugars, proline, and polyamines can be helpful in imparting drought and chilling tolerance in crop plants.

3.12.2.2 Application of Plant Growth Regulators

Plant growth hormones have a dynamic role for increasing the ability of plants to acclimatize against abiotic stress conditions. Auxins, gibberellins, cytokinins, ethylene, abscisic acid, salicylic acid, jasmonates, and brassinosteroids have been widely studied in plants under different abiotic stresses. These plant hormones are being used efficiently under stress conditions for enhancing crop production. Plants normally synthesize various types of phytohormones under individual and concurrent abiotic stresses depending upon the strength of their defense mechanism (Nadeem et al. 2016). Abscisic acid is synthesized in response to both drought and chilling and contribute in stress tolerance of crop plants (Liu et al. 1998). Abscisic acid, jasmonates, and salicylic acid often activate phosphoprotein cascade pathways that led to expression of genes associated with cold stress tolerance in plants (Kolaksazov et al. 2013). Greater chilling tolerance has been recorded in abscisic acid-treated maize (Xin and Li 1992), and levels of abscisic acid are increased in chilling-tolerant genotypes (Farooq et al. 2009a). Under water deficit and low temperature stress conditions, plant growth regulator including brassinosteroids significantly improved chlorophyll content and increased water potential in soybean (Zhang et al. 2004). Khan et al. (2015) reported that application of brassinosteroids led to considerable improvements in plant tolerance against chilling stress in terms of plant growth, photosynthesis, and antioxidant system in tomato plants. Reguera et al. (2013) stated that drought and low temperature tolerance in rice was concomitant with the accumulation of cytokinins. Drought and low temperature tolerance induced by cytokinins has also been observed in tobacco (Rivero et al. 2007) and wheat (Shang 2000). Kaya et al. (2006) reported that drought and low temperature tolerance can also be enhanced by the application of gibberellins in maize. Exogenous gibberellins application breaks the rosette by rapid enlargement of differentiated tissues under cold stress (Khan et al. 2017). Similarly, jasmotae has been reported to trigger the plant defense responses to abiotic stresses including drought and chilling (Pauwels et al. 2009; Seo et al. 2011). Exogenous application of methyl jasmonate improved chilling tolerance in wheat by modulating antioxidant defense system and production of soluble protein content (Qi et al. 2005). The relationship between enhanced ethylene levels and chilling tolerance in crop plants has also been reported by Chu and Lee (1989) and Zhao et al. (2014).

Salicylic acid is also considered as potential chemical compound for drought and chilling tolerance in several plant species (Farooq et al. 2009a, b). Salicylic acid application was found to be effective in alleviating the chilling (Miura and Tada 2014) and freezing injury (Tasgin et al. 2003). In maize, the harmful effects of low-temperature stress were decreased by the application of salicylic acid (Janda et al. 1999; Wang et al. 2012). Horváth et al. (2007) reported that significant improvement in performance of winter wheat was related with salicylic acid induced higher activities of catalase under low temperature stress. Exogenous auxin application improved wheat yield significantly under soil water deficit and low temperature stress conditions (Abdoli et al. 2013).

Non-hormonal growth regulators, e.g., triazoles, paklobutrazol, chlorocholinchloride, meflidid, and unikonazol (Lurie et al. 1994; Feng et al. 2003), are also used to enhance drought and chilling tolerance in crop plants (Feng et al. 2003; Lukatkin et al. 2012). Still and Pill (2003) observed that foliar applications of paclobutrazol remained effective in improving drought and low temperature tolerance in tomato seedlings. These evidences suggest the exogenous application of various plant growth regulators is a potent method to improve the drought and chilling tolerance in plants.

3.12.2.3 Use of Mineral Nutrients

Mineral nutrients play important role in improving plant tolerance against stress conditions (Marschner 1995; Waraich et al. 2012). To alleviate the adverse effects of abiotic stresses, nitrogen fertilization has been reported to have significant role in stress alleviation (Waraich et al. 2011a). Nitrogen as nitric oxide is a highly reactive that protects plant against stress conditions by acting as scavenger to reactive oxygen species (Wendehenne et al. 2001; Waraich et al. 2012). Under stress, phosphorus helps plants to maintain leaf water potential which, in turn, enhances stomatal conductivity and photosynthetic rates (Waraich et al. 2011b). Supplementation of potassium doses and its uptake in plant parts could be beneficial for obtaining reasonable yields under stress conditions (Valadabadi and Farahani 2010). Potassium-induced increases in proline and free sugar contents in rice were reported by Pandey et al. (2004) under soil water deficit conditions. Kafkafi (1990) reported that low temperature injuries can be avoided in carnation plants by supplementing potassium in irrigation water.

Calcium has been reported to enhance the concentrations of amino acids, polyamines (putrescine and spermidine), and chlorophyll content in red spruce under low temperature stress (Schaberg et al. 2011). Waraich et al. (2012) reported that calcium is an essential nutrient for stomatal closure in chilling-tolerant genotypes. Application of calcium in *Vicia faba* enhanced plant biomass, plant water relations, and chlorophyll content while reduced membrane leakage (Abdel-Basset 1998). Magnesium also improves root morphological characteristics which help to increase uptake of water and nutrients via roots, while boron nutrition enhances sugar transport in the plant body and improves seed germination and grain formation process (Waraich et al. 2012). Application of selenium was found to improve plant tolerance against chilling (Hawrylak-Nowak et al. 2010) and soil water deficit stress (Valadabadi et al. 2010).

3.12.2.4 Seed Priming

Seed priming is considered to improve the abiotic stress tolerance (including chilling and drought) in different plant species (Hasanuzzaman and Fujita 2011; Hussain et al. 2016; Wang et al. 2016; Samota et al. 2017). Seed priming promotes the germination-related metabolic functions without radicle protrusion from seeds (Hussain et al. 2016; Wang et al. 2016; Zheng et al. 2016). It is a useful technique to acclimatize against stress conditions (Samota et al. 2017) by modulating physio-biochemical processes in developing seedlings (Hasanuzzaman and Fujita 2011). Seed priming with plant hormones and other chemical compounds can play effective role in alleviating the adverse effects of drought and chilling stresses at earlier growth stages. Seed priming activated antioxidant defense system thus improved seed germination and early growth in tobacco seedlings (Xu et al. 2011). Seed priming with ascorbic acid, salicylic acid, and H_2O_2 improved seedling establishment under suboptimal temperature (Ahmad et al. 2014b). Hussain et al. (2016)

found that the seed priming with selenium and salicylic acid were effective in improving chilling tolerance in rice cultivars. Likewise, seed priming either with 2.5% K_2HPO_4 or 2.5% K_2HPO_4 + KNO_3 were observed to induce chilling tolerance at early growth stages than non-primed seeds (Farooq et al. 2009a). Seed priming with salicylic acid improved relative water and chlorophyll contents, antioxidant activities, plant biomass, and grain yield of wheat (Singh and Usha 2003), whereas selenium priming regulated activities/levels of enzymatic and non-enzymatic antioxidants in rapeseed under drought stress (Hasanuzzaman and Fujita 2011). The priming of tomato and bean seeds with 0.1–0.5 mM salicylic acid improved plant tolerance to drought and low temperature stresses (Senaratna et al. 2000).

3.13 Conclusion and Future Prospects

It is now evident that environmental stresses including low temperature stress lead to the considerable reduction in photosynthetic performance mediated through stress-induced stomatal or nonstomatal limitations. However, it is not easy to discriminate between the effects of these limitations on overall photosynthetic capacity of a plant. Certainly, it depends on the species to what extent the process of photosynthesis under stress conditions is controlled by stomatal or nonstomatal factors. Nevertheless, knowledge of the proportion of the stomatal and nonstomatal factors controlling the process of photosynthesis is vital for appointing the future research on photosynthesis. Salinity, drought, or temperatures modulate gas exchange characteristics, such as net photosynthetic rate and stomatal conductance. Although genetic differences in photosynthetic capacity exist at intraspecific and interspecific levels, *net photosynthetic rate* is considered as one of the potential, physiological, selection criteria for stress tolerance (Ashraf 2004). Since an unreliable association of the photosynthetic capacity with the degree of stress tolerance exists in many species, use of the photosynthetic capacity as a physiological marker will be realistic only in those species for which a positive relationship between the photosynthetic performance and the growth under stress conditions has been confirmed. Improvement in the photosynthetic performance of C3 plants to bring it at par with that of C4 plants has been an important goal of plant biologists. This could be achieved either by overexpressing the key C3 enzymes or by transferring the C4 enzymes to C3 plants. The production of transgenic lines overexpressing C3 or C4 enzymes has had limited success in enhancing photosynthetic performance in most transformed lines. It is not clear yet up to what extent C4 enzymes transferred to C3 plants take part in effective assimilation of CO_2 in C3 engineered plants.

It depends undoubtedly on a multitude of factors including the partitioning of introduced C4 enzymes within the leaf tissues and the interlinking of C4 enzymes with photosynthesis-related pathways functioning in C3 plants. In most of the studies on generations of transgenic lines made for enhanced photosynthetic performance that were published so far, a single C4 enzyme gene has been transferred, but in most cases this did not result in a major success in terms of enhanced photosynthetic capacity in transgenic C3 lines. Thus, a generation of transgenic C3 plants overexpressing multiple C4 enzymes is one of the premier themes of future research.

The elucidation of stress-induced signal transduction pathways is vital to enhance plant tolerance to different stresses. Undoubtedly, by integrating the advances in molecular genetics and cell biology, our knowledge of the signaling pathways has increased greatly over recent years, although it is still far from a full understanding to perception and signaling of environmental cues in plants. There is little information on the components involved in either the perception or signaling involved in stress responses. One of several promising advances made recently in stress biology, knowledge of long- and short-distance signaling is contemplated as of vital importance for understanding up- or downregulation of photosynthesis under various stresses, because signaling pathways involved in plant stress responses are interconnected at several points (Chaves et al. 2009). This calls for ample understanding of the signal transduction pathways induced by different stresses to improve plant tolerance to various stressful conditions. Efforts are currently underway in different laboratories all over the world to identify signaling components, such as transcription factors and protein kinases, particularly mitogen-activated protein kinases, involved in stress adaptation in plants. Thus, understanding the prospective association between the stress-induced protein kinase pathways and the genetic and epigenetic regulation of the gene expression in plants remains the important area of future research. Photosynthetic response to different types of stresses is quite intricate, because it entails the interaction of several restrictions occurring at different locations of the cell or the leaf, and at different phases of the plant growth and development. Furthermore, the duration and intensity of the stress can also significantly affect the photosynthetic capacity. Thus, the complete elucidation of the factors involved in the regulation of photosynthetic capacity under stressful environments would help devise appropriate strategies to grow plants successfully in stress-prone areas.

References

Abdel-Basset R (1998) Calcium channels and membrane disorders induced by drought stress in Vicia faba, plants supplemented with calcium. Acta Physiol Plant 20:149–153

Abdoli M, Saeidi M, Azhand M, Jalali-Honarmand S, Esfandiari E, Shekari F (2013) The effects of different levels of salinity and indole-3-acetic acid (IAA) on early growth and germination of wheat seedling. J Stress Physiol Biochem 9:329–338

Adams WW, Demmig-Adams B, Winter K (1990) Relative contributions of zeaxanthin-related and zeaxanthin unrelated types of 'high-energy-state' quenching of chlorophyll fluorescence in spinach leaves exposed to various environmental conditions. Plant Physiol 92:302–309

Adams WW, Demmig-Adams B (2004) Chlorophyll fluorescence as a tool to monitor plant response to the environment. In: Papageorgiou GC, Govindjee (eds) Chlorophyll a fluorescence: a signature of photosynthesis, advances in photosynthesis and respiration, vol 19. Springer, Dordrecht, pp 583–604

Adams WW, Zarter CR, Mueh KE, Amiard V, Demmig-Adams B (2006) Energy dissipation and photoinhibition: a continuum of photoprotection. In: Demmig-Adams B, Adams WW, Mattoo

AK (eds) Photoprotection, photoinhibition, gene regulation, and environment, advances in photosynthesis and respiration, vol 21. Springer, Dordrecht, pp 49–64

Adams WWIII, Muller O, Cohu CM, Demmig-Adams B (2013a) May photoinhibition be a consequence, rather than a cause, of limited plant productivity? Photosynthesis Res 117(1–3): 31–44

Adams HD, Germino MJ, Breshears DD, Barron-Gafford GA, Guardiola-Claramonte M, Zou CB, Huxman TB (2013b) Nonstructural leaf carbohydrate dynamics of Pinus edulis during drought-induced tree mortality reveal role for carbon metabolism in mortality mechanism. New Phytol 197(4):1142–1151

Adam S, Murthy SDS (2014) Effects of cold stress on photosynthesis of plants and possible protection mechanisms. In: Gaur RK, Sharma P (eds) Approaches to plant stress and their management. Springer, India, pp 219–226

Adams WW, Muller O, Cohu CM, Demmig-Adams B (2014) Photosystem II efficiency and non-photochemical quenching in the context of source-sink balance. In: Demmig-Adams B, Garab G, Adams WW, Govindjee (eds) Non-photochemical quenching and energy dissipation in plants, algae and cyanobacteria, advances in photosynthesis and respiration, vol 40. Springer, Dordrecht, pp 503–529

Aghaee A, Moradi F, Zare-maivan H, Zarinkamar F, Irandoost HP, Sharifi P (2011) Physiological responses of two rice (Oryza sativa L.) genotypes to chilling stress at seedling stage. Afr J Biotechnol 10:7617–7621

Ahmad P, Wani MR, Azooz MM, Tran LSP (2014a) Improvement of crops in the era of climatic changes. Springer, New York, NY

Ahmad I, Basra SMA, Wahid A (2014b) Exogenous application of ascorbic acid, salicylic acid and hydrogen peroxide improves the productivity of hybrid maize at low temperature stress. Int J Agric Biol 16(4):825–830

Alam B, Jacob J (2002) Overproduction of photosynthetic electrons is associated with chilling injury in green leaves. Photosynthetica 40(1):91–95

Alberdi M, Bravo LA, Gutiérrez A, Gidekel M, Corcuera LJ (2002) Ecophysiology of Antarctic vascular plants. Physiol Plant 115:479–486

Alexandratos N, Bruinsma J (2012) World Agriculture towards 2030/2050: The 2012 Revision; ESA Working Paper No. 12–03. FAO, Rome

Ali F, Bano A, Fazal A (2017) Recent methods of drought stress tolerance in plants. Plant Growth Regul 82:363–375

Allen RD (1995) Overexpression of chloroplastic Cu/Zn superoxide dismutase in plants. In: Gartland KMA, Davey MR (eds) Methods in molecular biology, vol 44. Humana Press Inc., Totowa, NJ, pp 309–323

Allen DJ, Ratner K, Giller Y-E, Gussakovsky EE, Shahak Y, Ort DR (2000) An overnight chill induces a delayed inhibition of photosynthesis at midday in mango (Mangifera indica L.). J Exp Bot 51(352):1893–1902

Allen DJ, Ort DR (2001) Impacts of chilling temperature on photosynthesis in warm-climate plants. Trends Plant Sci 6(1):36–42

Alscher RG, Donahue JL, Cramer CL (1997) Reactive oxygen species and antioxidants: relationships in green cells. Physiol Plant 100(2):224–233

Ameglio T, Cochard H, Ewers FW (2001) Stem diameter variation and cold hardiness in walnut trees. J Exp Bot 52:2135–2142

Amme S, Matros A, Schlesier B, Mock H-P (2006) Proteome analysis of cold stress response in Arabidopsis thaliana using DIGE-technology. J Exp Bot 57(7):1537–1546

Andrews JR, Fryer MJ, Baker NR (1995) Characterization of chilling effects on photosynthetic performance of maize crops during early-season growth using chlorophyll fluorescence. J Exp Bot 46(9):1195–1203

Anjum F, Yaseen M, Rasool E, Wahid A, Anjum S (2003) Water stress in barley (Hordeum vulgare L.). i. Effect on morphological characters. Pak J Agric Sci 40(1–2):45–49

Apel K, Hirt H (2004) Reactive oxygen species: metabolism, oxidative stress, and signal transduction. Annu Rev Plant Biol 55:373–399

Aguilera C, Stirling CM, Long SP (1999) Genotypic variation within Zea mays for susceptibility to and rate of recovery from chill-induced photoinhibition of photosynthesis. Physiol Plant 106(4): 429–436

Armstrong AF, Badger MR, Day DA, Barthet MM, Smith PMC, Millar AH, Whelan J, Atkin OK (2008) Dynamic changes in the mitochondrial electron transport chain underpinning cold acclimation of leaf respiration. Plant Cell Environ 31(8):1156–1169

Aro EM, Virgin I, Andersson B (1993) Photoinhibition of photosystem II. Inactivation, protein damage and turnover. Biochim Biophys Acta 1143(2):113–134

Asada K, Endo T, Mano J, Miyake C (1998) Molecular mechanisms for relaxation of and protection from light stress. In: Satoh K, Murata N (eds) Stress responses of photosynthetic organisms, pp 37–52

Ashraf M (2004) Some important physiological selection criteria for salt tolerance in plants. Flora 199(5):361–376

Athar HR, Ashraf M (2005) Photosynthesis under drought stress. In: Pessarakli M (ed) Handbook of photosynthesis. CRC Press/Taylor and Francis Group, New York, pp 793–804

Atkin OK, Tjoelker MG (2003) Thermal acclimation and the dynamic response of plant respiration to temperature. Trends Plant Sci 8(7):343–351

Atkin OK, Bruhn D, Hurry VM, Tjoelker MG (2005) Evans review no. 2: the hot and the cold: unravelling the variable response of plant respiration to temperature. Funct Plant Biol 32(2): 87–105

Atkin OK, Scheurwater I, Pons TL (2006) High thermal acclimation potential of both photosynthesis and respiration in two low land Plantago species in contrast to an alpine congeneric. Glob Chang Biol 12(3):500–515

Augustyniak A, Perlikowski D, Rapacz M, Koscielniak J, Kosmala A (2018) Insight into cellular proteome of Lolium multiflorum/Festuca arundinacea introgression forms to decipher crucial mechanisms of cold acclimation in forage grasses. Plant Sci 272:22–31

Badger MR, Björkman O, Armond PA (1982) An analysis of photosynthetic response and adaptation to temperature in higher plants: temperature acclimation in the desert evergreen. Nerium oleander L. Plant Cell Environ 5(1):85–99

Baena-Gonzalez E, Gray JC, Tyystjarvi E, Aro EM, Maenpaa P (2001) Abnormal regulation of photosynthetic electron transport in a chloroplast ycf9 inactivation mutant. J Biol Chem 276: 20795–20802

Bagnall D, Wolfe J, King RW (1983) Chill-induced wilting and hydraulic recovery in mung bean plants. Plant Cell Environ 6(6):457–464

Baker NR, Long SP, Ort DR (1988) Photosynthesis and temperature with particularly references to effects on quantum yield. In: Long SP, Wooward FI (eds) Plants and temperature. Company of Biologists, Cambridge, pp 347–375

Baker NR, Bradbury M, Farage PK, Ireland CR, Long SP (1989) Measurements of the quantum yield of carbon assimilation and chlorophyll fluorescence for assessment of photosynthetic performance of crops in the field. Philos Trans R Soc Lond Ser B Biol Sci 323(1216):295–308

Baker NR, Nie G-Y, Ortizlopez A, Ort DR, Long SP (1990) Analysis of chill-induced depressions of photosynthesis in maize. In: Baltscheffsky M (ed) Current research in photosynthesis, vol 4. Kluwer, Dordrecht, pp 565–572

Baker NR (1994) Low temperature stress and photosynthesis. In: Foyer C, Mullineaux P (eds) Photooxidative stresses on plants: causes and amelioration. CRC Press, Boca Raton, FL, pp 127–154

Ball M, Hodges V, Laughlin G (1991) Cold-induced photoinhibition limits regeneration of snow gum at tree-line. Funct Ecol 5:663–668

Banerjee A, Roychoudhury A (2015) WRKY proteins: signaling and regulation of expression during abiotic stress responses. Sci World J 2015:Article ID 807560

Banerjee A, Roychoudhury A (2016) Plant responses to light stress: oxidative damages, photoprotection and role of phytohormones. In: Ahammed GJ, Yu J-Q (eds) Plant hormones under challenging environmental factors, Springer, Netherlands, pp 181–213
Banerjee A, Wani SH, Roychoudhury A (2017) Epigenetic control of plant cold responses. Front Plant Sci 8:1643
Banerjee A, Roychoudhury A (2018a) Regulation of photosynthesis under salinity and drought stress. In: Singh VP, Singh S, Singh R, Prasad SM (eds) Environment and photosynthesis: a future prospect. Studium Press (India) Pvt. Ltd., New Delhi, pp 134–144
Banerjee A, Roychoudhury A (2018b) Abiotic stress, generation of reactive oxygen species, and their consequences: an overview. In: Singh VP, Singh S, Tripathi DK, Prasad SM, Chauhan DK (eds) Revisiting the role of reactive oxygen species (ROS) in plants: ROS boon or bane for plants? Wiley, Hoboken, NJ
Banerjee A, Roychoudhury A (2020) Gibberellic acid-priming promotes fluoride tolerance in a susceptible indica rice cultivar by regulating the antioxidant and phytohormone homeostasis. J Plant Growth Regul 39(4):1476–1487
Banon S, Fernandez JA, Franco JA, Torrecillas A, Alarcón JJ, Sánchez-Blanco MJ (2004) Effects of water stress and night temperature preconditioning on water relations and morphological and anatomical changes of Lotus creticus plants. Sci Hortic 101(3):333–342
Barrero-Gil J, Salinas J (2013) Post-translational regulation of cold acclimation response. Plant Sci 205-206:48–54
Basu S, Ramegowda V, Kumar A, Pereira A (2016) Plant adaptation to drought stress. F1000 Res 5: 1554
Battaglia M, Beadle C, Loughhead S (1996) Photosynthetic temperature responses of Eucalyptus globulus and Eucalyptus nitens. Tree Physiol 16(1–2):81–89
Beale CV, Long SP (1995) Can perennial C4 grasses attain high efficiencies of radiant energy conversion in cool climates? Plant Cell Environ 18(6):641–650
Beale CV, Bint DA, Long SP (1996) Leaf photosynthesis in the C4-grass Miscanthus x giganteus, growing in the cool temperate climate of southern England. J Exp Bot 47(295):267–273
Begonia GB, Begonia MT (2007) Plant photosynthetic production as controlled by leaf growth, phenology, and behavior. Photosynthetica 45(3):321–333
Berglund AH, Nilsson R, Liljenberg C (1999) Permeability of large unilamelar digalactosyldiacyl glycerol vesicle for protons and glucose-influence of α-tocopherl, β-carotene, zeaxanthin and cholesterol. Plant Physiol Biochem 37(3):179–186
Bernacchi CJ, Singsaas EL, Pimentel C, Portis AR, Long SP (2001) Improved temperature response functions for models of Rubisco-limited photosynthesis. Plant Cell Environ 24(2):253–259
Bernacchi CJ, Portis AR, Nakano H, von Caemmerer S, Long SP (2002) Temperature response of mesophyll conductance. Implications for the determination of rubisco enzyme kinetics and for limitations to photosynthesis in vivo. Plant Physiol 130(4):1992–1998
Bernacchi CJ, Pimentel C, Long SP (2003) In vivo temperature response functions of parameters required to model RuBPlimited photosynthesis. Plant Cell Environ 26(9):1419–1430
Berry J, Björkman O (1980) Photosynthetic response and adaptation to temperature in higher plants. Annu Rev Plant Physiol 31:491–543
Bilska-Kos A, Panek P, Szulc-Głaz A, Ochodzki P, Cisto A, Zebrowski J (2018) Chilling-induced physiological, anatomical and biochemical responses in the leaves of Miscanthus × giganteus and maize (Zea mays L.). J Plant Physiol 228:178–188
Biswal B, Raval MK, Biswal UC, Joshi P (2008) Response of photosynthetic organelles to abiotic stress: modulation by sulfur metabolism. In: Khan NA, Singh S, Umar S (eds) Sulfur assimilation and abiotic stress in plants. Springer-Verlag, Berlin, pp 167–191
Björkman O, Pearcy RW (1971) Effect of growth temperature on the temperature dependence of photosynthesis in vivo and on the CO2 fixation by carboxydismutase in vitro in C3 and C4 species. Carnegie Inst Wash Year Book 70:511–520
Boese SR, Huner NPA (1990) Effect of growth temperature and temperature shifts on spinach leaf morphology and photosynthesis. Plant Physiol 94:1830–1836

Boese SR, Wolfem DW, Melkonian JJ (1997) Elevated CO2 mitigates chilling-induced water stress and photosynthetic reduction during chilling. Plant Cell Environ 20(5):625–635

Bongi G, Long SP (1987) Light-dependent damage to photosynthesis in olive leaves during chilling and high temperature stress. Plant Cell Environ 10(3):241–250

Borawska-Jarmułowicz B, Mastalerczuk G, Pietkiewicz S, Kalaji MH (2014) Low temperature and hardening effects on photosynthetic apparatus efficiency and survival of forage grass varieties. Plant Soil Environ 60(4):177–183

Bota J, Flexas J, Medrano H (2004) Is photosynthesis limited by decreased rubisco activity and RuBP content under progressive water stress? New Phytol 162:671–681

Bousba R, Ykhlef N, Djekoun A (2009) Water use efficiency and flag leaf photosynthetic in response to water deficit of durum wheat (Triticum durum Desf.). World J Agr Sci 5:609–616

Bowler C, Van Montagu M, Inze D (1992) Superoxide dismutase and stress tolerance. Ann Rev Plant Physiol Plant Mol Biol 43:83–116

Bravo LA, Saavedra-Mella FA, Vera F, Guerra A, Cavieres LA, Ivanov AG, Huner NPA, Corcuera LJ (2007) Effect of cold acclimation on the photosynthetic performance of two ecotypes of Colobanthus quitensis (Kunth) Bartl. J Exp Bot 58(13):3581–3590

Bray EA, Bailey-Serres J, Weretilnyk E (2000) Responses to abiotic stresses. In: Gruissem W, Buchannan B, Jones R (eds) Biochemistry and molecular biology of plants. American Society of Plant Physiologists, Rockville, MD, pp 1158–1249

Bredenkamp GJ, Nie GY, Baker NR (1992) Perturbation of chloroplast development in maize by low growth temperature. Photosynthetica 27:401–411

Breidenbach RW, Rank DR, Fontana AJ, Hansen LD, Criddle RS (1990) Calorimetric determination of tissue responses to thermal extremes as a function of time and temperature. Thermochim Acta 172:179–186

Brestic M, Cornic G, Fryer MJ, Baker NR (1995) Does photorespiration protect the photosynthetic apparatus in French bean leaves from photoinhibition during drought stress? Planta 196(3): 450–457

Brini F (2017) Photosynthesis under stressful environmental conditions: existing challenges. In: Songh S (ed) Environment and photosynthesis: a future prospect, Chapter: 5. Studium Press (Pvt.), Limited, pp 68–91

Brooks A, Farquhar GD (1985) Effect of temperature on the CO2/O2 specificity of ribulose-1,5-bisphosphate carboxylase/oxygenase and the rate of respiration in the light. Planta 165(3): 397–406

Brock MT, Galen C (2005) Drought tolerance in the alpine dandelion, Taraxacum ceratophorum (Asteraceae), its exotic congenter T. officinale and interspecific hybrids under natural and experimental conditions. Amer J Bot 92:1311–1321

Brown RH, Bouton JH (1993) Physiology and genetics of interspecific hybrids between phytosynthetic type. Annu Rev Plant Physiol Plant Mol Biol 44:435–456

Brugnoli E, Scartazza A, De Tullio MC, Monteverdi MC, Lauteri M, Augusti A (1998) Zeaxanthin and non-photochemical quenching in sun and shade leaves of C3 and C4 plants. Physiol Plant 104:727–734

Buchanan BB (1980) The role of light in the regulation of chloroplast enzymes. Ann Rev Plant Physiol 31:341–374

Buetow DE (1962) Differential effects of temperature on the growth of Euglena gracilis. Exp Cell Res 27:137–142

Bullard MJ, Heath MC, Nixon PMI (1995) Shoot growth, radiation interception and dry matter production and partitioning during the establishment phase of Miscanthus sinensis 'Giganteus' grown at two densities in the U.K. Ann Appl Biol 126(2):365–378

Bunce JA (2000) Acclimation of photosynthesis to temperature in eight cool and warm climate herbaceous C3 species: temperature dependence of parameters of a biochemical photosynthesis model. Photosynthesis Res 63:59–67

Bunkelmann JR, Trelease RN (1996) Ascorbate peroxidase. A prominent membrane protein in oilseed glyoxyosomes. Plant Physiol 110:589–598

Byrd GT, Ort DR, Ogren WL (1995) The effects of chilling in the light on ribulose-1,5-bisphosphate carboxylase/ oxygenase activation in tomato (Lycopersicon esculentum Mill.). Plant Physiol 107:585–591

Cai C-T, Cai Z-Q, Yao T-Q, Qi X (2007) Vegetative growth and photosynthesis in coffee plants under different watering and fertilization managements in Yunnan. SW China Photosynth 45(3): 455–461

Calegario FF, Cosso RG, Fagian MM, Almeida FV, Jardim WF, Jezěk P, Arruda P, Vercesi AE (2003) Stimulation of potato tuber respiration by cold stress is associated with an increased capacity of both plant uncoupling mitochondrial protein (PUMP) and alternative oxidase. J Bioenerg Biomembr 35:211–220

Calzadilla PI, Maiale SJ, Ruiz OA, Escaray FJ (2016) Transcriptome response mediated by cold stress in lotus japonicus. Front Plant Sci 7:374

Calzadilla PI, Vilas JM, Escaray FJ, Unrein F, Carrasco P, Ruiz OA (2019) The increase of photosynthetic carbon assimilation as a mechanism of adaptation to low temperature in lotus japonicas. Sci Reporter 9:Article No. 863

Camejo D, Rodríguez P, Morales MA, Dell'Amico JM, Torrecillas A, Alarcón JJ (2005) High temperature effects on photosynthetic activity of two tomato cultivars with different heat susceptibility. J Plant Physiol 162:281–289

Campbell C, Atkinson L, Zaragoza-Castells J, Lundmark M, Atkin O, Hurry V (2007) Acclimation of photosynthesis and respiration is asynchronous in response to changes in temperature regardless of plant functional group. New Phytol 176(2):375–389

Campos PS, Quartin V, Ramalho JC, Nunes MA (2003) Electrolyte leakage and lipid degradation account for cold sensitivity in leaves of Coffea sp. plants. J Plant Physiol 160(3):283–292

Carey EV, Callaway RM, DeLucia EH (1997) Stem respiration of ponderosa pines grown in contrasting climates: implications for global climate change. Oecologia 111(1):19–25

Casanova-Katny M, Bravo LA, Molina Montenegro M, Corcuera LJ, Cavieres LA (2006) Photosynthetic performance of Colobanthus quitensis (Kunth) Bartl. (Caryophyllaceae) in a high-elevation site of the Andes of Central Chile. Rev Chil Hist Nat 79(1):41–53

Cavaco AM, Da Silva AB, Arrabaça MC (2003) Effects of long-term chilling on growth and photosynthesis of the C4 gramineae Paspalum dilatatum. Physiol Plant 119(1):87–96

Cavieres LA, Arroyo MTK (1999) Tasa de enfriamiento adiabá'tico del aire en el valle del río Molina, provincia de Santiago, Chile central (330 S). Revista Geográfica de Chile Terra Australis 44:79–86

Cechin I, Rossi SC, Oliveira VC, Fumis TF (2006) Photosynthetic responses and proline content of mature and young leaves of sunflower plants under water deficit. Photosynthetica 44:143–146

Celerin M, Gilpin AA, Schisler NJ, Ivanov AG, Miskiewicz E, Krol M, Laudenbach DE (1998) ClpB in a cyanobacterium: predicted structure, phylogenetic relationships, and regulation by light and temperature. J Bacteriol 180(19):5173–5182

Cen YP, Sage RF (2005) The regulation of rubisco activity in response to variation in temperature and atmospheric CO2 partial pressure in sweet potato. Plant Physiol 139:979–990

Centritto M, Brilli F, Fodale R, Loreto F (2011) Different sensitivity of isoprene emission, respiration and photosynthesis to high growth temperature coupled with drought stress in black poplar (Populus nigra) saplings. Tree Physiol 31(3):258–261

Chandra R, Pareak RP (1990) Influence the soil temperature on nodulation, nitrogen fixation and compactitiveness of chickpea(Cicer arietinum L) rhizobium spp. LegRes 13:95–99

Chandra S (2003) Effects of leaf age on transpiration and energy exchange of Ficus glomerata,a multipurpose tree species of central Himalayas. Physiol Mol Biol Plants 9:255–260

Chaves MM, Flexas J, Pinheiro C (2009) Photosynthesis under drought and salt stress: regulation mechanisms from whole plant to cell. Ann Bot 103:551–560

Chen Y, Wang X-M, Zhou L, Wang D, Qi YH, Jiang DA (2015) Rubisco activase is also a multiple responder to abiotic stresses in rice. PLoS One 10(10):e0140934

Cheng C, Yun KY, Ressom HW, Mohanty B, Bajic VB, Jia Y, Yun SJ, Reyes BG (2007) An early response regulatory cluster induced by low temperature and hydrogen peroxide in seedlings of chilling-tolerant japonica rice. BMC Genomics 8(1):175–192

Chinthapalli B, Murmu J, Raghavendra AS (2003) Dramatic difference in the responses of phosphoenolpyruvate carboxylase to temperature in leaves of C3 and C4 plants. J Exp Bot 54(383):707–714

Chu C, Lee TM (1989) The relationship between ethylene biosynthesis and chilling tolerance in seedlings of rice (Oryza sativa L.). Bot Bull Acad Sin 30:263–273

Cleland RE, Bendall DS (1992) Photosystem I cyclic electron transport: measurement of ferredoxin–plastoquinone reductase activity. Photosynth Res 34(3):409–418

Cohu CM, Muller O, Adams WW, Demmig-Adams B (2014) Leaf anatomical and photosynthetic acclimation to cool temperature and high light in two winter versus two summer annuals. Physiol Plant 152:164–173

Collatz GJ, Ribascarbo M, Berry JA (1992) Coupled photosynthesis–stomatal conductance model for leaves of C4 plants. Aust J Plant Physiol 19(5):519–538

Collier DE, Cummins WR (1990) The effect of low growth and measurement temperature on the respiratory properties of five temperate species. Ann Bot 65(5):533–538

Cossins AR (1994) Homeoviscous adaptation of biological membranes and its functional significance. In: Cossins AR (ed) Temperature adaptation of biological membranes. Portland Press, London, pp 63–76

Cowling SA, Sage RF (1998) Interactive effects of low atmospheric CO2 and elevated temperature on growth, photosynthesis and respiration in Phaseolus vulgaris. Plant Cell Environ 21(4): 427–435

Creissen G, Broadbent P, Stevens R, Wellbum AR, Mullineaux P, P. (1996) Manipulation of glutathione metabolism in transgenic plants. J Biochem 24(2):465–472

Crespi MD, Zabaleta EJ, Pontis HG, Salerno GL (1991) Sucrose synthase expression during cold acclimation in wheat. Plant Physiol 96(3):887–891

Cui S, Huang F, Wang J, Ma X, Cheng Y, Liu J (2005) A proteomic analysis of cold stress responses in rice seedlings. Proteomics 5(12):3162–3172

Cvetkovic J, Muller K, Baier M (2017) The effect of cold priming on the fitness of Arabidopsis thaliana accessions under natural and controlled conditions. Sci Rep 7:44055

Dahal K, Kane K, Gadapati W, Webb E, Savitch LV, Singh J, Sharma P, Sarhan F, Longstaffe FJ, Godzinski B, Huner NPA (2012) The effects of phenotypic plasticity on photosynthetic performance in winter rye, winter wheat, and Brassica napus. Physiol Plant 144(2):169–188

Dai X, Xu Y, Ma Q, Xu W, Wang T, Xue Y (2007) Overexpression of an R1R2R3 MYB gene, OsMYB3R-2, increases tolerance to freezing, drought, and salt stress in transgenic Arabidopsis. Plant Physiol 143(4):1739–1751

DaMatta FM, Maestri M, Mosquim PR, Barros RS (1997) Photosynthesis in coffee (Coffea arabica and and C. canephora) as affected by winter and summer conditions. Plant Sci 128(1):43–50

Davis AP, Govaerts R, Bridson DM, Stoffelen P (2006) An annotated taxonomic conspectus of the genus Coffea (Rubiaceae). Bot J Linnean Soc 152(4):465–512

Davison PA, Hunter CN, Horton P (2002) Overexpression of β-carotene hydroxylase enhances stress tolerance in Arabidopsis. Nature 418(6894):203–206

Day ME (2000) Influence of temperature and leaf-to-air vapor pressure deficit on net photosynthesis and stomatal conductance in red spruce (Picea rubens). Tree Physiol 20(1):57–63

DeRidder BP, Crafts-Brandner SJ (2008) Chilling stress response of postemergent cotton seedlings. Physiol Plant 134(3):430–439

Demmig-Adams B, Adams WWIII (1992a) Carotenoid composition in sun and shade leaves of plants with different life forms. Plant Cell Environ 15(4):411–419

Demmig-Adams B, Adams WWIII (1992b) Photoprotection and other responses of plants to high light stress. Annu Rev Plant Biol 43:599–626

Demmig-Adams B, Adams WW III, Logan BA, Verhoeven AS (1995) Xanthophyll cycle- dependent energy dissipation and flexible photosystem II efficiency in plants acclimated to light stress. Aust J Plant Physiol 22(2):249–260
Demmig-Adams B, Adams WW (1996) Xanthophyll cycle and light stress in nature: uniform response to excess direct sunlight among higher plant species. Planta 198(3):460–470
Dickman LT, McDowell NG, Sevanto S, Pangle RE, Pockman WT (2015) Carbohydrate dynamics and mortality in a piñon-juniper woodland under three future precipitation scenarios. Plant Cell Environ 38(4):729–739
Diédhiou CJ, Popova OV, Dietz KJ, Golldack D (2008) The SNF1-type serine-threonine protein kinase SAPK4 regulates stress-responsive gene expression in rice. BMC Plant Biol 8(1):49
Dwivedi P, Raghvendra AS (2004) Modulation of osmatic stress effects on photosynthesis and respiration by temperature in mesophyll protoplast of pea. Ind J Expt Biol 42(12):1208–1211
Dodd IC (2003) Hormonal interactions and stomatal responses. J Plant Growth Regul 22(1):32–46
Doubnerová V, Ryšlavá H (2011) What can enzymes of C4 photosynthesis do for C3 plants under stress? Plant Sci 180(4):575–583
Dreyer E, Le Roux X, Montpied P, Daudet FA, Masson F (2001) Temperature response of leaf photosynthetic capacity in seedlings from seven temperate tree species. Tree Physiol 21(4): 223–232
Du Y-C, Nose A, Wasano K (1999) Effects of chilling temperature on photosynthetic rates, photosynthetic enzyme activities and metabolite levels in leaves of three sugarcane species. Plant Cell Environ 22(3):317–324
Dulai S, Molnár I, Molnár-Láng M (2011) Changes of photosynthetic parameters in wheat / barley introgression lines during salt stress. Acta Biol Szeged 55(1):73–75
Duncan DR, Widholm JM (1991) Proline is not me primary determination of chilling tolerance induced by mannitol or abscisic acid in regenerable maize callus cultures. Plant Physiol 95(4): 1284–1287
Dwyer SA, Ghannoum O, Nicotra A, von Caemmerer S (2007) High temperature acclimation of C4 photosynthesis is linked to changes in photosynthetic biochemistry. Plant Cell Environ 30(1): 53–66
Eamus D (1987) Influence of preconditioning upon the changes in leaf conductance and leaf water potential of soybean, induced by chilling, water stress and abscisic acid. Aust J Plant Physiol 14(3):331–339
Ebrahim MKH, Vogg G, Osman MNEH, Komor E (1998) Photosynthetic performance and adaptation of sugarcane at suboptimal temperatures. J Plant Physiol 153(5–6):587–592
Edwards GE, Nakamoto H, Burnell JN, Hatch MD (1985) Pyruvate, Pidikinase and NADP- malate dehydrogenase in C4 photosynthesis: properties and mechanism of light/dark regulation. Annu Rev Plant Physiol 36:255–286
Edwards EA, Enard C, Creissen G, Mullineaux PM (1994) Synthesis and properties of glutathione reductase in stressed peas. Planta 192(1):137–143
Edwards NT, Hanson PJ (1996) Stem respiration in a closed-canopy upland oak forest. Tree Physiol 16(4):433–439
Ehleringer JR (1978) Implications of quantum yield differences on the distributions of C3 and C4 grasses. Oecologia 31(3):255–267
Ehleringer JR, Sage RF, Flanagan LB, Pearcy RW (1991) Climate change and the evolution of C4 photosynthesis. Trends Ecol Evol 6(3):95–99
Einset J, Winge P, Bones A (2007) ROS signaling pathways in chilling stress. Plant Signal Behav 2(5):365–367
El-Saht HM (1998) Responses to chilling stress in French bean seedlings: antioxidant compounds. Biologia Plant 41(3):395–402
Ensminger I, Busch F, Huner NPA (2006) Photostasis and cold acclimation: sensing low temperature through photosynthesis. Physiol Plant 126(1):28–44

Equiza MA, Miravea JP, Tognetti JA (2001) Morphological, anatomical and physiological responses related to differential shoot vs. root growth inhibition at low temperature in spring and winter wheat. Ann Bot 87:67–76

Fan J, Hu Z, Xie Y, Chan Z, Chen K, Amombo E, Chen L, Fu J (2015) Alleviation of cold damage to photosystem II and metabolisms by melatonin in Bermudagrass. Front Plant Sci 6:925

Farooq M, Basra SMA, Wahid A, Cheema ZA, Cheema MA, Khaliq A (2008) Physiological role of exogenously applied glycine betaine in improving drought tolerance of fine grain aromatic rice (Oryza sativa L.). J Agron Crop Sci 194:325–333

Farooq M, Aziz T, Wahid A, Lee DJ, Siddique KHM (2009a) Chilling tolerance in maize: agronomic and physiological approaches. Crop Pasture Sci 60:501–516

Farooq M, Wahid A, Kobayashi N, Fujita D, Basra SMA (2009b) Plant drought stress: effects, mechanisms and management. Agron Sustain Dev 29:185–212

Farooq M, Wahid A, Basra SMA, Din I (2009c) Improving water relations and gas exchange with brassinosteroids in rice under drought stress. J Agron Crop Sci 195:262–269

Farquhar GD, von Caemmerer S, Berry JA (1980) A biochemical model of photosynthetic CO2 assimilation in leaves of C3 species. Planta 149(1):78–90

Farquhar GD, von Caemmerer S (1982) Modelling of photosynthetic response to environmental conditions. In: Lange OL, Nobel PS, Osmond CB, Ziegler H (eds) physiological plant ecology II: water relations and carbon assimilation, Encyclopedia of plant physiology, vol 12B. Springer-Verlag, Berlin, pp 550–587

Farrell TC, Fox KM, William RL, Fukai S, Lewin LG (2006) Minimizing cold damage during reproductive development among temperate rice genotypes. II. Genotypic variation and flowering traits related to cold tolerance screening. Aust J Agric Res 57:89–100

Feng Z, Guo A, Feng Z (2003) Amelioration of chilling stress by triadimefon in cucumber seedlings. Plant Growth Regul 39:277–283

Feng L, Han Y, Liu G, An B, Yang J, Yang G, Li Y, Zhu Y (2007) Overexpression of sedoheptulose-1,7-bisphosphatase enhances photosynthesis and growth under salt stress in transgenic rice plants. Funct Plant Biol 34(9):822–834

Fiorani F, Umbach LA, Siedow NJ (2005) The AOX of plant mitochondria is involved in the acclimation of shoot growth at low temperature. A study of Arabidopsis AOX1 a transgenic plants. Plant Physiol 139:1795–1805

Fitter A, Hay RKM (2002) Environmental physiology of plants. Academic Press, San Diego, CA

Flanagan LB, Johnson BG (2005) Interacting effects of temperature, soil moisture and plant biomass production on ecosystem respiration in a northern temperate grassland. Agric For Meteorol 130(3–4):237–253

Flexas J, Badger M, Chow WS, Medrano H, Osmond CB (1999) Analysis of the relative increase in photosynthetic O2 uptake when photosynthesis in grapevine leaves is inhibited following low night temperatures and/or water stress. Plant Physiol 121(2):675–684

Flexas J, Bota J, Loreto F, Cornic G, Sharkey TD (2004) Diffusive and metabolic limitations to photosynthesis under drought and salinity in C3 plants. Plant Biol 6:1–11

Foyer CH (1993) Ascorbic acid. In: Alscher RG, Hess JL (eds) Antioxidants in higher plants. CRC Press, Boca Raton, FL, pp 31–58

Foyer CH, Leiandais M, Kunert KJ (1994) Photooxidative stress in plants. Physiol Plant 92(4): 616–617

Foyer CH (1996) Oxygen processing in photosynthesis. J Biochem 24(2):427–433

Foyer CH, Noctor G (2000) Oxygen processing in photosynthesis: regulation and signaling. New Phytol 146(3):359–388

Fracheboud Y, Haldimann P, Leipner J, Stamp P (1999) Chlorophyll fluorescence as a selection tool for cold tolerance of photosynthesis in maize (Zea mays L.). J Exp Bot 50(338):1533–1540

Frank HA, Young AJ, Britton G, Cogdell RJ, Guszecki WI (1999) In: Frank HA, Young AJ, Britton G, Cogdell RJ (eds) The photochemistry of carotenoids. Kluwer Academic, Dordrecht, pp 363–379

Franks PJ, Farquhar GD (2007) The mechanical diversity of stomata and its significant in gas-exchange control. Plant Physiol 143:78–87
Freschi L, Mercier H (2012) Connecting environmental stimuli and crassulacean acid metabolism expression: Phytohormones and other signaling molecules. Prog Bot 73:231–255
Fryer MJ, Oxborough K, Martin B, Ort DR, Baker NR (1995) Factors associated with depression of photosynthetic quantum efficiency in maize at low growth temperature. Plant Physiol 108:761–767
Fryer MJ, Andrews JR, Oxborough K, Blowers DA, Baker NR (1998) Relationship between CO2 assimilation, photosynthetic electron transport, and active O2 metabolism in leaves of maize in the field during periods of low temperature. Plant Physiol 116(2):571–580
Fu J, Huang B (2001) Involvement of antioxidants and lipid peroxidation in the adaptation of two cool-season grasses to localized drought stress. Environ Exp Bot 45:105–114
Fukayama H, Tsuchida H, Agarie S (2001) Significant accumulation of C4-specific pyruvate, orthophosphate dikinase in a C3 plant, rice. Plant Physiol 127:1136–1146
Furbank RT, Chitty JA, Jenkins CLD, Taylor WC, Trevanion SJ, von Caemmerer S, Ashton AR (1997) Genetic manipulation of key photosynthetic enzymes in the C4 plant Flaveria bidentis. Aust J Plant Physiol 24(4):477–485
Galmés J, Medrano H, Flexas J (2007) Photosynthetic limitations in response to water stress and recovery in Mediterranean plants with different growth forms. New Phytol 175:81–93
Garstka M, Venema JH, Runiak I, Gieczewska K, Rosiak M, Koziol-Lipinska J, Kierdaszuk B, Vredenberg WJ, Mostowska A (2007) Contrasting effect of dark-chilling on chloroplast structure and arrangement of chlorophyll-protein complexes in pea and tomato: plants with a different susceptibility to non-freezing temperature. Planta 226(5):1165–1181
Gecin MV (2013) Effects of cold stress on photosynthesis in barley. Thesis: University of Western Ontario
Geissler N, Hussin S, Koyro HW (2009) Interactive effects of NaCl salinity and elevated atmospheric CO2 concentration on growth, photosynthesis, water relations and chemical composition of the potential cash crop halophyte Aster tripolium L. Environ Exp Bot 65:220–231
Ghosh S, Bagchi S, Majumder AL (2001) Chloroplast fructose- 1,6-bisphosphatase from Oryza differs in salt tolerance property from the Porteresia enzyme and is protected by osmolytes. Plant Sci 160:1171–1181
Gianoli E, Inostroza P, Zuñiga-Feest A, Reyes-Diaz M, Cavieres LA, Bravo LA, Corcuera LJ (2004) Ecotypic differentiation in morphology and cold resistance in population of Colobanthus quitensis (Cariophyllaceae) from the Andes of Central Chile and maritime Antarctica. Arct Antarct Alp Res 36:484–489
Gill SS, Tuteja N (2010) Reactive oxygen species and antioxidant machinery in abiotic stress tolerance in crop plants. Plant Physiol Biochem 48(12):909–930
Gill SS, Khan NA, Tuteja N (2011) Differential cadmium stress tolerance in five indian mustard (Brassica juncea L.) cultivars: an evaluation of the role of antioxidant machinery. Plant Signal Behav 6(2):293–300
Gillham DJ, Dodge AD (1986) Hydrogen-peroxide scavenging systems within pea chloroplasts. Planta 167(2):246–251
Gilmour SJ, Sebolt AM, Salazar MP, Everard JD, Thomashow MF (2000) Over expression of the Arabidopsis CBF3 transcriptional activator mimics multiple biochemical changes associated with coldacclimation. Plant Physiol 124:1854–1186
Giri J (2011) Glycinebetaine and abiotic stress tolerance in plants. Plant Signal Behav 6:1746–1751
Gladstone J (1992) Viticulture and environment. Winetitles, Underdale
Goh CH, Ko SM, Koh S, Kim YJ, Bae HJ (2012) Photosynthesis and environments: photoinhibition and repair mechanisms in plants. J Plant Biol 55(2):93–101
Gombos Z, Vigh L (1986) Primary role of the cytoplasmic membrane in thermal acclimation evidenced in nitrate-starved cells of the blue-green alga, Anacystis nidulans. Plant Physiol 80(2):415–419

Gombos Z, Wada H, Hideg E, Murata N (1994a) The unsaturation of membrane lipids stabilizes photosynthesis against heat stress. Plant Physiol 104(2):563–567

Gombos Z, Wada H, Murata N (1994b) The recovery of photosynthesis from low temperature photoinhibition is accelerated by the unsaturation of membrane lipids: a mechanism of chilling tolerance. Proc Natl Acad Sci U S A 91(19):8787–8791

Gombos Z, Murata N (1998) Genetic engineering of the ultrastructure of membrane glycerolipids: effect on the ability of the photosynthetic mechinery to tolerate temperature stress. In: "Lipids in photosynthesis: structure, function and genetics", Advances in photosynthesis and respiration, vol 6. Springer, Dordrecht, pp 177–192

Gonzalez-Meler MA, Ribascarbo M, Giles L, Siedow JN (1999) The effect of growth and measurement temperature on the activity of the alternative respiratory pathway. Plant Physiol 120(3):765–772

Goulas E, Schubert M, Kieselbach T, Kleczkowski LA, Gardestrom P, Schroder W, Hurry V (2006) The chloroplast lumen and stromal proteomes of Arabidopsis thaliana show differential sensitivity to short- and long-term exposure to low temperature. Plant J 47:720–734

Grace SC, Logan BA (1996) Acclimation of foliar antioxidant systems to growth irradiance in three broadleaved evergreen species. Plant Physiol 112:1631–1640

Graham D, Patterson BD (1982) Responses of plants to low, non-freezing temperatures:proteins, metabolism, and acclimation. Ann Rev Plant Physiol 33:347–372

Greaves JA (1996) Improving suboptimal temperature tolerance in maize—the search for variation. J Exp Bot 47(3):307–323

Grimaud F, Renaut J, Dumont E, Sergeant K, Lucau-Danila A, Blervacq AS, Sellier H, Bahrman N, Lejeune-Hénaut I, Bruno D, Goulas E (2013) Exploring chloroplastic changes related to chilling and freezing tolerance during cold acclimation of pea (Pisum sativum L.). J Proteo 80:145–159

Gruszecki WI, Sielewiesiuk J (1991) Galactolipid multibilayers modified with xanthophylls: orientational and diffractometric studies. Biochim Biophys Acta 1069(1):21–26

Gruszecki WI, Grudzinski W, Banaszek-Glos A, Matula M, Kernen P, Krupa Z, Sielewiesiuk J (1999) Xanthophyll pigments in light-harvesting complex ii in monomolecular layers: localisation, energy transfer and orientation. Biochim Biophys Acta 1412(2):173–183

Guo Z, Ou W, Lu S, Zhong Q (2006) Differential responses of antioxidative system to chilling and drought in four rice cultivars differing in sensitivity. Plant Physiol Biochem 44(11–12):828–836

Gupta PK, Balyan HS, Gahlaut V (2017) QTL analysis for drought tolerance in wheat: present status and future possibilities. Agronomy 7:1–21

Gururani MA, Venkatesh J, Tran LSP (2015) Regulation of photosynthesis during abiotic stress-induced photoinhibition. Mol Plant 8(9):1304–1320

Gusta LV, Trischuk R, Weiser CJ (2005) Plant cold acclimation: the role of abscisic acid. J Plant Growth Regul 24:308–318

Guy L (1990) Cold acclimation and freezing stress tolerance: role of protein metabolism. Annu Rev Plant Physiol 41:187–223

Guy CL, Haskel D, Neven L, Klein P, Smelser C (1992a) Hydration state responsive proteins link cold and drought stress in spinch. Planta 188:265–270

Guy CL, Huber JLA, Huber SC (1992b) Sucrose phosphate synthase and sucrose accumulation at low temperature. Plant Physiol 100(1):502–508

Guy CL, Li QB (1998) The organization and evolution of the spinach stress 70 molecular chaperone gene family. Plant Cell 10:539–556

Guy C, Haskell D, Li QB (1998) Association of proteins with the stress 70 molecular chaperones at low temperature evidence for the existence of cold labile proteins in spinach. Cryobiology 36: 301–314

Guye MG, Wilson JM (1987) The effects of chilling and chill-hardening temperatures on stomatal behaviour in chill-sensitive species and cultivars. Plant Physiol Biochem 25:717–721

Hahn M, Walbot V (1989) Effects of cold-treatment on protein synthesis and mRNA levels in rice leaves. Plant Physiol 91(3):930–938

Hajihashemi S, Noedoost F, Genus JMC, Djalovic I, Siddique KHM (2018) Effect of cold stress on photosynthetic traits, carbohydrates, morphology, and anatomy in nine cultivars of Stevia rebaudiana. Front Plant Sci 9:1430

Häusler RE, Hirsch HJ, Kreuzaler F, Peterhänsel C (2002) Overexpression of C4 cycle enzymes in transgenic C3 plants: A biotechlogical approach to improve C3 photosynthesis. J Exp Bot 53(369):591–607

Hikosaka K, Ishikawa K, Borjigidai A, Muller O, Onoda Y (2006) Temperature acclimation of photosynthesis: mechanisms involved in the changes in temperature dependence of photosynthetic rate. J Exp Bot 57(2):291–302

Haldiman P (1998) Low growth temperature induced changes to pigment composition and photosynthesis in Zea mays genotypes differing in chilling sensitivity. Plant Cell Environ 21(2): 200–208

Hällgren JE, Strand M, Lundmark T (1991) Temperature stress. In: Raghavendra AS (ed) Physiology of trees. J. Wiley & Sons, New York, pp 301–335

Halliwell B, Gutteridge JMC (1989) Free radicals in biology and medicine. Clarendon Press, Oxford

Harb A, Krishnan A, Madana MR (2010) Molecular and physiological analysis of drought stress in Arabidopsis reveals early responses leading to acclimation in plant growth. Plant Physiol 154: 1254–1271

Harley PC, Sharkey TD (1991) An improved model of C3 photosynthesis at high CO2-reversed O2 sensitivity explained by lack of glycerate reentry into the chloroplast. Photosynth Res 27(3): 169–178

Harley PC, Thomas RB, Reynolds JF, Strain BR (1992) Modelling photosynthesis of cotton grown in elevated CO2. Plant Cell Environ 15(3):271–282

Harrison J, Tonkinson C, Eagles C, Foyer C (1997) Acclimation to freezing temperatures in perennial ryegrass (Lolium perenne). Acta Physiol Plant 19:505–515

Hasanuzzaman M, Fujita M (2011) Selenium pre-treatment up-regulates the antioxidant defense and methylglyoxal detoxification system and confers enhanced tolerance to drought stress in rapeseed seedlings. Biol Trace Elem Res 143:1758–1776

Hasanuzzaman M, Nahar K, Fujita M (2013) Extreme temperature responses, oxidative stress and antioxidant defense in plants. In: Vahdati K, Leslie C (eds) Plants, abiotic stress—plant responses application in agriculture. Intech Open, Rijeka, pp 169–205

Havaux M, Gruszecki WI (1993) Heat and light induced chlorophyll fluroscence changes in potato leaves containing high or low levels of carotenoid zeaxanthin: indication of a regulatory effectof the zeaxanthin on thylokoid membrane fluidity. Photochem Photobiol 58(4):607–614

Havaux M (1996) Short-term responses of photosystem I to heat stress - induction of a PS II independent electron transport through PS I fed by stromal components. Photosynth Res 47:85–97

Havaux M (1998) Carotenoids as membrane stabilizer in chloroplasts. Trends Plant Sci 3(4): 147–151

Havaux M, Niyogi K (1999) The violaxanthin cycle protects plants from photooxidative damage by more than one mechanism. Proc Natl Acad Sci U S A 96(15):8762–8767

Hawrylak-Nowak B, Matraszek R, Szymańska M (2010) Selenium modifies the effect of short-term chilling stress on cucumber plants. Biol Trace Elem Res 138:307–315

Heber U, Gerst U, Krieger A, Neimanis S, Kobayashi Y (1995) Coupled cyclic electron transport in intact chloroplasts and leaves of C3 plants – does it exist – if so, what is its function? Photosynth Res 46(1–2):269–275

Heidarvand L, Millar AH, Taylor NL (2017) Responses of the mitochondrial respiratory system to low temperature in plants. Crit Rev Plant Sci 36(4):217–240

Hendriks JHM, Kolbe A, Gibon Y, Stitt M, Geigenberger P (2003) ADP-glucose pyrophosphorylase is activated by post translational redox-modification in response to light and to sugars in leaves of Arabidopsis and other plant species. Plant Physiol 133:838–849

Hendrickson L, Ball MC, Wood JT, Chow WS, Furbank RT (2004a) Low temperature effect on photosynthesis anf growth of grapevine. Plant Cell Environ 27(7):795–809

Hendrickson L, Chow WS, Furbank RT (2004b) Low temperature effects on grape vine photosynthesis: the role of inorganic phosphate. Funct Plant Biol 31(8):789–801

Hikosaka K, Murakami A, Hirose T (1999) Balancing carboxylation and regeneration of ribulose-1,5-bisphosphate in leaf photosynthesis: temperature acclimation of an evergreen tree, Quercus myrsinaefolia. Plant Cell Environ 22(7):841–849

Hill KE, Guerin GR, Hill RS, Watling JR (2014) Temperature influences stomatal density and maximum potential water loss through stomata of Dodonaea viscosa subsp. Angustissima along a latitude gradient in southern Australia. Aust J Bot 62:657–665

Hirayama T, Shinozaki K (2010) Research on plant abiotic stress responses in the post-genome era: past present. Plant J 61:1041–1062

Hirschberg J, Chamovitz D (1994) Carotenoid in cyanobacteria. In: Bryant DA (ed) The molecular biology of Cyanobacteris. Kulwer Academic Publisher, Dordrecht, pp 559–579

Hodkinson TR, Renvoize SA (2001) Nomenclature of Miscanthus x giganteus (Poaceae). Kew Bull 56(3):759–760

Hoermiller II, Naegele T, Augustin H, Stutz S, Weckwerth W, Heyer AG (2017) Subcellular reprogramming of metabolism during cold acclimation in Arabidopsis thaliana. Plant Cell Environ 40(5):602–610

Holaday AS, Martindale W, Aired R, Brooks AL, Leegood RC (1992) Changes in activities of enzymes of carbon metabolism in leaves during exposure of plants to low temperature. Plant Physiol 98(3):1105–1114

Holaday AS, Mahan JR, Payton P (2016) Molecular biology and physiology effects of chilling temperatures on photosynthesis. J Cotton Sci 20:220–231

Hopkins WG (1999) The physiology of plants under stress. In: Introduction to plant physiology, 2nd edn. Wiley, New York, pp 451–475

Horváth E, Pál M, Szalai G, Páldi E, Janda T (2007) Exogenous 4-hydroxybenzoic acid and salicylic acid modulate the effect of short term drought and freezing stress on wheat plants. Biol Plant 51:480–487

Hoshida H, Tanaka Y, Hibino T, Hayashi Y, Tanaka A, Takabe T, Takabe T (2000) Enhanced tolerance to salt stress in transgenic rice that overexpresses chloroplast glutamine synthetase. Plant Mol Biol 43(1):103–111

Hu WH, Song XS, Shi K, Xia XJ, Zhou YH, Yu JQ (2008) Changes in electron transport, superoxide dismutase and ascorbate peroxidase isoenzymes in chloroplasts and mitochondria of cucumber leaves as influenced by chilling. Photosynthetica 46(4):581–588

Huang Q, Wang Y, Li B, Chang J, Chen M, Li K, Yan G, He G (2015) TaNAC29, a NAC transcription factor from wheat, enhances salt and drought tolerance in transgenic Arabidopsis. BMC Plant Biol 15(1):1–15

Hudspeth RL, Grula JW, Dai Z, Edwards GE, Ku MS (1992) Expression of maize phosphoenolpyruvate carboxylase in transgenic tobacco. Effects on biochemistry and physiology. Plant Physiol 98(2):458–464

Huner NPA (1985) Acclimation of winter rye to clod-hardening temperatures results in an increased capacity for photosynthetic electron transport. Can J Bot 63(3):506–511

Huner NPA, Oquist G, Hurry VM, Krol M, Falk S, Griffith M (1993) Photosynthesis, photoinhibition and low temperature acclimation in cold tolerant plants. Photosynthesis Res 37(1):19–39

Huner NPA, Öquist G, Sarhan F (1998) Energy balance and acclimation to light and cold. Trends Plant Sci 3:224–230

Hur J, Jung K-H, Lee C-H, An G (2004) Stress-inducible OsP5CS2 gene is essential for salt and cold tolerance in rice. Plant Sci 167:417–426

Hurry VM, Huner NPA (1991) Low growth temperature effects a differential inhibition of photosynthesis in spring and winter wheat. Plant Physiol 96(2):491–497

Hurry VM, Huner NPA (1992) Effect of cold hardening on sensitivity of winter and spring wheat leaves to short-term photoinhibition and recovery of photosynthesis. Plant Physiol 100(3): 1283–1290

Hurry VM, Malmberg G, Gardeström P, Öquist G (1994) Effect of short-term shift to low temperature and of long-term cold hardening on photosynthesis and Ribulose-1,5-bisphosphate carboxylase/oxygenase and sucrose phosphate synthase activity in leaves of winter rye (Secale cereal L.). Plant Physiol 106:983–990

Hurry VM, Strand C, Tobiaeson M, Gradestrom P, Oquist G (1995) Cold hardening of spring and winter wheat and rape results in differential effects on growth, carbon metabolism, and carbohydrate content. Plant Physiol 109(2):697–706

Hurry V, Strand A, Furbank R, Stitt M (2000) The role of inorganic phosphate in the development of freezing tolerance and the acclimatization of photosynthesis to low temperature is revealed by the pho mutants of Arabidopsis thaliana. Plant J 24(3):383–396

Hurry V, Druart N, Cavaco A, Gardestrom P, Strand A (2002) Photosynthesis at low temperatures A case study with Arabidopsis. In: Li PH, Palva ET (eds) Plant cold hardiness. Kluwer, Academic Plenum Publisher, New York, pp 161–179

Hussain S, Khan F, Cao W, Wu L, Geng M (2016) Seed priming alters the production and detoxification of reactive oxygen intermediates in rice seedlings grown under sub-optimal temperature and nutrient supply. Front Plant Sci 7:439

Hussain HA, Hussain S, Khaliq A, Ashraf U, Anjum SA, Men S, Wang L (2018) Chilling and drought stresses in crop plants: implications, cross talk, and potential management opportunities. Front Plant Sci 9:393

Inagaki M, Omori E, Kim JY, Komatsu Y, Scott G, Ray MK, Yamada G, Matsumoto K, Mishina Y, Ninomiya-Tsuji J (2008) TAK1-binding protein 1, TAB1, mediates osmotic stress-induced TAK1 activation but is dispensable for TAK1-mediated cytokine signaling. J Biol Chem 283: 33080–33086

IPCC-Intergovernmental Panel ob Climate Change (2001) Contribution of Working Group I to the Third Assessment Report of the Intergovernmental Panel on Climate Change. In: Houghton JT, Ding Y, Griggs DJ, Noguer M, van der Linden PJ, Dai X, Maskell K, Johnson CA (eds) Climate Change 2001: The Scientific Basis. Cambridge University Press, Cambridge, MA, p 881

IPCC-Intergovernmental Panel ob Climate Change (2007) Contribution of Working Groups I, II and III to the Fourth Assessment Report of the Intergovernmental Panel on Climate Change. In: Pachauri RK, Reisinger A (eds) Climate Change 2007: Synthesis Report.; Core Writing Team. IPCC, Geneva, p 104

IPCC-Intergovernmental Panel ob Climate Change (2014) Contribution of Working Group III to the Fifth Assessment Report of the Intergovernmental Panel on Climate Change. In: Edenhofer O, Pichs-Madruga R, Sokona Y, Farahani E, Kadner S, Seyboth K, Adler A, Baum I, Brunner S, Eickemeier P et al (eds) Climate Change 2014: Mitigation of Climate Change. Cambridge University Press, Cambridge

Islam MT (2011) Effect of temperature on photosynthesis, yield attributes and yield of aromatic rice genotypes. Int J Sustain Crop Prod 6:14–16

Ishikawa HA (1996) Ultrastructural features of chilling injury: injured cells and the early events during chilling of suspension cultured mung bean cells. Am J Bot 83(7):825–835

Ishikawa T, Sakai K, Yoshimura K, Takeda T, Shigeoka S (1996) cDNAs encoding stromal and thylakoid bound ascorbate peroxidase differing in the presence or absence of their 3′-coding regions. FEBS Lett 384(3):289–293

Ivanov AG, Rosso D, Savitch LV, Stachula P, Rosembert M, Oquist G, Hurry V, Hüner NPA (2012) Implications of alternative electron sinks in increased resistance of PSII and PSI photochemistry to high light stress in cold acclimated Arabidopsis thaliana. Photosynth Res 113(1–3):191–206

Izui K, Ishijima S, Yamaguchi Y, Murata T, Shigesada K, Sugiyama T, Katsuki H (1986) Changing and sequence analysis of cDNA encoding active phosphor enol pyruvate carboxylase of the C4-pathway from maize. Nuclic Acids Res 14:1615–1628

Jahnke LS, Hull MR, Long SP (1991) Chilling stress and oxygen metabolizing enzymes in Zea mays and Zea diploperennis. Plant Cell Environ 14(1):97–104

Janda T, Szalai G, Tari I, Paldi E (1999) Hydroponic treatment with salicylic acid decreases the effects of chilling injury in maize (Zea mays L.) plants. Planta 208:175–180

Janknecht R (2003) Regulation of the ER81 transcription factor and its coactivators by mitogen- and stress-activated protein kinase 1 (MSK1). Oncogene 22:746–755

Jarmuszkiewicz W, Sluse-Goffart CM, Vercesi AE, Sluse FE (2001) AOX and uncoupling protein: thermogenesis versus cell energy balance. Biosci Rep 21:213–222

Jha UC, Bohra A, Jha R (2017) Breeding approaches and genomics technologies to increase crop yield under low-temperature stress. Plant Cell Rep 36:1–35

Jia W, Zhang J (2008) Stomatal movements and long-distance signaling in plants. Plant Signal Behav 3:772–777

Jiang QW, Kiyoharu O, Ryozo I (2002) Two novel mitogen-activated protein signaling components, OsMEK1 and OsMAP1, are involved in a moderate low-temperature signaling pathway in Rice1. Plant Physiol 129:1880–1891

Johnson GN, Young AJ, Scholes JD, Horton P (1993) The dissipation of excess excitation energy in British plant species. Plant Cell Environ 16(6):673–679

Johnsen K, Maier C, Sanchez F, Anderson P, Butnor J, Waring R, Linder S (2007) Physiological girdling of pine trees via phloem chilling: proof of concept. Plant Cell Environ 30:128–134

Jonak C, Ökrész L, Bögre L, Hirt H (2002) Complexity, cross talk and integration of plant MAP kinase signalling. Curr Opin Plant Biol 5:415–424

Jongdee B, Fukai S, Cooper M (2002) Leaf water potential and osmotic adjustment as physiological traits to improve drought tolerance in rice. Field Crop Res 76:153–163

Jones PG, Inouye M (1994) The cold shock response—a hot topic. Mol Microbiol 11:811–818

Jones HG, Demmers-Derks HHWM (1999) Photoinhibition as a factor in altitudinal or latitudinal limits of species. Phyton 39(4):91–98

Jouve L, Engelmann F, Noirot M, Charrier A (1993) Evaluation of biochemical markers (sugar, proline, malonedialdehyde and ethylene) for cold sensitivity in microcuttings of two coffee species. Plant Sci 91(1):109–116

Jump AS, Hunt JM, Penuelas J (2006) Rapid climate change related growth decline at the southern range edge of Fagus sylvatica. Glob Chang Biol 12(11):2163–2174

Jung S, Steffen KL (1997) Influence of photosynthetic photon flux densities before and during long-term chilling on xanthophyll cycle and chlorophyll fluorescence quenching in leaves of tomato (Lycopersicon hirsutum). Physiol Plant 100:958–966

Kafkafi U (1990) The functions of plant K in overcoming environmental stress situations. In: Proceedings of the 22nd Colloquium. International Potash Institute, Bern, pp 81–93

Kaiser WM (1979) Reversible inhibition of the Calvin cycle and activation of oxidative pentose phosphate cycle in isolated intact chloroplasts by hydrogen peroxide. Planta 145(4):377–382

Kajala K, Brown NJ, Williams BP, Borrill P, Taylor LE, Hibberd JM (2012) Multiple Arabidopsis genes primed for recruitment into C4 photosynthesis. Plant J 69:47–56

Kalaji MH, Bosa K, Kościelniak J, Hossain Z (2011) Chlorophyll a fluorescence – A useful tool for the early detection of temperature stress in spring barley (Hordeum vulgare L.). OMICS 15:925–934

Kalisz A, Jezdinský A, Pokluda R, Sekara A, Grabowska A, Gil J (2016) Impacts of chilling on photosynthesis and chlorophyll pigment content in juvenile basil cultivars. Hort Environ Biotechnol 57:330–339

Kanda HO (1998) Studies on nucleic acid and protein metabolism in root relating to low temperature-tolerant elongation ability of fig leaf gourd for root stock. Bull Akita Prefect Coll Agric 24:29–62

Kant P, Kant S, Gordon M, Shaked R, Barak S (2007) STRS1 and STRS2, two DEAD- box RNA helicases that attenuate Arabidopsis responses to multiple abiotic stresses. Plant Physiol 145: 814–830

Karaba A, Dixit S, Greco R, Aharoni A, Trijatmiko KR, Marsch-Martinez N, Krishnan A, Nataraja KN, Udayakumar M, Pereira A (2007) Improvement of water use efficiency in rice by expression of HARDY, an Arabidopsis drought and salt tolerance gene. Proc Natl Acad Sci U S A 104: 15270–15275

Kathuria H, Giri J, Nataraja KN, Murata N, Udaykumar M, Tyagi AK (2009) Glycinebetaine-induced water-stress tolerance in codA-expressing transgenic indica rice is associated with up-regulation of several stress responsive genes. Plant Biotechnol J 7(6):512–526

Kattge J, Knorr W (2007) Temperature acclimation in a biochemical model of photosynthesis: a reanalysis of data from 36 species. Plant Cell Environ 30(9):1176–1190

Kaya MD, Okçub G, Ataka M, Çıkılıc Y, Kolsarıcıa Ö (2006) Seed treatments to overcome salt and drought stress during germination in sunflower (Helianthus annuus L.). Eur J Agron 24:291–295

Kelly GJ (2006) Photosynthesis, carbon metabolism: the Calvin cycle's golden jubilee. In: Kelly GJ, Latzko E (eds) Thirty years of photosynthesis 1974–2004. Springer, Heidelberg, pp 382–410

Khan TA, Fariduddin Q, Yusuf M (2015) Lycopersicon esculentum under low temperature stress: an approach toward enhanced antioxidants and yield. Environ Sci Pollut Res 22:14178–14188

Khan TA, Fariduddin Q, Yusuf M (2017) Low temperature stress: is phytohormone application is remedy? Environ Sci Pollut Res 24(27):21574–21590

Kiani SP, Talia P, Maury P, Grieu P, Heinz R, Perrault A, Nishinakamasu V, Hopp E, Gentzbittel L, Paniego N, Sarrafi A (2007) Genetic analysis of plant water status and osmotic adjustment in recombinant inbred lines of sunflower under two water treatments. Plant Sci 172(4):773–787

Kingston-Smith AH, Harbinson J, Williams J, Foyer CH (1997) Effect of chilling on carbon assimilation, enzyme activation, and photosynthetic electron transport in the absence of photoinhibition in maize leaves. Plant Physiol 114(3):1039–1046

Kingston-Smith AH, Harbinson J, Foyer CH (1999) Acclimation of photosynthesis, H2O2 content and antioxidants in maize (Zea mays) growth at sub-optimal temperatures. Plant Cell Environ 22(9):1071–1083

Kingston-Smith AH, Foyer CH (2000) Bundle-sheath proteins are more sensitive to oxidative damage than those of the mesophyll in maize leaves exposed to paraquat or low temperatures. J Exp Bot 51(342):123–130

Kirschbaum MUF (1999) Modelling forest growth and carbon storage in response to increasing CO2 and temperature. Tellus B 51(5):871–888

Kitroongruang N, Jodo S, Hisai J, Kato M (1992) Photosynthesis characteristics of melons grown under high temperatures. J Jpn Soc Hort Sci 61:107–114

Klikoff L (1966) Temperature dependence of the oxidative rates of mitochondria in Danthonia intermedia, Penstemon davidsonii and Sitanion hystrix. Bot Gaz 129(3):227–230

Knight H, Trewavas AJ, Knight MR (1996) Cold calcium signaling in Arabidopsis involves two cellular pools and a change in calcium signature after acclimation. Plant Cell 8(3):489–503

Knight H, Brandt S, Knight MR (1998) A history of drought stress alters calcium signaling pathways in Arabidopsis. Plant J 16(6):681–387

Koch KE (1996) Carbohydrate-modulated gene expression in plants. Ann Rev Plant Physiol Plant Mol Biol 47:509–540

Kogami H, Shono M, Koike T, Yanagiswa S, Izui K, Sentoku S, Tanifugi S, Uchimiya H (1994) Molecular and physiological evaluation of transgenic tobacco plants expressing a maize phosphoenolpyruvate carboxylase gene under the control of the cauliflower mosaic virus 35S promoter. Transgenic Res 3:287–296

Kolaksazov M, Laporte F, Ananieva K, Dobrev P, Herzog M, Ananiev ED (2013) Effect of chilling and freezing stresses on jasmonate content in Arabis alpina. Bulg J Agric Sci 19:15–17

Königer M, Winter K (1993) Reduction of photosynthesis in sun leaves of Gossypium hirsutum L. under conditions of high light intensities and suboptimal leaf temperatures. Agronomie 13(7): 659–668

Kornyeyev D, Logan BA, Allen RD, Holaday AS (2001) Enhanced photochemical light utilization and decreased chilling-induced photoinhibition of photosystem II in cotton overexpressing genes encoding chloroplast-targeted antioxidant enzymes. Physiol Plant 113(3):323–331

Kozaki A, Takeba G (1996) Photorespiration protects C3 plants from photooxidation. Nature 384: 557–560

Krall JP, Edwards GE (1993) PEP carboxylases from two C4 species of Panicum with markedly different susceptibilities to cold inactivation. Plant Cell Physiol 34(1):1–11

Kratsch HA, Wise RR (2000) The ultrastructure of chilling stress. Plant Cell Environ 23:337–350

Krause GH (1994a) Photoinhibition induced by low temperatures. In: Baker NR, Bowyer JR (eds) Photoinhibition of photosynthesis. From molecular mechanisms to the field. BIOS Scientific Publishers, Oxford, pp 331–348

Krause GH (1994b) The role of oxygen photoinhibition of photosynthesis. In: Foyer CH, Mullineaux PM (eds) Causes of photooxidative stress and amelioration of defence systems in plants. CRC Press, Boca Raton, FL, pp 43–76

Krol M, Huner NPA, Mac Intosh A (1988) Chloroplast biogenesis at cold hardening temperatures. Development of photosystem I and photosystem II activities in relation to pigment accumulation. Photosynth Res 14:97–112

Krol M, Ivanov AG, Jansson S, Kloppstech K, Huner NP (1999) Greening under high light or cold temperature affects the level of xanthophyll-cycle pigments, early light-inducible proteins, and light-harvesting polypeptides in wild-type barley and the Chlorina f2 mutant. Plant Physiol 120: 193–204

Krause GH (1994) Photoinhibition induced by low temperatures. In: Baker NR, Bowyer JR (eds) Photoinhibition of photosynthesis from molecular mechanism to the field. BiosScientific Publishers Ltd., Oxford, pp 301–348

Ku MSB, Agarie S, Nomura M, Fukuyama H, Tsuchida H, Ono K, Hirose S, Toki S, Miyao-Tokutomi M, Matsuoka M (1999) High-level expression of maize phosphoenolpyruvate carboxylase in transgenic rice plants. Nat Biotechnol 17:72–80

Kubien DS, von Caemmerer S, Furbank RT, Sage R (2003) C4 photosynthesis at low temperature. A study using transgenic plants with reduced amounts of rubisco. Plant Physiol 132(3): 1577–1585

Kubien DS, Sage RF (2004a) Dynamic photo-inhibition and carbon gain in a C4 and a C3 grass native to high latitudes. Plant Cell Environ 27(11):1424–1435

Kubien DS, Sage RF (2004b) Low-temperature photosynthetic performance of a C4 grass and a co-occurring C3 grass native to high latitudes. Plant Cell Environ 27(7):907–916

Kuczyńska P, Latowski D, Niczyporuk S, Olchawa-Pajor M, Jahns P, Gruszecki WI, Strzałka K (2012) Zeaxanthin epoxidation–an in vitro approach. Acta Biochim Polonica 59:105–107

Kumar A, Bernier J, Verulkar S, Lafitte HR, Atlin GN (2008) Breeding for drought tolerance: direct selection for yield, response to selection and use of drought-tolerant donors in upland and lowland-adapted populations. Field Crops Res 107:221–231

Kumar S, Singh B (2009) Effect of water stress on carbon isotope discrimination and rubisco activity in bread and durum wheat genotypes. Physiol Mol Biol Plants 15:281–286

Kumar A, Li C, Portis ARJ (2009) Arabidopsis thaliana expressing a thermostable chimeric rubisco activase exhibits enhanced growth and higher rates of photosynthesis at moderately high temperatures. Photosynth Res 100:143–153

Kumar M, Uniyal M, Kumar N, Kumar S, Gangwar R (2015) Conventional and molecular breeding for development of drought tolerant maize cultivars. J Crop Sci Tech 4:1–3

Kurek I, Chang TK, Bertain SM, Madrigal A, Liu L, Lassner MW, Zhu G (2007) Enhanced thermostability of Arabidopsis rubisco activase improves photosynthesis and growth rates under moderate heat stress. Plant Cell 19:3230–3241

Kyparissis A, Petropoulun Y, Manetas Y (1995) Summer survival of leaves in a soft-leaved shrub (Phlomis fruticosa L., Labiatae) under Mediterranean field conditions: avoidance of photoinhibitory damage through decreased chlorophyll contents. J Exp Bot 46(12):1825–1831

Labate CA, Leegood RC (1988) Limitation of photosynthesis by changes in temperature. Factors affecting the response of carbon dioxide assimilation to temperature in barley leaves. Planta 173(4):519–527
Labate CA, Adcock MD, Leegood RC (1990) Effect of temperature on the photosynthetic carbon assimilation in leaves of maize and barley. Planta 181(4):547–554
Laisk A, Edwards GE (1998) Oxygen and electron flow in C4 photosynthesis: Mehler reaction, photorespiration and CO2 concentration in the bundle sheath. Planta 205:632–645
Lang NT, Nha CT, Ha PTT, Buu CB (2013) Quantitative trait loci (QTLs) associated with drought tolerance in rice (Oryza sativa L.). Sabrao J Breed Genet 45:409–421
Larcher W (1995) Physiological plant ecology, 3rd edn. Springer, Berlin
Larcher W (2003) Physiological plant ecology, 4th edn. Springer, Berlin
Latowski D, Åkerlund HE, Strzalka K (2004) Violaxanthin deepoxidase, the xanthophylls cycle enzyme, requires lipid inverted hexagonal structures for its activity. Biochemistry 43:4417–4420
Laureau C, Bligny R, Streb P (2011) The significance of glutathione for photoprotection at contrasting temperatures in the alpine plant species Soldanella alpine and Ranunculus glacialis. Physiol Plant 143:246–260
Law RD, Crafts-Brandner SJ (1999) Inhibition and acclimation of photosynthesis to heat stress is closely correlated with activation of ribulose–1,5–bisphosphate carboxylase/oxygenase. Plant Physiol 120(1):173–181
Lawlor DW (2001) Photosynthesis, 3rd edn. Scientific Publishers Limited, Oxford
Lawlor DW (2002) Limitation to photosynthesis in water stressed leaves: stomata versus metabolism and the role of ATP. Ann Bot 89(7):1–15
Lawlor DW, Cornic G (2002) Photosynthetic carbon assimilation and associated metabolism in relation to water deficits in higher plants. Plant Cell Environ 25(2):275–294
Lawrence C, Holaday AS (2000) Effects of mild night chilling on respiration of expanding cotton leaves. Plant Sci 157(2):233–244
Lee AE, Staebler MA, Tollenaar M (2002) Genetic variation in physiological discriminators for cold tolerance—early autotrophic phase of maize development. Crop Sci 42(6):1919–1929
Lee Y-P, Kim SH, Bang JW, Lee HS, Kwak SS, Kwon SY (2007) Enhanced tolerance to oxidative stress in transgenic tobacco plants expressing three antioxidant enzymes in chloroplasts. Plant Cell Rep 26(5):591–598
Leegood RC, Walker DA (1983) The role of transmembrane solute flux in regulation of CO2 fixation in chloroplasts. Biochem Soc Trans 11(1):74–76
Leegood RC, Furbank RT (1986) Stimulation of photosynthesis by 2% O2 at low temperatures is restored by phosphate. Planta 168(1):84–93
Leegood RC, Edwards GE (1996) Carbon metabolism and photorespiration: temperature dependence in relation to other environmental factors. In: Baker NR (ed) Photosynthesis and the environment. Kluwer Academic Publishers, Dordrecht, pp 191–221
Leipner J, Fracheboud Y, Stamp P (1997) Acclimation by suboptimal growth temperature diminishes photooxidative damage in maize leaves. Plant Cell Environ 20(3):366–372
Leipner J, Basilides A, Stamp P, Fracheboud Y (2000) Hardly increased oxidative stress after exposure to low temperature in chillingacclimated and non-acclimated maize leaves. Plant Biol 2(2):243–251
Leroux O (2012) Collenchyma: a versatile mechanical tissue with dynamic cell walls. Ann Bot 110: 1083–1098
Leuning R (1995) A critical appraisal of a combined stomatal-photosynthesis model for C3 plants. Plant Cell Environ 18(4):339–355
Levitt J (1980) Responses of plants to environmental stresses. In: Chilling, freezing and high temperatures stresses, vol 1. Academic Press, New York, p 426
Lewis Smith RI (2003) The enigma of Colobanthus quitensis and Deschampsia antartica in Antarctica. In: Huiskes AHL, Gieskes WWC, Rozema J, Schorno RML, van der Vies SM, Wolff WJ (eds) Antarctic biology in a global context. Backhuys Publishers, Leiden, pp 234–239

Leuning R (2002) Temperature dependence of two parameters in a photosynthesis model. Plant Cell Environ 25(9):1205–1210

Li X-G, Wang X-M, Meng Q-W, Zou Q (2004) Factors limiting photosynthetic recovery in sweet pepper leaves after short-term chilling stress under low irradiance. Photosynthetica 42(2): 257–262

Li M, Liu H, Li L, Yi X, Zhu X (2007) Carbon isotope composition of plants along an altitudinal gradient and its relationship to environmental factors on the Qinghal-Tibet Plateau. Polish J Ecol 55:67–78

Li X, Cai J, Liu F, Dai T, Cao W, Jiang D (2014) Cold priming drives the sub-cellular antioxidant systems to protect photosynthetic electron transport against subsequent low temperature stress in winter wheat. Plant Physiol Biochem 82:34–43

Li X, Tan D-X, Jiang D, Liu F (2016) Melatonin enhances cold tolerance in drought-primed wild-type and abscisic acid-deficient mutant barley. J Pineal Res 61:328–339

Li NN, Yue C, Cao HL, Qian WJ, Hao XY, Wang YC, Wang L, Ding CQ, Wang XC, Yang YJ (2018a) Transcriptome sequencing dissection of the mechanisms underlying differential cold sensitivity in young and mature leaves of the tea plant (Camellia sinensis). J Plant Physiol 225: 144–155

Li S, Yang Y, Zhang Q, Liu N, Xu Q, Hu L (2018b) Differential physiological and metabolic response to low temperature in two zoysiagrass genotypes native to high and low latitude. PLoS One 13(6):e0198885

Lichtenthaler HK (1996) Vegetation stress: an introduction to the stress concept in plants. J Plant Physiol 148:4–14

Lichtenthaler HK (1998) The stress concept in plants: an introduction. In: Csermely P (ed) Stress of life: from molecules to man. Annals of New York Academy of Sciences, New York, NY, pp 187–198

Lin YS, Medlyn BE, Ellsworth DS (2012) Temperature responses of leaf net photosynthesis: the role of component processes. Tree Physiol 32(2):219–231

Liu Q, Kasuga M, Sakuma Y, Abe H, Miura S, Yamaguchi-Shinozaki Y, Shinozaki K (1998) Two transcription factors, DREB1 and DREB2, with an EREBP /AP2 DNA binding domain separate two cellular signal transduction pathways in drought- and low-temperature-responsive gene expression, respectively, in Arabidopsis. Plant Cell 10:1391–1406

Lloyd J, Farquhar GD (1994) 13C discrimination during CO2 assimilation by the terrestrial biosphere. Oecologia 99(3–4):201–215

Logan BA, Adams WW, Demmig-Adams B (2007) Avoiding common pitfalls of chlorophyll fluorescence analysis under field conditions. Funct Plant Biol 34:853–859

Logan BA, Demmig-Adams B, Adams WW, Bilger W (2014) Context, quantification, and measurement guide for non-photochemical quenching of chlorophyll fluorescence. In: Demmig-Adams B, Garab G, Adams WW, Govindjee (eds) Non-photochemical quenching and energy dissipation in plants, algae and cyanobacteria. Advances in photosynthesis and respiration, vol 40. Springer, Dordrecht, pp 187–201

Long SP, East TM, Baker NR (1983) Chilling damage to photosynthesis in young Zea mays. J Exp Bot 34:177–188

Long SP, Postl WF, Bolhar-Nordenkampf HR (1993) Quantum yields for uptake of carbon dioxide in C3 vascular plants of contrasting habitats and taxonomic groupings. Planta 189:226–234

Long SP, Humphries S, Falkowski PG (1994) Photoinhibition of photosynthesis in nature. Annu Rev Plant Physiol Plant Mol Biol 45:633–662

Long SP (1999) Environmental responses. In: Sage RF, Monson RK (eds) C4 plant biology. Academic Press, San Diego, CA, pp 215–249

Lovisolo C, Schubert A (1998) Effects of water stress on vessel size and xylem hydraulic conductivity in Vitis vinifera L. J Exp Bot 49:693–700

Lundmark M, Cavaco AM, Trevanion S, Hurry V (2006) Carbon partitioning and export in transgenic Arabidopsis thaliana with altered capacity for sucrose synthesis grown at low 3 temperature: a role for metabolite transporters. Plant Cell Environ 29(9):1703–1714

Lukatkin AS, Brazaityte A, Bobinas C, Duchovskis P (2012) Chilling injury in chilling-sensitive plants: a review. Agriculture 99:111–124
Lurie S, Ronen R, Lipsker Z, Aloni B (1994) Effects of paclobutrazol and chilling temperatures on lipids, antioxidants and ATPase activity of plasma membrane isolated from green bell pepper fruits. Physiol Plant 91:593–598
Lynch DV, Thompson GA Jr (1982) Low temperature-induced alterations in the chloroplast and microsomal membrane of Dunaliella salina. Plant Physiol 69:1369–1375
Lynch DV (1990) Chilling injury in plants: the relevance of membrane lipids. In: Katterman F (ed) Environmental injury to plants. Academic press, New York, pp 17–34
Ma Y-Z, Holt NF, Li X-P, Niyogi KK, Fleming GR (2003) Evidence for direct carotenoid involvement in the regulation of photosynthetic light harvesting. Proc Natl Acad Sci U S A 100(8):4377–4382
Mächler F, Schnyder H, Nosberger J (1984) Influence of inorganic phosphate on photosynthesis of wheat chloroplasts I. photosynthesis and assimilate export at 50C and 250C. J Exp Bot 35(153): 481–487
Mahajan S, Tuteja N (2005) Cold, salinity and drought stress: an overview. Arch Biochem Biophys 444:139–158
Maherali H, Reid CD, Polley HW, Johnson HB, Jachson RB (2002) Stomatal acclimation over a subambient to elevated CO2 gradient in a C3/C4 grassland. Plant Cell Environ 25(4):557–566
Makino A, Nakano H, Mae T (1994) Response of ribulose-1,5-bisphosphate carboxylase, cytochrome f, and sucrose synthesis enzymes in rice leaves to leaf nitrogen and their relationships to photosynthesis. Plant Physiol 105(1):173–179
Makino A (2011) Photosynthesis, grain yield, and nitrogen utilization in rice and wheat. Plant Physiol 155:125–129
Manivannan P, Jaleel CA, Sankar B, Kishorekumar A, Somasundaram R, Alagu Lakshmanan GM, Panneerselvam R (2007) Growth, biochemical modifications and proline metabolism in Helianthus annuus L. as induced by drought stress. Colloids Surf B Biointerfaces 59(2): 141–149
Mantovani A, Vieira RC (2000) Leaf micromorphology of Antarctic pearlwort Colobanthus quitensis (Kunth) Bartl. Polar Biol 23:531–538
Maroco JP, Ku MSB, Edwards GE (1997) Oxygen sensitivity of C4 photosynthesis: evidence from gas exchange and chlorophyll fluorescence analyses with different C4 subtypes. Plant Cell Environ 20(12):1525–1533
Maroco JP, Ku MSB, Edwards GE (2000) Utilization of O2 in the metabolic optimization of C4 photosynthesis. Plant Cell Environ 23(1):115–121
Maroco JP, Rodrigues ML, Lopes C, Chaves MM (2002) Limitations to leaf photosynthesis in field-grown grapevine under drought-metabolic and modelling approaches. Func Plant Biol 29:451–459
Maqbool MA, Aslam M, Ali H (2017) Breeding for improved drought tolerance in chickpea (Cicer arietinum L.). Plant Breed 136:300–318
Marschner H (1995) Mineral nutrition of higher plants, vol 2. Academic Press, London, p 889
Martin B, Ort DR, Boyer JS (1981) Impairment of photosynthesis by chilling-temperatures in tomato (Lycopersicon esculentum cultivar Rutgers). Plant Physiol 68(2):329–334
Martindale W, Leegood RC (1997) Acclimation of photosynthesis to low temperature in Spinacia oleracea L. I. Effects of acclimation on CO2 assimilation and carbon partitioning. J Exp Bot 48(10):1865–1872
Maruyama S, Yatomi M, Nakamura Y (1990) Response of rice leaves to low temperature: I. Changes in basic biochemical parameters. Plant Cell Physiol 31(3):303–309
Matsuba K, Imaizumi N, Kaneko S, Samejima M, Ohsugi R (1997) Photosynthetic responses to temperature of phosphoenolpyruvate carboxykinase type C4 species differing in cold sensitivity. Plant Cell Environ 20(2):268–274

Matsuo T, Graham D, Patterson BD, Hockley DB (1994) An electrophoretic method to detect cold-induced dissociation of proteins in crude extracts of higher plants. Anal Biochem 223(2): 181–184

Matsuoka M, Furbank R, Fukayama H, Miyao M (2001) Molecular engineering of C4 photosynthesis. Annu Rev Plant Physiol Plant Mol Biol 52:297–314

Maxwell K, Johnson GN (2000) Chlorophyll fluorescence—a practical guide. J Exp Bot 51(345): 659–668

Mawson BT, Svoboda J, Cummins RW (1986) Thermal acclimation of photosynthesis by the arctic plant Saxifraga cernua. Can J Bot 64(1):71–76

Mawson BT, Cummins RW (1989) Thermal acclimation of photosynthetic electron transport activity by thylakoid of Saxifraga cernua. Plant Physiol 89(1):325–332

McDowell N, Pockman WT, Allen CD, Breshears DD, Cobb N, Kolb T, Plaut J, Sperry J, West A, Williams DG, Yepez EA (2008) Mechanisms of plant survival and mortality during drought: why do some plants survive while others succumb to drought? New Phytol 178(4):719–739

McKersie BD, Bowley SR (1997) Active oxygen and freezing tolerance in transgenic plants. In: Li PH, Chen THH (eds) Plant cold hardiness. Springer, Boston, MA

McNulty AK, Cummins WR (1987) The relationship between respiration and temperature in leaves of the arctic plant Saxifraga cernua. Plant Cell Environ 10:319–325

Medici LO, Azevedo RA, Canellas LP, Machado AT, Pimentel C (2007) Stomatal conductance of maize under water and nitrogen deficits. Pesq Agropec Bras 42(4):599–601

Medlyn BE, Dreyer E, Ellsworth D, Forstreuter M, Harley PC, Kirschbaum MUF, Le Roux X, Montpeid P, Strassemeyer J, Walcroft A, Wang K, Loustau D (2002) Temperature response of parameters of a biochemically based model of photosynthesis. II. A review of experimental data. Plant Cell Environ 25(9):1167–1179

Medrano H, Escalona JM, Bota J, Gulías J, Flexas J (2002) Regulation of photosynthesis of C3 plants in response to progressive drought: the stomatal conductance as a reference parameter. Ann Bot 89:895–905

Meehl GA, Covey C, Delworth T, Latif M, McAvaney B, Mitchell JFB, Stouffer RJ, Taylor KE (2007) The WCRP CMIP3 multimodel dataset—A new era in climate change research. Bull Am Meteorol Soc 88:1383–1394

Melis A (1999) Photosystem-II damage and repair cycle in chloroplasts: what modulates the rate of photodamage in vivo? Trends Plant Sci 4:130–135

Melo LC, dos Santos JB, Ramalho MAP (1997) Choice of parents to obtain common bean (Phaseolus vulgaris) cultivars tolerant to low temperatures at the adult stage. Brazil J Genet 20(2):283–292

Medlyn BE, Dreyer E, Ellsworth D, Forstreuter M, Harley PC, Kirschbaum MUF, Le Roux X, Montpied P, Strasseneyer J, Walcroft A, Wang K (2002a) Temperature response of parameters of a biochemically based model of photosynthesis. II. A review of experimental data. Plant Cell Environ 25(9):1167–1179

Medlyn BE, Loustau D, Delzon S (2002b) Temperature response of parameters of a biochemically based model of photosynthesis. I. Seasonal changes in mature maritime pine (Pinus pinaster Ait.). Plant Cell Environ 25(9):1155–1165

Medlyn BE, Duursma RA, Eamus D, Ellsworth DS, Prentice C, Barton CVM, Crous KY, Angelis P, Freeman M, Wingate L (2011) Reconciling the optimal and empirical approaches to modelling stomatal conductance. Glob Change Biol 17(6):2134–2144

Menzel A, Sparks TH, Estrella N, Koch E, Aasa A, Ahas R, Alm-Kübler K, Bissolli P, Braslavská O, Briede A, Chmielewski FM, Crepinsek Z, Curnel Y, Dahl A, Defila C, Donnelly A, Filella Y, Måge F, Mestre A, Nordli Ø, Peñuelas J, Pirinen P, Remišová V, Scheifinger H, Striz M, van Vliet AJH, Wielgolaski FE, Zach S, Zust A (2006) European phenological response to climate change matches the warming pattern. Glob Chang Biol 12(10): 1969–1976

Mercado JA, Reid MS, Valpuesta V, Quesada MA (1997) Metabolic changes and susceptibility to chilling stress in Capsicum annuum plants grown at suboptimal temperature. Aust J Plant Physiol 24(6):759–767

Millenaar FF, Lambers H (2003) The AOX: in vivo regulation and function. Plant Biol 5:2–15

Miskiewicz E, Ivanov AG, Williams JP, Khan MU, Falk S, Huner NPA (2000) Photosynthetic accilation of the filamentous cyanobacterium, Plectonema baryanum, UTEX 485 to temperature and light. Plant Cell Physiol 41(6):767–775

Misra AN, Latowski D, Strzalka K (2006) The xanthophyll cycle activity in kidney bean and cabbage leaves under salinity stress. Russ J Plant Physiol 53:113–121

Mitchell RAC, Barber J (1986) Adaptation of photosynthetic electron-transport rate to growth temperature in pea. Planta 169(3):429–436

Miyao M (2003) Molecular evolution and genetic engineering of C4 photosynthetic enzymes. J Exp Bot 54:179–189

Miura K, Tada Y (2014) Regulation of water, salinity, and cold stress responses by salicylic acid. Fron Plant Sci 5:4

Mizoguchi T, Irie K, Hirayama T, Hayashida N, Yamaguchi-Shinozaki K, Katsumoto K, Shinozaki K (1996) A gene encoding a mitogen-activated protein kinase kinase kinase is induced simultaneously with genes for a mitogen-activated protein kinase and an S6 ribosomal protein kinase by touch, cold, and water stress in Arabidopsis thaliana. Proc Natl Acad Sci U S A 93(2):765–769

Møller IM (2001) Plant mitochondria and oxidative stress: electron transport, NADPH turnover, and metabolism of reactive oxygen species. Annu Rev Plant Physiol Plant Mol Biol 52:561–591

Møller IM, Jensen PE, Hansson A (2007) Oxidative modifications to cellular components in plants. Annu Rev Plant Biol 58:459–481

Moore AL, Carré JE, Affourtit C, Albury MS, Crichton PG, Kita K, Heathcote P (2008) Compelling EPR evidence that the alternative oxidase is a diiron carboxylate protein. Biochim Biophys Acta 1777(4):327–333

Moseki B (2004) Effects of low temperature stress on chloroplast development in a range of Zea mays cultivars. S Afr J Bot 70(5):726–729

Mott KA, Peak D (2010) Stomatal responses to humidity and temperature in darkness. Plant Cell Environ 33(2):1084–1090

Mouthon J, Daufresne M (2006) Effects of the 2003 heat wave and climatic warming on mollusc communities of the Saone: a large lowland river and of its two main tributaries (France). Glob Chang Biol 12(3):441–449

Mumm P, Wolf T, Fromm J, Roelfsema MR, Marten I (2011) Cell type-specific regulation of ion channels within the maize stomatal complex. Plant Cell Physiol 52:1365–1375

Munro KD, Hodges DM, DeLong JM, Forney CF, Kristie DN (2004) Low temperature effects on ubiquinone content, respiration rates and lipid peroxidation levels of etiolated seedlings of two differentially chilling-sensitive species. Physiol Plant 121(3):488–497

Murata N, Ishizaki-Nishizawa O, Higashi S, Hayashi H, Tasaka Y, Nishida I (1992) Genetically engineered alteration in the chilling sensitivity of plants. Nature 356:710–713

Murchie EH, Niyogi KK (2011) Manipulation of photoprotection to improve plant photosynthesis. Plant Physiol 155(1):86–92

Nadeem SM, Ahmad M, Zahir ZA, Kharal MA (2016) Role of phytohormones in stress tolerance of plants. In: Hakeem KR, Akhtar MS (eds) Plant, soil and microbes. Springer, Cham, pp 385–421

Naidu SL, Moose SP, Al-Shoabi KA, Raines CA, Long SP (2003) Cold-tolerance of C4 photosynthesis in Miscanthus x giganteus - effects of cold on photosynthetic enzymes. Plant Physiol 132(3):1688–1697

Naidu SL, Long SP (2004) Potential mechanisms of low-temperature tolerance of C4 photosynthesis in Miscanthus x giganteus: an in vivo analysis. Planta 220(1):145–155

Nakagami H, Pitzschke A, Hirt H (2005) Emerging MAP kinase pathways in plant stress signalling. Trends Plant Sci 10:339–346

Nie GY, Baker NR (1991) Modifications to the thylakoid composition during development of maize leaves at low temperatures. Plant Physiol 95:184–191

Nie GY, Long SP, Baker NR (1992) The effects of development at suboptimal growth temperatures on photosynthetic capacity and susceptibility to chilling-dependent photo inhibition in Zea mays. Physiol Plant 85(3):554–560

Nie GY, Hendrix DL, Weber AN, Kimball BA, Long SP (1995) Increased accumulation of carbohydrates and decreased photosynthetic gene transcript levels in wheat grown at an elevated CO2 concentration in the field. Plant Physiol 108:975–983

Nishida I, Murata N (1996) Chilling sensitivity in plants and cyanobacteria: the crucial contribution of membrane lipid. Annu Rev Plant Physiol Plant Mol Biol 47:541–568

Niyogi KK, Björkman O, Grossman AR (1997) Chlamydomonas xanthophyll cycle mutants identified by video imaging of chlorophyll fluorescence quenching. Plant Cell 9:1369–1380

Niyogi KK (2000) Safety valves for photosynthesis. Cur Opin Plant Biol 3(6):455–460

Noguchi K, Yamori W, Hikosaka K, Terashima I (2015) Homeostasis of the temperature sensitivity of respiration over a range of growth temperatures indicated by a modified Arrhenius model. New Phytol 207(1):34–42

Olien CR, Smith MN (1997) Ice adhesions in relation to freeze stress. Plant Physiol 60:499–503

Oliveira G, Penuelas J (2004) Effects of winter cold stress on photosynthesis and photochemical efficiency of PSII of the Mediterranean Citrus albidus L. and Quercus ilex L. Plant Ecol 175: 179–191

Ordentlich A, Linzer RA, Raskin I (1991) Alternative respiration and heat evolution in plants. Plant Physiol 97(4):1545–1550

Ortiz-Lopez A, Nie GY, Ort DR, Baker NR (1990) The involvement of photoinhibition of photosystem II and impaired membrane energization in the reduced quantum yield of carbon assimilation in chilled maize. Planta 181:78–84

Osmond CB, Björkman O, Anderson DJ (1980) Physiological processes in plant ecology. Towards a synthesis with atriplex. Springer-Verlag, Berlin

Osmond CB, Grace SC (1995) Perspectives on photoinhibition and photorespiration in the field: quintessential inefficiencies of the light and dark reactions of photosynthesis. J Exp Bot 47(4): 1351–1362

Ouk M, Basnayake J, Tsubo M, Fukai S, Fischer KS, Cooper M, Nesbitt H (2006) Use of drought response index for identification of drought tolerant genotypes in rainfed lowland rice. Field Crops Res 99(1):48–58

Öquist G, Hurry VM, Huner NPA (1993a) Low-temperature effects on photosynthesis and correlation with freezing tolerance in spring and winter cultivars of wheat and rye. Plant Physiol 101: 245–250

Öquist G, Hurry VM, Huner NPA (1993b) The temperature dependence of the redox state of Q (A) and susceptibility of photosynthesis to photoinhibition. Plant Physiol Biochem 31(5): 683–691

Ort DR, Baker NR (2002) A photoprotective role for O2 as an alternative electron sink in photosynthesis? Curr Opin Plant Biol 5(3):193–198

Pammenter NW, Loreto F, Sharkey TD (1993) End product feedback on the photosynthetic electron transport. Photosyn Res 35(1):5–15

Pan X, Lada RR, Caldwell CD, Falk KC (2011) Water-stress and N-nutrition effects on photosynthesis and growth of Brassica carinata. Photosynthetica 49:309–315

Pan J, Lin S, Woodbury NW (2012) Bacteriochlorophyll excitedstate quenching pathways in bacterial reaction centers with the primary donor oxidized. J Phys Chem B 116:2014–2022

Pandey R, Agarwal RM, Jeevaratnam K, Sharma GL (2004) Osmotic stress induced alterations in rice (Oryza sativa L.) and recovery. Plant Growth Regul 42:79–87

Pardossi A, Vernieri P, Tognoni F (1992) Involvement of abscisic acid in regulating water status in Phaseolus vulgaris L. during chilling. Plant Physiol 100:1243–1250

Paredes M, Quiles MJ (2015) The effects of cold stress on photosynthesis in hibiscus plants. PLoS One 10:e0137472

Parida AK, Dasgaonkar VS, Phalak MS, Umalkar GV, Aurangabadkar LP (2007) Alterations in photosynthetic pigments, protein, and osmotic components in cotton genotypes subjected to short-term drought stress followed by recovery. Plant Biotechnol Reports 1:37–48

Park EJ, Jeknic Z, Chen THH (2006) Exogenous glycine betaine application increases chilling tolerance in tomato plants. Plant Cell Physiol 47(6):706–714

Partelli FL, Vieira HD, Viana AP, Batista-Santos P, Rodrigues AP, Leitão AE, Ramalho JC (2009) Low temperature impact on photosynthetic parameters of coffee genotypes. Pesq Agropec Bras 44(11):1404–1415

Paul MJ, Driscoll SP, Lawlor DW (1992) Sink-regulation of photosynthesis in relation to temperature in sunflower and rape. J Exp Bot 43:147–153

Paul MJ, Foyer CH (2001) Sink regulation of photosynthesis. J Exp Bot 52(360):1383–1400

Pauwels L, Inzé D, Goossens A (2009) Jasmonate-inducible gene: what does it mean? Trend Plant Sci 14(2):87–91

Payton P, Allen RD, Trolinder N, Holaday AS (1997) Over-expression of chloroplast-targeted Mn superoxide dismutase in cotton (Gossypium hirsutum L., cv. Coker 312) does not alter the reduction of photosynthesis after short exposures to low temperature and high light intensity. Photosynthesis Res 52(3):233–244

Payton P, Webb R, Kornyeyev D, Allen R, Holaday AS (2001) Protecting cotton photosynthesis during moderate chilling at high light intensity by increasing chloroplastic antioxidant enzyme activity. J Exp Bot 52(365):2345–2354

Pearce RS (1999) Molecular analysis of acclimation to cold. Plant Grow Regul 29:47–76

Pearcy RW (1977a) Acclimation of photosynthetic and respiratory carbon dioxide exchange to growth temperature in Atriplex lentiformis (Torr.) Wats. Plant Physiol 59(5):795–799

Pearcy RW (1977b) Effects of growth temperature on the thermal stability of the photosynthetic apparatus of Atriplex lentiformis (Torr.) Wats. Plant Physiol 59(5):873–878

Perera NH, Hartmann E, Holaday AS (1995) Regulation of cotton photosynthesis during moderate chilling. Plant Sci 111:133–143

Pérez P, Morcuende R, del Molino IM, de la Puente LS, Martinez-Carrasco R (2001) Contrasting responses of photosynthesis and carbon metabolism to low temperatures in tall fescue and clovers. Physiol Plant 112:478–486

Pérez-Torres E, Dinamarca J, Bravo LA, Corcuera LJ (2004) Responses of Colobanthus quitensis (Kunth) Bartl. to high light and low temperature. Polar Biol 27:183–189

Peynevandi KM, Razavi SM, Zahri S (2018) The ameliorating effects of polyamine supplement on physiological and biochemical parameters of Stevia rebaudiana Bertoni under cold stress. Plant Prod Sci 21:123–131

Piao S, Ciais P, Friedlingstein P, Peylin P, Reichstein M, Luyssaert S, Margolis H, Fang J, Barr A, Chen A, Grelle A, Hollinger DY, Laurila T, Lindroth A, Richardson AD, Vesala T (2008) Net carbon dioxide losses of northern ecosystems in response to autumn warming. Nature 451(7174):49–52

Pietikäinen J, Pettersson M, Bååth E (2005) Comparison of temperature effects on soil respiration and bacterial and fungal growth rates. FEMS Microbiol Ecol 52(1):49–58

Pilon-Smits EAH, Ebskamp MJM, Paul MJ, Jeuken MJW, Weisbeek PJ, Smeekens SCM (1995) Improved performance of transgenic fructan-accumulating tobacco under drought stress. Plant Physiol 107(1):125–130

Pittermann J, Sage RF (2000) Photosynthetic performance at low temperature of Bouteloua gracilis lag., a high-altitude C4 grass from the Rocky Mountains, USA. Plant Cell Environ 23(8): 811–823

Pittermann J, Sage R (2001) The response of the high altitude C4 grass Muhlenbergia montana (Nutt.) A.S. Hitchc. To long- and short-term chilling. J Exp Bot 52(357):829–838

Plaxton WC (2004) Plant response to stress: biochemical adaptation to phosphate deficiency. In: Encyclopedia of Plant and Crop Science. Marcel Dekker, New York, pp 976–980

Poetsch W, Hermans J, Westhoff P (1991) Multiple cDNAs of phosphoenolpyruvate carboxylase in the C4 dicot Flaveria trinervia. FEBS Lett 292:133–136

Pollock CJ, Lloyed EJ (1987) The effect of low growth temperature upon stach, sucrose and fructose synthesis in leaves. Ann Bot 60:231–235

Powles SB, Björkman O (1982) Photoinhibition of photosynthesis: effect on chlorophyll fluorescence at 77K in intact leaves and in chloroplast membranes of Nerium oleander. Planta 156(2): 97–107

Prasad TK, Anderson MD, Martin BA, Stewart CR (1994) Evidence for chilling-induced oxidative stress in maize seedlings and a regulatore role for hydrogen peroxide. Plant Cell 6(1):65–74

Purvis AC, Shewfelt RL (1993) Does the alternative pathway ameliorate chilling injury in sensitive plant tissues? Physiol Plant 88(4):712–718

Qi F, Li J, Duan L, Li Z (2005) Inductions of coronatine and MeJA to low-temperature resistance of wheat seedlings. Acta Bot Boreal Occident Sin 26:1776–1780

Quartin VL, Ramalho JC, Campos PS, Nunes MA (2004) A importância da investigacao na cafercultura : o problema do frio. In: Tembo PJL, Neto JFC, Pomball AM (eds) Livro de Actas do 10 Coloquio angola – agricultura. Sociedade e Desenvolvimento Rural, Instituto Superior de Agronomia Press, Lisboa, pp 64–72

Raghavendra AS, Padmasree K (2003) Beneficial interactions of mitochondrial metabolism with photosynthetic carbon assimilation. Trends Plant Sci 8(11):546–553

Rahnama A, Poustini K, Tavakkol-Afshari R, Tavakoli A (2010) Growth and stomatal responses of bread wheat genotypes in tolerance to salt stress. Int J Biol Life Sci 6:216–221

Raines CA (2006) Transgenic approaches to manipulate the environmental responses of the C3 carbon fixation cycle. Plant Cell Environ 29:331–339

Raines CA (2011) Increasing photosynthetic carbon assimilation in C3 plants to improve crop yield: current and future strategies. Plant Physiol 155:36–42

Ramalho JC, Quartin V, Leitão E, Campos PS, Carelli MLC, Fahl JI (2003) A. Cold acclimation ability of photosynthesis among species of the tropical Coffea genus. Plant Biol 5(6):631–641

Rasion JK (1980) Effect of low temperature on respiration. In: Davies DD (ed) The biochemistry of plants: metabolism and respiration, vol 2. Academic Press, New York, pp 613–626

Reda F, Mandoura HMH (2011) Response of enzymes activities, photosynthetic pigments, proline to low or high temperature stressed wheat plant (Triticum aestivum L.) in the presence or absence of exogenous proline or cysteine. Int J Acad Res 3(4):108–115

Reddy AR, Chaitanya KV, Vivekanandan M (2004) Drought induced responses of photosynthesis and antioxidant metabolism in higher plants. J Plant Physiol 161:1189–1202

Reguera M, Peleg Z, Abdel-Tawab YM, Tumimbang EB, Delatorre CA, Blumwald E (2013) Stress induced cytokinin synthesis increases drought tolerance through the coordinated regulation of carbon and nitrogen assimilation in rice. Plant Physiol 163:1609–1622

Reyes L, Jennings PH (1997) Effects of chilling on respiration and induction of cyanide- resistant respiration in seedling roots of cucumber. J Am Soc Hortic Sci 122(2):190–194

Ribeiro RV, Machado EC, Oliveira RF, Pimentel C (2003) High temperature effects on the response of photosynthesis to light in sweet orange plants infected with Xylella fastidiosa. Braz J Plant Physiol 15(2):89–97

Ribeiro RV, Machado EC, Oliveira RF (2004) Growth– and leaf–temperature effects on photosynthesis of sweet orange seedlings infected with Xylella fastidiosa. Plant Pathol 53(3):334–340

Ribeiro RV, Machado EC, Oliveira RF (2006) Temperature response of photosynthesis and its interaction with light intensity in sweet orange leaf discs under non-photorespiratory condition. Ciência e Agrotecnologia 30(4):670–678

Rivero RM, Kojima M, Gepstein A, Sakakibara H, Mittler R, Gepstein S, Blumwald E (2007) Delayed leaf senescence induces extreme drought tolerance in a flowering plant. Proc Natl Acad Sci U S A 104(49):19631–19636

Rizza F, Pagani D, Stanca AM, Cattivelli L (2001) Use of chlorophyll fluorescence to evaluate the cold acclimation and freezing tolerance of winter and spring oats. Plant Breed 120:389–396

Rodriguez JL, Davies WJ (1982) The effects of temperature and ABA on stomata of Zea mays L. J Exp Bot 33(136):977–987

Rodriguez VM, Malvar RA, Butron A, Ordas A, Revilla P (2007) Maize populations as sources of favorable alleles to improve cold-tolerant hybrids. Crop Sci 47:1779–1786

Roháček K (2002) Chlorophyll fluorescence parameters: the definitions, photosynthetic meaning, and mutual relationships. Photosynthetica 40:13–29

Roychoudhury A, Banerjee A (2016) Endogenous glycine betaine accumulation mediates abiotic stress tolerance in plants. Trop Plant Res 3:105–111

Royer DL (2001) Stomatal density and stomatal index as indicators of paleoatmospheric CO2 concentration. Rev Palaeobot Palynol 114:1–28

Ruan C, Shao H, da Silva JAT (2012) A critical review on the improvement of photosynthetic carbon assimilation in C3 plants using genetic engineering. Crit Rev Biotechnol 32:1–21

Ruban AV, Pascal AA, Robert B, Horton P (2002) Activation of zeaxanthin is an obligatory event in the regulation of photosynthetic light harvesting. J Biol Chem 277:7785–7789

Ruelland E, Vaultier MN, Zachowski A, Hurry V (2009) Cold signalling and cold acclimation in plants. Adv Bot Res 49:35–150

Ruelland E, Zachowski A (2010) How plants sense temperature. Environ Exp Bot 69:225–232

Rumeau D, Peltier G, Cournac L (2007) Chlororespiration and cyclic electron flow around PSI during photosynthesis and plant stress response. Plant Cell Environ 30:1041–1051

Ruuska SA, Badger MR, Andrews TJ, von Caemmerer S (2000) Photosynthetic electronic sinks in transgenic tobacco with reduced amount of rubisco: little evidence for Mehler reaction. J Exp Bot 51(Suppl. 1):357–368

Sage RF, Sharkey TD (1987) Effect of temperature on the occurance of O2 and CO2 insensitive photosynthesis in field grown plants. Plant Physiol 84(3):658–664

Sage RF (1990) A model describing the regulation of ribulose-1,5- bisphosphate carboxylase, electron transport, and triose phosphate sse in response to light-intensity and CO2 in C3 plants. Plant Physiol 94(4):1728–1734

Sage RF, Sharkey TD, Seemann JR (1990) Regulation of ribulose-1,5-bisphosphate carboxylase in response to light intensity and CO2 in the C3 annuals Chenopodium album L. and Phaseolus vulgaris L. Plant Physiol 94(4):1735–1742

Sage RF (1999) Why C4 photosynthesis? In: Sage RF, Monson RK (eds) C4 plant biology. Academic Press, San Diego, ca, pp 3–16

Sage RF, Wedin DA, Li M (1999) The biogeography of C4 photosynthesis: patterns and controlling factors. In: Sage RF, Monson RK (eds) C4 plant biology. Academic Press, San Diego, ca, pp 313–373

Sage RF (2002) Variation in the kcat of rubisco in C3 and C4 plants and some implications for photosynthetic performance at high and low temperature. J Exp Bot 53(369):609–620

Sage RF, Kubien DS (2007) The temperature response of C3 and C4 photosynthesis. Plant Cell Environ 30(9):1086–1106

Sage RF, Zhu XL (2011) Exploiting the engine of C4 photosynthesis. J Exp Bot 62:2989–3000

Saibo NJM, Lourenço T, Oliveira MM (2009) Transcription factors and regulation of photosynthetic and related metabolism under environmental stresses. Ann Bot 103:609–623

Sales CRG, Ribeiro RV, Silveira JAG, Machado EC, Martins MO, Lagôa AMMA (2013) Superoxide dismutase and ascorbate peroxidase improve the recovery of photosynthesis in sugarcane plants subjected to water deficit and low substrate temperature. Plant Physiol Biochem 73:326–336

Salvucci ME, Crafts-Brandner SJ (2004a) Inhibition of photosynthesis by heat stress: the activation state of rubisco as a limiting factor in photosynthesis. Physiol Plant 120(2):179–186

Salvucci ME, Crafts-Brandner SJ (2004b) Relationship between the heat tolerance of photosynthesis and the thermal stability of rubisco activase in plants from contrasting thermal environments. Plant Physiol 134:1460–1470

Samota MK, Sasi M, Awana M, Yadav OP, Amitha MSV, Tyagi A, Kumar S, Singh A (2017) Elicitor-induced biochemical and molecular manifestations to improve drought tolerance in rice (Oryza sativa L.) through seed-priming. Front Plant Sci 8:934

Sassenrath GF, Ort DR (1990) The relationship between inhibition of photosynthesis at low temperature and the inhibition of photosynthesis after rewarming in chill-sensitive tomato. Plant Physiol Biochem 28(4):457–465

Sassenrath GF, Ort DR, Portis AR Jr (1990) Impaired reductive activation of stromal bisphosphatases in tomato leaves following low-temperature exposure at high light. Arch Biochem Biophys 282(2):302–308

Savitch LV, Gray GR, Huner NPA (1997) Feed back limited photosynthesis and regulation of sucrose-starch accumulation during cold acclimation and low-temperature stress in spring and winter wheat. Planta 201(1):18–26

Savitch LV, Leonardos ED, Krol M, Jansson S, Grodzinski B, Huner NPA, Öquist G (2002) Two different strategies for light utilization in photosynthesis in relation to growth and cold acclimation. Plant Cell Environ 25(6):761–771

Saravanavel R, Ranganathan R, Anantharaman P (2011) Effect of sodium chloride on photosynthetic pigments and photosynthetic characteristics of Avicennia officinalis seedlings. Recent Res Sci Technol 3:177–180

Sausen TL, Rosa LMG (2010) Growth and carbon assimilation limitations in Ricinus communis (Euphorbiaceae) under soil water stress conditions. Acta Bot Bras 24(3):648–654

Sawada S (1970) An ecophysiological analysis of the difference between the growth rates of young wheat seedlings grown in various seasons. J Fac Sci Univ Tokyo III 10(11/13):233–263

Schaberg PG, Minocha R, Long S, Halman JM, Hawley GJ, Eagar C (2011) Calcium addition at the Hubbard brook experimental forest increases the capacity for stress tolerance and carbon capture in red spruce (Picea rubens) trees during the cold season. Trees 25(6):1053–1061

Scheibe R (1990) Light/dark modulation: regulation of chloroplast metabolism in a new light. Botanica Acta 103(4):327–334

Schrader SM, Wise RR, Wacholtz WF, Ort DR, Sharkey TD (2004) Thylakoid membrane responses to moderately high leaf temperature in Pima cotton. Plant Cell Environ 27:725–735

Schreiber U, Neubauer C (1990) O2-dependent electron flow, membrane energization and the mechanism of nonphotochemical quenching of chlorophyll fluorescence. Photosynth Res 25: 279–293

Sen Gupta A, Heinen JL, Holady AS, Burke JJ, Allen RD (1993) Increased resistance to oxidative stress in transgenic plants that over-express chloroplastic cu/Zn superoxide dismutase. Proc Natl Acad Sci U S A 90(4):1629–1633

Senaratna T, Touchell D, Bunn E, Dixon K (2000) Acetyl salicylic acid (aspirin) and salicylic acid induce multiple stress tolerance in bean and tomato plants. Plant Growth Regul 30:157–161

Seo PJ, Kim MJ, Park JY, Kim SY, Jeon J, Lee YH, Kim J, Park CM (2010) Cold activation of a plasma membrane-tethered NAC transcription factor induces a pathogen resistance response in Arabidopsis. Plant J 61:661–671

Seo JS, Joo J, Kim MJ, Kim YK, Nahm BH, Song SI, Cheong JJ, Lee JS, Kim JK, Choi YD (2011) OsbHLH148, a basic helix-loop-helix protein, interacts with OsJAZ proteins in a jasmonate signaling pathway leading to drought tolerance in rice. Plant J 65:907–921

Sevanto S (2014) Phloem transport and drought. J Exp Bot 65(7):1751–1759

Sevanto S, McDowell NG, Dickman LT, Pangle R, Pockman WT (2014) How do trees die? A test of the hydraulic failure and carbon starvation hypotheses. Plant Cell Environ 37(1):153–161

Shah K, Kumar RG, Verma S, Dubey RS (2001) Effect of cadmium on lipid peroxidation. Superoxide anion generation and activity of antioxidant enzyme in growing rice seedlings. Plant Sci 161:1135–1144

Shah F, Huang J, Cui K, Nie L, Shah T, Chen C, Wang K (2011) Impact of high-temperature stress on rice plant and its traits related to tolerance. Agric Sci 149(05):545–556

Shang Z (2000) Effect of 6-BA and KT on photophosphorylation activity in wheat flag leaves under water stress. Acta Agric Boreal Sin 15:34–38

Shao HB, Liang ZS, Shao MA (2006) Osmotic regulation of 10 wheat (Triticum aestivum L.) genotypes at soil water deficits. Biointerfaces 47:132–139

Sharkey TD (1985a) O2-insensitive photosynthesis in C3 plants – its occurrence and a possible explanation. Plant Physiol 78(1):71–75
Sharkey TD (1985b) Photosynthesis in intact leaves of C3 plants – physics, physiology and rate limitation. Bot Rev 51(1):53–105
Sharkey TD (1988) Estimating the rate of photorespiration in leaves. Physiol Plant 73(1):147–152
Sharkey TD (1990) Feedback limitation of photosynthesuis and the physiological role of ribulose bisphosphate carboxylase carbamylation. Bot Mag Tokyo 2:87–105
Sharkey TD (2005) Effects of moderate heat stress on photosynthesis: importance of thylakoid reactions rubisco deactivation, reactive oxygen species, and thermotolerance provided by isoprene. Plant Cell Environ 28:269–277
Sharkey TD, Zhang R (2010) High temperature effects on electron and proton circuits of photosynthesis. J Integr Plant Biol 52:712–722
Shen B, Jenesen RG, Bohnert HJ (1997) Mannitol protects against oxidation by hydroxyl radicals. Plant Physiol 115:527–532
Sehrawat A, Abat JK, Deswal R (2013) RuBisCO depletion improved proteome coverage of cold responsive S-nitrosylated targets in Brassica juncea. Front Plant Sci 4:342
Shou H, Bordallo P, Wang K (2004) Expression of the Nicotiana protein kinase (NPK1) enhanced drought tolerance in transgenic maize. J Exp Bot 55:1013–1019
Siebke K, Ghannoum O, Conroy JP, Badger MR, von Caemmerer S (2003) Photosynthetic oxygen exchange in C4 grasses: the role of oxygen as electron acceptor. Plant Cell Environ 26(12): 1963–1972
Sieger SM, Kristensen BK, Robson CA, Amirsadeghi S, Eng EW, Abdel-Mesih A, Møller IM, Vanlerberghe GC (2005) The role of AOX in modulating carbon use efficiency and growth during macronutrient stress in tobacco cells. J Exp Bot 6:1499–1515
Sierra-Almeida A, Casanova-Katny MA, Bravo LA, Corcuera LJ, Cavieres LA (2007) Comparing the photosynthetic responses to temperature and light between Antarctic and Andean populations of Colobanthus quitensis (Caryophylaceae). Rev Chil Hist Nat 80(3):335–343
Silim SN, Ryan N, Kubien DS (2010) Temperature responses of photosynthesis and respiration in Populus balsamifera L.: acclimation versus adaptation. Photosynthesis Res 104:19–30
Silva EA, Damatta FM, Ducatti C, Regazzi AJ, Barros RS (2004) Seasonal changes in vegetative growth and photosynthesis of Arabica coffee trees. Field Crops Res 89(2–3):349–357
Singh B, Usha K (2003) Salicylic acid induced physiological and biochemical changes in wheat seedlings under water stress. Plant Growth Regul 39:137–141
Singh CR, Curtis C, Yamamoto Y, Hall NS, Kruse DS, He H, Hannig EM, Asano K (2005) Eukaryotic translation initiation factor 5 is critical for integrity of the scanning preinitiation complex and accurate control of GCN4 translation. Mol Cell Biol 25(13):5480–5491
Skulachev VP (1999) Anion carriers in fatty acid mediated physiological uncoupling. J Bioenerg Biomembr 31(5):431–445
Smirnoff N (1995) Antioxidant systems and plant response to the environment. In: Smirnoff N, Scientific BIOS (eds) Environment and plant metabolism. Publishers Ltd., Oxford, pp 217–242
Smith BN, Jones AR, Hanson LD, Criddle RS (1999) Growth, respiration rate, and efficiency response to temperature. In: Handbook of plant and crop stress, 2nd edn. Marcel Dekker, New York, pp 417–440
Somersalo S, Kyei-Boahen S, Pehu E (1996) Exogenous glycine betaine application as a possibility to increase low temperature tolerance of crop plants. Nordisk Jordbruksforsk 78:102
Song Y, Chen Q, Ci D, Shao X, Zhang D (2014) Effects of high temperature on photosynthesis and related gene expression in poplar. BMC Plant Biol 14:111
Songstad DD, Duncan DR, Widholm JM (1990) Proline and polyamine involvement in chilling tolerance of maize suspension cultures. J Exp Bot 41:289–294
Sonoike K (1996) Photoinhibition of photo system I – its physiological significance in the chilling sensitivity of plants. Plant Cell Physiol 37(3):239–247

Sonoike K (1998) Various aspects of inhibition of photosynthesis under light/chilling stress. Photoinhibition at chilling temperatures versus chilling damage in the light. J Plant Res 111(1101):121–129

Sonoike K (1999) The different roles of chilling temperatures in the photoinhibition of photo system I and photo system II. J Photochem Photobiol B Biol 48(2–3):136–141

Sowiński P, Rudzińska-Langwald A, Adamczyk J, Kubica I, Fronk J (2005) Recovery of maize seedling growth, development and photosynthetic efficiency after initial growth at low temperature. J Plant Physiol 162:67–80

Srivastava S, Fristensky B, Kav NNV (2004) Constitutive expression of a PR10 protein enhances the germination of Brassica napus under saline conditions. Plant Cell Physiol 45:1320–1324

Stachula P (2015) Short and long term low temperature response in Arabidopsis thaliana. Umeå Plant Science Centre, Department of Plant Physiology, Umeå University, Sweden

Steponkus PL, Uemura M, Webb MS (1993) A contrast of the cryostability of the plasma membrane of winter rye and spring oat-two species that widely differ in their freezing tolerance and plasma membrane lipid composition. In: Steponkus PL (ed) Advances in low-temperature biology, vol 2. JAI Press, London, pp 211–312

Steward CR, Martin BA, Reding L, Cerwick S (1990) Respiration and alternative oxidase in corn seedling tissues during germination at different temperatures. Plant Physiol 92(3):755–760

Stewart JJ, Demmig-Adams B, Cohu CM, Wenzl CA, Muller O, Adams WW (2016) Growth temperature impact on leaf form and function in Arabidopsis thaliana ecotypes from northern and southern Europe. Plant Cell Environ 39:1549–1558

Still JR, Pill WG (2003) Germination, emergence and seedlings growth of tomato and impatiens in response to seed treatment with paclobutrazol. Hortic Sci 38:1201–1204

Stirling CM, Nie GY, Aguilera C, Nugawela A, Long SP, Baker NR (1991) Photosynthetic productivity of an immature maize crop: changes in quantum yield of CO2 assimilation, conversion efficiency and thylakoid proteins. Plant Cell Environ 14:947–954

Stitt M, Hurry V (2002) A plant foe all seasons: alteration in photosynthetic carbon metabolism during cold acclimation in Arabidopsis. Curr Opin Plant Biol 5:199–206

Stockinger EJ, Gilmour SJ, Thomashow MF (1997) Arabidopsis thaliana CBF1 encodes an AP2 domain-containing transcriptional activator that binds to the C-repeat /DRE, a cis-acting DNA regulatory element that stimulates transcription in response to low temperature and water deficit. Proc Natl Acad Sci U S A 94:1035–1040

Strain BR, Higginbotham KO, Mulroy JC (1976) Temperature preconditioning and photosynthetic capacity of Pinus taeda L. Photosynthetica 10:47–53

Strand A, Hurry V, Gustafsson P, Gardeström P (1997) Development of Arabidopsis thaliana leaves at low temperatures releases the suppression of photosynthesis and photosynthetic gene expression despite the accumulation of soluble carbohydrates. Plant J 12(3):605–614

Strand Å, Hurry V, Henkes S, Huner N, Gustafsson P, Gardeström P, Stitt M (1999) Acclimation of Arabidopsis leaves developing at low temperature: increasing cytoplasmic volume accompanies increased activities of enzymes in the Calvin cycle and in the sucrose-biosynthesis pathway. Plant Physiol 119(4):1387–1397

Strand A, Foyer CH, Gustafsson P, Hurry V (2003) Increased expression of sucrosephosphate synthase in transgenic Arabidopsis thaliana results in improved photosynthetic performance and increased freezing tolerance al low temperatures. Plant Call Environ 26:523–535

Strauss AJ, Krüger GHJ, Strasser RJ, Heerden PDR (2006) Ranking of dark chilling tolerance in soybean genotypes probed by the chlorophyll a fluorescence transient O-J-I-P. Environ Exp Bot 56(2):147–157

Strauss AJ, Kruger GHJ, Strasser RJ, van Heerden PDR (2007) The role of low soil temperature in the inhibition of growth and PSII function during dark chilling in soybean genotypes of contrasting tolerance. Physiol Plant 131(1):89–105

Streb P, Josse EM, Gallouët E, Baptist F, Kuntz M, Cornic G (2005) Evidence for alternative electron sinks to photosynthetic carbon assimilation in the high mountain plant species Ranunculus glacialis. Plant Cell Environ 28:1123–1135

Strzalka K, Gruszecki WI (1994) Effect of beta-carotene on structural and dynamic properties of model phosphatidylcholine membranes. I. an epr spin label study. Biochim Biophys Acta 1194(1):138–142

Subczynski WK, Markowska E, Sielewiesiuk J (1991) Effect of polar carotenoids on the oxygen diffusion-concentration product in lipid bilayers. An EPR spin label study. Biochim Biophys Acta 1068(1):68–72

Suh JP, Jeung JU, Lee JI, Choi YH, Yea JD, Virk PS, Mackill DJ, Jena KK (2010) Identification and analysis of QTLs controlling cold tolerance at the reproductive stage and validation of effective QTLs in cold-tolerant genotypes of rice (Oryza sativa L.). Theor Appl Genet 120(5): 985–995

Sun JD, Edwards GE, Okita TW (1999) Feedback inhibition of photosynthesis in rice measured by O2 dependent transients. Photosynthesis Res 59:187–200

Suzuki N, Mittler R (2006) Reactive oxygen species and temperature stresses: a delicate balance between signalling and destruction. Physiol Plant 126(1):45–51

Suzuki K, Nagasuga K, Okada M (2008) The chilling injury induced by high root temperature in the leaves of rice seedlings. Plant Cell Physiol 49(3):433–442

Szalai G, Janda T, Bartok T, Paldi E (1997) Role of light in changes in free amino-acid and polyamine contents at chilling temperature in maize (Zea mays). Physiol Plant 101(2):434–438

Taiz L, Zeiger E (2010) Plant physiology, 5th edn. Sinauer Associates, Sunderland

Takeda Y, Ogawa T, Nakamura Y, Kasamo K, Sakata M, Ohta E (1995) 31P-NMR study of the physiological conditions in intact root cells of mung bean seedlings under low temperature stress. Plant Cell Physiol 36(5):865–871

Takeuchi Y, Akagi H, Kamasawa N, Osumi M, Honda H (2000) Aberrant chloroplasts in transgenic rice plants expressing a high level of maize NADP-dependent malic enzyme. Planta 211:65–274

Talts P, Pärnik T, Gardeström P, Keerberg O (2004) Respiratory acclimation in Arabidopsis thaliana leaves at low temperature. J Plant Physiol 161(5):573–579

Tamburino R, Vitale M, Ruggiero A, Sassi M, Sannino L, Arena S, Costa A, Batelli N, Zambrano N, Scaloni A, Grillo S, Scotti N (2017) Chloroplast proteome response to drought stress and recovery in tomato (Solanum lycopersicum L.). BMC Plant Biol 17(1):40

Tanaka Y, Sano T, Tamaoki M, Nakajima N, Kondo N, Hasezawa S (2005) Ethylene inhibits abscisic acid-induced stomatal closure in Arabidopsis. Plant Physiol 138(4):2337–2343

Tardy T, Havaux M (1997) Thylakoid membrane fluidity and thermostability during the operation of the xanthophyll cycle in higher-plant chloroplasts. Biochim Biophys Acta 1330(2):179–193

Tasgin E, Atici O, Nalbantoglu B (2003) Effects of salicylic acid and cold on freezing tolerance in winter wheat leaves. Plant Growth Regul 41:231–236

Taub D (2010) Effects of rising atmospheric concentrations of carbon dioxide on plants. Nat Educ Knowl 3(10):21

Teige M, Scheikl E, Eulgem T, Doczi R, Ichimura K, Shinozaki K, Dangl J, Hirt H (2004) The MKK2 pathway mediates cold and salt stress signaling in Arabidopsis. Mol Cell 15:141–152

Terashima I, Noguchi K, Itohnemoto T, Park YM, Kubo A, Tanaka K (1998) The cause of PSI photoinhibition at low temperatures in leaves of Cucumis sativus, a chilling-sensitive plant. Physiol Plant 103(3):295–303

Tewari AK, Tripathy BC (1998) Temperature-stress-induced impairment of chlorophyll biosynthetic reactions in cucumber and wheat. Plant Physiol 117(3):851–858

Thakur P, Kumar S, Malik JA, Berger JD, Nayyar H (2010) Cold stress effects on reproductive development in grain crops: an overview. Environ Exp Bot 67:429–443

Tikkanen M, Grieco M, Nurmi M, Rantala M, Soursa M, Aro EM (2012) Regulation of the photosynthetic apparatus under fluctuating growth light. Phil Trans R Soc B 367(1608): 3486–3493

Tjus SE (2001) Active oxygen produced during selective excitation of photosystem i is damaging not only to photosystem I, but also to photosystem II. Plant Physiol 125(4):2007–2015

Thomashow MF (2001) So what's new in the field of plant cold acclimation? Lots! Plant Physiol 125:89–93

Thomashow MF (2010) Molecular basis of plant cold acclimation: insights gained from studying the CBF cold response pathway. Plant Physiol 154(2):571–577

Toth S, Nagy V, Puthur JT, Kovács L, Garab G (2011) The physiological role of ascorbate as leaves photosystem II electron donor: protection against photoinactivation in heat-stressed leaves. Plant Physiol 156(1):382–392

Tsonev T, Velikova V, Georgieva K, Hyde PF, Jones HG (2003) Low temperature enhances photosynthetic down-regulation in french bean (Phaseolus vulgaris L.) plants. Ann Bot 91(3): 343–352

Tsuchida H, Tamai T, Fukayama H, Agarie S, Nomura M, Onodera H, Ono K, Nishizawa Y, Lee BH, Hirose H, Toki S, Ku MSB, Miyao M (2001) High level expression of C4-specific NADP-malic enzyme in leaves and impairment of photoautotrophic growth in a C3 plant, rice. Plant Cell Physiol 42(2):138–145

Uemura M, Steponkus PL (1997) Effect of cold acclimation on membrane lipid composition and freeze induced membrane destabilization, plant cold hardiness. In: Li PH, Chen THH (eds) Plant cold hardiness. Springer, Boston, MA

Usuda H, Ku MSB, Edwards GE (1985) Influence of light intensity during growth on photosynthesis and activity of several key photosynthetic enzymes in a C4 plant (Zea mays). Physiol Plant 63(1):65–70

Valadabadi SA, Farahani HA (2010) Studying the interactive effect of potassium application and individual field crops on root penetration under drought condition. Asian J Agric Sci 1:19–24

Valadabadi SA, Shiranirad AH, Farahani HA (2010) Ecophysiological influences of zeolite and selenium on water deficit stress tolerance in different rapeseed cultivars. J Ecol Nat Environ 2(8):154–159

Val J, Baker NR (1989) Light-dependent chilling effects on phosphorylation of thylakoid proteins and consequences for associated photochemical activities in maize. Physiol Plant 77(3): 420–426

Van Camp W, Capiau K, Van Montagu M, Inze D, Slooten L (1996) Enhancement of oxidative stress tolerance in transgenic tobacco plants overproducing Fe-superoxide dismutase in chloroplasts. Plant Physiol 112:1703–1714

Van den Ende W, Valluru R (2009) Sucrose, sucrosyl oligosaccharides, and oxidative stress: scavenging and salvaging? J Exp Bot 60(1):9–18

Van Heerden PDR, Kruger GHJ, Loveland JE, Parry MAJ, Foyer CH (2003) Dark chilling imposes metabolic restrictions on photosynthesis in soybean. Plant Cell Environ 26(2):323–337

Vanlerberghe GC, Vanlerberghe AE, Mclntosh L (1997) Molecular genetic evidence of the ability of AOX to support respiratory carbon metabolism. Plant Physiol 11:657–661

Várkonyi Z, Masamoto K, Debrebzeny M, Zsiros O, Ughy B, Gombos Z, Domonkos I, Farkas T, Wada H, Szalontai B (2002) Low-temperature-induced accumulation of xanthophylls and its structural consequences in the photosynthetic membranes of the cyanobaterium Cylindrospermosis raciborskii: an FTIR spectroscopic study. Proc Natl Acad Sci U S A 99(4):2410–2415

Varshney RK, Kudapa H, Roorkiwal M, Thudi M, Pandey MK, Saxena RK, Chamarthi SK, Mohan SM, Mallikarjuna N, Upadhyaya H, Gaur PM, Krishnamurthy L, Saxena KB, Nigam SN, Pande S (2012) Advances in genetics and molecular breeding of three legume crops of semi-arid tropics using, next-generation sequencing and high-throughput genotyping technologies. J Bio Sci 37(5):811–820

Velikova V, Sharkey TD, Loreto F (2012) Stabilization of thylakoid membranes in isoprene-emitting plants reduces formation of reactive oxygen species. Plant Signal Behav 7:139–141

Venema JH, Posthumus F, De Vries M, Van Hasselt PR (1999a) Differential response of domestic and wild Lycopersicon species to chilling under low light: growth, carbohydrate content, photosynthesis and the xanthophyll cycle. Physiol Plant 105:81–88

Venema JH, Posthumus F, van Hasselt PR (1999b) Impact of suboptimal temperature on growth, photosynthesis, leaf pigments and carbohydrates of domestic and high-altitude wild Lycopersicon species. J Plant Physiol 155:711–718

Vercesi AE (2001) The discovery of an uncoupling mitochondrial protein in plants. Biosci Rep 21: 195–200

Verulkar SB, Mandal NP, Dwivedi JL, Singh BN, Sinha PK, Mahato RN, Dongre P, Singh ON, Bose LK, Swain P, Robin S, Chandrababu R, Senthil S, Jain A, Shashidhar HE, Hittalmani S, Cruz CV, Paris T, Haefele S, Serraj R, Atlin G, Kumar A (2010) Breeding resilient and productive genotypes adapted to drought-prone rainfed ecosystem of India. Field Crop Res 117(2–3):197–208

Vezza ME, Llanes A, Travaglia C, Agostinia E, Talanoa MA (2018) Arsenic stress effects on root water absorption in soybean plants: physiological and morphological aspects. Plant Physiol Biochem 123:8–17

Vinocur B, Altman A (2005) Recent advances in engineering plant tolerance to abiotic stress: achievements and limitations. Curr Opin Biotechnol 16:123–132

Vogg G, Heim R, Hansen J, Schäfer C, Beck E (1998) Frost hardening and photosynthetic performance of scots pine (Pinus sylvestris L.) needles. I. Seasonal changes in the photosynthetic apparatus and its function. Planta 204:193–200

von Caemmerer S, Furbank RT (1999) Modeling C4 photosynthesis. In: Sage RF, Monson RK (eds) C4 plant biology. Academic Press, san Diego, CA, pp 173–211

von Caemmerer S (2000) Biochemical models of leaf photosynthesis. CSIRO Publishing, Collingwood VIC, pp 91–122

Voss I, Sunil B, Scheibe R, Raghavendra AS (2013) Emerging concept for the role of photorespiration as an important part of abiotic stress response. Plant Biol 15(4):713–722

Wise RR (1995) Chilling-enhanced photooxidation: the production, action and study of reactive oxygen species produced during chilling in the light. Photosyn Res 45:79–97

Wahid A, Rasul E (2005) Photosynthesis in leaf, stem, flower and fruit. In: Pessarakli M (ed) Handbook of photosynthesis, 2nd edn. CRC Press, Boca Raton, FL, pp 479–497

Walcroft AS, Whitehead D, Silvester WB, Kelliher FM (1997) The response of photosynthetic model parameters to temperature and nitrogen concentration in Pinus radiata D. Don Plant Cell Environ 20(11):1338–1348

Wang W, Vinocur B, Altman A (2003) Plant responses to drought, salinity and extreme temperatures: towards genetic engineering for stress tolerance. Planta 218:1–14

Wang LJ, Li SH (2006) Salicylic acid-induced heat or cold tolerance in relation to Ca2+ homeostasis and antioxidant systems in young grape plants. Plant Sci 170:685–694

Wang C, Ma XL, Hui Z, Wang W (2008) Glycine betaine improves thylakoid membrane function of tobacco leaves under low temperature stress. Photosynthetica 46(3):400–409

Wang LJ, Fan L, Loescher W, Duan W, Liu GJ, Cheng JS, Luo HB, Li SH (2010) Salicylic acid alleviates decreases in photosynthesis under heat stress and accelerates recovery in grapevine leaves. BMC Plant Biol 10:Article no. 34

Wang Y, Hu J, Qin G, Cui H, Wang Q (2012) Salicylic acid analogues with biological activity may induce chilling tolerance of maize (Zea mays) seeds. Botany 90:845–855

Wang W, Chen Q, Hussain S, Mei J, Dong H, Peng S, Huang J, Cui K, Nie L (2016) Pre- sowing seed treatments in direct-seeded early rice: consequences for emergence, seedling growth and associated metabolic events under chilling stress. Sci Rep 6:19637

Waraich EA, Ahmad R, Ashraf MY, Saifullah, Ahmad M (2011a) Improving agricultural water use efficiency by nutrient management in crop plants. Acta Agric Scand Sect B Plant Soil Sci 61: 291–304

Waraich EA, Ahmad R, Ashraf MY (2011b) Role of mineral nutrition in alleviation of drought stress in plants. Aust J Crop Sci 5:764–777

Waraich EA, Ahmad R, Halim A, Aziz T (2012) Alleviation of temperature stress by nutrient management in crop plants: a review. J Soil Sci Plant Nutr 12:221–244

Ward DA (1987) The temperature acclimation of photosynthesis to CO2 in Zea mays and its relationship to the activity of photosynthetic enzymes and the CO2-concentrating mechanism of C4 photosynthesis. Plant Cell Environ 10(5):407–411

Wanner LA, Junttila O (1999) Cold induced freezing in Arabidopsis. Plant Physiiol 120:391–400

Watanabe T, Kume T (2009) A general adaptation strategy for climate change impacts on paddy cultivation: special reference to the Japanese context. Paddy Water Environ 7:313–320

Wendehenne D, Pugin A, Klessig DF, Durner J (2001) Nitric oxide: comparative synthesis and signaling in animal and plant cells. Trend Plant Sci 6:177–183

Whitmarsh AJ, Davis RJ (2000) Regulation of transcription factor function by photo-respiration. Cell Mol Life Sci 57:1172–1183

Wiese C, Shi LB, Heber U (1998) Oxygen reduction in the Mehler reaction is insufficient to protect photosystems I and II of leaves against photoinactivation. Physiol Plant 102(3):437–446

Wilkinson S, Clephan AL, Davies WJ (2001) Rapid low temperature-induced stomatal closure occurs in cold-tolerant Commelina communis L. leaves but not in cold-sensitive Nicotiana rustica L. leaves, via a mechanism that involves apoplastic calcium but not abscisic acid. Plant Physiol 126:1566–1578

Williams JP, Kahn MU, Mitchell K, Johnson G (1988) The effect of temperature on the level and biosynthesis of unsaturated fatty acids in diglycerols of Brassica napus leaves. Plant Physiol 87: 904–910

Wilson KE, Ivanov AG, Öquist G, Grodzinski B, Sarhan F, Huner NPA (2006) Energy balance, organellar redox status, and acclimation to environmental stress. Can J Bot 84(9):1355–1370

Wingler A, Lea PJ, Quick WP, Leegood RC (2000) Photorespiration: metabolic pathways and their role in stress protection. Philos Trans R Soc Lond Ser B Biol Sci 355(1402):1517–1529

Winter K, Königer M (1991) Dry matter production and photosynthetic capacity in Gossypium hirsutum L. under conditions of slightly suboptimum leaf temperatures and high levels of irradiance. Oecologia 87(2):190–197

Wolfe DW (1991) Low temperature effects on early vegetative growth, leaf gas exchange and water potential of chilling-sensitive and chilling-tolerant crop species. Ann Bot 67(3):205–212

Woodall AA, Britton G, Jackson MJ (1997) Carotenoids and protection of phospholipids in solution or in liposomes against oxidation by peroxyl radicals: relationship between carotenoid structure and protective ability. Biochim Biophys Acta 1336(3):575–686

Wu H, Hou L-L, Zhou Y-E, Fan Z-C, Shi J-Y, Rouzi A, Zhang J-S (2012) Analysis and evaluation indicator selection of chilling tolerance of different cotton genotypes. Agric Sci Tech Hunan 13(11):2338–2346

Wu G, Liu H, Hua L, Luo Q, Lin Y, He P, Feng S, Liu J, Ye Q (2018) Differential responses of stomata and photosynthesis to elevated temperature in two co-occurring subtropical forest tree species. Front Plant Sci 9:467

Xiao B, Huang Y, Tang N, Xiong L (2007) Overexpression of a LEA gene in rice improves drought resistance under the field conditions. Theor Appl Genet 115:35–46

Xie XJ, Shen SHH, Li YX, Xao XY, Li BB, Xu DF (2011) Effect of photosynthetic characteristic and dry matter accumulation of rice under high temperature at heading stage. Afr J Agr Res 6: 1931–1940

Xin Z, Li PH (1992) Abscisic acid-induced chilling tolerance in maize suspension-cultured cells. Plant Physiol 99:707–711

Xin Z, Browse J (2001) Cold comfort farm: the acclimation of plants to freezing temperatures. Plant Cell Environ 23:893–902

Xing W, Rajashekar CB (2001) Glycinebetaine involvement in freezing tolerance and water stress is Arabidopsis thaliana. Environ Exp Bot 46:21–28

Xiong FS, Ruhland CT, Day TA (1999) Photosynthetic temperature response of the Antartic vascular plants Colobanthus quitensis and Deschampsia antartica. Physiol Plant 106:276–286

Xiong FS, Mueller EC, Day TA (2000) Photosynthetic and respiratory acclimation and growth response of Antarctic vascular plants to contrasting temperature regimes. Am J Bot 87:700–710

Xiong FS, Day TA (2001) Effect of solar ultraviolet-B radiation during springtime ozone depletion on photosynthesis and biomass production of Antarctic vascular plants. Plant Physiol 125:738–751

Xu S, Hu J, Li Y, Ma W, Zheng Y, Zhu S (2011) Chilling tolerance in Nicotiana tabacum induced by seed priming with putrescine. Plant Growth Regul 63:279–290

Xu Z, Jiang Y, Jia B, Zhou G (2016) Elevated-CO2 response of stomata and its dependence on environmental factors. Front Plant Sci 7:657
Yadegari LZ, Heidari R, Carapetian J (2008) Chilling pretreatment causes some changes in respiration, membrane permeability and some other factors in soybean seedlings. Res J Biol Sci 3(9):1054–1059
Yamada M, Morishita H, Urano K, Shiozaki N, Yamaguchi-Shinozaki K, Shinozaki K, Yoshiba Y (2005) Effects of free proline accumulation in petunias under drought stress. J Exp Bot 56: 1975–1981
Yamaguchi-Shinozaki K, Shinozaki K (2006) Transcriptional regulatory networks in cellular responses and tolerance to dehydration and cold stresses. Annu Rev Plant Biol 57:781–803
Yamasaki T, Yamakawa T, Yamane Y, Koike H, Satoh K, Katoh S (2002) Temperature acclimation of photosynthesis and related changes in photosystem II electron transport in winter wheat. Plant Physiol 128(3):1087–1097
Yamori W, Noguchi K, Hikosaka K, Terashima I (2009) Cold-tolerant crop species have greater temperature homeostasis of leaf respiration and photosynthesis than cold-sensitive species. Plant Cell Physiol 50(2):203–215
Yan W, Zhong Y, Shangguan Z (2017) Contrasting responses of leaf stomatal characteristics to climate change: a considerable challenge to predict carbon and water cycles. Glob Chang Biol 104:145–157
Yang XL, Xu H, Li D, Gao X, Li TL, Wang R (2018) Effect of melatonin priming on photosynthetic capacity of tomato leaves under low-temperature stress. Photosynthetica 56:884–892
Ye C, Fukai S, Godwin I, Reinke RB, Snell PB, Schiller J, Basnayake J (2009) Cold tolerance in rice varieties at different growth stages. Crop Past Sci 60:328–338
Yokoi S, Quintero FJ, Cubero B, Ruiz MT, Bressan RA, Hasegawa PM, Pardo JM (2002) Differential expression and function of Arabidopsis thaliana NHX Na+ /H+ antiporters in the salt stress response. Plant J 30(5):529–539
Yordanova R, Popova L (2007) Effect of exogenous treatment with salicylic acid on photosynthetic activity and antioxidant capacity of chilled wheat plants. Gen Appl Plant Physiol 33:155–170
Yoshida R, Kanno A, Sato T, Kaтeya T (1996) Cool temperature-induced chlorosis in rice plants. 1. Relationship between the induction and a disturbance of etioplast development. Plant Physiol 110(3):997–1005
Yoshimura K, Miyao K, Gaber A, Takeda T, Kanaboshi H, Miyasaka H, Shigeoka S (2004) Enhancement of stress tolerance in transgenic tobacco plants overexpressing Chlamydomonas glutathione peroxidase in chloroplasts or cytosol. Plant J 37(1):21–33
Zhang J, Kirkham MB (1996) Antioxidant response to drought in sunflower and sorghum seedlings. New Phytol 132:361–373
Zhang S, Klessig DF (2001) MAPK cascades in plant defense signaling. Trends Plant Sci 6:520–527
Zhang HX, Hodson JN, Williams JP, Blumwald E (2001) Engineering salt-tolerant brassica plants: characterization of yield and seed oil quality in transgenic plants with increased vacuolar sodium accumulation. Proc Natl Acad Sci U S A 98:12832–12836
Zhang JZ, Creelman RA, Zhu JK (2004) From laboratory to field. Using information from Arabidopsis to engineer salt, cold and drought tolerance in crops. Plant Physiol 135:615–621
Zhang S, Scheller HV (2004) Photoinhibition of photosystem I at chilling temperature and subsequent recovery in Arabidopsis thaliana. Plant Cell Physiol 45:1595–1602
Zhang C, Tian S (2009) Crucial contribution of membrane lipids unsaturation to acquisition of chilling-tolerance in peach fruit stored at 0°C. Food Chem 115:411–415
Zhang LT, Zhang SS, Gao HY, Xue ZC, Yang C, Meng XL, Meng QW (2011) Mitochondrial alternative oxidase pathway protects plants against photoinhibition by alleviating inhibition of the repair of photodamaged PSII through preventing formation of reactive oxygen species in Rumex K-1 leaves. Physiol Plant 143:396–407

Zhang K, Wang J, Lian L, Fan W, Guo N, Lv S (2012) Increased chilling tolerance following transfer of a betA gene enhancing glycinebetaine synthesis in cotton (Gossypium hirsutum L.). Plant Mol Biol Reprod 30(5):1158–1171

Zhang XD, Wang RP, Zhang FJ, Tao FQ, Li WQ (2013) Lipid profiling and tolerance to low-temperature stress in Thellungiella salsuginea in comparison with Arabidopsis thaliana. Biol Plant 57:149–153

Zhao M, Liu W, Xia X, Wang T, Zhang WH (2014) Cold acclimation-induced freezing tolerance of Medicago truncatula seedlings is negatively regulated by ethylene. Physiol Plant 152:115–129

Zauralov OA, Lukatkin AS (1997) Aftereffect of low temperatures on the respiration of heat-loving plants. Rus J Plant Physiol 44(5):640–644

Zheng Y, Xu M, Hou R, Shen R, Qiu S, Ouyang Z (2013) Effects of experimental warming on stomatal traits in leaves of maize (Zea mays L.). Ecol Evol 3:3095–3111

Zheng M, Tao Y, Hussain S, Jiang Q, Peng S, Huang J, Cui K, Nie L (2016) Seed priming in dry direct-seeded rice: consequences for emergence, seedling growth and associated metabolic events under drought stress. Plant Growth Regul 78(2):167–178

Zhou S, Chen FC, Nahashon S, Chen T (2006) Cloning and characterization of glycolate oxidase and NADH-dependent hydroxypyruvate reductase genes in Pachysandra terminalis. HortSci 41(5):1226–1230

Zhou X, Chen S, Wu H, Xu H (2017). Effects of cold stress on the photosynthesis and antioxidant system of rhododendron chrysanthemum pall. Preprints, 2017030131

Zhu XG, de Sturier E, Long SP (2007) Optimizing the distribution of resources between enzymes of carbon metabolism can dramatically increase photosynthetic rate: a numerical simulation using an evolutionary algorithim. Plant Physiol 145(2):513–526

Ziegler P (1998) Environmental stress and stress responses plants. In: The co-action between living system and the planet. University of Genava, Genava, pp 67–100

Živčák M, Brestič M, Olšovská K, Slamka P (2008) Performance index as a sensitive indicator of water stress in Triticum aestivum L. Plant Soil Environ 54(4):133–139

Low-Temperature Stress and Nitrogen Metabolism in Plants: A Review

4

Abstract

Cold stress can lead to major crop losses. In natural ecosystems, there are several inputs of soil nitrogen. Even though our atmosphere consists of nearly 80% of N_2, it is the limiting element in most ecosystems. Bio-available nitrogen can enter an ecosystem via atmospheric nitrogen deposition, mostly in the form of ammonium and nitrogen oxides. Larger input of nitrogen comes from nitrogen fixed by microorganisms, called diazotrophs. Nitrogen input from legumes can be a sustainable source of nitrogen in agricultural systems. The symbiotic microorganisms in the root nodules, the rhizobia, can take up gaseous di-nitrogen from the air and "fix" the nitrogen into molecules that can be assimilated by the plant. In return, the plant provides the rhizobia with a carbon source in the form of di-carboxylic acids. The enzyme responsible for the nitrogen fixation, nitrogenase, is irreversibly damaged when exposed to oxygen. The use of nitrogen from soil by plants involves several steps, including uptake, assimilation, translocation, and, when the plant is aging, recycling and remobilization. Although generally low, soil nitrogen availability can fluctuate greatly in both space and time due to factors such as precipitation, temperature, wind, soil type, and pH. Therefore, the preferred form in which nitrogen is taken up depends on plant adaptation to soil conditions. Generally, plants adapted to low pH and reducing soils as found in mature forests or arctic tundra tend to take up ammonium or amino acids, whereas plants adapted to higher pH and more aerobic soils prefer nitrate. Nitrate is easily transported in the plant and can be assimilated both in leaves and roots. The partitioning of nitrate reduction between roots and shoots may vary with plant species, age, and environmental factors. After nitrate reduction, nitrite is translocated to the chloroplast, where it is reduced to ammonium by the second enzyme of the pathway, the nitrite reductase. Ammonium, originating from nitrate reduction, and also from

A. Bhattacharya, *Physiological Processes in Plants Under Low Temperature Stress*, https://doi.org/10.1007/978-981-16-9037-2_4

photorespiration or amino acid recycling, is mainly assimilated in the plastid/chloroplast by the so-called GS/GOGAT.

Keywords

Nitrogen cycle · Biological nitrogen fixation · Factors affecting biological nitrogen fixation · Nitrate reduction · Ammonia assimilation · Nitrogen metabolism · Cold acclimation and freezing tolerance · Role of free amino acid · Role of γ-aminobutyric acid · Low-temperature stress and protein synthesis · Protein management opportunities

4.1 Low-Temperature Stress and Metabolic Response in Plants

Only one-third of the total land area on Earth is free of ice, and 42% of land experiences temperatures below −20 °C (Ramankutty et al. 2008). In such areas, plants require specialized mechanisms to survive exposure to low temperature. Cold stress can be classified as chilling (0–15 °C) and freezing (<0 °C) stresses. Generally, plants originating from temperate regions, such as spinach and *Arabidopsis*, exhibit a variable degree of chilling tolerance and can increase their freezing tolerance during exposure to chilling and nonfreezing temperatures. Plants often encounter a wide range of environmental stress conditions, such as drought, salt, and extreme temperatures. Climatic changes, especially in the fall and winter months, can adversely affect plant health and potentially lead to plant death. Cold acclimation or the exposure to nonfreezing temperatures prior to freezing stress is a process that plants use to acquire freezing tolerance and mitigate cold injury (Janská et al. 2010). During this process, plants redirect their metabolism to accumulate cryoprotectant compounds or stress-responsive elements. Fluctuating or increased temperatures prior to freezing can disrupt cold acclimation processes and accelerate injury through cellular dehydration, reduced metabolic functions, or cellular membrane rupture. Damage sustained due to the disruption of acclimation, or de-acclimation, can then inhibit photosynthesis and other metabolic processes, thus limiting or preventing growth and development upon warmer temperatures (Park et al. 2006). De-acclimation and freezing injury have been observed in many different plant species, including turfgrasses (Webster and Ebdon 2005; Espevig et al. 2011, 2014). Winter injury can greatly diminish turfgrass performance for many species, and reestablishment of damaged turf can be costly (Webster and Ebdon 2005). It is predicted that climate change will continue to occur at a dangerous rate, and rapid de-acclimation can potentially lead to the extinction of plant species, which makes it necessary to better understand the cold acclimation process (Root et al. 2003). Increased knowledge on the mechanisms in which C3 and C4 plants undergo cold acclimation can provide insight to aid in the development of more cold-tolerant species or provide improved management practices to reduce low-temperature injury. Low-temperature stress is one of the most important environmental factors that affect the growth and development of plants and limit the distribution of plant and crop productivity (Theocharis et al. 2012; Chen et al. 2014). Extreme climate events have become more frequent over the last two decades. The plant itself has evolved a series of protective mechanisms to cope with low temperature and to maintain a metabolic balance of substance (Zhu et al. 2015) and energy in response to an adverse environment (Wang et al. 2013). Moreover, symbiotic interactions may represent an effective strategy for improving tolerance to low-temperature stress in plants (Duhamel and Vandenkoornhuyse 2013).

Every plant has some requirement of some physical factors, viz., light, temperature, air, water, etc., and for some chemical factors, for example, inorganic nutrients and organic nutrients in optimum amount and in the available form at the right time as its nutrients for its proper growth and development. Any deviation from this optimum condition of any factor essential for its growth will lead to an aberrant change in physiological processes, and due to this, the plant body will experience tension and this state can be referred to as plant under stress. When this condition prevails for some time, stress causes injury with or without producing visible symptoms, which can be referred to as strain. The injury occurs as a result of aberrant metabolism and may be expressed as a reduction in growth, value, or death of two plants.

- Zero stress: In concept, this is the level of exposure to an environmental factor that leads to injury or to a reduction in growth, yield, or value.
- Temperature stress: Temperature on earth changes along with latitudinal and altitudinal clines, and also with the season, most plants are specially adopted to grow over a limited range of temperatures (Sonoike 1998). Extremes of temperatures limit plant survival and its reproduction, and therefore temperature is required as one of the strongest determinants of plants' growth and metabolism (Smith et al. 1999). The extremes of temperatures, either very high or very low, both cause severe changes in the plant metabolism machinery.

Both these conditions are stressful for the plant growth and disturb plants' physiology and biochemical process (Dwivedi and Raghvendra 2004). All plant species perform their metabolic arthritis and complete the life cycle under an optimum temperature range (Kumar et al. 2018). Among the various climatic factors vital for the growth of autotrophic plants, temperature is the crucial one, besides light, water, atmospheric CO_2, minerals, and oxygen. Temperature, in a true sense, dictates the rate at which metabolic activity proceeds in plants, and it must be present at a certain level of warmth in plants for productive metabolism (Ziegler 1998).

Any change in temperature affects plants greatly, so stress imposed by temperature has important implications for agriculture; for example, it has been reported that a 1 °C decrease in the world mean temperature. Temperature stress can be categorized into two, viz., low- and high-temperature stress. Low-temperature stress has been defined as the temperatures considerably lower than the optimal growth temperatures resulting in temperature stress in plants. In the majority of plants, this low temperature is usually below 1 °C; this can further be subdivided into two main categories, depending on whether this low temperature leads to the formation of ice within the tissues or not. Both these conditions cause water deficit inside the plant (Pearce 1999).

Low temperature and frost have a negative effect on the metabolism with a corresponding reduction in crop quality and quantity. Chilling resistance and freeing resistance are complex phenomenon, as most plants are capable of hardening to cold, i.e., acquisition of increased resistance upon exposure to low temperature. These results in irreversible and proper honor loss of cell proteins and disturbed plasma membrane integrity. Tropical plants are more susceptible to chilling than those growing in cold regions. When exposed to temperature higher than 12–13 °C, an altered nitrate uptake and its metabolism are observed under low temperature. There becomes reduced nitrogen uptake (Chandra and Pareak 1990), decreased nitrogen partitioning in young shoots, induced remobilization of nitrogen from older to younger leaves, and inhibition in nitrogen assimilation process at low temperatures. Low-temperature treatment modulates the effect of hyper-osmoticum stress on photosynthesis and respiration and results in increased participating of alternative pathway (Dwivedi and Raghvendra 2004). The crown of the plant, consisting of shoot apex and young leaves, is much colder hardy and can tolerate low temperatures better than the more mature older parts. Low-temperature stress also manifests increased oxidative damage in plants. There is enhancement of antioxidant defense mechanism of the plant under low-temperature stress.

As a major response, plant under low temperature stress shows enhanced activities and content of scavenging enzyme and antioxidant (Shah et al. 2001; Sylvie et al. 1997). As chilling and freezing temperatures cause water deficit inside the cells, the compatible solute mannitol, which alleviates water stress, has been shown to protect chloroplast phosphokinase from hydroxyl radical damage under stressful conditions (Shen et al. 1997). Plant accumulates certain compatible nitrogenous compounds specific to a particular stress. These compounds help to combat that particular stress condition (Choudhary and Singh 2000; Sakamoto and Murata 2000, 2001). The most common nitrogenous compounds that accumulate upon chilling injury are proline and glycine betaine (Borsani et al. 1999). In barley and radish, substantial amounts of proline, serine, glycine, and alkaline compounds accumulate on exposure to low temperatures (Nandwal et al. 2000). Low temperature causes reduced nitrate uptake by the roots (Dubey and Pessarakli 2002). Soybean plants when exposed to 15 °C for 4 days showed about a 52–61% decrease in nitrogen partitioning in the young shoots compared to plants grown at 25 °C. It is suggested that tolerance to low temperature can be increased by increasing nitrogen supply to young shoots. An enhancement in the level of soluble protein during the cold hardening was accompanied by increased activities of nitrate reeducates and glutamine synthetase in many plant species, including wheat, barley, sugar beet, rose, sunflower, etc. (Gusta et al. 1999). Nitrate reductase shows species-dependent activity when the plant is exposed to low temperature (Dubey and Pessarakli 2002). Since nitrate reductase is a substrate inducible enzyme and depends on the level of nitrates in the active pool, low-temperature stress along with low soil pH influenced nitrate reduction by significant reduction in nitrate reductase in actively growing plants (Gusta et al. 1999). Reduced nitrate uptake under low temperature would thus lead to a decrease in nitrate reductase activity.

There are certain types of protein that are either produced or induced under low temperature, and these proteins are named after the inducing agent, e.g., early dehydration inducible, low temperature induced, low temperature responsive, or cold responsive (Thomashow 1998). These compounds exhibit ability to arrest the growth of plant pathogens, and levels of certain low-temperature proteins have been shown to increase in winter barley following exposure to low temperature, i.e., above freezing hardening temperature (2 °C), both in the field and under controlled environment conditions (Gaudet et al. 2003; Pearce 1999).

Carbon and nitrogen metabolisms are fundamental processes in plants. Metabolic processes of nitrogen, including nitrogen uptake, transport, reduction and assimilation, and amino acid metabolism (Kusano et al. 2011; Luo et al. 2013), are affected by both symbiosis and low-temperature stress. Several studies have demonstrated that arbuscular mycorrhiza fungi can take up and transfer significant amounts of nitrogen (accounting for 20–50% of the total root nitrogen) to the host plant (Govindarajulu et al. 2005; Pérez-Tienda et al. 2012; Ngwene et al. 2013). Toussaint et al. (2004) demonstrated that the activities of key enzymes involved in nitrogen assimilation, such as glutamine synthetase (EC 6.3.1.2) and glutamate dehydrogenase (EC 1.4.1.3), in arbuscular mycorrhiza carrot roots were greater than those in nonarbuscular mycorrhiza plants. Liu et al. (2013a) reported that arbuscular mycorrhiza rice plants under low-temperature stress had higher nitrogen content and activities of nitrate reductase and glutamine synthetase than nonarbuscular mycorrhiza plants. However, little is known about the nitrogen metabolism of arbuscular mycorrhiza colonization association under low-temperature stress.

Carbon metabolism processes involve photosynthetic carbon assimilation, sucrose and starch metabolism, and carbohydrate transport and utilization. When plants are grown under low temperatures, a series of physiological processes associated with carbon metabolism are altered, such as inhibition of photosynthesis, changes in carbohydrate content, and carbohydrate metabolic enzyme activity (Theocharis et al. 2012). Zhu et al. (2010a) demonstrated that colonization with *Glomus etunicatum* increased the photosynthetic performance of maize seedlings under low-temperature stress. Under arbuscular mycorrhiza symbiosis, the sugar content was also altered under low-temperature stress (Zhu et al. 2010b; Liu et al. 2013a). Furthermore, Liu et al. (2013a) reported that arbuscular mycorrhiza rice had higher sucrose synthase (EC 2.4.1.13) and sucrose phosphate synthase (EC 2.4.1.14) activities than nonarbuscular mycorrhiza plants at low temperature. Thus, knowledge regarding the effects of arbuscular mycorrhiza on plant carbon metabolism under low-temperature stress is limited. It is well known that plant carbon and nitrogen metabolisms are tightly coordinated in nearly every biochemical pathway (Coruzzi and Bush 2001). Assimilation of nitrogen requires energy and carbon skeletons produced by carbon metabolism, and assimilation of photosynthetic carbon requires a large amount of nitrogen (Nunes-Nesi et al. 2010). In arbuscular mycorrhiza symbiosis, the transfer and metabolism of nitrogen are accompanied by the biosynthesis and flux of carbon from plants to arbuscular mycorrhiza fungi (Liu et al. 2013a). Fellbaum et al. (2012) demonstrated that the flux of carbon acts as a key trigger to regulate the uptake and transport of nitrogen in arbuscular mycorrhiza symbiosis. However, the interaction between carbon and nitrogen metabolisms in arbuscular mycorrhiza plants has not been illustrated to date, particularly at low temperature, although this knowledge is essential for understanding the regulatory mechanisms of carbon and nitrogen metabolism in arbuscular mycorrhiza symbiosis (Zhu et al. 2015).

4.2 Nitrogen Cycle

More than one-fourth of terrestrial surface is exposed to low temperature, one-third is represented by dry habitats including saline, alkaline environments, and dry sea beds. The main limitations of the denitrification in agriculture, including soils, availability of O_2, water, and NO_3^-, are coupled to interplays of temperature, pH, salinity, and severe desiccation in more unusual habitats. Denitrifying microorganisms can adapt to low or high pH, take advantage of sporadic rain events in deserts, maintain their activity at a temperature below the freezing point of water, survive multiple freeze-thaw cycles, and take advantage of environmental conditions in hypersaline environments. The nitrogen cycle sustains life by converting inert di-nitrogen gas (N_2) into fixed bio-available forms (e.g., ammonium, nitrate), as well as returning it via gases such as N_2 and nitrous oxide (N_2O) back into the atmosphere. Effects of long-term warming on key components of the carbon cycle, which is tightly coupled to the nitrogen cycle, have been highlighted, but how global warming might systematically affect the balance of the nitrogen cycle is still largely unknown. The effects of long-term warming on denitrification and nitrification were investigated using long-term, experimental mesocosm, allowing us to study the effect of warming on natural communities of bacteria involved in these processes. Denitrification activity responded to warming in the short-term in a predictable way; however, long-term moderate warming of 3–5 °C increased the specific activity of the sediment and had a pronounced effect on the ratio of N_2O to N_2. Molecular analysis of denitrifying communities in our long-term mesocosm experiment also suggested a profound alteration of the communities underlying these differences in the process. It has been indicated that we may see similar effects on the nitrogen cycle as we have previously determined in the carbon cycle, with the balance of nitrogen species consumed and created becoming out of balance.

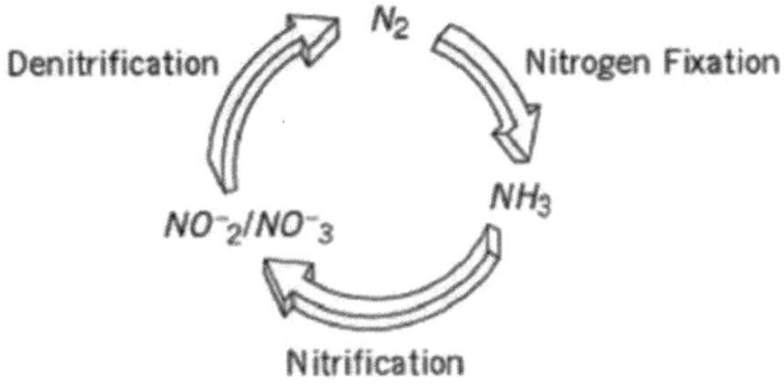

4.2.1 Nitrogen Fixation

With the exception of water-associated hydrogen and oxygen, all soil-derived nutrients (nitrogen, phosphorus, potassium, calcium, iron, magnesium, etc.) are introduced initially into terrestrial food webs through the weathering of soil minerals. While this does occur with nitrogen, however, soil mineral weathering contributes typically a minor addition that by itself would be insufficient to produce the productivity found in terrestrial plant communities. The principal nitrogen input to terrestrial ecosystems stems from the ability of certain bacteria to transform atmospheric nitrogen into ammonia, which the bacteria can then incorporate into organic nitrogen compounds. This nitrogen fixation occurs asymbiotically or symbiotically with plants and, in the case of lichens, with fungi. Free-living asymbiotic, nitrogen-fixing heterotrophs (including *Azotobacter and Clostridium*) use organic compounds as their energy source. The majority of these species require aerobic conditions, although some are facultative, surviving under limited oxygen levels. While the rate of asymbiotic nitrogen fixation is low per hectare, it is significant on a global basis. In addition, there exist free-living nitrogen fixing photoautotrophs: cyanobacteria, which are found in soil and aquatic systems. Other nitrogen fixers include those that are associated with grasses where they occur in the plant rhizosphere (the area of soil surrounding the plant root) and on leaf surfaces.

In symbiotic relationships between bacteria and plants, energy- and nutrient-containing organic compounds are supplied by the plant to the bacteria. This increases the efficiency of symbiotic bacterial nitrogen fixation and ammonia production. Any surplus ammonia is excreted by the bacterium and acquired by the host plant. Leguminous plants, including those used as food crops (e.g., soybeans, beans, and peas) and cover crops (e.g., clovers and alfalfa), are the plants best known for their symbiotic relationship with nitrogen-fixing bacteria. In these plants, species of *Rhizobium* infect the roots, triggering the formation of nodules that the bacteria then occupy. Symbiotic relationships also exist between nitrogen-fixing microbes and nonleguminous plant species. The nitrogen-fixing actinomycete, *Frankia*, forms symbiotic relationships with 170 known species within eight plant families. Globally, symbiotic nitrogen fixation represents 8.5% of the nitrogen that is taken up by plants. The addition of nitrogen through fertilizer applications is about 50% of this amount. Due to low efficiency in the transfer of fertilizer nitrogen to plants, however, this represents approximately 2% of the total plant nitrogen uptake (Paul and Clark 1989).

While the plant is surrounded by an atmosphere that is rich in nitrogen, plant uptake of nitrogen occurs through the roots, and both ammonia and nitrates are forms that are taken up. Within the plant, ammonia is used to form proteins, with energy being required for the necessary chemical transformations. Nitrate is used to form proteins as well, but first, it must be reduced to ammonia in a process that requires additional energy to be used by the plant. Consequently, plant uptake of ammonia is preferable to that of nitrate because it requires less energy expenditure by the plant to transform it into organic nitrogen forms. However, nitrate is more mobile than ammonia within the soil, which produces a greater delivery of it to the root. Within the plant, this nitrogen can be incorporated into proteins and nucleic acids, becoming part of cellular constituents. Parasitism and herbivory can then transfer these organic nitrogen compounds from the plant to the consumers of the food chain. Predation can transfer the nitrogen to higher trophic levels. At all levels within the food chain, through death, litter formation, or production of animal waste (urine or feces), nitrogen-containing organic compounds along with other organic molecules are transferred to the soil. Once there, the decomposer food web uses this material for its energy and nutrient (including nitrogen) needs.

4.2.2 Decomposition

Following the introduction of litter to the soil, amino sugars and nucleic acids in the litter can be used as energy sources or building blocks for other organic molecules by the soil decomposers. Consequently, nitrogen in these molecules may become part of the soil decomposer food web. Alternatively, nitrogen that is contained in these molecules may be transformed by the microbes to ammonia that can be released to the soil through a general process termed mineralization. This inorganic form of nitrogen can be taken up by plants and other microbes or immobilized by clay minerals. While the majority of soil organic nitrogen is present in decomposing plant and microbial matter, up to 40% occurs in by-products of decomposition that are somewhat resistant to further decomposition. Having been altered greatly by the complexities of decomposition, the primary biological source of this mixture of principally phenolic compounds is unrecognizable.

4.2.3 Nitrification

Nitrification consists of the stepwise transformation of ammonia to nitrite and then to nitrate by a group of chemoautotrophic bacteria. These bacteria, which are obligate aerobes, obtain their energy through the oxidation of ammonia or nitrite and their carbon from carbon dioxide or soil carbonates. Two groups of oxidizers are involved in nitrification: the nitrite formers include members of the *Nitrosoccus* and

Nitrobacter genera and the nitrate formers are *Nitrobacter*. In this latter oxidative transformation, oxygen is taken from water and not oxygen. Heterotrophic nitrification can also occur in soils but is considered insignificant in agricultural settings. A variety of environmental factors influences the rate of nitrification through effects on the activity of nitrifying bacteria. The optimum soil acidity range for nitrifying organisms is pH 6.6–8.0, with nitrifying rates dropping off on either side of this range. With nitrifying bacteria being obligate aerobes, oxygen availability in the soil is of great importance to their functioning. Consequently, soil moisture content exerts a strong influence on nitrification.

Nitrification rates are typically low below 5 °C and above 4 °C, and the optimum temperature range is 30–35 °C. In temperate regions, through a combination of the influences of moisture and temperature, soil nitrification is greatest in the spring and the fall (Paul and Clark 1989). Following the production and release of nitrate into the soil through nitrification, there are several things that can happen. Nitrate can be taken up by organisms to fill their nutritional need for nitrogen. Within organisms, multiple biochemical reactions of the assimilatory reduction biological pathway transform nitrate to ammonia that then can be processed into proteins. Unlike ammonia, nitrate is not readily adsorbed by clay minerals. As a consequence, it can be leached from the rooting zone in soils. This leaching reduces soil nitrogen availability to plants and can lead to the export of soil nitrogen to the ground water and, subsequently, to surface waters.

4.2.4 Denitrification

Nitrate, due to its high water-soluble characteristic, is possibly the major nitrogen contaminant in the water (Guo et al. 2013; Tang et al. 2010). High concentration nitrate could contribute to water eutrophication (Bhatnagar and Sillanpää 2011) and even impose a serious threat to human health, such as malformation, carcinoma, and mutation when the nitrate transformed into nitrosamines (Zhao et al. 2011). The World Health Organization (WHO) has recommended that the nitrate concentration in drinking water should be lower than 10 mg/L (Xing et al. 2016). The nitrate remediation, thereby, is a big challenge and has become a matter of great concern in recent years (He et al. 2018). Research studies have shown that the most common methods to remove nitrate from wastewater included biological denitrification with microorganisms and physicochemical reduction using ion exchange, electrodialysis, reverse osmosis, zero-valent iron, and zero-valent magnesium (Bhatnagar and Sillanpää 2011; Sierra-Alvarez et al. 2007). Several reports demonstrated that bioremediation has much higher efficiency, has lower cost, and is easier to implement compared with physicochemical methods for nitrate removal from wastewater (Leyva-Díaz et al. 2015; Shrimali and Singh 2001). Researchers discovered that biological denitrification is the most promising approach because the bacteria could reduce the nitrate to the harmless nitrogen gas (Matějů et al. 1992). Conversely, under anaerobic soil conditions, nitrate can be used as an electron acceptor during cellular respiration of certain bacteria. These denitrifying bacteria are principally heterotrophic; common forms include *Pseudomonas*, *Bacillus*, and *Alcaligenes*. In a series of reactions, ammonia and other forms of reduced nitrogen molecules can be produced. In enzymatic denitrification, nitrate is first transformed to nitrites and then to nitrous oxide and nitrogen. In dissimilatory reduction, nitrate is also used as an electron acceptor, and ammonia, but not gaseous nitrogen forms, is produced (Paul and Clark 1989).

As with both nitrogen fixation and nitrification, environmental conditions exert a significant influence on the rate of denitrification. Sufficient soil water is required to produce the necessary anaerobic soil conditions. The optimal range of soil pH for denitrifying bacteria is pH 6–8, and it can occur down to pH 4. As with other microbial reactions, the rate of denitrification increases exponentially with temperature increase and occurs from 5 to 7.5 °C. Lastly, a simple statistic can be used to emphasize the importance of this microbial process in agricultural soils. On average, 15% of the nitrogen in fertilizer is lost to the atmosphere through denitrification (Paul and Clark 1989).

4.2.5 Biogeochemical Linkage of Nitrogen and Carbon

While the above-mentioned bacterial processes exert a great influence on the soil chemistry of nitrogen, an understanding of the nitrogen cycle is incomplete if its relationship with carbon is not considered. The stoichiometric relationship between nitrogen and carbon in the organic molecules of living tissue is of great importance as a factor that influences the processes of decomposition and, consequently, the biological cycling of soil nitrogen and carbon. Although highly variable, the C:N ratio in trees, crops, and grasses averages 50:1 (Paul and Clark 1989). In plants, while this ratio for the cytoplasm is substantially lower, high amounts of carbon-rich, nitrogen-poor organic compounds of the cell wall dramatically increase the whole-plant C:N ratio to the high value just stated. This ratio in plant tissue has an important influence on food chain transfers. For proper health, herbivores require a minimum amount of nitrogen in their food. As a consequence, it benefits herbivores to be selective in their consumption of plant species or specific tissues that do not contain overly high C:N ratio compounds.

As mentioned, the C:N ratio of dead biomass also influences soil decomposition and the resultant mineralization of nitrogen. Agricultural soils have approximately 45% carbon and 6–9% nitrogen, which gives a C:N ratio of 10:1. Typical crop residues have 40% carbon and 1.6% nitrogen to produce a C:N ratio of 25:1. The C:N ratio of soil organisms is lower, ranging from 5:1 to 8:1. For the different groups of soil microorganisms, the C:N ratio also varies due to the relative proportions of carbon and nitrogen in the cytoplasm and cell wall components (peptidoglycan in bacteria and chitin in fungi). Fungi contain approximately 45% carbon and 3–10% nitrogen to yield a C:N ratio with a range of 4.5:1 to 15:1. For bacteria, the ratio ranges from 3:1 to 5:1 (Paul and Clark 1989).

In agricultural soils having the typical crop residue C:N ratio of 25:1, the nitrogen needs of the decomposer microbes can be met solely through the nitrogen that is available in the plant litter. Consequently, this produces no net mineralization or immobilization of nitrogen by the microbial food web. When the C:N ratio in litter is greater than 25:1, however, there is insufficient nitrogen to maintain optimal microbial growth. To meet this nitrogen need of the microbes, additional nitrogen may be acquired from soil inorganic nitrogen or from the use of soil organic matter (C:N of approximately 10:1). If this is the case, microbial growth and decomposition rates will not be limited by nitrogen availability. If this supplemental nitrogen is not available, the rate of decomposition will be reduced (Paul and Clark 1989).

4.2.6 Soil Organic Matter

Given the great importance of its influence on soil nitrogen dynamics and its role as a storage compartment of carbon in the global carbon cycle, it is desirable to explore further the organic matter component of soils. As a by-product of terrestrial food chains, dead organic matter accumulates in soils. This organic matter serves as a food source for the decomposer food web of the soil, as well as stores a substantial amount of the world's carbon.

Soil organic matter is also important for its capacity to store soil nutrients (including nitrogen) and to increase the water-holding capacity of soils. In addition, it influences soil structure. It helps form soil mineral grains into various sized aggregates that, in turn, produce open pathways, which facilitate soil aeration and water movement. Decomposition processes produce a wide variety of chemical forms of soil carbon. Some of these are fairly ephemeral, while others are relatively resistant to decomposition and consequently enhance the accumulation of carbon in the soils of most terrestrial ecosystems. This amount of carbon stored in the soil varies geographically due to the relative influences of climate on the processes of photosynthesis (production of organic matter) and cellular respiration (destruction of organic matter). The source of soil organic carbon is atmospheric CO_2 that enters terrestrial food chains through photosynthesis. This carbon can then be transferred to soils as dead organic matter in the form of plant litter, organic matter solubilized from plant tissues, food web wastes, and dead consumers within food webs. Organic matter is contributed to soils primarily as plant material in either dissolved or particulate forms. Particulate forms consist principally of carbohydrates (including cellulose, hemicellulose, and lignin) and proteins. The former constitutes 85% or more of a plant's biomass, and the latter constitutes a lesser amount (typically less than 15%) (Paul and Clark 1989). Once on or within the soil, this dissolved and particulate organic matter is modified by the decomposer food web. This produces a mixture of soil organic matter that includes these raw materials for the decomposer food web plus their decomposing residues, decomposition by-products, highly resistant humic matter, and even the microorganisms themselves.

Chemically, soil organic matter consists of a wide range of organic compounds within each of four categories: carbohydrates (of microbial origin, which constitute 10–20% of the soil organic matter); nitrogen-containing compounds (20%); aliphatic fatty acids, alkanes, and other related compounds (10–20%); and aromatic compounds, which make up the remainder of the material (Paul and Clark 1989). The latter group is quite resistant to decomposition and is the predominant constituent of soil humic matter. Humic matter is dark and spongy in texture and is composed of a mixture of degradation residues that are fairly resistant to degradation by decomposition, somewhat the end-product of decomposition. Inputs of dead plant matter to the soil occur at the soil surface from the above-ground plant parts and below the surface from plant roots. Decomposition occurs in both above- and below-ground locations, and there is a tendency for dissolved and fine particulate soil organic matter to be transferred from the soil surface down into the soil through the action of percolating soil water. As a consequence, soil organic matter accumulates both on the soil's mineral surface and in the upper portion of the mineral soil.

The amount of organic matter in the soil is influenced by the climatic conditions of moisture and temperature and various soil conditions, including nutrient availability, water holding capacity, and aeration. Moisture influences the rate of photosynthesis, and therefore the production of organic matter in terrestrial ecosystems. Seasonality in moisture availability can produce seasonality in photosynthesis. Water availability and, consequently, photosynthesis are decreased during dry seasons and during periods when soil water is frozen. Both water and temperature influence decomposition rates, and thus the accumulation of soil organic matter and the generation of CO_2. Being ectotherms, the metabolic rates of decomposers (and therefore the decomposition rate) increase with increasing temperatures to a point and then fall off when temperatures are hotter than optimal for the decomposers. Likewise, the amount of water in the soil can limit decomposition if the soil is too dry or too wet. Waterlogged soils produce anaerobic conditions that slow decomposition rates. Under these moisture conditions, facultative and obligate anaerobic microorganisms can be active and can produce CH_4.

Globally, total annual plant production is estimated at 40×10^9 Million tones (Paul and Clark 1989). Decomposition almost equals this amount, and consequently, there is a net accumulation of organic matter within soils. Geographic variations in climatic factors (water and temperature) have produced a geographic variation in the accumulation of soil organic matter, with some ecosystems having large quantities while others contain small amounts. For example, in grasslands, the annual dry season limits decomposition producing the accumulation of soil organic matter. Tropical wet forests have a greater production of biomass but lower soil carbon than temperate forests. Greater accumulation of soil organic matter in the latter forests is due to colder winter temperatures that limit decomposition. The highest accumulation of soil organic matter occurs in highly productive wetlands where the rate of decomposition is slowed due to O_2 limitation in the waterlogged soils.

4.2.7 Perturbations and Nitrogen and Carbon Biogeochemistry

Human perturbations of ecosystems alter environmental conditions that influence the microbial processes of the nitrogen and carbon biogeochemical cycles. Land-use changes that decrease plant cover (e.g., deforestation for development or agriculture) can reduce carbon storage in the soil through decreasing litter inputs to the soil and increasing the rate of decomposition in the soil through increased soil warming. Globally, deforestation and other land-use changes contribute approximately 1.5 Gt of carbon to the atmosphere each year (Paul and Clark 1989). To compensate for such losses in agricultural systems, animal manure and green manures (cover crops that are plowed under) provide supplemental additions of organic matter to the soil. Soil nitrogen is also impacted through such land-use changes. In agricultural soils, nitrogen is removed with harvested plant material and lost through soil erosion, leaching, and volatilization. To counter the losses, supplemental additions of inorganic N fertilizers (e.g., urea, nitrate, and ammonium) and organic nitrogen in the form of animal wastes can be added. When these additions are excessive, runoff and leaching can contaminate water bodies. Crop rotation, which incorporates the use of nitrogen-fixing plants (e.g., alfalfa), adds nitrogen to the soil.

Other farming practices alter the soil's aeration, water, and temperature regimes, which in turn can alter the rates of microbial processes leading to losses through nitrogen leaching and emissions of nitrogen, carbon dioxide, and methane. Irrigation can lead to soil water logging and associated changes in aeration status. The use of heavy mechanical farm equipment can compact the soil. The traditionally used plow, the moldboard plow, turns over the soil to the depth of the plow. This disrupts soil structure altering aeration and water retention, as well as exposes bare soil, making it susceptible to erosion. No-till approaches better maintain soil structure and decrease potential erosion but require the use of herbicides. While agricultural practices produce local alterations in the nitrogen and carbon biogeochemical cycles, cumulative land-use change over the Earth translates to global-scale disruptions of the cycles. With deforestation and other land-use change contribute significantly to increases in atmospheric CO_2, the potential for climate alteration at the global-scale increases. Global climate change has the potential to impact agricultural systems in different manners. The predicted increased temperature would likely increase the metabolic rates of soil microbes, thereby accelerating the rates of decomposition and nitrogen transformations. For some geographic regions, drier conditions are predicted, while for others, precipitation may increase. Such variation would produce quite different geographic effects on microbial activity and the biogeochemical cycles of nitrogen and carbon. This, in turn, would contribute to significant variability in the effects of climate change on agricultural productivity among different geographical regions.

4.2.8 Organic Nitrogen

Soil organic matter is the major storehouse of many plant nutrients in soils, including nitrogen, phosphorus, sulfur, calcium, and magnesium. Soil organic matter is composed of a stable material called humus, an easily decomposed material (litter), soil microbes, and some other organic molecules. Typically, humus will contain 45–55% carbon and about 5% nitrogen. In other words, soil humus typically has a carbon to nitrogen ratio (C:N ratio) of approximately 12:1 for surface soils and 8–10:1 for subsurface soils.

4.2.9 Mineralization

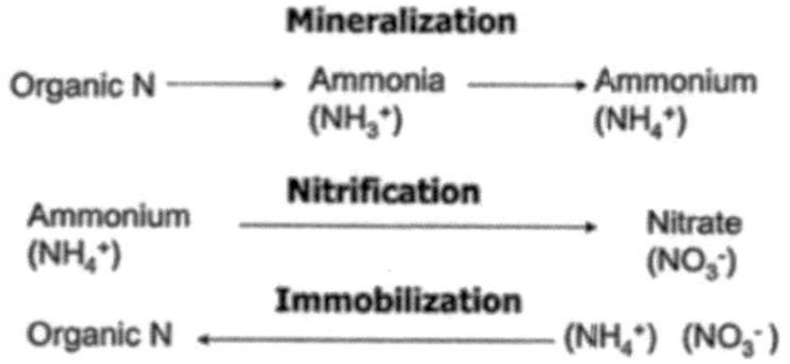

Nitrogen that is present in soil organic matter, crop residues, and manures is converted to the inorganic form by the process of mineralization. Initially, larger organic matter molecules are broken down into smaller ones, with soil microorganisms attacking these remaining materials by producing specific enzymes. The transformation of organic nitrogen to the ammonia and ammonium forms is referred to as ammonification. The resultant ammonium can then be transformed into the nitrate form by a process called nitrification. Since the decomposition process is carried out by living organisms, it is affected by several environmental variables, including soil moisture, temperature, pH, the C:N ratio, and the type of organic materials in the residue. Cotton and grain sorghum stalks and small grain residues are relatively high in carbon and low in nitrogen (having C:N ratios greater than 30: 1). Soil organisms may need additional nitrogen to decompose these residues. When the nitrogen supply is limited, soil microbes compete with plants for fertilizer nitrogen in a process called immobilization. When these microbes die, nitrogen tied up in the decomposition process again becomes available for crop use.

4.2.10 Biological Nitrogen Fixation

Approximately 80% of the Earth's atmosphere is nitrogen gas. Unfortunately, nitrogen gas is unusable by most living organisms. Plants, animals, and microorganisms can die of nitrogen deficiency, surrounded by nitrogen gas they cannot use. All organisms use the ammonia form of nitrogen to manufacture amino acids, proteins, nucleic acids, and other nitrogen-containing components necessary for life. Nitrogen fixation is a process by which molecular nitrogen in the air is converted into ammonia or related nitrogenous compounds in soil (Postgate 1998). Atmospheric nitrogen is molecular dinitrogen, a relatively nonreactive molecule that is metabolically useless to all but a few microorganisms. Biological nitrogen fixation converts nitrogen into ammonia, which is metabolized by most organisms. Nitrogen fixation is essential to life because fixed inorganic nitrogen compounds are required for the biosynthesis of all nitrogen-containing organic compounds, such as amino acids and proteins, nucleoside triphosphate, and nucleic acids. As part of the nitrogen cycle, it is essential for agriculture and the manufacture of fertilizer. It is also, indirectly, relevant to the manufacture of all nitrogen chemical compounds, which includes some explosives, pharmaceuticals, and dyes.

Biological nitrogen fixation is one of the important soil microbial activities, which is affected by all ongoing processes in the soil as well as other soil microorganisms. The biological nitrogen fixation process depends on the occurrence and survival of *Rhizobium* in soils and also on its efficiency (Adamovich and Klasens 2001). The rate of nitrogen fixation was affected by many different physiological and environmental factors in the soil, such as temperature, water holding capacity, water stress, salinity, nitrogen level, pH, and other nutrients (Dabessa et al. 2018). Many of these factors, including temperature, affect many aspects of nitrogen fixation and assimilation, as well as factors such as respiratory activity, gaseous diffusion, and solubility of dissolved gases, which ultimately affect plant growth (Coskan and Dogan 2011). There are several environmental factors affecting biological nitrogen fixation. Severe environmental conditions such as salinity, unfavorable soil pH, nutrient deficiency, mineral toxicity, extreme temperature conditions, low or extremely high levels of soil moisture, inadequate photosynthesis, and disease conditions can affect plant growth and development. As a result, even the persistent *Rhizobium* strains will not be able to perform root infection and nitrogen fixation in their full capacity (Panchali 2011). Temperature has a profound influence on nitrogen metabolism. Little activity is observed at low temperatures, and warming promotes the microbial nitrogen fixation and uptake of fixed gas (Saha et al. 2017). The plant nitrogenase activity reduces dramatically as a result of the formation of ineffective nodules at high temperatures (40 °C) (Hungria and Franco 1993; Hungria and Vargas 2000). In addition, the relative activity of the rhizobia is altered by temperature so that *Rhizobium* that is highly effective at a specified range of temperature is less active at another range of temperatures. For these reasons, greater nitrogen gains probably can be achieved by improvements in the heat resistance of the symbiosis. The optimum temperatures for growth in culture vary among strains and species; values between 27 and 39 °C have been noted. The maximum temperatures are generally 35–39 °C, but proliferation may take place up to 42 °C (Al-Falih 2002). Rhizobial survival in soil exposed to high temperature is greater in soil aggregates than in nonaggregated soil and is favored by dry rather than moist conditions (Zahran 1999).

Legumes hold a position of special significance within agricultural plants: their ability to fix atmospheric nitrogen through the presence of rhizobia in root nodules makes their growth theoretically independent both of soil nitrogen status and extraneous fertilizer nitrogen addition. Among them are important grain and forage crops that grow in tropical or temperate climatic zones, including soybean, beans, peas, lucerne, clovers, trefoils, and lupin (Sprent et al. 2010). Nitrogen fixation occurs when these plants are in the symbiotic state, and the agents of fixation are soil bacteria from the genera *Rhizobium*, *Bradyrhizobium*, *Sinorhizobium*, *Mesorhizobium*, *Azorhizobium*, and *Allorhizobium* as well as some more genera that lie phylogenetically outside this group. These include *Methylobacterium*, *Devosia*, *Ochrabactrum*, and *Phyllobacterium* in the α-Proteobacteria and *Burkholderia*, *Ralstonia, and Cupriavidus* from the β-Proteobacteria (Sprent et al. 2010). Strictly speaking, the term "rhizobia" should be limited to the traditional group of α-Proteobacteria listed above plus, possibly, *Agrobacterium* (Young et al. 2001; Willems 2006). Nitrogen fixation is carried out naturally in soil by microorganisms termed diazotrophs that include bacteria such as *Azotobacter* and *archaea*. Some nitrogen-fixing bacteria have symbiotic relationships with plant groups, especially legumes (Zahran 1999). All biological nitrogen fixation is affected by enzymes called nitrogenases (Wagner 2011). These enzymes contain iron, often with a second metal, usually molybdenum but sometimes vanadium. Biological nitrogen fixation occurs when atmospheric nitrogen is converted to ammonia by an enzyme called nitrogenase (Postgate 1998). The overall reaction for biological nitrogen fixation is

$$\mathbf{N_2 + 16ATP + 8e^- + 8H^+ \rightarrow 2NH_3 + H_2 + 16ADP + 16P_i}$$

The process is coupled to the hydrolysis of 16 equivalents of ATP and is accompanied by the co-formation of one equivalent of H_2 (Lee et al. 2014). The conversion of N_2 into ammonia occurs at a metal cluster called Fe-Mo cofactor. The mechanism proceeds via a series of protonation and reduction steps wherein the Fe-Mo active site hydrogenates the N_2 substrate (Hoffman et al. 2013). In free-living dizotrophs, nitrogenase-generated ammonia is assimilated into glutamate through the glutamine synthetase/glutamate synthase pathway. The microbial *nif* genes required for nitrogen fixation are widely distributed in diverse environments (Gaby and Buckley 2011; Hoppe et al. 2014). Nitrogenases are rapidly degraded by oxygen. For this reason, many bacteria cease production of the enzyme in the presence of oxygen. Many nitrogen-fixing organisms exist only in anaerobic conditions, respiring to draw down oxygen levels or binding the oxygen with a protein such as leghemoglobin (Postgate 1998).

Plants that contribute to nitrogen fixation include those of the legume family with taxa such as kudzu, clover, soybean, alfalfa, lupins, peanut, red gram, green gram, black gram, pea, etc. They contain and support symbiotic rhizobia bacteria within nodules in the root systems, producing nitrogen compounds that help the plant grow and compete with other plants (Kuypers et al. 2011). When the plant dies, the fixed nitrogen is released, making it available to other plants; this helps to fertilize the soil (Postgate 1998; Smil 2000). The great majority of legumes have this association, but a few genera, e.g., *Styphnolobium*, do not. In many traditional and organic farming practices, fields are rotated through various types of crops, which usually include one consisting mainly or entirely of clovers or buckwheat, often referred to as "green manure."

4.2.11 Nitrogen Fixation in Nonleguminous Plants

Biological nitrogen fixation reflects the activity of a phylogenetically diverse list of bacteria, archaea, and symbioses. Genetic and molecular techniques now provide efficient means to identify organisms with the potential to fix nitrogen, and the application of these techniques in an array of environments has broadened considerably our understanding of the suite of organisms that can carry out biological nitrogen fixation. In the past, most evaluations of terrestrial biological nitrogen fixation considered rhizobial and actinorhizal symbioses; free-living and symbiotic cyanobacterial fixers; and free-living and symbiotic (or associated) heterotrophic bacteria. More recent research has demonstrated that a dizzying array of microbes are capable of biological nitrogen fixation, including archaea as well as many previously undiscovered bacteria (Reed et al. 2011). Even the best-known system, the nodulated legume/rhizobial symbiosis, has been found to be much more diverse than previously understood, with new discoveries ranging from additional microbial partners (e.g., *Burkholderia*) that nodulate legumes and fix nitrogen effectively (Gyaneshwar et al. 2011) to a diversity of pathways by which nodules themselves are initiated and infected (Sprent 2007).

Nitrogen is an essential element in plant development and a limiting factor in plant growth. Only some prokaryotes are able to use atmospheric nitrogen through a process known as biological nitrogen fixation, which is the conversion of atmospheric nitrogen to ammonia, a form that can be used by plants (Lam et al. 1996; Franche et al. 2009). The bacteria responsible for nitrogen fixation are called diazotrophs; they encode nitrogenase, the enzyme complex that catalyzes the conversion of nitrogen gas to ammonia. The nitrogenase complex is highly conserved in free-living and symbiotic diazotrophs. Various types of associations/interactions occur between diazotrophs and their host plants (Santi et al. 2013). The highly specific and most efficient processes for nitrogen fixation involve the formation of root nodules on legumes and nonlegumes. The diazotrophic bacteria involved in these endosymbiotic interactions include rhizobia (Gram-negative) members of the alpha-subgroup of the phylum proteobacteria that associate with legumes (family Fabaceae) (Oldroyd and Downie 2008; Desbrosses and Stougaard 2011) and the nonlegume *Parasponia* species (family Cannabaceae), and *Frankia* sp. (Gram-positive) members of the actinomycete family that associate with a broad spectrum of plants belonging to eight families collectively called actinorhizal plants. In addition, nitrogen-fixing cyanobacteria (mainly *Nostoc* sp.) have also been found to colonize different plant organs, either intracellularly in the family Gunneraceae or extracellularly in liverworts, hornworts, *Azolla*, and *Cycadaceae*. In contrast with these symbioses, some diazotrophs, such as *Azospirillum* spp., *Azoarcus* spp., and *Herbaspirillum*, form associative and/or endophytic relationships with a wide variety of plant roots, including those of cereals. In all these associations and symbioses, for the host plants, the expected benefit of the interaction is the fixed nitrogen provided by the symbiotic partner, which, in return, receives reduced carbon and possibly all the other nutrients it requires. In addition, the symbiotic or endophytic plant structure colonized by the nitrogen-fixing microorganisms may provide the appropriate conditions for protecting the nitrogenase complex from oxygen exposure.

Other nitrogen-fixing families include the following:

- *Parasponia*, a tropical genus in the family Cannabaceae, which are able to interact with rhizobia and form nitrogen-fixing nodules (den Camp et al. 2010).
- Actinorhizal plants such as alder and Bayberry can form nitrogen-fixing nodules, thanks to a symbiotic association with *Frankia* bacteria. These plants belong to 25 genera (Dawson 2008) distributed across eight families.

A number of nonleguminous plants are recognized to have the ability to fix nitrogen either through exogenous or endogenous symbiosis with nitrogen-fixing microorganisms. This can be a potential source of nitrogen for agriculture and can be of greater economic importance. Some cereal crops of commercial importance like rice, wheat, maize, and millets are found to have an association with microorganisms that are capable of assimilating atmospheric nitrogen. Fixation of nitrogen by bacteria living in symbiosis with leguminous plants is a popular phenomenon. Various attempts have been made to extend the host range of *Rhizobium* from legumes to nonlegumes through plant genetic manipulation. The transfer of nif genes along with others for functional nitrogen fixation was considered to be the most suitable strategy to achieve symbiotic nitrogen fixation in nonlegumes (Saikia and Jain 2007). However, the highly complicated system of gene regulation posed great hindrance in achieving a functional nitrogen-fixing transgenic system. Genetic engineering through biotechnological means has seen little or no success in achieving the induction of symbiosis between cereals and diazotrophs. In rice, associative nitrogen fixation supplied 20–25% nitrogen of the total need, as reported in one of the studies conducted at International Rice Research Institute I, Philippines (Ladha et al. 1987). Rice seedlings when inoculated with some strains of *Burkholderia* spp. isolated from rice plants contributed a relatively higher level of nitrogen to rice via biological nitrogen fixation (Baldani et al. 2000). *Acetobacter diazotrophicus* could enhance rice growth, and growth promotion might be related to the transfer of biologically fixed nitrogen, although other factors such as auxin production could be involved (Sevilla and Kennedy 2000).

Inoculation of wheat with *Azospirillum brasilense* and *Azospirilium lipoferum* significantly increased the yields of foliage, grain, and branching of root hairs and gave varying degrees of acetylene reduction (Kennedy and Tchan 1992) from 48 ηmol ethylene h^{-1} $plant^{-1}$ to 10 ηmol ethylene h^{-1} $plant^{-1}$. Wheat para-nodules can be readily colonized externally and to some extent internally by the cyanobacterium *Nostoc* spp. strain 2S9B and that such co-existence provides favorable conditions for nitrogen fixation (Gantar and Elhai 1999). Acetylene reduction by maize inoculated with *Azospirillum* showed that carbon substrates and oxygen pressure have significant effects on it. Enhanced nitrogenase activity (acetylene reduction assay) was noticed in maize plants treated with *Azospirillum* either alone or with 2,4-D, but higher activity was noticed in the 2,4-D and *Azospirillum* treated plants (Saikia et al. 2004). The rate of activity varied greatly from 300 to 800 ηmol C_2H_4 h^{-1} g^{-1} dry wt of roots (Kennedy and Tchan 1992).

The ability to fix nitrogen is present in other families that belong to the orders Cucirbitales, Fagales, and Rosales, which together with the Fabales form a clade of eurosids. The ability to fix nitrogen is not universally present in these families. For example, of 122 Rosaceae genera, only 4 fix nitrogen. Fabales were the first lineage to branch off this nitrogen-fixing clade; thus, the ability to fix nitrogen may be plesiomorphic and subsequently lost in most descendants of the original nitrogen-fixing plant; however, it may be that the basic genetic and physiological requirements were present in an incipient state in the most recent common ancestors of all these plants, but only evolved to full function in some of them.

4.3 Factors Influencing Biological Nitrogen Fixation in Legumes

The complex process of legume biological nitrogen fixation is affected by environmental conditions such as temperature, water content, nitrogen concentration, root zone pH, plant nutrient status including carbon and nitrogen substrates in roots, and genetic variation in potential nitrogen fixation capacity. It is also affected by plant nutritional status such as phosphorus and potassium levels that control nodule growth and nitrogenase activity directly or indirectly (Havelka et al. 1982). In this section, we focus only on how legume biological nitrogen fixation responds to environmental conditions and plant carbon and nitrogen substrates with a view to improving models.

4.3.1 Temperature

Nodulation and nitrogen fixation are observed under a wide range of temperatures, and the optimum temperature is between 20 and 30 °C. Elevated temperatures may delay nodule initiation and development and interfere with nodule structure and functioning in temperate legumes, whereas in tropical legumes, nitrogen fixation efficiency is mainly affected. Furthermore, temperature changes affect the competitive ability of *Rhizobium* strains. Low temperatures reduce nodule formation and nitrogen fixation in temperate legumes; however, in the extreme environment of the high arctic, native legumes can nodulate and fix nitrogen at rates comparable to those observed with legumes in temperate climates, indicating that both the plants and their rhizobia have successfully adapted to arctic conditions. In addition to low temperatures, arctic legumes are exposed to a short growing season, a long photoperiod, low precipitation, and low soil nitrogen levels. In this review, we present results on a number of structural and physiological characteristics that allow arctic legumes to function in extreme environments (Bordeleau and Prevost 1994). Trolldenier (1982) studied the relation of nitrogenase activity (ethylene evolution) to soil temperature or incubation temperature of roots on two genera of swamp plants, viz., rice (*Oryza sativa*) and reed (*Phragmites communis*). For both intact rice plants and excised rice roots, the optimum temperature was 35 °C. On excised roots, nitrogenase activity responded more sensitivity to changes in temperature. In contrast to intact rice plants, no ethylene evolution occurred on excised roots at 17 and 44 °C. On reed roots, temperature optimum was between 26 and 30 °C, which is clearly lower than on rice (35 °C). The temperature range in which nitrogen fixation occurred was, however, similar to that of rice, although on a lower level. These results suggest a higher potential of the tropics for associative nitrogen fixation, while in cooler climates, the lower temperatures appear to be a major limiting factor.

Generally, soil temperature inhibits legume biological nitrogen fixation through its control on nodulation, nodule establishment, and nitrogenase activity when it is either too high or too low (Whitehead 1995). Therefore, minimum and maximum soil temperatures, and the range of temperature between these that are favorable for nitrogen fixation, could be used to define the response of nitrogen fixation to soil temperature. Soil temperature in the root zone is one of the controlling factors for nodulation and nodule establishment (Table 4.1). For example, the nodulation of arrow leaf clover (*Trifolium vesiculosum* Savi.) is accelerated at a root temperature of 25 °C compared with that growing at both 18 °C and 32 °C (Schomberg and Weaver 1992). However, the response of nodule establishment to soil temperature differs between species and varieties. In soybean (*Glycine max* (L.) Merr.), more nodules are produced in the early growth stage at 25 °C, while 20 °C is optimal for nodule size after nodule generation is completed compared to 15 °C and 30 °C (Lindemann and Ham 1979). In contrast, nodule establishment is enhanced with increasing temperature in the range of 10–35 °C for white clover (*Trifolium repens* L.) regardless of the varieties and the rhizobia strains (Whitehead 1995). Nitrogenase activity responds slightly differently to soil temperature between species. There are a large number of studies on the response of nitrogen fixation to temperature in legume crops. Minimum temperatures for nitrogen fixation differ among species from 2 to 10 °C, and normally tropical and subtropical legumes have higher minimum temperatures than temperate species. Nitrogenase activity is high around 12–35 °C and reaches maximum at 20–25 °C in most legumes (Fig. 4.1). Nitrogen fixation in subterranean clover (*Trifolium subterraneum* L.) is very active at a wide range of temperatures, from 5 to 30 °C, but declines dramatically with low temperature and almost stops at 2 °C. Generally, 35 °C or 40 °C is the upper limit of temperature for legume biological nitrogen fixation.

4.3.2 Soil Water Status

In a similar manner to soil temperature, soil water content in the root zone controls N fixation through nodule establishment and nodule activity, plus gas permeability (Sinclair et al. 1987). Soil water deficit inhibits nitrogen fixation (Whitehead 1995; Goh and Bruce 2005), and the inhibition is reinforced as drought stress becomes more intense (Albrecht et al. 1984). Nitrogen fixation by peanut (*Arachis hypogaea* L.) grown in soil at two-thirds of field water capacity is reduced by 18–40%

Table 4.1 Response of nitrogen fixation to soil temperature (°C) for legume species

	Species	Minimum	Optimum	Maximum	Reference
1.	*Medicago sativa* L.	2	20–25	–	Waughman (1977)
			35	40	Dart and Day (1971)
2.	*Trifolium vesiculosum*	–	25	–	Schomberg and Weaver (1992)
3.	*Medicago truncatula* Gaertn.	2	20	35	Dart and Day (1971)
4.	*Lupinus polyphyllus* Lindl.	1.5	25		Waughman (1977)
5.	*Lotus corniculatus* L.	10	25–27	35	Rao (1977)
6.	*Phaseolus vulgaris* L	–	–	35	Piha and Munns (1987)
				40	Michiels et al. (1994)
7.	*Vicia sativa* L.	2	20	40	Dart and Day (1971)
8.	*Vigna unguiculata* (L.) Walp	5	40	–	Dart and Day (1971)
		2	20–30	40	
9.	*Vicia faba* L.	5	20	–	Waughman (1977)
		5	15–25	–	Halliday (1975)
10.	*Pisum sativum* L.	0.5	25	–	Waughman (1977)
			20–26	–	Lie (1971)
11.	*Cyamopsis tetragonoloba*	–	–	37–40	Pate (1961)
12.	*Lupinus angustifolius* L.	10	20–30	–	Halliday (1975)
13.	*Arachis hypogaea* L	–	–	40	Kishinevsky et al. (1992)
14.	*Vicia atropurpurea* Desf.	–	24	–	Pate (1961)
		2	25	35	Dart and Day (1971)
15.	*Trifolium pratense* L.	–	12–26	–	Small and Joffe (1968)
		–	27	–	Kuo and Boersma (1971)
16.	*Glycine max L.*	–	20–25	–	Lindemann and Ham (1979)
		5	20–35	40	Dart and Day (1971)
		2	5–30	–	Dart and Day (1971)
17.	*Trifolium subterraneum L.*	5	12–32	–	Gibson (1971)
		–	15	–	Roughley and Dart (1969)
		5	13–26	–	Halliday and Pate (1976)
		–	26	–	Small and Joffe (1968)
18.	*Trifolium repens L.*	–	21	–	Masterson and Murphy (1976)
		9	–	–	Frame and Newbould (1986)

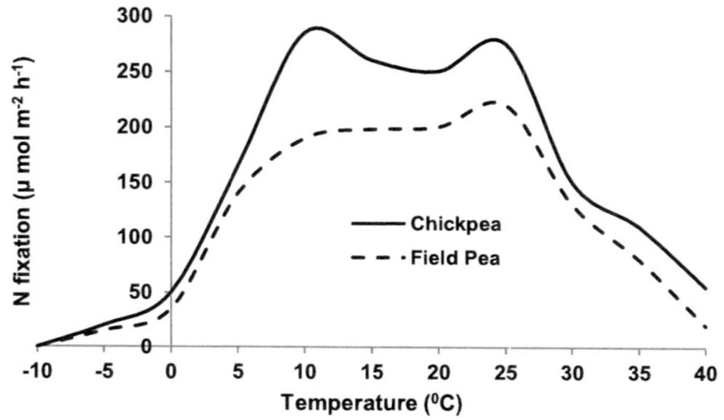

Fig. 4.1 Effect of temperature on nitrogen fixation (μ moles m^{-2} h^{-1}) in chickpea and field pea

compared to that in well-watered soil and by 44–69% when the soil water is one-third of field capacity (Pimratch et al. 2008). In addition, water-logging can seriously reduce nitrogen fixation through the depression of the establishment and activity of nodules (Havelka et al. 1982). It is not easy to quantify the relationship between nitrogen fixation and soil water status precisely due to the limitations of experimental controls, stress periods, and plant recovery (Ledgard and Steele 1992). Pimratch et al. (2008) tried to determine nitrogen fixation under drought stress with multiple-linear regression, but the correlation coefficients between the reduction of nitrogen fixation under drought compared to under field capacity and the soil water deficit were not statistically significant. A sigmoid function has been used to describe the response of N fixation by common bean (*Phaseolus vulgaris* L.), black gram (*Vigna mungo* (L.) Hepper), and cowpea (*Vigna unguiculata* (L.) Walp) to soil water stress, showing a sharper decline in nitrogen fixation as soil water stress becomes more severe (Serraj and Sinclair 1998; Sinclair et al. 1987).

4.3.3 Nitrogen Concentration in the Root Zone

It has been widely reported that soil mineral nitrogen in the root zone inhibits legume nodulation (Abdel Wahab et al. 1996), nodule establishment (Imsande 1986), and nitrogenase activity (Purcell and Sinclair 1990; Eaglesham 1989) as it costs less energy for legumes to take up nitrogen from soil than fix nitrogen biologically from the atmosphere (Cannell and Thornley 2000; Wood 1996). Normally, the severity of inhibition of nitrogen fixation by soil mineral nitrogen increases with soil mineral nitrogen content (Macduff et al. 1996; Lamb et al. 1995; Waterer and Vessey 1993a). It has been reported that a certain concentration of mineral nitrogen in the root zone, defined as "starter nitrogen," stimulates nodule establishment and nitrogen fixation compared to nonmineral nitrogen in some circumstances. And the concentrations of "starter nitrogen" that stimulate legume biological nitrogen fixation vary widely with cultivar and growth conditions but are normally less than 4 mM for ammonium and less than 2 mM for nitrate (Schomberg and Weaver 1992; Gulden and Vessey 1997; Gan et al. 2004). However, the time of external nitrogen application in relation to legume growth stage affects nodule growth and nitrogen fixation, and the later nitrogen is applied, the less nodule growth and nitrogen fixation is inhibited in field pea (*Pisum sativum* L.) (Waterer and Vessey 1993b). This is probably due to nodules being well established in the early stages before nitrogen is applied. Moreover, the inhibition of nitrogen fixation by nitrate was more severe than that of ammonia in white clover, field pea, and soybean (Svenning et al. 1996; Bollman and Vessey 2006; Gan et al. 2004) even though high amounts of either ammonia or nitrate retarded nitrogen fixation.

4.3.4 Carbon Demand for Fixation

Photosynthate partitioned to roots supports nodule growth, provides energy for nitrogen fixation, maintains a functional population of rhizobia, and allows the synthesis of amino compounds produced from nitrogen fixation (Minchin and Pate 1973; Layzell et al. 1979; King et al. 1986). Even though it is difficult to distinguish the proportion of CO_2 generated by nitrogen fixation from that generated by respiration for nodule growth and maintenance (Warembourg and Roumet 1989), the correlation between the rate of CO_2 produced from either nodulated roots or nodules and nitrogen fixation rate may be used to evaluate carbon consumption by nitrogen fixation (Lawrie and Wheeler 1973; Mahon 1977a, b; Warembourg and Roumet 1989). The carbon cost per unit fixed nitrogen (g carbon per g nitrogen fixed) varies widely with species, growth stage, and environmental conditions and ranges from 1.4 to 8.5 (Minchin and Pate 1973; Phillips 1980; Minchin et al. 1981; Sheehy 1987; Schulze et al. 1999). For example, it is reported that the carbon cost is 1.54 g carbon g nitrogen fixed in cowpea and 3.64 g carbon g nitrogen fixed in white lupin (*Lupinus albus* L.) from nodules during the early vegetative stage (Layzell et al. 1979, 1981) while it is 6.3–6.8 g carbon g nitrogen fixed for soybean, cowpea, and white clover from nodulated roots at periods of intense nitrogen fixation (Ryle et al. 1979). The carbon cost determined by CO_2 released from nodulated roots is generally higher than that from nodules as the former includes root respiration. The carbon cost of nitrogen fixation also varies with growth stages (Ryle et al. 1979; Twary and Heichel 1991), but it is a matter of debate that the carbon cost increases (Warembourg and Roumet 1989; Voisin et al. 2003) or decreases (Adgo and Schulze 2002) with the course of the legume life cycle. In addition, the strain of *Rhizobium* may affect the carbon cost significantly. For example, the carbon cost in alfalfa (*Medicago sativa* L.) nodules formed by strain P207 is on average 9.4 mol C mol N^{-1}, which is 59% higher than that of nodules formed by strain I02F51, 5.9 mol C molN^{-1} (Twary and Heichel 1991).

4.3.5 Seasonal Regulation of Biological Nitrogen Fixation

The rate of legume biological nitrogen fixation changes with physiological growth stages. It is low in the early growth stages while nodules are established (Lawrie and Wheeler 1973) and reaches a maximum value between early flowering and early seed-filling, depending on the species and growing conditions (Klucas 1974; Nelson et al. 1984; Jensen 1987). After the peak, nitrogen fixation decreases dramatically or even ceases during seed-filling (Herridge and Pate 1977; Beverly and Jarrell 1984; Sinclair et al. 1987), due to nodule senescence (Lawrie and Wheeler 1973) and poor carbon supply as a result of the strong demand for seed dry matter accumulation (Herridge and Pate 1977; Voisin et al. 2003, 2007).

4.3.6 Changes in Nitrogen Fixation with Plant Growth Stage

The quantitative effect of plant growth stage on the legume biological nitrogen fixation rate is considered in very few models, and in general, the process is stopped forcibly after the legume attains a certain growth stage. For example, nitrogen fixation ceases at the beginning of seed growth for cowpea and black gram but continues until the end of seed-filling for soybean in Sinclair's model (Sinclair et al. 1987), whereas it stops at physiological maturity in CROPGRO (Boote et al. 2008). A more specific function, similar to the temperature response function, is incorporated into EPIC and STICS to simulate the seasonal change of nitrogen fixation (Sharpley and Williams 1990; Bouniols et al. 1991; Cabelguenne et al. 1999).

4.3.7 Influence of Nitrate on Nodulation

There is information on the negative effect of nitrate on the nodulation induced by rhizobia (Giannakis et al. 1988; Alcantar-Gonzales et al. 1988). It was postulated that the presence of nitrate in the medium completely inhibited or drastically reduced the efficiency of infection in strains that possessed any nitrate reductase activity (Alcantar-Gonzales et al. 1988). However, it was shown that at least in some *Rhizobium* strains, the presence of nitrate did not affect this process negatively. Serrano and Chamber (1990) investigated three *Bradyrhizobium* sp. (*Lupinus*) strains differing in nitrate reductase activity: the first strain possessed constitutive nitrate reductase, the second inducible nitrate reductase, and the third did not express nitrate reductase activity. The results demonstrated that the strains with constitutive nitrate reductase and inducible nitrate reductase activity were capable of infecting roots sufficiently even at relatively high (12 mM) nitrate concentration, whereas the third strain completely lost the ability to form nodules. Moreover, at nitrate concentration of 1–2 mM, the strains with nitrate reductase activity (either constitutive or inducible) infected roots slightly more efficiently than in the absence of nitrate. The described "patterns" of infection ability resulted in the production of durable nodules, which was observed three weeks after. These results show that there is a possibility of the formation of symbiotic associations, in which the presence of nitrate does not inhibit nodulation. The effect of nitrate reductase activity on the efficiency of root infection might be explained by the removal of nitrate ions from the bacterial environment (O'Hara and Daniel 1985).

4.3.8 Influence of Nitrate on Nitrogenase Activity

A decrease in the efficiency of nitrogen fixation by bacteroids in the presence of nitrate was repeatedly reported (Chamber-Perez et al. 1997; Serrano and Chamber 1990; Arrese-Igor et al. 1997; Minchin et al. 1986a). Many hypotheses have been proposed to explain the mechanism of the nitrate effect. Most of the research concerning this problem approached it from the plant host's "viewpoint," overlooking the fact that the bacterial micro-symbiont is not entirely subordinated to plant metabolism. In recent years, several papers were published showing the other side of the symbiotic associations, and they reported new observations might be helpful in attempts at understanding the effect of bacteroid nitrate and nitrite reductases on nitrogenase activity. These results will be discussed in the final chapter. Hypotheses explain the decreased nitrogen fixation as the effect of decreased bacteroid respiration, caused by increased resistance to O_2 diffusion in the nodule cortex (Minchin et al. 1986a, b). The fast changes in gas diffusion resistance have been explained by various mechanisms: as changes in osmotic potential of nodule cells (Witty et al. 1986), filling of intercellular spaces in nodules by H_2O molecules (Witty et al. 1986; Walsh et al. 1989; Ismade 1991), or variable quantity of passing phloem sap (Layzell and Hunt 1990).

There are reports confirming the presumption that the presence of nitrate raises the gas diffusion resistance. In the presence of nitrate, both the energy cost of nitrogen fixation (expressed as the quantity of released CO_2 per mol of reduced C_2H_2) and the gas diffusion resistance increase, but the efficiency of bacteroid respiration (measured as μ mol CO_2 per gram dry mass per min) decreases. Some authors have suggested that the increase in gas diffusion resistance could be a secondary response to metabolic changes, as in the case of the effects caused by water stress and low temperature (Arrese-Igor et al. 1997).

4.4 Nitrogen Assimilation

Plants absorb nitrogen from the soil in the form of nitrate and ammonium. In aerobic soils where nitrification can occur, nitrate is usually the predominant form of available nitrogen that is absorbed (Xu et al. 2012). However, this is not always the case as ammonia can predominate in grasslands (Jackson et al. 1989) and in flooded, anaerobic soils like rice paddies (Tcherkez and Hodges 2007). Plant roots themselves can affect the abundance of various forms of nitrogen by changing the pH and secreting organic compounds or oxygen (Li et al. 2007). This influences microbial activities like the inter-conversion of various nitrogen species, the release of ammonia from organic matter in the soil, and the fixation of nitrogen by nonnodule-forming bacteria.

Ammonium ions are absorbed by the plant via ammonia transporters (structurally related membrane transport proteins called Amt transporters in bacteria and plants, methyl-ammonium/ammonium permeases in yeast). Nitrate is taken up by several nitrate transporters that use a proton gradient to power the transport (Sorgona et al. 2011; Tischner 2000). Nitrogen is transported from the root to the shoot via the xylem in the form of nitrate, dissolved ammonia, and amino acids. Usually (Scheurwater et al. 2002) (but not always, Stewart et al. 1986), most of the nitrate reduction is carried out in the shoots while the roots reduce only a small fraction of the absorbed nitrate to ammonia. Ammonia (both absorbed and synthesized) is incorporated into amino acids via the glutamine synthetase-glutamate synthase (GS-GOGAT) pathway (Masclaux-Daubresse et al. 2006). While nearly all (Kiyomiya et al. 2001) the ammonia in the root is usually incorporated into amino acids at the root itself, plants may transport significant amounts of ammonium ions in the xylem to be fixed in the shoots (Schjoerring et al. 2002). This may help avoid the transport of organic compounds down to the roots just to carry the nitrogen back as amino acids (Fig. 4.2).

4.4.1 Nitrate Reduction

Plants have a fundamental dependence on inorganic nitrogen, and 85–90 million metric tons of nitrogenous fertilizers are added to the soil worldwide annually (Good et al. 2004). The ability of a plant to capture nitrogen from the soil depends on soil type, environment, and species. It has been estimated that 50–70% of the nitrogen provided to the soil is lost (Hodge et al. 2000). The use of nitrogen by plants involves several steps, including uptake, assimilation, and translocation, and when the plant is aging, recycling and remobilization (Masclaux-Daubresse et al. 2010). Although generally low, soil nitrogen availability can fluctuate greatly in both space and time due to factors such as precipitation, temperature, wind, soil type, and pH. Therefore, the preferred form in which nitrogen is taken up depends on plant adaptation to soil conditions. Generally, plants adapted to low pH and reducing soils as found in mature forests or arctic tundra tend to take up ammonium or amino acids, whereas plants adapted to higher pH and more aerobic soils prefer nitrate (Maathuis 2009).

Heat stress can sharply decrease nitrate reductase activity and disrupt nitrate assimilation in plants (Klimenko et al. 2006). For instance, raising root-zone temperature leads to the excessive accumulation of nitrate in hydroponic lettuce (Van der Boon et al. 1990). Compared with relatively low soil temperature (10 °C), relatively high soil temperature (18 °C) enhances nitrate absorption and leads to the excessive accumulation of nitrate in the roots of radish (Sirtautas et al. 2011). Earlier studies have shown that a great portion of absorbed nitrate is assimilated in the roots under low temperatures (Calatayud et al. 2008). Under weak light conditions, low temperature not only led to marked decreases in the activities of

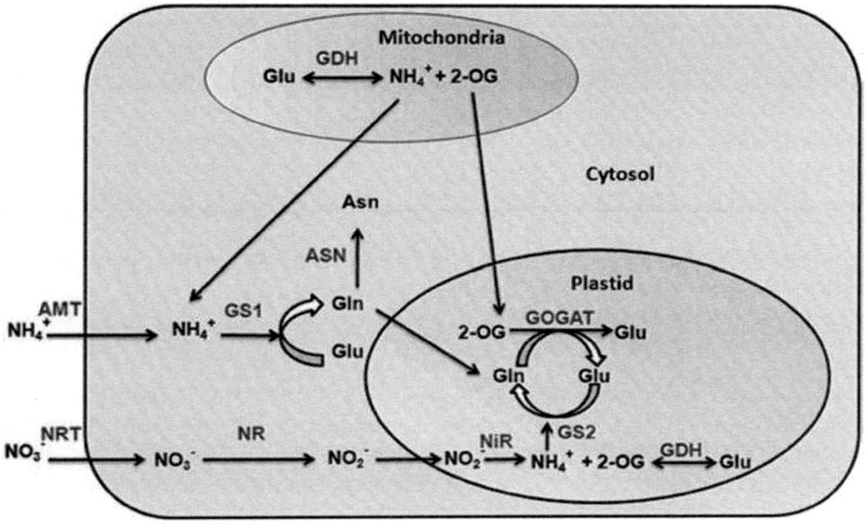

Fig. 4.2 Nitrogen uptake and assimilation process in plants. The uptake of nitrate (NO_3^-) and ammonium (NH_4^+) ions is mediated by nitrate and ammonium transporters, respectively. The NO_3 entered into the cell is reduced to nitrite ions (NO_2^-) by an enzyme nitrate reductase. The nitrite ion then moves to the plastid and is reduced to ammonium ion by the nitrite reductase enzyme. The ammonium is then incorporated into amino acids by glutamine synthetase and glutamate synthase via GS/GOGAT cycle. The ammonium ion transported by ammonium transporters directly enters into GS/GOGAT cycle. The two additional enzymes, glutamate dehydrogenase (GDH) and asparagine synthetase (ASN), also participate in ammonium assimilation. The GS, GDH, and ASN are the key enzymes involved in the synthesis of glutamine (Gln), glutamate (Glu), and asparagine (Asn)

nitrate reductase, glutamine-2-oxoglutarate aminotransferase (GOGAT), and glutamine synthetase (GS) but also resulted in a low nitrate accumulation in the leaves of tomatoes (Shu et al. 2016). This indicates that, in addition to the inhibition of nitrate reduction, low temperature impairs nitrate uptake and translocation in plants. The translocation of nitrates from the roots to the shoots is regulated by both environmental and intrinsic factors (Bian et al. 2020). Overnight low-temperature exposure inhibits photosynthesis in chilling-sensitive species, such as tomato and cucumber, by as much as 60%. It had been shown that low temperature stalled the endogenous rhythm controlling transcription of certain nuclear-encoded genes in chilling-sensitive plants causing the synthesis of the corresponding transcripts and proteins to be mistimed upon rewarming. The activity of nitrate reductase, the first and rate-limiting step in the assimilation of nitrate into amino acids in leaves, is subjected to a varied range of regulatory influences, including low temperature (Tucker and Ort 2002) and a robust circadian rhythm. Although nitrate reductase regulation is disrupted by low temperatures, the change is transient and does not alter the phase of the nitrate reductase endogenous rhythm following the chill. There is a temporary induction of *de novo* transcription of nitrate reductase causing an increase in both nitrate reductase protein and activity. This occurs regardless of the time in the

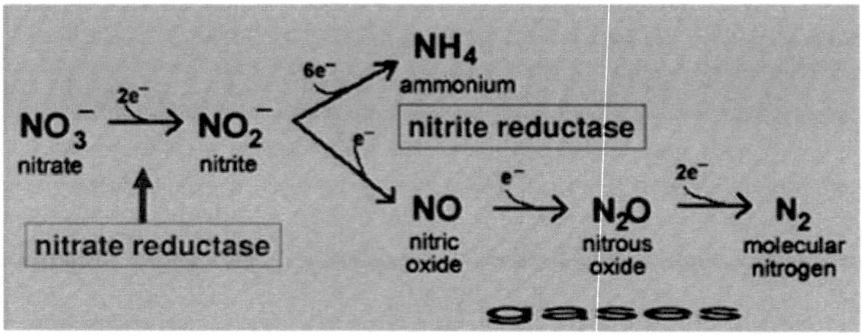

Fig. 4.3 Nitrate reduction and denitrification

circadian cycle that the chilling episode is initiated, thereby decoupling the normally closely coordinated processes of carbon and nitrogen assimilation. This decoupling would be expected to deplete cellular reductant and carbon skeleton reserves as well as allow accumulation of cytotoxic intermediates of nitrogen assimilation, thereby contributing to the low-temperature-induced disruption of metabolism that takes place in photosynthetic cells of chilling-sensitive plant species (Tucker and Ort 2002).

Plants require nitrogen from soil or from the atmosphere by symbiotic nitrogen fixation (Vance 1990). Soil-derived nitrogen, generally in the form of NO_3^-, is reduced to ammonia by a two-step process. Nitrate uptake occurs at the root level, and two nitrate transport systems have been shown to coexist in plants and to act coordinately to take up nitrate from the soil solution and distribute it within the whole plant (Daniel-Vedele et al. 1998; Tsay et al. 2007). Once taken up by root cells, nitrate must be transported across several cell membranes and distributed in various tissues. The nitrogen sources taken up by higher plants are nitrate or ammonium as inorganic nitrogen sources and amino acids under particular conditions of soil composition. Nitrogen assimilation requires the reduction of nitrate to ammonium, followed by ammonium assimilation into amino acids.

The first step, reduction of nitrate to nitrite (Fig. 4.3), is catalyzed by the plant nitrate reductase (Solomonson and Barber 1990), which is an inducible plant enzyme, depending on the availability of nitrate and light (Campbell 1999). Nitrate reduction takes place in both roots and shoots but is spatially separated between the cytoplasm where the reduction takes place and plastids/chloroplasts where nitrite reduction occurs. Nitrate reduction into nitrite is catalyzed in the cytosol by the enzyme nitrate reductase (Meyer and Stitt 2001). This enzyme is a homodimer, each monomer being associated with three prosthetic groups: flavin adenine dinucleotide (FAD), heme, and a molybdenum cofactor. Since 1993, considerable work has been done to characterize the nitrate reductase in different species (Meyer and Stitt 2001). Although the nitrate reductase enzyme is thought to be localized in the cytosol, an association with the plasma membrane has been found in corn roots and barley (Ward et al. 1989). In the regulation of nitrate reductase activity, both gene expression and post-translational processes are involved (Kaiser and Huber 2001). For a long time, it has been well known that nitrate inhibits nitrogenase activity in legume plant nodules. Many investigators have demonstrated the negative effect of nitrate on root infection by *Rhizobium* and on the ratio of the nodule dry mass to the whole plant mass. The effect of nitrate on the symbiosis between legumes and rhizobia can be divided into three major areas:

- The role of nitrate availability during root infection
- The relationship between nitrate availability and nitrogenase activity
- The effect of nitrate on the ratio of the nodule dry mass to the whole plant mass (Streeter and Wong 1988)

Aside from nitrogenase activity, in many symbiotic associations between legumes and rhizobia, the activity of nitrate reductase has also been observed, e.g., in root nodules of soybean (Heckmann and Drevon 1987; Hunter 1983) and in yellow lupin. It has been reported that nitrate reductase activity in nodules of yellow lupine is a few dozen times higher than in other parts of the plant (Polcyn and Luciñski 2001). Moreover, 97% of nodular nitrate reductase activity is localized in bacteroids (Polcyn and Luciñski 2001).

It is known that the concentration of free oxygen inside nodules is the major factor affecting nitrogenase activity (Layzell and Hunt 1990). Oxygen availability in the infected zone of root nodules is controlled by the plant host (Parsons and Day 1990; Ianetta et al. 1993; James et al. 1991; Drevon et al. 1995; Layzell and Hunt 1990; Minchin 1997). On the one hand, this is due to the presence of leghemoglobin; on the other hand, oxygen availability is limited by the diffusion resistance (Parsons and Day 1990; Becana and Klucas 1992; Chamber-Perez et al. 1997). The presence of nitrate may directly or indirectly influence the efficiency of gas diffusion resistance (Arrese-Igor et al. 1997).

There are several reports stating that enhancement of nitrogen fixation is due to the presence of active nitrate reductase in bacteroids. Symbiotic associations in which nitrate reductase function is complementary to that of nitrogenase are possible since bacteroids of many rhizobial strains are capable of performing dissimilatory nitrate reduction (Chamber-Perez et al. 1997; Serrano and Chamber 1990). The role of denitrification might be dual. On the one hand, it would remove toxic nitrite; on the other hand, it could supply ATP under oxygen deprivation during the first step of the denitrification process named nitrate respiration (Arrese-Igor et al. 1990; Garcia-Plazaola et al. 1993; O'Hara and Daniel 1985; Chamber-Perez and Serrano 1991). All these reports suggest that nitrogen fixation and denitrification, two processes that are antagonistic in the nitrogen cycle, may perform complementary functions in the cells of symbiotic rhizobia that would be important for survival of bacteria inside root nodules (Luciñski et al. 2002).

Nitrate reduction is carried out in two steps. Nitrate is first reduced to nitrite in the cytosol by nitrate reductase using NADH or NADPH (Tischner 2000). Nitrite is then reduced to ammonia in the chloroplasts (plastids in roots) by a ferredoxin-dependent nitrite reductase. In photosynthesizing tissues, it uses an isoform of ferredoxin (Fd1) that is reduced by photosystem I, while in the root, it uses a form of ferredoxin (Fd3) that has a less negative midpoint potential and can be reduced easily by NADPH (Hanke et al. 2004). In nonphotosynthesizing tissues, NADPH is generated by glycolysis and the pentose phosphate pathway.

4.4.1.1 Partitioning of Nitrate Reductase Activity Between Leaves and Roots

Nitrate is easily transported in the plant and can be assimilated both in leaves and roots. The partitioning of nitrate reduction between roots and shoots may vary with plant species, age, and environmental factors. In most herbaceous plants, nitrate reduction occurs predominantly in the leaves (Gojon et al. 1994), while studies on nitrate concentration in xylem sap exudates and in vivo and in vitro nitrate reductase activity showed that woody species reduce most of their nitrate in the roots (Cruz et al. 1991, 1993; Gojon et al. 1991; Lee and Titus 1992; Thomas and Hilker 2000).

Coffee plants present high potential for nitrate assimilation in leaves as well as in roots. Nevertheless, there is a lot of controversy about this subject. Some reports have shown higher nitrate reductase activity in leaves (Da Matta et al. 1999; Amaral et al. 2001), and others in roots (Carelli and Fahl 1991; Queiroz et al. 1991, 1993). It is generally accepted that the distribution of nitrate reduction between roots and shoots is mainly dependent on the ability of the root to export nitrate to the shoots (Gojon et al. 1994). Under high nitrate availability, roots of young coffee trees showed high capacity to export nitrate to the shoots. After supplying 15 mol.m^{-3} NO_3^- to nitrogen-starved intact plants, both roots and leaves accumulated similar nitrate levels; at the same time, nitrate reductase activity was rapidly induced in roots and leaves in a similar way (Carelli and Fahl 2006). Such evidence is consistent with the composition of nitrogenous compounds in the xylem sap exudates of coffee seedlings reported by Mazzafera and Gonçalves (1999). It was observed that seedlings fed with Hoagland's solution presented nitrate nitrogen as the most abundant nitrogen form in the xylem sap, representing 51.9% of the total sap nitrogen, whereas amino acids and ureides accounted for 41.2% and 6.6%, respectively (Mazzafera and Gonçalves 1999). Hence, it would appear that coffee plants have a high ability to partition absorbed nitrate and nitrate reductase activity between leaves and roots (Carelli and Fahl 2006). In this sense, coffee trees differ from some other woody species, in which the nitrate influx into the leaf may be a major factor limiting the leaf nitrate reductase activity (Cruz et al. 1993; Lee and Titus 1992; Gojon et al. 1994).

4.4.1.2 Effect of Environmental Factors on Nitrate Reduction

Nitrate accumulation in plants is affected greatly by environmental factors. Santamaria et al. (2001) observed an interaction between light intensity, nitrogen availability, and temperature on nitrate accumulation in rocket. Under conditions of low light availability, an increase in temperature increases the nitrate accumulation. Under high light intensity, an increase in temperature increases the nitrate content mainly when the nitrogen supply is high. Chadjaa et al. (2001) have studied the effect of artificial lighting in greenhouses on nitrate accumulation in lettuce. High-pressure sodium vapor lamps were more effective than metal halide lamps at increasing the nitrate reductase activity and reducing the nitrate accumulation. The effect of climate on nitrate accumulation was studied by Grzebelus and Baranski (2001), who found that nitrate content was less in the year that had a high rainfall. In warm and wet years, increased accumulation of nitrate is possible, regardless of whether the nitrogen originates from organic or mineral sources (Custic et al. 2003). Plant nitrate levels were influenced by weather conditions more significantly than by the form and application rates of fertilizers (Custic et al. 2003). Nitrate accumulation varies with the season (Vieira et al. 1998), being higher in autumn-winter than in spring (Santamaria et al. 1999). It is accepted that plants in winter are not able to use all the nitrogen available in the soil due to less favorable light and temperature conditions (Anjana et al. 2007).

4.4.1.3 Ammonium Depression of Nitrate Reduction

Linked to the multiphasic nature of nitrate transport is the possible differential effect of ammonium on the kinetics of nitrate assimilation. Growth in the presence of ammonium concentrations >1 μM usually represses the use of nitrate completely or nearly so (Dortch 1990; Flynn et al. 1997). If there is to be a significant use of nitrate under these conditions, it follows that repression cannot be complete, that nitrate is transported at a significant rate, and that the enzymes of nitrate assimilation are synthesized and operable. Maximum transport rates for ammonium and nitrate vary with the nitrogen status of cells, typically increasing with moderate nitrogen stress and then declining with extreme nitrogen stress. In the ammonium–nitrate interaction model, these relationships are described by mathematical functions relating maximum transport rates to cellular N:C (Flynn et al. 1999). In microalgae that attain higher growth rates and N:C when growing on ammonium, there is little or no residual capacity to use nitrate (Flynn 1999). On the other hand, for there to be a potential for cells to use nitrate when supplied with ammonium at >1 μM, the difference between the maximum rates of nitrate and ammonium transport cannot be too dissimilar during nitrogen replete growth at high N:C (Flynn et al. 1999). Furthermore, the significance of nitrate nutrition under such conditions will be enhanced if the capacity to transport nitrate is elevated at high nitrate concentrations by the presence of bi- or multiphasic transporters. It would be logical for such multiple transport systems for nitrate to have different control thresholds, different levels of control metabolites being required to halt nitrate transport. This could allow, in the presence of ammonium, the continued use of nitrate supplied at high concentration entering via the low-affinity system, but not of nitrate entering through the high-affinity system. The latter might be expected to be more costly to construct or operate (perhaps being a primary rather than secondary active transporter, Falkowski 1975a, b; Raven 1980) and thus subject to tighter regulation.

4.4.1.4 Effect of Tissue Age on Nitrate Assimilation

It is widely accepted that the ability of leaves and roots to reduce nitrate decreases with age (Carelli and Magalhães 1981; Kenis et al. 1992). Like in other species, leaf nitrate reductase activity in coffee plants is low during initial leaf expansion, reaches a peak in recently expanded leaves, and declines in the older ones (Fahl et al. 1992;

Queiroz et al. 1993; Carelli and Fahl 2005). However, this age-dependent nitrate reductase activity decrease in coffee leaves was not as evident as in annual plants, since in the latter case, very low nitrate reductase activity values were observed in older leaves (Carelli and Magalhães 1981). In the young coffee tree, a perennial evergreen species, older leaf nitrate reductase activity maintained nearly 70% of the values observed for recently expanded leaves, remaining nearly constant thereafter (Fahl et al. 1992; Carelli and Fahl 2005). In addition, total nitrogen and chlorophyll content increased as the leaf expanded, remaining constant in mature leaves. All these results indicate that coffee leaves are able to maintain their metabolic activities for a longer period of time (Fahl et al. 1992).

A decrease in nitrate reductase activity with leaf aging has been correlated with the tissue's ability to synthesize proteins (Travis and Key 1971). The lower nitrate reductase activity found in older leaves of corn is mainly due to a low level of nitrate reductase protein (Kenis et al. 1992), and it had been suggested that the decreased ability of older leaves to form an active nitrate reductase protein in response to its substrate may be due to a decrease in nitrate reductase mRNA abundance with age (Kenis et al. 1992). On the other hand, the finding that nitrate reductase synthesis is dependent on the photosynthesis suggests that the fall in nitrate reductase activity in older leaves could also be related to the lower photosynthetic capacity, and this could also be the case in coffee plants, where parallelism between NRA and photosynthesis is evident, in response to increasing leaf age (Fahl et al. 1992).

4.4.1.5 Effect of Low Temperature on Nitrate Reductase Activity

It has been shown that nitrate reductase in fact also displays a nitrite reductase activity, catalyzing NAD(P)H-dependent reduction of nitrite to nitrite oxide. Nitrate reductase is one of the enzymes that are capable of producing nitric oxide in plants. This activity is different from the plastidic nitrite reducing activity catalyzed by nitrite reductase, which reduces nitrite to ammonium. It is now clear that nitrate reductase can produce nitric oxide through its nitrate reductase–nitrite reductase activity. Nitric oxide is a diffusible, very reactive gas that is involved in the regulation of many processes in plants (Rockel et al. 2002; Wendehenne et al. 2004; Meyer et al. 2005).

In plants, it is known that nitrate is the primary factor regulating nitrate reductase induction (Cabello et al. 1998). Nitrate reductase activity is also regulated by a number of other factors, including growth hormones, reduced nitrogen metabolites, drought, temperature, and light (Foyer et al. 1998). It was shown that nitrate reductase activity is also regulated by cold (Vogel and Dawson 1991; Pietilainen et al. 1991; Atkin and Cummins 1994; Yaneva et al. 1996, 2002; Gao et al. 2000; Vasilieva et al. 2001; Tucker and Ort 2002; Vona et al. 2004; Rigano et al. 2006). In these studies at low temperatures in the short term, it was observed that nitrate reductase activity shows different responses in different plants. However, evidence for the effect of low temperature on the nitrate reductase activity in plants is ambiguous (Aydin and Nalbantoğlu 2011). It was reported that nitrate reductase activity was increased by cold in black alder leaves and roots, *Pinus sylvestris* needles, and tomato and winter wheat leaves (Vogel and Dawson 1991; Pietilainen et al. 1991; Yaneva et al. 1996, 2002; Tucker and Ort 2002) and was decreased by cold in arctic plant leaves and roots, cryophilic algae species, marine phytoplankton, and pea leaves (Atkin and Cummins 1994; Gao et al. 2000; Vasilieva et al. 2001; Vona et al. 2004; Rigano et al. 2006). In these short-term studies at low temperatures, different effects were observed on nitrate reductase activity. The nitrate reductase activity also shows a circadian rhythm in a normal diurnal pattern (Reins and Heldt 1992).

There are few studies done to show evidence related to nitrate reductase activity under cold stress. In a study, it was demonstrated that short-term low temperature causes a rapid activation of nitrate reductase in winter wheat leaves resulting from nitrate reductase protein dephosphorylation by protein phosphatase (Yaneva et al. 2002). In the other studies, temperature responses of nitrate reductase were studied in the marine phytoplankton, psychrophilic alga, and mesophilic alga in which cold decreases nitrate reductase activity and nitrate reductase protein values (Gao et al. 2000; Rigano et al. 2006). The results strongly suggested that structural modification (s) for cold adaptation affect thermodynamic properties of each of the functional domains within nitrate reductase holoenzyme (Rigano et al. 2006). It was also found that the flavin adenine dinucleotide (FAD) domain and electron transport among redox centers are sensitive to elevated temperatures (Gao et al. 2000). Aydin and Nalbantoğlu (2011) undertook a study on the effects of long-term cold (5/3 °C) treatments on the diurnal pattern of nitrate reductase activity in spinach (*Spinacia oleracea* L. cv. Gladiator) leaves. When the activities with cold treatment were compared to the control (20/15 °C), nitrate reductase activity was decreased by 5 °C in the light period and by 3 °C in the dark period. The cold treatment also affected the timing of circadian rhythm in the diurnal pattern of nitrate reductase activity.

4.4.1.6 Nitrate Accumulation as Influenced by Nitrate Distribution in Plants

In general, vegetables that are consumed with their roots, stems, and leaves have a high nitrate accumulation, whereas those with only fruits and melons as consumable parts have a low nitrate accumulation (Zhou et al. 2000). The nitrate content of various parts of a plant differs (Santamaria et al. 1999). Indeed, the vegetable organs can be listed by decreasing nitrate content as follows: petiole > leaf > stem > root > inflorescence > tuber > bulb > fruit > seed (Santamaria et al. 1999). In lettuce and "head chicory," inner leaves accumulate less nitrate than outer leaves, and in parsley and spinach, leaf blades accumulate less nitrate than petioles (Santamaria et al. 1999, 2001). Nitrate concentration in the petiole was more than double that in the lamina of rocket leaf (Elia et al. 2000); the difference was as high as 6.6 times greater in spinach (Anjana et al. 2006). Similarly, nitrate concentration in the petiole stems was higher than in the leaves, with the lowest in the roots in leafy vegetables (Chen et al. 2004). Chen et al. (2004) have also shown that nitrate supply has a significant effect on nitrate distribution, both in the metabolic pool and the storage pool of leaf blades. The comparison of the nitrate storage pool with the nitrate concentration in the whole leaf (nitrate metabolic pool + nitrate storage pool) suggests that about 90% of the nitrate accumulates in the storage pool. Functional diversity of nitrate compartmentation in the cells of various species has been discussed by Izmailov (2004). The concentration of nitrate in various plant parts could be a measure to evaluate the balance between adequate and excessive available nitrogen for optimum growth of plants during a growth season. Typically, however, this balance assumes two markedly different modes depending on the vegetable crop under consideration. Plants that develop fruit or storage organs, such as potatoes and tomatoes, usually have low nitrate nitrogen concentrations in petioles as the crop approaches the harvest. The decline in the petiolar nitrate concentration depends on the translocation of soluble nitrogen to the developing storage organ and on the gradual decrease in the available soil nitrogen. Vegetables that do not develop storage organs as the foodstuff have a different pattern of nitrate accumulation wherein nitrate often continues accumulating with the growing plant's age (Maynard et al. 1976). The findings of Anjana et al. (2006) were similar, showing that nitrate concentration in spinach was higher at later stages of plant growth. Therefore, the optimum physiological age for harvesting needs to be standardized for different leafy vegetable crops. Since nitrate continues to accumulate in spinach plants until maturity, the extent of nitrate intake by humans through spinach leaves may be reduced by harvesting the crop during the vegetative stage.

4.4.2 Ammonia Assimilation

After nitrate reduction, nitrite is translocated to the chloroplast, where it is reduced to ammonium by the second enzyme of the pathway, the nitrite reductase. Ammonium, originating from nitrate reduction, and also from photorespiration or amino acid recycling, is mainly assimilated in the plastid/chloroplast by the so-called GS/GOGAT (glutamine synthetase/glutamate oxo-glutarate aminotransferase) cycle (Lea and Miflin 1974; Lea and Forde 1994). The glutamine synthetase fixes ammonium on a glutamate molecule to form glutamine. This glutamine reacts subsequently with 2-oxoglutarate to form two molecules of glutamate, and this step is catalyzed by the glutamine 2-oxoglutarate aminotransferase (or glutamate synthase, GOGAT). The decameric structure of maize glutamate synthetase was described by Unno et al. (2006). It has been shown that there are two classes of nuclear genes that code for glutamate synthetase, the *GLN2* and *GLN1* genes. *GLN2*, present as a single nuclear gene in all the species studied so far, codes for the chloroplastic GS2, thought to be involved in the primary assimilation of ammonium coming from nitrate reduction in both C3 and C4 plants and in the re-assimilation of ammonium produced from photorespiration in C3 plants. The magnitude of the ammonium flux through the photorespiration pathway in the leaves of C3 plants was indeed estimated to exceed that produced from nitrate reduction by five- to tenfold (Keys et al. 1978).

Conversely, the *GLN1* gene family codes for cytosolic glutamate synthetase-1 isoforms, present in different organs such as roots or stems and thought to be involved in ammonium recycling during particular developmental steps such as leaf senescence and in glutamine synthesis for transport into the phloem sap

(Bernard and Habash 2009). Two different forms of glutamate synthase are present in plants: Fd-GOGAT and NADH-GOGAT use ferredoxin and NADH as the electron donors, respectively (Vanoni et al. 2005). Fd-GOGAT is predominantly localized in leaf chloroplasts, whereas NADH-GOGAT is primarily located in plastids of nonphotosynthetic tissues, such as roots, etiolated leaf tissues, and companion cells. The structural, mechanistic, and regulatory properties of GOGAT enzymes and their role in amino acid metabolism have been reviewed by Suzuki and Knaff (2005).

In addition to the GS/GOGAT cycle, three enzymes probably participate in ammonium assimilation. Cytosolic asparagine synthetase catalyzes the ATP-dependent transfer of the amido group of glutamine to a molecule of aspartate to generate glutamate and asparagine (Lam et al. 2003). Masclaux-Daubresse et al. (2006) provided evidence that asparagine synthetase can also use ammonia as a substrate. Indeed, providing $^{15}NH_4^+$ to leaf discs in the presence of azaserine, glutamate biosynthesis was blocked but not asparagine biosynthesis, thus showing that asparagine production is not aminotransferase dependent. Asparagine has a higher nitrogen/carbon ratio than glutamine and can be used as a long-range transport and storage compound, especially in legumes (Rochat and Boutin 1991; Lam et al. 2003). Asparagine synthetase could, in certain situations, compensate for the reduced glutamine synthetase-dependent ammonium assimilatory activity. Carbamoyl-phosphate synthase forms carbamoylphosphate, a precursor of citrulline and arginine, within plastids using bicarbonate, ATP, and ammonium or the amide group of glutamine (Masclaux-Daubresse et al. 2010). Finally, the mitochondrial NADH-glutamate dehydrogenase could alternatively incorporate ammonium into glutamate in response to high levels of ammonium under stress (Skopelitis et al. 2006). However, the major catalytic activity for glutamate dehydrogenase in plant cells has been reported to be glutamate deamination (Masclaux-Daubresse et al. 2006; Purnell and Botella 2007). Glutamate dehydrogenase activity was shown to be essential for plant survival in dark conditions (Miyashita and Good 2008).

Nitrate reductase, nitrite reductase, and glutamate synthase require reducing power as either NADH or ferredoxin, according to the enzyme. Glutamine synthetase and asparagine synthetase need ATP. In addition, carbon skeletons and especially keto-acids are essential to forming organic nitrogen as amino acids. The availability of carbon skeletons for ammonium condensation and the supply of ATP, ferredoxin, and NADH as products of photosynthesis, respiration, and photorespiration pathways are thus essential for nitrogen assimilation.

In the chloroplasts (Tcherkez and Hodges 2007), glutamine synthetase incorporates this ammonia as the amide group of glutamine using glutamate as a substrate. Glutamate synthase (Fd-GOGAT and NADH-GOGAT) transfers the amide group onto a 2-oxoglutarate molecule producing two glutamates. Further transaminations are carried out to make other amino acids (most commonly asparagines) from glutamine. While the enzyme glutamate dehydrogenase does not play a direct role in the assimilation, it protects the mitochondrial functions during periods of high nitrogen metabolism and takes part in nitrogen remobilization (Lea and Miffins 2003).

4.4.2.1 Effect of Low Temperature on Ammonia Assimilation

The overall trend of global climate change is warming; however, the accelerated melting and disappearance of Arctic Sea-Ice has also led to increasing local occurrences of abnormally low temperatures (Mori et al. 2014; Overland et al. 2015). Low-temperature stress damages plants by inducing oxidative stress, hindering photosynthesis, and indirectly inducing osmotic stress (Allen and Ort 2001; Beck et al. 2007). Increasing nitrogen deposition is another environmental problem that attracts global attention (Stevens et al. 2004; Galloway et al. 2008; Liu et al. 2013b). Excess nitrogen deposition leads to nutritional imbalance, cell membrane destruction, and oxidative stress in plants (Pearce and van der Wal 2002; Koranda et al. 2007). Nitrogen in wet deposition mainly occurs in the forms of NO^{-}_{3} and NH^{+}_{4}, but they have different effects on plants. On the one hand, as H^{+} is generated during NH^{+}_{4} assimilation and OH^{-} is generated during NO^{-}_{3} assimilation, NH^{+}_{4} is more apt to decrease tissue pH and disrupt membrane function (Pearce et al. 2003; Paulissen et al. 2004). On the other hand, NH^{+}_{4} is more readily available than NO^{-}_{3} for utilization by plants as less energy is needed during its assimilation (Flaig and Mohr 1992). Moreover, excess nitrogen deposition can reduce the resistance of plants to low-temperature stress (Fenn et al. 1998).

Mosses, a group of primitive land plants with small stature that mainly acquire their nutrients from air, are relatively sensitive to environmental change and, therefore, generally more vulnerable to global changes than vascular plants (Pearce and van der Wal 2002; Harmens et al. 2011; Schröder et al. 2014). A comprehensive understanding of the stress responses of mosses and the mechanisms underlying them will facilitate the accurate evaluation of the fate of mosses on the background of global change. Studies of the responses of mosses to low-temperature stress have primarily been limited to the model species *Physcomitrella patens* (Nagao et al. 2006; Bhyan et al. 2012), and little attention has been paid to other species or their ecological roles (Yin and Zhang 2016).

4.4.2.2 pH and Ionic Balance During Nitrogen Assimilation

Every nitrate ion reduced to ammonia produces one OH^{-} ion. To maintain a pH balance, the plant must either excrete it into the surrounding medium or neutralize it with organic acids. This results in the medium around the plant's roots becoming alkaline when they take up nitrate. To maintain ionic balance, every nitrate taken into the root must be accompanied by either the uptake of a cation or the excretion of an anion. Plants like tomatoes take up metal ions like K^{+}, Na^{+}, Ca^{2+}, and Mg^{2+} to exactly match every nitrate taken up and store these as the salts of organic acids like malate and oxalate (Kirkby and Knight 1977). Other plants like the soybean balance most of their nitrate intake with the excretion of OH^{-} or bi-carbonate (Bruno et al. 1988). Plants that reduce nitrates in the shoots and excrete alkali from their roots need to transport the alkali in an inert form from the shoots to the roots. To achieve this, they synthesize malic acid in the leaves from neutral precursors like carbohydrates. The potassium ions brought to the leaves along with the nitrate in the xylem are then sent along with the malate to the roots via the phloem. In the roots, the malate is consumed. When malate is converted back to malic acid prior to use, an OH^- is released and excreted ($RCOO^- + H_2O \rightarrow RCOOH + OH^-$). The potassium ions are then recirculated up the xylem with fresh nitrate. Thus, the plants avoid having to absorb and store excess salts and also transport the OH^- (Bruno et al. 1992). Plants like castor reduce a lot of nitrate in the root itself and excrete the resulting base. Some of the base produced in the shoots is transported to the roots as salts of organic acids, while a small amount of the carboxylates are just stored in the shoot itself (Susan and Raven 1987).

4.5 Cold Acclimation and Tolerance to Freezing Stress

In many overwintering temperate plant species, a period of low positive temperature will increase tolerance to a subsequent subzero (freezing) temperature. This process is known as cold or frost acclimation and is acknowledged to be complex, involving a number of biochemical and physiological changes. Cellular and metabolic changes that occur during cold acclimation include increased levels of sugars, soluble proteins, proline, and organic acids as well as the appearance of new isoforms of proteins and altered lipid membrane composition (Hughes and Dunn 1990). There is an increasing body of evidence to show that many of these biochemical and physiological changes are regulated by low temperature through changes in gene expression, and in recent years, a number of low temperature-responsive genes have been cloned from a range of both dicotyledon and monocotyledon species. The molecular basis of cold acclimation and acquired freezing tolerance in plants, mainly *Arabidopsis* and winter cereals, has been extensively studied. To adapt to cold stress during cold acclimation, gene expression is reprogrammed, and the metabolism is also modified (Chinnusamy et al. 2010). Cold response is a very complex trait involving many different metabolic pathways, gene regulations, and cell compartments (Hannah et al. 2005).

4.5.1 Cold Acclimation

Cold acclimation is a process that plants use to acquire low-temperature tolerance through the subjection to low nonfreezing temperatures before periods of freezing stress. For example, nonacclimated rye is killed by temperatures of −5 °C, but after exposure to low nonfreezing temperatures, it can survive to temperatures of −30 °C (Thomashow 1999). Plants have also been shown to benefit and gain tolerance from exposure to sub-zero temperatures before periods of freezing stress. Some plants need not only low temperatures but also short photoperiods during acclimation to obtain cold tolerance (Janská et al. 2010). Low-temperature tolerance gained from cold acclimation can be lost due to temperature or photoperiod increases, causing deacclimation and cold injury (Junttila and Kaurin 1990; Welling et al. 2002; Chang et al. 2015). During cold acclimation, plants redirect their metabolism toward the synthesis of cryoprotectant molecules like soluble sugars, sugar alcohols, and low-molecular-weight nitrogen compounds like proline or glycine betaine (Janská et al.

2010). The increase in freezing tolerance during cold acclimation can be due to changes in lipid composition for more unsaturated fatty acids, as well as the accumulation of sucrose or other sugars, which may also stabilize membranes (Anchordoguy et al. 1987). During cold acclimation, there is also an increase in cold-regulated proteins and other stress-induced proteins like dehydrins that are involved in stabilizing cell constituents. Also, during this process, antioxidant enzymes are induced that are involved in the scavenging of reactive oxygen species (Janská et al. 2010).

Cold acclimation results in many different metabolic changes, and it has been reported that increases in calcium, glycine betaine, abscisic acid, sucrose, and potassium are involved in the cold acclimation process and have been shown to induce cold tolerance in various plant species when applied exogenously (Shi et al. 2014; Xing and Rajashekar 2001; Zhang et al. 2006; Espevig et al. 2011). Calcium is an important secondary messenger that facilitates the initial plant response to cold stress through an influx of calcium from cellular storage to the cytoplasm, which induces the expression of cold-related genes by protein kinase activation (Janská et al. 2010; Knight et al. 1996). The exogenous application of calcium has been shown to improve chilling and freezing tolerance in bermudagrass (*Cynodon dactylon*) (Shi et al. 2014) and other plant species, such as chickpea (*Cicer arietinum*) (Nayyar et al. 2005) and alfalfa (*Medicago sativa*) (Monroy and Dhindsa 1995). Treatment with $CaCl_2$ is reported to alleviate oxidative damages from cold stress (Shi et al. 2014). Foliar $CaCl_2$ treatments promoted plant growth by increasing photosynthetic efficiency and nitrogen assimilation under cold stress (Shi et al. 2014). In addition, calcium treatments induce the accumulation of protective compounds in amino acids, such as proline, or soluble carbohydrates like sucrose, glucose, fructose, and trehalose (Nayyar et al. 2005). Exogenous calcium has been shown to provide protection to not only cold stress but to other abiotic stresses involving water deprivation, such as heat or salinity (Tan et al. 2011; Yin et al. 2015). An increase in cold tolerance is commonly associated with the capacity to increase protective carbohydrates, such as sucrose that is reported to protect plants from freeze-induced dehydration by decreasing ice formation through increasing the intracellular solute concentration (Espevig et al. 2011). This carbohydrate may help regulate carbohydrate metabolism and assimilate partitioning, impacting carbon and nitrogen allocation, which directly affects plant development and may serve roles in tolerance or recovery from stress periods (Espevig et al. 2011). In addition, sucrose has been shown to act as a source in cell signaling during stress, which induces the expression of various genes and plays roles in metabolite transport to aid in substrate recycling to sustain metabolic processes (Espevig et al. 2011). Foliar or exogenous treatments of sucrose have been reported to increase low-temperature tolerance in many plants, such as *Arabidopsis* (Rekarte-Cowie et al. 2008), cucumber (*Cucumis sativus*) (Cao et al. 2014), and stone fruits (*Prunus cerasus*) (Barraco et al. 2012) by activating antioxidants, stress-responsive genes, and other protective compounds like proline.

In *Arabidopsis thaliana*, cold acclimation is established through an intricate network of signaling pathways. The best characterized of them is mediated by a family of three AP2-type transcription factors named C-REPEAT/DRE BINDING FACTORs (CBF1 to CBF3) (Medina et al. 2011). These factors are estimated to control the induction of around 12% of the *Arabidopsis* cold-responsive genes (Fowler and Thomashow 2002). The expression of each *CBF*, in turn, is specifically induced by low temperature in a fast and transient manner (Gilmour et al. 1998; Liu et al. 1998; Medina et al. 1999), and their individual overexpression in *Arabidopsis* results in the constitutive expression of the target genes and increased freezing tolerance (Gilmour et al. 2000; Fowler and Thomashow 2002). As expected, considering their importance in cold acclimation, the expression of *CBFs* has been shown to be remarkably regulated (Medina et al. 2011). Another relevant signaling pathway involved in cold acclimation is mediated by abscisic acid. This hormone accumulates in response to low temperature (Lang et al. 1994), and exogenous abscisic acid application increases plant freezing tolerance (Gilmour and Thomashow 1991). Furthermore, mutants affected in abscisic acid biosynthesis and perception show altered tolerance to freezing temperatures before and after cold acclimation (Llorente et al. 2000). The emerging picture is that, in addition to abscisic acid, other hormones are also involved in cold acclimation. For example, ethylene accumulates in *Arabidopsis*, bean (*Phaseolus vulgaris*), rhododendron (*Rhododendron* spp.), winter rape (*Brassica napus*), tomato (*Solanum lycopersicum*), and winter rye (*Secale cereale*) after exposure to low temperature (Yu et al. 2001; Wang et al. 2012a). Moreover, the cold-induced accumulation of ethylene in winter rye promotes the synthesis of antifreezing proteins and the consequent increase in freezing tolerance (Yu et al. 2001). In addition, exogenous application of ethephon, an ethylene releaser, enhances rhododendron freezing tolerance (Harber and Fuchigami 1989), and overexpression in the sense or antisense orientation of *TERF2/ERF2*, a tomato ethylene inducible gene encoding a transcription factor, can increase or decrease freezing tolerance in tomato and tobacco (*Nicotiana tabacum*), respectively (Zhang and Huang 2010). In contrast with these data, Shi et al. (2012) proposed that ethylene acts as a negative regulator of freezing tolerance and cold acclimation in *Arabidopsis*, which indicates that the role of this hormone in the low-temperature response still needs further investigation (Catalá et al. 2014).

4.5.2 Physiological and Metabolic Responses to Low Temperature

Low temperature is one of the most severe weather events that destructively affect crop growth, quality, and yield (Taşgın et al. 2006; Hu et al. 2016). Many different metabolic and physiological changes occur in response to low temperature for cool season and warm season grass species. These changes include the alteration in membrane stability, osmotic adjustment, and accumulation of protective compounds like amino acids, sugars, carbohydrates, or antioxidant enzymes (Hoffman et al. 2010, 2014; Shi et al. 2014). Cold or chilling stress impairs the chloroplast microstructure, photosynthetic metabolism, and energy production (Bilska and Pawe 2010), directly leading to inhibited photosynthesis, which is a severe threat to crop production. Additionally, the efficiency of photosynthetic electron transport in plants is significantly decreased under chilling stress, resulting in a burst of reactive oxygen species that directly causes cellular oxidative damages and increases membrane rigidity (Lawlor and Tezara 2009; Chaves et al. 2009; Xie et al. 2009; Nakashima et al. 2007). Low temperature also disrupts the carbon reduction cycle and the control of stomatal conductance (Allen and Ort 2001). Low temperature impacts plant growth and development by limiting the rate of photosynthesis and contributing to the degradation of cell components like chlorophyll. Also, typical symptoms of freezing stress are intracellular or extracellular ice formation, which can lead to membrane damage or cellular dehydration (Shi et al. 2014). Low temperature can cause damage or denaturing to other important cellular components like proteins and enzymes and can lead to the production of reactive oxygen species (McKersie and Bowley 1997). Chlorophyll content decreases under cold stress, which might be because low temperature suppresses chlorophyll biosynthesis, probably by inhibiting the activities of chlorophyll biosynthetic enzymes (Nagata et al. 2007). An earlier study showed that the optimum temperature of divinyl reductase activity was 30 °C, low activity was observed at 10 °C, and no activity was found at 50 °C (Nagata et al. 2007). Zhao et al. (2020) observed that under low temperature, chlorophyll biosynthesis was inhibited and chlorophyll accumulation declined during greening. The activities of glutamate-1-semialdehyde transaminase, Mg-chelatase, and protochlorophyllide oxidoreductase were downregulated under low temperatures. Chloroplasts at 18 °C had loose granum lamellae, while the thylakoid and lamellar structures of grana could hardly develop at 12 °C after 48 h of greening. Furthermore, the accumulation of reactive oxygen species and malondialdehyde and cell death were all induced by low temperatures. Phenotypically, decreasing temperatures result in reduced leaf expansion, wilting, chlorosis, necrosis, and damage to tissues like reproductive organs (Mahajan and Tuteja 2005). These symptoms of cold or freezing stress dramatically affect turf quality and overall plant performance and can potentially lead to plant death, with a better understanding of the physiological and metabolic impacts of cold stress,

The basic structure and organization of most plant tissues and cells dictate to a large extent the site of initiation of ice crystallization, determining the course and location of cell-water freezing (Mazur 1963, 1969; Burke et al. 1976). In the freezing process, plant tissues are affected by the following facts:

- Plant cells are not immersed in an aqueous solution but are surrounded by a water-saturated environment.
- The amount of osmotically available extracellular water is small relative to the amount of water inside living cells (Wenkert 1980).
- Apoplastic solution has a much lower solute concentration than the cell (Wenkert 1980).
- Cellular water, because of the higher solute concentration, has a greater freezing point depression (Levitt 1972, 1980).
- A functionally intact cell membrane is an effective barrier to the propagation of ice crystals (Dowgert and Steponkus 1984; Steponkus 1984).
- Liquid water can freely move across the plasma membrane in either direction.

- The effectiveness of the cell membrane as a barrier to ice may vary with cold acclimation or temperature (Steponkus et al. 1982, 1983; Dowgert and Steponkus 1984).
- The presence of heterogeneous nucleators inside cells is minimized, excluded, or masked (Burke et al. 1976; Sakai and Larcher 1987).
- Many tissues and a portion of the internal extracellular volume is free air space, normally saturated with water vapor.

The other major factor in the freezing of plant cells is the rate of cooling (Steponkus et al. 1983). Crystallization of the water inside the cell may occur by internal nucleation or by penetration of an external ice crystal into the cell (Steponkus et al. 1983). For yeast cells and leaf protoplasts suspended in a liquid medium, the rate of cooling required for the formation of intracellular freezing must be, respectively, $\geq$10 °C/min (Mazur 1969) and 3–16 °C/min (Steponkus et al. 1982, 1983). In contrast, some circumstantial evidence suggests cooling rates of $\geq$3 °C/h may favor intracellular freezing in intact plants (Steffen et al. 1989).

In several plant species (e.g., *Lolium temulentum*), progressive changes in gene expression resulted from several days or longer exposure to low temperatures (e.g., 5 °C) (Karimzadeh et al. 2000), and these changes occurred in parallel with an increase in chilling or freezing tolerance (Ougham and Howarth 1988). Such low-temperature tolerance is often referred to as hardening; altered gene expression is probably an essential component of the hardening process (Whelan 1991; Fujita et al. 1992; Howarth and Ougham 1993). Cold-hardening enables an organism to tolerate sub-zero temperatures after an initial exposure to low but noninjurious suboptimal temperatures (Ougham and Howarth 1988; Whelan 1991; Howarth and Ougham 1993). The duration of low-temperature treatment required for changes in protein synthesis differs between species, as does the duration of cold required for the development of hardiness (Mohapatra et al. 1987a; Hughes and Dunn 1990, 1996). For example, one cultivar of alfalfa attained cold hardening after 2 days at 4 °C while the other cultivar required 17 days (Mohapatra et al. 1987a).

Tropical and subtropical plants are sensitive to chilling stress and lack the capacity of cold acclimation. However, temperate plants have the ability to tolerate freezing temperatures following a period of exposure to nonfreezing temperatures, which is termed cold acclimation (Chinnusamy et al. 2007). Temperate plants are tolerant to seasonal changes in temperature and can tolerate cold stress during early spring and winter. Many important crops, such as rice, corn, soybean, potato, cotton, and tomato, are chilling sensitive and incapable of cold acclimation. In contrast, some crops, such as oats, are chilling tolerant but freezing sensitive. On the other hand, barley, wheat, and rye are well adapted to freezing temperatures (Zhang et al. 2011). However, some plants such as *Arabidopsis*, winter wheat (*Triticum aestivum* L.), and barley (*Hordeum vulgare* L.) are not able to tolerate nonfreezing temperatures without cold acclimation (Zhao et al. 2015). Similarly, *Santalum album* and *Betula utilis* (Himalayan Birch), as dominant vegetations in cold environments (Eastern Nepal), are also sensitive to cold during the growing season (Zhang et al. 2017; Pandey et al. 2018).

Each plant has different enrichment pathways in different periods of cold stress, such as the amino sugar and nucleotide sugar metabolism pathway, alanine and protein export, and the aspartate and glutamate metabolisms that are highly enriched during the latter stages of cold stress periods in tea plants (Ritonga and Chen 2020). In contrast, cellular components, biological processes, and the molecular function category are highly enriched in the early periods of cold stress in tea plants. The same result was found using transcriptome analysis in which flavonoid biosynthesis, phagosome, plant hormone signal transduction, and fructose were highly enriched at the beginning of cold stress (Hao et al. 2018). In tomatoes, even though the expression of ethylene signaling genes decreased after cold stress, none of the protein genes decreased. This shows the different regulation of gene and protein levels (Mata et al. 2019). In addition, abscisic acid signaling pathway gene expression improved the cold hardiness in grapevine buds only during the cold acclimation period (Rubio and Pérez 2019).

In some species, cold treatment has no effect on protein synthesis. For example, in *Lolium temulentum*, the pattern of polypeptide synthesis in the leaf base did not alter significantly in response to an abrupt drop in temperature from 20 to 5 °C (Thomas and Stoddart 1984; Stoddart et al. 1986; Ougham 1987). In contrast, the tropical cereal, sorghum (*Sorghum bicolor*), grown initially at 35 °C, responded markedly to low temperatures. In one cultivar, more new patterns of cold-shock proteins were detectable at 5 °C than in the other and both exhibited heat shock protein synthesis at 45 °C, a temperature which for many species would be high enough to abolish all protein synthesis (Ougham and Howarth 1988). Perhaps, differential responses between species or varieties are partly explicable by differing levels of developmental regulation of the perception of cold. The ability of plant organs to withstand and respond to cold treatment is remarkable, and low-temperature effects on the induction of cold-shock proteins are well documented (Howarth and Ougham 1993).

The ability to acclimate to frost tolerance is under genetic control, and a low temperature of 6–2 °C, which will cold acclimate genetically competent plants, such as winter cereals, will damage chill-sensitive species, such as maize. However, not all chill-tolerant plants are able to acclimate for frost tolerance, and many spring cereal cultivars can be classified in this group. In addition to acclimation for frost tolerance, a low positive temperature treatment may also acclimate chill-tolerant plants for growth at low temperatures (Hughes and Pearce 1988). In species and cultivars that will cold acclimate, the low-temperature treatment may also alter the growth habit and vernalize the plants. Furthermore, there is often an interaction between low temperature and other environmental factors, particularly day length and water status (Pan et al. 1994).

A study of plant cold-hardiness and freezing-stress injury had two primary goals. The first was to describe the mechanism(s) during the freeze/thaw cycle that leads to cell injury and death (Steponkus 1984). It was believed that once the mechanisms of injury were understood, it should be a relatively straightforward matter to determine the cellular and biochemical alterations necessary for conferring a higher level of cold tolerance. Nevertheless, researchers have developed a general understanding of the basic elements of the formation of ice in plant tissues (the freezing and thawing of cellular water) (Mazur 1963, 1969). Studies of the process by which plants become tolerant to the stresses imposed by freezing and thawing have gained added importance in efforts to understand plant freezing-tolerance mechanisms (Levitt 1972, 1980; Grace 1987).

A large number of plant species display physiological or cellular perturbations when encountering low-temperature stress. Under low-temperature stress, plants need to maintain cell behavior and activity, and, in particular, the stability of the cell membrane and structure of the protein with biological activity, for survival in adverse environments (Chen et al. 2018). The exposure of plants to subzero temperature leads to ice formation in plant tissues (Puhakainen 2004). Higher concentrations of active ice nucleators in the apoplastic solution of plants lead to a higher freezing point. As a result, ice crystals first form in the extracellular space of plant cells. Ice formation outside cells reduces the water potential of the apoplastic solution, which leads to water flowing from the cells (Ritonga and Chen 2020). Therefore, freezing stress at the cellular level is often followed by dehydration stress. Ice crystals will lead to an increase in electrolyte leakage and membrane lipid phase changes. As the freezing continues, osmotic forces produce cellular dehydration, which facilitates the formation of intracellular ice crystals. At the extreme, ice crystals can puncture plant cells and lead to cytosol outflow and ultimately cause the plants to die (Zhang et al. 2011; Sun et al. 2019; Demidchik et al. 2014). Therefore, preventing the formation of intracellular ice crystals and avoiding the growth of ice crystals are important for plants to tolerate cold stress. The most popular approach used by plants to deal with low-temperature stress is cold acclimation, which allows plants to survive freezing via accumulation of cryoprotective polypeptides (e.g., *COR15a*) and osmolytes (e.g., soluble sugars and proline). It has been reported that sugar content in *Euphorbia resinifera*, *Echinocactus grusonii*, *Aloe vera*, *Crassula lacteal*, *Bryophyllum pinnatum*, *Yucca aloifolia*, and *Sansevieria trifasciata* is increased after cold stress (Khan et al. 2015). Cold adaptive plants always store more sugar (D-glucose, D-glucose 6-phosphate, amylose, starch, and maltose) in their underground tissues (Janská et al. 2010; Londo et al. 2018).

A study involving grapevine reveals that grapevine leaves have a watery appearance, indicating tissue damage and cell leakage after freezing treatment. After 4 days of recovery, damage was found in the leaves (Londo et al. 2018). Necrosis of plants is normally caused by overproduction of reactive oxygen species. The elevated H_2O_2 level in plants under low-temperature stress is the result of increased oxygenation reaction in the chloroplasts, which leads to an increased glycolate content. The glycolate is converted to glyoxylate in peroxisomes by glycolate oxidase, which is accompanied by the accumulation of H_2O_2. The physiological process of reactive oxygen species toxic concentration in plants could be relieved by developing a complicated and efficient reactive oxygen species scavenging and antioxidant defense system. Plants require the use of low reactive oxygen species concentrations as mediators for signal transduction. Nitric oxide is implicated in the response of plants to low-temperature stress. Abscisic acid, Ca^{2+}, and $H_2O_2^-$ associated nitric oxide are shared by signaling cold stress events (Ding et al. 2019). Hemoglobin is believed to be a modifier of low-temperature plant response through the transition of nitric oxide (Abiri et al. 2017). Moreover, an increased malondialdehyde content and Ca^{2+} in the cytosol characterized lipid degradation, while the activity of antioxidant enzymes, such as catalase, superoxide dismutase, ascorbate peroxidase, and peroxidase, was minimized (Wang et al. 2018; Sun et al. 2019; Beck et al. 2004; Longo et al. 2017). In addition, abscisic acid and reactive oxygen species can induce Ca^{2+} and will affect cold signaling (Chinnusamy et al. 2007).

After methyl jasmonate treatment in bell pepper (Wang et al. 2018) and ethylene treatment in *Arabidopsis* (Hao et al. 2017), bell pepper and *Arabidopsis* are more tolerant to cold stress and chilling stress. Their physiological function is changed, namely, *via* a decrease of malondialdehyde content and an increase of antioxidant enzyme activities, such as catalase, superoxide dismutase, and peroxidase. This proves that these plants have a high tolerance to cold stress (Sun et al. 2019). Other research about cold stress revealed that cold could induce the inhibition of CO_2 assimilation in *Zea mays* and was also related to a persistent depression of the photochemical efficiency of PS II (Savitch et al. 2011). Moreover, nitric oxide is known as a signal in an early transduction network. Cold triggers nitric oxide a few hours after exposure to low temperature in *Arabidopsis* (Cantrel et al. 2011). Arabidopsis hemoglobin 1 (*AHb1*) interferes with nitric oxide production before and after exposure to low temperature. However, *A. thaliana* over-expressing *AHb1* resists its capacity to produce nitric oxide during cold stress (Cantrel et al. 2011).

Nevertheless, lowering the temperature decreased the translocation of xylem nitrogen in *Secale cereale* and *Brassica napus* by about 60% and 30%, respectively. It is suggested that low temperatures will directly affect the absorption mechanism of nitrates and nitrogen accumulation in the roots. This is caused by an obstruction of the nitrogen xylem flow greater than the nitric oxide absorption (Laine et al. 1994). Low temperature also contributed to decreasing the chlorophyll content of rice. Putting rice at low temperatures modified the number of chloroplasts, the arrangement of grana (no normal stacked membranous structures), and the lamellar structures in chloroplasts.

4.5.3 Inducible Response

Freezing tolerance of vast numbers of temperate and alpine species is not static but varies seasonally and in response to short-term variations in temperature. For centuries it has been known that temperate perennials follow a seasonal rhythm of dormancy and freezing tolerance (Levitt 1972). Temperate woody perennials rely on photoperiodic cues to signal the end of the active growth season and initiation of the developmental and physiological changes necessary to promote cold acclimation and increase freezing tolerance (Van Huystee et al. 1967). Following initiation by photoperiodic responses, acquisition of additional freezing tolerance via cold acclimation generally requires exposure to low nonfreezing temperatures (Weiser 1970; Sakai and Yoshida 1968). Tolerance to the mildest form of freezing, frost, is typically minimal or nonexistent in spring and summer, but in fall, susceptibility to even severe freezing stress declines. In winter, tolerance or hardiness reaches its greatest level. This is evident by the fact that some species when fully acclimated are not injured by exposure to temperatures approaching absolute zero (0 °K) (Sakai and Sugawara 1973).

Trees at the seedling stage are physiologically and morphologically very sensitive to cold stress, which can produce cell damage and reduce growth or even the tree's chances of survival (Yildiz et al. 2014). Therefore, the ability to cope with low temperatures is an important parameter driving the distribution of trees such as conifers in different locations (Bannister and Neuner 2001). At the molecular level, cold stress alters membrane fluidity and consequently membrane permeability. Plants respond to cold stress by inducing physiological and molecular changes, including the plant metabolic profile. These changes may play an advantageous role in cell cryoprotection during cold stress or prior to freezing temperatures (Thomashow 1999). However, seedlings' resistance to cold temperatures as well as the metabolic phenotype is likely to differ according to the plant genetic structure, the genetic origin, or the provenance (Kreyling et al. 2010), and this difference may affect the plant's ability to adapt to cold stress (Alonso-Blanco et al. 2005). For instance, differential sensitivity to winter cold was reported by Colombo et al. (1995) in Douglas fir (*Pseudotsuga menziensii*) tissues from different seedling provenances. In a similar manner, provenance significantly affected stem and needle hardiness of Scotch pine (*Pinus sylvestris*) (Bannister and Neuner 2001). At the molecular level, there are several descriptions of genes whose role is important in cold tolerance or acclimation (Thomashow 1999) or even genes whose overexpression may increase cold tolerance (Porcel et al. 2018), but there are no descriptions in the literature of metabolomic or molecular changes affecting different provenances of *Pinus halepensis* (Aleppo pine), an important conifer species widely used for afforestation or reforestation programs under climates with dry, hot summers, and cold winters. This lack of information is limiting not only our basic knowledge on the species but the chances of success in afforestation and reforestation programs. Selection and use of provenances and genotypes with potential resistance to sudden changes in environmental conditions, including cold, is crucial. But for many provenances, we do not have any information. There are several references in the literature studying the effect of cold in *Pinus halepensis* (Pardos et al. 2003; Puértolas et al. 2005), but none of them consider the effect of cold at the molecular or metabolic level. One advantage of having this information is that if we identify the physiological or molecular profile of cold-tolerant and cold-sensitive species, we can have a valuable tool to select provenances with more chances to survive under cold environments, as a higher content of a given metabolite may correlate with higher chances of survival (Taibi et al. 2018).

4.5.4 Nitrogen and Amino Acid Metabolism

Nitrogen is critical for plant health due to its involvement in amino acids, proteins, chlorophyll, and many other metabolites (Stitt et al. 2002). Nitrogen metabolism interacts with many different plant processes involved in growth and development, like rubisco for light-harvesting complexes during photosynthesis. Nitrogen and carbon are closely linked by the interaction of nitrogen assimilation and carbon metabolism, but under periods of stress or carbon starvation, proteins and some amino acids are catabolized (Stitt et al. 2002). Amino acids are important metabolites that are crucial for the synthesis of proteins and play multiple functions in plant growth and stress response. Amino acids can be precursors for the synthesis of various metabolic factors like hormones, chlorophyll, cell wall, secondary metabolites, and other cell constituents (Less and Galili 2008). They are synthesized in many different mechanisms using their respective substrates, all requiring a source of nitrogen. Plants can readily take up nitrogen forms from the soil like ammonium or the air to lead to the production of certain amino acids. Commonly the synthesis of one amino acid can also come from the degradation of proteins or other amino acids (Singh 1998). These stages of amino acid metabolism can be observed during a plant's life cycle. During germination, storage proteins may be broken down into amino acids, which will aid in the production of proteins necessary for growth initiation. For actively growing tissues, amino acids are upregulated to provide substrates for proteins or provide signals for the production of hormones or other metabolites. In senescing tissues, amino acids are allocated to provide energy to stressed tissues (Singh 1998). It is evident that free amino acid pools are constantly changing with regard to environmental sensors or plant health status. Throughout a plant's life cycle or during periods of stress, there is an interplay between free amino acids, protein biosynthesis, metabolic functions, and signal transduction pathways (Keough 2016).

Adaptive mechanisms to stress allow for amino acids like proline and arginine to respond directly to environmental signals, while others are produced by protein breakdown (Less and Galili 2008). It has been reported that stress can induce molecules or amino acids into specific compounds to alleviate the effects of stress. For example, under salt stress, an accumulation of proline has been observed with less proline degradation and the breakdown of serine into glycine betaine (Singh 1998). During environmental stress, plants undergo a bulk amino acid and protein degradation, leading to an accumulation of free amino acids to be used as alternative substrates in various processes to mitigate the effects of stress (Hildebrandt et al. 2015). Also, the upregulation of nitrogen-rich amino acids, like proline, supplies an energy source for not only stress periods but also post-stress recovery (Chen et al. 1964). This continuous process of the allocation of amino acids, proteins synthesis and degradation, is crucial for plant development as well as response to abiotic stress.

4.5.5 Protein Metabolism

The metabolic responses of plants at low temperatures have been covered in the numerous major monographs and reviews dealing with cold stress (Graham and Patterson 1982; Levitt 1972, 1980; Li 1984). Attempts have been made to correlate metabolic and biochemical responses with cold tolerance. Steponkus (1984) performed correlative studies of biochemical changes but did not provide an understanding of how cold acclimation leads to increased freezing tolerance. Studies of cold acclimation had a focus on some of the more rapid physiological and molecular responses in plants subjected to low temperatures, and these studies revealed that plant and algal cells can rapidly begin to alter membrane lipid composition (Dickens and Thompson 1981; Lynch and Thompson 1984), RNA (Cattivelli and Bartels 1989), and protein content (Gilmour et al. 1988; Maimiroli et al. 1986; Yacoob and Filion 1986a, b) within hours of low-temperature exposure. These findings of rapid biochemical alterations in response to low temperature are in agreement with observations of the rapid induction of freezing tolerance at inductive temperatures and at noninductive temperatures by desiccation (Siminovitch and Cloutier 1982) and abscisic acid (Chen et al. 1983; Chen and Gusta 1983).

Wheat seedlings transferred from 25 to 5 °C, total soluble protein accumulation, in vivo protein synthesis and breakdown, in vitro protein breakdown, and SDS-PAGE profiles of proteinases in gelatine-containing gels were analyzed (Pinedo et al. 2000). Leaf protein content increased within a 7-day period (70% over the initial value) in plants exposed to 5 °C. The fast protein accumulation observed on days 0–2 was mainly attributed to a decreased breakdown. In further days, parallelly to a slowdown in the rate of protein accumulation, the leaf proteolytic activity increased. The incubation temperature also had an influence on the proteolytic activity: Q_{10} values for the 15–5 °C range were 80–200% higher than those observed for the 25–15 °C range. On the other hand, the in vivo protein synthesis capacity, at either 25 or 55 °C, was not significantly modified in cold-treated plants. In addition to the enhanced activities of two serine-proteinases (previously found in control plants by SDS-PAGE analysis), cold-treated plants displayed a new proteinase, which has not been detected so far.

4.5.6 Free Amino Acids

Ability of plants to counteract such freezing effect with erratic temperature patterns which compel premature deacclimation followed by a cold backlash (Arora 2013) has critical importance for their survival against spring frost.

Freezing stress at molecular levels also makes plants suffer biochemical strains such as excessive production of reactive oxygen species (ROS) during electron transport processes in photosynthesis and mitochondrial respiration. It also causes perturbation of amino acids, carbohydrates, and lipid metabolisms (Thomashow 1999).

Proline (Pro) is one of compatible solutes. It is widely known that Pro is accumulated in response to various abiotic stresses. It mainly serves as a mediator of osmotic adjustment, a chaperone in the cell, and a major component of cell wall structural proteins (Kavi Kishor and Sreenivasulu 2014).

Pro accumulation is known to confer tolerance in response to salt or temperature stress in different species (Chu et al. 1974; Bagdi and Shaw 2013). Recently, some studies have demonstrated that exogenous Pro can mitigate detrimental effects of salt and cadmium stresses (Heuer 2003; Hoque et al. 2007; Islam et al. 2009). On the other hand, Pro accumulation does not necessarily lead to dehydration or salt-stress tolerance in Arabidopsis thaliana (Liu and Zhu 1997; Nakashima et al. 1998). In our previous studies, contrary to our expectations largely based on published reports of Pro accumulation in response to cold stress in various herb-aceous species (Chu et al. 1974; Duncan and Widholm 1987; Savouré et al. 1997; Hur et al. 2004), Pro accumulation of peach shoots was likely a response to activation of metabolism associated with growth resumption in eco-dormant shoots at relatively high temperatures to be conducive for growth. Ability of plants to counteract such freezing effect with erratic temperature patterns which compel premature deacclimation followed by a cold backlash (Arora 2013) has critical importance for their survival against spring frost.

Plants subjected to stress show accumulation of proline and other amino acids. The role played by accumulated amino acids in plants varies from acting as an osmolyte, regulation of ion transport, modulating stomatal opening, and detoxification of heavy metals. Amino acids also affect the synthesis and activity of some enzymes, gene expression, and redox homeostasis (Rai 2002). The accumulation of amino acids is the most common response of plants to abiotic stresses (Ali et al. 2019), and in plant tissues exposed to such stresses, concentrations of amino acids may rise up to the milli molar range (Planchet et al. 2011; Singh et al. 1972). Amino acids have been shown to have multifarious functional roles in plant stress tolerance by acting as compatible osmolytes (Deivanai et al. 2011), being involved in pH regulation and in the detoxification of reactive oxygen species (Rizwan et al. 2016), and acting as a nitrogen and carbon reserve, mainly for the synthesis of specific enzymes and precursors of various secondary metabolites such as flavonoids and lignins, as well as being an available stock of amino acids, which are useful during recovery from stress. Amino acids also act as regulatory and signaling molecules (Munns 2002). These protecting roles of amino acids are also supported by the fact that stress-tolerant plants accumulate higher amino acids than sensitive plants and show a correlation between stress tolerance and amino acid levels. The increase in the level of some amino acids has been reported to lead to beneficial effects during stress acclimation. In some cases, amino acids accumulate as a sink of excess nitrogen, indicating a reduction in growth rate. In the latter case, the increase of amino acid concentrations indicates damage to cell functioning rather than an adaptive response to environmental stresses (Munns 2002). Amino acids are well-known biostimulants that have positive effects on plant growth and yield and significantly mitigate the injuries caused by abiotic stresses (Kowalczyk et al. 2008). Amino acids as constituents of a variety of proteins have the potential to alleviate abiotic stresses in crop plants (Sadak et al. 2015; Sadak and Abdelhamid 2015). In addition, metabolite analysis of the mitochondrial protein complex composition following oxidative stress suggested that enhancement of amino acid catabolism into the TCA cycle might compensate for a reduced electron supply from the TCA cycle (Obata et al. 2011). The metabolic connection of the amino acid pathway with the TCA cycle has also recently emerged from a dedicated bioinformatics approach, analyzing the network expression behavior of the entire set of *Arabidopsis* genes encoding metabolic enzymes and transcription factors to multiple abiotic and biotic stresses as well as to some nutritional and hormonal cues (Less et al. 2011). Plants have the ability to maintain the cellular content of specific amino acids for stress tolerance, but this is plant-species specific. Some high-yielding cultivars are an exception to that and are more tolerant through their exogenous use (Ali and Ashraf 2011) through varying modes. This alters the internal cellular metabolic activities in different ways. Application of amino acids as a foliar spray caused an increase in the content of total carbohydrates and polysaccharides of stressed and nonstressed plants, and similar results have been reported in other plant species (Nahed et al. 2010).

Duke et al. (1978) studied the free amino acid concentration in germinating soybean (Glycine max [L.] Merr. cv. Wells) seedlings at 10 and 23 °C, and it was observed that after 5 days of germination at 10 °C, glutamate and aspartate were in high concentration in both cotyledons and axes (38 and 24% of total free amino acids recovered, respectively), whereas the concentrations of their amide derivatives, asparagine and glutamine, were low in cotyledons (4.4%) and high in axes (21%). In contrast, after 5 days of germination at 23 °C, asparagine and glutamine accounted for 22 and 45% of total free amino acids in cotyledons and axes, respectively, and aspartate and glutamate concentrations were low. The activities of glutamine synthetase and asparagine synthetase were considerably lower in tissues from the 10 °C treatment than those from the 23 °C treatment. Aspartate and glutamate concentrations were nearly equal in all but one sample. Both glutamate oxaloacetate transaminase and glutamate dehydrogenase activities were much higher in axis tissues at 23 °C as compared to 10 °C. This indicates a difference in enzyme complement for glutamate dehydrogenase with the two treatments. Hydrolysis of free amino acid sample (basic fraction) aliquots showed large quantities of peptides in 23 °C-treated axes at 2 days, while few or no peptides were found in the 10 °C treatment. Amino acid residues most prevalent in peptides were aspartate, threonine, serine, glutamate, and glycine.

The ability of plants to counteract such freezing effect with erratic temperature patterns, which compel premature deacclimation followed by a cold backlash (Arora 2013), has critical importance for their survival against spring frost. Freezing stress at molecular levels also makes plants suffer biochemical strains such as excessive production of reactive oxygen species during electron transport processes in photosynthesis and mitochondrial respiration. It also causes perturbation of amino acids, carbohydrates, and lipid metabolisms (Thomashow 1999). Proline is one of the compatible solutes. It is widely known that proline is accumulated in response to various abiotic stresses. It mainly serves as a mediator of osmotic adjustment, a chaperone in the cell, and a major component of cell wall structural proteins (Kavi Kishor and Sreenivasulu 2014). Proline accumulation is known to confer tolerance in response to salt or temperature stress in different species (Chu et al. 1974; Bagdi and Shaw 2013). Studies have demonstrated that exogenous proline can mitigate detrimental effects of salt and cadmium stresses (Heuer 2003; Hoque et al. 2007; Islam et al. 2009). On the other hand, proline accumulation does not necessarily lead to dehydration or salt stress tolerance in *Arabidopsis thaliana* (Liu and Zhu 1997; Nakashima et al. 1998). Contrary to expectations on proline accumulation in response to cold stress in various herbaceous species (Chu et al. 1974; Duncan and Widholm 1987; Savouré et al. 1997; Hur et al. 2004), proline accumulation of peach shoots was likely a response to activation of metabolism associated with growth resumption in eco-dormant shoots at relatively high temperatures to be conducive for growth (Shin et al. 2016). Therefore, the relationship between proline accumulation and abiotic stress tolerance in plants does not seem to be universal. Furthermore, the critical correlation between proline synthesis and freezing stress has not been investigated yet. Amino acids including proline are precursors and constituents of proteins. They are also mobilized as precursors of other nitrogen compounds. Several studies have reported that free amino acids such as aspartic acid, serine, glycine, and proline are accumulated in response to drought and osmotic stresses in different species (Hanower and Brzozowska 1975; Yang et al. 2000; Di Martino et al. 2003). Although the correlation between proline and abiotic stresses has been of interest in many studies, correlations of other free amino acids with abiotic stresses have been less studied (Shin et al. 2018).

4.5.7 Role of γ-Aminobutyric Acid in Cold and Frost Tolerance

Among the numerous compounds that seem to serve as osmoprotectants in plants is the four-carbon, nonprotein amino acid γ-aminobutyric acid, which under stress conditions may represent a significant fraction of the free amino acid pool (Bown and Shelp 1989). Intracellular levels of γ-aminobutyric acid are typically low, but they rapidly increase in response to several abiotic stresses, such as hypoxia, drought, cold, heat shock, and mechanical stimulation (Bown and Shelp 1997). In this context, given also some peculiar properties like the zwitterionic form at neutral pH and high solubility, a role as a compatible osmolyte has been hypothesized (Shelp et al. 1999). γ-Aminobutyric acid is synthesized by a cytosol-localized glutamate decarboxylase, a Ca^{2+} calmodulin-dependent protein (Baum et al. 1993) that can thus be rapidly activated during various stress situations that elicit changes in cytosolic Ca^{2+} concentration (Snedden et al. 1996). Following irreversible glutamate decarboxylation, γ-aminobutyric acid may be further metabolized to succinic acid in the so-called γ-aminobutyric acid shunt, a short pathway bypassing two steps in the Krebs cycle. The γ-aminobutyric acid shunt takes place in the mitochondrion by means of two enzymes, a γ-aminobutyric acid transaminase using either α-ketoglutarate or pyruvate as amino acceptor, and a succinic semialdehyde dehydrogenase (Shelp et al. 1999). Although the γ-aminobutyric acid shunt is widely distributed in most prokaryotes and eukaryotes, it is only in animals that a major role for γ-aminobutyric acid, as the predominant inhibitory neurotransmitter of the brain, has been well established.

Some experimental evidence has been obtained suggesting that a possible involvement of γ-aminobutyric acid in several physiological processes in plants (Bouché and Fromm 2004). Because of its role in neurotransmission, γ-aminobutyric acid could be produced to deter insect feeding and interfere with their development (MacGregor et al. 2003). γ-Aminobutyric acid might function as an endogenous signaling molecule as well since there are indications that glutamate/γ-aminobutyric acid receptors do exist in plants as well (Kang and Turano 2003). γ-Aminobutyric acid levels increased under long-term conditions that limit glutamine production, reduce protein synthesis, or enhance protein degradation; thus, γ-aminobutyric acid might be a temporary store of nitrogen and play a role as a sensor of nitrogen status and C:N balance, or a long-distance interorgan signal molecule (Beuve et al. 2004). Moreover, because glutamate decarboxylase activity is low pH-dependent and consumes H^+, γ-aminobutyric acid synthesis might function as a pH-stat, being activated under conditions that cause cytosolic acidification (Crawford et al. 1994). On the other hand, Arabidopsis *pollen–pistil interaction2* mutant resulting in growth inhibition and misguidance of pollen tubes in pistils was shown to lack a functional γ-aminobutyric acid transaminase; thus, a gradient of γ-aminobutyric acid concentration seems essential for the growth and guidance of pollen tubes (Palanivelu et al. 2003). Moreover, succinic semialdehyde dehydrogenase was found to be negatively regulated by ATP and NADH, suggesting a control of the γ-aminobutyric acid shunt by mitochondrial energy charge and reducing potential (Breitkreuz and Shelp 1995). The disruption of the unique succinic semialdehyde dehydrogenase gene in *Arabidopsis thaliana* resulted in plants undergoing necrotic cell death caused by an abnormal accumulation of reactive oxygen species. Such behavior might rely upon the ability of the γ-aminobutyric acid shunt to supply NADH and/or succinate to mitochondrial metabolism under conditions that inhibit the TCA cycle, impair respiration, and enhance the production of reactive oxygen species (Bouché et al. 2003). In most cases, γ-aminobutyric acid accumulation can be simply associated with the cited physiological processes, and the speculated links are far from being definitely proved. Overall, the exact role of the γ-aminobutyric acid shunt in plants still awaits elucidation (Mazzucotelli et al. 2006).

4.5.8 Exogenous Application of Amino Acid Under Temperature Stress

The exogenous application of different antioxidants in plants might be helpful in minimizing the harmful effects of stress. Many substances are useful in the way that they provide crucial protection against oxidative damage and, many of them, when applied in small quantities, are also able to enhance the plants' ability to assimilate applied nutrients or provide benefits to plant development (Calvo et al. 2014). Biostimulating effect is widely described in many scientific papers on substances such as tocopherol, polyamines, L-ascorbate, glutathione, glycine betaine, putrescine, glycine, spermine, spermidine, salicylic acid, hydrogen peroxide, silicone, alginic acid, polyphenols, and trehalose (Bafeel and Ibrahim 2008; Winter et al. 2015; Ahmad et al. 2014a; Ali et al. 2013; Møller et al. 2007; Ashraf and Foolad 2007; Sotiropoulos et al. 2005; Hassanein et al. 2013; Venkatesh and Park 2014;

Yang et al. 2017; Xiaochuang et al. 2017; Waqas et al. 2017; Matysiak et al. 2018; Moradtalab et al. 2018).

One of the main categories of biostimulants is protein hydrolysates and other nitrogen-containing compounds. This group includes free amino acids. In general, protein oxidation by reactive oxygen species is irreversible, but this does not apply to proteins that are rich in sulfur-containing amino acids (arginine, histidine, proline, threonine, tryptophan). These amino acids, apart from the thiol group, also contain a carbonyl group, which inhibits the effects of the proteolytic attack (Ghezzi and Bonetto 2003). Thus, it has been demonstrated that exogenous application of certain amino acids allows for plants to survive unfavorable stress conditions and stimulates their further, undisturbed growth and development (Ali et al. 2013; Calvo et al. 2014; du Jardin 2015; Rai and Sharma 1989; Hassanein et al. 2009; Rafiee et al. 2013). Some studies have confirmed that, in plants, the most functionally diverse amino acid and a precursor for the biosynthesis of nitric oxide and polyamines is L-arginine (Nasibi et al. 2013; Liu et al. 2007; Nejadalimoradi et al. 2014). Polyamines and their precursor arginine have been implicated as vital modulators in a variety of processes in higher plants, such as growth, physiology, and development, as well as in plant response to various stress factors. It has also been recorded that both endogenous and exogenous arginine play a role in plant stress response (Liu et al. 2015; Winter et al. 2015; Qados 2010; Greaves 1996). Glycine is the simplest amino acid in nature and is one of the most abundant free amino acids in soil. It is more rapidly taken up by plants than other amino acids. However, there have been far fewer studies on the role of the exogenous application of glycine on plant's growth (Yang et al. 2017). Matysiak et al. (2020) tested ten different concentrations (from 0.006 to 9.0 mM) of amino acids, viz., L-arginine and L-glycine, on maize crop under constant temperature condition 20–22 °C/23–25 °C (night/day) and an fluctuating stress of rising and dropping temperatures between 12–15 °C (night) and 30–38 °C (day). A large discrepancy between the effects of the two amino acids was observed after foliar application. Under optimal thermal conditions, L-arginine increased the mass of shoots and roots by 55–59%. Under stress conditions, root mass was increased even by 100% when compared to the control plants. The best results were recorded at concentrations of 6 mM and 3 mM. Plants that were treated with glycine concentrations generally reached the height of untreated plants or less. It was shown that glycine applied at concentrations of 0.2–3 mM has a negative effect on the fresh mass of the crop.

Plant thermal acclimation mechanisms include the accumulation of compatible nitrogen-rich solutes, such as amino acids, that confer stress tolerance because they act as osmolytes (regulating ion transport and stomatal opening), they have enzyme protecting effect, and they influence gene expression and redox homeostasis (Botta 2013). This increase of free amino acids naturally originates from hydrolysis of functional proteins or by inhibition of new protein synthesis. However, external applications of amino acids have the advantage of avoiding protein breakdown and saving energy resources. In a study in a controlled environment on lettuce plants subjected to three different types of cold stress, treated plants have a higher fresh weight than control plants, exhibiting a higher stomatal conductance, which implies productive improvements. Also, heat stress tolerance was evaluated on ryegrass plants (*Lolium perenne* L.) under several temperatures regimes where lawn physiological and quality parameters were measured. In this case, at a high temperature (36 °C), ryegrass treated with enzyme hydrolyzed proteins shows a superior photosynthetic efficiency (Fv/Fm) and maintains higher levels of chlorophylls and carotenoids. These findings suggest that foliar application of enzyme hydrolyzed proteins has a similar effect to natural plant amino acids and promotes a better, more prompt crop recovery from temperature stress.

4.5.9 Enzyme Variation

Studies of enzymes from plants subjected to low temperatures have demonstrated changes in activity, freeze stability, and isozymic variation when compared to plants that were maintained at warm temperatures (Guy 1990). Using electrophoretic techniques to separate enzymes from nonhardened and hardened tissues, peroxidase activity was shown to be increased in hardened stems of four widely unrelated woody species (McCown et al. 1969). In three of the species, several peroxidase isozymes present in hardened tissues were not found in nonhardened tissues. However, in another study, no change in peroxidases, glucose-6-phosphate, 6-phosphogluconate, and malate dehydrogenases was observed in willow stem during de-acclimation (Hall et al. 1970). Differences were noted for lactate dehydrogenase, where the activity increased during deacclimation. Consistent with the above finding is the demonstration that invertase in wheat leaves at low temperature undergoes a shift from a lower-molecular-weight form to a higher-molecular-weight form (Roberts 1974). In studies with alfalfa, Krasnuk et al. (1975, 1976a, b) observed during winter increased activity of a number of dehydrogenases associated with respiratory pathways, including glucose-6-phosphate dehydrogenase, lactate, and isocitrate dehydrogenase (Krasnuk et al. 1976b). The increases in enzyme activities paralleled increases in soluble protein content during winter, suggesting that the increase in activity may result from higher amounts of enzyme. Also, in winter, new isozymic variants for ATPases, esterases, acid phosphatases, leucine aminopeptidases, peroxidases, and some dehydrogenases were present (Krasnuk et al. 1976b).

However, it is not clear whether the altered stability was an intrinsic property of the enzyme, was produced by new isozymic variants, or arose from some other factor present in the tissue extracts (Guy 1990). A study of glutathione reductase from the spinach has demonstrated, along with the appearance of additional isozymic forms in cold-acclimated tissue, increased activity, freezing stability, and altered kinetic behavior (Guy and Carter 1984). Glutathione reductase activity was decreased by freezing/thawing both in vitro and in vivo. However, the enzyme from cold-acclimated plants was less sensitive to freezing than its counterpart from nonacclimated plants. In contrast, ferredoxin NADP reductase from nonacclimated and cold-acclimated wheat showed identical kinetic parameters and freeze/thaw inactivation (Riov and Brown 1976), but activity was increased during cold acclimation. These studies illustrate the potential for alterations in enzymes in response to low-temperature exposure and the apparent selective basis on which such changes may occur.

4.5.10 Protein Content

Accumulation of soluble protein in cold-acclimated cortical bark cells of black locust trees was first correlated with freezing tolerance (Siminovitch and Briggs 1949). A study of the seasonal variation protein content and hardiness of the bark cells demonstrated the accumulation of soluble proteins in fall, which closely paralleled the induction of freezing tolerance (Siminovitch and Briggs 1953). Throughout the winter, the protein concentration remained high but declined rapidly during the breaking of dormancy and the resumption of growth in the spring. The decline of soluble protein again closely matched the loss of freezing tolerance. Studies with black locusts have verified these early findings (Siminovitch et al. 1968; Sakai and Sugawara 1973; Pomeroy et al. 1970; Brown and Bixby 1975). Freezing tolerance may not be explained as a simple function of the soluble protein concentration. There are many reasons why some plants might accumulate soluble proteins during cold acclimation, but with the exception of a protoplasmic augmentation hypothesis (Siminovitch et al. 1968), no clear mechanistic rationale for conferring greater freezing tolerance has been proposed for this hardening response. In temperate deciduous perennials like black locusts, the export of nitrogenous materials during senescence could provide the nitrogen source for the accumulation of proteins in the cortical cells of the living bark. Therefore, in black locusts, the accumulation of protein in fall could be more centrally related to the storage processes necessary in deciduous species than to freezing tolerance.

In the early studies of black locusts (Siminovitch and Briggs 1949), the soluble protein content was found to be increased during the period when the cells were most hardy, and most of the studies have confirmed the notion that cold-acclimated and freezing-tolerant plants contain new protein species not present in nonacclimated plants. When compared to nonacclimated plants, the shift in protein content in cold-acclimated tissues was not dramatic but subtle, involving mostly the appearance and disappearance of minor bands in gels. Unfortunately, several major obstacles have discouraged the pursuit of this line of study and obscured the significance of the cumulative findings.

4.5.11 Genetic Analysis of Cold Acclimation

The genetic analysis of cold acclimation has been carried out primarily in cereals. Winter hardiness of a cereal crop is composed of a number of interacting components, frost tolerance, vernalization requirement, and photoperiod sensitivity, each of which is complex. In the diploid cereal barley, it has been suggested that a change in the growth habit caused by the vernalization response can contribute directly to winter hardiness of the crop (Hughes and Dunn 1996); however, Doll et al. (1989) were able to separate vernalization requirement and good frost tolerance in a number of doubled haploid lines produced by crossing a winter, frost hardy cultivar with four spring, frost-sensitive cultivars. Takahashi and Yasuda (1970) concluded that there were three major genes for vernalization requirement located on barley chromosomes 4, 7, and 5. Hackett et al. (1992) mapped a quantitative trait locus for vernalization on barley chromosome 4, although this locus may strictly be a heading (flowering) date locus (Pan et al. 1994). A 74-point linkage analysis of a spring x winter barley cross by Pan et al. (1994) has identified significant quantitative trait loci for winter survival on chromosome 7.

The homeologous wheat chromosome for barley chromosome 7 is chromosome 5. In hexaploid wheat, five vernalization loci have been identified. Of the chromosomes identified as carrying genes involved in frost tolerance in wheat, those most frequently implicated are chromosomes 5A, 7A, 4D, and 5D (Sutka and Snape 1989). Sutka and Snape (1989) have reported a locus having a major effect on frost tolerance on chromosome 5A, which is closely linked to the recessive allele. Roberts (1990) identified two loci on wheat chromosome 5A, which are involved in frost tolerance. These genetic studies only reveal loci that are involved in the response of cereals to low temperatures, where the parents used in the cross contain different alleles that segregate in the progeny of the F_1 hybrid or alter the genotype of cytological substitution or addition lines. The power of this analysis to dissect the processes of frost acclimation depends, therefore, on the amount of variation available within the crop/species.

4.6 Protein Synthesis

4.6.1 Types of Proteins

Plants produce different types of proteins to cope with environmental factors like heat shock protein, cold shock protein, and proteins produced during drought stress and salinity as given below (Key et al. 1981).

4.6.1.1 Cold Shock Proteins

Plants vary in their "cold-hardiness" depending on the level of low-temperature stress they can tolerate. Many temperate agricultural crops, including tobacco, tomatoes, potatoes, corn, and apples, as well as certain ornamentals such as impatiens, begonias, and phlox, are considered "very tender" or "half-hardy" and sustain freezing damage between 0 and −4 °C. As such, sub-zero temperatures in early spring and late autumn not only shorten the growing season but also result in substantial crop loss. Damage to plant cells invariably results following the growth of ice crystals in the extracellular space (Guy 1990; Pearce and Ashworth 1992; Wisniewski et al. 1997). The growth of large ice crystals, at the expense of smaller ones, occurs through a thermodynamically favorable process known as ice recrystallization. Cells may sustain mechanical damage or cellular dehydration resulting from the sequestration of intracellular water caused by the osmotic imbalance imposed by the exclusion of solutes from recrystallized ice (Pearce 2001). This can have devastating effects on cell metabolism associated with protein inactivation or denaturation and an increase in reactive oxygen species (Thomashow 1998). If the dehydration is severe, the loss of cell volume may also facilitate the collapse of membrane structures resulting in cell lysis.

Ice-binding proteins are low-temperature-associated proteins found in certain cold-hardy animals and microbes as well as some freeze-tolerant plants. These proteins serve to control the growth of ice crystals and mitigate freezing damage. Plant ice-binding proteins appear to be particularly efficient at adsorbing ice crystals, thereby preventing the migration of water molecules at temperatures close to freezing or under freeze-thaw conditions. This property is known as ice recrystallization inhibition. Because of these functional properties, plant ice-binding proteins have also been variously known as antifreeze proteins. Plant ice-binding proteins have likely evolved independently many times (Gupta and Deswal 2012, 2014). This diversity has resulted in a multiplicity of other functions associated with biotic and abiotic stress responses, including, most relevantly, the inhibition of bacterial ice nucleation, which would otherwise result in freezing at high sub-zero temperatures (Tomalty and Walker 2014; Bredow 2017). Indeed, the wide distribution of ice-binding proteins across a variety of plant species points to their importance in freezing tolerance and also highlights their possible utility for the development of freeze-tolerant crops (Bredow and Walker 2017).

Cold shock response in *Escherichia coli* is triggered by a sudden temperature downshift, which causes a transient inhibition of synthesis of most cellular proteins, resulting in a growth lag period called the acclimation phase (Etchegaray and Inouya 1999). During this acclimation phase, at least 15 different cold shock proteins are significantly induced, and some of them are essential for cell growth at low temperatures (Jones et al. 1992; Jones and Inouya 1996; Teana et al. 1991). It has been shown that the induction of cold shock protein A is caused mainly by dramatic stabilization of its mRNA at low temperature (Brandi et al. 1996; Fang et al. 1997; Goldenberg et al. 1996). Bacteria respond to a rapid temperature drop (cold shock) by a transient induction of cold-induced proteins (Phadtare 2004), and the production of cold-induced proteins increases with the severity of the cold shock (Hébraud and Potier 1999). In *Escherichia coli*, numerous cold-induced proteins have been identified so far, including, cold shock protein family (Yamanaka et al. 1998) and RNA helicase cold shock domain (Charollais et al. 2004). Cold shock proteins counteract some harmful effects of temperature downshift and thus help the cells to adapt (Phadtare 2004). After the immediate cold shock response, the synthesis of cold shock proteins declines, and the synthesis of other proteins increases. This enables the cells to grow at a low temperature, although at a slower rate (Ermolenko and Makhatadze 2002). Cold shock proteins are known to be important during cold shock response; however, recent studies have shown that cold shock proteins might have a wider role in the stress tolerance of bacteria (Schmid et al. 2009; Duval et al. 2010; Loepfe et al. 2010; Michaux et al. 2012; Schärer et al. 2013; Wang et al. 2014; Derman et al. 2015).

Like heat shock protein, cold shock domain proteins are also found in higher plants that have vascular bundles or conducting tissues (like xylem) such as *Pinus sylevestris* and lower plants (nonvascular plants such as mosses and liverworts) (Karlson and Imai 2003). Thomashow (1998) stated that the synthesis of cold shock protein enhances cold tolerance. While sufficient research has been performed to characterize cold shock domain proteins in animals and bacteria, not much research has been conducted on plants for their roles. The first functionally characterized plant cold shock domain protein is the wheat cold shock protein (Karlson et al. 2002).

Cold shock proteins are small nucleic acid-binding proteins ranging from 65 to 75 amino acids in length (Graumann and Marahiel 1996; Czapski and Trun 2014), and they have been found in psychrophilic, mesophilic, thermophilic, and even hyperthermophilic bacteria (Phadtare 2004; Jin et al. 2014). Cold shock protein was first described in *E. coli* (Goldstein et al. 1990). Later it was found that *E. coli* cold shock protein family consists of nine homologous proteins named cold shock protein through cold shock proteins that share a 46–91% amino acid sequence similarity (Yamanaka et al. 1998). Structurally cold shock proteins are highly conserved; however, their thermostability varies (Lee et al. 2013; Jin et al. 2014). Cold shock protein A of the psychrotrophic *Listeria monocytogenes* has a melting temperature of 40 °C (Lee et al. 2013), whereas cold shock protein of thermophilic *Thermus aquaticus* has a more rigid structure and a melting temperature as high as 76 °C, suggesting that psychrophilic cold shock proteins need higher structural flexibility to accommodate nucleic acids upon cold shock (Jin et al. 2014). Keto-Timonen et al. (2016) have reviewed cold shock proteins from enteropathogenic bacteria, *Yersinia*.

4.6.1.2 Heat Shock Protein

These are a family of proteins that are produced by cells in response to exposure to stressful conditions. They were first described in relation to heat shock but are now known to also be expressed during other stresses, including exposure to cold, UV light, and during wound healing or tissue remodeling. Heat shock proteins are found in virtually all living organisms, from bacteria to humans. Heat shock proteins are named according to their molecular weight. For example, Hsp60, Hsp70, and Hsp90 refer to families of heat shock proteins on the order of 60, 70, and 90 kilo daltons (kDa) in size, respectively (reviewed by Bhattacharya 2019). Heat shock proteins play an important role in protein–protein interactions such as folding and assisting in the establishment of proper protein conformation (shape) and prevention of unwanted protein aggregation. By helping to stabilize partially unfolded proteins, heat shock proteins aid in transporting proteins across membranes within the cell. Some members of the heat shock protein family are expressed at low to moderate levels in *all* organisms because of their essential role in protein maintenance.

Heat shock proteins, or stress proteins, are highly conserved and present in all plants and animals (Vierling 1991). The heat shock response in plants is similar to that of other organisms. Trimerization and activation of heat shock factors regulate the induction of heat shock genes (Young 2010). Almost all kinds of stresses induce gene expression and synthesis of heat shock proteins in cells that are subjected to stress (Iqbal et al. 2010; Westerheide et al. 2009). In *Arabidopsis* and some other plant species, low temperature, osmotic, salinity, oxidative, desiccation, high-intensity irradiations, wounding, and heavy metal stresses were found to induce the synthesis of heat shock proteins (Swindell et al. 2007). Plants living in extreme environments require a cell physiology able to compensate for stressful conditions, and membranes and proteins are among the most affected cell components during stress (Huang and Xu 2008).

Heat shock proteins can be said to be proteins that are usually expressed when responding to stress conditions (Banilas et al. 2012; Saidi et al. 2009; Krivokapich et al. 2006; Tsan and Gao 2004; Gorski et al. 2003). Both at cellular and organismic level exposure, heat shock proteins are highly conserved (Saidi et al. 2009). For a cell to survive under stress, it is quite important to maintain the proteins in their functional conformation, and aggregation of nonnative proteins should be prevented. Abiotic stresses usually cause protein dysfunction (Wang et al. 2004a). The heat shock protein is synthesized in plants under heat stress conditions and is hence also called stress-induced protein. Heat shock protein responds to heat stress on a molecular level in plants (Gupta et al. 2010). According to De Maio (1999), the stress initiates gene expression and biosynthesis of heat shock proteins to cope with the effect of stress. The production of heat shock protein is common phenomenon in all organisms under heat stress conditions (Gupta et al. 2010). Heat shock proteins are generally categorized into five evolutionarily conserved families: HSP100, 90, 70, 60, and the small heat shock proteins (sHSPs) (Krishna 2003). Some of the heat shock proteins act as molecular chaperones. Most, but not all, heat shock proteins are molecular chaperones, which bind and stabilize proteins at intermediate stages of folding, assembly, degradation, and translocation across membranes (Bhattacharya 2018).

4.6.1.3 Antifreeze Proteins

Ice is a big problem for organisms that live in cold climates. Once the temperature dips below freezing, ice crystals steadily grow and burst cells. This danger, however, has not limited the spread of life on Earth to temperate regions. Organisms of all types—plants, animals, fungi, and bacteria—have developed ways to combat the deadly growth of ice crystals. In some cases, they pack their cells with small antifreeze compounds like sugars or glycerol. But in cases where extra help is needed, cells make specialized antifreeze proteins to protect themselves as the temperature drops.

Antifreeze proteins or ice structuring proteins refer to a class of polypeptides produced by certain animals, plants, fungi, and bacteria that permit their survival in temperatures below the freezing point of water. But the antifreeze proteins found in the plant cell are different from other organisms in many ways. These proteins are commonly found in freeze-resistant plants in cold regions (Griffith and Yaish 2004). Under freezing temperature, the process of crystallization begins in plants, which could be fatal for the growth of plants. Antifreeze proteins bind to ice crystals and inhibit the process of crystallization (Venketesh and Dayananda 2008). Antifreeze proteins bind to small ice crystals to inhibit the growth and recrystalization of ice that would otherwise be fatal (Goodsell 2009). There is also increasing evidence that antifreeze proteins interact with mammalian cell membranes to protect them from cold damage. This work suggests the involvement of antifreeze proteins in cold acclimatization (Fletcher et al. 2001). It has been noted that during cold acclimation antifreeze protein genes produce antifreeze proteins. These proteins are activated under low temperatures and enhance the production of antifreeze proteins (Teijo and Griffith 1999). The production of the antifreeze protein is a complicated and tissue-specific process. Nayyar (2003) reported that *Arabidopsis* has CBF/DREB1 proteins (act as a transcription factors), which control the expression of cold-induced genes that enhances resistances in plants against freezing.

4.6.1.4 Late Embryogenesis Abundant Proteins

These proteins are generally found in plants and in some bacteria and some invertebrates that protect against protein aggregation due to desiccation or osmotic stresses associated with low temperature (Goyal et al. 2005; Hundertmark and Hincha 2008; Liu et al. 2011). Late embryogenesis abundant proteins were initially discovered accumulating late in embryogenesis of cotton seeds (Dure III et al. 1981). Although abundant in seeds and pollens, late embryogenesis abundant proteins have been found to protect against desiccation, cold, or high salinity in a variety of organisms (Gal et al. 2004; Menze et al. 2009; Hundertmark and Hincha 2008). Late embryogenesis abundant proteins function by mechanisms that are distinct from those displayed by heat shock molecular chaperones (Goyal et al. 2005).

Late embryogenesis abundant proteins are synthesized in plants in response to desiccation in seed and leave tissues. Late embryogenesis abundant proteins belong to the glycine-rich protein family. According to Gal et al. (2004), the late embryogenesis abundant protein was first observed in seeds during the development of the embryo under drought stress. These proteins were first observed in cottonseed during late embryogenesis and are associated with desiccation (Gal et al. 2004). They also protect from freezing, protect other proteins' function, and also stabilize the membranes (Li et al. 2012).

4.6.1.5 Pathogenesis-Related Proteins

Plants also produce defensive proteins against the pathogenic attack, which are known as pathogenesis-related proteins. The pathogenic infection activates the gene that is coded for pathogenesis-related proteins, and these genes produce pathogenesis-related proteins. The main role of these proteins is the protection against the pathogen (Jwa and Hwang 2017). A few proteins help the immune system to break down the cell wall of a fungus or bacterium, while others spread signals of infection to nearby cells. Studies indicated that many proteins are found in the vine (climbing plant), which acts as a defensive protein against pathogens like thaumatin and chitinases like proteins (Van-Loon 1985). Plants produce different types of defensive proteins under pathogenic effects. During stress conditions, different types of genes are activated. Various types of the small RNAs (50–250 nucleotides length) play a significant role in plants' defense via the production of defensive proteins.

4.6.1.6 Seed Storage Proteins

The plant seed is not only an organ of propagation and dispersal but also the major plant tissue harvested by humankind. The amount of protein present in seeds varies from 10% (in cereals) to 40% (in certain legumes and oilseeds) of the dry weight, forming a major source of dietary protein. Although the vast majority of the individual proteins present in mature seeds have either metabolic or structural roles, all seeds also contain one or more groups of proteins that are present in high amounts and that serve to provide a store of amino acids for use during germination and seedling growth. These storage proteins are of particular importance because they determine not only the total protein content of the seed but also its quality for various end uses. For example, the low content of lysine, threonine, and tryptophan in various cereal seeds and of cysteine and methionine in legume seeds is due to the low proportions of these amino acids in the major storage proteins and may limit the nutritional quality of the seeds for monogastric animals. In the case of wheat, the storage proteins form the gluten fraction, whose properties are largely responsible for the ability to use wheat flour to make bread, other baked goods, and pasta. These properties are not shared by the storage proteins of other cereals. Despite wide variation in their detailed structures, all seed storage proteins have a number of common properties. First, they are synthesized at high levels in specific tissues and at certain stages of development. In fact, their synthesis is regulated by nutrition, and they act as a sink for surplus nitrogen. However, most also contain cysteine and methionine, and adequate sulfur is therefore also required for their synthesis. Many seeds contain separate groups of storage proteins, some of which are rich in sulfur amino acids and others of which are poor in them. The presence of these groups may allow the plant to maintain high levels of storage protein synthesis despite variations in sulfur availability. The strict tissue specificity of seed storage protein synthesis contrasts with that of tuber storage proteins, which may be synthesized in vegetative tissues under unusual conditions (for example, in vitro or after removal of tubers) (Shewry 1995). A second common property of seed storage proteins is their presence in the mature seed in discrete deposits called protein bodies, whose origin has been the subject of some dispute and may in fact vary both between and within species. Finally, all storage protein fractions are mixtures of components that exhibit polymorphism both within single genotypes and among genotypes of the same species. This polymorphism arises from the presence of multigene families and, in some cases, proteolytic processing and glycosylation. Seed storage proteins are biosynthesized in specific tissues and in a specific stage of growth. The synthesis of seed storage protein depends on nitrogen and acts as a sink for extra nitrogen. The seed storage proteins are classified into three groups, viz. albumins, prolamins, and globulins.

Protein is the macro-biomolecules that consist of a large number of polymers known as amino acids. Due to this reason, amino acids are also known as the building blocks of protein. These polymers contain hydrogen, carbon, nitrogen, oxygen atoms, and sulfur. A few proteins also have additional elements like phosphorus and iron. Protein is found in almost all cells of all organisms and is the most abundant molecule found in the living system apart from water. It is believed that about 100,000 different types of protein are found in organisms (Christopher 2004). These proteins are different from each other depending on their amino acid sequences. Therefore, proteins are a very significant group of molecules found in living systems. They are involved in many metabolic pathways occurring in plant cells, such as catalyzing reaction, replication of DNA, and transporting molecules inside and outside of the cell, and can be considered building blocks of cell membranes (Joshi 2018).

It is known that proteins consist of several amino acids, which differ from each other depending on the order and number of the amino acid, in molecular weight, and their function. The type of protein based on the number and arrangement of amino acids in polypeptide changes. Pierce et al. (2011) stated that the type, structure, and function of the protein happen as a result of translation of genetic code on mRNA. Each genetic code links to a specific amino acid, so each gene is coded for a specific number of polypeptides. Normally, proteins are synthesized in the ribosome and rough endoplasmic reticulum, but the production of proteins is regulated by the messenger RNA (mRNA). The messenger RNA carries messages in the form of a genetic code or codon. These codes have specific information for making a polypeptide chain by binding of different types for amino acids. According to Attardi (1967), mechanism of protein synthesis consists of two steps. The first step is the transcription of DNA (deoxyribonucleic acid) into RNA (ribonucleic acid), and the second step is the translation of RNA into protein. The synthesis of protein happens in ribosomes. The biosynthesis of protein starts at the amino-terminal of amino acids and continues to their carboxy-terminal.

4.6.2 Protein Synthesis

The genetic code, commonly known as codon (a group of three adjacent nucleotides in the mRNA), specifies the attachment of amino acids to growing peptide chain, called polypeptide chain (Attardi 1967). The sequence of amino acids in the polypeptide chain has all information necessary for generating the structure of protein molecules (Anderson et al. 1967). A similar mechanism of production of protein is found in plants. Protein biosynthesis plays a significant role in fulfilling the energy requirement for the growth of plants (Amthor 2000). The rate of production of proteins can be assessed by investigating the quantity of ribosomes loaded in polysomes. The environmental factors such as water deficit decrease the polysome loading (Beilharz and Preiss 2004). In nonstressed conditions, plants normally have high polysome loading in light and have low in the night (Pal et al. 2014). This approach specifies that plants produce maximum protein in the daytime while it is constrained in the night by the rate of starch degradation and availability of carbon and energy.

As early as 1970, Weiser (1970) proposed that cold acclimation of temperate woody perennials may require both transcriptional activation of a set of genes normally not expressed under nonacclimating conditions and the synthesis of new proteins during the development of maximal freezing tolerance. The following conclusions were drawn:

- Presumably, translation competent polysomes could be isolated from cold-acclimated tissues (Bixby and Brown 1975; Brown 1972).
- rRNA increased during cold acclimation (Sarhan and D'Aoust 1975).
- RNA polymerase activities were increased in response to low temperature (Sarhan and Chevrier 1985).
- Proteins accumulated during the fall in woody perennial bark cells (Sakai and Yoshida 1968; Siminovitch et al. 1968; Pomeroy et al. 1970), but little direct evidence existed to support the proposal that altered gene expression and synthesis of new proteins occurred during cold acclimation.

Almost simultaneously, it was shown that the RNA base composition in cold-acclimated tissues differed from that in nonacclimated tissues (Sarhan and D'Aoust 1975), and plants at low temperature synthesized two unique proteins in response to cold acclimation (Rochat and Therrien 1975). Freezing-tolerant winter wheat synthesized two hydrophilic high-molecular-weight proteins at 3 °C, while freezing-sensitive spring wheat was unable to make similar proteins (Rochat and Therrien 1975). Independent studies of wheat and/or rye subjected to cold acclimation (Perras and Sarhan 1989; Sarhan and Perras 1987) or desiccation treatments have yielded analogous results (Cloutier 1983, 1984).

There are reports stating that when exposed to low temperature, a new set of proteins are synthesized (Cooper and Ort 1988; Guy and Haskell 1987, 1988; Guy et al. 1985; Hahn and Walbot 1989; Johnson-Flanagan and Singh 1987; Kurkela et al. 1988; Laroche and Hopkins 1987; Maimiroli et al. 1986; Meza-Basso et al. 1986; Mohapatra et al. 1987a; Ougham 1987; Perras and Sarhan 1989; Robertson et al. 1987; Sarhan and Perras 1987; Tseng and Li 1987; Yacoob and Filion 1986a, b). Nearly all studies of protein synthesis at low temperatures have focused on a descriptive approach of the response. Out of this collective work, which includes species that range from chilling sensitive to freezing-tolerant, several features of the protein synthesis response at low temperature have emerged. The set of new proteins synthesized at low temperature does not appear to be highly conserved as those of heat shock are (Barnett et al. 1979). This was taken that protein synthesis at a low temperature of widely divergent species involves no similar proteins, but only that at present little or no evidence exists to support a high conservation of the response. That the set of proteins synthesized at low temperature appears not to be highly conserved as heat shock reflects on the diversity of cold-stress situations and adaptive responses (Grace 1987; Sakai and Larcher 1987; Woodward and Williams 1987). Also, the synthesis and concentration of housekeeping proteins at low temperature continue even under cold shock treatments (Gilmour et al. 1988; Guy and Haskell 1987; Guy et al. 1998; Maimiroli et al. 1986; Perras and Sarhan 1989; Yacoob and Filion 1986a, b). Support for the idea that synthesis of housekeeping proteins at low temperature continues comes from the apparent increase in rRNA (Laroche and Hopkins 1987) and protein synthesis capacity (Laroche and Hopkins 1987; Bixby and Brown 1975) in cold-acclimated tissues. At low temperatures, cold-acclimated tissues can synthesize protein faster than nonacclimated tissue (Guy et al. 1985). This would seem to be consistent with the need to maintain a basal level of cellular metabolism even at low temperatures. Therefore, the alterations in protein synthesis at low temperature were found to be subtler, less visually dramatic, and kinetically slower than those in response to many other environmental stresses. Low-temperature responses in protein synthesis are further distinguished from other stresses by the fact that many of the unique proteins are continuously synthesized at higher rates for extended periods. While synthesis of some of the new proteins at low temperature is stable, continuing for several days or weeks (Guy and Haskell 1987; Guy et al. 1985), that of others is transient (Cattivelli and Bartels 1989; Tseng and Li 1987). Time course studies have indicated that new proteins can appear within one day and perhaps as early as an hour following exposure to low temperature (Guy and Haskell 1987; Maimiroli et al. 1986). However, a few studies have correlated the changes in protein synthesis with the induction and loss of freezing tolerance. Time course studies with spinach have shown the synthesis of several high-molecular-weight proteins closely correlates with the induction and loss of freezing tolerance during cold acclimation and de-acclimation (Guy and Haskell 1987). During cold acclimation, the synthesis of these proteins was upregulated, while during de-acclimation, synthesis was downregulated. Such time course studies provide evidence to support the notion that new proteins synthesized at low temperature go beyond a role in the adjustment of metabolism to low temperature and that some may play a central role in freezing tolerance. Proteins synthesized at low temperature during cold stress could be related to the other temperature-responsive proteins of heat shock (Barnett et al. 1979; Key et al. 1981). Comparison of proteins synthesized during cold acclimation and heat shock suggests that the two responses elicit the synthesis of distinctly different proteins (Ougham 1987; Guy et al. 1985), although some were apparently electrophoretically similar (Guy and Haskell 1987; Yacoob and Filion 1986a, b). More specific protein blotting analyses of corn subjected to cold shock have demonstrated the accumulation of both low- and high-molecular-weight heat shock proteins (Yacoob and Filion 1986a, b). This has been partially confirmed by the finding of increased RNA levels for the 27,000 Mr (molecular weight) heat shock proteins in chilled soybean (Kuznetsov et al. 1987) and protein sequence homology of a 79,000 Mr protein accumulated in spinach during cold acclimation. While the heat shock response and cold acclimation may involve a common subset of "stress" proteins, it is reasonable to expect that heat and freezing tolerance involve different adaptive mechanisms that require their own unique stress proteins (Guy 1990).

The first direct evidence to support Weiser's (Weiser 1970) hypothesis of altered gene expression during cold acclimation was the observation that newly translatable mRNAs were induced in spinach leaves exposed to 5 °C (Guy et al. 1985). This finding also provided an explanation for the previous observation of the altered base composition of RNA during cold acclimation (Sarhan and D'Aoust 1975).

Subsequent studies have confirmed and extended these findings (Cattivelli and Bartels 1989; Gilmour et al. 1988; Guy and Haskell 1988; Hughes and Pearce 1988; Johnson-Flanagan and Singh 1987; Kushad and Yelenosky 1987; Meza-Basso et al. 1986; Mohapatra et al. 1987b; Schaffer and Fischer 1988; Tseng and Li 1987). In many studies, some of the mRNAs detected in cold-acclimated tissues encode proteins that are translated and accumulated at low temperature (Gilmour et al. 1988; Guy and Haskell 1987, 1988, 1989; Guy et al. 1985; Kurkela et al. 1988; Meza-Basso et al. 1986). In the few studies that have examined the time course of the changes in mRNAs, it is clear that within hours (Gilmour et al. 1988; Cattivelli and Bartels 1989) to one day (Guy and Haskell 1988) of low-temperature exposure, new mRNAs begin to appear in herbaceous species. This represents one of the fastest documented biochemical responses to low temperature in nature and again suggests a potential role of the synthesis of new proteins in tolerance mechanisms. However, many of the responses of temperate perennial plants to low temperature, at the mRNA level, may parallel that found in winter flounder, where antifreeze protein mRNA synthesis is preferentially regulated at low temperature following priming by endogenous clock mechanisms (Price et al. 1986).

Some mRNAs decline during exposure to low temperature. The mRNA level for the small subunit of ribulose bisphosphate carboxylase/oxygenase was decreased in *Brassica napus* (Meza-Basso et al. 1986) and rice (Hahn and Walbot 1989). In rice, the decline in the mRNA level also resulted in a lack of synthesis of the small subunit, while the large subunit synthesis was less affected at low temperatures (Hahn and Walbot 1989). This decline seemed to cause a loss in the coordinate regulation of the large and small subunit syntheses. In chilling-sensitive rice, transcript levels for some organelle- and nucleus-encoded genes were also suppressed at low temperatures (Hahn and Walbot 1989).

Partial cDNA clones for mRNAs accumulated during low-temperature exposure of chilling-sensitive tomato fruit and freezing-tolerant alfalfa have been isolated. Three cDNAs were isolated from chilled tomato fruit that recognizes distinct RNAs of 0.8, 1.2, and 1.9 kb (Schaffer and Fischer 1988). Time course RNA blot analyses have indicated rapid accumulation of the RNAs at temperatures below 16 °C, with the greatest accumulation observed at 4 °C. Sequence analysis of one of the cDNAs indicated strong homology with conserved domains of thiol protease genes. Schaffer and Fischer (1988) postulate that one function of the induced thiol protease in tomato fruit might be to degrade proteins denatured at low temperature. If they are correct, the appearance of this enzyme at low temperature in chilling-sensitive tissues might be an injury response to cold stress instead of an adaptive response that increases tolerance. In contrast, RNA blot analysis of rice showed the transcript of ubiquitin to decrease during low-temperature exposure (Hahn and Walbot 1989). The biological role of the other two genes remains unknown. The same situation holds true for the cold acclimation-specific cDNAs isolated from alfalfa (Mohapatra et al. 1989). The three cDNAs hybridize to transcripts ranging from 0.9 to 1.4 kb. RNA blot analyses have indicated that the transcripts' steady-state levels increase rapidly during cold acclimation but do not increase in response to abscisic acid administration, wounding, water stress, or heat shock. The relative levels of expression of the three genes seem to correspond to cultivar hardiness levels during cold acclimation. Since no one knows how many genes are activated at low temperatures, the two studies discussed above may only have scratched the surface. As more genes are cloned and characterized, a better understanding of both adjustment of metabolism to low temperature and cold tolerance mechanisms should emerge.

4.7 Chilling Stress and Potential Management Opportunities

Crops grown under open environments often pass through periods of abiotic stress during their life cycle. Such stresses many a time overlap so that the crop growth and productivity of the crops are adversely affected. Plants pass through a series of morphological, physiological, biochemical, and molecular changes in a quest to mitigate such adversities of the abiotic stresses. A lot of work reports the adaptive responses of crop plants to different abiotic stresses wherein emphasis has been laid on individual stress factors (Jongdee et al. 2002; Chinnusamy et al. 2007; Nejad et al. 2010; Basu et al. 2016; Wang et al. 2016; Anjum et al. 2017). Nonetheless, understanding the plant responses to combined stress factors is inevitable for enhancing the adaptation of plants under field conditions (Pandey et al. 2015). Crops experience periods of extremely low temperatures in many regions of the world (Ruelland et al. 2009; Wang et al. 2016) and are exposed to limited water availability owing to either drought or disturbed water movement and uptake under low ambient temperatures (Shinozaki et al. 2003; Zhang et al. 2004; Beck et al. 2007). Such an exposure of plants to chilling and drought simultaneously hampers plant growth and hence is detrimental to productivity (Tommasini et al. 2008). Studies show that primarily chilling and drought pose a similar impact on stomatal development and leaf growth; nevertheless, the mechanisms of drought-induced changes in some physiological processes are quite different than those induced by chilling (Deng et al. 2012). In general, chilling stress thermodynamically declines the kinetics of several physiological as well as metabolic processes occurring in plants (Ruelland et al. 2009; Hussain et al. 2016). It severely reduces the rate and uniformity of germination, hampers seedling vigor, and delays ontogenic plant development (Cruz and Milach 2004; Oliver et al. 2007), resulting in severe crop yield losses (Ruelland et al. 2009).

4.7.1 Plant Responses to Chilling and Drought Stresses

Suboptimal temperatures at any growth stage of crop plants evoke negative effects on crop growth and development. Such detrimental effects on plants depend on the magnitude of stress and its period and the plant growth stage. Such effects are elicited in terms of morphological, physiological, biochemical, and molecular processes in plants. Chilling stress is known to cause significant reductions in plant growth and development. Various studies have reported the negative effects of chilling stress on the growth of rice (Hussain et al. 2016; Wang et al. 2016), cotton

(Zhao et al. 2012), tomato (Starck et al. 2000), potato (Svensson et al. 2002), muskmelon (Wang et al. 2004b), and sugarcane (Thakur et al. 2010; Anjum et al. 2011; Zhu et al. 2013). Chilling stress may cause the decreased rate of both cell division and elongation, resulting in smaller leaf area (Ben-Haj-Salah and Tardieu 1995; Jouyban et al. 2013), while slow leaf initiation rate leads to a reduced number of leaves under periods of chilling stress (Lukatkin et al. 2012; Jouyban et al. 2013). Chilling stress restricts root growth by decreasing root length, biomass, and morphology (Cutforth et al. 1986), which ultimately reduces the volume of the root system for exploring the nutrients and water (Richner et al. 1996). The cultivated tomato genotype (*Solanum lycopersicum* L.) displays limited growth and development at temperatures under 12 °C (Hu et al. 2006). At temperatures between 0 and 12 °C, plants are damaged by chilling stress. The severity of damage is proportional to the length of time spent in this temperature range, with alterations in flowering (Adams et al. 2001), germination (Foolad and Lin 2000), cellular turgor (Bloom et al. 2004), nutrient absorption (Bloom et al. 1998), and photoinhibition of photosynthesis (Venema et al. 2000). Unlike cultivated tomatoes, wild tomato species such as *Solanum habrochaites*, *Solanum chilense*, and *Solanum peruvianum* recover rapidly after exposure to suboptimal temperatures, which is expressed through greater growth capacity and decreased inhibition of photosynthesis (Brüggemann and Linger 1994) or minor alterations in nitrogen metabolism (Jones et al. 1998). Wild species have been used for constructing genetic maps and identifying genes of agronomic importance. In addition, it is well established that freeze-induced membrane damage results primarily from the severe dehydration associated with freezing (Steponkus 1984; Steponkus et al. 1993). As temperatures drops below 0 °C, ice formation is generally initiated in the intercellular spaces due, in part, to the extracellular fluid having a higher freezing point (lower solute concentration) than the intracellular fluid. Because the chemical potential of ice is less than that of liquid water at a given temperature, the formation of extracellular ice results in a drop in water potential outside the cell. Consequently, there is a movement of unfrozen water down the chemical potential gradient from inside the cell to the intercellular spaces. At 10 °C, more than 90% of the osmotically active water typically moves out of the cells, and the osmotic potential of the remaining unfrozen intracellular and intercellular fluid is greater than 5 osmolar.

The accumulation of sucrose and other simple sugars that typically occurs with cold acclimation also seems likely to contribute to the stabilization of membranes as these molecules can protect membranes against freeze-induced damage in vitro (Strauss and Hauser 1986; Anchordoguy et al. 1987). In addition, there is emerging evidence that certain novel hydrophilic and late embryogenesis abundant polypeptides also participate in the stabilization of membranes against freeze-induced injury. These hydrophilic and late embryogenesis abundant polypeptides are predicted to contain regions capable of forming amphipathic α-helices, which are shown to have a strong effect on the intrinsic curvature of monolayers and their propensity to form hexagonal II phase. They are said to defer their formation at lower temperatures (Epand et al. 1995). Whether the regions predicted to form amphipathic α-helices actually form such structures is uncertain. An additional hypothesis suggests that the extensive water-binding capacity of these hydrophilic proteins might provide a protective environment in the proximity of membranes during freezing and result in membrane stabilization. Although freezing injury is thought to result primarily from membrane lesions caused by cellular dehydration, additional factors may also contribute to freezing-induced cellular damage. There is evidence that freeze-induced production of reactive oxygen species contributes to membrane damage and that intercellular ice can form adhesions with cell walls and membranes and cause cell rupture (Olien and Smith 1977). In addition, there is evidence that protein denaturation occurs in plants at low temperatures (Guy et al. 1998), which could potentially result in cellular damage. In these cases, the enhancement of antioxidative mechanisms (Aroca et al. 2003), increased levels of sugars in the apoplastic space (Livingston and Henson 1998), and the induction of genes encoding molecular chaperones (Guy and Li 1998), respectively, could have protective effects.

Suboptimal temperatures reduce the growth and metabolic activities, leading to a reduction in agronomic and yield attributes of the crops. Rizza et al. (1994) reported that chilling stress severely impaired barley productivity. Chilling stress in chickpea at the reproductive stage led to flower abortion and poor pod setting (Clarke and Siddique 2004; Nayyar et al. 2005). In temperate growing areas, cold temperature is responsible for 30–40% reduction in rice yields because chilling stress resulted in degeneration of spikelets, panicle deformation, and poor spikelet fertility (Andaya and Mackill 2003). In grain crops, low-temperature stress during reproductive development may induce flower abscission, ovule abortion, pollen sterility, pollen tube distortion, poor fruit set, and hence reduced final yields (Thakur et al. 2010). Whaley et al. (2004) found that chilling stress during stem elongation reduced the internode extension, caused spikelet death, and lowered biomass accumulation and grain yield in wheat. Low temperature at jointing and booting stages of wheat significantly decreased the number of productive tillers per plant (Li et al. 2015) and negatively affected the development of young spikes (Thakur et al. 2010).

4.7.2 Molecular Responses

Crop growth and productivity are negatively affected by molecular alterations under various environmental stresses. Chilling stress induces the expression of a number of genes in plants (Hussain et al. 2018). In *Arabidopsis*, expression of DREB1 (also called CBFs or the c-box binding factors) was affected in response to chilling stresses (Stockinger et al. 1997; Liu et al. 1998). Likewise, Yamaguchi-Shinozaki and Shinozaki (1994) identified the dehydration responsive element and is regulated in response to chilling stresses. Expression of homeodomain leucine zipper (4D-Zip) protein, which interacts with CaCBFIB (Mahajan and Tuteja 2005), was highly regulated by low-temperature stresses (Kim et al. 2004). Chinnusamy et al. (2007) stated that cold temperature affects membrane fluidity, disrupts nucleic acid and protein structures, hinders water and nutrient uptake, and causes significant changes in the plant transcriptome. It also drastically disturbs the cellular metabolism either by directly decreasing the rates of biochemical reactions or indirectly by gene expression reprogramming. Chilling stress rapidly increased the levels of cytoplasmic calcium in *Arabidopsis* (Polisensky and Braam 1996) and alfalfa (Monroy and Dhindsa 1995) due to influx of calcium from extracellular stores (Mahajan and Tuteja 2005). However, Jouyban et al. (2013) reported that response of plants to chilling stress was associated with changes in gene transcription of low-molecular-weight proteins. Differences in gene expression in response to stress also occur among plant parts. Kreps et al. (2002) found that leaves and roots exhibited differential gene expressions during cold acclimation, and 86% of the cold-induced genes were not shared between leaves and roots. Chilling stress also induces the accumulation of proline, which in turn induces the expression of genes having proline-responsive elements in their promoters (Satoh et al. 2002; Chinnusamy et al. 2007). It can be said that profound changes in gene expression occur under chilling stresses (Maqbool et al. 2017). Overall, the majority of the transcription factors were upregulated under chilling stress; a few transcription factors having a role in primary growth processes were downregulated (Bhargava and Sawant 2013).

Both cold-stress-induced transcripts and constitutively expressed transcripts need to be processed, exported to the cytoplasm, and kept in conformations that are competent for translation. RNA can fold into extensive secondary structures that could interfere with its function, and cold temperatures exacerbate this interference. In bacteria, nucleic-acid-binding cold shock proteins accumulate at cold temperatures and function as transcription antiterminators or translational enhancers by destabilizing RNA secondary structure (Jones and Inouye 1994). Some cold shock protein domain-containing proteins in plants are upregulated by cold stress and might function as RNA chaperones in the regulation of translation (Nakaminami et al. 2006; Kim et al. 2001). A different cold-responsive nucleic-acid-binding protein, a zinc finger-containing glycine-rich RNA-binding protein from *Arabidopsis* designated at RZ-1a, is also upregulated by cold stress, and genetic analysis supports its function in freezing tolerance (Kim et al. 2005). Another group of RNA chaperones, RNA helicases, are involved in every step of RNA metabolism. In cyanobacteria, a cold-induced DEAD-box RNA helicase was suggested to unwind cold-stabilized secondary structure in the 5′-untranslated region of RNA during cold stress (Yu and Owttrim 2000). Compared to other organisms, plants have the largest number of DEAD-box RNA helicase genes. One of these helicases, which is encoded by the *Arabidopsis* low expression of osmotically responsive gene 4, is essential for plant tolerance of chilling and freezing stress (Gong et al. 2002). Low expression of the osmotically responsive gene is required for the efficient export of RNA from the nucleus to the cytoplasm (Gong et al. 2005). The *Arabidopsis* nucleoporin AtNUP160 suppressor of auxin resistance1 (SAR1) also controls RNA export and is crucial for chilling and freezing tolerance (Dong et al. 2006). Both low expression of osmotically responsive genes and AtNUP160 proteins are enriched at the nuclear rim (Gong et al. 2002; Dong et al. 2006). Defects in the nucleocytoplasmic transport of RNA seem to affect cold tolerance preferentially, because the low expression of osmotically responsive gene 4 and AtNUP160 mutant plants do not have severe growth or developmental phenotypes, nor are they strongly altered in the tolerance of other abiotic stresses.

4.7.3 Management Strategies

Plants often develop a cascade of strategies through physiological, biochemical, and molecular modifications at the cell level to cope with chilling stress. Studies have shown that mechanisms of plant tolerance to chilling stress can be regulated through developing tolerant plant genotypes, genetic modifications, seed treatments, application of plant growth regulators and compatible solutes, and use of plant mineral nutrients.

4.7.3.1 Selection and Breeding Strategies

What genes have important roles in cold acclimation? What are their functions? How do plants sense low temperature and activate the cold-acclimation response? These are some of the key questions that investigators working in the field of cold acclimation are actively engaged in answering. Knowledge in these areas is not only important for an overall understanding of how plants sense and respond to changes in the environment but also has potential practical applications. Determining the nature of the genes and mechanisms responsible for freezing tolerance and the sensing and regulatory mechanisms that activate the cold-acclimation response provide the potential for new strategies to improve the freezing tolerance of agronomic plants (Jan et al. 2009). Such strategies would be highly significant as traditional plant breeding approaches have had limited success in improving freezing tolerance (Sarhan and Danyluk 1998). The freezing tolerance of wheat varieties and oilseed varieties developed so far, for instance, are only marginally better than those developed approximately a hundred years ago (Rapacz and Markowski 1999). Classical genetic studies have demonstrated that the ability of plants to cold acclimate is a quantitative trait involving the action of many genes with small additive effects (Thomashow 1990). Many approaches have been taken to determine the nature of genes with roles in freezing tolerance: the isolation and characterization of genes induced during cold acclimation; the isolation and characterization of mutants affected in freezing tolerance; and transcriptome analysis. Most molecular studies on plant responses to cold stress are focused on the mechanism of cold acclimation rather than on chilling tolerance. Nevertheless, evidence indicates that some of the molecular changes that occur during cold acclimation are also important for chilling tolerance (Gong et al. 2002; Dong et al. 2006). In other words, it appears that chilling tolerance that is exhibited by temperate plants is not entirely constitutive and that at least part of it is developed during exposure to chilling temperatures.

Integration of conventional, molecular, and omic-based techniques can successfully be employed to develop tolerant plant genotypes (Maqbool et al. 2017). Conventional breeding has been instrumental in developing chilling tolerant cultivars in various crops (Jha et al. 2017). Various rice varieties in Indonesia and Malaysia were released for chilling stress tolerance by conventional breeding methods (Ahmad et al. 2014b). Based on both open-air and controlled conditions, Rodriguez et al. (2007) reported that "EP80×Puenteareas" population in maize is an important source of low-temperature tolerance. Implementation of the modern omics approaches and identification of quantitative trait loci (QTLs)/genes for chilling tolerance can significantly support crop improvement strategies aimed to develop high yielding cultivars under low-temperature conditions (Ahmad et al. 2014a; Jha et al. 2017). In rice, selection using different parameters led to the development of low-temperature tolerant genotypes based on low spikelet sterility (Farrell et al. 2006; Ye et al. 2009). Suh et al. (2010) reported that phenotyping selection and simple sequence repeat (SSR) makers' identification methods are helpful in the screening of plant genotypes against chilling stress. They found three QTLs responsible for seed setting percentage under chilling stress by using a recombinant inbred line (RIL) population developed by tropical japonica × temperate japonica.

Nearly isogenic lines contain introgressions in a defined region of the genome and are obtained through backcrosses and marker-assisted selection, and near-isogenic lines (NILs) have been used to study and isolate genes responsible for different agronomic traits (Elizondo and Oyanedel 2010). NILs allow for studying the behavior of specific genomic regions, offering an alternative to classic strategies of mapping based on F2 populations. F2 mapping populations are ephemeral, while those of NILs are permanent. This facilitates the identification of QTLs in multiyear studies in various localities. However, it requires more time to form a NIL population than an F2. NILs express different characteristics among themselves through their allelic combination and are developed to study genes involved in morphogenesis and metabolic regulation, as well as to introduce new genetic variability in germplasm of interest (Monforte and Tanksley 2000).

4.7.3.2 Molecular and Functional Genomics Approaches

Plants respond to chilling stress through a series of mechanistic changes in morpho-physiological features (Shinozaki and Yamaguchi-Shinozaki 1996). Many genes that respond to low temperature at the transcriptional level have been previously identified (Thomashow 1994; Kasuga et al. 2004; Yadav 2010; Todaka et al. 2015; Ali et al. 2017; Kaur and Asthir 2017). Liu et al. (1998) stated that abscisic acid is generally produced under both drought and chilling stresses and has significant contributions to stress tolerance in plants. They further stated that overexpression of various regulatory elements induces drought and low-temperature tolerance in plants (Liu et al. 1998). Kasuga et al. (2004) reported that overexpression of DREB1A enhanced chilling tolerance in tobacco plants. Furthermore, genes encoding transcription factors were recognized better for improving stress tolerance in plants than various other genes responsible for cold tolerance in *Arabidopsis* and rice (Yadav 2010). Tuberosa et al. (2008) stated that identification of QTLs involved in chilling stress tolerance may directly relate to acquiring valuable genetic diversity in the physiological features in plants. Low-temperature restrictions have been overwhelmed by the support of cold-tolerant genes in genetically modified crops (Sanghera et al. 2011). Cold tolerance in tobacco can also be enhanced by the transgenic overexpression of the chloroplast omega-3 fatty acid desaturase gene

(Kodama et al. 1994). Upchurch (2008) also reported the involvement of omega-3 desaturases to establish chilling tolerance in plants. Liu et al. (2008) described that the overexpression of Lefad7 in transgenic tomatoes could enhance the chilling stress tolerance. Lukatkin et al. (2012) reported that plant response to low temperature is linked with the extent of change of gene transcription of low-molecular-weight proteins. Mahajan and Tuteja (2005) described that CBF binds with COR genes carrying CRT/DRE elements, and the overexpression of these COR genes has an imperative role for chilling tolerance and cold acclimation in plants (Thomashow 1994; Stockinger et al. 1997; Yadav 2010). These findings signify the roles of cold-inducible genes to protect plant cells against cold stress (Yadav 2010).

Some studies have shown that some genes are expressed by both drought and cold stress concurrently while some are only responsive to individual stress (Thomashow 1994; Liu et al. 1998; Kasuga et al. 2004; Sanghera et al. 2011). Overexpression of DREB1A has been found to have been linked with drought and chilling stress in wheat, groundnut, and tobacco plants (Kasuga et al. 2004; Pellegrineschi et al. 2004). In order to improve tolerance against chilling and drought stress, applications of various molecular and engineering strategies are needed. It is well evident that the integration of traditional and molecular breeding approaches with genetic engineering and marker-assisted selection may help to improve individual and concurrent environmental stresses in crop plants (Chaves and Oliveira 2004).

4.7.3.3 Agronomic and Physiological Measures

Plants respond to chilling stress via a series of agronomic and physiological processes. Exogenous application of compatible solutes, plant growth hormones, plant mineral nutrients, and seed priming can be effective in alleviating the adverse effects of drought and chilling stresses.

4.7.3.4 Application of Compatible Solutes

Compatible solutes protect the plants from osmotic stress by not posing any detrimental effects on enzymes, membranes, and other macromolecules even at higher concentrations (Cechin et al. 2006; Kiani et al. 2007). Compatible solutes include glycine betaine, soluble sugars, proline, sugar alcohols, trehalose, and some organic acids (Kiani et al. 2007). Many of the osmolytes are organic solutes in nature while some of them are essential ions such as K^+ (Yokoi et al. 2002). Among various osmolytes, proline has been found to have multiple roles in plant stress regulation and tolerance potential (Yamada et al. 2005; Farooq et al. 2009a, b). Proline and polyamines have been reported to impart chilling tolerance in maize (Songstad et al. 1990). Proline retains the potential to relieve low-temperature injury in cold-sensitive plants. Although most of the chilling-sensitive plants accumulate proline, the levels of accumulation are not enough to induce cold tolerance; therefore, exogenous application of proline can be a potential option to enhance its level in plants. The beneficial role of proline in augmenting drought tolerance in rice (Hsu et al. 2003) and pea cultivars (Alexieva et al. 2001) has been well reported. Soluble sugars and sugar alcohols serve as osomoregulators, signaling molecules, and cryoprotectants in plants (Ruelland et al. 2009) and help in cleansing stress-generated reactive oxygen species (Van den Ende and Valluru 2009). Chilling tolerance has been observed by an increase of total soluble sugars in many plant species (Farooq et al. 2009a; Wang et al. 2016). Soluble sugars reduce osmotic potential and enhance water absorption under water-limited conditions and thus improve stress tolerance in plants (Farooq et al. 2009b).

4.7.3.5 Application of Plant Growth Regulators

Plants normally synthesize various types of phytohormones under individual and concurrent abiotic stresses depending upon the strength of their defense mechanism (Nadeem et al. 2016). Abscisic acid is synthesized in response to both drought and chilling and contributes to stress tolerance of crop plants (Liu et al. 1998). Abscisic acid, jasmonate, and salicylic acid often activate phosphoprotein cascade pathways that lead to the expression of genes associated with cold stress tolerance in plants (Kolaksazov et al. 2013). Greater chilling tolerance has been recorded in abscisic acid-treated maize (Xin and Li 1992), and levels of abscisic acid are increased in chilling tolerant genotypes (Farooq et al. 2009a). Jasmonates have been reported to trigger the plant defense responses to abiotic stresses, including drought and chilling (Pauwels et al. 2009; Seo et al. 2011). Exogenous application of methyl jasmonate improved chilling tolerance in wheat by modulating antioxidant defense system and production of soluble protein content (Qi et al. 2005). The relationship between enhanced ethylene levels and chilling tolerance in crop plants has also been reported by Zhao et al. (2014).

Salicylic acid is also considered as a potential chemical compound for drought and chilling tolerance in several plant species (Farooq et al. 2009a, b). Salicylic acid application was found to be effective in alleviating the chilling (Miura and Tada 2014) and freezing injury (Tasgin et al. 2003). In maize, the harmful effects of low-temperature stress were decreased by the application of salicylic acid (Janda et al. 1999; Wang et al. 2012b). Exogenous application of glycine betaine plays critical roles in abiotic stress tolerance in plants by modifying antioxidant activities, membrane integrity, osmotic adjustment, and reactive oxygen species detoxification (Somersalo et al. 1996; Xing and Rajashekar 2001; Farooq et al. 2008). Xing and Rajashekar (2001) reported that glycine betaine accumulation was positively related to the cold stress tolerance in plants. Glycine betaine improved the growth of *Solanum tuberosum* plants under low-temperature conditions (Somersalo et al. 1996) and freezing tolerance in *Arabidopsis thaliana* (Xing and Rajashekar 2001). In short, the application of compatible solutes such as glycine betaine, soluble sugars, proline, and polyamines can be helpful in imparting chilling tolerance in crop plants.

4.7.3.6 Effect of Nitric Oxide

Plants' ability to tolerate below zero temperatures relies on complex and varied processes that involve the accumulation of endogenous as well as inducible components often regulated by environmental factors (Costa-Broseta et al. 2018). The endogenous components that favor constitutive freezing tolerance have been extensively studied and mainly refer to metabolites with osmoprotective activities to limit freeze-induced dehydration and to avoid ice nucleation inside the cells (Janská et al. 2010), with hormonal activities (Eremina et al. 2016) or with antioxidant functions (Winkel-Shirley 2002). Moreover, other regulatory molecules such as polyamines, lipids, reactive oxygen species, and nitric oxide have also been described to be involved in freezing tolerance (Cuevas et al. 2008; Chen and Thelen 2016; Takahashi et al. 2016; van Buer et al. 2016; Zhao et al. 2009; Puyaubert and Baudouin 2014).

Nitric oxide is endogenously produced in diverse living organisms and regulates a wide array of biological processes, including many responses of plants to environmental abiotic and biotic stresses (Siddiqui et al. 2011; Arasimowicz-Jelonek and Floryszak-Wieczorek 2014). In plants, nitric oxide is generated through both oxidative and reductive biosynthetic pathways, which are enhanced under stress (Gupta et al. 2011; Mur et al. 2013). Nitric oxide is a free radical that reacts with metals and reactive oxygen species (Thomas 2015), thus contributing to redox homeostasis and alleviating oxidative stress. It has also been reported that nitric oxide improves the antioxidant capacity of plants (Correa-Aragunde et al. 2015). However, nitric oxide can potentiate or attenuate oxidative stress in plants when it acts either in a chronic or acute mode (Groß et al. 2013). Besides redox-related functions, nitric oxide also has the potential of triggering post-translational modifications of many target proteins, which then display often altered function, activity, stability, and/or subcellular localization. Nitric oxide induces S-nitrosylation of cysteine and nitration of tyrosine (Astier and Lindermayr 2012), but also ubiquitylation of lysine and phosphorylation of serine, threonine, and tyrosine (Hess and Stamler 2012). In addition, many of those post-translational modifications also alter the stability of the target proteins through the regulation of proteolytic degradation (Guerra and Callis 2012).

4.7.4 Exogenous Diethyl Aminoethyl Hexanoate Ameliorates Low-Temperature Stress

Throughout the growing season, crops frequently suffer from various types of environmental stress. As one of the major abiotic stresses, low temperature negatively affects plant growth and development (Heidarvand et al. 2017). Brief exposure to low temperature may disrupt plant physiological processes, such as water status, photosynthesis, and nitrogen metabolism, but plants generally survive (Theocharis et al. 2012; Chen et al. 2014); prolonged exposure to low temperature may lead to plant necrosis or death. As a thermophilic C4 plant that originates from subtropical regions, maize growth is highly susceptible to low temperatures. Nitrogen metabolism, including nitrogen uptake, transport, reduction, and assimilation, as well as amino acid metabolism, is a fundamental process in plants (Bashri et al. 2018). Moreover, most plant stress-responsive physiological processes involve nitrogen metabolisms, such as enhanced nutrient uptake and transport, improved photosynthetic regulation, rapid synthesis of osmotic solutes, and structural alterations (Xie et al. 2019). Hence, nitrogen metabolism is extremely important for the growth and low temperature tolerance of plants. Chemical regulation is widely applied in agricultural production as a strategy to prevent or alleviate the adverse effects induced by abiotic stresses. Diethyl aminoethyl hexanoate, a plant growth regulator, is involved in the regulation of a wide range of metabolic and physiological responses of crop plants such as maize, cotton, soybean, peanut, tomato, and wheat (Wen et al. 2019; Zhou et al. 2018; Qi et al. 2013; Zhang et al. 2016). Exogenous diethyl aminoethyl hexanoate increases grain weight through involvement in the synthesis of sucrose and starch (Wen et al. 2019); promotes seed germination and seedling establishment by mediating fatty acid metabolism and glycometabolism (Zhou et al. 2018); enhances seedling growth through altered photosynthesis by accelerating chlorophyll biosynthesis and increasing the activities of phosphoenolpyruvate carboxylase and ribulose-1,5-bisphosphate carboxylase; and regulates hormone balance by enhancing the contents of auxin, zeatin riboside, and gibberellin but decreasing the content of abscisic acid (Qi et al. 2013). Exogenous diethyl aminoethyl hexanoate also has positive effects on the improvement of plant stress resistance, such as resistance to salinity stress and heavy metal stress (Zhang et al. 2016; Li et al. 2018). Despite accumulating research that has enriched our understanding of the improvement of growth and development following diethyl aminoethyl hexanoate application, the possible role of diethyl aminoethyl hexanoate in alleviating low-temperature stress has not yet been explored (Zhang et al. 2020).

References

Abdel Wahab AM, Zahran HH, Abd-Alla MH (1996) Root-hair infection and nodulation of four grain legumes as affected by the form and the application time of nitrogen fertilizer. Folia Microbiol 41(4):303–308

Abiri R, Azmi N, Maziah M, Norhana Z, Yusof B, Atabaki N, Sahebi M, Valdiani A, Kalhori N, Azizi P, Hanafi MM (2017) Role of ethylene and the APETALA 2/ethylene response factor superfamily in rice under various abiotic and biotic stress conditions. Environ Exp Bot 134:33–44

Adamovich A, Klasens V (2001) Symbiotically fixed nitrogen in forage legume–grass mixture. Grassland Science in Europe 6:12

Adams S, Cockshull K, Cave C (2001) Effect of temperature on the growth and development of tomato fruits. Ann Bot 88(5):1–9

Adgo E, Schulze J (2002) Nitrogen fixation and assimilation efficiency in Ethiopian and German pea varieties. Plant Soil 239(2):291–299

Ahmad I, Basra SMA, Wahid A (2014a) Exogenous application of ascorbic acid, salicylic acid and hydrogen peroxide improves the productivity of hybrid maize at low temperature stress. Int J Agric Biol 16(4):825–830

Ahmad P, Wani MR, Azooz MM, Tran LSP (2014b) Improvement of crops in the era of climatic changes. Springer, New York

Albrecht SL, Bennett JM, Boote KJ (1984) Relationship of nitrogenase activity to plant water stress in field-grown soybeans. Field Crops Res 8:61–71

Alcantar-Gonzales GM, Migianac-Maslow A, Champigny ML (1988) Effect of nitrate supply on energy balance and acetylene reduction and nitrate reductase activities of soybean root nodules infected with *Bradyrhizobium japonicum*. CR Acad Sci Paris 307:145–152

Alexieva V, Sergiev I, Mapelli S, Karanov E (2001) The effect of drought and ultraviolet radiation on growth and stress markers in pea and wheat. Plant Cell Environ 24(12):1337–1344

Al-Falih AMK (2002) Factors affecting the efficiency of symbiotic nitrogen fixation by *Rhizobium*. Pak J Biol Sci 5(11):1277–1293

Ali Q, Ashraf M (2011) Exogenously applied glycinebetaine enhances seed and seed oil quality of maize (*Zea mays* L.) under water deficit conditions. Environ Exp Bot 71(2):249–259

Ali Q, Anwar F, Ashraf M, Saari N, Perveen R (2013) Ameliorating effects of exogenously applied proline on seed composition, seed oil quality and oil antioxidant activity of maize (*Zea mays* L.) under drought stress. Int J Mol Sci 14(1):818–835

Ali F, Bano A, Fazal A (2017) Recent methods of drought stress tolerance in plants. Plant Growth Regul 82:363–375

Ali Q, Rehman H, Haider MZ, Shahid S, Aslam N, Shehzad F, Naseem J, Ashraf R, Ali S, Hussain SM (2019) Chapter 12. Role of amino acids in improving abiotic stress tolerance to plants. In: Hasanuzzaman M, Fujita M, Oku H, Islam MT (eds) Plant tolerance to environmental stress. Taylor & Francis Group, Abingdon, pp 175–204

Allen DJ, Ort DR (2001) Impacts of chilling temperatures on photosynthesis in warm-climate plants. Trends Plant Sci 6(1):36–42

Alonso-Blanco C, Gomez-Mena C, Llorente F, Koornneef M, Salinas J, Martínez-Zapater JM (2005) Genetic and molecular analyses of natural variation indicate CBF2 as a candidate gene for underlying a freezing tolerance quantitative trait locus in *Arabidopsis*. Plant Physiol 139(3): 1304–1312

Amaral JAT, Da Matta FM, Rena AB (2001) Effects of fruiting on the growth of arabica coffee trees as related to carbohydrate and nitrogen status and to nitrate reductase activity. Braz J Plant Physiol 13(1):66–74

Amthor JS (2000) Direct effect of elevated CO_2 on nocturnal in situ leaf respiration in nine temperate deciduous tree species is small. Tree Physiol 20(2):139–144

Anchordoguy TJ, Rudolph AS, Carpenter JF, Crowe JH (1987) Modes of interaction of cryoprotectants with membrane phospholipids during freezing. Cryobiology 24(4):324–331

Andaya VC, Mackill DJ (2003) Mapping of QTLs associated with cold tolerance during the vegetative stage in rice. J Exp Bot 54(392):2579–2585

Anderson J, Dahlberg JE, Bretscher MS, Revel M, Clark BFC (1967) GTP stimulated binding of initiator tRNA to ribosomes directed by f2 bacteriophage RNA. Nature 216(5120):1072–1076

Anjana A, Umar S, Iqbal M, Abrol YP (2006) Are nitrate concentrations in leafy vegetables within safe limits? In Proceedings of the Workshop on Nitrogen in Environment, Industry and Agriculture, New Delhi, India, pp 81–84

Anjana A, Umar S, Iqbal M (2007) Nitrate accumulation in plants, factors affecting the process, and human health implications. A review. Agron Sustain Dev 27(1):45–57

Anjum SA, Xie Y, Wang LC, Saleem MF, Man C, Lei W (2011) Morphological, physiological and biochemical responses of plants to drought stress. Afr J Agric Res 6(9):2026–2032

Anjum SA, Ashraf U, Tanveer M, Khan I, Hussain S, Shahzad B, Zohaib A, Abbas F, Saleem MF, Ali I, Wang LC (2017) Drought induced changes in growth, osmolyte accumulation and antioxidant metabolism of three maize hybrids. Front Plant Sci 8:69

Arasimowicz-Jelonek M, Floryszak-Wieczorek J (2014) Nitric oxide: an effective weapon of the plant or the pathogen? Mol Plant Pathol 15(4):406–416

Aroca R, Irigoyen JJ, Sánchez-Díaz M (2003) Drought enhances maize chilling tolrerance. II. Photosynthetic traits and protective mechanisms against oxidative stress. Physiol Plant 117(4):540–549

Arora R (2013) Physiological study of the components of winter-hardiness in Rhododendron: cold acclimation, deacclimation kinetics, and reacclimation ability. Acta Hortic 990:67–81

Arrese-Igor C, Garcia-Plazaola JI, Hernandez A, Aparicio-Tejo PM (1990) Effect of low nitrate supply to nodulated lucerne on time course of activities of enzymes involved in inorganic nitrogen metabolism. Physiol Plant 80(2):185–190

Arrese-Igor C, Minchin FR, Gordon AJ, Nath AK (1997) Possible causes of the physiological decline in soybean nitrogen fixation in the presence of nitrate. J Exp Bot 48(4):905–913

Ashraf M, Foolad M (2007) Roles of glycine betaine and proline in improving plant abiotic stress resistance. Environ Exp Bot 59(2):206–216

Astier J, Lindermayr C (2012) Nitric oxide-dependent posttranslational modification in plants: an update. Int J Mol Sci 13(11):15193–15208

Atkin OK, Cummins WR (1994) The effect of root temperature on the induction of nitrate reductase activities and nitrogen uptake rates in arctic plant species. Plant Soil 159(2):187–197

Attardi G (1967) The mechanism of protein synthesis. Ann Rev Microbiol 21:383–416

Aydin B, Nalbantoğlu B (2011) Effects of cold and salicylic acid treatments on nitrate reductase activity in spinach leaves. Turk J Biol 35:443–448

Bafeel SO, Ibrahim MM (2008) Antioxidants and accumulation of α-tocopherol induce chilling tolerance in *Medicago sativa*. Int J Agric Biol 10:593–598

Bagdi DL, Shaw BP (2013) Analysis of proline metabolic enzymes in *Oryza sativa* under NaCl stress. J Environ Biol 34(4):677–681

Baldani VLD, Baldani JI, Dobereiner J (2000) Inoculation of rice plants with the endophytic diazotrophs *Herbasprillum* seropedicae and *Burkholderia* spp. Biol Fertil Soils 30(5):485–491

Banilas G, Korkas E, Englezos V, Nisiotou AA, Hatzopoulos P (2012) Genome wide analysis of the heat shock protein 90 gene family in grapevine (*Vitis vinifera* L.). Aust J Grape Wine Res 18(1): 29–38

Bannister P, Neuner G (2001) Frost resistance and the distribution of conifers. In: Bigras FJ, Colombo SJ (eds) Conifer cold hardiness, Tree physiology, vol 1. Springer, Dordrecht, pp 3–21

Barnett T, Altschuler M, McDaniel CN, Mascarenhas JP (1979) Heat shock induced proteins in plant cells. Dev Genet 1(4):331–340

Barraco G, Chatelet P, Balsemin E, Decourcelle T, Sylvestre I, Engelmann F (2012) Cryopreservation of *Prunus cerasus* through vitrification and replacement of cold hardening with preculture on medium enriched with sucrose and/or glycerol. Sci Hort 148:104–108

Bashri G, Singh M, Mishra RK, Kumar J, Singh VP, Mohan S (2018) Kinetin regulates UV-B-induced damage to growth, photosystem II photochemistry, and nitrogen metabolism in tomato seedlings. J Plant Growth Regul 37(1):233–245

Basu S, Ramegowda V, Kumar A, Pereira A (2016) Plant adaptation to drought stress. F1000 Res 5: 1554

Baum G, Chen Y, Arazi T, Takatsuji H, Fromm H (1993) A plant glutamate decarboxylase containing a calmodulin binding domain: cloning, sequence, and functional analysis. J Biol Chem 268(26):19610–19617

Becana M, Klucas RV (1992) Oxidation and reduction of leghemoglobin in root nodules of leguminous plants. Plant Physiol 98(4):1217–1221

Beck EH, Heim R, Hansen J (2004) Plant resistance to cold stress: mechanisms and environmental signals triggering frost hardening and dehardening. J Biosci 29(4):449–459

Beck EH, Fettig S, Knake C, Hartig K, Bhattarai T (2007) Specific and unspecific responses of plants to cold and drought stress. J Biosci 32(3):501–510

Beilharz TH, Preiss T (2004) Widespread use of poly (A) tail length control to accentuate expression of the yeast transcriptome. RNA 13(7):982–997

Ben-Haj-Salah H, Tardieu F (1995) Temperature affects expansion rate of maize leaves without change in spatial distribution of cell length (analysis of the coordination between cell division and cell expansion). Plant Physiol 109(3):861–870

Bernard SM, Habash DZ (2009) The importance of cytosolic glutamine synthetase in nitrogen assimilation and recycling. New Phytol 182(3):608–620

Beuve N, Rispail N, Lainé P, Cliquet J-B, Ourry A, Le Deunff E (2004) Putative role of γ-aminobutyric acid (GABA) as a long-distance signal in up-regulation of nitrate uptake in *Brassica napus* L. Plant Cell Environ 27(8):1035–1046

Beverly RB, Jarrell WM (1984) Cowpea response to N form, rate, and timing of application. Agron J 76(4):663–668

Bhargava S, Sawant K (2013) Drought stress adaptation: metabolic adjustment and regulation of gene expression. Plant Breed 132(1):21–32

Bhatnagar A, Sillanpää M (2011) A review of emerging adsorbents for nitrate removal from water. Chem Eng J 168(2):493–504

Bhattacharya A (2018) Abiotic stress and heat shock proteins (HSPs) in plants. In: Abiotic stress and plant physiology, Metabolic activities, vol 1. New India Publishing Agency, New Delhi, pp 359–445

Bhattacharya A (2019) Effect of high temperature on protein metabolism. In: Effect of high temperature on crop productivity and metabolism of macro molecules. Elsevier, San Diego, pp 217–310

Bhyan SB, Minami A, Kaneko Y, Suzuki S, Arakawa K, Sakata Y, Takezawa D (2012) Cold acclimation in the moss *Physcomitrella patens* involves abscisic acid-dependent signaling. J Plant Physiol 169(2):137–145

Bian Z, Wang Y, Zhang X, Grundy S, Yang Q, Cheng R (2020) A review of environment effects on nitrate accumulation in leafy vegetables grown in controlled environments. Food 9(6):732

Bilska A, Pawe S (2010) Closure of plasmodesmata in maize (*Zea mays*) at low temperature: a new mechanism for inhibition of photosynthesis. Ann Bot 106(5):675–686

Bixby JA, Brown GN (1975) Ribosomal changes during induction of cold hardiness in black locust seedlings. Plant Physiol 56(5):617–621

Bloom AJ, Randall LB, Meyerhof PA, St. Clair, D.A. (1998) The chilling sensitivity of root ammonium influx in a cultivated and wild tomato. Plant Cell Environ 21(2):191–199

Bloom A, Zwieniecki M, Passioura J, Randall L, Holbrook N, St. Clair D (2004) Water relations under root chilling in a sensitive and tolerant tomato species. Plant Cell Environ 27(8):971–979

Bollman MI, Vessey JK (2006) Differential effects of nitrate and ammonium supply on nodule initiation, development, and distribution on roots of pea (*Pisum sativum*). Can J Bot 84(6): 893–903

Boote KJ, Hoogenboom G, Jones JW, Ingram KT (2008) Modeling nitrogen fixation and its relationship to nitrogen uptake in the CROPGRO model. In: Ma L, Ahuja LR, Bruulsem TW (eds) Modeling. CRC Press, Florence, pp 13–46

Bordeleau LM, Prevost D (1994) Nodulation and nitrogen fixation in extreme environments. Plant Soil 161:115–125

Borsani O, Diaz P, Monza J (1999) Proline is involved in water stress. J Plant Physiol 155(2): 269–273

Botta A (2013) Enhancing plant tolerance to temperature stress with amino acids: an approach to their mode of action. Acta Hortic 1009:29–35

Bouché N, Fromm H (2004) GABA in plants: just a metabolite? Trends Plant Sci 9(3):110–115

Bouché N, Fait A, Bouchez D, Moller SG, Fromm H (2003) Mitochondrial succinic semialdehyde dehydrogenase of the c-aminobutyrate shunt is required to restrict levels of reactive oxygen intermediates in plants. Proc Natl Acad Sci U S A 100(11):6843–6848

Bouniols A, Cabelguenne M, Jones CA, Chalamet A, Charpenteau JL, Marty JR (1991) Simulation of soybean nitrogen nutrition for a silty clay soil in southern France. Field Crops Res 26:19–34

Bown AW, Shelp BJ (1989) The metabolism and physiological roles of γ-aminobutyric acid. Biochemistry (Life Science Advances) 8:21–25

Bown AW, Shelp BJ (1997) The metabolism and functions of γ-aminobutyric acid. Plant Physiol 115(1):1–5

Brandi A, Pietroni P, Gualerzi CO, Pon CL (1996) Post-transcriptional regulation of CspA expression in *Escherichia coli*. Mol Microbiol 19(2):231–240

Bredow M (2017) Characterization and knockdown of a family of ice-binding proteins from freeze-tolerant grasses. PhD thesis, Queen's University; Canada.

Bredow M, Walker VK (2017) Ice-binding proteins in plants. Front Plant Sci 8:2153

Breitkreuz KE, Shelp BJ (1995) Subcellular distribution of the 4-aminobutyrate shunt in developing soybean cotyledon protoplasts. Plant Physiol 108(1):99–103

Brown GN (1972) Changes in ribosomal patterns and a related membrane fraction during induction of cold hardiness in *Mimosa* epicotyl tissues. Plant Cell Physiol 13(2):345–351

Brown GN, Bixby JA (1975) Soluble and insoluble protein patterns during induction of freezing tolerance in black locust seedlings. Physiol Plant 34(3):187–191

Brüggemann W, Linger P (1994) Long term chilling of young tomato plants under low light. IV. Differential responses of chlorophyll fluorescence quenching coefficients in *Lycopersicon* species of different chilling sensitivity. Plant Cell Physiol 35(4):585–591

Bruno T, Grignon N, Grignon C (1988) Charge balance of NO_3-fed soybean estimation of K+ and carboxylate regirculation. Plant Physiol 88(3):605–612

Bruno T, Muller B, Grignon C (1992) Effect of phloem translocated malate on NO_3-uptake by roots of intact soybean plants. Plant Physiol 99(3):1118–1123

Burke MJ, Gusta LV, Quamme HA, Weiser CJ, Li PH (1976) Freezing and injury in plants. Annu Rev Plant Physiol 27:507–528

Cabelguenne M, Debaeke P, Bouniols A (1999) EPIC phase, a version of the EPIC model simulating the effects of water and nitrogen stress on biomass and yield, taking account of developmental stages: validation on maize, sunflower, sorghum, soybean and winter wheat. Agric Syst 60(3):175–196

Cabello P, de la Haba P, Govzales-Fontes A, Maldonado JM (1998) Induction of nitrate reductase, nitrite reductase, and glutamine synthetase isoforms in sunflower cotyledons as affected by nitrate, light and plastid integrity. Protoplasma 201(1–2):1–7

Calatayud Á, Gorbe E, Roca D, Martínez PF (2008) Effect of two nutrient solution temperatures on nitrate uptake, nitrate reductase activity, NH_4^+ concentration and chlorophyll a fluorescence in rose plants. Environ Exp Bot 64(1):65–74

Calvo P, Nelson L, Kloepper JW (2014) Agricultural uses of plant biostimulants. Plant Soil 383(1–2):3–41

Campbell WH (1999) Nitrate reductase structure, function and regulation: bridging the gap between biochemistry and physiology. Annu Rev Plant Physiol Plant Mol Biol 50:277–303

Cannell MGR, Thornley JHM (2000) Modelling the components of plant respiration: some guiding principles. Ann Bot 85(1):45–54

Cantrel C, Vazquez T, Puyaubert J, Rezé N, Lesch M, Kaiser WM, Dutilleul C, Guillas I, Zachowski A, Baudouin E (2011) Nitric oxide participates in cold-responsive phosphosphingolipid formation and gene expression in *Arabidopsis thaliana*. New Phytol 189(2):415–427

Cao YY, Yang MT, Li X, Zhou ZQ, Wang XJ, Bai JG (2014) Exogenous sucrose increases chilling tolerance in cucumber seedlings by modulating antioxidant enzyme activity and regulating proline and soluble sugar contents. Sci Hortic 179:67–77

Carelli MLC, Fahl JI (1991) Distribuição da assimilação de nitrato e de matéria seca em plantas jovens de café cultivadas em diferentes níveis de nitrogênio. Bragantia 50(1):29–37

Carelli ML, Fahl JI (2005) Partitioning of nitrate reductase activity in *Coffea arabica* L. and its relation to carbon assimilation under different irradiance regimes. Braz J Plant Physiol 18(3): 397–406

Carelli MLC, Fahl JI (2006) Partitioning of nitrate reductase activity and its relation to carbon assimilation under different irradiance regimes. Braz J Plant Physiol 18(3):397–406

Carelli MLC, Magalhães AC (1981) Development of nitrate reductase activity in green tissues of soybean seedlings (*Glycine max* L. Merr.). Z Pflanzenphysiol 104(1):17–24

Catalá R, López-Cobollo R, Castellano MM, Angosto T, Alonso JM, Ecker JR, Salinasa J (2014) The Arabidopsis 14-3-3 protein rare cold inducible 1A links low-temperature response and ethylene biosynthesis to regulate freezing tolerance and cold acclimation. Plant Cell 26(8): 3326–3342

Cattivelli L, Bartels D (1989) Cold-induced mRNAs accumulate with different kinetics in barley coleoptiles. Planta 178(2):184–188

Cechin I, Rossi SC, Oliveira VC, Fumis TF (2006) Photosynthetic responses and proline content of mature and young leaves of sunflower plants under water deficit. Photosynthetica 44(1): 143–146

Chadjaa H, Vezina L-P, Dorais M, Gosselin A (2001) Effects of lighting on the growth, quality and primary nitrogen assimilation of greenhouse lettuce (*Lactuca sativa* L.). Acta Hortic 559:325–331

Chamber-Perez MA, Serrano A (1991) C-N balance, nitrate and nitrite reductases in *Bradyrhizobium sp. (Lupinus)* nodules exposed to nitrate, and the relationships to nitrogenase activities. J Plant Physiol 139(1):70–75

Chamber-Perez MA, Camacho-Martinez M, Soriano-Niebla JJ (1997) Nitrate-reductase activities of *Bradyrhizobium sp.* in tropical legumes: effects of nitrate on O_2 diffusion in nodules and carbon costs of N_2 fixation. J Plant Physiol 150(1–2):92–96

Chandra R, Pareak RP (1990) Influence the soil temperature on nodulation, nitrogen fixation and compactiveness of chickpea (*Cicer arietinum* L) *Rhizobium* spp. Leg Res 13:95–99

Chang CY, Unda F, Zubilewich A, Mansfield SD, Ensminger I (2015) Sensitivity of cold acclimation to elevated autumn temperature in field-grown *Pinus strobes* seedlings. Front Plant Sci 6:165

Charollais J, Dreyfus M, Iost I (2004) CsdA, a cold-shock RNA helicase from *Escherichia coli*, is involved in the biogenesis of 50S ribosomal subunit. Nucleic Acids Res 32(9):2751–2759

Chaves MM, Oliveira MM (2004) Mechanisms underlying plant resilience to water deficits: prospects for water-saving agriculture. J Exp Bot 55(407):2365–2384

Chaves MM, Flexas J, Pinheiro C (2009) Photosynthesis under drought and salt stress: regulation mechanisms from whole plant to cell. Ann Bot 103(4):551–560

Chen THH, Gusta LV (1983) Abscisic acid-induced freezing resistance in cultured plant cells. Plant Physiol 73(1):71–75

Chen M, Thelen JJ (2016) Acyl-lipid desaturase 1 primes cold acclimation response in *Arabidopsis*. Physiol Plant 158(1):11–22

Chen D, Kessler B, Monselise SP (1964) Studies on water regime and nitrogen metabolism of citrus seedlings grown under water stress. Plant Physiol 39(3):379–386

Chen THH, Li PH, Brenner ML (1983) Involvement of abscisic acid in potato cold acclimation. Plant Physiol 71(2):362–355

Chen B-M, Wang Z-H, Li S-X, Wang G-X, Song H-X, Wang X-N (2004) Effects of nitrate supply on plant growth, nitrate accumulation, metabolic nitrate concentration and nitrate reductase activity in three leafy vegetables. Plant Sci 167(3):635–643

Chen X, Song F, Liu F, Tian C, Liu S, Xu H, Zhu X (2014) Effect of different *Arbuscular mycorrhizal* fungi on growth and physiology of maize at ambient and low temperature regimes. Sci World J 2014:956141

Chen L, Zhao Y, Xu S, Zhang Z, Xu Y, Zhang J, Chong K (2018) Os MADS 57 together with Os TB 1 coordinates transcription of its target Os WRKY 94 and D14 to switch its organogenesis to defense for cold adaptation in rice. New Phytol 218(1):219–231

Chinnusamy V, Zhu J, Zhu JK (2007) Cold stress regulation of gene expression in plants. Trends Plant Sci 12(10):444–451

Chinnusamy V, Zhu JK, Sunkar R (2010) Gene regulation during cold stress acclimation in plants. Methods Mol Biol 639:39–55

Choudhary A, Singh RP (2000) Cadmium induced change in diamine oxidase activity and polyamine levels in *Vigna radiata* Wilcrez seedling. J Plant Physiol 156(5–6):704–710

Christopher MD (2004) Principles of protein folding, misfolding and aggregation. Semin Cell Dev Biol 15:3–16

Chu TM, Aspinall D, Paleg LG (1974) Stress metabolism. VI. Temperature stress and the accumulation of proline in barley and radish. Aust J Plant Physiol 191):87–97

Clarke HJ, Siddique KHM (2004) Response of chickpea genotypes to low temperature stress during reproductive development. Field Crops Res 90(2):323–334

Cloutier Y (1983) Changes in the electrophoretic patterns of the soluble proteins of winter wheat and rye following cold acclimation and desiccation stress. Plant Physiol 71(2):400–403

Cloutier Y (1984) Changes of protein patterns in winter rye following cold acclimation and desiccation stress. Can J Bot 62(2):366–371

Colombo SJ, Zhao S, Blumwald E (1995) Frost hardiness gradients in shoots and roots of *Picea mariana* seedlings. Scand J For Res 10(1):32–36

Cooper P, Ort DR (1988) Changes in protein synthesis induced in tomato by chilling. Plant Physiol 88(2):454–461

Correa-Aragunde N, Foresi N, Lamattina L (2015) Nitric oxide is a ubiquitous signal for maintaining redox balance in plant cells: regulation of ascorbate peroxidase as a case study. J Exp Bot 66(10):2913–2921

Coruzzi G, Bush DR (2001) Nitrogen and carbon nutrient and metabolite signaling in plants. Plant Physiol 125(1):61–64

Coskan A, Dogan K (2011) Symbiotic nitrogen fixation in soybean. In: Soybean physiology and biochemistry. INTECH Open Access Publisher, pp 167–182

Costa-Broseta A, Perea-Resa C, Castillo MC, Ruiz MF, Salinas J, León J (2018) Nitric oxide controls constitutive freezing tolerance in Arabidopsis by attenuating the levels of osmoprotectants, stress-related hormones and anthocyanins. Sci Rep 8(1):9268

Crawford LA, Bown AW, Breitkreuz KE, Guinel FC (1994) The synthesis of γ-aminobutyric acid in response to treatments reducing cytosolic pH. Plant Physiol 104(3):865–871

Cruz R, Milach S (2004) Cold tolerance at the germination stage of rice: methods of evaluation and characterization of genotypes. Sci Agric 61:1–8

Cruz CM, Soares MIM, Martins-Loução MA (1991) Nitrate reduction in seedlings of carob (*Ceratonia siliqua* L.). New Phytol 119(3):413–419

Cruz CM, Lips SH, Martins-Loução MA (1993) Nitrogen assimilation and transport in carrot plants. Physiol Plant 89(3):524–531

Cuevas JC, López-Cobollo R, Alcázar R, Zarza X, Koncz C, Altabella T, Salinas J, Tiburcio AF, Ferrando A (2008) Putrescine is involved in Arabidopsis freezing tolerance and cold acclimation by regulating abscisic acid levels in response to low temperature. Plant Physiol 148(2): 1094–1105

Custic M, Poljak M, Coga L, Cosic T, Toth N, Pecina M (2003) The influence of organic and mineral fertilization on nutrient status, nitrate accumulation, and yield of head chicory. Plant Soil Environ 49(5):218–222

Cutforth HW, Shaykewich CF, Cho CM (1986) Effect of soil water and temperature on corn (*Zea mays* L.) root growth during emergence. Can J Soil Sci 66(1):51–58

Czapski TR, Trun N (2014) Expression of csp gene in *E. coli* K-12 in defined rich and defined minimal media during normal growth, and after cold-shock. Gene 547(1):91–97

Da Matta FM, Amaral JAT, Rena AB (1999) Growth periodicity in trees of *Coffea arabica* L. in relation to nitrogen supply and nitrate reductase activity. Field Crops Res 60(3):223–229

Dabessa A, Abebe Z, Bekele S (2018) Limitations and strategies to enhance biological nitrogen fixation in sub-humid tropics of Western Ethiopia. J Agric Biotechnol Sustain Dev 10(7): 122–131

Daniel-Vedele F, Filleur S, Caboche M (1998) Nitrate transport: a key step in nitrate assimilation. Curr Opin Plant Biol 1(3):235–239

Dart P, Day J (1971) Effects of incubation temperature and oxygen tension on nitrogenase activity of legume root nodules. Plant Soil 35:167–184

Dawson JO (2008) Chapter 8. Ecology of actinorhizal plants. In: Nitrogen fixing actinorhizal symbiosis. Nitrogen fixation: origin, application and research progress. Springer, pp 199–234

De Maio A (1999) Heat shock proteins: facts, thoughts, and dreams. Shock 11(1):1–12

Deivanai S, Xavier R, Vinod V, Timalata K, Lim OF (2011) Role of exogenous proline in ameliorating salt stress at early stage in two rice cultivars. J Stress Physiol Biochem 7(4): 157–174

Demidchik V, Straltsova D, Medvedev SS, Pozhvanov GA, Sokolik A, Yurin V (2014) Stress-induced electrolyte leakage: the role of K^+-permeable channels and involvement in programmed cell death and metabolic adjustment. J Exp Bot 65(5):1259–1270 den Camp RO, Streng A, De Mita S, Cao Q, Polone E, Liu W, Ammiraju JSS, Kudrna D, Wing R, Untergasser A, Bisseling T, Geurts R (2010) LysM-type mycorrhizal receptor recruited for rhizobium symbiosis in non-legume Parasponia. Science 331(6019):909–912

Deng B, Du W, Liu C, Sun W, Tian S, Dong H (2012) Antioxidant response to drought, cold and nutrient stress in two ploidy levels of tobacco plants: low resource requirement confers polytolerance in polyploids? Plant Growth Regul 66(1):37–47

Derman Y, Söderholm H, Lindström M, Korkeala H (2015) Role of csp genes in NaCl, pH, and ethanol stress response and motility in *Clostridium botulinum* ATCC 3502. Food Microbiol 46: 463–470

Desbrosses GJ, Stougaard J (2011) Root nodulation: a paradigm for how plant-microbe symbiosis influences host developmental pathways. Cell Host Microbe 10(4):348–358

Di Martino C, Delfine S, Pizzuto R, Loreto F, Fuggi A (2003) Free amino acids and glycine betaine in leaf osmoregulation of spinach responding to increasing salt stress. New Phytol 158(3): 455–463

Dickens BF, Thompson GA (1981) Rapid membrane response during low temperature acclimation. Correlation of early changes in the physical properties and lipid composition of *Tetrahymena microsomal* membranes. Biochim Biophys Acta 644(2):211–218

Ding C, Lei L, Yao L, Wang L, Hao X, Li N, Wang Y, Yin P, Guo G, Yang Y (2019) The involvements of calcium-dependent protein kinases and catechins in tea plant [*Camellia sinensis* (L.) O. Kuntze] cold responses. Plant Physiol Biochem 143:190–202

Doll H, Haahr V, Segaard B (1989) Relationships between vernalization requirement and winter hardiness in doubled haploids of barley. Euphytica 42(3):209–213

Dong C-H, Hu X, Tang W, Zheng X, Kim YS, Lee B, Zhu J-K (2006) A putative *Arabidopsis* nucleoporin, AtNUP160, is critical for RNA export and required for plant tolerance to cold stress. Mol Cell Biol 26(24):9533–9543

Dortch Q (1990) The interaction between nitrate and ammonium in phytoplankton. Mar Ecol Prog Ser 61:183–201

Dowgert MF, Steponkus PL (1984) Behavior of the plasma membrane of isolated protoplasts during a freeze-thaw cycle. Plant Physiol 75(4):1139–1151

Drevon JJ, Derensard C, Iretki H, Payre H, Roy G, Serraj R (1995) La salinité abaisse la conductans des nodosités á la diffusion de l'oxygéne. In: Drevon JJ, editor. Facteurs Limitant la Fixation Symbiotique de l'Azote dans le Bassin Méditerranen. INRA, Les Colleques nr 77, Paris, pp 73–84

du Jardin P (2015) Plant biostimulants: definition, concept, main categories and regulation. Sci. Hortic 196:3–14

Dubey RS, Pessarakli M (2002) Physiological mechanisms of nitrogen absorption and assimilation in plants under stressful conditions. In: Handbook of plant and crop physiology, 2nd edn. Marcel Dekkar Inc, New York, pp 637–656

Duhamel M, Vandenkoornhuyse P (2013) Sustainable agriculture: possible trajectories from mutualistic symbiosis and plant neodomestication. Trends Plant Sci 18(11):597–600

Duke SH, Schrader LE, Miller MG (1978) Low temperature effects on soybean (*Glycine max* [L.] Merr. cv. Wells) free amino acid pools during germination. Plant Physiol 62(4):642–647

Duncan DR, Widholm JM (1987) Proline accumulation and its implication in cold tolerance of regenerable maize callus. Plant Physiol 83(3):703–708

Dure L III, Greenway SC, Galau GA (1981) Developmental biochemistry of cotton seed embryogenesis and germination changing messenger ribonucleic acid populations as shown by *in vitro* and *in vivo* protein synthesis. Biochemistry 20(14):4162–4168

Duval BD, Mathew A, Satola SW, Shafer WM (2010) Altered growth, pigmentation, and antimicrobial susceptibility properties of *Staphylococcus aureus* due to loss of the major cold shock gene cspB. Antimicrob Agents Chemother 54(6):2283–2290

Dwivedi P, Raghvendra AS (2004) Modulation of osmotic stress effects on photosynthesis and respiration by temperature in mesophyll protoplast of pea. Indian J Exp Biol 42(12):1208–1211

Eaglesham ARJ (1989) Nitrate inhibition of root-nodule symbiosis in doubly rooted soybean plants. Crop Sci 29(1):115–119

Elia A, Conversa G, Gonnella M (2000) Dosi di azoto, produzione e accumulo di nitrati di lattuga allevata in idrocoltura, Atti V Giornate Scientifiche SOI. Sirmione 1:229–230

Elizondo R, Oyanedel E (2010) Field testing of tomato chilling tolerance under varying light and temperature conditions. Chilean J Agric Res 70(4):552–558

Epand RM, Shai Y, Segrest JP, Anantharamaiah GM (1995) Mechanisms for the modulation of membrane bilayer properties by amphipathic helical peptides. Biopolymers 37(5):319–328

Eremina M, Rozhon W, Poppenberger B (2016) Hormonal control of cold stress responses in plants. Cell Mol Life Sci 73(4):797–810

Ermolenko DN, Makhatadze GI (2002) Bacterial cold-shock proteins. Cell Mol Life Sci 59:1902–1913

Espevig T, DaCosta M, Hoffman L, Aamlid TS, Ronsmo AM, Clarke B, Huang B (2011) Freezing tolerance and carbohydrate changes of two species during cold acclimation. Crop Sci 51(3): 1188–1197

Espevig T, Höglind M, Aamlid TS (2014) Dehardening resistance of six turfgrasses used on golf greens. Environ Exp Bot 106:182–188

Etchegaray JP, Inouya M (1999) CspA, CspB, and CspG, major cold shock proteins of *Escherichia coli*, are induced at low temperature under conditions that completely block protein synthesis. J Bacteriol 181(6):1827–1830

Fahl JI, Carelli MLC, Magalhães AC (1992) Asimilación de carbono y nitrógeno en hojas de café. Turrialba 42:523–527

Falkowski PG (1975a) Nitrate uptake in marine pehytoplankton: (nitrate, chloride)-activated adenosine triphosphatase from *Skeletonema costatum* (Bacillariophyceae). J Phycol 11(3): 323–326

Falkowski PG (1975b) Nitrate uptake in manne phytoplank-comparison of half saturation constants from seven species. Lirnnol Oceanogr 20(3):412–417

Fang L, Jiang W, Bae W, Inouye M (1997) Promoter-independent cold-shock induction of cspA and its derepression at 37 degrees C by mRNA stabilization. Mol Microbiol 23(2):355–364

Farooq M, Basra SMA, Wahid A, Cheema ZA, Cheema MA, Khaliq A (2008) Physiological role of exogenously applied glycine betaine in improving drought tolerance of fine grain aromatic rice (*Oryza sativa* L.). J Agron Crop Sci 194(5):325–333

Farooq M, Aziz T, Wahid A, Lee DJ, Siddique KHM (2009a) Chilling tolerance in maize: agronomic and physiological approaches. Crop Pasture Sci 60(6):501–516

Farooq M, Wahid A, Kobayashi N, Fujita D, Basra SMA (2009b) Plant drought stress: effects, mechanisms and management. Agron Sustain Dev 29(1):185–212

Farrell TC, Fox KM, William RL, Fukai S, Lewin LG (2006) Minimizing cold damage during reproductive development among temperate rice genotypes. II. Genotypic variation and flowering traits related to cold tolerance screening. Aust J Agric Res 57(1):89–100

Fellbaum CR, Gachomo EW, Beesetty Y, Choudhari S, Strahan GD, Pfeffer PE, Kiers ET, Bucking H (2012) Carbon availability triggers fungal nitrogen uptake and transport in arbuscular mycorrhizal symbiosis. Proc Natl Acad Sci U S A 109(7):2666–2671

Fenn ME, Poth MA, Aber JD, Baron JS, Bormann BT, Johnson DW, Lemly AD, McNutty SG, Ryan DF, Stottlemyer R (1998) Nitrogen excess in north American ecosystems: predisposing factors, ecosystem responses, and management strategies. Ecol Appl 8(3):706–733

Flaig H, Mohr H (1992) Assimilation of nitrate and ammonium by the Scots pine (*Pinus sylvestris*) seedling under conditions of high nitrogen supply. Physiol Plant 84(4):568–576

Fletcher GL, Hew CL, Davies PL (2001) Antifreeze protein of teleost fishes. Annu Rev Physiol 63: 359–390

Flynn KJ (1999) Nitrate transport and ammonium-nitrate interaction at high nitrate concentration and low temperature. Mar Ecol Prog Ser 187:283–287

Flynn KJ, Fasham MJR, Hipkin CR (1997) Modelling the interactions between ammonium and nitrate uptake in marine phytoplankton. Philos Trans R Soc Lond B Biol Sci 352(1361): 1625–1645

Flynn KJ, Page S, Wood G, Hipkin CR (1999) Variations in the maximum transport rates for ammonium and nitrate in the prymnesiophyte *Emiliania huxleyi* and the raphidophyte *Heterosigma carterae*. J Plankton Res 21(2):355–371

Foolad M, Lin G (2000) Relationship between cold tolerance during seed germination and vegetative growth in tomato: Germplasm evaluation. J Am Soc Hortic Sci 125(6):679–683

Fowler S, Thomashow MF (2002) Arabidopsis transcriptome profiling indicates that multiple regulatory pathways are activated during cold acclimation in addition to the CBF cold response pathway. Plant Cell 14(8):1675–1690

Foyer CH, Valadier MH, Migge A, Becker TW (1998) Drought-induced effects on nitrate reductase activity and mRNA and on the co-ordination of nitrogen and carbon metabolism in maize leaves. Plant Physiol 117(1):283–292

Frame J, Newbould P (1986) Agronomy of white clover. Adv Agron 40:1–88

Franche C, Lindström K, Elmerich C (2009) Nitrogen-fixing bacteria associated with leguminous and non-leguminous plants. Plant Soil 321(1–2):35–59

Fujita M, Kawada N, Tahir M (1992) Relationship between cold resistance, heading traits and ear primordia development of wheat cultivars. Euphytica 64(1–2):123–130

Gaby JC, Buckley DH (2011) A global concensus of nitrogenase diversity. Environ Microbiol 13(7):1790–1799

Gal TZ, Glaer I, Koltai H (2004) An LEA group 3 family member is involved in survival of *C. elegans* during exposure to stress. FEBS Lett 577(1–2):21–26

Galloway JN, Townseed AR, Erisman JW, Bekunda M, Cai ZC, Freney JR, Martinelli LA, Seitzinger SP, Sutton MA (2008) Transformation of the nitrogen cycle: recent trends, questions, and potential solutions. Science 320(5878):889–892

Gan Y, Stulen I, van Keulen H, Kuiper PJC (2004) Low concentrations of nitrate and ammonium stimulate nodulation and N_2 fixation while inhibiting specific nodulation (nodule DW g^{-1} root dry weight) and specific N_2 fixation (N_2 fixed g^{-1} root dry weight) in soybean. Plant Soil 258: 281–292

Gantar M, Elhai J (1999) Colonization of wheat para nodules by the nitrogen fixing cyanobacterium *Nostoc* spp. strain 2S9B. New Phytol 141(3):373–379

Gao Y, Smith GJ, Alberte RS (2000) Temperature dependence of nitrate reductase activity in marine phytoplankton: biochemical analysis and ecological implications. J Physiol 36(2): 304–313

Garcia-Plazaola JI, Beccerril JM, Arrese-Igor C, Hernandez A, Gonzales-Murua C, Aparicio-Tejo PM (1993) Denitrifying ability of thirteen *Rhizobium meliloti* strains. Plant Soil 149(1):43–50

Gaudet D, Loroche A, Frick M, Huel R, Puchelski B (2003) Cold induced expression of plant defense in and lipid transfer protein transcript in winter wheat. Physiol Plant 117(2):195–205

Ghezzi P, Bonetto V (2003) Redox proteomics: identification of oxidatively modified proteins. Proteomics 3(7):1145–1153

Giannakis C, Nicholas DJD, Wallace W (1988) Utilization of nitrate by bacteroids of *Bradyrhizobium japonicum* in the soybean root nodule. Planta 174(1):51–58

Gibson A (1971) Factors in the physical and biological environment affecting nodulation and nitrogen fixation by legumes. Plant Soil 35:139–152

Gilmour SJ, Thomashow MF (1991) Cold acclimation and cold-regulated gene expression in ABA mutants of *Arabidopsis thaliana*. Plant Mol Biol 17(6):1233–1240

Gilmour SJ, Hajela RK, Thomashow MF (1988) Cold acclimation in Arabidopsis thaliana. Plant Physiol 87(3):745–750

Gilmour SJ, Zarka DG, Stockinger EJ, Salazar MP, Houghton JM, Thomashow MF (1998) Low temperature regulation of the Arabidopsis CBF family of AP2 transcriptional activators as an early step in cold-induced COR gene expression. Plant J 16(4):433–442

Gilmour SJ, Sebolt AM, Salazar MP, Everard JD, Thomashow MF (2000) Overexpression of the Arabidopsis CBF3 transcriptional activator mimics multiple biochemical changes associated with cold acclimation. Plant Physiol 124(4):1854–1865

Goh KM, Bruce GE (2005) Comparison of biomass production and biological nitrogen fixation of multi-species pastures (mixed herb leys) with perennial ryegrass-white clover pasture with and without irrigation in Canterbury, New Zealand. Agric Ecosyst Environ 110(3–4):230–240

Gojon A, Bussi C, Grignon C, Salsac L (1991) Distribution of NO_3^- reduction between roots and shoots of peach-tree seedlings as affected by NO_3^- uptake rate. Physiol Plant 82(4):505–512

Gojon A, Plassard C, Bussi C (1994) Root/shoot distribution of NO_3^- assimilation in herbaceous and woody species. In: Garnier RJE (ed) A whole plant perspective on carbon-nitrogen interactions. SPB Academic Publishing, The Hague, pp 131–147

Goldenberg D, Azar I, Oppenheim AB (1996) Differential mRNA stability of the cspA gene in the cold-shock response of *Escherichia coli*. Mol Microbiol 19(2):241–248

Goldstein J, Pollitt NS, Inouye M (1990) Major cold shock protein of *Escherichia coli*. Proc Natl Acad Sci U S A 87:283–287

Gong Z, Lee H, Xiong L, Jagendorf A, Stevenson B, Zhu J-K (2002) RNA helicase-like protein as an early regulator of transcription factors for plant chilling and freezing tolerance. Proc Natl Acad Sci U S A 99(17):11507–11512

Gong Z, Dong C-H, Lee H, Zhu J, Xiong L, Gong D, Stevenson B, Zhu JK (2005) A DEAD box RNA helicase is essential for mRNA export and important for development and stress responses in *Arabidopsis*. Plant Cell 17(1):256–267

Good AG, Shrawat AK, Muench DG (2004) Can less yield more? Is reducing nutrient input into the environment compatible with maintaining crop production? Trends Plant Sci 9(12):597–605

Goodsell DS (2009) Molecule of the month: antifreeze proteins. The Scripps Research Institute and the RCSB PDB, Educational portal for learning macro-biomolecules

Gorski SM, Chittaranjan S, Pleasance ED, Freeman JD, Anderson CL, Varhol RJ, Coughlin SM, Zuvderduvn SD, Jones SJ, Marra MA (2003) A SAGE approach to discovery of genes involved in autophagic cell death. Curr Biol 13(4):358–363

Govindarajulu M, Pfeffer PE, Jin HR, Abubaker J, Douds DD, Allen JW, Bucking H, Lammers PJ, Shachar-Hill Y (2005) Nitrogen transfer in the arbuscular mycorrhizal symbiosis. Nature 435(7043):819–823

Goyal K, Walton LJ, Tunnacliffe A (2005) LEA proteins prevents protein aggregation due to water stress. Biochem J 388(1):151–157

Grace J (1987) Climatic tolerance and the distribution of plants. New Phytol 106(Suppl):113–130

Graham D, Patterson BD (1982) Responses of plants to low, nonfreezing temperature: proteins, metabolism, and acclimation. Annu Rev Plant Physiol 33:347–372

Graumann PL, Marahiel MA (1996) Some like it cold: response of microorganisms to cold shock. Arch Microbiol 166(5):293–300

Greaves JA (1996) Improving suboptimal temperature tolerance in maize-the search for variation. J Exp Bot 47:307–323

Griffith M, Yaish MW (2004) Antifreeze proteins in overwintering plants: a tale of two activities. Trends Plant Sci 9(8):399–405

Groß F, Durner J, Gaupels F (2013) Nitric oxide, antioxidants and prooxidants in plant defence responses. Front Plant Sci 4:419

Grzebelus D, Baranski R (2001) Identification of accessions showing low nitrate accumulation in a germplasm collection of garden beet. Acta Hortic 563:253–255

Guerra DD, Callis J (2012) Ubiquitin on the move: the ubiquitin modification system plays diverse roles in the regulation of endoplasmic reticulum- and plasma membrane-localized proteins. Plant Physiol 160(1):56–64

Gulden R, Vessey JK (1997) The stimulating effect of ammonium on nodulation in *Pisum sativum* L. is not long lived once ammonium supply is discontinued. Plant Soil 195:195–205

Guo L, Chen Q, Fang F, Hu Z, Wu J, Miao A, Xiao L, Chen X, Yang L (2013) Application potential of a newly isolated indigenous aerobic denitrifier for nitrate and ammonium removal of eutrophic lake water. Bioresour Technol 142:45–51

Gupta R, Deswal R (2012) Low temperature stress modulated secretome analysis and purification of an antifreeze protein from *Hippophae rhamnoides*, a Himalayan wonder plant. J Prot Res 11(5): 2684–2696

Gupta R, Deswal R (2014) Antifreeze proteins enable plants to survive in freezing conditions. J Biosci 39(5):931–944

Gupta SC, Sharma A, Mishra M, Mishra RK, Chowdhuri DK (2010) Heat shock proteins in toxicology: how close and how far? Life Sci 86(11–12):377–384

Gupta KJ, Fernie AR, Kaiser WM, van Dongen JT (2011) On the origins of nitric oxide. Trends Plant Sci 16(3):160–168

Gusta LV, O'Connor BJ, Lafond GL (1999) Phosphorus and nitrogen effect on the freezing tolerance of norstar winter wheat. Can J Plant Sci 79(2):191–195

Guy CL (1990) Cold acclimation and freezing stress tolerance: role of protein metabolism. Annu Rev Plant Physiol Plant Mol Biol 41:187–223

Guy CL, Carter JV (1984) Characterization of partially purified glutathione reductase' from cold hardened and non-hardened spinach leaf tissue. Cryobiology 21(4):454–464

Guy CL, Haskell D (1987) Induction of freezing tolerance in spinach is associated with the synthesis of cold acclimation induced proteins. Plant Physiol 84:872–878

Guy CL, Haskell D (1988) Detection of polypeptides associated with the cold acclimation process in spinach. Electrophoresis 9(11):787–796

Guy CL, Haskell D (1989) Preliminary characterization of high molecular mass proteins associated with cold acclimation in spinach. Plant Physiol Biochem 27:777–784

Guy C, Li QB (1998) The organization and evolution of the spinach stress 70 molecular chaperone gene family. Plant Cell 10(4):539–556

Guy CL, Niemi KJ, Brambl R (1985) Altered gene expression during cold acclimation of spinach. Proc Natl Acad Sci U S A 81:3673–3677

Guy C, Haskell D, Li QB (1998) Association of proteins with the stress 70 molecular chaperones at low temperature: evidence for the existence of cold labile proteins in spinach. Cryobiology 36(4):301–314

Gyaneshwar P, Hirsch AM, Moulin L, Chen W-M, Elliott GN, Bontemps C, de Los Santos PE, Gross E, Dos Reis FB, Sprent JL, Young JPW, James EK (2011) Legume- nodulating betaproteobacteria: diversity, host range, and future prospects. Mol Plant Microbe Interact 24(11):1276–1288

Hackett CA, Ellis RP, Forster BP, McNlcol JW, Macaulay M (1992) Statistical analysis of a linkage experiment in barley involving quantitative trait loci for height and ear-emergence time and two genetic markers on chromosome 4. Theor Appl Genet 85(1):120–126

Hahn M, Walbot V (1989) Effects of cold-treatment on protein synthesis and mRNA levels in rice leaves. Plant Physiol 91(3):930–938

Hall TC, McLeester RC, McCown BH, Beck GE (1970) Enzyme changes during deaccilmation of willow stem. Cryobiology 7(2):130–135

Halliday J (1975) An interpretation of seasonal and short term fluctuations in nitrogen fixation. PhD thesis, University of Western Australia, Perth

Halliday J, Pate JS (1976) Symbiotic nitrogen fixation by coralloid roots of the *Cycad Macrozamia riedlei*: physiological characteristics and ecological significance. Aust J Plant Physiol 3(3): 349–358

Hanke GT, Kimata-Ariga Y, Taniguchi I, Hase T (2004) A post genomic characterization of *Arabidopsis* ferredoxins. Plant Physiol 134(1):255–264

Hannah MA, Heyer AG, Hincha DK (2005) A global survey of gene regulation during cold acclimation in *Arabidopsis thaliana*. PLoS Genet 1:e26

Hanower P, Brzozowska J (1975) Effects of osmotic stress on composition of free amino acids in cotton leaves. Phytochemistry 14:1691–1694

Hao J, Yang J, Dong J, Fei SZ (2017) Characterization of BdCBF genes and genome-wide transcriptome profiling of BdCBF3-dependent and -independent cold stress responses in *Brachypodium distachyon*. Plant Sci 262:52–61

Hao X, Wang B, Wang L, Zeng J, Yang Y, Wang X (2018) Comprehensive transcriptome analysis reveals common and specific genes and pathways involved in cold acclimation and cold stress in tea plant leaves. Sci Hortic 240:354–368

Harber RM, Fuchigami LH (1989) Ethylene-induced stress resistance. In: Li PH (ed) Low temperature stress physiology in crops. CRC Press, Boca Raton, pp 81–89

Harmens H, Norris DA, Cooper DM, Mills G, Steinnes E, Kubin E, Thöni L, Abrol JR, Carballeira A, Coskun M, De Temmerman L, Froloval M, González-Miqueo L, Jeran Z, Leblond S, Liiv S, Maňkovská B, Pesch R, Poikolainen J, Rühling A, Santamaria JM, Simonéié P, Schröder W, Suchara I, Yurukova L, Zechmeister HG (2011) Nitrogen concentrations in mosses indicate the spatial distribution of atmospheric nitrogen deposition in Europe. Environ Pollut 159(10):2852–2860

Hassanein RA, Hassanein AA, Haider AS, Hashem HA (2009) Improving salt tolerance of *Zea mays* L. plants by presoaking their grains in glycine betaine. Aust J Basic Appl Sci 3(2): 928–942

Hassanein RA, El-Khawas SA, Ibrahim SK, El-Bassiouny HM, Mostafa HA, Abdel-Monem AA (2013) Improving the thermo tolerance of wheat plant by foliar application of arginine or putrescine. Pak J Bot 45(1):111–118

Havelka UD, Boyle MG, Hardy RWF (1982) Biological nitrogen fixation. In: Stevenson FJ (ed) Nitrogen in agricultural soils. ASA, Madison, pp 365–422

He T, Ye Q, Sun Q, Cai X, Ni J, Li Z, Xie D (2018) Removal of nitrate in simulated water at low temperature by a novel psychrotrophic and aerobic bacterium, *Pseudomonas taiwanensis* strain J. Int Biomed Res 2018:4984087

Hébraud M, Potier P (1999) Cold shock response and low temperature adaptation in psychrotrophic bacteria. J Mol Microbiol Biotechnol 1(2):211–219

Heckmann MO, Drevon JJ (1987) Nitrate metabolism in soybean root nodules. Physiol Plant 69(4): 721–725

Heidarvand L, Millar AH, Taylor NL (2017) Responses of the mitochondrial respiratory system to low temperature in plants. Crit Rev Plant Sci 36(4):1–24

Herridge DF, Pate JS (1977) Utilization of net photosynthate for nitrogen fixation and protein production in an annual legume. Plant Physiol 60(5):759–764

Hess DT, Stamler JS (2012) Regulation by S-nitrosylation of protein post-translational modification. J Biol Chem 287(7):4411–4418

Heuer B (2003) Influence of exogenous application of proline and glycinebetaine on growth of salt-stressed tomato plants. Plant Sci 165(4):693–699

Hildebrandt TM, Nunes Nesi A, Araujo WL, Braun HP (2015) Amino acid catabolism in plants. Mol Plant 8(11):1563–1579

Hodge A, Robinson D, Fitter A (2000) Are microorganisms more effective than plants at competing for nitrogen? Trends Plant Sci 5(7):304–308

Hoffman L, DaCosta M, Ebdon JS, Watkins W (2010) Physiological changes during cold acclimation of perennial ryegrass accessions differing in freeze tolerance. Crop Sci 50(3):1037–1047

Hoffman BM, Lukovanov D, Dean DR, Seefeldt LC (2013) Nitrogenase: a draft mechanism. Acc Chem Res 46(2):587–595

Hoffman L, DaCosta M, Bertrand A, Castonguay Y, Ebdon JS (2014) Comparative assessment of metabolic responses to cold acclimation and deacclimation in annual bluegrass and creeping bentgrass. Environ Exp Bot 106:197–206

Hoppe B, Kahl T, Karasch P, Wubet T, Bauhus J, Buscot F, Krüger D (2014) network analysis reveals ecological links between N-fixation bacteria and wood decaying fungi. PLOS One 9(2): e88141

Hoque MA, Okuma E, Banu MNA, Nakamura Y, Shimoishi Y, Murata Y (2007) Exogenous proline mitigates the detrimental effects of salt stress more than exogenous betaine by increasing antioxidant enzyme activities. J Plant Physiol 164(5):553–561
Howarth CJ, Ougham HJ (1993) Gene expression under temperature stress. New Phytol 125(1): 1–26
Hsu SY, Hsu YT, Kao CH (2003) The effect of polyethylene glycol on proline accumulation in rice leaves. Biol Plant 46:73–78
Hu WH, Zhou YD, Du YS, Xia XJ, Yu JQ (2006) Differential response of photosynthesis in greenhouse and field ecotypes of tomato to long-term chilling under low light. J Plant Physiol 163(12):1238–1246
Hu ZR, Fan JB, Xie Y, Amombo E, Liu A, Gitau MM, Khaldun ABM, Chen L, Fu JM (2016) Comparative photosynthetic and metabolic analyses reveal mechanism of improved cold stress tolerance in bermudagrass by exogenous melatonin. Plant Physiol Biochem 100:94–104
Huang B, Xu C (2008) Identification and characterization of proteins associated with plant tolerance to heat stress. J Integr Plant Biol 50(10):1230–1237
Hughes MA, Dunn MA (1990) The effect of temperature on plant growth and development. Biotechnol Genet Eng Rev 8(1):161–188
Hughes MA, Dunn MA (1996) The molecular biology of plant acclimation to low temperature. J Exp Bot 47(296):291–305
Hughes MA, Pearce RS (1988) Low temperature treatment of barley plants causes altered gene expression in shoot meristems. J Exp Bot 39(10):1461–1467
Hundertmark M, Hincha DK (2008) LEA (late embryogenesis abundant) proteins and their encoding sense in Arabidopsis thaliana. BMC Genomics 9:118
Hungria M, Franco AA (1993) Effects of high temperature on nodulation and nitrogen fixation by *Phaseolus vulgaris* L. Plant Soil 49(1):95–102
Hungria M, Vargas MA (2000) Environmental factors affecting N2 fixation in grain legumes in the tropics, with an emphasis on Brazil. Field Crops Res 65(2):151–164
Hunter WJ (1983) Soybean root and nodule nitrate reductase. Physiol Plant 59(3):471–475
Hur J, Jung KH, Lee CH, An G (2004) Stress-inducible OsP5CS2 gene is essential for salt and cold tolerance in rice. Plant Sci 167(3):417–426
Hussain S, Khan F, Cao W, Wu L, Geng M (2016) Seed priming alters the production and detoxification of reactive oxygen intermediates in rice seedlings grown under sub-optimal temperature and nutrient supply. Front Plant Sci 7:439
Hussain HA, Hussain S, Khaliq A, Ashraf U, Anjum SA, Men S, Wang L (2018) Chilling and drought stresses in crop plants: implications, cross talk, and potential management opportunities. Front Plant Sci 9:393
Ianetta PPM, de Lorenzo C, James EK, Fernandez-Pascual M, Sprent JI, Lucas MM, Witty JF, de Felipe MR, Minchin FR (1993) Oxygen diffusion in lupin nodules. I. Visualisation of diffusion barrier operation. J Exp Bot 44(9):1461–1467
Imsande J (1986) Inhibition of nodule development in soybean by nitrate or reduced nitrogen. J Exp Bot 37:348–355
Iqbal N, Farooq S, Arshad R, Hameed A (2010) Differential accumulation of high and low molecular weight heat shock proteins in Basmati rice (*Oryza sativa* L.) cultivars. Genet Resour Crop Evol 57(1):65–70
Islam MM, Hoque MA, Okuma E, Banu MNA, Shimoishi Y, Nakamura Y, Murata Y (2009) Exogenous proline and glycinebetaine increase antioxidant enzyme activities and confer tolerance to cadmium stress in cultured tobacco cells. J Plant Physiol 166(15):1587–1597
Ismade J (1991) Regulation of the nodule efficiency by the undisturbed soybean plant. J Exp Bot 42:687–691
Izmailov SF (2004) Saturation and utilization of nitrate pools in pea and sugar beet leaves. Russ J Plant Physiol 51(2):189–193
Jackson LE, Schimel JP, Firestone MK (1989) Short term partitioning of ammonium and nitrate between plants and microbes in annual grassland. Soil Biol Biochem 21(3):409–415

James EK, Sprent JI, Minchin FR, Brewin NJ (1991) Intracellular location of glycoprotein in soybean nodules: effect of altered rhizosphere oxygen concentration. Plant Cell Environ 14(5): 467–476

Jan N, Mahbiib-ul-Hussain, Andrabi KI (2009) Cold resistance in plants: a mystery unresolved. Mol Biol Genet 12(3)

Janda T, Szalai G, Tari I, Paldi E (1999) Hydroponic treatment with salicylic acid decreases the effects of chilling injury in maize (*Zea mays* L.) plants. Planta 208(2):175–180

Janská A, Marsik P, Zelenkova S, Ovesna J (2010) Cold stress and acclimation—what is important for metabolic adjustment? Plant Biol 12(3):395–405

Jensen ES (1987) Seasonal patterns of growth and nitrogen fixation in field-grown pea. Plant Soil 101:29–37

Jha UC, Bohra A, Jha R (2017) Breeding approaches and genomics technologies to increase crop yield under low-temperature stress. Plant Cell Rep 36(1):1–35

Jin B, Jeong KW, Kim Y (2014) Structure and flexibility of the thermophilic cold-shock protein of *Thermus aquaticus*. Biochem Biophys Res Commun 451(3):402–407

Johnson-Flanagan AM, Singh J (1987) Alteration of gene expression during the induction of freezing tolerance in *Brassica napus* suspension cultures. Plant Physiol 85(3):699–705

Jones PG, Inouya M (1996) RbfA, a 30S ribosomal binding factor, is a cold-shock protein whose absence triggers the cold-shock response. Mol Microbiol 21(6):1207–1218

Jones PG, Inouye M (1994) The cold-shock response—a hot topic. Mol Microbiol 11(5):811–818

Jones PG, Krah R, Tafuri SR, Wolffe AP (1992) DNA Gyrase, CS7.4, and the cold shock response in *Escherichia coli*. J Bacteriol 174(18):5798–5802

Jones T, Tucker D, Ort D (1998) Chilling delays circadian pattern of sucrose phosphate synthase and nitrate reductase activity in tomato. Plant Physiol 118(1):149–158

Jongdee B, Fukai S, Cooper M (2002) Leaf water potential and osmotic adjustment as physiological traits to improve drought tolerance in rice. Field Crops Res 76(2–3):153–163

Joshi R (2018) Biosynthesis of protein in plants under different environmental factors. J Med Plant Stud 6(2):261–264

Jouyban Z, Hasanzade R, Sharafi S (2013) Chilling stress in plants. Int J Agric Crop Sci 5:2961–2968

Junttila O, Kaurin Å (1990) Environmental control of cold acclimation in Salix pentandra. Scand J For Res 5(1–4):195–204

Jwa NS, Hwang BK (2017) Convergent evolution of pathogen effectors toward reactive oxygen species signaling networks in plants. Front Plant Sci 8:1687

Kaiser WM, Huber SC (2001) Post-translational regulation of nitrate reductase: mechanism, physiological relevance and environmental triggers. J Exp Bot 52(363):1981–1989

Kang J, Turano FJ (2003) The putative glutamate receptor 1.1 (AtGLR1.1) functions as a regulator of carbon and nitrogen metabolism in *Arabidopsis thaliana*. Proc Natl Acad Sci U S A 100(11): 6872–6877

Karimzadeh G, Francis D, Davies MS (2000) Low temperature-induced accumulation of protein is sustained both in root meristems and in callus in winter wheat but not in spring wheat. Ann Bot 85(6):769–777

Karlson D, Imai R (2003) Conservation of the cold shock domain protein family in plants. Plant Physiol 131(1):12–1510

Karlson D, Nakaminami K, Toyomasu T, Imai R (2002) A cold-regulated nucleic acid-binding protein of winter wheat shares a domain with bacterial cold shock proteins. J Biol Chem 277(38):35248–35256

Kasuga M, Miura S, Shinozaki K, Yamaguchi-Shinozaki K (2004) A combination of the Arabidopsis DREB1A gene and stress-inducible rd29A promoter improved drought-and low-temperature stress tolerance in tobacco by gene transfer. Plant Cell Physiol 45(3):346–350

Kaur G, Asthir B (2017) Molecular responses to drought stress in plants. Biol Plant 61(2):201–209

Kavi Kishor PB, Sreenivasulu N (2014) Is proline accumulation per se correlated with stress tolerance or is proline homeostasis a more critical issue? Plant Cell Environ 37(2):300–311

Kenis JD, Silvent ST, Luna CM, Campbell WH (1992) Induction of nitrate reductase in detached corn leaves: the effect of the age of the leaves. Physiol Plant 85(1):49–56

Kennedy IR, Tchan YT (1992) Biological nitrogen fixation in non-leguminous field crops: recent advances. Plant Soil 141:98–118

Keough JT (2016) Physiological and metabolic factors associated with cold and freezing tolerance and induced recovery in cool season and warm season grass species. MSc thesis in plant biology, New Brunswick

Keto-Timonen R, Hietala N, Palonen E, Hakakorpi A, Lindstrom M, Korkeala H (2016) Cold shock proteins: a minireview with special emphasis on csp-family of enteropathogenic Yersinia. Front Microbiol 7(3991):1151

Key JL, Lin CY, Chen YM (1981) Heat shock proteins of higher plants. Proc Natl Acad Sci U S A 78(6):3526–3530

Keys A, Bird I, Cornelius M, Lea P, Wallsgrove R, Miflin B (1978) Photorespiratory nitrogen cycle. Nature 275:741–743

Khan H, Shah SH, Uddin N, Azhar N, Asim M, Syed S, Ullah F, Tawab F, Inayat J (2015) Biochemical and physiological changes of different plants species in response to heat and cold stress. ARPN J Agric Biol Sci 10(6):213–216

Kiani SP, Talia P, Maury P, Grieu P, Heinz R, Perrault A, Nishinakamasu V, Hopp E, Gentzbittel L, Paniego N, Sarrafi A (2007) Genetic analysis of plant water status and osmotic adjustment in recombinant inbred lines of sunflower under two water treatments. Plant Sci 172(4):773–787

Kim JC, Lee SH, Cheong YH, Yoo C-M, Lee SI, Chun HJ, Yun D-J, Hong JC, Lee SY, Lim CO, Cho MJ (2001) A novel cold-inducible zinc finger protein from soybean, SCOF-1, enhances cold tolerance in transgenic plants. Plant J 25(3):247–259

Kim S, An CS, Hong YN, Lee KW (2004) Cold-inducible transcription factor, CaCBF, is associated with a homeodomain leucine zipper protein in hot pepper (*Capsicum annuum* L.). Mol Cell 18(3):300–308

Kim Y-O, Kim JS, Kang H (2005) Cold-inducible zinc finger-containing glycine-rich RNA-binding protein contributes to the enhancement of freezing tolerance in *Arabidopsis thaliana*. Plant J 42(6):890–900

King BJ, Layzell DB, Canvin DT (1986) The role of dark carbon dioxide fixation in root nodules of soybean. Plant Physiol 81:200–205

Kirkby EA, Knight AH (1977) Influence of the levels of nitrate nitrogen on ion uptake and assimilation, organic acid accumulation and cation-anion balance in whole tomato plants. Plant Physiol 60(3):349–353

Kishinevsky BD, Sen D, Weaver RW (1992) Effect of high root temperature on *Bradyrhizobium*-peanut symbiosis. Plant Soil 143:275–282

Kiyomiya S, Nakanishi H, Uchida H, Tsuji A, Nishiyama S, Futatsubashi M, Tsukada H, Ishioka NS, Watanabe S, Ito T, Mizuniwa C, Osa A, Matsuhashi S, Hashimoto S, Sekine T, Mori S (2001) Real time visualization of ^{13}N-translocation in rice under different environmental conditions using positron emitting tracer imaging system. Plant Physiol 124(4):1743–1753

Klimenko S, Peshkova A, Dorofeev NV (2006) Nitrate reductase activity during heat shock in winter wheat. J Stress Physiol Biochem 2(1):50–55

Klucas RV (1974) Studies on soybean nodule senescence. Plant Physiol 54:612–616

Knight H, Trewavas AJ, Knight MR (1996) Cold calcium signaling in Arabidopsis involves two cellular pools and a change in calcium signature after acclimation. Plant Cell 8(3):489–503

Kodama H, Hamada T, Horiguchi G, Nishimura M, Iba K (1994) Genetic enhancement of cold tolerance by expression of a gene for chloroplast [omega]-3 fatty acid desaturase in transgenic tobacco. Plant Physiol 105(2):601–605

Kolaksazov M, Laporte F, Ananieva K, Dobrev P, Herzog M, Ananiev ED (2013) Effect of chilling and freezing stresses on jasmonate content in *Arabis alpina*. Bulg J Agric Sci 19(2):15–17

Koranda M, Kerschbaum S, Wanek W, Zechmeister H, Richter A (2007) Physiological responses of bryophytes *Thuidium tamariscinum* and *Hylocomium splendens* to increased nitrogen deposition. Ann Bot 99(1):161–169

Kowalczyk K, Zielony T, Gajewski M (2008) Effect of aminoplant and Asahi on yield and quality of lettuce grown on rockwool. In: Dąbrow ZT, editor. Biostimulators in modern agriculture vegetable crops. Wieś Jutra, Warsaw, pp 35–43

Krasnuk M, Jung GA, Witham FH (1975) Electrophoretic studies of the relationship of peroxidases, polyphenol oxidase, and indoleacetic acid oxidase to cold tolerance of alfalfa. Cryobiology 12(1):62–80

Krasnuk M, Jung GA, Witham FH (1976a) Electrophoretic studies of several dehydrogenases in relation to cold tolerance of alfalfa. Cryobiology 13(2):375–393

Krasnuk M, Witham FH, Jung GA (1976b) Electrophoretic studies of several hydrolytic enzymes in relation to cold tolerance of alfalfa. Cryobiology 13(2):225–242

Kreps JA, Wu Y, Chang HS, Zhu T, Wang X, Harper J (2002) Transcriptome changes for Arabidopsis in response to salt, osmotic, and cold stress. Plant Physiol 130(4):2129–2141

Kreyling J, Wiesenberg GLB, Thiel D, Wohlfart C, Huber G, Walter J, Jentsch A, Konnert M, Beierkuhnlein C (2010) Cold hardiness of *Pinus nigra* Arnold as influenced by geographic origin, warming, and extreme summer drought. Environ Exp Bot 78:99–108

Krishna P (2003) Plant responses to heat stress. Topics Curr Genet 4:73–101

Krivokapich SJ, Molina V, Bergagna HFJ, Guarnera EA (2006) Epidemiological survey of *Trichinella* infection in domestic, synanthropic and sylvatic animals from Argentina. J Helminthol 80(3):267–269

Kumar R, Chaurasiya PC, Singh RN, Singh S (2018) A review report: low temperature stress for crop production. Int J Pure Appl Biosci 6(2):575–598

Kuo T, Boersma L (1971) Soil water suction and root temperature effects on nitrogen fixation in soybeans. Agron J 63(3):901–904

Kurkela S, Franck M, Heino P, Lang V, Palva ET (1988) Cold induced gene expression in *Arabidopsis thaliana* L. Plant Cell Rep 7:495–498

Kusano M, Fukushima A, Redestig H, Saito K (2011) Metabolomic approaches toward understanding nitrogen metabolism in plants. J Exp Bot 62(4):1439–1453

Kushad MM, Yelenosky G (1987) Evaluation of polyamine and proline levels during low temperature acclimation of citrus. Plant Physiol 84(3):692–695

Kuypers MMM, Marchant HK, Kartal B (2011) The microbial nitrogen-cycling network. Nat Rev Microbiol 1(1):1–14

Kuznetsov VV, Kimpel JA, Goekjian G, Key JL (1987) Elements of non specificity in responses on the plant genome to chilling and heat stress. Sov Plant Physiol 34:685–693

Ladha JK, Tirol-Padre A, Punzulan GC, Watanabe I (1987) Nitrogen fixing (C_2H_2- reducing) activity and plant growth characters of 16 wetland rice varieties. Soil Sci Plant Nutr 33(1–2): 187–200

Laine P, Bigot J, Ourry A, Boucaud J (1994) Effects of low temperature on nitrate uptake, and xylem and phloem flows of nitrogen, in *Secale cereale* L. and *Brassica napus* L. New Phytol 127(4):675–683

Lam HM, Coschigano KT, Oliveira IC, Melo-Oliveira R, Coruzzi GM (1996) The molecular-genetics of nitrogen assimilation into amino acids in higher plants. Annu Rev Plant Physiol Plant Mol Biol 47:569–593

Lam HM, Wong P, Chan HK, Yam K-M, Chen L, Chow C-M, Coruzzi GM (2003) Overexpression of the ASN1 gene enhances nitrogen status in seeds of Arabidopsis. Plant Physiol 132(2): 926–935

Lamb JFS, Barnes DK, Russelle MP, Vance CP, Heichel GH, Henjum KI (1995) Ineffectively and effectively nodulated alfalfas demonstrate biology nitrogen fixation continues with high nitrogen fertilization. Crop Sci 35:153–157

Lang V, Mantyla E, Welin B, Sundberg B, Palva ET (1994) Alterations in water status, endogenous abscisic acid content, and expression of rab18 gene during the development of freezing tolerance in Arabidopsis thaliana. Plant Physiol 104(4):1341–1349

Laroche A, Hopkins WG (1987) Polysomes from winter rye seedlings grown at low temperature I. Size class distribution, composition, and stability. Plant Physiol 85(3):648–854

Lawlor DW, Tezara W (2009) Causes of decreased photosynthetic rate and metabolic capacity in water-deficient leaf cells: a critical evaluation of mechanisms and integration of processes. Ann Bot 103(4):561–579

Lawrie AC, Wheeler CT (1973) The supply of photosynthetic assimilates to nodules of *Pisum Sativum* L. in relation to the fixation of nitrogen. New Phytol 72:1341–1348

Layzell DB, Hunt S (1990) Oxygen and the regulation of nitrogen fixation in legume nodules. Physiol Plant 80(2):322–327

Layzell DB, Rainbird RM, Atkins CA, Pate JS (1979) Economy of photosynthate use in nitrogen-fixing legume nodules: observations on two contrasting symbioses. Plant Physiol 64:888–891

Layzell DB, Pate JS, Atkins CA, Canvin DT (1981) Partitioning of carbon and nitrogen and the nutrition of root and shoot apex in a nodulated legume. Plant Physiol 67:30–36

Lea PJ, Forde BG (1994) The use of mutants and transgenic plants to study amino acid metabolism. Plant Cell Environ 17(5):541–556

Lea PJ, Miffins BJ (2003) Glutamate synthase and the synthesis of glutamate in plants. Plant Physiol Biochem 41(6–7):555–564

Lea P, Miflin B (1974) Alternative route for nitrogen assimilation in higher plants. Nature 251(5476):614–616

Ledgard SF, Steele KW (1992) Biological nitrogen fixation in mixed legume/grass Pastures. Plant Soil 141:137–153

Lee HJ, Titus JS (1992) Nitrogen accumulation and nitrate reductase activity in MM.106 apple trees as affected by nitrate supply. J Hortic Sci 67(2):273–281

Lee J, Jeong KW, Jin B, Ryu KS, Kim EH, Ahn JH, Kim Y (2013) Structural and dynamic features of cold-shock proteins of *Listeria monocytogenes*, a psychrophilic bacterium. Biochemistry 52(14):2492–2504

Lee CC, Ribbe MW, Hu Y (2014) Cleaving the N,N triple bond: The transformation of dinitrogen to ammonia by nitrogenases. In: Metal driven biogeochemistry of gaseous compounds in the environment. Metal ion in life science, vol. 14, pp 147–174

Less H, Galili G (2008) Principal transcriptional programs regulating plant amino acid metabolism in response to abiotic stresses. Plant Physiol 147(1):316–330

Less H, Angelovici R, Tzin V, Galili G (2011) Coordinated gene networks regulating Arabidopsis plant metabolism in response to various stresses and nutritional cues. Plant Cell 23(4): 1264–1271

Levitt J (1972) Responses of plants to environmental stresses. Academic, New York, p 697

Levitt J (1980) Responses of plants to environmental stresses. In: Chilling, freezing and high temperature stresses, vol 1, 2nd edn. Academic, New York, p 497

Leyva-Díaz JC, González-Martínez A, González-López L, Muñío MM, Poyatos JM (2015) Kinetic modeling and microbiological study of two-step nitrification in a membrane bioreactor and hybrid moving bed biofilm reactor-membrane bioreactor for wastewater treatment. Chem Eng J 259:692–702

Li PH (1984) Sub zero temperature stress physiology of herbaceous plants. Hortic Rev 6:373–416

Li YL, Fan XR, Shen QR (2007) The relationship between rhizosphere nitrification and nitrogen-use efficiency in rice plants. Plant Cell Environ 31(1):73–85

Li S, Chakraborty N, Borcar A, Menze MA, Toner M, Hand S (2012) Late embryogenesis abundant proteins protect human hepatoma cells during acute desiccation. Proc Natl Acad Sci U S A 109(51):20859–20864

Li XN, Pu HC, Liu F, Zhou Q, Cai J, Dai T, Cao W, Jiang D (2015) Winter wheat photosynthesis and grain yield responses to spring freeze. Agron J 107(3):1002–1010

Li Z, Zhang R, Zhang H (2018) Effects of plant growth regulators (DA-6 and 6-BA) and EDDS chelator on phytoextraction and detoxification of cadmium by *Amaranthus* hybridus Linn. Int J Phytoremediation 20(11):1121–1128

Lie TA (1971) Symbiotic nitrogen fixation under stress conditions. Plant Soil 35:117–127

Lindemann WC, Ham GE (1979) Soybean plant growth, nodulation, and nitrogen fixation as affected by root temperature. Soil Sci Soc Am J 43:1134–1137

Liu J, Zhu JK (1997) Proline accumulation and salt-stress induced gene expression in a salt-hypersensitive mutant of *Arabidopsis*. Plant Physiol 114(2):591–596
Liu Q, Kasuga M, Sakuma Y, Abe H, Miura S, Yamaguchi-Shinozaki K, Shinozaki K (1998) Two transcription factors, DREB1 and DREB2 with an EREBP /AP2 DNA binding domain separate two cellular signal transduction pathways in drought- and low-temperature- responsive gene expression, respectively, in Arabidopsis. Plant Cell 10(8):1391–1406
Liu JH, Kitashiba H, Wang J, Ban Y, Moriguchi T (2007) Polyamines and their ability to provide environmental stress tolerance to plants. Plant Biotechnol 24(1):117–126
Liu XY, Li B, Yang JH, Sui N, Yang XM, Meng QW (2008) Overexpression of tomato chloroplast omega-3 fatty acid desaturase gene alleviates the photo inhibition of photosystems 2 and 1 under chilling stress. Photosynthetica 46(2):185–192
Liu Y, Chakrabortee S, Li R, Zheng Y, Tunnacliffe A (2011) Both plant and animal LEA proteins acts as kinetic stabilizers of polyglutamine-dependent protein aggregation. FEBS Lett 585(4): 630–634
Liu ZL, Li YJ, Hou HY, Zhu XC, Rai V, He XY, Tian CJ (2013a) Differences in the arbuscular mycorrhizal fungi-improved rice resistance to low temperature at two N levels: aspects of N and C metabolism on the plant side. Plant Physiol Biochem 71:87–95
Liu XJ, Zhang Y, Han WX, Tang AH, Shen JL, Cui ZL, Vitousek P, Erisman JW, Goulding K, Christie P, Fangmaier A, Zhang F (2013b) Enhanced nitrogen deposition over China. Nature 494(7438):459–462
Liu JH, Wang W, Wu H, Gong X, Moriguchi T (2015) Polyamines function in stress tolerance: from synthesis to regulation. Front Plant Sci 6:827
Livingston DP, Henson CA (1998) Apoplastic sugars, fructans, fructan exohydrolase, and invertase in winter oat: responses to second-phase cold hardening. Plant Physiol 116(1):403–408
Llorente F, Oliveros JC, Martínez-Zapater JM, Salinas J (2000) A freezing-sensitive mutant of *Arabidopsis*, *frs1*, is a new *aba3* allele. Planta 211(5):648–655
Loepfe C, Raimann E, Stephan R, Tasara T (2010) Reduced host cell invasiveness and oxidative stress tolerance in double and triple csp gene family deletion mutants of *Listeria monocytogenes*. Foodborne Pathog Dis 7(7):775–783
Londo JP, Kovaleski AP, Lillis JA (2018) Divergence in the transcriptional landscape between low temperature and freeze shock in cultivated grapevine (Vitis vinifera). Hortic Res 5(1):10–24
Longo V, Kamran RV, Michaletti A, Toorchi M, Zolla L, Rinalducci S (2017) Proteomic and physiological response of spring barley leaves to cold stress. Int J Plant Biol Res 5(1):1–10
Luciñski R, Polcyn W, Ratajczak L (2002) Nitrate reduction and nitrogen fixation in symbiotic association *Rhizobium*—legumes. Acta Biochim Polonica 49(2):537–546
Lukatkin AS, Brazaityte A, Bobinas C, Duchovskis P (2012) Chilling injury in chilling-sensitive plants: a review. Agriculture 99(2):111–124
Luo J, Li H, Liu T, Polle A, Peng C, Luo Z (2013) Nitrogen metabolism of two contrasting poplar species during acclimation to limiting nitrogen availability. J Exp Bot 64(14):4207–4224
Lynch DV, Thompson GA (1984) Microsomal phospholipid molecular species alterations during low temperature acclimation in *Dunaliella*. Plant Physiol 74(2):193–197
Maathuis FJM (2009) Physiological functions of mineral nutrients. Curr Opin Plant Biol 12(3): 250–258
Macduff JH, Jarvis SC, Davidson IA (1996) Inhibition of N_2 fixation by white clover (*Trifolium repens* L.) at low concentrations of NO_3 in flowing solution culture. Plant Soil 180:287–295
MacGregor KB, Shelp BJ, Peiris S, Bown AW (2003) Overexpression of glutamate decarboxylase in transgenic tobacco plants deters feeding by phytophagous insect larvae. J Chem Ecol 29(9): 2177–2182
Mahajan S, Tuteja N (2005) Cold, salinity and drought stresses: an overview. Arch Biochem Biophys 444(2):139–158
Mahon JD (1977a) Respiration and the energy requirement for nitrogen fixation in nodulated pea roots. Plant Physiol 60:817–821

Mahon JD (1977b) Root and nodule respiration in relation to acetylene reduction in intact nodulated peas. Plant Physiol 60:812–816

Maimiroli N, Terzi V, Odoardi Stanca M, Lorenzoni C (1986) Protein synthesis during cold shock in barley tissues: comparison of two genotypes with winter and spring growth habit. Theor Appl Genet 73(2):190–196

Maqbool MA, Aslam M, Ali H (2017) Breeding for improved drought tolerance in chickpea (*Cicer arietinum* L.). Plant Breed 136(3):300–318

Masclaux-Daubresse C, Reisdorf-Cren M, Pageau K, Lelandais M, Grandjean O, Kronenberger J, Valadier MH, Feraud M, Jouglet T, Sizuki A (2006) Glutamine-synthetase- glutamate synthase pathway and glutamate dehydrogenase play distinct roles in the sink- source nitrogen cycle in tobacco. Plant Physiol 140(2):444–456

Masclaux-Daubresse C, Daniel-Vedele F, Dechorognat J, Chardon F, Gaufichon L, Suzuki A (2010) Nitrogen uptake, assimilation and remobilization in plants: challenges for sustainable and productive agriculture. Ann Bot 105(7):1141–1157

Masterson CL, Murphy PM (1976) Application of the acetylene reduction technique to the study of nitrogen fixation by white clover in the field. In: Nutman PS (ed) Symbiotic nitrogen fixation in plants. Cambridge University Press, London, pp 299–316

Mata CI, Hertog ML, Van Raemdonck G, Baggerman G, Tran D, Nicolai BM (2019) Omics analysis of the ethylene signal transduction in tomato as a function of storage temperature. Postharvest Biol Technol 155:1–10

Matějů V, Čižinská S, Krejčí J, Janoch T (1992) Biological water denitrification—a review. Enzyme Microb Technol 14(3):170–183

Matysiak K, Miziniak W, Kaczmarek S, Kierzek R (2018) Herbicides with natural and synthetic biostimulants in spring wheat. Cienc Rural 48(11):1–10

Matysiak K, Kierzek R, Siatkowski I, Kowalska J, Krawczyk R, Miniziak W (2020) Effect of exogenous application of amino acids l-arginine and glycine on maize under temperature stress. Agronomy 10(6):769

Maynard DN, Barker AV, Minotti PL, Peck NH (1976) Nitrate accumulation in vegetables. Adv Agron 28:71–118

Mazur P (1963) Kinetics of water loss from cells at subzero temperatures and the likelihood of intracellular freezing. J Genet Physiol 47(2):347–369

Mazur P (1969) Freezing injury in plants. Annu Rev Plant Physiol 20:419–448

Mazzafera P, Gonçalves KV (1999) Nitrogen compounds in the xylem sap of coffee. Phytochemistry 50(3):383–386

Mazzucotelli E, Tartari A, Cattivelli L, Forlani G (2006) Metabolism of γ-aminobutyric acid during cold acclimation and freezing and its relationship to frost tolerance in barley and wheat. J Exp Bot 57(14):3755–3766

McCown BH, McLeester RC, Beck GE, Hall TC (1969) Environment-induced changes in peroxidase zymograms in the stems of deciduous and evergreen plants. Cryobiology 5(6):410–412

McKersie BD, Bowley SR (1997) Active oxygen and freezing tolerance in transgenic plants. Plant Cod Hardness 68:203–214

Medina J, Bargues M, Terol J, Pérez-Alonso M, Salinas J (1999) The Arabidopsis CBF gene family is composed of three genes encoding AP2 domain-containing proteins whose expression is regulated by low temperature but not by abscisic acid or dehydration. Plant Physiol 119(2): 463–470

Medina J, Catalá R, Salinas J (2011) The CBFs: Three Arabidopsis transcription factors to cold acclimate. Plant Sci 180(1):3–11

Menze MA, Boswell L, Toner M, Hand SC (2009) Occurrence of mitochondria-targeted late embryogenesis abundant (LEA) gene in animals increases organelle resistance to water stress. J Biol Chem 284(16):10714–10719

Meyer C, Stitt M (2001) Nitrate reductase and signalling. In: Lea PJ, Morot-Gaudry J-F (eds) Plant nitrogen. Springer, New York, pp 37–59

Meyer C, Lea US, Provan F, Kaiser WM, Lillo C (2005) Is nitrate reductase a major player in the plant NO (nitric oxide) game? Photosynth Res 83(2):181–189
Meza-Basso L, Alberdi M, Raynal M, Ferrero-Cadinanos ML, Delseny M (1986) Changes in protein synthesis in rapeseed (*Brassica napus*) seedlings during a low temperature treatment. Plant Physiol 82(3):733–738
Michaux C, Martini C, Shioya K, Ahmed Lecheheb S, Budin-Verneuil A, Cosette P, Sanguinetti M, Hartke A, Verneuil N, Christphe Giardb J (2012) CspR, a cold shock RNA-binding protein involved in the long-term survival and the virulence of *Enterococcus faecalis*. J Bacteriol 194(24):6900–6908
Michiels J, Verreth C, Vanderleyden J (1994) Effects of temperature stress on bean-nodulating *Rhizobium* strains. Appl Environ Microbiol 60:1206–1212
Minchin FR (1997) Regulation of oxygen diffusion in legume nodules. Soil Biol Biochem 29(5–6): 881–888
Minchin FR, Pate JS (1973) The carbon balance of a legume and the functional economy of its root nodules. J Exp Bot 24:259–271
Minchin FR, Summerfield RJ, Hadley P, Roberts EH, Rawsthorne S (1981) Carbon and nitrogen nutrition of nodulated roots of grain legumes. Plant Cell Environ 4:5–26
Minchin FR, Minquez MI, Sheehy JE, Witty JF, Skot L (1986a) Relationships between nitrate and oxygen supply in symbiotic nitrogen fixation by white clover. J Exp Bot 37(8):1103–1113
Minchin FR, Sheehy JE, Witty JF (1986b) Further errors in acetylene reduction assay: effects of plant disturbance. J Exp Bot 37(10):1581–1591
Miura K, Tada Y (2014) Regulation of water, salinity, and cold stress responses by salicylic acid. Front Plant Sci 5:4
Miyashita Y, Good AG (2008) NAD(H)-dependent glutamate dehydrogenase is essential for the survival of Arabidopsis thaliana during dark-induced carbon starvation. J Exp Bot 59(3): 667–680
Mohapatra SS, Poole RJ, Dhindsa RS (1987a) Cold acclimation, freezing resistance and protein synthesis in alfalfa (*Medicago sativa* L. cv. Saranac). J Exp Bot 38(10):1697–1703
Mohapatra SS, Poole RJ, Dhindsa RS (1987b) Changes in protein patterns and translatable messenger RNA populations during cold acclimation of alfalfa. Plant Physiol 84(4):1172–1176
Mohapatra S, Wolfraim L, Poole RJ, Dhindsa RS (1989) Molecular cloning and relationship to freezing tolerance of cold-acclimation-specific genes of alfalfa. Plant Physiol 89(1):375–380
Møller IM, Jensen PE, Hansson A (2007) Oxidative modifications to cellular components in plants. Annu Rev Plant Biol 58:459–481
Monforte A, Tanksley S (2000) Development of a set of nearly isogenic and backcross recombinant inbred lines containing most of the *Lycopersicon hirsutum* genome in a *L. esculentum* genetic background: a tool for gene mapping and gene discovery. Genome 43(5):803–813
Monroy AF, Dhindsa RS (1995) Low-temperature signal transduction: induction of cold acclimation-specific genes of alfalfa by calcium at 25 degrees C. Plant Cell 7(3):321–331
Moradtalab N, Weinmann M, Walker F, Höglinger B, Ludewig U, Neumann G (2018) Silicon improves chilling tolerance during early growth of maize by effects on micronutrient homeostasis and hormonal balances. Front Plant Sci 9:420
Mori M, Watanabe M, Shiogama H, Inoue J, Kimoto M (2014) Robust Arctic sea-ice influence on the frequent Eurasian cold winters in past decades. Nat Geosci 7(12):869–874
Munns R (2002) Comparative physiology of salt and water stress. Plant Cell Environ 25(2): 239–250
Mur LA, Mandon J, Persijn S, Cristescu SM, Moshkov IE, Novikova GV, Hall MA, Harren FJN, Hebelstrup KM, Gupta KJ (2013) Nitric oxide in plants: an assessment of the current state of knowledge. AoB Plants 5:pls052
Nadeem SM, Ahmad M, Zahir ZA, Kharal MA (2016) Role of phytohormones in stress tolerance of plants. In: Hakeem KR, Akhtar MS (eds) Plant, soil and microbes. Springer International Publishing, Cham, pp 385–421

Nagao M, Oku K, Minami A, Mizuno K, Sakurai M, Arakawa K, Fujikawa S, Takezawa D (2006) Accumulation of the anderose in association with development of freezing tolerance in the moss *Physcomitrella patens*. Phytochemistry 67(7):702–709

Nagata N, Tanaka R, Tanaka A (2007) The major route for chlorophyll synthesis includes [3,8-divinyl]-chlorophyllide a reduction in Arabidopsis thaliana. Plant Cell Physiol 48(12): 1803–1808

Nahed G, Abdel Aziz AA, Mazher M, Farahat MM (2010) Response of vegetative growth and chemical constituents of *Thuja orientalis* L. plant to foliar application of different amino acids at Nubaria. J Am Sci 6(3):295–301

Nakaminami K, Karlson DT, Imai R (2006) Functional conservation of cold shock domains in bacteria and higher plants. Proc Natl Acad Sci U S A 103(26):10122–10127

Nakashima K, Satoh R, Kiyosue T, Yamaguchi-Shinozaki K, Shinozaki K (1998) A gene encoding proline dehydrogenase is not only induced by proline and hypo-osmolarity, but is also developmentally regulated in the reproductive organs of *Arabidopsis*. Plant Physiol 118(4):1233–1241

Nakashima K, Tran LS, Nguyen DV, Fujita M, Maruyama K, Todaka D, Ito Y, Hayashi N, Shinozaki K, Yamaguchi-Shinozaki K (2007) Functional analysis of a NAC-type transcription factor OsNAC6 involved in abiotic and biotic stress responsive gene expression in rice. Plant J 51(4):617–630

Nandwal AS, Godara M, Sheokhand S, Kamboj DV, Kundu BS, Kushad MS, Kumar B, Sharma SK (2000) Salinity induced change in plant water status, nodule functioning and ionic distribution in phenotypically differing ecotype of *Vigna radiata* L. J Plant Physiol 156(3):350–359

Nasibi F, Heidari T, Asrar Z, Mansoori H (2013) Effect of arginine pre-treatment on nickel accumulation and alleviation of the oxidative stress in *Hyoscyamus niger*. J Soil Sci Plant Nutr 13(3):680–689

Nayyar H (2003) Calcium as environmental sensor in plants. Curr Sci 84(7):893–902

Nayyar H, Bains T, Kumar S (2005) Low temperature induced floral abortion in chickpea: relationship to abscisic acid and cryoprotectants in reproductive organs. Environ Exp Bot 53(1):39–48

Nejad TS, Bakhshande A, Nasab SB, Payande K (2010) Effect of drought stress on corn root growth. Rep Opin 2(2):47–52

Nejadalimoradi H, Nasibi F, Kalantari KM, Zanganeh R (2014) Effect of seed priming with L-arginine and sodium nitroprusside on some physiological parameters and antioxidant enzymes of sunflower plants exposed to salt stress. Agric Commun 2(1):23–30

Nelson DR, Bellville RJ, Porter CA (1984) Role of nitrogen assimilation in seed development of soybean. Plant Physiol 74:128–133

Ngwene B, Gabriel E, George E (2013) Influence of different mineral nitrogen sources (NO^{3-}-N vs. NH^{4+} -N) on arbuscular mycorrhiza development and N transfer in a *Glomus intraradices*—cowpea symbiosis. Mycorrhiza 23:107–117

Nunes-Nesi A, Fernie AR, Stitt M (2010) Metabolic and signaling aspects underpinning the regulation of plant carbon nitrogen interactions. Mol Plant 3(6):973–996

O'Hara GW, Daniel RM (1985) Rhizobial denitrification: a review. Soil Biol Biochem 17(1):1–9

Obata T, Matthes A, Koszior S, Lehmann M, Araújo WL, Bock R, Sweetlove LJ, Fernie AR (2011) Alteration of mitochondrial protein complexes in relation to metabolic regulation under short-term oxidative stress in *Arabidopsis* seedlings. Phytochemistry 72(10):1081–1091

Oldroyd GE, Downie JA (2008) Coordinating nodule morphogenesis with rhizobial infection in legumes. Annu Rev Plant Biol 59:519–546

Olien CR, Smith MN (1977) Ice adhesions in relation to freeze stress. Plant Physiol 60(4):499–503

Oliver SN, Dennis ES, Dolferus R (2007) ABA regulates apoplastic sugar transport and is a potential signal for cold-induced pollen sterility in rice. Plant Cell Physiol 48(9):1319–1330

Ougham HJ (1987) Gene expression during leaf development in *Lolium temulentum*: patterns of protein synthesis in response to heat-shock and cold shock. Physiol Plant 70(3):479–484

Ougham HJ, Howarth CJ (1988) Temperature shock proteins in plants. In: Long SM, Woodward FI, editors. Plants and temperature, Sympo. Soc. Exp. Biol. Company of Biologists, Cambridge, pp 259–280

Overland J, Francis JA, Hall R, Hanna E, Kim S-J, Vihma T (2015) The melting Arctic and midlatitude weather patterns: are they connected? J Climate 28(20):7917–7932

Pal R, Fatima Z, Hameed S (2014) Efflux pumps in drug resistance of *Mycobacterium tuberculosis*: a panoramic view. Int J Curr Microbiol Appl Sci 3(8):528–546

Palanivelu R, Brass L, Edlund AF, Preuss D (2003) Pollen tube growth and guidance is regulated by POP2, an *Arabidopsis* gene that controls GABA levels. Cell 114(1):47–59

Pan A, Hayes PM, Chen F, Chen THH, Blake T, Wright S, Karsai I, Beds Z (1994) Genetic analysis of the components of winter hardiness in barley *(Hordeum vulgare* L.). Theor Appl Genet 89(7–8):900–910

Panchali K (2011) Symbiotic nitrogen fixation and seed development of genetically modified soybean in relation to Bradyrhizobium inoculation and nitrogen use under acidic and saline dykeland and soil conditions. MSc thesis, Dalhousie University, Halifax

Pandey P, Ramegowda V, Senthil-Kumar M (2015) Shared and unique responses of plants to multiple individual stresses and stress combinations: physiological and molecular mechanisms. Front Plant Sci 6:723

Pandey S, Carrer M, Castagneri D, Petit G (2018) Xylem anatomical responses to climate variability in Himalayan birch trees at one of the world's highest forest limit. Perspect Plant Ecol Evol Syst 33:34–41

Pardos M, Royo A, Gil L, Pardos JA (2003) Effect of nursery location and outplanting date on field performance of *Pinus halepensis* and *Quercus ilex* seedlings. Forestry 76(1):67–81

Park EJ, Jeknic Z, Chem THH (2006) Exogenous application of glycine betaine increases chilling tolerance in tomato plants. Plant Cell Physiol 47(6):706–714

Parsons R, Day DA (1990) Mechanisms of soybean nodule adaptation to different oxygen pressures. Plant Cell Environ 13(6):501–512

Pate JS (1961) Temperature characteristics of bacterial variation in legume symbiosis. Nature 192: 637–639

Paul EA, Clark FE (1989) Soil microbiology and biochemistry. Academic, San Diego, p 275

Paulissen MPCP, van der Ven PJM, Dees AJ, Bobbink R (2004) Differential effects of nitrate and ammonium on three fen bryophyte species in relation to pollutant nitrogen input. New Phytol 164(3):451–458

Pauwels L, Inzé D, Goossens A (2009) Jasmonate-inducible gene: what does it mean? Trend Plant Sci 14(2):87–91

Pearce RS (1999) Molecular analysis of acclimation to cold. Plant Growth Regul 29:47–76

Pearce RS (2001) Plant freezing and damage. Ann Bot 87(4):417–424

Pearce RS, Ashworth EN (1992) Cell shape and localization of ice in leaves or overwintering wheat during frost stress in the field. Planta 188(3):324–331

Pearce ISK, van der Wal R (2002) Effect of nitrogen deposition on growth and survival of montane Racomitrium lanuginosum heath. Biol Conserv 104(1):83–89

Pearce ISK, Woodin SJ, van der Wal R (2003) Physiological and growth responses of the montane bryophyte *Racomitrium lanuginosum* to atmospheric nitrogen deposition. New Phytol 160(1): 145–155

Pellegrineschi A, Reynolds M, Pacheco M, Brito RM, Almeraya R, Yamaguch-Shinozaki K, Hoisington D (2004) Stress induced expression in wheat of the *Arabidopsis thaliana* DREB1A gene delays water stress symptoms under greenhouse conditions. Genome 47(3): 493–500

Pérez-Tienda J, Valderas A, Camañes G, García-Agustín P, Ferrol N (2012) Kinetics of NH^{4+} uptake by the arbuscular mycorrhizal fungus *Rhizophagus irregularis*. Mycorrhiza 22(6): 485–491

Perras M, Sarhan F (1989) Synthesis of freezing tolerance proteins in leaves, crown, and roots during cold acclimation of wheat. Plant Physiol 89(2):577–585

Phadtare S (2004) Recent developments in bacterial cold-shock response. Curr Issues Mol Biol 6(2):125–136
Phillips AD (1980) Efficiency of symbiotic nitrogen fixation in legumes. Annu Rev Plant Biol 31: 29–49
Pierce BG, Hourai Y, Weng Z (2011) Accelerating protein docking in ZDOCK using an advanced 3D convolution library. PLoS One 6(9):e24657
Pietilainen P, Poikolainen J, Lahdesmaki P (1991) Long-term monitoring of nitrate reductase-activity in the needles of *Pinus sylvestris* in the context of environmental-temperature and ground frost as an indicator of nitrogen-balance in N Finland. Anna Bot Fennici 28(2):131–134
Piha MI, Munns DN (1987) Sensitivity of the common bean (*Phaseolus vulgaris*) symbiosis to high soil temperature. Plant Soil 98:183–194
Pimratch S, Jogloy S, Vorasoot N, Toomsan B, Patanothai A, Holbrook CC (2008) Relationship between biomass production and nitrogen fixation under drought-stress conditions in peanut genotypes with different levels of drought resistance. J Agron Crop Sci 194:15–25
Pinedo ML, Hernandez GF, Conde RD, Tognetti JA (2000) Effect of low temperature on the protein metabolism of wheat leaves. Biol Plant 43(3):363–367
Planchet E, Rannou O, Ricoult C, Boutet-Mercey S, Maia-Grondard A, Limami AM (2011) Nitrogen metabolism responses to water deficit act through both abscisic acid (ABA)- dependent and independent pathways in Medicago truncatula during post germination. J Exp Bot 62(2):605–615
Polcyn W, Luciñski R (2001) Functional similarities of nitrate reductase from yellow lupine bacteroids to bacterial denitrification system. J Plant Physiol 158(7):829–834
Polisensky DH, Braam J (1996) Cold-shock regulation of the Arabidopsis TCH genes and the effects of modulating intracellular calcium levels. Plant Physiol 111(4):1271–1279
Pomeroy MK, Siminovitch D, Wrightman F (1970) Seasonal biochemical changes in the living bark and needles of red pine (*Pinus resinosa*) in relation to adaptation to freezing. Can J Bot 48(5):953–967
Porcel R, Bustamante A, Ros R, Serrano R, Mulet JM (2018) BvCOLD1: a novel aquaporin from sugar beet (*Beta vulgaris L.*) involved in boron homeostasis and abiotic stress. Plant Cell Environ 41(12):1–14
Postgate J (1998) Nitrogen fixation, 3rd edn. Cambridge University Press, Cambridge
Price JL, Gourlie BB, Lin Y, Huang RCC (1986) Induction of winter flounder anti freeze protein messenger RNA at 4°C i*n vivo* and in vitro. Physiol Zool 59(6):679–695
Puértolas J, Gil L, Pardos JA (2005) Effects of nitrogen fertilization and temperature on frost hardiness of Aleppo pine (*Pinus halepensis* mill.) seedlings assessed by chlorophyll fluorescence. Forestry 78(5):501–511
Puhakainen T (2004) Short-day potentiation of low temperature-induced gene expression of a c-repeat-binding factor-controlled gene during cold acclimation in Silver Birch. Plant Physiol 136(4):4299–4307
Purcell LC, Sinclair TR (1990) Nitrogenase activity and nodule gas permeability response to rhizospheric NH_3 in soybean. Plant Physiol 92:268–272
Purnell M, Botella J (2007) Tobacco isozyme 1 of NAD(H)-dependent glutamate dehydrogenase catabolizes glutamate *in vivo*. Plant Physiol 143:530–539
Puyaubert J, Baudouin E (2014) New clues for a cold case: nitric oxide response to low temperature. Plant Cell Environ 37(12):2623–2630
Qados AMSA (2010) Effect of arginine on growth, nutrient composition, yield and nutritional value of mung bean plants grown under salinity stress. Nat Sci 8(7):30–42
Qi F, Li J, Duan L, Li Z (2005) Inductions of coronatine and MeJA to low-temperature resistance of wheat seedlings. Acta Bot Boreal Occident Sin 26(9):1776–1780
Qi R, Gu W, Zhang J, Hao L, Zhang M, Duan L, Li Z (2013) Exogenous diethyl aminoethyl hexanoate enhanced growth of corn and soybean seedlings through altered photosynthesis and phytohormone. Aust J Crop Sci 7(13):2021–2028

Queiroz CGS, Alves JD, Rena AB, Cordeiro AT (1991) Efeito do cloranfenicol, propanol, pH e temperatura sobre a atividade da redutase do nitrato em cafeeiros jovens. Rev Bras Bot 14(1): 73–77

Queiroz CGS, Rena AB, Cordeiro AT, Alves JD (1993) Distribuição da atividade da redutase do nitrato no cafeeiro: a importância da raiz. Rev Bras Bot 16(1):31–35

Rafiee H, Mehrafarin A, Qaderi A, Kalate JS, Naghdi BH (2013) Phytochemical, agrochemical and morphological responses of pot marigold (*Calendula officinalis* L.) to foliar application of bio-stimulators (bioactive amino acid compounds). J Med Plants 12(47):48–61

Rai VK (2002) Role of amino acids in plant responses to stresses. Biol Plant 45(4):481–487

Rai VK, Sharma UD (1989) Modulation of osmotic closure of stomata, stomatal resistance and K^+ fluxes by exogenous amino acids in *Vicia faba* leaves. Biochem Physiol Pllanzen 185(5–6): 369–376

Ramankutty N, Evan AT, Monfreda C, Foley JA (2008) Farming the planet: 1. Geographic distribution of global agricultural lands in the year 2000. Glob Biogeochem Cycles 22(1): GB1003

Rao VR (1977) Effect of root temperature on the infection processes and nodulation in *Lotus* and *Stylosanthes*. J Exp Bot 28:241–259

Rapacz M, Markowski A (1999) Winter hardiness, frost resistance and vernalization requirement of European winter oilseed rape (*Brassica napus* var. oleifera) cultivars within the last 20 years. J Agron Crop Sci 183(4):243–253

Raven J (1980) Nutrient transport in microalgae. Adv Microb Physiol 21:47–226

Reed SC, Cleveland CC, Townsend AR (2011) Functional ecology of free-living nitrogen fixation: a contemporary perspective. Annu Rev Ecol Evol Syst 42:489–512

Reins B, Heldt W (1992) Decrease of nitrate reductase activity in spinach leaves during a light- dark transition. Plant Physiol 98(2):573–577

Rekarte-Cowie I, Ebshish OS, Mohamed KS, Pearce RS (2008) Sucrose helps regulate cold acclimation of *Arabidopsis thaliana*. J Exp Bot 59(15):4205–4217

Richner W, Soldati A, Stamp P (1996) Shoot-to-root relations in field grown maize seelings. Agron J 88(1):56–61

Rigano VD, Vona V, Lobosco O, Carillo P, Lunn JE, Carfagna S, Esposito S, Caiazzo M, Rigano C (2006) Temperature dependence of nitrate reductase in the psychrophilic unicellular alga *Koliella antarctica* and the mesophilic alga *Chlorella sorokiniana*. Plant Cell Environ 29(7): 1400–1409

Riov J, Brown GN (1976) Comparative studies of activity and properties of ferredoxin-$NADP^+$ reductase during cold hardening of wheat. Can J Bot 54(16):1896–1902

Ritonga FN, Chen S (2020) Physiological and molecular mechanism involved in cold stress tolerance in plants. Plants 9(5):560

Rizwan M, Ali S, Adrees M, Rizvi H, Zia-ur-Rehman M, Hannan F, Qayyum MF, Hafeez F, Ok YS (2016) Cadmium stress in rice: toxic effects, tolerance mechanisms, and management: a critical review. Environ Sci Pollut Res Int 23(18):17859–17879

Rizza F, Crosatti C, Stanca AM, Cattivelli L (1994) Studies for assessing the influence of hardening on cold tolerance of barley genotypes. Euphytica 75(1):131–138

Roberts DWA (1974) The invertase complement of cold-hardy and cold sensitive wheat leaves. Can J Bot 53(13):1333–1337

Roberts DWA (1990) Identification of loci on chromosome 5A of wheat involved in control of cold hardiness, vernalization, leaf length, rosette growth habit, and height of hardened plants. Genome 33:247–259

Robertson AJ, Gusta LV, Reaney MJT, Ishikawa M (1987) Protein synthesis in bromegrass (*Bromus inermis* Leyss) cultured cells during the induction of frost tolerance by abscisic acid or low temperature. Plant Physiol 84(4):1331–1336

Rochat C, Boutin J-P (1991) Metabolism of phloem-borne amino acids in maternal tissues of fruit of nodulated or nitrate-fed pea plants (*Pisum sativum* L.). J Exp Bot 42(23):207–214

Rochat E, Therrien HP (1975) Etude des protéines des blés résistant, Kharkov, et sensible, Selkirk, au cours de l'endurcissement au froid. 1. Protéines solubles. Can J Bot 53:2411–2416

Rockel P, Strube F, Rockel A, Wildt J, Kaiser WM (2002) Regulation of nitric oxide (NO) production by plant nitrate reductase in vivo and in vitro. J Exp Bot 53(366):103–110

Rodriguez VM, Malvar RA, Butron A, Ordas A, Revilla P (2007) Maize populations as sources of favorable alleles to improve cold-tolerant hybrids. Crop Sci 47(5):1779–1786

Root TL, Price JT, Hall KR et al (2003) Fingerprints of global warming on wild animals and plants. Nature 421(6918):57–60

Roughley RJ, Dart PJ (1969) Reduction of acetylene by nodules of *Trifolium subterraneum* as affected by root temperature, *Rhizobium* strain and host cultivar. Arch Microbiol 69:171–179

Rubio S, Pérez FJ (2019) ABA and its signaling pathway are involved in the cold acclimation and deacclimation of grapevine buds. Sci Hortic 256:108565

Ruelland E, Vaultier MN, Zachowski A, Hurry V (2009) Cold signalling and cold acclimation in plants. Adv Bot Res 49:35–150

Ryle GJA, Powell CE, Gordon AJ (1979) The respiratory costs of nitrogen fixation in soyabean, cowpea, and white clover I. Nitrogen fixation and the respiration of the nodulated root. J Exp Bot 30:135–144

Sadak MS, Abdelhamid MT (2015) Influence of amino acids mixture application on some biochemical aspects, antioxidant enzymes and endogenous polyamines of *Vicia faba* plant grown under seawater salinity stress. Gesunde Pflanze 67:119–129

Sadak MSH, Abdelhamid MT, Schmidhalter U (2015) Effect of foliar application of amino acids on plant yield and some physiological parameters in bean plants irrigated with seawater. Acta Biol Colomb 20(1):141–152

Saha B, Saha S, Das A, Bhattacharyya PK, Basak N, Sinha AK, Poddar P (2017) Biological nitrogen fixation for sustainable agriculture. In: Agriculturally important microbes for sustainable agriculture. Springer, Singapore, pp 81–128

Saidi Y, Finka A, Muriset M, Bromberg Z, Weiss YG, Maathuis FJ, Goloubinoff P (2009) The heat shock response in moss plants is regulated by specific calcium- permeable channels in the plasma membrane. Plant Cell 21(9):2829–2843

Saikia SP, Jain V (2007) Biological nitrogen fixation with non-legumes: an achievable target or a dogma? Curr Sci 92(3):317–322

Saikia SP, Srivastava GC, Jain V (2004) Nodule-like structures induced on the roots of maize seedlings by the addition of synthetic auxin 2,4-D and its effects on growth and yield. Cereal Res Commun 32(1):83–89

Sakai A, Larcher W (1987) Frost survival of plants. Springer, Berlin, p 321

Sakai A, Sugawara Y (1973) Survival of popular callus at super-low temperatures after cold acclimation. Plant Cell Physiol 14:1201–1204

Sakai A, Yoshida S (1968) The role of sugar and related compounds in variations of freezing resistance. Cryobiology 5:160–174

Sakamoto A, Murata N (2000) Genetic engineering of glycine betaine synthesis in plants: currents status and implication for enhancement of stress tolerance. J Exp Bot 51(342):81–88

Sakamoto A, Murata N (2001) The use of bacterial choline oxidase, a glycine betaine synthesizing enzyme to create stress resistant transgenic plants. Plant Physiol 125(1):180–188

Sanghera GS, Wani SH, Hussain W, Singh NB (2011) Engineering cold stress tolerance in crop plants. Curr Genom 12(1):30–43

Santamaria P, Elia A, Serio F, Todaro E (1999) A survey of nitrate and oxalate content in retail fresh vegetables. J Sci Food Agric 79(13):1882–1888

Santamaria P, Elia A, Gonnella M, Parente A, Serio F (2001) Ways of reducing rocket salad nitrate content. Acta Hortic 548:529–537

Santi C, Bogusz D, Franche C (2013) Biological nitrogen fixation in non-legume plants. Ann Bot 111(5):743–767

Sarhan F, Chevrier N (1985) Regulation of RNA synthesis by DNA-dependent RNA polymerases and RNases during cold acclimation in winter and spring wheat. Plant Physiol 78(2):250–255

Sarhan F, D'Aoust MJ (1975) RNA synthesis in spring and winter wheat during cold acclimation. Physiol Plant 35(1):62–65
Sarhan F, Danyluk J (1998) Engineering cold-tolerant crops-throwing the master switch. Trends Plant Sci 3(8):289–290
Sarhan F, Perras M (1987) Accumulation of a high molecular weight protein during cold hardening of wheat (*Triticure aestivum* L.). Plant Cell Physiol 28:1173–1179
Satoh R, Nakashima K, Seki M, Shinozaki K, Shinozaki KY (2002) ACTCAT, a novel cis- acting element for proline- and hypoosmolarity-responsive expression of the ProDH gene encoding proline dehydrogenase in Arabidopsis. Plant Physiol 130(2):709–719
Savitch LV, Ivanov AG, Gudynaite-Savitch L, Huner NPA, Simmonds J (2011) Cold stress effects on PSI photochemistry in Zea mays: differential increase of FQR-dependent cyclic electron flow and functional implications. Plant Cell Physiol 52(6):1042–1054
Savouré A, Hua XJ, Bertauche N, van Montagu M, Verbruggen N (1997) Abscisic acid- independent and abscisic acid-dependent regulation of proline biosynthesis following cold and osmotic stresses in Arabidopsis thaliana. Mol Gen Genet 254(1):104–109
Schaffer MA, Fischer RL (1988) Analysis of mRNAs that accumulate in response to low temperature identifies a thiol protease gene in tomato. Plant Physiol 87(2):431–436
Schärer K, Stephan R, Tasara T (2013) Cold shock proteins contribute to the regulation of listeriolysin O production in *Listeria monocytogenes*. Foodborne Pathog Dis 10:1023–1029
Scheurwater I, Koren M, Lambers H, Atkin OK (2002) The contribution of roots and shoots to whole plant nitrate reduction in fast- and slow-growing grass species. J Exp Bot 53(374): 1635–1642
Schjoerring JK, Husted S, Mäck G, Mattsson M (2002) The regulation of ammonium translocation in plants. J Exp Bot 53(370):883–890
Schmid B, Klumpp J, Raimann E, Loessner MJ, Stephan R, Tasara T (2009) Role of cold shock proteins in growth of *Listeria monocytogenes* under cold and osmotic stress conditions. Appl Environ Microbiol 75(6):1621–1627
Schomberg HH, Weaver RW (1992) Nodulation, nitrogen fixation, and early growth of arrow leaf clover in response to root temperature and starter nitrogen. Agron J 84:1046–1050
Schröder W, Pesch R, Schönrock S, Harmens H, Mills G, Fagerli H (2014) Mapping correlations between nitrogen concentrations in atmospheric deposition and mosses for natural landscapes in Europe. Ecol Indic 36:563–571
Schulze J, Adgo E, Merbach W (1999) Carbon costs associated with N_2 fixation in *Vicia faba* L. and *Pisum sativum* l. over a 14-day period. Plant Biol 1:625–631
Seo JS, Joo J, Kim MJ, Kim YK, Nahm BH, Song SI, Cheong JJ, Lee JS, Kim LK, Choi YD (2011) OsbHLH148 a basic helix-loop-helix protein, interacts with OsJAZ proteins in a jasmonate signaling pathway leading to drought tolerance in rice. Plant J 65(6):907–921
Serraj R, Sinclair TR (1998) N_2 fixation response to drought in common bean (*Phaseolus vulgaris* L.). Ann Bot 82:229–234
Serrano A, Chamber M (1990) Nitrate reduction in *Bradyrhizobium sp (Lupinus)* strains and its effects on their symbiosis with *Lupinus luteus*. J Plant Physiol 136(2):240–246
Sevilla M, Kennedy C (2000) Colonization of rice and other cereals by *Acetobacter diazotrophicus*, an endophyte of sugarcane. In: Ladha JK, Reddy PM (eds) The quest for nitrogen fixation in rice. International Rice Research Institute, Manila, pp 151–165
Shah K, Kumar RG, Verma S, Dubey RS (2001) Effect of cadmium on lipid peroxidation, superoxide anion generation and activity of antioxidant enzyme in growing rice seedlings. Plant Sci 161(6):1135–1144
Sharpley AN, Williams JR (1990) EPIC-erosion/productivity impact calculator 1. Model documentation. USDA Tech. Bull. No. 1768, pp 1–235
Sheehy JE (1987) Photosynthesis and nitrogen fixation in legume plants. Crit Rev Plant Sci 5:121–159
Shelp BJ, Bown AW, McLean MD (1999) Metabolism and function of gamma-aminobutyric acid. Trends Plant Sci 4(11):446–452

Shen B, Jenesen RG, Bohnert HJ (1997) Mannitol protects against oxidation by hydroxyl radicals. Plant Physiol 115(2):527–532

Shewry PR (1995) Plant storage proteins. Biol Rev 70(3):375–426

Shi Y, Tian S, Hou L, Huang X, Zhang X, Guo H, Yang S (2012) Ethylene signaling negatively regulates freezing tolerance by repressing expression of CBF and type-A ARR genes in Arabidopsis. Plant Cell 24(6):2578–2595

Shi H, Ye T, Zhong B, Liu X, Chan Z (2014) Comparative proteomic and metabolomic analyses reveal mechanisms of improved cold stress tolerance in bermudagrass (*Cynodon dactylon* (L.) Pers.) by exogenous calcium. J Integr Plant Biol 56(11):1064–1079

Shin H, Oh S, Arora R, Kim D (2016) Proline accumulation in response to high temperature in winter-acclimated shoots of *Prunus persica*: a response associated with growth resumption or heat stress? Can J Plant Sci 96(4):630–638

Shin H, Oh S, Kim D, Hong JK, Yun JG, Lee SW, Son KH (2018) Induced freezing tolerance and free amino acids perturbation of spinach by exogenous proline. J Plant Biotechnol 45(4): 357–363

Shinozaki K, Yamaguchi-Shinozaki K (1996) Molecular responses to drought and cold stress. Curr Opin Biotechnol 7(2):161–167

Shinozaki K, Yamaguchi-Shinozaki K, Seki M (2003) Regulatory network of gene expression in the drought and cold stress responses. Curr Opin Plant Biol 6(5):410–417

Shrimali M, Singh KP (2001) New methods of nitrate removal from water. Environ Pollut 112(3): 351–359

Shu S, Tang Y, Yuan Y, Sun J, Zhong M, Guo S (2016) The role of 24- epibrassinolide in the regulation of photosynthetic characteristics and nitrogen metabolism of tomato seedlings under a combined low temperature and weak light stress. Plant Physiol Biochem 107:344–353

Siddiqui MH, Al-Whaibi MH, Basalah MO (2011) Role of nitric oxide in tolerance of plants to abiotic stress. Protoplasma 248(3):447–455

Sierra-Alvarez R, Beristain-Cardoso R, Salazar M, Gómez J, Razo-Flores E, Field JA (2007) Chemolithotrophic denitrification with elemental sulfur for groundwater treatment. Water Res 41(6):1253–1262

Siminovitch D, Briggs DR (1949) The chemistry of the living bark of the black locust tree in relation to frost hardiness. I. Seasonal variations in protein content. Arch Biochem 23(1):8–17

Siminovitch D, Briggs DR (1953) Studies on the chemistry of the living bark of the black locust tree in relation to frost hardiness. IV. Effects of ringing on translocation, protein synthesis, and the development of hardiness. Plant Physiol 28(2):177–200

Siminovitch D, Cloutier Y (1982) Twenty-four-hour induction of freezing and drought tolerance in plumules of winter rye seedlings by desiccation stress at room temperature in the dark. Plant Physiol 69:250–255

Siminovitch D, Rheaume B, Pomeroy K, Lepage M (1968) Phospholipid, protein, and nucleic acid increases in protoplasm and membrane structures associated with development of extreme freezing resistance in black locust tree cells. Cryobiology 5(3):202–225

Sinclair TR, Muchow RC, Ludlow MM, Leach GJ, Lawn RJ, Foale MA (1987) Field and model analysis of the effect of water deficits on carbon and nitrogen accumulation by soybean, cowpea and black gram. Field Crops Res 17:121–140

Singh BK (1998) In: Plant amino acid synthesis. Biochemistry and Biotechnology, Marcell and Dekker (Books in Soil, Plant and Environment), New York

Singh TN, Aspinall D, Paleg LG (1972) Proline accumulation and varietal adaptability to drought in barley; a potential metabolic measure of drought resistance. Nature 286(67):188–190

Sirtautas R, Samuoliene G, Brazaityte A, Duchovskis P (2011) Temperature and photoperiod effects on photosynthetics indices of radish (Raphanus sativa L.). Zemdirbyste 98(1):57–62

Skopelitis DS, Paranychianakis NV, Paschalidis KA, Pliakonis ED, Delis ID, Yakoumakis DI, Kouvarakis A, Papadakis AK, Stephanou EG, Roubelakis-Anfelakis KA (2006) Abiotic stress generates ROS that signal expression of anionic glutamate dehydrogenases to form glutamate for proline synthesis in tobacco and grapevine. Plant Cell 18(10):2767–2781

Small JGC, Joffe A (1968) Physiological studies on the genus *Trifolium* with special reference to the South African species. II. Influence of root temperature on growth, nodulation and symbiotic nitrogen fixation. S Afr J Agric Sci 11:41–56

Smil V (2000) In: Cycles of life. Scientific American Library, USA.

Smith BN, Jones AR, Hanson LD, Criddle RS (1999) Growth, respiration rate, and efficiency response to temperature. In: Handbook of plant and crop stress, 2nd edn. Marcel Dekker, New York, pp 417–440

Snedden WA, Koutsia N, Baum G, Fromm H (1996) Activation of a recombinant petunia glutamate decarboxylase by calcium/calmodulin or by a monoclonal antibody which recognizes the calmodulin binding domain. J Biol Chem 271(8):4148–4153

Solomonson LP, Barber MJ (1990) Assimilatory nitrate reductase: functional properties and regulation. Annu Rev Plant Physiol Plant Mol Biol 41:225–253

Somersalo S, Kyei-Boahen S, Pehu E (1996) Exogenous glycine betaine application as a possibility to increase low temperature tolerance of crop plants. Nordisk Jordbruksforsk 78:10

Songstad DD, Duncan DR, Widholm JM (1990) Proline and polyamine involvement in chilling tolerance of maize suspension cultures. J Exp Bot 41(224):289–294

Sonoike K (1998) Various aspects of inhibition of photosynthesis under light/chilling stress. Photoinhibition at chilling temperatures versus chilling damage in the light. J Plant Res 111(1101):121–129

Sorgona A, Lupini A, Mercati F, Di Dio L, Sunseri F, Abenavoli MR (2011) Nitrate uptake along the maize primary root: an integrated physiological and molecular approach. Plant Cell Environ 34(7):1127–1140

Sotiropoulos TE, Dimassi KN, Therios IN (2005) Effects of L-arginine and L-cysteine on growth, and chlorophyll and mineral contents of shoots of the apple rootstock EM 26 cultured in vitro. Biol Plant 49(3):443–445

Sprent JL (2007) Evolving ideas of legume evolution and diversity: a taxonomic perspective on the occurrence of nodulation. New Phytol 174:11–25

Sprent JI, Odee DW, Dakora FD (2010) African legumes: a vital but under-utilized resource. J Exp Bot 61(5):1257–1265

Starck Z, Niemyska B, Bogdan J, Tawalbeh RNA (2000) Response of tomato plants to chilling stress in association with nutrient or phosphorus starvation. Plant Soil 226:99–106

Steffen KL, Arora R, Palta JP (1989) Relative sensitivity of photosynthesis and respiration to freeze-thaw stress in herbaceous species. Plant Physiol 89(4):1372–1379

Steponkus PL (1984) Role of the plasma membrane in freezing injury and cold acclimation. Annu Rev Plant Physiol 35:543–584

Steponkus PL, Dowgert MF, Evans RY, Gordon-Kamm W (1982) Cryobiology of isolated protoplasts. Academic, New York, pp 459–474

Steponkus PL, Dowgert MF, Gordon-Kamm WJ (1983) Destabilization of the plasma membrane of isolated plant protoplasts during a freeze-thaw cycle: the influence of cold acclimation. Cryobiology 20(4):448–465

Steponkus PL, Uemura M, Webb MS (1993) A contrast of the cryostability of the plasma membrane of winter rye and spring oat-two species that widely differ in their freezing tolerance and plasma membrane lipid composition. In: Steponkus PL (ed) Advances in low-temperature biology, vol 2. JAI Press Ltd, London, pp 211–312

Stevens CJ, Dise NB, Mountford JO, Gowing DJ (2004) Impact of nitrogen deposition on the species richness of grasslands. Science 303(5665):1876–1879

Stewart GR, Popp M, Holzapfel I, Stewart JA, Dickie-Eskew ANN (1986) Localization of nitrate reduction in ferns and its relationship to environment and physiological characteristics. New Phytol 104(3):373–384

Stitt M, Müller C, Matt P, Gibon Y, Carillo P, Morcuende R, Scheible WR, Krapp A (2002) Carbon and nitrogen relationships and signaling: steps towards an integrated view of nitrogen metabolism. J Exp Bot 53(370):959–970

Stockinger EJ, Gilmour SJ, Thomashow MF (1997) *Arabidopsis thaliana* CBF1 encodes an AP2 domain-containing transcriptional activator that binds to the C-repeat /DRE, a cis-acting DNA regulatory element that stimulates transcription in response to low temperature and water deficit. Proc Natl Acad Sci U S A 94(3):1035–1040

Stoddart JL, Thomas H, Lloyd EJ, Pollock CJ (1986) The use of temperature-profiled transducer for the study of low-temperature growth in Gramineae. Planta 167(3):359–363

Strauss G, Hauser H (1986) Stabilization of lipid bilayer vesicles by sucrose during freezing. Proc Natl Acad Sci U S A 83(8):2422–2426

Streeter J, Wong PP (1988) Inhibition of legume nodul formation and N_2 fixation by nitrate. Plant Sci 7(1):1–23

Suh JP, Jeung JU, Lee JI, Choi YH, Yea JD, Virk PS, Mackill DJ, Jena KK (2010) Identification and analysis of QTLs controlling cold tolerance at the reproductive stage and validation of effective QTLs in cold-tolerant genotypes of rice (*Oryza sativa* L.). Theor Appl Genet 120(5): 985–995

Sun X, Zhu Z, Zhang L, Fang L, Zhang J, Wang Q, Li S, Liang Z, Xin H (2019) Overexpression of ethylene response factors VaERF080 and VaERF087 from *Vitis* amurensis enhances cold tolerance in *Arabidopsis*. Sci Hortic 243:320–326

Susan A, Raven JA (1987) Intercellular pH regulation in *Ricinus communis* grown with ammonia or nitrate as N source: the role of long distance transport. J Exp Bot 38(4):580–596

Sutka J, Snape JW (1989) Location of a gene for frost resistance on chromosome 5A of wheat. Euphvtica 42:41–44

Suzuki A, Knaff DB (2005) Glutamate synthase: structural, mechanistic and regulatory properties, and role in the amino acid metabolism. Photosynth Res 83(2):191–217

Svenning MM, Junttila O, Macduff JH (1996) Differential rates of inhibition of N_2 fixation by sustained low concentrations of NH^{+4} and NO^{-3} in northern ecotypes of white clover (*Trifolium repens* L.). J Exp Bot 47:729–738

Svensson J, Ismail A, Palva ET, Close TJ (2002) Dehydrins. In: Storey KB, Storey JB (eds) Cell and molecular responses to stress. Elsevier, Amsterdam, pp 155–171

Swindell WR, Huebner M, Weber AP (2007) Transcriptional profiling of Arabidopsis heat shock proteins and transcription factors reveals extensive overlap between heat and non-heat stress response pathways. BMC Genomics 8:125

Sylvie FM, Murchie E, Hirel B, Galteir N, Quick WP, Fayer CH (1997) Manipulation of pathways of sucrose biosynthesis and nitrogen assimilation in transformed plants to improve photosynthesis and productivity. In: A molecular approach to primary metabolism in higher plants. Taylar and Francis Ltd., Abingdon, pp 125–153

Taibi K, Del Campo AD, Vilagrosa A, Bellés JM, López-Gresa MP, López-Nicolás JM, Mulet JM (2018) Distinctive physiological and molecular responses to cold stress among cold-tolerant and cold-sensitive *Pinus halepensis* seed sources. BMC Plant Biol 18:236

Takahashi R, Yasuda S (1970) Genetics of earliness and growth habit in barley. In: Milan RA, editor. Barley genetics, vol. II, Proc. 2nd Intl. Barley Genet. Sympo., Washington State University Press, pp 388–408

Takahashi D, Kawamura Y, Uemura M (2016) Cold acclimation is accompanied by complex responses of glycosylphosphatidylinositol (GPI)-anchored proteins in *Arabidopsis*. J Exp Bot 67(17):5203–5215

Tan W, Meng QW, Brestic M, Olsovska K, Yang X (2011) Photosynthesis is improved by exogenous calcium in heat-stressed tobacco plants. J Plant Physiol 168(17):2063–2071

Tang S, Yang Q, Shang H, Sun T (2010) Removal of nitrate by autosulfurotrophic denitrifying bacteria: optimization, kinetics and thermodynamics study. Fresenius Environ Bull 19(12 B):3193–3198

Tasgin E, Atici O, Nalbantoglu B (2003) Effects of salicylic acid and cold on freezing tolerance in winter wheat leaves. Plant Growth Regul 41(3):231–236

Taşgın E, Atici O, Nalbantoğlu B, Popova LP (2006) Effects of salicylic acid and cold treatments on protein levels and on the activities of antioxidant enzymes in the apoplast of winter wheat leaves. Phytochemistry 67(7):710–715

Tcherkez G, Hodges M (2007) How stable isotopes may help to elucidate primary nitrogen metabolism and its interaction with (photo)respiration in C3 leaves. J Exp Bot 59(7):1685–1693

Teana L, Brandi A, Falconi M, Spurio R, Pon CL, Gualerzi CO (1991) Identification of a cold shock transcriptional enhancer of the Escherichia coli gene encoding nucleoid protein H- NS. Proc Natl Acad Sci U S A 88(23):10907–10911

Teijo MH, Griffith M (1999) Snow-mold induced apoplastic proteins in winter rye leaves lack antifreeze activity. Plant Physiol 121(2):665–674

Thakur P, Kumar S, Malik JA, Berger JD, Nayyar H (2010) An overview: cold stress effects on reproductive development in grain crops. Environ Exp Bot 67(3):429–443

Theocharis A, Clement C, Barka EA (2012) Physiological and molecular changes in plants grown at low temperatures. Planta 235(6):1091–1105

Thomas DD (2015) Breathing new life into nitric oxide signaling: a brief overview of the interplay between oxygen and nitric oxide. Redox Biol 5:225–233

Thomas FM, Hilker C (2000) Nitrate reduction in leaves and roots of young pedunculate oaks (*Quercus robur*) growing on different nitrate concentrations. Environ Exp Bot 43(1):19–32

Thomas H, Stoddart JL (1984) Kinetics of leaf growth in *Lolium temulentum* at optimal and chilling temperatures. Ann Bot 53(3):341–347

Thomashow MF (1990) Molecular genetics of cold acclimation in higher plants. Adv Genet 28:99–131

Thomashow MF (1994) *Arabidopsis thaliana* as a model for studying mechanisms of plant cold tolerance. In: Meyerowitz EM, Somerville CR (eds) Arabidopsis. Cold Spring Harbor Laboratory Press, New York, pp 807–834

Thomashow MF (1998) Role of cold-responsive genes in plant freezing tolerance. Plant Physiol 118(1):1–8

Thomashow MF (1999) Plant cold acclimation: freezing tolerance genes and regulatory mechanism. Annu Rev Plant Physiol Mol Biol 50:571–599

Tischner R (2000) Nitrate uptake and reduction in higher and lower plants. Plant Cell Environ 23(10):1005–1024

Todaka D, Shinozaki K, Yamaguchi-Shinozaki K (2015) Recent advances in the dissection of drought-stress regulatory networks and strategies for development of drought-tolerant transgenic rice plants. Front Plant Sci 6:84

Tomalty HE, Walker VK (2014) Perturbation of bacterial ice nucleation activity by a grass antifreeze protein. Biochem Biophys Res Commun 452(3):636–641

Tommasini L, Svensson JT, Rodriguez EM, Wahid A, Malatrasi M, Kato K, Wanamaker S, Resnik J, Close TJ (2008) Dehydrin gene expression provides an indicator of low temperature and drought stress: transcriptome-based analysis of Barley (*Hordeum vulgare* L.). Funct Integr Genomics 8(4):387–405

Toussaint JP, St-Arnaud M, Charest C (2004) Nitrogen transfer and assimilation between the arbuscular mycorrhizal fungus *Glomus intraradices* Schenck & Smith and Ri T-DNA roots *of Daucus carota* L. in an *in vitro* compartmented system. Can J Microbiol 50(4):251–260

Travis RL, Key JL (1971) Correlation between polyribosome level and the ability to induce nitrate reductase in dark grown corn seedlings. Plant Physiol 48(5):617–620

Trolldenier G (1982) Effect of soil temperature on nitrogen fixation on roots of rice and reed. Plant Soil 68(2):217–222

Tsan MF, Gao B (2004) Heat shock protein and innate immunity. Cell Mol Immunol 1(4):274–279

Tsay YF, Chiu CC, Tsai CB, Ho CH, Hsu PK (2007) Nitrate transporters and peptide transporters. FEBS Lett 581(12):2290–2300

Tseng MJ, Li PH (1987) Changes in nucleic acid and protein synthesis during induction of cold hardiness. In: Li PH, Sakai A (eds) Plant cold hardiness. Liss, New York, pp 1–27

Tuberosa R, Frascaroli E, Salvi S, Sanguineti MC, Conti S, Landi P (2008) QTLS for tolerance to abiotic stresses in maize: present status and prospects. Maydica 50(3):559–569
Tucker DE, Ort DR (2002) Low temperature induces expression of nitrate reductase in tomato that temporarily overrides circadian regulation of activity. Photosynth Res 72(3):285–293
Twary SN, Heichel GH (1991) Carbon costs of dinitrogen fixation associated with dry matter accumulation in alfalfa. Crop Sci 31(4):985–992
Unno H, Uchida T, Sugawara H, Kurisu G, Sugiyama T, Yamaya T, Sakakibara H, Hase T, Kusunoki M (2006) Atomic structure of plant glutamine synthetase: a key enzyme for plant productivity. J Biol Chem 281(39):29287–29296
Upchurch RG (2008) Fatty acid unsaturation, mobilization, and regulation in the response of plants to stress. Biotechnol Lett 30(6):967–977
van Buer J, Cvetkovic J, Baier M (2016) Cold regulation of plastid ascorbate peroxidases serves as a priming hub controlling ROS signaling in *Arabidopsis thaliana*. BMC Plant Biol 16(1):163
Van den Ende W, Valluru R (2009) Sucrose, sucrosyl oligosaccharides, and oxidative stress: scavenging and salvaging? J Exp Bot 60:9–18
Van der Boon J, Steenhuizen J, Steingrover EG (1990) Growth and nitrate concentration of lettuce as affected by total nitrogen and chloride concentration, NH_4/NO_3 ratio and temperature of the recirculating nutrient solution. J Hortic Sci 65(3):309–321
Van Huystee RB, Weiser CJ, Li PH (1967) Cold acclimation in *Cronus stolonifera* under natural and controlled photoperiod and temperature. Int J Plant Sci 128(3/4):200–205
Vance CP (1990) Symbiotic nitrogen fixation: recent genetic advances. In: Miflin BJ, Lea PJ (eds) The biochemistry of plants: intermediary nitrogen metabolism, vol 16. Academic, San Diego, pp 48–88
Van-Loon LC (1985) Pathogenesis-related proteins. Plant Mol Biol 4:111–116
Vanoni M, Dossena L, van den Heuvel R, Curti B (2005) Structure–function studies on the complex iron-sulfur flavoprotein glutamate synthase: the key enzyme of ammonia assimilation. Photosynth Res 83(2):219–238
Vasilieva GG, Mironova NV, Glyanko AK (2001) The low above-zero temperature effect in the zone of roots on nitrate reductase activity in pea organs in the process of vegetating. Turk J Bot 25:255–260
Venema J, Villerius L, Van Hasselt P (2000) Effect of acclimation to suboptimal temperature on chilling induced photodamage: comparison between a domestic and a high altitude wild *Lycopersicon* species. Plant Sci 152(2):153–163
Venkatesh J, Park SW (2014) Role of L-ascorbate in alleviating abiotic stresses in crop plants. Bot Stud 55(1):38
Venketesh S, Dayananda C (2008) Properties, potentials and prospects of antifreeze proteins. Crit Rev Biotechnol 28(1):57–82
Vieira IS, Vasconselos EP, Monteiro AA (1998) Nitrate accumulation, yield and leaf quality of turnip greens in response to nitrogen fertilisation. Nutr Cycl Agroecosys 51(3):249–258
Vierling E (1991) The roles of heat shock proteins in plants. Annu Rev Plant Physiol Plant Mol Biol 42(1):579–620
Vogel CS, Dawson JO (1991) Nitrate reductase activity, nitrogenase activity and photosynthesis of black alder exposed to chilling temperatures. Physiol Plant 82(4):551–558
Voisin A-S, Salon C, Jeudy C, Warembourg FR (2003) Symbiotic N2 fixation activity in relation to C economy of *Pisum sativum* L. as a function of plant phenology. J Exp Bot 54:2733–2744
Voisin A-S, Bourion V, Duc G, Salon C (2007) Using an ecophysiological analysis to dissect genetic variability and to propose an ideotype for nitrogen nutrition in pea. Ann Bot 100:1525–1536
Vona V, Rigano VDM, Lobosco O, Carfagna S, Esposito S, Rigano C (2004) Temperature responses of growth, photosynthesis, respiration and NADH: nitrate reductase in cryophilic and mesophilic algae. New Phytol 163(2):325–331
Wagner SC (2011) Biological nitrogen fixation. Nat Educ Know 3(10):15

Walsh KB, McCully ME, Canny MJ (1989) Vascular transport and soybean nodule function: nodule xylem is a blind alley, not a throughway. Plant Cell Environ 12(4):395–405
Wang W, Vinocur B, Shoseyov O, Altman A (2004a) Role of plant heat shock proteins and molecular chaperons in the abiotic stress response. Trend Plant Sci 9(5):1360–1385
Wang B, Vinocur O, Altman AS (2004b) Role of plant heat-shock proteins and molecular chaperones in the abiotic stress response. Trend Plant Sci 9(5):244–252
Wang H, Huang J, Liang X, Bi Y (2012a) Involvement of hydrogen peroxide, calcium, and ethylene in the induction of the alternative pathway in chilling-stressed Arabidopsis callus. Planta 235(1): 53–67
Wang Y, Hu J, Qin G, Cui H, Wang Q (2012b) Salicylic acid analogues with biological activity may induce chilling tolerance of maize (*Zea mays*) seeds. Botany 90(9):845–855
Wang X, Zhao Q, Ma C, Zhang Z, Cao H, Kong Y, Yue C, Hao X, Chen L, Ma J, Jin J, Li X, Yang Y (2013) Global transcriptome profiles of *Camellia sinensis* during cold acclimation. BMC Genomics 14:415
Wang Z, Wang S, Wu Q (2014) Cold shock protein A plays an important role in the stress adaptation and virulence of *Brucella melitensis*. FEMS Microbiol Lett 354(1):27–36
Wang W, Chen Q, Hussain S, Mei J, Dong H, Peng S, Huang J, Cui K, Nie L (2016) Pre-sowing seed treatments in direct-seeded early rice: consequences for emergence, seedling growth and associated metabolic events under chilling stress. Sci Rep 6:19637
Wang L, Ma R, Yin Y, Jiao Z (2018) Role of carbon ion beams irradiation in mitigating cold stress in *Arabidopsis thaliana*. Ecotoxicol Environ Saf 162:341–347
Waqas MA, Khan I, Akhter MJ, Noor MA, Ashraf U (2017) Exogenous application of plant growth regulators (PGRs) induces chilling tolerance in short-duration hybrid maize. Environ Sci Pollut Res 24(12):11459–11471
Ward M, Grimes H, Huffaker R (1989) Latent nitrate reductase activity is associated with the plasma membrane of corn roots. Planta 177(4):470–475
Warembourg FR, Roumet C (1989) Why and how to estimate the cost of symbiotic N_2 fixation? A progressive approach based on the use of ^{14}C and ^{15}N isotopes. Plant Soil 115:167–177
Waterer JG, Vessey JK (1993a) Effect of low static nitrate concentrations on mineral nitrogen uptake, nodulation, and nitrogen fixation in field pea. J Plant Nutr 16:1775–1789
Waterer JG, Vessey JK (1993b) Nodulation response of auto regulated or NH^{+4} –inhibited pea *(Pisum sativum)* after transfer to stimulatory (low) concentrations of NH^{+4}. Physiol Plant 88: 460–466
Waughman GJ (1977) The effect of temperature on nitrogenase activity. J Exp Bot 28:949–960
Webster DE, Ebdon JS (2005) Effects of nitrogen and potassium fertilization on perennial ryegrass cold tolerance during deacclimation in late winter and early spring. HortSci 40(3):842–849
Weiser CJ (1970) Cold resistance and injury in woody plants. Science 169(3952):1269–1278
Welling A, Moritz T, Palva ET, Junttila O (2002) Independent activation of cold acclimation by low temperature and short photoperiod in hybrid aspen. Plant Physiol 129(4):1633–1641
Wen D, Li Y, He L, Zhang C (2019) Transcriptome analysis reveals the mechanism by which spraying diethyl aminoethyl hexanoate after anthesis regulates wheat grain filling. BMC Plant Biol 19(1):327
Wendehenne D, Durner J, Klessig DF (2004) Nitric oxide: a new player in plant signalling and defence responses. Curr Opin Plant Biol 7(4):449–455
Wenkert W (1980) Measurement of tissue osmotic pressure. Plant Physiol 65(4):614–617
Westerheide SD, Anckar J, Stevens SM, Sistonen L, Morimoto RI (2009) Stress-inducible regulation of heat shock factor 1 by the deacetylase SIRT1. Science 323(5917):1063–1066
Whaley JM, Kirby EJM, Spink JH, Foulkes MJ, Sparkes DL (2004) Frost damage to winter wheat in the UK: the effect of plant population density. Eur J Agron 21(1):105–115
Whelan EDP (1991) Differential response to chilling injury of the group 6 chromosomes of cv. Chinese Spring wheat. Genome 34(1):144–150
Whitehead DC (1995) Legumes: biological nitrogen fixation and interaction with grasses. In: Whitehead DC (ed) Grassland nitrogen. CAB International, Wallingford, pp 36–57

Willems A (2006) The taxonomy of rhizobia: an overview. Plant Soil 287(1/2):3–14
Winkel-Shirley B (2002) Biosynthesis of flavonoids and effects of stress. Curr Opin Plant Biol 5(3): 218–223
Winter G, Todd CD, Trovato M, Forlani G, Funck D (2015) Physiological implications of arginine metabolism in plants. Front Plant Sci 6:534
Wisniewski M, Lindow SE, Ashworth EN (1997) Observations of ice nucleation and propagation in plants using infrared video thermography. Plant Physiol 113:327–334
Witty JF, Minchin FR, Skot L, Sheehy JE (1986) Nitrogen fixation and oxygen in legume root nodules. Oxfort Surveys Plant Mol Cell Biol 3:275–314
Wood M (1996) Nitrogen fixation: how much and at what cost? In: Younie D, editor. Legumes in sustainable farming systems. Occasional Symposium No. 30, British Grassland Society, Reading, pp 26–35
Woodward FI, Williams BG (1987) Climate and plant distribution at global and local scales. Vegetation 69(1/3):189–197
Xiaochuang C, Chu Z, Lianfeng Z, Junhua Z, Hussain S, Lianghuan W, Qianyu J (2017) Glycine increases cold tolerance in rice via the regulation of N uptake, physiological characteristics, and photosynthesis. Plant Physiol Biochem 112:251–260
Xie G, Kato H, Sasaki K, Imai R (2009) A cold-induced thioredoxin h of rice, OsTrx23, negatively regulates kinase activities of OsMPK3 and OsMPK6 in vitro. FEBS Lett 583(17):2734–2738
Xie TL, Gu WR, Li CF, Li J, Wei S (2019) Exogenous DCPTA increases the tolerance of maize seedlings to PEG-simulated drought by regulating nitrogen metabolism-related enzymes. Agronomy 9(11):676
Xin Z, Li PH (1992) Abscisic acid-induced chilling tolerance in maize suspension-cultured cells. Plant Physiol 99(2):707–711
Xing W, Rajashekar CB (2001) Glycine betaine involvement in freezing tolerance and water stress is *Arabidopsis thaliana*. Environ Exp Bot 46(1):21–28
Xing W, Li D, Li J, Hu Q, Deng S (2016) Nitrate removal and microbial analysis by combined micro-electrolysis and autotrophic denitrification. Bioresour Technol 211:240–247
Xu G, Fan X, Miller AJ (2012) Plant nitrogen assimilation and use efficiency. Annu Rev Plant Biol 63:153–182
Yacoob RK, Filion WG (1986a) The effects of cold-temperature stress on gene expression in maize. Biochem Cell Biol 65:112–119
Yacoob RK, Filion WG (1986b) Temperature-stress response in maize: a comparison of several cultivars. Can J Genet Cytol 28:1125–1135
Yadav SK (2010) Cold stress tolerance mechanisms in plants. A review. Agron Sustain Dev 30: 515–527
Yamada M, Morishita H, Urano K, Shiozaki N, Yamaguchi-Shinozaki K, Shinozaki K, Yoshiba Y (2005) Effects of free proline accumulation in petunias under drought stress. J Exp Bot 56(417): 1975–1981
Yamaguchi-Shinozaki K, Shinozaki K (1994) A novel cis-acting element in an Arabidopsis gene is involved in responsiveness to drought, low-temperature, or high-salt stress. Plant Cell 6(2): 251–264
Yamanaka K, Fang L, Inouye M (1998) The CspA family in *Escherichia coli*: multiple gene duplication for stress adaptation. Mol Microbiol 27(2):247–255
Yaneva I, Mack G, Vunkova-Radeva R, Tischner R (1996) Changes in nitrate reductase activity and the protective effect of molybdenum during cold stress in winter wheat grown on acid soil. J Plant Physiol 149(1–2):211–216
Yaneva IA, Hoffmann GW, Tischner R (2002) Nitrate reductase from winter wheat leaves is activated at low temperature via protein dephosphorylation. Physiol Plant 114(1):65–72
Yang CW, Lin CC, Kao CH (2000) Proline, ornithine, arginine and glutamic acid contents in detached rice leaves. Biol Plant 43(2):305–307

Yang X, Cui X, Zhao L, Guo D, Feng L, Wei S, Zhao C, Huang D (2017) Exogenous glycine nitrogen enhances accumulation of glycosylated flavonoids and antioxidant activity in lettuce (*Lactuca sativa* L.). Front Plant Sci 8:2098

Ye C, Fukai S, Godwin I, Reinke RB, Snell PB, Schiller J, Basnayake J (2009) Cold tolerance in rice varieties at different growth stages. Crop Past Sci 60(4):328–338

Yildiz D, Nzokou P, Deligoz A, Koc I, Genc M (2014) Chemical and physiological responses of four Turkish red pine (*Pinus brutia* Ten.) provenances to cold temperature treatments. Eur J Forest Res 133(5):809

Yin BF, Zhang YM (2016) Physiological regulation of *Syntrichia caninervis* Mitt in different microhabitats during periods of snow in the *Gurbantünggüt* desert, northwestern China. J Plant Physiol 194:13–22

Yin Y, Yang R, Han Y, Gu Z (2015) Comparative proteomic and physiological analyses reveal the protective effect of exogenous calcium on the germinating soybean response to salt stress. J Proteomics 113:110–126

Yokoi S, Quintero FJ, Cubero B, Ruiz MT, Bressan RA, Hasegawa PM, Pardo JM (2002) Differential expression and function of *Arabidopsis thaliana* NHX Na^+ /H^+ antiporters in the salt stress response. Plant J 30(5):529–539

Young JC (2010) Mechanisms of the Hsp70 chaperone system. Biochem Cell Biol 88(2):291–300

Young JM, Kuykendall LD, Martinez-Romero E, Kerr A, Sawada H (2001) A revision of Rhizobium Frank 1889, with an emended description of the genus, and the inclusion of all species of Agrobacterium Conn 1942 and Allorhizobium undicola de Lajudie et al. 1998 as new combinations: *Rhizobium radiobacter, R. rhizogenes, R. rubi, R. undicola* and *R. vitis*. Int J Syst Evol Microbiol 51(Pt 1):89–103

Yu E, Owttrim GW (2000) Characterization of the cold stress-induced cyanobacterial DEAD-box protein CrhC as an RNA helicase. Nucleic Acids Res 28(20):3926–3934

Yu XM, Griffith M, Wiseman SB (2001) Ethylene induces antifreeze activity in winter rye leaves. Plant Physiol 126(3):1232–1240

Zahran HH (1999) Rhizobium-legume symbiosis and nitrogen fixation under severe conditions and in an arid climate. Microbiol Mol Biol Rev 63(4):968–989

Zhang Z, Huang R (2010) Enhanced tolerance to freezing in tobacco and tomato overexpressing transcription factor TERF2/ LeERF2 is modulated by ethylene biosynthesis. Plant Mol Biol 73(3):241–249

Zhang JZ, Creelman RA, Zhu JK (2004) From laboratory to field. Using information from Arabidopsis to engineer salt, cold and drought tolerance in crops. Plant Physiol 135(2):615–621

Zhang X, Ervin EH, LaBranche AJ (2006) Metabolic defense responses of seeded Bermuda grass during acclimation to freezing stress. Crop Sci 46(6):2598–2605

Zhang F, Jiang Y, Bai L-P, Zhang L, Chen L-J, Li HG, Yin Y, Yan W-W, Yi Y, Guo Z-F (2011) The ICE-CBF-COR pathway in cold acclimation and AFPs in plants. Middle East J Sci Res 8(2):493–498

Zhang C, He P, Li Y, Li Y, Yao H, Duan J, Hu S, Zhou H, Li S (2016) Exogenous diethyl aminoethyl hexanoate, a plant growth regulator, highly improved the salinity tolerance of important medicinal plant *Cassia obtusifolia* L. J Plant Growth Regul 35:330–344

Zhang X, Da Silva JAT, Niu M, Li M, He C, Zhao J, Zeng S, Duan J, Ma G (2017) Physiological and transcriptomic analyses reveal a response mechanism to cold stress in *Santalum album* L. leaves. Sci Rep 7:42165

Zhang J, Li S, Cai Q, Wang Z, Cao J, Yu T, Xie T (2020) Exogenous diethyl aminoethyl hexanoate ameliorates low temperature stress by improving nitrogen metabolism in maize seedlings. PLoS One 15(4):e0232294

Zhao MG, Chen L, Zhang LL, Zhang WH (2009) Nitric reductase dependent nitric oxide production is involved in cold acclimation and freezing tolerance in Arabidopsis. Plant Physiol 151(2): 755–767

Zhao Y, Feng C, Wang Q, Yang Y, Zhang Z, Sugiura N (2011) Nitrate removal from groundwater by cooperating heterotrophic with autotrophic denitrification in a biofilm-electrode reactor. J Hazard Mater 192(3):1033–1039

Zhao J, Li S, Jiang T, Liu Z, Zhang W (2012) Chilling stress- the key predisposing factor for causing *Alternaria alternata* infection and leading to cotton (*Gossypium hirsutum* L.) leaf senescence. PLoS One 7(4):e36126

Zhao M, Liu W, Xia X, Wang T, Zhang WH (2014) Cold acclimation-induced freezing tolerance of Medicago truncatula seedlings is negatively regulated by ethylene. Physiol Plant 152(1): 115–129

Zhao C, Lang Z, Zhu JK (2015) Cold responsive gene transcription becomes more complex. Trends Plant Sci 20(8):466–468

Zhao Y, Han Q, Ding C, Huangm Y, Liao J, Chen T, Feng S, Zhou L, Zhang Z, Chen Y, Yuan S, Yuan M (2020) Effect of low temperature on chlorophyll biosynthesis and chloroplast biogenesis of rice seedlings during greening. Int J Mol Sci 21(4):1390

Zhou Z-Y, Wang M-J, Wang J-S (2000) Nitrate and nitrite contamination in vegetables in China. Food Rev Int 16(1):61–76

Zhou W, Chen F, Zhao S, Yang C, Meng Y, Shuai H, Luo X, Dai Y, Yin H, Du J, Liu J, Fan G, Liu W, yang, W. and Shu, K. (2018) DA-6 promotes germination and seedling establishment from aged soybean seeds by mediating fatty acid metabolism and glycometabolism. J Exp Bot 70(1):101–114

Zhu XC, Song FB, Xu HW (2010a) Arbuscular mycorrhizae improves low temperature stress in maize via alterations in host water status and photosynthesis. Plant Soil 331:129–137

Zhu XC, Song FB, Xu HW (2010b) Influence of arbuscular mycorrhizae on lipid peroxidation and antioxidant enzyme activity of maize plants under temperature stress. Mycorrhiza 20:325–332

Zhu JJ, Li YR, Liao JX (2013) Involvement of anthocyanins in the resistance to chilling- induced oxidative stress in *Saccharum officinarum* L. leaves. Plant Physiol Biochem 73:427–443

Zhu XC, Song FB, Liu FL, Liu SQ, Tian CJ (2015) Carbon and nitrogen metabolism in *Arbuscular* mycorrhizal maize plants under low-temperature stress. Crop Pasture Sci 66:62–70

Ziegler P (1998) Environmental stress and stress responses plants. In: The co-action between living system and the planet. University of Genava, Switzerland, pp 67–100

5 Lipid Metabolism in Plants Under Low-Temperature Stress: A Review

Abstract

Cold temperatures often severely restrict the growth, distribution, and productivity of plants. The freezing tolerance of plants from temperate climates can be improved by undergoing periods of cold acclimation. Lipids play an important role in the mechanism of frost or cold tolerance in plants. Lipids are widely distributed in the plant kingdom. They are found in abundance in reproductive tissues (e.g., seeds and fruits) of some higher plants where they form important reserve food material such cotyledon of sunflower, rape, peanuts, and almond; endosperm of castor beans, coconut plants, etc.; and mesocarp of avocado pears. Plant membrane lipids have the tendency to change from gel to liquid crystalline phase in response to low-temperature stress. This process is due to the increased level of lipid desaturation. The responsible components of this process are, among others, the fatty acid desaturases. Controlling the activity of these enzymes affects the amount of polyunsaturated fatty acids on the glycerol backbone and eventually controls the plant sensitivity to low-temperature stress. These metabolic processes trigger a series of changes at the transcriptional level, causing differential expression in genes. Numerous approaches toward this process from chemical to the advanced mass spectrometry were taken during the past decades. Metabolomics and transcriptomics seem to be the keys toward describing these complex mechanisms and providing the necessary understanding of lipid metabolic response to low-temperature stress.

Keywords

Role of lipids in plant · Effect of lipid metabolism to cold stress · Effects of cold stress on different plants · Role of lipid signaling in stress response · Role of free ftty acid in signaling · Role of lipid in mitigating cold stress · Mechanism of temperature adaptation · Cold stress symptoms · Sensing and strategies · Role of

A. Bhattacharya, *Physiological Processes in Plants Under Low Temperature Stress*,
https://doi.org/10.1007/978-981-16-9037-2_5

triglycerides to cold response · Glycerolipid response to freezing temperature · Lipid hydrolysis under low-temperature stress

5.1 Biological Role of Lipids in Plants

Lipids are major and vital cellular constituents because they provide the structural basis for cell membranes and provide an energy stock for metabolism. Increasing evidence has indicated that lipids are involved not only in the initiation of defense reactions, as signal mediators (Wang 2004; Wang et al. 2006a; Munnik and Testerink 2009), but also in the mitigation processes in response to stress in plant cells (Gaude et al. 2008; Nakamura et al. 2009; Moellering et al. 2010; Gasulla et al. 2013; Okazaki et al. 2013). The current changes in the global climate significantly affect agronomy (Karl et al. 2009), and thus studies on the stress responses of plants mediated by lipids are more important than ever. Signaling lipids include a wide range of lipid classes, such as lysophospholipid, fatty acid, phosphatidic acid, inositol phosphate, diacylglycerol, oxylipin, sphingolipid, and *N*-acylethanolamine (Wang 2004; Wang et al. 2006a; Kang et al. 2008; Munnik and Testerink 2009; Kilaru et al. 2011; Markham et al. 2013). These lipids are usually present in small quantities in tissues and are quickly synthesized from preexisting membrane lipids or biosynthetic intermediates of membrane lipids. During this process, lipid-hydrolytic enzymes such as phospholipases and esterase are critically important (Wang 2004; Wang et al. 2006a; Munnik and Testerink 2009). In plants, there are many types of lipases and esterases with different substrate preferences, distributions, and modes of induction following stress treatment (Okazaki and Saito 2014). There are also many genes encoding lipoxygenases and kinases that use lipids as substrates (Li-Beisson et al. 2013), suggesting a highly complex and flexible lipid-mediated signaling system that enables adaptation to various types of environmental stresses present in plants. In addition to their roles in signaling, lipids also play a role as mitigators to reduce the impact of stressors. For instance, membrane lipid remodeling occurs in response to stresses such as freezing, dehydration, and nutrition depletion (Gaude et al. 2008; Nakamura et al. 2009; Moellering et al. 2010; Gasulla et al. 2013; Okazaki et al. 2013), and this remodeling contributes to the maintenance of membrane properties affecting lipid dynamics, membrane integrity, and membrane-bound protein functions. Surface protection through wax and cutin deposition and suberization is also an essential anti-stress function of lipids (Samuels et al. 2008).

Lipids function as the structural components of cell membranes, which serve as permeable barriers to the external environment of cells. In plants, lipids play especially important roles as signaling and energy storage compounds. Plant lipids include triacylglycerols, phospholipids, galactolipids, and sphingolipids (Suh et al. 2015). Extracellular lipids are especially important in plants and are comprised of cutin, suberin, and waxes. Jasmonic acid is a particularly important signaling compound in plants (Li-Beisson et al. 2013). In particular, storage oils that accumulate in seeds or fruits as triacylglycerol have been widely used for industrial applications including in paints, soaps, cosmetics, detergents, and pharmaceuticals (Dyer et al. 2008). In the bioeconomy, plant storage oils are used as renewable for the production of biodiesel and nonpetroleum-based biomaterials (Durrett et al. 2008). Beyond bioenergy and industrial bioproducts, plant lipids are altered to improve the quantity and quality of oils for food and feed. Thus, our understanding about plant lipid chemistry and biosynthesis is necessary for downstream manipulation of lipids via biotechnology. This special issue focuses on current topics that are currently important for basic and applied plant lipid research.

While plant lipid research has been active for centuries (Ohlrogge and Browse 1995), our understanding about plant lipid biosynthesis and metabolism has more recently been advanced by the availability of the *Arabidopsis thaliana* genome sequence, followed by genomics of crop plants. Genome-scale research has greatly facilitated the identification of genes that underlie acyl lipid biosynthesis and metabolism (Li-Beisson et al. 2013). Even more recently, technological developments in lipidomics have allowed the analysis of the complete lipid profile within cells, tissues, organs, and organisms (Horn and Chapman 2012). McGlew et al. (2014) describe the integration of *Arabidopsis* mutant information within the ARALIP plant acyl lipid metabolism. This tool allows facile and rapid access to information about each mutant. In addition, there is a compendium of literature sources that describe the basis for lipid composition of relevant mutants. Both Parveez et al. (2014) and Goold et al. (2014) review cover storage and neutral lipids in oil palm and microalgae, respectively. Parveez et al. (2014) describe advances in biotechnology of oil palm that includes biochemical studies, gene and promoter isolation, vector construction for transformation, and genetic transformation technology. Goold et al. (2014) report the biogenesis, structure, and function of lipid droplets in the green microalga *Chlamydomonas reinhardtii* and model microalgae, such as species in *Haematococcus* and *Nannochloropsis*. Lee and Suh (2015), Vishwanath et al. (2014), and Molina and Kosma (2014) are related with the extracellular lipids, cuticular waxes, cutin, and suberins. Lee and Suh (2015) specifically describe recent advances in the understanding of the biological functions of genes involved in cuticular wax biosynthesis, transport, and the regulation of wax deposition from *Arabidopsis* and crop species. Vishwanath et al. (2014) focus on the transport of suberin components, polymer assembly, and the regulation of suberin deposition. Molina and Kosma (2014) outline our current understanding about HXXXD/BAHD acyltransferases that are involved in extracellular lipid biosynthesis. Lee et al. (2015) describe the metabolic engineering of plant seed oil for producing hydroxy fatty acid, which is an important industrial raw material. Morales-Cedillo et al. (2014) report on the function of membrane lipids on the activity of plasma membrane H+-ATPase. Kobayashi et al. (2014) describe the roles of anionic phosphatidylglycerol in photosynthetic electron transport in thylakoid membranes by characterization of a phosphatidylglycerol deficient *Arabidopsis* mutant (pgp1-2) under Pi-controlled conditions. Chen et al. (2015a) describe the production of high oleic linseed oil for oxidative stability of oils and industrial applications via RNAi-mediated gene silencing technology. Kuntam et al. (2015)

describe new fluorochromes for the investigation of dynamic oil bodies within living plant cells using fluorescence microscopy.

5.2 Lipids in Plants

Plant lipids as a group exhibit great structural diversity from simple lipids such as free fatty acids to complex lipids such as sphingolipids. This large diversity comprises a wide range of physical and chemical properties enabling their involvement in a wide variety of physiological processes. They constitute a varied group composed of different forms and structures, typically divided into eight main groups, fatty acids, glycerolipids, glycerol-phospholipids, sphingolipids, sterol lipids, prenol lipids, saccharolipids, and polyketides (Fahy et al. 2005). The unique physical properties of the different lipid species confer a multitude of biological functions involving energy storage and plant metabolism with active surface protection, structural, and signaling roles.

Fats or glycerides or acylglycerols are esters of high molecular weight fatty acids and trihydroxy alcohol, the glycerol. Most of the glycerides in nature have all three hydroxyl groups of glycerol esterified with fatty acids and are called as triglycerides (or "fats"). Monoglycerides and diglycerides having one and two of the three hydroxyl groups of glycerol esterified, respectively, do not occur in appreciable amounts in living organisms although they are important intermediate metabolites. In plants, the fatty acids usually consist of even number of carbon atoms. Fatty acids constituting the fats may be saturated or unsaturated. Tables 5.1 and 5.2 show major saturated and unsaturated fatty acids, respectively, in plants. The carbon atoms in fatty acids are usually numbered from the carboxylic carbon side. The carbon atom no. 2 which is adjacent to the carboxylic carbon is also called as a-carbon, while the next or no. 3 carbon is β-carbon and so on. The number written before the colon denotes the no. of carbon atoms in the fatty acid, while that after the colon is the

Table 5.1 Major saturated plant fatty acids

Common name	No. of C-atoms	Structure
Lauric acid	12	$CH_3\ (CH_2)_{10}\ COOH$
Myristic acid	14	$CH_3\ (CH_2)_{12}\ COOH$
Palmitic acid	16	$CH_3\ (CH_2)_{14}\ COOH$
Stearic acid	18	$CH_3\ (CH_2)_{16}\ COOH$
Arachidonic acid	20	$CH_3\ (CH_2)_{18}\ COOH$

Table 5.2 Major unsaturated plant fatty acids

Common name	No. of C-atoms	Structure
Palmitoleic acid	16:1 (*9c*)	$CH_3\ (CH_2)_5\ CH = CH\ (CH_2)_7\ COOH$
Oleic acid	18:1 (*9c*)	$CH_3\ (CH_2)_7\ CH = CH\ (CH_2)_7\ COOH$
Linoleic acid	18:1 (*9c, 12c*)	$CH_3\ (CH_2)_3\ (CH_2\ CH = CH)_2\ (CH_2)_7\ COOH$
α-Linolenic acid	18:3 (*9c 12c 15c*)	$CH_3\ (CH_2\ CH = CH)_3\ (CH_2)_7\ COOH$

number of double bonds present. Each of the numbers in parenthesis is the lowest numbered carbon atom of the two connected by a double bond, counting being made from the carboxyl carbon side. The letter c which follows immediately denotes that the double bond has cis configuration.

Some plant fatty acids have unusual structure in their molecules. Although these constitute majority of the fatty acids in plants, they are not widespread in plant kingdom and occur only in a few plant species. Triglycerides may be solid or liquid at ordinary room temperature and are termed as fats or oils, respectively. The fats are rich in saturated fatty acids while the oils are rich in unsaturated fatty acids. In plants, the triglycerides mostly occur as oils. Natural fats and oils also contain minor amounts of other lipids in addition to the glycerides.

The majority of plant lipids act as energy stores (e.g., triacylglycerols in seeds) or as plant membrane components (e.g., galactosylglycerides in chloroplast membranes and phosphoglycerides in non-chloroplast membranes). Furthermore, surface lipids such as waxes and cutin provide an impervious barrier on the plant surface to reduce water loss and to provide protection against plant pathogens and toxins (Hatfield et al. 2007). Lipids are found in relatively small amounts in temperate grasses and generally decrease as the plant matures (Frame and Laidlaw 2011). The cuticle is a barrier coating the outer surface of epidermal cells of organs of the aerial parts of the plants. It protects against water loss and various abiotic and biotic stress. The structure and properties of the cuticle have received increased attention in the past years, and a number of detailed reviews have been published (Goodwin and Jenks 2005; Muller and Riederer 2005; Reina-Pinto and Yephremov 2009; Schreiber 2010; Domínguez et al. 2011a, b; Nawrath et al. 2013; Yeats and Rose 2013). The cuticle is structurally diverse among species but exhibits the organization of a composite material consisting in cutin, a polyester that is partly covered and interspersed with waxes (epicuticular and intracuticular waxes). Lipids are found in relatively small amounts in temperate grasses and generally decrease as the plant matures (Frame and Laidlaw 2011). The epicuticular waxes and the cuticle with intracuticular waxes are referred to as the cuticle proper. The cuticle proper lies above a so-called cuticular layer made of cutin and polysaccharides that is closely associated with the cell wall of the underlying epidermis cell. The cutin polymer is typically made of esterified ω- and mid-chain hydroxy and epoxy C16 and C18 fatty acids and some glycerol (Heredia 2003). A number of recent reviews have been published that describe various aspects of the biological functions of the cuticle in relation to their physical and biochemical properties (Muller and Riederer 2005; Reina-Pinto and Yephremov 2009).

The term triglyceride is, however, frequently synonymous with the term fats, and thus the latter represents both fats and oils. Fats (triglycerides) are poorer in oxygen in comparison with carbon. Fats are insoluble in water; hence, it can be stored in the cell without disturbing their osmotic relations. Fats can be hydrolyzed by alkalies into glycerol and salt of fatty acids—a process called as saponification. Fats together with waxes constitute the "natural lipids." But while fats are fatty acid esters of trihydroxy alcohol, glycerol and the waxes are predominantly fatty acid esters of long-chain monohydric alcohol. Neutral lipids are readily soluble in nonpolar hydrocarbon solvents such as light petroleum and benzene.

Fats are widely distributed in plant kingdom. They are found in abundance in reproductive tissues (e.g., seeds and fruits) of some higher plants where they form important reserve food material such cotyledon of sunflower, rape (*Brassica napus*) peanuts (*Arachis hypogeal*), and almond (*Prunus amygdalus*); endosperm of castor beans, coconut plants, etc.; and mesocarp of avocado pears (*Persea anericana*). Some cereal such as wheat and barley (*Hordeum distichom*) which store more starch as chief reserve food in their endosperm rich fat content in the cells of their aleuron layer. Within the cells of the abovementioned tissues, the fats are stored in cell organelles called spherosomes (or oleosomes or oil bodies) which are extensively distributed throughout the cytoplasm. Fat is one of the three main macronutrients, along with carbohydrate and protein. Fats, also known as triglycerides, are esters of three fatty acid chains and the alcohol glycerol. The terms fats and oils are categorized according to the number and bonding of the carbon atoms in the aliphatic chain. Fats that are saturated fats have no double bonds between the carbons in the chain. Unsaturated fats have one or more double bonded carbons in the chain. The nomenclature is based on the nonacid (non-carbonyl) end of the chain. This end is called the omega (ω) end or the n-end. Thus α-linolenic acid is called a ω-3-fatty acid because the third carbon from that end is the first double bonded carbon in the chain counting from that end. Some oils and fats have multiple double bonds and are therefore called polysaturated fats. Unsaturated fats can be further divided into *cis*-fats, which are the most common in nature, and trans fats, which are rare in nature. Unsaturated fats can be altered by reaction with hydrogen affected by a catalyst. This action, called hydrogenation, tends to break all the double bonds and makes a fully saturated fat. To make vegetable shortening, then, liquid *cis*-unsaturated fats such as vegetable oils are hydrogenated to produce saturated fats, which have more desirable physical properties; e.g., they melt at a desirable temperature (30–40 °C) and store well, whereas polyunsaturated oils go rancid when they react with oxygen in the air. However, trans fats are generated during hydrogenation as contaminants created by an unwanted side reaction on the catalyst during partial hydrogenation.

Saturated fats can stack themselves in a closely packed arrangement, so they can solidify easily and are typically solid at room temperature. Fats serve both as energy sources for the body and as stores for energy in excess of what the body needs immediately. Each gram of fat when burned or metabolized releases about nine food calories (37 kJ = 8.8 kcal). Fats are broken down in the healthy body to release their constituents, glycerol and fatty acids. Glycerol itself can be converted to glucose by the liver and so become a source of energy. There are many different kinds of fats, but each is a variation on the same chemical structure. All fats are derivatives of fatty acids and glycerol. Most fats are glycerides, particularly triglycerides. One chain of fatty acid is bonded to each of the three –OH groups of the glycerol by the reaction of the carboxyl end of the fatty acid (–COOH) with the alcohol, i.e. three chains per molecule. Water is eliminated and the carbons are linked by an –O– bond through dehydration synthesis. The properties of any specific fat molecule depend on the particular fatty acids that constitute it. Fatty acids form a family of compounds that are composed of increasing numbers of carbon atoms linked into a zigzag chain (hydrogen atoms to the side). The more carbon atoms there are in any fatty acid, the longer its chain will be. Long chains are more susceptible to intermolecular forces of attraction (in this case, van der Waals forces), and so the longer ones melt at a higher temperature.

Fatty acid chains may also differ by length, often categorized as short to very long.

- Short-chain fatty acids (SCFA) are fatty acids with aliphatic tails of fewer than six carbons (i.e., butyric acid).
- Medium-chain fatty acids (MCFA) are fatty acids with aliphatic tails of 6–12 carbons, which can form medium-chain triglycerides.
- Long-chain fatty acids (LCFA) are fatty acids with aliphatic tails of 13–21 carbons.
- Very long-chain fatty acids (VLCFA) are fatty acids with aliphatic tails of 22 or more carbons.

Any of these aliphatic fatty acid chains may be glycerated, and the resultant fats may have tails of different lengths from very short to very long, e.g., cerotic acid or *hexacosanoic acid*, a 26-carbon long-chain saturated fatty acid. Long-chain fats are exemplified by tallow (lard) whose chains are 17 carbons long. Most fats found in food, whether vegetable or animal, are made up of medium- to long-chain fatty acids, usually of equal or nearly equal length. Many cell types can use either glucose or fatty acids for this energy.

In biology and biochemistry, a lipid is a biomolecule that is soluble in nonpolar solvents (McNaught and Wilkinson 1997). Nonpolar solvents are typically hydrocarbons used to dissolve other naturally occurring hydrocarbon lipid molecules that do not (or do not easily) dissolve in water, including fats, waxes, sterols, fat-soluble vitamins (such as vitamins A, D, E, and K), monoglycerides, diglycerides, triglycerides, phospholipids, and others. The main biological functions of lipids include storing energy, signaling, and acting as structural components of cell membranes (Fahy et al. 2009; Subramaniam et al. 2011). Lipids may be broadly defined as hydrophobic or amphiphilic small molecules; the amphiphilic nature of some lipids allows them to form structures such as vesicle liposomes or membranes in an aqueous environment. Biological lipids originate entirely or in part from two distinct types of biochemical subunits or "building-blocks": ketoacyl and isoprene groups. Using this approach, lipids may be divided into eight categories: fatty acids, glycerolipids, glycerophospholipids, sphingolipids, saccharolipids, and polyketides (derived from condensation of ketoacyl subunits) and sterol lipids and prenol lipids (derived from condensation of isoprene subunits).

5.3 Importance of Lipid Metabolism to Cold Stress

Within the broader context of how lipid composition affects the cell biology and physiology of plants, the question of how membrane unsaturation affects a plant's ability to tolerate low-temperature stresses has a long and distinguished history. A series of *Arabidopsis* mutants with defects in lipid metabolism has provided important examples of how membrane fatty acid composition can affect plant temperature responses, although it is not completely clear how these specialized examples relate to chilling and freezing tolerance in horticulture. In this review, we will first provide a general introduction to the lipid mutants and then consider in detail some of the studies that have demonstrated the requirements of polyunsaturated membranes for proper growth and function of plants at chilling temperatures (Tokuhisa et al. 1997).

Increases in the frequency, severity, and duration of temperature extremes are anticipated to be a frequent feature of our weather. Climate change is driving greater temperature fluctuations, resulting in frequent periods of cold temperatures. Susceptible crops exposed to cold temperatures have impaired growth and development, and this limits the use of valuable varieties and lowers yields. Therefore, it is important to understand the mechanisms underlying cold stress responses in the search for more tolerant crops and those that maintain yield under temperature stress. Sensing and reacting to cold stress is a complex process that appears to require a multitude of mechanisms and pathways. The strategies that plants utilize to endure stressful conditions are varied and involve a multitude of molecular, metabolic, and physiological adaptations. These strategies make changes that occur in the first instance to protect the plant, followed by cold acclimation enhancing the plant survival under the low-temperature stress (Fowler and Thomashow 2002; Thomashow 1999). Although many of these processes are mediated by transcription factors that activate stress-related gene expression, the response of plants to cold is not limited to the transcription network (Zhu et al. 2007a; Barrero-Gil and Salinas 2013). As a major component of plasma and endomembranes, lipids have a recognizable structural role in mitigating the impact of cold temperatures (Nishida and Murata 1996). Biological membranes have a crucial role in the protection, homeostasis, and metabolism of the cell. There phospholipids are combined with sphingolipids, sterols, and proteins that interact with and have active role in metabolism (Cacas et al. 2012). Indeed, cold temperature resilience is often defined by the ability of the cell membranes to adapt (Steponkus 1984; Takahashi et al. 2013; Uemura et al. 1995a).

Temperature may oscillate considerably during the rice growing season within this region and affect crop development unfavorably. Anticipation of sowing time has made cold temperature problem at the initial stages of development, an important issue that may affect establishment of adequate crop stands and cause chlorosis and plant death. Temperature is an abiotic stress factor to which knowledge of the physiological basis can be useful to define new strategies of selection for cold-tolerant genotypes. Throughout the growing season, crops frequently suffer from various types of environmental stress. As one of the major abiotic stresses, low temperature negatively affects plant growth and development (Heidarvand et al.

2017). Brief exposure to low temperature may disrupt plant physiological processes, such as water status, photosynthesis, and nitrogen metabolism, but plants generally survive (Theocharis et al. 2012a; Chen et al. 2014); prolonged exposure to low temperature may lead to plant necrosis or death. As a high temperature resistance C4 plant maize (*Zea mays* L.) growth is highly susceptible to low temperature. Nitrogen metabolism, including nitrogen uptake, transport, reduction, and assimilation as well as amino acid metabolism, is a fundamental process in plants (Bashri et al. 2018). Moreover, most plant stress-responsive physiological processes involve nitrogen metabolism, such as enhanced nutrient uptake and transport, improved photosynthetic regulation, rapid synthesis of osmotic solutes, and structural alterations (Xie et al. 2019). Hence, metabolic activity is extremely important for the growth and low-temperature tolerance of plants.

Environmental temperature is one of the key factors that affect the yield and quality of crop plants by influencing their distribution, major physiological and biochemical processes (Jumrani and Bhatia 2014; Pham et al. 2015; Souza Lucéia et al. 2016; Wang et al. 2016a; Sunoj et al. 2016; Yang et al. 2016; Xu et al. 2016). Temperature remarkably affects the growth and metabolism of plants. When exposed to different stresses, plants have different ways of dealing with it to compensate for their lack of mobility. Cold is an abiotic stress that has intensively affected plants because of its negative effect on growth and development which is related to yield productivity rates in cultivated crops. This is limiting factor for growing plants from warmer region in different climatic areas when the temperature is less suitable for cultivation. The plant response to cold stress varies from species to species and even within the same plant family. However, regardless of plant species complexity, one major common mechanism used by plant to deal with cold stress is changes in membrane lipid composition to protect its membrane stability and integrity. It was reported that there is always great accumulation of soluble sugars in response to cold stress (Deryabin et al. 2005) and even the major role that is played by light in this complex response mechanism (Zhu et al. 2007a). These metabolic processes trigger a series of changes in transcriptional level, determining differential expression of genes. It is believed that deciphering the mechanism of lipid of plants exposed to low temperature will help in finding the answer in cold stress that could be a major breakthrough for the future of agriculture (Badea and Basu 2009).

Every cellular compartment is affected by low-temperature exposure. Moreover, membrane bilayers themselves are directly affected by chilling and freezing exposure. Indeed, temperature-dependent changes in bilayer structure are believed to be essential to the perception and transduction of temperature information (Murata and Los 1997). Upon exposure to chilling, membrane bilayers undergo a reduction in fluidity. Membrane fluidity is used to describe molecular motion within the bilayer, with reduced fluidity being attributed to the unsaturation of membrane lipids (Murata and Los 1997). Loss of fluidity can result in a phenomenon known as leakage—the loss of essential solutes from the plant cell, indicating a loss of the membranes' semipermeable nature. However, when membrane fluidity is reduced, the altered physical state of the membrane may also regulate the conformation and activity of bound proteins, such as calcium ion channels (Los and Murata 2004). This has been proposed to enable the transduction of temperature perception to the nucleus via calcium ion influx (Ca^{2+}). Thus the plasma membrane itself it thought to act as a "biological thermometer," playing an integral role in the plant temperature-sensing machinery (Los and Murata 2004).

According to Wang et al. (2006a), temperature variation is one of the biggest environmental stresses to which plants may be submitted. Many species of temperate origin may develop tolerance when exposed to temperature change. This process is known as thermal adaptation that is associated to biochemical and physiological responses caused mainly by alterations in lipidic fluidity of membranes (Hur et al. 2004). Cold acclimation involves altered gene expression that affects membrane composition and accumulation of compatible solutes (Uemura et al. 2006). This is possible through the action of specific enzymes which are capable of altering the level of lipidic unsaturation of membranes. Therefore, fatty acid composition of the lipids that constitute the plant cell membranes is being studied as a key factor for cold sensitivity (Ito and Simpson 1996). Lipids of plant cell membranes are characterized by a high content of polyunsaturated fatty acids (Wang et al. 2006b) that consist of carboxylic acids with hydrocarbonated chains with a length between 4 and 36 Carbons (C4 to C36). In some, this chain is totally saturated (do not contain double bonds) and non-ramificated; in others, the chain contains one or more double bonds. The physical proprieties of fatty acids and of the substances that contain them are mainly determined by the length and degree of unsaturation of this hydrocarbonated chain. This nonpolar chain is responsible by the low solubility of fatty acids in water. Under temperature of 25 °C, C12 to C14 fatty acids have a cerous consistency, while unsaturated fatty acids of the same carbon length are liquid oily (Da Cruz et al. 2010).

When winter rape plants were transferred from favorable temperature conditions (25/20 °C day/night temperature) to 5 °C, the frost resistance of the leaves was increased, whereas the frost tolerance of the roots remained unaffected. This permitted an analysis of the changes in lipid and fatty acid composition both as related to functioning of the plant at low temperature alone (roots) and as related to adaptation to freezing and functioning at low temperature (leaves). Transfer of the plants to 5 °C leads to an increase in the level of linolenic acid in roots and leaves. This increase was most evident in the phosphatidyl choline and ethanolamine fractions of the leaves and in the neutral lipids and in an unidentified phospholipid from the roots. It was concluded that upon transfer of the plants to 5 °C, a general and nonspecific increase in linolenic acid level contributed to functioning of the rape plants at low temperature and that parallel but minor increases in linolenic acid level of digalactosyl diglyceride, phosphatidyl inositol, and the unknown phospholipid in roots and leaves could only contribute to low-temperature functioning in specific membrane enzyme locations. Combined adaptation of the leaves to freezing tolerance and low-temperature functioning was correlated with a higher level of phosphatidyl choline and ethanolamine, predominantly esterified with linolenic acid (Smolenska and Kuiper 1977).

It is known to great extent that one of the membrane lipid response actions when exposed to cold would be the increase in the unsaturation of fatty acids compared to normal condition (Tasseva et al. 2004a; De Palma et al. 2008). However, the ratio differs substantially from plant to plant as even the same plant can act differentially depending upon the way of exposure to low temperature (i.e., acclimated (3–-4 °C) vs. nonacclimated plants (22–24 °C)). Most of the research conducted on the impact of cold stress in roots and leaves. Studies have shown that low temperature strongly induces suberin layers in the roots of chilling-sensitive cucumber plants, while in contrast, low temperature produces a much weaker induction of suberin layers in the roots of the chilling-tolerant figleaf gourd (Lee et al. 2005a, b; Ahn et al. 1999). Lee et al. (2005c) examined the effect of low temperature on fatty acid unsaturation and lipoxygenase activity in cucumber and figleaf gourd. The double bond index demonstrated that membrane lipid unsaturation shows hyperbolic saturation curve in figleaf gourd roots while a biphasic response in cucumber roots to low temperature. In figleaf gourd, the hyperbolic response in the double bond index was primarily due to accumulation of linolenic acid. Chilling stress also significantly induced lipoxygenase activity in figleaf gourd roots. These results suggest that the degree of unsaturation of root plasma membrane lipids correlates positively with chilling tolerance. Therefore, studies that compare the effects of chilling on cucumber and figleaf gourd may provide broad insight into stress response mechanisms in chilling-sensitive and chilling-tolerant plants. Furthermore, these studies may provide important information regarding the relationship between lipid unsaturation and lipoxygenase function/activity and between lipoxygenase activity and water channeling during the response to chilling stress.

5.3.1 Role of Cytosolic Lipid Droplets in Stress Response

While only small numbers of lipid droplets are present in vegetative tissues (e.g., leaves) under non-limiting growth conditions (Pyc et al. 2017), they have been found to accumulate in vegetative tissues during senescence (Brocard et al. 2017), as well as under stresses. In *Arabidopsis*, three isoforms of small rubber particle proteins (SRPs) are primarily responsible for maintaining lipid droplet stability in leaves (Kim et al. 2016), and the overexpression or suppression of *SRP*s has been found to result in an increased or decreased number of lipd droplets, respectively, in leaf tissues. Since lipid droplets provide a scaffold for the binding of enzymes that function in stress tolerance and the provision of substrates for the biosynthesis of bioactive compounds, it is possible that the reduction in lipid droplets observed following the disruption of *SRP*s simply leads to the lack of an adequate binding surface for these lipid droplet-associated proteins (Park et al. 2018). Therefore, it stands to reason that the overexpression of *SRP*s and the corresponding increase in lipid droplet number in leaves could enhance the binding of lipid droplet-associated proteins and potentially ameliorate stress tolerance. While this has yet to be assessed, its further study is certainly warranted because it may provide another possible approach for improving resiliency to various types of stress.

Caleosin, which can sense alterations in Ca^{2+} concentrations that typically occur during stress-related signaling (Park et al. 2018; Purkrtova et al. 2008; Aubert et al. 2010), is an abundant lipid droplet-associated protein. In addition to their Ca^{2+} sensing ability, caleosins also possess peroxygenase activity, which is responsible for the epoxidation and hydroxylation of polyunsaturated fatty acids (Blée et al. 2012). Lipoxygenases play an important role in polyunsaturated fatty acid oxidation, the products of which act as substrates for downstream reactions that yield biologically active oxylipins (Prost et al. 2005) and are known to be involved in various developmental processes, including stress tolerance (Yang et al. 2012b).

5.3.2 Role of Triacylglycerol in Stress Response

Under various stress conditions, plant cells undergo substantial alterations to both plastidic and extraplastidic membrane lipid composition via lipid remodeling, which aids in the maintenance of membrane fluidity, stability, and integrity (Balogh et al. 2013). Environmental stress can induce the degradation of monogalactosyldiacylglycerol and chlorophyll in chloroplasts (Hölzl and Dörmann 2019), which results in the accumulation of toxic lipid intermediates (including diacylglycerol, free fatty acids, and phytyl) that can damage plant cells. In many cases, triacylglycerol appears to act as a transit pool to sequester some of these toxic intermediates, thus preventing cellular damage under stress conditions. In *Arabidopsis*, the plastoglobule-localized acyltransferases1 and plastoglobule-localized acyltransferases2 have been found to provide a detoxifying role by catalyzing the conversion of diacylglycerol into triacylglycerol and using acyl-CoA, galactolipids, and acyl carrier proteins as acyl donors. Meanwhile, plastoglobule-localized acyltransferases1 and plastoglobule-localized acyltransferases2 can also convert free fatty acids and phytyl into fatty acid phytyl ester as a means of detoxification (Lippold et al. 2012). While the expression levels of plastoglobule-localized acyltransferases1 and plastoglobule-localized acyltransferases2 have been found to increase under nitrogen starvation and senescence (Lippold et al. 2012), it remains to be determined whether they are also involved in the response to other types of stress.

As is the case with excessive levels of heat, cold stress can also severely restrict plant growth and decrease crop yields. Under freezing stress, plants accumulate diacylglycerol, largely through the activity of SENSITIVE TO FREEZING2 (SFR2), which encodes a galactolipid:galactolipid galactosyltransferase that is localized to the outer chloroplast envelope (Moellering et al. 2010). Freezing stress induces the formation of apoplastic ice and cellular dehydration. These both lead to membrane leakage, which results in the release of Mg^{2+} and cellular acidification (Barnes et al. 2016), as well as the consequential posttranslational activation of SFR2. Once activated, this enzyme transfers galactosyl groups from monogalactosyldiacylglycerol into galactolipid acceptors, leading to increased production of oligogalactolipids and diacylglycerol (Barnes et al. 2016). While oligogalactolipids can serve to enhance the stability of the chloroplast membrane

(Tan et al. 2018), diacylglycerol is released into the cytosol during membrane shrinkage (which is caused by freezing) (Tan et al. 2018; Guo et al. 2005), becoming available for conversion to phosphatidic acid via the catalytic action of the diacylglycerol kinase. Though the freezing-induced accumulation of phosphatidic acid occurs mainly through the action of diacylglycerol kinase, the hydrolysis of membrane phospholipids by phospholipase-D also contributes under these conditions (Arisz et al. 2013).

Growing evidence has suggested that triacylglycerol plays an important role in the ability of plants to withstand freezing stress by modulating the relative conversion of excess diacylglycerol into triacylglycerol instead of phosphatidic acid via the catalytic actions of diacylglycerol acyltransferase and diacylglycerol kinase (Lu et al. 2020). For example, the expression of *Arabidopsis* diacylglycerol acyltransferase1 is upregulated under low-temperature conditions, and dgat1 mutants display increased sensitivity to freezing stress compared to wild-type plants. In line with this, a comparative genomics study between freezing-sensitive and freezing-tolerant *Boechera stricta* lines indicated that the upregulation of diacylglycerol acyltransferase1 might provide a common mechanism for conferring freezing stress tolerance (Arisz et al. 2018), feasibly because the increased conversion of diacylglycerol to triacylglycerol would be associated with a decrease in diacylglycerol to phosphatidic acid reactions.

5.3.3 Lipid Antioxidant and Galactolipid Remodeling Under Low Temperature

The thylakoid membrane consists mostly of galactolipids (galactolipids are a type of glycolipid whose sugar group is galactose). They differ from glycosphingolipids in that they do not have nitrogen in their composition. They are the main part of plant membrane lipids where they substitute phospholipids to conserve phosphate for other essential processes—composed of 50% of monogalactosyldiacylglycerol and 26% digalactosyldiacylglycerol, phosphatidylglycerol, and sulfoquinovosyldiacylglycerol making up most of the remainder (Spicher et al. 2016). The abundance of galactolipids in the photosynthetic membranes suggests they have not only typical bilayer functions but also specific roles, such as the stabilization of photosynthetic complexes, membrane architecture (curvature), and thylakoid stack (grana) formation (Dörmann 2013). In addition to membrane lipids, the thylakoids contain embedded lipid antioxidants such as tocopherols, plastochromanol, as well as plastoquinone, which is better known as an electron transporting redox molecule (Munné-Bosch and Alegre 2002; Gruszka et al. 2008; Mène-Saffrané and DellaPenna 2010; Nowicka and Kruk 2012; Rastogi et al. 2014; Ksas et al. 2015).

Chloroplasts and photosynthetic membranes respond to environmental changes. They acclimate to intensity and quality of light, with changes in the structure of thylakoid membranes, the size of grana stacks, chlorophyll content, amount, and localization of light harvesting complexes (Kanervo et al. 1997; Rochaix 2007;

Lichtenthaler 2010; Rochaix et al. 2012). The modification of the saturation level of membrane lipids is an important acclimation strategy of plants in response to temperature changes, allowing to maintain parameters such as membrane fluidity and permeability in response to varying temperature (Zheng et al. 2011). It is also reported that reversible increase in the size of chloroplasts and number of plastoglobules occurs under moderate temperature stress (Zhang et al. 2010). Upon abiotic stress, such as high or low temperature, the homeostasis of reactive oxygen species metabolism in the plant cell is challenged (Lichtenthaler and Burkart 1999; Apel and Hirt 2004; Zhang et al. 2010). A protective system in part based on lipid soluble molecules is in place to protect plant cells and the photosynthetic membranes against the action of reactive oxygen species.

5.3.4 Low-Temperature Stress and Free Fatty Acids

At low temperatures, plant membranes undergo transition from a liquid crystalline to a gel-like phase with reduced fluidity and ion leakage, and deactivation of membrane proteins occurs (Upchurch 2008). Observations suggest that saturated phosphatidylglycerol content in chloroplast membranes may be related to the phase transition temperature and thus related to the low-temperature adaptability of plants (Iba 2002). The temperature of the phase transition in chilling-sensitive plants is higher than the phase transition temperature for chilling-resistant plants which suggests an association with the level of fatty acid saturation of phosphatidylglycerol in plastid membranes. In chilling-resistant plants, the amount of phosphatidylglycerol with saturated fatty acids (16:0, 18:0) is less than 20%, but chilling-sensitive plants contain 40% or more saturated phosphatidylglycerol (Murata et al. 1992; Wolter et al. 1992).

Maintenance of polyunsaturated fatty acid levels in chloroplast lipids has been shown to contribute to low-temperature survival and the normal formation of chloroplast membranes under chilling stress (Routaboul et al. 2000; Iba 2002). Both the *Arabidopsis* fad5 mutant, which lacks an active chloroplast x-9 fatty acid desaturase and accumulates high levels of palmitic acid (16:0), and the fad6 mutant, which lacks an active x-6 fatty acid desaturase and accumulates high levels of palmitoleic acid (16:1) and oleic acid (18:1), have reduced levels of polyunsaturated fatty acids in the chloroplast galactolipids. Seedlings of these mutants subjected to chilling stress became chlorotic and had half the chlorophyll content of wild type. In addition, the number of thylakoid membranes in these mutants decreased and their chloroplasts were smaller. The *Arabidopsis* fad2 mutant, deficient in an endoplasmic reticulum localized x-6 desaturase, had significantly decreased polyunsaturates in the extrachloroplast membrane lipids. Long-term culture of these mutant plants at low temperature (6 °C) eventually resulted in their withering and death, but not for wild-type plants (Miquel et al. 1993). It has been suggested that plants capable of cold acclimation accumulate polyunsaturates during cold stress. Cold acclimating potato (*Solanum commersonii*) was found to accumulate linoleic acid (18:2) in the membrane glycerolipids of the leaves, whereas commercial, nonacclimating potato (*Solanum tuberosum*) did not show this trait during cold stress (Vega et al. 2004).

Trienoic fatty acids, hexadecatrienoic acid (16:3) and linolenic acid (18:3), are the major polyunsaturated fatty acid species in membrane lipids. Increasing trienoic fatty acids in chloroplast membranes has been shown to enhance low-temperature tolerance in plants during the early growth stage (Iba 2002). Overexpressing *Arabidopsis* chloroplast x-3 desaturase genes FAD7 or FAD8 in tobacco increased trienoic fatty acids and decreased dienoic fatty acids, hexadecadienoic acid (16:2) and linoleic acid (18:2), in leaf tissue (Nishiuchi and Ida 1998). Comparison of wild type with transgenic tobacco showed that chilling tolerance was achieved in transgenic seedlings, but not in mature plants (Kodama et al. 1995). When the levels of trienoic fatty acids in tobacco phospholipids, the main constituent of extrachloroplastic membranes, were increased by overexpressing an endoplasmic reticulum localized x-3 fatty acid desaturase FAD3 gene, no significant differences were detected between wild type and transgenic tobacco for resistance to chilling and freezing (Hamada et al. 1998).

5.4 Effect of Low-Temperature Stress on Different Plants

It is known to a great extent that one of membrane lipid's response actions when exposed to cold would be the increase in the unsaturation of fatty acids as compared to unsaturated fatty acids under normal conditions (Tasseva et al. 2004a, b; De Palma et al. 2008). However, this ratio differs substantially from plant to plant, and even the same plant can act differently depending on the way of exposure to low temperature (i.e., acclimated (3–4 °C) vs. nonacclimated plants (22–24 °C)). Most of the research focused on the impact of cold on leaves and roots. In a study conducted by Makarenko et al. (2011) on fatty acid composition of lipids from the seedlings and roots of crops having different cold resistance and grown at 27 or 4 °C, biosynthesis of fatty acid in the lipids of seedlings and roots of cold-susceptible maize (*Zea mays* L.) at both growth temperatures was controlled by chloroplast ω6 desaturase and microsomal ω6 desaturase, respectively. The content of linoleic acid was 56.2% and 43.3% in the coleoptiles of maize seedlings grown at 4 and 27 °C, respectively, and in the roots it was 52.2% and 38.5%, respectively. The content of α-linolenic acid in the coleoptiles was 6.7–6.8% at both temperatures, while in the root lipids it was higher at low temperature (3.15% at 27 °C vs. 4.7% at 4 °C). Fatty acid biosynthesis in the seedling coleoptiles of wheat (*Triticum aestivum* L.) and Siberian wild rye (*Elymus sibiricus* L.) grown at low temperature was controlled by the chloroplast ω3 desaturase. A minor increase in the content of α-linolenic acid was observed at low temperature: 29.7–30.2% in wheat and 22.8–25.8% in wild rye. In the root tissues of these species, the biosynthesis of α-linolenic acid was controlled by the microsomal ω3 desaturase (Harwood 2005). The content of α-linolenic acid was higher at low temperature: in wheat it was 6.1% at 27 °C and 17.1% at 4 °C, while in Siberian wild rye, 7.1% and 12.0% at 4 and 27 °C, respectively.

5.4.1 General Effects of Chilling Temperature (Between 20 and 0 °C) on Plant Physiology

Photosynthesis is one of the central processes in plant physiology, and it is sensitive to chilling temperatures, which is why it is not surprising that chloroplasts are often first sites of ultrastructural chilling injury in the cell even before the physical symptoms are visible (Kimball and Salisbury 1973). These ultrastructural injuries affect other compartments as well and include swelling and disorganization of the chloroplasts and mitochondria, reduced size and number of starch granules, dilation of thylakoids and unstacking of grana, formation of small vesicles of chloroplast peripheral reticulum, lipid droplet accumulation in chloroplasts, and condensation of chromatin in the nucleus (Kratsch and Wise 2000). In addition, chilling reduces the fluidity of the membranes in the cell (Levitt 1980) causing, among other things, the inhibition of mitochondrial and chloroplastic electron transfer reactions (Huner et al. 1998). For example, the chloroplast electron transfer chain becomes over-reduced under chilling temperatures due to the combined effects of impaired mobility of plastoquinone within the thylakoid membrane and reduced utilization of NADPH by the Calvin cycle (Huner et al. 1998). Linked to both the inhibited metabolic activity and the impaired electron transfer reactions, chilling injury is often associated with the accumulation of reactive oxygen species (Prasad and Saradhi 2004). The activities of the reactive oxygen species scavenging enzymes will also be attenuated by low temperature, inhibiting the ability of the scavenging systems to counterbalance the increase in reactive oxygen species formation. Accumulation of reactive oxygen species has negative impacts on membranes (Allen and Ortm 2001) presented as electrolyte leakage. There is no ice formation at chilling temperatures, and chilling conditions can last from hours to days. The critical temperature varies depending on the plant species involved: for bananas it is 15 °C, rice 18 °C, beans (wilting) 10 °C, and wheat 4 °C (Levitt 1980). Chilling affects the entire internal environment of the cell, and chilling has been found to change the entire metabolic response of the cell, with some processes recovering quickly and others slowly. Chilling injury is therefore a result of the combined consequences of these different metabolic perturbations (Kratsch and Wise 2000).

5.4.2 General Effects of Freezing Temperature (Below 0 °C) on Plant Physiology

Exposure to freezing temperatures causes more severe effects to cells than chilling temperatures because freezing can trigger the formation of intracellular ice, which leads irreversibly to cell death. Freezing injury starts when ice is formed in the extracellular spaces of plant tissues. Ice formation is dependent on the presence of ice nucleators (organic or inorganic substances) and the rate at which the external temperature declines (Weiser 1970; Levitt 1980). During freezing, much of the cellular water moves out from the cell in response to the water potential gradient that is established between the extracellular ice crystal and the intracellular liquid water. This water loss causes cell shrinkage and dehydration. If cooling is sufficiently slow, the cell will dehydrate to the extent required to establish an equilibrium between the intra- and extracellular aqueous spaces, and intracellular freezing will not occur (nonlethal effect). If there are no ice nucleators present (no bacteria, dust, or any other particle), and water is very clean, it remains in a supercooled state above −38 °C, the temperature at which water self-nucleates (Thomashow 1998). Plant-produced ice nucleating agents have been found in the extracellular liquid of a number of species, including the annual winter rye (Brush et al. 1994) and perennial trees such as *Prunus* (Gross et al. 1988) and *Citrus* (Constantinidou and Menkissoglu 1992). There are also extrinsic nucleators that are usually present on plant surfaces during radiation frost. These include bacteria (Baca 1987; Lindow 1988), dust particles, and wind motion. As soon as membranes are affected by ice protrusion inside the cell, this is lethal and causes cell death. There are protective strategies that plants can deploy in order to minimize the action of freezing temperatures and ice formation. One of such strategy is the synthesis of antifreeze proteins. These proteins have two main functions: the first is to lower the temperature at which ice is formed, without influencing melting point of the solution (Duman 1994). The second is to inhibit the growth of ice crystals by binding to ice nuclei and preventing the small ice crystals from joining to form larger crystals, a process called recrystallization (Tursman and Duman 1995). Plant antifreeze proteins are secreted into the apoplast, and while they have been found in a number of freezing-tolerant monocotyledons (e.g., winter rye and winter wheat, winter barley), the synthesis of antifreeze proteins does not appear to be a general response of plants to low temperature, but rather it is specific for particular plants (Antikainen and Griffith 1997). Thus many cold-tolerant species are able to mount a defense against the physical presence of ice in their tissues, but there remain the damaging effects on plant physiology. The extent that plants experience such damage will depend on the type of plant and its developmental state. Information on the responses generated by different cell structures, such as the plasma membrane and the organelles (main focus in this thesis is related to plastids), during freezing improve our understanding of how cells can overcome or cope with extreme stress events such as ice formation in their tissues in order to survive.

Different crop species have different mechanism(s) for tolerance to low-temperature stress. The following are the tolerance mechanisms for some different crop species:

5.4.3 Rapeseed (*Brassica napus*)

Tasseva et al. (2004a, b) have studied the lipid changes in response to cold in the endoplasmic reticulum of rapeseed. They observed an increase in the level of unsaturation when the plants were grown at 4 °C. Choosing the endoplasmic reticulum for their analysis, they were hoping that it would provide more insight on the lipid metabolism, knowing that the endoplasmic reticulum is the site for lipid biosynthesis. They have studied the action of the desaturases and acyltransferases after cold stress. For the experiment, they studied frost-tolerant and frost-sensitive cultivars, respectively. They used the double bond index to measure the unsaturation levels of the lipids for analyzing their results. There were not major differences in lipid changes between the two cultivars with the exception of the increase (3%) in the very long-chain polyunsaturated acids in the sensitive plants at 4 °C. Overall, there was an important increase in linoleic (18:2) and alpha-linolenic (18:3) acids in both types of studied plants when exposed to low temperature. However, what is notable is the time of accumulation of these acids in the tested plants. The tolerant plants showed a very fast accumulation in unsaturated acids, after only a few days compared to the sensitive ones, conferring a more rapid adaptation to cold. This could lead to the conclusion that the tolerant plants have a better responding mechanism to reduce cold stress membrane damages. To better understand this process, Tasseva et al. (2004a, b) performed analysis at the transcriptional level, too. In this way, they found that there were changes in the transcript level of certain enzymes involved in lipid metabolism. All the enzymes did not show major difference in transcript level when compared to results from the sensitive plants. They concluded that this desaturase played a major role in the cold tolerance mechanism of membrane lipids.

Smolenska and Kuiper (1977) reported that when winter rape (*Brassica napus*) plants were transferred from favorable temperature conditions (25/20 °C day/night temperature) to 5 °C, the frost resistance of the leaves was increased, whereas the frost tolerance of the roots remained unaffected. It has been observed that the changes in lipid and fatty acid composition both as related to functioning of the plant at low temperature alone (roots) and as related to adaptation to freezing and functioning at low temperature (leaves), it was observed that transfer of the plants to 5 °C leads to an increase in the level of linolenic acid in roots and leaves. This increase was most evident in the phosphatidyl choline and ethanolamine fractions of the leaves and in the neutral lipids and in an unidentified phospholipid from the roots. It was concluded that upon transfer of the plants to 5 °C, a general and nonspecific increase in linolenic acid level contributed to functioning of the rape plants at low temperature; and parallel but minor increases in linolenic acid level of digalactosyl diglyceride, phosphatidyl inositol, and the unknown phospholipid in roots and leaves could only contribute to low-temperature functioning in specific membrane enzyme locations. Combined adaptation of the leaves to freezing tolerance and low-temperature functioning was correlated with a higher level of phosphatidyl choline and ethanolamine, predominantly esterified with linolenic acid.

Fatty acid composition is affected by environmental conditions, temperature being the main factor. Earlier it had been reported that for conventional double-low rapeseed varieties, low minimal temperatures during the 60 days following the onset of flowering were related to higher α-linolenic acid contents. Baux et al. (2008) reported that monitoring the fatty acid profile in low-linolenic varieties from the beginning of seed filling to full maturity showed that α-linolenic acid synthesis occurred mainly between 550 and 850 degree-days (base 0 °C) after the onset of flowering, that is, during the 20 first days of seed filling in Swiss conditions, i.e., 41–60 days after the onset of flowering. During this period, the determination coefficient of a second order regression between final α-linolenic acid (C18:3) content and minimal daily temperature was even better, with $R^2 = 0.87$. A significant positive relation was also found for the regression between minimal temperature and oleic acid (C18:1) content for the cultivar Splendor ($R^2 = 0.77$), but no correlation could evidence a relation between temperature and linoleic acid (C18:2) content.

5.4.4 Potato (*Solanum tuberosum*)

The implication of the stearoyl-acyl carrier protein desaturase has also been, recently, studied by De Palma et al. (2008) and the implication of this desaturase gene in changing the composition of the membrane lipids in potato plants. It has been suggested that the expression of the stearoyl-acyl carrier protein desaturase (w-9) gene in plants will increase the cold tolerance in plants due to the increased desaturation of the fatty acids and thus a better membrane control of damage at the membrane level. They have compared transgenic plants carrying w-9 gene and nontransgenic plants that were either cold-acclimated or nonacclimated. Plants with increased cold tolerance showed a substantial increase in linoleic acid (18:2) compared to the more sensitive plants that presented an increase in alpha-linolenic acid (18:3). It was suggested that the major component of this mechanism was the level of desaturases in the plants. However, upon fatty acid analysis of the plants, they observed that the extrachloroplastic membranes of the transgenic plants contained, among the normal fatty acids, a *cis*-vaccenic acid (18:1 Δ11) in phosphatidylcholine and phophatidylethanolamine which is unusual in plants (3%). It was hypothesized that the presence of this peculiar fatty acid is due to the excess of the stearoyl-acyl carrier protein desaturase as previously stated by Tasseva et al. (2004a, b). Overall, the membrane lipids upon stress have changed their composition showing increase (by 5–10%) in unsaturated fatty acids in potato. Potato plants were also used as a material to study cold stress and changes in membrane lipids. Deryabin et al. (2005) used transgenic plants that contained a yeast-derived invertase gene to study peroxidation of membrane lipids exposed to low temperatures. Invertases are enzymes that are involved in sugar metabolism that catalyzes the hydrolysis of sucrose into glucose and fructose. They focused on the importance of soluble sugar in lipid peroxidation due to low-temperature exposure. Lipid peroxidation has been measured as the accumulation of its secondary product malonic dialdehyde in potato leaves. Analyzing the plants after chilling (3 °C), they have found that there is a slight increase in malonic dialdehyde in both B33-inv plants (tolerant) and nontransgenic (sensitive) plants with a higher content (17%) in the sensitive plant. However, after decreasing the temperature (−1 °C), they observed an even higher increase (62%) for the sensitive compared to tolerant plants. They have concluded that this low increase in malonic dialdehyde for the transgenic plants is due to the invertase gene and that it plays an important role in tolerance to low temperatures.

5.4.5 Wheat (*Triticum aestivum*)

Bohn et al. (2007) studied the effect of cold on the plasma membrane lipids of winter wheat. It is known that the recovery of the plant after exposure to low temperature during winter is critical for its productivity. Their work focused on two different methods to induce cold, either by exposure to low temperature or by treatment with the hormone abscisic acid on both cold- and nonacclimated plants. The plant material chosen for testing had different levels of cold tolerance. Three major lipid groups were taken into consideration, such as free sterols, glycolipids, and phospholipids, when applying cold treatment. It was observed that the stress response mechanism is very complex. Moreover, it was observed that the way of inducing the stress plays an important role in the plant response. Plants from etiolated seedlings stressed by abscisic acid presented different protein ratios for almost all lipid classes, free sterols, phospholipids, glycolipids, and total lipids, while only classes for glycolipids were identified by tolerant plants derived from green seedlings. These are very important findings for biotechnology, knowing that hormone manipulation is considered one of the many options in genetic engineering. Regardless of the stress-inducing method, there have been notable changes in membrane lipids upon stress. The amount of free sterols increased in all plants, as well as the phospholipids, with the exception of the susceptible plants. Glycolipid amount showed a decrease for all tested plants. The amount of alpha-linolenic acid (18:3) doubled for the resistant plant in phosphatidylcholine and phosphatidylethanolamines. However, the amount of linoleic acid (18:2) increased in phospholipids for the same plants. This confirms the findings of De Palma et al. (2008) on potato tolerant plants. Analogous experiments with similar results involving the role of abscisic acid on the desaturation of fatty acids were performed by Bakht et al. (2006) on chickpea.

5.4.6 Rice (*Oryza sativa*)

Rice is another species that has been studied for low-temperature stress and its influence on membrane lipids. Zhu et al. (2007b) have studied the unsaturated levels of fatty acids in lipids of the chloroplastic thylakoid membrane. After studying three rice cultivars with different levels of resistance, they observed positive correlations between the unsaturated fatty acids (18:1, 18:2, and 18:3) and the electron transport activities, D1 protein (constitutes the core of the PSII reaction center) content and **Fv/Fm** (PSII variable to maximum florescence). Conversely, these correlations were negative for the saturated acids (16:0, 18:0). They concluded that the increase in unsaturation during low temperature definitely had a positive impact on repairing PSII from damage from chilling (11 °C) with an important role in preserving the integrity of the membranes. Studies on the changes in fatty acid composition and its association with low temperature stress in rice are rare, and the great majority of them are related to the types and percentages of fatty acids found in the cultivars (Juliano 1994; Wu et al. 2000; Kitta et al. 2005; Bravi et al. 2006; Deepa et al. 2008).

However, cold tolerance improvement via tissue culture was related to alterations in fatty acid composition, with an increase in fatty acid unsaturation (Bertin et al. 1998).

5.4.7 Sorghum (*Sorghum bicolor*)

When exposed to chilling, but nonfreezing, temperatures, plant cell membranes undergo a number of modifications in their lipid and fatty acid properties (Lyons et al. 1979). Thi et al. (1989) has pointed out that the most currently observed changes are an accumulation of phospholipids, particularly phosphatidylcholine, and an increase in the level of fatty acid unsaturation. Mahmood et al. (2008) reported that the total oil contents in the seeds of sorghum genotypes ranged from 5.0 to 8.2% (w/w). Furthermore, oleic acid (31.12–48.99%), palmitoleic acid (0.43–0.56%), linoleic acid (27.59–50.73%), linolenic acid (1.71–3.89%), stearic acid (1.09–2.59%), and palmitic acid (11.73–20.18%) were present in the seed oil of different sorghum varieties when analyzed by GC-MS. It was observed that in most of the varieties, polyunsaturated fatty acids were higher than monounsaturated fatty acids. The two atypical saturated fatty acids, octanedioic (C8:0) and azelaic acid (C9: 0), were found in some varieties. Sorghum are tropical plants, which show a chilling sensitivity differing from one species to another. Thus, *Sorghum bicolor* is sensitive to chilling, while *Sorghum halepense* could resist to temperatures lower than 5 °C (Bagnall 1979).

5.4.8 Soybean (*Glycine max*)

Soybean seeds contain about 21% oil, 40% protein, 34% carbohydrates, and 5% ash (Burton 1997), although cultivars with less than 18% oil or over 50% protein may be found (Yadav 1996). The oil fraction is essentially composed of triacylglycerols, and the composition and distribution of fatty acids in the triacylglycerol molecules determine the oil quality. Soybean oil, like most edible oils, is composed of palmitic (C16:0), stearic (C 18:0), oleic (C18:1), linoleic (C18:2), and linolenic (C18:3) acids (Lanna et al. 2005). Oleic, linoleic, and linolenic acids are 18 carbon unsaturated fatty acids, containing one, two, and three *cis*-double bonds interrupted by a methylene group, respectively. Soybean oil contains about 11% palmitic, 4% stearic, 24% oleic, 54% linoleic, and 7% linolenic acids (Kinney 1996). The quality of the oil fraction varies considerably among these sources, and it depends on the fatty acid composition and, specially, on the proportion of unsaturated fatty acids, mainly oleic, linoleic, and linolenic acids (Somerville and Browse 1991).

Most of the polyunsaturated fatty acid production in developing seeds occurs via desaturation of oleic and linoleic acids catalyzed by desaturases in the smooth endoplasmic reticulum (Ohlrogge et al. 1991; Somerville and Browse 1991; Arondel et al. 1992). The substrate for the desaturases is phosphatidylcholine (Slack et al. 1979; Browse and Slack 1981). The fatty acids linked to phosphatidylcholine, which may become unsaturated, can subsequently be incorporated into storage triacylglycerol molecules. The linoleic and linolenic acid levels in the oil depend on their biosynthesis rate and availabilities (Yadav 1996). The polyunsaturated fatty acids can become available through two distinct mechanisms:

- Reversible reaction by which cholinephosphotransferase (EC 2.7.8.2) converts phosphatidylcholine containing polyunsaturated fatty acids into diacylglycerols, which can be used for oil synthesis via diacylglycerol acyltransferase (Slack et al. 1983; Stymne and Stobart 1984).
- Reversible reaction by which acyl-CoA:lysophosphatidylcholine acyltransferase (EC 2.3.1.23) catalyzes the exchange of acyl groups, generally between oleoyl-CoA and a polyunsaturated acyl group linked to position 2 of phosphatidylcholine (Stymne et al. 1983; Griffiths et al. 1988).

According to Stymne et al. (1983), oleoyl-CoA was a preferential substrate in the acyl-CoA pool for lysophosphatidylcholine acyltransferase activity, while stearoyl-CoA was completely excluded. Griffiths et al. (1985) and Stobart and Stymne (1985) showed that during exchange of acyl groups, oleoyl-CoA enters position 2 of phosphatidylcholine, liberating linoleate which is preferentially used in the acylation of position 2 of glycerol-3-phosphate. Thus, this enzyme regulates the type of acyl groups constituting the triacylglycerols, which accumulate in developing seeds. According to Stobart et al. (1983), the exchange of acyl groups between acyl-CoA and phosphatidylcholine is a major step for regulating quality of polyunsaturated fatty acids in the acyl-CoA pool for oil synthesis in developing safflower seeds. Increase in the production of reactive oxygen species is a well-known mechanism for stress defense, and Tambussi et al. (2004) studied the amount of reactive oxygen species as an effect of membrane response to low temperature. Levels of malonic dialdehyde were calculated at low temperature and at room temperature. However, in contrast to other reports, there were no significant increases in malonic dialdehyde under cold stress.

5.4.9 Food Legumes

Several abiotic and biotic factors limit the production potential of legumes (Dita et al. 2006) with temperature stress as one of the most important (Kumar et al. 2010). Global climate change is resulting in extreme temperature (high or low) situations in different regions of the world (Porter and Semenov 2005), which affects the performance of winter as well as summer season legumes (Stoddard et al. 2006; Board and Kahlon 2011). Crop legumes are divided between two sister clades: the phaseoloid and the galegoid (Foyer et al. 2016). These clades represent temperately adapted long day plants (galegoid) and tropically adapted short day plants (phaseoloid). While studies of vernalization in legumes are comparatively limited, two models have emerged for each lineage—pea (*Pisum sativum*), representing temperate legumes, and soybean (*Glycine max*) representing tropical legumes (Weller and

Ortega 2015). Although the genes and gene families central to flowering time are largely conserved between legumes and *Arabidopsis thaliana*, there are numerous examples of gene duplication and loss. For example legumes lack the phyC clade of photoreceptor genes, having only three phytochromes for the detection of day length (phyA, phyB, and phyE) (Hecht et al. 2007). With regard to the flowering locus gene family, FT/TFL1, both clades show gene family expansion, with the galegoid clade showing two distinctly divergent sequences from *Arabidopsis thaliana*. Interestingly the FLC clade, shown to be critical in the *Arabidopsis thaliana* flowering response, is apparently absent among galegoid legumes and soybean. These data indicate that within the legume family there are fundamental differences with regard to the initiation of flowering and the perception of light and temperature. Each legume species has its own range of maximum and minimum temperatures, threshold temperature (Bhandari et al. 2017), and extreme variations in temperature will have serious repercussions on every stage of plant development (Zinn et al. 2010), resulting in severe loss of productivity. Faba bean (*Vicia* faba L.) is sensitive to water, cold, and heat stress. Studies evaluating winter hardiness and frost tolerance in faba bean genotypes have indicated that traits such as changes in fatty acid composition, ion leakage, and free proline content are strongly correlated with frost tolerance (Arbaoui et al. 2008; Link et al. 2010).

5.4.10 Adzuki Beans (*Vigna angularis*)

Kojima et al. (1998) studied the effect of low temperature on membrane lipids of Adzuki beans (*Vigna angularis* L.). It was observed that among other modifications, phosphatidylcholine, phosphatidylethanolamines and phosphatidylinositolof phospholipids, monogalatocyldiacyl-glycerol and digalatocyldiacyl-glycerol of glycolipids and free sterols. Acclimation at 3 °C of bean cell culture did not have the same growth rate as it as at 24 °C. Of the free sterol, sitosterols and stigmasterol present in culture at 3 °C were more distinct as compared to that of 24 °C. Moreover, Kojima et al. (1998) also reported increase in unsaturated fatty acids in cultures at 3 °C.

5.4.11 Strawberry (*Fragaria virginiana*)

When the lipid composition in the leaves of wild strawberry (*Fragaria virginiana* Duchesne) was analyzed (O'Neil et al. 1981) throughout an annual growth cycle in the field, cellular hardiness to temperature stress was assessed concomitantly by a solute leakage technique. Leaves were shown to be very sensitive to an applied temperature of −5 °C during the summer months but insensitive to a 35 °C treatment. This general pattern was also seen in young overwintering leaves but was reversed after a period of low-temperature hardening of these same leaves. Associated with cold hardening of the overwintering leaves was a twofold increase in the phospholipid content of the leaf membranes with a proportionately smaller increase in free sterols. The large increase in phospholipids presumably is due primarily to the proliferation of a sterol-poor membrane fraction, probably the endoplasmic reticulum. These quantitative changes in membrane material may be important in increasing freezing tolerance in the overwintering leaf cells by enhancing the overall capacity of the cell for plasma membrane and tonoplast extension through vesicle fusion using components from this endomembrane pool. Analysis of electron micrographs of hardened leaf cells showed an increase in vesiculated smooth endoplasmic reticulum and tonoplast membrane over nonhardened leaf cells, the latter resulting in an enhanced tonoplast surface area to vacuolar volume ratio. During this same period, no changes in the fatty acid or free sterol composition were detectable, suggesting that regulation of membrane fluidity via these components is not required for cold acclimation in this species. During aging and senescence of both the overwintering and the summer leaves, the cellular membranes remained functionally intact but became progressively more vulnerable to temperature stress. Free sterol content increased during this time. This feature may be related to the inability of the older leaves to withstand environmental stress. Increasing sensitivity of the cellular membranes to stress may, in turn, be causally related to the actual onset of senescence in these leaves, thus explaining why only the older leaves senesce when the plant is challenged by periodic environmental stress.

5.4.12 Sugarcane (*Saccharum officinarum*)

Sugarcanes (*Saccharum* spp. hybrids of *Saccharum officinarum* and *Saccharum spontaneum*) are tropical grasses widely grown in both hemispheres in over 120 countries worldwide. Sugarcane is considered a cold-sensitive plant (Tai and Lentini 1998) and occupies the tropical and subtropical areas where frost is not a usual phenomenon. However, when it occurs there are crop failures accompanied by reduced capacity for sucrose accumulation in the stalks (D'Hont et al. 2008). Cold damage of mature plants can kill regrowth of young plants and ratoons. The optimum temperature for sugarcane growth is about 35 °C, and temperatures below 20 °C significantly limit growth and yield (Moore 1987). Low-temperature stress, including temperatures above (chilling) and below 0 °C (freezing), causes severe losses in most crops (Xin and Browse 2000). The accumulation of ice in the intercellular spaces can potentially result in the physical disruption of cells and tissues. The injury also involves effects on membrane structure and function caused by dehydration (Uemura et al. 1995a).

Sun et al. (2016) subjected cold-resistant and cold-susceptible genotypes of sugarcane at 25 °C and 4 °C, respectively. The results showed that as the root growth was inhibited under cold stress, the average root length and root volume were decreased in the cold treatment as compared to the control. As the cold stress prolonged, the activities of root vigor were significantly decreased in both sugarcane varieties, but it decreased more obviously in cold-susceptible than in cold-resistant genotype. The contents of malondialdehyde, proline, soluble sugar, and soluble protein and the activities of peroxidase and superoxide dismutase were higher in cold treatments than in control.

5.4.13 Tobacco (*Nicotiana tabacum*)

Tobacco plants are a thermophilic crop plant originating from tropical region, and they are sensitive to the change of temperature (Yamori et al. 2010; Popov et al. 2013; Zhang et al. 2013). Long-term effects of different growth temperatures (18.5, 23.5, and 28.5 °C, daily average) on growth, development, and plastid pigment metabolism have been studied extensively in tobacco (Yang et al. 2018). It was reported that as compared with the plants grown under 23.5 °C, lower (18.5 °C) and higher (28.5 °C) growth temperature inhibited the growth of tobacco plants. In general, treatment with 28.5 °C accelerated the flowering and senescence of tobacco plants by enhancing the accumulation of $O^{\cdot -}_2$ and H_2O_2 in leaves, while exposing to 18.5 °C had the opposite effects. Treatment with 18.5 °C increased the content of chlorophylls and reduced the carotenoids in leaves. In contrast, treatment with 28.5 °C decreased the chlorophylls and increased the carotenoids. Moreover, both $O^{\cdot -}_2$ and H_2O_2 took part in the breakdown of chlorophylls in tobacco leaves to some extent. The results suggest that growth temperature could regulate growth, development, and plastid pigment metabolism, and 23.5 °C could be an optimal temperature for growth, development, and metabolism of plastid pigments of tobacco plants under the experimental conditions.

Changes in the fatty acid composition of leaf and root lipids of heat-loving tobacco (*Nicotiana tabacum* L., cv. Samsun) plants during low-temperature hardening (8 °C for 6 days) were studied (Popov et al. 2012). Hardening could improve leaf but not root cold tolerance. The relative content of polyunsaturated (18:2n-6 and 18:3n-3) fatty acids increased and the proportion of saturated and monounsaturated FAs decreased. In contrast, in the roots, hardening slightly increased the concentration of saturated fatty acids (16:0 and 18:0) and reduced the level of unsaturated fatty acids (18:1n-9, 18:2n-6, and 18:3n-3). At the same time, root lipids contained much C20–24 fatty acids, and their content increased during hardening. It was suggested that an increased fatty acid saturation and elevated proportion of C20–24 fatty acids in the root lipids resulting in the lower membrane fluidity could be a reason for incapability of heat-loving tobacco plant roots of hardening and plant death at the low-temperature stress.

5.4.14 Coffee (*Coffea arabica*)

Quality and content of the chemical composition of coffee beans widely vary based on the species of coffee, altitude, soil, daily temperature variations, and the place of growing (Bertrand et al. 2006; Villarreal et al. 2009; Mehari et al. 2016a, b, c). Coffee taste is different due to the presence of different volatile and nonvolatile chemical constituents. Fatty acids are found in crude oil in the form of glycerides, sometimes present as free fatty acids. Fatty acids are a member of the larger class of ubiquitous lipids. Lipids are the most important components of coffee beans (Alvarez and Rodríguez 2000). Coffee oil (lipids) mainly consists of triglycerides, sterols, tocopherols, and the coffee characteristic diterpenes and their esters with fatty acids (Wagemaker et al. 2011). Many fatty acids are present in the oil extract from arabica green coffee beans (Romano et al. 2014; Martin et al. 2001). Main fatty acids present in green coffee beans are linoleic and palmitic acids. Minor fatty acid compositions are myristic, palmitoleic, eicosenoic, behenic, linolenic, and arachidic acids (Martin et al. 2001). The fatty acid compositions are classified as saturated fatty acids, monounsaturated fatty acids, and polyunsaturated fatty acids (Dong et al. 2017). In general, the extracted coffee oil contained saturated fatty acids in the range of 49.4–59.2%, polyunsaturated fatty acids have the range 29.5–39.2%, and monounsaturated fatty acids are in the range of 4.30–9.69 (Hurtado-Benavides et al. 2016).

Fatty acids are an important factor to assess the quality of the coffee. It is affected by altitude, soil, and the place where coffee plants are grown. The effect of altitude of the coffee plants on the chemical composition of green coffee beans has been reported. Bertrand et al. (2006) have reported the comparison of bean biochemical composition and beverage quality of arabica hybrids involving Sudanese-Ethiopian origins with traditional varieties at various elevations in Central America. Tolessa et al. (2017) have studied the influence of growing altitude, shade, and harvest period on quality and biochemical composition of Ethiopian specialty coffee. Hagos et al. (2018) have studied the correlation between caffeine contents in green coffee beans and altitudes of the coffee plants grown in southwest Ethiopia. Mintesnot and Dechassa (2018) have studied the effect of altitude, shade, and processing methods on the quality and biochemical composition (caffeine, trigonelline, and chlorogenic acids) of green coffee beans in Ethiopia. Adem et al. (2018) have reported the effect of altitude on biochemical composition (caffeine, chlorogenic acids, and sucrose contents) and quality of green arabica coffee beans. Bodner et al. (2019) have reported the effect of harvesting altitude on the aroma released by coffee powder. Gebrekidan et al. (2019) have reported the influence of altitudes of coffee plants on the alkaloid contents of Ethiopian green coffee beans.

Campos et al. (2003) reported that cold-sensitive genotypes of coffee plants showed the highest percentage of malonic dialdehyde. It was observed that as the temperature decreases, the glycolipid ratio monogalactosyldiacylglycerol/digalactosyldiacylglycerol decreases especially in more tolerant genotypes. Increase in unsaturation was observed in all genotypes. However, the highest percent of alpha-linolenic acid (18:3) was observed in tolerant genotypes, and it was in contrast to the observations of Bohn et al. (2007) and Deryabin et al. (2005).

5.4.15 Cucumber (*Cucumis sativa*) and Fig Leaf Gourd (*Cucurbita ficifolia*)

Bulder et al. (1991) studied the total lipid, phospholipid, and fatty acid composition of leaf and root as well as leaf phosphatidyl glycerol fatty acid composition which were determined for two low temperature-tolerant inbred lines of cucumber (*Cucumis sativus* L.) and a control variety, all grown at two temperature regimes. It was observed that there were hardly any genotypic differences at either temperature regime. At low temperature, the degree of unsaturation of fatty acids increased. There was no correlation between fatty acid composition of phosphatidyl glycerol and genotypic differences in growth. However, at low temperature, the inbred lines reacted by lowering the amount of phosphatidyl glycerol without changing the proportion of relatively saturated phosphatidyl glycerol fatty acids, while the control variety reacted by lowering the proportion of relatively saturated phosphatidyl glycerol fatty acids at a stable phosphatidyl glycerol content. In general, no valid selection criterion based on lipid composition was found, indicating that lipid composition at present shows no promise as a suitable selection criterion for genetic adaptation to low-temperature conditions.

Lee et al. (2005c) studied the lipid unsaturation and lipoxygenase activity in after cold stress between cold-tolerant fig-leaf gourd (*Cucurbita ficifolia*) and sensitive species of cucumber (*Cucumis* sativus). When comparing for double bond index for several days, it was observed that the tolerant plants have very uniformly ascending scale and a low and a relatively constant value for electrolyte leakage. However, in sensitive plants, the double bond index values were quite variable and lowest value on second day and highest value on third day of low-temperature stress. It was concluded that a gradual increase in double bond index could be positively correlated with low-temperature tolerance. Also the ratio between C18 and C16 differed between plants.

5.4.16 Zoysiagrass

Zoysiagrass (*Zoysia* spp.) is one of the most widely used warm-season turfgrasses in home lawns, athletic fields, and parks because of its good performance to high temperature, water deficit, traffic tolerance, and low maintenance (Zhang et al. 2009a; Patton and Reicher 2007). Optimum temperature for the growth and development of zoysiagrass is from 25 to 30 °C. Thus, low-temperature stress is one of the essential environmental limiting factors for the geographic distribution and growing season of zoysiagrass in transitional and temperate regions (Pompeiano et al. 2015). Among all the warm-season turfgrasses, however, zoysiagrass is the most tolerant species in freezing tolerance than other ones, but the cold injuries during the winter season vary widely among zoysiagrass species and genotypes (Patton and Reicher 2007). Preliminary investigation in a field study showed that zoysiagrass native to high latitude exhibits great freezing tolerance than the ones native to low latitude. The physiological, biochemical, and molecular basis for those differences has only partially been explored. Li et al. (2018) explored chilling and freezing induced the upregulation of C-repeat binding factor (*ZjCBF*), late embryogenesis abundant (*ZjLEA3*), and dehydration-responsive element binding (*ZjDREB1*) transcription factors in two genotypes of zoysiagrass, whereas those genes exhibited higher expression levels in Latitude-40 (high latitude grass), particularly under freezing temperature. These results suggested that zoysiagrass native to higher latitude exhibiting higher freezing tolerance may attribute to the higher carbohydrates serving as energy reserves and stress protectants that stabilize cellular membranes. The phytohormones may serve signals in regulating plant growth, development, and adaptation to low temperature as well as inducing the upregulated *ZjCBF*, *ZjLEA3*, and *ZjDREB*1 expressions, thus resulting in a higher cold tolerance.

5.5 Roles of Lipid Signaling in Stress Responses

Whereas glycerolipids are known to be the major components of biological membranes, they also serve many other diverse functions such as the release of partially deacylated glycerolipids, lysophospholipids and free fatty acids, which have been implicated in cell signaling associated with various stress-related responses, growth and development (Wang 2004; Munnik and Testerink 2009). The enzyme phospholipase-A (PLA) is critically important as it is involved in the formation of both lysophospholipids and fatty acids. Phospholipase-A can be classified in terms of its reaction specificity: phospholipase-A-1 removes acyl moieties from the sn–1 position of phospholipids, and phospholipase-A-2 does so from the sn–2 position, whereas some phospholipase-A have hydrolytic activity at both positions. Lysophospholipids are usually present in trace quantities in plant tissues, but under stress conditions such as freezing, the level of these lipids substantially increases (Welti et al. 2002). The enhanced production of lysophospholipids and free fatty acids is also often evoked in plants by the application of some elicitors, which is sometimes accompanied by a transient increase in phospholipase-A activity (Roy et al. 1995; Chandra et al. 1996; Narvaez-Vasquez et al. 1999; Scherer et al. 2002).

There have been several reports suggesting physiological roles of lysophospholipids in stress responses and plant-environment interactions. The transient increase of lysophosphatidylcholine generated by phospholipase-A-2 at the plasma membrane has been implicated in the induced phytoalexin production in elicitor-treated cultured cells of *Eschscholzia californica* (Viehweger et al. 2002). Lysophosphatidylcholine and lysophosphatidylethanolamine also act as signal transducers, respectively, in arbuscular mycorrhizal symbiosis in *Solanum tuberosum* (Drissner et al. 2007) and pollen germination (Kim et al. 2011). Fatty acids have also been demonstrated to be involved in stress responses (Kachroo and Kachroo 2009). Oleic acid can function as a stimulator of the signaling enzyme phospholipase-D, which has an anticell death function (Zhang et al. 2003). Oleic acid also modulates levels of nitric oxide-associated protein, thereby regulating nitric oxide-mediated defense signaling in *Arabidopsis* (Mandal et al. 2012). Patatin-related phospholipase-A have been well studied with respect to their role in the production of fatty acid and their oxidized products, oxylipins, demonstrating the importance of fatty acid metabolites produced in response to drought and pathogen infection and in organ development (Yang et al. 2007, 2012a; Scherer et al. 2010; Li et al. 2011). Free fatty acids also regulate salt, drought, and heavy metal tolerance, as well as wound-induced responses and defenses against insect/herbivore feeding in plants (Upchurch 2008). Fatty acid derivatives are also involved in stress responses in plants. Azelaic acid, a saturated dicarboxylic acid containing nine carbon atoms, has been reported to prime plants in order to accumulate salicylic acid, a known defense signal for systemic acquired resistance upon pathogen infection (Jung et al. 2009). Lysophospholipids and free fatty acids are likely to be involved in the signaling of various stress responses and cell regulation. Thus, the suppression of the expression of particular phospholipase-A genes by reverse genetics or by the inhibition of phospholipase-A activities by inhibitors is a popular approach to investigate the role of these lipid species in vivo; however, plant phospholipase-A does not solely function as a hydrolytic enzyme of phospholipids. In *Arabidopsis*, transcription factor MYB30 and the secretory phospholipase-A-2α have been reported to physically interact in vivo, following MYB30-mediated translocation of phospholipase-A-2α from cytoplasmic vesicles to the plant cell nucleus (Froidure et al. 2010). Although the translocation of phospholipase-A-2α possibly modulates lysophospholipid and/or fatty acid-mediated signaling, this report indicates that the effect of phospholipase-A gene silencing in plants is not straightforward.

5.5.1 Role of Lipid as Signaling Molecule During Stress Response

Lipids are major and vital cellular constitutent because they provide the structural basis for cell membranes and provide an energy stock for metabolism. Increasing evidence has indicated that lipids are involved not only in the initiation of defense reactions, as signal mediators (Wang 2004; Wang et al. 2006a; Munnik and Testerink 2009), but also in the mitigation processes in response to stress in plant cells (Gaude et al. 2008; Nakamura et al. 2009; Moellering et al. 2010; Gasulla et al. 2013; Okazaki et al. 2013). Current changes in the global climate significantly affect agronomy (Karl et al. 2009), and thus studies on the stress responses of plants mediated by lipids are more important than ever.

Signaling lipids include a wide range of lipid classes, such as lysophospholipid, fatty acid, phosphatidic acid, inositol phosphate, diacylglycerol, oxylipin, sphingolipid and *N*–acylethanolamine (Wang 2004; Wang et al. 2006a; Kang et al. 2008; Munnik and Testerink 2009; Kilaru et al. 2011; Markham et al. 2013). These lipids are usually present in small quantities in tissues, and are quickly synthesized from pre-existing membrane lipids or biosynthetic intermediates of membrane lipids. In plants, there are many types of lipases and esterases with different substrate preferences, distributions and modes of induction following stress treatment. There are also many genes encoding lipoxygenases and kinases that use lipids as substrates (Li-Beisson et al. 2013), suggesting a highly complex and flexible lipid-mediated signaling system that enables adaptation to various types of environmental stresses is present in plants. In addition to their roles in signaling, lipids also play a role as mitigators to reduce the impact of stressors. For instance, membrane lipid remodeling occurs in response to stresses such as freezing, dehydration and nutrition depletion (Gaude et al. 2008; Nakamura et al. 2009; Moellering et al. 2010; Gasulla et al. 2013; Okazaki et al. 2013), and this remodeling contributes to the maintenance of membrane properties affecting lipid dynamics, membrane integrity and membrane-bound protein functions. Surface protection through wax and cutin deposition and suberization are also an essential anti-stress function of lipids (Kolattukudy 1981; Samuels et al. 2008).

5.5.2 Role of Phosphatidic Acid in Signaling

The roles of phosphatidic acid during various stress responses have been demonstrated through the study of knockout mutants of the *PLD* or *PLC* genes or by the application of a phospholipase-D inhibitor (*n*–butanol), which leads to an inactive phosphatidyl alcohol (Munnik et al. 1995; Hong et al. 2009). Thus far, phosphatidic acid and phospholipase-D have been shown to be involved in basal defense and nonhost resistance against pathogenic fungi (Pinosa et al. 2013), signal transduction of the salicylic acid-mediated signaling cascade (Kalachova et al. 2013), low-temperature stress response (Ruelland et al. 2002; Arisz et al. 2013), salt-stress response (Yu et al. 2010; McLoughlin et al. 2013), microtubule organization in response to salt stress (Zhang et al. 2012c), transduction of hydrogen peroxide signals through interaction with cytosolic glyceraldehyde-3-phosphate dehydrogenase (Guo et al. 2012b), suppression of cell death induced by hydrogen peroxide (Zhang et al. 2003), and response to abscisic acid, drought (Katagiri et al. 2005; Zhang et al. 2005; Hong et al. 2008; Uraji et al. 2012), and nitrogen signaling (Hong et al. 2009). Although it is sometimes difficult to discriminate the functions of phosphatidic acid as a signaling molecule and as a metabolic intermediate for other membrane lipids, phosphatidic acid and phospholipase-D are essential in the lipid remodeling that occurs during phosphorus starvation (Misson et al. 2005) and in normal root hair development, where phospholipase-D is directly regulated by the transcription factor GL2 (Ohashi et al. 2003).

As phosphatidic acid is implicated in various stress signaling pathways, there are several reports on downstream targets suggesting a relationship between phosphatidic acid and abscisic acid responses based on the investigation of abscisic acid-dependent stomatal closure. Phosphatidic acid also affects abscisic acid signaling through the activation of the kinase of long-chain bases of sphingolipid (SPHK) (Guo et al. 2011, 2012b). Activation of phospholipase-Dα1 requires SPHK activity; thus, SPHK and phospholipase-Dα1 are suggested to be co-dependent in the amplification of the response to abscisic acid (Guo et al. 2012b). Phosphatidic acid also regulates NADPH oxidase activity and the production of reactive oxygen species in abscisic acid-mediated stomatal closure in *Arabidopsis* (Zhang et al. 2009b). There are also many other phosphatidic acid binding proteins that are not described here (Janda et al. 2013).

5.5.3 Role of Free Fatty Acid in Signaling

Fatty acids have also been demonstrated to be involved in stress responses (Kachroo and Kachroo 2009). Oleic acid can function as a stimulator of the signaling enzyme phospholipase-D, which has an anticell death function (Zhang et al. 2003). Oleic acid also modulates levels of nitric oxide-associated protein, thereby regulating nitric oxide-mediated defense signaling in *Arabidopsis* (Mandal et al. 2012). Patatin-related phospholipase-A have been well studied with respect to their role in the production of fatty acid and their oxidized products, oxylipins, demonstrating the importance of fatty acid metabolites produced in response to drought and pathogen infection and in organ development (Yang et al. 2007, 2012a; Scherer et al. 2010; Li et al. 2011). Fatty acid derivatives are also involved in stress responses in plants. Azelaic acid, a saturated dicarboxylic acid containing nine carbon atoms, has been reported to prime plants in order to accumulate salicylic acid, a known defense signal for systemic acquired resistance upon pathogen infection (Jung et al. 2009).

Lysophospholipids and free fatty acids are likely to be involved in the signaling of various stress responses and cell regulation. Thus, the suppression of the expression of particular phospholipase-A genes by reverse genetics or by the inhibition of phospholipase-A activities by inhibitors is a popular approach to investigate the role of these lipid species in vivo; however, plant phospholipase-A does not solely function as a hydrolytic enzyme of phospholipids. Recently, in *Arabidopsis*, transcription factor MYB30 and the secretory phospholipase-A_2–α have been reported to physically interact in vivo, following MYB30-mediated translocation of phospholipase-A2α from cytoplasmic vesicles to the plant cell nucleus (Froidure et al. 2010). Although the translocation of phospholipase-A2α possibly modulates lysophospholipid and/or fatty acid-mediated signaling, this report indicates that the effect of phospholipase-A gene silencing in plants is not straightforward (Okazaki and Saito 2014).

5.5.4 Role of Sphingolipids in Signaling

Sphingolipids are a class of lipids containing a backbone of sphingoid bases. A set of aliphatic amino alcohols that includes sphingosine is referred to as a long-chain base. In plants, there are various types of sphingolipids, such as long-chain base, long-chain base phosphate, ceramide (*N*-acyl long-chain base), glucosylceramide, and glycosyl inositolphosphoceramide. Sphingolipids are known to play roles in various environmental stress responses (Berkey et al. 2012; Markham et al. 2013). It should be especially noted that sphingolipids are an essential component in the regulation of cell death and that the disruption of sphingolipid biosynthetic genes in plants often induces programmed cell death (Liang et al. 2003; Shi et al. 2007) and enhanced pathogen-induced hypersensitive responses (Liang et al. 2003; Wang et al. 2008; Gan et al. 2009). These reports suggest that elevated ceramide might be linked with enhanced cell death phenotypes through salicylic acid-dependent signaling pathways; however, the relationship among ceramide, salicylic acid, and enhanced cell death seems to be rather complex, as a hydroxylase-deficient mutant that leads to elevated ceramide and salicylic acid does not display a programmed cell deathlike phenotype (Konig et al. 2012), and a mutant that accumulates glycosphingolipid with an abnormal glycosylation pattern has a constitutive hypersensitive response phenotype without elevated ceramide accumulation (Mortimer et al. 2013). Thus, other metabolites such as long-chain base may be the cause of enhanced cell death (Shi et al. 2007; Markham et al. 2013). With respect to signaling, sphingolipid-mediated enhanced cell death is mediated by a mitogen-activated protein kinase (Saucedo-Garcia et al. 2011) and a Ca^{2+}-dependent protein kinase regulated by 14-3-3-proteins (Lachaud et al. 2013).

Another type of sphingolipid, the long-chain base phosphates, is implicated in signaling during drought (Ng et al. 2001; Coursol et al. 2003; Worrall et al. 2008). It has been reported that long-chain base kinase activity is transiently stimulated by abscisic acid and that heteromeric G proteins are the downstream component of long-chain base phosphate in the abscisic acid signaling pathway (Coursol et al. 2003; Worrall et al. 2008). There are reports on the activation of long-chain base kinase by phosphatidic acid, suggesting interplay between phospholipase-D-dependent phosphatidic acid signaling and sphingolipid-dependent signaling (Guo et al. 2012b). Long-chain base phosphates are also implicated in the cold-stress response (Cantrel et al. 2011; Dutilleul et al. 2012; Guillas et al. 2013).

5.5.5 The Signal Transduction of Membrane Lipids Under Cold Stress

Lipids also play essential roles other than as signal mediators. For instance, the well-defined relationship between membrane composition and chilling sensitivity has been verified by genetic modification that changed the saturation rate of a particular lipid class (Murata et al. 1992; Wolter et al. 1992; Nishida and Murata 1996). Recently, a detailed analysis of changes in glycerolipid metabolism in response to temperature change was also reported (Burgos et al. 2011; Degenkolbe et al. 2012). Some lipids, including oxylipin, function as protective compounds by having anti-microbial properties (Blee 2002; Prost et al. 2005; Nalam et al. 2012; Shimada et al. 2014). In addition to these roles, plant lipids also function as stress mitigators to reduce the severity of environmental stresses. For example, polyunsaturated fatty acids attenuate cell damage upon stress by scavenging reactive oxygen species (Méne-Saffrane et al. 2009). In the currently accepted model for temperature sensing, cold stress causes a change in membrane fluidity and rearrangement of the cytoskeleton, followed by an influx of calcium that triggers downstream responses to confer cold tolerance (Guo and Liu 2018). When cold signals are sensed by the plasma membrane, a series of signal transduction processes of membrane lipids are activated, leading to downstream actions to deliver the cold signal.

Phosphatidic acid is the precursor of phospholipids biosynthesis, acts as the main lipid signal in eukaryotes, binds to specific protein, and activates the mitogen-activated protein kinase signal pathway, Ca^{2+}-dependent protein kinase, NADPH oxidase, and ion channel (Testerink and Munnik 2011). The biosynthesis of phosphatidic acid involves two different pathways: one is the direct hydrolysis of phospholipids by phospholipase-D, and the other way is phospholipase-C that catalyzes the hydrolysis of poly-phosphatidylethanolamine together with diacylglycerol kinase and synthesis of phosphatidic acid from diacylglycerol (Arisz et al. 2009). Extensive research suggests that phosphatidic acid is involved in the processes of plant growth, differentiation, reproduction, hormone response, and signal transduction under various biological and abiotic stresses (Hou et al. 2016; Meringer et al. 2016). Under cold stress, the two pathways of phospholipase-D/diacylglycerol kinase and phospholipase-D are both responsive (Chen et al. 2015c). Phospholipase-D can catalyze the hydrolysis of phosphodiester bond and produce inositol triphosphate, diester glycerol, acetylcholine, and phosphatidic acid. As the second messengers in cells, inositol triphosphate, diester glycerol, phosphatidic acid, and acetylcholine can cause a series of secondary reactions by changing intracellular Ca^{2+} and protein kinase levels, thus completing the process of cell response to cold signals (Hong et al. 2016). Moreover, the activity of phospholipase-D is closely related to the response of plant to low temperature (Muzi et al. 2016); short-term chilling stress (0–180 min) causes rapid and transient increases in phospholipase-D activity of young leaves, while long-term chilling stress (24–36 h) causes significant decreases in phospholipase-D activity in young leaves and roots (Margutti et al. 2017). Furthermore, phospholipase-D is also involved in abscisic acid signal transduction under cold stress. Abscisic acid influences the activity of mitochondrial membrane—binding phospholipase-D in alpine ion mustard leaves through the mediation of Ca^{2+} under cold stress (He et al. 2017). As the initial enzyme of phospholipid degradation, phospholipase-D can accelerate the degradation of phospholipids, resulting in phosphatidic accumulation in the membrane.

The monogalactose diglyceride and digalactose diglyceride are vital components of chloroplast and thylakoid membrane lipids and are closely related to plant photosynthesis. They are synthesized by the catalysis of galactosylglycerol synthetase and digalactoglycerol synthase, respectively (Rocha et al. 2018). During freezing conditions, freezing-sensitive genotypes transfer galactosyl from monogalactose diglyceride to another monogalactose diglyceride and produces oligogalactosols, including galactosyl diacylglycerol and trigalactosyl diacylglycerol, leaving diacylglycerol as a by-product (Roston et al. 2014). The diacylglycerol is converted into triacylglycerol, and then triacylglycerol and oligogalactolipids derived from monoacyl diglyceride specifically increase in response to freezing (Vu et al. 2014). Therefore, the metabolic pathway of triacylglycerol is closely related to plant cold tolerance. Diacylglycerol acyltransferase is a rate-limiting enzyme in the Kennedy pathway, one of the biosynthesis pathways of triacylglycerol (Kennedy 1963). In peanut, *AhDGAT1-1* and *AhDGAT1-2* heterologous expression in a *Saccharomyces cerevisiae* triacylglycerol-deficient quadruple mutant could restore lipid body formation, synthesize triacylglycerol, and markedly accumulate higher levels of fatty acids (Peng et al. 2013; Tang et al. 2013; Chi et al. 2014). Studies have shown that diacylglycerol acyltransferase1 plays a role in adaptive responses to chilling injury in plants, which can modulate the production of triacylglycerol and phosphatidic acid that cooperate with diacylglycerol kinase (Chi et al. 2014; Chen et al. 2016; Yan et al. 2018). The expression of *diacylglycerol acyltransferase1*, *diacylglycerol kinase2*, *3*, and *5*, in *Arabidopsis thaliana* during cold stress can regulate the dynamic balance of diacylglycerol, triacylglycerol, and phosphatidic acid, thereby maintaining the integrity of membrane system and intracellular redox state (Tan et al. 2018).

Lipids also play essential roles other than as signal mediators. For instance, the well-defined relationship between membrane composition and chilling sensitivity has been verified by genetic modification that changed the saturation rate of a particular lipid class (Murata et al. 1992; Wolter et al. 1992; Nishida and Murata 1996). A detailed analysis of changes in glycerolipid metabolism in response to temperature change was also reported (Burgos et al. 2011; Degenkolbe et al. 2012). Some lipids, including oxylipin, function as protective compounds by having anti-microbial properties (Blee 2002; Prost et al. 2005; Nalam et al. 2012; Shimada et al. 2014). In addition to these roles, plant lipids also function as stress mitigators to reduce the severity of environmental stresses. For example, polyunsaturated fatty acids attenuate cell damage upon stress by scavenging reactive oxygen species (Méne-Saffrane et al. 2009). In this section, we will summarize recent reports on stress mitigation based on the plant lipid metabolism.

5.6 Role of Lipid in Mitigating the Low-Temperature Stress

The impact of cooling on plants depends on the chilling rate, exposure time, and other associated stresses (Buchanan et al. 2015). Depending on the tolerance of the species, the damage caused by low temperatures (chilling injury) may occur in a few hours or days (Larcher 2004), determining the degree of sensitivity of a given species. When there is a sudden cold, the injury can occur after a few minutes and is called a "cold shock" (Levitt 1980; Phadtare et al. 1999).

Cold tolerance in plants is an intricate quantitative trait that always occurs in combination or in succession, and it is not controlled by a single regulatory pathway or gene, making conventional breeding approaches for cold tolerance challenging (Kumar et al. 2015; Wang et al. 2017). With the development of biotechnology in agriculture, extensive and in-depth studies on the mechanism of cold tolerance in plants in terms of morphological, anatomical, physiological, biochemical, and molecular biology have been conducted. Lyons (1973) proposed that chilling damage initially occurs at the cellular and organ levels. The biomembrane system, including cell membrane, nuclear membrane, and organelle membrane, is the initial site of injury, particularly in terms of its structure, function, stability, and enzyme activity, thereby resulting in substantial metabolic imbalance, especially involving respiration and photosynthesis. These changes in turn affect the plant growth and development and eventually incur damages at the whole-plant level, leading to the occurrence of chilling damage (Zhang et al. 2019). Biomembrane is also the main repository of lipid for plants (Yu 2008), and fatty acid is the main component of biomembrane, which has been used as the primary index to evaluate peanut quality. Studies have further shown that chilling tolerance in peanut is closely correlated with the composition and structure of the membrane lipids, particularly the saturation of membrane fatty acids (Tang 2011). The complex physiological, biochemical, and molecular mechanisms between membrane lipid metabolism and cold tolerance are being continuously explored to improve cold tolerance.

5.6.1 Effects of Cold Stress on Membrane Permeability

The regulatory mechanism of biomembrane fluidity is one of the principal mechanisms that plants accommodate to changes in temperature conditions (Fig. 5.1), and it is affected by the distribution ratio of various lipids on the membrane and the unsaturation of the glycerol lipid group (Li et al. 2016; Barrero-Sicilia et al. 2017). When plants are subjected to cold stress, membrane lipids change from liquid crystal state to the gel state (Murata et al. 1982), causing electrolyte leakage and loss of balance of intracellular ions (Huang et al. 2015). Chilling injury symptoms include dehydration, wilt, chlorosis, and accelerated senescence (Upadhyaya et al. 2009). Generally, under cold stress, proline, soluble sugars, and soluble proteins accumulate in the cytosol of sensitive and tolerant cultivars. Furthermore, the increasing amplitude of these osmotic regulation substances in varieties with stronger cold tolerance is larger than cold-sensitive varieties. However, the content of these significantly decreases in the cytosol when peanut plants are subjected to unbearable chilling (Bai et al. 2018a; Kazemi-Shahandashti and Maali-Amiri 2018).

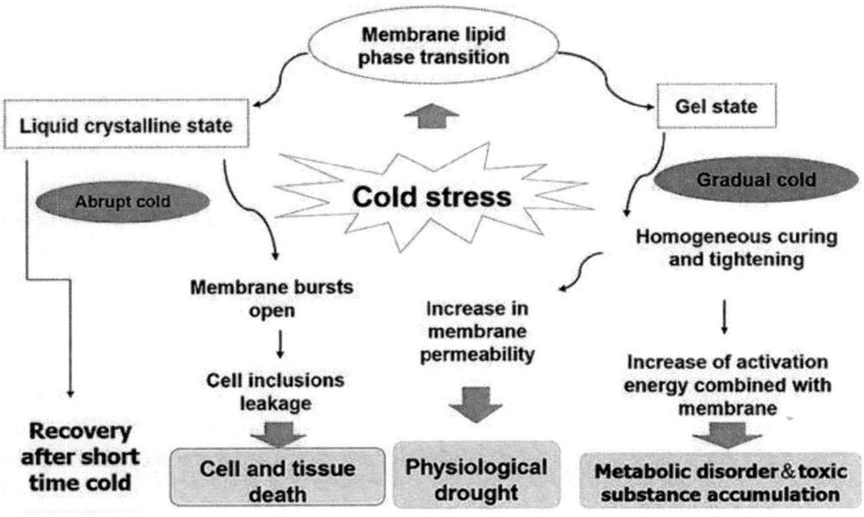

Fig. 5.1 Effect of cold stress on membrane permeability

5.6.2 Effects of Cold Stress on Membrane Lipid Peroxidation

The damage of the plant biomembrane system at low temperature is also related to membrane lipid peroxidation and protein destruction induced by reactive oxygen species (Grant et al. 2014). Membrane lipid peroxidation refers to a series of free radical reactions on double bonds of unsaturated fatty acid on the membrane, which is initiated by oxygen free radicals ($O_2^{-\cdot}$, H_2O_2, ·OH) on unsaturated fatty acids in lipids (Thomas et al. 2016). The structure of the membrane system is destroyed by reactive oxygen species, resulting in metabolic disorders, and massive accumulation of toxic substances in plants (Liu et al. 2013a; Choudhury et al. 2016; Huang et al. 2016). Malondialdehyde is the product of membrane lipid peroxidation that accumulates to higher concentrations in sensitive than tolerant genotypes (Iqbal et al. 2018a, b; Zhong et al. 2018). Cold-tolerant cultivars can resist external environmental stress by relying on the antioxidant enzyme system, which scavenges reactive oxygen species and superoxide anion free radicals produced in plant cells (Tian et al. 2015).

5.6.3 Effects of Cold Stress on Membrane Lipid Component

The main changes in lipid composition that occur in all microorganisms as a result of the lowering of growth temperature are in the fatty acyl (mainly C14-C18) composition of complex glycero-acyl lipids, which are the predominant membrane lipids in all microorganisms except archaea (Russell 2008). In comparison, the cold-dependent changes in head-group composition of lipids are much smaller and do not follow a set pattern (Russell 1989, 2003). The outer membrane of Gram-negative bacteria has lipopolysaccharide instead of phospholipid in the outer leaflet of the lipid bilayer. The lipopolysaccharide contains diagnostic hydroxyl (C12-C16) and hydroxyacyl fatty acids as well as the more usual saturated and monounsaturated fatty acids, and, although the outer membrane does not act as a semipermeable membrane, its fatty acid composition is thermally regulated, probably to modulate the function of porins and other proteins such as receptors (Kumar et al. 2002). Several types of anaerobic bacteria contain ether lipids with (C14-C18) alkyl or alkenyl chains, e.g., the plasmalogens in clostridia (Harwood and Russell 1984). Available data indicates that low temperature has similar effects on the alkyl/alkenyl and acyl compositions, which vary considerably between different phylogenetic groups of anaerobes (Goldfine 1984). In contrast to bacteria, the archaea contain ether-linked glycero-alkyl lipids that are based on C20 (or C25) phytanyl units instead of fatty acids. The focus of attention has been on (hyper) thermophilic archaea, in which the temperature-dependent changes in lipid composition are either much smaller or nonexistent compared with those in bacteria. Little is known about cold adaptation of membranes in psychrophilic archaea, even though they are abundant, particularly in cold oceanic waters and wet soils (Cavicchioli 2006). Eukaryotic microorganisms (yeasts, filamentous fungi, and algae) plus a few exceptional species of bacteria contain sterols that are phylogenetically specific, with ergosterol being the most common (Harwood and Russell 1984). The sterol composition does not usually change significantly with alterations of temperature: instead, the relative proportions of total sterol and phospholipid are a key feature of membrane fluidity regulation in the cold in eukaryotic microorganisms.

The membrane lipids of plants are mainly composed of phospholipids, which include phosphatidyl choline, phosphatidyl ethanolamine, phosphatidyl inositol, phosphatidyl glycerol, phosphatidic acid, glycolipids that consist of monogalactose diglyceride and digalactose diglyceride, and a small amount of sulfolipids and neutral lipids, such as cholesterol (Jouhet et al. 2004). Biomembrane is a dynamic equilibrium system that adaptively adjusts the internal composition based on changes in external temperature. Phospholipid content is positively correlated to cold tolerance in plants, and cold tolerance is weakened when phospholipid synthesis is blocked (Saita et al. 2016). The percentage of high-melting point molecules in total molecular species or saturated fatty acids in total fatty acids in phosphatidyl glycerol is significantly related to plant cold sensitivity, which is higher in cold-sensitive cultivars (Eriksson et al. 2011). The monogalactose diglyceride and digalactose diglyceride are important components of thylakoid membrane lipids, which are closely related to photosynthesis, and their contents also change dynamically at low temperature (Kobayashi 2016). Lipidomic analysis of maize leaves after cold treatment shows an increase in the phosphatidic acid and digalactose diglyceride, but a decrease in phosphatidyl choline and monogalactose diglyceride, resulting in enhanced turnover of phosphatidyl choline to phosphatidic acid which serves as precursors for galactolipid synthesis under low-temperature conditions (Gu et al. 2017).

Lipid transfer protein acts as a carrier for lipid transfer among different cell membranes, and changes in its activity can lead to alterations in membrane lipid composition and affect cold tolerance (Sun et al. 2015). Choi and Hwang (2015) reported that the *BLT101*-overexpressing transgenic wheat lines (BLT101ox) under cold stress loose less water and showed decreased expression of the genes induced by hormones (such as auxin and cytokinin) compared to nontransgenic plants. After prolonged cold treatment, BLT101ox leaves show normal phenotypes, whereas the nontransgenic plants display dehydrated and withered leaves. Nonspecific lipid transfer proteins, small molecular basic protein with abundant content, are responsible for the intermembrane transport of phospholipids by changing the composition of membrane lipids, participating in the biosynthesis of membranes, and transporting lipids among different organelles (Liu et al. 2015a).

5.6.4 Effect of Cold Stress on Membrane Remodeling

Lipid biosynthesis is composed of a multitude of processes that span different cellular organelles, such as plastids, the endoplasmic reticulum, and cytosol employing complex pathways and trafficking mechanisms. Fatty acid synthesis utilizes the carbon flux derived from photosynthesis in the form of pyruvate, and it is localized in the plastid, after which the fatty acyl chains are then channeled to produce more complex lipid molecules in the plastid or transported in the cytosol and endoplasmic reticulum. This is regulated according to the supply and demand for acyl chains, and its production is balanced according to requirements. The regulation of fatty acid synthesis and further modifications is complex (Ohlrogge and Jaworski 1997; Li-Beisson et al. 2013).

Cutin and suberin, present in plant cell walls, are complex fatty acid-derived polymers that confer chemical and physical barriers that protect cells from external pathogens and control the movement of gases, water, and solutes (Shepherd and Griffiths 2006). *Arabidopsis* sfr3 mutants which are sensitive to freezing present an alteration of cuticle permeability due to the variation of wax composition and morphology, in the cold and in flower development. This phenotype arises from a missense mutation of the ACC1 gene that encodes a cytoplasmic acetyl-CoA carboxylase involved in de novo fatty acid biosynthesis (Baud et al. 2003; Amid et al. 2012). Plasma membranes are considered as the primary barrier between the organism and external environment, being the first to experience the injurious effects of stress. Alterations in lipid composition/structure in the plasma membrane under environmental stresses are crucial to maintaining membrane stability and functions. In general, biological membranes have a crucial role not only in the protection, homeostasis, and metabolism of the cell but also in signal recognition and signaling cascades which are proposed as key regulatory processes under both optimum and stress conditions mediated by protein-lipid interactions (Xin and Browse 2000: Steponkus 1984).

5.6.5 Effects of Cold Stress on Membrane Lipid Unsaturation

Plants can modulate the stability and fluidity of membrane by changing the unsaturation of fatty acids in membrane lipids, which is of great significance for organisms to maintain normal photosynthesis and respiratory metabolism and resist cold stress (Mironov et al. 2012; Karabudak et al. 2014). In general, the content of unsaturated fatty acids in lipid membranes increases with decreasing temperature. In addition, compared to cold-sensitive cultivars, the content and the degree (number of double bonds) of unsaturated fatty acids in lipids are higher in cold-tolerant cultivars (Nejadsadeghi et al. 2015). The biomembrane of chilling-sensitive genotypes undergoes a phase transition from liquid crystal to gel even at room temperature due to high saturation of fatty acids, whereas cold-tolerant genotypes can keep the phase transition temperature lower than the cold treatment temperature, thus avoiding phase transition (Ianutsevich et al. 2016). The main fatty acids in various peanut cultivars are similar, including palmitic acid (16:0), stearic acid (18:0), oleic acid (18:1), linoleic acid (18:2), linolenic acid (18:3), and arachidic acid (20:0). However, their contents vary among cultivars after low-temperature treatment, i.e., contents of 18:1, 18:2, and 18:3 rapidly increase, whereas those of 16:0 and 18: 0 decrease (Tang 2011).

In plant cells, saturated fatty acids are synthesized by the type II fatty acid synthase system with the aid of an acyl carrier protein. The biosynthesis of unsaturated fatty acid is conducted by the desaturation of saturated fatty acids depending on two kinds of acyl-lipases, which include glycerol-3-phosphateacyl transferase that is responsible for the lipidization on the C-1 position of the glycerol skeleton and monoacyl-glycerol-3-phosphateacyl transferase that is responsible for the lipidization at the C-2 position, as well as various fatty acid desaturases (Klempova et al. 2013). Acyl carrier protein is a small, acidic protein that plays an essential role in fatty acid synthesis by elongating fatty acid chains. In peanut, *AhACP1*, *AhmtACP3*, *AhACP4*, and *AhACP5* have been identified and have been proven to be closely linked with plant cold tolerance (Wei 2012; Lei et al. 2014). The overexpression and antisense inhibition of *AhACP1* in transgenic tobacco could alter the content of total lipids and composition of fatty acid in leaves, leading to a significant increase or decrease in the content of C18:2 and C18:3, thereby becoming more tolerant or sensitive to cold stress, respectively (Bai et al. 2018b). The glycerol-3-phosphateacyl transferase is the first acyl-lipase in phosphatidyl glycerol biosynthesis that can transfer the aliphatic acyl to C-1 position of glycerol-3-phosphate to synthesize 1-acyl-glycerol-3-phosphoric acid. Cui et al. (2017) indicated that glycerol-3-phosphateacyl transferases from different chilling-tolerant varieties have different selectivities to acyl group substrates, i.e., cold-sensitive genotypes prefer C16:0, whereas chilling-tolerant genotypes have the same selectivity for C16:0 and C18:1. The expression of glycerol-3-phosphateacyl transferase under cold stress is closely correlated to cold tolerance (Li et al. 2018).

Freezing is a notorious stress affecting crop yields through the initial growth of extracellular ice formation, resulting in cellular dehydration and loss of membrane integrity. Oligogalactolipids were recently reported to be responsible for the mitigation of freezing stress (Moellering et al. 2010). The mutant sfr2 (sensitive freezing 2) of *Arabidopsis* was found to be highly sensitive to freezing that resulted in the rupture of chloroplasts (Warren et al. 1996; Thorlby et al. 2004). Detailed biochemical characterization revealed that SFR2 is a galactosyltransferase that progressively transfers a galactose moiety from a monogalactosyldiacylglycerol molecule to other galactolipids including monogalactosyldiacylglycerol, resulting in the formation of oligogalactolipids in the chloroplast envelope and the removal of monogalactosyldiacylglycerol from the envelope membrane, changing the ratio of bilayer- to non-bilayer-forming membrane lipids (Moellering et al. 2010). The importance of lipid metabolism in the protection against freezing was also suggested by the increased tolerance of an *Arabidopsis* mutant with a disruption in phospholipase-Dα (Welti et al. 2002). Regarding the mitigation of freezing stress, D8 unsaturation of long-chain base in sphingolipids is also required for the alleviation of damage caused by freezing (Chen et al. 2012).

5.7 Mechanism of Temperature Adaptation

Despite the predicted global warming scenario for the next decades, models of climate simulation reveal that cold events will persist in the current century and may be more intense than in the twentieth century (Kodra et al. 2011). These events occur in both tropical and temperate regions and usually have a short duration, from 24 h to a few days (Kodra et al. 2011; Marengo et al. 1997; Barroso et al. 1999; Petoukhov and Semenov 2010). Depending on the intensity, the temperature of these events may be low but above 0 °C that is called chilling, or if it is below 0 °C, it is defined as freezing, in which ice formation may occur within the tissues (Chen 1994; Puyaubert and Baudouin 2014).

Poikilotherms (an organism that cannot regulate its body temperature except by behavioral means such as basking or burrowing) which represent most of the organisms on Earth have to survive and adapt to environmental stresses. One of the most important environmental stressor is temperature change. Such alterations lead to changes in membrane lipid order and, in the case of elevated temperatures, may cause protein unfolding and denaturation. If these alterations are not corrected rapidly, then the function and, maybe, the survival of the stressed organism will be in danger. Lowering the ambient temperature of poikilotherms is likely to cause an increase in membrane lipid order. Because of the mixture of lipid classes and lipid molecular species in biological membranes, this will mean that one or more of these molecular species will be below its transition temperature. As soon as this phenomenon becomes significant, then phase separation of the affected lipids can take place. This heterogeneity of membrane structure is already seen, to an extent, with the so-called lipid rafts, but when it becomes quantitatively important, as a result of lower environmental temperatures, then impairment of membrane function will occur. So the first thing that a poikilotherm must do is to correct the increased average membrane lipid order (or to put it in common, but incorrect jargon, the reduced membrane "fluidity").

When the growth temperature of poikilotherms is lowered, a number of changes in their membrane lipids have been observed which are commensurate with attempt to maintain lipid order at physiologically advantageous values. These alterations could, therefore, be viewed as adaptation although, in some cases, they may be coincidental and driven by factors, such as oxygen availability which also changed as ambient temperatures change. Of the various alterations which have been observed, an increase in fatty acid unsaturation is the most common. Since this increase clearly has adaptive significance, we will concentrate for the rest of the mini-review on this aspect, although with examples of other potentially adaptive changes.

Low temperature is the primary determining factor limiting geographical distribution and growing season of plants (Anderson and Taliaferro 2000). The injury symptoms are often exhibited at temperatures below 12 °C for plant species native to tropical and subtropical areas (Esmaili and Salehi 2012). However, the injuries caused by low temperature are generally categorized into chilling stress (temperatures above 0 °C) and freezing stress (temperatures below 0 °C) (Huang et al. 2014). Chilling stress is often leading to growth and photosynthesis reduction, leaf wilting and chlorosis or even necrosis, cellular membrane damages, as well as oxidative stress in plants (Suzuki and Mittler 2006). In addition, temperature decline below 0 °C caused freezing stress, which results in the formation of ice crystals within the cell, mechanical damages, as well as metabolic dysfunction in plants (Xin and Browse 2000). However, plants have evolved complex mechanisms to tolerate chilling and freezing stresses, such as accumulation of carbohydrates and proteins, resulting in large quantities of soluble sugars, amino acids, and cold-induced stress-related proteins (Trischuk et al. 2014), as well as hormone homeostasis inducing the gene expression that functions to stabilize membranes against freezing-induced injury (Thomashow 1999; Fowler and Thomashow 2002).

When plants are exposed to low, nonfreezing temperatures, the degree of fatty acid unsaturation and the content of phospholipids typically increase. Such lipid changes enhance membrane fluidity and reduce the propensity of cellular membranes to undergo freezing-induced non-bilayer phase formation, thus enhancing membrane integrity (Uemura et al. 1995b; Thomashow 1999). To analyze changes in specific lipid molecular species of *Arabidopsis* as a result of low-temperature stress, lipid profiling was used (Welti et al. 2002). Phosphatidylcholine, phosphatidylethanolamine, phosphatidylglycerol, monogalactosyldiacylglycerol, and digalactosyldiacylglycerol species that contain two polyunsaturated acyl species, such as 36:5- linoleoyl/linolenoyl (18:2/18:3-) and 36:6- (di18:3-) phosphatidylcholine, 36:5- (18:2/18:3-) and 36:6- (di18:3-) phosphatidylethanolamine, 34:6- (18:3/hexadecatrienoyl (16:3)-) monogalactosyldiacylglycerol, and 36:6- (di18:3-) digalactosyldiacylglycerol, were shown to increase in plants during cold acclimation. At the same time, the level of more saturated species in these lipid classes, such as 36:2- (stearoyl (18:0)/ 18:2 and di-oleoyl (18:1)-) and 36:3- (18:1/18:2 and 18:0/18:3-) phosphatidylcholine, decreased. The data indicate an increase in desaturation activity and concomitant remodeling of both extraplastidic (phosphatidylcholine and phosphatidylethanolamine) and plastidic (monogalactosyldiacylglycerol, digalactosyldiacylglycerol, and phosphatidylglycerol) lipid species. On the other hand, the content of phosphatidylinositol molecular species was unchanged in response to cold exposure. Cold exposure did induce significant increases in specific molecular species of the lipid metabolites lysophosphatidylcholine, lysophosphatidylethanolamine, and phosphatidic acid. These data suggest that phospholipases, such as phospholipase-A and phospholipase-D, which produce lysophospholipids and GPA, respectively, are activated during cold acclimation. Phosphatidic acid and lysophospholipids are minor phospholipids with potential regulatory functions, such as activation of target signaling proteins, regulation of cytoskeletal organization, and regulation of ion channel function (Wang et al. 2006a, b, c). The observed changes in profiles imply that lipids play both structural and regulatory roles in plant adaptation and survival during low-temperature exposure (Welti et al. 2007).

Damage related to cold is the change in the cell membranes' fluidity. This damage can also be caused by heat; however, under cold exposure, membranes become rigid, while under heat they become more fluid. Chilling may cause a reduction in membrane fluidity due to the transition of the lipid components from a fluid-crystalline state to a solid-gel state, leading to dysfunctions such as increased permeability, water and intracellular solute loss, and inactivation of transport channels (Larcher 2004; Buchanan et al. 2015). The membrane lipids can move within one of the layers and rotate around their own axis, and these movements are thermodynamically directed and temperature dependent (Ruelland and Zachowski 2010). These movements are reduced when temperature decreases, making the membrane more rigid. In most chilling-sensitive plants, the temperature at which the transition phase begins is close to 10 °C. In addition, differences in the structure of membrane lipids can influence chilling tolerance. Sensitive plants usually have a higher content of saturated fatty acid residues in the lipids, while the more resistant ones show an increase in the desaturation of these molecules (Ahmad and Prasad 2012). The loss of membrane fluidity is one of the main signals for the activation of the cold response pathways and may lead to an enhanced reactive oxygen species and reactive nitrogen species synthesis, causing an imbalance between the production of these reactive species and the protection of the antioxidant defense system, generating injuries like lipid peroxidation (Repetto et al. 2012). Malondialdehyde is one of the main products of lipid peroxidation and is considered a good indicator of structural integrity of membranes exposed to low temperature (Jouve et al. 1993).

5.7.1 *cis*-Unsaturated Fatty Acids and Membrane Fluidity

Membrane lipids are usually in the liquid crystalline structure (Gunstone et al. 2007), and the melting behavior of the fatty acyl part of the molecule can be conveniently measured by methods such as differential scanning calorimetry. The resultant transient temperature allows the effect of introducing a *trans*- or *cis*-double bond into a saturated hydrocarbon chain to be assessed. In general, *trans*-double bonds have rather little effect, whereas the first *cis*-double bond causes a massive change in transient temperature (Table 5.3). Interestingly, although a second *cis*-double bond produces a further significant lowering of transient temperature, additional double bonds have little, if any, effect. This raises the question about the function of trienoic-hexaenoic fatty acids in biological membranes which is often assumed to relate to an ability of such acids to confer "fluidity." Clearly, if dienoic fatty acids have similar transient temperatures, then the functions of highly unsaturated fatty acids must lie elsewhere.

In reducing the transient temperature of a typical membrane lipid such as phosphatidylcholine, it is not necessary to have an unsaturated fatty acid at both the *sn*-1 and the *sn*-2 positions on the glycerol backbone. Moreover, it should also be born in mind that swapping the positions of attached fatty acids (remodeling) may also alter the transient temperature by 8–9 °C (Gurr et al. 2002). Such remodeling has been termed an emergency response because it does not need any net lipid synthesis. As expected from such data, when organisms modify the unsaturation of their membrane lipids, a clear correlation with fluidity can be measured. Thus, use of Fourier transform infrared (FTIR) spectroscopy showed that:

- Organisms, which respond to low temperatures by increasing fatty acid unsaturation, exhibited increased membrane disorder.

Table 5.3 Values of transient temperatures for gel to liquid crystalline phase transition of phosphatidylcholine

Lipids	Transient temperature °C
Saturated	
Dimyristoyl-PC (14:0, 14:0-PC)	24
Myristoyl, palmitoyl-PC (14:0, 16:0-PC)	36
Palmitoyl, myristoyl-PC (16:0, 14:0-PC)	27
Dipalmitoyl-PC (16:0, 16:0-PC)	41
Distearoyl-PC (18:0, 18:0-PC)	55
Unsaturated	
Palmitoyl, oleoyl-PC (16:0, 18:1-PC)	−1
Dioleoyl-PC (18:1, 18:1-PC)	−19
Stearoyl, oleoyl-PC (18:0, 18:1-PC)	−6
Stearoyl, linoleoyl-PC (18:0, 18:2-PC)	−16
Stearoyl, linolenoyl-PC (18:0, 18:3-PC)	−13
Stearoyl, arachidonyl-PC (18:0, 20:4-PC)	−13

- Mutants which lacked the required fatty acid desaturases (Fads) showed less adaptability (Szalontai et al. 2000).
- Modification of phospholipid synthesis to increase the incorporation of saturated fatty acids rendered the transgenic organism chilling-sensitive (Murata et al. 1992; Mikami and Murata 2003).

5.7.2 Low-Temperature Effect on Lipids of Cyanobacteria

Cyanobacteria, being prokaryotic, have a very simple membrane system which, for their photosynthetic cells, only includes the cytoplasmic and thylakoid membranes. Murata and coworkers have studied cyanobacteria such as *Anabaena variabilis* and *Synechocystis* in some detail. Earlier experiments (Mikami and Murata 2003) showed that cold induced an increase in fatty acid unsaturation mediated by the expression of genes for the desaturases concerned (Los et al. 1997). Furthermore, careful experiments where the fluidity of cyanobacterial membranes was manipulated established a clear relationship between membrane lipid order and expression of fatty acid desaturases (Vigh et al. 1993). In turn, it was also shown that disruption of desaturase genes and, hence, an inability to adjust membrane lipid unsaturation following cooling stress caused the transgenic organisms to be cold-sensitive (Murata and Wada 1995).

However, although membrane fluidity could, clearly, be a sensor for environmental cooling, the way in which this could impact on gene expression was not immediately obvious. Use of DNA microarrays identified 45 cold-inducible genes in *Synechocystis*. A complete analysis of the function importance of these genes has yet to be carried out (Mikami and Murata 2003). It was found that the cold sensor histidine kinase 33 regulated the expression of 28 of the above 45 genes (Mikami and Murata 2003). The mechanism of histidine kinase 33 activation has been examined, and cold stress suggested to result in dimerization and, hence, activation of proteins (Mikami et al. 2002). This activation could be mediated via two transmembrane domains that are sensitive to membrane fluidity (Mikami and Murata 2003). It is not surprising that a histidine kinase may be involved in transmitting the cold stress signal from the membrane to the gene expression machinery because such enzymes have a common role in mediating stress signals in prokaryotes. Indeed, it is of interest that another histidine kinase, *Des*K, was identified as a cold sensor that regulated the cold-inducible expression of a Δ5-desaturase in *Bacillus subtilis* (Aguilar et al. 2001).

The fatty acid desaturases in cyanobacteria work with complex lipids as substrates (Murata and Wada 1995). Upregulation of gene expression for the desaturases is likely to be a major factor in increasing production of unsaturated fatty acids at low temperatures. However, observations with the Δ12-desaturase from *Synechocystis* sp. PCC6803 suggest that its mRNA is much more stable at 22 °C compared to 34 °C [lifetimes of 20 and 2 min, respectively (Murata and Wada 1995)]. It should also be noted that low temperature induced transcription of only three of the four desaturase genes in *Synechocystis* and that *des*C was unaffected (Los et al. 1997).

5.7.3 Low-Temperature Effects on Lipids of Higher Plants

As in cyanobacteria, higher plants often respond to low-temperature stress by increasing membrane lipid unsaturation despite the generally high level of polyunsaturated fatty acids, especially in leaf tissues (Harwood 1994). However, certain molecular species are especially important, and a good correlation has been noted for many plants where chilling sensitivity is associated with significant amounts of 16:0/16:0- or 16:0/*trans*-16:1-phosphatidylglycerol species (Harwood 1998). Efforts to increase chilling tolerance in higher plants have been made either by the introduction of desaturase genes that can desaturate palmitate to palmitoleate (Gomboz and Murata 1998) or by modifying the synthetic pathway to phosphatidylglycerol where oleate is acylated (Murata et al. 1992). A number of subtleties in the response of different plants to low temperatures have also been revealed. There are several other ways organisms may change their membrane lipid composition. Two reports suggest that phospholipid class composition can be affected. Sugar bean roots responded to low temperatures by lowering the ratio of phosphatidylcholine/phosphatidylethanolamine (Lindberg et al. 2005). The amount of unsaturation in these phospholipids was also especially increased. Similarly, in *Brassica napus*, endoplasmic reticulum membranes of hypocotyls grown at low temperature (4 °C) were enriched in polyunsaturated fatty acids and phosphatidylethanolamine compared to 22 °C growth (Tasseva et al. 2004a). Part of these changes in phosphatidylcholine and phosphatidylethanolamine after cold stress may be related to the activity of phospholipases. In *Arabidopsis*, the most abundant phospholipase-Dα can lead to a decline in phosphatidylcholine rather than phosphatidylethanolamine on sublethal freezing where an increase in phosphatidic acid and lysophospholipids is seen (Welti et al. 2002). Indeed, a phospholipid-α-deficient *Arabidopsis* was prepared and showed improved freezing tolerance (associated with reduced reciprocal changes in phosphatidylcholine and phosphatidic acid).

Dyer et al. (2001) expressed the *Brassica napus FAD3* gene in yeast in order to better understand its regulation by low temperature. Although a shift in growth temperature from 30 to 20 °C was associated with an 8.5-fold increase in desaturase protein, there was no appreciable change in transcriptional activity (Dyer et al. 2001). This suggested posttranscriptional regulation as a main mechanism of control and agreed with previous data that there was no apparent increase in the rate of transcription or stability of *Fad* mRNA at lower temperatures (Heppard et al. 1996; Okuley et al. 1994) with the exception of the *Arabidopsis FAD8* gene (Gibson et al. 1994). The increase in *FAD3* desaturase protein observed in the yeast studies (Dyer et al. 2001) may be due to a reduction in proteinase degradation imparted through a reduction in membrane fluidity.

5.7.4 Low-Temperature Stress in Field and Forage Production

Field crops may be subject to low-temperature stress throughout their growing cycle, but particularly in spring and again in late summer when natural temperature variations will expose plants to conditions well below their temperature optima. Decreased growth rates with decreasing temperature may also be associated with differential activity of phosphoenolpyruvate carboxylase. This enzyme has no association with membranes or lipids, yet it has markedly lower activity at low temperatures than at high, and the level of activity is considerably reduced in chilling-susceptible as compared to chilling-tolerant crop species (Andrews 1987). Plants of a number of crop species may be protected against chilling injury by exposure to abscisic acid, applied before the low-temperature stress. Photosynthesis supports all plant growth, and it is well recognized that the process is highly temperature-sensitive, varying considerably between species and between groups of species from different ecological habitats (Berry and Bjorkrnan 1980; Ooquist 1983). In crop plants, there is considerable variability; corn and soybeans have higher temperature optima than wheat and barley (Woolhouse 1981). At 2 °C, a significant part of the seedling population forms no chloroplasts, and those that do have few thylakoids. Chilling susceptibility is directly related to the level of saturation of the fatty acids of this component (Murata et al. 1982). A number of workers (Woolhouse 1981; Miedema 1982) have emphasized that in normally adapted plants, it is leaf and stem growth that is most reduced by low temperature, and photosynthesis is less limited. The excess production of carbohydrate over that consumed during growth at low temperature is accumulated as long-chain polysaccharides of various types.

In field conditions, overwintering crop plants rarely survive and survival is dependent, or partially so, on a buffer of snow, which insulates the plants from natural air temperatures and allows them to benefit from the upward flux of heat from the soil. When this cover is reduced or destroyed, crop plants generally die. If the snow cover is increased as in the case of snow catching stubble management (Fowler 1983), the survival of crops will generally be increased. The effect of severe low temperatures on cold hardened tissues of plants is caused by lack of snow cover or removal of snow cover by wind, involve the transfer of intracellular water to extracellular ice by movement along the gradient of water vapor pressure. Ice forms between the cells in large relatively harmless glaciers. Cold hardening of plants before such severe low-temperature exposure increases tolerance to the stress.

5.7.5 How Low Temperature Is Perceived?

Rigidification of membrane lipids has been thought to be a general mechanism which could provide a simple way in which homeostatic regulation of membrane properties could be achieved (Guschina and Harwood 2006). Review provided by Vigh et al. (1998), the evidence for this is quite persuasive. However, different observations suggest that a very simple relationship with decreased membrane fluidity causing increased desaturase expression/activity is not tenable. Not all desaturases are induced by low temperature. In addition, a complication is that many membranes contain specific domains (such as lipid rafts) and this increases the complexity of any response to temperature. These aspects are reviewed at some length by Vigh et al. (2005). Moreover, the mediation of cold sensor proteins between a membrane environment and the expression machinery gives an opportunity for variable responses in different organisms and for different genes. In the same way as transcriptional changes to low temperatures have been examined in cyanobacteria by the use of arrays, large numbers of cold-inducible genes have been identified in plants (Seki et al. 2001; Fowler and Thomashow 2002). This information has led to suggestions as to how transduction of cold signals could be achieved (Mikami and Murata 2003). The possible involvement of the histidine kinase, *Hik33*, is in the low-temperature response of cyanobacteria. Detailed studies suggest that the two transmembrane domains of *Hik33* perceive a decrease in membrane fluidity and this allows its activation (through dimerization) (Mikami and Murata 2003). In turn, this would lead to differential expression of a number of genes. Alternatively, membrane fluidity could lead to alterations in stability of key enzymes such as membrane-associated *Fads*. Thus, the proposal of Dyer et al. (2001) was based on the observation that the stearoyl-CoA Δ9-desaturase in animals contains a 33-amino-acid segment at the N-terminus which facilitates proteolytic degradation. However, the protease responsible for the desaturase's turnover is also associated with the endoplasmic reticulum. Thus, any change in physical characteristics of a poikilotherm membrane might alter accessibility for proteinase attack and, hence, stabilize desaturase levels (Dyer et al. 2001). The local environment of the desaturase in the membrane could, of course, be affected not only by temperature but, if lipid rafts are present, then by other effectors also (Vigh et al. 2005).

Low temperature-induced gene expression can be described as transient or long term. Transient expression is generally considered up to 5 h of cold exposure, whereas long-term expression genes are transcribed and remain activated during the entire cold stress period. Transcriptome analyses in *Arabidopsis* and other economically significant species, such as rice and wheat, have yielded a multitude of potential candidate genes participating in the low-temperature stress response; however, some of their functions are still unknown (Kilian et al. 2007; Winfield et al. 2010; Zhang et al. 2012a). Extended analyses within the *Arabidopsis functional genomics* network project have allowed the detection of 24,000 protein-encoding genes via the Affymetrix ATH1 gene chip, enabling the identification of more than 2000 cold response genes (1335 upregulated and 1061 downregulated genes), of which more than 170 genes encode cold-induced transcription factors (Kilian et al. 2007). Although there are several families of stress-induced transcription factors, a significant portion of the cold-induced genes are activated by the abscisic acid-independent CBF/DREB1 (C-REPEAT-BINDING FACTOR/DEHYDRATION RESPONSIVE ELEMENT-BINDING FACTOR 1) regulon in *Arabidopsis* (Yamaguchi-Shinozaki and Shinozaki 1994; Vogel et al. 2005). The regulation can be up- or downstream of transcription, and it is activated by low temperatures through a complex network controlling the expression of many genes. Transcriptional regulation during the cold response of plants has been extensively reviewed (Chinnusamy et al. 2007, 2010; Thomashow 2010).

5.8 Cold Stress: Symptoms, Sensing, and Strategy

Plants are poikilothermic and sessile, as such they are highly influenced by the surrounding environment. Of the many abiotic challenges faced by plants, low temperature often defines the natural distribution of plant species and limits the survival range of many agricultural crops. Cold stress reduces plant growth and development, altering the physical and chemical composition of cell membranes, and causes electrolyte leakage, decreased protoplasmic streaming, and changes in cell metabolism (Thomashow 1999). Further responses to cold include changes in nucleic acid and protein synthesis, enzyme conformation, affinity, water and nutrient balance, and impairment of photosynthesis, specifically the downregulation and photodamage of photosystem II (PSII) (Chinnusamy et al. 2007). Despite the deleterious effects of suboptimal temperatures, plants can endure low nonfreezing temperatures, which increase resilience to stress and extend the temperature range in a process called cold acclimation (Thomashow 1999; Chinnusamy et al. 2010). Our understanding of how plants respond to cold temperatures has come through investigations of the regulatory network underlying cold acclimation. The process is complex and system-wide, incorporating turnover at the transcription, proteomic, and metabolic levels (Janska et al. 2010; Degenkolbe et al. 2012; Barrero-Gil and Salinas 2013). The changes in protein and lipid membrane composition help to restore metabolite homeostasis and are considered a mechanism by which cells sense cold temperatures (Takahashi et al. 2013). The fluid state of the plasma membrane is a structural and functional asset for its metabolic and physical role. The plasma membrane transitions from a fluid state at high temperature to a rigid gel phase when low temperatures are present. Low temperature-induced changes in the physical conformation of membranes are mainly due to the increased level of unsaturated lipids enhancing membrane fluidity and stabilization allowing cells to mechanically adapt to cold (Miquel et al. 1993; Chen and Thelen 2013). Sugars, nitrogen compounds, and proteins produced during cold acclimation have an active role in stabilizing both membrane lipids and proteins, conserving membrane integrity (Janska et al. 2010). Other sensing mechanisms, closely related to lipid turnover, but not fully understood, involve Ca^{2+} channels, kinases, and G-protein-associated receptors. Cold temperatures typically trigger a signaling cascade which includes $Ca^{2\pm}$ flux and $Ca^{2\pm}$-dependent protein kinase, mitogen-activated protein kinase-mediated cascades, and the consequent generation of lipid signaling molecules, e.g., phosphatidic acid, diacylglycerol, and inositol phosphate, which collectively contribute to responsive gene expression (Janska et al. 2010; Xiong et al. 2002; Hou et al. 2016).

Although there are a number of variables affecting the severity and time course of chilling injury (Kratsch and Wise 2000), general ultrastructural symptoms are largely similar across species (Table 5.4). They include swelling and disorganization of both chloroplasts and mitochondria, reduced size and number of starch granules, dilation of thylakoids and unstacking of grana, formation of small vesicles of chloroplast peripheral reticulum, lipid droplet accumulation in chloroplasts, and condensation of chromatin in the nucleus (Nessler and Wernsman 1980; Murphy and Wilson 1981; Leddet and Geneves 1982; Wise et al. 1983; Musser et al. 1984; Gemel et al. 1986; Ma et al. 1990; Ishikawa 1996; Yun et al. 1996). Cold hardening serves to protect the plant from subsequent seasonal freezing and is not generally considered to be injurious to the plant. It must also be noted that the inherent sensitivity of the plant to chilling affects the time course, extent, and symptomology of injury; therefore, wherever possible, species discussed in this review will be identified as well as their chilling sensitivity. Some plants, by nature, are more resistant to chilling stress than others. Generally speaking, the more sensitive a plant is to chilling, the sooner and more extensive are the ultrastructural changes (Nessler and Wernsman 1980; Wise and Naylor 1987; Ma et al. 1990). Some highly resistant plants may show no discernible effects from chilling, even for long periods of time, unless another stress factor is simultaneously imposed. Wise et al. (1983) with three species of differing chilling sensitivities to various combinations of chilling, irradiance, and water stress observed that the most chilling-sensitive species (*Gossypium hirsutum* L.) suffered damage within 24 h, with the response being total disruption of the chloroplast envelope and cell plasmolysis. A less chilling-sensitive plant (*Phaseolus vulgaris* L.) showed a delayed response (within 144 h) characterized by chloroplast swelling, thylakoid dilation, randomly tilted grana stacks, and accumulation of lipid droplets in the stroma. The most chilling-resistant of the three species (*Brassica oleracea* L.) showed no signs of chilling injury until exposed to prolonged chilling, light, and water stress simultaneously. Even then, plastids remained intact.

Table 5.4 Factors and symptoms observed in ultrastructural studies of chilling stress

Aspect of chilling stress	Symptom/observation	References
Chilling in dark	Starch is depleted. Extent of chloroplast swelling	Wise et al. (1983)
Chilling in light	Starch depletion, chloroplast swelling, formation of peripheral reticulum, and intraleaf light intensity gradient noted	Murphy and Wilson (1981); Musser et al. (1984); Terashima and Inoue (1985)
Duration of chilling	First stage: Distorted thylakoid membrane, decrease in starch, formation of peripheral reticulum	Van Hasselt (1974); Wise et al. (1983); Wu et al. (1997)
	Second stage: Disintegration of chloroplast envelop, lipid droplet accumulation	
	Third stage: Unstacking of grana, dilation of thylakoid intraspaces	
Acclimation to chilling	Smaller grana stacks, free sterol ratio in leaf cell membranes all change	Garber and Steponkus (1976); Huner et al. (1993); O'Neil et al. (1981)
Recovery from chilling	Ability to recover is species dependent	Murphy and Wilson (1981)
Inherent chilling sensitivity	Mitochondrial swelling, plasmolysis	Ishikawa (1996); Murphy and Wilson (1981); Nessler and Wernsman (1980); Wise and Naylor (1987); Wise et al. (1983)

The accumulation of abscisic acid in response to abiotic stress mediates signal transduction and gene expression. The role of abscisic acid in the cold stress response results from its function as a plant hormone, regulating growth, sugar content, leaf senescence, seed dormancy and germination, and cell division and elongation (Xue-Xuan et al. 2010). One contribution of abscisic acid to cold acclimation is its role in promoting phospholipid [inositol 1,4,5-triphosphate metabolism] (Lee et al. 1996) and the generation of second messenger (Xiong et al. 2002; Sanchez and Chua 2001). However, in the low-temperature response pathway, abscisic acid is only transiently accumulated at low levels (Xue-Xuan et al. 2010; Lang et al. 1994). Accordingly, changes in plant gene expression response to chilling are triggered by both abscisic acid-dependent and abscisic acid-independent mechanisms (Thomashow 1999; Xue-Xuan et al. 2010).

Membrane fluidity is susceptible to multiple stresses, especially extreme temperatures. However, both cold-induced rigidification and heat-induced fluidization are detrimental to membrane function, resulting in protein deactivation, electrolyte leakage, and even cytoskeleton destabilization (Sangwan et al. 2002; Liu et al. 2013b; He et al. 2018). In fact, being a thermodynamic property, membrane fluidity may serve as a sensor in temperature signaling (Ruelland and Zachowski 2010). Interestingly, the effects of cold and heat can be mimicked at 25 °C by dimethylsulfoxide and benzyl alcohol, respectively. In alfalfa (*Medicago sativa*), the two chemicals induced opposite changes in membrane fluidity coupled with cytoskeleton disassembly, which triggered Ca^{2+} signaling to activate two distinct mitogen-activated protein kinase cascades, leading to corresponding stress response (Sangwan et al. 2002).

5.8.1 Lipid Profiling Under Cold Stress

Changes in membrane lipid composition play multiple roles in plant adaptation and survival in the face of chilling and freezing damage (Wang et al. 2006b). Membrane lipids undergo substantial alterations when plants are exposed to low temperatures. The degree of fatty acid unsaturation and the content of phospholipids increase during cold acclimation. Such lipid changes enhance membrane fluidity and reduce the propensity of cellular membranes to undergo freezing-induced non-bilayer phase formation, thus enhancing membrane integrity and cellular functions (Uemura et al. 1995b). In addition, membrane lipids are rich sources for generating lipid mediators. Lipid mediators, such as phosphatidic acid, lysophospholipids, and free fatty acids, are involved in various cellular processes, including signal transduction, vesicular trafficking, and/or cytoskeletal reorganization, in plant stress responses (Wang 2004). Thus, membrane lipid changes play both structural and regulatory roles in plant adaptation and survival when subjected to low-temperature damage. Information on how lipid species change and how the alterations are generated will help us understand the functions of lipids and membranes in plant response to temperature stresses.

Lipid profiling has been applied to analyze changes in lipid molecular species of *Arabidopsis* under low-temperature stresses. Significant changes in membrane lipid molecular species occurred when *Arabidopsis* plants were cold-acclimated at 4 °C (Welti et al. 2002). Phosphatidylcholine, phosphatidylethanolamine, phosphatidylglycerol, monogalactosyldiacylglycerol, and digalactosyldiacylglycerol species that contain two polyunsaturated acyl species, such as 36:5- (18:3-18:2-) and 36:6- (di18:3-) phosphatidylcholine, 36:5- and 36:6-phosphoethanolamine, 34:6- (18:3-16:3) monogalactosyldiacylglycerol, and 36:6-digalactosyldiacylglycerol, increased, while the levels of more saturated species in these lipid classes, such as 36:2- (in which 18:0-18:2- combinations are more common than 18:1-18:1-) and 36:3- (in which 18:1-18:2- combinations are more common than 18:0-18:3-) phosphatidylcholine, decreased. By contrast, phosphatidylinositol molecular species levels were unchanged. Cold acclimation also induced significant increases in species of lipid metabolites lysophosphatidylcholine, lysophosphatidylethanolamine, and phosphatidic acid, suggesting that phospholipases, such as phospholipase-A and phospholipase-D, are activated during cold acclimation.

Freezing induced a decrease in total membrane lipid content, and the decrease resulted primarily from the loss of phosphatidylcholine, phosphatidylethanolamine, and phosphatidylglycerol (Welti et al. 2002). At −8 °C, a sublethal freezing temperature for cold-acclimated *Arabidopsis*, the amount of almost all molecular species of phosphatidylglycerol, phosphatidylcholine, and phosphatidylethanolamine decreased, with the 36:4 and 36:5 species contributing the most to the decline in phosphatidylethanolamine and phosphatidylcholine levels. The levels of lipid metabolites phosphatidic acid, lysophosphatidylcholine, and lysophosphatidylethanolamine increased dramatically. As during cold acclimation, the change in phosphatidylinositol molecular species was minimal, and some phosphatidylinositol species actually increased, monogalactosyldiacylglycerol levels tended to decline, but no net loss of digalactosyldiacylglycerol occurred in *Arabidopsis* exposed to −8 °C.

In plants, several lipolytic enzymatic activities have been described; these include phospholipase-D, phospholipase-A, phospholipase-C, nonspecific acyl hydrolase, and galactolipases. Phospholipase-D and phospholipase-C have been implicated in plant response to low temperatures (Ruelland et al. 2002; Welti et al. 2002; Li et al. 2004). The large decline in major membrane phospholipids, but not galactolipids, suggests that at the sublethal freezing temperature, phospholipases are activated to a greater extent than galactolipases. The low level of the phospholipase products, lysophospholipids, in comparison with the phospholipase-D product, phosphatidic acid, indicates that phospholipase-D is more active than phospholipase in plant tissues. In addition, comparison of the molecular species profile of phosphatidic acid with other lipid classes helps identify the substrates that give rise to phosphatidic acid. Upon freezing, most of the phosphatidic acid species that increased have the same acyl chains as the phosphatidylcholine species that decreased. These data, together with the analysis of phospholipase-Dα1-deficient mutant, lead to the conclusion that freezing-induced phosphatidic acid is derived primarily from phosphatidylcholine. However, upon freezing, plants gained one phosphatidic acid molecular species, 34:6 phosphatidic acid, that was absent from the plants before the freezing treatment. Moreover, 34:6 phosphatidic acid has the acyl composition, 18:3-16:3, which is found otherwise only in galactolipids, particularly in monogalactosyldiacylglycerol. This suggests that some phosphatidic acid is formed by phosphorylation of diacylglycerol produced by hydrolysis of monogalactosyldiacylglycerol.

5.8.2 Role of Unsaturated Fatty Acids to Low-Temperature Stress

Abiotic stressors include temperature (low or high), water (deficit or excess), ultraviolet, salt, heavy metals, and so on. High salinity, for example, can induce ionic toxicity, osmotic pressure, oxidative damage, and nutrient deficiency (Zhao et al. 2010; Feng et al. 2014a, b), thereby strongly suppressing the whole lifecycle of most plants (Guo et al. 2012a, 2015, 2018a). More seriously, in the context of climate change and soil salinization, cold/heat, drought, and salt usually arise together, posing an increasingly severe threat to plant subsistence and cultivation worldwide (Yuan et al. 2015; He et al. 2018). Unsaturated fatty acids are coming into the limelight as one of the general defense systems against various biotic and abiotic stresses (He et al. 2018). In higher plants, the most common unsaturated fatty acids are three 18-carbon (C18) species: 18:1 (oleate), 18:2 (linoleate), and 18:3

(α-linolenate), where m:n denotes a fatty acid with m carbon atoms and n *cis*-double bonds (He and Ding 2020; He et al. 2020). Unsaturated fatty acids have been implicated in stress defense as follows:

- Ingredients and modulators of cellular membranes in glycerolipids
- Reserve of carbon and energy in triacylglycerol
- Stocks of extracellular barrier constituents
- Precursors of various bioactive molecules
- Regulators of stress signaling

There are two pathways for membrane glycerolipid biogenesis: the "prokaryotic" one in plastids and the "eukaryotic" one in the endoplasmic reticulum. Both are based on sequential acylation of glycerol-3-phosphate, with phosphatidic acid and diacylglycerol being the intermediates. The first reaction is catalyzed by glycerol-3-phosphate acyltransferase, rendering it a key enzyme. With the emergence of substrate glycerolipids, C18 polyunsaturated fatty acids are then generated by the ω-6 and ω-3 pair of fatty acid desaturases (FADs), i.e., plastidial FAD6-FAD7/FAD8 or endoplasmic reticulum FAD2-FAD3 (Ohlrogge and Browse 1995; He et al. 2020).

In the eukaryotic pathway, phosphatidylcholine is the vector for fatty acid desaturation. However, nascent 18:1 exported out of plastids appears to prefer the acyl-editing way to be directly incorporated into phosphatidylcholine rather than the multistep de novo way (He et al. 2020). The exchanged acyl chains (e.g., 18:2) can then enter other metabolic fates, for example, the production of triacylglycerol prevailing in seed cells or cutin unlocked in epidermal cells. Of note, unlike phosphatidylcholine and triacylglycerol that are directly converted from diacylglycerol, cutin or suberin is assembled via monoacylglycerol (Beisson et al. 2012).

Further, C18 unsaturated fatty acids, esterified or released, can be modified into a great variety of bioactive molecules, such as jasmonic acid, azelaic acid, and 2-hydroxy-octadecatrienoic acid from oxidation. Nevertheless, when incorporated into triacylglycerol, their major fate is to retrograde into acetyl-CoA via β-oxidation in peroxisomes. Notably, fatty acids removed from membrane lipids for turnover will be deposited in triacylglycerol first (Fan et al. 2017).

5.8.3 Lipid Metabolism of Horticultural Products

Storage of horticultural products at low temperature is advantageous to preventing softening and prolonging the postharvest life; however, cold storage triggers the pedicle pitting and pericarp and pulp adhesion in postharvest, which also is accompanied with abnormal changes of stress-related enzymes, destruction of outer cell wall structure, and decrease of fiber filaments (Zhou et al. 2014a, b). The development of this cold injury reduces consumer acceptance of the fruit, thus limiting its storage life. And it has been confirmed that cold stress accelerated the energy consumption and membrane lipid peroxidation and affected normal active oxygen metabolism in blueberry. Chilling injury is one of the main problems affecting the market value of horticultural fruits and vegetables, which is also one of the principal abiotic stresses. Cold stress at low temperatures above zero (0–10 °C) and freezing stress at subzero temperatures lead to decreased cell membrane fluidity, water potential, and osmotic stress causing irreversible damage. In previous research, cold stress (4 °C) has been found to damage the cellular structure of peppers (*Capsicum annuum* L.) (Ge et al. 2019; Kong et al. 2018a; Purvis 2002), and the most typical cold injury symptom in peppers is water immersion sunken spots (Lim et al. 2007; Wang et al. 2016b); in banana, the peel would pit and discolor at low temperatures (Chen et al. 2008a); in peach, cold injury symptoms include lack of juiciness (Lurie and Crisosto 2005), poor flavor, and the brown-red discoloration of flesh (Crisosto et al. 1999); in mango, cold injury symptoms include scalding, softening, internal browning, and electrolyte leakage (Rámy et al. 2018); in litchi, fruit quality, membrane permeability, enzyme activities, and energy changed during cold storage (Liu et al. 2011); in pear, the aroma was less under or after long-term storage at 0 °C, seriously affecting its quality (Shi et al. 2018a, b; Zhou et al. 2014c); in harvested cucumber (*Cucumis sativus* L.) fruit, disorder characterized by surface pitting and dark watery patches was confirmed when held at 7 °C or below (Chen and Yang 2013). In conclusion, it has been identified when the fruits are subjected to cold stress, the cell structure changed, the membranes disrupted, the malondialdehyde content increased, and the reactive oxygen accumulated (Kratsch and Wise 2000; Singh et al. 2013), finally leading to the decline of fruit quality. Although the cold injury symptoms in different fruits and vegetables are distinct, the nutritional and economic values are adversely affected. And the mechanisms involved in the cold injury of blueberry are still not clear. Therefore, there also remains a need to study the chilling mechanism and develop more effective techniques for blueberry fruit storage and transport at low temperature (Zhang et al. 2020).

Horticultural products of tropical or subtropical origin are susceptible to chilling injury when exposed to chilling temperatures (Wang 1994). Cold injury is a physiological disorder, the symptoms of which involve browning, pitting, or abnormal ripening, thereby affecting negatively the horticultural products' commercial quality and storage life (Chen et al. 2008b). For example, when exposed to temperatures less than 13 °C, banana fruits are subjected to cold injury, displaying pitting on peel surface, disappearance of aroma, and abnormal ripening (Guo et al. 2018b). In the case of papaya, improper low temperatures can lead to water-soaked flesh and scalding skin (Zou et al. 2014). Cucumber is highly susceptible to cold injury, mainly developing into dehydration and large sunken areas on the surface (Wang et al. 2018). Therefore, chilling damage has a negative effect on the quality of horticultural products and thus minimizing their commercial profits. It is generally accepted that damage to the membrane is the primary event in chilling injury. The cell membrane acts as a barrier to protect the cell from injury and perturbation, providing a constant environment for intracellular bioprocesses. Once the membranes disintegrate under cold stress, multiple metabolic processes are affected, which then causes ion leakage, insufficient energy, and overproduction of reactive oxygen species, thereby resulting in membrane disruption and cell death (Patel et al. 2016). For example, ultrastructure analysis indicated that membrane integrity and function were destroyed in chilled blueberries, leading to cell death and finally pitting surface (Wang et al. 2019). The composition of the membrane lipids manifests themselves in different forms in their physical state at low temperature, and high levels of unsaturated fatty acids are beneficial to sustaining normal function of the membrane (Mendoza 2014). When subjected to nonfreezing low temperature, cellular membrane in chilling-sensitive plants undergoes a rapid physical change from a flexible liquid crystalline to a solid-gel structure. Because of the altered physical state, the membrane fails to keep a constant milieu for biochemical process, which leads to cellular metabolism disorders and appearance of cold injury (Liang et al. 2020). Early studies have reported the lipid changes in mitochondrial membranes isolated from chilling-injured fruits, but none of the mitochondrial membranes showed phase transitions from chilling-resistant plants (Raison et al. 1971). Also, mitochondrial membranes from different varieties of apple were approved to experience lipid phase transitions at 30–10 °C (McGlasson and Raison 1973). Avocado fruits stored in cold temperatures display particle-depleted regions in chilling-injured fruits, but no change occurs when they are kept at 20 °C.

Whether the temperature causes the membrane to undergo a phase change or phase separation is still unclear, and any significant observations of the temperature-induced membrane physical state are yet shorted. Nonetheless, numerous studies implicated that the plasma membrane is the primary site of cold injury (Lyons 1973; Platt-Aloia and Thomson 1987; Parkin et al. 1989; Marangoni et al. 1996). The composition of the membrane, especially the type of lipids, plays a determining effect on its physical characteristics and stability. Many studies have found that compared with the cold-sensitive plants, the percentage of unsaturated fatty acids in cold-resistant plants is higher. For example, "Qingzhong" fruits (a cold-resistant loquat cultivar) have a higher content of unsaturated fatty acids than those in "Fuyang" fruit (a cold-sensitive loquat cultivar) (Pareek et al. 2014). Intermittent warming attenuated the cold injury of peppers largely because of efficiently increasing the unsaturated fatty acids' content (Liu et al. 2015b). This might be attributed to the easier solidification of saturated fatty acids than of the unsaturated fatty acids at low temperature, which can enhance membrane fluidity at a critical temperature. It seems possible that the more the contents of unsaturated fatty acids the membrane contains, the more flexible the membrane is exposed to cold stress. Reduction of the levels of phosphatidic acid and diacylglycerol may contribute to the enhanced cold injury tolerance of the mutant plants (Chen et al. 2015d; Tan et al. 2018). Likewise, monogalactosyldiacylglycerol and phosphatidylcholine were sharply decreased during the cold injury development in pepper (Kong et al. 2018a). Together, these results suggest the high sensitivity of lipids in response to cold temperature. In addition to lipids, other membrane compositions such as membrane proteins are also thought to make an effect on its physical inflection (Yoshida et al. 1986). Overall, alterations of plasma membrane composition play a decisive role in its flexibility and function and hence chilling tolerance.

5.8.4 Characterization of Cold Stress Responses

Cell membrane stiffness, protein complex instability, and photosynthesis weakness are detected under chilling stress, and much more serious injuries occur under freezing stress conditions. Many physiology and biochemistry processes can be influenced by cold stress, which is dependent on the plant species, stage of plant development, and length of the stress period. Many plant species, including *Brassica napus* L., have the ability to acclimate to the cold. However, this ability can be increased after exposure to chilling stress, and the capacity of acclimation varies greatly among species (Jian et al. 2020). Cold acclimation mechanisms have been fully explored and have expanded our understanding of *Arabidopsis thaliana.* Numerous cold-regulated genes have been identified in diverse species, including *Arabidopsis thaliana* (Zhao et al. 2009), *Oryza sativa* (Chen et al. 2003; Maruyama et al. 2014; He et al. 2016; Xiaochuang et al. 2017), *Zea mays* (Bilska-Kos et al. 2016; Riva-Roveda et al. 2016), *Prunus persica* (Sanhueza et al. 2015; Bustamante et al. 2016), and many other species (Khodakovskaya et al. 2006; Kosova et al. 2012; Tian et al. 2013; Zhao et al. 2014; Liu et al. 2017), using RNA sequencing technology. All of these findings have expanded our understanding of molecular mechanisms in response to cold stress tolerance (Chinnusamy et al. 2007). Nevertheless, many cold stress phenotypes cannot be explained by the functions of a single gene, highlighting the need to identify additional regulatory networks and pathways, including key metabolites.

In addition to the very sophisticated gene regulatory networks, plants have evolved various complex metabolite response patterns in the low-temperature stress response (Kaplan et al. 2004). Increasing evidence has shown that gene expression and biochemical pathways are altered under cold stress conditions. Sugars, lipids, and amino acids, for example, have been widely reported to be involved in the cold stress response, not only for protein synthesis process but as precursors of other key metabolites in the response to cold stress (Cook et al. 2004). By integrating transcriptome and metabolome data, a broader and more comprehensive understanding of genes and metabolites' co-regulation of low-temperature stress in *Arabidopsis* has been performed (Kaplan et al. 2007). In rice, genes encoding proteins involved in sucrose-starch metabolism and the glyoxylate cycle were upregulated, and these changes were correlated with the accumulation of glucose, fructose, and sucrose in rice after exposure to cold or dehydration, as revealed by joint metabolite, phytohormone, and gene transcript analyses (Maruyama et al. 2014). In *Arabidopsis thaliana*, flavonoids are regarded as determinants of freezing tolerance and cold acclimation (Schulz et al. 2016). Carboxylates, amino acids, carbohydrates, and sugar alcohols were also made contributing to reduce freezing tolerance in the pgm mutant (Hoermiller et al. 2017).

5.9 Metabolic Changes During Cold Stress and Cold Acclimation

When exposed to low temperatures, plants respond with changes in their pattern of gene expression and protein products. Different studies have indicated that the membrane systems of the cell are the primary site of freezing injury in plants (Levitt 1980; Steponkus 1984). In addition, it is well established that freeze-induced membrane damage results primarily from the severe dehydration associated with freezing (Steponkus 1984; Steponkus et al. 1993). Many species of tropical or subtropical origin are injured or killed by nonfreezing low temperatures and exhibit various symptoms of chilling injury such as chlorosis, necrosis, or growth retardation. In contrast, chilling-tolerant species are able to grow at such cold temperatures.

Multiple forms of membrane damage can occur as a consequence of freeze-induced cellular dehydration including expansion-induced lysis, lamellar-to-hexagonal II phase transitions, and fracture jump lesions (Steponkus et al. 1993). Thus, a key function of cold acclimation should be to stabilize membranes against freezing injury. Indeed, cold acclimation prevents expansion-induced lyses and the formation of hexagonal II phase lipids in rye and other plants (Steponkus et al. 1993). This ability to adapt has an impact on the distribution and survival of the plant and on crop yields. Multiple mechanisms appear to be involved in this stabilization. The best documented are changes in lipid composition (Steponkus et al. 1993). Secondly, temperature-induced change in membrane fluidity is another consequence in plants during temperature stresses and might represent a potential site of perception and/or injury (Horvath et al. 1998; Orvar et al. 2000). Adaptation of living cells to chilling temperatures is a function of alteration in the membrane lipid composition by increased fatty acid unsaturation. Genetically engineered tobacco plants overexpressing chloroplast glycerol-3-phosphate acyltransferase (GPAT) gene (involved in phosphatidyl glycerol fatty acid desaturation) from squash (*Cucurbita maxima*) and *Arabidopsis thaliana* showed an increase in the number of unsaturated fatty acids and a corresponding decrease in the chilling sensitivity. At low temperature, greater membrane lipid unsaturation appears to be crucial for optimum membrane function. An *Arabidopsis* fatty acid biosynthesis (*FAB1*) mutant with more saturated membranes showed decreased quantum efficiency of photosystem II (PSII), chlorophyll content, and the amount of chloroplast glycerolipids after prolonged exposure to low temperature (Wu et al. 1997). A triple mutant fatty acid desaturation (*fad3-2 fad7-2 fad8*) devoid of trienoic fatty acids (18:3 or 16:3) produced a phenotype similar to *FAB1*, when plants were subjected to prolonged low-temperature exposure (Routaboul et al. 2000). Similarly, *fad5* and *fad6* mutants with more saturated membranes became chlorotic and showed growth retardation during low-temperature incubation (Hugly and Somerville 1992). In addition to membrane unsaturation, it appears that lipid asymmetry in the membrane also contributes to membrane physical structure at low temperature (Gomes et al. 2000). The accumulation of sucrose and other simple sugars that typically occurs with cold acclimation also seems likely to contribute to the stabilization of membranes as these molecules can protect membranes against freeze-induced damage in vitro (Strauss and Hauser 1986; Anchordoguy et al. 1987).

The acquisition of freezing tolerance is called cold acclimation, which is generally initiated after a period of decreases in above 0 °C temperatures and photoperiod (Levitt 1980). Growth inhibition resulting from low temperature reduced the capacity for energy utilization that leads to feedback inhibition of photosynthesis (Huang et al. 2014). Differences in the capacity to minimize photoinhibition and recover photosynthesis during cold acclimation have been shown to contribute to intra- and interspecific differences in freezing tolerance (Pocock et al. 2001). Carbohydrates are synthesized from a complex series of reactions involving photosynthesis in plants (Huang et al. 2014). Changes in the three major types of water-soluble carbohydrates, monosaccharides (glucose and fructose), disaccharides (sucrose, trehalose), and fructans, have often been reported in plants subjected to low temperature (Theocharis et al. 2012a). Those sugars may serve as a typical osmoprotectant, cellular membrane stabilizer, scavengers of reactive oxygen species, and signaling molecules in plants to tolerant chilling and freezing stress (Theocharis et al. 2012a). Many studies have shown that higher accumulation of sugars or individual sugar fractions such as starch (Zhang et al. 2006), fructans (Dionne et al. 2001), and sucrose (Zhang et al. 2006) during cold acclimation is associated with great cold tolerance in turfgrass (Cai et al. 2004; Hoffman et al. 2010). Sucrose consists of the monosaccharides glucose and fructose and is intimately involved in growth, development, storage, signaling, and stress acclimation (Janska et al. 2010). Leaf starch potentially contributed carbon to carbohydrate accumulation in response to low temperature (Sicher 2011). Fructans often accumulated during cold acclimation in cereals and grasses, which serve to stabilize cell membranes (Livingston III and Heyer 2009). Furthermore, trehalose, a nonreducing disaccharide of glucose, has been implicated in modulating levels of freezing tolerance and may also be involved in starch accumulation (Theocharis et al. 2012b).

Plants have the ability to acclimate to changing environmental conditions. However, long-term climate change driven by rising temperatures may have a deleterious impact on plant physiology and in turn negative effects on crop yields. Upon abiotic stress, such as high or low temperature and high light, the homeostasis of reactive oxygen species metabolism in the plant cell is challenged (Lichtenthaler and Burkart 1999; Apel and Hirt 2004; Zhang et al. 2010). A protective system in part based on lipid soluble molecules is in place to protect plant cells and the photosynthetic membranes against the action of reactive oxygen species (Spicher et al. 2016). Carotenoids, such as β-carotene, lutein, neoxanthin, and three xanthophyll cycle carotenoids, zeaxanthin, violaxanthin, and antheraxanthin, besides structural stabilization and light harvesting roles, play important photoprotective roles in scavenging of singlet oxygen species and quenching of chlorophyll triplet states and in excess energy dissipation by non-photochemical quenching that implicates xanthophylls and lutein (Choudhury and Behera 2001; Gruszecki and Strzalka 2005; Shumskaya and Wurtzel 2013). Another category of molecules, the prenylquinones (including tocopherols, plastoquinone, plastochromanol) are lipid soluble compounds that function as membrane-protective antioxidant molecules. Tocopherols have widely been described as antioxidants against high-light triggered oxidative stress (Havaux et al. 2005; Krieger-Liszkay and Trebst 2006; Kobayashi and DellaPenna 2008).

It has long been suspected and is now well accepted that temperature acclimation results from a complex process involving a number of physiological and biochemical changes, including changes in membrane structure and function, tissue water content, global gene expression, protein, lipid, and primary and secondary metabolite composition (Gilmour et al. 2000; Shinozaki and Dennis 2003). Advances in genome sequencing and global gene expression analysis techniques have further established the multigenic quality of environmental stress responses and the complex nature of temperature acclimation (Seki et al. 2001; Fowler and Thomashow 2002; Kreps et al. 2002). Literally hundreds of genes have been linked with environmental stress responses. By contrast, less is known about stress responses of plants at the metabolite and metabolome level (Cook et al. 2004; Rizhsky et al. 2004). Global metabolite profiling analysis holds the promise to permit simultaneous monitoring of precursors, intermediates, and products of metabolic pathways. It is a discovery tool that can detect and monitor unidentified mass spectral tags as well as identified metabolites that play important roles in metabolism and physiology and, in the context of this work, stress tolerance. Metabolite profiling analysis was performed using gas chromatography-mass spectrometry to determine similarities and differences in temporal metabolite responses and to identify novel compounds that exhibit temperature-specific responses during the induction of acquired thermotolerance in response to heat shock and during induction of acquired freezing tolerance in response to cold shock. Metabolite profiling has revealed that cold shock influenced metabolism more profoundly than heat shock. However, the majority of heat shock responses were shared with cold shock, uncovering a novel relationship between heat shock and cold shock responses not previously known (Kaplan et al. 2004).

A key aspect of plant adaptation to temperature stress concerns modulation of membrane fluidity (Murata and Los 1997; Iba 2002). Membrane fluidity is affected by two key aspects of lipid composition:

- The degree of unsaturation in the fatty acyl groups of glycerolipid molecules
- The relative proportions of various lipid classes in the lipid bilayer (Iba 2002; Murakami et al. 2000; Murata et al. 1992)

A comprehensive survey was conducted on glycerolipid metabolite as well as enzyme gene transcript changes under temperature stress in three plant species, *Arabidopsis thaliana* (*Arabidopsis*), *Atriplex lentiformis* (saltbush), and *Triticum aestivum* L. (wheat) (Li et al. 2015). Results show that glycerolipid pathway adjustment between the chloroplast and cytosolic compartments is fundamental to the modulation of membrane lipid attributes pertinent to temperature stress adaptation.

Dynamic changes in the membrane lipid composition of imbibing seeds are required to maintain the structural equilibrium of membranes necessary for proper seed germination. At low temperatures, imbibition of cold water by the seeds causes lipids with high-melting temperature (because of high saturation in the fatty acid chains) to solidify and become separated within the membrane. This results in the loss of fluidity and selective permeability that disrupts membrane reorganization, thereby exacerbating cytoplasmic leakage (Yu et al. 2015; Liu et al. 2002; Noblet et al. 2017; Płażek et al. 2018). Unless the structural integrity of the membranes is reestablished, the excessive solute leakage can lead to chilling injuries resulting in metabolic dysfunction, abnormal root tip development, and even embryo death. Remodeling cell membrane fluidity is an adaptive response of plants to cold stress and is achieved by increasing the unsaturated fatty acid composition of membrane lipids (Płażek et al. 2018; Upchurch 2008).

These modifications are mediated in part by the activities of fatty acid desaturases (Miquel and Browse 1994; Vega et al. 2004; Hernández et al. 2019). The *cis*-double bonds created by the desaturases prevent the fatty acid molecules from packing tightly together, thereby keeping the membranes fluid and allowing membrane-bound proteins to move through (Rikin et al. 1993). Under cold stress, membrane flexibility through increased lipid unsaturation allows the maintenance of a homeostatic environment that is integral to the functioning of the cell (Upchurch 2008). The importance of increasing the levels of polyunsaturated fatty acids, via the desaturation of fatty acids such as palmitic, stearic, and oleic acids, toward enhancing the ability of higher plants to adapt to decreases in temperature has been illustrated in several studies. For example, transgenic tobacco plants overexpressing an *Arabidopsis thaliana* ω-fatty acid desaturase gene exhibited lesser chilling injuries compared to the wild-type when subjected to 15 °C. The enhanced chilling tolerance of the transformed plants was attributed to the increased production of linolenic acid (18:3) in the leaves in response to cold (Kodama et al. 1995). In rice, overexpression of the Δ-12 fatty acid desaturase that catalyzes the conversion of oleic acid to linoleic acid resulted in the accumulation of the latter in the leaf tissues. The higher production of linoleic acid in the plant tissues enhanced the chilling tolerance of the transgenic plants at the vegetative and reproductive stages upon exposure to 10 °C for 8 days and to 20 °C for 2 days, respectively (Shi et al. 2012). Studies in olives also showed that exposure of fruits to a suboptimal temperature of 15 °C increased the expression of stearoyl-acyl carrier protein desaturase, concomitantly increasing the unsaturated fatty acid contents in the microsomal membrane lipids of the mesocarp of a cold-tolerant cultivar. Increase in both stearoyl-acyl carrier protein desaturase transcripts and unsaturated fatty acid levels in the cold-tolerant but not the cold-sensitive cultivar underscores the role of membrane unsaturation in the ability of the plants to adjust to low temperatures (Hernández et al. 2019). Aside from enhancing cold tolerance of plants in vivo, modulation of membrane lipid composition also provides an adaptive mechanism for seeds to germinate at low temperatures (Shi et al. 2012; Noblet et al. 2017). In corn, cold germination screening of two hybrids at 10 °C resulted in higher electrolyte leakage in the cold-sensitive than in the cold-tolerant hybrid. Phospholipidomic assays to profile changes in the lipid membrane composition of the two hybrids during seed imbibition at suboptimal temperature revealed a significant accumulation in the di-unsaturated linoleic acid (C18:2) fraction, accompanied by a substantial decrease in the monounsaturated oleic acid (C18:1) fraction in the cold-tolerant hybrid. Conversely, the proportion of the saturated stearic acid (C18:0) and oleic acid (C18:1) fractions increased, whereas linoleic acid (C18:2) content decreased in the seeds of the cold-sensitive hybrid (Noblet et al. 2017). Transgenic rice seeds engineered to overproduce linoleic acid also showed higher germination rate at 4 °C compared to the wild-type seeds. Biochemical assays revealed a significant reduction in the electrolyte leakage of the transgenic seeds germinated at 4 °C compared to the seeds of the untransformed plants germinated at the same suboptimal temperature (Shi et al. 2012). Transgenic canola seeds producing three times more stearic acid compared to conventional varieties because of the knockdown expression of stearoyl-acyl desaturase demonstrate in reverse the importance of lipid unsaturation in the ability of seeds to germinate under low temperature. An increase by 30% in the saturated stearic acid levels and a 40% increase in the proportion of the total saturated fatty acids of the transgenic seeds relative to the nontransgenic seeds resulted in poor germination even at an ambient temperature of 25 °C. The incorporation of higher levels of saturates in the membrane lipids is presumed to have contributed to the rigidity of the cell membrane during imbibition, ultimately resulting in deficient germination (Thompson and Li 1997). Extensive studies on cell membrane remodeling show that the incorporation of unsaturated fatty acids in the membrane lipids during cold water imbibition effectively increases cell membrane fluidity. During cold water imbibition, the more flexible membranes are expected to reorganize faster from the rigid, gel state (hexagonal configuration) to a crystalline-liquid state (lamellar configuration). This significantly minimizes the electrolyte leakage during imbibition and rapidly restores cellular function, thereby facilitating faster and higher seed germination even under cold conditions (Sanchez et al. 2019).

5.9.1 Role of Triacylglycerol in Plant Stress Response

At low temperatures, imbibition of cold water by the seeds causes lipids with high-melting temperature, because of high saturation in the fatty acid chains, to solidify and become separated within the membrane. This results in the loss of fluidity and selective permeability that disrupts membrane reorganization, thereby exacerbating cytoplasmic leakage (Kunst et al. 1988; Mendes et al. 2013; Higashi et al. 2018; Fan et al. 2013; Mueller et al. 2017). Unless the structural integrity of the membranes is reestablished, the excessive solute leakage can lead to chilling injuries resulting in metabolic dysfunction, abnormal root tip development, and even embryo death. Remodeling cell membrane fluidity is an adaptive response of plants to cold stress and is achieved by increasing the unsaturated fatty acid composition of membrane lipids (Mueller et al. 2017; Moellering et al. 2010). These modifications are mediated in part by the activities of fatty acid desaturases (Barnes et al. 2016; Tan et al. 2018; Guo et al. 2005). The *cis*-double bonds created by the desaturases prevent the fatty acid molecules from packing tightly together, thereby keeping the membranes fluid and allowing membrane-bound proteins to move through (Arisz et al. 2013). Under cold stress, membrane flexibility through increased lipid unsaturation allows the maintenance of a homeostatic environment that is integral to the functioning of the cell (Moellering et al. 2010). The importance of increasing the levels of polyunsaturated fatty acids, via the desaturation of fatty acids such as palmitic, stearic, and oleic acids, toward enhancing the ability of higher plants to adapt to decreases in temperature has been illustrated in several studies. For example, transgenic tobacco plants overexpressing an *Arabidopsis thaliana* ω-fatty acid desaturase gene (FAD7) exhibited lesser chilling injuries compared to the wild-type when subjected to 15 °C. The enhanced chilling tolerance of the transformed plants was attributed to the increased production of linolenic acid (18:3) in the leaves in response to cold (Zhang et al. 2009b). In rice, overexpression of the Δ-12 fatty acid desaturase that catalyzes the conversion of oleic acid to linoleic acid resulted in the accumulation of the latter in the leaf tissues. The higher production of linoleic acid in the plant tissues enhanced the chilling tolerance of the transgenic plants at the vegetative and reproductive stages upon exposure to 10 °C for 8 days and to 20 °C for 2 days, respectively (Baxter et al. 2014). Studies with olives also showed that exposure of fruits to a suboptimal temperature of 15 °C increased the expression of stearoyl-acyl carrier protein desaturase (SAD), concomitantly increasing the unsaturated fatty acid contents in the microsomal membrane lipids of the mesocarp of a cold-tolerant cultivar. Increase in both stearoyl-acyl carrier protein desaturase transcripts and unsaturated fatty acid levels in the cold-tolerant but not the cold-sensitive cultivar underscores the role of membrane unsaturation in the ability of the plants to adjust to low temperatures (Guo et al. 2005). Aside from enhancing cold tolerance of plants in vivo, modulation of membrane lipid composition also provides an adaptive mechanism for seeds to germinate at low temperatures (Fan et al. 2013; Baxter et al. 2014). In corn, cold germination screening of two hybrids at 10 °C resulted in higher electrolyte leakage in the cold-sensitive than in the cold-tolerant hybrid. Transgenic rice seeds engineered to overproduce linoleic acid also showed higher germination rate at 4 °C compared to the wild-type seeds. Biochemical assays revealed a significant reduction in the electrolyte leakage of the transgenic seeds germinated at 4 °C compared to the seeds of the untransformed plants germinated at the same suboptimal temperature (Baxter et al. 2014).

In plant cells, the cytosolic and chloroplast compartment each possesses its own set of glycerolipid assembly pathway and generates glycerolipid species of defined molecular characteristics (Browse and Somerville 1991; Ohlrogge and Browse 1995; Hölzl et al. 2009). The lipid pathway located in the chloroplast, which traces its origin to symbiogenesis, is called the prokaryotic pathway. The cytosolic endoplasmic reticulum lipid pathway, on the other hand, is known as the eukaryotic pathway. The two pathways are separated by barriers of the chloroplast envelope, but they are by no means insulated from each other. In fact, a significant portion of the glycerolipid moieties required for the biogenesis of photosynthetic membrane systems in the chloroplast is supplied by the eukaryotic pathway. The relative contributions of the two pathways, however, are very different among plant species (Mongrand et al. 1998). In species like *Arabidopsis*, about half of the glycerolipid moieties in chloroplast come from the eukaryotic pathway (Ohlrogge and Browse 1995). In species, such as wheat, construction of photosynthesis apparatus relies almost entirely on the eukaryotic pathway, while the contribution of prokaryotic pathway is limited to the production of phosphatidylglycerol only (Heinz and Roughan 1983; Arunga and Morrison 1971). Despite the divergent pattern of relevant inputs from the two pathways, at the core is a basic requirement of glycerolipid channeling from the endoplasmic reticulum to chloroplasts in all plant species.

Changes in the proportion of glycerolipid classes under temperature stress have been previously reported in different plant species. The most common change is the adjustment between monogalactosyldiacylglycerol and digalactosyldiacylglycerol, and it has been proposed that this metabolic change aids in survival at suboptimal temperatures (Chen et al. 2006). A greater proportion of digalactosyldiacylglycerol has been observed at high temperatures in *Arabidopsis*, while low temperature induced the synthesis of monogalactosyldiacylglycerol in *Brassica napus* (Chen et al. 2006; Johnson and Williams 1989). In addition to the alterations in the proportion of glycerolipid species, changes were also observed in the same lipid species with different fatty acid combinations. In *Brassica napus*, the major molecular species of monogalactosyldiacylglycerol is a mixture of *sn-1* 18:3/*sn-2* 16:3 and *sn-1* 18:3/*sn-2* 18:3 molecule species (Ohlrogge and Browse 1995; Hölzl et al. 2009; Mongrand et al. 1998; Heinz and Roughan 1983). An increase in monogalactosyldiacylglycerol (18:3/16:3) relative to monogalactosyldiacylglycerol (18:3/18:3) molecule species was observed in *Brassica napus* leaves grown at low temperature (Johnson and Williams 1989). In contrast, a decrease in monogalactosyldiacylglycerol (18:3/16:3) relative to monogalactosyldiacylglycerol (18:3/18:3) was observed in *Arabidopsis* grown at high temperatures (Chen et al. 2006).

Rebalancing of the two glycerolipid pathways influences both the distribution of glycerolipid classes of different head groups and the degree of fatty acid desaturation in glycerolipid molecules. Due to substrate specificity of the plastidic lysophosphatidic acyltransferase, the prokaryotic pathway produces glycerolipids with C16 fatty acyl moiety at the *sn-2* position of the glycerol backbone and hence a molecular configuration of C34 (C18/C16, *sn-1*/*sn-2*) (Ohlrogge and Browse 1995; Heinz and Roughan 1983). The *sn-2* C16 fatty acids in the monogalactosyldiacylglycerol originating from the prokaryotic pathway are subsequently desaturated to form C16:3 fatty acids, leading to the formation of the highly unsaturated C34:6 (18:3/16:3) monogalactosyldiacylglycerol. The eukaryotic pathway generates glycerolipids with only C18 fatty acid at the *sn-2* and produces C36 (C18/C18) and C34 (C16:0/C18). Since the *sn-1* 16:0 does not undergo any further desaturation, the eukaryotic C34:3 (16:0/18:3) has a level of desaturation much lower than that of the prokaryotic C34:6 (18:3/16:3). Through comprehensive lipid profiling and lipidomics analysis, it was shown that high temperatures repress the input of the prokaryotic pathway and promote the channeling of eukaryotic lipids to the chloroplast, thereby reducing potential of fatty acid desaturation in the photosynthesis apparatus. It was also observed that there was a differential channeling of eukaryotic glycerolipid species from the endoplasmic reticulum to the chloroplast: C34 diacylglycerol moieties of 16:0/C18 (*sn-1*/*sn-2*) from the endoplasmic reticulum were preferentially transported to the chloroplast over that of C36 (C18/C18) under high temperature (Li et al. 2016). These two layers of metabolic adjustments collectively reduce the overall degree of fatty acid desaturation in chloroplast lipids, particularly that of monogalactosyldiacylglycerol, at higher temperature. In wheat, where chloroplast lipid synthesis is dependent completely on imports from the endoplasmic reticulum, a differential channeling of C34 (C16/C18) DAG versus C36 (C18/C18) diacylglycerol was also observed. Under low temperature, there was a reduction of C34 diacylglycerol with a corresponding increase of C36 diacylglycerol (Li et al. 2016). This metabolic alteration increased the overall potential of glycerolipid unsaturation in the chloroplast, i.e., increased level of C36 (18:3/18:3). The California desert plant *Atriplex lentiformis* presents an exceptional case, which shuts off its prokaryotic pathways at high temperature and falls back on the eukaryotic pathway completely for chloroplast lipid synthesis (Pearcy 1978). This metabolic rebalancing allows the endoplasmic reticulum to deliver more C34 diacylglycerol (16:0/C18), thereby reducing the degree of lipid desaturation under extreme heat. These three species demonstrate the essential role of glycerolipid pathway adjustment during adaptation to temperature stress.

Land plants living in restricted spaces usually encounter environmental stresses that affect their growth and productivity. Among these stressful conditions, cold is one of the major constraining factors that limit the geographical distribution and agricultural productivity of crops (Shi et al. 2015). Tobacco (*Nicotiana tabacum* L.), a *Solanaceae* species, originates from tropical and subtropical America but is now commercially cultivated worldwide due to its high economic value. In addition, it is commonly used as an excellent model plant species for studies on physiological traits and molecular functions. However, tobacco is sensitive to cold stress (Zhao et al. 2009). Cold acclimation is recognized as a typical process by which many temperate plants develop chilling and freezing tolerance (Thomashow 1999). Hence, exploring the molecular mechanism of cold acclimation in tobacco is of great value for breeding cold-tolerant varieties for agricultural production. Numerous studies have shown that cold acclimation is closely associated with complex changes in the regulation of gene expression, accumulation of protective metabolites, and maintenance of metabolic homeostasis (Joosen et al. 2006; Yue et al. 2015; Chen et al. 2015b). Many efforts in cold acclimation studies of various plant species have suggested that cold induces extensive recombination of gene transcription (Yang et al. 2019; Byun et al. 2018; Jeon and Kim 2013). Presently, the product of the best known cold-responsive gene involved in the cold acclimation signaling pathway is the ICE1-CBF-COR module (Xu et al. 2020). In this model, *Arabidopsis* plants exposed to cold stress induce the expression of C-repeat binding factors/dehydration-responsive element binding factors, which can bind to the promoter of cold-responsive genes to regulate their transcription (Cook et al. 2004; Chinnusamy et al. 2003). Despite these valuable insights, the whole understanding of plants during cold acclimation in terms of the potential links between gene expression and metabolite regulatory networks remains largely unknown.

Modern metabolomics techniques enable the monitoring of low-molecular-weight metabolites to gain a comprehensive understanding of biological processes during exposure to various stresses (Fiehn et al. 2000; Zhang et al. 2012b). The identification of general cold responses has generated metabolomic data from plants growing during cold acclimation. For example, the accumulation of sugars and the synthesis of amino acids are well recognized as essential strategies during cold acclimation not only because both sugars and amino acids are signaling molecules but also because they are precursors of large amounts of metabolites with multiple stress functions (Cook et al. 2004; Krasensky and Jonak 2012). These technological advances from the combination of transcriptomic and metabolomic analyses have provided several possible approaches for dissecting molecular features involved in the readjustment of gene expression and metabolic pathways. Using an integrated genetic and metabolite analyses, Jia et al. detected several genes and metabolites that were specifically modulated in response to various stages of drought stress in *Astragalus membranaceus* (Jia et al. 2016). Moreover, Moschen et al. (2017) integrated transcriptomic and metabolic data to reveal alterations in the levels of hub transcription factors and metabolites involved in the regulation of drought stress tolerance in sunflower (Moschen et al. 2017).

5.9.2 Effect of Exogenous Lipids on Cold Acclimation and Fatty Composition

Theoretically, exogenous lipids can be used for (a) synthesizing intracellular compounds and obtaining energy and (b) the biosynthesis of both lycopene and cellular lipids because they use acetyl-CoA (resulting from lipid degradation) as a common precursor substance (Lingran et al. 2019). The catabolism of lipids promotes glucose toward the hexose monophosphate pathway due to intracellular energy excess, hence increasing the production of NADPH. Thus, cellular lipid synthesis can be stimulated because the biosynthesis and desaturases of both carotenoids and aliphatic chains require NADPH as cofactors (Beyer et al. 1994; Garba et al. 2017; Miziorko 2011). However, the effects of exogenous lipids on intracellular lipids and the relationship between cellular fatty acid composition and lycopene production are not very clear.

Several modes of change of cellular fatty acids at varying environmental temperatures have been found in fungi, yeasts, and bacteria, such as the degree of fatty acid unsaturation, fatty acid chain length, and cellular fatty acid content. The main change reflects the degree of unsaturation of fatty acyl chains through the control of fatty acid desaturation both at the level of transcription and posttranscriptional regulation (Dyer et al. 2001; Gargano et al. 1995; Vigh et al. 2005). Remodeling of membrane composition and functionality to maintain proper cellular function is widely accepted and termed homeoviscous adaptation (Cossins 1994). In addition to phospholipids, *Blakeslea trispora* also reserves large amounts of neutral lipids in lipid bodies where lipophilic carotenoids are deposited. It has been shown that hypothermic conditions can result in (a) an increase in the degree of desaturation of fatty acids in the lipid bilayer and (b) changes in the qualitative and quantitative composition of neutral lipids (Feofilova et al. 2000). Feofilova et al. (2005) showed that the degree of desaturation of fatty acids of both (+) and (−) strains of *Blakeslea trispora increased* in the lipid bilayer, although the (−) strain lacks linolenic acid.

Plant oils are known to improve the lycopene production and biomass of *Blakeslea trispora*. However, their effects on intracellular fatty acid composition and the selection basis of exogenous plant oils are not very clear. In this work, the lipid composition and lycopene production of *Blakeslea trispora* in response to different exogenous plant oils and temperature variation were investigated. The aims were to reveal the effects of several common plant oils on intracellular lipids and the relationship between cellular fatty acid composition and lycopene production. Based on these findings, lycopene production could be improved by altering the intracellular fatty acid composition of *Blakeslea trispora*, such as by adding suitable plant oils, changing the fermentation temperature, and adjusting other culture conditions.

5.10 Lipid Biosynthesis Under Low-Temperature Stress in Plants

Within the broader context of how lipid composition affects the cell biology and physiology of plants, the question of how membrane unsaturation affects a plant's ability to tolerate low-temperature stresses has a long and distinguished history (Tokuhisa et al. 1997). A series of *Arabidopsis* mutants with defects in lipid metabolism has provided important examples of how membrane fatty acid composition can affect plant temperature responses. A central dilemma in membrane biology is the need to explain lipid diversity. Despite the involvement of particular lipids in processes such as signal transduction and protein attachment, the dominant view of a membrane is still the fluid mosaic model of Singer and Nicolson (1972) in which lipids simply compose the fluid, semipermeable bilayer that is the matrix for the functional membrane proteins. In principle, this bulk lipid role could be satisfied by a single unsaturated species of phosphatidylcholine. However, the number of chemically different membrane lipids is very much larger than the number of individual amino acids or the number of different carbohydrate monomers. Each membrane of the cell has a characteristic and distinct complement of lipid types, and within a single membrane each class of lipids has a distinct fatty acid composition. It is obvious, then, that such simple models are unsatisfactory and that new hypotheses are needed to relate membrane structure to cell and organism function.

The chemistry of membrane lipid synthesis in plant cells is similar to animals, but some differences occur in both molecular mechanisms and cellular organization (Browse and Somerville 1991). The plant fatty acid synthase is located, essentially exclusively, in the chloroplasts (or other plastids) and is composed of eight individual proteins (Shimakata and Stumpf 1982). Most of the 16:0-ACP' produced by the fatty acid synthase is elongated to 18:0-ACP and then desaturated in the chloroplast stroma by a soluble desaturase (McKeon and Stumpf 1982; Shanklin and Somerville 1991), so that 16:0- and 18:I-ACP are the primary products of chloroplast fatty acid synthesis. These products are used by two quite distinct pathways for the biosynthesis of membrane glycerolipids and the associated production of polyunsaturated fatty acids. The "prokaryotic" pathway (Roughan and Slack 1982) located in the plastid inner envelope uses 18:1 acyl carrier protein and 16:0 acyl carrier proteins for the sequential acylation of glycerol-3-phosphate. Because of the substrate specificities of the plastid acyltransferases, the phosphatidic acid formed has 18:I at the sn-J position and 16:0 at the sn-2 position. The phosphatidic acid is used for the synthesis of phosphatidylglycerol or is converted to diacylglycerol by a phosphatidic acid phosphatase located in the chloroplast inner envelope membrane. This diacylglycerol pool can act as a precursor for the synthesis of the other major thylakoid lipids like monogalactosyldiacylglycerol, digalactosyldiacylglycerol, and sulfoquinovosyldiacylglycerol (Joyard et al. 1993). The "eukaryotic" pathway begins with the hydrolysis of 16:0- and 18:I-ACPs and the export of these fatty acids to the cytoplasm as CoA thioesters. The CoA esters are used for the synthesis of 16:0118:1- and 18:11, 18:I phosphatidic acid in the endoplasmic reticulum. This phosphatidic acid gives rise to the phospholipids such as phosphatidic choline, phosphatidylethanolamine, and phosphatidylinositol that are characteristic of the various extrachloroplast membranes (Moore 1982). In addition, however, the diacylglycerol moiety of phosphatidylcholine is returned to the chloroplast envelope where it enters the diacylglycerol pool and contributes to the synthesis of thylakoid lipids (Browse et al. 1986). Initially, all the major glycerolipids are synthesized using only 16:0 and 18:1 acyl groups. Subsequent desaturation of the lipids involves membrane-bound desaturases of the chloroplast (Slack et al. 1976; Heinz 1993) and the endoplasmic reticulum. Biochemical characterization of these desaturases by traditional approaches has been very limited, because it has not been possible to solubilize and purify them. Instead, the increased understanding of the reactions and regulation of chloroplast lipid synthesis and desaturation has come from the isolation and characterization of four classes of *Arabidopsis* mutants, each one deficient in a specific desaturation step (Browse and Somerville 1991; Miquel and Browse 1992).

In plants, de novo fatty acid biosynthesis mainly takes place in the plastidial compartment (Ohlrogge and Browse 1995) from acetyl-CoA, which is a direct product of photosynthesis. Plastid pyruvate dehydrogenase (EC 1.2.4.1) is the main route for a rapid and stable supply of acetyl-CoA through its action on pyruvate (resulting from glycolysis or the pentose phosphate pathway). Another possible source is the import and activation of free acetate by acetyl-CoA synthase (ACS, EC 6.2.1.1) (Lin and Oliver 2008). The major product of fatty acid synthase is palmitic acid, except for the elongation of palmitic acid and the desaturation of stearic acid which take place in the chloroplast. Other changes (elongation, desaturation, hydroxylation, and epoxidation) occur mainly in the endoplasmic reticulum (Fatiha 2019). Two enzyme systems are required for fatty acid formation: acetyl-CoA carboxylase (ACCase, EC 6.4.1.2) of which two forms have been identified in plants (Li-Beisson et al. 2010) and fatty acid synthase which is a multienzyme complex present in the stroma of chloroplasts (Roughan and Ohlrogge 1996). Owing to the substrate specificity of the chloroplast lysophosphatidic acyltransferase, the prokaryotic pathway in the chloroplast generates glycerolipids with a C16 fatty acyl moiety at the *sn-2* position. The eukaryotic pathway, by contrast, produces glycerolipid molecules that have only C18 fatty acid at *sn-2*. The two pathways are compartmentalized by membrane barriers but coordinate closely to meet the overall demand of cellular membrane biogenesis and maintenance (Kunst et al. 1988; Ohlrogge and Browse 1995). However, there is considerable flexibility in the relative contributions of the two pathways to overall glycerolipid production. In species such as *Arabidopsis thaliana* and spinach (*Spinacia oleracea*), biosynthesis of major chloroplast lipids, particularly monogalactosyldiacylglycerol and digalactosyldiacylglycerol, draws nearly equal contribution from the two pathways (Ohlrogge and Browse 1995). The *sn-2* C16 fatty acyl moiety in monogalactosyldiacylglycerol and digalactosyldiacylglycerol produced through the prokaryotic pathway undergoes a series of desaturations, leading to the formation of 16:3 fatty acid. Plant species that rely on a significant contribution from the prokaryotic pathway for chloroplast lipid biosynthesis are therefore called 16:3 plants. By contrast, plants such as pea (*Pisum sativum*) and wheat (*Triticum aestivum*), wherein, besides phosphatidylglycerol, the biosynthesis of chloroplast glycerolipids is almost entirely dependent on the eukaryotic pathway, are classified as 18:3 plants (Heinz and Roughan 1983; Browse et al. 1986; Ohlrogge and Browse 1995).

In addition to the soluble acyl-ACP desaturases, the fatty acids synthesized in the chloroplast (palmitate, stearate, and oleate) are desaturated by membrane-bound desaturases that utilize complex lipid substrates such as phosphatidylcholine in the endoplasmic reticulum or monogalactosyldiacylglycerol in the plastid (Harwood 1996). These fatty acids were used for the synthesis of glycerolipids by two distinct metabolic pathways (prokaryotic and eukaryotic pathway) and in different cellular compartments (plastids and endoplasmic reticulum) (Roughan and Slack 1982). The importance of both biosynthetic pathways depends on the plant species. In the photosynthetically active "C18:3" plants, only the extrachloroplast galactolipid pathway is functional; in the case of "C16:3" plants, the two pathways coexist, and their importance differs according to the species and the conditions of the environment.

5.10.1 Temperature-Induced Lipid Pathway Adjustments in Plants

Studies show that membrane lipid biosynthesis is robust and well maintained even in the face of genetic lesions of major metabolic steps (Browse et al. 1986; Johnson and Williams 1989; Brockman et al. 1990; Somerville and Browse 1991; Härtel et al. 2000). For example, the *Arabidopsis act1* mutant, deficient in the activity of the chloroplast glycerol-3-phosphate acyltransferase, is essentially devoid of 16:3-containing monogalactosyldiacylglycerol from the prokaryotic pathway. However, the proportions of various classes of glycerolipids in *act1* were similar to those of the wild-type due to an apparent increase in the trafficking of endoplasmic reticulum lipids to chloroplasts (Kunst et al. 1988). A notable case illustrating the biological significance of a highly coordinated and flexible pathway adjustment is that of metabolic response to phosphate deprivation in *Arabidopsis* (Benning et al. 2006; Benning 2009). Under phosphate starvation, phospholipids are replaced by galactolipids in not only the thylakoid membrane but also the extraplastidic membranes (Härtel et al. 2000), thereby reducing the need for phosphorus consumption for membrane lipid biosynthesis. Proportional changes of glycerolipid species have also been documented in plants subjected to temperature fluctuations (Pearcy 1978; Johnson and Williams 1989; Harwood 1991, 1994; Chen et al. 2006). An increased proportion of digalactosyldiacylglycerol was observed at high temperature in *Arabidopsis* (Chen et al. 2006), whereas biosynthesis of monogalactosyldiacylglycerol in canola (*Brassica napus*) was induced by low temperature (Johnson and Williams 1989).

Modular regulations of the two glycerolipid pathways in response to metabolic perturbations are evident not only in glycerolipid profiles but also at the transcript level of many enzymatic components associated with the two pathways (Shen et al. 2010). Now it has been documented that strictly coordinated gene expression at the biochemical pathway level in parallel with changes of specific glycerolipid pools was reported in *Arabidopsis* in response to light and temperature stimuli (Szymanski et al. 2014). Nonetheless, how glycerolipid composition is mechanistically modulated between the two glycerolipid pathways and precisely what factors are involved in this process remain largely unknown. In this study, we explored the significance of glycerolipid pathway balance in temperature-induced lipid compositional changes in three plant species that have distinctive modes of lipid pathway interactions. In *Arabidopsis*, a typical 16:3 plant, low temperature induces an augmented prokaryotic pathway, whereas high temperature enhances the eukaryotic pathway. *Atriplex lentiformis* switches lipid profile from 16:3 to 18:3 when it encounters heat and reduces its overall lipid desaturation through drastically increasing the contribution of the eukaryotic pathway. In wheat, which is an 18:3 plant, low temperature causes a reduced contribution from the endoplasmic reticulum pathway to chloroplast lipid biosynthesis. In sync with the metabolic changes is coordinated expression of glycerolipid pathway genes at the transcript level. It has been demonstrated that in *Arabidopsis thaliana*, a 16:3 plant, the chloroplast pathway is upregulated in response to low temperature, whereas high temperature promotes the eukaryotic pathway. Operating under a similar mechanistic framework, *Atriplex lentiformis* at high temperature drastically increases the contribution of the eukaryotic pathway and correspondingly suppresses the prokaryotic pathway, resulting in the switch of lipid profile from 16:3 to 18:3. In wheat (*Triticum aestivum*), an 18:3 plant, low temperature also influences the channeling of glycerolipids from the endoplasmic reticulum to chloroplast. Li et al. (2015) have demonstrated differential trafficking of diacylglycerol moieties from the endoplasmic reticulum to chloroplast was uncovered in three plant species as another layer of metabolic adaptation under temperature stress.

As major components of plant membranes, lipids create hydrophobic barriers that separate cells from their environment and prevent diffusion of organelle contents in and out of a cell. Lipids are composed of a hydrophilic, polar head attached to a glycerol backbone and a hydrophobic tail composed of two fatty acids (Dörmann 2013; Dörmann and Benning 2002). Membrane lipids in plants are synthesized via two distinct pathways, i.e., the prokaryotic and the eukaryotic pathways. The prokaryotic pathway occurs in the chloroplasts and produces the major plastidic lipids such as monogalactosyldiacylglycerol, digalactosyldiacylglycerol, trigalactosyl-diacylglycerol, and sulfoquinovosyl diacylglycerol, otherwise known as sulfolipids. In contrast, the eukaryotic pathway occurs in the endoplasmic reticulum and synthesizes phospholipids including phosphatidic acid, phosphatidylcholine, phosphatidylethanolamine, phosphatidylglycerol, phosphatidylinositol, and phosphatidylserine (Nakamura 2017; Yu et al. 2015). In both the prokaryotic and eukaryotic pathways, membrane lipid biosynthesis is initiated by the formation of phosphatidic acid, which is then utilized to produce plastidic lipids or phospholipids, depending on the site of biosynthesis. Phosphatidic acids produced in the chloroplasts can also be converted to diacylglycerol which then serves as a precursor for plastidic lipid synthesis. During the lifecycle of plants, an active lipid exchange between the chloroplast and endoplasmic reticulum occurs via the import of the diacylglycerol moiety of phosphatidylcholine from the endoplasmic reticulum to the chloroplast envelope where it contributes to the diacylglycerol pool used to synthesize plastidic lipids (Nakamura 2017; Yu et al. 2015; Bates et al. 2013). Besides their function in membrane lipid synthesis, phosphatidic acid and phosphatidylcholine also serve as intermediaries in storage lipid metabolism. The biosynthesis of fatty acids that make up storage lipids follows the same pathway as those that comprise membrane lipids using many similar precursors (Bates et al. 2013; Murphy 1994; Guschina et al. 2014; Gasulla et al. 2013). They are packaged in the seeds primarily as triacylglycerols which are composed of a glycerol backbone esterified with three fatty acids in saturated and/or unsaturated forms (Huang 1992; Ohlrogge et al. 2015).

Fatty acids are the major components of both membrane and storage lipids. They are made up of carboxylic acids with highly reduced hydrocarbon chains. The saturated fatty acids contain carbon molecules linked by single bonds, whereas the unsaturated fatty acids consist of carbon chains that are linked by one or multiple *cis*-double bonds (Poirier et al. 2006). The most common fatty acids in both membrane and storage lipids contain 16 or 18 carbons that are either saturated or unsaturated. All unsaturated fatty acid species are derived by the sequential unsaturation of the fully saturated species, i.e., palmitic acid (C16:0) and stearic acid (C18:0). This chemical reaction is catalyzed by the activity of various fatty acid desaturase enzymes. Desaturases convert single bonds at specific positions in the fatty acyl chains to double bonds by the removal of two hydrogen atoms (Murphy 1994, 1999; Los and Murata 2002; Ohlrogge and Browse 1995; Bates et al. 2013; Guschina et al. 2014).

Triacylglycerol is the major component of vegetative oils and is composed of a glycerol backbone bearing three esterified fatty acids. In plants, triacylglycerol is mainly stored as a high-energy storage compound within lipid droplets in seeds or fruits (Xu and Shanklin 2016). Vegetative tissues such as leaves and stems, on the other hand, are typically regarded as non-lipid-storing tissues, at least under favorable environmental conditions, due to their limited contents of triacylglycerol. However, small amounts of triacylglycerol do exist in these tissues (Xu and Shanklin 2016), where they have been suggested to play a role in the sequestration of toxic lipid intermediates following membrane breakdown and light-induced stomatal opening (Lin and Oliver 2008; McLachlan et al. 2016). Over recent years, a growing number of studies have also linked vegetative triacylglycerol production to plant stress response, and the precise mechanisms behind such a function are in the process of being unraveled (Yang and Benning 2018).

5.10.2 Glycerolipid Response to Cold and Freezing Temperature

Low-temperature injury is one of the most important factors causing substantial agricultural losses. Glycerolipids are the principal constituent of cellular membranes; remodeling of glycerolipids plays important roles in temperature adaptation in plants (Zheng et al. 2016). Temperate plants can endure freezing stress, but even chilling at above-zero temperatures can induce death in tropical species. Freezing-induced injury to plants differs from that associated with chilling, because of the presence of ice inside plant tissues. As a result, there are major differences in the physiological, biochemical, and molecular responses of plants to freezing and chilling stresses (Ploschuk et al. 2014; Grimaud et al. 2013; Chinnusamy et al. 2007; Sung et al. 2003). For example, chilling tolerance in pea (*Pisum sativum*) was suggested to be related to an increase in proteins that are involved in osmotic adjustment and antioxidative responses; however, freezing tolerance seemed to depend on proteins that maintain the stability of the photosystems as well as the capacity for photosynthesis (Grimaud et al. 2013).

Membranes are integral to the structure and function of all cells and are very sensitive to environmental changes. The responses of membranes to temperature changes have been investigated extensively, and results indicate that membrane damage is the major cause of low temperature-induced injury in plants (Steponkus 1984; Hincha et al. 1987; Krause et al. 1988). Phospholipids and galactolipids are the main glycerolipids and are, respectively, the major constituents of plasma and chloroplast membranes. Regulation of the level of saturation and the constitution of glycerolipids is important for plants to maintain the integrity and fluidity of their membranes under temperature stresses (Nishida and Murata 1996; Wang et al. 2006a, b, c; Li et al. 2008). After introduction into a warmer lowland area from an alpine region, the level of saturation in all measured phospholipids of *Meconopsis racemosa* leaves increased (Zheng et al. 2014). Changing the level of saturation of glycerolipids has been identified as a strategy used by plants to cope with both long-term high and low temperatures; in contrast, upon short-term temperature stresses, the level of saturation of glycerolipids was maintained (Zheng et al. 2011). In addition to reducing the level of saturation of glycerolipids, adjustment of the composition of lipids to balance the ratio of bilayer- to non-bilayer-forming membrane is also important for plant responses to low temperatures. At low temperatures, the proportion of phospholipids tends to increase (Li et al. 2008; Uemura et al. 1995a). Glycerolipids harboring large polar head groups, such as digalactosyldiacylglycerol and phosphatidylcholine, are bilayer lipids that may increase membrane stability. In contrast, monogalactosyldiacylglycerol and phosphatidylethanolamine, which have relatively small head groups, show a higher propensity for transition to non-bilayer HII-type structures (Hazel and Williams 1990; Moellering et al. 2010). Increases in the ratio of bilayer- to non-bilayer-forming membrane lipids, for example, ratio between phosphatidylcholine and phosphatidylethanolamine and between digalactosyldiacylglycerol and monogalactosyldiacylglycerol, have been observed in plants responding to low temperatures (Moellering et al. 2010; Welti et al. 2002). Leaves of *Brassica napus* showed a tendency to synthesize more desaturated phosphatidylcholine and monogalactosyldiacylglycerol at low temperatures (Tasseva et al. 2004a). Plants that have different kinds of monogalactosyldiacylglycerol molecules have different pathways for synthesizing lipids: plants that have only 36:6 (total acyl carbons/total double bonds) monogalactosyldiacylglycerol molecules are 18:3 plants, which only have the eukaryotic-type lipid synthesis pathway; plants that contain both 34:6 and 36:6 monogalactosyldiacylglycerol molecules are 16:3 plants, which have both prokaryotic- and eukaryotic-type lipid synthesis pathways (Buchanan et al. 2002). However, little is known about the relationships between the different pathways for synthesizing lipids in plants that are tolerant to low temperatures.

Most studies on changes in the lipidome in response to low temperatures used the model plant *Arabidopsis* and investigated its freezing tolerance (Wang et al. 2006b; Li et al. 2008; Moellering et al. 2010; Degenkolbe et al. 2012). However, freezing-induced membrane damage may differ from that associated with chilling, because cells can suffer different types of injury. The lipidome responses of plants upon chilling-induced injury or even death have scarcely been studied, and only Li et al. (2014) have focused on this issue, studying the *Jatropha curcas* lipidome response to chilling treatment. Rice is a chilling-sensitive plant that dies after the temperature falls to 4 °C (Shinkawa et al. 2013). In contrast, after acclimation to cold at 4 °C, *Arabidopsis* can tolerate freezing temperatures as low as −6 °C (Li et al. 2004).

5.10.3 Triacylglycerol Biosynthesis

Triacylglycerols are neutral lipids and are the major component of oilseed oil. These storage lipids represent the main source of carbon and energy mobilized during germination. Other tissues can also accumulate triacylglycerols, such as senescence leaves or pollen grains (Kaup et al. 2002; Kim et al. 2002). Their biosynthesis occurs at the endoplasmic reticulum membrane during the storage accumulation phase after embryogenesis. The triacylglycerol results from the esterification at the sn-3 position of diacylglycerol of fatty acid from the pool of cytosolic acyl-CoA by the action of diacylglycerol acyltransferase (diacylglycerol acyltransferase, EC 2.3.1.20) or phospholipid:diacylglycerol acyltransferase (phospholipid:diacylglycerol acyltransferase, EC 2.3.1.158) (Zhang et al. 2015). The acyl-CoAs can be released from phosphatidylcholine after desaturation with lysophosphatidylcholine acyltransferase and reincorporated into the diacylglycerol triacylglycerol chain. This allows for a renewal of the fatty acid composition of triacylglycerols (Battey et al. 1989).

Seed triacylglycerols often contain unusual fatty acids, and more than 300 different fatty acids are known to occur in seed triacylglycerol. Chain length may range from less than 8 to over 22 carbons. The position and number of double bonds may also be unusual, and hydroxy, epoxy, or other functional groups can modify the acyl chain. The synthesis of these unusual fatty acids involves just one additional or alternative enzymatic step from primary lipid metabolism. All the enzymes identified to date that are involved in unusual fatty acid biosynthesis are structurally related to enzymes of primary lipid metabolism. Many of the unusual fatty acids are found in taxonomically dispersed families implying that the recruitment of enzymes for the synthesis of these unusual fatty acids might have occurred a number of independent times during angiosperm evolution. Plants that synthesize unusual monounsaturated fatty acids have an additional desaturase, which is closely related to the Δ9-desaturase but introduces a double bond at a different location on the acyl carrier protein. In coriander (*Coriandrum sativum*), petroselinic acid is synthesized by a desaturase that introduces a double bond between carbons 4 and 5 of a C16 acyl carrier protein (Δ4-desaturase). This fatty acid is then extended by two carbons and cleaved from acyl carrier protein to produce the free fatty acid. These last two steps are thought to require a specialized condensing enzyme and a specialized acyl carrier protein thioesterase (Grosbois 1971). Plants that synthesize medium-chain fatty acids have several thioesterases. Indeed, plants that produce seeds with high concentrations of 8–14 carbon atoms, like *Cuphea lanceolata* rich in decanoic acid (C10: 0) and *Umbellularia californica* rich in laurate (C12: 0), contain specific thioesterase for medium fatty acid chains. By removing acyl groups from acyl carrier proteins prematurely, the medium-chain thioesterases simultaneously prevent their further elongation and release them for triacylglycerol synthesis outside the plastids (Pollard et al. 1990).

Seeds of *Ricinus communis* L. are the source of castor oil, used for the production of high-quality lubricants due to its high proportion of the unusual fatty acid ricinoleic acid. Castor bean seed oil contains 90% of the unusual hydroxy fatty acid. Castor bean seeds contain an oleate hydroxylase which is structurally similar to extraplastidial membrane-bound Δ12-desaturases (fatty acid desaturase2), and only four amino acid substitutions are needed to convert an 18:1-desaturase into an 18:1-hydroxylase (Van de Loo et al. 1995). The synthesis of these fatty acids is thought to take place on the endoplasmic reticulum and uses fatty acids esterified to the major membrane lipid phosphatidylcholine as a substrate. Borage (*Borago officinalis* L.) seeds and evening primrose (*Oenothera biennis* L.) seeds are rich in γ-linolenic acid (Δ6, 9, 12), respectively, from 22 to 25% and from 8 to 10%, an essential fatty acid. Its synthesis takes place in the endoplasmic reticulum during the formation of the seed. The precursor is a linoleoyl-phosphatidylcholine, and the desaturation is catalyzed by a D6 desaturase (Galle et al. 1992). Very long-chain fatty acids (AGTLCs, containing more than 18 carbons) are used in the biosynthesis of many lipids involved in seed storage and waxes. Very long-chain fatty acids (VLCFAs) are synthesized in the following by-products of elongation of a C18 fatty acyl precursor by two carbons originating from malonyl CoA. Each elongation step requires four enzymatic reactions: condensation between an acyl precursor and malonyl-CoA, followed by a reduction, dehydration, and another reduction. In higher plants, triacylglycerol can be synthesized stored either in the cytosol (within lipid droplets) or in the chloroplast (within plastoglobules), termed cytosolic triacylglycerol and chloroplastic triacylglycerol, respectively (Huang 2018; Chapman et al. 2012). The major route of cytosolic triacylglycerol biosynthesis requires de novo fatty acid biosynthesis and triacylglycerol assembly, both of which take place in distinct cellular compartments (Chapman and Ohlrogge 2012). Chloroplasts are the primary site of fatty acid biosynthesis, which begins with the conversion of pyruvate to acetyl-CoA. Fatty acyl chains of 16 and 18 carbons in length are generated from acetyl-CoA and are then transported out of the plastid and activated into acyl-CoA for triacylglycerol biosynthesis (Shockey et al. 2002). The assembly of cytosolic triacylglycerol occurs on the endoplasmic reticulum via the Kennedy pathway using glycerol-3-phosphate produced during glycolysis as the carbon backbone and acyl-CoA as the acyl donor (Xu and Shanklin 2016). In this pathway, glycerol-3-phosphate is first acylated at the sn-1 position by sn-glycerol-3-phosphate acyltransferase to produce lysophosphatidic acid (Shockey et al. 2016; Singer et al. 2016), which is then acylated at the sn-2 position through the action of acyl-CoA/lysophosphatidic acid acyltransferase to yield phosphatidic acid (Kim et al. 2005). Mg^{2+}-dependent phosphatidic acid phosphatases then convert the phosphatidic acid into sn-1,2-diacylglycerol (Pascual and Carman 2013). The final step is catalyzed by diacylglycerol acyltransferase, which acylates the sn-3 position of sn-1,2-diacylglycerol to yield triacylglycerol (Ichihara et al. 1988). De novo synthesized 18:1-diacylglycerol can also be channeled into 18:1-phosphatidylcholine through the catalytic action of cytidine 5′-diphosphocholine-choline: diacylglycerol choline phosphotransferase (Tasseva et al. 2004b) and phosphatidyl-choline:diacylglycerol choline phosphotransferase (Lu et al. 2009). Phosphatidylcholine is regarded as the major site for the modification of fatty acids, where 18:1- phosphatidylcholine is converted into polyunsaturated phosphatidylcholine modified fatty acid via the acyl-editing cycle, which involves deacylation and re-acylation but does not generate new phosphatidylcholine (Bates and Browse 2012). In addition, plastid 18:1 acyl-CoA can also be directly incorporated into polyunsaturated phosphatidic modified fatty acid through the activity of lysophosphatidylcholine acyltransferase (Pascual and Carman 2013; Xu et al. 2012). Phospholipid diacylglycerol acyltransferase is then able to use polyunsaturated phosphatidic modified fatty acid as an acyl donor to catalyze the acylation of diacylglycerol into triacylglycerol, producing lysophosphatidylcholine as a by-product. This route is termed the acyl-CoA-independent pathway, which differs from the Kennedy pathway (Dahlqvist et al. 2000).

5.11 Lipid Hydrolysis Under Low-Temperature Stress

Low-temperature damage, also associated with freezing injury, can arise from changes in the intercellular water temperature which affects water potential through the plasma membrane, affecting also its own conformation causing structural damage and hence increasing ion leakage. Formation of non-bilayer lipid structures in both plasma and chloroplast membranes, such as the inverted hexagonal (HII)-type structures, can occur under cold temperatures causing lesions in the membrane. Under such conditions, changes in the lipid composition, and membrane repair systems, mediated by many lipid pathway enzymes, are activated in order to maintain its structure, keeping membrane integrity and avoiding damage (Steponkus 1984). Plants often respond to low-temperature stress by increasing membrane lipid unsaturation, chain shortening, altering lipid class composition, and/or changing the lipid/protein ratio (Takahashi et al. 2013).

The chilling response of plants is complex and based on the interplay of two important metabolic processes—lipolytic degradation of membrane lipids and a set of oxidative reactions leading to lipid peroxidation and membrane damage evoked in chilling-sensitive plants subjected to low temperature and light. The effects of chilling of detached leaves and intact plants differ and are often neglected during experiments. In closely related species, the activity of several constitutive enzymes (i.e., superoxide dismutase, ascorbate peroxidase, and glutathione reductase) appears to be higher in chilling-tolerant than in chilling-sensitive species, while in several native, closely related chilling-sensitive species, lipid acyl hydrolase (galactolipase) activity is higher than in chilling-tolerant species. Moreover, in chilling-insensitive plants, galactolipase activity is very low and is neither activated by detachment of leaves nor under stress conditions in growing plants. Dark and low-temperature treatments of detached leaves of chilling-sensitive species and post-chilling recovery of growing plants in the light activate galactolipase, which is responsible for the release of free fatty acids, the main substrates of peroxidation by lipoxygenase and free radicals. In several chilling-sensitive species, increased galactolipase activity is an important factor contributing to chilling susceptibility. Thus, it seems likely that enhancement of chilling tolerance may be achieved by genetically suppressing galactolipase in order to reduce both the degradation of chloroplast lipids and the level of released free fatty acid, thereby avoiding the deleterious action of their peroxidation products on plant tissues (Kaniuga 2008).

Plants deposit a lipidic layer composed of waxes and cutin on their surface, which serves as an impermeable barrier protecting them from excessive water loss, pathogens, and toxins (Nawrath et al. 2013; Serrano et al. 2014). The storage lipids are represented mainly by triacylglycerols and to a much lesser degree by sterol esters, whereas lipid composition of membranes is more complex and differs between cell compartments, cell types, and organisms (Bouvier-Navé et al. 2010; Horn et al. 2011; Goold et al. 2014; Shimada et al. 2019). Broadly, galactolipids are much more abundant in plastidial membranes, whereas non-plastidial membranes are composed mostly of phospholipids (Li-Beisson et al. 2013). Lipid droplets have been identified in cells of diverse organisms, including bacteria (Zhang et al. 2017), yeast (Grillitsch et al. 2011), algae (Zienkiewicz and Zienkiewicz 2020), plants (Tzen et al. 1993), nematodes (O'Rourke et al. 2009), and mammals (Birsoy et al. 2013). Lipid droplets were considered as a simple triacylglycerol storage compartment (Zienkiewicz and Zienkiewicz 2020), but intense studies revealed that lipid droplets represent highly dynamic structures, involved in a number of cellular processes, like regulation of energy homeostasis, remodeling of membranes, and signaling (Chitraju et al. 2017; Welte and Gould 2017; Yang and Benning 2018; Fernandez-Santos et al. 2020).

Triacylglycerol lipases play one of the most essential roles in lipid droplet degradation in plants and algae. These enzymes hydrolyze triacylglycerols stored in lipid droplets leading to the release of fatty acids, diacylglycerol, monoacylglycerols, and glycerol (Graham 2008; Kelly and Feussner 2016). Glycerol is subsequently phosphorylated and oxygenated and enters into glycolysis, whereas fatty acids undergo *β*-oxidation into acetyl-CoAs (Eastmond and Graham 2001). In plants (Graham 2008; Pracharoenwattana et al. 2010; Rinaldi et al. 2016) and green algae (Kong et al. 2018b, c), β-oxidation the latter process takes place in peroxisomes (glyoxysomes), meanwhile in diatoms, both mitochondrial and peroxisomal fatty acid *β*-oxidations are tough to occur (Chauton et al. 2013; Jallet et al. 2020). So far, the triacylglycerol lipases directly involved in lipid droplet mobilization are best characterized in plants, mainly in germinating seeds and growing seedlings of *Arabidopsis thaliana* (Eastmond 2006; Quettier and Eastmond 2009; Kelly et al. 2011).

Certain molecular species are especially important, and a good correlation has been noted for many plants where chilling sensitivity is associated with significant amounts of 16:0/16:0- or 16:0/*trans*-16:1-phosphatidylglycerol species (Murata and Yamaya 1984). Chilling tolerance in higher plants has been made possible by the introduction of desaturase genes (Ishizaki-Nishizawa 1996; Murata et al. 1992; Wada et al. 1990) and the desaturation of palmitate to palmitoleate or by modifying the biosynthetic pathway to phosphatidylglycerol where oleate (rather than palmitate) is acylated (Gao et al. 2015). Apart from fatty acid unsaturation, phospholipid class composition can be affected, e.g., sugar bean roots responded to low temperatures by lowering the ratio of phosphatidylcholine to phosphatidylethanolamine. Changes in phosphatidylcholine and phosphatidylethanolamine after cold stress may be related to the activity of phospholipases. Phospholipase-Dα can lead to a decline in phosphatidylcholine rather than phosphatidylethanolamine on sublethal freezing where an increase in phosphatidic acid and lysophospholipid is seen (Guschina and Harwood 2006). Phospholipase-D is activated by cold and increases the production of the signaling molecule phosphatidic acid (Ruelland et al. 2002); however, it is also proposed to have a mechanical role as a membrane anchor (Bargmann and Munnik 2006). Belonging to the same family of enzymes, phospholipase-D from δ class can bind microtubules and interact with the cytoskeleton stabilizing the membrane (Gardiner 2001). In *Arabidopsis*, phospholipase-Dδ mutants showed impaired freezing tolerance and reduced phosphatidic acid levels, whereas overexpression of this gene has the opposite effect (Li 2004). The regulation of phospholipase-Dδ and phospholipase-Dα expression is driven by ACYL CoA-BINDING PROTEINS (ACBP6, ACBP1), which influence the phosphatidic acid signaling cascade and increase freezing tolerance (Chen et al. 2008b; Du 2010).

Levels of free fatty acids is major determinant of oil quality. Minimizing acidity, i.e., free fatty acids in the oil palm mesocarp, is an important requisite for producing good quality of palm oil. The mesocarp of the ripe oil palm fruit accumulates large amounts of palm oil, an economically desirable product. The level of free fatty acids is an important determinant of oil quality, and little is known of the factors affecting this parameter in the oil palm fruit. Evidence is presented here for the presence of an active endogenous lipase in the oil palm mesocarp. Lipase activity was located in the oil body fraction. Maximum in vitro activity was observed at 18 °C, but using an in vivo assay, maximum activity was obtained at 5 °C. Low-temperature activation was less pronounced in a water-saturated environment.

References

Adem MW, Meulenaer BD, Duchateau L, Boeckx P (2018) Effect of altitude on biochemical composition and quality of green arabica coffee beans can be affected by shade and postharvest processing method. Food Res Int 105:278–285

Aguilar PS, Hernandez-Arriaga AM, Cybulski LE, Erazo AC, de Mendoza D (2001) Molecular basis of thermosensing: a two-component signal transduction thermometer in *Bacillus subtilis*. EMBO J 20:1681–1691

Ahmad P, Prasad MNV (2012) Environmental adaptations and stress tolerance of plants in the era of climate change. Springer, Dordrecht

Ahn SJ, Im YJ, Chung GC, Cho BH, Suh SR (1999) Physiological responses of grafted-cucumber leaves and rootstock roots affected by low root temperature. Sci Hortic 81:397–408

Allen DJ, Ortm DR (2001) Impacts of chilling temperatures on photosynthesis in warm-climate plants. Trends Plant Sci 6(1):36–42

Alvarez AMR, Rodríguez MLG (2000) Lipids in pharamaceutical and cosmetic preparations. Grasas Aceites 51:74–96

Amid A, Lyctovchenko A, Fernie AR, Warren G, Thorlby GJ (2012) The sensitive to freezing3 mutation of *Arabidopsis thaliana* is a cold-sensitive allele of homomeric acetyl-CoA carboxylase that results in cold-induced cuticle deficiencies. J Exp Bot 63(14):5289–5299

Anchordoguy TJ, Rudolph AS, Carpenter JF, Crowe JH (1987) Modes of interaction of cryoprotectants with membrane phospholipids during freezing. Cryobiology 24(4):324–331

Anderson JA, Taliaferro CM (2000) Freeze tolerance of seed-producing turf Bermuda grasses. Crop Sci 42:190–192

Andrews CJ (1987) Low-temperature stress in field and forage crop production—an overview. Can J Plant Sci 67:1121–1133

Antikainen M, Griffith M (1997) Antifreeze protein accumulation in freezing-tolerant cereals. Physiol Plant 99(3):423–432

Apel K, Hirt H (2004) Reactive oxygen species: metabolism, oxidative stress, and signal transduction. Annu Rev Plant Biol 55:373–399

Arbaoui M, Balko C, Link W (2008) Study of faba bean (*Vicia faba* L.) winter-hardiness and development of screening methods. Field Crop Res 106(1):60–67

Arisz SA, Christa T, Teun M (2009) Plant PA signaling via diacylglycerol kinase. Biochim Biophys Acta 1791(9):869–875

Arisz SA, van Wijk R, Roels W, Zhu JK, Haring MA, Munnik T (2013) Rapid phosphatidic acid accumulation in response to low temperature stress in Arabidopsis is generated through diacylglycerol kinase. Front Plant Sci 4:1–15

Arisz SA, Heo JY, Koevoets IT, Zhao T, van Egmond P, Meyer JA, Zeng W, Niu X, Wang B, Mitchell-Oids T, Schranz ME, Testerink C (2018) Diacylglycerol acyltransferase 1 contributes to freezing tolerance. Plant Physiol 177(4):1410–1424

Arondel V, Lemieux B, Hwang I, Gibson S, Goodeman HM, Somerville CR (1992) Map-based cloning of a gene controlling w-3 fatty acid desaturation in *Arabidopsis*. Science 258:1353–1354

Arunga RO, Morrison WR (1971) The structural analysis of wheat flour glycerolipids. Lipids 6(10): 768–776

Aubert Y, Vile D, Pervent M, Aldon D, Ranty B, Simonneau T, Vavasseur A, Galaud JP (2010) RD20, a stress-inducible caleosin, participates in stomatal control, transpiration and drought tolerance in *Arabidopsis thaliana*. Plant Cell Physiol 51(12):1975–1987

Baca S (1987) Variability in ice nucleation strains of pseudomonas syringae isolated from diseased woody plants in pacific northwest nurseries. Plant Dis 71(5):412

Badea C, Basu S (2009) Effect of low temperature on metabolism of membrane lipids in plants and associated gene expression. Plant Omics 2(2):78–84

Bagnall D (1979) Low temperature responses of three Sorghum species. In: Lyons JM, Graham D, Raison JK (eds) Low temperature stress in crop plants; the role of the membrane. Academic, New York, pp 67–80

Bai DM, Xue YY, Zhao JJ, Huang L, Tian YX, Quan BQ, Fang JH (2018a) Identification of old—tolerance during germination stage and genetic diversity of SSR markers in peanut landraces of Shanxi province. Acta Agron Sin 44(10):1459–1467

Bai H, Su LC, Hu B, Ling L (2018b) Expression of AhDREB1, an AP2/ERF transcription factor gene from peanut, is affected by histone acetylation and increases abscisic acid sensitivity and tolerance to osmotic stress in *Arabidopsis*. Int J Mol Sci 19(5):e1441

Bakht J, Bano A, Dominy P (2006) The role of abscisic acid and low temperature in chickpea (*Cicer arietinum*) cold tolerance. II. Effects on plasma membrane structure and function. J Exp Bot 57(14):3707–3715

Balogh G, Péter M, Glatz A, Gombos I, Török Z, Horváth I, Harwood JL, Vígh L (2013) Key role of lipids in heat stress management. FEBS Lett 587(13):1970–1980

Bargmann BO, Munnik T (2006) The role of phospholipase D in plant stress responses. Curr Opin Plant Biol 9(5):515–522

Barnes AC, Benning C, Roston RL (2016) Chloroplast membrane remodeling during freezing stress is accompanied by cytoplasmic acidification activating SENSITIVE TO FREEZING2. Plant Physiol 171(3):2140–2149

Barrero-Gil J, Salinas J (2013) Post-translational regulation of cold acclimation response. Plant Sci 205–206:48–54

Barrero-Sicilia C, Silvestre S, Haslam RP, Michaelson LV (2017) Lipid remodelling: unravelling the response to cold stress in *Arabidopsis* and its extremophile relative *Eutrema salsugineum*. Plant Sci 263:194–200

Barroso JB, Corpas FJ, Carreras A, Sandalio LM, Valderrama R, Palma JM, Lupiáñez JA, del Río LA (1999) Localization of nitric-oxide synthase in plant peroxisomes. J Biol Chem 274(51): 36729–36733

Bashri G, Singh M, Mishra RK, Kumar J, Singh VP, Prasad SM (2018) Kinetin regulates uv-b-induced damage to growth, photosystem ii photochemistry, and nitrogen metabolism in tomato seedlings. J Plant Growth Regul 37(1):233–245

Bates PD, Browse J (2012) The significance of different diacylgycerol synthesis pathways on plant oil composition and bioengineering. Front Plant Sci 3:1–11

Bates PD, Stymne S, Ohlrogge J (2013) Biochemical pathways in seed oil synthesis. Curr Opin Plant Biol 16(3):358–364

Battey JF, Schmid KM, Ohlrogge JB (1989) Genetic engineering for plant oils: potential and limitations. Trends Biotechnol 7(5):122–125

Baud S, Guyon V, Kronenberger J, Wuilléme S, Miquel M, Caboche M, Lepiniec L, Rochat C (2003) Multifunctional acetyl-CoA carboxylase 1 is essential for very long chain fatty acid elongation and embryo development in *Arabidopsis*. Plant J 33(1):75–86

Baux A, Hebeisen T, Pellet D (2008) Effects of minimal temperatures on low-linolenic rapeseed oil fatty-acid composition. Eur J Agron 29(2–3):102–107

Baxter A, Mittler R, Suzuki N (2014) ROS as key players in plant stress signalling. J Exp Bot 65(5): 1229–1240

Beisson F, Li-Beisson Y, Pollard M (2012) Solving the puzzles of cutin and suberin polymer biosynthesis. Curr Opin Plant Biol 15(3):329–337

Benning C (2009) Mechanisms of lipid transport involved in organelle biogenesis in plant cells. Annu Rev Cell Dev Biol 25:71–91

Benning C, Xu C, Awai K (2006) Non-vesicular and vesicular lipid trafficking involving plastids. Curr Opin Plant Biol 9(3):241–247

Berkey R, Bendigeri D, Xiao S (2012) Sphingolipids and plant defense/disease: the "death" connection and beyond. Front Plant Sci 3:68

Berry J, Bjorkrnan O (1980) Photosynthetic response and adaptation to temperature in higher plants. Annu Rev Plant Physiol 31:491–543

Bertin P, Bullens P, Bouharmont J, Kinet J-M (1998) Somaclonal variation and chilling tolerance improvement in rice: changes in fatty acid composition. Plant Growth Regul 24(1):31–41

Bertrand B, Vaast P, Alpizar E, Etienne H, Davrieux F, Charmetant P (2006) Comparison of bean biochemical composition and beverage quality of Arabica hybrids involving Sudanese-Ethiopian origins with traditional varieties at various elevations in Central America. Tree Physiol 26:1239–1248

Beyer P, Nievelstein V, Albabili S, Bonk M, Kleinig H (1994) Biochemical aspects of carotene desaturation and cyclization in chromoplast membranes from *Narcissus pseudonarcissus*. Pure Appl Chem 66(5):1047–1056

Bhandari K, Sharma KD, Rao BH, Siddique KHM, Gaur P, Agrawal SK, Nair RM, Nayyar H (2017) Temperature sensitivity of food legumes: a physiological insight. Acta Physiol Plant 39(3):68

Bilska-Kos A, Szczepanik J, Sowinski P (2016) Cold induced changes in the water balance affect immunocytolocalization pattern of one of the aquaporins in the vascular system in the leaves of maize (*Zea mays* L.). J Plant Physiol 205:75–79

Birsoy K, Festuccia WT, Laplante M (2013) A comparative perspective on lipid storage in animals. J Cell Sci 126(Pt. 7):1541–1552

Blee E (2002) Impact of phyto-oxylipins in plant defense. Trends Plant Sci 7(7):315–322

Blée E, Flenet M, Boachon B, Fauconnier ML (2012) A non-canonical caleosin from Arabidopsis efficiently epoxidizes physiological unsaturated fatty acids with complete stereoselectivity. FEBS J 279(20):3981–3995

Board JE, Kahlon CS (2011) Soybean yield formation: what controls it and how it can be improved? In: El-Shemy HA (ed) Soybean physiology and biochemistry. InTech Open Access, Rijeka, pp 1–36

Bodner M, Morozova K, Kruathongsri P, Thakeow P, Scampicchio M (2019) Effect of harvesting altitude, fermentation time and roasting degree on the aroma released by coffee powder monitored by proton transfer reaction mass spectrometry. Eur Food Res Technol 245(4): 1499–1506

Bohn M, Lüthje S, Sperling P, Heinz E, Dörffling K (2007) Plasma membrane lipid alteration induced by cold acclimation and abscisic acid treatment of winter wheat seedling differing in frost resistance. J Plant Physiol 164(2):146–156

Bouvier-Navé P, Berna A, Noiriel A, Compagnon V, Carlsson AS, Banas A, Stymne S, Schaller H (2010) Involvement of the phospholipid sterol acyltransferase1 in plant sterol homeostasis and leaf senescence. Plant Physiol 152(1):107–119

Bravi E, Pereetti G, Montanari L (2006) Fatty acids by high-performace liquid chromatography and evaporative light-scattering detector. J Chromatogr 1134(1–2):210–214

Brocard L, Immel F, Coulon D, Esnay N, Tuphile K, Pascal S, Claverol S, Fouillen L, Bessoule J-J, Bréhélin C (2017) Proteomic analysis of lipid droplets from *Arabidopsis* aging leaves brings new insight into their biogenesis and functions. Front Plant Sci 8:894

Brockman JA, Norman HA, Hilderbrand DF (1990) Effects of temperature, light and a chemical modulator on linolenate biosyntheis in mutant and wild type *Arabidopsis cali*. Phytochemistry 29(5):1447–1453

Browse JA, Slack CR (1981) Catalase stimulates linoleate desaturase activity in microsomes from developing linseed cotyledons. FEBS Lett 131:111–114

Browse J, Somerville C (1991) Glycerolipids synthesis: biochemistry and regulation. Annu Rev Plant Physiol Plant Mol Biol 42:467–506

Browse J, Warwick N, Somerville CR, Slack CR (1986) Fluxes through the prokaryotic and eukaryotic pathways of lipid synthesis in the '16:3' plant *Arabidopsis thaliana*. Biochem J 235(1):25–31

Brush RA, Griffith M, Mlynarz A (1994) Characterization and quantification of intrinsic ice nucleators in winter rye (*Secale cereale*) leaves. Plant Physiol 104(2):725–735

Buchanan B, Gruissem W, Jones R (2002) Biochemistry and molecular biology of plants. Wiley, Rockville, pp 456–527

Buchanan BB, Gruissem W, Jones RL (2015) Biochemistry & molecular biology of plants, 2nd edn. Wiley Blackwell, West Sussex, pp 456–527

Bulder HAM, Speek EJ, Van Hasselt PR, Kupier PJC (1991) Growth temperature and lipid composition of cucumber genotypes differing in adaptation to low energy conditions. J Plant Physiol 138(6):655–660

Burgos A, Szymanski J, Seiwert B, Degenkolbe T, Hannah MA, Giavalisco P, Willmitzer L (2011) Analysis of short-term changes in the Arabidopsis thaliana glycerolipidome in response to temperature and light. Plant J 66(4):656–668

Burton JW (1997) Soybean (Glycine max (L.) Merr.). Field Crops Res 53:171–186

Bustamante CA, Monti LL, Gabilondo J, Scossa F, Valentini G, Budde CO, Lara MV, Fernie AR, Drincovich MF (2016) Differential metabolic rearrangements after cold storage are correlated with chilling injury resistance of peach fruits. Front Plant Sci 7(499):1478

Byun MY, Cui LH, Lee J, Park H, Lee A, Kim WT, Lee H (2018) Identification of rice genes associated with enhanced cold tolerance by comparative transcriptome analysis with two transgenic rice plants overexpressing DaCBF4 or DaCBF7, isolated from antarctic flowering plant *Deschampsia antarctica*. Front Plant Sci 9:601

Cacas JL, Furt F, Le Guédard M, Schmitter JM, Buré C, Gerbeau-Pissot P, Moreau P, Bessoule JJ, Simon-Plas F, Mongrand S (2012) Lipids of plant membrane rafts. Prog Lipid Res 51(3): 272–299

Cai Q, Wang S, Cui Z, Sun J, Ishii Y (2004) Changes in freezing tolerance and its relationship with the contents of carbohydrates and proline in overwintering centipedegrass (*Eremochloa ophiuroides* (Munro) Hack.). Plant Prod Sci 7(4):421–426

Campos PS, Quartin V, Ramalho JC, Nunes MA (2003) Electrolyte leakage and lipid degradation account for cold sensitivity in leaves of *Coffea* sp. plants. J Plant Physiol 160(3):283–292

Cantrel C, Vazquez T, Puyaubert J, Reze N, Lesch M, Kaiser WM, Dutilleul C, Guillas I, Zachowski A, Baudouin E (2011) Nitric oxide participates in cold-responsive phosphosphingolipid formation and gene expression in *Arabidopsis thaliana*. New Phytol 189(2):415–427

Cavicchioli R (2006) Cold-adapted archaea. Nat Rev Microbiol 4(5):331–343

Chandra S, Heinstein PF, Low PS (1996) Activation of phospholipase A by plant defense elicitors. Plant Physiol 110(3):979–986

Chapman KD, Ohlrogge JB (2012) Compartmentation of triacylglycerol accumulation in plants. J Biol Chem 287(4):2288–2294

Chapman KD, Dyer JM, Mullen RT (2012) Biogenesis and functions of lipid droplets in plants: thematic review series: lipid droplet synthesis and metabolism: from yeast to man. J Lipid Res 53(2):215–226

Chauton MS, Winge P, Brembu T, Vadstein O, Bones AM (2013) Gene regulation of carbon fixation, storage, and utilization in the diatom *Phaeodactylum tricornatum* acclimated to light/ dark cycles. Plant Physiol 161(2):1034–1048

Chen THH (1994) Plant adaptation to low temperature stress. Can J Plant Pathol 16(3):231–236

Chen MJ, Thelen JJ (2013) ACYL-LIPID DESATURASE2 is required for chilling and freezing tolerance in *Arabidopsis*. Plant Cell 25(4):1430–1444

Chen B, Yang H (2013) 6-Benzylaminopurine alleviates chilling injury of postharvest cucumber fruit through modulating antioxidant system and energy status. J Sci Food Agric 93(8): 1915–1921

Chen J-Q, Dong Y, Wang Y-J, Liu Q, Zhang J-S, Chen S-Y (2003) An AP2/EREBP-type transcription-factor gene from rice is cold-inducible and encodes a nuclear-localized protein. Theor Appl Genet 107(6):972–979

Chen J, Burke JJ, Xin Z, Xu C, Velten J (2006) Characterization of *Arabidopsis* thermosensitive mutant atts02 reveals an important role for galactolipids in thermotolerance. Plant Cell Environ 29(7):1437–1448

Chen JY, He LH, Jiang YM, Wang Y, Joyce DC, Ji ZL, Lu WJ (2008a) Role of phenylalanine ammonia-lyase in heat pretreatment-induced chilling tolerance in banana fruit. Physiol Plant 132(3):318–328

Chen QF, Xiao S, Chye ML (2008b) Overexpression of the *Arabidopsis* 10-kilodalton acyl-coenzyme A-binding protein ACBP6 enhances freezing tolerance. Plant Physiol 148(1): 304–315

Chen M, Markham JE, Cahoon EB (2012) Sphingolipid Δ8 unsaturation is important for glucosylceramide biosynthesis and low-temperature performance in *Arabidopsis*. Plant J 69(5):769–781

Chen X, Song F, Liu F, Tian C, Liu S, Xu H, Zhu X (2014) Effect of different arbuscular mycorrhizal fungi on growth and physiology of maize at ambient and low temperature regimes. Sci World J 2014:1–7

Chen Y, Zhou X-R, Zhang Z-J, Dribnenki P, Singh S, Green A (2015a) Development of high oleic oil crop platform in flax through RNAi-mediated multiple FAD2 gene silencing. Plant Cell Rep 34(4):643–653

Chen J, Han G, Shang C, Li J, Zhang H, Liu F, Wang J, Liu H, Zhang Y (2015b) Proteomic analyses reveal differences in cold acclimation mechanisms in freezing-tolerant and freezing-sensitive cultivars of alfalfa. Front Plant Sci 6:105

Chen Y, Cui Q, Xu Y, Yang S, Ming G, Wang Y (2015c) Effects of tung oilseed FAD2 and DGAT2 genes on unsaturated fatty acid accumulation in *Rhodotorula glutinis* and *Arabidopsis t haliana*. Mol Gen Genomics 290(4):1605–1613

Chen QF, Xu L, Tan WJ, Chen L, Qi H, Xie LJ, Chen MX, Liu BY, Yu LJ, Yao N, Zhang JH, Shu W, Xiao S (2015d) Disruption of the Arabidopsis defense regulator genes SAG101, EDS1, and PAD4 confers enhanced freezing tolerance. Mol Plant 8(10):1536–1549

Chen BB, Wang J, Zhang G, Liu J, Manan S, Hu H, Zhao J (2016) Two types of soybean diacylglycerol acyltransferases are differentially involved in triacylglycerol biosynthesis and response to environmental stresses and hormones. Sci Rep 6:2854

Chi X, Hu R, Zhang X, Chen M, Chen N, Pan L, Wang T, Wang M, Yang Z, Wang Q, Yu S (2014) Cloning and functional analysis of three diacylglycerol acyltransferase genes from peanut (*Arachis hypogaea* L.). PLoS One 9(9):e105834

Chinnusamy V, Ohta M, Kanrar S, Lee B-H, Hong X, Agarwal M, Zhu J-K (2003) ICE1: a regulator of cold-induced transcriptome and freezing tolerance in Arabidopsis. Genes Dev 17(8):1043–1054

Chinnusamy V, Zhu J, Zhu JK (2007) Cold stress regulation of gene expression in plants. Trends Plant Sci 12(10):444–451

Chinnusamy V, Zhu JK, Sunkar R (2010) Gene regulation during cold stress acclimation in plants. Methods Mol Biol 639:39–55

Chitraju C, Mejhert N, Haas JT, Diaz-Ramirez LG, Grueter CA, Imbriglio JE, Pinto S, Koliwad SK, Walther TC, Farese RV Jr (2017) Triglyceride synthesis by DGAT1 protects adipocytes from lipid-induced ER stress during lipolysis. Cell Metab 26(2):407–418

Choi C, Hwang CH (2015) The barley lipid transfer protein, BLT101, enhances cold tolerance in wheat under cold stress. Plant Biotechnol Rep 9(4):197–207

Choudhury NK, Behera RK (2001) Photoinhibition of photosynthesis: role of carotenoids in photoprotection of chloroplast constituents. Photosynthetica 39(4):481–488

Choudhury FK, Rivero RM, Blumwald E, Mittler R (2016) Reactive oxygen species, abiotic stress and stress combination. Plant J 90(5):856–867

Constantinidou HA, Menkissoglu O (1992) Characteristics and importance of heterogeneous ice nuclei associated with citrus fruits. J Exp Bot 43(4):585–591

Cook D, Fowler S, Fiehn O, Thomashow MF (2004) A prominent role for the CBF cold response pathway in configuring the low-temperature metabolome of *Arabidopsis*. Proc Natl Acad Sci U S A 101(42):15243–15248

Cossins A (1994) Homeoviscous adaptation of biological membranes and its functional significance. Temperature adaptation of biological membranes. Portland Press, London, pp 63–76

Coursol S, Fan LM, Le Stunff H, Spiegel S, Gilroy S, Assmann SM (2003) Sphingolipid signalling in Arabidopsis guard cells involves heterotrimeric G proteins. Nature 423(6940):651–654

Crisosto CH, Mitchell FG, Ju Z (1999) Susceptibility to chilling injury of peach nectarine and plum cultivars grown in California. HortScience 34(6):1116–1118

Cui LB, Zhu L, Jiang LX (2017) The effect of ABA on biosynthesis of fatty acids and storage proteins and the relevant mechanism in cruciferae oilseed. J Agric Biotechnol 25(7):1059–1071

D'Hont A, Souza GM, Menossi M, Vincentz M, Van-Slyuis M-A, Glaszmann JC, Ulian E (2008) Sugarcane: a major source of sweetness, alcohol and bioenergy. In: Moore PH, Ming R (eds) Genomics of tropical crop plants. Springer, New York, pp 483–513

Da Cruz RP, Golombieski JI, Bazana MT, Cabreira C, Silveira TF, da Silva LP (2010) Alterations in fatty acid composition due to cold exposure at the vegetative stage in rice. Braz J Plant Physiol 22(3):199–207

Dahlqvist A, Stahl U, Lenman M, Banas A, Lee M, Sandager L, Ronne H, Stymne S (2000) Phospholipid:diacylglycerol acyltransferase: an enzyme that catalyzes the acyl-CoA-independent formation of triacylglycerol in yeast and plants. Proc Natl Acad Sci U S A 97(12): 6487–6492

De Palma M, Grillo S, Massarelli I, Costa A, Balogh G, Vigh L, Leone A (2008) Regulation of desaturase gene expression, changes in lipid composition and freezing tolerance in potato plants. Mol Breed 21:15–26

Deepa G, Singh V, Naidu KA (2008) Nutrient composition and physicochemical properties of Indian medicinal rice—Njavara. Food Chem 106(1):165–171

Degenkolbe T, Giavalisco P, Zuther E, Seiwert B, Hincha DK, Willmitzer L (2012) Differential remodeling of the lipidome during cold acclimation in natural accessions of *Arabidopsis thaliana*. Plant J 72(6):972–982

Deryabin AN, Dubinina IM, Burakhanova EA, Astakhova NV, Subel'nikova EP, Trunova TI (2005) Influence of yeast derived invertase gene expression in potato plants on membrane lipid peroxidation at low temperature. J Therm Biol 30:73–77

Dionne J, Castonguay Y, Nadeau P, Desjardins Y (2001) Freezing tolerance and carbohydrate changes during cold acclimation of green-type annual bluegrass (*Poa annua* L.) ecotypes. Crop Sci 41:443–451

Dita MA, Rispail N, Prats E, Rubiales D, Singh KB (2006) Biotechnology approaches to overcome biotic and abiotic stress constraints in legumes. Euphytica 147(1):1–24

Domínguez E, Cuartero J, Heredia A (2011a) An overview on plant cuticle biomechanics. Plant Sci 181(2):77–84

Domínguez E, Heredia-Guerrero JA, Heredia A (2011b) The biophysical design of plant cuticles: an overview. New Phytol 189(4):938–949

Dong W, Hu R, Chu Z, Zhao J, Tan L (2017) Effect of different drying techniques on bioactive components, fatty acid composition, and volatile profile of robusta coffee beans. Food Chem 234:121–130

Dörmann P (2013) Galactolipids in plant membranes. In: eLS. Wiley, Chichester, pp 1–6

Dörmann P, Benning C (2002) Galactolipids rule in seed plants. Trends Plant Sci 7(3):112–118

Drissner D, Kunze G, Callewaert N, Gehrig P, Tamasloukht M, Boller T, Felix G, Amrhein N, Bucher M (2007) Lyso-phosphatidylcholine is a signal in the arbuscular mycorrhizal symbiosis. Science 318(5848):265–268

Du ZY (2010) Depletion of the membrane-associated acyl-coenzyme A-binding protein ACBP1 enhances the ability of cold acclimation in *Arabidopsis*. Plant Physiol 152(3):1585–1597

Duman JG (1994) Purification and characterization of a thermal hysteresis protein from a plant, the bittersweet nightshade *Solanum dulcamara*. Biochim Biophys Acta 1206(1):129–135

Durrett TP, Benning C, Ohlrogge J (2008) Plant triacylglycerols as feedstocks for the production of biofuels. Plant J 54(4):593–607

Dutilleul C, Benhassaine-Kesri G, Demandre C, Reze N, Launay A, Pelletier S, Renou JP, Zachowski A, Baudouin E, Guillas I (2012) Phytosphingosine-phosphate is a signal for AtMPK6 activation and Arabidopsis response to chilling. New Phytol 194(1):181–191

Dyer JM, Chapital DC, Cary JW, Pepperman AB (2001) Chilling-sensitive, post-transcriptional regulation of a plant fatty acid desaturase expressed in *Yeast*. Biochem Biophys Res Commun 282(4):1019–1025

Dyer JM, Stymne S, Green AG, Carlsson AS (2008) High-value oils from plants. Plant J 54(4): 640–654

Eastmond PJ (2006) SUGAR-DEPENDENT1 encodes a patatin domain triacylglycerol lipase that initiates storage oil breakdown in germinating Arabidopsis seeds. Plant Cell 18(3):665–675

Eastmond PJ, Graham IA (2001) Re-examining the role of the glyoxylate cycle in oilseeds. Trends Plant Sci 6(2):72–78

Eriksson SK, Michael K, Jan P, Gerhard G, Pia H (2011) Tunable membrane binding of the intrinsically disordered dehydrin Lti30, a cold-induced plant stress protein. Plant Cell 23(6): 2391–2404

Esmaili S, Salehi H (2012) Effects of temperature and photoperiod on postponing bermudagrass (*Cynodon dactylon* [L.] Pers.) turf dormancy. J Plant Physiol 169:851–858

Fahy E, Subramanian S, Murphy RC, Nishijima M, Raetz CRH, Shimizu T, Spener F, van Meer G, Wakelam MJ, Dennis EA (2005) Update of lipid maps comprehensive classification system for lipids. J Lipid Res 46(5):839–861

Fahy E, Subramaniam S, Murphy RC, Nishijima M, Raetz CR, Shimizu T, Spener F, van Meer G, Wakelam MJ, Dennis EA (2009) Update of the Lipid Maps comprehensive classification system for lipids. J Lipid Res 50(Suppl S1):S9–S14

Fan J, Yan C, Xu C (2013) Phospholipid:diacylglycerol acyltransferase-mediated triacylglycerol biosynthesis is crucial for protection against fatty acid-induced cell death in growing tissues of Arabidopsis. Plant J 76(6):930–942

Fan J, Yu L, Xu C (2017) A central role for triacylgly cerol in membrane lipid breakdown, fatty acid β-oxidation, and plant survival under extended darkness. Plant Physiol 174(3):1517–1530

Fatiha AID (2019) Chater 5: Plant lipid metabolism. In: Advances in lipid metabolism. V.B. Rdrigo IntechOpen

Feng ZT, Deng YQ, Fan H, Sun QJ, Sui N, Wang BS (2014a) Effects of NaCl stress on the growth and photosynthetic characteristics of *Ulmus pumila* L. seedlings in sand culture. Photosynthetica 52(2):313–320

Feng ZT, Sun QJ, Deng YQ, Sun SF, Zhang JG, Wang BS (2014b) Study on pathway and characteristics of ion secretion of salt glands of *Limonium bicolor*. Acta Physiol Plant 36(10): 2729–2741

Feofilova E, Tereshina V, Memorskaya A, Khokhlova N (2000) Different mechanisms of the biochemical adaptation of mycelial fungi to temperature stress: changes in the lipid composition. Microbiology 69(5):509–515

Feofilova E, Tereshina V, Memorskaya A (2005) Biochemical mechanisms of temperature adaptation in the (+) and (−) strains of *Blakeslea trispora*. Microbiology 74(6):650–654

Fernandez-Santos R, Izquierdo Y, Lopez A, Muniz L, Martinez M, Cascon T, Hamberg M, Castresana C (2020) Protein profiles of lipid droplets during the hypersensitive defense response of Arabidopsis against Pseudomonas infection. Plant Cell Physiol 61(6):1144–1157

Fiehn O, Kopka J, Trethewey RN, Willmitzer L (2000) Identification of uncommon plant metabolites based on calculation of elemental compositions using gas chromatography and quadrupole mass spectrometry. Anal Chem 72(15):3573–3580

Fowler DB (1983) Effect of management practices on winter survival and yield of winter wheat produced in regions with harsh winter climates. In: Fowler DB, Gusta LV, Slinkard AE, Hobin BA (eds) Proceedings Western Canada winter wheat conference. University of Saskatchewan, Saskatoon, pp 238–282

Fowler S, Thomashow MF (2002) *Arabidopsis* transcriptome profiling indicates that multiple regulatory pathways are activated during cold acclimation in addition to the CBF cold response pathway. Plant Cell 14(8):1675–1690

Foyer CH, Lam HM, Nguyen HT, Siddique KHM, Varshney RK, Colmer TD, Cowling W, Bramley H, Mori TA, Hodgson JM, Cooper JW, Miller AJ, Kunert K, Vorster J, Cullis C, Ozga JA, Wahlqvist ML, Liang Y, Shou H, Shi K, Yu J, Fodor N, Kaiser BN, Wong FL, Valliyodan B, Considine MJ (2016) Neglecting legumes has compromised human health and sustainable food production. Nat Plants 2(8):1–10

Frame J, Laidlaw AS (2011) Improved grassland management. The Crowood Press, Marlborough

Froidure S, Canonne J, Daniel X, Jauneau A, Briere C, Roby D, Rivas S (2010) AtsPLA2-alpha nuclear relocalization by the *Arabidopsis* transcription factor AtMYB30 leads to repression of the plant defense response. Proc Natl Acad Sci U S A 107(34):15281–15286

Galle AM, Demandre C, Guerche P, Joseph M, Dubacq JP, Mazliak P et al (1992) Biosynthesis of γ-linolenic acid in cotyledons and microsomal preparations of developing seeds of of common borage (*Borage officinalis*). In: Chérif A et al (eds) Metabolism, structure and utilization of plant lipids. Centre National Pédagogique, Tunis, pp 185–188

Gan Y, Zhang L, Zhang Z, Dong S, Li J, Wang Y, Zheng X (2009) The LCB2 subunit of the sphingolip biosynthesis enzyme serine palmitoyltransferase can function as an attenuator of the hypersensitive response and Bax-induced cell death. New Phytol 181(1):127–146

Gao J, Wallis JG, Browse J (2015) Mutations in the prokaryotic pathway rescue the fatty acid biosynthesis1 mutant in the cold. Plant Physiol 169(1):442–452

Garba L, Ali MSM, Oslan SN, Rahman RNZRBA (2017) Review on fatty acid desaturases and their roles in temperature acclimatisation. J Appl Sci 17(6):282–295

Garber M, Steponkus PL (1976) Alterations in chloroplast thylakoids during cold acclimation. Plant Physiol 57(5):681–686

Gardiner JC (2001) A 90-kD phospholipase D from tobacco binds to microtubules and the plasma membrane. Plant Cell 13(9):2143–2158

Gargano S, Dilallo G, Kobayashi GS, Maresca B (1995) A temperature-sensitive strain of *Histoplasma capsulatum* has an altered Δ 9-fatty acid desaturase geneacid desaturase gene. Lipids 30(10):899–906

Gasulla F, Vom Dorp K, Dombrink I, Zahringer U, Gisch N, Dörmann P, Bartels D (2013) The role of lipid metabolism in the acquisition of desiccation tolerance in *Craterostigma plantagineum*: a comparative approach. Plant J 75(5):726–741

Gaude N, Nakamura Y, Scheible WR, Ohta H, Dörmann P (2008) Phospholipase C5 NPC5 is involved in galactolipid accumulation during phosphate limitation in leaves of *Arabidopsis*. Plant J 56(1):28–39

Ge WY, Kong XM, Zhao YB, Wei BD, Zhou Q, Ji SJ (2019) Insights into the metabolism of membrane lipid fatty acids associated with chilling injury in post-harvest bell peppers. Food Chem 295:26–35

Gebrekidan M, Redi-Abshiro M, Chandravanshi BS, Ele E, Mohammed AM, Mamo H (2019) Influence of altitudes of coffee plants on the alkaloids contents of green coffee beans. Chem Int 5(4):247–257

Gemel J, Golinowski W, Kaniuga Z (1986) Low-temperature induced changes in chloroplast ultrastructure in relation to changes of Hill reaction activity, manganese and free fatty acid levels in chloroplasts of chilling-sensitive and chilling-resistant plants. Acta Physiol Plant 8: 135–143

Gibson S, Arondel V, Iba K, Somerville C (1994) Cloning of a temperature-regulated gene encoding a chloroplast omega-3 desaturase from *Arabidopsis thaliana*. Plant Physiol 106(4): 1615–1621

Gilmour SJ, Sebolt AM, Salazar MP, Everard JD, Thomashow MF (2000) Overexpression of the Arabidopsis CBF3 transcriptional activator mimics multiple biochemical changes associated with cold acclimation. Plant Physiol 124(4):1854–1865

Goldfine H (1984) The control of membrane fluidity in plasmalogen-containing anaerobic bacteria. In: Kates M, Manson LA (eds) Membrane fluidity. Biomembranes, vol 12. Plenum, New York, pp 349–377

Gomboz Z, Murata N (1998) Genetic engineering of the unsaturation of membrane glycerolipids: effects on the ability of the photosynthetic machinery to tolerate temperature stress. In: Siegenthaler P-A, Murata N (eds) Lipids in photosynthesis: structure function and genetics. Kluwer, Dordrecht, pp 249–262

Gomes E, Jakobsen MK, Axelsen KB, Geisler M, Palmgreen MG (2000) Chilling tolerance in *Arabidopsis* involves ALA1, a member of a new family of putative aminophospholipid translocases. Plant Cell 12(12):2441–2453

Goodwin SM, Jenks MA (2005) Plant cuticle function as a barrier to water loss. In: Jenks MA, Hasegawa PM (eds) Plant abiotic stress. Blackwell Publishing Inc., Oxford, pp 14–36

Goold H, Beisson F, Peltier G, Li-Beisson Y (2014) Microalgal lipid droplets: composition, diversity, biogenesis and functions. Plant Cell Rep 34(4):545–555

Graham IA (2008) Seed storage oil mobilization. Annu Rev Plant Biol 59:115–142

Grant OM, Brennan D, Mellisho P, Salas CD, Dix CJ (2014) Impact of enhanced capacity to scavenge reactive oxygen species on cold tolerance of tobacco. Int J Plant Sci 175(5):544–554

Griffiths G, Stobart AK, Stymne S (1985) The acylation of *sn*-glycerol 3-phosphate and the metabolism of phosphatidate in microsomal preparations from the developing cotyledons of safflower (*Carthamus tinctorius* L.) seed. Biochem J 230:379–388

Griffiths G, Stymne S, Stobart AK (1988) The utilisation of fatty-acid substrates in triacylglycerol biosynthesis by tissue-slices of developing safflower (*Carthamus tinctorius* L.) and sunflower (*Helianthus annuus* L.) cotyledons. Planta 173:309–316

Grillitsch K, Connerth M, Kofeler H, Arrey TN, Rietschel B, Wagner B, Karas M, Daum G (2011) Lipid particles/droplets of the yeast *Saccharomyces cerevisiae* revisited: lipidome meets proteome. Biochim Biophys Acta 1811(12):1165–1176

Grimaud F, Renaut J, Dumont E, Sergeant K, Lucau-Danila A, Blervacq AS, Sellier H, Bahrman N, Lejeune-Hénaut I, Delbreil B, Goulas E (2013) Exploring chloroplastic changes related to chilling and freezing tolerance during cold acclimation of pea (*Pisum sativum* L.). J Proteomics 80:145–159

Grosbois M (1971) Biosynthèse des acides gras au cours du développement du fruit et de la graine du lierre. Phytochemistry 10(6):1261–1273

Gross DC, Proebsting EL, Maccrindle-Zimmerman H (1988) Development, distribution, and characteristics of intrinsic, nonbacterial ice nuclei in prunus wood. Plant Physiol 88(3):915–922

Gruszecki WI, Strzalka K (2005) Carotenoids as modulators of lipid membrane physical properties. Biochim Biophys Acta 1740(2):108–115

Gruszka J, Pawlak A, Kruk J (2008) Tocochromanols, plastoquinol, and other biological prenyllipids as singlet oxygen quenchers-determination of singlet oxygen quenching rate constants and oxidation products. Free Radic Biol Med 45(6):920–928

Gu Y, He L, Zhao C, Wang F, Yan B, Gao Y, Li Z, Yang K, Xu J (2017) Biochemical and transcriptional regulation of membrane lipid metabolism in maize leaves under low temperature. Front Plant Sci 8:2053

Guillas I, Guellim A, Reze N, Baudouin E (2013) Long chain base changes triggered by a short exposure of Arabidopsis to low temperature are altered by AHb1 non-symbiotic haemoglobin overexpression. Plant Physiol Biochem 63:191–195

Gunstone FD, Dijkstra AF, Harwood JL (2007) In: Gunstone FD (ed) Lipid handbook, 3rd edn. Taylor & Francis, London

Guo X, Liu D (2018) Cold signaling in plants: insights into mechanisms and regulation. J Integr Plant Biol 60(9):745–756

Guo J, Zhang Z, Bi Y, Yang W, Xu Y, Zhang L (2005) Decreased stability of photosystem I in dgd1 mutant of *Arabidopsis thaliana*. FEBS Lett 579(17):3619–3624

Guo L, Mishra G, Taylor K, Wang X (2011) Phosphatidic acid binds and stimulates Arabidopsis sphingosine kinases. J Biol Chem 286(15):13336–13345

Guo YH, Jia WJ, Song J, Wang DA, Chen M, Wang BS (2012a) *Thellungilla halophila* is more adaptive to salinity than *Arabidopsis thaliana* at stages of seed germination and seedling establishment. Acta Physiol Plant 34(4):1287–1294

Guo L, Mishra G, Markham JE, Li M, Tawfall A, Welti R, Wang X (2012b) Connections between sphingosine kinase and phospholipase-D in the abscisic acid signaling pathway in Arabidopsis. J Biol Chem 287(11):8286–8296

Guo JR, Suo SS, Wang BS (2015) Sodium chloride improves seed vigour of the euhalophyte *Suaeda salsa*. Seed Sci Res 25(3):335–344

Guo YF, Zhang YL, Shan W, Cai YJ, Liang SM, Chen JY, Lu WJ, Kuang JF (2018a) Identification of two transcriptional activators MabZIP4/5 in controlling aroma biosynthetic genes during banana ripening. J Agric Food Chem 66(24):6142–6150

Guo JR, Li YD, Han GL, Song J, Wang BS (2018b) NaCl markedly improved the reproductive capacity of the euhalophyte *Suaeda salsa*. Funct Plant Biol 45(3):350–361

Gurr MI, Harwood JL, Frayn KN (2002) Lipid biochemistry, 5th edn. Blackwells, Oxford

Guschina IA, Harwood JL (2006) Mechanisms of temperature adaptation in poikolitherns. FEBS Lett 580(23):5477–5483

Guschina IA, Everard JD, Kinney AJ, Quant PA, Harwood JL (2014) Studies on the regulation of lipid biosynthesis in plants: application of control analysis to soybean. Biochim Biophys Acta Biomembr 1838(6):1488–1500

Hagos M, Redi-Abshiro M, Chandravanshi BS, Ele E, Mohammed AM, Mamo H (2018) Correlation between caffeine contents in green coffee beans and altitudes of the coffee plants grown in southwest Ethiopia using ultraviolet-visible spectrophotometry. Bull Chem Soc Ethiop 32(1): 13–25

Hamada T, Kodama H, Takeshita K, Utsumi H, Iba K (1998) Characterization of transgenic tobacco with an increased α-linolenic acid level. Plant Physiol 118(2):591–598

Härtel H, Dörmann P, Benning C (2000) DGD1-independent biosynthesis of extraplastidic galactolipids after phosphate deprivation in *Arabidopsis*. Proc Natl Acad Sci U S A 97(19): 10649–10654

Harwood J (1991) Strategies for coping with low environmental temperatures. Trends Biochem Sci 16(4):126–127

Harwood JL (1994) Environmental factors affecting lipid metabolism. Prog Lipid Res 33:193–202

Harwood JL (1996) Fatty acid biosynthesis. In: Murphy DJ (ed) Plant lipids: biology, utilisation and manipulation. Blackwell Publishing, Oxford, pp 27–66

Harwood JL (1998) Involvement of chloroplast lipids in the reaction of plants submitted to stress. In: Siegenthaler P-A, Murata N (eds) Lipids in photosynthesis: structure function and genetics. Kluwer, Dordrecht, pp 287–302

Harwood JL (2005) Fatty acid biosynthesis. In: Murphy DJ (ed) Plant lipids: biology, utilisation and manipulation. Blackwell Publishing, Oxford, pp 27–66

Harwood JL, Russell NJ (1984) Lipids in plants and microbes. George Allen and Unwin, London

Hatfield RD, Jung HG, Broderick G, Jenkins TC (2007) Nutritional chemistry of forages. In: Barnes RF, Nelson JF, Collins M (eds) Forage—science of grassland agriculture. Blackwell Publishing, Oxford, pp 467–486

Havaux M, Eymery F, Porfirova S, Rey P, Dörmann P (2005) Vitamin E protects against photoinhibition and photooxidative stress in *Arabidopsis thaliana*. Plant Cell 17:3451–3469

Hazel JR, Williams EE (1990) The role of alterations in membrane lipid compositon in enabling physiological adaptation of organisms to their physical environment. Prog Lipid Res 29(3): 167–227

He M, Ding NZ (2020) Plant unsaturated fatty acids: multiple roles in stress response. Front Plant Sci 11:562785

He YA, Li YP, Cui LX, Xie LX, Zheng CK, Zhou GH, Zhou JJ, Xie XZ (2016) Phytochrome B negatively affects cold tolerance by regulating OsDREB1 gene expression through phytochrome interacting factor-like protein OsPIL16 in rice. Front Plant Sci 7:1963

He W, Zhang X, Yang P, Qiu Y, Wang X, Yang N (2017) Effects of abscisic acid on mitochondrial membrane phospholipase-D activity under low temperature stress. Guangxi Plants 37:742–748

He M, He CQ, Ding NZ (2018) Abiotic stresses: general defenses of land plants and chances for engineering multistress tolerance. Front Plant Sci 9:1771

He M, Qin CX, Wang X, Ding NZ (2020) Plant unsaturated fatty acids: biosynthesis and regulation. Front Plant Sci 11:390

Hecht V, Knowles CL, Vander Schoor JK, Liew LC, Jones SE, Lambert MJM, Weller JL (2007) Pea LATE BLOOMER1 is a GIGANTEA ortholog with roles in photoperiodic flowering, deetiolation, and transcriptional regulation of circadian clock gene homologs. Plant Physiol 144(2):648–661

Heidarvand L, Millar AH, Taylor NL (2017) Responses of the mitochondrial respiratory system to low temperature in plants. Crit Rev Plant Sci 36(4):1–24

Heinz E (1993) Biosynthesis of polyunsaturated fatty acids. In: Moore TS Jr (ed) Lipid metabolism in plants. CRC Press, Boca Raton, pp 33–89

Heinz E, Roughan PG (1983) Similarities and differences in lipid metabolism of chloroplasts isolated from 18:3 and 16:3 plants. Plant Physiol 72(2):273–279

Heppard EP, Kinney AJ, Stecca KL, Miao GH (1996) Developmental and growth temperature regulation of two different microsomal omega-6 desaturase genes in soybeans. Plant Physiol 110:311–319

Heredia A (2003) Biophysical and biochemical characteristics of cutin, a plant barrier biopolymer. Biochim Biophys Acta 1620:1–7

Hernández ML, Sicardo MD, Alfonso M, Martínez-Rivas JM (2019) Transcriptional regulation of stearoyl-acyl carrier protein desaturase genes in response to abiotic stresses leads to changes in the unsaturated fatty acids composition of olive mesocarp. Front Plant Sci 10:251

Higashi Y, Okazaki Y, Takano K, Myouga F, Shinozaki K, Knoch E, Fukushima A, Saito K (2018) HEAT INDUCIBLE LIPASE1 remodels chloroplastic monogalactosyl-diacylglycerol by liberating α-linolenic acid in Arabidopsis leaves under heat stress. Plant Cell 30(8):1887–1905

Hincha DK, Höfner R, Schwab KB, Heber U, Schmitt JM (1987) Membrane rupture is the common cause of damage to chloroplast membranes in leaves injured by freezing or excessive wilting. Plant Physiol 83(2):251–253

Hoermiller II, Naegele T, Augustin H, Stutz S, Weckwerth W, Heyer AG (2017) Subcellular reprogramming of metabolism during cold acclimation in *Arabidopsis thaliana*. Plant Cell Environ 40(5):602–610

Hoffman L, DaCosta M, Ebdon JS, Watkins E (2010) Physiological changes during cold acclimation of perennial ryegrass accessions differing in freeze tolerance. Crop Sci 50:1037–1047

Hölzl G, Dörmann P (2019) Chloroplast lipids and their biosynthesis. Annu Rev Plant Biol 70:51–81

Hölzl G, Witt S, Gaude N, Melzer M, Schöttler MA, Dörmann P (2009) The role of diglycosyl lipids in photosynthesis and membrane lipid homeostasis in *Arabidopsis*. Plant Physiol 150(3): 1147–1159

Hong Y, Zheng S, Wang X (2008) Dual functions of phospholipase-Dalpha1 in plant response to drought. Mol Plant 1(2):262–269

Hong Y, Devaiah SP, Bahn SC, Thamasandra BN, Li M, Welti R, Wang X (2009) Phospholipase-D epsilon and phosphatidic acid enhance Arabidopsis nitrogen signaling and growth. Plant J 58(3): 376–387

Hong Y, Zhao J, Guo L, Kim SC, Deng X, Wang G, Zhang G, Li M, Wang X (2016) Plant phospholipases D and C and their diverse functions in stress responses. Prog Lipid Res 62:55–74

Horn PJ, Chapman KD (2012) Lipidomics in tissues, cells and subcellular compartments. Plant J 70(1):69–80

Horn PJ, Ledbetter NR, James CN, Hoffman WD, Case CR, Verbeck GF, Chapman KD (2011) Visualization of lipid droplet composition by direct organelle mass spectrometry. J Biol Chem 286(5):3298–3306

Horvath I, Glatz A, Varvasovszki V, Torok Z, Pali T, Balogh G, Kovacs E, Nadasdi L, Benko S, Joo F, Vigh L (1998) Membrane physical state controls the signaling mechanism of the heat shock response in Synechocystis PCC 6803: identification of hsp17 as a 'fluidity gene'. Proc Natl Acad Sci U S A 95(7):3513–3518

Hou Q, Ufer G, Bartels D (2016) Lipid signalling in plant responses to abiotic stress. Plant Cell Environ 39(5):1029–1048

Huang A (1992) Oil bodies and oleosins in seeds. Annu Rev Plant Biol 43:177–200

Huang AHC (2018) Plant lipid droplets and their associated proteins: potential for rapid advances. Plant Physiol 176(3):1894–1918

Huang BR, DaCosta M, Jiang YW (2014) Research advances in mechanisms of turfgrass tolerance to abiotic stresses: from physiology to molecular biology. Crit Rev Plant Sci 33(2–3):141–189

Huang X, Chen MH, Yang LT, Li YR, Wu JM (2015) Effects of exogenous abscisic acid on cell membrane and endogenous hormone contents in leaves of sugarcane seedlings under cold stress. Sugar Tech 17:59–64

Huang S, Van AO, Schwarzl AM, Belt K, Millar AH (2016) Roles of mitochondrial reactive oxygen species in cellular signaling and stress response in plants. Plant Physiol 171(3): 1551–1559

Hugly S, Somerville CA (1992) Role for membrane lipid polyunsaturation in chloroplast biogenesis at low temperature. Plant Physiol 99(1):97–202

Huner NP, Oquist G, Hurry VM, Krol M, Falk S, Griffith M (1993) Photosynthesis, photoinhibition and low temperature acclimation in cold tolerant plants. Photosynth Res 37(1):19–39

Huner NPA, Öquist G, Sarhan F (1998) Energy balance and acclimation to light and cold. Trends Plant Sci 3(6):224–230

Hur J, Jung K-H, Lee C-H, An G (2004) Stress-inducible OsP5CS2 gene is essential for salt and cold toleranc in rice. Plant Sci 167(3):417–426

Hurtado-Benavides A, Dorado DA, Sánchez-Camargo ADP (2016) Study of the fatty acid profile and the aroma composition of oil obtained from roasted Colombian coffee beans by supercritical fluid extraction. J Supercrit Fluids 113:44–52

Ianutsevich EA, Danilova OA, Groza NV, Tereshina VM (2016) Membrane lipids and cyosol carbohydrates in *Aspergillus niger*, under osmotic, oxidative, and cold impact. Microbiology 85(3):302–310

Iba K (2002) Acclimative response to temperature stress in higher plants: approaches of gene engineering for temperature tolerance. Annu Rev Plant Biol 53:225–245

Ichihara K, Takahashi T, Fujii S (1988) Diacylglycerol acyltransferase in maturing safflower seeds: its influences on the fatty acid composition of triacylglycerol and on the rate of triacylglycerol synthesis. Biochim Biophys Acta 958(1):125–129

Iqbal H, Yaning C, Waqas M, Rehman H, Shareef M, Iqbal S (2018a) Hydrogen peroxide application improves quinoa performance by affecting physiological and biochemical mechanisms under water-deficit conditions. J Agron Crop Sci 204(6):541–553

Iqbal H, Yaning C, Waqas M, Shareef M, Raza ST (2018b) Differential response of quinoa genotypes to drought and foliage- applied H_2O_2 in relation to oxidative damage, osmotic adjustment and antioxidant capacity. Ecotoxicol Environ Saf 164:344–354

Ishikawa HA (1996) Ultrastructural features of chilling injury: injured cells and the early events during chilling of suspension-cultured mung bean cells. Am J Bot 83(7):825–835

Ishizaki-Nishizawa O (1996) Low-temperature resistance of higher plants is significantly enhanced by a nonspecific cyanobacterial desaturase. Nat Biotechnol 14(8):1003–1006

Ito MK, Simpson KL (1996) The biosynthesis of ω-3 fatty acids from 18:2ω in *Artemia* spp. Comp Biochem Physiol B Biochem Mol Biol 115(1):69–76

Jallet D, Xing D, Hughes A, Moosburner M, Simmons MP, Allen AE, Peers G (2020) Mitochondrial fatty acid β-oxidation is required for storage-lipid catabolism in a marine diatom. New Phytol 228(3):946–958

Janda M, Planchais S, Djafi N, Martinec J, Burketova L, Valentova O, Zachowski A, Ruelland E (2013) Phosphoglycerolipids are master players in plant hormone signal transduction. Plant Cell Rep 32(6):839–851

Janska A, Marsík P, Zelenková S, Ovesná J (2010) Cold stress and acclimation—what is important for metabolic adjustment? Plant Biol 12(3):395–405

Jeon J, Kim J (2013) Cold stress signaling networks in *Arabidopsis*. J Plant Biol 56:69–76

Jia X, Sun C, Zuo Y, Li G, Li G, Ren L, Chen G (2016) Integrating transcriptomics and metabolomics to characterise the response of Astragalus membranaceus Bge. var. mongolicus (Bge.) to progressive drought stress. BMC Genomics 17:188

Jian H, Xie L, Wang Y, Cao Y, Wan M, Lv D, Li J, Lu K, Xu X, Lin L (2020) Characterization of cold stress responses in different rapeseed ecotypes based on metabolomics and transcriptomics analyses. PeerJ 8:e8704

Johnson G, Williams JP (1989) Effect of growth temperature on the biosynthesis of chloroplastic galactosyldiacylglycerol molecular species in *Brassica napus* leaves. Plant Physiol 91(3): 924–929

Joosen RVL, Lammers M, Balk PA, Brønnum P, Konings MCJM, Perks M, Stattin E, van Wordragen M, Lonneke A, van der Geest HM (2006) Correlating gene expression to physiological parameters and environmental conditions during cold acclimation of *Pinus sylvestris*, identification of molecular markers using cDNA microarrays. Tree Physiol 26(10):1297–1313

Jouhet J, Maréchal E, Baldan B, Bligny R, Joyard J, Block MA (2004) Phosphate deprivation induces transfer of DGDG galactolipid from chloroplast to mitochondria. J Cell Biol 167(5): 863–874

Jouve L, Engelmann F, Noirot M, Charrier A (1993) Evaluation of biochemical markers (sugar, proline, malonedialdehyde and ethylene) for cold sensitivity in micro cuttings of two coffee species. Plant Sci 91(1):109–116

Joyard J, Block MA, Malberbe A, Marachal E, Douce R (1993) Origin and synthesis of galactolipid and sulfo-lipid headgroups. In: Moore TS Jr (ed) Lipid metabolism in plants. CRC Press, Boca Raton, pp 231–258

Juliano BO (1994) Polysaccharides, proteins and lipids of rice. In: Juliano BO (ed) Rice chemistry and technology. American Association of Cereal Chemists, St. Paul, pp 59–174

Jumrani K, Bhatia VS (2014) Impacts of elevate temperature on growth and yield of chickpea (*Cicer arietinum* L.). Field Crop Res 164:90–97

Jung HW, Tschaplinski TJ, Wang L, Glazebrook J, Greenberg JT (2009) Priming in systemic plant immunity. Science 324(5923):89–91

Kachroo A, Kachroo P (2009) Fatty acid-derived signals in plant defense. Annu Rev Phytopathol 47:153–176

Kalachova T, Iakovenko O, Kretinin S, Kravets V (2013) Involvement of phospholipase D and NADPH-oxidase in salicylic acid signaling cascade. Plant Physiol Biochem 66:127–133

Kanervo E, Tasaka Y, Murata N, Aro EM (1997) Membrane lipid unsaturation modulate processing of the photosystem II reaction-center protein D1 at low temperatures. Plant Physiol 114(3): 841–849

Kang L, Wang YS, Uppalapati SR, Wang K, Tang Y, Vadapalli V, Venables BJ, Chapman KD, Blancaflor EB, Mysore KS (2008) Overexpression of a fatty acid amide hydrolase compromises innate immunity in *Arabidopsis*. Plant J 56(2):336–349

Kaniuga Z (2008) Chilling response of plants: importance of galactolipase, free fatty acids and free radicals. Plant Biol 10(2):171–184

Kaplan F, Kopka J, Haskell DW, Zhao W, Schiller KC, Gatzke N, Sung DY, Guy CL (2004) Exploring the temperature-stress metabolome of *Arabidopsis*. Plant Physiol 136(4):4159–4168

Kaplan F, Kopka J, Sung DY, Zhao W, Popp M, Porat R, Guy CL (2007) Transcript and etabolite profiling during cold acclimation of Arabidopsis reveals an intricate relationship of cold-regulated gene expression with modifications in metabolite content. Plant J 50(6):967–981

Karabudak T, Bor M, Ozdemir F, Turkan I (2014) Glycine betaine protects tomato (*Solanum lycopersicum*) plants at low temperature by inducing fatty acid desaturase7 and lipoxygenase gene expression. Mol Biol Rep 41(3):1401–1410

Karl TR, Melillo JM, Peterson TC (2009) Global climate change impacts in the United States. Cambridge University Press, New York

Katagiri T, Ishiyama K, Kato T, Tabata S, Kobayashi M, Shinozaki K (2005) An important role of phosphatidic acid in ABA signaling during germination in *Arabidopsis thaliana*. Plant J 43(1): 107–117

Kaup MT, Froese CD, Thompson JE (2002) A role for diacylglycerol acyltransferase during leaf senescence. Plant Physiol 129(4):1616–1626

Kazemi-Shahandashti SS, Maali-Amiri R (2018) Global insights of protein responses to cold stress in plants: signaling, defence, and degradation. J Plant Physiol 226:123–135

Kelly AA, Feussner I (2016) Oil is on the agenda: lipid turnover in higher plants. Biochim Biophys Acta 1861(9 Pt B):1253–1268

Kelly AA, Quettier AL, Shaw E, Eastmond PJ (2011) Seed storage oil mobilization is important but not essential for germination or seedling establishment in Arabidopsis. Plant Physiol 157(2): 868–875

Kennedy JF (1963) The mechanics of dunes and antidunes in erodible-bed channels. J Fluid Mech 16(4):521–544

Khodakovskaya M, McAvoy R, Peters J, Wu H, Li Y (2006) Enhanced cold tolerance in transgenic tobacco expressing a chloroplast ω-3 fatty acid desaturase gene under the control of a cold-inducible promoter. Planta 223(5):1090–1100

Kilaru A, Herrfurth C, Keereetaweep J, Hornung E, Venables BJ, Feussner I, Chapman KD (2011) Lipoxygenase-mediated oxidation of polyunsaturated N-acylethanolamines in *Arabidopsis*. J Biol Chem 286(17):15205–15214

Kilian J, Whitehead D, Horak J, Wanke D, Wein S, Batistic O, D'Angelo C, Bornberg-Bauer E, Kudla J, Harter K (2007) The AtGenExpress global stress expression data set: protocols, evaluation and model data analysis of UV-B light, drought and cold stress responses. Plant J 50(2):347–363

Kim HU, Hsieh K, Ratnayake C, Huang AHC (2002) A novel group of oleosins is present inside the pollen of Arabidopsis. J Biol Chem 277(25):22677–22684

Kim HU, Li Y, Huang AHC (2005) Ubiquitous and endoplasmic reticulum-located lysophosphatidyl acyltransferase, LPAT2, is essential for female but not male gametophyte development in *Arabidopsis*. Plant Cell 17(4):1073–1089

Kim HJ, Ok SH, Bahn SC, Jang J, Oh SA, Park SK, Twell D, Ryu SB, Shin JS (2011) Endoplasmic reticulum- and Golgi-localized phospholipase A2 plays critical roles in Arabidopsis pollen development and germination. Plant Cell 23(1):94–110

Kim EY, Park KY, Seo YS, Kim WT (2016) *Arabidopsis* small rubber particle protein homolog srps play dual roles as positive factors for tissue growth and development and in drought stress responses. Plant Physiol 170(4):2494–2510

Kimball SL, Salisbury FB (1973) Ultrastructural changes of plants exposed to low temperatures. Am J Bot 60(10):1028

Kinney AJ (1996) Development of genetically engineered soybean oils for food applications. J Food Lipids 3:273–292

Kitta K, Ebihara M, Iizuka T, Yoshikawa R, Isshiki K, Kawamoto S (2005) Variations in lipid content and fatty acid composition of major non-glutinous rice cultivars in Japan. J Food Compos Anal 18(4):269–278

Klempova T, Mihalik D, Certik M (2013) Characterization of membrane-bound fatty acid desaturases. Gen Physiol Biophys 32(4):639–645

Kobayashi K (2016) Role of membrane glycerolipids in photosynthesis, thylakoid biogenesis and chloroplast development. J Plant Res 129(4):565–580

Kobayashi N, DellaPenna D (2008) Tocopherol metabolism, oxidation and recycling under high light stress in *Arabidopsis*. Plant J 55(4):607–618

Kobayashi K, Fujii S, Sato M, Toyooka K, Wada H (2014) Specific role of phosphatidylglycerol and functional overlaps with other thylakoid lipids in Arabidopsis chloroplast biogenesis. Plant Cell Rep 34(4):631–642

Kodama H, Horiguchi G, Nishiuchi T, Nishimura M, Iba K (1995) Fatty acid desaturation during chilling acclimation is one of the factors involved in conferring low-temperature tolerance to young tobacco leaves. Plant Physiol 107(4):1177–1185

Kodra E, Steinhaeuser K, Ganguly AR (2011) Persisting cold extremes under 21st-century warming scenarios. Geophys Res Lett 38(8):L08705

Kojima M, Suzuki H, Ohnishi M, Ito S (1998) Effect of growth temperature on lipids of Adzuki bean cells. Phytochemistry 47:1483–1487

Kolattukudy PE (1981) Structure, biosynthesis, and biodegradation of cutin and suberin. Annu Rev Plant Physiol Plant Mol Biol 32:539–567

Kong XM, Wei B, Gao Z, Zhou Y, Shi F, Zhou X, Zhou Q, Ji S (2018a) Changes in membrane lipid composition and function accompanying chilling injury in bell peppers. Plant Cell Physiol 59(1):167–178

Kong F, Burlacot A, Liang Y, Légeret B, Alseekh S, Brotman Y, Fernie AR, Krieger-Liszkay A, Beisson F, Peltier G, Li-Beisson Y (2018b) Interorganelle communication: peroxisomal MALATE DEHYDROGENASE2 connects lipid catabolism to photosynthesis through redox coupling in *Chlamydomonas*. Plant Cell 30(8):1824–1847

Kong F, Romero IT, Warakanont J, Li-Beisson Y (2018c) Lipid catabolism in microalgae. New Phytol 218(4):1340–1348

Konig S, Feussner K, Schwarz M, Kaever A, Iven T, Landesfeind M, Ternes P, Karlovsky P, Lipka V, Feussner I (2012) Arabidopsis mutants of sphingolipid fatty acid alpha-hydroxylases accumulate ceramides and salicylates. New Phytol 196(4):1086–1097

Kosova K, Prasil IT, Vitamvas P, Dobrev P, Motyka V, Flokova K, Novak O, Tureckova V, Rolcik J, Pesek B, Travnickova A, Gaudinova A, Galiba G, Janda T, Vlasakova E, Prasilova P, Vankova R (2012) Complex phytohormone responses during the cold acclimation of two wheat cultivars differing in cold tolerance, winter Samanta and spring Sandra. J Plant Physiol 169(6): 567–576

Krasensky J, Jonak C (2012) Drought, salt, and temperature stress-induced metabolic rearrangements and regulatory networks. J Exp Bot 63(4):1593–1608

Kratsch HA, Wise RR (2000) The ultrastructure of chilling stress. Plant Cell Environ 23(4): 337–350

Krause GH, Grafflage S, Rumich-Bayer S, Somersalo S (1988) Effects of freezing on plant mesophyll cells. Symp Soc Exp Biol 42:311–327

Kreps JA, Wu Y, Chang H-S, Zhu T, Wang X, Harper JF (2002) Transcriptome changes for Arabidopsis in response to salt, osmotic and cold stress. Plant Physiol 130(4):2129–2141

Krieger-Liszkay A, Trebst A (2006) Tocopherol is the scavenger of singlet oxygen produced by the triplet states of chlorophyll in the PSII reaction centre. J Exp Bot 57(8):1677–1684

Ksas B, Becuwe N, Chevalier A, Havaux M (2015) Plant tolerance to excess light energy and photooxidative damage relies on plastoquinone biosynthesis. Sci Rep 5:10919

Kumar GS, Jagannadham MV, Ray MK (2002) Low-temperature-induced changes in composition and fluidity of lipopolysaccharides in the Antarctic psychrotrophic bacterium. *Pseudomonas syringae*. J Bacteriol 184:6746–6749

Kumar S, Nayyar H, Bhanwara RK, Upadhyaya HD (2010) Chilling stress effects on reproductive biology of chickpea. J SAT Agric Res 8:1–14

Kumar D, Datta R, Hazra S, Sultana A, Mukhopadhyay R, Chattopadhyay S (2015) Transcriptomic profiling of *Arabidopsis thaliana* mutant pad2.1 in response to combined cold and osmotic stress. PLoS One 10(3):e0122690

Kunst L, Browse J, Somerville C (1988) Altered regulation of lipid biosynthesis in a mutant of Arabidopsis deficient in chloroplast glycerol-3-phosphate acyltransferase activity. Proc Natl Acad Sci U S A 85(12):4143–4147

Kuntam S, Puskás LG, Ayaydin F (2015) Characterization of a new class of blue-fluorescent lipid droplet markers for live cell imaging in plants. Plant Cell Rep 34(4):655–665

Lachaud C, Prigent E, Thuleau P, Grat S, Da Silva D, Briere C, Mazars C, Cotelle V (2013) 14-3-3-regulated Ca(2+)-dependent protein kinase CPK3 is required for sphingolipid-induced cell death in Arabidopsis. Cell Death Differ 20(2):209–217

Lang V, Mantyla E, Welin B, Sundberg B, Palva ET (1994) Alterations in water status: endogenous abscisic-acid content, and expression of rab18 gene during the development of freezing tolerance in *Arabidopsis thaliana*. Plant Physiol 104(4):1341–1349

Lanna AC, José IC, de Almeida Oliveira MG, Barros EG, Moreira MA (2005) Effect of temperature on polyunsaturated fatty acid accumulation in soybean seeds. Braz J Plant Physiol 17(2): 213–222

Larcher W (2004) Ecofisiologia vegetal. RiMa, São Carlos

Leddet PC, Geneves L (1982) Effects of low temperatures (+6 degrees C and +2 degrees C) on the fine structure of mitochondria and chloroplasts in the *Ephedra* tissues, cultivated in vitro. Ann Sci Nat 13:27–49

Lee SB, Suh MC (2015) Advances in the understanding of cuticular waxes in *Arabidopsis thaliana* and crop species. Plant Cell Rep 34(4):557–572

Lee YS, Choi YB, Suh S, Lee J, Assmann SM, Joe CO, Kelleher JF, Crain RC (1996) Abscisic acid-induced phosphoinositide turnover in guard cell protoplasts of *Vicia faba*. Plant Physiol 110(3): 987–996

Lee SH, Chung GC, Steudle S (2005a) Gating of aquaporins by low temperature in roots of chilling-sensitive cucumber and -tolerant figleaf gourd. J Exp Bot 56:985–995

Lee SH, Chung GC, Steudle E (2005b) Low temperature and mechanical stresses differently gate aquaporins of root cortical cells of chilling-sensitive cucumber and figleaf gourd. Plant Cell Environ 28(9):1191–1202

Lee SH, Ahn SJ, Im YJ, Cho K, Chung G-C, Cho B-H, Han O (2005c) Differential impact of low temperature on fatty acid unsaturation and lipoxygenase activity in figleaf gourd and cucumber roots. Biochem Biophys Res Commun 330(4):1194–1198

Lee K-R, Chen GQ, Kim HU (2015) Current progress towards the metabolic engineering of plant seed oil for hydroxy fatty acids production. Plant Cell Rep 34(4):603–615

Lei S, Tang GY, Xu PL, Zhao XB, Liu ZJ (2014) Cloning and analysis of 5′ flanking regions of *Arachisis hypogaea* L. genes encoding plastidial acyl carrier protein. Acta Agron Sin 40(3): 381–389

Levitt J (1980) Chilling, freezing, and high temperature stress. In: Responses of plants to environmental stresses, vol 1. Academic, New York

Li WQ (2004) The plasma membrane-bound phospholipase D delta enhances freezing tolerance in *Arabidopsis thaliana*. Nat Biotechnol 22(4):427–433

Li W, Li M, Zhang W, Welti R, Wang X (2004) The plasma membrane-bound phospholipase Dδ enhances freezing tolerance in *Arabidopsis thaliana*. Nat Biotechnol 22(4):427–433

Li W, Wang R, Li M, Li L, Wang C, Welti R, Wang X (2008) Differential degradation of extraplastidic and plastidic lipids during freezing and post-freezing recovery in *Arabidopsis thaliana*. J Biol Chem 283(1):461–468

Li M, Bahn SC, Guo L, Musgrave W, Berg H, Welti R, Wang X (2011) Patatin-related phospholipase pPLAIIIbeta-induced changes in lipid metabolism alter cellulose content and cell elongation in Arabidopsis. Plant Cell 23(3):1107–1123

Li ZG, Zeng HZ, Ao PX, Gong M (2014) Lipid response to short-term chilling shock and long-term chill hardening in *Jatropha curcas* L. seedlings. Acta Physiol Plant 36(10):2803–2814

Li Q, Zheng Q, Shen W, Cram D, Fowler DB, Wei Y, Zou J (2015) Understanding the biochemical basis of temperature-induced lipid pathway adjustments in plants. Plant Cell 27(1):86–103

Li Q, Shen W, Zheng Q, Fowler DB, Zou J (2016) Adjustments of lipid pathways in plant adaptation to temperature stress. Plant Signal Behav 11(1):e1058461

Li X, Liu P, Yang P, Fan C, Sun X, Zhang J (2018) Characterization of the glycerol-3-phosphate acyltransferase gene and its real-time expression under cold stress in *Paeonia lactiflora* pall. PLoS One 13(8):e0202168

Liang H, Yao N, Song JT, Luo S, Lu H, Greenberg JT (2003) Ceramides modulate programmed cell death in plants. Genes Dev 17(21):2636–2641

Liang SM, Kuang JF, Ji SJ, Chen QF, Deng W, Min T, Shan W, Chen JY, Lu WJ (2020) The membrane lipid metabolism in horticultural products suffering chilling injury. Food Qual Saf 4: 9–14

Li-Beisson Y, Shorrosh B, Beisson F, Andersson MX, Arondel V, Bates PD, Baud S, Bird D, Debono A, Durrett TP, Franke RB, Graham IA, Katayama K, Kelly AA, Larson T, Markham JE, Miquel M, Molina I, Nishida I, Rowland O, Samuels L, Schmid KM, Wada H, Welti R, Xu C, Zallot R, Ohlrogge J (2010) Acyl-lipid metabolism. Arabidopsis eBook 8:e0133

Li-Beisson Y, Shorrosh B, Beisson F, Andersson MX, Arondel V, Bates PD, Baud S, Bird D, DeBono A, Durrett TP, Franke RB, Graham IA, Katayama K, Kelly AA, Larson T, Markham JE, Miquel M, Molina I, Nishida I, Rowland O, Samuels L, Schmid KM, Wada H, Welti R, Xu C, Zallot R, Ohlrogge J (2013) Acyl-lipid metabolism. Arabidopsis Book 11:e0161

Lichtenthaler HK (2010) The non-mevalonate DOXP/MEP (deoxyxylulose 5-phosphate/ methylerythritol 4-phosphate) pathway of chloroplast isoprenoid and pigment biosynthesis. In: Rebeiz CA, Benning C, Bohnert HJ, Daniell H, Hoober JK, Lichtenthaler HK et al (eds) Advances in photosynthesis and respiration, vol 31. The chloroplast—basics and applications. Springer, Dordrecht, pp 95–118

Lichtenthaler HK, Burkart S (1999) Photosynthesis and high light stress. Bulg J Plant Physiol 25(3–4):3–16

Lim CS, Kang SM, Cho JL (2007) Bell pepper (*Capsicum annuum* L,) fruits are susceptible to chilling injury at the breaker stage of ripeness. HortScience 42(7):1659–1664

Lin W, Oliver DJ (2008) Role of triacylglycerols in leaves. Plant Sci 175:233–237

Lindberg S, Banas A, Stymne S (2005) Effects of different cultivation temperatures on plasma membrane ATPase activity and lipid composition of sugar beet roots. Plant Physiol Biochem 43: 261–268

Lindow SE (1988) Lack of correlation of in vitro antibiosis with antagonism of ice nucleation active bacteria on leaf surfaces by non-ice nucleation active bacteria. Phytopathology 78(4):444

Lingran F, Qiang W, Xiaobin Y, Kwame F (2019) Effects of exogenous lipids and cold acclimation on lycopene production and fatty acid composition in *Blakeslea trispora*. AMB Express 9:162

Link W, Balko C, Stoddard FL (2010) Winter hardiness in faba bean: physiology and breeding. Field Crop Res 115(3):287–296

Lippold F, vom Dorp K, Abraham M, Hölzl G, Wewer V, Yilmaz JL, Lager I, Montandon C, Besagni C, Kessler F, Stymne S, Dörmann P (2012) Fatty acid phytyl ester synthesis in chloroplasts of *Arabidopsis*. Plant Cell 24(5):2001–2014

Liu Q, Singh SP, Green AG (2002) High-stearic and high-oleic cottonseed oils produced by hairpin RNA-mediated post-transcriptional gene silencing. Plant Physiol 129(4):1732–1743

Liu H, Song L, You Y, Li YB, Duan XW, Jiang YM, Lu WJ (2011) Cold storage duration affects litchi fruit quality membrane permeability enzyme activities and energy charge during shelf time at ambient temperature. Postharvest Biol Technol 60(1):24–30

Liu YF, Han XR, Zhan XM, Yang JF, Wang YZ, Song QB, Chen X (2013a) Regulation of calcium on peanut photosynthesis under low night temperature stress. J Integr Agric 12(12):2172–2178

Liu W, Ji SX, Fang XL, Wang QG, Li Z, Yao F, Hou LL, Dai S (2013b) Protein kinase LTRPK1 influences cold adaptation and microtubule stability in rice. J Plant Growth Regul 32(3): 483–490

Liu F, Zhang X, Lu C, Zeng X, Li Y, Fu D, Wu G (2015a) Non-specific lipid transfer proteins in plants: presenting new advances and an integrated functional analysis. J Exp Bot 66(19): 5663–5681

Liu L, Wei Y, Shi F, Liu C, Liu X, Ji S (2015b) Intermittent warming improves postharvest quality of bell peppers and reduces chilling injury. Postharvest Biol Technol 101:18–21

Liu X, Shi W, Yin W, Wang J (2017) Distinct cold responsiveness of a StInvInh2 gene promoter in transgenic potato tubers with contrasting resistance to cold-induced sweetening. Plant Physiol Biochem 111:77–84

Livingston HD III, Heyer A (2009) Fructan and its relationship to abiotic stress tolerance in plants. Cell Mol Life Sci 66:2007–2023

Los DA, Murata N (2002) Sensing and responses to low temperature in cyanobacteria. In: Storey KB, Storey JM (eds) Sensing, signalling and cell adaptation; Storey, vol 3. Elsevier Science B. V., Amsterdam, pp 139–153

Los DA, Murata N (2004) Membrane fluidity and its roles in the perception of environmental signals. Biochim Biophys Acta Biomembr 1666(1–2):142–157

Los DA, Ray MK, Murata N (1997) Differences in the control of the temperature-dependent expression of four genes for desaturases in Synechocystis sp. PCC6803. Mol Microbiol 25: 1167–1175

Lu C, Xin Z, Ren Z, Miquel M, Browse J (2009) An enzyme regulating triacylglycerol composition is encoded by the ROD1 gene of *Arabidopsis*. Proc Natl Acad Sci U S A 106(44):18837–18842

Lu J, Xu Y, Wang J, Singer SD, Chen D (2020) The role of triacylglycerol in plant stress response. Plants 9(4):472

Lurie S, Crisosto CH (2005) Chilling injury in peach and nectarine. Postharvest Biol Technol 37(3): 195–208

Lyons JM (1973) Chilling injury in plants. Annu Rev Plant Physiol 24:445–466

Lyons JM, Graham D, Raison JK (1979) Low temperature stress in crop plants; The role of the membrane. Academic, New York

Ma S, Lin C, Chen Y (1990) Comparative studies of chilling stress on alterations of chloroplast ultrastructure and protein synthesis in the leaves of chilling-sensitive (mungbean) and -insensitive (pea) seedlings. Bot Bull Acad Sin 31:263–272

Mahmood S, Erdogan Orhan I, Ahsan Z, Erdem SA, Gulfraz M (2008) Fatty acid composition of seed oil of different *Sorghum bicolor* varieties. Food Chem 109(4):855–859

Makarenko SP, Dudareva LV, Katyshev AI, Konenkina TA, Nazarova AV, Rudikovskaya EG, Sokolova NA, Chernikova VV, Konstantinov YM (2011) The effect of low temperatures on fatty acid composition of crops with different cold resistance. Biochem (Moscow) Suppl Ser A Membr Cell Biol 5(1):64–69

Mandal MK, Chandra-Shekara AC, Jeong RD, Yu K, Zhu S, Chanda B, Navarre D, Kachroo A, Kachroo P (2012) Oleic acid-dependent modulation of NITRIC OXIDE ASSOCIATED1 protein levels regulates nitric oxide-mediated defense signaling in *Arabidopsis*. Plant Cell 24(4):1654–1674

Marangoni AG, Palma T, Stanley DW (1996) Membrane effects in postharvest physiology. Postharvest Biol Technol 7:193–217

Marengo J, Cornejo A, Satyamurty P, Nobre C, Sea W (1997) Cold surges in tropical and extratropical South America: the strong event in June 1994. Mon Weather Rev 125(11): 2759–2786

Margutti MP, Reyna M, Meringer MV, Racagni GE, Villasuso AL (2017) Lipid signalling mediated by PLD/PA modulates proline and H_2O_2 levels in barley seedlings exposed to short- and long-term chilling stress. Plant Physiol Biochem 113:149–160

Markham JE, Lynch DV, Napier JA, Dunn TM, Cahoon EB (2013) Plant sphingolipids: function follows form. Curr Opin Plant Biol 16(3):350–357

Martin MX, Pablos F, González AG, Valdenebro MX, León-Camacho M (2001) Fatty acid profiles as discriminant parameters for coffee varieties differentiation. Talanta 54(2):291–297

Maruyama K, Urano K, Yoshiwara K, Morishita Y, Sakurai N, Suzuki H, Kojima M, Sakakibara H, Shibata D, Saito K, Shinozaki K, Yamaguchi-Shinozaki K (2014) Integrated analysis of the effects of cold and dehydration on rice metabolites, phytohormones, and gene transcripts. Plant Physiol 164(4):1759–1771

McGlasson WB, Raison JK (1973) Occurrence of a temperature-induced phase transition in mitochondria isolated from apple fruit. Plant Physiol 52(4):390–392

McGlew K, Shaw V, Zhang M, Kim RJ, Yang W, Shorrosh B, Suh MC, Ohlrogge J (2014) An annotated database of *Arabidopsis* mutants of acyl lipid metabolism. Plant Cell Rep 34(4): 519–532

McKeon TA, Stumpf PK (1982) Purification and characterization of the stearoyl-acyl carrier protein desaturase and the acyl-acyl carrier protein thioesterase from maturing seeds of safflower. J Biol Chem 257(20):12141–12147

McLachlan DH, Lan J, Geilfus CM, Dodd AN, Larson T, Baker A, Hõrak H, Kollist H, He Z, Graham I, Mickelbart MV, Hetherinhton AM (2016) The breakdown of stored triacylglycerols is required during light-induced stomatal opening. Curr Biol 26(5):707–712

McLoughlin F, Arisz SA, Dekker HL, Kramer G, de Koster CG, Haring MA, Munnik T, Testerink C (2013) Identification of novel candidate phosphatidic acid-binding proteins involved in the salt-stress response of *Arabidopsis thaliana* roots. Biochem J 450(3):573–581

McNaught AD, Wilkinson A (1997) Lipids. In: McNaught AD, Wilkinson A (eds) Compendium of chemical terminology (the gold book), 2nd edn. Blackwell Scientific Publication, Oxford

Mehari B, Redi-Abshiro M, Chandravanshi BS, Combrinck S, Atlabachew M, McCrindle R (2016a) Profiling of phenolic compounds using UPLC–MS for determining the geographical origin of green coffee beans from Ethiopia. J Food Compos Anal 45:16–25

Mehari B, Redi-Abshiro M, Chandravanshi BS, Atlabachew M, Combrinck S, McCrindle R (2016b) Simultaneous determination of alkaloids in green coffee beans from Ethiopia: chemometric evaluation of geographical origin. Food Anal Methods 9:1627–1637

Mehari B, Redi-Abshiro M, Chandravanshi BS, Combrinck S, McCrindle R (2016c) Characterization of the cultivation region of Ethiopian coffee by elemental analysis. Anal Lett 49:2474–2489

Mendes A, Kelly AA, van Erp H, Shaw E, Powers SJ, Kurup S, Eastmond PJ (2013) bZIP67 regulates the omega-3 fatty acid content of Arabidopsis seed oil by activating fatty acid DESATURASE3. Plant Cell 25(8):3104–3116

Mendoza D (2014) Temperature sensing by membranes. Annu Rev Microbiol 68:101–116

Mène-Saffrané L, DellaPenna D (2010) Biosynthesis, regulation and functions of tocochromanols in plants. Plant Physiol Biochem 48(5):301–309

Méne-Saffrane L, Dubugnon L, Chetelat A, Stolz S, Gouhier-Darimont C, Farmer EE (2009) Nonenzymatic oxidation of trienoic fatty acids contributes to reactive oxygen species management in *Arabidopsis*. J Biol Chem 284(3):1702–1708

Meringer MV, Villasuso AL, Margutti MP, Usorach J, Pasquare SJ, Giusto NM, Machado EE, Racagni GE (2016) Saline and osmotic stresses stimulate PLD/diacylglycerol kinase activities and increase the level of phosphatidic acid and proline in barley roots. Environ Exp Bot 128:69–78

Miedema P (1982) The effects of low temperature on *Zea mays*. Adv Agron 35:93–128

Mikami K, Murata N (2003) Membrane fluidity and the perception of environmental signals in cyanobacteria and plants. Prog Lipid Res 42:572–643

Mikami K, Kanesaki Y, Suzuki I, Murata N (2002) The histidine kinase *Hik33* perceives osmotic stress and cold stress in *Synechocystis* sp. PCC6803. Mol Microbiol 46:905–915

Mintesnot A, Dechassa N (2018) Effect of altitude, shade, and processing methods on the quality and biochemical composition of green coffee beans in Ethiopia. East Afr J Sci 12(2):87–100

Miquel M, Browse J (1992) *Arabidopsis* mutants deficient in polyunsaturated fatty acid synthesis. Biochemical and genetic characterization of a plant oleoyl-phosphatidylcholine desaturase. J Biol Chem 267(3):1502–1509

Miquel MF, Browse JA (1994) High-oleate oilseeds fail to develop at low temperature. Plant Physiol 106(2):421–427

Miquel M, James D Jr, Dooner H, Browse J (1993) *Arabidopsis* requires olyunsaturated lipids for low-temperature survival. Proc Natl Acad Sci U S A 90(13):6208–6212

Mironov KS, Sidorov RA, Trofimova MS, Bedbenov VS, Tsydendambaev VD, Allakhverdiev SI, Los DA (2012) Light-dependent cold-induced fatty acid unsaturation, changes in membrane fluidity, and alterations in gene expression in *Synechocystis*. Biochim Biophys Acta Bioenerg 1817(8):1352–1359

Misson J, Raghothama KG, Jain A, Jouhet J, Block MA, Bligny R, Ortet P, Creff A, Somerville S, Rolland N, Patrick D, Nancy P, Herrera-Estrella L, Nussaume L, Thibaud M (2005) A genome-wide transcriptional analysis using *Arabidopsis thaliana* Affymetrix gene chips determined plant responses to phosphate deprivation. Proc Natl Acad Sci U S A 102(33):11934–11939

Miziorko HM (2011) Enzymes of the mevalonate pathway of isoprenoid biosynthesis. Arch Biochem Biophys 505(2):131–143

Moellering ER, Muthan B, Benning C (2010) Freezing tolerance in plants requires lipid remodeling at the outer chloroplast membrane. Science 330(6001):226–228

Molina I, Kosma D (2014) Role of HXXXD-motif/BAHD acyltransferases in the biosynthesis of extracellular lipids. Plant Cell Rep 34(4):587–601

Mongrand S, Bessoule J, Cabantous F, Cassagne C (1998) The $C_{16:3}$/$C_{18:3}$ fatty acid balance in photosynthetic tissues from 468 plant species. Phytochemistry 49(4):1049–1064

Moore TS (1982) Phospholipid biosynthesis. Ann Rev Plant Physiol 33:235–259

Moore PH (1987) Breeding for stress resistance. In: Heinz DL (ed) Sugarcane improvement through breeding. Elsevier, Amsterdam, pp 503–542

Morales-Cedillo F, González-Solís A, Gutiérrez-Angoa L, Cano-Ramírez DL, Gavilanes-Ruiz M (2014) Plant lipid environment and membrane enzymes: the case of the plasma membrane H^+-ATPase. Plant Cell Rep 34(4):617–629

Mortimer JC, Yu X, Albrecht S, Sicilia F, Huichalaf M, Ampuero D, Michaelson LV, Murphy AM, Matsunaga T, Kurz S, Stephens E, Baldwin TC, Ishii T, Napier JA, Weber APM, Handford MG, Dupree P (2013) Abnormal glycosphingolipid mannosylation triggers salicylic acid-mediated responses in Arabidopsis. Plant Cell 25(5):1881–1894

Moschen S, Di Rienzo JA, Higgins J, Tohge T, Watanabe M, González S, Rivarola M, García-García F, Dopazo J, Hopp HE, Hoefgen R, Fernie AR, Paniego N, Fernández P, Heinz RA (2017) Integration of transcriptomic and metabolic data reveals hub transcription factors involved in drought stress response in sunflower (*Helianthus annuus* L.). Plant Mol Biol 94(4–5):549–564

Mueller SP, Unger M, Guender L, Fekete A, Mueller MJ (2017) Phospholipid: diacylglycerol acyltransferase-mediated triacylglyerol synthesis augments basal thermotolerance. Plant Physiol 175:486–497

Muller C, Riederer M (2005) Plant surface properties in chemical ecology. J Chem Ecol 31:2621–2651

Munné-Bosch S, Alegre L (2002) The function of tocopherols and tocotrienols in plants. Crit Rev Plant Sci 21(2):31–57

Munnik T, Testerink C (2009) Plant phospholipid signaling: "in a nutshell". J Lipid Res 50(Suppl): S260–S265

Munnik T, Arisz SA, De Vrije T, Musgrave A (1995) G protein activation stimulates phospholipase-D signaling in plants. Plant Cell 7(12):2197–2210

Murakami Y, Tsuyama M, Kobayashi Y, Kodama H, Iba K (2000) Trienoic fatty acids and plant tolerance of high temperature. Science 287(5452):476–479

Murata N, Los DA (1997) Membrane fluidity and temperature perception. Plant Physiol 115(3): 875–879

Murata N, Wada H (1995) Acyl-lipid desaturases and their importance in the tolerance and acclimatization to cold of cyanobacteria. Biochem J 308:1–8

Murata N, Yamaya J (1984) Temperature-dependent phase behavior of phosphatidyl-glycerols from chilling-sensitive and chilling-resistant plants. Plant Physiol 74(4):1016–10124

Murata N, Sato N, Takahashi N, Hamasaki Y (1982) Compositions and positional distributions of fatty acids in phospholipids from leaves of chilling sensitive and chilling resistant plants. Plant Cell Physiol 23(6):1071–1079

Murata N, Ishizaki-Nishizawa O, Higashii S, Hayashi H, Tasaka Y, Nishida I (1992) Genetically engineered alteration in the chilling sensitivity of plants. Nature 356(6371):710–713

Murphy DJ (1994) Biogenesis, function and biotechnology of plant storage lipids. Prog Lipid Res 33:71–85

Murphy DJ (1999) Production of novel oils in plants. Curr Opin Biotechnol 10(2):175–180

Murphy C, Wilson JM (1981) Ultrastructural features of chilling-injury in *Episcia reptans*. Plant Cell Environ 4(3):261–265

Musser RL, Thomas SA, Wise RR, Peeler T, Naylor AW (1984) Chloroplast ultrastructure, chlorophyll fluorescence, and pigment composition in chilling-stressed soybeans. Plant Physiol 74(4):749–754

Muzi C, Camoni L, Visconti S, Aducci P (2016) Cold stress affects H+-ATPase and phospholipase-D activity in *Arabidopsis*. Plant Physiol Biochem 108:328–336

Nakamura Y (2017) Plant phospholipid diversity: emerging functions in metabolism and protein–lipid interactions. Trends Plant Sci 22(12):1027–1040

Nakamura Y, Koizumi R, Shui G, Shimojima M, Wenk MR, Ito T, Ohta H (2009) Arabidopsis lipins mediate eukaryotic pathway of lipid metabolism and cope critically with phosphate starvation. Proc Natl Acad Sci U S A 106(49):20978–20983

Nalam VJ, Keeretaweep J, Sarowar S, Shah J (2012) Root-derived oxylipins promote green peach aphid performance on *Arabidopsis* foliage. Plant Cell 24(4):1643–1653

Narvaez-Vasquez J, Florin-Christensen J, Ryan CA (1999) Positional specificity of a phospholipase A activity induced by wounding, systemin, and oligosaccharide elicitors in tomato leaves. Plant Cell 11(11):2249–2260

Nawrath C, Schreiber L, Franke RB, Geldner N, Reina-Pinto JJ, Kunst L (2013) Apoplastic diffusion barriers in *Arabidopsis*. Arabidopsis Book 11:e0167

Nejadsadeghi L, Reza MA, Hassan Z, Sanaz R, Behzad S (2015) Membrane fatty acid compositions and cold-induced responses in tetraploid and hexaploid wheats. Mol Biol Rep 42(2):363–372

Nessler CL, Wernsman EA (1980) Ultrastructural observations of extranuclear temperature sensitive lethality in *Nicotiana tabacum* L. Bot Gazette 141(1):9–14

Ng CK, Carr K, McAinsh MR, Powell B, Hetherington AM (2001) Drought-induced guard cell signal transduction involves sphingosine-1-phosphate. Nature 410(6828):596–599

Nishida I, Murata N (1996) Chilling sensitivity in plants and cyanobacteria: the crucial contribution of membrane lipids. Annu Rev Plant Physiol Plant Mol Biol 47:541–568

Nishiuchi T, Ida K (1998) Role of plastid ω-3 fatty acid desaturases in defence response of higher plants. J Plant Res 111:481–486

Noblet A, Leymarie J, Bailly C (2017) Chilling temperature remodels phospholipidome of *Zea mays* seeds during imbibition. Sci Rep 7(1):8886

Nowicka B, Kruk J (2012) Plastoquinol is more active than α-tocopherol in singlet oxygen scavenging during high light stress of *Chlamydomonas reinhardtii*. Biochim Biophys Acta 1817(3):389–394

O'Neil SD, Priestley DA, Chabot BF (1981) Temperature and aging effects on leaf membranes of a cold hardy perennial, *Fragaria virginiana*. Plant Physiol 68(6):1409–1415

O'Rourke EJ, Soukas AA, Carr CE, Ruvkun G (2009) *C. elegans* major fats are stored in vesicles distinct from lysosome-related organelles. Cell Metab 10(5):430–435

Ohashi Y, Oka A, Rodrigues-Pousada R, Possenti M, Ruberti I, Morelli G, Aoyama T (2003) Modulation of phospholipid signaling by GLABRA2 in root-hair pattern formation. Science 300(5624):1427–1430

Ohlrogge J, Browse J (1995) Lipid biosynthesis. Plant Cell 7:957–970

Ohlrogge JB, Jaworski JG (1997) Regulation of fatty acid synthesis. Annu Rev Plant Physiol Plant Mol Biol 48:109–136

Ohlrogge JB, Browse J, Somerville CR (1991) The genetics of plant lipids. Biochim Biophys Acta 1082:1–26

Ohlrogge J, Browse J, Jaworski J, Somerville C (2015) Lipids. In: Buchanan BB, Druissem W, Jones R (eds) Biochemistry and molecular biology of plants, 2nd edn. Wiley, West Sussex, pp 336–400

Okazaki Y, Saito K (2014) Roles of lipids as signaling molecules and mitigators during stress response in plant. Plant J 79(4):584–596

Okazaki Y, Otsuki H, Narisawa T, Kobayashi M, Sawai S, Kamide Y, Kusano M, Aoki T, Hirai MY, Saito K (2013) A new class of plant lipid is essential for protection against phosphorus depletion. Nat Commun 4:1510

Okuley J, Lightner J, Feldmann K, Yadav N, Lark E, Browse J (1994) Arabidopsis FAD2 gene encodes the enzyme that is essential for polyunsaturated lipid synthesis. Plant Cell 6:147–158

Ooquist G (1983) Effect of low temperature on photosynthesis. Plant Cell Environ 6:281–300

Orvar BL, Sangwan V, Omann F, Dhindsa RS (2000) Early steps in cold sensing by plant cells: the role of actin cytoskeleton and membrane fluidity. Plant J 23(6):785–794

Pareek S, Benkeblia N, Janick J, Cao S, Yahia EM (2014) Postharvest physiology and technology of loquat (*Eriobotrya japonica* Lindl.) fruit. J Sci Food Agric 94(8):1495–1504

Park KY, Kim WT, Kim EY (2018) The proper localization of RESPONSIVE TO DESICCATION 20 in lipid droplets depends on their biogenesis induced by STRESS-RELATED PROTEINS in vegetative tissues. Biochem Biophys Res Commun 495(2):1885–1889

Parkin KL, Marangoni A, Jackman RL, Yada RY, Stanley DW (1989) Chilling injury. A review of possible mechanisms. J Food Biochem 13(2):127–153

Parveez GKA, Rasid OA, Masani MYA, Sambanthamurthi R (2014) Biotechnology of oil palm: strategies towards manipulation of lipid content and composition. Plant Cell Rep 34(4):533–543

Pascual F, Carman GM (2013) Phosphatidate phosphatase, a key regulator of lipid homeostasis. Biochim Biophys Acta Mol Cell Biol Lipids 1831(3):514–522

Patel B, Tandel YN, Patel AH, Patel BL (2016) Chilling injury in tropical and subtropical fruits: a cold storage problem and its remedies: a review. Int J Sci Environ Technol 5(4):1882–1887

Patton AJ, Reicher ZJ (2007) *Zoysiagrass* species and genotypes differ in their winter injury and freeze tolerance. Crop Sci 47:1619–1627

Pearcy RW (1978) Effects of growth temperature on the fatty acid composition of the leaf lipids in *Atriplex lentiformis* (Torr.) Wats. Plant Physiol 61(4):484–486

Peng Z, Li L, Yang L, Zhang B, Chen G, Bi Y (2013) Overexpression of peanut diacylglycerol acyltransferase 2 in *Escherichia coli*. PLoS One 8(4):e61363

Petoukhov V, Semenov VA (2010) A link between reduced Barents-Kara sea ice and cold winter extremes over northern continents. J Geophys Res Atmos 115(21):1–11

Phadtare S, Alsina J, Inouye M (1999) Cold-shock response and cold-shock proteins. Curr Opin Microbiol 2(2):175–180

Pham VT, Herero M, Hormaza JI (2015) Effect of temperature on pollen germination and pollen tube growth in longan (*Dimocarpus longan* Lour.). Sci Hortic 197:470–475

Pinosa F, Buhot N, Kwaaitaal M, Fahlberg P, Thordal-Christensen H, Ellerstrom M, Andersson MX (2013) Arabidopsis phospholipase-Ddelta is involved in basal defense and non-host resistance to powdery mildew fungi. Plant Physiol 163(2):896–906

Platt-Aloia KA, Thomson WW (1987) Freeze fracture evidence for lateral phase separations in the plasmalemma of chilling-injured avocado fruit. Protoplasma 136:71–80

Płażek A, Dubert F, Kopeć P, Dziurka M, Kalandyk A, Pastuszak J, Wolko B (2018) Seed hydropriming and smoke water significantly improve low-temperature germination of *Lupinus angustifolius* L. Int J Mol Sci 19(4):992

Ploschuk EL, Bado LA, Salinas M, Wassner DF, Windauer LB, Insausti P (2014) Photosynthesis and fluorescence responses of *Jatropha curcas* to chilling and freezing stress during early vegetative stages. Environ Exp Bot 102:18–26

Pocock TH, Hurry VM, Savitch LV, Huner NPA (2001) Susceptibility to low-temperature photoinhibition and the acquisition of freezing tolerance in winter and spring wheat: the role of growth temperature and irradiance. Physiol Plant 113(4):499–506

Poirier Y, Antonenkov VD, Glumoff T, Hiltunen JK (2006) Peroxisomal β-oxidation—a metabolic pathway with multiple functions. Biochim Biophys Acta, Mol Cell Res 1763(12):1413–1426

Pollard MA, Anderson L, Fan C, Hawkins DJ, Davies HM (1990) A specific acyiACP thioesterase implicated in laurate production in immature cotyledons of *Umbellularia californica*. In: Quinn

PJ, Harwood JL (eds) Plant lipid biochemistry, structure and utilization. Portland, London, pp 163–165

Pompeiano A, Caturegli L, Grossi N, Volterrani M, Guglielminetti L (2015) Carbohydrate metabolism during wintering period in four *Zoysiagrass* genotypes. Plant Prod Sci 18(1):43–51

Popov VN, Antipina OV, Pchelkin VP, Tsydendambaev VD (2012) Changes in the content and composition of lipid fatty acids in tobacco leaves and roots at low-temperature hardening. Russ J Plant Physiol 59(2):177–182

Popov VN, Antipina OV, Burakhanova EA (2013) Involvement of cell-wall invertase in low-temperature hardening of tobacco plants. Russ J Plant Physiol 2:221–226

Porter JR, Semenov MA (2005) Crop responses to climatic variation. Philos Trans R Soc Lond B Biol Sci 360(1463):2021–2035

Pracharoenwattana I, Zhou W, Smith SM (2010) Fatty acid beta-oxidation in germinating Arabidopsis seeds is supported by peroxisomal hydroxypyruvate reductase when malate dehydrogenase is absent. Plant Mol Biol 72(1–2):101–109

Prasad KVSK, Saradhi PP (2004) Enhanced tolerance to photoinhibition in transgenic plants through targeting of glycinebetaine biosynthesis into the chloroplasts. Plant Sci 166(5): 1197–1212

Prost I, Dhondt S, Rothe G, Vicente J, Rodriguez MJ, Kift N, Carbonne F, Griffiths G, Esquerré-Tugayé M-T, Rosahl S, Castresana C, Hamberg M, Fournier J (2005) Evaluation of the antimicrobial activities of plant oxylipins supports their involvement in defense against pathogens. Plant Physiol 139(4):1902–1913

Purkrtova Z, Le Bon C, Kralova B, Ropers MH, Anton M, Chardot T (2008) Caleosin of Arabidopsis thaliana: effect of calcium on functional and structural properties. J Agric Food Chem 56(23):11217–11224

Purvis AC (2002) Diphenylamine reduces chilling injury of green bell pepper fruit. Postharvest Biol Technol 25(1):41–48

Puyaubert J, Baudouin E (2014) New clues for a cold case: nitric oxide response to low temperature. Plant Cell Environ 37(12):2623–2630

Pyc M, Cai Y, Greer MS, Yurchenko O, Chapman KD, Dyer JM, Mullen RT (2017) Turning over a new leaf in lipid droplet biology. Trends Plant Sci 22:596–609

Quettier A-L, Eastmond PJ (2009) Storage oil hydrolysis during early seedling growth. Plant Physiol Biochem 47(6):485–490

Raison JK, Lyons JM, Mehlhorn RJ, Keith AD (1971) Temperature induced phase changes in mitochondrial membranes detected by spin labeling. J Biol Chem 246(12):4036–4040

Rámy R, Mathieu L, Claudie DM (2018) Antioxidant and enzymatic responses to oxidative stress induced by cold temperature storage and ripening in mango (*Mangifera indica* L, cv, 'Cogshall') in relation to carotenoid content. J Plant Physiol 224–225:75–85

Rastogi A, Yadav DK, Szymañska R, Kruk J, Sedláøová M, Pospíšil P (2014) Singlet oxygen scavenging activity of tocopherol and plastochromanol in *Arabidopsis thaliana*: relevance to photooxidative stress. Plant Cell Environ 37(2):392–401

Reina-Pinto JJ, Yephremov A (2009) Surface lipids and plant defenses. Plant Physiol Biochem 47: 540–549

Repetto M, Semprine J, Boveris A (2012) Lipid peroxidation: chemical mechanism, biological implications and analytical determination. In: Catala A (ed) Lipid peroxidation. InTech, Rijeka, pp 3–30

Rikin A, Dillwith JW, Bergman DK (1993) Correlation between the circadian rhythm of resistance to extreme temperatures and changes in fatty acid composition in cotton seedlings. Plant Physiol 101(1):31–36

Rinaldi MA, Patel AB, Park J, Lee K, Strader LC, Bartel B (2016) The Roles of β-oxidation and cofactor homeostasis in peroxisome distribution and function in *Arabidopsis thaliana*. Genetics 204(3):1089–1115

Riva-Roveda L, Escale B, Giauffret C, Perilleux C (2016) Maize plants can enter a standby mode to cope with chilling stress. BMC Plant Biol 16(1):212

Rizhsky L, Liang H, Shuman J, Shulaev V, Davletova S, Mittler R (2004) When defense pathways collide. The response of Arabidopsis to a combination of drought and heat stress. Plant Physiol 134(4):1683–1696

Rocha J, Nitenberg M, Agnes GE, Juliette J, Eric M, Block MA, Breton C (2018) Do galactolipid synthases play a key role in the biogenesis of chloroplast membranes of higher plants? Front Plant Sci 9:126

Rochaix J-D (2007) Role of thylakoid protein kinases in photosynthetic acclimation. FEBS Lett 581(15):2768–2775

Rochaix J-D, Lemeille S, Shapiguzov A, Samol I, Fucile G, Willig A, Goldschmidt-Clermont M (2012) Protein kinases and phosphatases involved in the acclimation of the photosynthetic apparatus to a changing light environment. Philos Trans R Soc Lond Ser B Biol Sci 367(1608): 3466–3474

Romano R, Santini A, Grottaglie LL, Manzo N, Visconti A, Ritieni A (2014) Identification markers based on fatty acid composition to differentiate between roasted Arabica and Canephora (Robusta) coffee varieties in mixtures. J Food Compos Anal 35(1):1–9

Roston RL, Wang K, Kuhn LA, Christoph B (2014) Structural determinants allowing transferase activity in sensitive to freezing 2, classified as a family I glycosyl hydrolase. J Biol Chem 289(38):26089–26106

Roughan PG, Ohlrogge JB (1996) Evidence that isolated chloroplasts contain an integrated lipid-synthesizing assembly that channels acetate into long-chain fatty acids. Plant Physiol 110(4): 1239–1247

Roughan PG, Slack CR (1982) Cellular organization of glycerolipid metabolism. Annu Rev Plant Physiol 33:97–123

Routaboul JM, Fischer SF, Browse J (2000) Trienoic fatty acids are required to maintain chloroplast function at low temperatures. Plant Physiol 124(4):1697–1705

Roy S, Pouenat ML, Caumont C, Cariven C, Prevost MC, Esquerretugaye MT (1995) Phospholipase-activity and phospholipid patterns in tobacco cells treated with fungal elicitor. Plant Sci 107(1):17–25

Ruelland E, Zachowski A (2010) How plants sense temperature. Environ Exp Bot 69(3):225–232

Ruelland E, Cantrel C, Gawer M, Kader JC, Zachowski A (2002) Activation of phospholipases C and D is an early response to a cold exposure in Arabidopsis suspension cells. Plant Physiol 130(2):999–1007

Russell NJ (1989) Functions of lipids: structural roles and membrane functions. In: Ratledge C, Wilkinson SG (eds) Microbia lipids, vol 2. Academic, London, pp 279–365

Russell NJ (2003) Psychrophily and resistance to low temperatures. In: Encyclopaedia of life support systems. EOLSS Publishers Co Ltd. Published electronically, Contribution number 6–73-03–00

Russell NJ (2008) Membrane components and cold sensing. In: Margesin R, Schinner F, Marx JC, Gerday C (eds) Psychrophiles: from biodiversity to biotechnology. Springer, Berlin, Heidelberg, pp 177–190

Saita E, Albanesi D, Mendoza D (2016) Sensing membrane thickness: lessons learned from cold stress. Biochim Biophys Acta 1861(8 Pt B):837–846

Samuels L, Kunst L, Jetter R (2008) Sealing plant surfaces: cuticular wax formation by epidermal cells. Annu Rev Plant Biol 59:683–707

Sanchez JP, Chua NH (2001) *Arabidopsis* PLC1 is required for secondary responses to abscisic acid signals. Plant Cell 13(5):1143–1154

Sanchez J, Mangat PK, Angeles-Shim RB (2019) Weathering the cold: modifying membrane and storage fatty acid composition of seeds to improve cold germination ability in upland cotton (*Gossypium hirsutum* L.). Agronomy 9(11):684

Sangwan V, Örvar BL, Beyerly J, Hirt H, Dhindsa RS (2002) Opposite changes in membrane fluidity mimic cold and heat stress activation of distinct plant MAP kinase pathways. Plant J 31(5):629–638

Sanhueza D, Vizoso P, Balic I, Campos-Vargas R, Meneses C (2015) Transcriptomic analysis of fruit stored under cold conditions using controlled atmosphere in *Prunus persica* cv. "Red pearl". Front Plant Sci 6:788

Saucedo-Garcia M, Guevara-Garcia A, Gonzalez-Solis A, Cruz-Garcia F, Vázquez-Santana S, Markham JE, Lozano-Rosas MG, Dietrich CR, Ramos-Vega M, Cahoon EB, Gavilanes-Ruiz M (2011) MPK6, sphinganine and the LCB2a gene from serine palmitoyl-transferase are required in the signaling pathway that mediates cell death induced by long chain bases in Arabidopsis. New Phytol 191(4):943–957

Scherer GFE, Paul RU, Holk A, Martinec J (2002) Down-regulation by elicitors of phosphatidylcholine-hydrolyzing phospholipase C and up-regulation of phospholipase A in plant cells. Biochem Biophys Res Commun 293(2):766–770

Scherer GF, Ryu SB, Wang X, Matos AR, Heitz T (2010) Patatin-related phospholipase A: nomenclature, subfamilies and functions in plants. Trends Plant Sci 15(12):693–700

Schreiber L (2010) Transport barriers made of cutin, suberin and associated waxes. Trends Plant Sci 15:546–553

Schulz E, Tohge T, Zuther E, Fernie AR, Hincha DK (2016) Flavonoids are determinants of freezing tolerance and cold acclimation in *Arabidopsis thaliana*. Sci Rep 6(1):34027

Seki M, Narusaka M, Abe H, Kusuga M, Yamaguchi Shinozaki K, Carninci P, Hayashizaki Y, Shinozaki K (2001) Monitoring the expression pattern of 1300 Arabidopsis genes under drought and cold stresses by using a full-length cDNA microarray. Plant Cell 13(1):61–72

Serrano M, Coluccia F, Torres M, L'Haridon F, Metraux J-P (2014) The cuticle and plant defense to pathogens. Front Plant Sci 5:274

Shanklin J, Somerville CR (1991) Stearoyl-acyl-carrier-protein desaturase from higher plants is structurally unrelated to the animal and fungal homologs. Proc Natl Acad Sci U S A 88(6): 2510–2514

Shen W, Li JQ, Dauk M, Huang Y, Periappuram C, Wei Y, Zou J (2010) Metabolic and transcriptional responses of glycerolipid pathways to a perturbation of glycerol 3-phosphate metabolism in *Arabidopsis*. J Biol Chem 285(30):22957–22965

Shepherd T, Griffiths DW (2006) The effects of stress on plant cuticular waxes. New Phytol 171(3): 469–499

Shi L, Bielawski J, Mu J, Dong H, Teng C, Zhang J, Yang X, Tomishige N, Hanada K, Hannun YA, Zuo J (2007) Involvement of sphingoid bases in mediating reactive oxygen intermediate production and programmed cell death in Arabidopsis. Cell Res 17(12):1030–1040

Shi J, Cao Y, Fan X, Li M, Wang Y, Ming F (2012) A rice microsomal delta-12 fatty acid desaturase can enhance resistance to cold stress in yeast and *Oryza sativa*. Mol Breed 29(3):743–757

Shi YT, Ding YL, Yang SH (2015) Cold signal transduction and its interplay with phytohormones during cold acclimation. Plant Cell Physiol 56:7–15

Shi F, Zhou X, Zhou Q, Tan Z, Yao MM, Ji SJ (2018a) Membrane lipid metabolism changes and 31 aroma ester loss in low temperature stored 'Nanguo' pears. Food Chem 245:446–453

Shi F, Zhou X, Zhou Q, Tan Z, Yao MM, Ji SJ (2018b) Transcriptome analyses provide new possible mechanisms of aroma ester weakening of 'Nanguo' pear after cold storage. Sci Hortic 237:247–256

Shimada TL, Takano Y, Shimada T, Fujiwara M, Fukao Y, Mori M, Okazaki Y, Saito K, Sasaki R, Aoki K, Hara-Nishimura I (2014) Leaf oil body functions as a subcellular factory for the production of a phytoalexin in *Arabidopsis*. Plant Physiol 164(1):105–118

Shimada TL, Shimada T, Okazaki Y, Higashi Y, Saito K, Kuwata K, Oyama K, Kato M, Ueda H, Nakano A, Ueda T, Takano Y, Hara-Nishimura I (2019) HIGH STEROL ESTER 1 is a key factor in plant sterol homeostasis. Nat Plants 5:1154–1166

Shimakata T, Stumpf PK (1982) Isolation and function of spinach leaf ß-ketoacyl-[acyl-carrier-protein] syn-thases. Proc Natl Acad Sci U S A 79(19):5808–5812

Shinkawa R, Morishita A, Amikura K, Machida R, Murakawa H, Kuchitsu K, Ishikawa M (2013) Abscisic acid induced freezing tolerance in chilling-sensitive suspension cultures and seedlings of rice. BMC Res Notes 6:351

Shinozaki K, Dennis ES (2003) Cell signalling and gene regulation: global analyses of signal transduction and gene expression profiles. Curr Opin Plant Biol 6(5):405–409

Shockey JM, Fulda MS, Browse JA (2002) Arabidopsis contains nine long-chain acyl-coenzyme A synthetase genes that participate in fatty acid and glycerolipid metabolism. Plant Physiol 129(4): 1710–1722

Shockey J, Regmi A, Cotton K, Adhikari N, Browse J, Bates PD (2016) Identification of Arabidopsis GPAT9 (At5g60620) as an essential gene involved in triacylglycerol biosynthesis. Plant Physiol 170(1):163–179

Shumskaya M, Wurtzel ET (2013) The carotenoid biosynthetic pathway: thinking in all dimensions. Plant Sci 208:58–63

Sicher R (2011) Carbon partitioning and the impact of starch deficiency on the initial response of *Arabidopsis* to chilling temperatures. Plant Sci 181:167–176

Singer SJ, Nicolson GL (1972) The fluid mosaic model of the structure of cell membranes. Science 175(4023):720–731

Singer SD, Chen G, Mietkiewska E, Tomasi P, Jayawardhane K, Dyer JM, Weselake RJ (2016) Arabidopsis GPAT9 contributes to synthesis of intracellular glycerolipids but not surface lipids. J Exp Bot 67(15):4627–4638

Singh Z, Singh RK, Sane VA, Nath P (2013) Mango postharvest biology and biotechnology. Crit Rev Plant Sci 32:217–236

Slack CR, Roughan PG, Terpstra J (1976) Some properties of a microsomal oleate desaturase from leaves. Biochem J 155(1):71–80

Slack CR, Roughan PG, Browse J (1979) Evidence for an oleoyl phosphatidylcholine desaturase in microsomal preparations from cotyledons of safflower (*Carthamus tinctorius* L.) seed. Biochem J 179:649–656

Slack CR, Campbell LC, Browse JA, Roughan G (1983) Some evidence for the reversibility of the cholinephophotransferase-catalysed reaction in developing linseed cotyledons in vivo. Biochim Biophys Acta 754:10–20

Smolenska G, Kuiper PJC (1977) Effect of low temperature upon lipid and fatty acid composition of roots and leaves of winter rape plants. Physiol Plant 41(1):29–35

Somerville C, Browse J (1991) Plant lipids: metabolism, mutants, and membranes. Science 252(5002):80–87

Souza Lucéia F, Gasparetto Bruno F, Lopes Rodrigo R, Brrros Ingrid BI (2016) Temperature requirements for seed germination of *Pereskia aculeata* and *Pereskia grandifolia*. J Therm Biol 57:6–10

Spicher L, Glauser G, Kessler F (2016) Lipid antioxidant and galactolipid remodeling under temperature stress in tomato plants. Front Plant Sci 7:167

Steponkus PL (1984) Role of the plasma-membrane in freezing-injury and cold-acclimation. Annu Rev Plant Physiol Plant Mol Biol 35:543–584

Steponkus PL, Uemura M, Webb MS (1993) A contrast of the cryostability of the plasma membrane of winter rye and spring oat-two species that widely differ in their freezing tolerance and plasma membrane lipid composition. In: Steponkus PL (ed) Advances in low temperature biology, vol 2. JAI Press, London, pp 211–312

Stobart AK, Stymne S (1985) The regulation of the fatty-acid composition of the triacylglycerols im microsomal preparations from avocado mesocarp and the developing cotyledons of safflower. Planta 163:119–125

Stobart A, Stymne S, Glad G (1983) The synthesis of linoleate and phosphatidic acid and its relationship to oil production in the microsomes of developing seeds of safflower (*Carthamus tinctorius* L. var. GILA). Biochim Biophys Acta 754:292–297

Stoddard FL, Balko C, Erskine W, Khan HR, Link W, Sarker A (2006) Screening techniques and sources of resistance to abiotic stresses in cool-season food legumes. Euphytica 147(1):167–186

Strauss G, Hauser H (1986) Stabilization of lipid bilayer vesicles by sucrose during freezing. Proc Natl Acad Sci U S A 83(8):422–2426

Stymne S, Stobart AK (1984) Evidence for the reversibility of the acyl-CoA: lysophosphatidylcholine acyltransferase in microsomal preparations from developing safflower (*Carthamus tinctorius* L.) cotyledons and rat liver. Biochem J 223:305–314

Stymne S, Stobart AK, Glad G (1983) The role of the acyl-CoA pool in the synthesis of polyunsaturated 18-carbon fatty acids and triacylglycerol production in the microsomes of developing safflowers seeds. Biochim Biophys Acta 752:198–208

Subramaniam S, Fahy E, Gupta S, Sud M, Byrnes RW, Cotter D, Dinasarapu AR, Maurya MR (2011) Bioinformatics and system biology of the lipidome. Chem Rev 111(10):6452–6590

Suh MC, Hahne G, Liu JR, Stewart CN Jr (2015) Plant lipid biology and biotechnology. Plant Cell Rep 34:517–518

Sun W, Li Y, Zhao Y, Zhang H (2015) The TsnsLTP4, a nonspecific lipid transfer protein involved in wax deposition and stress tolerance. Plant Mol Biol Rep 33(4):962–974

Sun B, Liu GL, Phan TT, Yang LT, Li YR, Xing YX (2016) Effects of cold stress on root growth and physiological metabolisms in seedlings of different sugarcane varieties. Sugar Tech 19: 165–175

Sung DY, Kaplan F, Lee KJ, Guym CL (2003) Acquired tolerance to temperature extremes. Trends Plant Sci 8(4):179–187

Sunoj VSJ, Shroyer KJ, Jagadish SVK, Prasad PV (2016) Diurnal temperature amplitude alters physiological and growth response of maize (*Zea mays* L.) during the vegetative stage. Environ Exp Bot 130:113–121

Suzuki N, Mittler R (2006) Reactive oxygen species and temperature stresses: a delicate balance between signaling and destruction. Physiol Plant 126:45–51

Szalontai B, Nishiyama Y, Gombos Z, Murata N (2000) Membrane dynamics as seen by Fourier transform infrared spectroscopy in a cyanobacterium, *Synechocystis* PCC 6803. The effects of lipid unsaturation and the protein-to-lipid ratio. Biochim Biophys Acta 1509(1–2):409–419

Szymanski J, Brotman Y, Willmitzer L, Cuadros-Inostroza Á (2014) Linking gene expression and membrane lipid composition of *Arabidopsis*. Plant Cell 26(3):915–928

Tai PYP, Lentini RS (1998) Freeze damage of Florida sugarcane. In: Anderson DL (ed) Sugarcane handbook. Florida Cooperative Extension, Gainesville

Takahashi D, Li B, Nakayama T, Kawamura Y, Uemura M (2013) Plant plasma membrane proteomics for improving cold tolerance. Front Plant Sci 4:90

Tambussi EA, Bartoli CG, Guriamet JJ, Beltrano J, Araus JL (2004) Oxidative stress and photodamage at low temperature to soybean (*Glycine max* L. Merr.) leave. Plant Sci 167:19–26

Tan WJ, Yang YC, Zhou Y, Huang LP, Xu L, Chen QF, Yu LJ, Xiao S (2018) Diacylglycerol acyltransferase and diacylglycerol kinase modulate triacylglycerol and phosphatidic acid production in the plant response to freezing stress. Plant Physiol 177(3):1303–1318

Tang YY (2011) Screening of peanut genotypes for low temperature tolerance and identification of low temperature responsive genes. Ocean University of China, Qingdao

Tang G, Liu Z, Xu P, Zhao X, Shan L (2013) Cloning and functional analysis of diacylglycerol acyltransferase gene in *Arachis hypogaea*. Acta Botan Boreali-Occiden Sin 33:857–863

Tasseva G, Davy de Virville J, Cantrel C, Moreau F, Zachowski A (2004a) Changes in the endoplasmic reticulum lipid properties in response to low temperature in *Brassica napus*. Plant Physiol Biochem 42(10):811–822

Tasseva G, Richard L, Zachowski A (2004b) Regulation of phosphatidylcholine biosynthesis under salt stress involves choline kinases in *Arabidopsis thaliana*. FEBS Lett 566(1–3):115–120

Terashima I, Inoue Y (1985) Vertical gradients in photosynthetic properties of spinach chloroplasts dependent on intra-leaf light environment. Plant Cell Physiol 26(4):781–785

Testerink C, Munnik T (2011) Molecular, cellular, and physiological responses to phosphatidic acid formation in plants. J Exp Bot 62(7):2349–2361

Theocharis A, Clément C, Barka EA (2012a) Physiological and molecular changes in plants grown at low temperatures. Planta 235:1091–1105

Theocharis A, Bordiec S, Fernandez O, Paquis S, Dhondt-Cordelier S, Baillieul F, Clément C, Barka EA (2012b) *Burkholderia phytofirmans* PsJN primes *Vitis vinifera* L. and confers a better tolerance to low nonfreezing temperatures. Mol Plant-Microbe Interact 25(2):241–249

Thi ATP, Sidibé-Andrieu MD, Zuily-Fodil Y, da Silva JV (1989) Effects of low temperature on lipid and fatty acid composition of sorghum roots and shoots. In: Biacs PA, Gruiz K, Kremmer T (eds) Biological role of plant lipids. Springer, Boston

Thomas AH, Catala A, Vignoni M (2016) Soybean phosphatidylcholine liposomes as model membranes to study lipid peroxidation photoinduced by pterin. Biochim Biophys Acta Biomembr 1858(1):139–145

Thomashow MF (1998) Role of cold-responsive genes in plant freezing tolerance. Plant Physiol 118(1):1–8

Thomashow MF (1999) Plant cold acclimation: freezing tolerance genes and regulatory mechanisms. Annu Rev Plant Physiol Plant Mol Biol 50:571–599

Thomashow MF (2010) Molecular basis of plant cold acclimation: insights gained from studying the CBF cold response pathway. Plant Physiol 154(2):571–577

Thompson GA, Li CY (1997) Altered fatty acid composition of membrane lipids in seeds and seedling tissues of high-saturate canolas. In: Williams JP, Khan MU, Lem NW (eds) Physiology, biochemistry, and molecular biology of plant lipids. Kluwer Academic, Dordrecht, pp 313–315

Thorlby G, Fourrier N, Warren G (2004) The SENSITIVE TO FREEZING2 gene, required for freezing tolerance in Arabidopsis thaliana, encodes a beta-glucosidase. Plant Cell 16:2192–2203

Tian D-Q, Pan X-Y, Yu Y-M, Wang W-Y, Zhang F, Ge Y-Y, Shen X-L, Shen F-Q, Liu X-J (2013) De novo characterization of the Anthurium transcriptome and analysis of its digital gene expression under cold stress. BMC Genomics 14(1):827

Tian X, Liu Y, Huang Z, Duan H, Tong J, He X (2015) Comparative proteomic analysis of seedling leaves of cold-tolerant and -sensitive spring soybean cultivars. Mol Biol Rep 42:581–601

Tokuhisa J, Wu J, Miquel M, Xin Z, Browse J (1997) Investigating the role of lipid metabolism in chilling and freezing tolerance. In: Li PH, Chen THH (eds) Plant cold hardiness. Springer, Boston, pp 153–169

Tolessa K, D'heer J, Duchateau L, Boeckx P (2017) Influence of growing altitude, shade and harvest period on quality and biochemical composition of Ethiopian specialty coffee. J Sci Food Agric 97(9):2849–2857

Trischuk RG, Schilling BS, Low NH, Gray GR, Gusta LV (2014) Cold acclimation, de-acclimation and re-acclimation of spring canola, winter canola and winter wheat: the role of carbohydrates, cold-induced stress proteins and vernalization. Environ Exp Bot 106:156–163

Tursman D, Duman JG (1995) Cryoprotective effects of thermal hysteresis protein on survivorship of frozen gut cells from the freeze-tolerant centipedeLithobius forficatus. J Exp Zool 272(4): 249–257

Tzen JTC, Cao Y, Laurent P, Ratnayake C, Huang AHC (1993) Lipids, proteins, and structure of seed oil bodies from diverse species. Plant Physiol 101(1):267–276

Uemura M, Joseph RA, Steponkus PL (1995a) Cold-Acclimation of *Arabidopsis-Thaliana*—effect on plasma-membrane lipid-composition and freeze-induced lesions. Plant Physiol 109(1):15–30

Uemura M, Joseph RA, Steponkus PL (1995b) Effect of cold acclimation on the lipid composition of the inner and outer membrane of the chloroplast envelope isolated from rye leaves. Plant Physiol 114(4):1493–1500

Uemura M, Tominaga Y, Nakagawara C, Shigematsu S, Minami A, Kawamura Y (2006) Responses of the plasma membrane to low temperatures. Physiol Plant 126(1):81–89

Upadhyaya HD, Reddy LJ, Dwivedi SL, Gowda CLL, Singh S (2009) Phenotypic diversity in cold-tolerant peanut (*Arachis hypogaea* L.) germplasm. Euphytica 165:279–291

Upchurch RG (2008) Fatty acid unsaturation, mobilization, and regulation in the response of plants to stress. Biotechnol Lett 30(6):967–977

Uraji M, Katagiri T, Okuma E, Ye W, Hossain MA, Masuda C, Miura A, Nakamura Y, Mori IC, Shinozaki K, Murata Y (2012) Cooperative function of PLDdelta and PLDalpha1 in abscisic acid-induced stomatal closure in Arabidopsis. Plant Physiol 159(1):450–460

Van de Loo FJ, Turner PBS, Somerville C (1995) An oleate 12 hydroxylase from *Ricinus communis* L. is a fatty acyl desaturase homolog (ricinoleic acid/castor/FAH12/ transgenic plants). Proc Natl Acad Sci U S A 92(15):6743–6747

Van Hasselt P (1974) Photo-oxidative damage to the ultrastructure of *Cucumis* chloroplasts during chilling. Proc Konikl Nederl Akademie Van Wetenschappen Amsterdam Ser C 77:50–56

Vega SE, del Rio AH, Bamberg JB, Palta JO (2004) Evidence for the up-regulation of stearoyl-ACP (Δ9) desaturase gene expression during cold acclimation. Am J Potato Res 81(2):125–135

Viehweger K, Dordschbal B, Roos W (2002) Elicitor-activated phospholipase A2 generates lysophosphatidylcholines that mobilize the vacuolar H+ pool for pH signaling via the activation of Na+-dependent proton fluxes. Plant Cell 14(7):1509–1525

Vigh L, Los DA, Horvath I, Murata N (1993) The primary signal in the biological perception of temperature: Pd-catalyzed hydrogenation of membrane lipids stimulated the expression of the *des*A gene in *Synechocystis* PCC6803. Proc Natl Acad Sci U S A 90:9090–9094

Vigh L, Maresca B, Harwood JL (1998) Does the membrane's physical state control the expression of heat shock and other genes? Trends Biochem Sci 23:369–374

Vigh L, Escriba PV, Sonnleitner A, Sonnleitner M, Piotto S, Maresca B, Horváth I, Harwood JL (2005) The significance of lipid composition for membrane activity: new concepts and ways of assessing function. Prog Lipid Res 44(5):303–344

Villarreal D, Laffargue A, Posada H, Bertrand B, Lashermes P, Dussert S (2009) Genotypic and environmental effects on coffee (*Coffea arabica* L.) bean fatty acid profile: impact on variety and origin chemometric determination. J Agric Food Chem 57:11321–11327

Vishwanath SJ, Delude C, Domergue F, Rowland O (2014) Suberin: biosynthesis, regulation, and polymer assembly of a protective extracellular barrier. Plant Cell Rep 34(4):573–586

Vogel JT, Zarka DG, van Buskirk HA, Fowler SG, Thomashow MF (2005) Roles of the CBF2 and ZAT12 transcription factors in configuring the low temperature transcriptome of *Arabidopsis*. Plant J 41(2):195–211

Vu HS, Shiva S, Hall AS, Welti R (2014) A lipidomic approach to identify cold-induced changes in Arabidopsis membrane lipid composition. Methods Mol Biol 1166:199–215

Wada H, Gombos Z, Murata N (1990) Enhancement of chilling tolerance of a cyanobacterium by genetic manipulation of fatty acid desaturation. Nature 347(6289):200–203

Wagemaker TAL, Carvalho CRL, Maia NB, Baggio SR, Filho OG (2011) Sun protection factor, content, and composition of the lipid fraction of green coffee beans. Ind Crop Prod 33:469–473

Wang CY (1994) Chilling injury of tropical horticultural commodities. HortScience 29:986–988

Wang X (2004) Lipid signaling. Curr Opin Plant Biol 7(3):329–336

Wang X, Devaiah SP, Zhang W, Welti R (2006a) Signaling functions of phosphatidic acid. Prog Lipid Res 45(3):250–278

Wang X, Li W, Li M, Welti R (2006b) Profiling lipid changes in plant response to low temperature. Physiol Plant 126(1):90–96

Wang J, Ming F, Pittman J, Han Y, Hu J, Guo B, Shen D (2006c) Characterization of a rice (*Oryza sativa* L.) gene encoding a temperature-dependent chloroplast ω-3 fatty acid desaturase. Biochem Biophys Res Commun 340(4):1209–1216

Wang W, Yang X, Tangchaiburana S, Ndeh R, Markham JE, Tsegaye JF, Dunn TM, Wang GL, Bellizzi M, Parsons JF, Morrissey D, Bravo JE, Lynch DV, Xiao S (2008) An inositolphosphorylceramide synthase is involved in regulation of plant programmed cell death associated with defense in Arabidopsis. Plant Cell 20(11):3163–3179

Wang PC, Mo BT, Long ZF, Fan SQ, Wang HH (2016a) Factors affecting seed germination and emergence of *Sophora davidii*. Ind Crop Prod 87:261–265

Wang Q, Ding T, Zuo JH, Gao LP, Fan LL (2016b) Amelioration of postharvest chilling injury in sweet pepper by glycine betaine. Postharvest Biol Technol 112:114–120

Wang DZ, Jin YN, Ding XH, Wang WJ, Zhai SS, Bai LP, Guo ZF (2017) Gene regulation and signal transduction in the ICE-CBF-COR signaling pathway during cold stress in plants. Biochemistry 82(10):1444–1462

Wang J, Jiang Y, Li G, Lv M, Zhou X, Zhou Q, Fu W, Zhang L, Chen Y, Ji S (2018) Effect of low temperature storage on energy and lipid metabolisms accompanying peel browning of 'Nanguo' pears during shelf life. Postharvest Biol Technol 139:75–81

Wang Y, Ji S, Dai H, Kong X, Hao J, Wang S, Zhou X, Wei B, Cheng S, Zhou Q (2019) Changes in membrane lipid metabolism accompany pitting in blueberry during refrigeration and subsequent storage at room temperature. Front Plant Sci 10:829

Warren G, McKown R, Marin AL, Teutonico R (1996) Isolation of mutations affecting the development of freezing tolerance in *Arabidopsis thaliana* L. Heynh. Plant Physiol 111(4): 1011–1019

Wei L (2012) Cloning and functional analysis of ARACHIDONYL carrier protein gene in peanut. Shandong Normal University, Jinan

Weiser CJ (1970) Cold resistance and injury in woody plants: knowledge of hardy plant adaptations to freezing stress may help us to reduce winter damage. Science 169(3952):1269–1278

Weller JL, Ortega R (2015) Genetic control of flowering time in legumes. Front Plant Sci 6:207

Welte MA, Gould AP (2017) Lipid droplet functions beyond energy storage. Biochim Biophys Acta Mol Cell Biol Lipids 1862(10 Pt B):1260–1272

Welti R, Li W, Li M, Sang Y, Biesiada H, Zhou H-E, Rajashekar CB, Williams TD, Wang X (2002) Profiling membrane lipids in plant stress responses. Role of phospholipase-D alpha in freezing-induced lipid changes in *Arabidopsis*. J Biol Chem 277(35):31994–32002

Welti R, Shah J, Li W, Li M, Chen J, Burke JJ, Fauconnier ML, Chapman K, Chye ML, Wang X (2007) Plant lipidomics: discerning biological function by profiling plant complex lipids using mass spectrometry. Front Biosci 12:2494–2506

Winfield MO, Lu C, Wilson ID, Coghill JA, Edwards KJ (2010) Plant responses to cold: transcriptome analysis of wheat. Plant Biotechnol J 8(7):749–771

Wise RR, Naylor AW (1987) Chilling-enhanced photooxidation. The peroxidative destruction of lipids during chilling injury to photosynthesis and ultrastructure. Plant Physiol 83(2):272–277

Wise RR, McWilliam J, Naylor AW (1983) A comparative study of low temperature induced ultrastructural alterations of three species with differing chilling sensitivities. Plant Cell Environ 6(7):525–535

Wolter FP, Schmidt R, Heinz E (1992) Chilling sensitivity of *Arabidopsis thaliana* with genetically engineered membrane lipids. EMBO J 11(13):4685–4692

Woolhouse HW (1981) Crop physiology in relation to agricultural production: the genetic link. In: Johnson CB (ed) Physiological processes limiting plant productivity. Butterworths, London, pp 1–21

Worrall D, Liang YK, Alvarez S, Holroyd GH, Spiegel S, Panagopulos M, Gray JE, Hetherington AM (2008) Involvement of sphingosine kinase in plant cell signaling. Plant J 56(1):64–72

Wu J, Lightner J, Waewick N, Browse J (1997) Low-temperature damage and subsequent recovery of fab1 mutant *Arabidopsis* exposed to 28°C. Plant Physiol 113(2):347–356

Wu J, Hwang I-T, Hatzios KK (2000) Effects of chloroacetanilide herbicides on membrana fatty acid desaturation and lipid composition in rice, maize, and sorghum. Pestic Biochem Physiol 66(3):161–169

Xiaochuang C, Chu Z, Lianfeng Z, Junhua Z, Hussain S, Lianghuan W, Qianyu J (2017) Glycine increases cold tolerance in rice via the regulation of N uptake, physiological characteristics, and photosynthesis. Plant Physiol Biochem 112:251–260

Xie TL, Gu WR, Li CF, Li J, Wei S (2019) Exogenous DCPTA increases the tolerance of maize seedlings to PEG-simulated drought by regulating nitrogen metabolism related enzymes. Agronomy 9(11):676

Xin Z, Browse J (2000) Cold comfort farm: the acclimation of plants to freezing temperatures. Plant Cell Environ 23(9):893–902

Xiong LM, Schumaker KS, Zhu JK (2002) Cell signaling during cold, drought, and salt stress. Plant Cell 14(Suppl):S165–S183

Xu C, Shanklin J (2016) Triacylglycerol metabolism, function, and accumulation in plant vegetative tissues. Annu Rev Plant Biol 67:179–206

Xu J, Carlsson AS, Francis T, Zhang M, Hoffman T, Giblin ME, Taylor DC (2012) Triacyglycerol synthesis by PDAT1 in the absence of DGAT1 activity is dependent on re-acylation of LPC by LPCAT2. BMC Plant Biol 12(1):4

Xu GL, Singh SK, Reddy VR, Barnaby JY (2016) Soybean grown under elevated CO_2 benefits more under low temperature than high temperature stress: varying response of photosynthetic limitations, leaf metabolites, growth, and seed yield. J Plant Physiol 205:20–32

Xu J, Chen Z, Wang F, Jia W, Xu Z (2020) Combined transcriptomic and metabolomic analyses uncover rearranged gene expression and metabolite metabolism in tobacco during cold acclimation. Sci Rep 10:5242

Xue-Xuan X, Shao H, Yuan-Yuan M, Xu G, Jun-Na S, Dong-Gang G, Cheng-Jiang R (2010) Biotechnological implications from abscisic acid (ABA) roles in cold stress and leaf senescence as an important signal for improving plant sustainable survival under abiotic-stressed conditions. Crit Rev Biotechnol 30(3):222–230

Yadav NS (1996) Genetic modification of soybean oil quality. In: Verma DPS, Shoemaker RC (eds) Soybean genetics, molecular biology and biotechnology. CAB International, Wallingford, pp 127–188

Yamaguchi-Shinozaki K, Shinozaki K (1994) A novel cis-Acting element in an *Arabidopsis* gene is involved in responsiveness to drought, low-temperature, or high-Salt stress. Plant Cell 6(2): 251–264

Yamori W, Evans JR, Von CS (2010) Effects of growth and measurement light intensities on temperature dependence of CO_2 assimilation rate in tobacco leaves. Plant Cell Environ 3:332

Yan B, Xu X, Gu Y, Zhao Y, Zhao X, He L, Zhao C, Li Z, Xu J (2018) Genome-wide characterization and expression profiling of diacylglycerol acyltransferase genes from maize. Genome 61(10):735–743

Yang Y, Benning C (2018) Functions of triacylglycerols during plant development and stress. Curr Opin Biotechnol 49:191–198

Yang W, Devaiah SP, Pan X, Isaac G, Welti R, Wang X (2007) AtPLAI is an acyl hydrolase involved in basal jasmonic acid production and Arabidopsis resistance to *Botrytis cinerea*. J Biol Chem 282(25):18116–18128

Yang WY, Zheng Y, Bahn SC, Pana XQ, Lia MY, Vud HS, Rothd MR, Scheud B, Weltid R, Honge YY, Wanga XM (2012a) The patatin-containing phospholipase A pPLAIIalpha modulates oxylipin formation and water loss in *Arabidopsis thaliana*. Mol Plant 5(2):452–460

Yang XY, Jiang WJ, Yu HJ (2012b) The expression profiling of the lipoxygenase (LOX) family genes during fruit development, abiotic stress and hormonal treatments in cucumber (*Cucumis sativus* L.). Int J Mol Sci 13(2):2481–2500

Yang DL, Sun P, Li MF (2016) Chilling temperature stimulates growth, gene over-expression and podophyllotoxin biosynthesis in *Podophyllum hexandrum* royle. Plant Physiol Biochem 107: 197–203

Yang LY, Yang SL, Li JY, Ma JH, Pang T, Zou CM, He B, Gong M (2018) Effects of different growth temperatures on growth, development, and plastid pigments metabolism of tobacco (*Nicotiana tabacum* L.) plants. Bot Stud 59:5

Yang XY, Zhao T, Rao P, Gao K, Yang X, Chen Z, An X (2019) Transcriptome profiling of *Populus tomentosa* under cold stress. Ind Crop Prod 135(1):283–293

Yeats TH, Rose JKC (2013) The formation and function of plant cuticles. Plant Physiol 163(1): 5–20

Yoshida S, Kawata T, Uemura M, Niki T (1986) Properties of plasma membrane isolated from chilling-sensitive etiolated seedlings of *Vigna radiata* L. Plant Physiol 80(1):152–160

Yu S (2008) Cloning and expression analysis of the key enzymes in fatty acid metabolism of peanut. Nanjing Agricultural University, Nanjing

Yu L, Nie J, Cao C, Jin Y, Yan M, Wang F, Liu J, Xiao Y, Liang Y, Zhang W (2010) Phosphatidic acid mediates salt stress response by regulation of MPK6 on *Arabidopsis thaliana*. New Phytol 188(3):762–773

Yu X, Li A, Li W (2015) How membranes organize during seed germination: three patterns of dynamic lipid remodelling define chilling resistance and affect plastid biogenesis. Plant Cell Environ 38(7):1391–1403

Yuan F, Lyu MJA, Leng BY, Zheng GY, Feng ZT, Li PH, Zhu XG, Wang BS (2015) Comparative transcriptome analysis of developmental stages of the Limonium bicolor leaf generates insights into salt gland differentiation. Plant Cell Environ 38(8):1637–1657

Yue C, Cao H-L, Wang L, Zhou Y-H, Huang Y-T, Hao X-Y, Wang Y-C, Wang B, Yang Y-J, Wang Y-C (2015) Effects of cold acclimation on sugar metabolism and sugar-related gene expression in tea plant during the winter season. Plant Mol Biol 88(6):591–608

Yun JG, Hayashi T, Yazawa S, Katoh T, Yasuda Y (1996) Acute morphological changes of palisade cells of *Saintpaulia* leaves induced by a rapid temperature drop. J Plant Res 109: 339–342

Zhang W, Wang C, Qin C, Wood T, Olafsdottir G, Welti R, Wang X (2003) The oleate-stimulated phospholipase-D, PLDdelta, and phosphatidic acid decrease H_2O_2-induced cell death in Arabidopsis. Plant Cell 15(10):2285–2295

Zhang W, Yu L, Zhang Y, Wang X (2005) Phospholipase-D in the signaling networks of plant response to abscisic acid and reactive oxygen species. Biochim Biophys Acta 1736:1–9

Zhang XZ, Ervin EH, LaBranche AJ (2006) Metabolic defense responses of seeded bermudagrass during acclimation to freezing stress. Crop Sci 46:2598–2605

Zhang XZ, Ervin EH, Evanylo GK, Haering KC (2009a) Impact of biosolids on hormone metabolism in drought-stresses tall fescue. Crop Sci 49:364–370

Zhang Y, Zhu H, Zhang Q, Li M, Yan M, Wang R, Wang L, Welti R, Zhang W, Wang X (2009b) Phospholipase-Dα1 and phosphatidic acid regulate NADPH oxidase activity and production of reactive oxygen species in ABA-mediated stomatal closure in *Arabidopsis*. Plant Cell 21(8): 2357–2377

Zhang R, Wise RR, Struck KR, Sharkey TD (2010) Moderate heat stress of *Ariodopsis thaliana* leaves causes chloroplast swelling and plastoglobule formation. Photosynth Res 105(2): 123–134

Zhang AH, Sun H, Wang P, Han Y, Wang XJ (2012a) Modern analytical techniques in metabolomics analysis. Analyst 137:293–300

Zhang T, Zhao X, Wang W, Pan Y, Huang L, Liu X, Zong Y, Zhu L, Yang D, Fu B (2012b) Comparative transcriptome profiling of chilling stress responsiveness in two contrasting rice genotypes. PLoS One 7(8):e43274

Zhang Q, Lin F, Mao T, Nie J, Yan M, Yuan M, Zhang W (2012c) Phosphatidic acid regulates microtubule organization by interacting with MAP65-1 in response to salt stress in Arabidopsis. Plant Cell 24:4555–4576

Zhang W, Huang W, Yang Q, Zhang S, Hu H (2013) Effects of growth temperature on the selection flow for photorespiration in leaves of tobacco grown in the field. Physiol Plant 1:141–150

Zhang Y, Maximova SN, Guiltinan MJ (2015) Characterization of a stearoyl-acyl carrier protein desaturase gene family from chocolate tree *Theobroma cacao* L. Front Plant Sci 6:239

Zhang C, Yang L, Ding Y, Wang Y, Lan L, Ma Q, Chi X, Wei P, Zhao Y, Steinbüchel A, Zhang H, Liu P (2017) Bacterial lipid droplets bind to DNA via an intermediary protein that enhances survival under stress. Nat Commun 8:15979

Zhang H, Dong J, Zhao X, Zhang Y, Ren J, Xing L, Jiang C, Wang X, Wang J, Zhao S, Yu H (2019) Research progress in membrane lipid metabolism and molecular mechanism in peanut cold tolerance. Front Plant Sci 10:838

Zhang F, Ji S, Wei B, Cheng S, Wang Y, Hao J, Wang S, Zhou O (2020) Transcriptome analysis of postharvest blueberries (*Vaccinium corymbosum* 'Duke') in response to cold stress. BMC Plant Biol 20(1):80

Zhao L, Liu F, Xu W, Di C, Zhou S, Xue Y, Yu J, Su Z (2009) Increased expression of OsSPX1 enhances cold/subfreezing tolerance in tobacco and *Arabidopsis thaliana*. Plant Biotechnol J 7(6):550–561

Zhao KF, Song J, Fan H, Zhou S, Zhao M (2010) Growth response to ionic and osmotic stress of NaCl in salt-tolerant and salt-sensitive maize. J Integr Plant Biol 52(5):468–475

Zhao M, Liu W, Xia X, Wang T, Zhang W-H (2014) Cold acclimation-induced freezing tolerance of Medicago truncatula seedlings is negatively regulated by ethylene. Physiol Plant 152(1): 115–129

Zheng G, Tian B, Zhang F, Tao F, Li W (2011) Plant adaptation to frequent alterations between high and low temperatures: remodelling of membrane lipids and maintenance of unsaturation levels. Plant Cell Environ 34(9):1431–1442

Zheng G, Tian B, Li W (2014) Membrane lipid remodelling of *Meconopsis racemosa* after its introduction into lowlands from an alpine environment. PLoS One 9(9):e106614

Zheng G, Li L, Li W (2016) Glycerolipidome responses to freezing- and chilling-induced injuries: examples in *Arabidopsis* and rice. BMC Plant Biol 16:70

Zhong P, Liu J, Wang J, Chang B (2018) Physiological responses and cold resistance evaluation of peanut under low-temperature stress. J Nucl Agric Sci 32:1195–1202

Zhou Q, Zhang CL, Cheng SC, Wei BD, Liu XY, Ji SJ (2014a) Changes in energy metabolism accompanying pitting in blueberries stored at low temperature. Food Chem 164:493–501

Zhou Q, Ma C, Cheng SH, Wei BD, Liu XY, Ji SJ (2014b) Changes in antioxidative metabolism accompanying pitting development in stored blueberry fruit. Postharvest Biol Technol 88:88–95

Zhou X, Dong L, Li R, Zhou Q, Wang JW, Ji SJ (2014c) Low temperature conditioning prevents loss of aroma-related esters from 'Nanguo' pears during ripening at room temperature. Postharvest Biol Technol 100:23–32

Zhu J, Dong CH, Zhu JK (2007a) Interplay between cold-responsive gene regulation: metabolism and RNA processing during plant cold acclimation. Curr Opin Plant Biol 10(3):290–295

Zhu S-Q, Yu C-M, Liu X-Y, Ji B-H, Jiao D-M (2007b) Changes in unsaturated levels of fatty acids in thylakoid PS II membrane lipid during chilling induced resistance in rice. J Integr Plant Biol 49:463–471

Zienkiewicz K, Zienkiewicz A (2020) Degradation of lipid droplets in plants and algae-right time, many paths, one goal. Front Plant Sci 11:579019

Zinn KE, Tunc-Ozdemir M, Harper JF (2010) Temperature stress and plant sexual reproduction: uncovering the weakest links. J Exp Bot 61(7):1959–1968

Zou Y, Zhang L, Rao S, Zhu X, Ye L, Chen W, Li X (2014) The relationship between the expression of ethylenerelated genes and papaya fruit ripening disorder caused by chilling injury. PLoS One 9(12):e116002

Plant Growth Hormones in Plants under Low-Temperature Stress: A Review

6

Abstract

Many factors can cause and affect cell growth in the plant such as external (environmental) and internal factors; one of the most important internal factors is plant growth hormones. Many hormones required for cell growth, such as auxins, gibberellins, brassinosteroids, ethylene, jasmonates, salicylic acid, strigolactones, and cytokinins are able to accelerate or promote growth, but some hormone-like abscisic acid has an adverse effect on growth which increases seed dormancy by inhibiting cell growth. Also, plant hormones are able to break down dormancy for many plants and can alleviate abiotic stress (salinity, extreme temperatures, drought, etc.) which led to enhanced germination and improved growth for many plants, whether naturally occurring in the plant or by adding it to the plant in its artificial form or in the form of bio- or nano-fertilization in order to increase the productivity and improve its efficiency under extreme conditions. The plant growth regulators modulate plant growth and development and mediate responses to both biotic and abiotic stresses. The plant growth hormones are used commercially in agriculture. The approach of the application of plant growth hormones in horticulture shows good results. Chilling stress is known to reduce leaf size, stem extension, and root proliferation, disturb plant water relations, and impede nutrient uptake. Chilling stress in maize is a complex phenomenon with physiological and biochemical responses at both cellular and whole-organ level. CO_2 assimilation by leaves is reduced mainly due to membrane damage, photoinhibition, and disturbed activity of various enzymes. Enhanced metabolite flux through the photorespiratory pathway increases the oxidative load on tissues as both processes generate reactive oxygen species. Injury caused by reactive oxygen species to macromolecules under chilling stress is one of the major deterrents to growth. Plant growth hormones such as salicylic acid, gibberellic acid, and abscisic acid modulate the response of maize to chilling stress. Polyamines and several enzymes act as antioxidants and reduce the adverse

A. Bhattacharya, *Physiological Processes in Plants Under Low Temperature Stress*,
https://doi.org/10.1007/978-981-16-9037-2_6

effects of chilling stress. Chilling tolerance in maize can be managed through the development and selection of chilling-tolerant genotypes by breeding and genomic approaches. Agronomic approaches such as exogenous application of growth hormones and osmoprotectants to seeds or plants, and early vigor, can also aid in chilling tolerance.

Keywords

Importance of plant growth hormones · Low temperature and hormonal activity · Role of growth hormones under low temperature · Role of growth hormones under cold tolerance · ABA and low-temperature stress · Role of IAA in low-temperature stress · Brassinosteroids and low-temperature stress · Salicylic acid and low-temperature stress · Gibberellin and low-temperature stress · Cytokinin and low-temperature stress · Jasmonic acid and low-temperature stress · Thiourea and low-temperature stress · Metabolic activity of growth hormones for stress tolerance · Crop improvement under low temperature through growth hormones

6.1 Plant Growth Hormones

We all know that plants need light, water, oxygen, and nutrition to grow and develop. All these qualify as extrinsic factors. While extrinsic factors are important, there are intrinsic factors which have influence on plant growth and development. They can be intracellular genes or intercellular chemicals. These chemicals are called plant growth regulators. Plant hormones (also known as phytohormones) are not nutrients, but chemicals and not all plant cells respond to hormones, but those cells that do are programmed to respond at specific points in their growth cycle. The greatest effects occur at specific stages during the cell's life, with diminished effects occurring before or after this period (Swarup et al. 2007). Plants respond to external stimuli, sense changes in their environment, and their biological clock works in a precise manner. Many deciduous plants shed their leaves in a particular season or develop flowers and fruits during a specific time of the year. All these and many more rather intriguing responses of plants indicate that they have unique mechanisms to sense and communicate with the environment they live in. In higher plants, regulation and coordination of metabolism, growth, and morphogenesis depend on the transport of various chemical signals from one part to another. It was hypothesized that certain chemical messengers are responsible for the initiation and growth of different plant organs.

Plants need hormones at very specific times during plant growth and at specific locations. They also need to disengage the effects that hormones have when they are no longer needed. The production of hormones occurs very often at sites of active growth within the meristems, before the cells have fully differentiated. After production, they have sometimes moved to other parts of the plant, where they cause an immediate effect; or they can be stored in cells to be released later. Plants can also break down hormones chemically, effectively destroying them. Plant hormones frequently regulate the concentrations of other plant hormones (Campbell and Reece 2002; Davies 2010). Plant hormones are signal molecules, produced within plants that occur in extremely low concentrations. Plant hormones control all aspects of plant growth and development, from embryogenesis (Méndez-Hernández et al. 2019) to the regulation of organ size, pathogen defense (Shigenaga and Argueso 2016; Bürger and Chory 2019), stress tolerance (Ku et al. 2018; Ullah et al. 2018) and reproductive development (Pierre-Jetome et al. 2018). Each plant cell is capable of producing hormones, and these plant hormones shape the plant, affecting seed growth, time of flowering, the sex of flowers, senescence of leaves, and fruits. They affect which tissues grow upward and which grow downward, leaf formation and stem growth, fruit development and ripening, plant longevity, and even plant death. Hormones are vital to plant growth, and, lacking them, plants would be mostly a mass of undifferentiated cells. So they are also known as growth factors or growth hormones. Based on their action, plant growth hormones were broadly classified as follows:

- Plant Growth Promoters—They promote cell division, cell enlargement, flowering, fruiting, and seed formation. Examples are auxins, gibberellins, and cytokinins.
- Plant Growth Inhibitors—These chemicals inhibit growth and promote dormancy and abscission in plants. An example is an abscisic acid.

Growth hormones occur across the plant kingdom, and even in algae, where they have similar functions to those seen in higher plant (Tarakhovskaya et al. 2007). Some growth hormones also occur in microorganisms, such as unicellular fungi and bacteria; however in these cases they do not play a hormonal role and can better be regarded as secondary metabolites (Rademacher 1994). Plant growth regulators can be of a diverse chemical composition such as gases (ethylene), terpenes (gibberellic acid), or carotenoid derivatives (abscisic acid). Hormones have since long been defined as biochemicals produced in specific tissues which are transported to some distant cells to trigger response at low concentrations.

Since 1930, plant growth regulators have been a systematic use in different agriculture practices. For instance, controlling of vegetative growth, decreasing susceptibility toward biotic and abiotic stress, reducing the risk of lodging in field crops, breaking of bud dormancy, controlling of fruit set, improvement of fruit quality, acceleration or delay of fruit ripening for optimal harvesting, optimization of fruit storage and ripening, improved morphological structure, increases in yield, and modification of plant constituents (Alcázar et al. 2010; Choudhury et al. 2013; Rademacher 2015; Fahad et al. 2016; Bergstrand 2017). The plant growth regulators refer to artificially produced substances which in very low quantities normally act at sites other than the place of production and control different physiological processes that modulate plant growth and development. It is a structurally unrelated collection of small molecules derived from various essential metabolic pathways (Santner et al. 2009; Davies 2010).

These compounds are important regulators of plant growth and mediate responses to both biotic and abiotic stresses. Peleg and Blumwald (2011) expressed that the plant hormones play central roles in the ability of plants to adapt to changing environments, by mediating growth, development, nutrient allocation, and source/sink transitions. Although, abscisic acid is the most studied stress-responsive hormone, the role of cytokinins, brassinosteroids, and auxins during environmental stress is less studied. Recent evidence indicated that plant hormones are involved in multiple processes. Crosstalk between the different plant hormones results in synergetic or antagonist interactions that play crucial roles in response of plants to abiotic stress. The characterization of the molecular mechanisms regulating hormone synthesis, signaling, and action is facilitating the modification of hormone biosynthetic pathways for the generation of transgenic crop plants with enhanced abiotic stress tolerance in field (Alcázar et al. 2010; Peleg and Blumwald 2011; Verma et al. 2016).

Plant hormones are not nutrients, but chemicals that in small amounts promote and influence the growth (Öpik et al. 2005), development, and differentiation of cells and tissues. The biosynthesis of plant hormones within plant tissues is often diffuse and not always localized. Plants utilize simple chemicals as hormones, which move more easily through their tissues. They are often produced and used on a local basis within the plant body. Plant cells produce hormones that affect even different regions of the cell producing the hormone. Hormones are transported within the plant by utilizing four types of movements. For localized movement, cytoplasmic streaming within cells and slow diffusion of ions and molecules between cells are utilized. Vascular tissues are used to move hormones from one part of the plant to another; these include sieve tubes or phloem that moves sugars from the leaves to the roots and flowers and xylem that moves water and mineral solutes from the roots to the foliage. Not all plant cells respond to hormones, but those cells that do are programmed to respond at specific points in their growth cycle. The greatest effects occur at specific stages during the cell's life, with diminished effects occurring before or after this period. Plants need hormones at very specific times during plant growth and at specific locations. They also need to disengage the effects that hormones have when they are no longer needed. The production of hormones occurs very often at sites of active growth within the meristems, before cells have fully differentiated. After production, they are sometimes moved to other parts of the plant, where they cause an immediate effect; or they can be stored in cells to be released later. Plants use different pathways to regulate internal hormone quantities and moderate their effects; they can regulate the amount of chemicals used to biosynthesize hormones. Plants can also break down hormones chemically, effectively destroying them. Plant hormones frequently regulate the concentrations of other plant hormones (Swarup et al. 2007). Plants also move hormones around the plant diluting their concentrations. The concentration of hormones required for plant responses is very low (10^{-6} to 10^{-5} mol/l).

Abscisic acid regulates many important developmental processes and induced tolerance to different stresses including drought, soil salinity, and low temperature (Giraudat et al. 1994). Abscisic acid plays a direct role in mediating the photosynthesis and respiration in leaves and also inhibition of lateral root development (Zhou et al. 1998). Abscisic acid has been postulated to regulate the formation of xanthophylls in an inducible manner (Zhou and Leul 1998). Treatment with abscisic acid stimulated the accumulation of anthocyanin and phenolics and increased ethylene production in maize leaves (Parry 1993). Paclobutrazol is a triazolic group of fungicide which has plant growth-regulating properties. The growth properties of paclobutrazol are mediated by changes in balance of important plant hormones including the gibberellins, abscisic acid, and cytokinin (Fletcher and Hofstra 1990). Paclobutrazol treatment increased the chlorophyll and carotenoid content in corn seedlings (Pinhero and Fletcher 1994), *Lolium perenne* (Hampton and Hebblethwaite 2006), *Brassica carinata* (Setia et al. 1995), and *Dianthus caryophyllus* (Banon et al. 2002); enlarged chloroplasts were observed in barley (Khalil and Rahman 1995). Increased chlorophyll and carotenoid contents were in tomato (Berova and Zlatev 2000; Still and Pill 2004) and wheat (Berova et al. 2002) under triazole treatments.

6.2 Classes of Plant Growth Hormones

Plant growth-regulating substances, plant hormones, or simply phytohormones are compounds produced naturally by plants that participate in control of plant growth; also they are versatile chemical regulators of plant growth. When these substances are produced synthetically, they are called plant growth regulators (Santner et al. 2009; Davies 2010; Rademacher 2015). Plant growth and development require the integration of a variety of environmental and endogenous signals that, together with the intrinsic genetic program, determine plant form. Central to this process are plant growth regulators. The principal plant growth regulators include classical groups or others compounds. Among them are mentioned indole-3-acetic acid, indole-3-butyric acid, 2,4-dichlorophenoxyacetic acid, 1-naphthylacetic acid, 6-benzylaminopurine, benzyladenine, gibberellic acid, ethylene or ethephon, abscisic acid, jasmonate or methyl jasmonates, chlormequat chloride, as well as brassinosteroids, salicylic acid, nitric oxide, strigolactone, and polyamines (like putrescine, spermidine, and spermine), and other molecules such as α-tocopherol, thidiazuron, phenylureas, and triazoles. The nature, occurrence, transport, and effects of each plant growth regulators are different according to their characteristics (Chiwocha et al. 2009; Davies 2010; Ullah et al. 2012; Rademacher 2015; Fahad et al. 2016). Thiourea, a synthetic compound containing nitrogen (as NH_2) and sulfur (as SH), is an important plant growth regulator, which influences plant growth particularly under stress conditions (Garg et al. 2006). Since the discovery of the biological role of thiourea (Jocelyn 1972), it has only been employed to break seed dormancy and improve germination (Esashi et al. 1977). However, increasing research interest in thiourea has recently revealed several beneficial impacts on plant biology (Waqas et al. 2019).

In general, there are five major classes of plant growth regulators (Fig. 6.1), and each is grouped together into one of these classes based on their structural

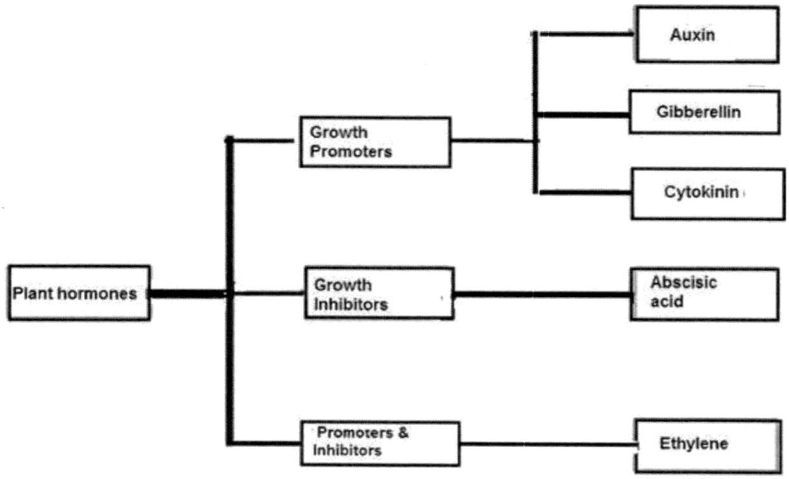

Fig. 6.1 Major classes of plant growth regulators

similarities and physiological responses. These include the auxins and gibberellins that stimulate predominantly cell elongation; cytokinins, the purine bases that stimulate cell division; ethylene, the olefinic gaseous molecule that regulates among other plant event fruit ripening; and abscisic acid, the sesquiterpene that regulates senescence and abscission of plant parts and helps in maintenance of plant water relations. Besides these five classes of naturally existing plant growth hormones, the steroid hormones brassinosteroids and polyamines are also reported to exhibit growth-regulating activities in plants and are the topics of extensive researches. While these natural plant growth hormones are distinctive both in chemical characteristics and in exhibiting characteristic responses, each of the plant growth hormones has the potential to alter almost all the aspects of plant growth and development to confer stress tolerance. It is documented that the induction in stress tolerance is induced by the manipulations in the concentrations of endogenous plant growth hormones under stress conditions by helping plants in many ways. Major classes of plant growth hormones have been reviewed extensively by Bhattacharya (2019, 2021).

6.2.1 Abscisic Acid

Abscisic acid is one of the most important plant growth inhibitor. This class of plant growth regulator is composed of one chemical compound normally produced in the leaves of plants, originating from chloroplasts, especially when plants are under stress. In general, it acts as an inhibitory chemical compound that affects bud growth, and seed and bud dormancy. Abscisic acid is an important chemical for plant responses to a range of abiotic stresses, including drought, temperature, and salinity

(Keskin et al. 2010; Verslues and Bray 2006). A dynamic balance between its biosynthesis and degradation, sensitized by developmental and environmental factors, determines the amount of available abscisic acid (Cutler and Krochko 1999). The functions of abscisic acid in plants are multiple. High cellular abscisic acid facilitates modifications in stomatal functioning, root hydraulic conductivity, photosynthesis, biomass allocation between roots and shoots, plant water relations, osmolyte production, and synthesis of stress-responsive proteins and genes to confer stress tolerance (Finkelstein et al. 2002, 2008; Hetherington 2001; Kim et al. 2010; Hoth et al. 2002; Seki et al. 2002). Abscisic acid is one of the most important stress hormones and participates in various crucial physiological processes during the plant life cycle, including stress responses, development, and reproduction (Dong et al. 2015; Zhu 2002). It mediates changes within the apical meristem, causing bud dormancy and the alteration of the last set of leaves into protective bud covers. From the experiments involving radio-labeled $^{18}O_2$, molecular genetic analysis of auxotrophs, and biochemical studies, the pathways for abscisic acid biosynthesis are identified. Abscisic acid is biosynthesized in cytosol through the carotenoid biosynthetic pathway (Milborrow 2001). Abscisic acid is biosynthesized from the C15 compound, farnesyl pyrophosphate, and a C40 carotenoid, involving isopentenyl pyrophosphate. Abscisic acid exists in all parts of the plant, and its concentration within any tissue seems to mediate its effects and function as a hormone (Kermode 2005). Stress from water or predation affects abscisic acid production and catabolism rates, mediating another cascade of effects that trigger specific responses from targeted cells.

6.2.2 Auxins

Auxins (indole acetic acid, IAA) play essential roles in diverse developmental events like root growth, vascular tissue differentiation, auxiliary bud formation, apical dominance, and flower organ development (Zhao 2010) and respond to abiotic stress (Bhattacharya 2019). These are compounds that positively influence cell enlargement, bud formation, and root initiation. They also promote the production of other hormones, and in conjunction with cytokinins, they control the growth of stems, roots, and fruits and convert stems into flowers (Osborne and McManus 2005). Auxins affect cell elongation by altering cell wall plasticity, stimulate cambium to divide, and in stems cause secondary xylem to differentiate. Auxins act to inhibit the growth of buds lower down the stems and also to promote lateral and adventitious root development and growth. Leaf abscission is initiated by the growing point of a plant ceasing to produce auxins. Auxins in seeds regulate specific protein synthesis (Walz et al. 2002) as they develop within the flower after pollination causing the flower to develop a fruit to contain the developing seeds. Auxins are toxic to plants in large concentrations; they are most toxic to dicots and less so to monocots. Auxins, especially 1-naphthaleneacetic acid and indole-3-butyric acid, are also commonly applied to stimulate root growth when taking cuttings of plants. The most common auxin found in plants is indole-3-acetic acid. Because of this property, synthetic auxin herbicides including 2,4-D (2,4-dichlorophenoxyacetic) and 2,4,5-T have been developed and used for weed control. The concentrations and ratios of indole-3-acetic acid and its derivatives in plant tissues and auxin homeostasis are regulated by its degradation, conjugation to amino acids, and transport. And these processes are sensitive to abiotic stresses in plants.

The auxin levels are modified in plant by two possible mechanisms under stress: one by changes in expression of auxin polar transporter gene and the other by inhibitions in polar transport by certain compounds accumulated in response to stress (Potters et al. 2009). Besides, auxin metabolism is also modulated by oxidative degradation of auxin through peroxidases induction under abiotic stress and also reactive oxygen species generated during stress (Kovtun et al. 2000). Upreti et al. (2012) reported that maintenance of high auxin in the reproductive organs of capsicum cultivars is responsible for lower abscission and high-temperature stress tolerance. Cold stress and auxin contents are potentially linked (Wyatt et al. 2002), as cold stress inhibits the root gravity response and also auxin. Shibasaki et al. (2009) suggested that auxin concentrations are regulated by changes in auxin transport in plants under cold stress (Shibasaki et al. 2009). Du et al. (2012) documented that the increased tolerance to cold stress is due to the combined effects of auxin and abscisic acid, as abscisic acid or carotenoid-deficient mutants show reduced auxin content and exhibit increased cold resistance (Du et al. 2013a). For abiotic stress response, Du et al. (2012, 2013a) found that the change in auxin homeostasis could influence abscisic acid synthesis and that the balance of auxin and abscisic acid homeostasis played a crucial role in diverse stress responses. In this regard, Taniguchi et al. (2010) suggested abscisic acid involvement in hydrotropic root responses through modulation in auxin.

6.2.3 Gibberellins

Gibberellins regulate various developmental processes, including stem elongation, germination, dormancy, flowering, flower development, and leaf and fruit senescence (Hedden and Sponsel 2015). The synthesis of gibberellins is strongly upregulated in seeds at germination and its presence is required for germination to occur. In seedlings as well as in adults plants, it strongly promote cell elongation, the transition between vegetative and reproductive growth and for pollen function during fertilization (Tsai et al. 1997). Gibberellins are involved in the natural process of breaking dormancy and other aspects of germination. Gibberellins in the seed embryo are believed to signal starch hydrolysis through inducing the synthesis of the enzyme α-amylase in the aleurone cells. In the model for gibberellin-induced production of α-amylase, it is demonstrated that gibberellins produced in the scutellum diffuse to the aleurone cells, where they stimulate the secretion of α-amylase (Campbell and Reece 2002). Gibberellins are produced in greater mass when the plant is exposed to cold temperatures. Its hormone binds to a receptor, and calcium activates the protein calmodulin, and the complex binds to DNA, producing an enzyme to stimulate growth in the embryo.

The primary mechanism by which these chemicals exert influence is through inhibition of the biosynthesis of gibberellins (Rademacher 2000) and application of gibberellin to retardant treated plants and to gibberellin-deficient mutants (Vettak-korumakankav et al. 1999). Radi et al. (2006) showed that the presoaking wheat seeds in gibberellic acid increased the germination potential especially at moderate salinization levels. Between gibberellin levels and the acquisition of stress protection, an intimate relationship exists (Vettak-korumakankav et al. 1999). In sugarcane, supplementing gibberellin as foliar application plays an important role on imparting salt tolerance in terms of enhancing nutrient uptake, as well as the morphological and physiological aspects (Shomeili et al. 2011). Ashraf et al. (2002) showed that gibberellin application increased the nutrient uptake, plant height, leaf area, and yield of wheat under saline conditions.

6.2.4 Cytokinins

Cytokinins are important growth promontory compounds involved in seed germination, morphogenesis, chloroplast biogenesis, maintenance of assimilate mobilization, fruit and leaf abscission, and regulation of stomatal functioning and root-to-shoot communication in the plant experiencing stress. Cytokinins are N6-substituted adenosine compounds with branched five-carbon side chain (zeatin and isopentenyl adenine). Cytokinins are a group of chemicals that influence cell division and shoot formation. They were called kinins in the past when the first cytokinins were isolated from yeast cells. They also help delay senescence of tissues, are responsible for mediating auxin transport throughout the plant, and affect internodal length and leaf growth. Cytokinins are involved primarily in cell growth and differentiation but also affect apical dominance, axillary bud growth, and leaf senescence. There are two types of cytokinins: adenine-type cytokinins represented by kinetin, zeatin, and 6-benzylaminopurine and phenylurea-type cytokinins like diphenylurea and thidiazuron (Aina et al. 2012). Cytokinins are synthesized in roots (Campbell et al. 2008). Cambium and other actively dividing tissues also synthesize cytokinins (Chen et al. 1985). Cytokinins participate in local and long-distance signaling, with the same transport mechanism as purines and nucleosides (Sakakibara 2006). Typically, cytokinins are transported in the xylem (Campbell et al. 2008). Plants experiencing abiotic stress tend to decline cytokinin concentration (Naqvi 1999; Pospisilova et al. 2000). Upon stress release, the cytokinin restores their normal levels, and response is fast, and species and cultivar dependent. Zhou et al. (2004) reported that the drought stress changes in cytokinin levels in the xylem sap of apple trunk, and the changes depended upon drought cycles. Upreti et al. (1998) in French bean and Upreti and Murti (2004a) in onion found profound decline in cytokinins as a result of drought stress with extent of decline dependent upon stress severity, growth stage, and cultivars. Satisha et al. (2005) witnessed a decline in cytokinins in the grape genotypes under soil moisture-deficit conditions. Further, the stress-induced decline in cytokinins depends upon the age of the plant organ, as the young leaves of stressed plants witness greater decline in cytokinins than the old leaves (Upreti and Murti 2004b). Root nodulation is sensitive to drought stress in French bean plants. Stomata play an important role in the control of water transpiration and gas exchange in plant leaves, and cytokinins are important in keeping stomata open by regulating abscisic acid action (Veselova et al. 2006).

6.2.5 Ethylene

Ethylene is a gas and its effectiveness as a plant hormone is dependent on its rate of production versus its rate of escaping into the atmosphere. Ethylene is produced at a faster rate in rapidly growing and dividing cells, especially in darkness. New growth and newly germinated seedlings produce more ethylene than can escape the plant, which leads to elevated amounts of ethylene, inhibiting leaf expansion. Ethylene governs the growth, development, and senescence of leaves, flowers, and fruits (Pierik et al. 2006; Masood et al. 2012; Nazar et al. 2014). It may also promote, inhibit, or induce senescence depending upon the optimal or suboptimal ethylene levels (Khan 2005; Pierik et al. 2006).

Abiotic stress conditions surge ethylene production by inducing 1-aminocyclopropane-1-carboxylic acid synthase and 1-aminocyclopropane-1-carboxylic acid oxidase activities. Drought stress enhances ethylene in French bean (Upreti et al. 1998), orange (Ben-Yehoshua and Aloni 1974), avocado (Adato and Gazit 1974), *Vicia faba* (El-Beltagy and Hall 1974), and many other plant species (Narayana et al. 1991). The increase in ethylene under stress is of adaptive significance as it helps plants to cope up stress by reducing water loss through increased senescence of fruits/leaves and reduced growth. The magnitude of ethylene increase under stress depends upon growth stage, stress intensity, and stress duration (Upreti et al. 1998, 2000), and higher stress levels tend to reduce ethylene concentration. In pineapple, drought stress has no effect on flower induction and produced significantly less ethylene and had lower 1-aminocyclopropane-1-carboxylic acid oxidase activity in leaf and stem tissues than the control plants (Min and Bartholomew 2005). Habben et al. (2014) reported that transgenic field-grown plants with downregulated 1-aminocyclopropane-1-carboxylic acid oxidase synthase enzyme activity yield higher under drought stress. Salt stress positively influences ethylene biosynthesis, which helps in promoting salt tolerance by enhancing Na/K homeostasis (Lockhart 2013).

6.2.6 Brassinosteroid

Brassinosteroids are a class of polyhydroxysteroids, the only example of steroid-based hormones in plants. Brassinosteroids inhibit root growth and leaf abscission. Brassinosteroid was the first identified brassinosteroid and was isolated from extracts of rapeseed (*Brassica napus*) pollen in 1979. Brassinosteroids play crucial roles in diverse aspects of plant biology, including cell elongation, cell division, root growth, photomorphogenesis, stomatal and vascular differentiation, seed germination, immunity, and reproduction (Gudesblat and Russinova 2011; Wei and Li 2016). These plant hormones regulate a broad range of processes in plant development and responses to environmental stresses, and their analogs have been shown to bring substantial increases in grain yield, depending on growth status. Numerous studies indicate brassinosteroid potential in enhancing ability of plant to cope with drought stress, salt stress, and high- and low-temperature stresses (Fariduddin et al. 2014). The growth regulatory and stress protection capability of brassinosteroids are linked to their action on metabolic processes associated with photosynthesis and nucleic acid and protein biosynthesis (Fariduddin et al. 2014).

It has been suggested that brassinosteroids regulate stress response by a complex sequence of biochemical reactions, such as activation or suppression of key enzymatic reactions, induction of protein synthesis, and the production of various chemical defense compounds (Bajguz and Hayat 2009). Most of the research on the response action of brassinosteroids to stress factors is made employing their exogenous application. Upreti and Murti (2004c) reported that application of epibrassinolide or homobrassinolide prior to drought stress results in increased root nodulation in French bean, and the response is mediated through induction in cytokinin synthesis and nitrogenase activity. Moreover, epibrassinolide is relatively more effective than homobrassinolide in rendering such response. Kagale et al. (2007) reported improved drought tolerance in *Brassica napus* seedlings treated with epibrassinolide, and the improved tolerance is due to reduction in reactive oxygen species, induction in antioxidative enzyme activities, and antioxidant contents (Zhang et al. 2008b; Li et al. 2012). Zhang et al. (2008b) also witnessed improvement in photosynthesis by regulating ribulose-1,5-bisphosphate carboxylase/oxygenase activity and sugar accumulation following brassinosteroid treatment in stressed plants.

Chilling tolerance in plants is influenced by brassinosteroids. The report of Huang et al. (2006) showed that epibrassinolide application upregulated 17 proteins which were downregulated by chilling to confer chilling tolerance. Chilling tolerance is also attributed to increase in membrane permeability (Janeczko et al. 2007), high pigment accumulation, and upregulation in cold-related genes (Kagale et al. 2007) besides high activation of Rubisco and expression of photosynthetic genes (Xia et al. 2009). Thus, the physiological responses of brassinosteroids though are variable due to complexities in the molecular mechanism of their action; their potential in improving abiotic stress tolerance has immense utility in managing plant responses to stress. For widening the scope in this area and effectively harnessing the benefits from brassinosteroid research, more investigations are needed on mechanisms by which brassinosteroids confer stress tolerance.

6.2.7 Jasmonates

Jasmonic acid is derived from α-linolenic acid. Jasmonates and its active derivatives have a significant role in regulating stress responses of plants to various biotic and abiotic stresses. Jasmonates are also involved in root growth, fruit ripening, tendril coiling, and viable pollen production (Delker et al. 2006; Wasternack 2007). Synthesis of jasmonates occurs in plastid, peroxisome, and cytosol. The most active jasmonate in plants is jasmonic acid. In addition to their role in defense, jasmonates are also believed to play roles in seed germination, the storage of protein in seeds, and root growth (Browse 2005). Although jasmonate regulates many different processes in the plant, its role in wound response is best understood. For example, in the tomato, wounding produces defense molecules that inhibit leaf digestion in the insect's gut. Another indirect result of jasmonate signaling is the volatile emission of jasmonate-derived compounds. Following its role in defense, jasmonates have also been implicated in cell death and leaf senescence.

6.2.8 Salicylic Acid

Salicylic acid is involved in the response of plants to abiotic stress, particularly soil water deficit, temperature, heavy metal, and osmotic stress (Rivas-San and Plasencia 2011). Salicylic acid is a phenolic compound which, despite its broad distribution in plants, has basal levels differing widely among species, with up to 100-fold differences having been recorded (Raskin et al. 1990). This disparity can be observed within members of the same family. For example, in the Solanaceae, whereas tobacco (*Nicotiana tabacum*) contains low basal levels of salicylic acid [<100 ng g^{-1} fresh weight] in leaves (Malamy et al. 1992), potato (*Solanum tuberosum*) might contain up to 10 μg of total salicylic acid g^{-1} fresh weight (Navarre and Mayo 2004). In the model plant *Arabidopsis thaliana*, basal levels of total salicylic acid range from 0.250 μg to 1 μg g^{-1} fresh weight (Wildermuth et al. 2001; Brodersen et al. 2005). Salicylic acid has been recognized as a regulatory signal mediating plant response to abiotic stresses such as chilling (Kang and Saltveit 2002), heat (Larkindale et al. 2005), and osmotic stress (Borsani et al. 2001). There is evidence that salicylic acid regulates processes such as seed germination, vegetative growth, photosynthesis, respiration, thermogenesis, flower formation, seed production, senescence, and a type of cell death that is not associated with the hypersensitive response. In addition, salicylic acid could contribute to maintaining cellular redox homeostasis through the regulation of antioxidant enzyme activity (Slaymaker et al. 2002) and induction of the alternative respiratory pathway (Moore et al. 2002).

6.2.9 Strigolactones

Strigolactones are known to stimulate the germination of arbuscular mycorrhiza spores (Besserer et al. 2006). Since they produce this effect at extremely low concentrations, it has been proposed that the mechanism of activation must be a signaling pathway (Besserer et al. 2006). When applied in terminal buds of stem, strigolactone can block the expression of transport proteins required to move auxin across the buds; these proteins are denominated PIN1 (Shinohara et al. 2013). Thus, it was not surprising that when analyzing strigolactone-deficient mutants, they were found to present an overexpression of PIN1 protein, which facilitates the transport of auxin in the terminal buds; auxin prevented the mitotic activity of these buds, stopping the plant to initiate secondary growth and branching (Shinohara et al. 2013).

6.2.10 Polyamines

Polyamines are low-molecular-weight aliphatic nitrogenous bases containing two or more amino groups and have potent biological activity (Xu et al. 2009; Vuosku et al. 2018). They are produced by organisms during metabolism and are present in almost all cells of both eukaryotic and prokaryotic (Chen et al. 2019). Because they play important roles in diverse plant growth and developmental processes and in environmental stress responses, they are considered as a new kind of plant growth regulators (Xu et al. 2014). In higher plants, polyamines are mainly present in their free form. Putrescine, spermidine, and spermine are the main polyamines in plants, and they are involved in the regulation of diverse physiological processes (Xu et al. 2014; Mustafavi et al. 2018), such as flower development, embryogenesis, organogenesis (Xu 2015), senescence, and fruit maturation and development. They are also involved in responses to biotic and abiotic stresses (Vuosku et al. 2012; de Oliveira et al. 2016; Reis et al. 2016; Mustafavi et al. 2018).

Exogenous polyamines were shown to accelerate the process of flower bud differentiation, and high polyamine contents in apical buds were beneficial for the initiation and maintenance of flower bud differentiation in *Chrysanthemum* (Xu 2015). In *Arabidopsis*, polyamines were found to be more abundant in flowers than in any other organ, and the addition of exogenous polyamines to poorly flowering plants significantly promoted their flowering response (Applewhite et al. 2010). Polyamines can bind to the phospholipid site of the cell membrane to prevent cytolysis and improve cold resistance (Li and He 2012). However, there are several different viewpoints on the relationship between putrescine and plant chilling stress (Wu and Yuan 2008). When sweet pepper and zucchini fruits were stored at chilling temperature, the putrescine content increased exponentially, accompanied by chilling damage. Storage under CO_2-modified atmosphere reduced the extent of cold damage and inhibited the accumulation of putrescine, suggesting that putrescine accumulated as a result of chilling stress (Serrano et al. 1997, 1998). In contrast, Roy and Wu (2002) proposed that putrescine accumulation caused chilling damage, and increased Spm may be a defense response to cold damage. They found that the putrescine, spermidine, and spermine contents gradually increased in loquat fruit stored at low temperatures. The application of exogenous spermidine maintained high levels of endogenous spermidine, and spermine, inhibiting putrescine accumulation and reducing chilling damage (Zhen et al. 2000; Roy and Wu 2001). Another opinion was that putrescine may accumulate as a defense response of plants to chilling damage, because putrescine accumulation was found to be positively correlated with the cold resistance of plants (Wang et al. 2003b).

6.3 Importance and Use of Plant Growth Hormones

Plant growth and productivity are adversely affected by various abiotic stresses. Plants are frequently exposed to a plethora of stress conditions such as low temperature, salt, drought, flooding, heat, and oxidative stress. Various anthropogenic activities have accentuated the existing stress factors. To provide tolerance against environmental stresses, plants are equipped with several built-in physiological and biochemical mechanisms occurring at cellular level. An understanding of processes linked to these mechanisms is vital in optimizing the crop growth and productivity under stress conditions. The cellular changes induced by low temperature include excess accumulation of toxic compounds, especially reactive oxygen species. The end result of reactive oxygen species accumulation is oxidative stress (Mittler 2002; Yin et al. 2008; Suzuki and Mittler 2006). Low-temperature conditions can create an imbalance between light absorption and light use by inhibiting the activity of the Calvin-Benson cycle. Enhanced photosynthetic electron flux to O_2 and over-reduction of the respiratory electron transport chain can also result in reactive oxygen species accumulation during chilling which causes oxidative stress (Hu et al. 2008). Plants have evolved a variety of responses for extreme temperatures, those that minimize damages and ensure the maintenance of cellular homeostasis (Kotak et al. 2007). A considerable amount of works have explored that there is a direct link between reactive oxygen species scavenging and plant stress tolerance under temperature extremes (Suzuki and Mittler 2006). Thus, the improvement of temperature stress tolerance is often related to enhanced activities of enzymes involved in antioxidant systems of plants. Plants exposed to extreme temperatures use several nonenzymatic and enzymatic antioxidants to cope with the harmful effects of oxidative stress; higher activities of antioxidant defense enzymes are correlated with higher stress tolerance. Different plant studies have revealed that enhancing antioxidant defense confers stress tolerance to low-temperature stress (Almeselmani et al. 2006, 2009; Babu and Devraj 2008; Huang and Guo 2005).

Cold stress is a significant abiotic stress which affects growth and development of crops, leading to loss of strength and lesions on the surface. These symptoms are triggered by changes in the physical and chemical organization of cell membranes, among other metabolic procedures (Barrero-Sicilia et al. 2017). Climate change controls greater changes in temperature, leading in frequent cold periods. Susceptible plants with cold temperatures have reduced growth and growth, restricted use of precious varieties, and reduced yields. Plants use separate strategies to cope with stressful conditions and integrate a variety of physiological, metabolic, and molecular adaptations. These methods initially generate modifications to safeguard the plant, followed by cold acclimation, which increases the survival of the plant under cold stress (Flowers 2004). While a lot of these mechanisms are facilitated by transcription factors that stimulate gene expression associated to stress, the transcription network is not restricted to the reaction of plants to cold (Zhu et al. 2007; Barrero-Sicilia et al. 2017). As a foremost element of plasma and endomembranes, lipids play a significant organizational role in mitigating the effects of

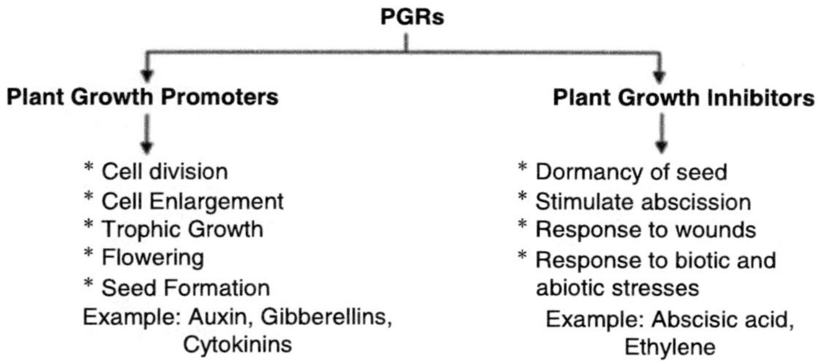

Fig. 6.2 Effects of plant growth regulators of different aspects of plant growth

cold temperatures (Nishida and Murata 1996). Cold stress decreases plant growth and development that affects the physical and chemical structure of the cell membrane, causes leakage of electrolytes, and reduces protoplasmic streaming and changes in the metabolism of cell (Thomashow 1999; Chen and Thelen 2013). Additional cold reactions comprise changes in nucleic acid and protein synthesis, water and nutrient equilibrium, enzyme affinity, and conformation and deficiency in photosynthesis, specifically downregulation and photodamage of photosystem II (Takahashi et al. 2013).

Changes in the structure of proteins and lipid membranes restore homeostasis of metabolites and are regarded a mechanism by which cells feel cold temperatures. For its metabolic and physical function, the liquid state of the plasma membrane is a structural and functional asset. When low temperatures are present, the plasma membrane transitions from a liquid state at elevated temperatures into a stiff gel stage (Miquel et al. 1993). Low temperature-mediated changes in the physical conformation of the membrane are mainly because of enhanced levels of unsaturated lipids, which increase the fluidity and stability of the membrane, enabling cells to adapt mechanically to cold (Janská et al. 2010).

Different aspects of plant growth are either positively or negatively affected by plant growth regulators (Fig. 6.2), and subsequently the yield is affected. Plant growth and crop yield are majorly affected by cold, drought, salt, and heavy metals. Abiotic stress impacts plants to molecular levels from morphological levels and is visible at all phases of plant development where drought occurs (Fahad et al. 2017). There are three significant stages of plant, vegetative development, pre-anthesis, and terminal phase, that are impacted by the drought (Shavrukov et al. 2017). Physiological reactions to stress include wilting of the leaf, abscission of the leaf, decreased leaf region, and decreased water loss through transpiration (Fghire et al. 2015). Abiotic stress resistance contains escape-avoidance and tolerance mechanisms. Detrimental impacts of stress can be decreased by osmotic modification, which helps with an active accumulation of solutes in the cytoplasm to maintain cell water balance (Shanker and Shanker 2016).

The use and application of plant growth regulators in agriculture have several practical examples. The plant growth regulators in this context are employed commercially. Besides, plant tissue culture is traditional practice and has relevant importance. The commercial uses of auxins include prevention of fruit and leaf drop, promotion of flowering, thinning of fruit, induction of parthenocarpic fruit development, and rooting of cuttings for plant propagation, among others (Taiz and Zeiger 2010). Considerable success had been obtained in the application of plant growth regulators in some process of plant development such as flowering and fruit development as well as ripening, harvesting, and post-harvesting of fruits and vegetables. Sajid et al. (2009) reported the effects of foliar application of plant growth regulators and nutrients for improvement of lily flowers (*Lilium* sp.). Tropical trees are generally induced to flower through environmental cues, whereas floral initiation of temperate deciduous trees is often autonomous. In the tropical evergreen tree mango (*Mangifera indica* L.), cool temperature is the only factor known to induce flowering, but does not ensure floral initiation will occur because there are important interactions with vegetative growth involvement of plant hormones (Bangerth 2009). In horticultural trees there are four major pathways to flowering. Those have been characterized in *Arabidopsis*, including environmental induction through photoperiod and temperature, autonomous floral initiation, and regulation by gibberellins and nitrate used (Marín et al. 2011).

The temperate deciduous apple tree (*Malus domestica* Borkh.) flowers autonomously, with floral initiation dependent on aspects of vegetative development in the growing season before anthesis. In general, for floral initiation of horticultural trees, the effect of the environment, its interactions with vegetative growth, and the roles of plant growth regulators and plant carbohydrates play a crucial role (Wilkie et al. 2008). Tang et al. (2009) observed effects of exogenous application of plant growth regulators on the development of ovule and subsequent embryo rescue of stenospermic grape (*Vitis vinifera* L. cv.), Centennial Seedless, Thompson Seedless, and Crimson Seedless sprayed by chlormequat chloride, benzyladenine, ethylene, and putrescine.

Cotton (*Gossypium hirsutum* L.) grows best under warm temperatures and high light intensity, is somewhat drought tolerant, and often continues or resumes growth late in the season. Because of these growth habits, alterations in growth and development of the crop are often agronomically desirable. These alterations may be accomplished with plant growth regulators (Cothren and Oosterhuis 2010). In seedless grapes (*Vitis* spp.), Bhat et al. (2011) found that the application of ethylene and brassinosteroid along with gibberellic acid increases the leaf number, leaf area, and leaf dry matter and indirectly influences the fruit yield and quality. The effects of nitrogen fertilization and chlormequat have been studied in other culture such us wheat (*Triticum aestivum* L.) by Shekoofa and Emam (2008). Tiwari et al. (2011) reported about the effect of gibberellic acid and other plant growth regulators as naphthalene acid on hybrid rice (*Oryza sativa* L.) seed production. The influence of exogenous applications of plant growth regulators on fruit quality of young 'Kinnow' mandarin (*Citrus nobilis* × *Citrus deliciosa*) trees was assayed by Khalid et al. (2012). Benzyladenine and kinetin at 20 ppm were applied at flowering stage, and benzyladenine; kinetin at 10, 20, and 30 ppm; and gibberellin at 10 ppm were applied at fruit setting stage. This study revealed that irrespective of shelf duration, the fruit harvested from young trees treated with 10 ppm gibberellin, 30 ppm benzyladenine, and kinetin at fruit setting stage showed significantly higher juice contents and lower rag contents in comparison with control.

Other plant growth regulators also have an important role in plant growth as shown by Fahad et al. (2016). The authors demonstrated that the exogenous application of plant growth regulators (ascorbic acid, α-tocopherol, brassinosteroids, methyl jasmonate, and triazoles) in four different combinations increases thermal stability and better rice performance under high-temperature stress. The plant growth regulators increased the biosynthesis of metabolites as evidence of stress tolerance. Besides, chemical plant growth regulators are used in the production of ornamental potted and bedding plants. In recent years, the interest in climatic growth regulation has been resurrected, not least due to the coming phaseout of the plant growth regulator, chlormequat chloride. Policy makers in many countries are aiming toward reducing the use of agrochemicals. At the same time, there is a strong demand from consumers for products produced without chemicals. Global environmental change presents both old and new challenges as immediate challenge is to adapt plant, animal, and food systems to changing temperature, nutrient, and water conditions (Fresco 2009; Bergstrand 2017).

The effect of different plant growth regulators (30 ppm of napthyl acetic acid, 30 ppm of gibberellin, and 30 ppm of 2,4-D) on growth and yield of tomato (*Solanum lycopersicum* L.) was studied by Uddain et al. (2009) and reported that there were an increase in plant height, number of leaves/plant, number of branches per plant, number of flowers cluster, number of flower cluster per plant, number of flowers per plant, number of fruits cluster, number of fruits per plant, and average weight of tomato fruit per plant (Dias 2019). Sajid et al. (2009) worked with foliar application of either gibberellin (20 ppm) or a nutrient solution [having 19 ppm of KNO_3, 16.5 ppm of NH_4NO_3, 1.7 ppm of KH_2PO_4, 4.4 ppm of $CaCl_2$, and 3.7 ppm of $MgSO_4$] or both on lily plants. Foliar application of gibberellin and nutrient application were found to have improved the productivity and quality of lily cut flowers and may be exploited commercially in horticulture industry to fetch premium price for cut flowers. Meijón et al. (2009) studied the improvement of compactness and floral quality in azalea by means of application of plant growth regulators (daminozide, paclobutrazol, and chlormequat chloride). The results showed that daminozide and paclobutrazol are the best options to control vegetative development and to promote the flowering of azalea japonica in a cold and humid zone such as Asturias. However, daminozide treatment induces floral deformation in one of the tested cultivars.

In the same way, Shirzad et al. (2012) demonstrated the influence of plant growth regulators (indole butyric acid, gibberellin, and ethylene) and planting method on growth and yield in oil pumpkin (*Cucurbita pepo*). Their results showed that plant growth regulators and planting method had significant effects on vegetative, flowering, and yield characteristics including leaf area, number of male and female flowers per plant, number of fruit by plant, fruit fresh weight, seed length and width, number of seed per fruit, seed yield, percentage of seed oil, and oil yield.

The effects of plant growth regulators also have been studied in plant abiotic stress. Ullah et al. (2012) showed the effects of plant growth regulators (as salicylic acid and putrescine) on growth and oil quality of canola (*Brassica napus* L.). The salicylic acid was effective to reduce the drought-induced accumulation of glucosinolates and erucic acid in canola oil. It is inferred that salicylic acid is economical and environment-friendly alternative and can be implicated to improve the plant growth and oil quality of *Brassica napus*. Another important aspect in the plant growth regulator applications is the determination of their residues in agriculture-derived foods. Multi-residue determination of plant growth regulators (i.e., chlormequat, mepiquat, paclobutrazol, uniconazole, ethephon, and flumetralin) in apples (*Malus domestica* Borkh.) and tomatoes by Xue et al. (2011) revealed that the limits of quantification were lower than their maximum residue limits. The effect of plant regulators on vegetables has been studied with satisfactory results in different plant species, for instance, *Ipomoea batatas* (L.) Lam. (Alam et al. 2010), ridge gourd (*Luffa acutangula* L. Roxb) (Hilli et al. 2010), potato (*Solanum tuberosum* L.) (Hartmann et al. 2011), lettuce (*Lactuca sativa* L.) (Mohebodini et al. 2011), onion (*Allium cepa* L.), garlic (*Allium sativum* L.) (Ouzounidou et al. 2011), pumpkin (Shirzad et al. 2012), and tomato (*Solanum lycopersicum* L.) (Choudhury et al. 2013).

6.3.1 In Vitro Plant Tissue Culture

In plant tissue culture, the use of plant growth regulator is an important practice, regardless of the plant species, for example, in rapid in vitro micropropagation of sugarcane (*Saccharum officinarum* L.) through callus culture (Behera and Sahoo 2009) and in the study of hormonal regulation of branching in grasses especially with auxin, cytokinin, and strigolactones, among others (McSteen 2009). Idowu et al. (2009) showed the importance of the use of plant growth regulators and tissue culture techniques to horticulture, the achievements and limitations of tissue culture, and some insights into current and possible future developments. On the other hand, Perera et al. (2009) evaluated the effect of plant growth regulators on ovary culture of coconut (*Cocos nucifera* L.) and found consistent callogenesis was obtained by culturing unfertilized ovaries in medium containing 100 μM 2,4-D and 0.1% activated charcoal. Callus formation was improved by application of 9 μM thidiazuron. Embryogenic calli were subcultured onto somatic embryogenesis induction medium containing 66 μM 2,4-D. Stunted growth was observed in the somatic embryos after subculture in medium containing abscisic acid. Maturation of somatic embryos could be achieved in medium without growth regulators. Conversion of somatic embryos was induced by adding gibberellin to conversion medium containing 5 μM benzyladenine, while 2-isopentenyladenine increased the frequency of plant regeneration. Plant regeneration of *Brassica oleracea* subsp. *italica* (broccoli) is affected by plant growth regulators such as 6-benzylaminopurine and napthyl acetic acid. The highest percentage of shoot tip explant producing shoot (100%) and highest number of shoot produced per shoot tip explant (3.76) were recorded with 5 ppm 6-benzylaminopurine. For rooting of shoots, napthyl acetic acid, indole-3-acetic acid, and indole butyric acid at 0, 0.2 ppm, and 1 ppm were applied. The highest percentage of shoots with roots (100%) and highest mean number of roots produced per shoot (6.5) occurred on medium with 0.2 mg ppm indole butyric acid (Ravanfar et al. 2009). The effect of plant growth regulators (napthyl acetic acid, 2,4-D, benzyladenine, and thidiazuron) on callus induction and plant regeneration in tuber segment culture of potato cultivar 'Diamant' worked by Elaleem et al. (2009) demonstrated that the best degree for callus formation (6.0) was obtained on MS medium supplemented with 2,4-D alone at 3.0 ppm or 2,4-D in combination with benzyladenine both at 2.0 ppm.

Garcia et al. (2011) on in vitro study related to the influence of explant type, plant growth regulators (napthyl acetic acid, picloram, and 2,4-D), salt composition of basal medium, and light on callogenesis and regeneration in *Passiflora suberosa* L. (Passifloraceae) checked that the direct organogenesis in this species was obtained through shoot development from internodal segments in the presence of benzyadenine. Indirect organogenesis was achieved from all explant types on media supplemented with benzyladenine, used alone or in combination with napthyl acetic acid. Alizadeh et al. (2012) verified that culture establishment increased using either 6-benzylaminopurine.

6.4 Physiological Responses in Plants to Low Temperature

Temperate plants are tolerant to seasonal changes in temperature and can tolerate cold stress during early spring and winter. Many important crops, such as rice, corn, soybean, potato, cotton, and tomato, are chilling-sensitive and incapable of cold acclimation. In contrast, some crops, such as oats, are chilling tolerant but freezing sensitive. However, some plants such as *Arabidopsis*, winter wheat (*Triticum aestivum* L.), and barley (*Hordeum vulgare* L.) are not able to tolerate non-freezing temperatures without cold acclimation (Zhao et al. 2015). Similarly, *Santalum album* and *Betula utilis* (Himalayan birch), as dominant vegetations in cold environments of Eastern Nepal, are also sensitive to cold during the growing season (Zhang et al. 2017; Pandey et al. 2018).

Each plant has different enrichment pathways in different periods of cold stress, such as the amino sugar and nucleotide sugar metabolism pathway, alanine and protein export, and the aspartate and glutamate metabolism are highly enriched during the latter stages of cold stress periods in tea plants. In contrast, cellular components, biological processes, and the molecular function category are highly enriched in the early periods of cold stress in tea plants. The same result was found using transcriptome analysis in which flavonoid biosynthesis, phagosome, plant hormone signal transduction, and fructose were highly enriched at the beginning of cold stress (Hao et al. 2018). In tomato, even though the expression of ethylene signaling genes decreased after cold stress, none of the protein genes decreased. This shows the different regulation of gene and protein levels (Mata et al. 2019). In addition, abscisic acid signaling pathway gene expression improved the cold hardiness in grapevine buds only during the cold acclimation period (Rubio and Pérez 2019). Furthermore, the overexpression of certain genes enhanced cold tolerance by increasing endogenous jasmonate levels under cold stress. However, they could not interact with each other directly, indicating they interact indirectly to achieve cold tolerance (Zheng et al. 2018). In banana and *Zoysia japonica*, improved tolerance to multilevel stresses, such as cold, salt, and drought, was obtained through incorporation of cold-tolerant aquaporin genes (Xu et al. 2020). These results show that every gene has a different mechanism in different plants to achieve abiotic or biotic stress tolerance. Due to these differences of signal transduction pathways and metabolisms, the mechanisms related to cold stress are complex (Tolosa and Zhang 2020). The methods plants use to deal with adverse environmental stress, including stress avoidance and stress tolerance, have been examined in the literature (Puijalon et al. 2011; Gnida 2015). Studies of low-temperature stress and cold acclimation based on the model plant *Arabidopsis* contribute substantially to understanding the molecular mechanisms of cold acclimation (Liu and Zhou 2018). In addition, numerous studies have been carried out in other plant species to reveal the molecular mechanism and gene regulatory networks (Ritonga and Chen 2020).

Low temperature is the primary determining factor limiting geographical distribution and growing season of plants (Anderson and Taliaferro 2000). The injury symptoms are often exhibited at temperatures below 12 °C for plant species native to tropical and subtropical areas (Esmaili and Salehi 2012). However, the injuries caused by low temperature are generally categorized into chilling stress (temperatures above 0 °C) and freezing stress (temperatures below 0 °C) (Huang et al. 2014). Chilling stress is often leading to growth and photosynthesis reduction, leaf wilting and chlorosis or even necrosis, cellular membrane damages, as well as oxidative stress in plants (Suzuki and Mittler 2006). In addition, temperatures declined below 0 °C caused freezing stress, which result in the formation of ice crystals within the cell, mechanical damages, as well as metabolic dysfunction in plants (Xin and Browse 2000). However, plants have evolved complex mechanisms to tolerate chilling and freezing stresses, such as accumulation of carbohydrates and proteins, resulting in large quantities of soluble sugars, amino acids, and cold-induced stress-related proteins (Trischuk et al. 2014), as well as hormone homeostasis inducing the gene expression that functions to stabilize membranes against freezing-induced injury (Thomashow 1999; Fowler and Thomashow 2002).

Cold, chilling, and heat stress during critical stages of plant development have strong impacts on developmental cascades of crop plants (Zhou et al. 2018). Temperature stress reduces the plant's optimal biochemical and physiological functioning by modulating molecular mechanisms (Djanaguiraman et al. 2018; Muhlemann et al. 2018; Takahashi and Shinozaki 2019). For instance, jatropha (*Jatropha curcas*) is an important bioenergy crop, but its ability to produce biofuel is often restricted under cold stress (Wang et al. 2013b). However, thiourea (1.3 mM) seed priming enhanced cold tolerance by decreasing leaf senescence and maintaining the relative water content in jatropha plants grown at 4 °C (Yadav et al. 2012).

Low-temperature stress not only decreases grain yield but also affects crop grain quality (Dreccer et al. 2018). However, thiourea (2.6 mM) foliar spray applied at the six-leaf and tasseling stages was effective in improving grain yield and the quality of maize (*Zea mays*). These improvements were attributed to better growth, and higher chlorophyll content, rate of photosynthesis, and dry matter accumulation under chilling stress (Waqas et al. 2017). In bell pepper (*Capsicum annuum*), thiourea primed seed had an improved germination rate under cold stress (15 °C), and subsequent seedlings from primed seed exhibited significantly better survival rates under cold stress (4 °C) (Yadav et al. 2011).

Plants in their physical environment face several types of variation. Plants are affected to function in an oscillating environment, and normal external changes are countered by internal changes without any harm to growth or development. The possibility of abiotic or environmental stress is to cause physical harm to the plant due to serious or chronic adverse environmental circumstances. Any adverse influence of inanimate factors on living beings in a fixed setting is described as abiotic stress. To substantially impact the organism's demographic output or individual physiology, the nonliving factor must alter the surrounding beyond its ordinary variation range. Due to the continuous climate change and environmental deterioration induced by human activity, physical surrounding stress has become a key threat to food security. Water deficit stress, salt stress, imbalances in nutrients (including mineral toxicity and deficiencies), and temperature extremes are significant environmental limitations on productivity of crops all over the world (Fahad et al. 2017).

Exposure to sub-optimum temperature is a major challenge to plant growth and development. Cold injury is a physiological damage caused by temperatures between 0 and 16 °C (Lyons 1973). Chilling sensitivity is manifested by susceptible plants whose germination, growth, development of reproductive organs, and post-harvest longevity are restricted within a range of chilling temperatures. Chilling temperature effects on plants lead to a reduction or complete crop failure due to either direct damage or delayed maturation. Even a small drop in temperature, causing no visible damage to chilling-sensitive plants, caused to up to 50% reduction in their productivity (Jouyban et al. 2013). Plants can grow only in species-specific temperature range. Cold stress has strong limiting effect with regard to geographical distribution, agronomic yield, and product quality (Bracale and Corragio 2003) of a crop species. However, chilling temperatures may differ for different species. For instance, temperate plants may be more chilling tolerant than the tropical plants. Cold stress adversely affects plant growth and development throughout the ontogeny of the plant (Foolad and Lin 2000). Seeds are often the first to encounter stress during a plant's life cycle. Seedlings appear to be more susceptible to chilling than plants at advanced stages of development (Lyons 1973). Under field conditions, low soil temperatures lead to poor seedling establishment with consequences for both crop survival and yield. At later stages, cold stress results in reduced plant growth and development, poor flower development and fruit set, and substantial reduction in fruit yield.

Survival and geographical spreading of plants are also greatly affected by low temperatures. Significant loss of crop due to reduced plant growth and crop efficiency is usually caused by cold stress (Browse and Xin 2001). Cell and tissue dehydration and cellular water crystallization are caused by cold stress, thereby reducing plant growth and productivity. Reduced membrane conductivity, increased water viscosity, and hydro active stomata closure are inhibited resulting in water stress and increased leakage of electrolytes at low temperatures (Pearce 2001). It also delays metabolism, dissipates energy, and causes free radicals to form as a result of oxidative stress (Farooq et al. 2009a). Climate change is one of the most serious problems facing the world today. The Intergovernmental Panel on Climate Change (IPCC) predicts that during the next decades, CO_2 concentration and average temperature will increase, the precipitation will increase in high latitudes, and decrease in most subtropical regions, as well as there will be increases in extreme events, including heat waves, storms, and floods (IPCC 2007). The temperatures considerably lower than the optimal growth temperatures result in two temperature stress in plant. In majority of plants, this temperature is usually below 10 °C; the low temperature, i.e., lower than 10 °C, can be further subdivided into two main categories. This is decided by the fact whether there is formation of ice within the tissues or none. Both this condition also causes water deficit inside the plant (Pearce 1999). Low temperature and frost have a negative effect on the metabolism with a corresponding reduction in crop quality and quantity. Low temperature and frost have a negative effect on the metabolism with a corresponding reduction in crop quality and quantity. Chilling and freezing resistance are complex phenomenon, as most plants are capable of hardening to cold, i.e., acquisition of increased resistance upon exposure to low temperature. These result in irreversible and proper honor loss of cell proteins and disturbed plasma membrane integrity. When exposed to temperature higher than 12–13 °C, tropical plants are more susceptible to chilling than those growing in cold regions. Many tropical and subtropical plants undergo certain adaptive changes which are commonly referred to as cold acclimation, freezing tolerance, or winter hardening.

The survival of plants at low temperature involves two distinct mechanisms, tolerance and cold acclimation. The tolerance to cooling or low temperatures (0° to 15 °C) can be defined as the intrinsic ability of plants to survive under these conditions without injury or damage (Somerville 1995). Acclimatization at suboptimal temperatures is characterized by the increased potential of plants to tolerate changes generated in the physicochemical composition of the cells, producing osmolytes such as soluble sugars and alcohols, as well as the nitrogenous compounds proline and glycine betaine (Peng et al. 2008; Thomashow 1999). Moreover, the accumulation of osmolytes and reactive oxygen species induced by stress caused by this environmental condition leads to the induction of cold-regulated proteins, dehydrins, as well as transcription factors which contribute to stress tolerance through the abscisic acid-dependent pathway (Agarwal et al. 2006; Uno et al. 2000; Hong et al. 2000). These stress-induced proteins lead to the removal of reactive oxygen species, the stabilization of membrane phospholipids, and eventually maintenance of ionic homeostasis (Agarwal et al. 2006). Innate tolerance at low temperatures is recognized as a complex response that involves the regulation of metabolic and morphological adaptive pathways, including proteins involved in signaling, gene expression, and control of repair to stress damage.

It had been stated that the membrane systems of the cell are primary sites of chilling injury in plants (Steponkus 1984). Lipids in the plasma membrane are made up of two kinds of fatty acids: unsaturated and saturated fatty acids. The relative proportion of these two types of fatty acids in the lipids of the plasma membrane determines the fluidity of the membrane (Uemura and Steponkus 1994). Cold-sensitive plants usually have a higher proportion of saturated fatty acids in their plasma membrane. On the contrary, cold-resistant plants have a higher proportion of unsaturated fatty acids and hence a lower transition temperature. Pereira da Cruz et al. (2010) observed higher linolenic acid and palmitic acid contents in cold-tolerant genotypes than in the cold-sensitive genotypes. The presence of a gel phase domain in the membrane bilayer prevents proper functioning of integral membrane proteins and does not maintain an effective permeability barrier. Under extensive cold events, loss of water through membranes leads to an irreversible damage. Chilling temperatures have been reported to lower expression and activity of cytochrome oxidase in the mitochondrial inner membrane of a chilling susceptible genotype of maize (Prasad et al. 1994b). Through the primary effect on membrane lipids, chilling stress causes a cascade of secondary events including loss of turgor, leakage of cytoplasmic solutes, lack of energy metabolism, disruption of photosystems, cell autolysis, and cell death (Lyons and Raison 1970). Increase in permeability and leakage of cell solutes is one of the responses which can occur at an early or late stage of the chilling exposures. Jatimliansky et al. (2004) showed that electrolyte leakage was higher in chilled than control maize plants. Inhibition of photosynthesis is an early response to low temperature. The response of photosynthesis to chilling includes changes in pigment composition (Haldimann 1998), decreased quantum efficiency (Fryer et al. 1995), modified thylakoid membranes, and impaired chloroplastic development (Nie et al. 1995). Photosynthetic rate and chlorophyll content decreased significantly under low temperature in grafted squash seedlings (Chen et al. 2000). In maize, a reduction in chloroplast development (Moseki and Musil 2004b) and photosynthetic yield (Moseki and Musil 2004a) has been observed when subjected to low growth temperatures. Inhibition of photosynthesis occurs via inhibition of electron transport chain. Guo-li and Zhen-fei (2005) found dramatic decline in photosynthetic rates in cold-tolerant as well as cold-sensitive cultivar of rice during chilling. Low temperature slows all the metabolic reactions, but in terms of photosynthesis, those involved in CO_2 fixation and regulation of stomatal aperture are particularly sensitive. CO_2 assimilation by leaves is reduced mainly due to membrane damage, photoinhibition, and disturbed activity of various enzymes (Farooq et al. 2009a). Chilling affects both the maximal catalytic activities and the activation states of photosynthetic enzymes. Inhibition of activity of some photosynthetic enzymes, e.g., ribulose-1,5-bisphosphate carboxylase/oxygenase (Rubisco) or fructose-1,6-bisphosphatase, is observed at low temperatures (Kingston-Smith et al. 1997), resulting in slowing down of carbon assimilation. A gradual temperature fall disturbs the water balance in tissues and, consequently, a series of the metabolic reactions. Water stress is the primary basis for the changes caused by chilling and other abiotic stresses (Bohnert et al. 1995). Chilling of sensitive plants affects all components of water regime and causes loss of water, resulting in strong wilting (Bloom et al. 2004). It is based on the two main factors: rapid decline in the ability of roots to absorb water and transport it to the shoots (Bolger et al. 1992a, 1992b) and reduced ability to close stomata in response to subsequent water deficit (Bloom et al. 2004). Chilling of maize seedlings to 5.5 °C for 24 h resulted in decreased root hydraulic conductance (Melkonian et al. 2004). Roots reduce their growth and capacity for water and mineral uptake under mild cold stress (Stamp et al. 1997). Uptake of potassium and phosphorus by maize roots is strongly dependent on temperature (Bravo and Uribe 1981). Shoot growth of maize seedlings at suboptimal root zone temperatures was limited by reduced supply of nutrients through the roots and by the direct influence of temperature on shoot meristems (Hund et al. 2007). Low root zone temperatures reduced nutrient acquisition by decreasing root growth (Pritchard et al. 1990) and uptake rates of mineral nutrients (Bravo and Uribe 1981). Chilling treatments greatly inhibited sorghum growth and nitrogen uptake during chilling exposure (Ercoli et al. 2004). Chilling, in general, impairs respiratory activity, the cytochrome pathway of electron transport, and ATPase activity. In maize, populations with high germination levels have higher fluidity of mitochondrial membranes and higher activity of cytochrome C oxidase relative to populations with low germination levels (Santis et al. 1999). Reports are there for an increase in respiratory activity during chilling, and prolonged elevation of the respiration rate after cold exposure may indicate irreversible metabolic dysfunction and accumulation of incompletely oxidized intermediates (Yadegari et al. 2008). Cold-tolerant crop species have greater temperature homeostasis of leaf respiration than cold-sensitive species (Yamori et al. 2009). Chilling reduces the cytochrome path of the electron transport in seedlings (Prasad et al. 1994b) and enhances alternative respiratory pathways (Ribascarbo et al. 2000). These alternative pathways induced by chilling caused a decrease in superoxide generated in mitochondria (Hu et al. 2008). Low temperature is known to induce oxidative stress in tissues leading to generation of reactive oxygen species including superoxide radicals, hydrogen peroxide, and hydroxyl radicals. Reactive oxygen species are believed to be responsible for cold injury in plants (Prasad et al. 1994a). Most subcellular compartments are potential sites of reactive oxygen species production, and they include chloroplasts, mitochondria, peroxisomes, plasma membranes, and endoplasmic reticulum (Srivalli et al. 2003). Reactive oxygen species-induced cell death can result from oxidative processes such as membrane lipid peroxidation, protein oxidation, enzyme inhibition, and DNA and RNA damages (Mittler 2002). In maize seedlings, lipid peroxidation was increased by about twofold in non-acclimated seedlings compared to control or acclimated during 4 °C stress and recovery (Prasad 1996).

6.4.1 Classification of Plants under Low Temperature

Plants can be classified into three broad classes in accordance to their reaction(s) to different ranges of low temperature:

- **Chilling-sensitive plants**: This group of plants is seriously injured by temperature above 0 °C but below 15 °C.
- **Chilling-resistant plants**: This group of plants is able to tolerate low temperature but is seriously injured when ice starts to form in tissues.
- **Frost-resistant plants**: This group of plants can tolerate exposure to very low temperatures (−50 °C to −100 °C) even when immersed in liquid nitrogen.

Some of the common symptoms of chilling stress are rapid wilting followed by water-soaked patches which develop into sunken pits that reflect cell tissue collapse. Following warming, the sunken pits usually dry up, leaving necrotic patches of tissues on the leaf surface. Plants may develop physiological disorders when exposed to low but non-freezing temperatures. Symptoms of chilling injury can differ widely between species but usually develop rapidly in plants native to tropical and subtropical climates and almost imperceptibly slowly in plants originating in cool-temperate climates. Within the range of chilling temperatures, that is, from the temperature of the freezing point of the plant tissue up to about 13 °C, the rate at which chilling injury develops intensifies with decreasing temperature and increasing duration. Chilling-sensitive foliage plants such as *Episcia* spp. which are native to the Amazon basin are killed within 30 min of being exposed to 1 °C. Survival of such highly sensitive plants necessitated year-round protection from cold. In modern-day agriculture, many plants are often cultivated outside their customary microclimate. For example, avocado (*Persea americana* Mill.) was taken from tropical highlands in Mexico and is now grown in the temperate North Island of New Zealand and in inland Australia where nights are cold. When the new location has temperature minima below those of the region in which the plant evolved, problems of chilling injury can arise. With annual crops, the time of greatest risk is likely to be early in a growth season and especially during seedling establishment. Chilling injury to seedlings can show up as necrotic lesions on the young roots and shoots, with slowed growth and increased susceptibility to disease attack and even death. Crops adversely affected by low temperatures during the establishment period take longer to mature, and this in turn may mean that they are at risk to chilling temperatures toward the end of the growing season.

Chilling injury is the damage of chilling-sensitive plant species during storage at temperatures above the freezing point of tissues but lower than 15 °C. Chilling-sensitive plants are seriously injured in a temperature range of 15 °C and 0 °C. The ability of plants in a vegetative state to survive the action of chilling temperatures without harm to the future growth and development is called cold resistance. In turn, chilling-sensitive plants are sensitive to chilling, and after prolonged storage in these temperatures, external symptoms of injury are developed, and death of the organism occurs (Table 6.1). Plants, which have the visual injuries at temperatures above

Table 6.1 Chilling-sensitive crops and their symptoms

Crop	Lowest safe temperature (°C)	Chilling injury symptoms
Asparagus	0–2	Dull, gray green limp tips
Beans (snap)	7	Pitting and russeting
Cucumber	7	Pitting, water soaked, decay
Brinjal	7	Surface scaled, *Alternaria* rot, seed blacking
Ladyfinger	7	Discoloration, water-soaked areas, pitting, decay
Pepper	7	Pitting, *Alternaria* rot, seed blacking
Potato	2	Mahogany browning, sweetening
Pumpkin	10	Decay, specially *Alternaria* blight
Squash	10	Decay, specially *Alternaria* blight
Sweet potato	10	Decay, pitting, internal discoloration
Tomato (ripe)	7–10	Water soaking, softening, decay
Tomato (mature green)	13	Poor color when ripe, *Alternaria* blight

15 °C, are called "very sensitive to chilling." A number of tropical or subtropical plants, such as rice, maize, tomato, cucumber, cotton, soybeans, etc., when introduced in the higher latitudes have not acquired substantial resistance to chilling, despite the long history of cultivation in temperate regions (Wilson 1985). Chilling temperature effects on plants in temperate climates lead to a reduction or complete crop failure due to either direct damage or delayed maturation. Even a small drop in temperature, causing no visible damage to chilling-sensitive plants, caused to up to 50% reduction in their productivity.

Cold stress occurs when plants are subjected to very low temperatures. Cold stress is a major abiotic stress that reduces productivity of crops by affecting quality and life after harvest. Cold stress impacts all cellular function characteristics in crops. Cold stress causes injury to plants by changes in the membrane structure and decrease in the protoplasmic streaming, electrolyte leakage, and plasmolysis which lead to cellular damage. The metabolism of the cells is damaged by an increase or decrease in respiration rate, and depending on the intensity of the stress, synthesis of abnormal metabolites occurs due to abnormal anaerobic respiration. Due to cellular damage and altered metabolism, there are reduced plant growth, abnormal ripening of fruits, internal discoloration (vascular browning), and increased susceptibility to decay, and also these cause the death of the plant (Devasirvatham and Tan 2018). Low temperature or cold stress often affects plant growth and crop productivity and leads to substantial crop losses (Xin and Browse 2000; Sanghera et al. 2011). Chilling stress results from temperatures cool enough to produce injury without forming ice crystals in plant tissues, whereas freezing stress results in ice formation within plant tissues. Plants differ in their tolerance to chilling (0–15 °C) and freezing (<0 °C) temperatures. Both chilling and freezing stresses are together termed low temperature or cold stress: the damage due to cold stress can range from chilling injury and freezing injury to suffocation and heaving. In general, plants from temperate climatic regions are considered to be chilling tolerant to variable degrees, and their freezing tolerance can be increased by exposing to cold, but non-freezing, temperatures; this process is known as cold acclimation. However, generally the plants of tropical and subtropical origins are sensitive to chilling stress and lack this mechanism of cold acclimation (Sanghera et al. 2011). Low temperature may affect several aspects of crop growth, viz., survival, cell division, photosynthesis, water transport, growth, and finally crop yield.

6.5 Role of Different Growth Hormones on Growth and Yield

Chilling stress, damage caused by low but above freezing temperatures, has been recognized as a unique environmental impact on crop plant growth. Susceptibility to chilling injury imposes late plantings or prevents the cultivation of many crops in regions where temperatures can drop below the optimal growth temperatures for individual plant species. Early planting could contribute to a longer growing season, a more effective utilization of late spring and early summer rainfall, and to an enhanced yield potential (Ercoli et al. 2004).

Plants exposed to chilling temperatures suffer injury, exhibiting reduction or stop of growth. This phenomenon is initially reversible, but ultimately irreversible and can result in cell death. Chilling adversely affects a wide range of processes, including crop germination and establishment, photosynthesis, flowering, plant growth, and grain yield (Ting et al. 1991; Staggenborg and Vanderlip 1996; Verheul et al. 1996). The degree of chilling injury can vary with plant species, stage of crop development, and the conditions of irradiance and mineral nutrition during chilling stress. In chilling-sensitive species, it has been recognized that chilling causes greater injury under illuminated conditions, owing to photooxidative damage determined by the increase of the formation of harmful active oxygen species (Wise 1995; Aroca et al. 2001). The degree of dysfunction observed after chilling of tomato plants depended on mineral nutrition, affecting the production of root-borne phytohormones (Starck et al. 1997), and plants grown with limited supply of mineral nutrients were more susceptible to chilling (Starck et al. 2000). In spinach, nitrogen deficiency greatly decreased the rate of acclimation to low temperature (Martindale and Leegood 1997).

Chilling injury is time dependent and changes in physiological activity preceding the development of visual symptoms of injury. At chilling temperatures, the rates of respiration and ion leakage increase, and photosynthetic activity and carbohydrate metabolism decrease (White and Schmidt 1989; Sonoike 1999). Chilling temperatures cause starch accumulation in chloroplasts of some plant species, inhibiting the photosynthetic process because of feedback inhibition of photosynthetic enzymes (Rogers et al. 1977). Photoinhibition of photosynthesis is enhanced at chilling temperatures, and one of the major reasons for this enhancement is the decreased utilization of excitation energy in carbon metabolism at chilling temperature (Sonoike 1999).

Chilling damage depends not only on conditions during chilling but also on the environment before and after chilling, such as temperature, light conditions, and water status of the plants prior to chilling and the light conditions after chilling (Starck et al. 1994; Szalai et al. 1996; Smeets and Wehner 1997). The effects of chilling stress on sorghum have been studied in relation to chilling-induced declines in photosynthetic capacity. Reduced growth may result from more direct effects of temperature on growing tissues or from the reduction of photosynthate flow through the phloem. According to Wardlaw and Bagnall (1981), although low temperatures may reduce the carrying capacity of the phloem of sorghum, this is unlikely to be an important factor in regulating plant growth at low temperatures. Lang and Minchin (1986) in sequential chill-warm cycles demonstrated that the translocation process was sensitive only to the first cycle, subsequent cycles having no effect owing to sorghum adaptation mechanisms.

In wheat, booting to anthesis is very critical period for yield determination, and exposure to low temperature at this period significantly deterioration in the wheat florets occurs, and pollen development enters the meiosis stage, and the most cold-sensitive period (Slafer et al. 2001). Low temperatures induce abnormal pollen and ovule development, resulting in a reduction in anther size and fertile pollen grains, thereby decreasing the rate of seed setting (Thakur et al. 2010). In rice, cold damage during the reproductive stage results in spikelet degradation and an increase in sterile spikelets (Wang et al. 2013a). In the early stages of pollen development, anthers are the strongest metabolic pool, and large amounts of sugars are therefore transported to the anthers to support development (Castro and Clément 2007). Serrago et al. (2008) revealed a positive correlation between basal spikelet gestation floret and the dry weight, as well as carbohydrate content at flowering, suggesting that the supply of assimilates determines the number of fertile florets. Abnormal glucose metabolism, a lack of starch accumulation in the pollen grains, and abnormal degradation of the tapetum are thought to be the main causes of floret abortion (Ji et al. 2013; Niu et al. 2013). Low-temperature stress also has a significant effect on sugar metabolism (Wanner and Junttila 1999). In order to enhance membrane stability, plants accumulate a large amount of carbohydrates under low temperature, inducing stomatal closure and photosynthetic electron transport damage and inhibiting carbon assimilation (Li et al. 2014). Low-temperature stress also reduces the supply of soluble carbohydrates to reproductive tissues, leading to nutrient deficiencies in the tapetum and endosperm (Nayyar et al. 2005).

Plant growth regulators play important roles in the enhancement of the growth and productivity of the most economic crops. Studies were carried out to investigate the effect of indole-3-acetic acid (100, 200, and 400 ppm), naphthalene acetic acid (10, 20, and 40 ppm), and kinetin (25, 50, and 100 ppm) on *Lupinus termis* plant growth, yield, and seed chemical composition as well as hormone contents compared with untreated plants. The obtained results indicated that the highest number of branches was obtained with kinetin treatment, especially at 100 ppm, while the maximum values of both number and dry weight of pods were obtained as a result of foliar application with kinetin at 50 ppm. Spraying the lupin plants with bioregulators increased significantly the seed number per pod and per plant, and

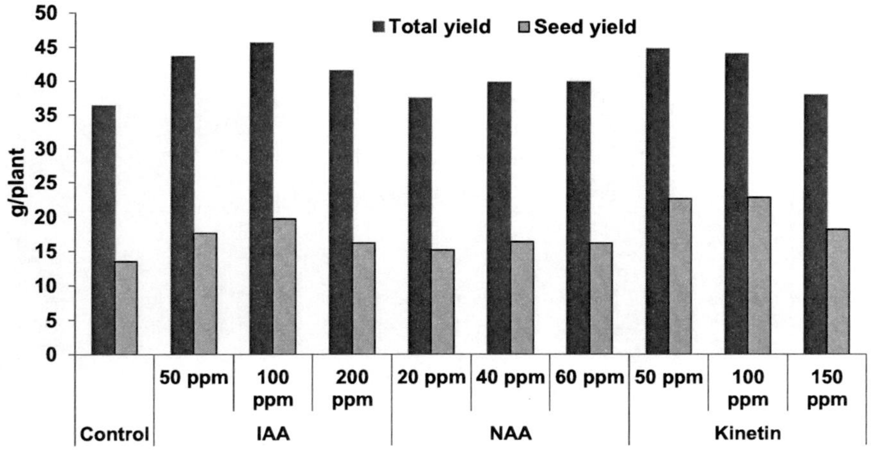

Fig. 6.3. Effect of foliar application of plant growth regulator on total dry matter yield (g/plant) and seed yield (g/plant) in desi chickpea

seed yield/plant except those plants treated with indole-3-acetic acid at 100 ppm. Applications of indole-3-acetic acid, naphthalene acetic acid, and kinetin treatments had no significant effect on crop index, harvest index, and shelling percentage. The maximum seed yield was recorded with the application of kinetin at 50 ppm followed by kinetin at 100 ppm as well as indole-3-acetic acid at 100 ppm. High levels of gibberellins and indole-3-acetic acid were observed in seeds produced from plants treated by indole-3-acetic acid relative to those treated by naphthalene acetic acid or kinetin. Spraying the lupin plants with the bioregulators indole-3-acetic acid, naphthalene acetic acid, and kinetin caused an increase in protein content and total carbohydrates and decrease of oil percent in the lupin seeds (El-Saeid et al. 2011). Foliar application of different concentration of indole-3-acetic acid, naphthalene acetic acid, and kinetin has been reported (Vijaylaxmi unpublished data) to interact with total dry matter yield as well as seed yield in desi chickpea as compared to control (Fig. 6.3). All the treatments of plant growth regulators had higher total dry matter as well as seed yield. Total dry matter yield was recorded to be higher in 100 ppm of indole-3-acetic acid and kinetin. Among hormonal treatment, seed yield was highest under kinetin treatment. The study was further extended to look into causes of the differences in yield, and it was revealed that the number of branches and number of pods of desi chickpea interacted with the different concentrations of various plant growth hormones (Fig. 6.4).

Gromadzinski et al. (1990) reported that plant growth regulators resulted in significant increases in pod number, shortening of offshoots, and seed yields of yellow- and narrow-leaved lupin (*Lupinus luteus* and *L. angustifolius*, resp) plants but only when applied at the correct rate and date. Prusinski et al. (1999) reported that sprayed plants of traditional yellow lupin cultivar with varied doses of the

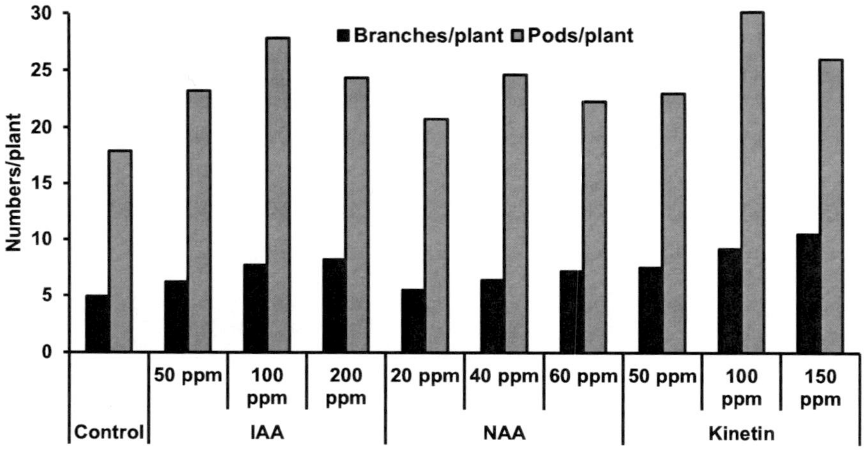

Fig. 6.4. Effect of foliar application of plant growth regulator on number of branches/plant and number of pods/plant in desi chickpea

growth regulators indole-3-butyric acid and 6-benzylaminopurine limited the unfavorable abscission of generative organs, which resulted in a significantly higher seed yield, than the control. Emery et al. (2000) reviewed the literature and stated that a wide range of environmental factors can influence the extent of "flower shedding," and as a consequence the yield potential of many pulses may not be realized.

Theories as to the cause include hypotheses based on competition for nutrients. They add that at different times during fruit development, 18 different forms of cytokinin were detected in the component tissues (Emery et al. 2000). El-Saeid et al. (2010) reported that when cowpea (*Vigna sinensis* L.) plants were sprayed with indoleacetic acid at the rate of 25 and 50 ppm, there are increased number of leaves, shoot dry weight, and number of produced flowers per plant and consequently significantly increased number and weight of pods and seeds per plant. Meanwhile 50 and 100 ppm indoleacetic acid significantly decreased the number of flowers abscised from cowpea plant. Seymour (1996) reported that timing of application was critical, with application 6 weeks before flowering giving the most consistent positive responses. El-Bassiouny and Shukry (2001) reported that foliar application of indole-3-acetic acid at three concentrations (12.5, 25, and 50 ppm) induced increments of the plant height, fresh and dry weights, number of branches and number of leaves per cowpea plant, as well as yield components (pods per plant, seeds per pod, weight of seeds per pod and per plant, and weight of seeds/ha). Khalil et al. (2006) reported that kinetin treatments decreased stem height of lentil plant, but the treatment induced significant increase in the number of leaves and branches per plant, shoot dry weight, and number of flowers produced per plant. Meanwhile, all treatment decreased the percentage of abscised flowers and increased number and weight of pods and seeds per plant.

Plant endogenous hormones act as signaling factors during low-temperature stress, with abscisic acid and gibberellic acid playing an important role in cold resistance (Sun et al. 2009). Increased endogenous abscisic acid levels induce upregulation of stress response genes, thereby enhancing cold tolerance (Yang et al. 2011). Oliver et al. (2007) further revealed that low-temperature stress induced an increase in abscisic acid accumulation in rice, causing a decrease in Inv activity in the cell wall of the tapetum, which is the main cause of pollen abortion. Gibberellin also plays an important role in initiating meiosis and development of the tapetum in pollen mother cells. In rice pollen, for example, high-temperature stress was found to cause a decrease in gibberellins and a significant increase in the floret abortion rate (Tang et al. 2008). A series of studies has shown that the photosynthetic rate of functional wheat leaves decreases significantly under low-temperature stress at booting (Zhang et al. 2015, 2016), with an accumulation of carbohydrates (Zeng et al. 2011) and changes in endogenous hormone contents and antioxidant enzyme activity (Sun et al. 2009; Li et al. 2014). However, few studies have documented the effects of low temperature at booting on young wheat ears. Phytohormone content assays showed an increase in the content of abscisic acid in young wheat ears but a decrease in the content of auxin and gibberellins. The grain number per spike and 1000-grain weight also showed a downward trend following low-temperature stress. Overall, these findings suggest that low temperature at booting induces abscisic acid accumulation in winter wheat, altering the activity of the enzymes involved in sucrose metabolism, which leads to an accumulation of sucrose in the young ears, thereby having a negative effect on wheat production (Zhang et al. 2019).

Salicylic acid and other salicylates, thiourea and other thiol compounds, and potassium nitrate and other nitrites are known to affect various physiological and biochemical activities of plants and may play a key role in regulating plant growth and productivity. Studies have provided conflicting evidences regarding role of salicylic acid in flowering. However, some demonstrations argue the role of salicylic acid in flowering. In salicylic acid-deficient *Arabidopsis*, initiation of flowering failed when irradiated with UVC (100–280 nm), and substantial flowering occurred when grown under non-stress condition than wild-type plants (Shulaev et al. 1995). It is well known that salicylic acid application induces flowering in a number of plants (Cleland 1974; Cleland and Ajami 1974). However, the rigorous mechanism of flower-inducing property of salicylic acid has not been explored. Kumar et al. (2000) studied the cumulative effect of salicylic acid in combination with gibberellin, kinetin, napthyl acetic acid, ethrel, and chlormequat chloride and found a synergistic effect between salicylic acid and gibberellin on flowering as compared to other combinations of hormones. It was reported that the dry matter accumulation was significantly enhanced in *Brassica juncea*, with foliar application of salicylic acid. However, higher concentrations of salicylic acid had an inhibitory effect. In sugarcane gibberellin increased internodal elongation, while glyphosate, 2-chloroethylphosphonic acid, and other regulators increase the deposition of sucrose, diquat and 2-chloroethylphosphonic acid inhibit flowering, and paraquat desiccates leaves just prior to harvest to facilitate leaf removal or burning. Gibberellin increased the rate of cell division and stimulation of vegetative growth (Arteca

1996). Leaf area was found to increase significantly after potassium nitrate application probably due to promoting role of potassium in plant growth.

Ethylene modulates plant growth and development under normal conditions and is also a key feature in the response of plants to a wide range of stresses. Ethylene is synthesized in plants in response to various stresses typically in response to the presence of metals, organic and inorganic chemicals, cold or heat stress, drought or flood, ultraviolet light, insect and nematode damage, phyto- and pathogens (both fungi and bacteria), and mechanical wounding (Glick 2014). A small fraction ethylene synthesis, which consumes the existing pool of 1-aminocyclopropane-1-carboxylate (precursor of ethylene) in stressed plant tissues, is believed to be responsible for initiating the transcription of genes that encode plant defensive/protective proteins (Robison et al. 2001). However, higher level of ethylene production following synthesis by the plant of additional 1-aminocyclopropane-1-carboxylate in response to a stress is generally injurious to plant growth and usually involved in initiating senescence, chlorosis, and leaf abscission (Glick 2014). There has been a major increase in the utilization of brassinosteroids in agricultural applications as a mean to boost crop productivity and stress tolerance. Crosstalk and interactions between brassinosteroids and other plant growth regulators occur through either the modification or intersection of their primary signaling cascades and function to regulate a large and diverse array of biological processes (Choudhary et al. 2012). Abscisic acid regulates both stress- and non-stress-related plant responses by acting as a signal molecule. Interactions between brassinosteroids and abscisic acid regulate the expression of many genes that govern several biological processes, such as seed germination, stomatal closure, and plant responses to environmental stresses. Interaction between soil bacterium may be negative, positive, or neutral, and sometimes the effect of soil bacterium changes according to the conditions of soil.

Sweet cherry trees exhibit strong apical dominance, producing vigorous, upright shoot growth with limited lateral branching, particularly in young trees (Elfving and Visser 2006; Jacyna 2002; Jacyna and Puchała 2004; Jacyna et al. 2005). The interaction of endogenous auxins and cytokinins is thought to play a role in the control by apical dominance of lateral bud activity in the spring and during shoot development (Theron et al. 1987). Pruning can interrupt the hormonally based apical dominance control system and encourage buds that otherwise might remain quiescent to form shoots (Elfving and Visser 2007). For this reason, commonly used training systems for sweet cherries rely on pruning as a principal means for stimulation of lateral branching during canopy development (Long 2001, 2005; Long et al. 2005; Nugent et al. 2001). However, pruning removes buds that might otherwise have developed into shoots or spurs and delays cropping (Robinson et al. 2005). Therefore, alternative means for stimulation of bud growth have been examined in apple and, to a lesser extent, in sweet cherry. Scoring (a circumferential knife cut through the phloem tissue around a shoot or stem) and notching (removal of a small piece of phloem above a bud) have been used to interfere with hormone movement and to promote shoot development in apple (Cook and Strydom 2000; Greene and Autio 1994; Niu et al. 1998; Ouellette et al. 1996). Bud removal (disbudding) can affect and may improve shoot growth in apple and other tree fruit species (Peano et al. 2005) and may show promise for shoot growth stimulation in sweet cherry as well (Perry 1999; Robinson et al. 2005). Very limited information suggests scoring or notching in the spring may positively affect lateral shoot development in sweet cherry (Long 2005; Long et al. 2005).

6.6 Effect of Low Temperature on the Activity of Plant Growth Hormone

Temperature stress is becoming the major concern for plant scientists worldwide due to the frequent climatic change. That adversely influences the sustainable agricultural practices around the globe (Watanabe and Kume 2009; Shah et al. 2011). Therefore, low-temperature stress is an inevitable environmental factor that extensively affects plant growth and development. Environmental stresses can reduce growth, quality, yield, and productivity (Thakur et al. 2010; Zhang et al. 2013a), which cause significant crop losses (Xin and Browse 2000; Wang et al. 2003a; Vinocur and Altman 2005). Low-temperature stress decreases capacity and efficiency of photosynthesis through changes in gas exchange, pigment content, and chloroplast development and also declines chlorophyll fluorescence (Farooq et al. 2009d). There are many studies which indicated that cellular membranes are the primary site of chilling injury in plants (Thomashow 2001; Zhang and Tian 2009). It is damaged in two ways, viz., disruption of protein lipid structure, protein denaturation, and leakage of solutes that indicate the membrane instability. Due to low-temperature stress, the fatty acids become unsaturated, and the ratio of lipid-protein in the membrane is altered, ultimately affecting the membrane fluidity as well as structure (Wang and Li 2006). The membrane stability also leads to leakage of free electrons and generation of reactive oxygen species. Differential activation of antioxidant system is also assumed to confer tolerance to chilling stress in various crop plants (Khan et al. 2017).

Cold and chilling stress during critical stages of plant development have strong impacts on developmental cascades of crop plants (Zhou et al. 2018). Temperature stress reduces the plant's optimal biochemical and physiological functioning by modulating molecular mechanisms (Djanaguiraman et al. 2018; Muhlemann et al. 2018; Takahashi and Shinozaki 2019). For instance, jatropha (*Jatropha curcas*) is an important bioenergy crop, but its ability to produce biofuel is often restricted under cold stress (Wang et al. 2013b). However, thiourea (1.3 mM) seed priming enhanced cold tolerance by decreasing leaf senescence and maintaining the relative water content in jatropha plants grown at 4 °C (Yadav et al. 2012). Low-temperature stress not only decreases grain yield but also affects crop grain quality (Dreccer et al. 2018). However, thiourea (2.6 mM) foliar spray applied at the six-leaf and tasseling stages was effective in improving grain yield and the quality of maize (*Zea mays*). These improvements were attributed to better growth, and higher chlorophyll content, rate of photosynthesis, and dry matter accumulation under chilling stress (Waqas et al. 2017). In bell pepper (*Capsicum annuum*), thiourea primed seed had an improved germination rate under cold stress (15 °C), and subsequent seedlings from primed seed exhibited significantly better survival rates under cold stress (4 °C) (Yadav et al. 2011). Stress tolerance mechanism in plants is regulated via transcriptional activation or repression. Evidence suggests that thiourea-induced cold tolerance is linked with the increased expression of genes involved in transcriptional regulation, osmotic adjustment, antioxidant defense (Cu/Zn superoxide dismutase), and metabolite biosynthesis (Patade et al. 2012).

6.6.1 Abscisic Acid and Low-Temperature Stress

Chilling stress causes dehydration, mainly due to reduction in water uptake by roots, and induces closure of stomata. This adaptive process is controlled by abscisic acid, which acts as messenger in the regulation of the plant's water status (Mahajan and Tuteja 2005). Stress-responsive genes can be expressed either through an abscisic acid-independent or abscisic acid-dependent pathway (Chinnusamy et al. 2004). Abscisic acid is a phytohormone critical for plant growth and development and plays an important role in integrating various stress signals and controlling a number of downstream stress responses. Abscisic acid is also involved in various physiological processes of plant development like flowering, seed development, synthesis of storage proteins and lipids, embryo morphogenesis, dormancy, stomatal opening and senescence triggering and remobilization (Nambara and Marion-Poll 2005; Tripathi and Tuteja 2007; Cutler et al. 2010), as well as induction of dehydration tolerance genes (Xiong et al. 2001). Several stressors including low temperature trigger synthesis of abscisic acid which plays a key role in modulating the response of plants under low-temperature stress. Induced synthesis of abscisic acid leads to the closure of stomata, thereby reducing water loss via transpiration (Stoll et al. 2000; Wilkinson and Davies 2009). In several plant species, exogenous treatment of abscisic acid could substitute for low-temperature treatment and induce freezing tolerance in cell suspension and callus cultures (Xue-Xuan et al. 2010). It was shown that abscisic acid biosynthesis is an important factor for tolerance of tomato plants to suboptimal root zone temperature (Ntatsi et al. 2013). Plants are frequently subjected to localized root chilling especially early in spring after cold nights when the air temperature increases much quicker than the soil temperature creating a steep temperature gradient between shoots and roots. Such conditions often occur in wheat fields of temperate climates. The responses of stomata to a decrease in air temperature have been studied intensively by Wilkinson et al. (2001). However, the reaction of a plant to a decrease in root zone temperature might be quite different. Accumulation of abscisic acid in shoots of plants was reported shortly after cooling their root to 12 °C (Smith and Dale 1988), which is likely to be important for the control of stomatal conductance. In the field root zone temperatures may be significantly lower. Thus the aim of the present work was to study the effect of root cooling down to 6–7 °C on the content of hormones (abscisic acid, indole-3-acetic acid, and cytokinins) in shoots, roots, and root exudate of wheat seedlings in relation to changes in root temperature-dependent water relations. Indole-3-acetic acid and cytokinins have been analyzed because they are believed to act as antagonists of abscisic acid (Mansfield and McAinsh 1995). The involvement of cytokinins in reaction to low root zone temperature was reported in earlier publications (Kudoyarova et al. 1998).

In some cases, the degree of freezing tolerance induced by abscisic acid treatment was much higher than that induced by low temperature (Keith and McKersie 1986). Some plant species capable of cold hardening such as wheat, spinach, and potato had increased endogenous level of abscisic acid during low-temperature acclimation (Taylor et al. 1990). It was reported that abscisic acid accumulation was correlated with the genotype of wheat cultivar against freezing tolerance (Galiba et al. 1993). High concentrations of glucose can induce abscisic acid biosynthesis and transcription of gene and its expression. It is also reported that changes of concentrations of abscisic acid could mediate some sugar signals (Arroyo et al. 2003), which help in the regulation of abscisic acid biosynthesis (Leon and Sheen 2003). High concentrations of glucose lead to increase in the concentrations of abscisic acid within cells (Arenas-Huertero et al. 2000) with simultaneous overexpression of genes related to abscisic acid biosynthesis (Cheng et al. 2002). The role of abscisic acid has been well established under stress condition due to high temperature (Bajguz 2009; Lu et al. 2009), salinity and draught, or low-temperature stress. Abscisic acid enhanced antioxidant defense by decreasing the accumulation of reactive oxygen species caused by low temperatures (Liu et al. 2011). Abscisic acid and brassinosteroids interact in many processes, such as seed germination, regulation of gene expression, and signal output (Nemhauser et al. 2006; Finkelstein et al. 2008; Zhang et al. 2009a). Various researches showed that the endogenous level of abscisic acid content was enhanced by brassinosteroid in *Chlorella vulgaris* cells subjected to short-term heat stress (Bajguz 2009). In maize leaves, nitric oxide has also been shown to mediate brassinosteroid-induced abscisic acid biosynthesis involved in tolerance to oxidative stress (Zhang et al. 2011).

6.6.2 Role of Auxins under Cold Stress

Auxins are widely recognized as a key growth regulator involved in defense responses against biotic and abiotic stresses (Wolters and Jurgens 2009). Indole-3-acetic acid regulates plant growth particularly by stimulating cell elongation and is more intricately involved in development and growth of roots. Auxins also play central role in hormonal crosstalk, which modulate various developmental processes (Chandler 2009; Depuydt and Hardtke 2011) like cell division and differentiation, fruit development, formation of roots from cuttings, and inhibition of lateral branching and leaf fall. It plays important role in regulating all aspect of growth and development of plants, but its role in low-temperature stress is yet to be dissected in detail. It has been observed that gravitropic response of inflorescence in *Arabidopsis* was inhibited at 4 °C, but this response was restored at room temperature (Wyatt et al. 2002; Nadella et al. 2006). Moreover, exposure of *Arabidopsis* roots to low temperature prevents gravity responses (Rahman 2013). Auxins also interact with brassinolide that regulates cell elongation in shoot and root and tropic responses (Hardtke et al. 2007). Jasmonic acid and auxin act antagonistically in regulation of coleoptile elongation, whereas they act synergistically to regulate root growth, lateral root development, and flowering (Chandler 2009). Moreover, it has been reported that interaction of indole-3-acetic acid and abscisic acid regulates formation of embryonic axis (Belin et al. 2009) and lateral root development (Shkolnik-Inbar and Bar-Zvi 2010). Auxin and ethylene also interact both in synergistic and antagonistic modes for regulating various developmental processes, such as apical hook formation, root and shoot elongation, root and shoot gravitropism, lateral root development, phototropism, and leaf abscission (Muday et al. 2012).

6.6.3 Brassinosteroids and Low-Temperature Stress

Brassinosteroids, steroidal plant hormones, have also been implicated to confer tolerance in plants against temperature stress. Brassinosteroids could be exogenously applied through seed soaking, root treatment, and/or foliar spray. Of these, seed soaking and foliar spray treatments are most commonly used modes of application. Application of 24-epibrassinolide prior to cold treatment reduced the ion leakage in chilling-exposed rape plants (Janeczko et al. 2007), whereas it increased the antioxidant defense (and also osmoregulation) in chilling-stressed young grapevines (Xi et al. 2013). The roles of brassinosteroids in low-temperature stress are summarized in Table 6.2. Application of brassinosteroids improved plant tolerance to low-temperature stress (Khan et al. 2015b) and enhanced the plant growth, photosynthesis, and antioxidant system of tomato under low-temperature conditions (Khan et al. 2015b). It was revealed that application of brassinosteroids further enhanced the activities of antioxidant enzymes under low-temperature stress (Ahammed et al. 2012; Khan et al. 2015b). The enhanced activities of the antioxidative enzymes as a result of brassinosteroid applications could be due to increased de novo synthesis or activation of the enzymes, which is mediated through transcription and/or translation of specific genes to gain tolerance (Bajguz 2000). Brassinosteroids have been implicated in a wide range of physiological and molecular responses in plants, like cell elongation and cell division in stem, and inhibit ions of root growth, promotion of xylem differentiation, and abscission of plant organs (Nemhauser et al. 2004).

6.6.4 Salicylic Acid and Low Temperature

Salicylic acid is a monohydroxybenzoic acid, a type of phenolic acid and beta hydroxy acid which functions as a plant hormone. Various studies have shown that salicylic acid has been implicated for improving plant tolerance to environmental stress such as chilling, heat, salt, drought, and ultraviolet light (He et al. 2005b; Ogawa et al. 2005). On the basis of physiological, molecular studies, it was found that salicylic acid application inhibited electrolyte leakage, reduced the accumulation

Table 6.2 Effect of brassinosteroid applications on plant responses to low-temperature stress in different plant species

Plant species	Low-temp stress	Mode of application	Metabolic response	References
Cucumis sativus	10/8 and 5/3 °C	Foliar spray	Improved PS-related attribute and enhancing CAT, POX, and SOD	Fariduddin et al. (2011)
Cucumis sativus	10/8 °C for 24 h	Foliar spray	Protected PS apparatus and antioxidant capacity	Jiang et al. (2013)
Brassica juncea	4 °C	Foliar spray	Alleviated the toxic effect of H_2O_2 through activities of CAT, APX, and SOD	Kumar et al. (2010)
Vitis vinifera	4 and 0 °C	Foliar spray	Increased antioxidant defense and osmoregulation	Xi et al. (2013)
Solanum lycopersicum	1 °C for 21 days	Fruits soaked in BR	Inhibited activity of phospholipase and lipoxygenase in fruits	Aghdam and Mohammadkhani (2014)
Capsicum annum	18 days stored at 3 °C	Fruits soaked in BR	Increased activities of antioxidant enzymes like CAT, POX, APX, and GR	Wang et al. (2012)
Chorispora bungeana	15/5 °C	Foliar spray	Activated chilling-induced oxidative damage through antioxidant defense system and decreased the levels of active oxygen species	Liu et al. (2009)
Brassica napus	2 °C	Through injection	Reduced the leakage of ion	Janeczko et al. (2007)
Prunus persica	1 °C for 4 weeks	Fruits were immersed	Increase in electrolyte leakage, total phenolic content	Gao et al. (2016)
Citrus sinensis	3 °C for 5 months	Fruits were immersed	Stimulation of antioxidant enzymes, decreased lipid peroxidation	Ghorbani and Pakkish (2014)
Lycopersicon esculentum	10/3, 12/7, 20/14, 25/18 °C	Foliar spray	Enhanced PS and antioxidant enzymes such as CAT, POX, and SOD	Khan et al. (2015a, b)

BR brassinosteroids, *PS* photosynthesis, *CAT* catalase, *POX* peroxidase, *APX* ascorbate peroxidase, *GR* glutathione reductase, *SOD* superoxide dismutase

of active oxygen species, induced the expression of selected genes, increased the activities of cell protective enzymes, and enhanced efficiency of photosynthesis during chilling treatment (Kang et al. 2004). In young maize plants, salicylic acid application decreased harmful effect of low-temperature stress (Janda et al. 1999). The foliar application of salicylic acid (0.5 mM) on chilling-sensitive banana or root treatment for 1 day caused an activation of superoxide dismutase, catalase, and ascorbate peroxidase activities at 5 °C (Kang et al. 2003a). Seeds of bean soaked in

0.1 mM of salicylic acid significantly enhanced the percent germination, germination rate under optimum and low-temperature stress (Gharib and Hegazi 2010). Additionally, seed priming with salicylic acid improved seedling emergence, root and shoot length, and fresh and dry mass of seedling and leaf compared to the control both at optimum and chilling temperatures in maize (Farooq et al. 2008c). Pre-sowing treatment of wheat seeds with salicylic acid and the resulting seedlings had improved growth of shoots and roots, as evident from the increase in both their masses and lengths (Rakhmankulova et al. 2010). Moreover, foliar application of salicylic acid or aspirin improved net photosynthetic rate, transpiration rate, and stomatal conductance in both soybean and maize (Khan et al. 2003). Salicylic acid has been used to regulate photosynthetic activities of *Cucumis sativus* seedlings and also showed tolerance under low temperature and light intensity. Furthermore, in bean and tomato, both 0.5 and 0.1 mM concentrations of acetyl salicylic acid and salicylic acid proved effective not only against heat and drought stress but also against low-temperature stress (Senaratna et al. 2000). The fact that seed imbibition with acetyl salicylic acid and salicylic acid confers stress tolerance in plants which is more consistent with a signaling role of these molecules leads to the expression of tolerance rather than a direct effect (Senaratna et al. 2000). Wang and Li (2006) also showed that application of salicylic acid to grape plants exposed to low-temperature stress led to a decrease in the rates of lipid peroxidation and electrolyte leakage and induced cold tolerance. Furthermore, after cold stress, increased cytosolic Ca^{2+} in salicylic acid-treated mesophyll cells was pumped back to vacuoles or intercellular spaces, which maintained the structure of chloroplasts (Wang and Li 2006). Moreover, other studies confirmed the role of salicylic acid in chilling tolerance in tomato (Ding et al. 2002), winter wheat (Tasgin et al. 2006), banana (Kang et al. 2003a), *Brassica juncea* (Setia et al. 2006), radish (Biao 2006), *Zoysia matrella* (Wang et al. 2009), eggplant (Chen et al. 2011), and barley (Mutlu et al. 2013).

6.6.5 Gibberellic Acids and Low Temperature

Gibberellic acids are mostly involved in plant growth and development process. It controls seed germination, leaf expansion, stem elongation, and flowering in many plant species (Magome et al. 2004). Gibberellic acid plays an important role in low-temperature stress. Foliar spray of gibberellic acid (500 mg l^{-1}) enhanced bolting and flowering of the plants subjected to low temperature of 6 °C for 5 weeks in *Limonium gmelinii* plants (Guda et al. 2000). In addition to this, when plants are exposed to low temperature, root mannitol and starch declined, and sucrose enhanced in *Delphinium* (Ogasawara et al. 2001). The higher starch and mannitol contents in gibberellic acid-treated seedlings indicated that the gibberellic acid application inhibits starch and mannitol utilization to sucrose (Ogasawara et al. 2001). These outcomes suggested that exogenous application of gibberellic acid breaks the rosette by means of rapid enlargement of differentiated tissues in cold. Chemical inhibitors of gibberellic acid biosynthesis, such as uniconazole and paclobutrazol, limit germination (Jacobsen and Olszewski 1993). These findings suggested that de novo biosynthesis of gibberellic acid is essential for seed germination in *Arabidopsis thaliana* (Hedden and Kamiya 1997). It was revealed that bioactive gibberellic acid has been increased just before protrusion of radicle in germinating *Arabidopsis thaliana* seeds (Ogawa et al. 2003). Moreover, the effect of low temperature on gibberellic acid content in germinating seeds has also been examined in *Pyrus malus*, *Corylus avellana*, and *Arabidopsis thaliana* seeds (Derkx et al. 1994). These studies indicated that gibberellic acid is more abundantly found in cold-treated seeds than non-cold-treated. Furthermore, DELLA proteins, a family of putative transcriptional regulators that inhibit the cell proliferation and expansion that drives the growth of plant organs, are stabilized in the lack of bioactive gibberellic acid, which leads to reduction in plant growth and confers tolerance to numerous environmental stresses (Fleet and Sun 2005; Davière et al. 2008; Achard et al. 2009). Treatment of gibberellic acid triggers degradation of DELLA proteins by the ubiquitin-proteasome system and results in the loss of stress tolerance (Davière et al. 2008; Achard et al. 2009). It was also found that application of the gibberellic acid suppressed low-temperature-induced male sterility in the modern rice cultivars, viz., Sasanishiki and Nipponbare (Sakata et al. 2014).

6.6.6 Cytokinins and Low-Temperature Stress

Cytokinins are classes of plant growth hormone that promote a numerous developmental and physiological processes, including cell proliferation, cell division, root and leaf differentiation, chloroplast biogenesis, and the suppression of leaf senescence (Kieber and Schaller 2014). Low temperatures employ contrary effects on plant growth and development and limit the geographic distribution of plants. Plants respond and acclimatize to low temperature via modifications in various features of physiological processes, including gene expression (Shinozaki et al. 2003). Low temperature restricts plant growth of *Arabidopsis thaliana* by extended interval of cell cycle. During the low-temperature stress, mutant plants with higher cytokinin level as well as wild type treated with exogenous cytokinins showed better relative growth rates than wild-type plants due to an increase in the cell number (Xia et al. 2009). When *Zea mays* seedlings were treated with 4 °C in the night plants formed smaller leaves, accompanied by decrease in cell number, longer cell cycles but cell cycle were unaffected (Rymen et al. 2007). In addition to this, cytokinins have an ability to improve plant yield under chilling conditions. Further study confirmed that low temperature induced a reduction in the content of monogalactosyl-diacylglycerol, a non-bilayer-forming lipid in thylakoid membranes of *Arabidopsis thaliana* (Yamaryo et al. 2003; Qi et al. 2004). During cold treatment, the content of cytokinins was reduced because of induced activity of cytokinin oxidase (Veselova et al. 2005) accompanied by our expression of genes related to cold acclimation mediated by increased level of abscisic acid (Xiong et al. 2001). These results suggested that low temperature could modify phytohormone level, delay cell cycle process, and activate cold acclimation signaling pathways.

6.6.7 Jasmonic Acid and Low-Temperature Stress

The plant hormone jasmonic acid regulates plant response to abiotic stress, such as low temperature, as well as biotic stress (Wasternack 2007; Pauwels and Goossens 2011; Li et al. 2015). Jasmonic acid is directly involved in a number of physiological processes like vegetative growth, cell cycle regulation, stamen and trichome development, senescence, regulation of anthocyanin biosynthesis, fruit ripening, and cell cycle regulation (Pauwels et al. 2008; Reinbothe et al. 2009; Yoshida et al. 2009). The exogenous application of methyl jasmonate improved cold tolerance and also enhanced antioxidant and soluble protein content in wheat (Qi et al. 2005). In addition, plant treated with methyl jasmonate significantly improved endogenous jasmonic acid accumulation and encouraged low-temperature tolerance in banana fruit during cold storage (Zhao et al. 2013). Likewise, other physiological studies revealed that exogenously applied methyl jasmonate enhanced the cold tolerance in peach and pomegranate fruits during low-temperature storage (Feng et al. 2003; Zolfagharinasab and Hadian 2007). Xu et al. (2016) suggested that cold motivated the germination of dormant seeds and stimulated a transient rise in jasmonic acid content in wheat, demonstrating involvement of jasmonic acid in release of dormancy via cold stratification (Xu et al. 2016). Moreover, molecular studies also showed that the jasmonic acid signaling pathway plays a significant role in the initial cold response by modifying the C-repeat binding factor signaling pathway (Hu et al. 2013). Furthermore, earlier findings indicated that the ratio of unsaturated/saturated fatty acid in methyl jasmonate-treated fruit was considerably higher than that in control fruit (Cao et al. 2009). These results suggest that the decrease in chilling injury by methyl jasmonate treatment could have improved antioxidant activity and maintained higher ratio of unsaturated/saturated fatty acid (Cao et al. 2009).

6.6.8 Thiourea and Low-Temperature Stress

Thiourea supplementation has also been reported to mitigate the adverse impacts of heat stress in crop plants (Asthir et al. 2013; Khanna et al. 2017). Seed pre-treatment (6.6 mM) and foliar spray of thiourea (6.6 mM) at the anthesis stage were effective in improving tolerance against terminal heat by lowering oxidative damage (cell membrane injury, thiobarbituric acid reactive substances, and H_2O_2 contents) and increasing the antioxidant potential in wheat under heat stress (29.4 °C) at the grain-filling stage (Asthir et al. 2013). DNA profiling using random amplified polymorphic DNA (RAPD) after thiourea application indicated significant variations in sunflower plants. More than 87% of DNA bands were polymorphic and identified positive markers for thermotolerance. The identified positive markers could be used to identify genes associated with heat tolerance in marker-assisted breeding (Waqas et al. 2019). Prolific root systems and optimal grain filling are crucial to improve stress tolerance and enhance final yield (Palta and Turner 2018). Thiourea (10 mM) increased root growth and grain-filling rate to impart heat tolerance in bread wheat

(Freeha et al. 2011). Application of thiourea increased potassium uptake and reduced abscisic acid biosynthesis in chickpea under heat stress (Aldasoro et al. 1981).

6.6.8.1 Influence of Thiourea on Other Plant Hormones

Thiourea can modulate numerous aspects of growth and development of agronomic and horticultural crops under normal conditions as well as stressed environments through interactions with other phytohormones (Abdelkader et al. 2012; Patel et al. 2016; Srivastava et al. 2017). Interactions between thiourea and other phytohormones, such as salicylic acid (Abdelkader et al. 2012), benzoic acid and polyamines (Hassanein et al. 2015), abscisic acid, auxin, and jasmonates (Srivastava et al. 2017), are already well known. There are two modes of thiourea interaction with phytohormones and other plant growth regulators. Either exogenous application of thiourea enhances the endogenous levels of phytohormones, such as abscisic acid, or thiourea interacts positively when applied in combination with other plant growth hormones, and phytohormones lead to improved performance of stressed plants. However, optimized dosage is essential. For example, a thiourea-mediated rise in the endogenous level of salicylic acid contributed to reversing the impacts of mild and high-temperature stress in sunflower (Akladious 2014). Moreover, the combined application of salicylic acid and thiourea interacted positively in terms of improved yield and nutritional contents of grains in drought-stressed wheat by tightly controlling oxidative damage (Hassanein et al. 2015). Benzoic acids are vital structural elements for numerous essential plant hormones (Widhalm and Dudareva 2015). Externally applied benzoic acid and thiourea increased wheat crop productivity and grain nutrition by improving photosynthetic pigments and maintaining flag leaf greenery at critical growth stages (Amin et al. 2016). Gooseberry (*Emblica officinalis* Gaertn.) shrubs often face heavy fruit drop, which significantly reduces yield. Thiourea in combination with gibberellic acid dramatically enhanced fruit quality parameters (ascorbic acid, sugar contents, and acidity) and fruit setting (Anshuman and Singh 2015). Thiourea application also influenced the biosynthesis of numerous hormones including an increase in the level of cytokinins (zeatin, zeatin riboside), gibberellins, and indole-3-acetic acid and decreased the growth inhibitory hormone abscisic acid under both normal and stressed conditions (Abdelkader et al. 2012).

6.7 Role of Growth Hormones in Plants under Low Temperature

Environmental stresses, including drought, salinity, high light, and extreme temperatures, influence plant growth and productivity. These abiotic stresses result in reductions in growth, stomatal and nonstomatal limitations on photosynthesis, and alterations in both hormonal balance and reduction/oxidation processes, potentially leading to enhanced lipid peroxidation, protein oxidation, and DNA damage (Munns and Tester 2008; Mittler et al. 2011; Munné-Bosch et al. 2013). Plant responses and adaptation to abiotic stresses are controlled by molecular signal transduction cascades. In these cascades, plant hormones as a part of the signal network function as central integrators that link and reprogram the complex stress-adaptive signaling cascades (Ma et al. 2006; Golldack et al. 2014). Several plant hormones, such as ethylene (Zhao and Schaller 2004; Cheng et al. 2013; Müller and Munné-Bosch 2015), abscisic acid (Wu et al. 2007), jasmonates (Cela et al. 2011; Cheng et al. 2013), salicylic acid (Jayakannan et al. 2015), gibberellins (Magome et al. 2008), cytokinins (Wu et al. 2014), auxin (He et al. 2005a), and brassinosteroids (Divi et al. 2010), have been reported to be involved in stress signaling. Despite its simple two-carbon structure, the plant hormone ethylene serves as a key mediator of biotic and abiotic stress factors. Transcription factors control the majority of stress response genes and of more than 1800 transcription factors identified in *Arabidopsis thaliana* (Riechmann et al. 2000; Gong et al. 2004). Ethylene response factor superfamily plays a pivotal role in adaptation to biotic and abiotic stresses, such as those caused by pathogens, wounding, cold and heat stress, UV light, drought, and salinity (Mizoi et al. 2012). A genome-wide analysis of plant species has identified, for instance, the following numbers of members of the ethylene response factor superfamily: *Arabidopsis*, 147 (Nakano et al. 2006); *Populus* spp., 200 (Zhuang et al. 2008); *Brassica* spp., 291 (Song et al. 2013); *Citrus* spp., 108 (Ito et al. 2014); *Vitis* spp., 149 (Licausi et al. 2010); *Solanum* spp., 155 (Charfeddine et al. 2015); and *Oryza* spp., 180 (Nakano et al. 2006; Sharoni et al. 2011).

6.7.1 Effect on Plant Architecture

Multiple studies in *Arabidopsis* have implicated plant hormones in the temperature regulation of plant architecture. When grown at low (<4 °C) temperatures, reductions in plant size are accompanied by the expression of genes involved in cold acclimation. These include the C-repeat binding factor regulon, a suite of genes involved in protecting plants against freezing damage through the synthesis of membrane-stabilizing proteins and accumulation of compatible solutes (Thomashow 1999). Plants constitutively expressing C-repeat binding factor transcription factors display both enhanced freezing tolerance and a severe reduction in growth (Gilmour et al. 2004). This growth retardation has recently been shown to involve C-repeat binding factor-mediated stabilization of the DELLA family of growth repressing proteins, through reductions in the level of gibberellins (Achard et al. 2008). Gibberellins regulate growth through targeting DELLA proteins for proteasome-mediated degradation (Dill et al. 2004; Fu et al. 2004). The C-repeat binding factor-mediated reduction in bioactive gibberellins therefore ensures coordination of cold acclimation with plant growth at low temperatures (Patel and Franklin 2009).

The plant hormone salicylic acid has also been implicated in the retardation of plant growth during chilling. This hormone performs an important role in plant pathogen defenses (Kunkel and Brooks 2002) but is also involved in plant responses to a variety of abiotic stresses and accumulates at low temperatures (Scott et al. 2006). Transgenic plants expressing the salicylic acid hydroxylase transgene have been shown to display enhanced cell expansion and a 2.7-fold increase in biomass when compared to wild-type controls grown for 2 months at 5 °C (Scott et al. 2006). These plants contained reduced amounts of salicylic acid suggesting a link between salicylic acid accumulation and reduced plant size. The enhanced cell expansion of salicylic acid hydroxylase plants at 4 °C has recently been shown to coincide with elevated expression of the cell cycle gene and increased endoreduplication (Xia et al. 2009). This study also reports a role for cytokinin in mediating plant growth rates at low temperatures. Mutants containing elevated cytokinin levels were shown to display enhanced cell division at 4 °C. No alterations in expression of C-repeat binding factor were recorded in salicylic acid hydroxylase or plants with elevated cytokinin level at low temperatures, suggesting the effects of salicylic acid and cytokinin on temperature-regulated growth to be independent of C-repeat binding factor regulon (Xia et al. 2009).

The hormone auxin has a fundamental role in temperature-dependent elongation growth. High temperature (28 °C/29 °C)-mediated increases in *Arabidopsis* hypocotyl elongation have been shown to correlate with elevated levels of free auxin and auxin-induced gene expression (Gray et al. 1998; Koini et al. 2009). Higher levels of free auxin were also recorded in *Arabidopsis* flower stems grown in positive difference between day and night temperature conditions when compared to plants grown in a negative difference between day and night temperature regime (Thingnaes et al. 2003). Elongation responses to reduced red/far-red in shade avoidance have been demonstrated to involve auxin synthesis via a tryptophan-dependent pathway (Tao et al. 2008).

6.8 Mechanism of Cold Tolerance and Role of Growth Hormones

Chilling resistance or tolerance can be defined as the ability of the chilling-sensitive plant to withstand the strain of the low-temperature stress (Li et al. 2018). The acquisition of freezing tolerance is called cold acclimation, which is generally initiated after a period of decreases in above 0 °C temperatures and photoperiod (Levitt 1980). The growth inhibition resulting from low temperature reduced the capacity for energy utilization that leads to feedback inhibition of photosynthesis (Huang et al. 2014). Differences in the capacity to minimize photoinhibition and recover photosynthesis during cold acclimation have been shown to contribute to intra- and interspecific differences in freezing tolerance (Pocock et al. 2001). Carbohydrates are synthesized from a complex series of reactions involving photosynthesis in plants (Huang et al. 2014). Changes in the three major types of water-soluble carbohydrates monosaccharides (glucose and fructose), disaccharides (sucrose, trehalose), and fructans have often been reported in plants subjected to low temperature (Theocharis et al. 2012a). Those sugars may serve as a typical osmoprotectant, cellular membrane stabilizer, scavengers of reactive oxygen species, and signaling molecules in plants to tolerate chilling and freezing stress (Theocharis et al. 2012a). Many studies have showed that higher accumulation of sugars or individual sugar fractions such as starch (Zhang et al. 2006a), fructans

(Dionne et al. 2001), and sucrose (Zhang et al. 2006a) during cold acclimation are associated with great cold tolerance in turfgrass (Cai et al. 2004; Hoffman et al. 2010). Sucrose consists of the monosaccharides glucose and fructose, and its intimate involvement in growth, development, storage, signaling, and stress acclimation were established (Janská et al. 2010). Leaf starch potentially contributed carbon to carbohydrate accumulation in response to low temperature (Sicher 2011). Fructans often accumulated during cold acclimation in cereals and grasses, which serve to stabilize cell membranes (Livingston and Heyer 2009). Furthermore, trehalose, a nonreducing disaccharide of glucose, has been implicated in modulating levels of freezing tolerance and may also be involved in starch accumulation (Theocharis et al. 2012b).

Plants resort to many adaptive strategies in response to different abiotic stresses such as high salt, dehydration, cold, heat, and excessive osmotic pressure which ultimately affect the plant growth and productivity (Yancey et al. 1982). Against these stresses, plants adapt themselves by different mechanisms including change in morphological and developmental pattern as well as physiological and biochemical processes (Bohnert et al. 1995). It is the ability of the plants to undergo such changes that accounts for major differences in chilling susceptibilities between species and protect against irreversible chilling-induced injury. Chilling sensitivity or tolerance was linked with the extent of lipid unsaturation in some plant species (Riken et al. 1993). A downward shift in the temperature generally increases the degree of unsaturation of membrane lipids which compensates for the decrease in the fluidity of membrane lipids. The increase in the degree of unsaturation of membrane lipids is also correlated with the sustained activity of membrane-bound enzymes at lower temperature. Leaves of transgenic tobacco plants with decreased levels of fatty acid unsaturation in phosphatidylglycerol exhibited a slightly lower level of the steady-state oxidation of the photosystem I than wild-type plants at low temperature (Ivanov et al. 2012).

Another mechanism involves cellular defense against membrane lipid peroxidation caused by chill-induced increase in the generation of reactive oxygen species (Prasad et al. 1994b). These reactive oxygen species are kept under control by various antioxidants (α-tocopherol, ascorbate, etc.) and antioxidant enzymes (superoxide dismutase, catalase, peroxidase, glutathione reductase, etc.) (Bowler et al. 1992). Lower concentrations of antioxidant compounds (ascorbate, glutathione, and β-carotene) were found in chilling-sensitive lines than chilling-tolerant lines in maize (Hodges et al. 1996). In rice also, activities of antioxidant enzymes, viz., superoxide dismutase, catalase, ascorbate peroxidase, and glutathione reductase, were higher in chilling-tolerant cultivar than in chilling-sensitive (Huang and Guo 2005). These results indicated that higher activities of defense enzymes and higher contents of antioxidants under stress were associated with tolerance to chilling. Reactive oxygen species production in mitochondria and chloroplast can be decreased by the alternative channeling of group of enzymes called alternative oxidases. Alternative oxidase decreases reactive oxygen species production by two mechanisms: they prevent electrons from reducing O_2 into superoxide radical, and they reduce the overall level of O_2, the substrate for reactive oxygen species production in the organelles. In the mitochondria of preemergent maize seedlings, the cytochrome pathway was induced in both acclimated and non-acclimated mitochondria at the end of the 4° C stress period (Prasad 2000). Another mechanism of chilling tolerance involves the induction of molecular chaperones by low temperature.

In response to cold and other osmotic stresses, plants accumulate a range of compatible solutes including carbohydrates, such as sucrose, raffinose, sorbitol and trehalose glucosides, raffinose, and arabinoxylans, and amino acids such as glutamic acid, alanine, glycine, proline, serine, polyamines, and betaines (Theocharis et al. 2012a). These different molecules, which are often degraded once the stress has passed, are referred to as osmolytes, osmoprotectants, or compatible solutes. Sugars play multiple roles in low-temperature tolerance. As typical compatible osmolytes, they contribute to the preservation of water within plant cells (Ruelland et al. 2009). Sugars might protect plant cell membranes during cold-induced dehydration, replacing water molecules in establishing hydrogen bonds with lipid molecules (Ruelland et al. 2009). The most abundantly accumulated free sugar in response to low temperature is sucrose (Kaurin et al. 1981). During cold acclimation, levels of soluble sugars increased in both genotypes of *Arabidopsis*, but raffinose accumulation discriminated the more tolerant one (Klotke et al. 2004).

Paldi (2002) suggested the role of some nitrogen-containing compounds (glycinebetaine, agmatine, putrescine, and proline) in chilling tolerance of maize. Proline content increased significantly in sunflower under moisture stress (Deshmukh and Srivastava 1982), in barley under salinity stress (El-Tayeb et al. 2005), and in maize under chilling stress (Chen and Li 2002). Low-temperature stress has been reported to cause significant increases in the levels of the soluble nonenzymatic antioxidants ascorbate and glutathione (Airaki et al. 2011). Low temperature induces abscisic acid accumulation providing a mechanism for chilling resistance. Higher abscisic acid levels were accumulated and maintained in tolerant cultivars of rice (Liu et al. 2003). Likewise in maize, under field conditions, chilling-tolerant genotypes accumulated higher amounts of abscisic acid during chilling periods than chilling-sensitive genotypes (Janowiak et al. 2003). Exogenous application of abscisic acid improved germination (Basra et al. 1989), increased protein content (Stupnikova et al. 2001), and decreased leaf water deficit (Aroca et al. 2003) in maize. Plant acclimation to low temperature depends on changes in the expression of specific genes encoding products that confer increased cold tolerance (Doherty et al. 2009). The process involves modifying preexisting proteins and up- or downregulating gene expression and protein synthesis. Several studies have suggested that activity of cold−/chilling-induced genes may facilitate the metabolic changes that confer low-temperature tolerance (Doherty et al. 2009). They may also be involved in the signal transduction of the stress response (Thomashow 2010). Yakovlev and Borovsky (2003) suggested the accumulation of proteins named as cold shock proteins during low-temperature stress. In *Arabidopsis*, cold acclimation induced the accumulation of four major mRNAs which code for highly hydrophobic proteins similar to cryoprotectant proteins in spinach. Several cold-regulated genes whose expression was changed during freezing acclimation (Thomashow 2010) have also been isolated and characterized. Hiilovaara-Teijo et al. (1999) suggested the accumulation of antifreeze proteins in winter rye plants during cold acclimation. Some genes expressed differentially during chilling acclimation, designated as chilling acclimation-responsive genes, have been isolated from various plants, including maize (Anderson et al. 1994a).

Plant hormones may serve as signals in regulating plant growth, development, and adaptation to environmental stress (Srivastava 2002). For abscisic acid, they can affect the diverse processes such as leaf senescence, seed dormancy and germination, as well as cell division and elongation (Finkelstein and Gibson 2002). In addition, it has been also reported to be involved in the responses to abiotic stresses including chilling and freezing in plants (Finkelstein and Gibson 2002). Cytokinins are essentially involved in various plant developmental processes, including cell division and enlargement, chloroplast biogenesis, nutrient mobilization, and leaf senescence (Ryu and Cho 2009), as well as facilitate the responses to delay both stomatal closure and leaf senescence under abiotic stresses (Ryu and Cho 2009). Auxin can promote root initiation and also delay plant senescence (Zhang et al. 2009b). Gibberellins are involved in plant growth and development such as leaf expansion, stem elongation, and flowering (Ryu and Cho 2009). Jasmonic acid is a critical signaling molecule for diverse developmental processes and defense responses in plants (Llanes et al. 2016) and has been shown to induce a response similar to that of abscisic acid for alleviating chilling injury (Hu et al. 2013). Salicylic acid has also been demonstrated to be involved in the responses of plants to a broad range of abiotic stresses, including low temperature (Khan et al. 2015a).

Cold acclimation and the acquisition of freezing tolerance in plants is highly dependent upon the induction of stress-related genes, such as the dehydration-responsive elements or C-repeat binding factor genes (*CBF/DREB1*) as well as the cold-induced stress proteins such as late embryogenesis abundant proteins (Hundertmark and Hincha 2008; Thomashow 1999). Low temperature upregulated the expression of *CBF/DREB1* which in turn induces the expression of *cold-regulated* (*COR*) genes to enhance freezing tolerance (Kurepin et al. 2011). The accumulation of late embryogenesis abundant proteins functions in direct protection from freezing-induced cellular dehydration and mechanical damage (Theocharis et al. 2012a).

Plant's stress tolerance can be improved with an exogenous use of stress alleviating chemicals (Wahid and Shabbir 2005; Wahid et al. 2007; Farooq et al. 2009b). Plant growth regulators are powerful tools for maximizing yield and quality and increasing net income to farmers. Plant growth regulators have been used from many years to alter the behavior of agricultural and horticultural crops for the economic benefit of the grower. Control of vegetative vigor, stimulation of flowering, regulation of crop load, reduction of fruit drop, and delay or stimulation of fruit maturity and ripening are best examples of regulation with the exogenous applications. Novel plant growth regulators with possible benefits for fruit growers are continually being made available by industries. In addition, research is in progress to find new uses for growth regulator products that have been made available for specific uses.

Every aspect of plant growth and development is controlled by plant hormones, and these serve as key integrators of exogenous (environmental) and endogenous (developmental) cues. The classes of phytohormones are auxins, gibberellins, cytokinins, ethylene, abscisic acid, brassinosteroids, salicylic acid, and jasmonates, with strigolactones representing a relatively new addition (Stamm et al. 2011). Stress may induce common responses such as enhancement of plant hormones. For instance, wounding can induce increased production of ethylene, auxin, and abscisic acid. Many kinds of stresses, including water, salt, and temperatures, induce abscisic acid synthesis and therefore abscisic acid is considered as a plant stress hormone. It regulates several important aspects of plant growth and development. Recent studies have demonstrated a pivotal role for abscisic acid in modulation at the gene level of adaptive responses for plants in adverse environmental conditions. Abscisic acid is also involved in several other physiological processes such as stomatal closure, embryo morphogenesis, development of seeds, synthesis of storage proteins and lipids, germination, leaf senescence, and defense against pathogens. Nevertheless, abscisic acid acts as a mediator in controlling adaptive plant responses to environmental stresses. In several instances, it has been implicated in signal transduction at the single-cell level. Other than plant hormones, exogenous plant growth regulators such as inorganic and organic chemicals and booster's application will also have strong impact on plant adaptation to abiotic stress either independently or synergistically with one another (Srivastava et al. 2008).

Many plant growth regulators influence with the signaling pathways of one another, thereby promoting plant tolerance to abiotic stress. The coordination of the responses triggered by the multiple stimuli is controlled by a network of intricate signal transduction pathways. This network has many signaling components and a small number of highly interconnected components which are central for the functioning of the network (Hetherington and Woodward 2003). For instance, in many studies nitric oxide has been found to mediate the action of stomatal closure of abscisic acid (Garcia-Mata and Lamattina 2007). Plant growth regulators also promote the uptake and metabolism of nutrient elements as well. Among the stress alleviating compounds, thiourea is one of the important molecules with two functional groups: "thiol" is important to oxidative stress response, and "imino" partly fulfills the nitrogen requirement. Apart from thiol compounds, other chemicals including plant growth regulators, 6-benzyladenine, prohexadione calcium, N-2-chloro-4-pyridinyl-N-phenylurea, 2,4-dichlorophenoxyacetic acid, 3,5,6-trichloro-2-pyridyloxyacetic acid, and aminoethoxyvinyl glycine, dithiothreitol, potassium nitrate, thiourea, salicylic acid, and silicon products were used, to a small extent, to screen for their ability to increase agricultural yield without compromising quality (Minhas et al. 2017).

Exogenous application of thiourea (e.g., as seed priming, foliar spray, medium supplementation, soil application) stimulates defense mechanisms in plants under abiotic stress (Akladious 2014; Srivastava et al. 2017; Seleiman and Kheir 2018). Thiourea application modulates key physiological events and mechanisms, including photosynthesis, nitrogen metabolism, proline metabolism, antioxidant defense systems, and plant water relations during different plant developmental stages (Kaya et al. 2015; Vineeth et al. 2016; Wakchaure et al. 2018; Kaya et al. 2019). Thiourea application upregulates the expression of genes involved in encoding reactive oxygen species activated ion channels, antioxidants, regulation of redox state, aquaporins, osmotic adjustment, metabolite biosynthesis, calcium signaling, and hormonal regulation, such as catalase and cytochrome P450 (Srivastava et al. 2009, 2011; Demidchik et al. 2010; Patade et al. 2012). The application of thiourea also modulates posttranscriptional regulation to enhance the expression of defense-related genes by the synchronization of microRNAs and hormones (Srivastava et al. 2017). Plant responses to thiourea vary by plant species and depend on the concentration applied. For instance, in potato (*Solanum tuberosum*), a lower concentration of thiourea (250, 500 mM) increased carotenoid, ascorbic acid, polyphenol, and nitrate contents, while a higher concentration (1000 mM) significantly decreased nitrate content (Mani et al. 2012). However, in lentil (*Lens culinaris*) grown under rainfed conditions, a higher concentration of thiourea (13 mM) significantly enhanced seed yield, quality, and economic returns of the crop when compared with a low concentration (6.5 mM) of thiourea (Premaradhya et al. 2018). High doses (such as 60 mM) can cause more severe injury to wheat (*Triticum aestivum*) plants compared with low doses (20 mM and 40 mM) (Suryavanshi et al. 2016). Besides the dose applied, plant species, growth stage, climatic conditions, treated plant organ, duration of treatment, and application method also affect the ability of thiourea to alleviate abiotic stress damage (Kaya et al. 2013; Waqas et al. 2017; Seleiman and Kheir 2018).

In the current scenario, various kinds of plant growth regulators such as thiourea, salicylic acid, silicon, nitric oxide, hydrogen peroxide, brassinosteroid, polyamines, etc. have been tested for enhancing the plant stress tolerance as well as the crop yield (Jisha et al. 2013; Srivastava et al. 2010a). Most of these plant growth regulator treatments are applied either for seed priming or as foliar spray. Seed priming is a pre-sowing treatment that partially hydrates seeds without allowing the radicle emergence. The seed priming may be induced alone or in combination with the foliar application of plant growth regulators at the time or early flowering or grain-filling stage. The mechanism of seed priming mediated action is not well understood, but it may act in two ways. First, seed priming sets in motion germination-related activities (e.g., respiration, endosperm weakening, gene transcription and translation, etc.) that facilitate the transition of quiescent dry seeds into germinating state and lead to improved germination potential. Second, priming imposes abiotic stress on seeds that represses radicle protrusion but stimulates stress responses, e.g., accumulation of late embryogenesis abundant, potentially inducing cross-tolerance (Chen and Arora 2013). These two strategies together also impose a "memory" and may mediate a greater stress tolerance in the subsequent generation (Pastor et al. 2013). The specific influence of exogenous plant growth regulators on crop plants has already been integrated into the crop production systems, e.g., the use of chlormequat chloride in cereal crop growth, ethephon for influencing the development and maturity of various crops, and mepiquat chloride in cotton. Several such plant growth regulators also may be useful in overcoming the production constraints. Foliar application of synthetic auxin, and naphthal acetic acid has been used to enhance apple flower (McArtney et al. 2007), ethephon has also enhanced flowering in bearing apple trees. The plant growth regulator prohexadione-calcium, an inhibitor of gibberellic acid biosynthesis, reduces shoot extension in several fruit tree species (Rademacher et al. 2006). Exogenous cytokinin application of benzyladenine has effectively improved lateral branching of nursery trees (Hrotkó et al. 1999) by limiting blindwood and improves canopy development.

Stressful environments interrupt the balance between generation and utilization of reactive oxygen species leading to toxicity by enhancing production of reactive oxygen species such as superoxide radicals (O_2^-), hydrogen peroxide (H_2O_2), hydroxyl radicals (OH^-), etc. in plants, thereby creating a state of oxidative stress in them (Panda et al. 2003a, 2003b). This increased reactive oxygen species level in plants causes oxidative damage to biomolecules such as lipids, proteins, and nucleic acids, thus altering the redox homeostasis (Smirnoff 1993; Gille and Singler 1995; Srivastava et al. 2010a). When applied exogenously at suitable concentrations, plant growth regulators enhance the efficiency of antioxidant system, upregulate osmolytes, and enhance the expression of stress-responsive genes in plants (Srivastava et al. 2010b; Knorzer et al. 1999), and it was also proposed plant growth regulators for sustainable agriculture by integrating redox signaling as a possible unifying mechanism (Srivastava et al. 2009).

Plant growth hormones interact with specific target tissues to cause physiological responses, such as growth and development, and, thus, affect plant's ability to respond to its environment (Farooq et al. 2015; Ratnakumar et al. 2015, 2016; Srivastava et al. 2010b). Each response is often the result of two or more hormones acting together. Among them, many hormones can be synthesized in the laboratory, therefore increasing the quantity of hormones available for commercial applications. For instance, gibberellin is widely used as plant growth hormones. Likewise, salicylic acid is another important plant hormone that regulates a wide range of metabolic and physiological responses in plants such as seed germination, seedling establishment, cell growth, respiration, stomatal closure, senescence-associated gene expression, responses to abiotic stresses, basal thermotolerance, nodulation in legumes, and fruit yield. Hence they are used as plant growth hormone for enhancing plant growth and yield (Vlot et al. 2009) and for particular stress alleviation. However, judicious use of salicylic acid dose is required as the excess concentration proves toxic for the plant growth (Donovan et al. 2013). Exogenous application of cytokinin, preferably kinetin, to the foliage of plants has also been shown to increase crop productivity. The application of plant sterols and steroid hormones such as brassinosteroids is essential for plant growth, reproduction, and responses to various abiotic and biotic stresses. The use of these brassinosteroids (Vriet et al. 2012) has also been proposed as a promising strategy for crop improvement.

When the temperature is below 15 °C, maize growth is strongly restricted (Miedema 1982), in a manner affecting seed germination and possibly resulting in growth inhibition, chlorosis, wilting, necrosis, and even death of the seedlings (Janowiak et al. 2002). At this temperature, in addition to damage to the photosynthetic systems of the leaves, complex physiological and biochemical changes in the plants themselves can occur (Bajguz and Hayat 2009), resulting in a decline in biomass and yield, which affects maize yield (Louarn et al. 2008). Brassinosteroids are an important class of polyhydroxylated sterols found in the plant kingdom and are widely used in agriculture to increase production (Sun et al. 2020) and enhance plant resistance to environmental stress (Krishna 2003). Since Hamada (1986) first reported that brassinosteroid has an ameliorating effect on plant abiotic stress, many researchers have conducted extensive research on the role of brassinosteroid (Fariduddin et al. 2013; Vardhini 2017). Brassinosteroids control responses to stresses such as heat, cold, and drought stress (Nolan et al. 2017a). The relationship between brassinosteroids and stress responses is complex. Although the application of brassinosteroids promotes tolerance to several stresses (Kagale et al. 2007; Bajguz and Hayat 2009; Yuan et al. 2010; Anjum et al. 2011; Divi et al. 2016), analysis of brassinosteroid-deficient and brassinosteroid-insensitive mutants revealed that impairment of the brassinosteroid pathway is often associated with increased survival in the face of stresses such as drought stress (Feng et al. 2015; Northey et al. 2016; Nolan et al. 2017b). Studies have affirmed that multiple aspects of the brassinosteroids signaling pathway interface with stress responses but suggest that the outcome depends on the particular components and tissues that are affected (Nolan et al. 2020). Brassinosteroids can induce plant tolerance to abiotic stresses such as high temperature, low temperature (Yu et al. 2002; Zhang et al. 2013b), drought (Tanveer et al. 2019), salinity (Ding et al. 2012), and heavy metals (Bajguz 2010). It has a positive contribution to the cellular biochemical, physiological, and morphological processes of higher plants, especially affecting cell division, antioxidant metabolism, gas exchange, and the plant growth rate (Yuan et al. 2012).

6.9 Metabolic Effects of Growth Hormones for Abiotic Stress Tolerance

Investigating how abiotic stresses affect plant growth and development at the physiological, biochemical, and molecular levels is critical to increasing the productivity of crops, because stresses cause widespread crop losses throughout the world (Kazan 2015). Environmental factors imposing stress on plants, including drought, salinity, heat, chilling, freezing, ozone, pathogens, and UV radiation, are the major environmental cues that limit crop productivity. The period and development of stress, stages of the plant, and biotic and abiotic factors may influence the stress response (Feller and Vaseva 2014). Some crops may be affected at an early stage but recover and finally survive. Susceptibility or resistance to stresses may differ markedly among species or genotypes of crops. Various plant hormones have shown positive plant-protective functions against abiotic stresses, and attempts have been made to assign specific hormone(s) for specific stresses as well as combinations of stresses, as some hormones have shown multiple stress resistance functions (Wani et al. 2016)..

Plants must regulate their growth and development to respond to various internal and external stimuli (Wolters and Jurgens 2009). Phytohormones, a diverse group of signaling molecules found in small quantities in cells, mediate these responses. Their pivotal roles in promoting plant acclimation to ever-changing environments by mediating growth, development, source/sink transitions, and nutrient allocation have been well established (Fahad et al. 2015a). Phytohormones are considered the most important endogenous substances for modulating physiological and molecular responses, a critical requirement for plant survival as sessile organisms (Fahad et al. 2015a). Phytohormones act either at their site of synthesis or elsewhere in plants following their transport (Peleg and Blumwald 2011). Phytohormones are of key importance in plant development and plastic growth. They include auxin, cytokinins, abscisic acid, ethylene, gibberellins, salicylic acid, brassinosteroids, and jasmonates. The strigolactone is relatively new phytohormone.

6.9.1 Abscisic Acid

Abscisic acid owes its name to its role in abscission of plant leaves and is perhaps the most studied phytohormone for its response and distinct role in plant adaptation to abiotic stresses and is accordingly termed a "stress hormone." It is an isoprenoid plant hormone produced in the plastidial 2-C-methyl-D-erythritol 4-phosphate pathway. Abscisic acid plays an influential role during several plant physiological processes and developmental stages including seed dormancy and development, stomatal opening, embryo morphogenesis, and synthesis of storage proteins and lipids (Sreenivasulu et al. 2010). Abscisic acid is considered an essential messenger in the adaptive response of plants to abiotic stress, and its role in stress tolerance has received much attention. In response to environmental stresses, endogenous abscisic acid levels increase rapidly, activating specific signaling pathways and modifying gene expression levels (O'Brien and Benkova 2013). Nemhauser et al. (2006) have reported that abscisic acid transcriptionally regulates up to 10% of protein-encoding genes. Abscisic acid also acts as an internal signal enabling plants to survive under adverse environmental conditions (Keskin et al. 2010).

Under water-deficit conditions, abscisic acid plays a vital role in providing plants the ability to signal to their shoots that they are experiencing stressful conditions around the roots, eventually resulting in water-saving antitranspirant activity, notably stomatal closure and reduced leaf expansion (Wilkinson et al. 2012). Abscisic acid is also involved in robust root growth and other architectural modifications under drought stress (Giuliani et al. 2005) and nitrogen deficiency (Zhang et al. 2007). Abscisic acid regulates the expression of numerous stress-responsive genes and in the synthesis of late embryogenesis abundant proteins, dehydrins, and other protective proteins (Verslues et al. 2006; Sreenivasulu et al. 2012). Abscisic acid regulates the processes involved in cell turgor maintenance and synthesis of osmoprotectants and antioxidant enzymes conferring desiccation tolerance (Chaves et al. 2003). Zhang et al. (2006b) reported a proportional increase in abscisic acid concentration upon exposure of plants to salinity.

6.9.2 Auxins

Although auxin has been studied for over 100 years, its biosynthesis, transport, and signaling pathways are still not clear (Ke et al. 2015). However, some interconnecting pathways have been proposed so far for biosynthesis of auxin in plants, including four tryptophan-dependent and one tryptophan-independent pathways (Mano and Nemoto 2012). Indole-3-acetic acid is one of the most multi-functional phytohormones and is vital not only for plant growth and development but also for governing and/or coordinating plant growth under stress conditions (Kazan 2013). The presence of an auxin biosynthesis, signaling, and transport apparatus in single-celled green algae is clear evidence of the evolutionary role played by auxin during the adaptation of plants to diverse land environments (De Smet et al. 2010). Though there has been a recent upsurge in our understanding of auxin regulation of plant growth and development, its role as a regulator of stress response is still little understood (Kazan 2013).

Interestingly, there is growing evidence that indole-3-acetic acid plays an integral part in plant adaptation to salinity stress (Fahad et al. 2015a; Iqbal et al. 2014). It increases root and shoot growth of plants growing under salinity or heavy metal stresses (Sheng and Xia 2006; Egamberdieva 2009). Salinity reduced indole-3-acetic acid levels in maize plants, but salicylic acid application effectively increased them (Fahad and Bano 2012), indicating that hormonal balance and crosstalk are critical to signal perception, transduction, and mediation of stress response (Fahad et al. 2015a). Auxin stimulates the transcription of a large number of genes called primary auxin response genes, and these genes have been identified and characterized in several plant species including rice, *Arabidopsis*, and soybean (Javid et al. 2011). Auxin is regarded as an influential constituent of defense responses via regulation of numerous genes and mediation of crosstalk between abiotic and biotic stress responses (Fahad et al. 2015b). However, identification of novel genes involved in stress responses may prove to be a vital target for engineering abiotic stress tolerance in principal crops.

6.9.3 Cytokinins

Cytokinins are important growth promontory compounds involved in seed germination, morphogenesis, chloroplast biogenesis, maintenance of assimilate mobilization, fruit and leaf abscission, and the regulation of stomatal functioning and root-to-shoot communication in the plant experiencing stress. These are synthesized primarily in roots (Binns 1994), besides the shoot apex and other plant tissues. These play influential roles in many plant growth and developmental processes and are considered as master regulators during plant growth and development (Kang et al. 2012; Nishiyama et al. 2011). Alteration of endogenous levels of cytokinins in response to stress indicates their involvement in abiotic stress (O'Brien and Benkova 2013), including drought (Kang et al. 2012) and salinity (Nishiyama et al. 2011). Mutants and transgenic cells/tissues with altered activity of cytokinin metabolic enzymes or perception machinery points toward their crucial involvement in several crop traits including productivity and increased stress tolerance (Zalabák et al. 2013).

Although plant responses to cytokinins have been evaluated most often via their external application, stressful conditions are also known to enhance their endogenous levels via uptake and enhanced biosynthesis (Pospíšilová 2003a). In contrast to the abscisic acid inhibition of seed germination, they also release seeds from dormancy (Fahad et al. 2015b). Cytokinins are often considered abscisic acid antagonists (Pospíšilová 2003b). In water-stressed plants, decreased cytokinin content and accumulation of abscisic acid lead to an increased abscisic acid/cytokinin ratio. The reduced cytokinin levels enhance apical dominance, which, together with the abscisic acid regulation of stomatal aperture, aids in adaptation to drought stress (O'Brien and Benkova 2013).

Cytokinins are N6-substituted adenosine compounds with branched five-carbon side chain (zeatin and isopentenyl adenine). The cytokinins are biosynthesized following two pathways: de novo biosynthetic pathway (Chen and Melitz 1979) and t-RNA pathway (Skoog and Armstrong 1970). Most of the biologically active cytokinins are biosynthesized by de novo biosynthetic pathway. The formation of N6-(Δ2-isopentenyl) adenosine-50-mono phosphate from Δ2-isopentenyl pyrophosphate and adenosine-50-monophosphate through catalytic reactions by isopentenyl transferase is the important step in cytokinin biosynthesis (Mok and Mok 2001). In the other pathway, t-RNA is degraded and isomerized to cis-zeatin by cis-trans isomerase (Mok and Mok 2001). The cytokinin levels are regulated through conversions to riboside derivatives and N- and O-linked glycoside derivatives (Murti and Upreti 2000; Zazimalova et al. 1999).

The cytokinin decline is presumed as the outcome of either reduction in cytokinin biosynthesis or their enhanced degradation or even both. Upreti and Murti (1999) reported that the stress-induced decline in root nodulation is linked to decline in cytokinins in roots/nodules. Stoll et al. (2000) in grapevine found that the partial root drying had negative influence on the xylem cytokinin concentration concomitant with distinct increase in xylem sap pH. Cytokinin increased abiotic stress resistance at low cytokinin levels employing mutants lacking the functional cytokinin receptor (Jeon et al. 2010; Kang et al. 2012). Cytokinins are important in keeping stomata open by regulating abscisic acid action (Veselova et al. 2006). However, there are reports regarding cytokinins antagonizing many physiological processes like stomatal closure, leaf senescence, and leaf expansion mediated by abscisic acid (Cowan et al. 1999). Both naturally occurring and synthetic cytokinins increase transpiration rate and increase in stomatal aperture (Incoll et al. 1990; Incoll and Jewer 1987). The stomatal responses of cytokinins are found variable. Blackman and Davies (1983) revealed that zeatin alone did not affect stomatal opening, but partially reverse abscisic acid-induced stomatal closure. In contrast, zeatin riboside or kinetin decreased stomatal opening and had no effect on abscisic acid-induced stomatal closure.

6.9.4 Ethylene

Ethylene is the simplest olefinic gaseous hormone which regulates a wide range of plant developmental processes such as pollination, seed germination, abscission and senescence, flowering, fruit ripening, root formation, and gravitropism. It is biosynthesized by the conversion of methionine to ethylene. Ethylene, a gaseous phytohormone, is involved in several phases of plant growth and development, notably fruit ripening, flower senescence, and leaf and petal abscission, besides being an essential regulator of stress responses (Gamalero and Glick 2012; Groen and Whiteman 2014). It is biosynthesized from methionine via S-adenosyl-L-methionine and the cyclic nonprotein amino acid 1-aminocyclopropane-1-carboxylic acid. 1-Aminocyclopropane-1-carboxylic acid synthase converts S-adenosyl-L-methionine to 1-aminocyclopropane-1-carboxylic acid, while 1-aminocyclopropane-1-carboxylic acid oxidase catalyzes the conversion of 1-aminocyclopropane-1-carboxylic acid to ethylene. Abiotic stresses including low temperature and salinity alter endogenous ethylene levels in plants. Enhanced tolerance was accordingly achieved with higher ethylene concentrations (Shi et al. 2012). Ethylene also plays a major role in the defense response of plants to heat stress (Larkindale et al. 2005). Environmental stress induces ethylene accumulation which increases plant survival chances under these adverse conditions (Gamalero and Glick 2012). Ethylene has been proposed to function via modulation of gene expression considered as the effectors of ethylene signal (Klay et al. 2014). Ethylene in combination with other phytohormones such as jasmonate and salicylic acid often acts cooperatively. These are considered the main players involved in regulating plant defense against pests and pathogens (Li et al. 2013). The biosynthesis, transport, and accumulation of these hormones trigger a cascade of signaling pathways involved in plant defense (Matilla-Vazquez and Matilla 2014). As concluded by Yin et al. (2015), ethylene and abscisic acid seem to act synergistically or antagonistically to control plant growth and development.

Increase in ethylene under stress is of adaptive significance as it helps plants to cope with stress by reducing water loss through increased senescence of fruits/leaves and reduced growth. The magnitude of ethylene increase under stress depends upon growth stage, stress intensity, and stress duration (Upreti et al. 1998, 2000), and higher stress levels tend to reduce ethylene concentration. Yang et al. (2013) illustrated a key role for ethylene in salt tolerance by relating its ability to retain potassium rather than decreasing sodium in roots and shoots. Cold stress alters ethylene levels in plants and the enhanced ethylene level contributes in cold tolerance (Machacckova et al. 1989; Ciardi et al. 1997; Zhao et al. 2014). Zhao et al. (2009) in tomato suggested positive relationship between ethylene and freezing tolerance from the negative effects of 1-methylcyclopropene, an ethylene biosynthesis inhibitor on freezing tolerance. Further support for ethylene in cold tolerance is evident from the study of Lockhart (2013) that the ethylene biosynthesis inhibitor 1-methylcyclopropene reduces cold tolerance in tomato, whereas ethephon enhances cold tolerance in tomato. Hoffman et al. (1983) showed that abscisic acid interacts with ethylene metabolism. Ethylene exerts responses through modulation of gene expression function at transcriptional level by ethylene response factor, by regulating gene expression under abiotic stress conditions (Zhang et al. 2008a; Hussain et al. 2011). A number of cold-inducible genes have been identified, such as late embryogenesis abundant proteins and enzymes for sugar metabolism and fatty acid synthesis (Fowler and Thomashow 2002).

6.9.5 Gibberellins

The gibberellins are a large group of tetracyclic diterpenoid carboxylic acids, but only a few of them function as growth hormones in higher plants, predominant forms being GA1 and GA4 (Sponsel and Hedden 2004). The gibberellins show positive effects on seed germination, leaf expansion, stem elongation, flower and trichome initiation, and flower and fruit development (Yamaguchi 2008). They are essential for plants throughout their life cycle for growth-stimulatory functions. They also promote developmental phase transitions (Colebrook et al. 2014). Interestingly, there is increasing evidence for their vital roles in abiotic stress response and adaptation (Colebrook et al. 2014). Experiments have been performed to investigate the role of gibberellins in osmotic stress response in *Arabidopsis thaliana* seedlings (Skirycz et al. 2011; Claeys et al. 2012). Gibberellins are known to interact with all other phytohormones in numerous developmental and stimulus-response processes (Munteanu et al. 2014). The interactions between gibberellins and ethylene include both negative and positive mutual regulation depending on the tissue and signaling context (Munteanu et al. 2014). Radi et al. (2006) showed that the presoaking wheat seeds in gibberellic acid increased the germination potential especially at moderate salinization levels. Between gibberellin levels and the acquisition of stress protection, an intimate relationship exists (Vettak-korumakankav et al. 1999). In sugarcane, supplementing gibberellins as foliar application plays an important role on imparting salt tolerance in terms of enhancing nutrient uptake, as well as the morphological and physiological aspects (Shomeili et al. 2011). Ashraf et al. (2002) showed that gibberellin application increased the nutrient uptake, plant height, leaf area, and yield of wheat under saline conditions. Evidence also shows the involvement of gibberellin in relieving NaCl-induced growth inhibition in rice (Wen et al. 2010).

The precise mechanisms by which gibberellin is linked to stress tolerance are less understood. One possible mechanism is via its possible interactions with other phytohormones. In this context, the auxin is expected to promote gibberellin biosynthesis (Wolbang et al. 2004). Similarly, gibberellic acid application enhances the catabolism of abscisic acid (Gonai et al. 2004). Application of paclobutrazol and application of gibberellins to the dwarf phenotype reversed the inherent stress tolerance. Reversal of the dwarf phenotype by specific gibberellins suggests that the interconversion of gibberellins is compromised. Still and Pill (2003) found that foliar applications or seed priming with paclobutrazol is effective in improving drought tolerance in tomato seedlings. Likewise paclobutrazol improves plant water status in strawberry (Navarro et al. 2007) and peach (George and Nissen

1992). The treatment of rice plants with gibberellic acid during submergence promoted elongation growth and compromised survival, indicating that gibberellins negatively impact tolerance to prolonged submergence (Das et al. 2005). Conversely, treatment of non-tolerant cultivars with gibberellin biosynthesis inhibitor, paclobutrazol, restricts underwater elongation and enhanced submergence survival (Bahadur 2018).

6.9.6 Brassinosteroids

Brassinosteroids comprise a relatively new group of polyhydroxy steroidal plant hormones with strong growth and development-promoting potential. Brassinosteroids are polyhydroxylated steroidal phytohormones that are structurally related to animal steroid hormones and possess distinct growth-promoting properties (Bishop and Yokota 2001; Clouse and Sasse 1998). These are considered as phytohormones due to inheriting pleiotropic effects, which influence diverse range of developmental and physiological processes including the promotion of cell elongation and cell division, photomorphogenesis, rhizogenesis, senescence and abscission, xylem differentiation, seed germination, and fruit ripening (Clouse and Sasse 1998). They were first isolated and characterized in pollen of the rape plant (*Brassica napus*). More than 70 brassinosteroids have been isolated from plants. However, brassinolide, 28-homobrassinolide, and 24-epibrassinolide constitute the three most bioactive brassinosteroids and are widely used in physiological and experimental studies (Vardhini et al. 2006). They are present in almost every part of plants including pollen, flower buds, fruits, seeds, vascular cambium, leaves, shoots, and roots (Bajguz and Hayat 2009). They play a critical role in numerous developmental processes such as stem and root growth, floral initiation, and development of flowers and fruits (Bajguz and Hayat 2009). However, recent findings suggest stress impact mitigating roles of brassinosteroids and associated compounds in various plants subjected to various abiotic stresses. These abiotic stresses are high temperature (Janeczko et al. 2011), chilling (Wang et al. 2014), soil salinity (Abbas et al. 2013), light (Kurepin et al. 2012), drought (Mahesh et al. 2013), flooding (Liang and Liang 2009), metals/metalloids (Bajguz 2010), and organic pollutants (Ahammed et al. 2012). Research has shown tremendous potential of brassinosteroids and associated compounds in the modulation of components of antioxidant defense system in response to and to counteract the abiotic stress-induced oxidative burst, reviewed by Vardhini and Anjum (2015).

It has been suggested that brassinosteroids regulate stress response by a complex sequence of biochemical reactions, such as activation or suppression of key enzymatic reactions, induction of protein synthesis, and the production of various chemical defense compounds (Bajguz and Hayat 2009). Kagale et al. (2007) reported improved drought tolerance in *Brassica napus* seedlings treated with epibrassinolide, and the improved tolerance is due to reduction in reactive oxygen species, induction in antioxidative enzyme activities, and antioxidant contents (Zhang et al. 2008b; Li et al. 2012). Zhang et al. (2008b) also witnessed improvement in photosynthesis by regulating ribulose-1,5-bisphosphate carboxylase/oxygenase activity and sugar accumulation following brassinosteroid treatment in stressed plants.

Singh and Shono (2005) found epibrassinolide-treated tomato plants as more tolerant to high temperature than untreated plants as a result of high accumulation of heat shock proteins and improvement in photosynthetic efficiency. Epibrassinolide treatment to tomato plants prior to high-temperature exposure protects Rubisco enzyme and RuBP regeneration under heat stress in order to provide better protection against high-temperature stress (Ogweno et al. 2008). Chilling tolerance in plants is influenced by brassinosteroids. The report of Huang et al. (2006) showed that epibrassinolide application upregulated 17 proteins which were downregulated by chilling to confer chilling tolerance. Chilling tolerance is also attributed to increase in membrane permeability (Janeczko et al. 2007), high pigment accumulation, and upregulation in cold-related genes (Kagale et al. 2007) besides high activation of Rubisco and expression of photosynthetic genes (Xia et al. 2009). Thus, the physiological responses of brassinosteroids though are variable due to complexities in the molecular mechanism of their action; their potential in improving abiotic stress tolerance has immense utility in managing plant responses to stress.

6.9.7 Jasmonates

The cyclopentanone phytohormones derived from the metabolism of membrane fatty acids including primarily methyl jasmonate and its free acid jasmonic acid are collectively called jasmonates and are widespread in the plant kingdom. These multifunctional compounds are involved in crucial processes associated with plant development and survival including reproductive processes, flowering, fruiting, senescence, secondary metabolism, and direct and indirect defense responses (Seo et al. 2001; Fahad et al. 2015c). Jasmonate is the most abundant, best known, and best characterized of the plant growth hormones. In addition to developmental functions of plants, jasmonate activates plant defense responses to pathogenic attack as well as environmental stresses including drought, salinity, and low temperature (Pauwels et al. 2008; Seo et al. 2011). Jasmonates are vital signaling molecules induced by various environmental stresses including salinity (Pauwels et al. 2008), drought (Seo et al. 2011; Du et al. 2013b), and UV irradiation (Demkura et al. 2010). They have great potential to mitigate an array of threatening environmental stresses (Dar et al. 2015). The exogenous application of methyl jasmonate effectively reduced salinity stress symptoms in soybean seedlings (Yoon et al. 2009). Remarkably, endogenous levels of jasmonate increased in rice roots under salinity stress and reported to counteract the deleterious effects of salinity stress (Wang et al. 2001). Jasmonate applications alleviate heavy metal stress in plants by activating the antioxidant machinery (Yan et al. 2013). Methyl jasmonate confers tolerance in *Arabidopsis thaliana* plants against copper and cadmium stress via accumulation of phytochelatins (Maksymiec et al. 2007).

Jasmonates are biosynthesized by the sequential action of enzymes present in plastid, peroxisome, and cytoplasm (Feussner and Wasternack 2002). Jasmonate biosynthesis is initiated by the release of α-linolenic acid (18:3) from chloroplast membranes by phospholipase 1 to generate jasmonate substrate (Vick and Zimmerman 1983). Jasmonate goes through several biochemical modifications (Sembdner and Parthier 1993; Koch et al. 1997; Seo et al. 2001; Staswick and Tiryaki 2004; Swiatek et al. 2004). Jasmonate and methyl jasmonate have biological activity in plants (Wasternack 2007; Fonseca et al. 2009). Among various environmental perturbations, cold stress or low temperature stress limits plant performance and geographical distribution. Cold stress can be categorized into chilling (0–15 °C) and freezing that causes mayhem in tropical and subtropical plants by inducing chlorosis, necrosis, membrane damage, changes in cytoplasm viscosity, and changes in enzyme activities (Ruelland and Zachowski 2010). All these physiological and biochemical changes elicit a cascade of events that cause changes in gene expression pattern and protein products and thus in due course induce plant species to adopt stratagems to tolerate low non-freezing temperatures and complete their life cycle, an experience known as cold acclimation response. Tropical and subtropical fruits like mango, avocado, papaya, etc. exhibit symptoms due to chilling injuries such as browning discoloration and off-flavor in the fruit. Studies have specified the role of methyl jasmonate in alleviating chilling injury by inducing the production of cryoprotective agents, proteinase inhibitors, polyamines, abscisic acid, lower activity of lipoxygenases, and antioxidants (Wang and Buta 1994; González-Aguilar et al. 2000; Cao et al. 2009; Zhao et al. 2013). Role of methyl jasmonate in response to freezing tolerance has also been studied in rice seedlings. It has been reported that rather than treating rice with methyl jasmonate during or after stress imposition, methyl jasmonate treatment before chilling remarkably enhances the survival ratio of chilled rice seedlings (Lee et al. 1997). Furthermore, it was observed that methyl jasmonate maintained the well-watered status of chilled plants by preventing stomatal opening and enhancing hydrolytic conductivity.

6.9.8 Salicylic Acid

Salicylic acid is an endogenous plant growth regulator and reportedly has a wide range of metabolic and physiological responses that affect the growth and development of plants (Hayat et al. 2010). Salicylic acid an active phenolic compound plays an important role in providing resistance to plants against pathogens. It takes part in the defense mechanism of plants in response to abiotic stresses such as low- and high-temperature stress (Horváth et al. 2007a; Yuan and Lin 2008; Miura and Tada 2014). Salicylic acid is a plant hormone which plays a crucial role in the regulation of physiological and molecular mechanisms to acclimatize plants in extreme environmental conditions and is believed to have a role in plant's response to abiotic stress (Horváth et al. 2007b; Yuan and Lin 2008; Bari and Jones 2009; Okrent et al. 2009). Salicylic acid and other phenolic compounds (benzoic acid, acetyl salicylic acid, 5-sulfosalicylic acid) have been found to increase cold tolerance of plants when applied exogenously (Horváth et al. 2007b; Okrent et al. 2009). Salicylic acid is a vital signal molecule for modulating plant response to various abiotic stresses (Hayat et al. 2010). Salicylic acid-induced alleviation of cold stress has been reported in tomato (Meena et al. 2017; Orabi et al. 2015; Khan et al. 2015b), cucumber (Zhang et al. 2003), squash (Mady 2014), and watermelon (Sayyari et al. 2013). Salicylic acid-induced alleviation of low-temperature stress has been reported in vegetables. On the basis of different studies, it has been concluded that salicylic acid enhanced the low-temperature stress tolerance significantly and increased plant growth, photosynthetic pigments, accumulations of osmoprotectants, and activity of antioxidant enzymes. It also increased production and quality of vegetables under low-temperature stress (Meena and Kaur 2017).

Salicylic acid is a powerful tool in regulating cold stress tolerance and is involved in the regulation of pathogenesis-associated protein expression (Miura and Tada 2014). In addition to defense responses, it plays a vital role in the regulation of plant growth, ripening and development, as well as responses to abiotic stresses (Rivas-San and Plasencia 2011; Hara et al. 2012; Sharma and Laxmi 2015). The synthesis of salicylic acid occurs via two pathways: the isochorismate and the phenylalanine ammonia-lyase pathway. Of these, the major pathway is the isochorismate pathway in *Nicotiana benthamiana* (Uppalapati et al. 2007) and tomato (Catinot et al. 2008). An interesting and noteworthy general belief is that low concentrations of salicylic acid enhance the antioxidant capacity of plants, but high concentrations of salicylic acid cause cell death or susceptibility to abiotic stresses (Jumali et al. 2011). Most genes that respond positively to acute salicylic acid treatment are associated with stress and signaling pathways that eventually led to cell death. Salicylic acid consists of genes encoding chaperones, heat shock proteins, antioxidants, and genes involved in the biosynthesis of secondary metabolites, such as sinapyl alcohol dehydrogenase, cinnamyl alcohol dehydrogenase, and cytochrome P450 (Jumali et al. 2011).

Salicylic acid is involved in plant response to abiotic stresses such as drought (Miura et al. 2013), salinity (Fahad and Bano 2012; Khodary 2004), chilling (Yang et al. 2012a), and heat (Fayez and Bazaid 2003). Salicylic acid along with abscisic acid is involved in the regulation of drought response (Miura and Tada 2014). Drought stress induced a fivefold increase in the endogenous levels of salicylic acid in *Phillyrea angustifolia* (Munne-Bosch and Penuelas 2003). The salicylic acid content in barley roots was increased approximately twofold by water deficit (Bandurska and Stroinski 2005). The salicylic acid inducible genes PR1 and PR2 (pathogenesis-related genes) are induced by drought stress (Miura et al. 2013). However, the detailed molecular mechanisms and role of salicylic acid in abiotic stress tolerance remain largely unknown, and more comprehensive investigations are needed in this direction.

Several studies have been reported about various beneficial effects of salicylic acid on plants under abiotic stresses (Miura and Tada 2014; Pál et al. 2014). Sayyari et al. (2013) reported that foliar application of salicylic acid increased cold tolerance of cucumber. In tomato and bean plants, the lower concentration of salicylic acid and acetyl salicylic acid, at 0.1 and 0.5 mM, has been found more effective against low-temperature stress (Orabi et al. 2015; Senaratna et al. 2003). Acetyl salicylic acid is a close analog of salicylic acid and when applied exogenously is converted to salicylic acid spontaneously, having similar effects to salicylic acid in plant defense processes (Beckers and Spoel 2006; Hayat et al. 2007; Scotter et al. 2007). Orabi et al. (2015) reported that salicylic acid led to enhancement of endogenous growth regulators under low temperature. Salicylic acid enhanced plant growth and cell division via regulation of other hormones (Zarghami et al. 2014) and mitigating abiotic stresses by increasing the growth-regulating hormones like auxins and cytokinins (Shakirova et al. 2003). Moreover, the application of salicylic acid increases the content of auxin and abscisic acid and prevents the reduction of cytokinin under drought and salinity stress conditions, thereby producing a higher total biomass and seed vigor index (Shakirova et al. 2003; Kabiri et al. 2012).

Exogenous application of salicylic acid on *Hordeum vulgare* genotypes resulted in cold tolerance by enhancing antioxidant enzymes, ice nucleation activity, and the patterns of apoplastic proteins (Mutlu et al. 2013). Accumulation of endogenous free salicylic acid and glucosyl salicylic acid has been reported during chilling in *Arabidopsis* shoots, wheat, and grape berry (Scott et al. 2004; Wan et al. 2009; Kosová et al. 2012). However, it has been observed that concentration and duration of applied salicylic acid greatly influence its utility. High concentration and continual application of salicylic acid decreases the cold tolerance capacity of plants as observed in some *Arabidopsis* mutants in which salicylic acid is over-accumulated, exhibiting a dwarf phenotype and freezing sensitivity (Scott et al. 2004; Miura et al. 2010). Siboza et al. (2014) reported that combined treatment of salicylic acid and methyl jasmonate reduced accumulation of reactive oxygen species and lipid peroxidation and also increased chilling tolerance in lemon during cold storage by increasing the synthesis of total phenolics and phenylalanine ammonia lyase and inhibiting the activity of polyphenol oxidase and peroxidase. Induction of phenylalanine ammonia lyase activity has been considered a good marker of chilling injury (Martínez-Téllez and Lafuente 1993; Sanchez-Ballesta et al. 2000). Kim et al. (2013) reported upregulation of cold-inducible genes by *CAMTA* TFs to the tune of 15%. It has been reported that salicylic acid does not contribute to freezing tolerance but genes induced by salicylic acid at low temperature increase the resistance to pathogen attack.

6.9.9 Strigolactones

Strigolactones were initially known as host-derived germination stimulants for parasitic weeds from the genera *Striga* and *Orobanche* and *Phelipanche* species (Xie et al. 2010; Ruyter-Spira et al. 2013). Later on, strigolactones were found to be host-detection and hyphal-branching signals for arbuscular mycorrhizal fungi (Akiyama et al. 2005; Besserer et al. 2006). In addition to their original, dual role as signaling molecules in the rhizosphere, strigolactones were further demonstrated to be a new class of branch-inhibiting phytohormones, which regulate the overall architecture of land plants (Gomez-Roldan et al. 2008; Umehara et al. 2008). Strigolactones constitute a small class of carotenoid-derived compounds, and by now more than 20 strigolactone- and strigolactone-like compounds have been identified so far in the root exudates of several plant species. They all share a conserved tricyclic lactone structure containing rings referred to as ABC rings, linked via an enol ether bridge to an invariable α,β-unsaturated furanone moiety named D ring. Several types of strigolactones can be synthesized by single plant species, whereas mixtures of different types and quantities of strigolactone molecules are produced in intraspecific varieties (Xie et al. 2010; Yoneyama et al. 2013). Although they are produced and exuded in small amounts primarily in roots, other plant parts can also synthesize them (Koltai and Beveridge 2013). A comparative study involving wild-type and mutant *Arabidopsis* plants indicates their role in the development of root system architecture (Ruyter-Spira et al. 2011). Application of GR24, a synthetic and biologically active strigolactone (Gomez-Roldan et al. 2008; Umehara et al. 2008), repressed lateral root formation in wild-type seedlings and strigolactone synthesis mutants (max3 and max4) but not in the strigolactone-response mutant (max2), suggesting the MAX2-dependent negative effect of strigolactone on lateral root formation (Ruyter-Spira et al. 2011; Kapulnik et al. 2011). Apparently, strigolactones are involved in plant responses to environmental stimuli from their early evolution. In higher plants, they participate in both shoot and root architecture in response to nutritional conditions (Kapulnik and Koltai 2014). Strigolactones also act as signaling molecules for plant interactions with microbes. They stimulate nodulation in the legume-rhizobium interaction process (Soto et al. 2010; Foo and Davies 2011). Overall, it can be concluded that strigolactones constitute an important group of signaling molecules and are key regulators of plants' developmental adaptations to changing environmental conditions. They have the potential to be used in agriculture for various purposes including as inducers of suicidal seed germination of parasitic plants (Vurro and Yoneyama 2012).

Detailed analysis of the mutants revealed roles of strigolactones in root growth and development, leaf shape and senescence, internode elongation, secondary growth, and drought stress responses (Brewer et al. 2013; de Saint Germain et al. 2013; Ha et al. 2014). In addition to these major roles, strigolactones are also involved in signaling pathways in promoting seed germination in crop plants (Pepperman and Bradow 1988). In lower plants strigolactones promote rhizoid elongation in moss, liverworts, and stoneworts of which only liverworts show mycorrhizal symbiosis (Delaux et al. 2012). Role for strigolactones has also been recognized during various abiotic and biotic stress conditions making them suitable target for generating genetically engineered crop plants with improved yield (Pandey et al. 2016).

6.9.10 Polyamines

Polyamines are important growth regulatory polycationic molecules known to be involved in a wide range of developmental events including organogenesis, embryogenesis, floral initiation and development, senescence, fruit development and ripening, and root development (Galston and Kaur-Sawhney 1995; Kumar et al. 1997; Kakkar et al. 2000). These are biosynthesized by decarboxylation of amino acids, ornithine, or arginine in the reaction catalyzed by enzymes ornithine decarboxylase and arginine decarboxylase, leading to the formation of putrescine, which by subsequent additions of aminopropyl moiety produces spermidine and spermine, respectively. The dynamics of polyamine metabolism is complex due to coexistence of degradation and conjugation pathways of transport and uptake mechanisms (Martin-Tanguy 2001). Molecular studies reveal that the polyamines are involved in signal transduction pathway through effects on calcium fluxes (Thomas et al. 1993) and interaction with certain transcriptional factors (Wang et al. 1999) and protein kinases (Datta et al. 1987). Polyamines are also found to play an important role in conferring tolerance against drought, salinity, flooding, heat stress, and chilling stress in plants (Gill and Tuteja 2010). The stress-induced increase in polyamines is a reflection of upward regulation of enzymes associated with their biosynthesis and release from polyamine conjugates (Gupta et al. 2013). In general, plants experiencing abiotic stresses tend to increase polyamine levels, and increase in polyamines helps in the regulation of plant tolerance to stress. Generally, a polyamine response due to stress depends on species, besides type and concentration of polyamines (Ali et al. 2009). Application of exogenous polyamine is efficient in manipulating the levels of endogenous polyamine during stress, and the stress protection effect is rendered through the maintenance of membrane integrity, regulation of gene expression for the synthesis of osmotically active solutes, reduction in reactive oxygen species production, and controlling of the accumulation of Na + and Cl^- ion in different organs.

Upreti and Murti (2005) reported cultivar difference in polyamine content changes in French bean cultivars under drought stress conditions. Moreover, the stress response on individual polyamine varied with stress duration. The putrescine which increased initially with stress declined under severe stress regimes. In contrast, spermidine levels consistently declined and spermine levels progressively increased with stress. The spermine level under stress was related with abscisic acid and stress tolerance of cultivars. High-temperature stress alters polyamine balance, which helps in providing thermotolerance to plant by stabilizing membrane structural integrity and functionality (Edreva et al. 1998).

6.10 Strategies for Crop Improvement through Growth Hormones under Low-Temperature Stress

Unlike other environmental stresses such as salinity, drought, and high temperatures, studies on the management of crops for chilling tolerance are relatively scarce. Exogenous application of glycinebetaine and proline (Ashraf and Foolad 2007) and hormones (Basra et al. 2006) is well documented, but there is still room for improvement. Plant growth and development are precisely regulated by coordination of both exogenous (environmental) and endogenous signals (Peleg and Blumwald 2011; Santner et al. 2009). External environmental factors such as light, temperature, moisture, and atmospheric carbon dioxide at optimum levels are essential for normal metabolic processes in plants. Among those various environmental factors, temperature is of great significance in the regulation of plant phenological development (Bahuguna and Jagadish 2015). Phytohormones are the endogenous messenger molecules that play a critical role in plant growth, development, and responses to various stresses. A number of previous reports provided solid evidence that phytohormones are actively involved in the response of plants to temperature stress (Ahammed et al. 2014; Peleg and Blumwald 2011; Xia et al. 2015). Moreover, majority of the phytohormones provide physiological protection against temperature stress. In addition to the individual role of a plant hormone, crosstalk between multiple hormones precisely coordinates plant defense response to heat stress. Although significant advancement has been achieved in the molecular mechanisms of heat perception, the mechanisms that control phytohormone-mediated responses to heat stress still remain largely unknown (Ahammed et al. 2016).

Mechanisms of tolerance or sensitivity to a particular stress depend on the genetic makeup of the plant species and the initiation of physiological processes within a plant or plant part (Dubey 1999). To minimize losses due to chilling, development of special techniques or modification of plant structures according to environmental conditions favorable to the plant is imperative. These techniques include developing resistance in the plant species itself or applying chemicals or growth regulators, which help to inhibit abnormal physiological processes within the plant. Studies show that the mechanism of chilling tolerance in plants can be initiated by exogenous application of chemicals, salt, or growth regulators.

6.10.1 Compatible Organic Solutes

One of the most common stress responses in plants is overproduction of different types of compatible organic solutes (Serraj and Sinclair 2002). Compatible solutes are low-molecular-weight, highly soluble compounds that are usually nontoxic at higher cellular concentrations. Generally, they protect plants from stress through different courses including contribution to cellular osmotic adjustment, detoxification of reactive oxygen species, protection of membrane integrity, and stabilization of enzymes/proteins (Bohnert and Jensen 1996). In response to chilling stress, tolerant plants synthesize certain molecules with cryoprotective roles (Sakamoto and Murata 2002). These compounds are low in molecular weight, highly soluble, and generally nontoxic even at high cytosolic concentrations. Other organic compounds such as alcohols increase chilling tolerance of sensitive plants by increasing fluidity of their membranes, thereby retarding a phase transition at chilling temperatures. These compounds protect plants from stress through different means such as contribution toward osmotic adjustment, detoxification of reactive oxygen species, and stabilization of membranes and native structures of enzymes and proteins (Ashraf and Foolad 2007).

Glycine betaine is a quaternary ammonium compound and has been proposed to be an effective cryoprotectant as it protects the cold-stressed plants from photoinhibition-induced damage to photosynthetic machinery (Giri 2011) by protecting the various components of photosystem II (Kurepin et al. 2017) and also as a molecular chaperone. Glycine betaine is one of the several such compatible solutes that has osmoprotection function and is known to protect proteins and enzyme activities and even stabilize membranes during freezing (Rhodes and Hanson 1993). It may play an important role in inducing freezing tolerance during the cold acclimation process. The accumulation of endogenous glycinebetaine was closely related to development of freezing tolerance. Accumulation of glycinebetaine may play a role in the induction of freezing tolerance during cold acclimation. The cryoprotective effects of glycinebetaine appear to come from its compatibility with macromolecular structure and function. It has been suggested that glycinebetaine can help stabilize protein tertiary structure and prevent or even reverse the disruption of the tertiary structure of proteins caused by noncompatible solutes (Bateman et al. 1992). Exogenously supplied glycine betaine improved the performance of tomato plants by Karabudak et al. (2014). Similarly, glycine betaine-supplemented sweet pepper plants also performed better than untreated one under cold stress (Wang et al. 2016) which corroborated similar reports on wheat (Zhang and Huang 2010) and other grain crops (Thakur et al. 2010).

Trehalose, also known as tremalose or mycose, a nonreducing disaccharide, is widely spread in biological systems, bacteria, yeast, fungi, lower and higher plants, as well as insects and invertebrates (Elbein et al. 2003), and its function is associated with tolerance against multiple abiotic stresses. It is an alpha-linked disaccharide and has been proposed to be playing protective roles under various abiotic stresses including cold and freezing (Iordachescu and Imai 2011). Trehalose has multiple functions, and some of them are species specific. In microorganisms trehalose appears to act as an energy source, during certain stages of development such as spore germination (Elbein et al. 2003). In anhydrobiotic organisms, trehalose is known to accumulate to high concentrations to survive complete dehydration (Drennan et al. 1993), by preserving the membranes during drought period (Crowe et al. 1984). In *Escherichia coli*, trehalose protects against cold stress, presumably by stabilizing cell membranes and preventing protein denaturation, whereas in yeast it plays a role in osmotic (Hounsa et al. 1998), heat, and desiccation tolerance (Hottiger et al. 1987), and it may act as a free radical scavenger (Benaroudj et al. 2001). There are numerous reports regarding trehalose levels going up under low temperatures in *Arabidopsis* (Veyres et al. 2008) and rice (Li et al. 2011). Thus, exogenously supplied trehalose can well protect not only cold-stressed plants as proved in *Arabidopsis* (Bae et al. 2005) but also yeast (Hirasawa et al. 2001) and even *Cryptococcus laurentii and Rhodotorula glutinis (*Li et al. 2008*)*. The protective effects of trehalose have been credited to its membrane stabilizing ability and depression in phase transition temperature of biomembranes and remain amorphous even under completely dehydrated conditions (Sundaramurthi et al. 2010).

Proline has been suggested to play multiple roles in plant stress tolerance. Proline is a well-established protective molecule and has been reported to impart protection against cold stress to many plants by not only maintaining its osmolarity but also acting as molecular chaperone, thereby stabilizing the structure and hence functioning of crucial proteins and enzymes (Diehl et al. 2013). It also protects the plant by maintaining the double-stranded structure of genetic material, i.e., DNA (Law and Jacobsen 2010), and by upregulating the antioxidative stress machinery (Rejeb et al. 2014).Thus, if exogenously supplied, proline can well mitigate the harmful effects of cold stress as reported by Kumar et al. (2011) in cold-stressed chickpea. Proline-supplemented chickpea plants exhibited improved pollen viability, pollen germination, and pollen tube growth along with better flower and pod retention. Similarly, proline conferred appreciable cold tolerance to rapeseed plants as well (Jonytiene et al. 2012).

Proline also acts as a mediator of osmotic adjustment, a stabilizer of proteins and membranes (Arakawa and Timasheff 1983; Rudolph et al. 1986), an inducer of osmotic stress-related genes, a scavenger of reactive oxygen species (Saradhi et al. 1995), and a readily available source of nitrogen and carbon (Brugière et al. 1999). In higher plants, proline can be synthesized via the glutamate and ornithine pathways (Delauney and Verma 1993). The former is considered a major pathway under osmotic stress and nitrogen limitation (Peng et al. 1996), whereas the latter is a route when the level of available nitrogen is high (Delauney and Verma 1993). Glutamine synthetase, a key enzyme involved in the assimilation of inorganic nitrogen into organic forms, also appears to play a major role in regulating proline accumulation under osmotic stress (Brugière et al. 1999). Most chilling-sensitive plants, which also accumulate proline under chilling, did not acquire improved cold tolerance (Chu et al. 1978). It appears that proline may have potential for alleviating chilling injury in chilling-sensitive plants, but for some reason this system failed under natural conditions.

The protective molecules don't seem to accumulate in sufficient amounts, so the exogenous treatment has been proposed, or the plants may be even genetically modified to over-accumulate specific molecule which in turn can well impart cold tolerance to the targeted plant. Numerous such transgenic plants have been raised and summarized (Table 6.1).

Transgenic plants	Biomolecules accumulated	References
Arabidopsis thaliana	Proline	Zhu et al. (2012)
Oryza sativa	Proline	Yang et al. (2012b)
Nicotiana tabacum	Proline	Pan et al. (2012)
Brassica napus	Proline	Gomaa et al. (2012)
Brassica campestris	Glycine betaine	Wang et al. (2010)
Triticum aestivum	Glycine betaine	Zhang and Huang (2010)
Ipomoea batatas	Glycine betaine	Fan et al. (2012)
Xenopus laevis	Trehalose	Li et al. (2009)

(continued)

Transgenic plants	Biomolecules accumulated	References
Oryza sativus	Trehalose	Hossain et al. (2010)
Nicotiana tabacum	Trehalose	Yang et al. (2010)
Hordeum vulgare	Trehalose	Ligaba et al. (2011)
Arabidopsis thaliana	Trehalose	Kang et al. (2011)
Oryza sativa	Trehalose	Li et al. (2011)

6.10.2 Plant Growth Regulators

Plant growth hormones have a dynamic role for increasing the ability of plants to acclimatize against abiotic stress conditions. Auxins, gibberellins, cytokinins, ethylene, abscisic acid, salicylic acid, jasmonates, and brassinosteroids have been widely studied in plants under different abiotic stresses. These plant hormones are being used efficiently under stress conditions for enhancing crop production (Hussain et al. 2018). Plants normally synthesize various types of phytohormones under individual and concurrent abiotic stresses depending upon the strength of their defense mechanism (Nadeem et al. 2016). Abscisic acid is synthesized in response to chilling as well as under soil water deficit and contributes in stress tolerance of crop plants (Liu et al. 1998). Abscisic acid, jasmonate, and salicylic acid often activate phosphoprotein cascade pathways that led to expression of genes associated with cold stress tolerance in plants (Kolaksazov et al. 2013). Greater chilling tolerance has been recorded in abscisic acid-treated maize (Xin and Li 1992), and levels of abscisic acid are increased in chilling-tolerant genotypes (Farooq et al. 2009c). Khan et al. (2015b) reported that application of brassinosteroids led to considerable improvements in plant tolerance against chilling stress in terms of plant growth, photosynthesis, and antioxidant system in tomato plants. Exogenous gibberellin application breaks the rosette by rapid enlargement of differentiated tissues under cold stress (Khan et al. 2017). Similarly, jasmonate has been reported to trigger the plant defense responses to abiotic stresses including chilling (Pauwels et al. 2009; Seo et al. 2011). Exogenous application of methyl jasmonate improved chilling tolerance in wheat by modulating antioxidant defense system and production of soluble protein content (Qi et al. 2005). The relationship between enhanced ethylene levels and chilling tolerance in crop plants has also been reported by Chu and Lee (1989) and Zhao et al. (2014). Salicylic acid is also considered as potential chemical compound for drought and chilling tolerance in several plant species (Farooq et al. 2009a, c). Salicylic acid application was found to be effective in alleviating the chilling (Miura and Tada 2014) and freezing injury (Tasgin et al. 2003). In maize, the harmful effects of low-temperature stress were decreased by the application of salicylic acid (Janda et al. 1999; Wang et al. 2012).

Endogenous abscisic acid levels in chilling-sensitive plants increase upon exposure to chilling stress (Anderson et al. 1994b; Janowiak and Dörffling 1996). Increased chilling tolerance has been observed even in abscisic acid-treated cell suspension cultures of maize (Xin and Li 1992). The protective function of abscisic acid seems to be manifold: stabilization of membranes and protection against oxidative stress (Prasad et al. 1994b); stabilization of water status by increasing root hydraulic conductivity and closing stomata (Markhart 1984); induction of abscisic acid-responsive genes (Anderson et al. 1994b); change of calcium signature in subcellular localization; and further mechanisms (Li et al. 1997). Chilling-tolerant maize genotypes accumulate abscisic acid faster and in higher amounts than chilling-sensitive genotypes when exposed to chilling stress (Ristic et al. 1998; Janowiak et al. 2002). Among others, there are relatively few reports regarding polyamine involvement in chilling stress (Bouchereau et al. 1999). During chilling, spermidine content in leaves markedly increased in cold-tolerant cucumber cultivars, but not in sensitive ones, while putrescine and spermine did not change. Numerous studies reported that salicylic acid improves plant resistance to various abiotic stresses including ultraviolet light, drought, salt, and temperature extremes (Yalpani et al. 1994; He et al. 2005b; Ogawa et al. 2005; Horváth et al. 2007a; Farooq et al. 2008c).

Salicylic acid plays crucial roles in response to external stimulation and by activating defense system in plants. Activation of phospholipase D is an early response to low temperature, involved in the accumulation of free salicylic acid and the development of thermotolerance induced by low-temperature acclimation in grape berries (Wan et al. 2009). Salicylic acid acts as an endogenous phytohormone from phenolic compound family, having the capacity of antioxidant protection system, and regulates different physiological and biochemical processes in plant such as activity of photosynthesis pigments, maintenance of tissue water contents and reduced membrane permeability, adjustment of the activity of antioxidant enzymes, stomatal conductivity, and tolerance to various abiotic stresses (Hayat et al. 2010; Kabiri et al. 2012; Farooq et al. 2008c; Carvalho et al. 2011). Furthermore, salicylic acid treatment maintains indole-3-acetic acid and cytokinin levels in the plant tissues, which enhance the cell division and dry weight as reported by Sakhabutdinova et al. (2003). Miura and Tada (2014) observed that the effects of salicylic acid on the physiological and biochemical processes of plants depend on the concentration and period of application, type of plant, the stage of growth, and environmental conditions (Orabi et al. 2015). Several workers have reported that low concentration of salicylic acid may increase the activation of antioxidants (Farooq et al. 2008a) and tolerance to various abiotic stresses; similar reports have been made by Senaratna et al. (2003) that treatment at low concentrations salicylic acid of (0.1 and 0.5 mM) promoted tolerance to chilling stress in bean and tomato. But high concentrations of salicylic acid may cause negative results or susceptibility to abiotic stresses (Hara et al. 2012). The exogenous application of salicylic acid (0.5 mM) enhanced the low-temperature stress tolerance of cucumber (Kang and Saltveit 2002) and potato (Mora-Herrera et al. 2005). Similarly, Kang et al. (2003b) reported that the foliar application of salicylic acid at 0.5 mM on the leaves of banana enhanced the chilling tolerance.

6.10.3 Exogenous Use of Organic Compounds

Like many other abiotic stresses, chilling results in a range of morphological and biochemical changes in growing plants. Many strategies have been used to induce chilling tolerance in plants, some of which are detailed below. Unlike other environmental stresses such as salinity, drought, and high temperatures, studies on the management of crops for chilling tolerance are relatively scarce. Exogenous application of glycinebetaine and proline (Ashraf and Foolad 2007) and hormones (Basra et al. 2006) is well documented, but there is still room for improvement. Certain maize genotypes, including many inbreds, are unable to synthesize glycinebetaine because they are defective in the first step of betaine biosynthesis (Rhodes et al. 1989). In those that can synthesize glycinebetaine, synthetic capacity is low. For example, stressed maize plants only accumulated 2–5 mmol/g fresh weight glycinebetaine (Rhodes et al. 1987, 1989), which was about five- to tenfold lower than stress-tolerant plants. Glycinebetaine is one of the several such compatible compounds that have a cryoprotective role. It helps to stabilize protein tertiary structure and prevent or reverse disruption of the tertiary structure of proteins caused by noncompatible solutes (Bateman et al. 1992) and even stabilizes membranes during freezing (Rhodes and Hanson 1993). Lower concentrations (below 100 mM) of glycinebetaine are more effective in preventing inactivation of thylakoids during freezing (Coughlan and Heber 1982).

Proline application experiments indicate that an intracellular proline level higher than 2 mmol/g fresh weight before chilling is needed to reduce chilling-enhanced lipid peroxidation and improve chilling tolerance. Total proline accumulation in cells treated with and without abscisic acid during chilling was potentially sufficient to improve chilling tolerance, but proline leakage rendered control cells unable to benefit from endogenous synthesis of proline in relation to the alleviation of chilling injury. Proline participation in improving chilling tolerance in abscisic acid-treated maize cells was evident from the following: (1) the inhibition of proline accumulation by L-methionine-D,L-sulfoximine (an inhibitor of glutamine synthetase)-reduced abscisic acid improved chilling tolerance, and (2) the addition of glutamine to the medium prevented L-methionine-D,L-sulfoximine-induced reduction in chilling tolerance (Farooq et al. 2009c).

The level of putrescine and spermidine contents inside the plant increased, while spermine contents decreased, which were helpful in chilling tolerance (Mónika et al. 2001). Polyamine pre-treatment of a cold-sensitive cultivar alleviated chilling injury in different species (Groppa and Benavides 2008) including maize (Cao et al. 2008). Salicylic acid is another important compound for chilling tolerance in several plant species including maize. It plays an important role in a complex signaling pathway induced by many biotic and abiotic stresses. Farooq et al. (2008c) reported that seed priming with salicylic acid improved chilling tolerance in hybrid maize. Exogenous salicylic acid increased chilling tolerance of maize plants (Janda et al. 1999). Exposing seedling radicles to aqueous solutions of 0.5 m salicylic acid for 24 h before chilling at 2.5 °C for 1–4 days reduced chilling-induced increases in electrolyte leakage from maize and rice leaves, and cucumber hypocotyls, but not from their radicles. Salicylic acid treatments that induced chilling tolerance in the aerial portion of seedlings did not induce chilling tolerance in radicles, even though salicylic acid treatments were applied to radicles. A direct physiological effect of salicylic acid is the alteration of antioxidant enzyme activities in vivo. Certain enzymes (glutathione peroxidase and glutathione reductase) were activated by salicylic acid treatment, while catalase was inhibited (Janda et al. 1999). Catalase seems to be a key enzyme in salicylic acid-induced stress tolerance. In maize scutella, 1 mM salicylic acid caused a slight inhibition (9%) in catalase activity, while 5 mM caused a 35% decrease (Guan and Scandalios 1995). Among other effects, 24 h hydroponic treatment of 2-week-old maize plants with 0.5 mM salicylic acid increased tolerance to chilling-induced increases in electrolyte leakage and changes in chlorophyll fluorescence (Janda et al. 1999).

Production of abscisic acid in leaves has been related to chilling tolerance in hybrid maize (Aroca et al. 2003). The application of abscisic acid to chilling-sensitive plants before, during, or even shortly after a low-temperature treatment has been shown to protect from chilling injury (Lee et al. 1993; Anderson et al. 1994b; Prasad et al. 1994b). When chilling-sensitive plants are exposed to chilling temperatures, endogenous abscisic acid levels increase (Daie et al. 1981). The comparative effectiveness of abscisic acid and calcium chloride during cold stress in maize in terms of antioxidant enzyme activity suggested that abscisic acid at lower temperatures was more protective against cold stress (Khorshidi and Nojavan 2006). Likewise, even abscisic acid-treated cell suspension cultures of maize displayed increased chilling tolerance (Xin and Li 1992). The protective function of abscisic acid against chilling stress seems to be manifold and may include stabilization of membranes and protection against oxidative stress (Prasad et al. 1994b); stabilization of water status by increasing root hydraulic conductivity and closing stomata (Markhart 1984); induction of abscisic acid-responsive genes (Anderson et al. 1994b); changes in calcium signature in subcellular localization; and further mechanisms (Li et al. 1997).

6.10.4 Agronomic Management

There are postharvest techniques that improve germination or seedling growth or facilitate delivery of seeds and other material at the time of sowing. This term is used in industry, and more recent additions to the scientific literature describe seed treatments after harvest (Taylor et al. 1998; Farooq et al. 2006a, b, c, d, 2008b; Wahid et al. 2008). The aim of these treatments is to reduce time between planting and emergence to protect seeds from biotic and abiotic stresses during seedling establishment and to synchronize emergence, resulting in a uniform stand and improved yield.

Maize seed treatment with 100 mg/L glycinebetaine significantly improved germination rate, root and shoot lengths, seedling fresh and dry weights, leaf and root scores, relative water content, soluble sugars, a-amylase activity, and antioxidants compared with untreated seeds under optimal and stress conditions (Farooq et al. 2008b). Khorshidi and Nojavan (2006) compared the efficacy of abscisic acid and calcium chloride-treated seeds against cold stress in maize on the basis of antioxidant activities. In this case, the role of abscisic acid at lower temperatures was more protective against cold stress than calcium chloride in maize. In another study, treatment of seeds with osmotic solution (2.5% potassium chloride), plant growth regulators (indole acetic acid, cytokinin, ethephon, and gibberellic acid), and water suggested that only one treatment of gibberellic acid improved seedling vigor and grain yield compared with the control treatment (Subedi and Ma 2005). Use of 3 mM aminotriazole for 18 h inhibited catalase activity and initiated oxidative stress in maize seedlings at low temperatures. This was due to a twofold increase in protein carbonyl content, an indication of oxidative damage, in non-acclimated and aminotriazole-treated acclimated seedlings, suggesting that catalase and other antioxidant enzymes play a major role in inducing chilling tolerance in preemergent maize seedlings (Prasad 1997).

Seed priming with $CaCl_2$ and KCl has also been found to improve chilling tolerance in different plant species including maize (Farooq et al. 2008a, d, e). Osmopriming of maize seeds using polyethylene glycol or different potassium salts accelerated germination when chilled (10 °C). Seeds primed in solutions of either 2.5% K_2HPO_4 or 2.5% K_2HPO_4 + KNO_3 were more chilling tolerant than untreated seeds, and the effect was largely retained after seeds were dried. Embryo phospholipid fractions and sterols increased during salt priming, and the proportion of phospholipid which was diphosphatidylglycerol also increased. Increased diphosphatidylglycerol content of primed embryos was suggested to be due to enhanced internal organization of mitochondrial membranes, while the benefit of osmotic priming may be partly due to an increased potential for adenine triphosphate accumulation (Basra et al. 1988).

Sowing time may be changed to avoid the chilling stress depending upon the local environment. For example, to avoid an early-season stress, sowing may be delayed, while for the terminal chilling stress, early sowing may be preferred. Light and tactical application of irrigation may also aid to alleviate the adverse effects of chilling stress in maize. Mulch is a protective covering placed around plants to heat or cool the root zone, to prevent the evaporation of moisture, limit the growth of weeds, and prevent winter damage in perennial/biennial crops. Different organic and inorganic mulches can be used to save the maize plant from chilling injury. Soil management practices have the potential to change the characteristics of the soil surface, to create a favorable thermal environment for plants. Several studies have reported that the soil thermal conditions under mulch are different from those of non-mulched soil. Soil temperature increases under mulched conditions compared with unmulched soils during chilling stress (Bristow 1988). Ramakrishnan et al. (2006) reported that polythene mulch may increase the soil temperature by ~6 °C at 0.05 m depth and by 4 °C at 0.10 m depth.

1-Naphthaleneacetic Acid

1-Naphthaleneacetic acid is a synthetic plant growth hormone in the auxin family and is an ingredient in many commercial plant rooting horticultural products; it is a rooting agent and used for the vegetative propagation of plants from stem and leaf cuttings. It is also used for plant tissue culture.

The hormone 1-naphthaleneacetic acid does not occur naturally, and, like all auxins, is toxic to plants at high concentrations. 1-Naphthaleneacetic acid is widely used in agriculture for various purposes. It is considered to be only slightly toxic, but when at higher concentrations, it can be toxic to animals. This was shown when tested on rats via oral ingestion at 1000–5900 mg/kg. 1-Naphthaleneacetic acid has been shown to greatly increase cellulose fiber formation in plants when paired with gibberellic acid. Because it is in the auxin family, it has also been understood to prevent premature dropping and thinning of fruits from stems. It is applied after blossom fertilization. Increased amounts can actually have negative effects, however, and cause growth inhibition to the development of plant crops. It has been used on many different crops including apples, olives, oranges, potatoes, and various other hanging fruits. In order for it to obtain its desired effects, it must be applied in concentrations ranging from 20 to 100 μg/ml (Navalón et al. 1997). 1-Naphthaleneacetic acid present in the environment undergoes oxidation reactions with hydroxyl radicals and sulfate radicals. In micropropagation of various plants, 1-naphthaleneacetic acid is typically added to a medium containing nutrients essential to the plants' survival. It is added to help induce root formation in various plant types. It can also be applied by spraying it onto plants and which is typical in agricultural use. It is prohibited in many areas to use it in high concentrations due to the health concerns toward humans and other animals

References

Abbas S, Latif HH, Elsherbiny EA (2013) Effect of 24-epibrassinolide on the physiological and genetic changes on two varieties of pepper under salt stress conditions. Pak J Bot 45(4): 1273–1284

Abdelkader AF, Hassanein RA, Ali H (2012) Studies on effects of salicylic acid and thiourea on biochemical activities and yield production in wheat (*Triticum aestivum* var. Gimaza 9) plants grown under drought stress. African J Biotechnol 11(64):12728–12739

Achard P, Gong F, Cheminant S, Alioua M, Hedden P, Genschik P (2008) The cold-inducible CBF1 factor-dependent signaling pathway modulates the accumulation of the growth-repressing DELLA proteins via its effect on gibberellin metabolism. Plant Cell 20(8):2117–2129

Achard P, Gusti A, Cheminant S, Alioua M, Dhondt S, Coppens F, Beemster GT, Genschik P (2009) Gibberellin signaling controls cell proliferation rate in Arabidopsis. Curr Biol 19(14): 1188–1193

Adato I, Gazit S (1974) Water-deficit, ethylene production, and ripening in avocado fruits. Plant Physiol 53(1):45–46

Agarwal PK, Agarwal P, Reddy MK, Sopory SK (2006) Role of DREB transcription factors in abiotic and biotic stress tolerance in plants. Plant Cell Rep 25(12):1263–1274

Aghdam MS, Mohammadkhani N (2014) Enhancement of chilling stress tolerance of tomato fruit by postharvest brassinolide treatment. Food Bioprocess Technol 7(3):909–914

Ahammed G, Xia X, Li X, Shi K, Yu J, Zhou Y (2014) Role of brassinosteroid in plant adaptation to abiotic stresses and its interplay with other hormones. Curr Protein Pept Sci 16(5):462–473

Ahammed GJ, Li X, Zhou J, Zhou YH, Yu JQ (2016) Role of plant hormones in plant adaptation to heat stress. In: Plant hormones under challenging environmental factors, chapter 1. Springer Science + Buiseness Media, Dordrecht, pp 1–21

Ahammed GJ, Yuan HL, Ogweno JO, Zhou YH, Xia XJ, Mao WH, Shi K, Yu JQ (2012) Brassinosteroid alleviates phenanthrene and pyrene phytotoxicity by increasing detoxification activity and photosynthesis in tomato. Chemosphere 86(5):546–555

Aina O, Quesenberry K, Gallo M (2012) Thidiazuron induced tissue culture regeneration from quartered seed explants of Arachis paraguariensis. Crop Sci 52(3):1076–1083

Airaki M, Leterrier M, Mateos RM, Valderrama R, Chaki M, Barroso JB, Del Rio LA, Palma JM, Corpas FJ (2011) Metabolism of reactive oxygen species and reactive nitrogen species in pepper (*Capsicum annuum* L.) plants under low temperature stress. Plant Cell Environ 35(2):281–295

Akiyama K, Matsuzaki K, Hayashi H (2005) Plant sesquiterpenes induce hyphal branching in arbuscular mycorrhizal fungi. Nature 435(7043):824–827

Akladious SA (2014) Influence of thiourea application on some physiological and molecular criteria of sunflower (*Helianthus annuus L.*) plants under conditions of heat stress. Protoplasma 251(3): 625–638

Alam I, Sharmin SA, Naher K, Alam J, Anisuzzaman M, Alam MF (2010) Effect of growth regulators on meristem culture and plantlet establishment in sweet potato (*Ipomoea batatas* L Lam.). Plant Omics 3(2):35

Alcázar R, Altabella T, Marco F, Bortolotti C, Reymond M, Koncz C, Tiburcio AF (2010) Polyamines: molecules with regulatory functions in plant abiotic stress tolerance. Planta 231(6):1237–1249

Aldasoro JJ, Matilla A, Nicolás G (1981) Effect of ABA, fusicoccin and thiourea on germination and K+ and glucose uptake in chick-pea seeds at different temperatures. Physiol Plant 53(2): 139–145

Ali RM, Abbas HM, Kamal RK (2009) The effects of treatment with polyamines on dry matter and some metabolites in salinity-stressed chamomile and sweet marjoram seedlings. Plant Soil Environ 55(11):477–483

Alizadeh M, Singh SK, Patel VB (2012) Comparative performance of *in vitro* multiplication in four grape (*Vitis* spp.) rootstock genotypes. Int J Plant Prod 4(1):41–50

Almeselmani M, Deshmukh PS, Sairam RK (2009) High temperature stress tolerance in wheat genotypes: role of antioxidant defence enzymes. Acta Agron Hungarica 57(1):1–14

Almeselmani M, Deshmukh PS, Sairam RK, Kushwaha SR, Singh TP (2006) Protective role of antioxidant enzymes under high temperature stress. Plant Sci 171(3):382–388

Amin AA, El-Kader AAA, Abouziena HF, El-Awadi M, Gharib FAE (2016) Effects of benzoic acid and thiourea on growth and productivity of wheat (*Triticum aestivum* L.) plants. Int Sci Res J 72(4):1032–1037

Anderson BE, Ward JM, Schoeder JI (1994a) Evidence for extracellular reception site for abscisic acid in *Commelina* guard cell. Plant Physiol 104(4):1177–1183

Anderson JA, Taliaferro CM (2000) Freeze tolerance of seed-producing turf bermudagrasses. Crop Sci 42(1):190–192

Anderson MD, Prasad TK, Martin BA, Steward CR (1994b) Differential gene expression in chilling-acclimated maize seedlings and evidence for the involvement of abscisic acid in chilling tolerance. Plant Physiol 105(1):331339

Anjum SA, Wang LC, Farooq M, Hussain M, Xue LL, Zou CM (2011) Brassinolide application improves the drought tolerance in maize through modulation of enzymatic antioxidants and leaf gas exchange. J Agron Crop Sci 197(3):177–185

Anshuman S, Singh HK (2015) Application of plant growth regulators to improve fruit yield and quality in Indian gooseberry (*Emblica officinalis* Gaertn.). J Agric Search 2(1):20–23

Applewhite PB, Kaur-Sawhney R, Galston AW (2010) A role for spermidine in the bolting and flowering of *Arabidopsis*. Physiol Plant 108(3):314–320

Arakawa T, Timasheff SN (1983) Preferential interactions of proteins with solvent components in aqueous amino acid solutions. Arch Biochem Biophys 224(1):169–177

Arenas-Huertero F, Arroyo A, Zhou L, Sheen J, Leon P (2000) Analysis of *Arabidopsis* glucose insensitive mutants, gin5 and gin6, reveals a central role of the plant hormone ABA in the regulation of plant vegetative development by sugar. Genes Dev 146(1):2085–2096

Aroca R, Irigoyen JJ, Sanchez-Diaz M (2001) Photosynthetic characteristics and protective mechanisms against oxidative stress during chilling and subsequent recovery in two maize varieties differing in chilling sensitivity. Plant Sci 161(4):719–726

Aroca R, Vernieri P, Irigoyen JJ, Sancher-Diaz M, Tognoni F, Pardossi A (2003) Involvement of abscisic acid in leaf and root of maize (*Zea mays* L.) in avoiding chilling-induced water stress. Plant Sci 165(3):671–679

Arroyo A, Bossi F, Finkelstein RR, Leon P (2003) Three genes that affect sugar sensing (abscisic acid insensitive 4, abscisic acid insensitive 5, and constitutive triple response 1) are differentially regulated by glucose in *Arabidopsis*. Plant Physiol 1336(1):231–242

Arteca RN (1996) *Plant growth substances: principles and applications*. Chapman and Hall Inc., New York, NY, USA

Ashraf M, Foolad MR (2007) Roles of glycinebetaine and proline in improving plant abiotic stress resistance. Environ Exp Bot 59(2):206–216

Ashraf M, Karim F, Rasul E (2002) Interactive effects of gibberellic acid (GA3) and salt stress on growth, ion accumulation and photosynthetic capacity of two spring wheat (*Triticum aestivum* L.) cultivars differing in salt tolerance. Plant Growth Regul 36(1):49–59

Asthir B, Thapar R, Farooq M, Bains NS (2013) Exogenous application of thiourea improves the performance of late sown wheat by inducing terminal heat resistance. Int J Agric Biol 15(6): 1337–1342

Babu NR, Devraj VR (2008) High temperature and salt stress response in French bean (*Phaseolus vulgaris*). Aust J Crop Sci 2(2):40–48

Bae H, Herman E, Sicher R (2005) Exogenous trehalose promotes non-structural carbohydrate accumulation and induces chemical detoxification and stress response proteins in *Arabidopsis thaliana* grown in liquid culture. Plant Sci 168(5):1293–1301

Bahadur R (2018) Role of plant growth regulators under abiotic stress. In: Bhattacharya A (ed) *Abiotic stress and plant physiology . Vol. 2. Productivity*. New India Publishing Agency, New Delhi, pp 253–328

Bahuguna RN, Jagadish KSV (2015) Temperature regulation of plant phenological development. Environ Exp Bot 111:83–90

Bajguz A (2000) Effect of brassinosteroids on nucleic acid and protein content in cultured cell of *Chlorella vulgaris*. Plant Physiol Biochem 38(3):209–215

Bajguz A (2009) Brassinosteroid enhanced the level of abscisic acid in *Chlorella vulgaris* subjected to short-term heat stress. J Plant Physiol 166(8):882–886

Bajguz A (2010) An enhancing effect of exogenous brassinolide on the growth and antioxidant activity in *Chlorella vulgaris* cultures under heavy metals stress. Environ Exp Bot 68(2): 175–179

Bajguz A, Hayat S (2009) Effects of brassinosteroids on the plant responses to environmental stresses. Plant Physiol Biochem 47(1):1–8

Bandurska H, Stroinski A (2005) The effect of salicylic acid on barley response to water deficit. Acta Physiol Plant 27(3):379–386

Bangerth KF (2009) Floral induction in mature, perennial angiosperm fruit trees: similarities and discrepancies with annual-biennial plants and the involvement of plant hormones. Sci Hortic 122(2):153–163

Banon S, Gonzalez A, Cano EA, Franco JA, Fernandez JA (2002) Growth, development and colour response of potted *Dianthus caryophyllus* cv. Mondriaan to paclobutrazol treatment. *Sci Hortic* 94(3–4):371–377

Bari R, Jones JDG (2009) Role of plant hormones in plant defense response. Plant Mol Biol 69(4): 473–488

Barrero-Sicilia C, Silvestre S, Haslam RP, Michaelson LV (2017) Lipid remodelling: unravelling the response to cold stress in Arabidopsis and its extremophile relative *Eutrema salsugineum*. Plant Sci 263:194–200

Basra AS, Bedi S, Malik CP (1988) Pre-sowing hydration of maize seeds for stimulation of low-temperature germination and its effects on phospholipid changes in the embryos. Curr Sci 57(24):1340–1342

Basra AS, Dhillon R, Malik CP (1989) Influence of seed pre-treatments with plant growth regulators on metabolic alterations of germinating maize embryos under stressing temperature regimes. Ann Bot 64(1):37–41

Basra SMA, Farooq M, Wahid A, Khan MB (2006) Rice seed invigoration by hormonal and vitamin priming. Seed Sci Technol 34(3):753–758

Bateman JB, Evans GF, Brown PR, Gabriel C, Grant EH (1992) Dielectric properties of the system bovine albumin urea betaine in aqueous solution. Phys Med Biol 37(1):175–182

Beckers GJM, Spoel SH (2006) Fine-tuning plant defense signalling: salicylate versus jasmonate. Plant Biol 8(1):1–10

Behera KK, Sahoo S (2009) Rapid in vitro micropropagation of sugarcane (*Saccharum officinarum* L. cv Nayana) through callus culture. Nat Sci 7(4):1–10

Belin C, Megies C, Hauserova E, Lopez-Molina L (2009) Abscisic acid represses growth of the Arabidopsis embryonic axis after germination by enhancing auxin signaling. Plant Cell 21(8): 2253–2268

Benaroudj N, Lee DH, Goldberg AL (2001) Trehalose accumulation during cellular stress protects cells and cellular proteins from damage by oxygen radicals. J Biol Chem 276(26):24261–24267

Ben-Yehoshua S, Aloni B (1974) Effect of water stress on ethylene production by detached leaves of Valencia orange (*Citrus sinensis* Osbeck). Plant Physiol 53(6):863–865

Bergstrand KJI (2017) Methods for growth regulation of greenhouse produced ornamental pot- and bedding plants–a current review. Folia Hortic 29(1):63–74

Berova M, Zlatev Z (2000) Physiological response and yield of paclobutrazol treated tomato plants (*Lycopersicon esculentum*). Plant Growth Regul 30(2):117–123

Berova M, Zlatev Z, Stoeva N (2002) Effect on Paclobutrazol on wheat seedlings under low temperature stress. Blug J Plant Physiol 28(1–2):75–84

Besserer A, Puech-Pagès V, Kiefer P, Gomez-Roldan V, Jauneau A, Roy S, Portais JC, Roux C, Bécard G, Séjalon-Delmas N (2006) Strigolactones stimulate arbuscular mycorrhizal fungi by activating mitochondria. PLoS Biol 4(7):e226

Bhat ZA, Rashid R, Bhat JA (2011) Effect of plant growth regulators on leaf number, leaf area and leaf dry matter in grape. Notulae Sci Biol 3(1):87–90

Bhattacharya A (2019) Effect of high temperature stress on the metabolism of growth regulators. In: Effect of high temperature on crop productivity and metabolism of macro molecules. Elsevier, Academic Press, USA, pp 485–590

Bhattacharya A (2021) Role of plant hormones during soil water deficit. A review. In: Bhattacharya A (ed) *Soil water deficit and physiological issues in plants*. Springer, pp 489–584

Biao WN (2006) Alleviating effects of exogenous salicylic acid (SA) on chilling stress in radish. J Southwest Agric Univ 28:782–785

Binns A (1994) Cytokinin accumulation and action: biochemical, genetic and molecular approaches. Annu Rev Plant Physiol Plant Mol Biol 45:173–196

Bishop GJ, Yokota T (2001) Plants steroid hormones, brassinosteroids: current highlights of molecular aspects on their synthesis/metabolism, transport, perception and response. Plant Cell Physiol 42(2):114–120

Blackman PG, Davies WJ (1983) The effects of cytokinins and ABA on stomatal behavior of maize and commelina. J Exp Bot 34(12):1619–1626

Bloom AJ, Zwieniecki MA, Passioura JB, Randall LB, Holbrook NM, Clair DA, St. (2004) Water relations under root chilling in a sensitive and tolerant tomato species. Plant Cell Environ 27(8): 971–979

Bohnert HJ, Jensen RG (1996) Metabolic engineering for increased salt tolerance the next step. Aust J Plant Physiol 23(5):661–666

Bohnert HJ, Nelson DE, Jensen RG (1995) Adaptations to environmental stresses. Plant Cell 7(7): 1099–1111

Bolger TP, Upchurch DR, McMichael BL (1992a) Temperature effect on cotton root hydraulic conductance. Environ Exp Bot 32(1):49–54

Bolger TP, Upchurch DR, McMichael BL (1992b) Temperature effects on cotton root hydraulic conductance. Environmental induced water flow in plants. Plant Physiol 57:738–739

Borsani O, Valpuesta V, Botella A (2001) Evidence for a role of salicylic acid in the oxidative damage generated by NaCl and osmotic stress in Arabidopsis seedlings. Plant Physiol 126(3): 1024–1030

Bouchereau A, Aziz A, Larher F, Martin-Tanguy J (1999) Polyamines and environmental challenges: recent development. Plant Sci 140(2):103–125

Bowler C, Van Montagu M, Inze D (1992) Superoxide dismutase and stress tolerance. Ann Rev Plant Physiol Mol Biol 43:83–116

Bracale M, Corragio I (2003) Chilling and freezing stresses in plants: cellular responses and molecular strategies for adaptation. In: di Toppi LS, Pawlik-Skowronska B (eds) Abiotic stresses in plants. Springer, Dordrecht, pp 23–51

Bravo FP, Uribe EG (1981) Temperature dependence of the concentration kinetics of absorption of phosphate and potassium in corn roots. Plant Physiol 67(4):815–819

Brewer PB, Koltai H, Beveridge CA (2013) Diverse roles of strigolactones in plant development. Mol Plant 6(1):18–28

Bristow KL (1988) The role of mulch and its architecture in modifying soil temperature. Aust J Soil Res 26(2):269–280

Brodersen P, Malinovsky FG, Hématy K, Newman MA, Mundy J (2005) The role of salicylic acid in the induction of cell death in Arabidopsis acd11. Plant Physiol 138(2):1037–1045

Browse J (2005) Jasmonate: an oxylipin signal with many roles in plants. Vitam Horm 72:431–456

Browse J, Xin Z (2001) Temperature sensing and cold acclimation. Curr Opin Plant Biol 4(3): 241–246

Brugière N, Dubois F, Limami AM, Lelandais M, Roux Y, Sangwan RS, Hirel B (1999) Glutamine synthetase in the phloem plays a major role in controlling proline production. Plant Cell 11(10): 1995–2011

Bürger M, Chory J (2019) Stressed about hormones: how plants orchestrate immunity. Cell Host Microbe 26(2):163–172

Cai Q, Wang S, Cui Z, Sun J, Ishii Y (2004) Changes in freezing tolerance and its relationship with the contents of carbohydrates and proline in overwintering centipedegrass (*Eremochloa ophiuroides* (Munro) hack.). Plant Prod Sci 7(4):421–426

Campbell NA, Reece JB (2002) Biology, 6th edn. Benjamin Cummings, San Francisco

Campbell NA, Reece JB, Urry LA, Cain ML, Wasserman SA, Minorsky PV, Jackson RB (2008) Biology, 8th edn. Pearson, Benjamin Cummings, San Francisco, pp 827–830

Cao DD, Hu J, Gao CH, Guan YJ, Zhang S, Xiao JF (2008) Chilling tolerance of maize (*Zea mays* L.) can be improved by seed soaking in putrescine. Seed Sci Technol 36(1):191–197

Cao S, Zheng Y, Wang K, Jin P, Rui H (2009) Methyl jasmonate reduces chilling injury and enhances antioxidant enzyme activity in postharvest loquat fruit. Food Chem 115(4):1458–1463

Carvalho RF, Piotto FA, Schmidt D, Peters LP, Monterio CC, Azevedo RA (2011) Seed priming with hormones dose not alleviate induced oxidative stress in maize seedling subjected to salt stress. Sci Agric 68(5):598–602

Castro AJ, Clément C (2007) Sucrose and starch catabolism in the anther of *Lilium* during its development: a comparative study among the anther wall, locular fluid and microspore/pollen fractions. Planta 225(6):1573–1582

Catinot J, Buchala A, Abou-Mansour E, Métraux JP (2008) Salicylic acid production in response to biotic and abiotic stress depends on isochorismate in *Nicotiana benthamiana*. FEBS Lett 582(4): 473–478

Cela J, Chang C, Munné-Bosch S (2011) Accumulation of γ- rather than α-tocopherol alters ethylene signaling gene expression in the vte4 mutant of *Arabidopsis thaliana*. Plant Cell Physiol 52(8):1389–1400

Chandler JW (2009) Auxin as compere in plant hormone crosstalk. Planta 231(1):1–12

Charfeddine M, Saïdi MN, Charfeddine F, Hammami A, Bouzid RG (2015) Genome-wide analysis and expression profiling of the ERF transcription factor family in potato (*Solanum tuberosum L.*). Mol Biotechnol 57(4):348–358

Chaves M, Moroco J, Pereira JS (2003) Understanding plant responses to drought-from genes to the whole plant. Funct Plant Biol 30(3):239–264

Chen CM, Ertl JR, Leisner SM, Chang CC (1985) Localization of cytokinin biosynthetic sites in pea plants and carrot roots. Plant Physiol 78(3):510–513

Chen CM, Melitz D (1979) Cytokinin biosynthesis in a cell free system from cytokinin autotrophic tobacco tissue cultures. FEBS Lett 107(1):15–20

Chen D, Shao Q, Yim L, Younis A, Zheng B (2019) Polyamine function in plants: metabolism, regulation on development, and roles in abiotic stress responses. Front Plant Sci 9:45

Chen GL, Nie LC, Li JW, Xiao K (2000) Effect of low temperature stress on photosynthetic characteristics of grafted squash seedlings. Acta Agric Shanghai 16(1):42–45

Chen K, Arora R (2013) Priming memory invokes seed stress-tolerance. Environ Exp Bot 94:33–45

Chen M, Thelen JJ (2013) ACYL-LIPID DESATURASE2 is required for chilling and freezing tolerance in arabidopsis. Plant Cell 25(4):1430–1444

Chen S, Zimei L, Cui J, Jiangang D, Xia X, Liu D, Yu J (2011) Alleviation of chilling-induced oxidative damage by salicylic acid pretreatment and related gene expression in eggplant seedlings. Plant Growth Regul 65(1):101–108

Chen WP, Li PH (2002) Membrane stabilization by abscisic acid under cold aids proline in alleviating chilling injury in maize (*Zea mays* L.) cultured cells. Plant Cell Environ 25(8): 955–962

Cheng MC, Liao PM, Kuo WW, Lin TP (2013) The Arabidopsis ETHYLENE RESPONSE FACTOR1 regulates abiotic stress-responsive gene expression by binding to different cis-acting elements in response to different stress signals. Plant Physiol 162(3):1566–1582

Cheng WH, Endo A, Zhou L, Penney J, Chen HC, Arroyo A, Leon P, Nambara E, Asami T, Seo M, Koshiba T, Sheen J (2002) A unique short-chain dehydrogenase/reductase in Arabidopsis glucose signaling and abscisic acid biosynthesis and functions. Plant Cell 14(11):2723–2743

Chinnusamy V, Schumaker K, Zhu JK (2004) Molecular genetic perspectives on cross-talk and specificity in abiotic stress signalling in plants. J Exp Bot 55(395):225–236

Chiwocha SD, Dixon KW, Flematti GR, Ghisalberti EL, Merritt DJ, Nelson DC, Riseborough JAM, Smith SM, Stevens JC (2009) Karrikins: a new family of plant growth regulators in smoke. Plant Sci 177(4):252–256

Choi H-I, Hong J-H, Ha J-O, Kang J-Y, Kim SY (2000) ABFs, a family of ABA responsive element binding factors. J Biol Chem 275(3):1723–1730

Choudhary SP, Yu JQ, Yamaguchi-Shinozaki K, Shinozaki K, Tran LSP (2012) Benefits of brassinosteroid crosstalk. Trends Plant Sci 17(10):594–605

Choudhury S, Islam N, Sarkar MD, Ali MA (2013) Growth and yield of summer tomato as influenced by plant growth regulators. Int J Sustain Agric 5(1):25–28

Chu C, Lee TM (1989) The relationship between ethylene biosynthesis and chilling tolerance in seedlings of rice (*Oryza sativa* L.). Bot Bull Acad Sin 30:263–273

Chu TM, Jusaitis M, Aspinall D, Paleg LG (1978) Accumulation of free proline at low temperatures. Physiol Plant 43(3):254–260

Ciardi JA, Deikman J, Orzolek MD (1997) Increased ethylene synthesis enhances chilling tolerance in tomato. Physiol Plant 101(2):333–340

Claeys H, Skirycz A, Maleux K, Inze D (2012) DELLA signaling mediates stress-induced cell differentiation in Arabidopsis leaves through modulation of anaphase-promoting complex / cyclosome activity. Plant Physiol 159(2):739–747

Cleland CF (1974) Isolation of flower-inducing and flower-inhibitory factors from aphid honeydew. Plant Physiol 54(6):899–903

Cleland CF, Ajami A (1974) Identification of the flower-inducing factor isolated from aphid honeydew as being salicylic acid. Plant Physiol 54(6):904–906

Clouse SD, Sasse JM (1998) Brassinosteroids: essential regulators of plant growth and development. Annu Rev Plant Physiol Plant Mol Biol 49:427–445
Colebrook EH, Thomas SG, Phillips AL, Hedden P (2014) The role of gibberellin signaling in plant responses to abiotic stress. J Exp Biol 217(1):67–75
Cook NC, Strydom DK (2000) The south African high density system. Acta Hortic 513:321–327
Cothren JT, Oosterhuis D (2010) Use of growth regulators in cotton production. In: Stewart JM, Oosterhuis DM, Heitholt JJ, Mauney JR (eds) Physiology of cotton. Springer, Dordrecht, pp 289–303
Coughlan SJ, Heber U (1982) The role of glycinebetaine in the protection of spinach thylakoids against freezing stress. Planta 156(1):62–69
Cowan AK, Cairns ALP, Bartels-Rahm B (1999) Regulation of abscisic acid metabolism: towards a metabolic basis of abscisic acid-cytokinin antagonism. J Exp Bot 50(334):595–603
Crowe JH, Crowe LM, Chapman D (1984) Preservation of membranes in anhydrobiotic organisms: the role of trehalose. Science 223(4637):701–703
Cutler AJ, Krochko JE (1999) Formation and breakdown of ABA. Trends Plant Sci 4(12):472–478
Cutler SR, Rodriguez PL, Finkelstein RR, Abrams SR (2010) Abscisic acid: emergence of a core signaling network. Annu Rev Plant Biol 61:651–679
Daie J, Campbell WF, Seeley SS (1981) Temperature-stress-induced production of abscisic acid and dihydrophaseic acid in warm- and cool-season crops. J Am Soc Hortic Sci 106(1):11–13
Dar TA, Uddin M, Khan MMA, Hakeem KR, Jaleel H (2015) Jasmonates counter plant stress: a review. Environ Exp Bot 115:49–57
Das KK, Sarkar RK, Ismail AM (2005) Elongation ability and non-structural carbohydrate levels in relation to submergence tolerance in rice. Plant Sci 168(1):131–136
Datta N, Schell MB, Roux SJ (1987) Stimulation of a nuclear NH kinase from pea plumules and its role in the phosphorylation of a nuclear polypeptide. Plant Physiol 84:1397–1401
Davière JM, de Lucas M, Prat S (2008) Transcriptional factor interaction: a central step in DELLA function. Curr Opin Genet Dev 18(4):295–303
Davies PJ (2010) The plant hormones: their nature, occurrence, and functions. In: Davies PJ (ed) Plant hormones: physiology, biochemistry and molecular Biology. Springer, Dordrecht, pp 1–15
de Oliveira LF, Elbl P, Navarro BV, Al E (2016) Elucidation of the polyamine biosynthesis pathway during Brazilian pine (*Araucaria angustifolia*) seed development. Tree Physiol 37(1):116–130
de Saint Germain A, Ligerot Y, Dun EA, Pillot JP, Ross JJ, Beveridge CA, Rameau C (2013) Strigolactones stimulate internode elongation independently of gibberellins. Plant Physiol 163(2):1012–1025
De Smet I, Voss U, Lau S, Wilson M, Shao N, Timme RE, Swarup R, Kerr I, Hodgman C, Bock R, Bennet M, Jurgens G, Beeckman T (2010) Unraveling the evolution of auxin signaling. Plant Physiol 155(1):209–221
Delauney AJ, Verma DPS (1993) Proline biosynthesis and osmoregulation in plants. Plant J 4(2): 215–223
Delaux PM, Xie X, Timme RE, Puech-Pages V, Dunand C, Lecompte E, Delwiche CF, Yoneyama K, Bécard G, Séjalon-Delmas N (2012) Origin of strigolactones in the green lineage. New Phytol 195(4):857–871
Delker C, Stenzel I, Hause B, Miersch O, Feussner I, Wasternack C (2006) Jasmonate biosynthesis in *Arabidopsis thaliana*—enzymes, products, regulation. Plant Biol 8(3):297–306
Demidchik V, Cuin TA, Svistunenko D, Smith SJ, Miller AJ, Shabala S, Shokolik A, Yurin V (2010) *Arabidopsis* root K^+-efflux conductance activated by hydroxyl radicals: single-channel properties, genetic basis and involvement in stress-induced cell death. J Cell Sci 123(9): 1468–1479
Demkura PV, Abdala G, Baldwin IT, Ballare CL (2010) Jasmonate-dependent and-independent pathways mediate specific effects of solar ultraviolet B radiation on leaf phenolics and anti herbivore defense. Plant Physiol 152(2):1084–1095

Depuydt S, Hardtke CS (2011) Hormone signalling crosstalk in plant growth regulation. Curr Biol 21(9):R365–R373

Derkx MPM, Vermeer E, Karssen CM (1994) Gibberellins in seeds of *Arabidopsis thaliana*: biological activities, identification and effects of light and chilling on endogenous levels. Plant Growth Regul 15(3):223–234

Deshmukh PS, Srivastava GC (1982) Variation in proline accumulation in sunflower genotypes under moisture stress conditions. Ind J Plant Physiol 25(4):396–399

Devasirvatham V, Tan D (2018) Impact of high temperature and drought stresses on chickpea production. Agronomy 8(8):145

Dias JPT (2019) Plant growth regulators in horticulture: practices and perspectives. Biotecnología Vegetal 19(1):3–14

Diehl RC, Guinn EJ, Capp MW, Tsodikov OV, Record MT Jr (2013) Quantifying additive interactions of the osmolyte proline with individual functional groups of proteins: comparisons with urea and glycine betaine, interpretation of m-values. Biochemistry 52(35):5997–6010

Dill A, Thomas SG, Hu J, Steber CM, Sun TP (2004) The Arabidopsis F-box protein SLEEPY targets gibberellins signaling repressors for gibberellins-induced degradation. Plant Cell 16(6): 1392–1405

Ding CK, Wang CY, Gross KC, Smith DL (2002) Jasmonate and salicylate induce the expression of pathogenesis-related-protein genes and increase resistance to chilling injury in tomato fruit. Planta 214(6):895–901

Ding H-D, Zhu X-H, Zhu Z-W, Yang S-J, Zha D-S, Wu X-X (2012) Amelioration of salt-induced oxidative stress in eggplant by application of 24-epibrassinolide. Biol Plant 56(4):767–770

Dionne J, Castonguay Y, Nadeau P, Desjardins Y (2001) Freezing tolerance and carbohydrate changes during cold acclimation of green-type annual bluegrass (*Poa annua* L.) ecotypes. Crop Sci 41(2):443–451

Divi UK, Rahman T, Krishna P (2010) Brassinosteroids-mediated stress tolerance in Arabidopsis shows interactions with abscisic acid, ethylene and salicylic acid pathways. *BMC Plant Biol* 10: article no. 151

Divi UK, Rahman T, Krishna P (2016) Gene expression and functional analyses in brassinosteroid-mediated stress tolerance. Plant Biotechnol J 14(1):419–432

Djanaguiraman M, Perumal R, Jagadish SVK, Ciampitti IA, Welti R, Prasad PVV (2018) Sensitivity of sorghum pollen and pistil to high-temperature stress. Plant Cell Environ 41(5):1065–1082

Doherty CJ, Van Buskirk HA, Myers SJ, Thomashow MF (2009) Roles for Arabidopsis CAMTA transcription factors in coldregulated gene expression and freezing tolerance. Plant Cell 21(3): 972–984

Dong T, Park Y, Hwang I (2015) Abscisic acid: biosynthesis, inactivation, homoeostasis and signalling. Essays Biochem 58:29–48

Donovan MP, Nabity PD, Delucia EH (2013) Salicylic acid-mediated reductions in yield in *Nicotiana attenuate* challenged by aphid herbivory. Arthropod Plant Interact 7(1):45–52

Dreccer MF, Fainges J, Whish J, Ogbonnaya FC, Sadras VO (2018) Comparison of sensitive stages of wheat, barley, canola, chickpea and field pea to temperature and water stress across Australia. Agric For Meteorol 248:275–294

Drennan PM, Smith MT, Goldsworth D, van Staden J (1993) The occurrence of trehalose in the leaves of the desiccation-tolerant angiosperm *Myrothamnus flabellifolius* Welw. J Plant Physiol 142(4):493–496

Du H, Liu HB, Xiong LZ (2013a) Endogenous auxin and jasmonic acid levels are differentially modulated by abiotic stresses in rice. Front Plant Sci 4:397

Du H, Wu N, Chang Y, Li X, Xiao J, Xiong L (2013b) Carotenoid deficiency impairs ABA and IAA biosynthesis and differentially affects drought and cold tolerance in rice. Plant Mol Biol 83(4–5):475–488

Du H, Wu N, Fu J, Wang S, Li X, Xiao J, Xiong L (2012) A GH3 family member, OsGH3-2, modulates auxin and abscisic acid levels and differentially affects drought and cold tolerance in rice. J Exp Bot 63(18):6467–6480

Dubey RS (1999) Protein synthesis by plants under stressful conditions. In: Pessarakli M (ed) Handbook of plant and crop stress. Marcel Dekker, Inc, New York, pp 365–397

Edreva A, Yordanov I, Kardjieva R, Gesheva E (1998) Heat shock responses of bean plants: involvement of free radicals, antioxidants and free radical/active oxygen scavenging systems. Biol Plant 41(2):185–191

Egamberdieva D (2009) Alleviation of salt stress by plant growth regulators and IAA producing bacteria in wheat. Acta Physiol Plant 31(4):861–864

Elaleem KGA, Modawi RS, Khalafalla MM (2009) Effect of plant growth regulators on callus induction and plant regeneration in tuber segment culture of potato (*Solanum tuberosum* L.) cultivar Diamant. Afr J Biotechnol 8(11):2529–2534

El-Bassiouny HMS, Shukry WM (2001) Cowpea growth pattern, metabolism and yield in response to IAA and biofertilizers under drought conditions. Egypt J Biol 3(1):117–129

Elbein AD, Pan YT, Pastuszak I, Carroll D (2003) New insights on trehalose: a multifunctional molecule. Glycobiology 13(4):17–27

El-Beltagy AS, Hall MA (1974) Effect of water stress upon endogenous ethylene levels in *Vicia faba*. New Phytol 73:47–60

Elfving DC, Visser DB (2006) Cyclanilide induces lateral branching in sweet cherry trees. HortScience 41(1):149–153

Elfving DC, Visser DB (2007) Improving the efficacy of cytokinin applications for stimulation of lateral branch development in young sweet cherry trees in the orchard. HortSci 42(2):251–256

El-Saeid HM, Abou-Hussein SD, El-Tohamy WA (2010) Growth characters, yield and endogenous hormones of cowpea plants in response to IAA application. Res J Agric Biol Sci 6(1):27–31

El-Saeid HM, Abouziena HF, AbdAlla MSA (2011) Effect of some bioregulators on white lupine (*Lupinus termis*) seed yield and its components and on endogenous hormones content in seeds. *Electronic J Polish Agric Univ* 14(4):article no. 2

El-Tayeb MA, El-ENany A, wahab, and Ahmed, N.L. (2005) Salicylic acid-induced adaptive response to copper stress in sunflower (*Helianthus annuus* L.). Plant Growth Regul 50(2): 191–199

Emery RJ, Ma Q, Atkins CA (2000) The forms and sources of cytokinins in developing white lupine seeds and fruits. Plant Physiol 123(4):1593–1604

Ercoli L, Mariotti M, Masoni A, Arduini I (2004) Growth responses of sorghum plants to chilling temperature and duration of exposure. Eur J Agron 2(1):93–103

Esashi Y, Katoh H, Leopold AC (1977) Dormancy and impotency of cocklebur seeds: IV. Effects of gibberellic acid, benzyladenine, thiourea, and potassium nitrate on the growth of embryonic axis and cotyledon segments. Plant Physiol 59(2):117–121

Esmaili S, Salehi H (2012) Effects of temperature and photoperiod on postponing bermudagrass (*Cynodon dactylon* [L.] Pers.) turf dormancy. J Plant Physiol 169(9):851–858

Fahad S, Bajwa AA, Nazir U, Anjum SA, Farooq A, Zohaib A, Sadia S, Nasim W, Adkins S, Saud S, Ihsan MZ, Alharby H, Wu C, Wang D, Huang, J. (2017) Crop production under drought and heat stress: plant responses and management options. Front Plant Sci, 29: 8

Fahad S, Bano A (2012) Effect of salicylic acid on physiological and biochemical characterization of maize grown in saline area. Pak J Bot 44(4):1433–1438

Fahad S, Hussain S, Bano A, Saud S, Hassan S, Shan D, Khan FA, Khan F, Chen YT, Wu C, Tabassum MA, Chun MX, Afzal M, Jan A, Jan MT, Huang JL (2015a) Potential role of phytohormones and plant growth-promoting rhizobacteria in abiotic stresses: consequences for changing environment. Environ Sci Pollut Res 22(7):4907–4921

Fahad S, Hussain S, Matloob A, Khan FA, Khaliq A, Saud S, Hassan S, Shan D, Khan F, Ullah N, Faiq M, Khan MR, Tareen AK, Khan A, Ullah A, Ullah N, Huang JL (2015b) Phytohormones and plant responses to salinity stress: a review. Plant Growth Regul 75(2):391–404

Fahad S, Hussain S, Saud S, Khan F, Hassan S, Nasim W, Arif M, Wang F, Huang J (2016) Exogenously applied plant growth regulators affect heat-stressed rice ollens. J Agron Crop Sci 202(2):139–150

Fahad S, Nie L, Chen Y, Wu C, Xiong D, Saud S, Hongyan L, Cui K, Huang J (2015c) Crop plant hormones and environmental stress. In: *Sustainable agricultural reviews*.", 15. Springer, Cham, pp 371–400

Fan W, Zhang M, Zhang H, Zhang P (2012) Improved tolerance to various abiotic stresses in transgenic sweet potato (*Ipomoea batatas*) expressing spinach betaine aldehyde dehydrogenase. PLoS One 7:e37344

Fariduddin Q, Yusuf M, Ahmad I, Ahmad A (2013) Brassinosteroids and their role in response of plants to abiotic stresses. Boil Plant 58(1):9–17

Fariduddin Q, Yusuf M, Ahmad I, Ahmad A (2014) Brassinosteroids and their role in response of plants to abiotic stresses. Biol Plant 58:9–17

Fariduddin Q, Yusuf M, Chalkoo S, Hayat S, Ahmad A (2011) 28- Homobrassinolide improves growth and photosynthesis in *Cucumis sativus* L. through an enhanced antioxidant system in the presence of chilling stress. Photosynthetica 49(1):55–64

Farooq M, Aziz T, Basra SMA, Cheema MA, Rehman H (2008c) Chilling tolerance in hybrid maize induced by seed priming with salicylic acid. J Agron Crop Sci 194(2):161–168

Farooq M, Aziz T, Basra SMA, Wahid A, Khaliq A, Cheema MA (2008d) Exploring the role of calcium to improve chilling tolerance in hybrid maze. J Agron Crop Sci 194(5):350–359

Farooq M, Aziz T, Cheema ZA, Khaliq A, Hussain M (2008e) Activation of antioxidant system by KCl treatments improves the chilling tolerance in hybrid maize. J Agron Crop Sci 194(6): 438–448

Farooq M, Aziz T, Hussain M, Rehman H, Jabran K, Khan MB (2008b) Glycinebetaine improves chilling tolerance in hybrid maize. J Agron Crop Sci 194:152–160

Farooq M, Basra SMA, Hafeez K (2006c) Seed invigoration by osmohardening in fine and coarse rice. Seed Sci Technol 34(1):181–187

Farooq M, Basra SMA, Khalid M, Tabassum R, Mehmood T (2006d) Nutrient homeostasis, reserves metabolism and seedling vigor as affected by seed priming in coarse rice. Can J Bot 84(8):1196–1202

Farooq M, Basra SMA, Rehman H, Saleem BA (2008a) Seed priming enhances the performance of late sown wheat (*Triticum aestivum* L.) by improving the chilling tolerance. *J Agron Crop Sci* 194(1):55–60

Farooq M, Basra SMA, Tabassum R, Afzal I (2006a) Enhancing the performance of direct seeded fine rice by seed priming. Plant Prod Sci 9(4):446–456

Farooq M, Basra SMA, Wahid A (2006b) Priming of field-sown rice seed enhances germination, seedling establishment, allometry and yield. Plant Growth Regul 49(2):285–294

Farooq M, Wahid A, Basra SMA, Din I (2009d) Improving water relations and gas exchange with brassinosteroids in rice under drought stress. J Agron Crop Sci 195(4):262–269

Farooq M, Wahid A, Kobayashi N, Fujita D, Basra SMA (2009a) Plant drought stress: effects, mechanisms and management. Agron Sustain Dev 29(1):185–212

Farooq M, Wahid A, Lee DJ (2009b) Exogenously applied polyamines increase drought tolerance of rice by improving leaf water status, photosynthesis and membrane properties. Acta Physiol Plant 31(5):937–945

Farooq MA, Saqib ZA, Akhtar J, Bakhat HF, Ratna-Kumar P, Dietz K (2015) Protective role of silicon (Si) against combined stress of salinity and boron (B) toxicity by improving antioxidant enzymes activity in rice. SILICON 11(4):2193–2197

Farooq MAE, Aziz TA, Abdul Wahid B, Lee DJC, Siddique KHM (2009c) Chilling tolerance in maize: agronomic and physiological approaches. Crop Pasture Sci 60:501–516

Fayez KA, Bazaid SA (2003) Improving drought and salinity tolerance in barley by application of salicylic acid and potassium nitrate. J Saudi Soc Agric Sci 13(1):45–55

Feller U, Vaseva II (2014) Extreme climate events: impacts of drought and high temperature on physiological processes in agronomically important plants. Front Environ Sci 39(2):1–17

Feng L, Zhao YH, Wang F, Zhang L (2003) Study on the effect of MeJA treatment for cold storage of peach. Food Sci 9:135–139

Feng Y, Yin Y, Fei S (2015) Down-regulation of BdBRI1, a putative brassinosteroid receptor gene produces a dwarf phenotype with enhanced drought tolerance in *Brachypodium distachyon*. Plant Sci 234:163–173

Feussner I, Wasternack C (2002) The lipoxygenase pathway. Annu Rev Plant Biol 53:275–297

Fghire R, Anaya F, Ali OI, Benlhabib O, Ragab R, Wahbi S (2015) Physiological and photosynthetic response of quinoa to drought stress. Chilean J Agric Res 75(2):174–183

Finkelstein R, Gampala SSI, Rock CD (2002) Abscisic acid signaling in seeds and seedling. Plant Cell 14(Suppl):e515–e545

Finkelstein RR, Ariizumi T, Steber C (2008) Molecular aspects of seed dormancy. Ann Rev Plant Biol 59:387–415

Finkelstein RR, Gibson SI (2002) ABA and sugar interactions regulating development. Cross-talk or voices in a crowd? Curr Opin Plant Biol 5(1):26–32

Fleet CM, Sun TP (2005) A DELLAcate balance: the role of gibberellin in plant morphogenesis. Curr Opin Plant Biol 8(1):77–85

Fletcher RA, Hofstra G (1990) Improvement of uniconazole induced protection in wheat seedlings. J Plant Growth Regul 9(4):207–217

Flowers TJ (2004) Improving crop salt tolerance. J Exp Bot 55(396):307–319

Fonseca S, Chico JM, Solano R (2009) The jasmonate pathway: the ligand, the receptor and the core signalling module. Curr Opin Plant Biol 12(5):539–547

Foo E, Davies NW (2011) Strigolactones promote nodulation in pea. Planta 234(5):1073–1081

Foolad MR, Lin GY (2000) Relationship between cold tolerance during seed germination and vegetative growth in tomato: germplasm evaluation. J Am Soc Hortic Sci 125(6):679–683

Fowler S, Thomashow MF (2002) Arabidopsis transcriptome profiling indicates that multiple regulatory pathways are activated during cold acclimation in addition to the CBF cold response pathway. Plant Cell 14(8):1675–1690

Freeha A, Abdul W, Muhammad F, Farrukh J (2011) Potential of foliar applied thiourea in improving salt and high temperature tolerance of bread wheat (Triticum aestivum). Int J Agric Biol 13(2):251–256

Fresco LO (2009) Challenges for food system adaptation today and tomorrow. Environ Sci Policy 12(4):378–385

Fryer MJ, Oxborough K, Martin B, Ort DR, Baker NR (1995) Factors associated with depression of photosynthetic quantum efficiency in maize at low growth temperature. Plant Physiol 108(2): 761–767

Fu X, Richards DE, Fleck B, Xie D, Burton N, Harberd NP (2004) The Arabidopsis mutant sleepygar2-1 protein promotes plant growth by increasing the affinity of the SCFSLY1 E3 ubiquitin ligase for DELLA protein substrates. Plant Cell 16(6):1406–1418

Galiba G, Tuberosa R, Kocsy G, Sutka J (1993) Involvement of chromosome 5A and chromosome 5D in cold-induced abscisic acid accumulation in and frost tolerance of wheat calli. Plant Breed 110(3):237–242

Galston AW, Kaur-Sawhney R (1995) Polyamines as endogenous growth regulators. In: Plant hormones. Springer, Dordrecht, pp 158–178

Gamalero E, Glick BR (2012) Ethylene and abiotic stress tolerance in plants. In: Ahmed P, Prasad MNV (eds) Environmental adaptations and stress tolerance of plants in the era of climate change. Springer, New York, pp 395–412

Gao H, Zhang ZK, Lv XG, Cheng N, Peng BZ, Cao W (2016) Effect of 24-epibrassinolide on chilling injury of peach fruit in relation to phenolic and proline metabolisms. Post Biol Technol 111:390–397

Garcia R, Pacheco G, Falcão E, Borges G, Mansur E (2011) Influence of type of explant, plant growth regulators, salt composition of basal medium, and light on callogenesis and regeneration in *Passiflora suberosa* L. (Passifloraceae). Plant Cell Tissue Organ Culture 106(1):47–54

Garcia-Mata C, Lamattina L (2007) Abscisic acid (ABA) inhibits light-induced stomatal opening through calcium-and nitric oxide-mediated signaling pathways. Nitric Oxide 17(3–4):143–151

Garg BK, Burman U, Kathju S (2006) Influence of thiourea on photosynthesis, nitrogen metabolism and yield of clusterbean (*Cyamopsis tetragonoloba* (L.) Taub.) under rainfed conditions of Indian arid zone. Plant Growth Regul 48(3):237–245

George AP, Nissen RJ (1992) Effects of water stress, nitrogen and paclobutrazol on flowering, yield and fruit quality of the low-chill peach cultivar Flordaprince. Sci Hortic 49(3–4):197–209

Gharib FA, Hegazi AZ (2010) Salicylic acid ameliorates germination, seedling growth, phytohormone and enzymes activity in bean (*Phaseolus vulgaris* L.) under cold stress. J Am Sci 6:675–683

Ghorbani B, Pakkish Z (2014) Brassinosteroid enhances cold stress tolerance of Washington navel orange (Citrus sinensis L.) fruit by regulating antioxidant enzymes during storage. Agric Conspec Sci 79(2):109–114

Gill SS, Tuteja N (2010) Polyamines and abiotic stress tolerance in plants. Plant Signal Behav 5(1): 26–33

Gille G, Singler K (1995) Oxidative stress in living cells. Folia Microbiol 40(2):131–152

Gilmour SJ, Fowler SG, Thomashow MF (2004) Arabidopsis transcriptional activators CBF1, CBF2 and CBF3 have matching functional activities. Plant Mol Biol 54(5):767–781

Giraudat J, Parcy F, Bertauche N, Gosti F, Leung J (1994) Current advances in abscisic acid action and signaling. Plant Mol Biol 26(5):1557–1577

Giri J (2011) Glycine betaine and abiotic stress tolerance in plants. Plant Signal Behav 6(11): 1746–1751

Giuliani S, Sanguineti MC, Tuberosa R, Bellotti M, Salvi S, Landi P (2005) Root-ABA1 a major constitutive QTL affects maize root architecture and leaf ABA concentration at different water regimes. J Exp Bot 56(422):3061–3070

Glick BR (2014) Bacteria with ACC deaminase can promote plant growth and help to feed the world. Microbiol Res 169(1):30–39

Golldack D, Li C, Mohan H, Probst N (2014) Tolerance to drought and salt stress in plants: unraveling the signaling networks. Front Plant Sci 5:151

Gomaa AM, Raldugina GN, Burmistrova NA, Radionov NV, Kuznetsov VV (2012) Response of transgenic rape plants bearing the Osmyb4 gene from rice encoding a trans-factor to low above-zero temperature. Russ J Plant Physiol 59(1):105–114

Gomez-Roldan V, Fermas S, Brewer PB, Puech-Pages V, Dun EA, Pillot JP, Letisse F, Matusova R, Danoun S, Portais JC, Bouwmeester H, Becard G, Beveridge CA, Rameau C, Rochange SF (2008) Strigolactone inhibition of shoot branching. Nature 455(7210):189–194

Gonai T, Kawahara S, Tougou M, Satoh S, Hashiba T, Hirai N, Kawaide H, Kamiya Y, Yoshioka T (2004) Abscisic acid in the thermo inhibition of lettuce seed germination and enhancement of its catabolism by gibberellin. J Exp Bot 55(394):111–118

Gong W, Shen YP, Ma LG, Pan Y, Du YL, Wang DH, Yang JY, Hu LD, Liu XF, Dong CX, Ma L, Chen YH, Yang XY, Gao Y, Zhu D, Tan X, Mu JY, Zhang DB, Liu YL, Dinesh-Kuamr SP, Li Y, Wang XP, Gu HY, Qu LJ, Bai SN, Lu YT, Li JY, Zhao JD, Zuo J, Huang H, Deng XW, Zhu YX (2004) Genome-wide ORFeome cloning and analysis of Arabidopsis transcription factor genes. Plant Physiol 135(2):773–782

González-Aguilar GA, Fortizm J, Cruz R, Baez R, Wang CY (2000) Methyl jasmonate reduces chilling injury and maintains postharvest quality of mango fruit. J Agric Food Chem 48:515–519

Gray WM, Ostin A, Sandberg G, Romano CP, Estelle M (1998) High temperature promotes auxin-mediated hypocotyl elongation in Arabidopsis. Proc Natl Acad Sci U S A 95(12):7197–7202

Greene DW, Autio WP (1994) Notching techniques increase branching of young apple trees. J Am Soc Hort Sci 119:678–682

Groen SC, Whiteman NK (2014) The evolution of ethylene signaling in plant chemical ecology. J Chem Ecol 40(7):700–716

Gromadzinski A, Ciesielski F, Mrówczynski M (1990) Influence of plant growth regulators on lupin yield. Materiały Sesji Instytutu Ochrony Roślin 30(2):65–70

Groppa MD, Benavides MP (2008) Polyamines and abiotic stress: recent advances. Amino Acids 34(1):35–45

Guan L, Scandalios JG (1995) Developmentally related responses of maize catalase genes to salicylic acid. Proc Natl Acad Sci U S A 92(3):5930–5934

Guda CD, Scordo E, Allera C, Farina E (2000) Effects of low temperatures and gibberellic acid on flowering of *Limonium gmelinii.* Acta Hortic 541:323–326

Gudesblat GE, Russinova E (2011) Plants grow on brassinosteroids. Curr Opin Plant Biol 14(5): 530–537

Guo-li W, Zhen-fei G (2005) Effects of chilling stress on photosynthetic rate and chlorophyll fluorescence parameter in seedlings of two rice cultivars differing in cold tolerance. Rice Sci 12(3):187–191

Gupta K, Dey A, Gupta B (2013) Plant polyamines in abiotic stress responses. Acta Physiol Plant 35:2015–2036

Ha CV, Leyva-González MA, Osakabe Y, Tran UT, Nishiyama R, Watanabe Y, Tanaka M, Seki M, Yamaguchi S, Dong NV, Yamaguchi-Shinozjaki K, Shinozaki K, Herra-Estrella L, Phan Tran LS (2014) Positive regulatory role of strigolactone in plant responses to drought and salt stress. Proc Natl Acad Sci U S A 111(2):851–856

Habben JE, Bao X, Bate NJ, DeBruin JL, Dolan D, Hasegawa D, Helentjaris TG, Lafitte RH, Lovan N, Mo H, Reimann K, Schussler JR (2014) Transgenic alteration of ethylene biosynthesis increases grain yield in maize under field drought-stress conditions. Plant Biotechnol J 12(6): 685–693

Haldimann P (1998) Low growth temperature-induced changes to pigment composition and photosynthesis in *Zea mays* L. genotypes differing in chilling sensitivity. Plant Cell Environ 21:200–208

Hamada K (1986) Brassinolide: some effects for crop cultivation. In, "*Proceedings of the international seminar plant growth regulation*", Tokyo, Japan, pp. 188–196

Hampton J, Hebblethwaite PD (2006) Effects of growth regulator paclobutrazol (PP 333) on the growth, development and yield of *Lolium perenne* grown for seed. Grass Forage Sci 40(1): 93–101

Hao X, Wang B, Wang L, Zeng J, Yang Y, Wang X (2018) Comprehensive transcriptome analysis reveals common and specific genes and pathways involved in cold acclimation and cold stress in tea plant leaves. Sci Hortic 240(9):354–368

Hara M, Furukawa J, Sato A, Mizoguchi T, Miura K (2012) Abiotic stress and role of salicylic acid in plants. In: Parvaiz A, Prasad MNV (eds) Abiotic stress responses in plants. Springer, New York, pp 235–251

Hardtke CS, Dorcey E, Osmont KS, Sibout R (2007) Phytohormone collaboration: zooming in on auxin–brassinosteroid interactions. Trends Cell Biol 17(10):485–492

Hartmann A, Senning M, Hedden P, Sonnewald U, Sonnewald S (2011) Reactivation of meristem activity and sprout growth in potato tubers require both cytokinin and gibberellin. Plant Physiol 155(2):776–796

Hassanein RA, Amin ABAES, Rashad ESM, Ali H (2015) Effect of thiourea and salicylic acid on antioxidant defense of wheat plants under drought stress. Int J ChemTech Res 7(1):346–354

Hayat Q, Hayat S, Irfan M, Ahmad A (2010) Effect of exogenous salicylic acid under changing environment: A review. Environ Exp Bot 68(1):14–25

Hayat S, Ali B, Ahmad A (2007) Salicylic acid- plant hormone. Springer, Dordrecht

He XJ, Mu RL, Cao WH, Zhang ZG, Zhang JS, Chen SY (2005a) AtNAC2, a transcription factor downstream of ethylene and auxin signaling pathways, is involved in salt stress response and lateral root development. Plant J 44(6):903–916

He YL, Liu YL, Cao WX, Huai MF, Xu BG, Huang BR (2005b) Effects of salicylic acid on heat tolerance associated with antioxidant metabolism in Kentucky bluegrass. Crop Sci 45(3): 988–995

Hedden P, Kamiya Y (1997) Gibberellin biosynthesis: enzymes, genes and their regulation. Annu Rev Plant Physiol Plant Mol Biol 48:431–460

Hedden P, Sponsel V (2015) A century of gibberellin research. J Plant Growth Regul 34(4): 740–760

Hetherington AM (2001) Guard cell signaling. Cell 107(6):711–714

Hetherington AM, Woodward FI (2003) The role of stomata in sensing and driving environmental change. Nature 424(6951):901–908

Hiilovaara-Teijo M, Hannukkala A, Griffith M, Yu XM, Pihakaski MK (1999) Snow-mold induced apoplastic proteins in winter rye leaves lack antifreeze activity. Plant Physiol 121(2):665–673

Hilli JS, Vyakarnahal BS, Biradar DP, Ravi H (2010) Effect of growth regulators and stages of spray on growth, fruit set and seed yield of ridgegourd (*Luffa acutangula* L. Roxb). Karnataka J Agric Sci 23(2):239–242

Hirasawa R, Yokoigawa K, Isobe Y, Kawai H (2001) Improving the freeze tolerance of baker's yeast by loading with trehalose. Biosci Biotechnol Biochem 65(3():522–526

Hodges DM, Andrews CJ, Johnson DA, Hamilton RI (1996) Antioxidant compound responses to chilling stress in differentially sensitive inbred maize lines. Physiol Plant 98(4):685–692

Hoffman L, DaCosta M, Ebdon JS, Watkins E (2010) Physiological changes during cold acclimation of perennial ryegrass accessions differing in freeze tolerance. Crop Sci 50(3):1037–1047

Hoffman NF, Liu Y, Yang SF (1983) Changes in 1-(malonylamino)cyclopropane-1-carboxylic acid content in wilted wheat leaves in relation to their ethylene production rates and 1-aminocyclopropane-1-carboxylic acid content. Planta 157(6):518–523

Horváth E, Szalai G, Janda T (2007b) Induction of abiotic stress tolerance by salicylic acid signaling. J Plant Growth Regul 26(3):290–300

Horváth E, Szalai G, Pál M, Páldi E, Janda T (2007a) Differences between the catalase isozymes of maize (*Zea mays* L.) in respect of inhibition by various phenolic compounds. Acta Biol Szegediensis 46(3–4):33–34

Hossain MA, Cho JI, Han M, Ahn CH, Jeon JS, An G, Park PB (2010) The ABRE-binding bZIP transcription factor OsABF2 is a positive regulator of abiotic stress and ABA signaling in rice. J Plant Physiol 167(17):1512–1520

Hoth S, Morgante M, Sanchez JP, Hamfley MK, Tingey SV, Chua NH (2002) Genome wide gene expression profiling in Arabidopsis thaliana reveals new targets of abscisic acid and largely impaired gene regulation in the abi1-1 mutant. Cell Sci 115(Pt 24):489–490

Hottiger T, Boller T, Wiemken A (1987) Rapid changes of heat and desiccation content correlated with changes of trehalose content in Saccharomyces cerevisiae cells subjected to temperature shifts. FEBS Lett 220(10):1223–1229

Hounsa CG, Brandt EV, Thevelein J, Hohmann S, Prior BA (1998) Role of trehalose in survival of *Saccharomyces cerevisiae* under osmotic stress. Microbiology 144(3):671–680

Hrotkó K, Magyar L, Öri B (1999) Improved feathering on one-year-old 'Germersdorfi FL 45' sweet cherry trees in the nursery. Gartenbauwissenschaft 64(2):75–78

Hu WH, Song XS, Shi K, Xia XJ, Zhou YH, Yu JQ (2008) Changes in electron transport, superoxide ismutase and ascorbate peroxidase isoenzymes in chloroplasts and mitochondria of cucumber leaves as influenced by chilling. Photosynthetica 46(4):581–588

Hu Y, Jiang L, Wang F, Yu D (2013) Jasmonate regulates the INDUCER OF CBF EXPRESSION-C-REPEAT BINDING FACTOR/DRE BINDING FACTOR1 cascade and freezing tolerance in *Arabidopsis*. Plant Cell 25(8):2907–2924

Huang BR, DaCosta M, Jiang YW (2014) Research advances in mechanisms of turfgrass tolerance to abiotic stresses: from physiology to molecular biology. Crit Rev Plant Sci 33(2–3):141–189

Huang LF, Zheng JH, Zhang YY, Hu WH, Mao WH, Zhou YH, Yu JQ (2006) Diurnal variation in gas exchange, chlorophyll fluorescence, quenching and light allocation in soybean leaves: the case for mid-day depression in CO_2 assimilation. Sci Hortic 110(2):214–218

Huang M, Guo Z (2005) Responses of antioxidant system to chilling stress in two rice cultivars differing in sensitivity. Biol Plant 49(1):81–84

Hund A, Richner W, Soldati A, Fracheboud Y, Stamp P (2007) Root morphology and photosynthetic performance of maize inbred lines at low temperature. Eur J Agron 27(1):52–61

Hundertmark M, Hincha DK (2008) LEA (late embryogenesis abundant) proteins and their encoding genes in *Arabidopsis thaliana*. BMC Genomics 9:118

Hussain HA, Hussain S, Khaliq A, Ashraf U, Anjum SA, Men S, Wang L (2018) Chilling and drought stresses in crop plants: implications, cross talk, and potential management opportunities. Front Plant Sci 9:393

Hussain SS, Kayani MA, Amjad M (2011) Transcription factors as tools to engineer enhanced drought tolerance in plants. Biotechnol Prog 27(2):297–306

Idowu PE, Ibitoye DO, Ademoyegun OT (2009) Tissue culture as a plant production technique for horticultural crops. Afr J Biotechnol 8(16):3782–3788

Incoll LD, Jewer PC (1987) Cytokinins and the water relations of whole plants. In: *Cytokinins—plant hormones in search of a role. Monograph 14*. British Plant Growth Regulator Group, pp 85–97

Incoll LD, Ray JP, Jewer PC (1990) Do cytokinins act as root to shoot signals? In: Davies WJ, Jeffcoat B (eds) Importance of root to shoot communication in the response to environmental stress. Br Soc Plant Growth Regulators, Bristol, pp 185–197

Iordachescu M, Imai R (2011) Trehalose and abiotic stress in biological systems. In Tech, Croatia, pp 215–234

IPCC (2007) *Climate change 2007: the physical science basis. Contribution of working group I to the fourth assessment report of the intergovernmental panel on climate change*. Cambridge University Press, Cambridge

Iqbal N, Umar S, Khan NA, Khan MIR (2014) A new perspective of phytohormones in salinity tolerance: regulation of proline metabolism. Environ Exp Bot 100:34–42

Ito TM, Polido PB, Rampim MC, Kaschuk G, Souza SGH (2014) Genome-wide identification and phylogenetic analysis of the AP2/ERF gene superfamily in sweet orange (*Citrus sinensis*). Genet Mol Res 13(3):7839–7851

Ivanov AG, Allakhverdiev SI, Huner NPA, Murata N (2012) Genetic decrease in fatty acid unsaturation of phosphatidylglycerol increased photoinhibition of photosystem I at low temperature in tobacco leaves. Biochim Biophys Acta 1817(8):1374–1379

Jacobsen SE, Olszewski NE (1993) Mutations at the SPINDLY locus of Arabidopsis alter gibberellin signal transduction. Plant Cell 5(8):887–896

Jacyna T (2002) Factors influencing lateral-branch formation in woody plants. Acta Agrobot 55(2): 5–25

Jacyna T, Barnard J, Andersen RL (2005) Natural growth habit of sweet cherry maiden trees. J Am Pom Soc 59(2):104–110

Jacyna T, Puchała A (2004) Application of environment friendly branch promoting substances to advance sweet cherry tree canopy development in the orchard. J Fruit Ornament Plant Res 12: 177–182

Janda T, Szalai G, Tari I, Paldi E (1999) Hydroponic treatment with salicylic acid decreases the effect of chilling injury in maize (*Zea mays* L.) plants. Planta 208(2):175–180

Janeczko A, Gullner G, Skoczowski A, Dubert F, Barna B (2007) Effects of brassinosteroid infiltration prior to cold treatment on ion leakage and pigment contents in rape leaves. Biol Plant 51(2):355–358

Janeczko A, Oklestkova J, Pociecha E, Koscielniak J, Mirek M (2011) Physiological effects and transport of 24- epibrassinolide in heat-stressed barley. Acta Physiol Plant 33(4):1249–1259

Janowiak F, Dörffling K (1996) Chilling of maize seedlings: changes in water status and abscisic acid content in ten genotypes differing in chilling tolerance. J Plant Physiol 147(5):582–588

Janowiak F, Luck E, Dorffling K (2003) Chilling tolerance of maize seedlings in the field during cold periods in spring is related to chilling-induced increase in abscisic acid level. J Agron Crop Sci 189(3):156–161

Janowiak F, Maas B, Dörlling K (2002) Importance of abscisic acid for chilling tolerance of maize seedlings. J Plant Physiol 159(6):635–643

Janská A, Maršík P, Zelenková S, Ovesná J (2010) Cold stress and acclimation—what is important for metabolic adjustment? Plant Biol 12(3):395–405

Jatimliansky JR, Garcia MD, Molina MC (2004) Response to chilling of *Zea mays, Tripsacum dactyloides* and their hybrid. Biol Plant 48(4):561–567

Javid MG, Sorooshzadeh A, Moradi F, Sanavy SAMM, Allahdadi I (2011) The role of phytohormones in alleviating salt stress in crop plants. Aust J Crop Sci 5(6):726–734

Jayakannan M, Bose J, Babourina O, Shabala S, Massart A, Poschenrieder C, Rengel Z (2015) The NPR1-dependent salicylic acid signalling pathway is pivotal for enhanced salt and oxidative stress tolerance in Arabidopsis. J Exp Bot 66(7):1865–1875

Jeon J, Kim NY, Kim S, Kang NY, Novak O, Ku SJ, Cho C, Lee DJ, Lee EJ, Strnad M, Kim J (2010) A subset of cytokinin two-component signaling system plays a role in cold temperature stress response in Arabidopsis. J Biol Chem 285(30):23371–23386

Ji CH, Li HY, Chen LB, Xie M, Wang FP, Chen Y, Liu YG (2013) A novel rice bHLH transcription factor. DTD, acts coordinately with TDR in controlling tapetum function and pollen development. Mol Plant 6(5):1715–1718

Jiang YP, Huang LF, Cheng F, Zhou YH, Xia XJ, Mao WH, Shi K, Yu JQ (2013) Brassinosteroids accelerate recovery of photosynthetic apparatus from cold stress by balancing the electron partitioning, carboxylation and redox homeostasis in cucumber. Physiol Plant 148(1):133–145

Jisha KC, Vijaya Kumari K, Puthur JT (2013) Seed priming for abiotic stress tolerance: an overview. Acta Physiol Plant 35(5):1381–1396

Jocelyn PC (1972) *Biochemistry of SH group; the occurrence, chemical properties metabolism and biological function of thiol and Disulphides*. Academic Press, London, pp 57–122

Jonytiene V, Burbulis N, Kupriene R, Blinstrubiene A (2012) Effect of exogenous proline and de-acclimation treatment on cold tolerance in *Brassica napus* shoots cultured *in-vitro*. J Food Agric Environ 10(1):327–330

Jouyban Z, Hasanzade R, Sharafi S (2013) Chilling stress in plants. Int J Agric Crop Sci 5(24): 2961–2968

Jumali SS, Said IM, Ismail I, Zainal Z (2011) Genes induced by high concentration of salicylic acid in *Mitragyna speciosa*. Aust J Crop Sci 5(3):296–303

Jutsz AM, Gnida A (2015) Mechanisms of stress avoidance and tolerance by plants used in phytoremediation of heavy metals. Arch Environ Prot 41(4):104–114

Kabiri R, Farahbakhsh H, Nasibi F (2012) Effect of drought and its interaction with salicylic acid on black cumin germination and seedling growth. World Appl Sci J 18(4):520–527

Kagale S, Divi UK, Krochko JE, Keller WA, Krishna P (2007) Brassinosteroid confers tolerance in *Arabidopsis thaliana* and *Brassica napus* to a range of abiotic stresses. Planta 225(2):353–364

Kakkar RK, Nagar PK, Ahuja PS, Rai VK (2000) Polyamines and plant morphogenesis. Biol Plant 43(1):1–11

Kang GZ, Wang CH, Sun GC, Wang ZX (2003b) Salicylic acid changes activities of H_2O_2-metabolizing enzymes and increases the chilling tolerance of banana seedlings. Environ Exp Bot 50(1):9–15

Kang GZ, Wang ZX, Sun GC (2003a) Participation of H_2O_2 in enhancement of cold chilling by salicylic acid in banana seedlings. Acta Bot Sin 45(5):567–573

Kang GZ, Zhu GH, Peng XX, Sun GC, Wang ZX (2004) Isolations of salicylic acid-induced genes in chilling-stressed banana seedling leaves using mRNA differential display. J Plant Physiol Mol Biol 30(2):225–228

Kang HG, Kim J, Kim B, Jeong H, Choi SH, Kim EK, Lee HY, Lim PO (2011) Over-expression of FTL1/DDF1, an AP2 transcription factor, enhances tolerance to cold, drought, and heat stresses in *Arabidopsis thaliana*. Plant Sci 180(4):634–641

Kang HM, Saltveit ME (2002) Chilling tolerance of maize, cucumber and rice seedling leaves and roots are differentially affected by salicylic acid. Physiol Plant 115(4):571–576

Kang NY, Cho C, Kim NY, Kim J (2012) Cytokinin receptor-dependent and receptor-independent path ways in the dehydration response of *Arabidopsis thaliana*. J Plant Physiol 169(14): 1382–1391

Kapulnik K, Delaux PM, Resnick N, Mayzlish-Gati E, Wininger S, Bhattacharya C, Séjalon-Delmas N, Combier JP, Bécard G, Belausov E, Beeckman T, Dor E, Hershenhorn J, Koltai H

(2011) Strigolactones affect lateral root formation and root-hair elongation in Arabidopsis. Planta 233(1):209–216
Kapulnik Y, Koltai H (2014) Strigolactone involvement in root development, response to abiotic stress, and interactions with the biotic soil environment. Plant Physiol 166(2):560–569
Karabudak T, Bor M, Özdemir F, Türkan I (2014) Glycine betaine protects tomato (*Solanum lycopersicum*) plants at low temperature by inducing fatty acid desaturase 7 and lipoxygenase gene expression. Mol Biol Rep 41(3):1401–1410
Kaurin A, Juntilla O, Hansen J (1981) Seasonal changes in frost hardiness in cloudberry (*Rubus chamacmorus*) in relation to carbohydrate content with special reference to sucrose. Physiol Plant 52(2):310–314
Kaya C, Ashraf M, Sönmez O (2015) Promotive effect of exogenously applied thiourea on key physiological parameters and oxidative defense mechanism in salt-stressed *Zea mays* L. plants. Turk J Bot 39:786–795
Kaya C, Aydemir S, Ashraf M, Dikilitas M (2013) Exogenous application of mannitol and thiourea regulates plant growth and oxidative stress responses in salt-stressed maize (*Zea mays* L.). J Plant Interact 8(3):234–241
Kaya C, Sarioğlu A, Akram NA, Ashraf M (2019) Thiourea-mediated nitric oxide production enhances tolerance to boron toxicity by reducing oxidative stress in bread wheat (*Triticum aestivum* L.) and durum wheat (*Triticum durum* Desf.) plants. J Plant Growth Regul 38(3): 1094–1109
Kazan K (2013) Auxin and the integration of environmental signals into plant root development. Ann Bot 112:1655–1665
Kazan K (2015) Diverse roles of jasmonates and ethylene in abiotic stress tolerance. Trends Plant Sci 20(4):219–229
Ke Q, Wang Z, Ji CY, Jeong JC, Lee HS, Li H, Xu B, Deng X, Kwak SS (2015) Transgenic poplar expressing Arabidopsis YUCCA6 exhibits auxin-overproduction phenotypes and increased tolerance to abiotic stress. Plant Physiol Biochem 94:19–27
Keith CN, McKersie BD (1986) The effect of abscisic acid on the freezing tolerance of callus cultures of *Lotus corniculatus* L. Plant Physiol 80(3):766–770
Kermode AR (2005) Role of abscisic acid in seed dormancy. J Plant Growth Regul 24:319–344
Keskin BC, Sarikaya AT, Yuksel B, Memon AR (2010) Abscisic acid regulated gene expression in bread wheat. Aust J Crop Sci 4(8):617–625
Khalid S, Malik AU, Khan AS, Jamil A (2012) Influence of exogenous applications of plant growth regulators on fruit quality of young 'kinnow' mandarin (*Citrus nobilis* × *Citrus deliciosa*) trees. Int J Agric Biol 14(2):229–234
Khalil IA, Rahman H (1995) Effect of paclobutrazol on growth, chloroplast pigment and sterol biosynthesis in maize (*Zea mays* L.). Plant Sci 105(1):15–21
Khalil S, El-Saeid HM, Shalaby M (2006) The role of kinetin in flower abscission and yield of lentil plant. J Appl Sci Res 2(9):587–591
Khan MI, Fatma M, Per TS, Anjum NA, Khan NA (2015a) Salicylic acid-induced abiotic stress tolerance and underlying mechanisms in plants. Front Plant Sci 6:462
Khan NA (2005) The influence of exogenous ethylene on growth and photosynthesis of mustard (*Brassica juncea*) following defoliation. Sci Hortic 105(4):499–505
Khan TA, Fariduddin Q, Yusuf M (2015b) *Lycopersicon esculentum* under low temperature stress: an approach toward enhanced antioxidants and yield. Environ Sci Pollut Res 22(18): 14178–14188
Khan TA, Fariduddin Q, Yusuf M (2017) Low-temperature stress: is phytohormones application a remedy? Environ Sci Pollut Res 24(27):21574–21590
Khan W, Prithiviraj B, Smith DL (2003) Photosynthetic responses of corn and soybean to foliar application of salicylates. J Plant Physiol 160(5):485–492
Khanna P, Kaur K, Gupta AK (2017) Root biomass partitioning, differential antioxidant system and thiourea spray are responsible for heat tolerance in spring maize. Proc Natl Acad Sci India Sect B Biol Sci 87(2):351–359

Khodary SEA (2004) Effect of salicylic acid on growth, photosynthesis and carbohydrate metabolism in salt stressed maize plants. Int J Agric Biol 6(1):5–8

Khorshidi M, Nojavan AM (2006) The effect of abscisic acid and $CaCl_2$ on the activities of antioxidant enzymes under cold stress in maize seedling under dark. Pak J Biol Sci 9(1):54–59

Kieber JJ, Schaller GE (2014) Cytokinins. *Arabidopsis Book* 12:e0168

Kim TH, Bohmer M, Nishilmura N, Schorder JI (2010) Guard cell signal transduction network: advances in understanding abscisic acid, CO_2 and ca^{+2} signalling. Ann Rev Plant Biol 61:561–569

Kim Y, Park S, Gilmour SJ, Thomashow MF (2013) Roles of *CAMTA* transcription factors and salicylic acid in configuring the low-temperature transcriptome and freezing tolerance of Arabidopsis. Plant J 75(3):364–376

Kingston-Smith AH, Harbinson J, Williams J, Foyer CH (1997) Effect of chilling on carbon assimilation, enzyme activation, and photosynthetic electron transport in the absence of photoinhibition in maize leaves. Plant Physiol 114(3):1039–1046

Klay I, Pirrello J, Riahi I, Bernadac A, Cherif A, Bouzayen M, Bouzid S (2014) Ethylene response factor Sl-ERF.B.3 is responsive to abiotic stresses and mediates salt and cold stress response regulation in tomato. *Sci World J* 2014:167681

Klotke J, Kopka J, Gatzke N, Heyer AG (2004) Impact of soluble sugar concentrations on the acquisition of freezing tolerance in accessions of *Arabidopsis thaliana* with contrasting cold acclimation. Plant Cell Environ 27(11):1395–1404

Knorzer OC, Lederer B, Durner J, Boger P (1999) Antioxidative defense activation in soybean cells. Physiol Plant 107(3):294–302

Koch T, Bandemer K, Boland W (1997) Biosynthesis of cis-jasmone: a pathway for the inactivation and the disposal of the plant stress hormone jasmonic acid to the gas phase? Helv Chim Acta 80(3):838–850

Koini MA, Alvey L, Allen T, Tilley CA, Harberd NP, Whitelam GC, Franklin KA (2009) High temperature-mediated adaptations in plant architecture require the bHLH transcription factor PIF4. Curr Biol 19(5):408–413

Kolaksazov M, Laporte F, Ananieva K, Dobrev P, Herzog M, Ananiev ED (2013) Effect of chilling and freezing stresses on jasmonate content in *Arabis alpina*. Bulg J Agric Sci 19(2):15–17

Koltai H, Beveridge CA (2013) Strigolactones and the coordinated development of shoot and root. In: Baluska F (ed) Long-distance systemic signaling and communication in plants. Springer, Berlin, pp 189–204

Kosová K, Prášil IT, Vítámvás P, Dobrev P, Motyka V, Floková K, Novák O, Turečková V, Rolčik J, Pešek B, Trávničková A, Gaudinová A, Galiba G, Janda T, Vlasáková E, Prášilová P, Vankova R (2012) Complex phytohormone responses during the cold acclimation of two wheat cultivars differing in cold tolerance, winter Samanta and spring Sandra. J Plant Physiol 169(6): 567–576

Kotak S, Larkindale J, Lee U, Von Koskull-döring P, Vierling E, Scharf KD (2007) Complexity of the heat stress response in plants. Curr Opin Plant Biol 10(3):310–316

Kovtun Y, Chiu WL, Tena G, Sheen J (2000) Functional analysis of oxidative stress-activated mitogen-activated protein kinase cascade in plants. *Proc Natl Acad Sci U S A* 97(6):2940–2945

Krishna P (2003) Brassinosteroid-mediated stress responses. J Plant Growth Regul 22(4):289–297

Ku YS, Sintaha M, Cheung MY, Lam HM (2018) Plant hormones signaling cross talks between biotic and abiotic responses. Int Mol Sci 19(10):3206

Kudoyarova GR, Farhutdinov RG, Mitrichenko AN, Teplova IR, Dedov AV, Veselov SU, Kulaeva ON (1998) Fast changes in growth rate and cytokinin content of the shoot following rapid cooling of roots of wheat seedlings. Plant Growth Regul 26(2):105–108

Kumar A, Altabella T, Taylor MA, Tiburcio AF (1997) Recent advances in polyamine research. Trends Plant Sci 2:124–130

Kumar M, Sirhindi G, Bhardwaj R, Kumar S, Jain G (2010) Effect of exogenous H_2O_2 on antioxidant enzymes of *Brassica juncea* L. seedlings in relation to 24-epibrassinolide under chilling stress. Ind J Biochem Biophys 47(6):378–382

Kumar P, Lakshmi NJ, Mani VP (2000) Interactive effects of salicylic acid and phytohormones on photosynthesis and grain yield of soybean (*Glycine max* L. Merrill). Physiol Mol Biol Plants 6(2):179–186

Kumar S, Malik J, Thakur P, Kaistha S, Sharma KD, Upadhyaya HD, Berger JD, Nayyar H (2011) Growth and metabolic responses of contrasting chickpea (*Cicer arietinum*) genotypes to chilling stress at reproductive phase. Acta Physiol Plant 33(3):779–787

Kunkel BN, Brooks DM (2002) Crosstalk between signalling pathways in pathogen defense. Curr Opin Plant Biol 5(4):325–331

Kurepin LV, Ivanov A, Zaman M, Pharis RP, Hurry V, Hüner NP (2017) Interaction of glycine betaine and plant hormones: protection of the photosynthetic apparatus during abiotic stress. In: Hou HJM, Najafpour MM, Moore GF, Allakhverdiev SI (eds) *Photosynthesis, structure, mechanism and application, chapter 9*. Springer international publishing AG, pp 185–202

Kurepin LV, Joo SH, Kim SK, Pharis RP, Back TG (2012) Interaction of brassinosteroids with light quality and plant hormones in regulating shoot growth of young sunflower and Arabidopsis seedlings. J Plant Growth Regul 31(2):156–164

Kurepin LV, Walton LJ, Pharis RP, Emery RJN, Reid DM (2011) Interactions of temperature and light quality on phytohormone-mediated elongation of *Helianthus annuus* hypocotyls. Plant Growth Regul 64(2):147–154

Lang A, Minchin PEH (1986) Phylogenetic distribution and mechanism of translocation inhibition by chilling. J Exp Bot 37(3):389–398

Larkindale J, Hall DJ, Knight MR, Vierling E (2005) Heat stress phenotypes of Arabidopsis mutants implicate multiple signaling pathways in the acquisition of thermo-tolerance. Plant Physiol 138(2):882–897

Law JA, Jacobsen SE (2010) Establishing, maintaining and modifying DNA methylation patterns in plants and animals. Nat Rev Genet 11(3):204–220

Lee TM, Lur HS, Chu C (1993) Role of abscisic acid in chilling tolerance of rice (*Oryza sativa* L.) seedlings. I. Endogenous abscisic acid levels. Plant Cell Environ 16(5):481–490

Lee TM, Lur HS, Chu C (1997) Role of abscisic acid in chilling tolerance of rice (*Oryza sativa* L.) seedlings: II. Modulation of free polyamine levels. Plant Sci 126(1):1–10

Leon P, Sheen J (2003) Sugar and hormone connections. Trends Plant Sci 86(1):110–116

Levitt J (1980) Responses of plants to environmental stresses. Vol 1: water, radiation. In: Salt and other stresses. Academic Press, New York

Li BQ, Zhou ZW, Tian SP (2008) Combined effects of endo- and exogenous trehalose on stress tolerance and biocontrol efficacy of two antagonistic yeasts. Biol Control 46(2):187–193

Li DD, Tai FJ, Zhang ZT, Li Y, Zheng Y, Wu YF, Li XB (2009) A cotton gene encodes a tonoplast aquaporin that is involved in cell tolerance to cold stress. Gene 438(1–2):26–32

Li HW, Zang BS, Deng XW, Wang XP (2011) Over-expression of the trehalose-6-phosphate synthase gene *OsTPS1* enhances abiotic stress tolerance in rice. Planta 234(5):1007–1018

Li PH, Chen WP, Jian L, Xin Z (1997) Abscisic acid-induced chilling tolerance in maize. In: Li PH, Chen THH (eds) *Plant cold hardiness. Molecular Biology, biochemistry and physiology*. Plenum Press, New York, London, pp 215–223

Li R, Afsheen S, Xin Z, Han X, Lou Y (2013) *OsNPR1* negatively regulates herbivore-induced JA and ethylene signaling and plant resistance to a chewing herbivore in rice. Physiol Plant 147(3): 340–351

Li R, Zhang J, Li J, Zhou G, Wang Q, Bian W, Erb M, Lou Y (2015) Prioritizing plant defence over growth through WRKY regulation facilitates infestation by non-target herbivores. elife 4: e04805

Li S, Yang Y, Zhang Q, Liu N, Xu Q, Hu L (2018) Differential physiological and metabolic response to low temperature in two zoysiagrass genotypes native to high and low latitude. PLoS One 13(6):e0198885

Li XN, Cai J, Liu FL, Dai TB, Cao WX, Jiang D (2014) Physiological, proteomic and transcriptional responses of wheat to combination of drought or waterlogging with late spring low temperature. Funct Plant Biol 41(7):690–703

Li Y, He J (2012) Advance in metabolism and response to stress of polyamines in plant. Acta Agric Boreali Sinica 27:240–245

Li YH, Liu YJ, Xu XL, Jin M, An LZ, Zhang H (2012) Effect of 24-epibrassinolide on drought stress-induced changes in *Chorispora bungeana*. Biol Plant 56(1):192–196

Liang JQ, Liang Y (2009) Effects of plant growth substances on water-logging resistance of oilseed rape seedling. J Southwest China Norm Univ (Nat Sci Ed.) 34:58–62. (in Chinese with English abstract)

Licausi F, Giorgi FM, Zenoni S, Osti F, Pezzotti M, Perata P (2010) Genomic and transcriptomic analysis of the AP2/ERF superfamily in Vitis vinifera. *BMC Genom* 11:article no. 719

Ligaba A, Katsuhara M, Shibasaka M, Djira G (2011) Abiotic stresses modulate expression of major intrinsic proteins in barley (*Hordeum vulgare*). CR Biol 334(2):127–139

Liu C, Zhao L, Yu G (2011) The dominant glutamic acid metabolic flux to produce gamma-amino butyric acid over proline in *Nicotiana tabacum* leaves under water stress relates to its significant role in antioxidant activity. J Integr Plant Biol 53(8):608–618

Liu CL, Chen HP, Liu EE, Peg XX, Lu SY, Guo ZF (2003) Multiple tolerance of rice to abiotic stresses and its relationship with ABA accumulation. Acta Agron Sin 29(5):725–729

Liu Q, Kasuga M, Sakuma Y, Abe H, Miura S, Yamaguchi-Shinozaki Y, Shinozaki K (1998) Two transcription factors, DREB1 and DREB2 with an EREBP /AP2 DNA binding domain separate two cellular signal transduction pathways in drought- and low-temperature-responsive gene expression, respectively, in Arabidopsis. *Plant Cell* 10(8):1391–1406

Liu Y, Zhou J (2018) MAPping kinase regulation of ICE1 in freezing tolerance. Trends Plant Sci 23(2):91–93

Liu YJ, Zhao ZG, Si J, Di CX, Han J, An LZ (2009) Brassinosteroids alleviate chilling-induced oxidative damage by enhancing antioxidant defense system in suspension cultured cells of *Chorispora bungeana*. Plant Growth Regul 59(3):207–214

Livingston HDIII, Heyer A (2009) Fructan and its relationship to abiotic stress tolerance in plants. Cell Mol Life Sci 66(13):2007–2023

Llanes A, Andrade A, Alenabo S, Luna V (2016) Alterations of endogenous hormonal levels in plants under drought and salinity. Am J Plant Sci 7(09):1357–1371

Lockhart J (2013) Salt of the earth: ethylene promotes salt tolerance by enhancing Na/K homeostasis. Plant Cell 25(9):3150

Long L (2001) Cherry training systems: selection and development. Pacific northwest publication PNW 543. Oregon State Univ, Corevalis, Ore. Extension

Long L (2005) Sweet cherry training systems: current practices in North America and Europe. Compact Fruit Tree 38:20–23

Long LE, Facteau T, Nuñez-Elisea R, Cahn H (2005) Developments in high density cherries in the USA. Acta Hortic 667:303–309

Louarn G, Chenu K, Fournier C, Andrieu B, Giauffret C (2008) Relative contributions of light interception and radiation use efficiency to the reduction of maize productivity under cold temperatures. Funct Plant Boil 35(10):885–899

Lu S, Su W, Li H, Guo Z (2009) Abscisic acid improves drought tolerance of triploid bermudagrass and involves H_2O_2- and NO-induced antioxidant enzyme activities. Plant Physiol Biochem 47(2):132–138

Lyons JM (1973) Chilling injury in plants. Ann Rev Plant Physiol 24:445–466

Lyons JM, Raison JK (1970) Oxidative activity of mitochondria isolated from plant tissues sensitive and resistant to chilling injury. Plant Physiol 45(4):386–389

Ma S, Gong Q, Bohnert HJ (2006) Dissecting salt stress pathways. J Exp Bot 57(5):1097–1107

Machacckova I, Hanisova A, Krekule J (1989) Levels of ethylene, ACC, MACC, ABA and proline as indicators of cold hardening and frost resistance in winter wheat. Physiol Plant 76(4): 603–607

Mady MA (2014) Inducing cold tolerability in squash (*Cucurbita pepo* l.) plant by using salicylic acid and chelated calcium application. Int J Agric Sci Res 4(4):9–14

Magome H, Yamaguchi S, Hanada A, Kamiya Y, Oda K (2008) The DDF1 transcriptional activator upregulates expression of a gibberellin-deactivating gene, GA2ox7, under high-salinity stress in *Arabidopsis*. Plant J 56(4):613–626

Magome H, Yamaguchi S, Hanada A, Kamiya Y, Odadoi K (2004) Dwarf and delayed-flowering 1, a novel Arabidopsis mutant deficient in gibberellins biosynthesis because of over expression of a putative AP2 transcription factor. Plant J 37(5):720–729

Mahajan SH, Tuteja N (2005) Cold, salinity and drought stresses: an overview. Arch Biochem Biophys 444(2):139–158

Mahesh K, Parshavaneni B, Ramakrishna B, Rao SSR (2013) Effect of brassinosteroids on germination and seedling growth of radish (*Raphanus sativus* L.) under PEG-6000 induced water stress. Am J Plant Sci 4(12):2305–2313

Maksymiec W, Wojcik M, Krupa Z (2007) Variation in oxidative stress and photochemical activity in *Arabidopsis thaliana* leaves subjected to cadmium and excess copper in the presence or absence of jasmonate and ascorbate. Chemosphere 66(3):421–427

Malamy J, Hennig J, Klessig DF (1992) Temperature-dependent induction of salicylic acid and its conjugates during the resistance response to tobacco mosaic virus infection. Plant Cell 4(3): 359–366

Mani F, Bettaieb T, Zheni K, Doudech N, Hannachi C (2012) Effect of hydrogen peroxide and thiourea on fluorescence and tuberization of potato (*Solanum tuberosum* L.). *J Stress Physiol Biochem* 8(3):61–71

Mano Y, Nemoto K (2012) The pathway of auxin biosynthesis in plants. J Exp Bot 63(8): 2853–2872

Mansfield TA, McAinsh MR (1995) Hormones as regulators of water balance. In: Davies PJ (ed) Plant hormones. Kluwer Academic Publishers, Dordrecht –Berlin/London, pp 598–616

Marín IC, Loef I, Bartetzko L, Searle I, Coupland G, Stitt M, Osuna D (2011) Nitrate regulates floral induction in Arabidopsis, acting independently of light, gibberellin and autonomous pathways. Planta 233(3):539–552

Markhart AH (1984) Amelioration of chilling-induced water stress by abscisic acid-induced changes in root hydraulic conductance. Plant Physiol 74(1):81–83

Martindale W, Leegood RC (1997) Acclimation of photosynthesis to low temperature in *Spinacia oleracea* L. II. Effects of nitrogen supply. J Exp Bot 48(10):1873–1880

Martínez-Téllez MA, Lafuente MT (1993) Chilling-induced changes in phenylalanine ammonia-lyase, peroxidase, and polyphenol oxidase activities in citrus flavedo tissue. Acta Hortic 343: 257–263

Martin-Tanguy J (2001) Metabolism and function of polyamines in plants: recent development (new approaches). Plant Growth Regul 34:135–148

Masood A, Iqbal N, Khan NA (2012) Role of ethylene in alleviation of cadmium-induced photosynthetic capacity inhibition by Sulphur in mustard. Plant Cell Environ 35(3):524–533

Mata CI, Hertog ML, Van Raemdonck G, Baggerman G, Tran D, Nicolai BM (2019) Omics analysis of the ethylene signal transduction in tomato as a function of storage temperature. Postharvest Biol Technol 155:1–10

Matilla-Vazquez MA, Matilla AJ (2014) Ethylene: role in plants under environmental stress. In: Ahmad P, Wani MR (eds) *Physiological mechanisms and adaptation strategies in plants under changing environment*, vol 2. Springer Science + Business Media, New York, pp 189–222

McArtney S, Obermiller JD, Green A (2007) Prohexadione-ca reduces russet and does not negate the efficacy of GA4+7 sprays for russet control on 'Golden Delicious' apples. HortScience 42(1): 550–554

McSteen P (2009) Hormonal regulation of branching in grasses. Plant Physiol 149(1):46–55

Meena YK, Kaur N (2017) Salicylic acid enhanced low temperature stress tolerance in vegetables: A review. Chem Sci Rev Lett 6(24):2428–2435

Meena YK, Khurana DS, Kaur N, Singh K (2017) Phenolic compounds enhanced low temperature stress tolerance in tomato (*Solanum lycopersicum* L.). Curr J App Sci Technol 20(3):1–9

Meijón M, Rodriguez RC, Cañal MJ, Feito I (2009) Improvement of compactness and floral quality in azalea by means of application of plant growth regulators. Sci Hortic 119(2):169–176
Melkonian J, Yu LX, Setter TL (2004) Chilling responses of maize (*Zea mays* L.) seedlings: root hydraulic conductance, abscisic acid and stomatal conductance. J Exp Bot 55(403):1751–1760
Méndez-Hernández HA, Ledezma-Rodríguez M, Avilez-Montalvo RN, Juátez-Gómez YL, Skeete A, Avilez-Montalvo J, De-la-Peña C, Loyola-Vargas VM (2019) Signaling overview of plant somatic embryogenesis. Front Plant Sci 10:77
Miedema P (1982) The effects of low temperature on *Zea mays*. Adv Agron 35:93–128
Milborrow BV (2001) The pathway of biosynthesis of abscisic acid in vascular plants: a review of the present state of knowledge of ABA biosynthesis. J Exp Bot 52(359):1145–1164
Min XJ, Bartholomew DP (2005) Effects of flooding and drought on ethylene metabolism, titratable acidity and fruiting of pineapple. Acta Hortic 666:135–148
Minhas PS, Rane J, Pasala RK (2017) Abiotic stress management for resilient agriculture. Springer, New Delhi
Miquel M, James D, Dooner H, Browse J (1993) Arabidopsis requires polyunsaturated lipids for low temperature survival. Proc Natl Acad Sci U S A 90(13):6208–6212
Mittler R (2002) Oxidative stress, antioxidants and stress tolerance. Trends Plant Sci 7(9):405–410
Mittler R, Vanderauwera S, Suzuki N, Miller G, Tognetti VB, Vandepoele K, Gollery M, Shulaev V, Van Breusegem F (2011) ROS signaling: the new wave? Trends Plant Sci 16(6): 300–309
Miura K, Lee J, Miura T, Hasegawa PM (2010) SIZ1 controls cell growth and plant development in *Arabidopsis* through salicylic acid. Plant Cell Physiol 51(1):103–113
Miura K, Okamoto H, Okuma E, Shiba H, Kamada H, Hasegawa PM, Murata Y (2013) SIZ1 deficiency causes reduced stomatal aperture and enhanced drought tolerance via controlling salicylic acid-induced accumulation of reactive oxygen species in Arabidopsis. Plant J 73(1): 91–104
Miura K, Tada YY (2014) Regulation of water, salinity and cold stress responses by salicylic acid. Front Plant Sci 5(4):4
Mizoi J, Shinozaki K, Yamaguchi-Shinozaki K (2012) AP2/ERF family transcription factors in plant abiotic stress responses. Biochim Biophys Acta 1819(2):86–96
Mohebodini M, Mokhtar JJ, Mahboudi F, Alizadeh H (2011) Effects of genotype, explant age and growth regulators on callus induction and direct shoot regeneration of lettuce (*Lactuca sativa* L.). *Aus J Crop Sci* 5(1):92–95
Mok DW, Mok MC (2001) Cytokinin metabolism and action. Annu Rev Plant Physiol Plant Mol Biol 52:89–118
Mónika M, Janda T, Horvath E, Paldi E, Szalai G (2001) Exogenous salicylic acid increases polyamine content but may decrease drought tolerance in maize. Plant Sci 162(4):569–574
Moore AL, Albury MS, Crichton PG, Affourtit C (2002) Function of the alternative oxidase: is it still a scavenger? Trends Plant Sci 7(11):478–481
Mora-Herrera ME, López-Delgado H, Castillo-Morales A, Foyer CH (2005) Salicylic acid and H_2O_2 function by independent pathways in the induction of freezing tolerance in potato. Physiol Plant 125(4):430–440
Moseki B, Musil CF (2004a) Characterization of low temperature stress effects on photosynthetic performance of maize cultivars using chlorophyll fluorescence. S Afr J Bot 70(5):730–733
Moseki B, Musil CF (2004b) Effects of low temperature stress on chloroplast development in a range of *Zea mays* cultivars. S Afr J Bot 70(5):726–729
Muday GK, Rahman A, Binder BM (2012) Auxin and ethylene: collaborators or competitors? Trends Plant Sci 17(4):181–195
Muhlemann JK, Younts TLB, Muday GK (2018) Flavonols control pollen tube growth and integrity by regulating ROS homeostasis during high-temperature stress. *Proc Natl Acad Sci U S A* 115(47):E11188–E11197
Müller M, Munné-Bosch S (2015) Ethylene response factors: A key regulatory hub in hormone and stress signaling. Plant Physiol 169(1):32–41

Munne-Bosch S, Penuelas J (2003) Photo and antioxidative protection and a role for salicylic acid during drought and recovery in field-grown *Phillyrea angustifolia* plants. Planta 217(5): 758–766

Munné-Bosch S, Queval G, Foyer CH (2013) The impact of global change factors on redox signaling underpinning stress tolerance. Plant Physiol 16(1):5–19

Munns R, Tester M (2008) Mechanisms of salinity tolerance. Annu Rev Plant Biol 59:651–681

Munteanu V, Gordeev V, Martea R, Duca M (2014) Effect of gibberellin cross talk with other phytohormones on cellular growth and mitosis to endoreduplication transition. Int J Adv Res Biol Sci 1(6):136–153

Murti GSR, Upreti KK (2000) Plant hormones. In: Hemantaranjan A (ed) Advances in plant physiology, vol 3. Scientific Publishers, Jodhpur, pp 109–148

Mustafavi SH, Badi HN, Sekara A, Al E (2018) Polyamines and their possible mechanisms involved in plant physiological processes and elicitation of secondary metabolites. Acta Physiol Plant 40(102):102

Mutlu S, Karadağoğlu Ö, Atici Ö, Nalbantoğlu B (2013) Protective role of salicylic acid applied before cold stress on antioxidative system and protein patterns in barley apoplast. Biol Plant 57(3):507–513

Nadeem SM, Ahmad M, Zahir ZA, Kharal MA (2016) Role of phytohormones in stress tolerance of plants. In: Hakeem KR, Akhtar MS (eds) Plant, soil and microbes. Springer International Publishing, Cham, pp 385–421

Nadella V, Shipp MJ, Muday GK, Wyatt SE (2006) Evidence for altered polar and lateral auxin transport in the gravity persistent signal (gps) mutants of *Arabidopsis*. Plant Cell Environ 29(4): 682–690

Nakano T, Suzuki K, Fujimura T, Shinshi H (2006) Genome-wide analysis of the ERF gene family in *Arabidopsis* and rice. Plant Physiol 140(2):411–432

Nambara E, Marion-Poll A (2005) Abscisic acid biosynthesis and catabolism. Annu Rev Plant Biol 56:165–185

Naqvi SSM (1999) Plant hormones and stress phenomenon. In: Pessarakli M (ed) Handbook of plant and crop stress. Marcel-Dekker, New York, pp 709–730

Narayana I, Lalonde S, Saini HS (1991) Water-stress-induced ethylene production in wheat: a fact or artifact? Plant Physiol 96(2):406–410

Navalón A, Blanc R, Vilchez JL (1997) Determination of 1-naphthylacetic acid in commercial formulations and natural waters by solid-phase spectrofluorimetry. *Mikcrochim Acta* **126**:33–38

Navarre DA, Mayo D (2004) Differential characteristics of salicylic acid-mediated signaling in potato. Physiol Mol Plant Pathol 64(4):179–188

Navarro A, Sanchez-Blanco MJ, Banon S (2007) Influence of paclobutrazol on water consumption and plant performance of *Arbutus unedo* seedlings. Sci Hortic 111(2):133–139

Nayyar H, Bains TS, Kumar S, Kaur G (2005) Chilling effects during seed filling on accumulation of seed reserves and yield of chickpea. J Sci Food Agric 85(11):1925–1930

Nazar R, Khan MIR, Iqbal N, Masood A, Khan NA (2014) Involvement of ethylene in reversal of salt-inhibited photosynthesis by sulfur in mustard. Physiol Plant 152(2):331–344

Nemhauser JL, Hong F, Chory J (2006) Different plant hormones regulate similar processes through largely non overlapping transcriptional responses. Cell 126(3):467–475

Nemhauser JL, Mockler TC, Chory J (2004) Interdependency of brassinosteroid and auxin signaling in *Arabidopsis*. PLoS Biol 2(9):1460–1471

Nie GY, Robertson EJ, Fryer MJ, Leech RM, Bake NR (1995) Response of the photosynthetic apparatus in maize leaves grown at low temperature on transfer to normal temperature. Plant Cell Environ 18(1):1–12

Nishida I, Murata N (1996) Chilling sensitivity in plants and cyanobacteria: the crucial contribution of membrane lipids. Ann Rev Plant Physiol Plant Mol Biol 47(1):541–568

Nishiyama R, Watanabe Y, Fujita Y, Tien LD, Kojima M, Werner T, Vankova R, Yamaguchi-Shinozaki K, Shinozaki K, Kakimoto T, Sakakibara H, Schmuelling T, Lam-Son PT (2011) Analysis of cytokinin mutants and regulation of cytokinin metabolic genes reveals important

regulatory roles of cytokinins in drought, salt and abscisic acid responses, and abscisic acid biosynthesis. Plant Cell 23(6):2169–2183
Niu NN, Liang WQ, Yang XJ, Jin WL, Wilson ZA, Hu JP, Zhang D (2013) EAT1 promotes tapetal cell death by regulating aspartic proteases during male reproductive development in rice. Nat Commun 4:1445
Niu ZM, Chen MK, Sun JB, Lu J, Xu YM (1998) Effects of bud notching on the endogenous hormone contents and shoot sprouting of apple trees. J Fruit Sci 15:198–202
Nolan T, Chen J, Yin Y (2017a) Cross-talk of brassinosteroid signaling in controlling growth and stress responses. Biochem J 474(16):2641–2661
Nolan TM, Brennan B, Yang M, Chen J, Zhang M, Li Z, Wang X, Bassham DC, Walley J, Yin Y (2017b) Selective autophagy of BES1 mediated by DSK2 balances plant growth and survival. Dev Cell 41(1):33–46
Nolan TM, Vukasinovic N, Liu D, Russinova E, Yin Y (2020) Brassinosteroids: multidimensional regulators of plant growth, development, and stress responses. Plant Cell 32(2):295–318
Northey JG, Liang S, Jamshed M, Deb S, Foo E, Reid JB, McCourt P, Samuel MA (2016) Farnesylation mediates brassinosteroid biosynthesis to regulate abscisic acid responses. Nat Plants 2(8):16114
Ntatsi G, Savvas D, Druege U, Schwarz D (2013) Contribution of phytohormones in alleviating the impact of sub-optimal temperature stress on grafted tomato. Sci Hortic 149:28–38
Nugent J, Lang G, Shane W, Perry R (2001) Dwarf rootstocks for sweet cherries have benefits. Fruit Growers News 40:28
O'Brien JA, Benkova E (2013) Cytokinin cross-talking during biotic and abiotic stress responses. Front Plant Sci 4:451
Ogasawara N, Hiramasu T, Ishiyama K, Fushimi H, Suzuki H, Takagi H (2001) Effects of gibberellic acid and temperature on growth and root carbohydrates of *delphinium* seedlings. Plant Growth Regul 33(3):181–187
Ogawa D, Nakajima N, Sano T, Tamaoki M, Aono M, Kwbo A, Kanna M, Ioki M, Kamada H, Saji H (2005) Salicylic acid accumulation under O_3 exposure is regulated by ethylene in tobacco plants. Plant Cell Physiol 46(7):1062–1072
Ogawa M, Hanada A, Yamauchi Y, Kuwahara A, Kamiya Y, Yamaguchi S (2003) Gibberellin biosynthesis and response during Arabidopsis seed germination. Plant Cell 15(7):1591–1604
Ogweno JO, Song XS, Shi K, Hu WH, Mao WH, Zhou YH, Yu JQ, Nogues S (2008) Brassinosteroids alleviate heat-induced inhibition of photosynthesis by increasing carboxylation efficiency and enhancing antioxidant systems in *Lycopersicon esculentum*. J Plant Growth Regul 27(1):49–57
Okrent RA, Brooks MD, Wildermuth MC (2009) Arabidopsis GH3.12 (PBS3) conjugates amino acids to 4-substituted benzoates and is inhibited by salicylate. J Biol Chem 284(15):9742–9754
Oliver SN, Dennis ES, Dolferus R (2007) ABA regulates apoplastic sugar transport and is a potential signal for cold-induced pollen sterility in rice. Plant Cell Physiol 48(9):1319–1330
Öpik H, Rolfe SA, Willis AJ, Street HE (2005) *The physiology of flowering plants*, 4th edn. Cambridge University Press, p 191
Orabi SA, Dawood MG, Saleem SR (2015) Comparative study between the physiological role of hydrogen peroxide and salicylic acid in alleviating low temperature stress on tomato plants grown under sandponic culture. Sci Agric 9(1):49–59
Osborne DJ, McManus MT (2005) Hormones, signals and target cells in plant development. Cambridge University Press, p 158
Ouellette DR, Unrath CR, Young E (1996) Manual and chemical branch inducement in fall- and spring-planted 'empire' apple on two rootstocks. HortScience 31(1):82–88
Ouzounidou G, Giannakoula A, Asfi M, Ilias I (2011) Differential responses of onion and garlic against plant growth regulators. Pak J Bot 43(4):2051–2057
Pál M, Kovács V, Szalai G, Soós V, Ma X, Liu H, Mei H, Janda T (2014) Salicylic acid and abiotic stress responses in rice. J Agro Crop Sci 200(1):1–11

Paldi E (2002) Role of some N-containing compounds in chilling tolerance of maize. Acta Biol Szeged 46(3–4):99–100

Palta JA, Turner NC (2018) Crop root system traits cannot be seen as a silver bullet delivering drought resistance. Plant Soil 439:1–13

Pan J, Zhang M, Kong X, Xing X, Liu Y, Zhou Y, Liu Y, Sun L, Li D (2012) ZmMPK17, a novel maize group D MAP kinase gene, is involved in multiple stress responses. Planta 235(4): 661–676

Panda SK, Chaudhary I, Khan MH (2003a) Heavy metal induced lipid peroxidation and affects antioxidants in wheat leaves. Biol Plant 46(2):289–294

Panda SK, Sinha LB, Khan MH (2003b) Does aluminium phyto-toxicity induce oxidative stress in green gram (*Vigna radiate*). Bulg J Plant Physiol 29(1–2):77–86

Pandey A, Sharma M, Pandey GK (2016) Emerging roles of strigolactones in plant responses to stress and development. Front Plant Sci 7:434

Pandey S, Carrer M, Castagneri D, Petit G (2018) Xylem anatomical responses to climate variability in Himalayan birch trees at one of the world's highest forest limits. Perspect Plant Ecol Evol Syst 33:34–41

Parry AD (1993) Abscisic acid metabolism. Methods Plant Biochem 9:381–402

Pastor V, Luna E, Mauch-Mani B, Ton J, Flors V (2013) Primed plants do not forget. Environ Exp Bot 94:45–56

Patade VY, Khatri D, Manoj K, Kumari M, Ahmed Z (2012) Cold tolerance in thiourea primed capsicum seedlings is associated with transcript regulation of stress responsive genes. Mol Biol Rep 39(12):10603–10613

Patel AH, Tandel YN, Patel RR, Bhatt AH, Patel BA (2016) Effect of nutrients and thiourea on economics of mango (*Mangifera indica* L.) cv. Kesar. *Int J Sci Enviorn Technol* 5(5):490–492

Patel D, Franklin KA (2009) Temperature-regulation of plant architecture. Plant Signal Behav 4(7): 577–579

Pauwels L, Goossens A (2011) The JAZ proteins: a crucial interface in the jasmonate signaling cascade. Plant Cell 23(9):3089–3100

Pauwels L, Inze D, Goossens A (2009) Jasmonate-inducible gene: what does it mean? Trends Plant Sci 14(2):87–91

Pauwels L, Morreel K, De Witte E, Lammertyn F, Van Montagu M, Boerjan W (2008) Mapping methyl jasmonate-mediated transcriptional reprogramming of metabolism and cell cycle progression in cultured *Arabidopsis* cells. Proc Natl Acad Sci U S A 105(4):1380–1385

Peano C, Giacalone G, Chiabarndo V, la Iacona T (2005) Potatura lunga e diradamento delle gemme nel controllo vegeto-produttivo del melo. Riv Frutt Ortofloricoltura 67:38–41

Pearce R (2001) Plant freezing and damage. Ann Bot 87(4):417–424

Pearce RS (1999) Molecular analysis of acclimation to cold. Plant Growth Regul 29:47–76

Peleg Z, Blumwald E (2011) Hormone balance and abiotic stress tolerance in crop plants. Curr Opin Plant Biol 14(3):290–295

Peng Y, Arora R, Li G, Wang X, Fessehaie A (2008) Rhododendron catawbiense plasma membrane intrinsic proteins are aquaporins, and their over-expression compromises constitutive freezing tolerance and cold acclimation ability of transgenic Arabidopsis plants. Plant Cell Environ 31(3):1275–1289

Peng Z, Lu Q, Verma DP (1996) Reciprocal regulation of delta 1-pyrroline-5-carboxylate synthetase and proline dehydrogenase genes controls proline levels during and after osmotic stress in plants. Mol Gen Genet 253(3):334–341

Pepperman AB, Bradow JM (1988) Strigol analogs as germination regulators in weed and crop seed. Weed Sci 36(6):719–725

Pereira da Cruz R, Golombieski JI, Bazana MT, Cabreira C, Silveira TF, Picolli da Silva L (2010) Alterations in fatty acid composition due to cold exposure at the vegetative stage in rice. Braz J Plant Physiol 22(3):199–207

Perera PI, Vidhanaarachchi VRM, Gunathilake TR, Yakandawala DMD, Hocher V, Verdeil JL, Weerakoon LK (2009) Effect of plant growth regulators on ovary culture of coconut (*Cocos nucifera* L.). Plant Cell Tissue Organ Culture 99(1):73–81
Perry R (1999) Debudding: clothespins critical to developing dwarf sweet cherries. Fruit Growers News 38:34–35
Pierik R, Tholen D, Poorter H, Visser EJ, Voesenek LACJ (2006) The Janus face of ethylene: growth inhibition and stimulation. Trends Plant Sci 11(4):176–183
Pierre-Jetome E, Drapek C, Benfey PN (2018) Regulation of division and differentiation of plant stem cells. Annu Rev Cell Dev Biol 34:269–310
Pinhero RG, Fletcher RA (1994) Paclobutrazol and ancymidol protect corn seedlings from high and low temperature stresses. Plant Growth Regul 15(1):47–53
Pocock TH, Hurry VM, Savitch LV, Huner NPA (2001) Susceptibility to low-temperature photoinhibition and the acquisition of freezing tolerance in winter and spring wheat: the role of growth temperature and irradiance. Physiol Plant 113(4):499–506
Pospíšilová J (2003a) Interaction of cytokinins and abscisic acid during regulation of stomatal opening in bean leaves. Photosynthetica 41(1):49–56
Pospíšilová J (2003b) Participation of phytohormones in the stomatal regulation of gas exchange during water stress. Biol Plant 46(4):491–506
Pospisilova J, Synkova H, Rulcova J (2000) Cytokinin and water stress. Biol Plant 43(3):321–328
Potters G, Pasternak TP, Guisez Y, Jansen MAK (2009) Different stresses, similar morphogenic responses: integrating a plethora of pathways. Plant Cell Environ 32(2):158–169
Prasad TK (1996) Mechanisms of chilling-induced oxidative stress injury and tolerance: changes in antioxidant system, oxidation of proteins and lipids and protease activities. Plant J 10(6): 1017–1026
Prasad TK (1997) Role of catalase in inducing chilling tolerance in pre-emergence maize seedlings. Plant Physiol 114(4):1369–1376
Prasad TK (2000) The role of oxidative stress in mediating the mechanisms of chilling injury and tolerance in maize. In: Basra AS (ed) Crop responses and adaptations to temperature stress. Food Products Press, Ney York, pp 1–52
Prasad TK, Anderson MD, Stewart CR (1994a) Evidence for chilling induced oxidative stress in maize seedlings and a regulatory role for hydrogen peroxide. Plant Cell 6(1):65–74
Prasad TK, Anderson MD, Stewart CR (1994b) Acclimation, hydrogen peroxide and abscisic acid protect mitochondria against irreversible chilling injury in maize seedlings. Plant Physiol 105(2):619–627
Premaradhya N, Shashidhar KS, Samuel Jeberson RK (2018) Effect and profitability of foliar application of thiourea on growth and yield attributes of lentil (*Lens culinaris* L.) under Manipur conditions of north-east, India. Int J Curr Microbiol App Sci 7(05):1040–1050
Pritchard J, Barlow PW, Adams JS, Tomos AD (1990) Biophysics of the inhibition of the growth of maize roots by lowered temperature. Plant Physiol 93(1):222–230
Prusinski J, Borowska M and Kaszkowiak E (1999) Effect of growth regulator, N-fertility, and micronutrients on seed yield and yield components in yellow lupin. Proc. of the 9th International Lupin Conference, Lupin, an ancient crop for the new millennium, Klink/Müritz, Germany, pp. 230–234
Puijalon S, Bouma TJ, Douady CJ, Groenendael JV, Anten NPR, Martel E, Bornette G (2011) Plant resistance to mechanical stress: evidence of an avoidance—tolerance trade-off. New Phytol 191(4):1141–1149
Qi F, Li J, Duan L, Li Z (2005) Inductions of coronatine and MeJA to low-temperature resistance of wheat seedlings. Acta Bot Boreal Occident Sin 26:1776–1780
Qi YH, Yamauchi Y, Ling JQ, Kawano N, Li DB, Tanaka K (2004) Cloning of a putative monogalactosyldiacylglycerol synthase gene from rice (*Oryza sativa* L.) plants and its expression in response to submergence and other stresses. Planta 219(3):450–458
Rademacher W (1994) Gibberellin formation in microorganisms. Plant Growth Regul 15(3): 303–314

Rademacher W (2000) Growth retardants: effects on gibberellin biosynthesis and other metabolic pathways. Annu Rev Plant Physiol Plant Mol Biol 51:501–531

Rademacher W (2015) Plant growth regulators: backgrounds and uses in plant production. J Plant Growth Regul 34(4):845–872

Rademacher W, Spinelli F, Costa G (2006) Prohexadione-ca: modes of action of a multi-functional plant bioregulator for fruit trees. Acta Hortic 727:97–106

Radi A, Lange T, Niki T, Koshioka M, Lange MJ (2006) Ectopic expression of pumpkin gibberellin oxidases alters gibberellins biosynthesis and development of transgenic *Arabidopsis* plants. Plant Physiol 140(2):528–536

Rahman A (2013) Auxin: a regulator of cold stress response. Physiol Plant 147(1):28–35

Rakhmankulova ZF, Fedyaev VV, Rakhmatulina SR, Ivanov CP, Gilvanova IR, Usmanov IY (2010) The effect of wheat seed presowing treatment with salicylic acid on its endogenous content, activities of respiratory pathways, and plant antioxidant status. Russ J Plant Physiol 57(6):778–783

Ramakrishnan A, Tam HM, Wani SP, Long TD (2006) Effect of mulch on soil temperature, moisture, weed infestation and yield of groundnut in northern Vietnam. Field Crops Res 95(2–3):115–125

Raskin I, Skubatz H, Tang W, Meeuse BJD (1990) Salicylic acid levels in thermogenic and non-thermogenic plants. Ann Bot 66(4):376–373

Ratnakumar P, Deokate PP, Rane J, Jain N, Kumar V, Drick V, Minhas PS (2016) Effect of ortho-silicic acid exogenous application on wheat (*Triticum aestivum*) under drought. J Func Environ Bot 6(1):34–42

Ratnakumar P, Singh Y, Sureshkumar P, Bhagat K, Nangare DD, Taware PB, Minhas PS (2015) Ray of hope after hailstorm: a delightful boon by bioregulator- ortho silicic acid. ICAR Newsl:2–3

Ravanfar SA, Aziz MA, Kadir MA, Rashid AA, Sirchi MHT (2009) Plant regeneration of Brassica oleracea subsp. italica (broccoli) cv Green marvel as affected by plant growth regulators. Afr J Biotechnol 8(11):2523–2528

Reinbothe C, Springer A, Samol I, Reinbothe S (2009) Plant oxylipins: role of jasmonic acid during programmed cell death, defence and leaf senescence. FEBS J 276(17):4666–4681

Reis RS, Vale EM, Heringer AS, Al E (2016) Putrescine induces somatic embryo development and proteomic changes in embryogenic callus of sugarcane. J Proteome 130:170–179

Rejeb KB, Abdelly C, Savouré A (2014) How reactive oxygen species and proline face stress together. Plant Physiol Biochem 80:278–284

Rhodes D, Hanson AD (1993) Quaternary ammonium and tertiary sulfonium compounds in higher plants. Ann Rev Plant Physiol Plant Mol Biol 44:357–384

Rhodes D, Rich PJ, Brunk DG, Ju GC, Rhodes JC, Pauly MH, Hansen LA (1989) Development of two isogenic sweet corn hybrids differing for glycinebetaine content. Plant Physiol 91(3): 1112–1121

Rhodes D, Rich PJ, Myers AC, Reuter CC, Jamieson GC (1987) Determination of betaines by fast atom bombardment mass spectrometry: identification of glycinebetaine deficient genotypes of *Zea mays*. Plant Physiol 84(3):781–788

Ribascarbo M, Aroca R, Gonzalez-Meler MA, Irigoyen JJ, Sanchezdiaz M (2000) The electron partitioning between the cytochrome and alternative respiratory pathways during chilling recovery in 2 cultivars of maize differing in chilling sensitivity. Plant Physiol 122(1):199–204

Riechmann JL, Heard J, Martin G, Reuber L, Jiang C, Keddie J, Adam L, Pineda O, Ratcliffe OJ, Samaha RR, Creelman R, Pilgrim M, Broun P, Zhang JZ, Ghandehari D, Sherman BK, Yu G (2000) Arabidopsis transcription factors: genome-wide comparative analysis among eukaryotes. Science 290(5499):2105–2110

Riken A, Dilwith JW, Bergman DK (1993) Correlation between the circadian rhythm of resistance to extreme temperatures and changes in fatty acid composition in cotton seedlings. Plant Physiol 101(1):31–36

Ristic Z, Yang G, Sterzinger A, Zhang L (1998) Higher chilling tolerance in maize is not always related to the ability for greater and faster abscisic acid accumulation. J Plant Physiol 153(1–2): 154–162

Ritonga FN, Chen S (2020) Physiological and molecular mechanism involved in cold stress tolerance in plants. Plant 9(5):560

Rivas-San VM, Plasencia J (2011) Salicylic acid beyond defense: its role in plant growth and development. J Exp Bot 62(10):3321–3338

Robinson T, Hoying S, Andersen R (2005) Management of high-density sweet-cherry orchards. NY Fruit Q 13:24–27

Robison MM, Shah S, Tamot B, Pauls KP, Moffatt BA, Glick BR (2001) Reduced symptoms of Verticillium wilt in transgenic tomato expressing a bacterial ACC deaminase. Mol Plant Pathol 2(3):135–145

Rogers RA, Dunn JH, Nelson CJ (1977) Photosynthesis and cold hardening in zoysia and bermudagrass. Crop Sci 17(5):727–732

Roy M, Wu R (2001) Arginine decarboxylase transgene expression and analysis of environmental stress tolerance in transgenic rice. Plant Sci 160(5):869–875

Roy M, Wu R (2002) Overexpression of S-adenosylmethionine decarboxylase gene in rice increases polyamine level and enhances sodium chloride-stress tolerance. Plant Sci 163(5): 987–992

Rubio S, Pérez FJ (2019) ABA and its signaling pathway are involved in the cold acclimation and deacclimation of grapevine buds. Sci Hortic 256:108565

Rudolph AS, Crowe JH, Crowe LM (1986) Effects of three stabilizing agents—proline, betaine, and trehalose–on membrane phospholipids. Arch Biochem Biophys 245(1):134–143

Ruelland E, Vaultier MN, Zachowski A, Hurry V, Kader JC, Delseny M (2009) Cold signalling and cold acclimation in plants. Adv Bot Res 49:35–150

Ruelland E, Zachowski A (2010) How plants sense temperature. Environ Exp Bot 69(3):225–232

Ruyter-Spira C, Al-Babili S, van der Krol S, Bouwmeester H (2013) The biology of Strigolactones. Trends Plant Sci 18(2):72–83

Ruyter-Spira C, Kohlen W, Charnikhova T, van Zeijl A, van Bezouwen L, de Ruijter N, Cardoso C, Lopez-Raez JA, Matusova R, Bours R, Verstappen F, Bouwmeester H (2011) Physiological effects of the synthetic strigolactone analog GR24 on root system architecture in Arabidopsis: another belowground role for strigolactones? Plant Physiol 155(2):721–734

Rymen B, Fiorani F, Kartal F, Vandepoele K, Inze D, Beemster GT (2007) Cold nights impair leaf growth and cell cycle progression in maize through transcriptional changes of cell cycle genes. Plant Physiol 143(3):1429–1438

Ryu H, Cho YG (2009) Plant hormones in salt stress tolerance. J Plant Biol 58(3):147–155

Sajid GM, Kaukab M, Ahmad Z (2009) Foliar application of plant growth regulators (PGRs) and nutrients for improvement of lily flowers. Pak J Bot 41(1):233–237

Sakakibara H (2006) Cytokinins: activity, biosynthesis, and translocation. Annu Rev Plant Biol 57: 431–449

Sakamoto A, Murata N (2002) The role of glycinebetaine in the protection of plants from stress clues from transgenic plants. Plant Cell Environ 25(2):163–171

Sakata T, Oda S, Tsunaga Y, Shomura H, Kawagishi-Kobayashi M, Aya K, Saeki K, Endo T, Nagano K, Kojima M, Sakakibara H, Watanabe M, Matsuoka M, Higashitani A (2014) Reduction of gibberellin by low temperature disrupts pollen development in rice. Plant Physiol 164(4):2011–2019

Sakhabutdinova AR, Fatkhutdinova DR, Bezrukova MV, Shakirova FM (2003) Salicylic acid prevents the damaging action of stress factors on wheat plants. Bulg J Plant Physiol., Special Issue:314–319

Sanchez-Ballesta MT, Lafuentea MT, Zacarias L, Granell A (2000) Involvement of phenylalanine ammonia-lyase in the response of *fortune mandarin* fruits to cold temperature. Physiol Plant 108(4):382–389

Sanghera GS, Wani SH, Hussain W, Singh NB (2011) Engineering cold stress tolerance in crop plants. Curr Genomics 12(1):30–43

Santis A, Landi P, Genchi G (1999) Changes of mitochondrial properties in maize seedlings associated with selection for germination at low temperature, fatty acid composition, cytochrome C oxidase and adenine nucleotide translocase activities. Plant Physiol 119(2):743–754

Santner A, Calderon-Villalobos LI, Estelle M (2009) Plant hormones are versatile chemical regulators of plant growth. Nat Chem Biol 5(5):301–307

Saradhi PP, Alia AS, Prasad KVSK (1995) Proline accumulates in plants exposed to UV radiation and protects them against UV induced peroxidation. Biochem Biophys Res Comm 209(1):1–5

Satisha J, Prakash GS, Murti GSR, Upreti KK (2005) Response of grape genotypes to water deficit: root, shoot growth and endogenous hormones. Ind J Plant Physiol 10(3):225–230

Sayyari M, Ghanbari F, Fatahi S, Bavandpour F (2013) Chilling tolerance improving of watermelon seedling by salicylic acid seed and foliar application. Not Sci Biol 5(1):67–73

Scott IM, Clarke SM, Wood JE, Mur LA (2004) Salicylate accumulation inhibits growth at chilling temperature in Arabidopsis. Plant Physiol 135(2):1040–1049

Scott IM, Clarke SM, Wood JE, Mur LAJ (2006) Salicylate accumulation inhibits growth at chilling temperatures in Arabidopsis. Plant Physiol 135(2):1040–1049

Scotter MJ, Roberts DPT, Wilson LA, Howard FAC, Davis J, Mansell N (2007) Free salicylic acid and acetyl salicylic acid content of foods using gas chromatography–mass spectrometry. Food Chem 105(1):273–279

Seki M, Ishida J, Narusaka M, Fujita M, Nanjo T, Umezawa T, Kamiya A, Nakajima M, Enju A, Sakurai T, Satou M, Akiyama K, Yamaguchi-Shinozaki K, Carninci P, Kawai J, Hayashizaki Y, Shinozaki K (2002) Monitoring the expression pattern of ca 7000. Arabidopsis genes under ABA treatment using full length cDNA microarray. Funct Integr Genomics 2(6):282–291

Seleiman MF, Kheir AMS (2018) Saline soil properties, quality and productivity of wheat grown with bagasse ash and thiourea in different climatic zones. Chemosphere 193:538–546

Sembdner G, Parthier B (1993) The biochemistry and the physiological and molecular actions of jasmonates. Annu Rev Plant Physiol Plant Mol Biol 44:569–589

Senaratna T, Merritt D, Dixon K, Bunn E, Touchell D, Sivasithamparam K (2003) Benzoic acid may act as the functional group in salicylic acid and derivatives in the induction of multiple stress tolerance in plants. Plant Growth Regul 39(1):77–81

Senaratna T, Touchell D, Bunn T, Dixon K (2000) Acetyl salicylic acid (aspirin) and salicylic acid induce multiple stress tolerance in bean and tomato plants. Plant Growth Regul 30(2):157–161

Seo HS, Song JT, Cheong JJ, Lee YH, Lee YW, Hwang I, Lee JS, Yang DC (2001) Jasmonic acid carboxyl methyltransferase: a key enzyme for jasmonate regulated plant responses. Proc Natl Acad Sci USA 98(8):4788–4793

Seo JS, Joo J, Kim MJ, Kim YK, Nahm BH, Song SI, Cheong JJ, Lee JS, Kim JK, Choi YD (2011) OsbHLH148, a basic helix-loop-helix protein, interacts with OsJAZ proteins in a jasmonate signaling pathway leading to drought tolerance in rice. Plant J 65(6):907–921

Serrago AR, Miralles DJ, Slafer GA (2008) Floret fertility in wheat as affected by photoperiod during stem elongation and removal of spikelets at booting. Eur J Agron 28(3):301–308

Serraj R, Sinclair TR (2002) Osmolyte accumulation can it really help increase crop yield under drought conditions. Plant Cell Environ 25(2):333–341

Serrano M, Mc M-M, Mt P (1997) Modified atmosphere packaging minimizes increases in putrescine and abscisic acid levels caused by chilling injury in pepper fruit. J Agric Food Chem 45(5):1668–1672

Serrano M, Pretel MT, Martinezmadrid MC (1998) CO_2 treatment of zucchini squash reduces chilling-induced physiological changes. J Agric Food Chem 46(7):2465–2468

Setia RC, Bhathal G, Setia N (1995) Influence of paclobutrazol on growth and yield of *Brassica carinata* A Br. Plant Growth Regul 16(2):121–127

Setia RC, Dhanda A, Setia N (2006) Chilling tolerance and related antioxidant enzyme activities induced by salicylic acid in leaves of *Brassica juncea* (L.) Czern and Coss. Environ Ecol 24: 886–889

Zhang Z, Huang R (2010) Enhanced tolerance to freezing in tobacco and tomato overexpressing transcription factor TERF2/LeERF2 is modulated by ethylene biosynthesis. Plant Mol Biol 73: 241–249

Seymour M (1996) Response of narrow-leafed lupine (*Lupinus angustifolius* L.) to foliar application of growth regulators in Western Australia. Aust J Exp Agric 36(4):473–478

Shah F, Huang J, Cui K, Nie L, Shah T, Chen C, Wang K (2011) Impact of high-temperature stress on rice plant and its traits related to tolerance. Agric Sci 149(5):545–556

Shakirova FM, Sakhabutdinova AR, Bezrukova MV, Fatkhutdinova RA, Fatkhutdinova DR (2003) Changes in the hormonal status of wheat seedlings induced by salicylic acid and salinity. Plant Sci 164(3):317–322

Shanker AK, Shanker C (2016) Abiotic and biotic stress in plants—recent advances and future perspectives. Intech Open, Rijeka, Rijeka

Shavrukov Y, Kurishbayev A, Jatayev S, Shvidchenko V, Zotova L, Koekemoer F, de Groot S, Soole K, Langridge P (2017) Early flowering as a drought escape mechanism in plants: how can it aid wheat production? Front Plant Sci 17:8

Shekoofa A, Emam Y (2008) Effects of nitrogen fertilization and plant growth regulators (PGRs) on yield of wheat (*Triticum aestivum* L.) cv. Shiraz. *J Agric Sci Technol* 10:101–108

Shi Y, Tian S, Hou L, Huang X, Zhang X, Guo H, Yang S (2012) Ethylene signaling negatively regulates freezing tolerance by repressing expression of CBF and type-A ARR genes in Arabidopsis. Plant Cell 24(6):2578–2595

Shinohara N, Taylor C, Leyse O (2013) Strigolactone can promote or inhibit shoot branching by triggering rapid depletion of the auxin efflux protein pin1 from the plasma membrane. PLoS Biol 11(1):e1001474

Shkolnik-Inbar D, Bar-Zvi D (2010) ABI4 mediates abscisic acid and cytokinin inhibition of lateral root formation by reducing polar auxin transport in *Arabidopsis*. Plant Cell 22(11):3560–3573

Sicher R (2011) Carbon partitioning and the impact of starch deficiency on the initial response of Arabidopsis to chilling temperatures. Plant Sci 181(2):167–176

Sharma M, Laxmi A (2015) Jasmonates: emerging players in controlling temperature stress tolerance. Front Plant Sci 6:1129

Sharoni AM, Nuruzzaman M, Satoh K, Shimizu T, Kondoh H, Sasaya T, Choi IR, Omura T, Kikuchi S (2011) Gene structures, classification and expression models of the AP2/EREBP transcription factor family in rice. Plant Cell Physiol 52(2):344–360

Sheng XF, Xia JJ (2006) Improvement of rape (*Brassica napus*) plant growth and cadmium uptake by cadmium-resistant bacteria. Chemosphere 64(6):1036–1042

Shibasaki K, Uemura M, Tsurumi S, Rahman A (2009) Auxin response in Arabidopsis under cold stress: underlying molecular mechanisms. Plant Cell 21(12):3823–3838

Shigenaga AM, Argueso CT (2016) No hormone to rule them all: interactions of plant hormones during the responses of plants to pathogens. Semin Cell Dev Biol 56:174–189

Shinozaki K, Yamaguchi-Shinozaki K, Seki M (2003) Regulatory network of gene expression in the drought and cold stress responses. Curr Opin Plant Biol 6(5):410–417

Shirzad S, Arooie H, Azizi M (2012) Influence of plant growth regulators (PGRs) and planting method on growth and yield in oil pumpkin (*Cucurbita pepo* var. styriaca). Notulae Sci Biol 4(2):101–107

Shomeili M, Nabipour M, Meskarbashee M, Memari HR (2011) Effects of gibberellic acid on sugarcane plants exposed to salinity under a hydroponic system. African J Plant Sci 5(10): 609–616

Shulaev V, León J, Raskin I (1995) Is salicylic acid a translocated signal of systemic acquired resistance in tobacco? Plant Cell 7(10):1691–1701

Siboza XI, Bertling I, Odindo AO (2014) Salicylic acid and methyl jasmonate improve chilling tolerance in cold-stored lemon fruit (*Citrus Limon*). J Plant Physiol 171(18):1722–1731

Singh I, Shono M (2005) Physiological and molecular effects of 24-epibrassinolide, a brassinosteroid on thermotolerance of tomato. Plant Growth Regul 47(2):111–119

Skirycz A, Claeys H, De Bodt S, Oikawa A, Shinoda S, Andriankaja M, Maleux K, Eloy NB, Coppens F, Yoo SD, Saito K, Inze D (2011) Pause-and-stop: the effects of osmotic stress on cell proliferation during early leaf development in Arabidopsis and a role for ethylene signaling in cell cycle arrest. Plant Cell 23(5):1876–1888

Skoog F, Armstrong DJ (1970) Cytokinins. Ann Rev Plant Physiol 21:359–384

Slafer GA, Abeledo LG, Miralles DJ, Gonzalez FG, Whitechurch EM (2001) Photoperiod sensitivity during stem elongation as an avenue to raise potential yield in wheat. Euphytica 119(1): 191–197

Slaymaker DH, Navarre DA, Clark D, del Pozo O, Martin GB, Klessig DF (2002) The tobacco salicylic acid-binding protein 3 (SABP3) is the chloroplast carbonic anhydrase, which exhibits antioxidant capacity and plays a role in the hypersensitive response. Proc Natl Acad Sci U S A 99(18):11640–11645

Smeets L, Wehner TC (1997) Environmental effects on genetic variation of chilling resistance in cucumber. Euphytica 97(2):217–225

Smirnoff N (1993) The role of active oxygen in response of plants to water deficit and desiccation. New Phytol 125(1):27–58

Smith PG, Dale JE (1988) The effects of root cooling and excision treatments on the growth of primary leaves of *Phaseolus vulgaris* L. rapid and reversible increases in abscisic acid content. New Phytol 110(3):293–300

Somerville C (1995) Direct tests of the role of membrane lipid composition in low-temperature induced photoinhibition and chilling sensitivity in plants and cyanobacteria. Proc Natl Acad Sci U S A 92(14):6215–6218

Song X, Li Y, Hou X (2013) Genome-wide analysis of the AP2/ERF transcription factor superfamily in Chinese cabbage (*Brassica rapa* ssp. pekinensis). *BMC Geno* **14**:article no. 573

Sonoike K (1999) The different roles of chilling temperatures in the photoinhibition of photosystem I and photosystem II. J Photochem Photobiol B Biol 48(2–3):136–141

Soto MJ, Fernandez-Aparicio MN, Castellanos-Morales V, Garcia-Garrido JM, Ocampo JA, Delgado MJ, Vierheilig H (2010) First indications for the involvement of strigolactones on nodule formation in alfalfa (*Medicago sativa*). Soil Biol Biochem 42(2):383–385

Sponsel VM, Hedden P (2004) Gibberellin, biosynthesis and inactivation. In: Davies PJ (ed) Plant hormones biosynthesis, signal transduction, action! Springer, Dordrecht, pp 63–94

Sreenivasulu N, Radchuk V, Alawady A, Borisjuk L, Weier D, Staroske N, Fuchs J, Miersch O, Strickert M, Usadel B, Wobus U, Grimm B, Weber H, Weschke W (2010) De-regulation of abscisic acid contents causes abnormal endosperm development in the barley mutant seg8. Plant J 64(4):589–603

Sreenivasulu N, Harshavardhan VT, Govind G, Seiler C, Kohli A (2012) Contrapuntal role of ABA: does it mediate stress tolerance or plant growth retardation under long-term drought stress? Gene 506(2):265–273

Srivalli B, Chinnusamy V, Khanna-Chopra R (2003) Antioxidant defense response to abiotic stresses in plants. J Plant Biol 30:121–139

Srivastava LM (2002) Plant growth and development: hormones and environment. Academic Press, San Diego, CA

Srivastava AK, Nathawat NS, Ramaswamy NK, Sahu MP, Singh G (2008) Evidence for thiol-induced enhanced in situ trans-location of 14C-sucrose from source to sink in *Brassica juncea*. Environ Exp Bot 64(3):250–255

Srivastava AK, Ramaswamy NK, Mukopadhayaya R, Jincy MGC, D'souza SF (2009) Thiourea modulates the expression and activity profile of mtATPase under salinity stress in seed of *Brassica juncea*. Ann Bot 103(3):403–410

Srivastava AK, Ramaswamy NK, Suprasanna P, D'souza SF (2010a) Genome-wide analysis of thiourea-modulated salinity stress-responsive transcripts in seeds of *Brassica juncea*: identification of signalling and effector components of stress tolerance. Ann Bot 106(5):663–674

Srivastava AK, Suprasanna P, Srivastava S, D'souza SF (2010b) Thiourea mediated regulation in the expression profile of aquaporins and its impact on water homeostasis under salinity stress in *Brassica juncea* roots. Plant Sci 178(6):517–522

Srivastava AK, Srivastava S, Penna S, D'Souza SF (2011) Thiourea orchestrates regulation of redox state and antioxidant responses to reduce the NaCl-induced oxidative damage in Indian mustard (*Brassica juncea* (L.) Czern.). Plant Physiol Biochem 49(6):676–686

Srivastava AK, Sablok G, Hackenberg M, Deshpande U, Suprasanna P (2017) Thiourea priming enhances salt tolerance through co-ordinated regulation of microRNAs and hormones in *Brassica juncea*. Sci Rep 7:45490

Staggenborg SA, Vanderlip RL (1996) Sorghum grain yield reductions caused by duration and timing of freezing temperatures. Agron J 88(3):473–477

Stamm P, Ramamoorthy R, Kumar PP (2011) Feeding the extra billions: strategies to improve crops and enhance future food security. Plant Biotechnol Rep 5(2):107–120

Stamp P, Feil B, Schortemeyer M, Richner W (1997) Responses of roots to low temperatures and nitrogen forms. In: Anderson HM, Barlow PW, Clarkson DT, Jackson MB, Shewry PR (eds) *Plant roots—from cell to system*, vol 73. Springer, Dordrecht, pp 143–154developments in plant and soil sciences

Starck Z, Choluj D, Niemyska B (1994) Effect of preceding temperature and subsequent condition on response of tomato plants to chilling. Acta Physiol Plant 16:329–336

Starck Z, Bogdan J, Choluj D (1997) Effect of phosphorus and potassium supply on tomato plant response to chilling stress. In: *Crop development for the cool and wet regions of Europe*, vol 814. COST, pp 49–56

Starck Z, Niemyska B, Bogdan J, Akour Tawalbeh RN (2000) Response of tomato plants to chilling stress in association with nutrient or phosphorus starvation. Plant Soil 226(1):99–106

Staswick PE, Tiryaki I (2004) The oxylipin signal jasmonic acid is activated by an enzyme that conjugates it to isoleucine in *Arabidopsis*. Plant Cell 16(8):2117–2127

Steponkus PL (1984) Role of plasma membrane in freezing injury and cold acclimation. Annu Rev Plant Physiol 35:543–584

Still JR, Pill WG (2003) Germination, emergence and seedlings growth of tomato and impatiens in response to seed treatment with paclobutrazol. Hortic Sci 38(6):1201–1204

Still JR, Pill WG (2004) Growth and stress tolerance of tomato seedlings (*Lycopersicon esculentum* mill.) in response to seed treatment with paclobutrazol. Hortic Sci 79(2):197–203

Stoll M, Loveys B, Dry P (2000) Hormonal changes induced by partial root zone drying of irrigated grapevine. J Exp Bot 51(350):1627–1634

Stupnikova IV, Borovskii GB, Voinikov VK (2001) ABA-induction of cold hardy state and heat stable COR proteins in maize seedlings and other cereals. Maize Genet Cooperation Newsl 75: 29–30

Subedi KD, Ma BL (2005) Seed priming does not improve corn yield in a humid temperate environment. Agron J 97(1):211–218

Sun XC, Hu CX, Tan QL, Liu JS, Liu HE (2009) Effects of molybdenum on expression of cold-responsive genes in abscisic acid (ABA)-dependent and ABA-independent pathways in winter wheat under low-temperature stress. Ann Bot 104(2):345–356

Sun Y, He Y, Irfan AR, Liu X, Yu Q, Zhang Q, Yang D (2020) Exogenous brassinolide enhances the growth and cold resistance of maize (*Zea mays* L.) seedlings under chilling stress. Agron 10(4):488

Sundaramurthi P, Patapoff TW, Suryanarayanan R (2010) Crystallization of trehalose in frozen solutions and its phase behavior during drying. Pharma Res 27(11):2374–2383

Suryavanshi P, Buttar GS, Brar AS (2016) Effect of osmoprotectants on performance of wheat (*Triticum aestivum*) under terminal heat stress condition of north-West India. Indian J Agric Sci 86:1037–1042

Suzuki N, Mittler R (2006) Reactive oxygen species and temperature stresses: a delicate balance between signaling and destruction. Physiol Plant 126(1):45–51

Swarup R, Perry P, Hagenbeek D, van der Straeten D, Beemster GTS, Sandberg G, Bhalerao R, Ljung K, Bennett MJ (2007) Ethylene upregulates auxin biosynthesis in Arabidopsis seedlings to enhance inhibition of root cell elongation. Plant Cell 19(7):20186–22196

Swiatek A, Van Dongen W, Esmans EL, Van-Onckelen H (2004) Metabolic fate of jasmonates in tobacco bright yellow-2 cells. Plant Physiol 135(1):161–172

Szalai G, Janda T, Paldi E, Szigeti Z (1996) Role of light in the development of post-chilling symptoms in maize. J Plant Physiol 148(3–4):378–383

Taiz l, Zeiger E (2010) Plant physiology. Sinauer associates, Inc, Sunderland, pp 866–867

Takahashi D, Li B, Nakayama T, Kawamura Y, Uemura M (2013) Plant plasma membrane proteomics for improving cold tolerance. Front Plant Sci 4:90

Takahashi F, Shinozaki K (2019) Long-distance signaling in plant stress response. Curr Opin Plant Biol 47(2):106–111

Tang RS, Zheng JC, Jin ZQ, Zhang DD, Huang YH, Chen LG (2008) Possible correlation between high temperature-induced floret sterility and endogenous levels of IAA, GA3 and ABA in rice (*Oryza sativa* L.). Plant Growth Regul 54(1):37–43

Tang D, Wang Y, Cai J, Zhao R (2009) Effects of exogenous application of plant growth regulators on the development of ovule and subsequent embryo rescue of stenospermic grape (*Vitis vinifera* L.). Sci Hortic 120(1):51–57

Taniguchi YY, Taniguchi M, Tsuge T, Oka A, Aoyama T (2010) Involvement of *Arabidopsis thaliana* phospholipase Dzeta 2 in root hydrotropism through the suppression of root gravitropism. Planta 231(2):491–497

Tanveer M, Shahzad B, Sharma A, Khan EA (2019) 24-Epibrassinolide application in plants: An implication for improving drought stress tolerance in plants. Plant Physiol Biochem 135:295–303

Tao Y, Ferrer J-L, Ljung K, Pojer F, Hong F, Long JA, Li L, Moreno JJ, Bowman ME, Ivans LJ, Cheng Y, Lim J, Zhao Y, Ballaré CL, Sandberg G, Noel JP, Chory J (2008) Rapid synthesis of auxin via a new tryptophan-dependent pathway is required for shade avoidance in plants. Cell 133(1):164–176

Tarakhovskaya ER, Maslov Y, Shishova MF (2007) Phytohormones in algae. Rus J Plant Physiol 54(2):163–170

Tasgin E, Atici O, Nalbantoglu B (2003) Effects of salicylic acid and cold on freezing tolerance in winter wheat leaves. Plant Growth Regul 41(3):231–236

Tasgin E, Atici O, Nalbantoglu B, Popova LP (2006) Effects of salicylic acid and cold treatments on protein levels and on the activities of antioxidant enzymes in the apoplast of winter wheat leaves. Phytochemistry 67(7):710–715

Taylor J, Bhalla M, Robertson J, Piening L (1990) Cytokinins and abscisic acid in hardening winter wheat. Can J Bot 68(7):1597–1601

Taylor AG, Allen PS, Bennett MA, Bradford KJ, Burris JS, Misra MK (1998) Seed enhancements. *Seed Sci Res* 8(2):245–256

Thakur P, Kumar S, Malik JA, Berger JD, Nayyar H (2010) Cold stress effects on reproductive development in grain crops: An overview. Environ Exp Bot 67(3):429–443

Theocharis A, Clément C, Ait Barka E (2012a) Physiological and molecular changes in plants grown at low temperatures. Planta 235(6):1091–1105

Theocharis A, Bordiec S, Fernandez O, Paquis S, Dhondt-Cordelier S, Baillieul F, Clément C, Barka EA (2012b) *Burkholderia phytofirmans* PsJN primes *Vitis vinifera* L. and confers a better tolerance to low nonfreezing temperatures. Mol Plant-Microbe Interact 25(2):241–249

Theron KI, Jacobs G, Strydom DK (1987) Correlative inhibition of axillary buds in apple nursery trees in relation to node position, defoliation, and Promalin application. J Am Soc Hortic Sci 112(5):732–734

Thingnaes E, Torre S, Ernstsen A, Moe R (2003) Day and night temperature responses in Arabidopsis: effects on gibberellin and auxin content, cell size, morphology and flowering time. Ann Bot 92(4):601–612

Thomas T, Gunnia UB, Yurkow EJ, Seibold JR, Thomas TJ (1993) Inhibition of calcium signalling in murine splenocytes by polyamines: differential effects of CD4 and CD8 T-cells. Biochem J 291(Pt 2):375–381

Thomashow MF (1999) Plant cold acclimation: freezing tolerance genes and regulatory mechanisms. Annu Rev Plant Physiol Plant Mol Biol 50:571–599

Thomashow MF (2001) So what's new in the field of plant cold acclimation? Lots! Plant Physiol 125(1):89–93

Thomashow MF (2010) Molecular basis of plant cold acclimation: insights gained from studying the CBF cold response pathway. Plant Physiol 154(2):571–577

Ting CS, Owens TG, Wolfe DW (1991) Seedling growth and chilling stress effects on photosynthesis in chilling-sensitive and chilling-tolerant cultivars of *Zea mays*. J Plant Physiol 137(5): 559–564

Tiwari DK, Pandey P, Giri SP, Dwivedi JL (2011) Effect of GA3 and other plant growth regulators on hybrid rice seed production. Asian J Plant Sci 10(2):133–139

Tolosa LN, Zhang Z (2020) The role of major transcription factors in Solanaceous food crops under different stress conditions: current and future perspectives. Plan Theory 9(1):56

Tripathi SK, Tuteja N (2007) Integrated signaling in flower senescence. Plant Signal Behav 2(6): 437–445

Trischuk RG, Schilling BS, Low NH, Gray GR, Gusta LV (2014) Cold acclimation, de-acclimation and re-acclimation of spring canola, winter canola and winter wheat: the role of carbohydrates, cold-induced stress proteins and vernalization. Environ Exp Bot 106:156–163

Tsai FY, Lin CC, Kao CH (1997) A comparative study of the effects of abscisic acid and methyl jasmonate on seedling growth of rice. Plant Growth Regul 21(1):37–42

Uddain J, Hossain KA, Mostafa MG, Rahman MJ (2009) Effect of different plant growth regulators on growth and yield of tomato. Int J Sustain Agric 1(3):58–63

Uemura M, Steponkus PL (1994) A contrast of the plasma membrane lipid composition of oat and rye leaves in relation to freezing tolerance. Plant Physiol 104(2):479–496

Ullah F, Bano A, Nosheen A (2012) Effects of plant growth regulators on growth and oil quality of canola (*Brassica napus* L.) under drought stress. Pak J Bot 44(6):1873–1880

Ullah A, Manghwar H, Shaban M, Khan AH, Akbar A, Ali U, Ali E, Fahad S (2018) Phytohormones enhanced drought tolerance in plants: a coping strategy. Environ Sci Pollut Res Int 25(33):33103–33118

Umehara M, Hanada A, Yoshida S, Akiyama K, Arite T, Takeda-Kamiya N, Magome H, Kamiya Y, Shirasu K, Yoneyama K, Kyozuka J, Yamaguchi S (2008) Inhibition of shoot branching by new terpenoid plant hormones. Nature 455(7210):195–200

Uno Y, Furihata T, Abe H, Yoshida R, Shinozaki K, Yamaguchi-Shinozaki K (2000) Arabidopsis basic leucine zipper transcription factors involved in an abscisic acid-dependent signal transduction pathway under drought and high-salinity conditions. Proc Natl Acad Sci U S A 97(21): 11632–11637

Uppalapati SR, Ishiga Y, Wangdi T, Kunkel BN, Anand A, Mysore KS, Bender CL (2007) The phytotoxin coronatine contributes to pathogen fitness and is required for suppression of salicylic acid accumulation in tomato inoculated with pseudomonas syringae pv tomato DC3000. Mol Plant-Microbe Interact 20(8):955–965

Upreti KK, Murti GSR, Bhatt RM (1998) Response of French bean cultivars to water deficits: changes in endogenous hormones, proline and chlorophyll. Biol Plant 40(3):381–388

Upreti KK, Murti GSR (1999) Water stress induced changes in root nodulation and cytokinin levels in French bean. Ind J Plant Physiol 26:1–4

Upreti KK, Murti GSR, Bhatt RM (2000) Response of pea cultivars to water stress: changes in morphophysiological characters, endogenous hormones and yield. Veg Sci 27(1):57–61

Upreti KK, Murti GSR (2004a) Changes in electrolyte leakage, chlorophyll concentration, endogenous hormones and bulb weight in onion in response to water stress. Trop Agric 81(2):127–132

Upreti KK, Murti GSR (2004b) Leaf growth and endogenous hormones under water stress and stress recovery in French bean. J Plant Biol 31:61–64

Upreti KK, Murti GSR (2004c) Effects of brassinosteroids on growth, nodulation, phytohormones content and nitrogenase activity in French bean under water stress. Biol Plant 48(3):407–411

Upreti KK, Murti GSR (2005) Water stress induced changes in common polyamines and abscisic acid in French bean. Ind J Plant Physiol 10(2):145–150

Upreti KK, Srinivasa Rao NK, Jayaram HL (2012) Floral abscission in capsicum under high temperature: role of endogenous hormones and polyamines. Ind J Plant Physiol 17(3):207–214

Vardhini CV, Anuradha S, Rao SSR (2006) Brassinosteroids: a great potential to improve crop productivity. Ind J Plant Physiol 11:1–12

Vardhini BV, Anjum NA (2015) Brassinosteroids make plant life easier under abiotic stresses mainly by modulating major components of antioxidant defense system. Front Environ Sci 2:1–16

Vardhini BV (2017) Modifications of morphological and anatomical characteristics of plants by application of brassinosteroids under various abiotic stress conditions—A review. Plant Gene 11(Part B):70–89

Verheul MJ, Picatto C, Stamp P (1996) Growth and development of maize (*Zea mays* L.) seedlings under chilling conditions in the field. Eur J Agron 5(1–2):31–43

Verma V, Ravindran P, Kumar PP (2016) Plant hormone-mediated regulation of stress responses. BMC Plant Biol 16(1):86

Verslues PE, Bray EA (2006) Role of abscisic acid (ABA) and *Arabidopsis thaliana* ABA-insensitive loci in low water potential induced ABA and Proline accumulation. J Exp Bot 57(1):201–212

Verslues PE, Agarwal M, Katiyar-Agarwal S, Zhu J, Zhu JK (2006) Methods and concepts in quantifying resistance to drought, salt and freezing, abiotic stresses that affect plant water status. Plant J 45(4):523–539

Veselova SV, Farhutdinov RG, Veselov SY, Kudoyarova GR, Veselov DS, Hartung W (2005) The effect of root cooling on hormone content, leaf conductance and root hydraulic conductivity of durum wheat seedlings (*Triticum durum* L.). J Plant Physiol 162:21–26

Veselova RG, Farkhutdinov D, Veselov S, Kudoyarova GR (2006) Role of cytokinins in the regulation of stomatal conductance of wheat seedlings under conditions of rapidly changing local temperature. Russ J Plant Physiol 53(6):756–761

Vettak-korumakankav NN, Falk D, Saxena P, Fletcher RA (1999) A crucial role for gibberellins in stress protection of plants. Plant Cell Physiol 40(5):542–548

Veyres N, Aono M, Sangwan-Norreel BS, Sangwan RS (2008) The Arabidopsis sweetie mutant is affected in carbohydrate metabolism and defective in the control of growth, development and senescence. Plant J 55(4):665–686

Vick BA, Zimmerman DC (1983) The biosynthesis of jasmonic acid: a physiological role for plant lipoxygenase. Biochem Biophys Res Commun 111(2):470–477

Vineeth TV, Kumar P, Krishna GK (2016) Bioregulators protected photosynthetic machinery by inducing expression of photorespiratory genes under water stress in chickpea. Photosynthetica 54(2):234–242

Vinocur B, Altman A (2005) Recent advances in engineering plant tolerance to abiotic stress: achievements and limitations. Curr Opin Biotechnol 16(2):123–132

Vlot AC, Dempsey DMA, Klessig DF (2009) Salicylic acid, a multifaceted hormone to combat disease. Annu Rev Phytopathol 47:177–206

Vriet C, Russinova E, Reuzeau C (2012) Boosting crop yields with plant steroids. Plant Cell 24(3): 842–857

Vuosku J, Suorsa M, Ruottinen M, Sutela S, Muilu-Mäkelä R, Julkunen-Tiitto R (2012) Polyamine metabolism during exponential growth transition in scots pine embryogenic cell culture. Tree Physiol 32(10):1274–1287

Vuosku J, Karppinen K, Muilu-Mäkelä R, Kusano T, Sagor GHM, Avia K (2018) Scots pine aminopropyltransferases shed new light on evolution of the polyamine biosynthesis pathway in seed plants. Ann Bot 121(6):1243–1256

Vurro M, Yoneyama K (2012) Strigolactones-intriguing biologically active compounds: perspectives for deciphering their biological role and for proposing practical application. Pest Manag Sci 68(5):664–668
Wahid A, Shabbir A (2005) Induction of heat stress tolerance in barley seedlings by pre-sowing seed treatment with glycine betaine. Plant Growth Regul 46(2):133–141
Wahid A, Parveen M, Gelani S, Basra SMA (2007) Pretreatment of seeds with H_2O_2 improves salt tolerance of wheat seedling by alleviation of oxidative damage and expression of stress proteins. J Plant Physiol 164(3):283–294
Wahid A, Sehar S, Perveen M, Gelani S, Basra SMA, Farooq M (2008) Seed pretreatment with hydrogen peroxide improves heat tolerance in maize at germination and seedling growth stages. Seed Sci Technol 36(3):633–645
Wakchaure GC, Minhas PS, Meena KK, Singh NP, Hegade PM, Sorty AM (2018) Growth, bulb yield, water productivity and quality of onion (*Allium cepa* L.) as affected by deficit irrigation regimes and exogenous application of plant bio-regulators. Agric Water Manag 199(3):1–10
Walz A, Park S, Slovin JP, Ludwig-Müller J, Momonoki YS, Cohen JD (2002) A gene encoding a protein modified by the phytohormones indoleacetic acid. Proc Natl Acad Sci U S A 99(3): 1718–1723
Wan SB, Tian L, Tian RR, Pan QH, Zhan JC, Wen PF, Chen JY, Zhang P, Wang W, Huang WD (2009) Involvement of phospholipase D in the low temperature acclimation induced thermotolerance in grape berry. Plant Physiol Biochem 47(6):504–510
Wang CY, Buta G (1994) Methyl jasmonate reduces chilling injury in *Cucurbita pepo* through its regulation of abscisic and polyamine levels. Environ Exp Bot 34(4):427–432
Wang Y, Devereux W, Stewart TM, Casero RA (1999) Cloning and characterization of polyamine-modulated factor-1, a transcriptional cofactor that regulators the transcription of the spermidine/spermine N1-acetyl transferase gene. J Biol Chem 274(31):22095–22101
Wang Y, Mopper S, Hasentein KH (2001) Effects of salinity on endogenous ABA, IAA, JA, and SA in *Iris hexagona*. J Chem Ecol 27(2):327–342
Wang W, Vinocur B, Altman A (2003a) Plant responses to drought, salinity and extreme temperatures: towards genetic engineering for stress tolerance. Planta 218(1):1–14
Wang Y, Lu W, Zhang Z (2003b) ABA and putrescine treatment alleviate the chilling damage of banana fruit. J Plant Physiol Mol Biol 29(6):549–554
Wang LJ, Li SH (2006) Salicylic acid-induced heat or cold tolerance in relation to Ca^{2+} homeostasis and antioxidant systems in young grape plants. Plant Sci 170(4):685–694
Wang Y, Yang ZM, Zhang QF, Li JL (2009) Enhanced chilling tolerance in *Zoysia matrella* by pre-treatment with salicylic acid, calcium chloride, hydrogen peroxide or 6-benzylaminopurine. Biol Plant 53(1):179–182
Wang QB, Xu W, Xue QZ, Su WA (2010) Transgenic *Brassica chinensis* plants expressing a bacterial codA gene exhibit enhanced tolerance to extreme temperature and high salinity. J Zhejiang Univ Sci B 11(11):851–861
Wang Y, Hu J, Qin G, Cui H, Wang Q (2012) Salicylic acid analogues with biological activity may induce chilling tolerance of maize (*Zea mays*) seeds. Botany 90(9):845–855
Wang J, Lin X, Sun Q, Jena KK (2013a) Evaluation of cold tolerance for japonica rice varieties from different country. Adv J Food Sci Technol 5(1):54–56
Wang H, Zou Z, Wang S, Gong M (2013b) Global analysis of transcriptome responses and gene expression profiles to cold stress of *Jatropha curcas* L. PLoS One 8(12):e82817
Wang XH, Shu C, Li HU, Hu XQ, Wang YX (2014) Effects of 0.01% brassinolide solution application on yield of rice and its resistance to autumn low-temperature damage. Acta Agric Jiangxi 26:36–38. (in Chinese with English abstract)
Wang Q, Ding T, Zuo J, Gao L, Fan L (2016) Amelioration of postharvest chilling injury in sweet pepper by glycine betaine. Postharvest Biol Technol 112:114–120
Wani SH, Kumar V, Shriram V, Sha SK (2016) Phytohormones and their metabolic engineering for abiotic stress tolerance in crop plants. Crop J 4(3):162

Wanner LA, Junttila O (1999) Cold-induced freezing tolerance in *Arabidopsis*. Plant Physiol 120(2):391–400
Waqas MA, Khan I, Akhter MJ, Noor MA, Ashraf U (2017) Exogenous application of plant growth regulators (PGRs) induces chilling tolerance in short-duration hybrid maize. Environ Sci Pollut Res 24(12):11459–11471
Waqas MA, Kaya C, Riaz A, Farooq M, Nawaz I, Wilkes A, Li Y (2019) Potential mechanisms of abiotic stress tolerance in crop plants induced by thiourea. Front Plant Sci 10:13–36
Wardlaw IF, Bagnall D (1981) Phloem transport and the regulation of growth of *Sorghum bicolor* (Moench) at low temperatures. Plant Physiol 68(2):411–414
Wasternack C (2007) Jasmonates: an update on biosynthesis, signal transduction and action in plant stress response, growth and development. Ann Bot 100(4):681–697
Watanabe T, Kume T (2009) A general adaptation strategy for climate change impacts on paddy cultivation: special reference to the Japanese context. Paddy Water Environ 7(4):313–320
Wei Z, Li J (2016) Brassinosteroids regulate root growth, development, and symbiosis. Mol Plant 9(1):86–100
Wen F, Zhang Z, Bai T, Xu Q, Pan Y (2010) Proteomics reveals the effects of gibberellic acid (GA3) on salt-stressed rice (*Oryza sativa* L.) shoots. Plant Sci 178(2):170–175
White RH, Schmidt RE (1989) Bermudagrass response to chilling temperatures as influenced by iron and benzyladenine. Crop Sci 29(3):768–773
Widhalm JR, Dudareva N (2015) A familiar ring to it: biosynthesis of plant benzoic acids. Mol Plant 8(1):83–97
Wildermuth MC, Dewdney J, Wu G, Ausubel FM (2001) Isochorismate synthase is required to synthesize salicylic acid for plant defense. Nature 414(6863):562–565
Wilkie JD, Sedgley M, Olesen T (2008) Regulation of floral initiation in horticultural trees. J Exp Bot 59(12):3215–3228
Wilkinson S, Clephan AL, Davies WJ (2001) Rapid low temperature-induced stomatal closure occurs in cold-tolerant *Commelina communis* leaves but not in cold-sensitive tobacco leaves, via a mechanism that involves apoplastic calcium but not abscisic acid. Plant Physiol 126(4): 1566–1578
Wilkinson S, Davies WJ (2009) Ozone suppresses soil drying- and abscisic acid (ABA)-induced stomatal closure via an ethylene dependent mechanism. Plant Cell Environ 32(8):949–959
Wilkinson S, Kudoyarova GR, Veselov DS, Arkhipova TN, Davies WJ (2012) Plant hormone interactions: innovative targets for crop breeding and management. J Exp Bot 63(9):3499–3509
Wise RR (1995) Chilling-enhanced photooxidation: the production, action and study of reactive oxygen species produced during chilling in the light. Photosynth Res 45(2):79–97
Wilson JM (1985) Economic importance of chilling injury. Outlook Agric 14(4):197–204
Wolbang CM, Chandler PM, Smith JJ, Ross JJ (2004) Auxin from the developing inflorescence is required for the biosynthesis of active gibberellins in barley stems. Plant Physiol 134(2): 769–776
Wolters H, Jurgens G (2009) Survival of the flexible: hormonal growth control and adaptation in plant development. Nat Rev Genet 10(5):305–317
Wu L, Chen X, Ren H, Zhang Z, Zhang H, Wang J, Wang XC, Huang R (2007) ERF protein JERF1 that transcriptionally modulates the expression of abscisic acid biosynthesis-related gene enhances the tolerance under salinity and cold in tobacco. Planta 226(4):815–825
Wu M, Yuan L (2008) Research progress in the relationship between polyamine and plant resistance. J Anhui Agric Sci 36:3516–3518
Wu X, He J, Chen J, Yang S, Zha D (2014) Alleviation of exogenous 6-benzyladenine on two genotypes of eggplant (*Solanum melongena* mill.) growth under salt stress. Protoplasma 251(1): 169–176
Wyatt SE, Rashotte AM, Shipp MJ, Robertson D, Muday GK (2002) Mutations in the gravity persistence signal loci in *Arabidopsis* disrupt the perception and/or signal transduction of gravitropic stimuli. Plant Physiol 130(3):1426–1435

Xi Z, Wang Z, Fang Y, Hu Z, Hu Y, Deng M (2013) Effects of 24- epibrassinolide on antioxidation defense and osmoregulation systems of young grapevines (*Vitis vinifera* L.) under chilling stress. Plant Growth Regul 71(1):57–65

Xia JC, Zhao H, Liu WZ, Li LG, He YK (2009) Role of cytokinin and salicylic acid in plant growth at low temperatures. Plant Growth Regul 57(3):211–221

Xia XJ, Zhou YH, Shi K, Zhou J, Foyer CH, Yu JQ (2015) Interplay between reactive oxygen species and hormones in the control of plant development and stress tolerance. J Exp Bot 66(10):2839–2856

Xie X, Yoneyama K, Yoneyama K (2010) The strigolactone story. Annu Rev Phytopathol 48:93–117

Xin Z, Li PH (1992) Abscisic acid-induced chilling tolerance in maize suspension-cultured cells. Plant Physiol 99(2):707–711

Xin Z, Browse J (2000) Cold comfort farm: the acclimation of plants to freezing temperatures. Plant Cell Environ 23(9):893–902

Xiong L, Ishitani M, Lee H, Zhu JK (2001) The Arabidopsis LOS5/ABA3 locus encodes a molybdenum cofactor sulfurase and modulates cold stress- and osmotic stress-responsive gene expression. Plant Cell 13(9):2063–2083

Xu C, Wu X, Zhang H (2009) Impact of D-Arg on drought resistance and endogenous polyamines in mycorrhizal *Pinus massoniana*. J Nanjing For Univ 33:019–023

Xu L, Xing ST, Sun X (2014) Effects of polyamines on hormones contents and the relationship with the flower bud differentiation in chrysanthemum. Plant Physiol J 50(8):1195–1202

Xu L (2015) The effect of polyamineon flower bud differentiation and bud germination of chrysanthemum. Shandong Agric Univ:31–36

Xu Q, Truong TT, Barrero JM, Jacobsen JV, Hocart CH, Gubler F (2016) A role for jasmonates in the release of dormancy by cold stratification in wheat. J Exp Bot 67(11):3497–3508

Xu Y, Hu W, Liu J, Song S, Hou X, Jia C, Li J, Miao H, Wang Z, Tie W (2020) An aquaporin gene MaPIP2-7 is involved in tolerance to drought, cold and salt stresses in transgenic banana (*Musa acuminata* L.). Plant Physiol Biochem 147:66–76

Xue J, Wang S, You X, Dong J, Han L, Liu F (2011) Multi residue determination of plant growth regulators in apples and tomatoes by liquid chromatography/tandem mass spectrometry. Rapid Commun Mass Spectrom 25(21):3289–3297

Xue-Xuan X, Hong-Bo S, Yuan-Yuan M, Gang X, Jun-Na S, Dong-Gang G, Cheng-Jiang R (2010) Biotechnological implications from abscisic acid (ABA) roles in cold stress and leaf senescence as an important signal for improving plant sustainable survival under abiotic-stressed conditions. Crit Rev Biotechnol 30(3):222–230

Yadav PV, Kumari M, Ahmed Z (2011) Seed priming mediated germination improvement and tolerance to subsequent exposure to cold and salt stress in capsicum. Res J Seed Sci 4(3): 125–136

Yadav PV, Kumari M, Meher LC, Arif M, Ahmed Z (2012) Chemical seed priming as an efficient approach for developing cold tolerance in *Jatropha*. J Crop Improv 26(1):140–149

Yadegari LZ, Heidari R, Carapetian J (2008) Chilling pretreatment causes some changes in respiration, membrane permeability and some other factors in soybean seedlings. Res J Biol Sci 3(9):1054–1059

Yakovlev AY, Borovsky GV (2003) Low- temperature protective proteins of fungi and plants. Mikologiyai-Fitopatologiya 37(6):87–93

Yalpani N, Enyedi AJ, Leon J, Raskin I (1994) Ultraviolet light and ozone stimulates accumulation of salicylic acid, pathogen-related proteins and virus resistance in tobacco. Planta 193(3): 372–376

Yamaguchi S (2008) Gibberellin metabolism and its regulation. Annu Rev Plant Physiol 59:225–251

Yamaryo Y, Kanai D, Awai K, Shimojima M, Masuda T, Shimada H, Takamiya K, Ohta H (2003) Light and cytokinin play a co-operative role in MGDG synthesis in greening cucumber cotyledons. Plant Cell Physiol 44(8):844–855

Yamori W, Noguchi K, Hikosaka K, Terashima I (2009) Cold-tolerant crop species have greater temperature homeostasis of leaf respiration and photosynthesis than cold-sensitive species. Plant Cell Physiol 50(2):203–215

Yan Z, Chen J, Li X (2013) Methyl jasmonate as modulator of cd toxicity in *Capsicum frutescens* var. fasciculatum seedlings. Ecotoxicol Environ Saf 98:203–209

Yancey PH, Clark ME, Hand SC, Bowlis RD, Somero GN (1982) Living with water stress: evolution of osmolyte system. Science 217(4566):1214–1222

Yang JS, Wang R, Meng JJ, Bi YP, Xu PL, Guo F, Wan SB, He QH, Li XG (2010) Over-expression of *Arabidopsis* CBF1 gene in transgenic tobacco alleviates photo-inhibition of PSII and PSI during chilling stress under low irradiance. J Plant Physiol 167(7):534–539

Yang W, Liu XD, Chi XJ, Wu CA, Li YZ, Song LL, Liu XM, Wang YF, Wang FW, Zhang C, Liu Y, Jong JM, Li HY (2011) Dwarf apple MbDREB1 enhances plant tolerance to low temperature, drought, and salt stress via both ABA-dependent and ABA-independent pathways. Planta 233(2):219–229

Yang Z, Cao S, Zheng Y, Jiang Y (2012a) Combined salicylic acid and ultrasound treatments for reducing the chilling injury on peach fruit. J Agric Food Chem 60:1209–1212

Yang A, Dai XY, Zhang WH (2012b) A R2R3-type MYB gene, OsMYB2, is involved in salt, cold, and dehydration tolerance in rice. J Exp Bot 63(7):2541–2556

Yang DL, Yang Y, He Z (2013) Role of plant hormones and their cross-talks in rice immunity. Mol Plant 6(3):675–685

Yin H, Chen QM, Yi MF (2008) Effects of short-term heat stress on oxidative damage and responses of antioxidant system in *Lilium longiflorum*. Plant Growth Regul 54(1):45–54

Yin CC, Ma B, Collinge DP, Pogson BJ, He SJ, Xiong Q, Duan KX, Chen H, Yang C, Lu X, Wang YQ, Zhang YK, Chu CC, Sun XH, Fang S, Chu JF, Lu TG, Chen SY, Zhang JS (2015) Ethylene responses in rice roots and coleoptiles are differentially regulated by a carotenoid isomerase-mediated abscisic acid pathway. Plant Cell 27(4):1061–1081

Yoneyama K, Kisugi T, Xie X, Yoneyama K (2013) Chemistry of strigolactones: why and how do plants produce so many strigolactones? In: de Bruijn FJ (ed) *Molecular microbial ecology of the rhizosphere, vol.1 & 2*. John Wiley & Sons Ltd, Hoboken, NJ. 2013, pp 373–379

Yoon JY, Hamayun M, Lee SK, Lee IJ (2009) Methyl jasmonate alleviated salinity stress in soybean. J Crop Sci Biotechnol 12(2):63–68

Yoshida Y, Sano R, Wada T, Takabayashi J, Okada K (2009) Jasmonic acid control of GLABRA3 links inducible defense and trichome patterning in *Arabidopsis*. Development 136(6): 1039–1048

Yu JQ, Zhou Y, Ye SF, Huang LF (2002) 24-epibrassinolide and abscisic acid protect cucumber seedlings from chilling injury. J Hortic Sci Biotechnol 77(4):470–473

Yuan S, Lin HH (2008) Role of salicylic acid in plant abiotic stress. Zeitschrift fur Naturforschung C Biosci 63(5–6):313–320

Yuan GF, Jia CG, Li Z, Sun B, Zhang LP, Liu N, Wang QM (2010) Effect of brassinosteroids on drought resistance and abscisic acid concentration in tomato under water stress. Sci Hortic (Amsterdam) 126(2):103–108

Yuan L, Shu S, Sun J, Guo S, Tezuka T (2012) Effects of 24-epibrassinolide on the photosynthetic characteristics, antioxidant system, and chloroplast ultrastructure in Cucumis sativus L. under ca $(NO_3)_2$ stress. Photosynth Res 112(3):205–214

Zalabák D, Pospíšilová H, Šmehilová M, Mrízová K, Frébort I, Galuszka P (2013) Genetic engineering of cytokinin metabolism: prospective way to improve agricultural traits of crop plants. Biotechnol Adv 31(1):97–117

Zarghami M, Shoor M, Ganjali A, Moshtaghi N, Tehranifar A (2014) Effect of salicylic acid on morphological and ornamental characteristics of petunia hybrida at drought stress. Ind J Fund Appl Life Sci 4(3):523–532

Zazimalova E, Kaminek M, Brezinova A, Motyka V (1999) Control of cytokinin biosynthesis and metabolism. In: Hooykaas PJJ, Hall MA, Libbenga KR (eds) Biochemistry and molecular biology of plant hormones. Elsevier, Amsterdam, pp 141–160

Zeng Y, Yu J, Cang J, Liu LJ, Mu YC, Wang J, Zhang D (2011) Detection of sugar accumulation and expression levels of correlative key enzymes in winter wheat (*Triticum aestivum*) at low temperatures. Biosci Biotechnol Biochem 75(4):681–687

Zhang Y, Chen K, Zhang S, Ferguson I (2003) The role of salicylic acid in postharvest ripening of kiwifruit. Postharvest Biol Technol 28(1):67–74

Zhang XZ, Ervin EH, LaBranche AJ (2006a) Metabolic defense responses of seeded bermudagrass during acclimation to freezing stress. Crop Sci 46(6):2598–2605

Zhang J, Jia W, Yang J, Ismail AM (2006b) Role of ABA in integrating plant responses to drought and salt stresses. Field Crops Res 97(1):111–119

Zhang S, Hu J, Zhang Y, Xie XJ, Knapp A (2007) Seed priming with brassinolide improves lucerne (*Medicago sativa* L.) seed germination and seedling growth in relation to physiological changes under salinity stress. Aust J Agric Res 58(8):811–815

Zhang M, Zhai Z, Tian X, Duan L, Li Z (2008a) Brassinolide alleviated the adverse effect of water deficits on photosynthesis and the antioxidant of soybean (*Glycine max* L.). Plant Growth Regul 56(3):257–264

Zhang X, Wollenweber B, Jiang D, Liu F, Zhao J (2008b) Water deficit and heat shock effects on photosynthesis of a transgenic *Arabidopsis thaliana* constitutively expressing ABP9, a bZIP transcription factor. J Exp Bot 59(4):839–840

Zhang C, Tian S (2009) Crucial contribution of membrane lipids unsaturation to acquisition of chilling-tolerance in peach fruit stored at 0°C. Food Chem 115(2):411–415

Zhang S, Yang C, Peng J, Sun S, Wang X (2009a) GASA5, a regulator of flowering time and stem growth in *Arabidopsis thaliana*. Plant Mol Biol 69(6):745–759

Zhang XZ, Ervin EH, Evanylo GK, Haering KC (2009b) Impact of biosolids on hormone metabolism in drought-stresses tall fescue. Crop Sci 49(5):364–370

Zhang A, Zhang J, Zhang J, Ye N, Zhang H, Tan M, Jiang M (2011) Nitric oxide mediates brassinosteroid-induced ABA biosynthesis involved in oxidative stress tolerance in maize leaves. Plant Cell Physiol 52(1):181–192

Zhang XD, Wang RP, Zhang FJ, Tao FQ, Li WQ (2013a) Lipid profiling and tolerance to low-temperature stress in *Thellungiella salsuginea* in comparison with *Arabidopsis thaliana*. Biol Plant 57(1):149–153

Zhang YP, Zhu XH, Ding HD, Yang SJ, Chen YY (2013b) Foliar application of 24-epibrassinolide alleviates high-temperature-induced inhibition of photosynthesis in seedlings of two melon cultivars. Photosynthetica 51(3):341–349

Zhang WJ, Huang ZL, Wang Q, Guan YN (2015) Effects of low temperature on leaf anatomy and photosynthetic performance in different genotypes of wheat following a rice crop. Int J Agric Biol 17(6):1165–1171

Zhang B, Jia D, Gao ZQ, Dong Q, He LH (2016) Physiological responses to low temperature in spring and winter wheat varieties. J Sci Food Agric 96(6):1967–1973

Zhang X, Da Silva JAT, Niu M, Li M, He C, Zhao J, Zeng S, Duan J, Ma G (2017) Physiological and transcriptomic analyses reveal a response mechanism to cold stress in *Santalum album* L. leaves. Sci Rep 7:42165

Zhang W, Wang J, Huang Z, Mi L, Xu K, Wu J, Fan Y, Ma S, Jiang D (2019) Effects of low temperature at booting stage on sucrose metabolism and endogenous hormone contents in winter wheat spikelet. Front Plant Sci 10:498

Zhao XC, Schaller GE (2004) Effect of salt and osmotic stress upon expression of the ethylene receptor ETR1 in *Arabidopsis thaliana*. FEBS Lett 562(1–3):189–192

Zhao D, Shen L, Fan B, Yu B, Zheng Y, Lv S, Sheng J (2009) Ethylene and cold participate in the regulation of LeCBF1 gene expression in postharvest tomato fruits. FEBS Lett 583(20): 3329–3334

Zhao Y (2010) Auxin biosynthesis and its role in plant development. Annu Rev Plant Biol 61:49–64

Zhao M, Wang J, Shan W, Fan J, Kuang J, Wu K, Li X, Chen W, He F, Chen J, Lu W (2013) Induction of jasmonate signalling regulators MaMYC2 and their physical interactions with

MaICE1 in methyl jasmonate-induced chilling tolerance in banana fruit. Plant Cell Environ 36(1):30–51
Zhao M, Liu W, Xia X, Wang T, Zhang WH (2014) Cold acclimation-induced freezing tolerance of Medicago truncatula seedlings is negatively regulated by ethylene. Physiol Plant 152(1): 115–129
Zhao C, Lang Z, Zhu JK (2015) Cold responsive gene transcription becomes more complex. Trends Plant Sci 20(8):466–468
Zhen Y, Li S, Su X (2000) Polyamine changes and chilling injury in cold-stored loquat fruits. J Integr Plant Biol 42(8):824–827
Zheng Y, Luo L, Wei J, Chen Q, Yang Y, Hu X, Kong X (2018) The glutamate receptors AtGLR1. 2 and AtGLR1. 3 increase cold tolerance by regulating jasmonate signaling in *Arabidopsis thaliana*. Biochem Biophys Res Commun 506(4):895–900
Zhou WJ, Leul M (1998) Uniconazole-induced alleviation of freezing injury in relation to changes in hormonal balance, enzyme activities and lipid peroxidation in winter rape. Plant Growth Regul 26(1):41–47
Zhou L, Jang JC, Jones TL, Sheen S (1998) Glucose and ethylene trans location crosstalk revealed by an Arabidopsis glucose—insensitive mutant. Proc Natl Acad Sci U S A 95(17):10294–10299
Zhou R, Cutler AJ, Ambrose SJ, Galka MM, Nelson KM, Squires TM, Loewen MK, Ashok S, Jadhav AS, Ross ARS, Taylor DC, Suzanne R, Abrams SR (2004) A new abscisic acid catabolic pathway. Plant Physiol 134(1):361–369
Zhou R, Hyldgaard B, Yu X, Rosenqvist E, Ugarte RM, Yu S, Wu Z, Ottosen CO, Zhao T (2018) Phenotyping of faba beans (*Vicia faba* L.) under cold and heat stresses using chlorophyll fluorescence. Euphytica 214(4):68
Zhu JK (2002) Salt and drought stress signal transduction in plants. Annu Rev Plant Biol 3:247–273
Zhu J, Dong C-H, Zhu J-K (2007) Interplay between cold-responsive gene regulation, metabolism and RNA processing during plant cold acclimation. Curr Opin Plant Biol 10(3):290–295
Zhu B, Peng RH, Xiong AS, Xu J, Fu XY, Zhao W, Jin XF, Meng XR, Gao J, Cai R, Yao QH (2012) Transformation with a gene for myo-inositol *O*-methyl transferase enhances the cold tolerance of *Arabidopsis thaliana*. Biol Plant 56(1):135–139
Zhuang J, Cai B, Peng RH, Zhu B, Jin XF, Xue Y, Gao F, Fu XY, Tian YS, Zhao W, Qiao YS, Zhang Z, Xiong AS, Yao QH (2008) Genome-wide analysis of the AP2/ERF gene family in *Populus trichocarpa*. Biochem Biophys Res Commun 371(3):468–474
Zolfagharinasab R, Hadian J (2007) Influence of methyl jasmonate on inducing chilling tolerance in pomegranate fruits (*Malas save*). Pak J Biol Sci 10(4):612–616

7 Effect of Low Temperature on Dry Matter, Partitioning, and Seed Yield: A Review

Abstract

Abiotic stresses adversely affect growth and productivity and trigger a series of morphological, physiological, biochemical, and molecular changes in plants. Drought, temperature extremes, and saline soils are the most common abiotic stresses that plant encounter. The efficiency with which crops produce grain from dry matter is the first determinant of harvest index. In this context, cereals, which have the highest carbohydrate content, tend to have the highest harvest indices. The high-energy cost of lipid production makes species with a high oil content (canola, soybean, peanut) inefficient at producing grain from assimilate, with high protein-seeded species (pea, lentil, chickpea) being intermediate. For most grain crops, harvest index rarely exceeds 0.5, excepting where fallen leaves are not taken into account in which case apparent harvest index may exceed 0.6.

Temperature can influence economic yield through its components, with variation in relative magnitude of effect on them. Extremes of temperature which reduces the reproductive sink through gametogenesis can substantially reduce harvest index. Different studies clearly show that temperatures exceeding the limits of adaptation substantially influence the metabolism, viability, physiology, and yield of many plants. Plants exposed to extreme temperatures often show a common response in the form of oxidative stress. However, the extent of damage caused by extreme temperatures depends greatly on the duration of the adverse temperature, the genotypes of the exposed plants, and their stage of growth. There is ample need to develop extreme temperature tolerance in crop plants by exploring suitable strategies. However, plant adaptation to either high temperature or low temperature is a multigenic response which is very complex in nature.

A. Bhattacharya, *Physiological Processes in Plants Under Low Temperature Stress*, https://doi.org/10.1007/978-981-16-9037-2_7

Keywords

Dry matter partitioning and yield · Carbon partitioning and acclimation to stress · Factors affecting harvest index · GxE interaction and harvest index · Increasing harvest index through management · Temperature effect on harvest index · Dry matter harvest index and yields in different crops · Low temperature and yield · Factors for yield gaps · Traits associated with seed yield · Climate change and seed yield · Temperature stress and seed yield · Physiological avenues for enhancing yield · Engineering for cold stress tolerance in plants

The growing world population together with the lack of expansion, or even reduction, of available arable lands needed to maintain agricultural sustainability (Cassman et al. 2003) implies that the relative importance of plant breeding to raise crop yield potential and adaptiveness is now greater than in the past (Slafer et al. 1999; Araus et al. 2002). This is not a minor expectation. Although there is a rising demand for irrigation and chemical fertilizers, the technological progress made in cereal cultivation over recent decades has led to a decline in the cost of cereal production per unit of output. Without this growth in productivity, many developing countries would have been forced to further extend cultivation into marginal lands, thus aggravating the dilemma of how to sustain natural resource bases (IRRI 1996). In the case of wheat, yield has been genetically enhanced virtually all over the world (Calderini et al. 1999). The magnitude of the improvement has depended upon the environmental conditions of the region. However, when wheat yields are expressed as a percentage of the mean yield of the trial in which they were assessed, the data appear to converge into a single trend. Abiotic stress factors frequently constrain the growth and productivity of major crop species such as cereals. The single greatest abiotic stress factor that limits crop growth worldwide is water availability (Araus et al. 2002). While genetic increases in yield potential are best expressed in optimum environments, they are also associated with enhanced yields under drought (Trethowan et al. 2002; Araus et al. 2002; Slafer and Araus 2007) and nitrogen deficiency (Ortiz-Monasterio et al. 1997; Abeledo et al. 2003). These gains are especially relevant given that further large increases in the area under irrigation are not expected, and land deterioration associated with intensive agriculture threatens those areas already irrigated (Araus 2004).

7.1 Dry Matter Partitioning

The ultimate objective of crop cultivation is to get maximum grain yield with available resources. In order to achieve potential grain yield, it is necessary to convert or transform more biomass into grain. Therefore, understanding of physiological processes involved in dry matter partitioning to grains is vital for improved grain yield and also for crop improvement. The relative amounts of recently assimilated carbon that are distributed in structures that contribute to growth in a photosynthetic capacity are particularly critical (Bhattacharya 2018). Assimilated carbon distributed to support leaf growth increases the capacity to gather light and carry out photosynthesis, thus compounding the effect of carbon assimilation on growth. So long as other essential resources such as nitrogen, phosphorus, and water are in sufficient supply, distributing a constant proportion of newly assimilated carbon to shoots produces an exponential increase in plant growth rate. However, leaves depend on other organs, and so sufficient assimilated carbon must be invested in root growth and other non-photosynthetic structures that supply the supporting resources. Ideally, homeostatic regulation of partitioning results in a balance between the growth of leaves and these non-photosynthetic organs so that availability of carbon and the other essential nutrients becomes co-limiting for each plant organ. This capacity for adaptive change in form represents a unique feature of plants. An important effecter of the process appears to be the responsiveness of plant gene expression to signals of altered resource availability. One effecter, changing sugar levels, can provide important information on overall status of assimilated carbon supplies or mobilizable reserves or both (Koch 1996).

Annual agricultural crop ecosystems are highly dynamic and provide different challenges compared to perennial forest and other perennial woody vegetation systems, especially in Australia where the climate is more variable than otherwise apparently similar climates (Nicholls et al. 1997). While growing crops initially represent a sink for atmospheric CO_2, the ultimate fate of assimilated carbon depends on the nature and use of crop products, the management of residues remaining after product harvest, the ability of a soil to accumulate and protect crop residue carbon against mineralization, and subsequent climatic conditions. Most approaches to carbon accounting in agriculture use measurements of harvested yield and estimates of the proportion of total plant carbon harvested to calculate the amount of carbon retained in the form of crop residues at any given location. Calculating the proportion of total plant carbon retained requires the use of defined values for the harvest index, the ratio of carbon allocation to shoots and roots, and the carbon contents of harvested products and residues. Bolinder et al. (1997), Gregory et al. (1996), Jackson et al. (1996a), and Robinson (2004) have reviewed the shoot to root ratios and carbon accounting. In agronomic terms, harvest index is calculated as the ratio of harvested product to total aboveground biological yield, which is an economic harvest index, and is usually calculated from unit area yield and dry matter data (machine-harvested yield/total plant dry weight at maturity). On a physiological basis, harvest index is a measure of the relative investment of plant resources in reproductive parts, and this is usually determined on a per plant basis and may include pod walls and other nonseed components (Huhn 1990a, b). Donald and Hamblin (1976), Evans (1993), and Sinclair (1998) have reported various ways to calculate harvest index, but none of these reviews were set within the context of agricultural carbon accounting, yet the harvest index data they reported have been relied upon to provide single harvest index values for individual crops that have been applied across regions, seasons (Bolinder et al. 2007; Johnson et al. 2006), and continents (Bondeau et al. 2007; Lobell et al. 2002). Crop harvest index refers to the aboveground portion of the crop, but for carbon-accounting purposes, it would also be necessary to account for the allocation of assimilated carbon to root dry matter. Comparisons among crop species reveal that the energy or carbon cost of producing seeds, the portion of the crop most often harvested, varies enormously. The quantity of seed that a given crop species can produce from a given amount of photosynthate (energy) is constrained by differences in plant biochemistry (Unkovich et al. 2010).

Environmental stress is a *state* of an organism in which a significant deviation from the conditions favorable for life elicits changes that can decrease or disrupt processes at all functional levels (Larcher 1995). Because a progression of stress responses is elicited, environmental stress is an important factor affecting regulation of carbon partitioning. Adaptation is the genetic potential of a plant determining the specific stress responses that are available for adjusting to stress. Acclimation is the response to the state of environmental stress that actually ameliorates the effects caused by the stress factors. The response to stress begins with an initial alarm phase during which structural damage and functional destabilization due to the stress reaction begin to develop. During the restoration phase that follows, available adaptive mechanisms initiate responses that repair damage and restore functions to a greater or lesser degree. If the stress continues during the resistance phase, acclimation brings about increased resistance or hardening. The genetically based adaptive program specifies both the development of constitutive adaptive mechanisms that determine the initial responses of the plant to acute stress reactions and the development of additional adaptive mechanisms in response to chronic stress reactions. Generally, acclimation alters carbon partitioning in ways that ameliorate the stress (Geiger and Servaites 1991).

Low temperatures have an effect on mineral nutrition of plants as absorption of ions by roots is difficult, as well as their movement in the aboveground parts of plants (Lukatkin et al. 2012). Chilling of plants leads to a decrease in the activity of nitrate reductase, reduction in the nitrogen incorporation in the amino acids and proteins, and a drop in the proportion of organic phosphorus and an increase in inorganic phosphorus content (Zia et al. 1994), which is a consequence of a breach of phosphorylation and enhanced decomposition of organic phosphorus compounds. Mechanisms to reduce the absorption of nutrients by chilling temperatures include depression of respiration and/or oxidative phosphorylation, impairing enzymatic transport systems associated with conformational protein changes in membranes, changes in membrane potential, reducing the supply of ATP to hydrogen ion transporting ATPase, as well as lowering the permeability coefficients for ions (Clarkson et al. 1988). Minorsky (1985) proposed a hypothesis to explain most of the secondary effects of chilling shock, which suddenly increases (by one to two orders) in cytosolic calcium ion concentration. It is assumed that the rapid increase in cytosolic calcium ion concentration due to chilling may serve as the primary physiological signal of cold exposure. It was shown that changes in intracellular calcium compartmentation in chilled plants, leading to an increase in cytosolic calcium ion concentration, stop cytoplasmic streaming and affect the subcellular structures (Woods et al. 1984). There is evidence that input of $^{45}Ca^{2+}$ in maize root cells increased by 20–25% at a temperature of 2 °C (Zocchi and Hanson 1982). Changes in cytosolic calcium ion concentration trigger cascade reactions in the cell, which leads to numerous disturbances at all levels of an organization. Jouyban et al. (2013) have shown that chilling induces abrupt reduction of calcium ATPase activity, which pumps out calcium ion in apoplast and/or in intracellular depots. So, this enhances the cytosolic calcium ion concentration. During the growth of maize seedlings on nutrient media with different calcium status, more intense chilling injury was observed at reduced or enhanced calcium ion doses in comparison with optimal dose (Lukatkin and Isaikina 1997). It has been argued that oxidative deficit that occurs during cooling of chilling-sensitive plants plays a leading role in the transduction of chilling injury (Hu et al. 2008). It was shown that concentration of cytosolic calcium ion changes is intimately connected to an oxidative deficit. Oxidative deficit causes an immediate increase in cytosolic calcium (Price et al. 1994), acting the same as chilling shock (Knight et al. 1996).

7.2 Dry Matter Partitioning and Yield in Cereals

Increases in the yield of cereal variety over the past half century have been associated with rising values in dry matter partitioning coefficients (harvest index) with little increase in biomass, although no plant breeder has purposely sought this increase (Donald and Hamblin 1976). In Australia, the improvement in harvest index of cereals is largely associated with early flowering and a greater duration of reproductive growth (Loss and Siddique 1994). Biomass and harvest indices in *Vicia faba* (Thompson and Taylor 1977) and *Cicer arietinum* (Thompson et al. 1997) are often lower than cereals and other pulse crops in Mediterranean-type environment. It has been shown that limited photosynthetic capacity in low-temperature stressed plants is often associated with limited capacity for carbon partitioning into the sucrose and starch biosynthesis and as a result triose phosphate accumulation and limited phosphate recycling (Huner et al. 1998; Savitch et al. 1997). However, cold acclimation results in a species-dependent adjustment of CO_2 assimilation rates often associated with upregulation of Calvin cycle enzymes and stimulation of sucrose biosynthesis (Savitch et al. 1997). It has been shown that cold acclimation of *Arabidopsis thaliana* as well results in an adjustment of sucrose and starch biosynthetic pathways (Strand et al. 1998). The suppression of photosynthesis is released upon leaf development at low growth temperature at the level of LHC (chlorophyll a and b binding protein) gene expression and Calvin cycle gene expression (Strand et al. 1999). Despite some similarities between cereals and *Arabidopsis* in biochemical mechanisms of acclimation to low temperature, cold acclimation of *Arabidopsis* was associated with higher ATP level and ATP/ADP ratio and higher glucose-6-phosphate/fructose-6-phosphate ratio when compared to controls maintained at 23 °C (Strand et al. 1997). It appears that starch biosynthesis might be. the major pathway of carbon partitioning in *Arabidopsis*. Moreover, the limited capacity to adjust starch biosynthesis upon cold acclimation rather than the adjustment of Calvin cycle enzymes might limit photosynthetic performance of cold-acclimated *Arabidopsis* plants. It might be suggested that mechanisms of upregulation of photochemical efficiency at low temperature in *Arabidopsis* plants differ from the one in cereals.

Harvest index depends on the relative duration of vegetative and reproductive phases, the proportion of dry matter assimilated during vegetative phase, and the amount of assimilate remobilization from vegetative to reproductive organs. Much of the advances in cereal yield have been from improvement in harvest index rather than in total crop dry matter (Donald and Hamblin 1976). Seed yield itself can be partitioned into the number of pods per unit area (or the product of plant density and the number of pods per plant), the number of seeds per pod, and mean seed weight. Of these yield components, the number of seeds per pod and mean seed weight are relatively stable, while seed yield is most highly correlated to the number of pods in chickpea (Siddique and Sedgley 1985) and many other legumes.

The key physiological constraint to potential productivity is the relatively low harvest index, at least as grown in traditional production systems. Harvest index of most pigeon pea is very low as compared with such species as soybean and groundnut and that much of the genetic advance made in latter species has been due to improvement in harvest index. It has been said that low harvest index in pigeon pea as compared to soybean is due to the following:

1. Relatively low nitrogen concentration in pigeon pea seeds as compared to soybean seeds.
2. Negligible oil content as compared to soybean. The energy concentration of pigeon pea seeds is relatively lower than that of soybean.
3. Loss of biomass (carbon and nitrogen) through senescent leaves, petioles, and stem in pigeon pea plants would be sufficient to sustain much greater seed yield without any need for improved nitrogen and carbon productivity.

In the short term, the manipulation of photothermal sensitivity offers the most powerful tool for improving harvest index. There are several successful applications exemplifying this approach (Byth et al. 1981; Wallis et al. 1981; Chauhan 1987), which rallies on the fact that under strong inductive photothermal conditions, time to flower is shortened and plants are less vegetative and less prone to loss of biomass through senescence of shaded leaves, branch, etc. In large part, although not always (Chauhan 1987) reduced biomass per plant is compensated by greater harvest index and the use of dense stands. Nevertheless, higher seed yields were recorded for pigeon pea from this approach (Whiteman et al. 1985). The fact that harvest index of pigeon pea is influenced by photothermal regime, questions were raised about its utility as selection criteria in a breeding program. Most obvious of these is that unless genotypes are tested under the same conditions of location/sowing date, the effect of photothermal regime will confound and probably obscure any inherent genotypic variations for the trait. Also, the sensitivity of harvest index to environmental stresses such as water deficit and low-temperature stress raises a specific problem in selection. Often, it has been questioned whether most pulses produce enough dry matter so as to give yield comparable to cereals. In many instances more than 14 t/ha dry matter production e.g., in pigeon pea, has been recorded (Khanna-Chopra et al.

1979; Narayanan and Sheldrake 1976). However, the fact remains that this dry matter production capacity is poorer than the cereals grown in the same season (Sinha et al. 1988).

Besides transport of recently fixed carbon and nitrogen to the developing seeds, pre-flowering stored reserves of carbon and nitrogen are reported to be an important assimilate source for seed filling, particularly when plants are subjected to water deficit. Generally in pulses, 2–42% of seed dry matter has been reported to be contributed by the remobilization from vegetative tissues (Bushby and Lawn 1992). In chickpea, a decrease of dry matter in the stem, leaf, and pod walls of field-grown plant suggests that dry matter (carbon, nitrogen, and others) is important as an alternative assimilate source to current photosynthates (Leport et al. 1999). It was estimated that one-third of the pod dry matter is derived from the remobilization of dry matter from vegetative tissues in chickpea (Khanna-Chopra and Sinha 1982). The studies conducted by Davies et al. (2000) showed that carbon and nitrogen assimilated prior to podding can supplement to the filling seeds in both well-watered and water-stressed chickpea. Thus, remobilization of pre-podding nitrogen is an essential source of nitrogen for seed filling irrespective of the environmental stresses.

Rice (*Oryza sativa* L.) is the staple food of mankind and provides 35–60% of the dietary calories consumed by three billion people, making it inarguably the most important crop worldwide (Confalonieri and Bocchi 2005). The demand for increasing rice production is because the population of traditional rice-producing countries will require 70% more rice by 2025 (IRRI 1995; Swaminathan 2007). The yield of rice is an integrated result of various processes, including canopy photosynthesis, conversion of assimilates to biomass, and partitioning of assimilates to grains (Wu et al. 1998; Ying et al. 1998). Studies on zinc and phosphorus interaction and their impact on dry matter accumulation and partitioning into leaf, culm, and panicle at different growth stages and harvest index of lowland rice have been made. It has been postulated that there is no difference in the dry matter accumulation and partitioning into various plant parts at different growth stages (tillering, heading, and physiological maturity), and the harvest index of lowland rice genotypes when applied with variable rates of phosphorus and zinc. Amanullah and Imanullah (2016) observed that the growing hybrid rice with application of phosphorus at 120 kg/ha along with 15 kg/ha of zinc not only increases total dry matter accumulation and partitioned greater amounts into the reproductive plant parts (panicles) but also results in higher harvest index.

7.3 Dry Matter Partitioning and Yield in Grain Legumes

Since domestication, grain legumes were subjected to poor agronomical (suboptimal) conditions, and this has led to the development of indeterminate plant types with poor harvest index; these traits are suitable for survival rather than productivity. These traits of legumes have been overexploited with the result that not much effort has been made in enhancing their genetic yield potential under high input situations. Various biotic and abiotic stresses as well as climatic changes to which these crops are generally exposed affect their yielding ability, and often the harvested yields are less than their maximum attainable yield levels. Pulse crops can be categorized into cool-season (dry bean, lentil, lupins, and chickpea) and warm-season (common bean, soybean, pigeon pea, mung bean, urad beans, and cowpea) crops based primarily on their ability to emerge in cool-season conditions and on frost tolerance (Miller et al. 2002). Minimum temperature for seed germination and crop growth differ among pulse crops, with soybean having a base temperature near 10 °C compared to base temperature near 0 °C for chickpea, dry beans, and lentil (Ney and Turc 1993). Consequently, soybean typically requires a relative later seeding date, mid-May to early June, to reduce the risk of frost injury. Chickpea, dry beans, and lentil tolerate a moderate degree of frost, 2–18 °C, depending upon the cultivars, degree of acclimation, and plant growth stage (Welbaun et al. 1997; Srinivasan et al. 1998). On the Canadian Prairies, daytime temperature for best growth of chickpea ranges from 21 to about 30 °C, whereas temperature range for best growth of field pea is 13–23 °C (Soltani et al. 2006; Wang et al. 2006). Temperature exceeding 30–32 °C limits yields of chickpea by hastening maturity and/or decreasing seeds/plant and seed weight (Wang et al. 2006). Chickpea will tolerate higher temperature than field pea during flowering; temperature >27 °C will often decrease flower numbers and flowering duration (Hnatowich 2000; Hawthorne et al. 2003). Lentil has poor tolerance for higher temperatures, especially at flowering and pod set (Erskine et al. 1994).

Food legumes are still relatively minor crops despite their role as a source of protein in the diet of predominantly vegetarian populace and their importance as components of animal feed and a major source of biological nitrogen fixation in a cropping system. An important explanation of the apparent low status of food legumes, despite their tropical and subtropical adaptation, is the tendencies to underestimate the role of these crops in farming systems. They are often grown as a secondary crop in arid and semiarid regions under poor edaphic and receding soil moisture conditions. Most food legumes have a long history of domestication almost as long as cereals, and during this time, they have been subjected to conscious and unconscious selections for better adaptation to environmental condition as well as better seed yield. Larger seed size, lack of germination inhibitors, non-shattering pods, and freedom from toxic compounds are all retrospective evidences of modification by man (Zohary and Hopf 1973). Despite this, yields of food legumes fall well short of the yields obtained for the major cereals like wheat, maize, rice, sorghum, etc. Although simple dry matter underestimates the dietary contribution of legumes by 25–30%, because of their high caloric contents of protein, it is of interest to explore the reason(s) for the apparent differences in the genetic yield potentials of these two important crop families.

Prevailing environmental factors of a sowing season may impose limitations on growth and processes of yield determination in crops (Agele et al. 2017). The effects of environmental alterations due to changing temperature, light intensity, water availability, and vapor pressure deficit on plant growth rates, dry matter production, and tissue water relations have been reported (Monteith 1981; Thompson et al. 1999). Gillett et al. (2001) and Vega et al. (2000) reported the effects of temperature on dry matter partitioning, while Duncan et al. (1978) and Sato et al. (2002) noted strong cultivar by environment interactions on plant growth rate. Although the partitioning of biomass to fruits is important to yield in crops, its physiological regulation is poorly understood (Hadley et al. 1983; Vega et al. 2000). A simple physiological model developed by Duncan et al. (1978) can be adapted to overcome the constraints imposed by the prevailing environmental factors of the growing seasons on crop growth and yield. This proposition emphasizes the role of carbon as an energy source and its adequacy to integrate effects of other resources (Agele and Olabomi 2010; Vega et al. 2000). Other sets of functions are used to relate dry matter partitioning, yield, crop growth, or development rate in crops to environmental factors (Gillett et al. 2001; De Koning 1994; Heuvelink 1995; Sakata and Higashitani 2008). Heuvelink (1995) and Agele and Olabomi (2010) reported that information on the dynamics of dry matter production and distribution extended with detailed climatic data are still very scarce for tropical crops. Charles-Edwards (1984), Egli and Zhen-Wen (1991), and Ball et al. (2000) proposed that each reproductive structure requires a minimum rate of assimilate supply during its initial development to sustain growth.

Pod yields of cowpea (*Vigna unguiculata* L. Walp) are attributed to a number of factors (Agele and Agbi 2012; Ntare et al. 1998; Ntare and Williams 1998; Purseglove 1988). In crops, the prevailing environmental factors of a sowing season may impose limitations on growth and processes of yield determination. It is hypothesized that crop growth, dry matter partitioning to seed, and yield of cowpea may differ between sowing seasons. The concept of thermal times describes crop development (McMaster and Wilhelm 1997). The response of crop development to temperature regimes is important to the description of the rate of progression toward key events, for example, onset of flowering and harvest index (Bell and Wright 1998). Accumulated heat unit (thermal time) is calculated from temperature coefficients for individual crops. The general cowpea breeding scheme includes evaluating a number of genotypes at various stages and testing selected ones at several locations and seasons across the years. Yield is a complex quantitative character and is generally influenced by environment, climate, and soil fluctuations; hence, the selection for superior genotypes based on yield at a single location may not be very effective (Shrestha et al. 2012).

7.4 Dry Matter Partitioning and Yield in Grain Oilseed Crops

According to Morgan et al. (2007), oilseed crops are in many ways more like a weed than a crop plant, partly because it has an indeterminate growth habit which produces many unproductive branches, flowers, and pods. This results in wasted resources and poor canopy architecture. As a result, harvest index in oilseed crops is usually less than 25% which is very low when compared to other crop species such as wheat which has about 50%. Little, however, is known about how architectural traits function in determining the oilseed crop canopy or how they are controlled genetically. Harvest index is difficult to study because it is the end product of many individual, component traits and is determined by the interactions of many genes and the variation resulting from the effects of the environment. *Brassica napus* is the most important oilseed crop grown in temperate agricultural regions, but despite extensive breeding efforts, it remains an inefficient plant in several aspects of its growth. The indeterminate growth habit of this crop leads to the production of infertile or poorly productive branches with large numbers of unproductive flowers and pods. These characters, among others, such as pod shattering before harvest, have led to oilseed rape being described as possessing "many characteristics associated more with weeds than a crop plant" (Thompson and Hughes 1986). This inefficiency is compounded by the fact that, in order to achieve high yields, it is necessary to have high input of nitrogen fertilizer which accounts for about 60% of the energy input into growing the crop (Kasterine and Batchelor 1998). This makes it only marginally useful as a crop when grown for alternative uses such as biodiesel production. This wasteful use of resources needs to be addressed to make this crop fit for the twenty-first century. An indication of the inefficiency of oilseed rape is seen in measurements of harvest index where typical values in the United Kingdom are between 20 and 25% (compared to 50% and higher for wheat). Harvest index is both physiologically and genetically complex and is determined by the actions of many individual traits, each of which is likely to be quantitative and also environmentally sensitive.

Hanway and Weber (1971c) reported in soybean average total dry matter accumulation of 9680 kg/ha for a yield of 2983 kg/ha with a peak accumulation rate of approximately 8–149 kg/ha/day (Hanway and Weber 1971b). At maturity, this yield represents around 29% of the dry matter partitioned to the seed, with the remaining allocated to the leaves (28%), stems (17%), petioles (15%), and pods (11%). Greater seed yields over the past half century can simply be explained as the product of increased total plant dry matter, greater harvest index, or both through better management practices, like irrigation (Frederick et al. 1991); physiological changes, like a longer reproductive growth period (Rowntree et al. 2014); and genetic improvement through breeding (Rincker et al. 2014). Multiple reports point to both increased total dry matter and a greater harvest index for new genetics within current production systems (Sadler and Karlen 1995; Sinclair 1998; Morrison et al. 1999; Kumudini et al. 2001; Koester et al. 2014; Rincker et al. 2014). Koester et al. (2014) found that harvest index increased by 0.0022% each year between 1923 and 2007 for commercially available varieties of soybean. Greater dry matter partitioning to the seed (high harvest index) may also affect partitioning to other plant parts. Bender et al. (2015) reported an average total dry matter accumulation of 9775 kg/ha for a yield of 3480 kg/ha, resulting in a harvest index of 37% within a modern production environment, with a peak accumulation rate of around 162 kg/ha/day. The higher harvest index reported by Bender et al. (2015) compared with Hanway and Weber (1971c) (29%) altered dry matter partitioning within the leaves (16%), stems and petioles (33%), and pods (14%) drastically, which likely affects nitrogen uptake, partitioning, and removal patterns (Sinclair 1998). Earlier studies on soybean dry matter and nitrogen uptake, partitioning, and removal have reported results at single yield levels (Bender et al. 2015; Hanway and Weber 1971a, b, c) and thus critically limited their widespread applicability, because seed yields vary significantly both spatially and temporally soybean production regions (Gasper et al. 2017).

Sunflower (*Helianthus annuus* L.) is one of the major edible oilseed crops of India after groundnut and rape seed-mustard accounting for nearly 6.5% of the total oilseed production of the country (Anonymous 1994). Several studies have addressed the issue of low productivity in sunflower, and attempts have been made to develop varieties/hybrids with wide adaptability to biotic and abiotic stresses. However, an important approach to identify the factors limiting the productivity lies in understanding the physiology of the plant and the possible constraints that limit seed yield. Despite such a high net assimilation rate in sunflower, the total biomass accumulated over the crop growth period is considerably less compared with many other crops of similar duration (Nanja Reddy et al. 1994). This is because of poor leaf area index values in sunflower. It has been argued that substantial improvement in the productivity could be obtained through improving the photosynthetic area. Several approaches including genetic and agronomic have also been proposed (Shivaram 1986; Nanja Reddy et al. 1994). According to Nanja Reddy et al. (2003), the medium duration types of sunflower (100–110 days) had high harvest index compared with early or long-duration varieties. Genotypes which had low partitioning of dry matter to stem plus thalamus had high harvest index. Sunflower genotypes which accumulated high biomass during post-flowering stages of development also showed high harvest index and seed yield. In a few genotypes, remobilization of photosynthates from vegetative plant parts to the seed resulted in high harvest index and seed yield. Among the sink characters, the number of seed per head, test weight, and seed density (weight/volume) also contributed to achieving high harvest index values. Identification and selection of genotypes based on these criteria will increase the production further. Since large amount of biomass is still locked up in the vegetative plant parts, any practice to manipulate the mobilization of photosynthates from vegetative parts to head also improves harvest index and thus seed yield. In our experiment, we observed that foliar application of boron nutrient and application of growth regulators to the head improved the translocation of photosynthates to the head and thus increased the harvest index and seed yield.

Papantoniou et al. (2013) in a field trial with four oilseed rape (*Brassica napus* L.) cultivars under different growing degree-days (581, 1247, 1609, 2749) showed that pod development occurred in a period when nitrogen accumulation by oilseed rape plants had stopped, leading to high values of contribution of pre-anthesis nitrogen accumulation to pod nitrogen content. Results indicated that hot and dry weather post-anthesis reduced dramatically the net assimilation rates; thus, translocation of pre-anthesis assimilates was crucial for pod development. The results demonstrate that variation in weather conditions between growing seasons is one of the main causes of seasonal variation in oilseed rape productivity under Mediterranean conditions.

7.5 Carbon Partitioning and Acclimation to Environmental Stress

A rapid shift from warm growth conditions to chilling temperatures (typically between 5 °C and 10 °C) leads to a pronounced inhibition of photosynthetic carbon fixation. In chilling-sensitive C3 plants such as tomato and bean, this inhibition is associated with the inactivation of regulatory thioredoxin-activated enzymes of the Calvin cycle (Sassenrath et al. 1991; Holaday et al. 1992; Brüggemann et al. 1994). Because the shift to low temperature is also associated with a strong activation of NADP-malate dehydrogenase in bean (Holaday et al. 1992), it is not due to a chill-induced lack of stromal reducing equivalents. In tomato the chill-induced reduction in photosynthesis is also associated with irreversible inactivation of Rubisco and loss of Rubisco protein (Brüggemann et al. 1992). Low temperature leads to a selective inhibition of end-product synthesis. This results in an accumulation of phosphorylated metabolites and a short-term phosphate limitation of photosynthesis (Leegood and Furbank 1986; Sharkey et al. 1986).

Chilling inhibits phloem export, leading to a rapid accumulation of soluble sugars and repressing photosynthetic gene expression (Krapp and Stitt 1995). Strand et al. (1997) showed this effect recently for *Arabidopsis* during short-term exposure to low temperatures. The inhibition of carbon fixation by low temperature may be followed by photoinhibition (Hurry and Huner 1992). Chilling also interrupts the circadian-regulated transcription of several nuclear-encoded photosynthetic genes in both chilling-sensitive tomato (Martino-Catt and Ort 1992) and chilling-tolerant *Arabidopsis* (Kreps and Simon 1997). These data show that short-term exposure of chilling-sensitive and chilling-tolerant species to low temperatures results in feedback-mediated downregulation of photosynthesis and photosynthetic gene expression, perturbations to circadian-regulated processes, and, in some instances, irreversible damage to photosynthetic proteins. In contrast, when leaves of cold-hardy herbaceous plants develop at low growth temperatures (5 °C), they show a remarkable recovery of photosynthetic capacity. Even in the short term, low temperatures lead to an increase in the activation state of the Calvin cycle enzymes in chilling-tolerant C3 plants such as spinach, winter rye, and winter oilseed rape (Holaday et al. 1992; Hurry et al. 1994, 1995), which is the opposite of the response in cold-sensitive species. When *Arabidopsis* leaves develop at low temperature, they no longer show the suppression of transcript levels for photosynthetic proteins that is found after shifting warm-grown leaves to low temperatures (Strand et al. 1997). As a result, leaves that have developed at low temperature contain higher activities of selected enzymes of the photosynthetic carbon reduction cycle (Hurry et al. 1994, 1995). This reversal of the sink regulation of photosynthetic gene expression (Jang et al. 1997) occurs even though these leaves produce and maintain large pools of soluble sugars.

Leaves of *Arabidopsis* that develop at 5 °C greatly increase their expression of sucrose-6-phosphate synthase and cytosolic fructose-1,6-bisphosphatase in the sucrose synthesis pathway in the cytosol but not in the starch synthesis pathway in the plastid (Strand et al. 1997). Marked increases in the sucrose-6-phoshphte synthase activity have also been reported for spinach (Guy et al. 1992; Holaday et al. 1992), for winter cultivars of oilseed rape and wheat (Hurry et al. 1995), and for rye (Hurry et al. 1994) after prolonged exposure to cold. The increased expression and activity of sucrose synthesis enzymes correlates with an accumulation of sugars in leaves at low temperature. There is no direct evidence that the increased sugar levels are due to increased synthesis and not just to a passive response to the inhibition of growth and phloem transport. It is also not clear whether the increased activities of the enzymes leading to sucrose merely counteract the direct inhibitory effect of low temperature or whether they actually lead to increased sucrose synthesis. Studies of *Arabidopsis* and other cold-tolerant herbaceous plants indicate that growth at low temperature leads to a developmental reprogramming of carbon metabolism that involves changes in gene expression and enzyme activity and possibly also changes of phosphate compartmentation. The resulting recovery of the photosynthetic rate and the preferential accumulation of soluble sugars could be an essential element for acclimation to low growth temperatures (Strand et al. 1999).

Another marked short-term effect found after shifting warm-grown plants to low temperatures is the accumulation of large pools of phosphorylated intermediates (Hurry et al. 1994; Strand et al. 1997), thought to reflect the inhibition of end-product synthesis at low temperatures. It has been suggested that this accumulation of phosphorylated intermediates contributes to the feedback downregulation of photosynthesis, because the cytosolic and chloroplastic pools of inorganic phosphorus become depleted to the point where photophosphorylation is inorganic phosphorus limited (Leegood and Furbank 1986; Labate and Leegood 1988; Stitt and Grosse 1988). Although this may be true for leaves exposed to short-term (minutes to hours) drops in temperature, it is unclear whether inorganic phosphorus limitation persists in the long term (days) in plants acclimating to low temperature. The subcellular compartmentation of inorganic phosphorus is tightly regulated, and 85–95% of the inorganic phosphorus may be located in the vacuole in well-fertilized plants (Bieleski and Ferguson 1983). Prolonged exposure to low-temperature-induced inorganic phosphorus deficiency might therefore be buffered by the release of inorganic phosphorus from the vacuole (Woodrow et al. 1984; Mimura et al. 1990), as occurs when choline feeding depletes inorganic phosphorus in the cytosol (Bligny et al. 1990).

7.6 Factors Affecting Dry Matter Partitioning

There are some factors which cause the low harvest index and ultimately lead to the low production of a crop. Factors that influence crop harvest index include the energy and protein content of seeds, long-term breeding achievements, and extreme (either hot or cold) temperatures during crop reproductive development (Sakata et al. 2000). Crop husbandry can also influence harvest index, especially delayed sowing, which shortens the length of the vegetative phase and increases harvest index. For wheat, and perhaps some other C3 cereals, excess nitrogen can enhance the allocation of photosynthate to structural carbon, which cannot be mobilized to grain later, resulting in a decrease in harvest index. Evidence for the balance between pre- and post-anthesis water use of field-grown crops having a significant influence on crop harvest index is equivocal. It has been stated that environment rather than genotype is seen as the major determinant of harvest index for most field crops for a given site year genotype (Turner et al. 1999). Frost and cold temperatures at flowering, or high temperatures during flowering or seed development, can contribute to substantial variation in harvest index. The balance between pre- and post-anthesis water use is not yet clearly implicated in variation in harvest index of field crops, but in some regions, vigorous growth early due to high soil nitrogen status, perhaps in conjunction with available water, reduces harvest index of wheat, and possibly barley. Field data on seasonal water use and harvest index for other crops are scant.

7.6.1 Energy Investment in Seeds, Fruits, and Storage Organs

Comparisons among crop species reveal that the energy or carbon cost of producing seeds, the portion of the crop most often harvested, varies enormously. The quantity of seed that a given crop species can produce from a given amount of photosynthate (energy) is constrained by differences in plant biochemistry. Unkovich et al. (2010) estimated productivity and nitrogen requirement of different crops from available data across the literature. The calculations were based on the following assumptions (Penning de Vries et al. 1974; Sinclair and De Wit 1975):

- The principal energy costs are in synthesis of carbohydrates (CHO), proteins, and lipids, the equivalent glucose costs of which are protein from NH_3, 1.62 g glucose/g protein; protein from NO_3, 2.48 g glucose/g protein; carbohydrate, 1.21 g/g; lipids, 3.03 g/g; lignin, 2.12 g/g; and organic acid, 0.91 g/g (Penning de Vries et al. 1983).
- Nitrogen for protein is acquired as nitrate for nonlegume crops (2.48 g glucose required/g nitrogen), and for legumes, the cost of symbiotic nitrogen fixation is the same as for ammonium assimilation (1.62 g glucose required/g nitrogen) (Pate and Layzell 1990).
- Nitrogen requirement approximates mg protein nitrogen in seed/total glucose consumed (mg/g), and protein is 15% nitrogen.

The estimated productivity and nitrogen requirement data vary slightly from those of Sinclair and De Wit (1975). Nevertheless, the rankings among crop species are valid. On average, C3 cereal crops turn out to be the most efficient in their conversion of photosynthate to grain with a biomass productivity averaging 0.73, due to the low lipid and protein contents of their seed. Barley is highlighted as being the most efficient. Flood and Martin (2001) noted a negative relationship between grain protein content and harvest index in wheat. Legumes tend to have a higher carbon cost for seed production, averaging 0.61 g seed dry matter/g glucose invested. Peanut stands out as having a very high cost of seed production (0.43 g/

g) due to high seed oil content (45%). This is similar to a range of other high oil species such as cotton, canola, and sunflower.

On a biochemical basis, one would thus expect cereals to have higher harvest indices than grain legumes, which in turn would be higher than the other species with an oil-rich seed. Compared to noncereal crops, the plasticity in grain protein concentration of wheat (Flood and Martin 2001), and other cereals, may reduce variations in harvest index. In contrast, lupin (*Lupinus angustifolius*) seed, for example, tends to have much more stable protein content (Duthion and Pigeaire 1993), and grain nitrogen concentration may dictate grain yield more so than for cereal crops, although nitrogen supply such as in nitrogen-fixing legumes should be non-limiting. Manufacture of seeds and other reproductive structures comes at a significant energy cost, both in terms of storage and biosynthesis, especially where seeds contain lipids and proteins, and thus seed biology impinges directly on the amount of seed that can be produced for a given amount of photosynthate.

7.6.2 Breeding

The values of harvest index for modern varieties of most intensively cultivated grain crops fall within the range of 0.4–0.6. Variation between varieties of the same species is illustrated by trends in the harvest indices of old, outclassed, and recent varieties of temperate and Mediterranean wheat and barley (compared under uniform conditions); this shows a progressive increase throughout the present century, although improvement has been much slower in some countries (Fig. 7.1). In most cases, the improvement in harvest index has been a consequence of increased grain population density coupled with stable individual grain weight. The high heritability of harvest index is explored by examining its response to variation in environmental factors (fertilization, population density, application of growth regulators) in the

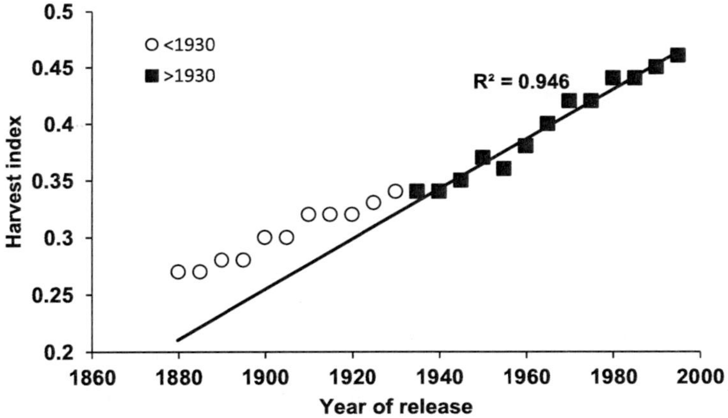

Fig. 7.1 Changes in harvest index of crops during pre- and post-periods of 1930

absence of severe stress. Harvest index of many pulse species and varieties tends to be low because selection has been for some yield in all seasons. Extension of the harvest index concept to express the partitioning of mineral nutrients as well as dry matter (e.g., the nitrogen harvest index) has provided a range of responses whose implications for production and breeding remain to be explored. It has been said that even though the principal cereal crops appear to be approaching the upper limit of harvest index, and future yield gains will have to be sought by increased biomass production, there will still be a need for the concept of harvest index as a tool in interpreting crop response to different environments and climatic change (Hay 1995).

Harvest index of crop plants has increased over time due to breeding for higher yield (Calderini et al. 1999; Whitehead et al. 2000), and lately, specifically for harvest index (Sinclair 1998). Shorter-statured, modern crop cultivars have higher harvest indices than their taller forebears, although total dry matter production is most often very similar (Evans 1993). A recurring theme in harvest index gain has also been a shortening of the vegetative phase of crop plants, providing a proportionately longer grain-filling period (Lopez Pereira et al. 2000). A similar effect can often be achieved by delayed sowing (Gomez-Macpherson and Richards 1995). These increases in harvest index of crops over time may be an unrecognized source of error where carbon accounting is conducted over extended historical time periods (Adger and Subak 1996) in which significant shifts in harvest index have occurred and crop residue returns are calculated from measured grain yields using a constant harvest index.

7.6.3 Determinacy

Most cereal crop cultivars are determinate, flowering within a very small window of time and then entering a grain-filling period such that vegetative and reproductive growths are separated in time. Determinate crops generally have higher harvest indices as most crop resources are diverted to grain production once flowering has commenced. In contrast, for indeterminate crops, such as most legumes and oilseeds, concurrent vegetative and reproductive growth is maintained, resulting in competition for resources between vegetative and reproductive sinks (Pate and Armstrong 1996), and flowering occurs over an extended period during which flowers and pods may be subjected to a range of climatic stresses. Harvest index could thus be more variable in indeterminate than determinate crops. In cases where both determinate and indeterminate cultivars are available (e.g., faba bean, soybean), the determinate varieties tend to have higher harvest indices (De Costa et al. 1997). While determinate crops are also thought to have less variable harvest indices, we cannot find sufficient data to reliably assess this. While flowering over an extended period in indeterminate crops means that developing flowers and pods are subject to a broader range of climatic conditions, some of which may cause abortion of reproductive parts, it also means that aborted flowers or seeds may be compensated for by later formed flowers.

7.6.4 Effects of Pests and Diseases

The improvement and application of pest and disease models to analyze and predict yield losses including those due to climate change is still a challenge for the scientific community (Donatelli et al. 2017). Quantifying the impacts of plant pests and diseases on crop performances represents one of the most important research questions for agricultural simulation modeling (Newman et al. 2003; Savary et al. 2006; Esker et al. 2012; Whish et al. 2015). The widely accepted categorization of yield levels incorporates the crop genetics among the factors defining potential yield, and groups the water and nitrogen stress as limiting factors to attainable yield. The classification of yield levels constitutes the basis to guide strategic decisions in the development and application of cropping system models (Jagtap et al. 1999; Cheero-Nayamuth et al. 2000; Abeledo et al. 2008), including the quantification and modeling of yield losses (Savary et al. 2006; Esker et al. 2012). Crop harvest indices can be reduced by pests and diseases through impacts on flowers or seed and pod development, or indirectly through destruction of grain. Diseases and pests can reduce kernel number or size, for example, Hoffman and Kolb (1998) showed that harvest index in wheat was correlated with the incidence of bean yellow mosaic virus, and McKirdy et al. (2002) showed that this virus caused shriveled grain in wheat, but less so in oat. In Australia, pea weevil (*Bruchus pisorum*) and other insect pests can inflict significant yield losses on grain legume crops (Hardie et al. 1995), significantly reducing economic yield and harvest index.

7.6.5 Temperature Extremes During Flowering and Grain Filling

For most plants, reproductive development is sensitive to a wide range of environmental impacts, with many grain crops having a preferred temperature range for flower and seed development. Large variations in temperature outside of this range may cause problems with flower or seed growth. Damage can be caused by freezing injury or due to temperatures just above freezing point. A definition of absolute values of these is problematic at the field level as there will be some differences among crop species, cultivars and age, and interactions with local landscape factors. For chickpea, Croser et al. (2003) defined freezing temperatures as <-1.5 °C and "chilling temperatures" between -1.5 and 15 °C. For other crops, the freezing temperatures are likely to be similar, but the range of chilling sensitivity will vary considerably regardless of crop species. Chilling temperature effects will be much more common than freezing temperatures, where seed or flower development is constrained due to specific effects on reproductive apparatus, and harvest index is likely to be significantly reduced, especially for determinate crops. The simplest example of a temperature extreme is frost, with freezing injury causing flower or seed abortion. This is mostly managed by combining sowing time with cultivar such that crop anthesis occurs after the last heavy frost (Boer et al. 1993). This approach is mostly successful, but severe frosts after anthesis still occur on occasion and cause reductions in harvest index and grain quality (Allen et al. 2001). With a detailed

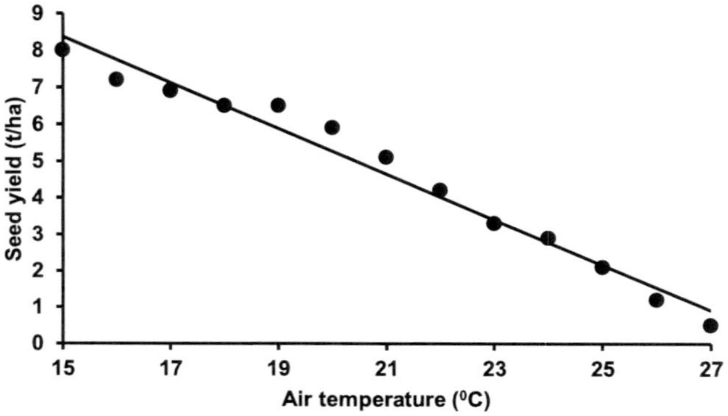

Fig. 7.2 Average maximum air temperature during grain-filling period

knowledge of crop sowing date and cultivar, the likelihood of frost damage can be estimated if combined with local historical weather measurements (Robertson et al. 2001), but the actual impact on yield or harvest index cannot be modeled as the actual damage to grain caused by frosts cannot be accurately predicted within a crop, region, or field. Frost events may not have a major impact on harvest index due to flexibility in sowing dates, the range of crop cultivars available, the relatively low incidence of frost occurrence, and its localized nature, except in exceptional ($\geq$1/20) years.

Other temperature stresses on reproductive performance of crops also affect harvest index. Working with canola in low rainfall, Si and Walton (2004) found that high post-anthesis rainfall and low post-anthesis temperature correlated with higher yield. Exposing pod to temperatures of 34–38 °C for 2 days is known to cause reduced seed weight by 8–12% as compared to 20 °C (Reader et al. 1997) and occurs in about 1 year in three in the main lupin-growing region, but impacts on crop harvest index were not measured. While one might expect pod abortion to result in a lower harvest index, in the case of lupin, an indeterminate crop, later formed pods can compensate to some extent (Seymour 1996), but at the field level, this is likely to be the exception rather than the rule (Dracup 1994). High post-anthesis temperatures can also reduce grain yield (Fig. 7.2) and quality in wheat, but with substantial differences in sensitivity among cultivars (Stone and Nicholas 1995).

7.6.6 Pre-/Post-anthesis Water Use

Total water use (evapotranspiration) may not always be the principal limitation to grain yield for rainfed grain crops (Passioura 2006), but the seasonal distribution of rainfall and plant available soil water relative to current crop demand and crop development is often thought to be critical (Angus and van Herwaarden 2001). In Australia, this has been most extensively studied in wheat (van Herwaarden et al. 1998a), but the effect of timing of rainfall has also been examined in field

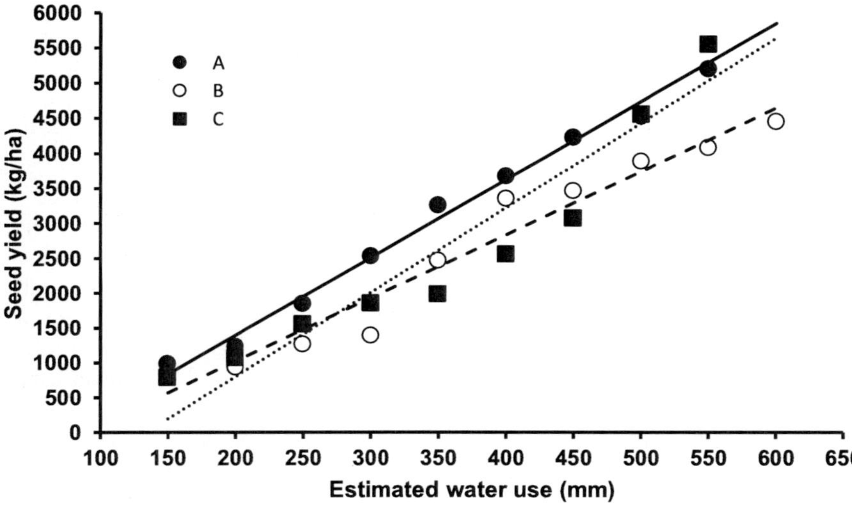

Fig. 7.3 Effect of total water use (mm) by plants of *Cicer arietinum* (A), *Pisum sativum* (B), and *Vicia faba* (C) on seed yield (kg/ha)

experiments with barley (Goyne et al. 1993), lupin (Palta et al. 2003), and canola (Wright et al. 1992). Harvest index decreased in the order irrigated > rainfed > rain excluded. Thus, having more water available to canola plants post-anthesis increased harvest index. This relationship was shown quite eloquently for wheat in the glasshouse experiments of Passioura (1977). Harvest index as well as seed yield was found to be independent of total water use (Fig. 7.3). For rainfed field crops, it is difficult to manage water supply to match current crop demand. The balance between pre- and post-anthesis growth will be a function of available soil water (rainfall soil type soil evaporation) and the rate it is used by the crop. Pre-anthesis transpiration can be increased by promoting more rapid development of crop leaf area. This can be done with the application of fertilizers, especially nitrogen, and with crop rotations which reduce root diseases and promote faster root growth. Together these could lead to faster leaf area development and earlier exhaustion of soil water. Field experiments employing irrigation and rainout shelters (Kirkegaard et al. 2007) examined seasonal water use and yield of wheat. In one experiment, additional post-anthesis water use of around 10 mm increased grain yield but had no impact on harvest index, while in a second experiment where a severe water stress was imposed, extraction of an additional 30 mm of water increased harvest index from 0.31 to 0.38. Averaging across site year may mask variations in fractional post-anthesis water use and harvest index due to crop sequence (Lenssen et al. 2007; O'Connell et al. 2002). Wahbi and Gregory (1989) presented a negative relationship between the percentage of total water use (20–35%) after anthesis and harvest index of barley grown in Syria and concluded that water use during grain filling was a poor indicator of harvest index.

7.6.7 Water Use, Soil Mineral Nitrogen, and Harvest Index

The field studies of van Herwaarden et al. (1998a) provide some evidence for there being a relationship between fractional water use after anthesis and harvest index; however, this dataset is confounded by interactions with nitrogen fertilizer applications. Indeed nitrogen fertilizer rates may have an influence on harvest index that does not operate through seasonal water use patterns. The work of van Herwaarden et al. (1998a, b) indicates that high soil nitrogen fertility might result in a decrease in the deposition water-soluble carbohydrates in wheat stems, tillers, and leaves and an increase in structural carbon, leading to less mobilization of vegetative carbon to grain and a lowering of harvest index. Examining wheat crops, Angus and van Herwaarden (2001) concluded that vigorous pre-anthesis growth can increase plant transpiration, at the expense of soil evaporation, and lead to a higher carbon fixation. However, structural carbohydrate laid down earlier in the growing season is not readily mobilized to the grain. In irrigation experiments with wheat receiving 44 kg N/ha (Kirkegaard et al. 2007), a greater fraction of stem water-soluble carbohydrates were mobilized to grain in more water-stressed plants, but this did not result in increased harvest index. In a second experiment, there were no significant differences in water-soluble carbohydrates between irrigation treatments. Replotting the harvest index dataset of Cantero-Martinez et al. (1995) on the basis of available nitrogen shows a weak negative relationship between available nitrogen at sowing and the harvest index of wheat, whereas no relationship was apparent for fractional water use after anthesis and crop harvest index in this same dataset. A similar relationship is indicated by Adcock (2006). Here harvest index of wheat in 2 years was lower after legumes (vetch or medic) than after nonlegumes (canola, barley, or wheat), although there were minimal differences in the fraction of water used after anthesis between rotations within a given year. The timing of seasonal water use was governed by absolute water availability (rainfall) and not through deferral of water use within the crop-growing season. In this study, soil mineral nitrogen was higher when the preceding crop was a legume (133–225 kg N/ha) than cereal or canola (81–149 kg N/ha), and this may have contributed to the differences in harvest index rather than seasonal water use patterns by the crops. In general lower soil fertility, lower nitrogen fertilizer applications resulting from a greater uncertainty of responses to nitrogen fertilizers due to climate variability, and cultivars with greater concentrations of water-soluble carbohydrates (van Herwaarden and Richards 2002), may reduce the likelihood of poor mobilization of stored carbohydrates to grain in the western wheat belt. In a pot study on barley, Fathi et al. (1997) found that reductions in grain yield and increases in grain protein due to post-anthesis water stress were greater when plants were grown under high nitrogen supply than when the supply of nitrogen was limited. This result is in concordance with the aforementioned studies on wheat.

7.6.8 Hormonal Control of Photosynthate Partitioning

Seed expansion and filling include all processes involved in formation and structural development of mature seed (Locascio et al. 2014). Seed-making process is highly coordinated and complex, requiring hormonal regulation (Ochatt 2011), and a constant exchange of signals from-and-to maternal tissues, and between cotyledon/ endosperm and embryo (Locascio et al. 2014). The continual interaction among these three constituents of seed ensures synchronized seed development (Locascio et al. 2014). Many reports showed substantial changes in hormone content during seed development and filling. Auxin is linked with maize grain filling through controlling cell wall-associated invertase enzyme activity. The increase in auxin during seed filling interval accompanied with increased grain-filling rate. The increased auxin is known to enhance sink capacity through enlargement of cell and increased nutrient assimilation (Kong et al. 2015). Abscisic acid, auxin, and cytokinins play a major role in grain filling through source photosynthate/nutrient remobilization and grain development in cereals (Yu et al. 2015). Notably, the source activity and sink strength are synchronized and can be altered by hormones and external/environmental stimulus. The grain filling is closely associated with senescence and senescence-related hormone abscisic acid, affecting the time for nutrient mobilization and grain filling in barley, wheat, rice, and sorghum (Yu et al. 2015). In barley, abscisic acid can induce the orchestration of gene expression of senescence-related genes (Yang et al. 2003a). Such activities are also in action during drought stress to accelerate seed filling. In rice and wheat, water stress at grain-filling stage caused reduced photosynthetic activity and induced accelerated leaf senescence, resulting in shortened grain-filling period due to enhanced abscisic acid level and remobilization of carbon pool from stem to leaves (Yang et al. 2003a; Yang and Zhang 2006). Drought-tolerant plants exhibit delayed leaf senescence; thus, cultivars with stay-green trait have advantage over terminal drought (Jordan et al. 2012), which is beneficial to sustain seed filling. Early onset of leaf senescence causes inadequate nutrient supply to grains during grain filling, whereas late initiation of nutrient mobilization may not support rapid developing seed. This synchronized developmental program is known to involve abscisic acid-inducible NAC-transcription factor, which regulates the expression of genes involved in mineral nutrient remobilization from leaf to grain, chlorophyll degradation, and leaf senescence in rice (Uauy et al. 2006). A reduced expression of this abscisic acid regulated NAC-transcription factor caused enhancement in yield in the transgenic rice, plausibly due to prolonged supply of photosynthates/nutrients into the developing grains through fine-tuned abscisic acid biosynthesis pathway. Cytokinins also play a key role in grain filling by initiating rapid cell division of endosperm cells (Kong et al. 2015). The increased level of cytokinins probably enhances sink strength by up-regulating cell division-related genes via sugar signaling, which involves enhanced phloem unloading and sugar import to endospermic cells through a cell wall-associated enzyme invertase (Rijavec et al. 2009). Overproduction of cytokinin delays early senescence through reduced proteolytic activity and inhibition of nitrogen remobilization, which increases life span and confers drought tolerance in crop plants (Gregersen et al. 2013). Thus, seed filling is determined by diverse plant hormones having specific functions, and their interplay might be vital in regulating various processes related to accumulation of seed reserves (Ochatt 2015).

Signaling in plants is well established and regulatory mechanisms appear to be highly sophisticated involving long-distance electrical signals and intercellular transport of macromolecules that help to optimize energy use and regulate plant development and defense pathways (Brenner et al. 2006), and are thus obvious targets for genetic modification. In wheat, photoperiod has been shown to influence the number of fertile florets (Miralles et al. 2000) in relation to spike growth (González et al. 2005a) probably by regulating sugar supply leading to programmed cell death (Ghiglione et al. 2008). When temperature is high around anthesis, yield losses can be very significant (Semenov and Shewry 2010). Here, stressed ethylene appears to be involved in signaling leading to grain abortion (Hays et al. 2007). Ethylene may also be involved in the stress response of growth, development, and functioning of plants under ozone stress (Wilkinson and Davies 2009), and abscisic acid (ABA)/ethylene ratios can be important in determining spikelet fertility (maize, Wang et al. 2002; rice, Yang et al. 2006) and the rate of grain filling in wheat (Yang et al. 2006). The photo-environment and the nutrient status of plants, both of which significantly impact plant growth, development, and grain yield, can do so through modification of hormone signal transduction chains (Pierik et al. 2004). Several studies have characterized post-anthesis changes in hormone concentration in wheat grains (Yang et al. 2006; Gutam et al. 2008), but there is little information on the physiological roles of these changes in hormone content of developing wheat grains in response to environmental factors. Post-anthesis application of abscisic acid and/or ethylene biosynthesis inhibitors directly to developing grains increased the ratio of abscisic acid to ethylene and increased the grain-filling rate (Yang et al. 2006). This observation, and that excessive ethylene production results in wheat grain abortion under high-temperature stress (Hays et al. 2007), suggests that grain abscisic acid accumulation in wheat is a desirable trait and also in rice. Further investigation is important here, however, as there is a clear indication that accumulation of abscisic acid in developing grains of maize can result in grain abortion (Wang et al. 2002).

7.6.9 Plant Growth and Carbon Partitioning

Regulation of carbon partitioning during developmental transitions or in response to stress is important to avoid a reduction in the supply of carbon to plant structures that could disrupt balanced growth (Geiger et al. 1992). When plants are subjected to drought stress, regulation of carbon partitioning helps the plants to acclimate (Meyer and Boyer 1981), thereby sustaining balanced growth of the various plant parts (Geiger and Servaites 1994; Geiger et al. 1992). Regulating the distribution of assimilated carbon regulation of assimilate partitioning between shoot growth and root growth maintains the balance between photosynthetic productivity and water and mineral uptake, thereby maintaining allometric and functional balance.

Regulation of genetic development helps avert damage that could result if growth were adjusted mainly by direct competition among plant parts. Programmed developmental changes associated with the transition from vegetative to reproductive stage in *Phaseolus vulgaris* seem successfully to prevent local shortages of assimilated carbon and functional imbalance (Geiger and Shieh 1988). The developmental events that initiate the transition from vegetative to reproductive state occur at a critical stage about 10 days after first anthesis (Geiger and Shieh 1988). Prior to this time, leaves are initiated at intervals of about 1.5 days, but about 10 days after the start of anthesis leaf initiation stops (Geiger et al. 1989). By this time, most of the flowers that will produce the final reproductive growth have set and flowering is coming to an end. Although the relative growth rate of fruits is over four times that of the vegetative organs at this stage, reproductive organs make up only about 10% of plant dry mass. The combination of vegetative and reproductive developmental events at this stage leads to a dramatic change in the pattern of carbon partitioning at about 20 days after anthesis (Geiger and Shieh 1988).

Regulating carbon partitioning to restore and maintain functional and allometric balance throughout the plant probably involves signals among the parts. For instance, the roots of most species must receive at least part of their reduced nitrogen from shoots by phloem translocation (Cooper and Clarkson 1989). These authors suggest that there is a single regulatory pool of amino-nitrogen common to roots and shoots and that this pool is cycled via the xylem and phloem. Various molecular species involved in this cycle are well suited to serve as signal molecules, activating gene expression in target organs. This is also supported by the responsiveness of many genes to glutamine, and includes sugar-responsive genes such as nitrate reductase (Stitt et al. 1995). Still, other carbohydrate-modulated genes such as that encoding the Rubisco small subunit do not respond to such changes.

7.6.10 Effects of Nitrogen Partitioning

It is important to consider nitrogen partitioning within strategies to optimize assimilate partitioning, as increasing biomass implies greater crop nitrogen uptake and/or nitrogen use efficiency (grain yield/nitrogen available from fertilizer and/or soil). Optimizing canopy nitrogen dynamics could provide an additional route to increase canopy photosynthesis and biomass by prolonging green area in the post-anthesis period (Parry et al. 2010). Nitrogen dynamics are an important factor in the maintenance of green leaf area in sorghum, with the onset and rate of leaf senescence among three hybrid lines in Australia explained by a supply-demand framework for nitrogen dynamics, in which individual grain nitrogen demand was sink determined and was initially met through nitrogen translocation from the stem and rachis, and then if these nitrogen pools were insufficient, from leaf nitrogen translocation (van Oosterom et al. 2010a, b). Genetic variation in the stay-green trait has been reported in bread wheat (Silva et al. 2000; Verma et al. 2004; Foulkes et al. 2007).

7.7 Exploring G × E Interaction for Increasing Harvest Index in Crops

Environmental factors are important determinants of harvest index and include seasonal pattern of water supply and extreme temperatures during crop reproductive development (Porker et al. 2020). One plausible explanation for a reduced harvest index in early sown slow developing cultivars is related to pattern of water use. In glasshouse experiments, the ratio of pre- and post-anthesis water use has been demonstrated to be strongly related to harvest index (Passioura 1977). When established early, winter wheats have a long duration of pre-anthesis growth, and in water-limited environments can use too much water before anthesis such that HI is low compared with spring wheats established later (Gomez-Macpherson and Richards 1995). Deferred water use trades off against the higher dry matter accumulation of early established winter cultivars which drives high grain number and produces water-soluble carbohydrates that can be translocated to grain, such that yield of early established winter wheats is at least equivalent to faster spring cultivars established later.

The second explanation is that increased plant height and more leaves lead to competition for carbohydrates between the developing spike and elongating stem of early sown crops (Gomez-Macpherson and Richards 1995). Genetic improvement is one approach to remove these limitations. Since the Green Revolution, improvement in wheat grain yield in many environments has been due to an increase in the number of grains per unit area and harvest index (Hay 1995; Sayre et al. 1997; Shearman et al. 2005a, b; Flohr et al. 2018). Increasing the sink size or improved partitioning to spike growth may provide a pathway for the improvement of the harvest index of wheat (Foulkes et al. 2010). While there may be potential to further adjust phasic development to alter the timing and duration of spike development, a study by Flohr et al. (2018) showed that harvest index improvement was not due to phase duration and suggests other factors are involved such as partitioning traits. Variation in grain growth rates under different air temperatures has been documented (Fig. 7.4). Earlier studies in many crops have outlined any genetic or management solution that could either increase the grain growth rate or enhance the remobilization of assimilates from vegetative tissues to grains after anthesis usually leads to a higher harvest index within a crop (Yang et al. 2000; Kemanian et al. 2007; Fletcher and Jamieson 2009; Zhang et al. 2012).

Crop management also has the capacity to modify the pattern of water use and biomass partitioning. There are many examples where variation in harvest index is mainly attributed to differences in crop management (Kemanian et al. 2007; Peltonen-Sainio et al. 2008). While delayed sowing is a management factor known to increase harvest index, this is counterproductive for slow developing cultivars as they will flower outside of the optimal window, produce less dry matter, and have reduced yield potential. Management factors that can improve harvest index and yield from an early sowing date have not yet been explored.

Reducing plant density has also been proposed as a way of reducing early dry matter accumulation and improving harvest index in early established slow

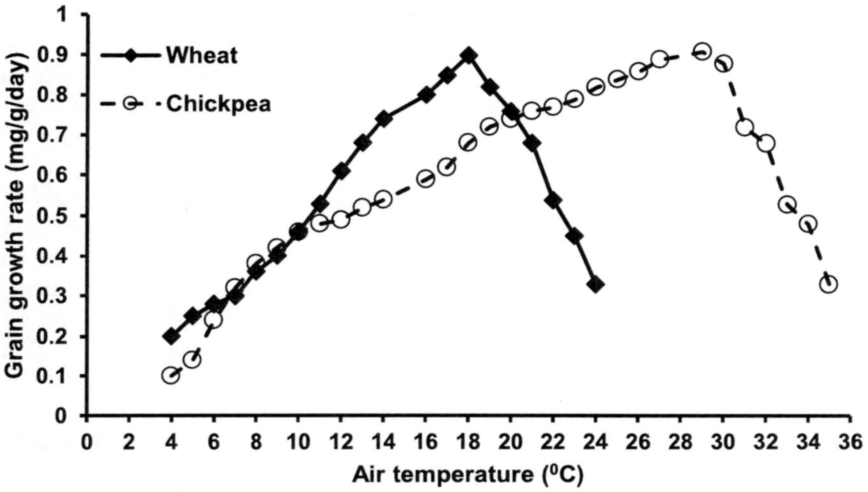

Fig. 7.4 Changes in grain growth rates in wheat and chickpea under different air temperatures

developing cultivars (Kirkegaard et al. 2014). Early sown winter cultivars provide additional farming system benefits; a longer vegetative period means more dry matter is accumulated for forage, and there is more time available for stock to graze before the onset of the reproductive phase (Bell et al. 2015). It is often thought that defoliation could be used as a management technique to reduce vegetative growth and early water use. While this would imply an increase in harvest index, it also trades off against reduced dry matter accumulation by anthesis as shown by Kirkegaard et al. (2015). There have been some measured instances of deferral of water use assisting recovery of wheat yield (Virgona et al. 2006; Harrison et al. 2010), but Harrison et al. (2011) found that defoliation on average reduces grain yield most likely due to reduced dry matter accumulation.

Oversupply of nitrogen early in crop development stimulates vigorous vegetative growth and can lead to water deficit in later reproductive phases. Excessive vegetative growth induced by excessive nitrogen is commonly known to lower grain yield and harvest index in water-limited environments and is associated with reduced post-anthesis carbon assimilation in response to a lack of soil water (van Herwaarden et al. 1998b). The water deficit also contributes to poor re-translocation of pre-anthesis reserves. Deferring nitrogen inputs until after the start of stem elongation in winter types therefore has the potential to conserve water use and manage early biomass production.

Trait-based physiological breeding has been used successfully to improve abiotic stress adaptation of wheat, including higher transpiration efficiency, greater early vigor, reduced tillering, and dehydration avoidance (Richards et al. 2002; Reynolds and Tuberosa 2008), and is well suited to the current objectives in which a number of complex traits (spike fertility, potential grain size, and lodging resistance) must be combined. In wheat, grain yield improvement has been highly associated with grain number per unit area (Canevara et al. 1994; Sayre et al. 1997; Brancourt-Hulmel et al. 2003; Shearman et al. 2005a, b; Peltonen-Sainio et al. 2007). Evidence suggests that grain sink strength remains a critical yield-limiting factor (Slafer and Savin 1994; Abbate et al. 1998; Miralles et al. 2000; Borras et al. 2004; Miralles and Slafer 2007) and that sink capacity will need to be improved if improvements in biomass and radiation-use efficiency are to be fully exploited as increased harvestable yield (Reynolds et al. 2001, 2005a, 2009; Acreche and Slafer 2009). Increased partitioning of assimilates to the developing spike and grain has historically had the single greatest impact on improving yield potential in wheat (Calderini et al. 1999), not only under high-yielding conditions (Brancourt-Hulmel et al. 2003) but also under stressful conditions such as those of Mediterranean regions (Loss and Siddique 1994; Acreche et al. 2008). Although greater biomass appears to be contributing to genetic yield progress in modern cultivars released since about 1990 (Donmez et al. 2001; Shearman et al. 2005a, b), there are clear indications that yield of modern wheats is still more sink than source limited during grain filling (Borras et al. 2004). A mirror image of this sink limitation is that the current or improved photosynthetic capacity during grain filling may not be fully exploited if it is not matched by adequate spike fertility (Reynolds et al. 2005a). For example, photosynthesis during the post-anthesis period (at both leaf and canopy levels) seems to be responsive to increases in strength of the sink, via source/sink manipulation treatments imposed around anthesis, even in modern cultivars with high grain numbers (Reynolds et al. 2005a; Acreche and Slafer 2009). Therefore, for improvements in photosynthesis to result in additional crop yield, extra assimilates must be partitioned to developing spikes and grains, and/or potential grain size must be increased to accommodate the extra assimilate (Foulkes et al. 2011). Although to date improvement of yield potential has been based mainly on yield per se, there is strong evidence that understanding traits at the physiological level will help to identify trait interactions and indirect selection criteria that could accelerate breeding progress (Jackson et al. 1996b; Araus et al. 2002; Slafer 2003; Reynolds and Borlaug 2006; Reynolds et al. 2009). An understanding of traits at the physiological level will help to predict synergies (between new environments and breeding, between agronomy and breeding, and between biomass and partitioning traits).

Yield increases associated with the introduction of semidwarf cultivars during the Green Revolution of the 1960s and 1970s were associated with reduction in stature, reduction in lodging, and also reduced competition from the growing stem resulting in greater spike growth, more grains per unit area, and higher harvest index (Fischer and Stockman 1986). Despite a hypothetical limit to harvest index of ~0.62 in wheat (Austin 1980), comparisons of genetic progress in harvest index over time indicate no systematic improvement since the early 1990s from values of ~50% in spring wheat (Sayre et al. 1997; Reynolds et al. 1999) and 55% in winter wheat (Shearman et al. 2005a, b). Future yield gains in modern semidwarfs through harvest index will require further gains in grain number. These may be achieved by increased partitioning to spikes during stem elongation, but if increased assimilate is not available for this, care must be taken that decreased stem partitioning does not decrease lodging resistance or potential grain size. The genetic and physiological interdependence of these traits will need to be established for easier deployment in breeding and to optimize any trade-offs.

7.8 Management Techniques to Increase Harvest Index

The challenge to produce more food under increasing water scarcity has led to the notion that crop water productivity (economic yield over the amount of water consumed) needs to increase (Kijne et al. 2002, 2003). However, how to increase water productivity is still being debated (Kijne et al. 2003; Belder et al. 2007). In cereals and at the crop level, it is proposed that water productivity can be defined as the ratio of grain yield over amount of water transpired (WP_T) (Bouman 2007). As the grain yield is the product of harvest index and total aboveground biomass, the WP_T in rice could be expressed as (Yang and Zhang 2010):

$$WP_T = Y/T = \mathrm{HI} \times B/T$$

where WP_T is the grain yield per unit water transpired (kg grain kg^{-1} water), *Y* is the grain yield (kg), *T* is the amount of water transpired (kg), HI is the harvest index (kg kg^{-1}), and *B* is the aboveground biomass (kg).

The fraction *B*/*T* is sometimes known as transpiration efficiency. The harvest index is the grain yield over total aboveground biomass. The grain yield and water productivity would be improved by either an increase in transpiration efficiency or an increase in harvest index. However, the ratio of biomass production over transpiration (*B*/*T*) has been shown to be fairly constant for a given species in a given climate (Ehlers and Goss 2003), and can be selected for in plant breeding (Bouman 2007). Plant biomass production is linearly coupled with the amount of water transpired, and a higher water-use efficiency is often a trade-off against lower biomass production (Zhang and Yang 2004). In agriculture, many ways of conserving water have been investigated, and techniques such as alternate partial root zone irrigation, deficit irrigation, and drip irrigation have shown that water-use efficiency can be enhanced (Kang et al. 2000; Tabbal et al. 2002; Li et al. 2010). In general, these techniques are a trade-off: a lower yield for a higher water-use efficiency (Zhang and Yang 2004). On the other hand, harvest index has been shown to be a variable factor in crop production. Variations in harvest index within a crop are mainly attributed to differences in crop management (Yang et al. 2000; Guo et al. 2004; Kemanian et al. 2007; D'Andrea et al. 2008; Peltonen-Sainio et al. 2008). A water and/or nitrogen management system that could increase growth rate during grain growth and/or enhance the remobilization of assimilates from vegetative tissues to grains during the grain-filling period usually leads to a higher harvest index within a crop (Xue et al. 2006; Bueno and Lafarge 2009; Fletcher and Jamieson 2009; Ju et al. 2009). In many situations, harvest index is closely associated with water-use efficiency and grain yield in wheat (*Triticum aestivum* L.) and rice (Yang et al. 2000, 2007; Zhang et al. 2008a, b). The question arises as to

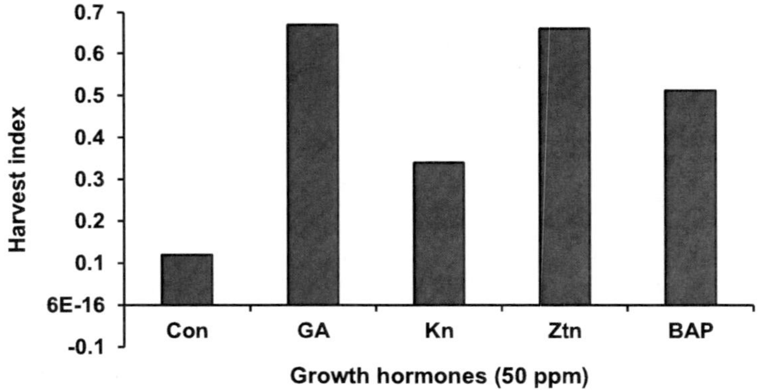

Fig. 7.5 Effect of 50 ppm of plant growth hormones (GA, gibberellins; Kn, kinetin; Ztn, zeatin; BAP, benzylaminopurine) in comparison to control (Con) on harvest index of rice crop

whether it is possible to achieve the dual goal of increasing food production and saving water through the manipulation of harvest index in crop production. Use of plant growth hormones to increase harvest index has been reported (Fig. 7.5). It was found that application of 50 ppm gibberellins increases the harvest index of rice crop.

Grain filling is the final stage of growth in cereals when fertilized ovaries develop into caryopses and depends on carbon from two resources: current assimilates and assimilates redistributed from reserve pools in vegetative tissues either pre- or post-anthesis (Samonte et al. 2001). The contribution of reserved assimilates in culms and leaf sheaths of rice plants is estimated at 10–40% of the final yield, depending on the cultivar and the environmental conditions (Takai et al. 2005). Remobilization of reserves to the grain is critical for grain yield if the plants are subjected to water stress or if the yield potential is largely based on the high biomass accumulation (Asseng and van Herwaarden 2003; Plaut et al. 2004). Remobilization and transfer of the stored assimilates in vegetative tissues to the grain in monocarpic plants such as rice and wheat require the initiation of whole plant senescence (Gan and Amasino 1997; Noodén et al. 1997).

Delayed whole plant senescence results in much nonstructural carbohydrate left in the straw and leads to a low harvest index. Usually, water stress at grain-filling time induces early senescence and shortens the grain-filling period but increases the remobilization of assimilates from the straw to grains (Asseng and van Herwaarden 2003; Plaut et al. 2004). It has been found that a controlled soil drying or a moderate soil drying, namely, plants can rehydrate overnight, and photosynthesis is not severely inhibited imposed at mid- and late grain-filling stages (from 9 days post-anthesis until maturity), can greatly enhance assimilate remobilization from vegetative tissues to grains and also accelerate the grain-filling rate. Such a controlled soil drying does not necessarily reduce grain yield even when plants are grown under normal nitrogen conditions. Furthermore, in cases where plant senescence is unfavorably delayed, such as by the heavy use of nitrogen, the gain from enhanced remobilization and accelerated grain-filling rate can outweigh the loss of reduced photosynthesis and a shortened grain-filling period, leading to an increased grain yield and higher harvest index and water-use efficicency.

The mechanism by which post-anthesis-controlled soil drying enhances the utilization of pre-stored assimilates is not fully understood. Many processes are likely to be involved, including the hydrolysis of stored carbohydrate, phloem loading, long-distance translocation, and phloem unloading into the kernels. It was observed that, in rice stems, both α- and β-amylase activities were enhanced by the soil drying treatment, with the former enhanced more than the latter, and significantly correlated with the concentrations of soluble sugars in the stems. The other two possible starch-breaking enzymes, α-glocosidase and starch phosphorylase, showed no significant differences in the activities between well-watered and soil drying treatments. Soil drying also increased the sucrose-phosphate synthase activity that is responsible for sucrose production (Yang et al. 2001). In grains, the activities of four enzymes involved in sucrose-to-starch conversion, sucrose synthase, ADP-glucose pyrophosphorylase, starch synthase, and starch branching enzyme, were significantly enhanced by soil drying, and positively correlated with the starch accumulation rate in grains (Yang et al. 2003b). These results suggest that, in the source (stems), enhanced activities of α-amylase and sucrose-phosphate synthase contribute to the fast hydrolysis of starch and increased carbon remobilization, and in the sink side (grains), an increased grain-filling rate is mainly attributed to the enhanced sink activity by regulating key enzymes involved in sucrose-to-starch conversion, when subjected to a mild soil drying during the grain-filling period.

7.9 Carbon Partitioning in Different Plant Parts

There is little knowledge of processes involved in carbon partitioning to the various organs of a plant (Whisler et al. 1986). Assimilate partitioning is thus the main weak point of current crop simulation models. Usually, carbon is translocated from a mature leaf to the nearest actively growing region, but this general pattern can be changed if leaves are removed or shaded or depending on sink requirements (Shibles et al. 1987). In pea, the presence of a sink and a source on the same reproductive node led some authors to study the possible preferential relationship between them.

At a node scale, the assimilates allocated to the pods come mainly from the leaf of the node, and to a lesser extent from the hull of the node (Flinn and Pate 1970). However, translocation of assimilates between nodes can occur (Harvey 1973; Szynkier 1974; Jeuffroy and Warembourg 1991). In pea and soybean, during the reproductive phase, carbon is preferentially exported to pods on the same node and, secondarily, to pods' two nodes below or above (Shibles et al. 1987), according to the vascularization of the plant. In a ^{14}C-labeling experiment, with plants growing under two different radiation intensities, Jeuffroy and Warembourg (1991) showed that there was no priority among the sinks in the plant during the period of seed formation. When the amount of assimilates produced by the whole plant was severely reduced, the amount of ^{14}C allocated to each individual organ was proportionally reduced. This led to a model for assimilate partitioning between the organs on a stem, at different dates during this period. In the model, the assimilates produced by the stem during a time interval are allocated to each pod in proportion to its demand. This demand is linked to the initial amount of dry matter of the pod before its final stage in seed abortion, and to the seed number of the pod.

Partitioning between reproductive and vegetative organs is dependent on the plant growth during the two preceding phyllochrons (Jeuffroy 1994a). This model was tested in various field conditions (Jeuffroy and Devienne 1995), with good accordance between observed and simulated growth of the different organs, i.e., pods of each node and vegetative parts. However, in situations with water stress, the growth rate of seeds which had begun to fill before the beginning of the stress remained stable, while abortions occurred on pods which had not yet reached this stage (New et al. 1994). This suggests that filling pods have priority over young pods in assimilate partitioning. In soybean, defoliation reducing the supply of assimilates did not also modify seed growth after the beginning of the linear growth phase of the seed (Munier-Jolain 1994). This observation could be used in a general model of assimilate partitioning.

At the pod scale, seed number can be predicted from pod growth dynamics during the first 120°-days after anthesis (Jeuffroy and Chabanet 1994). Thus, the pattern of assimilate partitioning and the relationship between early pod growth rate and seed number per pod have been used to construct a model for seed number production on each node of a pea stem (Jeuffroy 1992, 1994b). The inputs of the model are the development parameters (as presented above), the growth rate of the whole stem during the period of seed formation, and the climatic data (mean daily temperature). When tested in various field conditions (Jeuffroy 1994b), this model gave a good account of the variability in seed number partitioning among the nodes of the stems. This model requires measurements of the assimilate demand of the filling seeds. In soybean (Munier-Jolain 1994) and pea (Munier-Jolain and Ney 1995), the potential seed growth rate is determined by seed cell number, which is fixed at the beginning of seed filling. Thus, if the cell number is low at this stage, the accumulation of reserves in the seed until physiological maturity can be limited even if assimilate sources are large. The relationship between seed growth rate and cell number does not seem to differ between genotypes, even with substantial differences in individual seed weight.

Sucrose is the main form of translocated sugar in most plants. Its synthesis is catalyzed by sucrose-6-phosphate synthase, and its degradation is catalyzed by sucrose synthase or invertase (Preiss 1982). A major control point for the partitioning of photosynthate between sucrose and starch in the leaves is sucrose-6-phosphate synthase (Huber and Huber 1992). This postulate was confirmed when overexpression of maize sucrose-6-phosphate synthase in tomato leaves was shown to increase foliar sucrose concentration but reduce starch (Galtier et al. 1993). Some species, such as soybean, have been reported to have endogenous rhythms of sucrose-6-phosphate synthase activity (Kerr et al. 1985), whereas the enzyme appears to respond directly to the light cycle in other species (Kalt-Torres et al.

1987). Environmental factors such as water stress and temperature influence sucrose-6-phosphate synthase activity. Sucrose-6-phosphate synthase activity in mature leaves was reduced in plants subjected to water stress (Vassey et al. 1991; Castrillo 1992). This decrease in sucrose-6-phosphate synthase activity was attributed to inhibition of photosynthesis as a result of stomatal closure (Vassey et al. 1991). However, pigeon pea and soybean plants exposed to water stress had increased sucrose-6-phosphate synthase activity (Cheikh and Brenner 1992; Keller and Ludlow 1993). Soybean leaves had lower sucrose-6-phosphate synthase activity under cool conditions (18/14 °C) than under warm conditions (26/22 °C; Rufty et al. 1985). Spinach plants acclimated to cold temperature (10 °C) had higher sucrose-6-phosphate synthase activity than plants grown at 24 °C (Holaday et al. 1992). Khayat and Zieslin (1987) exposed rose plants to 12 and 18 °C night temperatures and found higher sucrose-6-phosphate synthase activity in shoots exposed to the higher night temperature. Sucrose synthase is considered the main catalytic enzyme for sucrose cleavage in many tissues, and its activity can be used as a biochemical marker for sink strength in plants (Sung et al. 1989; Xu et al. 1989). Growing tubers have high activity of sucrose synthase and ADP-glucose phosphorylase that declines as tubers mature (Sowokinos 1976). Heat stress alters carbohydrate partitioning in potato (*Solanum tuberosum* L.) plants from tubers to shoots and reduces overall plant yield (Wolf et al. 1990). Apart from source limitation, the capacity of the reproductive sinks to utilize the incoming assimilates is also affected under drought stress and may also play a role in regulating reproductive abortion (Zinselmeier et al. 1999). A reduced acid invertase activity can arrest the development of reproductive tissues due to improper phloem unloading (Goetz et al. 2001).

7.10 Temperature Effects on Dry Matter Partitioning

Temperature stress is becoming the major concern for plant scientists worldwide due to change in climate. The difficulty of climate change is further added considering its precisely projecting potential agricultural impacts (Watanabe and Kume 2009; Shah et al. 2011). Temperature stress has devastating effects on plant growth and metabolism, as these processes have optimum temperature limits in every plant species. It has been shown that nighttime temperature has got negative effect on grain dry matter accumulation as well as on harvest index in lentil (Fig. 7.6). Global climate change is making high temperature a critical factor for plant growth and productivity; high temperature is now considered to be one of the major abiotic stresses for restricting crop production (Hasanuzzaman et al. 2012). The US Environmental Protection Agency (EPA) indicated that global temperatures have risen during the last 30 years, and it was mentioned that the decade from 2000 to 2009 was the warmest ever recorded.

Low temperature or cold stress is another major environmental factor that often affects plant growth and crop productivity and leads to substantial crop losses (Xin and Browse 2001; Sanghera et al. 2011). Chilling stress results from temperatures cool enough to produce injury without forming ice crystals in plant tissues, whereas

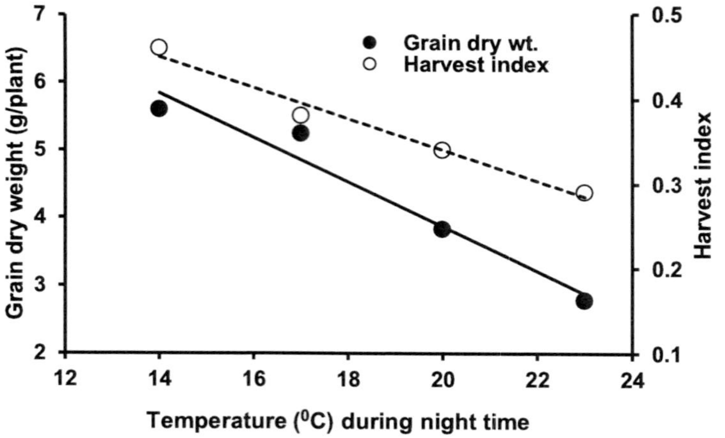

Fig. 7.6 Effect of nighttime temperature on dry matter accumulation (g/plant) and harvest index in lentil

freezing stress results in ice formation within plant tissues. Plants differ in their tolerance to chilling (0–15 °C) and freezing (<0 °C) temperatures. Both chilling and freezing stresses are together termed low temperature or cold stress: the damage due to cold stress can range from chilling injury and freezing injury to suffocation and heaving. In general, plants from temperate climatic regions are considered to be chilling-tolerant to variable degrees, and their freezing tolerance can be increased by exposing to cold, but non-freezing, temperatures; this process is known as cold acclimation. However, generally, the plants of tropical and subtropical origins are sensitive to chilling stress and lack this mechanism of cold acclimation (Sanghera et al. 2011). Low temperature may affect several aspects of crop growth, viz., survival, cell division, photosynthesis, water transport, growth, and finally crop yield.

For spring wheat and barley, an air temperature of about 20–25 °C is considered to be optimum for growth and development (Acevedo et al. 2002; Hakim et al. 2012; Hossain et al. 2012a, b and c). Optimal crop growth requires a non-limiting supply of water, nutrients, and radiation; as temperatures rise, the demand for growth resources increases due to higher rates of metabolism, development, and evapotranspiration (Rawson 1988). When growth resources are limited by heat stress, the size of plant organs, such as leaves, tillers, and spikes, is reduced (Martiniello and Teixeira da Silva 2011; Hossain et al. 2012a, b, c). The apparent sensitivity of metabolic processes to heat stress in the field (Reynolds et al. 2000), coupled with the reduced length of the life cycle at high temperatures (Hakim et al. 2012; Hossain et al. 2012a, b, c), results in low grain yield with lower total plant biomass in hot environments.

7.10.1 Low temperature (Chilling) Stress

Chilling stress is usually limited to plants native to or growing in tropical or subtropical regions of the world. The temperature range for chilling stress ranges from just above freezing to 15–20 °C. Plants vary greatly relative to their sensitivity to chilling stress. Chilling-sensitive plants have been defined as plants that are killed or injured by temperature above the freezing point of the tissues up to 15–20 °C (Graham and Patterson 1982). Chilling-resistant plants are those able to grow at temperature near 0 °C. However, considerable variability exists among plants relative not only to genetic factors but to stages of development, metabolic status (dormancy or active growth), and conditions under which plants are growing during and after the chilling episode. Generally, plants are more sensitive under nondormant conditions (high metabolic activity), during younger stages of development, during the day or under high-light intensities, under drought stress (although drought hardening can increase chilling tolerance), and when nutrients are limiting (particularly potassium, which is involved with osmotic adjustment).

Chilling stress can affect proteins both qualitatively and quantitatively. Generally, soluble proteins increase and proteins that have enzymatic functions can either be up- or downregulated, depending on specific function. The following are the enzymes' activities and processes that are affected by chilling stress (Table 7.1):

Generally, enzymes are more liable to high-temperature stress than they are to low-temperature stress. However, complex enzymes possessing subunits such as phosphate dikinase and phosphofructokinase involved in the carbon fixation reactions in C4 plants and glycolysis, respectively, are inactivated by chilling temperatures as a result of converting tetramers from diammers. Another enzyme critical osmotic relation in plant, potassium-mediated ATPase, is affected by chilling stress. Associated with reduced activity of this protein is leakage of K^+ from cells exposed to chilling stress. It appears that this process is reversible if the stress is not prolonged or too severe (Palta and Weiss 1993).

7.10.2 Low Root Temperature and Dry Matter Production

During the vegetative period, a lower flooded water temperature reduces the rate of tillering, delays leaf number, and reduces leaf elongation (Shimono et al. 2002). Since the low temperature of flooded water greatly affects biomass and grain yield of rice, studies have been conducted to identify the underlying mechanisms (Shimono et al. 2002). However, it is still unclear whether these mechanisms were caused by the integrated effects of low temperature on the root zone and shoot meristem near the soil surface or the effect on the root system (root temperature) alone, because in many studies flooded water was cooled in the paddy field accompanied with a decrease in air temperature (Nagasuga et al. 2011). Leaf growth, which is important for dry matter production, is strongly affected by environmental factors such as temperature (Shimono et al. 2004). Reduction of dry matter production by low temperature of flood water might be a direct effect of low temperature on the shoot

Table 7.1 Enzymes or processes up- or downregulated in plants as a result of chilling stress

Enzymes or processes upregulated	Enzymes or processes downregulated
(a) Phenolic synthesis	(a) Carbon fixation reactions
PAL (phenylalanine ammonium synthase)	NADP-malate dehydrogenase (C_4 reactions, high light)
CQT (hydroxycinnamoyl CoA quinate hydroxycinnamoyl transferase)	Pyruvate P_i dikinase (C_4 reactions, high light)
(b) Respiratory enzymes	(3) PEP carboxylase (C_4 reactions)
(1) Glycolysis	(4) Rubisco (dark reaction)
(2) TCA cycle	(5) Fructose-1,6-bisphosphatase (dark reaction)
(3) PPP (pentose phosphate pathway)	(b) Light reaction
(c) Cryoprotectants	(1) NADP reductase
(1) Invertase (starch and glucose)	(2) Plastocyanin
(2) Proline	(3) CF1
(3) Putrescine	(4) Ca^{2+} ATPase
(4) Betaine	(c) Respiration
(d) Antioxidants	NADP-malate dehydrogenase (TCA cycle)
(1) Glutathione	(2) 3-PGAL-dehydrogenase (glycolysis)
(2) Ascorbate	(3) Phosphofructokinase (glycolysis)
(e) Membrane lipids	(4) 6-Phosphogluconate dehydrogenase (PPP)
(1) Desaturases	(d) Nitrogen metabolism
(2) Acetyl-CoA carboxylase	Glutamate dehydrogenase (NH_4 to α-keto glutarate)
(f) Nonenzymtic proteins	(2) Aspartic β-semialdehyde dehydrogenase (lysine and alanine)
(1) Ice nucleators	(e) Starch metabolism
(2) Cryoprotective proteins	(1) Fructose-1,6-bisphosphatase
(g) Invertase (decreased inhibitor activity)	(2) Starch synthase
	(f) Antioxidants
	SOD (superoxide dismutase loss of Zn and Cu)
	(2) Catalase
	(3) Ascorbate peroxidase

growth. On the other hand, Shimono et al. (2004) indicated that the water content of plant and bleeding from the plant after shoot excision were lower under a low-flooded water temperature condition, suggesting that the low temperature of flooded water inhibits water uptake by the root and thereby reduces dry matter production. It is important to evaluate the effect of a low root temperature on water uptake separately from that on other processes to better understand the mechanism of rice plant resistance to low temperature. It has been shown that low root temperature inhibits water uptake by roots (Kramer and Boyer 1995). Since the water movement throughout a plant is strongly regulated by that in the root system, the inhibition of water uptake by low root temperature should affect plant water content and, as a result, dry matter production. Water uptake is controlled by both the water permeability of the root surface and the root volume. Low root temperature inhibits root growth, although there are wide differences among species in the optimal temperature for root growth (Kramer and Boyer 1995). However, there are few studies on the relationship between low root temperature and root volume in rice plants. On the other hand, Murai-Hatano et al. (2008) reported that root hydraulic conductivity in rice, which relates to its permeability, decreased with decreasing root temperature and that hydraulic conductivity decreased dramatically at temperatures below 15 °C. Such phenomenon has been reported in several other plant species (Clarkson 1976; Markhart et al. 1979; Wan et al. 2001; Lee et al. 2004). It is necessary to clarify which of these factors is mainly affected by a low root temperature.

7.11 Dry Matter, Grain Yield, and Harvest Index Values for Crops

Lower limits for crop dry matter, grain yield, and harvest index would all be near zero for rainfed crops, due to occasional drought or other catastrophic crop failure (e.g., pests, disease, weeds, frost). For some crops, zero and very low grain yields have been reported, and although zero yield crops would by default have zero harvest indices, this was not specifically reported for these particular crops. Across the dataset, mean harvest index for wheat and barley (0.37) was the same as that averaged for all grain legumes (0.37, lupin, field pea, faba bean, lentil, chickpea, and vetch), while oat harvest index was lowest (0.21) followed by canola (0.28). The low harvest index of oat may reflect its primary role as a forage crop and a low intensity of breeding for grain yield, and for canola the heavy energy (oil) investment in seeds. Apparently, high harvest index value for faba bean may partly be due to noncollection of fallen leaves in harvest index estimations.

The upper limits of crop growth are a function of crop leaf area development, incident radiation, and radiation-use efficiency (Sinclair 1994), assuming that water, nutrients, weeds, and other pests are not limiting. Maximum economic yields are then a function of harvest index. Relatively simple crop models have been effective in accurately estimating maximum potential yields and maximum achieved yields in a given environment (Sinclair 1993). Establishment of maxima is useful from the point of view of setting upper limits for management and for models, and perhaps for filtering out unrealistic or poor quality data. Upper limits for harvest index in wheat and barley have been suggested to be approximately 0.6, based on likely assimilate partitioning and the need for stem strength to prevent lodging, etc. (Slafer et al. 1999). However, many of the harvest index values above 0.5 reported in the literature may be due to a reporting of apparent rather than actual harvest indices for the reasons outlined earlier.

7.11.1 Wheat

Harvest index values in wheat ranged from 0.04 to 0.59. Maximum growth, yield, and harvest index of wheat were examined in a series of experiments on irrigated wheat during 1983–1985 (Stapper and Fischer 1990a, b, c), and combinations of nitrogen, sowing rate, seeding geometry, and sowing date were employed. Within the treatments and years, harvest index varied from 0.28 to 0.51. Maximum potential grain yield in this environment with extant material was considered to be 9–10 t/ha at 12% moisture. In the work of Mason and Fischer (1986), at one site in southern NSW, harvest index in wheat varied from 0.25 to 0.42 across 3 years and six tillage treatments. Averaging across data source site year provided 194 estimates of harvest index for rainfed wheat, a mean of 0.36. More than 50% of site year harvest index estimates fell within the 0.3–0.4 range.

7.11.2 Barley

Average dry matter (8.7 t/ha) and grain yield (2.6 t/ha) of barley were higher than those obtained for wheat. Harvest index averaged 0.38 tends to be higher than for wheat, perhaps a consequence of its lower grain nitrogen content, with 70% of records being between 0.35 and 0.45 (0.275–0.475). However, it would thus appear to be less variable than harvest index in wheat. There are few specific studies of harvest index in barley in Australia, averaging by source site year 0.38.

7.11.3 Sorghum

Mean dry matter yield for rainfed sorghum was 6.8 t/ha and 11.4 t/ha for irrigated crops. Grain yield varied from 0.8 to 10.0 t/ha for rainfed crops and 2.7–12.4 t/ha for irrigated crops. Harvest index for rainfed crops ranged from 0.14 to 0.70, and for irrigated crops 0.26–0.52.

7.11.4 Rice

Rice crop grain yields in Australia (14 t/ha) are among the highest in the world (Sheehy et al. 2004). Crop dry matter yield peaked at 31.6 t/ha (Ockerby and Fukai 2001), or in a more traditional rice-growing region 27.5 t/ha (Horie et al. 1997). In rice, harvest index ranges from 0.01 (Williams and Angus 1994) to 0.71 (Dunn and Beecher 1994). The low protein and lipid content of rice may serve to increase its harvest index. All rice grown is irrigated, and it has been shown that water deficits can be managed to increase harvest index, especially for high nitrogen crops (Zhang and Yang 2004). Low temperatures during reproductive development can also have a substantial impact on harvest index, but this is controlled to some extent by judicious floodwater management which can affect increases in canopy temperatures

(Williams and Angus 1994). Experimental data showed that a combination of temperature and plant nitrogen content (kg/ha) at microspore development could be used to model harvest index quite effectively. Harvest index values >0.6 have been reported in the literature, for example, 0.47–0.71 (Dunn and Beecher 1994) and 0.5–0.6 (Williams and Angus 1994), but these were more likely to be extreme examples of what can be induced experimentally rather than what is likely in commercial crops.

7.11.5 Maize

The mean values were 19.3 t/ha (dry matter yield), 5.5 t/ha (grain yield), and 0.49 (harvest index), including both rainfed and irrigated crops. The mean harvest index was very close to that observed in the United States (Prince et al. 2001). Maximum maize dry matter yield was 28.0 t/ha and grain yield 20.5 t/ha (Birch et al. 2006). Maize grown in temperate environments tends to have a higher harvest index than maize grown in tropical environments due to a shorter grain-filling period in the latter, caused by water limitation.

7.11.6 Chickpea

Mean dry matter production of chickpea is 3.8 t/ha and grain yield 1.1 t/ha. The maximum dry matter yield recorded was 10.7 t/ha for an irrigated crop at Merredin (Leport et al. 1999), with an accompanying grain yield of 4.2 t/ha. For rainfed crops, the maxima were very similar at 10.2 t DM/ha (Berger et al. 2004). Harvest index of chickpea ranged from 0.19 to 0.52, depending on seasonal water supply under controlled conditions. Ganguly and Bhattacharya (2000) have analyzed the changes in physiochemical traits in developing chickpea seeds under normal and late seeding conditions and reported that carbon/nitrogen (C/N) ratio had a negative effect on harvest index during early and late seed development stages. However, C/N ratio had a positive effect on seed dry matter at various seed developmental stages.

7.11.7 Field Pea

Mean dry matter production of field pea is 4.7 t/ha and grain yield 1.5 t/ha. Maxima were 11.6 t/ha for dry matter (Armstrong and Pate 1994a) and 4.5 t/ha for grain yield (McMurray 2005). These are well below yields attained for the crop in Europe (Jensen et al. 2004). For harvest index, absolute minima of 0.06 and maxima of 0.58 (French and Ewing 1989) were recorded. Armstrong and Pate (1994b) showed that harvest index of semidwarf types was higher than taller, conventional cultivars, as has been shown for other crop species. Harvest indices of up to 0.6 were reported for their glasshouse-grown plants.

7.11.8 Faba Bean

Dry matter of faba bean averages 5 t/ha and grain yield 1.7 t/ha, with absolute maxima of 10.7 t/ha for DM and 5.7 t/ha for grain (Loss and Siddique 1997). As for field pea, harvest index in faba bean is quite variable. The relatively high mean of 0.46 may be more apparent than real, with possible noncollection of fallen leaves skewing the values to the high end (De Costa et al. 1997). For faba bean, harvest index decreased with increasing plant density across a range of sowing dates (Adisarwanto and Knight 1997). This resulted from fewer pods/plant at higher sowing densities. L'opez-Bellido et al. (2005) reviewed harvest index in faba bean in relation to plant density and presented values ranging from 0.25 to 0.52. In a series of experiments with determinate and indeterminate faba bean and irrigation in the United Kingdom (De Costa et al. 1997), harvest index ranged from 0.34 to 0.68, for both irrigated, rainfed, and rain excluded treatments. Across genotypes and water stress, there was no trend in harvest index excepting that where water was provided continuously through trickle irrigation, harvest index was lower than all other treatments. Higher availability of water increased assimilate partitioning to vegetative matter.

7.11.9 Lentil

Maximum dry matter yield of 9.3 t/ha was reported (Siddique et al. 1998), and grain yield 4.1 t/ha. Studying a wide range of germplasms from across the globe, from wild types through old and modern cultivars, Whitehead et al. (2000) found harvest index in lentil to reach 0.52, including fallen leaves, etc. Maximum dry matter production was 18.3 t/ha, and crop residues of 14.1 t/ha were recorded across all the cultivars studied, and seed yield >4.2 t/ha. In a glasshouse study (Shrestha et al. 2006), harvest index ranged from 0.36 to 0.45 for water-stressed lentil to 0.53–0.59 for well-watered lentil.

7.11.10 Lupin

Mean values for lupin (*Lupinus polyphyllus*) were 5.2 t/ha for dry matter yield, 1.6 t/ha for grain yield, and 0.30 for harvest index. Maximum dry matter yield of 12.6 t/ha for irrigated lupin (Dracup et al. 1998) and 11.1 t/ha for rainfed lupin at Eradu have been reported. The grain yield maxima of 4.4 t/ha was recorded at Cressy, Victoria, Australia (Bishop and Mendham 1996). Harvest index of lupin ranged from 0.04 to 0.50, with an average of 0.30. However, the upper limit of 0.5 is used in the modeling of Farre et al. (2004); they included roots in their harvest index calculation. Large variation in lupin harvest index has been suggested (Dracup et al. 1998). Delane et al. (1988) concluded that "overall, results suggest that very poor pod setting on the main stem and low harvest index are not as widespread a problem as previously thought." In the studies of Delane et al. (1988), across 38 sites, crop dry matter explained 67–83% of yield variation of three genotypes, indicating that factors influencing pod set probably account for no more than 20% of yield variation. Using supplementary irrigation, Dracup et al. (1998) showed that terminal drought can have a substantial influence on yield and harvest index of lupin. Harvest index increased from 0.14 in unirrigated plants to 0.22 with the addition of approximately 190 mm water. Additional water during pod fill was seen as key to increased harvest index. In lupin, mild transient water stress may also increase harvest index by redirecting plant resources from vegetative growth to seed fill (French and Turner 1991). In lupin, up to 94% of flowers and pods may be aborted (Atkins and Pigeaire 1993); however, keeping flowers on such plants does not necessarily lead to increased seed yield (Palta and Ludwig 1996), and seed filling rather than increased pod number would appear to be more likely to increase lupin harvest index and yield (Palta et al. 2003). Increasing carbon supply during pod fill (Palta and Ludwig 2000) increased seed yield but not harvest index as concurrent vegetative growth maintained its share of the carbon sink.

7.11.11 *Brassica* Crops

Dry matter production of canola (*Brassica napus* L.) averages 6.8 t/ha and grain yield 1.6 t/ha. The maximum rainfed grain yield of 5.16 t/ha (Zhang and Evans 2004) was very similar to the irrigated yield of 5.2 t/ha (Wright et al. 1992). Maximum dry matter production (15 t/ha) was recorded in a rainfed crop (Ward et al. 2002). Harvest index for canola averaged 0.28, ranging from 0.04 (Thomas 2005) to 0.41 (Hirth et al. 2001). In field experiments examining canola response to nitrogen (Hocking et al. 1997), harvest index varied from 0.27 to 0.30 at one site, and from 0.29 to 0.34 at another. Robertson et al. (2004) examined harvest index in canola in response to sowing date and found that, aside from very early or very late (after mid-June) sowings, harvest index was relatively stable and yields principally a function of dry matter accumulation. Variation in harvest index of canola was less than that of most other field crops examined.

Chaturvedi et al. (1988) have pointed out that considerable variability existed in leaf area; number of primary, secondary, and tertiary branches; and length of main raceme in Indian mustard (*Brassica juncea* L.) Leaf area development of different stages of growth showed important bearing on yield. In other crops also, the growth analysis characters like leaf area index and leaf area duration had positive correlation with yield (Singh et al. 1985). Harvest index, biological yield, and test weight are also positively correlated with economic yield. They are supposed to be most important determinants of seed yield in these species. Yield per plant which has been affected by different accumulative characters possesses greater variability in different groups of cultivars.

7.11.12 Sunflower

Records of maximum dry matter yield and grain yield for rainfed sunflower (*Helianthus annuus* L.) crops were 15.6 and 5.7 t/ha, respectively. For an irrigated crop sown at optimum time and provided with luxuriant nutrients, total dry matter production was 17.1 t/ha at anthesis and 24.2 t/ha at maturity. Grain yield (0% moisture) was 6.02 t/ha and harvest index 0.4. Average yields for irrigated crops were 1.8–3.0 t/ha (Bange et al. 1977). Connor et al. (1985) studied variation in harvest index as a function of the timing and quantity of irrigation. Harvest index increased where water was predominantly applied post-anthesis, with post-anthesis irrigated crops having a harvest index of 0.53, as compared with unirrigated crop harvest index of 0.38. In India, Nanja Reddy et al. (2003) reported harvest index in sunflower to vary between 0.14 and 0.44 across genotypes.

7.12 Seed Yield

Several factors pose significant risk to farms leading to yield reduction when they are not correctly monitored and well managed. These factors can be grouped into three categories which are technological, biological, and environmental (Metclfe and Elkins 1980). Grain yield can be determined from the components of the biological (biomass) yield and harvest index (Passioura 1977), and the potential yield of bread wheat has increased due to increases in the harvest index (Curtis 2002). Furthermore, high temperature and soil water deficit decrease the biological yield and grain yield, although they do not significantly affect the harvest index (Mondal et al. 2015). Under high-temperature conditions, grain yield is related to the harvest index across bread wheat genotypes (Zhong-hu and Rajaram 1994). However, the importance of the biological yield and harvest index on the grain yield, in different wheat species with diverse phenotypes in rainfed areas of the Mediterranean zone, is not well known (Kobata et al. 2017). The past 50 years have seen seed yield increases for the world's major crops of 1 to 3% per year (Bruins 2009), due largely to genetic gains obtained with plant breeding (Le Buanec 2009; Ceccarelli et al. 2010), but also to improved agronomic inputs and specialized management systems (Bruins 2009). Globally, abiotic stresses drastically affect crop productivity leading to substantial yield losses. According to the IPCC (2014) report, the decline in food productivity and quality, primarily due to extreme temperatures and water deficit conditions, poses a serious threat to agriculture (IPCC 2014; Zandalinas et al. 2018). Therefore, under the changing climate, minimizing agricultural losses caused by these stresses has become a major challenge and has created a global concern to assure food security (Anjum et al. 2011). The effects of climate change are exacerbated by the continuous decline in availability and productivity of agricultural land (Zandalinas et al. 2018). The world is facing a gradual rise in heat wave frequency leading to warm days and nights, which is projected to exceed 2 °C by the end of the twenty-first century (IPCC 2014). Abiotic stresses markedly affect the reproductive development of various crops and, ultimately, reduce the final economic yields. While the

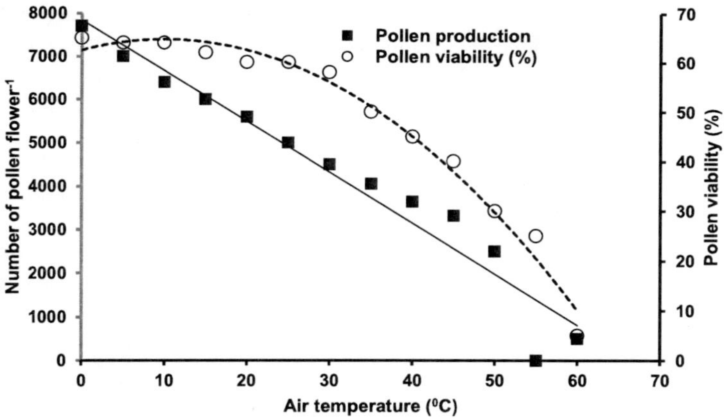

Fig. 7.7 Effect of air temperature on pollen production and percent pollen viability in pigeon pea (*Cajanus cajan* L.) crop

presence of abiotic stresses at any growth stage can affect crop yield, the seed filling stage is crucial for determining average seed weight, seed composition, and, therefore, the final quantitative and qualitative yield (Çakir 2004; Otegui and Slafer 2004; Prasad et al. 2017).

Stresses disrupt germination, vegetative growth, tiller production, dry matter partitioning, reproductive organ development, reproductive processes (Boyer and Westgate 2004; Prasad et al. 2011), grain filling (Sehgal et al. 2017), and grain quality (Gooding et al. 2003; Britz et al. 2007). Reproductive processes and grain filling are more sensitive to both these stresses, and have optimum and ceiling temperatures that are relatively lower than those for vegetative growth and development phases. For instance, exposure of wheat to short episodes (2–5 days) of heat stress (>24 °C) at reproductive stage (at start of heading) resulted in substantial damage to florets' fertility, while mean daily temperature of 35° caused total failures. More sensitivity of reproductive stage to abiotic stress, compared to vegetative stage, is mainly attributed to damage to male components, which are severely impacted as a result of disruption of developmental as well as functional aspects, such as sucrose and starch accumulation in pollen grains (Sita et al. 2017). It has been shown that the pollen production and percent pollen viability in pigeon pea (*Cajanus cajan* L.) plant are very much affected by air temperature (Fig. 7.7). The leaves show more resilience at reproductive ceiling temperatures, probably due to some effective mechanisms.

Seed filling is a crucial growth stage for all crops, which involves mobilization and transport processes required for importing various constituents, and many biochemical processes for the synthesis of proteins, carbohydrates, and lipids in the developing seeds (Barnabás et al. 2008; Prasad et al. 2008; Awasthi et al. 2014; Farooq et al. 2017a). Seed filling processes and the accumulation of reserves in the developing and maturing seeds are highly sensitive to environmental changes, which influence the qualitative and quantitative traits of the final yield (Yang and Zhang 2006). Abiotic stresses can hinder the accumulation of various seed constituents, primarily starch and proteins (Behboudian et al. 2001; Asthir et al. 2012; Farooq et al. 2017a, b) through inhibiting the enzymatic processes of synthesis of starch (Ahmadi and Baker 2001) and proteins (Triboï et al. 2003).

7.13 Low Temperature and Seed Yield

Low temperatures are the cause of enormous agricultural losses, especially in subtropical and temperate grain crops. The reproductive period is a vital phase in the life cycle of all annual plants, and metabolism during this phase ultimately determines crop yield. Plants exposed to cold temperature during reproduction show reduced metabolic rates leading to low yields. Low temperature induces flower abortion and pollen and ovule infertility, causes breakdown of fertilization, and affects seed filling, leading to low seed set and ultimately low grain yield. Physiological and metabolic bases of cold stress, including the role of phytohormones, specially abscisic acid and sugar-induced regulation are reviewed, and future research directions investigating molecular and metabolomic changes during the development of the gynoecium during cold stress have been suggested (Thakur et al. 2010).

Cold or chilling stress experiences by plants from 0 to 15 °C (Yadav 2010) leads to major crop losses. Various types of crops in tropical or subtropical origin are injured or killed by non-freezing low temperatures, and exhibit different symptoms such as poor germination, stunted seedlings, chlorosis, or growth retardation, reduced leaf expansion, and wilting and necrosis. In general, plants respond with changes in their pattern of gene expression and protein synthesis when exposed to low temperatures (Sanghera et al. 2011). In general, plants from temperate climatic regions are considered to be chilling-tolerant with variable degree compared to tropical and subtropical crops, and can increase their freezing tolerance by cold acclimation (Levitt 1980).

Cold stress is a common problem in rice cultivation and affects global production as a crucial factor (Zhou et al. 2012). Rice is a cold-sensitive plant that originated from tropical or subtropical zones. When low temperature occurs during the reproductive stages, it can cause serious yield and yield component losses (Farrell et al. 2006a, b). The optimum temperature for rice cultivation is between 25 °C and 35 °C, and in temperate regions, rice growth is impressed by limited period that favors its growth (Los Reyes et al. 2003). Exposure to cold temperature affects all phonological stages of rice and lowers grain production and yield too. Low temperature in vegetative stage can cause slow growth and reduce seedling vigor (Ali et al. 2006) and low number of seedlings, reduce tillering (Shimono et al. 2004), increase plant mortality (Baruah et al. 2009; Fujino et al. 2004; Farrell et al. 2006a, b), increase the growth period (Alvarado and Hernaiz 2007), and in reproductive stage, can cause to produce panicle sterility and lower grain production and yield (Shimono et al. 2004). Critical stages for cold damage include germination, booting, flowering, and filling stages (Satake 1976). Since the most sensitive stage for cold harm is the flowering stage, which occurs 10–12 days prior to heading (Ghadirnezhad and Fallah 2014), it would be better to evaluate the effect of cold stress on yield and yield components in flowering stages.

Winter wheat plants increase in freeze tolerance by acclimating to low temperatures. As temperature drops below 15 °C, winter wheat plants increase cellular solute content to tolerate temperatures as low as −20 °C (Pukacki and McKersie 1990; Vágújfalvi et al. 1999). The plants eventually stop growth and go dormant over winter months. During extended midwinter warm-ups, wheat plants may resume growth and tiller, then reacclimate to cold temperatures, and reenter dormancy (Prášil et al. 2004). However, once the vernalization requirement has been met, plants transition to reproductive growth and can no longer cold acclimate (Asseng et al. 2011; Prášil et al. 2004). When plants enter reproductive growth, freeze tolerance quickly declines, leaving plants susceptible to freeze events (Alt et al. 2020). A common practice among plant breeders is to adapt a cultivar's vernalization requirements to specific growing regions, delaying the reproductive growth stages to beyond the spring freeze-free dates (Zheng et al. 2015). However, extended midwinter warm-ups have accelerated wheat development, exposing wheat that has transitioned to reproductive growth stages to late spring freeze events (Gu et al. 2008). This extended period of above-average temperatures initiated reproductive growth in many crops that were later injured or killed by freezing temperatures (Grimm et al. 2013). Wheat cultivars with low vernalization requirements have earlier flowering and seed fill to avoid summer heat and drought stress (Asseng et al. 2011; Chenu et al. 2013). Wheat freeze tolerance primarily depends on the growth stage of the plant (Barlow et al. 2015; Fuller et al. 2007). Freeze injury to the original tillers or spike can significantly reduce seed yield as lower yielding, regenerative (new) tillers are formed after the freeze event (Ji et al. 2017a; Shroyer et al. 1995; Wu et al. 2014). Freeze damage can injure developing spikes and damage the flag leaf (Rawson and Evans 1971). Freeze injury can cause sterility, embryo death, or complete loss of the spike (Cromey et al. 1998; Shroyer et al. 1995).

7.14 Factors for Yield Gaps

The factors underlying the yield gaps are numerous and complex. They relate to inherent differences in crop architecture and developmental pattern which allow intercepting more light and remaining insensitive to day length and sustain a much higher concentration of inflorescence and grain per unit of land area in cereals than legumes. The genetic manipulation is only one aspect of crop improvement and may not be the most appropriate means for resolution of the primary limits of productivity for adaptation. For these cases, where the differences in plant performance relate to mainly one specific limitation, such as disease, the strategy of crop improvement is relatively clear. More commonly, however, agricultural environment imposed quite complex challenges with many factors bearing simultaneously, and there may be range responses by different genotypes to these changes. In these situations, relative differences among genotypes become important. The nature of the differences and of the appropriate resolution become less clear, and the objective designed for alternative strategies of crop improvement becomes more complex. Climate is the major factor of the environment conditioning the regional and seasonal adaptation and yield of the crop plant. Improvement in climatic adaptation is fundamental to the process of crop improvement in a broader sense, viz., the manipulation of both genotypes and environments to respectively maximize genetic potential and minimize environmental constraints to the expression of the potential. The key climatic factors influencing the food legumes are insolation, temperature, day length, and water availability.

Grain yield can be improved either by increasing biomass production or by converting more biomass to grain yield, which is a higher harvest index (Otegui and Salfer 2000). The harvest index is defined as the ratio of grain yield to total crop biomass (Hay and Porter 2006):

$$\text{Grain yield} = \text{total biomass} \times \text{harvest index} \ldots \quad (7.1)$$

Therefore, grain yield is proportional to harvest index (Donald and Hamblin 1976). Since the Green Revolution, improved harvest index has been the main determinant of yield improvement of cereal crops (Sinha et al. 1982), as crop biomass has not increased substantially (Hay and Gilbert 2001). Increased harvest index was principally associated with a reduction in plant stature. Although harvest index is a high heritable character (Hay 1995), it is associated with environmental conditions and phenology (Hammer and Broad 2003). The role of environmental factors may become more prominent under stress conditions (Hay and Gilbert 2001). In this context, there are two determinants of harvest index that can be genetically manipulated to maximize harvest index to achieve higher grain yield (Richards et al. 2002). The first one is the potential yield of a crop in a given environment and is independent of drought. The second one is drought-dependent harvest index and depends largely on water availability during grain filling. The harvest index of modern cereal varieties falls within the range of 0.4–0.6 (Hay 1995). Although harvest index of cereal crops is close to the upper limit under optimum conditions, a better understanding of the physiological traits associated with higher harvest index under rainfed conditions could potentially increase grain yield under drought stress, which is known to reduce biomass and harvest index.

7.15 Physiological Traits Associated with Improvement in Seed Yield

Research into the changes in physiological traits associated with genetic gains in yield potential is essential to improve our understanding of yield-limiting factors and to base future breeding strategies on. Cultivars released in distinct eras have been compared for yield and morphophysiological determinants. They have been grown simultaneously under fixed conditions, thereby removing the effects of varying management practices on yield (Slafer et al. 1994) and allowing the comparison of the physiological bases underlying the differences in yield capacity. Most of the traits identified in retrospective analyses are constitutive in nature; that is to say, they are expressed in the absence of stress (Calderini et al. 1999). These attributes are divided into four categories: (1) time to flowering and plant height, (2) biomass production and partitioning, (3) main yield components, and (4) cross-category interactions.

7.15.1 Time to Flower and Plant Height

The timing of flowering is a major trait related to the adaptation of cultivars to particular areas, thus determining crop performance under prevalent field conditions (Passioura 1996, 2002; Richards 1996a; Slafer and Araus 1998). This explains why time to flowering (phenological adjustment) is one of the first attributes optimized by breeding programs (Slafer 2003). Consistent changes in this trait are therefore to be expected only in regions in which the lengthening or shortening of the growing season has advantages for adaptation regardless of the cultivars released earlier. A scenario frequently reported in the literature, where the manipulation of time to heading may strongly affect adaptation, is that of regions characterized by a Mediterranean climate, i.e., a hot dry summer and humid temperate winter (Loss and Siddique 1994). In these environments, the vegetative and early reproductive phases of a crop frequently occur under adequate water availability. However, as the season progresses, drought becomes more intense and frequent water stress occurs during the late reproductive phases. After anthesis, grains fill under severe water and heat stress. The analysis of long-term trends in time to flowering for cultivars released in different eras reveals that, in most cases, there are few or no changes in this trait in regions with climates that differed from the Mediterranean, while reducing time to flowering has been a successful strategy when breeding for environments characterized by terminal stress factors (Álvaro et al. 2008a, b), unless the original landraces on which breeding started to improve yields already had a time to flowering conferring adaptation (Acreche et al. 2008). This strategy has achieved good results because earliness is probably the most effective means to increase yield in regions where drought commonly occurs during grain filling (Passioura 1996; Slafer and Whitechurch 2001). In addition to an adaptive advantage of a reduction in crop duration, this decrease has permitted augmented cropping intensity for cereals such as rice, and has also allowed land to be used for noncereal crop cultivation in cereal-based farming systems (IRRI 1996).

Breeders have always selected for reduced height when the initial elite material exceeds a threshold. For plant heights above this threshold (70–100 cm; Miralles and Slafer 1995a), there is no gain in biomass, while there is a proportional reduction in the harvest index (Royo et al. 2007). Below this threshold, further gain in the harvest index does not compensate for the loss in biomass because of extremely poor radiation distribution within the canopy and consequent reductions in radiation-use efficiency (Miralles and Slafer 1997). Thus, reducing height to an optimum range increases yield potential (a similar biomass more efficiently partitioned to grains) and simultaneously reduces the risk of lodging. Consequently, plant height was reduced universally through wheat breeding during the twentieth century (Calderini et al. 1999), and still other studies (Donmez et al. 2001; Brancourt-Hulmel et al. 2003; Jiang et al. 2003; Ramdani et al. 2003; Royo et al. 2007).

7.15.2 Dry Matter Production and Its Partitioning

The degree of partitioning of aboveground dry matter to grain yield, i.e., harvest index, is genetically complex. Since the Green Revolution, yield potential has continued to increase mainly due to improvements in harvest index (Reynolds et al. 1999; Abeledo et al. 2002). Despite a hypothetical limit to harvest index of ~0.62 in wheat (Austin 1980), there has been no systematic progress since the early 1990s from values of ~0.50–0.55. Austin (1980) predicted a maximum theoretical harvest index of 0.62 in winter wheat based on an extrapolation from the mean value (0.49) observed for the four most modern winter wheat cultivars. This assumed no change in aboveground dry matter and leaf lamina (physiological); and chaff (structural) dry matter cannot be decreased, stem and leaf sheath dry matter can be reduced by 50%, and chaff dry matter can be increased pro rata to accommodate extra grain. Investigation on breeding progress in winter wheat in the United Kingdom demonstrated that biomass increased by 17% among seven semidwarf cultivars released from 1980 to 1995 (Shearman 2001; Shearman et al. 2005a, b) over the 15-year period. Moreover, the ratio of chaff to grain decreased by 38 mg chaff g grain^{-1} over the 15-year period. Extrapolating from the mean value of the four most modern cultivars characterized by Austin et al. (1989) and making the same assumptions as Austin (1980) except that (1) aboveground biomass may be increased by 10% and (2) the ratio of chaff to grain may be decreased from the value assumed by Austin (1980) of 202 mg chaff g grain^{-1} to that of the most recent cultivar characterized by Shearman et al. (2005a, b) of 195 mg chaff g grain^{-1}, the theoretical maximum HI increases to 0.63. Assuming that the ratio of chaff to grain may be decreased by a further 10% to 175 mg chaff g grain^{-1}, and holding all the other assumptions the same, harvest index could theoretically be increased to 0.64. Strategies for optimizing assimilate partitioning to raise HI toward this theoretical maximum value of ~0.64 are now considered. Increased partitioning to grains could theoretically be increased by reducing competition from alternative sinks, especially during stem elongation when grain number is determined (Fischer 1985; Kirby 1988). These competing sinks include roots, leaves, stems (structural and soluble carbohydrate), and infertile tillers. The genetic range for partitioning to the respective plant organs at anthesis reported in the literature was recently reviewed by Reynolds et al. (2009) and concluded that the range for spikes was 0.12–0.29, for leaf lamina 0.19–0.31, and for stems and leaf sheaths 0.48–0.63 as a proportion of aboveground biomass. The dry matter lost in infertile tillers was 0.02–0.15 as a proportion of aboveground biomass (Berry et al. 2003), and the dry matter in roots was 0.06–0.17 as a proportion of total biomass. These data suggest the possibility for increasing partitioning to spikes at the expense of other plant organs. Some caution is required when suggesting decreasing root partitioning as a strategy to increase spike partitioning index (spike dry matter/aboveground dry matter) at anthesis, since future genetic gains in yield seem likely to depend on raising biomass which, in turn, may require increased access to water and nutrients. Reducing partitioning to the leaf lamina may also have negative effects in the context of improving photosynthetic capacity. However, decreasing leaf partitioning may be possible if combined with agronomic compensation by increasing plant density. On the other hand, reduced partitioning to the stem and infertile shoots may not have negative effects, although decreasing stem partitioning should take account of any effects of compacting the leaves of the canopy and decreasing the vertical light distribution and radiation-use efficiency.

As yield was greatly increased by genetic improvement during the last century and grain yield is simply a fraction of the dry matter accumulated by the crop during the growing season, one might expect a trend to have emerged for increased biomass with the release of newer higher-yielding cultivars. However, for small grain cereals, such as wheat, in almost all cases in which the physiological bases of yield improvements have been analyzed, final biomass is not associated with the year the cultivars were released with only a few exceptions. Thus, for example, recent genetic gains in grain yield in the United Kingdom have resulted from a combination of improved growth rate in the pre-anthesis period, which has led to increases in the number of grains' perm (Austin 1980), and a larger source for grain filling through greater soluble carbohydrate reserves in the stem (Shearman et al. 2005a, b). Under Mediterranean conditions, genetic gains in durum wheat have been associated with an increased contribution of pre-anthesis assimilates to grain filling and greater dry matter translocation efficiency (Tambussi et al. 2005a; Álvaro et al. 2008b). While the lack of a consistent increase in biomass through breeding implies that neither leaf photosynthesis (Austin 1980; Austin et al. 1989) nor radiation-use efficiency (Calderini et al. 1997) is related to past yield increases, in a few cases, positive relationships between maximum leaf photosynthesis and the year cultivars were released have been reported (Fischer et al. 1998; Jiang et al. 2003). However, these differences have not been associated with increases in biomass production. Looking in more detail at post-anthesis, it seems, however, that modern cultivars are characterized by higher radiation-use efficiency than their predecessors. This was the pattern observed in the only retrospective analysis of this characteristic that we are aware of (Calderini et al. 1997). This increased radiation-use efficiency may be simply a consequence of the increased demand for a larger sink in modern cultivars compared with their older counterparts. These kinds of results are in line with evidence from near-isogenic lines for semi-dwarfism (Miralles and Slafer 1997) and with the observation that the number of grains that grow after anthesis enhances photosynthetic efficiency (Richards 1996b). Also, it has recently been reported that increased post-anthesis biomass can be achieved by increased sink strength through positive feedback to photosynthesis (Reynolds et al. 2004). Greater stomatal conductance seems to be responsible, at least in part, for the higher photosynthetic rates (Fischer et al. 1998; Araus et al. 2002). Therefore, the genetic gains in grain yield in most countries have been virtually entirely due to modifications in the harvest index, and are probably related to reductions in plant height. In some studies, biomass partitioning was analyzed before maturity in cultivars released in different eras, and the partitioning effects observed at maturity, as differences in the harvest index, were already established by anthesis as variations in the spike-to-stem ratio (Slafer and Andrade 1993). A critical issue in this regard is that the harvest index in most modern wheat cultivars is close to its biological maximum, 60% (Austin 1980).

The case of maize shows some basic differences to that of small grain cereals. Maize grain yield in the United States has increased by about 2% per year from the start of the large-scale adoption of hybrids by maize growers in the late 1930s until the first decade of the twenty-first century, and about 75% of the yield improvement has been attributed to genetic gain and the rest to improved agronomic practices (Tollenaar and Lee 2006). Two important physiological processes appear to be involved in this increase in tolerance:

- Sustained leaf photosynthesis during grain filling, which contributes to increases in dry matter accumulation
- Increase in kernel number due to higher partitioning to the kernels during the sensitive period of kernel number determination

7.15.3 Main Yield Components

The two main yield components are the number of grains per m^2 and the averaged individual grain weight. Most experiments analyzing the effects of genetic improvement on yield have considered these components. The vast majority of the studies performed found that, while selecting for higher-yielding cultivars, wheat breeders have consistently increased the number of grains per unit land area and produced either no trend or even a trend toward slightly reduced individual grain weight (Donmez et al. 2001; García del Moral et al. 2002; Brancourt-Hulmel et al. 2003; Royo et al. 2005). However, earlier studies have reported a slight increase in this component during the twentieth century (Cox et al. 1988) or part of it (Calderini et al. 1995).

The increased number of grains per unit area of modern cultivars compared to their predecessors is attributed to greater survival of floret primordia (Álvaro et al. 2008a), as the number of potential florets per spike is apparently quite similar (Slafer and Andrade 1993; Miralles et al. 2002). Thus, the higher survival of floret primordia (a process that occurs during the last half of stem elongation) is the most important factor leading to higher yield potential in modern cultivars (Slafer et al. 1994). This finding is consistent with the observation that semi-dwarfing genes, which contribute significantly to enhanced yields in many breeding programs throughout the world, increase the number of grains per unit land area by augmenting the survival of floret primordia (Álvaro et al. 2008a).

7.15.4 Associations Between Yield Attributes

The genetic improvement of yield can be understood more mechanistically by inspecting and interpreting the relationships between the main attributes of growth and by partitioning the yield components described above. Studies on the mechanistic bases by which breeding has successfully increased yield may shed light on potential future alternatives. Genetic improvements in grain yield are almost unequivocally due to gains in the number of grains per unit land area, with no gains and even slight losses in the average weight of the grains, probably as a consequence of the smaller cells in the pericarp induced by insensitivity to gibberellins (Miralles and Slafer 1995b) and the increased number of grains of small potential size when grain number is improved (Acreche and Slafer 2006). This physiological basis of yield improvement in wheat is so strong that even under conditions characterized by terminal stress grain yield improvement has been associated with increases in grain number per unit land area (Acreche et al. 2008), mostly due to increases in the number of grains per spike (Royo et al. 2007). Consequently, a negative relationship between grain number per unit land area and the average weight of these grains has frequently been found (Slafer 1996). In fact, this negative relationship is also common when yield is increased by management practices. These observations are generally attributed to greater competition among grains as the number of grains m^{-2} increases. This is because, although wheat breeding has reduced the degree of post-anthesis sink limitation to yield (Kruk et al. 1997), photosynthetic capacity during grain filling together with the pre-anthesis assimilate reserves is in excess of the demands of the growing grains during post-anthesis (Borras et al. 2004; Reynolds et al. 2004) even under Mediterranean conditions (Cartelle et al. 2006). The high source to sink ratio may be responsible for a downregulation of photosynthetic capacity during grain filling. Reynolds et al. (2005b) suggested that radiation-use efficiency could be increased by increasing sink strength (i.e., more grains per unit land area), which is in line with evidence that cultivars with improved sink strength may possess higher photosynthetic rates in post-anthesis (Calderini et al. 1997). Further, Uddling et al. (2008) showed that this high source-sink ratio more than offsets the direct stimulating effect of elevated CO_2 on post-anthesis photosynthesis. It was reported that decreasing post-anthesis sink strength resulted in a reduction in both canopy radiation-use efficiency and leaf photosynthesis in wheat (Acreche et al. 2007).

With the above statements, it can be said that with only a few exceptions, crop yield is limited by sink strength during grain filling and that further increases in yield depend upon increases in sink strength after anthesis (to either further increase grain number per unit land area or to boost the potential size of the individual grains). In cereals, the number of grains per unit land area is determined by various factors, including plants per unit land area, spikes per plant, spikelets per spike, and grains per spikelet. These factors are sensitive to growth conditions from sowing to anthesis (Slafer and Rawson 1994). However, the number of grains m^{-2} is far more sensitive to changes in growth partitioning over a short window of time immediately before anthesis (coinciding with stem elongation) than to any changes in growth occurring before this time (Demotes-Mainard and Jeuffroy 2001). Plant breeders have increased grains m^{-2} precisely through the same mechanism: crop growth has not clearly and systematically been affected by breeding for higher yield, but there has been a consistent trend toward reduced plant height (a trait determined during stem elongation), thereby leading to enhanced partitioning toward the growing spikes, which in turn results in an increased spike dry matter associated with more grains per unit land area (Slafer and Andrade 1993). Studies demonstrating a consistent increase in yield caused by *Rht* genes in a wide variety of conditions also report the relevance of this mechanism (Royo et al. 2007). In this context, the positive relationship between the number of grains m^{-2} and the harvest index, found almost without exception when comparing modern and old wheat cultivars, can be explained. The number of grains is simply a reflection of changes in partitioning that occur before anthesis, thereby resulting in lower vegetative biomass.

7.15.5 Yield Potential Versus Stress Adaptation: GE Interaction

The limitations of traditional breeding are more evident when selecting for stress (e.g., drought, high as well as low temperature) adaptation because of GE interactions and higher within-site variability, which also diminishes heritability (Araus et al. 2002). Thus, although selecting for yield per se, the targeted environment may be sensitive when stress is uniformly severe (Ceccarelli and Grando 1996). This situation is rare in the field. For example, wheat cultivars tested in a particular set of stressful conditions may not show the same performance in another set (Cooper et al. 1997). Moreover, a crossover effect in the yield of genotypes of high and low yield potential when regressed against the environmental index over a wide range of conditions is not frequently found unless the conditions are too extreme. This observation suggests that, in general, genotypes selected under high-yielding environments perform better than those with lower yield potential when grown in a wide range of yielding environments (Araus et al. 2002; Slafer and Araus 2007). Review of breeding progress pointed out that selection for high yield in stress-free conditions has, to a certain extent, indirectly improved yield in many water-limiting conditions (Cattivelli et al. 2008). A clear example is provided by the comparison of two barley cultivars released at different times over recent decades (Tambussi et al. 2005b).

For maize, there is evidence that supports higher tolerance of low resource availability in newer hybrids bred for high yield potential. These hybrids perform better than older ones under stress as a result of parental line involvement which is associated with greater tolerance to high plant density (Tollenaar and Lee 2006). In fact, plant water deficit occurs more readily at high rather than at low crop density, and adaptation to high plant density involves resistance to drought stress when moisture is limited (Tollenaar and Wu 1999). Anthesis to silking interval (ASI) under drought has become shorter in modern hybrids, and selection has possibly led to an increase in the growth of spikelets and ears and a reduction in final spikelet number (Bänziger et al. 2000). Moreover, "stay green" or the reduction in the rate of leaf senescence during grain filling is one of the traits that most visually differentiates older from newer hybrids (Duvick et al. 2004). Changes in constitutive traits, such as plant phenology, are also involved in the distinct responses to limited resources. Older hybrids show a greater yield loss, in part, because they extract most of the plant-available water before entering the critical flowering period (Campos et al. 2004).

7.16 Changing Climate and Seed Yield

The impact of climate change in our agricultural systems is undoubtable. For example, drought followed by intense rain can increase the flooding potential, thereby creating conditions that favor fungal infestations of leaves, roots, and tuber crops. In addition, reduction of bees' density due to global climate change has led to local extinction of several plant species (Ajetomobi 2016). The production of enough food to match population growth while preserving the environment is a key challenge, especially in the face of climate change. This chapter will review factors affecting yields of crops and provide some strategies to overcome yield loss while preserving the environment. Climate change has resulted in a continuous increase in the frequency, intensity, and duration of extreme low-temperature events and poses a serious threat to agricultural production. Yield has been described as the combination of distinct independent processes (e.g., agronomical or physiological yield components) (Araus et al. 2002; Reynolds and Tuberosa 2008). Another approach involves retrospective studies, comparing genotypes released in different eras as a result of previous breeding programs (Araus et al. 2002; Sanguineti et al. 2006). Grain yield can be expressed as the integrated response of distinct plant processes to a limited resource such as radiation or water. This response involves the following two main steps: the production of photoassimilates and their further transformation into an economic (usually harvestable) component. An additional factor to consider is the phenological stage of the plant when the limited resource acts.

The impacts of changing climate and climate variability on seed production of arable crops such as rice, soybean, cotton, and peanut have been recently reviewed (Singh et al. 2013). Some authors (Nelson et al. 2009; Challinor et al. 2014) concluded that the overall impact of climate change on agriculture is expected to be negative, threatening global food security, whereas others (Jaggard et al. 2010) considered that global crop production could increase by around 50% by 2050 without extra land. However, the latter authors did stress that a large proportion of these yield increases will need to be delivered via plant breeders through the release of new cultivars adapted to cope with a changing environment (Ceccarelli et al. 2010; Ainsworth et al. 2008a, b; Ainsworth and Ort 2010). Enrichment of atmospheric CO_2 is likely to increase seed yields of C3 arable crops (e.g., wheat) by between 11 and 30% but leave yields of C4 crops (e.g., maize) unchanged (Jaggard et al. 2010; Hatfield et al. 2011). Yield increases are usually because of a greater number of seeds per unit area (Qaderi and Reid 2005; Baker et al. 1989; Prasad et al.

2003) as a result of increased resource use efficiency for radiation, water, and nitrogen (Olesen and Bindi 2002). Elevated CO_2, through decreasing stomatal conductance, increases water-use efficiency (Singh et al. 2013). In water stress conditions, this may allow plants to retain more seeds, but yields will always be lower compared to those in non-resource stress environments (Singh et al. 2013). Sometimes there is also an increase in seed mass (Vanaja et al. 2007), although the effects of elevated CO_2 on seed mass are highly variable (Hampton et al. 2013). Elevated temperature accelerates various aspects of plant metabolism and changes the rate of organ development, respiration, senescence, and source-sink relationships (Farrar and Williams 1991). Most arable crop species are vulnerable to heat stress both pre- and post-fertilization, and any exposure to temperatures above the optimum for the species can result in significant seed yield losses (Dornbos and Mullen 1991; Wang et al. 2006; Madan et al. 2012) with the extent of the reduction in individual seed yield components depending on the timing and duration of the temperature stress experienced. Temperature increases, which reduce seed number, accelerate seed growth rate, reduce the duration of seed filling, and sometimes reduce seed mass (Singh et al. 2013), can, therefore, negate any positive seed yield responses to increased CO_2.

Global warming has markedly shifted the distribution of temperature variability and extremes and precipitation patterns (Trnka et al. 2014). These shifts have resulted in more frequent cold stress events (including chilling stress and freezing stress) during the wheat-growing season (Gu et al. 2008; Lobell et al. 2015), with consequent negative effects on wheat production (Fuller et al. 2007; Venzhik et al. 2011). Moreover, global warming has accelerated the process of wheat growth and development, resulting in significant advancement of the low-temperature-sensitive stage of wheat, thereby increasing the probability of cold or frost injury in wheat (Ji et al. 2017a). Study showed that during the last several decades, low temperature has seriously affected wheat production in China (Zhong et al. 2008), the United States (Gu et al. 2008), Europe (Peings et al. 2013; Trnka et al. 2014), and Australia (Barlow et al. 2015; Crimp et al. 2016). However, due to the increasing global mean temperature during the last several decades, most studies have been concerned with the effects of heat stress on crop production (Gourdji et al. 2013; Liu et al. 2014; Rahmstorf and Coumou 2011). Low-temperature injury to crops, which includes both cold and freezing injuries, has usually been ignored (Ding et al. 2010; Rigby and Porporato 2008).

Leaf photosynthesis and biomass accumulation, which are the primary sources of grain yield, are the two most sensitive crop growth processes affected by cold stress (Sassenrath and Ort 1990). During the vegetative period, cold stress usually exacerbates the balance between the source of energy and the metabolic sink, which inhibits the photosynthesis rate and results in reduced leaf area (Allen and Ort 2001; Paul and Foyer 2001). A reduction in both photosynthesis rate and leaf area index will reduce the biomass accumulation and, consequently, grain yield (Subedi et al. 1998). Leonardos et al. (2003) reported that the plants in the cold condition had a lower relative growth rate and smaller leaf area and biomass but greater specific leaf weight. Fuller et al. (2007) reported that the flag leaves of two wheat cultivars were damaged under the treatment at −3 °C for 2 h, with damage increasing as the low temperature declined to −5 °C and −7 °C, resulting in 10–100% grain yield loss. Venzhik et al. (2011) showed that photosynthesis rate of the first leaf decreased by 18% after 5 h of low-temperature treatment, and the maximum quantum efficiency of photosystem II decreased by 8% after 24 h of treatment. Li et al. (2015) reported that 8 °C lower than ambient temperature for 5 days at jointing significantly decreased gas exchange rates and the maximum quantum efficiency of photosystem II in wheat leaves, causing a 5–14% reduction in grain yield. Studies have primarily focused on the effects of low temperature on photosynthesis and yield (Fuller et al. 2007; Subedi et al. 2010; Venzhik et al. 2011), with less focus on biomass production and the relative importance of biomass production to grain yield. Additionally, studies have primarily investigated the effect of either low-temperature level or low-temperature duration alone on wheat production (Al-Issawi et al. 2013; Whaley et al. 2004), whereas the combined effects of low-temperature level and duration have rarely been reported. Furthermore, the temperature regimes in most previous studies were based on fixed day/night temperature treatments (Li et al. 2015; Venzhik et al. 2011), which are not consistent with the natural daily temperature dynamics. In the actual environment, the impact of low temperature on crop growth and development is often affected by the combination of low-temperature intensity and duration under the daily temperature variation pattern (Shimono et al. 2007).

Few studies on the climate change and vegetable seed production have been reported. Carrot (*Daucus carota* L.) vegetative growth is increased by elevated CO_2 (Idso and Kimball 1989; Mortensen 1994; Wheeler et al. 1994) although the effect is temperature dependent. At 11 to 12 °C, elevated CO_2 had no effect on carrot growth, but at mean air temperatures of >25 °C, vegetative biomass was nearly doubled (Idso and Kimball 1989). However, there have been no reports on the effects of elevated CO_2 on carrot reproductive growth. Elevated CO_2 more than doubled radish (*Raphanus sativus* L.) vegetative dry matter (Usuda and Shimogawara 1998), but the response was temperature dependent, increasing as air temperature increased (Idso and Kimball 1989). In wild radish (*Raphanus raphanistrum*), elevated CO_2 increased flower and seed production (Miller et al. 1994), but another report found that in the same species, elevated CO_2 reduced flower and seed numbers (Case et al. 1998). Increasing mean air temperature from 20° to 26 °C reduced carrot vegetative biomass, main stem height, and days to first flowering. Umbels per plant were increased at the higher temperature, but the number of flowers per umbel was reduced, as was the fertility of the male flowers (Quagliotti 1967). When temperature was increased from 20 °C day/10 °C night to 30 °C day/20 °C night, a 13–20% reduction in mean carrot seed mass was found (Gray et al. 1988), which was associated with a reduction in seed filling time rather than any difference in seed relative growth rate. Daytime temperatures of 20–30 °C caused devernalization in radish (Sheen 2000), and a similar response has been recorded in vegetable brassicas (Weibe et al. 1992; Zenkteler et al. 2012). Increasing temperature from 17 to 26 °C significantly reduced radish pod number and seeds per pod (El Murabaa 1957), and a reduction in cabbage (*Brassica oleracea* L.) seed yield was associated with increasing temperature during pod setting (Kumar et al. 2009). Although *Brassica napus* L. (canola) is not a vegetable brassica, the negative effects of a 5 °C temperature increase on seed yield could not be compensated by elevated CO_2 (Frenck et al. 2011). It is likely that this response would also occur in vegetable brassicas.

Cold stress, which includes both chilling and freezing injuries, modifies gene expression and plant metabolism with consequent effects on many biological functions (Janmohammadi et al. 2015). There is a disruption of plasmatic membrane (Steponkus 1984), changes of protein synthesis (Duman and Wisniewski 2014), and limited breathing, photosynthesis, and carbon fixation (Liu et al. 2012). Freeze damage is not caused by the actual freeze temperature but by the formation of ice crystals (Beck et al. 2007). The formation of ice crystals in the extracellular space thus causes the drainage of water from protoplast into the extracellular space and the subsequent dehydration of cells. Although dehydration causes the fading of tissues, it prevents the adverse formation of ice in the protoplasts (Scott 2008). Despite this, the dehydration of protoplasts remains the main cause of freeze damage. The two distinct strategies taken by plants to combat low-temperature stress are avoidance and tolerance. Stress avoidance entails preventing the freezing of sensitive tissues. A more elaborate avoidance strategy involves supercooling, in which endogenous ice nucleation is prevented by inhibiting the formation of ice nucleators (Janská et al. 2010). Balbuena et al. (2011) studied sunflower (*Helianthus annuus* L.) cold acclimation changes in freezing; susceptible and tolerant lines were investigated using a label-free comparative proteomic approach. They stated that cold-responsive proteins were mostly involved in metabolism, protein synthesis, and energy and defense processes and that tolerant lines have different proteome responses to cold acclimation. Tetreault et al. (2016) examined variation in cold acclimation capacity and freezing tolerance among three natural populations of the perennial sunflower species. Freezing tolerance was the highest in plants from the northernmost latitude under both non-cold-acclimated and cold-acclimated experimental conditions. The plants from all populations retained the ability to increase freezing tolerance through the process of cold acclimation (Hniličková et al. 2017).

Accumulated cold degree-days (°C days), which include both duration and intensity of low temperature, have been used to quantify the effects of low temperature on the growth, development, and yield formation in rice (Roel et al. 2005; Shimono et al. 2005) and wheat (Ji et al. 2017a). Godwin et al. (1994) introduced accumulated cold degree-days into the CERES-Rice model to improve the simulation accuracy under low-temperature environments, suggesting that accumulated cold degree-days could effectively quantify the chilling injury effect in an extensive field experiment. Shimono (2011) employed accumulated cold degree-days to quantify the effects of low temperature on rice yield in northern Japan during the past 50 years and found that for each 1 °C·days increase in accumulated cold degree-days, rice yield was predicted to decline by 0.86%. Ji et al. (2017a) used accumulated cold degree-days to analyze the relationship between low-temperature stress and grain yield and yield components of wheat and observed significant negative linear relationships between accumulated cold degree-days and grain yield per plant, spike number per plant, grain number per spike, and 1000-grain weight. Although several studies have quantified the effects of low temperature on the yield of rice and wheat with accumulated cold degree-days, few similar studies have been conducted on wheat biomass production (Liu et al. 2019).

In rice, the most sensitive stage of low temperature is the booting stage. The percentage of fertile spikelets has been used as effective parameter of cold tolerance in rice. Low temperature at booting stage causes anther injury and degeneration of young microspores, resulting in high spikelet sterility and reduced rice yield. Cold temperature during reproductive stage leads to seed sterility. The fertilization stage ranging from pollen maturation to the completion of fertilization is sensitive to low temperature (Loitongbam et al. 2017). Ghadirnezhad and Fallah (2014) reported that low temperature had significant effects on all characters, such as number of panicle, length of panicle, and number of empty, full, and total grains; as a result, yield was significantly reduced. Many species of tropical and subtropical origin are injured or killed by non-freezing temperature. In contrast, chilling-tolerant species are able to grow at such low temperatures (Sanghera et al. 2011). Ye et al. (2009) reported yield losses due to cold stress are the result of incomplete pollen formation and subsequent floret sterility. Perveen et al. (2013) found that concentration of two photosynthetic pigments was negatively affected by low-temperature stress. Suzuki et al. (2008) reported that the effect of cold stress at reproductive stage of plant delays heading and results in pollen sterility, which is thought to be one of the key factors responsible for reduction in seed yield of the crop. Arshad et al. (2017) reported that cold stress during reproductive phase in rice results in the following:

- Delayed heading, due to its effect on anther respiration, which increases sucrose accumulation, protein denaturation, and asparagine levels, and decreases proline accumulation
- Pollen sterility owing to tapetal hypertrophy and related nutrient imbalances
- Reduced activity of cell wall-bound invertase in the tapetum of rice anthers
- Impaired fertilization due to inhibited anther dehiscence, stigma receptivity, and ability of the pollen tube to germinate through the style toward the ovary
- Floret sterility, which increases grain abortion, restricts grain size, and thus reduces grain yield

Due to its tropical origin, castor (*Ricinus communis* L.) is a heat-loving plant, and cool temperatures during seed filling have been reported to reduce seed yield. Severino and Auld (2013) found that 74% of the seeds reached maturity during 42 days of warmer air temperature (from 88 to 130 days after planting), while only 8% of the seeds reached maturity during 44 days of cooler air temperature. Many seeds (18% of the seed yield) were still immature when the killing frost occurred. Temperature is an important environmental factor influencing the rate of plant growth and development. For that reason, plant growth and development are more associated with the thermal time than with chronological time. The base temperature can assume a different value for each physiological process, and an adequate calculation of the degree-days depends on an accurate determination of the base temperature. The impact of air temperature on seed growth rate and the limited influence of the plant's physiological condition is well documented in the literature for crops such as maize (*Zea mays*), wheat (*Triticum aestivum*), soybean (*Glycine max* L.), and rice (*Oryza sativa* L.) (Soffield et al. 1977; Yoshida and Hara 1977; Egli and Wardlaw 1980; Tollenaar and Bruulsema 1988; Egli 1998; Tacarindua et al. 2012). The development of castor seed is similar to most species and occurs in three phases: cell division, seed filling, and desiccation (Severino and Auld 2014). The phase of cell division (Phase I) begins with ovule fertilization and ends when the seed has the maximum number of cells. During this phase, there is negligible accumulation of reserves (oil, protein, carbohydrates, and minerals), the seed grows close to its final size, and the water content is around 900 g kg^{-1}. The phase of seed filling (Phase II) is characterized by a linear increase in the dry weight due to accumulation of reserves. Usually, the Phase II does not overlap the cell division phase and it ends when the seed reaches the physiological maturity. In the desiccation phase (Phase III), the seed loses vascular connection with the mother-plant and passively loses water at a rate dependent upon environmental conditions (Egli 1998; Munier-Jolain and Ney 1998). The physiological conditions of the mother plant influence the number of seeds and the seed characteristics that are defined during the Phase I. However, once the Phase II begins, the mother plant plays the important role of supplying the seed with assimilates, but it has little control on the seed growth rate. For instance, during Phase I, the mother plant controls the abortion rate and the number of cells in the seed, which are determined by sucrose availability. Nevertheless, during the Phase II, there is no abortion, and seed growth is primarily controlled by the number of cells that was previously differentiated during Phase I (Munier-Jolain and Ney 1998). Thus, the seed growth during Phase II is influenced by environmental condition rather than by the mother plant (Egli 1998). Only under extreme conditions, such as severe drought, high temperatures, or pest attack, can seed growth during Phase II be significantly influenced by the physiological condition of the mother plant.

Temperature stress (cold, heat) during reproductive development is one of the serious constraints to the productivity of grain legumes as their cultivation is expanding to warmer environments and temperature variability is increasing due to climate change. Grain legumes exposed to temperature below 10–15 °C or above 30 °C show flower abortion, pollen and ovule infertility, impaired fertilization, and reduced seed filling, leading to substantial reduction in grain yield. For managing these effects of temperature extremes, it is important to improve the resistance of grain legumes by using improved breeding and genetic engineering tools. In this review article, the impact of both high- and low-temperature stresses on different phases of the reproductive stage, from meiosis to grain filling, and the sensitivity of different reproductive organs to temperature extremes are discussed. The review also covers the management options to improve resistance to temperature stress in grain legumes (Gogoi et al. 2018). Nayyar et al. (2005) pointed out that chilling (<15 °C) during the reproductive phase of chickpea leads to abortion of flowers and pods, infertile pods, smaller seeds, and reduced seed yields. Relative to control plants (17/28 °C mean minimum/maximum temperature), those subjected to cold stress (5/13 °C mean minimum/maximum temperature) showed a marked increase in electrolyte leakage, while cellular respiration, chlorophyll content, relative leaf water content, and rate and duration of seed filling decreased significantly. In cold-stressed plants, seed number per 100 pods, seed weight per plant, average seed weight, and average seed size decreased by 35, 43, 41, and 24%, respectively. Seed reserves of starch, protein, and fat decreased by 34, 33, and 43%, respectively, while total soluble sugars increased twofold. The accumulation of storage proteins such as globulins and albumins was inhibited to a greater extent than that of prolamins and glutelins. Most of the amino acids decreased as a result of stress, while some such as proline and glutamic acid increased significantly.

7.17 Effect of Temperature Stress on Seed Yield

While temperature and other abiotic stresses are clearly limiting factors for crops cultivated on marginal lands, crop productivity everywhere is often at the mercy of random environmental fluctuations. Current speculation about global climate change is that most agricultural regions will experience more extreme environmental fluctuations (Solomon et al. 2007). Hot or cold temperature stresses can be detrimental to all phases of plant development. As the majority of our food supply is a product of sexual reproduction in flowering plants, understanding how different plants cope with stress during their reproductive (gametophytic) phase is critical to managing the future of agricultural productivity. During the short time surrounding fertilization, even a single hot day or cold night can be fatal to reproductive success for many plant species. While temperature stress has been extensively studied (Iba 2002; Yamaguchi-Shinozaki and Shinozaki 2006; Chinnusamy et al. 2007; Kotak et al. 2007; Wahid et al. 2007), most of the literature emphasizes insights from experimentally accessible tissues, such as leaves and roots. By comparison, studies on sexual reproduction are often more difficult because gamete development and fertilization are complex processes that occur during a short window of time, and are predominantly hidden within tissues of the flower. Nevertheless, sexual reproduction has been long recognized as being highly stress-sensitive, with reproductive stress tolerance often a limiting trait in crop plant productivity (Herrero and Johnson 1980), and this has been extensively reviewed (Barnabás et al. 2008; Hedhly et al. 2008; Thakur et al. 2010).

The impact of temperature stress is a complex function of intensity, duration, and rate of temperature change (Wahid et al. 2007; Thakur et al. 2010). Both hot and cold stresses can alter multiple aspects of cellular physiology. For example, temperature can dramatically change membrane fluidity, nucleic acid and protein structures, as well as metabolite and osmolyte concentrations (Wang et al. 2003; Howarth 2005; Chinnusamy et al. 2007). Plants have evolved complex mechanisms to cope with daily changes in the environment. At a molecular level, this is illustrated by the thousands of transcriptional changes observed in seedlings, leaves, roots, and pollen as plants reprogram cellular processes to adapt to hot or cold temperatures (Fowler and Thomashow 2002; Kreps et al. 2002; Busch et al. 2005; Larkindale and Vierling 2008; Frank et al. 2009). Both hot and cold temperature stresses induce the production of reactive species of oxygen, which at elevated concentrations will result in oxidative damage and, potentially, cell death (Apel and Hirt 2004).

The gametophytic phase in flowering plants is often highly sensitive to hot or cold temperature stresses, with even a single hot day or cold night sometimes being fatal to reproductive success (Zinn et al. 2010). Temperature stress can sometimes have different effects on male and female structures, thereby creating asynchrony between male and female reproductive development (Herrero 2003; Hedhly et al. 2008). In maize (*Zea mays*), floral asynchrony is a significant problem under conditions of combined stress from heat and water deficit (Barnabás et al. 2008). In addition, high-temperature stress can shorten the period of time in which the stigmas in the flowers are receptive to pollen, and thereby decrease the chances for a successful fertilization. For example, the stigmas in peach (*Prunus persica* L.) at 30 °C lost their ability to support pollen germination after 3 days, whereas at 20 °C they were viable for 8 days (Hedhly et al. 2005). More is known of the effects of temperature stress on male reproductive structures (Barnabás et al. 2008; Thakur et al. 2010). For example, in wheat, heat stress during the period of microspore meiosis can induce tapetum degradation. This degradation of the nutritive tissues of the tapetum leads to pollen sterility. High temperatures cause poor anther dehiscence characterized by tight closure of the locules, which was shown to reduce pollen dispersal in rice and tomato (*Solanum lycopersicum*) (Matsui and Omasa 2002; Sato et al. 2002).

Post-flowering conditions affect yield mainly by influencing ovary or seed abortion, or by changing seed filling duration or seed filling rate. In general, seed filling duration is more plastic than seed filling rate. In spite of the importance of individual seed size, yield variation is more often due to changes in seed number per unit of ground area than to individual seed mass at maturity. Warming hastens crop development and therefore shortens the seed filling duration. Seed number can be reduced via the direct effects of high temperature on reproduction, particularly pollen formation and function. Because warming speeds reproductive development, the seeds that do develop are often small. Although seed filling rate is sometimes stimulated by warming, this effect often does not fully compensate for shortened seed filling duration (Thuzar et al. 2010).

Pollen development, fertilization, and asynchrony of stamen and gynoecium's development are sensitive to temperatures during flowering (Prasad et al. 1999; Croser et al. 2003; Boote et al. 2005). The loss of pollen or stigma viability under high-temperature stress might be the primary reason for the lowered number of seeds produced in the legume (Srinivasan et al. 1998; Davies et al. 1999; Hall 2004). Significant negative correlation between pollen production and temperature was found in groundnut (Prasad et al. 1999). Lower seed yield at high temperature under both ambient and elevated concentrations of CO_2 conditions was shown to be due to decreased pollen viability in groundnut and bean (Prasad et al. 2002, 2003; Kakani et al. 2002). Pollen sterility and pollen production at high temperatures may also be associated with early degeneration of the tapetal layer of pollen (Porch and Jahn 2001). The exact physiological reasons of pollen viability loss are not clearly known and need further investigation (Sailaja et al. 2005). Reproductive growth leading to seed yield is often depressed by the same increase in temperature that enhances vegetative growth and development. Flower initiation was reduced by temperature >32 °C and seed formation was delayed by 40/30 °C (Thomas et al. 2003). When soybean plants were exposed to temperature of 35 °C for 10 h during the day, yield reductions of about 27% were measured (Gibson and Mullen 1996). Hence, it is essential to protect crop yield from higher and more frequent episodes of extremely higher temperature both in current and future climates (Salem et al. 2007). Gan et al. (2004) found that the seed yield of canola decreases by 15% when high-temperature stress was applied before flowering, whereas the yield reduction was 58% when the stress was delayed to the period of flowering, and further to 77% when the stress was delayed to the pod developmental stage. Physiologically, the high-temperature stress during reproductive development may have affected flower abortion, sequent sink site, and later pod abscission, resulting in a decreased number of seeds per plant (Duthion and Pigeaire 1991). Also, high-temperature stress during reproductive development may have negatively affected cell expansion, cotyledon cell number, and thus seed filling rate, resulting in the lowered weight per seed (Munier-Jolain and Ney 1998).

Cold temperatures can induce pollen sterility, which may be due to disruption of sugar metabolism in the tapetum, ultimately abolishing starch accumulation (energy reserves) in the pollen grains (Oliver et al. 2005). Evidence suggests that cold-induced disruption of anther sugar transport and corresponding pollen sterility is signaled by abscisic acid, in part by downregulating expression of cell wall invertase and monosaccharide transporters (Oliver et al. 2007). Heat stress reduces carbohydrate deposition in pollen grains (Jain et al. 2007). The preexisting carbohydrate reserves within the pollen grains fuel tube growth, but pollen later switch to using carbohydrates provided by the transmitting tract of the style (Herrero and Arbeloa 1989). Cotton (*Gossypium hirsutum*) flowers exposed to moderately high temperatures showed reduced carbohydrate reserves (particularly sucrose) and ATP production in the pistils, which correlated with a heat stress-induced decline in net photosynthesis (Snider et al. 2009).

Rapeseed (*Brassica napus* L.) growth is controlled from the seedling emergence to flowering by photothermal factors and from flowering to maturity by temperature (Nanda et al. 1996). Because of the crop sensitivity in the early stages of development, the seed yield of canola is often limited by a poor plant establishment (Yang et al. 2014). From the beginning of the leaf development to the end of stem elongation, the phenological development is controlled by the temperature, vernalization, and photoperiod (Böttcher et al. 2016). Increased temperatures and photoperiod stop the vernalization period as the day length increases (Hoffmann et al. 2015). When temperatures stay continuously above 0 °C, rapeseed grows rapidly and produces most of the aboveground biomass over a period of several weeks (Marjanović-Jeromela et al. 2019). Optimum temperatures for photosynthetic, vegetative, and reproductive processes vary from 21 to 25 °C (Deligios et al. 2013). Flowering is the most critical stage of rapeseed development, because of the decrease in the total leaf area and reduced photosynthesis (Gabrielle et al. 1998; Robelin and Triboi 1983). Frost at the beginning of the flowering increases branching, and the flowering lasts longer, resulting in an increased number of poorly filled pods, formed on the lower branches (Lardon and Triboi-Blondel 1995; Balodis and Gaile 2016). Rapeseed growth stages are used to define the main components in yield estimation models (Weymann et al. 2015). Each of them can be affected by yield-limiting factors, of which climatic conditions are the most important ones. Temperature, irradiation, and precipitation directly and indirectly affect and determine the yield (Takashima et al. 2013). Unfavorable weather conditions may lead to a low sink capacity, which is not necessarily yield-limiting because of the compensatory effect between the yield components (Weymann et al. 2015). In the regions with a warmer climate, each 1 °C increase in the ambient temperature results in shorter growth periods. However, the production of the highest seed and oil yields requires rather limited day/night average temperature ranges (20/15 °C and 15/13 °C) (Kuo et al. 1980; Hocking and Stapper 2001). During the seed development, the temperature has a significant effect on the yield (Balodis and Gaile 2016). It affects the duration and rate of assimilation and therefore determines the assimilate availability for seed filling. The yield potential may be determined until the end of flowering, but whether this yield potential is realized or not mainly depends on the temperature and water availability in the subsequent growth phases (Weymann et al. 2015).

7.18 Physiological Avenues for Increasing Yield Potential

Retrospective studies on wheat indicate that the improvement in yield has more often been associated with augmented partitioning of biomass to the grain than with enhanced overall biomass (Sayre et al. 1997; Calderini et al. 1999). Thus, plant height is usually negatively related to the harvest index (Royo et al. 2007). However, there is a minimum height below which yield limitation is evident (Slafer et al. 2005). Moreover, given that the harvest index is estimated to have an upper limit of just over 60% (Austin 1980), and that this limit is already being approached (Shearman et al. 2005a, b), it is becoming more important than ever to unveil the physiological and genetic bases of radiation-use efficiency and biomass determination if increased yields are to be achieved (Araus et al. 2003).

7.18.1 Increasing Radiation-Use Efficiency

Radiation-use efficiency is a basic parameter used in analyzing crop growth based on resource capture and use efficiency (Stockle and Kemanian 2009). It is used to evaluate the relative performance of crops under different locations and management practices (Monteith 1977; Sinclair and Muchow 1999). A reduction in radiation-use efficiency has commonly been observed during pod-filling in all species. In addition to the reduction of radiation-use efficiency, low temperature and water stress also affect foliage expansion and eventually radiation capture (Jamieson et al. 1995; Stone et al. 2001; Inman-Bamber 2004).

Ribulose carboxylase/oxygenase (Rubisco) enzyme, which catalyzes the assimilation of CO_2 by the carboxylation of ribulose biphosphate in the Calvin cycle, is the most obvious target for attempts to improve photosynthetic rate and thus radiation-use efficiency. Moreover, since Rubisco exhibits a slow catalysis, large amounts are needed to sustain high photosynthetic rates, therefore negatively affecting nitrogen use efficiency. Rubisco also catalyzes the wasteful oxygenation of ribulose bisphosphate, initiating the process of photorespiration and the consequent release of CO_2, NH_3, and energy expenditure. The dual specificity of Rubisco seriously impacts upon water-use efficiency. Thus, stomatal closure leads to a lower internal CO_2/O_2, making carbon assimilation less efficient; conversely, the unrestricted exchange of gases through fully open stomates would exacerbate water loss. Therefore, high temperature and even more drought (the combination of water stress and high temperature) may further increase oxygenase activity of Rubisco relative to the carboxylation of RuBP and therefore photorespiration (Long et al. 2006; Parry et al. 2007).

Among environmental factors which affect the radiation-use efficiency, temperature is one of them. Low temperatures reduce radiation-use efficiency for different types of crops (Andrade et al. 1992; Bell et al. 1992). For example, Andrade et al. (1992) showed that the temperature from 15 to 18 °C during the vegetative period caused low radiation-use efficiency in maize. Also, high temperature (increased respiration sometimes above the rate of photosynthesis) had negative effects on crops (Monneveux et al. 2003). In an earlier study, decreased precipitation as combined with high temperature showed a considerable negative effect on pasture production in southern Mongolia (Munkhtsetseg et al. 2007). Under field conditions, the development of crops depends on the capacity of the canopy to intercept incident radiation and convert it into new biomass (Bhattacharya and Vijaylaxmi 2010; Bhattacharya 2019). The amount of incident radiation intercepted depends on the leaf area index and canopy orientation in space. The fraction of the incident photosynthetically active radiation that is absorbed by a canopy depends on the leaf area index and the arrangement of the leaves of the crop (Plenet et al. 2000). When there is no environmentally induced stress from water or air temperature, inadequate nutrition, pests, and diseases, the quantity of biomass produced by a growing crop is correlated to the amount of solar radiation or photosynthetically active radiation intercepted by the green leaf area of a crop during its growth cycle (Monteith 1977).

Except maize and sorghum, major crops like wheat, rice, barley, or soybean have C3 photosynthetic metabolism. There are different avenues to improve assimilation of CO_2 subsequently decreasing photorespiratory losses. While one strategy would be to introduce forms of Rubisco with higher relative specificity for CO_2, an alternative exists to insert the carbon concentration mechanisms of C4 plants. Such approach would be of greatest benefit under drought or other conditions that lead to stomatal closure and low intercellular CO_2 (von Caemmerer 2003). Other targets for the improvement of photosynthetic rate impact upon the capacity for RuBP regeneration and include other Calvin cycle enzymes, principally sedoheptulose-bisphosphatase, and components of electron transport and light capture by the photosystems (Parry et al. 2007, 2008). To date, no substantial increase in the carboxylase relative to oxygenase activity or catalytic rate has yet been achieved by the characterization of site-directed mutants and chimeric enzymes for cyanobacterial, algal, and bacterial forms of Rubisco (Parry et al. 2003a, 2007, 2008). Alternatively, natural variation, although comparatively small (Galmes et al. 2005), can be exploited to improve Rubisco specificity and thus CO_2 assimilation in crop plants (Parry et al. 2007). Models indicate that replacing native wheat Rubisco with that from *Limonium gibertii* will give a 12% increase in net photosynthesis and thus greatly increased yield. Although much progress has been made to develop the tools for introducing genes into the nuclear genome of elite cultivars of wheat (Shewry and Jones 2005), transformation of the plastid genome is still slow and limited to model species (Maliga 2004). Furthermore, the additional complexities of assembling copies of the large and small polypeptide subunits of Rubisco into a functional holoenzyme in vivo (requiring high levels of expression, interaction with chaperonins, posttranslational modification, and interaction with Rubisco activase) remain a major challenge (Parry et al. 2007).

An alternative approach to overcoming the limitations of Rubisco in C3 crops is to increase the CO_2 concentration at the catalytic site. For example, the International Rice Research Institute (IRRI) has been developing a research program on rice on this topic for more than a decade. However, the complexities of C4 "Kranz" anatomy (Araus et al. 1991) and the segregation of photosynthetic metabolism between the mesophyll and bundle sheath have prevented the introduction of such a system into C3 plants. The discovery of CO_2 pumps that operate within a single cell of some species (e.g., *Hydrilla verticillata*) has led to a number of biotechnological approaches introducing a single cell CO_2 pump into C3 crop plants such as rice and wheat (Leegood 2002). However, this approach still requires the coordinated, targeted expression of many transgenes and the assembly of gene products at the correct cellular and subcellular location, together with the appropriate regulation of the component enzymes and transporters. This has yet to be achieved (Leegood 2002) and, even if successful, additional structural changes may be required to maximize yield potential. The wild plant where this mechanism has been discovered is characterized by a very low rate of growth, and in fact this mechanism has developed to allow this species to tolerate extreme growing conditions in its natural environment.

Biotechnological approaches are also being explored to introduce bacterial enzymes to short-circuit the photorespiratory pathway, therefore eliminating the photorespiratory release of NH_3 (Parry et al. 2003b). A pathway which converts chloroplastic glycolate directly to glycerate without involvement of NH_3 has been introduced into *Arabidopsis* and releases CO_2 in the chloroplast. This is consistent with an increase in photosynthesis partly due to an increased CO_2 at the catalytic site of Rubisco (Kebeish et al. 2007). Other approaches to improve photosynthesis include preventing the decrease in Rubisco activity particularly under high temperature, by increasing concentration and thermal stability of Rubisco activase (Salvucci et al. 2001) and from altering the abundance of the Rubisco by targeting their synthesis or degradation (Andralojc et al. 2002; Parry et al. 2008).

Phosphoenolpyruvate carboxylase, which serves as the actual CO_2 pump of the C4 pathway, is specifically expressed in the mesophyll cells of C4 leaves. Therefore, regulation of phosphoenolpyruvate carboxylase expression is another potential target to increase in photosynthesis.

7.18.2 Biomass Increase

Moderate increases in biomass have started to be reported in spring wheat (Reynolds et al. 1999) and winter bread wheat (Shearman et al. 2005a, b). In some cases, these increases are associated with a small rise in the assimilation rate during the spike growth stage, and a much larger increase in photosynthetic rate during grain filling (Reynolds et al. 2005a). Further experiments, in which grain number was increased artificially in elite lines with a brief light treatment during the rapid spike growth stage, show that these lines have a photosynthetic capacity that exceeds that required to fill the grains they would normally set (Reynolds et al. 2005a). Nevertheless, in maize, gains in yield potential have been accompanied by clear increases in total biomass, and the harvest index has remained steady (Tollenaar and Lee 2006). The higher dry matter accumulation in newer than in older maize hybrids during the grain-filling period can be attributed, in part, to a longer duration of the grain-filling period in the former (Tollenaar and Lee 2006). However, the silking dates as well as relative maturity do not differ between modern and old hybrids (Cavalieri and Smith 1985), which suggest that heterosis has not been responsible for this genetic gain (Tollenaar et al. 2004).

7.18.3 Phenological Adjustment

An alternative approach to further increase the number of grains per unit land area might be to lengthen the stem elongation phase (Slafer and Whitechurch 2001). The hypothesis is that a longer stem elongation phase results in greater crop growth during this phase, higher spike dry matter at anthesis, and subsequently more grains being filled. This hypothesis would be viable only when the length of the stem elongation phase can be manipulated and when the expected higher biomass accumulated during this phase is not counterbalanced by reduced partitioning to the spikes. These manipulations might involve genes responsible for sensitivity to photoperiod or for earliness per se. The hypothetical involvement of photoperiod genes is supported by the fact that the duration of stem elongation is regulated by photoperiod response (Miralles and Richards 2000; González et al. 2002, 2003a), while that of earliness per se genes is backed by the existence of genetic variation in minimum duration of stem elongation and the intrinsic earliness of that phase. Briefly, there is clear genetic variation in the duration of stem elongation, even when holding constant the length of the period to anthesis (Kernich et al. 1997; Whitechurch et al. 2007). At least part of this variation may be due to sensitivity to photoperiod (Whitechurch et al. 2007). This sensitivity varies between phenological phases (González et al. 2005a). Exposing plots to photoperiod extensions during the stem elongation phase (natural day length was maintained before this phase) altered the duration of the phase, which was associated with changes in the number of fertile florets and grains as a result of modifications in spike dry matter at anthesis (González et al. 2003a, b, 2005b, c). These observations suggest that the mechanism by which photoperiod alters the final number of grains per m^2 is the same as that determined by radiation interception during stem elongation (González et al. 2005b). Therefore, isolating and subsequently manipulating the genetic control sensitivity to photoperiod (in this example, or genetic control of earliness per se) during stem elongation might be an effective avenue for increasing yield.

7.18.4 Plant Responses to Low Temperature

Cold or low-temperature stress comprises chilling (<20 °C) and freezing temperatures (<0 °C) that hamper the plant growth and development in many ways. Chilling-sensitive plants exposed to low temperatures usually show water-stress symptoms due to decreased root hydraulic conductance and leaf water and turgor potentials (Aroca et al. 2003). Cold stress effects on crop plant have been studied since a long time in many economically important crops (Farooq et al. 2009; Nahar et al. 2009a, b; Zhou-fei et al. 2009), among which some are chilling sensitive and unable to survive cold temperatures. Low temperature affects the plants in every stage of life starting from germination up to maturity.

7.18.5 Reproductive Development Under Low Temperature

Reproductive phases of the life cycle of plants are more vulnerable to cold stress (Nishiyama 1995). The reproductive phase facing low-temperature stress is influenced diversely in its different sub-phases. During the development of male gametophyte, low temperature causes disruption of meiosis, tapetal hypertrophy, stunted development of pollen grain, anther protein degradation, pollen sterility, and pollen tube deformation (Saini et al. 1983). In female gametophyte development, its effects are characterized by reduced style and ovary length, disruption of meiosis, reduced stigma receptivity, callose deposition in style, damage to embryo sac components, and arrest of the fertilization process. At flowering, low temperature may cause delayed flowering, bud abscission, and sterile or distorted flowers, while at grain filling, the source-sink relation is altered, kernel filling rate is reduced, and ultimately small sized, unfilled, or aborted seeds are produced (Jiang et al. 2002; Thakur et al. 2010). Flower buds of *Simmondsia chinensis* can be damaged or killed by temperatures from −2° to −5.5 °C (Nelson and Palzkill 1993). Farrell et al. (2006a, b) observed that cold water significantly reduced panicle emergence in rice. Low temperatures affect not only normal heading but also panicle exsertion and prevent the normal elongation of internodes of rice. Cold stress (11 °C) caused abnormality in panicle initiation process, delayed heading, incomplete panicle emergence, sterile and malformed spikelet, and spikelet degeneration symptom in rice genotypes (Nahar et al. 2009a). In canola, low but non-freezing temperatures prior to flowering slow the rate of plant development, delayed flowering, slow the rate of flower opening, and reduce the amount of pollen shed, and in severe frost, the pod abortion is evident in many species (Miller et al. 2001). Effect of low temperature prevailing at the early stages of plant growth is sometimes associated with the premature flowering in Chinese cabbage (Kalisz and Cebula 2001).

7.18.6 Yield Components and Yield Under Low Temperature

Reproductive phase products are the key components of economic yield, and hence low-temperature stress during the reproductive phase has significant economic and social consequences. All the adverse effects of cold stress ultimately lower the yield of crop. Low-temperature-induced yield reduction is a common phenomenon in many crops (Nahar et al. 2009a, b; Kalbarczyk 2009; Riaz-ud-din et al. 2010). Low temperature often causes flower abortion, pollen and ovule infertility, breakdown of fertilization, poor seed filling, and decreases in seed setting which ultimately reduce the grain yield (Thakur et al. 2010). In *Oryza sativa*, low temperatures are responsible for 30–40% yield reduction in temperate-growing areas (Andaya and Mackill 2003). It was observed that about 16 and 37% yield reduction in the rice due to late sowing mediated low-temperature stress (Nahar et al. 2009a). In another report, it was observed that low-temperature stress near about 11 °C caused yield reduction in maximum genotypes and only 23 genotypes were screened out among the 244 genotypes considering their better yielding ability under low temperature (Nahar et al. 2009b). The reduction in yield in *Cucumis sativus* L. was from 15 up to over 18% due to low soil temperature (Korkmaz and Dufault 2002). In *Cucumis melo* L., total yields decreased linearly for cold stress (21 and 32 h) which accounted for 10% lower than normal condition (Korkmaz and Dufault 2002). Frost prevailing just after flowering can result in yield reduction and the quality or grade loss. Premature flowering or bolting has occurred due to low-temperature stress which has economic importance for the Chinese cabbage industry because advanced flower stalk development results in an unmarketable head (Kalisz and Cebula 2001). In *B. napus* and *B. rapa*, various abnormal structures were observed like reduced diameter and extensive white patches, white reticulation, red-brown pigmentation, folded seed, extensively shriveled seed, etc., which are the causes for reduced market value of this crop (Angadi et al. 2000).

7.18.7 Traits for Tolerance to Abiotic Stress

Substantial improvements have been made in low-temperature tolerance through conventional breeding, more enhanced by the use of molecular markers associated with the trait. Although many genes have been identified as putative cold-tolerance genes, it would appear that none of these has yet demonstrated sufficient protection to merit commercialization in large-scale crop production. However, despite the trends of higher temperatures in many regions, protection against the devastating effects of low temperatures, particularly during the sensitive phases of seedling growth and crop maturation, remains an important focus area for crop improvement.

Water-use efficiency has some elements in common with drought tolerance, in that it can be seen as a mechanism whereby a plant can avoid drought by depleting soil moisture more slowly. However, the physiological mechanisms of the two traits are likely different, as is their place in agricultural cropping systems. Water-use efficiency is being recognized as a critically important trait in areas where crop production relies on dwindling supplies of subsurface irrigation water or where there is competition for water between urban and agricultural demands. Approaches to enhancing water-use efficiency center on altering the ratio of CO_2 assimilated per unit of water transpired (Chaerle et al. 2005). In a study, an *Arabidopsis* transcription factor, *HRD*, was shown to increase water-use efficiency in rice (Karaba et al. 2007). The transgenic plants in this study exhibited higher shoot biomass under well-irrigated conditions and more root biomass under drought stress. Water-use efficiency was also linked to changes in leaf morphology and mesophyll-cell structure that contributed to higher rates of photosynthesis. It is very likely that we will see considerably more effort in this area of research in the coming years as concerns over water availability increase.

7.18.8 Combination of Stress

Abiotic stresses often occur in combination. For example, heat and drought stress frequently occur simultaneously. Mittler (2006) has emphasized the importance of studying plant responses to combinations of stresses rather than stresses in isolation, since a plant's response and coping mechanism may differ according to which stresses or combination of stresses it is experiencing. Some stress protection mechanisms appear to confer tolerance of multiple stresses, for example, through effects on energy balance or detoxification of reactive oxygen species generated upon exposure to stress. Downregulation of poly (ADP-ribose) polymerase in *Arabidopsis* and canola increased tolerance of heat, drought, and high light (De Block et al. 2005). While this was originally believed to be due to reduced NAD^+ depletion and reduced ATP consumption, subsequent research indicated that reducing poly (ADP-ribose) polymerase expression also induced abscisic acid production, which, in turn, led to induction of a wider range of stress tolerance genes (Vanderauwera et al. 2007). This may provide a common mechanism of tolerance of multiple stresses. In a similar way, different stresses that result in oxygen-radical production (e.g., ozone, high temperature) may be mitigated by common mechanisms that detoxify these radicals.

Soil water deficit, salinity, and temperature are the most complex stress tolerances to breed for as the type (combinations of heat and drought or sodicity and salinity), timing in relation to plant growth stage, and intensity of stress can all vary considerably. There is no unified abiotic stress resistance mechanism for drought at the level of the whole plant or the single gene (Blum 2004). The traits associated with avoidance and tolerance can be constitutive (differing between genotypes) or adaptive (vary with the stage of the life cycle). Drought avoidance and drought tolerance involve different mechanisms and processes, and phenology is the single most important factor influencing whether a plant avoids drought. Drought stress is highly variable in its timing, duration, and severity, and this results in high environmental variation and GxE variation. The whole plant response to stress is complex because it is determined by component traits that interact and differ in their individual responses to the intensity and duration of water deficits and temperature. The use of managed stress environments can be very effective in breeding for drought tolerance; however, it is important to apply sufficient drought stress intensity to maximize $G \times E$ (Bänziger et al. 2006).

Clearly, stress tolerance is complex, and no single approach can provide a solution to the multiple environmental stresses that a plant might experience during its life cycle. Further evaluation of novel germplasm and putative stress-related genes under field conditions will be required to develop a robust set of new germplasm better adapted to withstand the stresses a plant may be exposed to.

7.19 Engineering for Cold Stress Tolerance in Crop Plants

Abiotic stresses adversely affect growth and productivity and trigger a series of morphological, physiological, biochemical, and molecular changes in plants. Cold stress is a major environmental factor that limits the agricultural productivity of plants in hilly areas. Plants respond and adapt to this stress to survive under stress conditions at the molecular and cellular levels as well as at the physiological and biochemical levels. However, expression of a variety of genes is induced by different stresses in diverse plants (Sanghera et al. 2011).

Low temperature often affects plant growth and crop productivity, which causes significant crop losses (Xin and Browse 2001). Plants differ in their tolerance to chilling (0–15 °C) and freezing (<0 °C) temperatures. In general, plants from temperate climatic regions are considered to be chilling-tolerant with variable degree, and can increase their freezing tolerance by being exposed to chilling, non-freezing temperatures, a process known as cold acclimation (Levitt 1980), which is associated with biochemical and physiological changes (Shinozaki and Yamaguchi-Shinozaki 1996; Thomashow 1998; Gilmour et al. 2000) and ultimately showed marked changes in gene expression, biomembrane lipid composition, and small molecule accumulation (Yamaguchi-Shinozaki and Shinozaki 2006). Besides, plants of tropical and subtropical origins are sensitive to chilling stress and lack the mechanism of cold acclimation. Low-temperature resistance in plants is a very complex trait, involving many different metabolic pathways and cell compartments (Hannah et al. 2005). Conventional breeding methods have met with limited success in improving the cold tolerance of important crop plants involving interspecific or intergeneric hybridization. Besides, in vitro-induced variations have also been applied to improve the abiotic stress tolerance of various crop plants but without much success. The conventional breeding approaches are limited by the complexity of stress tolerance traits, low genetic variance of yield components under stress condition, and lack of efficient selection criteria. It is important, therefore, to look for alternative strategies to develop cold stress-tolerant crops.

Low temperature can greatly restrict the growth and development of rice. The rice seedlings show growth retardation and lamina wrap, and part of blade even died under the condition of low temperature. In order to get more information about cold stress responses in rice, two-dimensional electrophoresis and bioinformatics analysis of mass spectrometry were used to preliminary survey the cold tolerance of cold-sensitive and cold resistance variety under cold stress (Ji et al. 2017). With analysis of bioinformation, the proteins were found involve in many aspects of rice development. The largest category of proteins is functioning on metabolism. By comparing the proteins from the two varieties, it was revealed that most proteins from cold-sensitive were downregulated and were upregulated in cold resistance genotypes. The results showed that the membrane composition and structure were damaged, metabolism changed dramatically, and rice defense system was activated under the cold stimulation. Fifty-nine proteins related to the resistance of cold stress were identified in our study, and we have investigated and classified all of their biological functions. The importance of our study are providing some conduct for the research of rice resistant to cold stress, supporting auxiliary technique for rice varieties, and widening the search field of cold tolerance in plants.

Renaut et al. (2004) studied the effect of cold acclimation on primary metabolism of poplar (*Populus tremula* L.) and has reported that during acclimation, the evolution of soluble carbohydrate contents was followed in the leaves. Sucrose, glucose, fructose, and trehalose accumulated rapidly under chilling conditions, while raffinose content increased after 1 week at 4 °C. Proteomic analyses, by bidimensional electrophoresis, performed during this stage revealed that a large number of proteins had higher expression, while much less proteins disappeared or had a lower abundance. Among the accumulating or appearing proteins proposed, about a third presented similarities with chaperone-like proteins (heat shock proteins, chaperonins). In addition, dehydrins and other late embryogenesis abundant proteins, i.e., stress-responsive proteins, detoxifying enzymes, and proteins involved in stress signaling and transduction pathways, were also activated or newly synthesized. Finally, cold exposure induced a decrease in the candidate proteins involved in cell wall or energy production.

Biotechnology offers new strategies that can be used to develop transgenic crop plants with improved tolerance to cold stress. Rapid advance in recombinant DNA technology and development of precise and efficient gene transfer protocols have resulted in efficient transformation and generation of transgenic lines in a number of crop species (Wani et al. 2008; Gosal et al. 2009; Wani and Gosal 2011). A number of genes have been isolated and characterized that are responsive to freezing stress. Many studies have suggested that cold-regulated gene expression is critical in plants for both chilling tolerance (Hsieh et al. 2002) and cold acclimation (Knight et al. 1999; Tamminen et al. 2001). Advent of molecular tools has made it possible to select directly at the gene label without waiting for the phenotype to show up. Transgenic approach is being pursued actively throughout the world to improve traits including tolerance to biotic and abiotic stresses in a number of crops (Ashraf et al. 2008). As compared to other stresses, plant responses to cold stress are complex, so the prospects of improving cold tolerance in crops seem not to be very bright. Despite this, efforts have been made during the last two decades to generate transgenic lines of different crops, which have shown improved tolerance to cold stress. Therefore, it is important to use most appropriate tools that help in reaching the goals. The genotype designed should be better than the available ones and must reach the farmers. An attempt has been made in this article to review the various mechanisms and genes involved in cold acclimatization and the possibilities where transgenic technology has been explored for breeding cold tolerance in crop plants.

7.19.1 Role of Proteomics in Crop Stress Management

Plant acclimation to stress is associated with profound changes in proteome composition. Since proteins are directly involved in plant stress response, proteomics studies can significantly contribute to unravel the possible relationships between protein abundance and plant stress acclimation. In this review, proteomics studies dealing with plant response to a broad range of abiotic stress factors—cold, heat, drought, waterlogging, salinity, ozone treatment, hypoxia and anoxia, herbicide treatments, inadequate or excessive light conditions, disbalances in mineral nutrition, enhanced concentrations of heavy metals, radioactivity, and mechanical wounding—are discussed. Most studies have been carried out on model plants *Arabidopsis thaliana* and rice due to large protein sequence databases available; however, the variety of plant species used for proteomics analyses is rapidly increasing. Protein response pathways shared by different plant species under various stress conditions (glycolytic pathway, enzymes of ascorbate-glutathione cycle, accumulation of late embryogenesis abundant proteins) as well as pathways unique to a given stress are discussed. Results from proteomics studies are interpreted with respect to physiological factors determining plant stress response (Kosova et al. 2011).

Of various abiotic stresses, drought, salinity, temperature (freezing/heat), light intensity, and heavy metal contamination are the most prevalent that considerably retard not only plant production but also the quality of crops (Ahmad et al. 2008, 2010, 2012; Ashraf et al. 2011; Ashraf 2014; Qadir et al. 2014). Stress as it is understood today is a factor that alters normal functioning of a number of mechanisms in an organism. During a stress, these mechanisms are upregulated at various levels of molecular, morphological, and physio-biochemical responses. As the stress comes under control, homeostasis is reestablished. In plants under stress conditions, signaling of kinase cascades, ion channels, accumulation of reactive oxygen species, and hormones are activated (Ahmad et al. 2010, 2012, 2013; Ashraf 2014; Rejeb et al. 2014; Ziogas et al. 2015; Molassiotis et al. 2016).

Temperature is a main environmental factor which affects growth, productivity, and distribution of plants (Shah et al. 2011; Bita and Gerats 2013; Zia et al. 2014). The phenomenon of exposing plants to temperatures from 0 to 15 °C (non-freezing temperatures) is called cold stress or chilling stress (Renaut et al. 2004). The increased freezing tolerance by plants under low temperature is called cold acclimation (Renaut et al. 2004; Timperio et al. 2008; Kosova et al. 2011; Miura and Furumoto 2013). It includes different changes in protein and gene expression as well as metabolites (Renaut et al. 2006; Miura and Furumoto 2013). Kawamura and Uemura (2003) studied plasma membrane proteome of *Arabidopsis* under low temperature. They detected 38 proteins, 27 of which were soluble, whereas 15 insoluble. Imin et al. (2004) reported 70 proteins including 12 new (47 up- and 11 downregulated) at pH between 4.0 and 7.0 in rice anthers exposed to cold stress (12 °C) for 48 h. A proteomic work on mitochondria of pea (*Pisum sativum*) is reported by Taylor et al. (2005). Twenty out of 33 proteins appear in response to cold stress at 4 °C for 1.5 days. Cheng et al. (2009) in their study on proteomics of soybean seeds subjected to cold stress (4 °C) reported 40 proteins (25 up- and 15 downregulated). These proteins are involved in cell growth/division, storage, defense of cell, energy protein synthesis, transcription, and transport (Ahmad et al. 2016).

LB-a is one of the freezing stress-responsive low abundant proteins and identified as Hsp70 (heat shock protein of 70 kD) which decreased in rice plants exposed to low-temperature stress (Hashimoto and Komatsu 2007; Bashir et al. 2010). This decrease in LB-a might be attributed to cold-induced chloroplast degradation (Bashir et al. 2010). The proteomic analysis of plants under low temperature indicated an increase in enzymes involved in reactive oxygen species scavenging, e.g., Imin et al. (2004) reported high accumulation of different isoforms of ascorbate peroxidase in trinucleate rice pollen under cold stress. Degand et al. (2009) showed that Cu/Zn-superoxide dismutase abundantly increased in chicory roots under cold stress. While Kosmala et al. (2009) detected an enhancement in enzymes which take part in ascorbate and glutathione metabolism in *Festuca pratensis* plants.

Freezing temperature induces desiccation, and imbalances plasma membranes leading to formation of inverted hexagonal phase membrane structure (Sung et al. 2003; Timperio et al. 2008). Timperio et al. (2008) reported that anti-freezing proteins play a significant role in maintaining plant growth against freezing injury. Anti-freezing proteins are similar to the pathogenesis-related protein involved in eliminating freezing stress and inducing disease resistance (Timperio et al. 2008). Winter cereal such as winter rye and wheat accumulated anti-freezing proteins in their apoplast to tolerate freezing stress (Marentes et al. 1993). Xu and Huang (2012) reported a significant increase in relative abundance of antioxidant-related proteins in low-temperature-tolerant wheat compared to those in sensitive. They also reported that carbohydrate metabolism related proteins. Xuan et al. (2013) studied the proteomics of *Zoysia japonica* and *Zoysia matrella* under cold stress. They showed that 700 proteins were resolved on 2-DE gels, but only 70 proteins were considerably overexpressed. They suggested that of all identified proteins, 45 proteins were participated in cellular metabolic processes. The cold-responsive proteins have been associated with the biosynthesis of carbohydrates, proteins, and nucleotides, reactive oxygen species scavenging, proteolysis, protein folding, and energy storage.

Proteome analysis performed in the vegetal kingdom has provided new dimensions to assess the changes in protein types and their expression levels under abiotic stress. As reported in this review, specific and novel proteins, protein-protein interactions, and posttranslational modifications have been identified, which play a role in signal transduction, antioxidative defense, anti-freezing, heat shock, metal binding, etc. However, besides specific protein production, plants respond to various stresses in a similar manner by producing heat shock proteins, indicating a similarity in the plant's adaptive mechanisms; in plants, more than in animals, heat shock proteins protect cells against many stresses. A relationship between reactive oxygen species and heat shock proteins also seems to exist, corroborating the hypothesis that during the course of evolution, plants were able to achieve a high degree of control over reactive oxygen species toxicity and are now using reactive oxygen species as signaling molecules to induce heat shock proteins (Timperio et al. 2008).

7.19.2 Chilling Acclimation

Tropical plants can be injured by above-zero chilling temperature; chilling-injured leaves may become purple or reddish and in some cases wilt (Costello et al. 2003). Chilling responses of tropical plants have been documented: an oxidative signaling regulatory network triggers an early response to chilling stress in japonica rice (Yun et al. 2010), and exogenous abscisic acid can induce freezing tolerance in chilling-sensitive rice seedlings (Shinkawa et al. 2013). Further, global expression profiling of chilling-induced genes in the chilling-tolerant japonica rice suggests a role of abscisic acid signaling in chilling tolerance (Chawade et al. 2013).

Chilling response and chilling acclimation may alter gene regulatory circuitry (Sunkar et al. 2007). However, the scale and mechanism of reprogramming of gene regulation as well as the molecular components involved remain to be investigated in important tropical crops, such as cassava. Moreover, although small noncoding RNAs (sncRNAs), particularly microRNAs (miRNAs), have been recognized as essential posttranscriptional gene regulators in plant development and stress responses (Sunkar et al. 2007; Sunkar and Zhu 2004; Zhu et al. 2008), their functions in chilling acclimation have not yet been well documented. Cassava (*Manihot* esculenta) is remarkably tolerant to drought and low-fertility soils. However, it is sensitive to low temperature, and chilling injury often occurs in spring planting and autumn harvest seasons. In this genome-wide study, we analyzed transcriptome variations of cassava plants in response to chilling and during chilling acclimation, aiming at elucidating gene regulatory networks underlying chilling acclimation.

References

Abbate PE, Andrade DH, Lazaro L, Baraitti JH, Beradocco HG, Inza VH, Marturano F (1998) Grain yield in recent Argentine wheat cultivars. Crop Sci 38(5):1203–1209

Abeledo LG, Calderini DF, Slafer GA, Slafer GA, Molina-Cano JL, Savin R, Araus JL (2002) Physiological changes associated with genetic improvement of grain yield in barley. In: Barley science: recent advances from molecular biology to agronomy of yield and quality. Food Product Press, New York, pp 361–386

Abeledo LG, Calderini DF, Slafer GA (2003) Genetic improvement of yield responsiveness to nitrogen fertilization and its physiological determinants in barley. Euphytica 133(3):291–298

Abeledo LG, Savin R, Slafer GA (2008) Wheat productivity in the Mediterranean Ebro Valley: analyzing the gap between attainable and potential yield with a simulation model. Eur J Agron 28(4):541–550

Acevedo E, Silva P, Silva H (2002) Wheat growth and physiology. In: Bread wheat: improvement and production, Ed. B.C. Curtis, FAO Plant Production and Protection Series No. 30. Rome, Italy, p 567

Acreche M, Slafer GA (2006) Grain weight response to increases in number of grains in wheat in a Mediterranean area. Field Crops Res 98(1):52–59

Acreche M, Briceno-Felix G, Martín Sóánchez JA, Slafer GA (2007) Radiation use efficiency and photosynthesis rate as affected by sink-strength in Mediterranean wheat. In: Proceedings of the 17th XVII conference of the spanish society of plant physiology, Alcalá de Henares

Acreche MM, Briceno-Félix G, Martín Sánchez JA, Slafer GA (2008) Physiological bases of genetic gains in Mediterranean bread wheat yield in Spain. Eur J Agron 28(3):162–170

Acreche M, Slafer GA (2009) Grain weight, radiation interception and use efficiency as affected by sink-strength in Mediterranean wheats released from 1940 to 2005. Field Crops Res 110(2): 98–105

Adcock D (2006) Soil water and nitrogen dynamics of farming systems on the upper Eyre Peninsula, South Australia. PhD, University of Adelaide, Adelaide

Adger WN, Subak S (1996) Estimating above-ground carbon fluxes from UK agricultural land. Geogr J 162(2):191–204

Adisarwanto T, Knight R (1997) Effect of sowing date and plant density on yield and yield components in the faba bean. Aust J Agric Res 48(8):1161–1168

Agele SO, Olabomi A (2010) Growth, fruit set efficiency and incidence of Blossom End Rot (BER) in *Capsicum* species grown in the rainy and dry seasons in a humid zone of Nigeria. Appl Trop Agric 15:112–124

Agele SO, Agbi MA (2012) Growth and yield responses of selected cowpea (*Vigna unguiculata* L. Walp) cultivars to weather and soil factors of the sowing seasons. Res Rev Bio Sci 7:265–276

Agele SO, Oyewusi IK, Aiyelari OP, Famuwagun IB (2017) Dry matter production, biomass partitioning and seed setting efficiencies in early- and late-rainy season cowpea in the rainforest agroecology of South-West Nigeria. Int J Bot 13:115–125

Ahmadi A, Baker DA (2001) The effect of water stress on the sucrose to starch pathway. Plant Growth Regul 35:81–91

Ahmad P, Sarwat M, Sharma S (2008) Reactive oxygen species, antioxidants and signaling in plants. J Plant Biol 51(3):167–173

Ahmad P, Jaleel CA, Salem MA, Nabi G, Sharma S (2010) Roles of enzymatic and non-enzymatic antioxidants in plants during abiotic stress. Crit Rev Biotechnol 30(3):161–175

Ahmad P, Ashraf M, Younis M, Hu X, Kumar A, Akram NA, Al-Qurainy F (2012) Role of transgenic plants in agriculture and biopharming. Biotechnol Adv 30(3):524–540

Ahmad F, Rafii MY, Ismail MR, Juraimi AS, Rahim HA, Asfaliza R, Latif MA (2013) Waterlogging tolerance of crops: breeding, mechanism of tolerance, molecular approaches, and future prospects. Biomed Res Int 2013:963525

Ahmad P, Abdel Latef AAH, Rasool S, Akram NA, Ashraf M, Gucel S (2016) Role of Proteomics in crop stress tolerance. Front Plant Sci 7(8):1336

Ainsworth EA, Ort DR (2010) How do we improve crop production in a warming world? Plant Physiol 154(2):526–530

Ainsworth EA, Rogers A, Leakey ADB (2008a) Targets for crop biotechnology in a future high-CO2 and high-O2 world. Plant Physiol 147(1):13–19

Ainsworth EA, Beier C, Calfapietra C, Ceulemans R, Durand-Tardif M, Farquhar GD, Godbold DL, Hendrey GR, Hickler T, Kaduk J, Karnosky DF, Kimball BA, Körner C, Koornneef M, Lafarge T, Leakey ADB, Lewin KF, Long SP, Manderscheid R, McNeill DL, Mies TA, Miglietta F, Morgan JA, Nagy J, Norby RJ, Norton RM, Percy KE, Rogers A, Soussana JF, Stitt M, Weigel HJ, White JW (2008b) Next generation of elevated CO2 experiments with crops: a critical investment for feeding the future world. Plant Cell Environ 31(9):1317–1324

Ajetomobi J (2016) Effects of weather extremes on crop yields in Nigeria. Afr J Food Agric Nutri Develop 16(4):11168–11184

Al-Issawi M, Rihan HZ, El-Sarkassy N, Fuller MP (2013) Frost hardiness expression and characterisation in wheat at ear emergence. J Agron Crop Sci 199:66–74

Alt AS, Linfsay AJ, Sulc RM, Lindsay LE (2020) Effect of temperature on survival and yield components of field-acclimated soft red winter wheat. Crop Sci 60:475–484

Ali MG, Naylor REL, Matthews S (2006) Distinguishing the effects of genotype and seed physiological age on low temperature tolerance of rice (*Oryza sativa* L.). Exp Agric 42(3): 337–349

Allen DJ, Ort DR (2001) Impacts of chilling temperatures on photosynthesis in warm-climate plants. Trends Plant Sci 6:36–42

Allen HM, Pumpa JK, Batten GD (2001) Effect of frost on the quality of samples of Janz wheat. Aust J Exp Agric 41:641–647

Alvarado R, Hernaiz S (2007) Antecedentes generals sobre el arroz en Chile. In: Alvarado R (ed) Arroz Manejo Tecnológico, vol 162. Boletín INIA, p 179

Álvaro F, Isidro J, Villegas D, García del Moral LF, Royo C (2008a) Old and modern durum wheat varieties from Italy and Spain differ in spike components. Field Crop Res 106(1):86–93

Álvaro F, Isidro J, Villegas D, García del Moral LF, Royo C (2008b) Breeding effects on grain filling, biomass partitioning and remobilization in Mediterranean durum wheat. Agron J 100(2): 361–370

Amanullah, Imanullah (2016) Dry matter partitioning and harvest index differ in rice genotypes with variable rates of phosphorus and zinc nutrition. Rice Sci 23(2):78–87

Andaya VC, Mackill DJ (2003) QTLs conferring cold tolerance at the booting stage of rice. Theo Appl Genet 106(6):1084–1109

Andrade HH, Uhart SA, Arguissain GG, Ruiz RA (1992) Radiation use efficiency of maize grown in a cool area. Field Crops Res 28:345–354

Andralojc PJ, Keys AJ, Kossmann J, Parry MAJ (2002) Elucidating the biosynthesis of 2-carboxyarabinitol 1-phosphate through reduced expression of chloroplastic fructose-1,6-bisphosphate phosphatase and radiotracer studies with 14CO2. Proc Nat Acad Sci 99:4742–4747

Angadi SV, Cutforth HW, Miller PR, McConkey BG, Entz MH, Brandt SA, Volkmar KM (2000) Response of three Brassica species to high temperature stress during reproductive growth. Can J Pl Sci 80:693–701

Angus J, van Herwaarden A (2001) Increasing water use and water use efficiency in dryland wheat. Agron J 93(2):290–298

Anjum SA, Wang LC, Farooq M, Hussain M, Xue LL, Zou CM (2011) Brassinolide application improves the drought tolerance in maize through modulation of enzymatic antioxidants and leaf gas exchange. J Agron Crop Sci 197:177–185

Anonymous (1994) Survey of Indian Agriculture. The Hindu Publication

Apel K, Hirt H (2004) Reactive oxygen species: metabolism, oxidative stress, and signal transduction. Annu Rev Plant Biol 55:373–399

Arshad MS, Farooq M, Asch F, Krishna JSV, Prasad PVV, Siddique KHM (2017) Thermal stress impact reproductive development and grain yield in rice. Plant Phiol Biochem 115:57–72

Armstrong EL, Pate JS (1994a) The field pea crop in S.W. Australia I. Patterns of biomass production and photosynthetic performance in genotypes of contrasting morphology. Aust J Agric Res 45:1347–1362

Armstrong EL, Pate JS (1994b) The field pea crop in S.W. Australia II. Effects of contrasting morphology and environment on reproductive performance. Aust J Agric Res 45:1363–1378

Araus JL, Brown HR, Byrd GT, Serret MD (1991) Comparative effects of growth irradiance on photosynthesis and leaf anatomy of *Flaveria brownie* (C4 like), *Flaveria linearis* (C3C4) and their F1 hybrid. Planta 183:497–504

Araus JL, Slafer GA, Reynolds MP, Royo C (2002) Plant breeding and water relations in C3 cereals: what should we breed for? Ann Bot 89:925–940

Araus JL, Slafer GA, Reynolds MP, Royo C (2003) Physiology of yield and adaptation in wheat and barley breeding. In: Nguyen HT, Blum A (eds) Physiology and biotechnology integration for plant breeding. Part physiological basis of yield and environmental adaptation. Marcel Dekker, Inc., New York, Basel, pp 1–49

Araus JL (2004) The problem of sustainable water use in the Mediterranean and research requirements for agriculture. Ann Appl Biol 144:229–272

Aroca R, Vernieri P, Irigoyen JJ, Sánchez-díaz M, Tognoni F, Pardossi A (2003) Involvement of abscisic acid in leaf and root of maize (Zea mays L.) in avoiding chilling induced water stress. Plant Sci 165:671–679

Ashraf M, Athar HR, Harris PJC, Kwon TR (2008) Some prospective strategies for improving crop salt tolerance. Adv Agron 97:45–110

Ashraf M, Akram NA, Al-Qurainy F, Foolad MR (2011) Drought tolerance: roles of organic osmolytes, growth regulators and mineral nutrients. Adv Agron 111:249–296

Ashraf M (2014) Stress-induced changes in wheat grain composition and quality. Crit Rev Food Sci Nutr 54(12):1576–1583

Asseng S, van Herwaarden AF (2003) Analysis of the benefits to wheat yield from assimilates stored prior to grain filling in a range of environments. Plant Soil 256(1):217–219

Asseng S, Foster IA, Turner NC (2011) The impact of temperature variability on wheat yields. Glob Chang Biol 17:997–1012

Asthir B, Koundal A, Bains NS (2012) Putrescine modulates antioxidant defense response in wheat under high temperature stress. Biol Plant 56:757–761

Atkins C, Pigeaire A (1993) Application of cytokinins to flowers to increase pod set in *Lupinus angustifolius* L. Aust J Agric Res 44:1799–1819

Austin RB (1980) Physiological limitations to cereals yields and ways of reducing them by breeding. In: Hurd RG, Biscoe PV, Dennis C (eds) Opportunities for increasing crop yields. Association of Applied Biology, Pitman, Boston, pp 3–19

Austin RB, Ford MA, Morgan CL (1989) Genetic improvement in the yield of winter wheat: a further evaluation. J Agr Sci 112:295–301

Awasthi R, Kaushal N, Vadez V, Turner NC, Berger J, Siddique KHM, Nayyar H (2014) Individual and combined effects of transient drought and heat stress on carbon assimilation and seed filling in chickpea. Funct Plant Biol 41(11):1148–1167

Baker HG, Allen LH Jr, Boote KJ, Jones P, Jones JW (1989) Response of soybean to air temperature and carbon dioxide concentration. Crop Sci 29(1):98–105

Balbuena TS, Salas JJ, Martínez-Force E, Garcés R, Thelen JJ (2011) Proteome analysis of cold acclimation in sunflower. J Proteome Res 10:2330–2346

Ball RA, Purcell LC, Vories ED (2000) Short-season soybean yield compensation in response to population and water regime. Crop Sci 40(4):1070–1078

Balodis O, Gaile Z (2016) Sowing date and rate effect on winter oilseed rape (*Brassica napus* L.) yield components' formation. Proc Latv Acad Sci Sect B 70(6):384–392

Bange M, Hammer G, Rickert K (1977) Environmental control of potential yield of sunflower in the subtropics. Aust J Agric Res 48(2):231–240

Bänziger M, Setimela PS, Hodson D, Vivek B (2006) Breeding for improved abiotic stress tolerance in maize adapted to southern Africa. Agric Water Manag 80(1-3):212–224

Barlow KM, Christy BP, O'Leary GJ, Riffkin PA, Nuttall JG (2015) Simulating the impact of extreme heat and frost events on wheat crop production: a review. Field Crops Res 171:109–119

Barnabás B, Jäger K, Fehér A (2008) The effect of drought and heat stress on reproductive processes in cereals. Plant Cell Environ 31:11–38

Baruah AR, Ishigo-Oka N, Adachi M, Oguma Y, Tokizono Y, Onishi K, Sano Y (2009) Cold tolerance at the early growth stage in wild and cultivated rice. Euphytica 165(3):459–470

Bashir H, Qureshi MI, Muneer S, Ahmad J, Zolla L (2010) Proteomic approaches to map thylakoid proteins and study differential protein expression under various abiotic stresses. In: Proceedings of International Conference of Biology, Biochemistry and Biotechnology (ICBBB), World Academy of Science, Engineering & Technology, Rome, pp 28–30

Beck EH, Fettiq S, Knake C, Hartiq K, Bhattarai T (2007) Specific and unspecific responses of plants to cold and drought stress. J Biosci 32(3):501–510

Behboudian MH, Ma Q, Turner NC, Palta JA (2001) Reactions of chickpea to water stress: yield and seed composition. J Sci Food Agric 81:1288–1291

Belder P, Bouman BAM, Spiertz JHJ (2007) Exploring options for water saving in lowland rice using a modeling approach. Agric Syst 92(1-3):91–114

Bell MJ, Wright GC, Hammer GL (1992) Night temperature affects radiation-use efficiency in peanut. Crop Sci 32(6):1329–1335

Bell MJ, Wright GC (1998) Groundnut growth and development in contrasting environments. 1. Growth and plant density responses. Exp Agric 34(1):99–112

Bell LW, Lilley JM, Hunt JR, Kirkegaard JA (2015) Optimising grain yield and grazing potential of crops across Australia's high-rainfall zone: a simulation analysis. 1. Wheat. Crop Pasture Sci 66(3):332–348

Bender RR, Haegele JW, Below FE (2015) Nutrient uptake, partitioning, and remobilization in modern soybean varieties. Agron J 107(2):563–573

Bänziger M, Edmeades GO, Beck D, Bellon M (2000) Breeding for drought and nitrogen stress tolerance in maize: from theory to practice. In: CIMMYT, Mexico, DF, p 68

Berger J, Turner N, Siddique KHM, Knights E, Brinsmead R, Mock I, Edmonson C, Khan T (2004) Genotype by environment studies across Australia reveal the importance of phenology for chickpea (*Cicer arietinum* L.) improvement. Aust J Agric Res 55(10):1071–1084

Berry PM, Spink JH, Foulkes MJ, Wade A (2003) Quantifying the contributions and losses of dry matter from non-surviving shoots in four cultivars of winter wheat. Field Crops Res 80(2): 111–121

Bhattacharya A, Vijaylaxmi (2010) Physiological responses of grain legumes to stress environments. In: Yadav SS, McNeil DL, Roberts R, Sharanagouda AP (eds) Climate change and management of cool season grain legume crops. Springer, Dordrecht, pp 35–86

Bhattacharya A (2018) Dry matter partitioning under abiotic stress. In: Bhattacharya A (ed) Abiotic stress and plant physiology, Productivity, vol 2. New India Publishing Agency, New Delhi, India, pp 419–476

Bhattacharya A (2019) Radiation use efficiency under different climatic conditions. In: Changing climate and resource use efficiency in plants. Elsevier, Academic Press, New York, pp 51–109

Bieleski RL, Ferguson IB (1983) Physiology and metabolism of phosphate and its compounds. In: Läuchli A, Bieleski RL (eds) Encyclopedia of plant physiology, New Series, Vol 15A. Springer-Verlag, Berlin, pp 422–449

Birch C, McLean G, Sawers A (2006) Analysis of high yielding maize production—a study based on a commercial crop. In: Humphreys E, O'Keefe K, Hutchins N, Gill R (eds) "Water to Gold", 6th Triennial Conference, Maize Association of Australia, Griffith, pp 55–65

Bishop AC, Mendham NJ (1996) Quantifying the yield-density relationship for narrow-leafed lupin (*Lupinus angustifolius*) in Tasmania. In: Asghar M (ed) 8th Australian Agronomy Conference, Australian Society of Agronomy, pp 100–103

Bita C, Gerats T (2013) Plant tolerance to high temperature in a changing environment: scientific fundamentals and production of heat stress tolerant crops. Front Plant Sci 4:273

Bligny R, Gardeström P, Roby C, Douce R (1990) 31P NMR studies of spinach leaves and their chloroplasts. J Biol Chem 265:1319–1326

Blum A (2004) The physiological foundation of crop breeding for stress environments. In: Proc. World Rice Research Conf., Tsukuba, Japan, November 2004, IRRI, Manila, The Philippines, pp 456–458

Boer R, Campbell LC, Fletcher DJ (1993) Characteristics of frost in a major wheat-growing region of Australia. Aust J Agric Res 44(8):1731–1743

Bolinder MA, Angers DA, Dubuc JP (1997) Estimating shoot to root ratios and annual carbon inputs in soils for cereal crops. Agric Ecosyst Environ 63(1):61–66

Bolinder MA, Janzen HH, Gregorich EG, Angers DA, VandenBygaart AJ (2007) An approach for estimating net primary productivity and annual carbon inputs to soil for common agricultural crops in Canada. Agric Ecosyst Environ 118(1-4):29–42

Bondeau A, Smith PC, Zaehle S, Schaphoff S, Lucht W, Cramer W, Gerten D, Lotze-Campen H, Muller C, Reichstein M, Smith B (2007) Modelling the role of agriculture for the 20th century global terrestrial carbon balance. Glob Change Biol 13(3):679–706

Boote KJ, Aleen LH, Prasad PVV, Baker JT, Gesch RW, Snyder AM, Pan D, Thomas JMG (2005) Elevated temperature and carbon dioxide impacts on pollination, reproductive growth, and yield of several globally important crops. J Agric Metero 60(5):469–474

Borras L, Slafer GA, Otegui ME (2004) Seed dry weight response to source–sink manipulations in wheat, maize and soybean: a quantitative reappraisal. Field Crop Res 86(2-3):131–146

Böttcher U, Rampin E, Hartmann K, Zanetti F, Flenet F, Morison M, Kage H (2016) A phenological model of winter oilseed rape according to the BBCH scale. Crop Pasture Sci 67(3):345–358

Bouman B (2007) A conceptual framework for the improvement of crop water productivity AR different spatial scale. Agric Syst 9(1):43–60

Boyer IS, Westgate ME (2004) Grain yield with limited water. J Exp Bot 55(407):2385–2394

Brancourt-Hulmel M, Doussinault G, Lecomte C, B'erard P, Le Buanec B, Trottet M (2003) Genetic improvement of agronomic traits of winter wheat cultivars released in France from 1946 to 1992. Crop Sci 43:37–45

Brenner ED, Stahlberg R, Mancuso S, Vivanco J, Baluska F, Van Volkenburgh E (2006) Plant neurobiology: an integrated view of plant signaling. Trends Plant Sci 11(8):413–419

Britz SJ, Prasad PVV, Moreau RA, Allen LH Jr, Kremer DF, Boote KJ (2007) Influence of growth temperature on amounts of tocopherols, tocotrienols and γ-oryzanol in brown rice. J Agric Food Chem 55:7559–7565

Bruins M (2009) The evolution and contribution of plant breeding to global agriculture. In: Proceedings of the second world seed conference, Rome, pp 18–31

Brüggemann W, van der Kooij TAW, van Hasselt PR (1992) Long-term chilling of young tomato plants under low light and subsequent recovery. II. Chlorophyll fluorescence, carbon metabolism and activity of ribulose-1,5-diphosphate carboxylase/oxygenase. Planta 186(2):179–187

Brüggemann W, Klaucke S, Maas-Kantelm K (1994) Long-term chilling of young tomato plants under low light. V. Kinetic and molecular properties of two key enzymes of the Calvin cycle in *Lycopersicon esculentum* Mill. and *L. peruvianum* Mill. Planta 194(2):160–168

Bueno CS, Lafarge T (2009) Higher crop performance of rice hybrids tah of elite inbreds of tropics. Hybrids accumulates more biomass during each phonological phase. Field Crops Res 112(2): 229–237

Busch W, Wunderlich M, Schoffl F (2005) Identification of novel heat shock factor-dependent genes and biochemical pathways in *Arabidopsis thaliana*. The Plant J 41(1):1–14

Bushby HVA, Lawn RJ (1992) Accumulation and partitioning of dry matter by contrasting genotype of mungbean (*Vigna radiata* (L.) Wilczek). Aust J Agric Res 43:1609–1628

Byth DE, Wallis ES, Saxena KB (1981) Adaptation and breeding strategies for pigeonpea. In: Proceedings of International Workshop on Pigeonpea, ICRISAT, Hyderabad, pp 450–465

Çakir R (2004) Effect of water stress at different development stages on vegetative and reproductive growth of corn. Field Crops Res 89:1–16

Cantero-Martinez C, O'Leary G, Connor D (1995) Stubble retention and nitrogen fertilisation in a fallow-wheat rainfed cropping system. 1. Soil water and nitrogen conservation, crop growth and yield. Soil Till Res 34(2):79–94

Calderini DF, Dreccer MF, Slafer GA (1995) Genetic improvement in wheat yield and associated traits. A re-examination of previous results and the latest trends. Plant Breed 114(2):108–112

Calderini DF, Dreccer MF, Slafer GA (1997) Consequences of breeding on biomass, radiation interception and radiation use efficiency in wheat. Field Crops Res 52(3):271–281

Calderini DF, Reynolds MP, Slafer GA (1999) Genetic gains in wheat yield and main physiological changes associated with them during the 20th century. In: Satorre EH, Slafer GA (eds) Wheat: ecology and physiology of yield determination. Food Product Press, New York, pp 351–377

Campos H, Cooper M, Habben JE, Edmeades GO, Schussler JR (2004) Improving drought tolerance in maize: a view from industry. Field Crops Res 90(1):19–34

Canevara MG, Romani M, Corbellini M, Perenzin M, Borghi B (1994) Evolutionary trends in morphological, physiological, agronomical and qualitative traits of *Triticum aestivum* L. cultivars bred in Italy since 1900. Eur J Agron 3(3):175–185

Cartelle J, Pedró A, Savin R, Slafer GA (2006) Grain weight responses to post-anthesis spikelet-trimming in an old and a modern wheat under Mediterranean conditions. Eur J Agron 25(4): 365–371

Case AL, Curtis PS, Snow AA (1998) Heritable variation in stomatal response to elevated CO2 in wild radish, *Raphanus raphanistrum* (Brassicaceae). Am J Bot 85(2):253–258

Cassman KG, Dobermann A, Walters DT, Yang H (2003) Meeting cereal demand while protecting natural resources and improving environmental quality. Annu Rev Environ Resourc 28:315–358

Castrillo M (1992) Sucrose metabolism in bean plants under water deficit. J Exp Bot 43(12): 1557–1561

Cattivelli L, Rizza F, Badek F-W, Mazzucotelli E, Mastrangelo AM, Francia E, Marè C, Tondelli, Stanca AM (2008) Drought tolerance improvement in crop plants: an integrated view from breeding to genomics. Field Crop Res 115(1):1–4

Cavalieri AJ, Smith OS (1985) Grain filling and field drying of a set of maize hybrids released from 1930 to 1982. Crop Sci 25(5):856–860

Ceccarelli S, Grando S (1996) Drought as a challenge for the plant breeder. Plant Growth Regul 20(2):149–155

Ceccarelli S, Grando S, Maatougui M, Michael M, Slash M, Haghparast R, Rahmanian M, Taheri A, Al-Yassin A, Benbelkacem A, Labidi M, Mimoun H, Nachit MN (2010) Plant breeding and climate changes. J Agric Sci 148(6):627–637

Chaerle L, Saibo N, van der Straeten D (2005) Tuning the pores: towards engineering plants for improved water use efficiency. Trends in Biotechnol 23:308–315

Challinor AJ, Watson J, Lobell DB, Howden SM, Smith DR, Chhetri N (2014) A meta-analysis of crop yield under climate change and adaptation. Nat Clim Chang 4(4):287–291

Chaturvedi GS, Singh BB, Prasad R, Chauhan YS, Padmakar. (1988) Physiological analysis of yield in Indian mustard (Brassica juncea L.) under irrigated condition. Ind J Plant Physiol 31: 38–44

Charles-Edwards DA (1984) On the ordered development of plants 1. An hypothesis. Ann Bot 53(5):699–707

Chauhan YS (1987) Screening for tolerance to salinity and water logging: case study with pigeonpea and chickpea. In: Adaptation of chickpea and pigeonpea to abiotic stress: Proceedings of Consultant's Workshop, ICRISAT, Hyderabad, pp 93–103

Chawade A, Lindlöf A, Olsson B, Olsson O (2013) Global expression profiling of low temperature induced genes in the chilling tolerant japonica rice Jumli Marshi. PLoS One 8:e81729

Cheng Z, Tang Y, Chen Y, Kim S, Liu H, Li SSC, Gu W, Zhao Y (2009) Molecular characterization of propionyllysines in non-histone proteins. Mol Cell Proteomics 8(1):45–52

Chenu K, Deihimfard R, Chapman SC (2013) Large scale characterization of drought pattern: a continent-wide modelling approach applied to the Australian wheatbelt-Spatial and temporal trends. New Phytol 198(3):801–820

Cheikh N, Brenner ML (1992) Regulation of key enzymes of sucrose biosynthesis in soybean leaves. Plant Physiol 100:1230–1237

Cheero-Nayamuth FC, Robertson MJ, Wegener MK, Nayamuth ARH (2000) Using a simulation model to assess potential and attainable sugar cane yield in Mauritius. Field Crop Res 66(3): 225–243

Chinnusamy V, Zhu J, Zhu JK (2007) Cold stress regulation of gene expression in plants. Trends in Plant Sci 12(10):444–451

Clarkson DT (1976) The influence of temperature on the exudation of xylem sap from detached root systems of rye (*Secale cereale*) and barley (*Hordeum vulgare*). Planta 132(3):297–304

Clarkson DT, Earnshaw MJ, White PA, Cooper HD (1988) Temperature dependent factors influencing nutrient uptake: an analysis of responses at different levels of organization. In: Plant and Temperature: Symp. Soc. Expt. Biol. Cam, pp 281–309

Confalonieri R, Bocchi S (2005) Evaluation of cropsyst for simulating the yield of flooded rice in northern Italy. Eur J Agron 23(5):315–332

Connor D, Jones T, Palta J (1985) Response of sunflower to strategies of irrigation I. Growth, yield and the efficiency of water use. Field Crops Res 10:15–36

Cooper HD, Clarkson DT (1989) Cycling of amino nitrogen and other nutrients between shoots and roots in cereals-a possible mechanism integrating shoot and root in the regulation of nutrient uptake. J Exp Bot 40(7):753–762

Cooper M, Stucker RE, DeLacy IH, Hatch BD (1997) Wheat breeding nurseries, target environments and indirect selection for grain yield. Crop Sci 37(4):1168–1176

Costello LR, Perry EJ, Matheny NP, Henry JM, Geisel PM (2003) Abiotic disorders of landscape plants-a diagnostic guide, vol 3420. University of California Division of Agriculture and Natural Resources, pp 133–138

Cox TS, Shroyer RJ, BenHui L, Sears RG, Martin TJ (1988) Genetic improvement in agronomic traits of hard red winter wheat cultivars from 1919 to 1987. Crop Sci 28(5):756–760

Crimp S, Zheng B, Khimashia N, Gobbett D, Chapman S, Howden SM, Nicholls N (2016) Recent changes in southern Australian frost occurrence: implications for wheat production risk. Crop Pasture Sci 67(8):801–811

Cromey MG, Wright DSC, Boddington HJ (1998) Effects of frost during grain filling on wheat yield and grain structure. NZ J Crop Hortic Sci 26:279–290

Croser JS, Clarke HJ, Siddique KHM, Khan TN (2003) Low temperature stress: implications for chickpea (Cicer arietinum L.) in a short season Mediterranean type environment. Crit Rev Plant Sci 22(2):185–219

Curtis BC (2002) Wheat in the world. In: Curtis BC, Rajaram S, Gomez Macpherson H (eds) Bread wheat improvement and production. Food and Agriculture Organization, Rome, p 544

D'Andrea KE, Otegui ME, de la Vega AJ (2008) Multi-attribute responses of maize inbred lines across managed environments. Euphytica 162(3):381–394

Davies SL, Turner NC, Siddique KHM, Plummer JA, Leport L (1999) Seed growth of desi and kabuli chickpea (*Cicer arietinum* L) in a short-season Mediterranean-type of environment. Aus J Exp Agric 39(2):181–188

Davies SL, Turner NC, Palta JA, Siddique KHM, Plummer JA (2000) Remobilization of carbon and nitrogen supports seed filling in chickpea subjected to water deficit. Aust J Agric Res 51(7): 855–866

De Block M, Verduyn C, De Brouwer D, Cornelissen M (2005) Poly (ADP-ribose) polymerase in plants affects energy homeostasis, cell death and stress tolerance. Plant J 41(1):95–106

De Costa WAJM, Dennett MD, Ratnaweera U, Nyalemegbe K (1997) Effects of different water regimes on field-grown determinate and indeterminate faba bean (*Vicia faba* L.). II. Yield, yield components and harvest index. Field Crops Res 52(1-2):169–178

De Koning ANM (1994) Development and dry matter distribution in glasshouse tomato: a quantitative approach. Ph.D. Thesis, Agricultural University, Wageningen, The Netherlands
Degand H, Faber AM, Dauchot N, Mingeot D, Watillon B, Van Cutsem P, Morsomme P, Boutry M (2009) Proteomic analysis of chicory root identifies proteins typically involved in cold acclimation. Proteomics 9(10):2903–2907
Delane RJ, Matthews JC, Hamblin J (1988) Effect of genotype and environment on harvest index of L. angustifolius. In: Donald CM, Hamblin J (eds) Proceedings 5th international lupin conference, Poznan, Poland, p 527
Deligios PA, Farci R, Sulas L, Hoogenboom G, Ledda L (2013) Predicting growth and yield of winter rapeseed in a Mediterranean environment: model adaptation at a field scale. Field Crops Res 144:100–112
Demotes-Mainard S, Jeuffroy MH (2001) Partitioning of dry matter and nitrogen to spike throughout the spike growth period in wheat crops subjected to nitrogen deficiency. Field Crops Res 70(2):153–165
Ding T, Qian W, Yan Z (2010) Changes in hot days and heat waves in China during 1961-2007. Int J Climatol 30(10):1452–1462
Dornbos DLJ, Mullen RE (1991) Influence of stress during soybean seed fill on seed weight, germination and seedling growth rate. Can J Plant Sci 71:373–383
Donald CM, Hamblin J (1976) The biological yield and harvest index of cereals as agronomic and plant breeding criteria. Adv Agron 28:361–405
Donatelli M, Magarey RD, Bregaglio S, Whish JPM, Savary S (2017) Modelling the impacts of pests and diseases on agricultural systems. Agric Syst 155:213–224
Donmez E, Sears RG, Shroyer JP, Paulsen GM (2001) Genetic gain in yield attributes of winter wheat in the Great Plains. Crop Sci 41(5):1412–1419
Dracup M (1994) Physiological aspects to lupin improvement for water-limited Mediterranean environments. In: Dracup M, Palta J (eds) First Australian Lupin Technical Symposium. Agriculture Western Australia, Perth, pp 140–151
Dracup M, Reader M, Palta J (1998) Variation in yield of narrow-leafed lupin caused by terminal drought. Aust J Agric Res 49:799–810
Duman JG, Wisniewski MJ (2014) The use of antifreeze proteins for frost protection in sensitive crop plants. Environ Exp Bot 106:60–69
Duncan WG, McCloud DE, McGraw RL, Boote KJ (1978) Physiological aspects of peanut yield improvement. Crop Sci 18(6):1015–1020
Dunn BW, Beecher HG (1994) Green manuring legume pasture for aerial-sown rice. Aust J Exp Agric 34(7):967–975
Duthion C, Pigeaire A (1991) Seed lengths corresponding to the final stage in seed abortion of three grain legumes. Crop Sci 31(6):1579–1583
Duthion C, Pigeaire A (1993) Stability of nitrogen concentration of inoculated white lupins during pod development. Ann Bot 72(1):55–61
Duvick DN, Smith JCS, Cooper M (2004) Long-term selection in a commercial hybrid maize breeding program. Plant Breed Rev 24:109–151
Egli DB, Wardlaw IF (1980) Temperature response of seed growth characteristics of soybean. Agron J 72(3):560–564
Egli DB, Zhen-Wen Y (1991) Crop growth rate and seeds per unit area in soybean. Crop Sci 31(2): 439–442
Egli DB (1998) Seed biology and the yield of grains. CAB International, New York
Ehlers W, Goss M (2003) Water dynamics in plant production. CABI Publishing, Wallingford, UK, pp 297–305
El Murabaa AIM (1957) Effects of high temperature on incompatibility in radish. Euphytica 6:268–270
Erskine W, Tufail M, Russel A, Tyagi MC, Rahman MM, Saxena MC (1994) Current and future strategies in breeding lentil for resistance to abiotic and abiotic stresses. Euphytica 73(1): 127–135

Esker PD, Savary S, McRoberts N (2012) Crop loss analysis and global food supply: focusing now on required harvests. CAB Rev 7:52

Evans LT (1993) Crop evolution, adaptation and yield. Cambridge University Press, Cambridge

Farrar JF, Williams ML (1991) The effects of increased atmospheric carbon-dioxide and temperature on carbon partitioning, source-sink relations and respiration. Plant Cell Environ 14(8): 819–830

Farrell TC, Fox KM, Williams RL, Fukai S, Lewin LG (2006a) Minimising cold damage during reproductive development among temperate rice genotypes. II. Genotypic variation and flowering traits related to cold tolerance screening. Aus J Agric Res 57(1):89–100

Farooq M, Aziz T, Wahid A, Lee DJ, Siddique KHM (2009) Chilling tolerance in maize agronomic and physiological approaches. Crop and Pasture Science 60(6):510–516

Farooq M, Gogoi N, Barthakur S, Baroowa B, Bharadwaj N, Alghamdi SS, Siddiqui KHM (2017a) Drought stress in grain legumes during reproduction and grain filling. J Agron Crop Sci 203(2): 81–102

Farooq M, Nadeem F, Gogoi N, Ullah A, Alghamdi SS, Nayyar H, Siddiqui KHM (2017b) Heat stress in grain legumes during reproductive and grain-filling phases. Crop Pasture Sci 68(11): 985–1005

Farre I, Robertson M, Asseng S, French R, Dracup M (2004) Simulating lupin development, growth, and yield in a Mediterranean environment. Aust J Agric Res 55(8):863–877

Farrell TC, Fox KM, Williams RL, Fukai S, Lewin LG (2006b) Minimising cold damage during reproductive development among temperate rice genotypes. II. Genotypic variation and flowering traits related to cold tolerance screening. Aust J Agric Res 57(1):89–100

Fathi G, McDonald G, Lance R (1997) Effects of post-anthesis water stress on the yield and grain protein concentration of barley grown at two levels of nitrogen. Aust J Agric Res 4891:67–80

Fischer RA (1985) Number of kernels in wheat crops and the influence of solar radiation and temperature. J Agric Sci 105(2):447–461

Fischer RA, Stockman YM (1986) Increase kernel number in Norin 10-derived dwarf wheat: evaluation of the cause. Aus J Plant Physiol 13(6):767–784

Fischer RA, Rees D, Sayre KD, Lu Z-M, Condon AG, Larqu'e-Saavedra, A. (1998) Wheat yield progress associated with higher stomatal conductance and photosynthetic rate, and cooler canopies. Crop Sci 38(6):1467–1475

Fletcher AL, Jamieson PD (2009) Causes of variation in the rate of increase of wheat harvest index. Field Crops Res 113(3):268–273

Flinn AM, Pate JS (1970) A quantitative study of carbon transfer from pod and subtending leaf to the ripening seeds of the field pea (*Pisum arvense* L.). J Exp Bot 21(1):71–82

Flohr BM, Hunt JR, Kirkegaard JA, Evans JR, Swan A, Rheinheimer B (2018) Genetic gains in NSW wheat cultivars from 1901 to 2014 as revealed from synchronous flowering during the optimum period. Eur J Agron 98:1–13

Flood R, Martin P (2001) Nitrogen accumulation and distribution at anthesis and maturity in ten wheats grown at three sites in north-western Victoria. Aust J Exp Agric 41(4):533–540

Foulkes MJ, Sylvester-Bradley R, Weightman R, Snape J (2007) Identifying physiological traits associated with improved drought resistance in winter wheat. Field Crops Res 103(1):11–24

Foulkes MJ, Slafer GA, Davies WJ, Berry PM, Sylvester-Badley R, Martre P, Calderini DF, Griffiths S, Reynolds MP (2010) Raising yield potential of wheat. III. Optimizing partitioning to grain while maintaining lodging resistance. J Exp Bot 62(2):469–486

Foulkes MJ, Slafer GA, Davies WJ, Berry PM, Sylvester-Bradley R, Martre P, Calderini DF, Griffith S, Reynolds MP (2011) Raising yield potential of wheat. III. Optimizing partitioning to grain while maintaining lodging resistance. J Exp Bot 62(2):469–486

Fowler S, Thomashow MF (2002) Arabidopsis transcriptome profiling indicates that multiple regulatory pathways are activated during cold acclimation in addition to the CBF cold response pathway. Plant Cell 14(8):1675–1690

Frank G, Pressman E, Ophir R, Althan L, Shaked R, Freedman M, Shen S, Firon N (2009) Transcriptional profiling of maturing tomato (*Solanum lycopersicum* L.) microspores reveals

the involvement of heat shock proteins, ROS scavengers, hormones, and sugars in the heat stress response. J Exp Bot 60(13):3891–3908
French RJ, Ewing MA (1989) Soil type influences the relative yields of different cereals and crop legumes in the Western Australian wheatbelt. Aust J Exp Agric 29(6):829–835
Frederick JR, Woolley JT, Hesketh JD, Peters DB (1991) Seed yield and agronomic traits of old and modern soybean cultivars under irrigation and soil water-deficit. Field Crops Res 27(1-2):71–82
French RJ, Turner NC (1991) Water deficit changes dry matter partitioning and seed yield in narrow leaved lupin (*Lupinus angustifolius* L.). Aust J Agric Res 42(3):471–484
Frenck G, van der Linden L, Mikkelsen TN, Brix H, Jorgensen RB (2011) Increased [CO2] does not compensate for negative effects on yield caused by higher temperatures and [O3] in *Brassica napus* L. Eur J Agron 35(3):127–134
Fujino K, Sekiguchi H, Sato T, Kiuchi K, Nonoue Y, Takeuchi Y, Ando T, Lim SY, Yano M (2004) Mapping of quantitative trait loci controlling low temperature germiability in rice (*Oryza* sativa L.). Theor Appl Genet 108(5):794–799
Fuller MP, Fuller AM, Kaniouras S, Christophers J, Fredericks T (2007) The freezing characteristics of wheat at ear emergence. Eu J Agron 26:435–441
Gabrielle B, Denoroy P, Gosse G, Justes E, Andersen MN (1998) A model of leaf area development and senescence of winter oilseed rape. Field Crops Res 57(2):209–222
Galmes J, Flexas J, Keys AJ, Cifre J, Mitchell RAC, Madgwick PJ, Haslam RP, Medrano H, Parry MAJ (2005) Rubisco specificity factor tends to be larger in plant species from drier habitats and with persistent leaves. Plant Cell Environ 28(5):571–579
Galtier N, Foyer CH, Huber J, Voelker TA, Huber SC (1993) Effect of elevated sucrose phosphate synthase activity on photosynthesis, assimilate partitioning, and growth in tomato (*Lycopersicon esculentum* vai-UC82B). Plant Physiol 101(2):535–543
Gan S, Amasino RM (1997) Making sense of senescence. Plant Physiol 113(2):313–319
Gan Y, Angadi SV, Cutforth HW, Potts D, Angadi VV, Mc Donald CL (2004) Canola and mustard response to short period of high temperature and water stress at different developmental stages. Can J Plant Sci 84(3):697–704
Ganguly SB, Bhattacharya A (2000) Harvest index in late sown chickpea: effect of physio-chemical traits in developing grains. Legume Res 23(4):222–226
García del Moral LF, Roca LF, Ramdani A, Rharrabti Y, Villegas D, Royo C (2002) Genetic improvement effects on durum wheat yield and yield components in the Mediterranean region. In: Villalobos FJ, Testi L (eds) Proceedings VII congress of the European Society for Agronomy, C'ordoba, Spain, pp 91–92
Gasper AP, Laboski CAM, Naeve SN, Conley SL (2017) Dry matter and nitrogen uptake, partitioning, and removal across a wide range of soybean seed yield levels. Crop Sci 57(4): 2170–2182
Geiger DR, Shieh W-J (1988) Analysing partitioning of recently fixed and of reserve carbon in reproductive *Phaseolus vulgaris*, L. plants. Plant Cell Environ 11(8):777–783
Geiger DR, Shieh WJ, Saluke RM (1989) Carbon partitioning among leaves, fruits and seeds during development of *Phaseolus vulgaris* L. Plant Physiol 91(1):291–297
Geiger DR, Servaites JC (1991) Allocation of recently fixed and reserve carbon in relation to stress. In: Mooney HA, Winner W, Pell EJ (eds) Integrated response of plants to stress. Academic Press, New York, pp 103–127
Geiger DR, Servaites JC, Shieh W-J (1992) Balance in the source-sink system: a factor in crop productivity. In: Baker NR, Thomas H (eds) Crop photosynthesis: spatial and temporal determinants. Elsevier Science Publishers BV, Amsterdam, pp 155–176
Geiger DR, Servaites JC (1994) Dynamics of self-regulation of photosynthetic carbon metabolism. Plant Physiol Biochem 32:173–183
Ghadirnezhad R, Fallah A (2014) Temperature effect on yield and yield components of different rice cultivars in flowering stage. Int J Agron 2014:1–4

Ghiglione HO, Gonzalez FG, Serrago R, Maldonado SB, Chilcott C, Curá JA, Miralles DJ, Zhu T, Casal JJ (2008) Autophagy regulated by day length determines the number of fertile florets in wheat. Plant J 55(6):1010–1024

Gilmour SJ, Sebolt AM, Salazar MP, Everard JD, Thomashow MF (2000) Overexpression of the *Arabidopsis* CBF3 transcriptional activator mimics multiple biochemical changes associated with cold acclimation. Plant Physiol 124(4):1854–1865

Gibson LR, Mullen RE (1996) Soybean seed quality reductions by high day and night temperature. Crop Sci 36(6):1615–1619

Gillett AG, Crout NMJ, Stokes DT, Sylvester-Bradley R, Scott RK (2001) An approach to modelling the effect of environmental and physiological factors upon biomass accumulation in winter wheat. J Agric Sci 136:369–381

Godwin DC, Meyer WS, Singh U (1994) Simulation of the effect of chilling injury and nitrogen supply on floret fertility and yield in rice. Aust J Exp Agric 34(7):921–926

Goetz M, Godt DE, Guivarc'h, A., Kahmann, U., Chriqui, D. and Roitsch, T. (2001) Induction of male sterility in plants by metabolic engineering of the carbohydrate supply. Proc Natl Acad Sci U S A 98(11):6522–6527

Gogoi N, Farooq M, Barthakur S, Baroowa B, Paul S, Bharadwaj N, Ramanjulu S (2018) Thermal stress impacts on reproductive development and grain yield in grain legumes. J. Plant Biol 61(5):265–291

Gooding MJ, Ellis RH, Shewry PR, Schofield JD (2003) Effects of restricted water availability and increased temperature on the grain filling, drying and quality of winter wheat. J Cereal Sci 37(3): 295–309

Gomez-Macpherson H, Richards RA (1995) Effect of sowing time on yield and agronomic characteristics of wheat in south-eastern Australia. Aust J Agric Res 46(7):1381–1399

González FG, Slafer GA, Miralles DF (2002) Vernalization and photoperiod responses in wheat reproductive phases. Field Crops Res 74(2-3):183–195

González FG, Slafer GA, Miralles DJ (2003a) Floret development and spike growth as affected by photoperiod during stem elongation in wheat. Field Crops Res 81(1):29–38

González FG, Slafer GA, Miralles DJ (2003b) Grain and floret number in response to photoperiod during stem elongation in fully and slightly vernalised wheats. Field Crops Res 81(1):17–27

González FG, Slafer GA, Miralles DJ (2005a) Pre-anthesis development and number of fertile florets in wheat as affected by photoperiod sensitivity genes *Ppd-D1* and *Ppd-B1*. Euphytica 146(3):253–269

González FG, Slafer GA, Miralles DJ (2005b) Photoperiod during stem elongation in wheat: is its impact on fertile floret and grain number determination similar to that of radiation? Funct Plant Biol 32(3):181–188

González FG, Slafer GA, Miralles DJ (2005c) Floret development and survival in wheat plants exposed to contrasting photoperiod and radiation environments during stem elongation. Funct Plant Biol 32(3):189–197

Gosal SS, Wani SH, Kang MS (2009) Biotechnology and drought tolerance. J Crop Improv 23(1): 19–54

Gourdji SM, Sibley AM, Lobell DB (2013) Global crop exposure to critical high temperatures in the reproductive period: historical trends and future projections. Environ Res Lett 8:Article 024041

Goyne P, Milroy S, Lilley J, Hare J (1993) Radiation interception, radiation use efficiency and growth of barley cultivars. Aust J Agric Res 44(6):1351–1366

Graham D, Patterson BD (1982) Responses of plants to low, nonfreezing temperatures: proteins, metabolism and acclimation. Annu Rev Plant Physiol 33:347–372

Gray D, Steckel JRA, Dearman J, Brocklehurst PA (1988) Some effects of temperature during seed development on carrot (*Daucus carota*) seed growth and quality. Ann Appl Biol 112(2): 367–376

Gregersen PL, Culetic A, Boschian L, Krupinska K (2013) Plant senescence and crop productivity. Plant Mol Biol 82(6):603–622

Gregory PJ, Palta JA, Batts GR (1996) Root systems and root: mass ratio-carbon allocation under current and projected atmospheric conditions in arable crops. Plant Soil 187(2):221–228

Grimm NB, Chapin FS, Bierwagen B, Gonzalez P, Groffman PM, Luo Y, Schimel J (2013) The impacts of climate change on ecosystem structure and function. Front Ecol Environ 11(9): 474–482

Gu L, Hanson PJ, Post WM, Kaiser DP, Yang B, Nemani R, Pattardy SG, Meyers T (2008) The 2007 eastern US spring freeze: increased cold damage in a warming world? Biosci 58(3): 253–262

Guo QF, Wang QC, Wang LM (2004) Maize production in China. Shanghai Science & Technology Press, Shanghai, pp 117–167

Gutam S, Nath V, Srivastava GC (2008) Endogenous hormonal content during grain development in hexapoloid and tetraploid wheat. Bangladesh J Agric Sci 33(3):493–502

Guy CL, Huber JLA, Huber SC (1992) Sucrose phosphate synthase and sucrose accumulation at low temperature. Plant Physiol 100(1):502–508

Hadley P, Roberts EH, Summerfield RJ, Minchin FR (1983) A quantitative model of reproductive development in cowpea [*Vigna unguiculata* (L) Walp.] in relation to photoperiod and temperature and implications for screening germplasm. Ann Bot 51(4):531–543

Hakim MA, Hossain A, Teixeira da Silva JA, Zvolinsky VP, Khan MM (2012) Yield, protein and starch content of 20 wheat (*Triticum aestivum* L.) genotypes exposed to high temperature under late sowing conditions. J Sci Res 4(2):477–489

Hall AE (2004) Breeding for adaptation to drought and heat in cowpea. Eur J Agron 21(4):447–454

Hammer GL, Broad IJ (2003) Genotype and environment effects on dynamics of harvest index during grain filling in sorghum. Agron J 95(1):199–206

Hampton JG, Boelt B, Rolston MP, Chastain TG (2013) Effects of elevated CO2 and temperature on seed quality. J Agric Sci 151(2):154–162

Hanway JJ, Weber CR (1971a) Accumulation of N, P, and K by soybean [*Glycine max* (L) Merrill] plants. Agron J 63(3):406–408

Hanway JJ, Weber CR (1971b) Dry Matter accumulation in eight Soybean (*Glycine max* (L) Merrill) varieties. Agron J 63(2):227–230

Hanway JJ, Weber CR (1971c) Dry matter accumulation in soybean [*Glycine max* (L.) Merrill] plants as influenced by N, P, and K fertilization. Agron J 63(2):263–266

Hannah MA, Heyer AG, Hincha DK (2005) A global survey of gene regulation during cold acclimation in *Arabidopsis* thaliana. PLoS Genet 1(2):e26

Hardie DC, Baker GJ, Marshall DR (1995) Field screening of *Pisum* accessions to evaluate their susceptibility to the pea weevil (*Coleoptera*: Bruchidae). Euphytica 84(2):155–161

Harrison MT, Kelman WM, Moore AD, Evans JR (2010) Grazing winter wheat relieves plant water stress and transiently enhances photosynthesis. Funct Plant Biol 37(8):726–736

Harrison MT, Evans JR, Dove H, Moore AD (2011) Dual-purpose cereals: can the relative influences of management and environment on crop recovery and grain yield be dissected? Crop Pasture Sci 62(11):930–946

Harvey DM (1973) The translocation of 14C photosynthate in *Pisum sativum* L. Ann Bot 37(152): 787–794

Hasanuzzaman M, Hossain MA, Teixeira da Silva JA, Fujita M (2012) Plant response and tolerance to abiotic oxidative stress: antioxidant defenses is a key factor. In: Venkateswarlu B, Shanker AK, Shanker C, Maheswari M (eds) Crop stress and management: prospective and Strategies. Springer, Berlin, pp 291–315

Hashimoto M, Komatsu S (2007) Proteomics analysis of rice seedling during cold stress. Proteomics 7(8):1293–1302

Hatfield JL, Boote KJ, Kimball BA, Ziska LH, Izaurralde RC, Ort D, Thomson AM, Wolfe D (2011) Climate impacts on agriculture: implications for crop production. Agron J 103(2): 351–370

Hawthorne W, Hannay J, Heinjus D (2003) Growing chickpeas. Govt. of South Australia, Department of Primary Industries and Resources

Hay RKM (1995) Harvest index: a review of its use in plant breeding and crop physiology. Ann Appl Biol 126(1):197–216

Hay RKM, Gilbert RA (2001) Variation in the harvest index of tropical maize: evaluation of recent evidence from Mexico and Malawi. Ann Appl Biol 138:103–109

Hay R, Porter J (2006) The physiology of crop yield, 2nd edn. Blackwell Publishing, UK

Hays DB, Do JH, Mason RE, Morgan G, Finlayson SE (2007) Heat stress induced ethylene production in developing wheat grains induces kernel abortion and increased maturation in a susceptible cultivar. Plant Sci 172(6):1113–1123

Hedhly A, Hormaza JI, Herrero M (2005) The effect of temperature on pollen germination pollen tube growth, and stigmatic receptivity in peach. Plant Biol 7(5):476–483

Hedhly A, Hormaza JI, Herrero M (2008) Global warming and plant sexual reproduction. Trends Plant Sci 14(1):30–36

Herrero MP, Johnson RR (1980) High-temperature stress and pollen viability of maize. Crop Sci 20(6):796–800

Herrero M, Arbeloa A (1989) Influence of the pistil on pollen-tube kinetics in peach (*Prunus persica*). Am J Bot 76(10):1441–1447

Herrero M (2003) Male and female synchrony and the regulation of mating in flowering plants. Philos Transac Royal Soc London Series B-Biol Sci 358(1434):1019–1024

Heuvelink E (1995) Growth, development and yield of a tomato crop: periodic destructive measurements in a greenhouse. Scient Hortic 61:77–99

Hnatowich G (2000) Pulse production manual. Saskatchewan Pulse Growes, Saskatoon, SK, Canada

Hniličková H, Hejnák V, Němcová L, Martinková J, Skalický M, Hnilička F, Grieu P (2017) Effect of freezing temperature on physiological traits in sunflower. Plant Soil Environ 63(8):375–380

Hirth J, Haines P, Ridley AM, Wilson K (2001) Lucerne in crop rotations on the Riverine Plains. 2. Biomass and grain yields, water use efficiency, soil nitrogen, and profitability. Aust J Agric Res 52:279–293

Hocking P, Randall P, DeMarco D (1997) The response of dryland canola to nitrogen fertilizer: partitioning and mobilization of dry matter and nitrogen, and nitrogen effects on yield components. Field Crops Res 54:201–220

Hocking PJ, Stapper M (2001) Effects of sowing time and nitrogen fertiliser on canola and wheat, and nitrogen fertiliser on Indian mustard. II. Nitrogen concentrations, N accumulation, and N fertiliser use efficiency. Aust J Agric Res 52(6):635–644

Hoffman TK, Kolb FL (1998) Effects of barley yellow dwarf virus on yield and yield components of drilled winter wheat. Plant Dis 82(6):620–624

Hoffmann MP, Jacobs A, Whitbread AM (2015) Crop modeling based analysis of site-specific production limitations of winter oilseed rape in northern Germany. Field Crops Res 178:49–62

Holaday AS, Martindale W, Alred R, Brooks AL, Leegood RC (1992) Changes in activities of enzymes of carbon metabolism in leaves during exposure of plants to low temperature. Plant Physiol 98(3):1105–1114

Horie T, Ohnishi M, Angus JF, Lewin LG, Tsukaguchi T, Matano T (1997) Physiological characteristics of high-yielding rice inferred from cross-location experiments. Field Crops Res 52(1-2):55–67

Hossain A, Teixeira da Silva JA, Lozovskaya MV, Zvolinsky VP (2012a) The effect of high temperature stress on the phenology, growth and yield of five wheat (*Triticum aestivum* l.) enotypes. Asian Aus J Plant Sci Biotech 6(1):14–23

Hossain A, Lozovskaya MV, Zvolinsky VP, Teixeira da Silva JA (2012b) Effect of soil and climatic conditions on yield-related components performance of spring wheat (*Triticum aestivum* L.) varieties in the northern Bangladesh. Nat Sci J Fund Appl Sci 2(39):69–78

Hossain A, Lozovskaya MV, Zvolinsky VP, Teixeira da Silva JA (2012c) Effect of soil and climatic conditions on phenology of spring wheat varieties in the northern Bangladesh. Nat Sci J Fund Appl Sci 2(39):78–86

Howarth CJ (2005) Genetic improvements of tolerance to high temperature. In: Ashraf M, Harris PJC (eds) Abiotic stresses: plant resistance through breeding and molecular approaches. Haworth Press Inc, New York, pp 277–300

Hsieh TH, Lee JT, Yang PT, Chiu LH, Charng Y, Wang YC, Chan MT (2002) Heterology expression of the *Arabidopsis* C-repeat/dehydration response element binding factor 1 gene confers elevated tolerance to chilling and oxidative stresses in transgenic tomato. Plant Physiol 129(3):1086–1094

Hu WH, Song XS, Shi K, Xia XJ, Zhou YH, Yu JQ (2008) Changes in electron transport, superoxide dismutase and ascorbate peroxidase isoenzymes in chloroplasts and mitochondria of cucumber leaves as influenced by chilling. Photosynthetica 46(4):581–588

Huber SC, Huber JL (1992) Role of sucrose-phosphate synthase in sucrose metabolism in leaves. Plant Physiol 99(4):1275–1278

Huhn M (1990a) Comments on the calculation of mean harvest indices. J Agron Crop Sci 165(2-3): 86–93

Huhn M (1990b) On the variability of harvest indices. J Agron Crop Sci 164(4):271–281

Huner NPA, Oquist G, Sarhan F (1998) Energy balance and acclimation to light and cold. Trends Plant Sci 3:224–230

Hurry VM, Huner NPA (1992) Effect of cold hardening on sensitivity of winter and spring wheat leaves to short-term photoinhibition and recovery of photosynthesis. Plant Physiol 100(3): 1283–1290

Hurry VM, Malmberg G, Gardeström P, Öquist G (1994) Effects of a short-term shift to low temperature and of long-term cold hardening on photosynthesis and ribulose 1,5-bisphosphate carboxylase/oxygenase and sucrose phosphate synthase activity in leaves of winter rye (*Secale cereale* L.). Plant Physiol 106(3):983–990

Hurry VM, Strand Å, Tobiæson M, Gardeström P, Öquist G (1995) Cold hardening of spring and winter wheat and rape results in differential effects on growth, carbon metabolism, and carbohydrate content. Plant Physiol 109(2):697–706

Iba K (2002) Acclimative response to temperature stress in higher plants: approaches of gene engineering for temperature tolerance. Annu RevPlant Biol 53:225–245

Idso SB, Kimball BA (1989) Growth response of carrot and radish to atmospheric CO2 enrichment. Environ Exp Bot 29(2):135–139

Imin N, Kerim T, Rolfe BG, Weinman JJ (2004) Effect of early cold stress on the maturation of rice anthers. Proteomics 4(7):1873–1882

Inman-Bamber NG (2004) Sugarcane water stress criteria for irrigation and drying off. Field Crops Res 89(1):107–122

IPCC (2014) Climate change synthesis report contribution of working groups I. II and III to the fifth assessment report of the intergovernmental panel on climate change. IPCC, Geneva, p 151

IRRI (1995) Rice Facts. International Rice Research Institute, Los Baños, the Philippines

IRRI (1996) IRRI toward 2020. International Rice Research Institute, Los Baños, p 43

Jackson RB, Canadell JJ, Mooney HO, Schulze E (1996a) A global analysis of root distributions for terrestrial biomes. Oecologia 108(3):389–411

Jackson P, Robertson M, Cooper M, Hammer G (1996b) The role of physiological understanding in plant breeding: from a breeding prospective. Field Crops Res 49(1):11–37

Jain M, Prasad PVV, Boote KJ, Hartwell AL, Chourey PS (2007) Effects of season-long high temperature growth conditions on sugar-to-starch metabolism in developing microspores of grain sorghum (*Sorghum bicolor* L. Moench). Planta 227(1):67–79

Jaggard KW, Qi A, Ober ES (2010) Possible change to arable crop yields by 2050. Philos Trans R Soc B Biol Sci 365(1554):2835–2851

Jagtap SS, Abamu FJ, Kling JG (1999) Long-term assessment of nitrogen and variety technologies on attainable maize yields in Nigeria using CERES-maize. Agric Syst 60(2):77–86

Jamieson PD, Martin RJ, Francis GS, Wilson DR (1995) Drought effects on biomass production and radiation-use efficiency in barley. Field Crops Res 43:77–86

Janmohammadi M, Zolla L, Rinalducci S (2015) Low temperature tolerance in plants: changes at the protein level. Phytochemistry 117:76–89

Jang J-C, León P, Zhou L, Sheen J (1997) Hexokinase as a sugar sensor in plants. Plant Cell 9(1): 5–19

Janská A, Maršík P, Zelenková S, Ovesná J (2010) Cold stress and acclimation—what is important for metabolic adjustment? Plant Biol 12(3):395–405

Jensen CR, Joernsgaard B, Andersen MN, Christiansen JL, Mogensen VO, Friis P, Petersen CT (2004) The effect of lupins as compared with peas and oats on the yield of the subsequent winter barley crop. Eur J Agron 20(4):405–418

Jeuffroy M-H, Warembourg FR (1991) Carbon transfer and partitioning between vegetative and reproductive organs in *Pisum sativum* L. Plant Physiol 97(1):440–448

Jeuffroy M-H (1992) P'tipois: a simulation model for pea seed number. In: Scaife A (ed) Proceedings of 2nd congress of European Society for Agronomy, Warwick, pp 90–91

Jeuffroy M-H (1994a) Les rbgles de partage des assimilats. In: INRA, ITCF and UNIP (eds) Agrophysiologie du Pois Protragineux, Applications h la Production Agricole, pp 49–73

Jeuffroy M-H (1994b) Le nombre de graines par tige. In: INRA, ITCF and UNIP (eds) Agrophysiologie du Pois Protragineux, Applications h la Production Agricole, pp 93–110

Jeuffroy M-H, Chabanet C (1994) A model to predict seed number per pod from early pod growth rate in pea (*Pisum sativum* L.). J Exp Bot 45(275):709–715

Jeuffroy M-H, Devienne F (1995) A simulation model for assimilate partitioning between pods in pea (*Pisum sativum* L.) during the period of seed set; validation in field conditions. Field Crops Res 41(2):79–89

Ji H, Xiao L, Xia Y, Song H, Liu B, Tang L, Liu L (2017a) Effects of jointing and booting low temperature stresses on grain yield and yield components in wheat. Agric Forest Meteor 243: 33–42

Ji L, Zhou P, Zhu Y, Liu F, Li R, Qui Y (2017) Proteomic analysis of rice seedlings under cold Stress. Protein J 36(4):299–307

Jiang QW, Kiyoharu O, Ryozo I (2002) Two novel mitogen-activated protein signaling components, OsMEK1 and OsMAP1, are involved in a moderate low-temperature signaling pathway in rice. Plant Physiol 129(4):1880–1891

Jiang GM, Sun JZ, Liu HQ, Qu CM, Wang KJ, Guo RJ, Bai KZ, Gao LM, Kuang TY (2003) Changes in the rate of photosynthesis accompanying the yield increase in wheat cultivars released in the past 50 years. J Plant Res 116(5):347–354

Johnson JMF, Allmaras RR, Reicosky DC (2006) Estimating source carbon from crop residues, roots and rhizodeposits using the National grain-yield database. Agron J 98(3):622–636

Jordan DR, Hunt CH, Cruickshank AW, Borrell AK, Henzell RG (2012) The relationship between the staygreen trait and grain yield in the elite Sorghum hybrids grown in a range of environments. Crop Sci 52(3):1153–1161

Jouyban Z, Hasanzade R, Sharafi S (2013) Chilling deficit in plants. Int J Agri Crop Sci 5(24): 2961–2968

Ju J, Yamamoto Y, Wang YL, Shan YH, Dong GC, Miyazaki A, Yoshida T (2009) Genotypic differences in dry matter accumulation, nitrogen use efficiency and harvest index in recombinant inbred lines of rice under hydroponic culture. Plant Prod Sci 12(2):208–216

Kalbarczyk R (2009) Potential reduction in cucumber yield (*Cucumis sativus* l.) in Poland caused by unfavourable thermal conditions of soil. Acta Scientiarum Polonorum Hortorum Cultus 8: 45–58

Kakani VG, Prasad PVV, Craufurd PQ, Wheeler TR (2002) Response of *in vitro* pollen germination and pollen tube growth of groundnut (*Arachis hypogaea* L.) genotype to temperature. Plant Cell Environ 25(12):1651–1661

Kalisz A, Cebula S (2001) Direct plant covering and soil mulching in the spring production of some Chinese cabbage cultivars. Effect of temperature on premature bolting. Folia Hortic 13(2): 13–22

Kalt-Torres W, Kerr PS, Usuda H, Huber SC (1987) Diurnal changes in maize leaf photosynthesis: 1. Carbon exchange rate, assimilation export rate and enzyme activities. Plant Physiol 83(2): 283–288

Kang S, Shi W, Zhang J (2000) An improved water-use efficiency for maize grown under regulated deficit irrigation. Field Crops Res 67(3):207–214

Karaba A, Dixit S, Greco R, Aharoni A, Trijatmiko KR, Marsch-Mrtinez N, Krishnan A, Nataraja KN, Udayakumar M, Pereira A (2007) Improvement of water use efficiency in rice expression of *HARDY*, an *Arabidopsis* drought and salt tolerance gene. Proc Natl Acad Sci U S A 104(39): 15270–15275

Kasterine A, Batchelor S (1998) Biodiesel—environmental impact and potential for niche markets. Research Review OS12, HGCA, London

Kawamura Y, Uemura M (2003) Mass spectrometric approach for identifying putative plasma membrane proteins of *Arabidopsis* leaves associated with cold acclimation. Plant J 36(2): 141–154

Kebeish R, Niessen M, Thiruveedhi K, Bari R, Hirsch HJ, Rosenkranz R, Stabler N, Schonfeld B, Kreuzaler F, Peterhansel C (2007) Chloroplastic photorespiratory bypass increases photosynthesis and biomass production in Arabidopsis thaliana. Nat Biotechnol 25(5):593–599

Keller F, Ludlow MM (1993) Carbohydrate metabolism in drought-stressed leaves of pigeonpea (*Cajanus cajun*). J Exp Bot 44(8):1351–1359

Kemanian AR, Stöckle CO, Huggins DR, Viega LM (2007) A simple method to estimate harvest index in grain crops. Field Crops Res 103(3):208–216

Kernich GC, Halloran GM, Flood RG (1997) Variation in duration of pre-anthesis phases of development in barley (*Hordeum vulgare*). Aust J Agric Res 48(1):59–66

Kerr PS, Rufty TW, Huber SC (1985) Endogenous rhythms in photosynthesis, sucrose-phosphate synthase activity, and stomatal resistance in leaves of soybean *Glycine max* [L.] Merr. Plant Physiol 77(2):275–280

Khanna-Chopra R, Koundal KR, Sinha SK (1979) Response of pigeonpea to water availability. In: Symposium on "Legumes in the Tropics", Selangor, Malaysia

Khanna-Chopra R, Sinha SK (1982) Photosynthesis rate and photosynthetic carboxylases during growth and development in bengal gram cultivar. Photosynthetica 16(4):509–513

Khayat E, Zieslin N (1987) Effect of night temperature on the activity of sucrose phosphate synthase, acid invertase, and sucrose synthase in source and sink tissues of *Rosa hybrida* cv Golden Times. Plant Physiol 84(2):447–449

Kijne JW, Barker R, Molden D (2002) Water productivity in agriculture: limits and opportunities for improvement. CABI Publishing, Wallingford, UK

Kijne JW, Tuong TP, Bennett J, Bouman BAM, Oweis T (2003) Ensuring food security via crop water productivity improvement. In: Background papers for the challenge program for food and water, Colonbo, Sri Lanka: CGIAR-IWMI, pp 1–42

Kirby EJM (1988) Analysis of leaf, stem and ear growth in wheat from terminal spikelet initiation to anthesis. Field Crops Res 18:127–140

Kirkegaard JA, Lilley JM, Howe GN, Graham JM (2007) Impact of subsoil water use on wheat yield. Aust J Agric Res 58:303–315

Kirkegaard JA, Hunt JR, Mcbeath TM, Lilley JM, Moore A, Verburg K, Robertson M, Oliver Y, Ward PR, Milroy S, Whitbread AM (2014) Improving water productivity in the Australian Grains industry—a nationally coordinated approach. Crop Pasture Sci 65(7):583–601

Kirkegaard JA, Lilley JM, Hunt JR, Sprague SJ, Ytting NK, Rasmussen IS, Graham JM (2015) Effect of defoliation by grazing or shoot removal on the root growth of field-grown wheat (*Triticum aestivum* L.). Crop Pasture Sci 66(4):249–259

Knight H, Trewavas AJ, Knight MR (1996) Cold calcium signaling in Arabidopsis involves two cellular pools and a change in calcium signature after acclimation. Plant Cell 8(3):489–503

Knight H, Veale EL, Warren GJ, Knight MR (1999) The sfr6 mutation in *Arabidopsis* suppresses low temperature induction of genes dependent on the CRT/DRE sequence motif. Plant Cell 11(5):875–886

Kobata T, Koç M, Baruntçular C, Tanno K, Inagaki M (2017) Harvest index is a critical factor influencing the grain yield of diverse wheat species under rain-fed conditions in the Mediterranean zone of southeastern Turkey and northern Syria. Plant Prod Sci 21(2):71–82

Koester RP, Skoneczka JA, Diers BW, Ainsworth EA (2014) Historic gains in soybean seed yield are driven by linear increases in light interception, energy conversion, and partitioning efficiencies. J Exp Bot 65(12):3311–3321

Kosmala A, Bocian A, Rapacz M, Jurczyk B, Zwierzykowski Z (2009) Identification of leaf proteins differentially accumulated during cold acclimation between *Festuca pratensis* plants with distinct levels of frost tolerance. J Exp Bot 60(12):3595–3609

Kosova K, Vítámvás P, Prášil PT, Renaut J (2011) Plant proteome changes under abiotic stress-contribution of proteomics studies to understanding plant stress response. J Proteome 74(8): 1301–1322

Kramer PJ, Boyer JS (1995) Water relations of plant and soils. Academic Press, San Diego

Krapp A, Stitt M (1995) An evaluation of direct and indirect mechanisms for the "sink-regulation" of photosynthesis in spinach: changes in gas exchange, carbohydrates, metabolites, enzyme activities and steady-state transcript levels after cold-girdling source leaves. Planta 195:313–323

Kreps JA, Simon AE (1997) Environmental and genetic effects on circadian clock-regulated gene expression in Arabidopsis. Plant Cell 9(3):297–304

Kreps JA, Wu YJ, Chang HS, Zhu T, Wang X, Harper JF (2002) Transcriptome changes for Arabidopsis in response to salt, osmotic, and cold stress. Plant Physiol 130(4):2129–2141

Koch KE (1996) Carbohydrate-modulated gene expression in plants. Ann. Rev Plant Physiol Plant Mol Biol 47:509–540

Kong L, Guo H, Sun M (2015) Signal transduction during wheat grain development. Planta 241(4): 789–801

Korkmaz A, Dufault RJ (2002) Short-term cyclic cold temperature stress on water melon yield. Hort Sci 37(3):487–489

Kotak S, Larkindale J, Lee U, von Koskull-Doring P, Vierling E, Scharf KD (2007) Complexity of the heat stress response in plants. Cur Opin Plant Biol 10(3):310–316

Kruk B, Calderini DF, Slafer GA (1997) Source-sink ratios in modern and old wheat cultivars. J Agric Sci (Cambridge) 128:273–281

Kumar PR, Yadav SK, Sharma SR, Lal SK, Jha DN (2009) Impact of climate change on seed production of cabbage in North Western Himilayas. World J Agric Sci 5(1):18–26

Kumudini S, Hume DJ, Chu G (2001) Genetic improvement in short season soybeans. Crop Sci 41(2):391–398

Kuo NC, Chen C, Cheng HH (1980) The effect of different temperatures on agronomic characters seed yield, oil content and fatty acid composition of rapeseed. Mem Coll Agric Natl Taiwan Univ 20:45–53

Labate CA, Leegood RC (1988) Limitation of photosynthesis by changes in temperature: factors affecting the response of carbon dioxide assimilation to temperature in barley leaves. Planta 173(4):519–527

Larcher W (1995) Plants under stress. In: Physiological plant ecology. Ecophysiology and Stress Physiology of Functional Groups, 3rd edn. Springer, Berlin

Lardon A, Triboi-Blondel AM (1995) Cold and freeze stress at flowering—effects on seed yield in winter rapeseed. Field Crops Res 44(2-3):95–101

Larkindale J, Vierling E (2008) Core genome responses involved in acclimation to high temperature. Plant Physiol 146(2):748–761

Le Buanec B (2009) Opening address. In: Proceedings of the second world seed conference, Rome, pp 13–16

Lee SH, Singh AP, Chung GC, Ahn SJ, Noh EK, Steudle E (2004) Exposure of roots of cucumber (*Cucumis sativus*) to low temperature severely reduces root pressure, hydraulic conductivity and active transport of nutrients. Physiol Plant 120(3):413–420

Leegood RC, Furbank RT (1986) Stimulation of photosynthesis by 2% oxygen at low temperatures is restored by phosphate. Planta 168(1):84–93

Leegood RC (2002) C4 photosynthesis: principles of CO2 concentration and prospects for its introduction into C3 plants. J Exp Bot 53(369):581–590

Lenssen AW, Johnson GD, Carlson GR (2007) Cropping sequence and tillage system influences annual crop production and water use in semiarid Montana. Field Crops Res 100(1):32–43

Leonardos ED, Savitch LV, Huner NP, Oquist G, Grodzinski B (2003) Daily photosynthetic and C-export patterns in winter wheat leaves during cold stress and acclimation. Physiol Plant 117(4):521–531

Leport L, Turner NC, French RJ, Barr MD, Duda R, Davies SL, Tennant D, Siddique KHM (1999) Physiological responses of chickpea genotypes to terminal drought in a Mediterranean environment. Eue J Agron 11(3-4):279–291

Levitt J (1980) Responses of plants to environmental stress. In: Chilling, freezing, and high temperature stress, vol 1. Academic Press, New York

Li F, Yu JM, Nong ML, Kang SZ, Zhang JH (2010) Partial rootzone irrigation enhanced soil enzyme activities and water use of maize under different ratios of inorganic to organic nitrogen fertilizers. Agric Water Manag 97(2):231–239

Li X, Pu H, Liu F, Zhou Q, Cai J, Dai T, Cao W, Jiang D (2015) Winter wheat photosynthesis and grain yield responses to spring freeze. Agron J 107(3):1002–1010

Liu YF, Qi MF, Li TL (2012) Photosynthesis, photoinhibition, and antioxidant system in tomato leaves stressed by low night temperature and their subsequent recovery. Plant Sci 196:8–17

Liu B, Liu L, Tian L, Cao W, Zhu Y, Asseng S (2014) Post-heading heat stress and yield impact in winter wheat of China. Glob Chang Biol 20(2):372–381

Liu L, Ji H, An J, Shi K, Ma J, Liu B, Tang L, Cao W, Zhu Y (2019) Response of biomass accumulation in wheat to low-temperature stress at jointing and booting stages. Environ Exp Bot 157:46–57

Lobell DB, Hicke JA, Asner GP, Field CB, Tucker CJ, Los SO (2002) Satellite estimates of productivity and light use efficiency in United States agriculture, 1982–98. Glob Change Biol 8(8):722–735

Lobell DB, Hammer GL, Chenu K, Zheng B, McLean G, Chapman SC (2015) The shifting influence of drought and heat stress for crops in northeast Australia. Glob Chang Biol 21(11): 4115–4127

Locascio A, Roig-Villanova I, Bernardi J, Varotto S (2014) Current perspectives on the hormonal control of seed development in Arabidopsis and maize: a focus on auxin. Front Plant Sci 5:412

Loitongbam B, Chandra K, Bisen P, Motasara N, Thakur P, Kulhari S (2017) Breeding for low temperature stress tolerance in reproductive stage of rice (*Oryza sativa* L.). Int J Bio-resource Stress Mange 8(6):857–862

Long SP, Zhu X, Naidu SL, Ort DR (2006) Can improvement in photosynthesis increase crop yields? Plant Cell Environ 29(3):315–330

Lopez Pereira M, Trapani N, Sadras VO (2000) Genetic improvement of sunflower in Argentina between 1930 and 1995: Part III. Dry matter partitioning and grain composition. Field Crops Res 67(3):215–221

L'opez-Bellido F, L'opez-Bellido L, L'opez-Bellido RJ (2005) Competition, growth and yield of faba bean (*Vicia faba* L.). Eur J Agron 23(4):359–378

Los Reyes BGD, Myers SJ, McGrath JM (2003) Differential induction of glyoxylate cycle enzymes by stress as a marker for seedling vigor in sugar beet (*Beta vulgaris*). Mol Gen Genomics 269(5):692–698

Loss SP, Siddique KHM (1994) Morphological and physiological traits associated with wheat yield increases in Mediterranean environments. Adv Agron 52:229–276

Loss SP, Siddique KHM (1997) Adaptation of faba bean (*Vicia faba* L.) to dryland Mediterranean-type environments I. Seed yield and yield components. Field Crops Res 52(1-2):17–28

Lukatkin AS, Isaikina EE (1997) Calcium status and chilling injury in maize seedlings. Rus J Plant Physiol 44(3):339–342

Lukatkin AS, Brazaityte A, Bobinas C, Duchovskis P (2012) Chilling injury in chilling-sensitive plants: a review. Agriculture 99(2):111–124

Madan P, Jagadish SVK, Craufurd PQ, Fitzgerald M, Lafarge T, Wheeler TR (2012) Effect of elevated CO2 and high temperature on seed-set and grain quality of rice. J Environ Bot 63(10): 3843–3852
Marentes E, Griffith M, Mlynarz A, Brush RA (1993) Proteins accumulate in the apoplast of winter rye leaves during cold acclimation. Physiol Plant 87(4):499–507
Marjanović-Jeromela A, Terzić S, Jankulovska M, Zorić M, Kondić-Špika A, Jocković M, Hristov N, Crnobarac J, Nagl N (2019) Dissection of year related climatic variables and their effect on winter rapeseed (*Brassica napus* L.) development and yield. Agron 9(9):517
Maliga P (2004) Plastid transformation in higher plants. Annu Rev Plant Biol 55:289–313
Markhart AH, Fiscuc EL, Naylor AW, Kramer PJ (1979) Effect of temperature on water and ion transport in soybean and broccoli systems. Plant Physiol 64(1):83–87
Martino-Catt S, Ort DR (1992) Low temperature interrupts circadian regulation of transcriptional activity in chilling-sensitive plants. Proc Natl Acad Sci U S A 89(9):3731–3735
Martiniello P, Teixeira da Silva JA (2011) Physiological and bio–agronomical aspects involved in growth and yield components of cultivated forage species in Mediterranean Environments: a review. Eur J Plant Sci Biotech 5(Special Issue 2):64–98
Mason I, Fischer RA (1986) Tillage practices and the growth and yield of wheat in southern New South Wales: Lockhart, in a 450 mm rainfall region. Aust J Exp Agric 26(4):457–468
Matsui T, Omasa K (2002) Rice (Oryza sativa L.) cultivars tolerant to high temperature at flowering: anther characteristics. Ann Bot 89:683–687
McKirdy SJ, Jones RAC, Nutter FW (2002) Quantification of yield losses caused by barley yellow dwarf virus in wheat and oats. Plant Dis 86(7):769–773
McMaster GS, Wilhelm WW (1997) Growing degree-days: one equation, two interpretations. Agric For Meteorol 87(4):291–300
McMurray L (2005) Filed pea trials. In: Crop Variety Evaluation 2004. South Australian Research and Development Institute, Adelaide
Metclfe DS, Elkins DM (1980) Crop production: principles and practices. Macmillan Publishing Co., Inc., New York
Meyer RF, Boyer JS (1981) Osmoregulation, solute distribution, and growth in soybean seedlings having low water potentials. Planta 151(5):482–489
Miller AS, Curtis PS, Snow AA (1994) Genotype-specific effects of elevated CO2 on fecundity in wild radish (*Raphanus raphanistrum*). Oecologia 97(1):100–105
Miller P, Lanier W, Brandt S (2001) Using growing degree days to predict plant stages. Montana State University Extension Service
Miller PR, McConkey BG, Clawton GW, Brandt SA, Staricka JA, Jhonston AM, Lafond G, Schatz BG, Baltensperger DD, Neill K (2002) Pulse crop adaptation in Northern Grait Plain. Agron J 94(2):261–272
Mimura T, Dietz K-J, Kaiser W, Schramm MJ, Kaiser G, Heber U (1990) Phosphate transport across biomembranes and cytosolic homeostasis in barley leaves. Planta 180(2):139–146
Minorsky PV (1985) A heuristic hypothesis of chilling injury in plants: a role for calcium as the primary physiological transducer of injury. Plant Cell Environ 8(2):75–94
Miralles DJ, Slafer GA (1995a) Yield, biomass and yield components in dwarf, semi dwarf and tall isogenic liens of spring wheat under recommended and late sowing dates. Plant Breed 114(5): 392–396
Miralles DJ, Slafer GA (1995b) Individual grain weight responses to genetic reduction in culm length in wheat as affected by source-sink manipulations. Field Crop Res 43(2-3):55–66
Miralles DJ, Slafer GA (1997) Radiation interception and radiation use efficiency of near isogenic wheat lines with different height. Euphytica 97(2):201–208
Miralles DJ, Richards RA (2000) Response of leaf and tiller emergence and primordium initiation in wheat and barley to interchanged photoperiod. Ann Bot 85(5):655–663
Miralles DJ, Richards RA, Slafer GA (2000) Duration of stem elongation period influences the number of fertile florets in wheat and barley. Aus J Plant Physiol 27(10):931–940

Miralles DJ, Rharrabti Y, Royo C, Villegas D, García del Moral LF (2002) Grain setting strategies of Mediterranean durum wheat cultivars released in different periods (1900–2000). In: Association of applied biologists conference 'Genotype to Phenotype: Narrowing the Gaps.' Royal Agricultural College, Cirencester, UK

Miralles DJ, Slafer GA (2007) Sink limitations to yield in wheat: how could it be reduced? J Agric Sci 145(2):139–150

Mittler R (2006) Abiotic stress, the field environment and stress combination. Trends in Plant Sci. 11(1):15–19

Miura K, Furumoto T (2013) Cold signaling and cold response in plants. Int J Mol Sci 14(3): 5312–5337

Molassiotis A, Job D, Ziogas V, Tanou G (2016) Citrus plants: a model system for unlocking the secrets of NO and ROS-inspired priming against salinity and drought. Front Plant Sci 7:229

Mondal S, Singh RP, Huerta-Espin J, Kehel Z, Autrique E (2015) Characterization of heat- and drought-stress tolerance in high-yielding spring wheat. Crop Sci 55(4):1552–1562

Monneveux P, Pastenes C, Reynolds MP (2003) Limitations to photosynthesis under light and heat stress in three high-yielding wheat genotypes. J Plant Physiol 160(6):657–666

Monteith JL (1977) Climate and the efficiency of crop production in Britain. Philos Trans R Soc Lond B 281(980):277–294

Monteith JL (1981) Chapter 2: Does light limit crop production? In: Johnson CB (ed) Physiological processes limiting plant productivity. Butterworths, London, UK., pp 23–38

Morgan C, Wells R, Bilsborrow PE, Miro B, Nightingale M, Bancroft I (2007) Improving harvest index in oilseed rape (Brassica napus) through modifying canopy architecture. In: 12th International Rapeseed Congress, vol 3. Wuhan, China, Conference, pp 53–55

Morrison MJ, Voldeng HD, Cober ER (1999) Physiological changes from fifty-eight years of genetic improvement of season soybean cultivars in Canada. Agron J 91(4):685–689

Mortensen LM (1994) Effects of elevated CO2 concentration on growth and yield of eight vegetable species in a cool climate. Sci Hortic 58(3):177–185

Munier-Jolain NG (1994) Etude de la variabilit6 du poids individuel des graines du soja de type indéterminé (*Glycine max*. L. Merrill, cv. Maple Arrow). Influence de l'appadtion srquentielle des organes reproducteurs. Thse de Doctorat, INA-PG, Paris, 85p

Munier-Jolain NG, Ney B (1995) Variability of seed growth rate for various pea cultivars (*Pisum sativum* L.) in relation to their seed cell number. In: Proceedings of 2nd European Conference on Grain Legumes, Copenhagen, pp 32–33

Munier-Jolain NG, Ney B (1998) Seed growth rate in grain legumes II. Seed growth rate depends on cotyledon cell number. J Exp Bot 49(329):1971–1976

Munkhtsetseg E, Kimura R, Wang J, Shinoda M (2007) Pasture yield response to precipitation and high temperature in Mongolia. J Arid Environ 70(1):94–110

Murai-Hatano M, Kuwagata T, Sakurai J, Nonami H, Ahamed A, Nagasuga K, Matsunami T, Fukushi K, Maeshima M, Okada M (2008) Effect of low root temperature on hydraulic conductivity of rice plants and the possible role aquaporins. Plant Cell Physiol 49(9):1294–1305

Nagasuga K, Murai-Hatano M, Kuwagata T (2011) Effect of low root temperature on dry matter production and root water uptake in rice plants. Plant Prod Sci 14(1):22–29

Nahar K, Hasanuzzaman M, Majumder RR (2009a) Effect of low temperature stress in transplanted Aman Rice varieties mediated by different transplanting dates. Acad J Plant Sci 2(3):132–138

Nahar K, Biswas JK, Shamsuzzaman AMM, Hasanuzzaman M, Barman HN (2009b) Screening of indica rice (*Oryza sativa* L.) genotypes against low temperature stress. Bot Res Int 2:295–303

Nanda R, Bhargava SC, Tomar DPS, Rawson HM (1996) Phenological development of Brassica campestris, Brassica juncea, Brassica napus and Brassica carinata grown in controlled environments and from 14 sowing dates in the field. Field Crops Res 46(1–3):93–103

Nanja Reddy YA, Sheshshayee MS, Uma Shaanker R, Virupakshappa K, Prasad TG (1994) Selection for high canopy assimilation rate is a good strategy to increase productivity in sunflower. Helia 17:45–52

Nanja Reddy YA, Uma Shaanker R, Prasad TG, Udaya Kumar M (2003) Physiological approaches to improving harvest index and productivity in sunflower. Helia 26:81–90

Narayanan A, Sheldrake AR (1976) Pulse Physiology Annual Report 1975-1976. I, Pigeonpea Physiology. ICRISAT, Hyderabad

Nayyar H, Bains TS, Kumar S, Kaur G (2005) Chilling effects during seed filling on accumulation of seed reserves and yield of chickpea. J Sci Food Agric 85(11):1925–1930

Nelson JM, Palzkill DA (1993) In: Janick J, Simon JE (eds) Irrigation effects on growth, cold tolerance of flower buds and seed yield of jojoba "new Crop". Wiley, New York, pp 360–362

Nelson GC, Rosegrant MW, Koo J, Robertson R, Sulser T, Zhu T, Ringler C, Msangi S, Palazzo A, Batka M, Magalhaes M, Valmonte-Santos R, Ewing M, Lee D (2009) Climate change: impact on agriculture and costs of adaptation. International Food Policy Research Institute, Washington, DC

Ney B, Turc O (1993) Heat-unit-based description of the reproductive development of pea. Crop Sci 33(3):510–514

New B, Duthion C, Turc O (1994) Phenological response of pea to water stress during reproductive development. Crop Sci 34(1):141–146

Newman JA, Gibson DJ, Parsons AJ, Thornley JHM (2003) How predictable are aphid population responses to elevated CO2. J Anim Ecol 72(4):556–566

Nicholls N, Drosdowsky W, Lavery B (1997) Australian rainfall variability and change. Weather 52:66–71

Nishiyama I (1995) Damage due to extreme temperatures. In: Matsuo T, Kamazawa K, Ishii R, Ishihara H, Hirata H (eds) Science of the rice plant. Food and Agriculture Policy Research, Tokyo, Japan, pp 769–812

Noodén LD, Guiamet JJ, John I (1997) Senescence mechanism. Physiol Plant 101(4):746–753

Ntare BR, Williams JH (1998) Heritability and genotype x environment interaction for yield and components of a yield model in segregating populations of groundnut under semi-arid conditions. Afr Crop Sci J 6(2):119–127

Ntare BR, Williams JH, Ndunguru BJ (1998) Effects of seasonal variation in temperature and cultivar on yield and yield determination of irrigated groundnut (*Arachis hypogaea*) during the dry season in the Sahel of West Africa. J Agric Sci 131(4):439–448

Ochatt SJ (2011) Immature seeds and embryos of *Medicago truncatula* cultured in vitro. In: Thorpe TA, Yeung EC (eds) Plant embryo culture: methods and protocols; methods in molecular biology, Springer protocols, vol 710. Humana Press, New York, NY, pp 39–52

Ochatt SJ (2015) Agroecological impact of an in vitro biotechnology approach of embryo development and seed filling in legumes. Agron Sustain Dev 35(2):535–552

Ockerby SE, Fukai S (2001) The management of rice grown on raised beds with continuous furrow irrigation. Field Crops Res 69(3):215–226

O'Connell M, Connor D, O'Leary G (2002) Crop growth, yield and water use in long fallow and continuous cropping sequences in the Victorian mallee. Aust J Exp Agric 42(7):971–983

Olesen JE, Bindi M (2002) Co-sequences of climate change for European agricultural productivity, land use and policy. Eur J Agron 16(4):239–262

Oliver SN, Van Dongen JT, Alfred SC, Mamun EA, Zhao X, Saini HS, Fernandes SF, Blanchard CL, Sutton BG, Geigenberger P, Dennis ES, Dolferus R (2005) Cold-induced repression of the rice anther-specific cell wall invertase gene *OSINV4* is correlated with sucrose accumulation and pollen sterility. Plant Cell Environ 28(12):1534–1551

Oliver SN, Dennis ES, Dolferus R (2007) ABA regulates apoplastic sugar transport and is a potential signal for cold-induced pollen sterility in rice. Plant Cell Physiol 48(9):1319–1330

Ortiz-Monasterio RJI, Sayre KD, Rajaram S, McMahon M (1997) Genetic progress in wheat yield and nitrogen use efficiency under four N rates. Crop Sci 37(3):898–904

Otegui ME, Salfer DGA (2000) Physiological basis for maize improvement. Harworth Press, Binghamton, New York, p 23

Otegui ME, Slafer GA (2004) Increasing cereal yield potential by modifying developmental traits. In: Proceedings of the 4th International Crop Science Congress, Btisbane, pp 1–11

Palta JP, Weiss LS (1993) Ice formation and freezing injury: an overview on the survival mechanism and molecular aspects of injury and cold acclimation in herbaceous plants. In: Li PH, Christersson L (eds) Advances in cold hardiness. CRC Press, Fl, Boca Raton, pp 143–176

Palta J, Ludwig C (1996) Pod set and seed yield as affected by cytokinin application and terminal drought in narrow-leafed lupin. Aust J Agric Res 48(1):81–90

Palta J, Ludwig C (2000) Elevated CO2 during pod filling increased seed yield but not harvest index in indeterminate narrow-leafed lupin. Aust J Agric Res 51(2):279–286

Palta J, Turner NC, French RJ, Buirchell B (2003) Selection for high lupin yield under terminal drought. In: Agribusiness crop updates 2003. Perth, Agriculture WA

Papantoniou AN, Tsialtas JT, Papakosta DK (2013) Dry matter and nitrogen partitioning and translocation in winter oilseed rape (*Brassica napus* L.) grown under rainfed Mediterranean conditions. Crop Pasture Sci 64(2):115–122

Parry MAJ, Andralojc PJ, Mitchell RAC, Madgwick PJ, Keys AJ (2003a) Manipulation of Rubisco: its Amount, activity, function and regulation. J Exp Bot 54(386):1321–1333

Parry MAJ, Madgwick PJ, Keys AJ, Carvahlo JFC, Lea PJ, Beale MH, Ward JL (2003b) Short circuiting photorespiration. Abstracts ASPB Annual Meeting Plant Biology 2003, Abs. 18002

Parry MAJ, Madgwick PJ, Carvahlo JFC, Andralojc PJ (2007) Prospects for increasing photosynthesis by overcoming the limitations of Rubisco. J Agric Sci 145(1):31–43

Parry MAJ, Keys AJ, Madgwick PJ, Carmo-Silva AE, Andralojc PJ (2008) Rubisco regulation: role for inhibitors. J Exp Bot 59(7):1569–1580

Parry, M., Reynolds, M., Salvucci, M., Raines, C.A., Andralojc, PJ., Zhu, X.G., Price, G.D., Condon, A.G. and Furbank, R.T. 2010. Raising yield potential of wheat: (II) increasing photosynthetic capacity and efficiency. J Exp Bot, 62(2): 453-467

Passioura JB (1977) Grain yield, harvest index, and water use of wheat. J Aust Inst Agric Sci 43(3-4):117–121

Passioura JB (1996) Drought and drought tolerance. Plant Growth Regul 20:79–83

Passioura JB (2002) Environmental biology and crop improvement. Funct Plant Biol 29:537–546

Passioura JB (2006) Increasing crop productivity when water is scarce—from breeding to field management. Agric Water Manag 80(1-3):176–196

Pate JS, Layzell DB (1990) Energetics and costs of nitrogen assimilation. In: Mifflin B, Lea P (eds) The biochemistry of plants, vol 16. Academic Press, pp 1–42

Pate JS, Armstrong EL (1996) Pea. In: Zamski E, Schaffer A (eds) Photoassimilate distribution in plants and crops and Source-Sink Relationships. Marcel and Dekker Inc, New York, pp 625–642

Paul MJ, Foyer CH (2001) Sink regulation of photosynthesis. J Exp Bot 52(360):1383–1400

Peings Y, Cattiaux J, Douville H (2013) Evaluation and response of winter cold spells over western Europe in CMIP5 models. Clim Dyn 41(11-12):3025–3037

Penning de Vries FWT, Brunsting AHM, Van Laar HH (1974) Products, requirements and efficiency of biosynthesis a quantitative approach. J Theor Biol 45(2):339–377

Penning de Vries FWT, Van Laar HH, Chardon M (1983) Bioenergetics of growth of seeds, fruits and storage organs. In: Smith WH (ed) Potential productivity of field crops under different environments. International Rice Research Institute, Los Banõs, Philippines, pp 37–59

Perveen S, Shinwari KI, Jan M, Mallok I, Rehman S, Khan MA, Jameel M (2013) Low temperature stress induced changes in biochemical parameters, protein banding pattern and expression of Zat12 and Myb genes in rice seedlings. J Stress Physiol Biochem 9(4):193–206

Pierik R, Whitelam GC, Voesenek LACJ, de Kroon H, Visser EJW (2004) Canopy studies on ethylene-insensitive tobacco identify ethylene as a novel element in blue light and plant-plant signalling. Plant J 38(2):310–319

Peltonen-Sainio P, Kangas A, Salo Y, Jauhiainen L (2007) Grain number dominates grain weight in temperate cereal yield determination: evidence based on 30 years of multi-location trials. Field Crops Res 100(2-3):179–188

Peltonen-Sainio P, Muurinen S, Rajala A, Jauhiainen L (2008) Variation in harvest index of modern spring barley, oat and wheat cultivars adapted to northern growing conditions. J Agric Sci 146(1):35–47

Plaut Z, Butow BJ, Blumenthal CS, Wrigley CW (2004) Transport of dry matte into developing wheat kernels and its contribution to grain yield under post-anthesis water deficit and elevated temperature. Field Crops Res 86(2-3):185–198

Plenet D, Mollier A, Pellerin S (2000) Growth analysis of maize field crops under phosphorus deficiency. II. Radiation-use efficiency, biomass accumulation and yield components. Plant Soil 224(2):259–272

Porch TG, Jahn M (2001) Effect of high temperature stress on microsporogenesis in heat-sensitive and heat-tolerant of Phaseolus vulgaris. Plant Cell Environ 24(7):723–731

Porker K, Straight M, Hunt JR (2020) Evaluation of G × E × M interactions to increase harvest index and yield of early sown wheat. Front Plant Sci 11:994

Prasad PVV, Craufurd PQ, Summerfield RJ (1999) Fruit number in relation to pollen production and viability in groundnut exposed to short episodes of heat stress. Ann Bot 84(3):381–386

Prasad PVV, Boote KJ, Allen LH Jr, Thomas JMG (2002) Effects of elevated temperature and carbon dioxide on seed-set and yield of kidney bean (*Phaseolus vulgaris* L.). Glob Chang Biol 8(8):710–721

Prasad PVV, Boote KJ, Hartwell L, Allen JR, Thomas JMG (2003) Super-optimal temperatures are detrimental to peanut (*Arachis hypogaea* L.) reproductive processes and yield at both ambient and elevated carbon dioxide. Glob Chang Biol 9(12):1775–1787

Prasad PVV, Pisipati SR, Mutava RN, Tuinstra MR (2008) Sensitivity of grain sorghum to high temperature stress during reproductive development. Crop Sci 48(5):1911–1917

Prasad PVV, Pisipati SR, Momcilovic I, Ristic Z (2011) Independent and combined effects of high temperature and drought stress during grain filling on plant yield and chloroplast protein synthesis elongation factor (EF-Tu) expression in spring wheat. J Agron Crop Sci 197(6): 430–441

Prasad PVV, Bheemanahalli R, Jagdish SK (2017) Field crops and the fear of heat stress-opportunities, challenges and future directions. Field Crops Res 200:114–121

Prášil IT, Prášilová P, Pánková K (2004) Relationships among vernalization, shoot apex development and frost tolerance in wheat. Ann Bot 94(3):413–418

Preiss J (1982) Regulation of the biosynthesis and degradation of starch. Annu Rev Plant Physiol 33:431–454

Price AH, Taylor A, Ripley SJ, Griffiths A, Trewavas AJ, Knight MR (1994) Oxidative signals in tobacco increase cytosolic calcium. Plant Cell 6(9):1301–1310

Prince S, Haskett J, Steininger M, Strand H, Wright R (2001) Net primary production of the US mid-West croplands from agricultural harvest yield data. Ecol Appl 11(4):1194–1205

Pukacki PM, McKersie BD (1990) Supercooling and ice nucleation events in the crown of winter wheat seedlings. Can J Plant Sci 70:1179–1182

Purseglove JW (1988) Tropical crops: dicotyledons. Longman, London, p 232

Qaderi MM, Reid DM (2005) Growth and physiological responses of canola (*Brassica napus*) to UV-B and CO2 under controlled environment conditions. Physiol Plant 125(2):247–259

Qadir S, Jamshieed S, Rasool S, Ashraf M, Akram NA, Ahmad P (2014) Modulation of plant growth and metabolism in cadmium-enriched environments. Arch Environ Contam Toxicol 229:51–88

Quagliotti L (1967) Effects of different temperatures on stalk development, flowering habit, and sex expression in the carrot *Daucus carota* L. Euphytica 16:83–103

Rahmstorf S, Coumou D (2011) Increase of extreme events in a warming world. Proc Natl Acad Sci U S A 108:17905–17909

Ramdani A, Martos V, Villegas D, García del Moral LF, Rharrabti Y, Royo C (2003) Changes on some agronomic and physiological traits of durum wheat cultivars released in Italy and Spain throughout the 20th century. In: Pogna NE, Roman'o M, Pogna EA, Galterio G (eds)

Proceedings of the 10th international wheat genetics symposium, vol 1. Paestum, Italy. pp 163–166
Rawson HM, Evans LT (1971) The contribution of stem reserves to grain development in a range of wheat cultivars of different height. Aus J Agric Res 22:851–863
Rawson HM (1988) Effect of high temperatures on the development and yield of wheat and practices to reduce deleterious effects. In: Klatt AR (ed) Wheat production constraints in tropical environments. D.F, CIMMYT, Mexico, pp 44–62
Reader MA, Dracup M, Atkins CA (1997) Transient high temperatures during seed growth in narrow-leafed lupin (*Lupinus angustifolius* L.) II. Injuriously high pod temperatures are likely in Western Australia. Aust J Agric Res 48(8):1179–1186
Rejeb IB, Pastor V, Mauch-Mani B (2014) Plant responses to simultaneous biotic and abiotic stress: molecular mechanisms. Planta 3(4):458–475
Renaut J, Lutts S, Hoffmann L, Hausman JF (2004) Responses of poplar to chilling temperature: proteomic and physiological aspects. Plant Biol 6(1):81–90
Renaut J, Hausman J, Wisniewski ME (2006) Proteomics and low-temperature studies: bridging the gap between gene expression and metabolism. Physiol Plant 126(1):97–109
Reynolds MP, Sayre KD, Rajaram S (1999) Physiological and genetic changes in irrigated wheat in the post Green Revolution period and approaches for meeting projected global demand. Crop Sci 39:1611–1621
Reynolds MP, Delgado MI, Gutiérrez-Rodríguez M, Larqué-Saavedra A (2000) Photosynthesis of wheat in a warm, irrigated environment. I: Genetic diversity and crop productivity. Field Crops Res 66:37–50
Reynolds MP, Calderini DF, Condon AG (2001) Physiological basis of yield gains in wheat associated with the LR19 translocation from *Agropyron elongatum*. Euphytica 119(1):137–141
Reynolds M, Condon AG, Rebetzke GJ, Richards RA (2004) Evidence for excess photosynthetic capacity and sink-limitation to yield and biomass in elite spring wheat. In: Fischer T, Turner N, Angus J, McIntyre L, Robertson M, Borrell A, Lloyd D (eds) New directions for a diverse planet: Proceedings of the 4th International Crop Science Congress, The Regional Institute Ltd., Gosford, NSW, Australia
Reynolds MP, Pellegrineschi A, Skovmand B (2005a) Sink limitation to yield and biomass: a summary of some investigations in spring wheat. Ann Appl Biol 146(1):39–49
Reynolds MP, Mujeeb-Kazi A, Sawkins M (2005b) Prospects for utilizing plant adaptive mechanisms to improve wheat and other crops in drought-and salinity-prone environments. Ann Appl Biol 146(2):239–259
Reynolds MP, Borlaug NE (2006) Impacts of breeding on international collaborative wheat improvement. J Agric Sci 144(1):3–17
Reynolds M, Tuberosa R (2008) Translational research impacting on crop productivity in drought-prone environments. Curr Opin Plant Biol 11(2):171–179
Reynolds MP, Foulkes MJ, Slafer GA, Berry PM, Parry MAJ, Snape JW, Angus WJ (2009) Raising yield potential in wheat. J Exp Bot 60(7):1899–1918
Riaz-ud-din G, Subhani GM, Ahmad N, Hussain M, Rehman AU (2010) Effect of temperature on development and grain formation in spring wheat. Pak J Bot 42(2):889–906
Richards RA (1996a) Defining selection criteria to improve yield under drought. Plant Growth Regul 20:157–166
Richards RA (1996b) Increasing yield potential in wheat—source and sink limitations. In: Reynolds MP, Rajaram S, McNab A (eds) Increasing yield potential in wheat: breaking the barriers, CIMMYT, Mexico DF, pp 134-149
Richards RA, Rebetzke GJ, Condon AG, van Herwaarden AF (2002) Breeding opportunities for increasing the efficiency of water use and crop yield in temperate cereals. Crop Sci 42(1): 111–121
Rigby JR, Porporato A (2008) Spring frost risk in a changing climate. Geophys Res Lett 35(1): 150–152

Rijavec T, Kovač M, Kladnik A, Chourey PS, Dermastia M (2009) A comparative study on the role of cytokinins in caryopsis development in the *maize miniature1* seed mutant and its wild type. J Integr Plant Biol 51(9):840–849

Rincker K, Nelson R, Specht J, Sleper D, Cary T, Cianzio SR, Casteel S, Conley S, Chen P, Davis V, Fox C, Graef G, Godsey C, Holshouser D, Jiang GL, Kantartzi SK, Kenworthy W, Lee C, Mian R, McHale L, Naeve S, Orf J, Poysa V, Schapaugh W, Shannon G, Uniatowski R, Wang D, Diers B (2014) Genetic improvement of U.S. soybean in maturity groups II, III, and IV. Crop Sci 54(4):1419–1432

Robelin M, Triboi AM (1983) Assimilation netted'une culture de colza d'hiver au cours du cycle de vegetation sous l'influence de l'environment climatique de la densite du peuplementet de la fertilization azotee. In: Proceedings of the 6th international rape seed conference, France, pp 98–103

Robertson MJ, Holland JF, Cawley S, Bambach R, Cocks B, Watkinson A (2001) Phenology of canola cultivars in the northern region and implications for frost risk. In: Science and technology: delivering results for agriculture. Proc. 10th Australian Agronomy Conference, Australian Society of Agronomy, Hobart

Robertson MJ, Holland JF, Bambach R (2004) Response of canola and Indian mustard to sowing date in the grain belt of north-eastern Australia. Aust J Exp Agric 44(1):43–52

Robinson D (2004) Scaling the depths: below-ground allocation in plants, forests and biomes. Funct Ecol 18(2):290–295

Roel A, Mutters RG, Eckert JW, Plant RE (2005) Effect of low water temperature on rice yield in California. Agron J 97(3):943–948

Rowntree SC, Suhre JJ, Weidenbenner NH, Wilson EW, Davis VM, Naeve SL, Diers BW, Esker P, Conley SP (2014) Physiological and phenological responses of historical soybean cultivar releases to earlier planting. Crop Sci 54(2):804–816

Royo C, García del Moral LF, Slafer G, Nachit MN, Araus JL (2005) Selection tools for improving yield-associated physiological traits. In: Royo C, Nachit MN, Di Fonzo N, Araus JL, Pfeiffer WH, Slafer GA (eds) Durum wheat breeding: current approaches and future strategies. Haworth Press, New York, pp 563–598

Royo C, Alvaro F, Martos V, Ramdani A, Isidro J, Villegas D, García del Moral LF (2007) Genetic changes in durum wheat yield components and associated traits in Italy and Spain during the 20th century. Euphytica 155:259–270

Rufty TW, Huber SC, Kerr PS (1985) Association between sucrose-phosphate synthase activity in leaves and plant growth rate in response to altered aerial temperature. Plant Sci 39(1):7–12

Sadler EJ, Karlen DL (1995) Aerial dry matter and nutrient accumulation comparisons among five soybean experiments. Commun Soil Sci Plant Anal 26(19-20):3145–3163

Sailaja K, Reddy KR, Reddy VR, Kakani VG, Zhao D (2005) Interactive effects of carbon dioxide, temperature, and UV-B radiation on soybean (*Glycine max* L.) flower and pollen morphology, pollen production, germination, and tube length. J Exp Bot 56(412):725–736

Saini HS, Sedgley M, Aspinall D (1983) Effect of heat-stress during floral development on pollen-tube growth and ovary anatomy in wheat (*Triticum aestivum* L.). Aust J Plant Physiol 10(2): 137–144

Sakata T, Takahashi H, Nishiyama I, Higashitani A (2000) Effects of high temperature on the development of pollen mother cells and microspores in barley *Hordeum vulgare* L. J Plant Res 113(4):395–402

Sakata T, Higashitani A (2008) Male sterility accompanied with abnormal anther development in plants-genes and environmental stresses with special reference to high temperature injury. Int J Plant Dev Biol 2(1):42–51

Salem MA, Kakani VG, Koti S, Reddy KR (2007) Pollen-based screening of soybean genotype for high temperatures. Crop Sci 47(1):219–231

Salvucci ME, Osteryoung KW, Crafts-Brandner SJ, Vierling E (2001) Exceptional sensitivity of Rubisco activase to thermal denaturation in vitro and in vivo. Plant Physiol 127(3):1053–1064

Samonte SO, Wilson LT, McClung AM, Tarpley L (2001) Seasonal dynamics of non-structural carbohydrate partitioning in 15 diverse rice genotypes. Crop Sci 41(3):902–909

Sanghera GS, Wani SH, Hussain W, Singh NB (2011) Engineering cold stress tolerance in crop plants. Curr Genomics 1200(12):30–43

Sanguineti MC, Duvick DN, Smith SP, Landi P, Tuberosa R (2006) Effects of long-term selection on seedling traits and ABA accumulation in commercial maize hybrids. Maydica 51:329–338

Sassenrath GF, Ort DR (1990) The relationship between inhibition of photosynthesis at low temperature and the inhibition of photosynthesis after rewarming in chill-sensitive tomato. Plant Physiol Biochem 28(4):457–465

Sassenrath GF, Ort DR, Portis AR (1991) Impaired reductive activation of stromal bisphosphatase in tomato leaves following low-temperature exposure at high light. Arch Biochem Biophys 282(2):302–308

Satake T (1976) Determination of the most sensitive stage to sterile type cool injury in rice Plants. Bull Hokkaido Nat Agric Exp Stat 113:1–33

Sato S, Peet MM, Thomas JF (2002) Determining critical pre- and post-anthesis periods and physiological processes in *Lycopersicon esculentum* Mill. exposed to moderately elevated temperatures. J Exp Bot 53(371):1187–1195

Savary S, Teng PS, Willocquet L, Nutter FW Jr (2006) Quantification and modeling of crop losses: a review of purposes. Annu Rev Phytopathol 44:89–112

Savitch LV, Gray GR, Huner NPA (1997) Feedback-limited photosynthesis and regulation of sucrose-starch accumulation during cold acclimation and low-temperature stress in a spring and winter wheat. Planta 201(1):18–26

Sayre KD, Rajaram S, Fischer RA (1997) Yield potential progress in short bread wheat in northwest Mexico. Crop Sci 37(1):36–42

Scott P (2008) Physiology and behaviour. Wiley, New York, p 305

Sehgal A, Sita K, Kumar J, Kumar S, Singh S, Siddique KHM, Nayyar H (2017) Effects of drought, heat and their interaction on the growth, yield and photosynthetic function of lentil (*Lens culinaris* Medikus) genotypes varying in heat and drought sensitivity. Front Plant Sci 8:1776

Semenov MA, Shewry PR (2010) Modelling predicts that heat stress and not drought will limit wheat yield in Europe. Sci Rep 1:66

Severino LS, Auld DL (2013) Seed yield and yield components of castor influenced by irrigation. Indust Crop Prod 49:52–60

Severino LS, Auld DL (2014) Study on the effect of air temperature on seed development and determination of the base temperature for seed growth in castor (*Ricinus communis* L.). Aust J Crop Sci 8(2):290–295

Seymour M (1996) Response of narrow-leafed lupin (*Lupinus angustifolius* L.) to foliar application of growth regulators in Western Australia. Aust J Exp Agric 36(4):473–478

Shah F, Huang J, Cui K, Nie L, Shah T, Chen C, Wang K (2011) Impact of high temperature stress on rice plant and its traits related to tolerance. J Agric Sci 149(5):545–556

Sharkey TD, Stitt M, Heineke D, Gerhardt R, Raschke K, Heldt HW (1986) Limitation of photosynthesis by carbon metabolism. II. O2-insensitive CO2 uptake results from limitations of triose phosphate utilization. Plant Physiol 81(4):1123–1129

Shearman VJ, Sylvester-Bradley R, Scott RK, Foulkes MJ (2005a) Physiological processes associated with wheat yield progress in the UK. Crop Sci 45(1):175–185

Sheehy JE, Peng S, Dobermann A, Mitchell PL, Ferrer A, Yang J, Zou Y, Zhong X, Huang J (2004) Fantastic yields in the system of rice intensification: fact or fallacy? Field Crops Res 88(1):1–8

Sheen TF (2000) Effect of day and night temperature variation and of high temperature on devernalization in radish. Acta Hortic 514:157–162

Shewry PR, Jones HD (2005) Transgenic wheat: where do we stand after the first 12 years? Ann Appl Biol 147(1):1–14

Shibles R, Secor J, Ford DM (1987) Carbon assimilation and metabolism. In: Soybeans: improvement, production, and uses, ASA-CSSA-SSSA, pp 535–588

Shimono H, Hasegawa T, Fujimura S, Iwama K (2002) Response of leaf photosynthesis and leaf water status to low water temperature at different growth stages. Field Crops Res 89(1):71–83
Shimono H, Hasegawa T, Fujimura S, Iwama K (2004) Responses of leaf photosynthesis and plant water status to low water temperature at different growth stages. Field Crops Res 89(1):71–83
Shimono H, Hasegawa T, Moriyama M, Fujimura S, Nagata T (2005) Modeling spikelet sterility induced by low temperature in rice. Agron J 97(6):1524–1536
Shimono H, Okada M, Kanda E, Arakawa I (2007) Low temperature-induced sterility in rice: evidence for the effects of temperature before panicle initiation. Field Crops Res 101(2): 221–231
Shimono H (2011) Earlier rice phenology as a result of climate change can increase the risk of cold damage during reproductive growth in northern Japan. Agric Ecosyst Environ 144(1):201–207
Shinozaki K, Yamaguchi-Shinozaki K (1996) Molecular response to drought and cold stress. Curr Opin Plant Biol 7(2):161–167
Shivaram B (1986) Physiological reasons for variations in productivity in sunflower genotypes under different plant populations. M.Sc. (Agri.) Thesis, Univ. Agric. Sci., Bangalore, India
Shrestha R, Turner NC, Siddique KHM, Turner D, Speijers J (2006) A water deficit during pod development in lentils reduces flower and pod numbers but not seed size. Aust J Agric Res 57(4):427–438
Shrestha S, Asch F, Dusserre J, Ramanantsoanirina A, Brueck H (2012) Climate effects on yield components as affected by genotypic responses to variable environmental conditions in upland rice systems at different altitudes. Field Crops Res 134:216–228
Shroyer J, Mikesell M, Paulsen G (1995) Spring freeze injury to Kansas wheat. Kansas State Univ. Ext. Serv
Si P, Walton G (2004) Determinants of oil concentration and seed yield in canola and Indian mustard in the lower rainfall areas of Western Australia. Aust J Agric Res 55(3):367–377
Siddique KHM, Sedgley RH (1985) The effect of reduced branching and water use of chickpea (Cicer arietinum L.) in a Mediterranean type environment. Field Crop Res 12:251–269
Siddique KHM, Loss SP, Regan KL, Pritchard DL (1998) Adaptation of lentil (*Lens culinaris* Medik) to short season Mediterranean-type environments: response to sowing rates. Aust J Agric Res 49(7):1057–1066
Silva SA, de Carvalho FIF, Caetano VR, Oliveira AC, Coimbra JLM, Vasconcellos NJS, Lorencetti C (2000) Genetic basis of stay-green trait in bread wheat. J New Seeds 2(2):55–68
Sinclair T, De Wit CT (1975) Photosynthate and nitrogen requirements for seed production by various crops. Science 189(4202):565–567
Sinclair T (1993) Crop yield potential and fairy tales. In: Buxton D, Shibles R, Forsberg R, Blad B, Asay K, Paulsen G, Wilson R (eds) International Crop Science 1. Crop Science Society America, Madison, pp 707–711
Sinclair T (1994) Limits to crop yield? In: Boote K, Bennett JM, Sinclair T, Paulsen GM (eds) Physiology and determination of crop yield. ASA, CSSA, SSSA, Madison, pp 509–532
Sinclair T (1998) Historical changes in harvest index and crop nitrogen accumulation. Crop Sci 38(3):638–643
Sinclair TR, Muchow RC (1999) Radiation use efficiency. Adv Agron 65:215–265
Singh DP, Singh P, Kumar A, Sharma HC (1985) Transpirational cooling as a screening technique for drought tolerance in oilseed Brassicas. Ann Bot 56:815–820
Singh RP, Prasad PVV, Reddy KR (2013) Impacts of changing climate and climate variability on seed production and seed industry. Adv Agron 118:49–110
Sinha SK, Bhargava SC, Goel A (1982) Energy as the basis of harvest index. J Agric Sci (Cambridge) 99:237–238
Sinha SK, Bhargava SC, Baldev B (1988) Physiological aspects of pulse crops. In: Baldev B, Ramanujam S, Jain HK (eds) Pulse Crops. Oxford & IBH Publishing Company, New Delhi, pp 421–455
Shearman VJ (2001) Changes in the yield limiting processes associated with the yield improvement of wheat. PhD Thesis, University of Nottingham, UK, p 238

Shearman VJ, Sylvester-Bradley R, Foulkes MJ (2005b) Physiological processes associated with wheat yield progress in the UK. Crop Sci 45(1):175–185

Shinkawa R, Morishita A, Amikura K, Machida R, Murakawa H, Kuchitsu K, Ishikawa M (2013) Abscisic acid induced freezing tolerance in chilling-sensitive suspension cultures and seedlings of rice. BMC Res Notes 6:351

Sita K, Sehgal A, Kumar J, Kumar S, Singh S, Siddique KHM, Nayyar H (2017) Identification of high-temperature tolerant lentil (*Lens culinaris* Medik.) genotypes through leaf and pollen traits. Front Plant Sci 8:744

Slafer GA, Andrade FH (1993) Physiological attributes related to the generation of grain yield in bread wheat cultivars released at different years. Field Crop Res 31(3-4):351–367

Slafer GA, Rawson HM (1994) Sensitivity of wheat phasic development to major environmental factors: a re-examination of some assumptions made by physiologists and modellers. Aust J Plant Physiol 21(4):393–426

Slafer GA, Savin R (1994) Grain mass change in a semi-dwarf and a standard height wheat cultivar under different sink-source relationships. Field Crops Res 37(1):39–49

Slafer GA, Calderini DF, Miralles DJ, Dreccer MF (1994) Preanthesis shading effects on the number of grains of three bread wheat cultivars of different potential number of grains. Field Crops Res 36(1):31–39

Slafer GA (1996) Differences in phasic development rate amongst wheat cultivars independent of responses to photoperiod and vernalization. A viewpoint of the intrinsic earliness hypothesis. J Agr Sci (Cambridge) 126(4):403–419

Slafer GA, Araus JL (1998) Improving wheat responses to abiotic stresses. In: Slinkard AE (ed) Proceedings of the 9th international wheat genetics symposium, Vol 1. University of Saskatchewan Extension Press, Saskatoon, pp 201–213

Slafer GA, Araus JL, Richards RA (1999) Physiological traits for future breeding to increase wheat the yield. In: Satorre EH, Slafer GA (eds) Wheat: ecology and physiology of yield determination. Food Product Press, New York, pp 379–415

Slafer GA, Whitechurch EM (2001) Manipulating wheat development to improve adaptation and to search for alternative opportunities to increase yield potential. In: Reynolds MP, Ortiz-Monasterio JI, McNab A (eds) Application of physiology in wheat breeding, CIMMYT, México, DF, pp 160–170

Slafer GA (2003) Genetic basis of yield as viewed from a crop physiologist's perspective. Ann Appl Biol 142(2):117–128

Slafer GA, Araus JL, Royo C, García Del Moral LF (2005) Promising eco-physiological traits for genetic improvement of cereal yields in Mediterranean environments. Ann Appl Biol 146(1): 61–70

Slafer GA, Araus JL (2007) Physiological traits for improving wheat yield under a wide range of conditions. In: Spiertz JHJ, Struik PC, van Laar HH (eds) Scale and complexity in plant systems research: gene-plant-crop relations. Springer, Dordrecht, pp 145–154

Snider JL, Oosterhuis DM, Skulman BW, Kawakami EM (2009) Heat stress-induced limitations to reproductive success in Gossypium hirsutum. Physiol Plant 137(2):125–138

Soffield I, Evans LT, Cook MG, Wardlaw IF (1977) Factors influencing the rate and duration of grain filling in wheat. Aust J Plant Physiol 4(5):785–797

Solomon S, Qin D, Manning M, Marquis M, Averyt K, Tignor MMB, Miller HL, Cheng Z (2007) Climate change 2007: the physical science basis. Contribution of working group I to the fourth assessment report of the intergovernmental panel on climate change. Cambridge University Press, New York

Soltani A, Hammer GL, Torabi B, Robertson MJ, Zeinali F (2006) Modeling chickpea growth and development: phenological development. Field Crops Res 99(1):1–13

Sowokinos JR (1976) Pyrophosphorylases in *Solanum tuberosum*. I. Changes in ADP-glucose pyrophosphorylase activities associated with starch biosynthesis during tuberization, maturation, and storage of potatoes. Plant Physiol 57(1):63–68

Srinivasan A, Johansen C, Saxena NP (1998) Cold tolerance during early reproductive growth of chickpea (*Cicer arietinum* L); Characterization of stress and genetic variation in pod set. Field Crops Res 57(2):181–193

Stapper M, Fischer RA (1990a) Genotype, sowing date and plant spacing influence on high-yielding irrigated wheat in southern New South Wales. I. Phasic development, canopy growth and spike production. Aust J Agric Res 41(6):997–1019

Stapper M, Fischer RA (1990b) Genotype, sowing date and plant spacing influence on high-yielding irrigated wheat in southern New South Wales. II. Growth, yield and nitrogen use. Aust J Agric Res 41(6):1021–1041

Stapper M, Fischer RA (1990c) Genotype, sowing date and plant spacing influence on high-yielding irrigated wheat in southern New South Wales. III. Potential yields and optimum flowering dates. Aust J Agric Res 41(6):1041–1056

Steponkus PL (1984) Role of the plasma membrane in freezing injury and cold acclimation. Annu Rev Plant Physiol 35(1):543–584

Stitt M, Krapp A, Klein D, Roper-Schwarz U, Paul M (1995) Do carbohydrates regulate photosynthesis and allocation by altering gene expression? In: Lucas WJ, Madore MA (eds) Carbon partitioning and source-sink interactions in plants. American Society of Plant Physiology, Rockville, MD, pp 68–77

Stockle CO, Kemanian AR (2009) Crop radiation capture and use efficiency: a framework for crop growth analysis. In: Sadras V, Calderini D (eds) Crop physiology. Academic Press, San Diego, pp 145–170

Stone P, Nicholas M (1995) A survey of the effects of high temperature during grain filling on yield and quality of 75 wheat cultivars. Aust J Agric Res 46(3):475–492

Stone PJ, Wilson DR, Reid JB, Gillespie RN (2001) Water deficit effects on sweet corn. I. Water use, radiation use efficiency, growth, and yield. Aust J Agric Res 52(1):103–113

Strand Å, Hurry V, Gustafsson P, Gardeström P (1997) Development of *Arabidopsis thaliana* leaves at low temperature releases the suppression of photosynthesis and photosynthetic gene expression despite the accumulation of soluble carbohydrates. Plant J 12(3):605–614

Strand A, Foyer CH, Gustafsson P, Gardeström. And Hurry, V. (1998) Altering flux through the sucrose biosynthesis pathway in transgenic *Arabidopsis thaliana* modifies photosynthetic acclimation at low temperatures and the development of freezing tolerance. Plant Cell Environ 26(4):523–535

Strand Å, Hurry V, Henkes S, Huner N, Gustafsson P, Gardeström. and Stitt, M. (1999) Acclimation of Arabidopsis leaves developing at low temperatures. increasing cytoplasmic volume accompanies increased activities of enzymes in the Calvin Cycle and in the sucrose-biosynthesis pathway. Plant Physiol 119(4):1387–1398

Stitt M, Grosse H (1988) Interactions between sucrose synthesis and CO2 fixation. IV. Temperature-dependent adjustment of the relation between sucrose synthesis and CO2 fixation. J Plant Physiol 133(4):392–400

Subedi KD, Gregory PJ, Summerfield RJ, Gooding MJ (1998) Cold temperatures and boron deficiency caused grain set failure in spring wheat (*Triticum aestivum* L.). Field Crops Res 57(3):277–288

Subedi KD, Gooding MJ, Gregory PJ (2010) Cultivar variation in boron accumulation and grain set in wheat under the influence of cold temperature. Ann Appl Biol 138(1):97–101

Sung SS, Xu D-P, Black CC (1989) Identification of actively filling sucrose sinks. Plant Physiol 89(4):1117–1121

Sung DY, Kaplan F, Lee KJ, Guy CL (2003) Acquired tolerance to temperature extremes. Trend Plant Sci 8(4):179–187

Sunkar R, Zhu J-K (2004) Novel and stress-regulated microRNAs and other small RNAs from Arabidopsis. Plant Cell 16(8):2001–2019

Sunkar R, Chinnusamy V, Zhu J, Zhu J-K (2007) Small RNAs as big players in plant abiotic stress responses and nutrient deprivation. Trends Plant Sci 12(7):301–309

Suzuki K, Nagasuga K, Okada M (2008) Chilling injury induced by high root temperature in the leaves of rice seedlings. Plant Cell Physiol 49(3):433–442

Swaminathan MS (2007) Can science and technology feed the world in 2025? Field Crops Res 104(1):3–9

Szynkier K (1974) The effect of removing of supply or sink organs on the distribution of assimilate in two vrieties of garden pea (*Pisum sativum* L.). Act Agric Scand 4:7–12

Tabbal DF, Bouman BAM, Bhuiyan SI, Sibayan EB, Sattar MA (2002) On-farm strategies for reducing water input in irrigated rice: case studies in the Philippines. Agric Water Manag 56(2): 93–112

Tacarindua CRP, Shiraiwa T, Homma K, Kumagai E, Sameshima R (2012) The response of soybean seed growth characteristics to increased temperature under near-field conditions in a temperature gradient chamber. Field Crops Res 131:26–31

Takai T, Fukuta Y, Shirawa T, Horie T (2005) Time-related mapping of quantitative trait loci controlling grain-filling in rice (*Oryza sativa* L.). J Exp Bot 56(418):2107–2118

Takashima NE, Rondanini DP, Puhl LE, Iralles DL (2013) Environmental factors affecting yield variability in spring and winter rapeseed genotypes cultivated in the southeastern Argentine Pampas. Eur J Agron 48:88–100

Tambussi EA, Noguès S, Araus JL (2005a) Ear of durum wheat under water stress: water relations and photosynthetic metabolism. Planta 221(3):446–458

Tambussi EA, Noguès S, Ferrio JP, Voltas J, Araus JL (2005b) Does a higher yield potential improve barley performance under Mediterranean conditions? A case study. Field Crop Res 91(2-3):149–160

Tamminen I, Mäkelä P, Heino P, Palva ET (2001) Ectopic expression of *ABI3* gene enhances freezing tolerance in response to abscisic acid and low temperature in *Arabidopsis thaliana*. Plant J 25(1):1–8

Taylor NL, Heazlewood JL, Day DA, Millar AH (2005) Differential impact of environmental stresses on the pea mitochondrial proteome. Mol Cell Proteomics 4(8):1122–1133

Tetreault HM, Kawakami T, Ungerer MC, Levy C (2016) Low temperature tolerance in the perennial sunflower *Helianthus maximiliani*. Am Midl Nat 175(1):91–102

Thakur P, Kumar S, Malik JA, Berger JD, Nayyar H (2010) Cold stress effects on reproductive development in grain crops: a review. Env Expt Bot 67(3):429–443

Thomashow MF (1998) Role of cold-responsive genes in plant freezing tolerance. Plant Physiol 118(1):1–8

Timperio AM, Egidi MG, Zolla L (2008) Proteomics applied on plant abiotic stresses: role of heat shock proteins (HSP). J Proteome 71(4):391–411

Thomas JMG, Boote KJ, Allen LH Jr, Gallo-Meagher M, Davis JM (2003) Elevated temperature and carbon dioxide effects on soybean seed germination and transcript abundance. Crop Sci 43(4):1548–1557

Thomas G (2005) Grain yield and protein content for cropping systems at Dunkerry south trial 1996–2001.CSIRO, Toowoomba. Vol 2006

Thompson R, Taylor H (1977) Yield components and cultivar sowing date and density in field beans (*Vicia faba* L.). Ann Appl Biol 86(2):313–320

Thompson KF, Hughes WG (1986) Breeding and varieties. In: Scarisbrick DH, Daniels RW (eds) Oilseed rape. Collins, London, pp 32–82

Thompson BD, Siddique KHM, Barr MD, Wilson JM (1997) Grain legume species in low rain fall Mediterranean type environment. I. Phenology and seed yield. Field Crop Res 54(2-3):173–187

Thompson DS, Smith PW, Davies WJ, Ho LC (1999) Interactions between environment, fruit water relations and fruit growth. Acta Hortic 487:65–70

Thuzar M, Putch AB, Abdullah NAP, Lassim MBM, Jusoff K (2010) The effects of temperature stress on the quality and yield of soyabean [(*Glycine max* L.) Merrill.]. J Agric Sci 2(1):172–179

Tollenaar M, Bruulsema TW (1988) Effects of temperature on rate and duration of kernel dry matter accumulation of maize. Can J Plant Sci 68(4):935–940

Tollenaar M, Wu J (1999) Yield improvement in temperate maize is attributed to greater water stress tolerance. Crop Sci 39(6):1597–1604
Tollenaar M, Ahmadzadeh A, Lee EA (2004) Physiological basis for grain yield improvement in maize. Crop Sci 44(Supplement 3):2086–2094
Tollenaar M, Lee EA (2006) Dissection of physiological processes underlying grain yield in maize by examining genetic improvement and heterosis. Maydica 51:399–408
Trethowan RM, van Ginkel M, Rajaram S (2002) Progress in breeding for yield and adaptation in global drought affected environments. Crop Sci 42(5):1441–1446
Triboï E, Martre P, Triboï-Blondel AM (2003) Environmentally-induced changes of protein composition for developing grains of wheat are related to changes in total protein content. J Exp Bot 54(388):1731–1742
Trnka M, Rötter RP, Ruiz-Ramos M, Kersebaum KC, Olesen JE, Žalud Z, Semenov MA (2014) Adverse weather conditions for European wheat production will become more frequent with climate change. Nat Clim Chang 4(7):637–643
Turner N, Wright G, Siddique KHM (1999) Adaptation of grain legumes (pulses) to water-limited environments. Adv Agron 71:193–229
Uauy C, Distelfeld A, Fahima T, Blechl A, Dubcovsky J (2006) A NAC Gene regulating senescence improves grain protein, zinc, and iron content in wheat. Science 314(5803):1298–1301
Uddling J, Gelang-Alfredsson J, Karlsson PE, Selldén G, Pleijel H (2008) Source–sink balance of wheat determines responsiveness of grain production to increased [CO2] and water supply. Agric Ecosys Environ 127(3-4):215–222
Unkovich M, Baldock J, Forbed M (2010) Variability in harvest index of grain crops and potential significance for carbon accounting: examples from Australian agriculture. In: Sparks DL (ed) Advances in agronomy, vol 105. Academic Press, Burlington, pp 173–219
Usuda H, Shimogawara K (1998) The effects of increased atmospheric carbon dioxide on growth, carbohydrates and photosynthesis in radish, *Raphanus sativus*. Plant Cell Physiol 39(1):1–7
Vágújfalvi A, Kerepesi I, Galiba G, Tischner T, Sutka J (1999) Frost hardiness depending on carbohydrate changes during cold acclimation in wheat. Plant Sci 144(2):85–92
Vanaja M, Reddy PR, Lakshmi NJ, Maheswari M, Vagheera P, Ratnakumar P, Jyothi M, Yadav SK, Venkateswarulu B (2007) Effect of elevated CO2 concentration on growth and yield of blackgram (*Vigna mungo* L. Hepper)—a rainfed pulse crop. Plant Soil Environ 53(2):81–88
Vanderauwera S, De Block M, van der Steene N, van der Cotte B, Metxlaff M, van Breusegem F (2007) Silencing of poly (ADP-ribose) polymerase in plants alters abiotic stress signal transduction. Proc Natl Acad Sci U S A 104(38):15150–15155
van Herwaarden A, Farquhar G, Angus J, Richards RA, Howe G (1998a) Haying-off', the negative grain yield response of dryland wheat to nitrogen fertiliser. I. Biomass, grain yield and water use. Aust J Agric Res 49(7):1067–1081
van Herwaarden A, Richards RA, Angus J, Ferquhar G (1998b) Haying-off, the negative grain yield response of dry land wheat to nitrogen fertilizer. II. Carbohydrate and protein dynamics. Aust J Agric Res 49(7):1083–1093
van Herwaarden A, Richards R (2002) Water-soluble carbohydrate accumulation in stems is related to breeding progress in Australian wheats. In: McComb J (ed) Plant Breeding for the 11th Millennium; 12th Australian Plant Breeding Conference, pp 878–882
van Oosterom EJ, Borrell AK, Chapman SC, Broad IJ, Hammer GL (2010a) Functional dynamics of the nitrogen balance of sorghum: I. N demand of vegetative plant parts. Field Crops Res 115(1):19–28
van Oosterom EJ, Borrell AK, Chapman SC, Broad IJ, Hammer GL (2010b) Functional dynamics of the nitrogen balance of sorghum. II. Grain filling period. Field Crops Res 115(1):29–48
Vassey TL, Quick WP, Sharkey TD, Stitt M (1991) Water stress, carbon dioxide, and light effects on sucrose-phosphate synthase activity in Phaseolus vulgaris. Physiol Plant 81(1):37–44
Vega CRC, Sadras VO, Andrade FH, Uhart SA (2000) Reproductive allometry in soybean, maize and sunflower. Ann Bot 85(4):461–468

Venzhik YV, Titov AF, Talanova VV, Frolova SA, Talanov AV, Nazarkina YA (2011) Influence of lowered temperature on the resistance and functional activity of the photosynthetic apparatus of wheat plants. Biol Bull 38(2):132–137

Verma V, Foulkes MJ, Caligari P, Sylvester-Bradley R, Snape J (2004) Mapping quantitative trait loci for flag leaf senescence as a yield determinant in winter wheat under optimal and drought-stressed environments. Euphytica 135(3):255–263

Virgona JM, Gummer FAJ, Angus JF (2006) Effects of grazing on wheat growth, yield, development, water use, and nitrogen use. Aust J Agric Res 57(12):1307–1319

von Caemmerer S (2003) C4 photosynthesis in a single C3 cell is theoretically inefficient but may ameliorate internal CO2 diffusion limitations of C3 leaves. Plant Cell Environ 26(8):1191–1197

Wahbi A, Gregory PJ (1989) Genotypic differences in root and shoot growth of barley (*Hordeum vulgare*) II. Field studies of growth and water use of crops grown in northern Syria. Aust J Exp Agric 25:389–399

Wahid A, Gelani S, Ashraf M, Foolad MR (2007) Heat tolerance in plants: an overview. Environ Exp Bot 61(3):199–223

Wallis ES, Byth DE, Whiteman PC (1981) Mechanized dry seed production of pigeonpea. In: Proceedings of International Workshop on Pigeonpea, Vol 1, ICRISAT, Hyderabad, pp 51–60

Wan X, Zwiazek JJ, Lieffers VJ, Landhausser SM (2001) Hydraulic conductance in aspen (*Populus tremuloides*) seedlings exposed to low root temperature. Tree Physiol 21(10):691–696

Wani SH, Sandhu JS, Gosal SS (2008) Genetic engineering of crop plants for abiotic stress tolerance. In: Malik CP, Kaur B, Wadhwani C (eds) Advanced topics in plant biotechnology and plant biology. MD Publications, New Delhi, pp 149–183

Wani SH, Gosal SS (2011) Introduction of OsglyII gene into Indica rice through particle bombardment for increased salinity tolerance. Biol Plant 55(3):536–540

Wang Z, Mambelli S, Setter TL (2002) Abscisic acid catabolism in maize kernels in response to water deficit at early endosperm development. Ann Bot 90(5):623–630

Wang WX, Vinocur B, Altman A (2003) Plant responses to drought, salinity and extreme temperatures: towards genetic engineering for stress tolerance. Planta 218(1):1–14

Wang J, Gan YT, Clarke F, McDonald CL (2006) Response of chickpea yield to high temperature stress during reproductive development. Crop Sci 46(5):2171–2178

Watanabe T, Kume T (2009) A general adaptation strategy for climate change impact on paddy cultivation: special reference to the Japanese context. Paddy Water Environ 7(4):313–320

Ward PR, Dunin FX, Micin SF (2002) Water use and root growth by annual and perennial pastures and subsequent crops in a phase rotation. Agric Water Manag 53(1-3):83–97

Weibe JH, Habegger R, Liebig HP (1992) Quantification of vernalization and devernalisation effects for kohl rabi (*Brassica oleracea* convar. acephala var. gongylodes L.). Sci Hortic 50(1-2):11–20

Welbaun GE, Bian D, Hill DR, Grayson RL, Gunatilaka MK (1997) Freezing tolerance, protein composition, and abscisic acid localization and content of pea epicotyl, shoot, and root tissues in response to temperature and water stress. J Exp Bot 48(3):643–654

Weymann W, Böttcher U, Sieling K, Kage H (2015) Effects of weather conditions during different growth phases on yield formation of winter oilseed rape. Field Crops Res 173:41–48

Whaley JM, Kirby EJM, Spink JH, Foulkes MJ, Sparkes DL (2004) Frost damage to winter wheat in the UK: the effect of plant population density. Eur J Agron 21(1):105–115

Wheeler TR, Morison JIL, Ellis RH, Hadley P (1994) The effects of CO2, temperature and their interaction on the growth and yield of carrot (*Daucus carota* L.). Plant Cell Environ 17(12): 1275–1284

Whish JPM, Herrmann NI, White NA, Moore AD, Kriticos DJ (2015) Integrating pest population models with biophysical crop models to better represent the farming system. Environ Model Softw 72:418–425

Whisler FD, Acock B, Baker DN, Fye RE, Hodges HF, Lambert JR, Lemmon HE, McKinion JM, Reddy VR (1986) Crop simulation models in agronomic systems. Adv Agron 40:141–208

Whitechurch EM, Slafer GA, Miralles DJ (2007) Variability in the duration of stem elongation in wheat and barley. J Agron Crop Sci 193(2):138–145
Whitehead SJ, Summerfield RJ, Muehlbauer FJ, Coyneb CJ, Ellis RH, Wheeler TR (2000) Crop improvement and the accumulation and partitioning of biomass and nitrogen in lentil. Crop Sci 40(1):110–120
Whiteman PC, Byth DE, Wallis ES (1985) Pigeonpea (Cajanus cajan (L.) Millsp.). In: Summerfield RJ, Roberts EH (eds) Grain legume crops. Collins Professional and Technical Books, London, pp 685–698
Williams RL, Angus J (1994) Deep floodwater protects high-nitrogen rice crops from low-temperature damage. Aust J Exp Agric 34(7):927–932
Wilkinson S, Davies WJ (2009) Ozone suppresses soil drying-and abscisic acid (ABA)-induced stomatal closure via an ethylene dependent mechanism. Plant Cell Environ 32(8):949–959
Wolf S, Marani A, Rudich J (1990) Effects of temperature and photoperiod on assimilate partitioning in potato plants. Ann Botany 66(5):513–520
Woods CM, Reid MS, Patterson BD (1984) Response to chilling deficit in plant cells. I. Changes in cyclosis and cytoplasmic structure. Protoplasma 121(1):8–16
Woodrow IE, Ellis JR, Jellings A, Foyer CH (1984) Compartmentation and fluxes of inorganic phosphate in photosynthetic cells. Planta 161(6):525–530
Wright P, Morgan J, Cass A (1992) Effects of water stress on yield and the components of yield in canola and Indian mustard. In: 6th Australian Agronomy Conference, Australian Society of Agronomy, Armidale, p 617
Wu G, Wilson LT, McClung AM (1998) Contribution of rice tillers to dry matter accumulation and yield. Agron J 90(3):317–323
Wu YF, Zhong XL, Hu X, Ren DC, Lv GH, Wei CY, Song JQ (2014) Frost affects grain yield components in winter wheat. NZ J Crop Hortic Sci 42(3):194–204
Xin Z, Browse J (2001) Cold comfort farm: the acclimation of plants to freezing temperatures. Plant Cell Environ 23(9):893–902
Xu D-P, Sun SS, Black CC (1989) Sucrose metabolism in lima bean seeds. Plant Physiol 89(4): 1106–1116
Xu C, Huang B (2012) Comparative analysis of proteomic responses to single and simultaneous drought and heat stress for two *Kentucky bluegrass* cultivars. Crop Sci 52(3):1246–1260
Xuan J, Song Y, Zhang H, Liu J, Guo Z, Hua Y (2013) Comparative proteomic analysis of the stolon cold stress response between the C4 perennial grass species *Zoysia japonica* and *Zoysia matrella*. PLoS One 8:e75705
Xue QW, Zhu ZX, Musick JT, Stewart BA, Dusek DA (2006) Physiological mechanisms contributing to the increased water-use efficiency in winter wheat under deficit irrigation. J Plant Physiol 163(2):154–164
Yadav SK (2010) Old stress tolerance mechanisms in plants. A review. Agron Sust Develop 30(3): 515–527
Yamaguchi-Shinozaki K, Shinozaki K (2006) Transcriptional regulatory networks in cellular responses and tolerance to dehydration and cold stresses. Annu Rev Plant Biol 57:781–803
Yang J, Zhang J, Huang Z, Zhu Q, Wang L (2000) Remobilization of carbon reserves is improved by controlled soil-drying during grain filling of wheat. Crop Sci 40(6):1645–1655
Yang J, Zhang J, Wang Z, Zhu Q (2001) Activities of starch hydrolytic enzymes and sucrose-phosphate synthase in the stems of rice subjected to water stress during grain filling. J Exp Bot 52(364):2169–2179
Yang JC, Zhang JH, Wang ZQ, Zhu QS, Liu LJ (2003a) Involvement of abscisic acid and cytokinins in the senescence and remobilization of carbon reserves in wheat subjected to water stress during grain filling. Plant Cell Environ 26(10):1621–1631
Yang J, Zhang J, Wang Z, Zhu Q, Liu L (2003b) Activities of enzymes involved in sucrose-to-starch metabolism in rice grains subjected to water stress during grain filling. Field Crops Res 81(1):69–81
Yang J, Zhang J (2006) Grain filling of cereal under soil drying. New Phytol 169(2):223–236

Yang J, Zhang J, Wang Z, Liu K, Wang P (2006) Post-anthesis development of inferior and superior spikelets in rice in relation to abscisic acid and ethylene. J Exp Bot 57(1):149–160

Yang J, Liu K, Wang Z, Du Y, Zhang J (2007) Water-saving and high-yielding irrigation for lowland rice by controlling limiting values of soil water potential. J Integra Plant Biol 49(10): 1445–1454

Yang J, Zhang J (2010) Crop management techniques to enhance harvest index in rice. J Exp Bot 61(12):3177–3189

Yang C, Gan Y, Harker KN, Kutcher HR, Gulden R, Irvine B, May WE (2014) Up to 32% yield increase with optimized spatial patterns of canola plant establishment in western Canada. Agron Sustain Dev 34(4):793–801

Ye H, Du H, Tang N, Li X, Xiong L (2009) Identification and expression profiling analysis of TIFY family genes involved in stress and phytochrome responses in rice. Plant Mol Biol 71(3): 291–305

Ying JF, Peng SB, He QR, Yang H, Yang CD, Visperas RM, Cassman KG (1998) Comparison of high-yield rice in tropical and subtropical environments: I. Determinants of grain and dry matter yields. Field Crops Res 57(1):71–84

Yoshida S, Hara T (1977) Effects of air temperature and light on grain filling of an indica and a japonica rice (*Oryza sativa* L.) under controlled environmental conditions. Soil Sci Plant Nut 23(1):93–107

Yun K-Y, Park MR, Mohanty B, Herath V, Xu F, Mauleon R, Wijaya E, Bajic VB, Bruskiewich R, de Los Reyes BG (2010) Transcriptional regulatory network triggered by oxidative signals configures the early response mechanisms of japonica rice to chilling stress. BMC Plant Biol 10: 16

Yu SM, Lo SF, Ho THD (2015) Source–sink communication: regulated by hormone, nutrient, and stress cross-signaling. Trends Plant Sci 20(12):844–857

Zandalinas SI, Mittler R, Balfag6n D, Arbona V, G6mez-Cadenas A (2018) Plant adaptations to the combination of drought and high temperatures. Physiol Plant 162(1):2–12

Zenkteler E, Samordakiewicz S, Kaluzewicz A, Knaflewski M (2012) Effect of devernalization on the transition from vegetative to prefloral phase of the broccoli (*Brassica oleracea* var. italica cv. 'Fiesta') shoot meristem. Acta Agrobot 65(1):29–36

Zhang X, Evans PM (2004) Grain yield production in relation to plant growth of wheat and canola following clover pastures in southern Victoria. Aust J Exp Agric 44(10):1003–1012

Zhang J, Yang J (2004) Crop yield and water use efficiency: a case study in rice. In: Bacon MA (ed) Water use efficiency in plant biology. Blackwell Publishing/CRC Press, Oxford

Zhang H, Zhang S, Zhang J, Yang J, Wang Z (2008a) Postanthesis moderate wetting drying improves both quality and quantity of rice yield. Agron J 100(3):726–734

Zhang Z, Zhang S, Yang J, Zhang J (2008b) Yield, grain quality and water use efficiency of rice under non-flooded mulching cultivation. Field Crops Res 108(1):71–81

Zhang H, Turner NC, Poole ML (2012) Increasing the harvest index of wheat in the high rainfall zones of southern Australia. Field Crops Res 129:111123

Zheng B, Chapman SC, Christopher JT, Frederiks TM, Chenu K (2015) Frost trends and their estimated impact on yield in the Australian wheatbelt. J Exp Bot 66(12):3611–3623

Zhong X, Mei X, Li Y, Yoshida H, Zhao P, Wang X, Han L, Hu X, Huang S, Huang J, Sun Z (2008) Changes in frost resistance of wheat young ears with development during jointing stage. J Agron Crop Sci 194(5):343–349

Zhou L, Zeng Y, Hu G, Pan Y, Yang S, You A, Zhang H, Li J, Li Z (2012) Characterization and identification of cold tolerant near-isogenic lines in rice. Breeding Sci 62(2):196–201

Zhong-hu Z, Rajaram S (1994) Differential responses of bread wheat characters to high temperature. Euphytica 72(3):197–203

Zhou-fei W, Jian-fei W, Fu-hua W, Yong-mei B, Yun-yu W, Hong-sheng Z (2009) Genetic control of germination ability under cold stress in rice. Rice Sci 16(3):173–180

Zhu Q-H, Spriggs A, Matthew L, Fan L, Kennedy G, Gubler F, Helliwell C (2008) A diverse set of microRNAs and microRNA-like small RNAs in developing rice grains. Genome Res 18:1456–1465

Zia MS, Salim M, Aslam M, Gill MA, Rahmatullah. (1994) Effect of low temperature of irrigation water on rice growth and nutrient uptake. The J Agric Sci 173(1):22–31

Zia MA, Ashraf M, Akram NA, Hussain M, Al-Qurainy F (2014) Heat-induced regulation in antioxidant defense system in wheat (*Triticum aestivum* L.). Pak J Bot 46(3):957–971

Zinn KE, Tunc-Ozdemir M, Harper JF (2010) Temperature stress and plant sexual reproduction: uncovering weakest link. J Exp Bot 61(7):1959–1968

Zinselmeier C, Jeong BR, Boyer JS (1999) Starch and the control of kernel number in maize at low water potentials. Plant Physiol 121(1):25–35

Ziogas V, Tanou G, Belghazi M, Filippou P, Fotopoulos V, Grigorios D, Molassiotis (2015) Roles of sodium hydrosulfide and sodium nitroprusside as priming molecules during drought acclimation in citrus plants. Plant Mol Biol 89(4-5):433–450

Zocchi G, Hanson JB (1982) Calcium influx into corn roots as a result of cold shock. Plant Physiol 70(1):318–319

Zohary D, Hopf M (1973) Domestication of pulses in the old world. Science 182(4115):887–894